ADVANCES IN MATERIALS SCIENCE AND ENGINEERING

PROCEEDINGS OF THE 7TH ANNUAL INTERNATIONAL WORKSHOP ON MATERIALS SCIENCE AND ENGINEERING, (IWMSE 2021), CHANGSHA, HUNAN, CHINA, 21–23 MAY 2021

Advances in Materials Science and Engineering

Edited by

Domenico Lombardo
Consiglio Nazionale delle Ricerche, Messina, Italy

Ke Wang
East China University of Technology, Nanchang City, China

CRC Press
Taylor & Francis Group
Boca Raton London New York Leiden

CRC Press is an imprint of the
Taylor & Francis Group, an **informa** business

A BALKEMA BOOK

CRC Press/Balkema is an imprint of the Taylor & Francis Group, an informa business

© 2022 selection and editorial matter, the Editors; individual chapters, the contributors

Typeset by MPS Limited, Chennai, India

All rights reserved. No part of this publication or the information contained herein may be reproduced, stored in a retrieval system, or transmitted in any form or by any means, electronic, mechanical, by photocopying, recording or otherwise, without written prior permission from the publisher.

Although all care is taken to ensure integrity and the quality of this publication and the information herein, no responsibility is assumed by the publishers nor the author for any damage to the property or persons as a result of operation or use of this publication and/or the information contained herein.

Library of Congress Cataloging-in-Publication Data

A catalog record has been requested for this book

Published by: CRC Press/Balkema
Schipholweg 107C, 2316 XC Leiden, The Netherlands
e-mail: enquiries@taylorandfrancis.com
www.routledge.com – www.taylorandfrancis.com

ISBN: 978-1-032-12707-1 (Hbk)
ISBN: 978-1-032-12712-5 (Pbk)
ISBN: 978-1-003-22585-0 (ebk)
DOI: 10.1201/9781003225850

Table of contents

Preface	xiii
Committees	xv
Editors	xvii
Invited Talk	xix
Keynote Speaker	xxi

Chapter 1: Various Materials Properties, Processing, and Manufactures

Preparation and ferroelectric properties of strontium doped BaTiO3 nanotubes Y.N. Sun, M.E. Ma, H. Li, T.Y. Yu & X.Y. Deng	3
Removal of wastewater-dissolved heavy metals by Na-carboxylate polyarylene ether sulfone Y.N. Wang, J.R. Zhang, Z.H. Lu, Y. Fu & B.D. Wei	9
MgO-based desulfurizer for SO_2 removal of sintering flue gas: performance and mechanism W. Cao, W.J. Zhang & Q. Wu	15
Effect of Mo and B on the structure and properties of high chromium cast iron surfacing metal Q.Y. Zhao, F.L. Tantai, H.F. Tian, C. Wang, W. Wang, Q.L. Hou & J.S. Sun	20
Effects of green and black tea on total phenolic content and textural properties of bread Z.L. Ma, N. Duan, D.X. Zhang, P. Wu, Y.H. Guo & D.Y. Li	26
Connecting technology of vertical components in fabricated concrete buildings: Current status and analysis G.Z. Lu, Z.J. Wang, Z.L. Guo, Z.S. Zhang & Y. Zhang	31
Design of intermolecular charge-transfer fluorescent probes and determination of the content of drugs C.F. Jin, L.S. Yu, Y. Wang, Y.Q. Cheng & Y.G. Lv	41
Optimization design and fatigue life analysis of damping conical rubber spring for rail vehicle R.G. Zhao, X.Q. Yang, Y.J. Huang, W.H. Liu, X. Zhou, Y.L. Liu & D.H. Ye	47
Experimental investigation on physical and mechanical properties of aluminium alloy 7050-T7451 after burning B. Zhao, J.W. Wu, Q. Chen & B.W. Qi	57
Dehydration kinetics of $Mg(OH)_2$ for thermochemical energy storage via model-free and model-fitting methods D.Y. Wang, Q.P. Wu, M.T. Li, E.X. Ren, Z.Y. Chang & L. Zhu	64
Straw-based boards and panels: Experiment, setbacks, and underlying factors M.X. Xie	70

Design of a distributed Bragg reflector of GaN-based light emitting diodes flip chips 75
J.W. Wang, Y.C. Xu, Q. Zhou, G.Y. Chen, M.M. Wang & P.P. Wu

Application technology of thin layer repair materials for cement concrete pavement 84
Y. Liu & X.J. Huang

Study on microstructure of new nickel-based powder superalloy 90
G.H. Yang, G.F. Tian, X.L. Zhang, Z. Liu, Z. Ji, X.Q. Yan & C.C. Jia

Study on preparation and characteristics of Ni-Cr-W alloy laser cladding coating 96
L. Chen, Z.D. Liu, B.K. Li, S.Y. Guo, J.X. Li & Y. Li

Effects of nano-fillers on the surface structure and properties of expoxy resin modified by polyhedral oligomeric silsesquioxane 103
Y.H. Fang & P. Wang

Effect of K_2CO_3 adding on properties of SiC porous ceramics 108
T.Y. Yu, H. Li, M.E. Ma, Y.N. Sun & X.Y. Deng

Effects of TiC and rare earth on the microstructure and performance of LASER cladding Mo_2FeB_2 113
Z.X. Zhang, H.F. Tian, F.L. Tantai, H. Liu, W. Wang & J.S. Sun

The effects of carbon and nitrogen sources on biosurfactant fermentation by *Ochrobactrum intermedium* XY-1 119
Y.M. Sun, H. You, W.M. Si, L. Liu, X. Xu & X.Z. Li

Study on single-pass forming process parameters of laser direct deposition of Fe55 alloy 124
G.L. Yin, S.Y. Chen, J. Liang, T. Cui, C.S. Liu & M. Wang

Preparation and performances of SiO_2 aerogel heat-insulation coatings 130
G.Z. Lu, Z.J. Wang, Z.S. Zhang & Z.F. Yin

Optimization design for the free surface parameters of a spherical hinge suspender connecting locomotive structural frame and axle box 137
R.G. Zhao, W.H. Liu, N. Ji, Y.J. Huang, X.Q. Yang, X. Zhou & M. Zaheer

Using boron-doped diamond sensor to detect heavy metal in water 146
L. Pang, X. Yu, X. Qian, T.Y. Yan, L.Q. Chen & H.Z. Li

Pore defects and process control of pure molybdenum using wire arc additive manufacturing 152
Y.A. Qiao, J.C. Wang, S.Y. Tang & C.M. Liu

Optimization and test evaluation of vibration control for an airborne radar frequency synthesizer 161
C. Peng & S.G. Yang

Catalytic hydroisomerization of n-Hexadecane over Pd/SAPO-41 bifunctional catalysts: Effect of molar ratio of ethylene glycol to water on their catalytic performance 170
J. Yang, X.M. Wei & X.F. Bai

Study on raw materials and structural performance of coral concrete composite structure 177
Y. Gao & Y.J. Lu

Research on creep-fatigue lifetime of GH4133B superalloy at elevated temperature used in turbine disk of aero-engine 186
R.G. Zhao, X. Zhou, Y.F. Liu, X.Q. Yang, W.H. Liu, Y.L. Liu & D.H. Ye

Experimental research on mechanical performance of sulphoaluminate cement rapid repair material *Y. Liu, X.J. Huang & Z.Q. Hou*	193
In situ fatigue of new nickel-based powder superalloy at high temperature *X.L. Zhang, G.F. Tian, G.H. Yang, Z. Liu, Z. Ji, X.Q. Yan & C.C. Jia*	199
Analysis of electrical contact failure of new beryllium bronze contact parts *Z.G. Kong & Y.C. Zhang*	204
Preparation and properties of laser cladding Cu-W alloy coating on pure copper electrical contacts *B.K. Li, Z.D. Liu, S.Y. Guo, L. Chen, J.X. Li & H.R. Ma*	212
Effect of laser energy density on microstructure and properties of SLM 24CrNiMoY alloy steel *M. Sun, S.Y. Chen, M.W. Wei, X.W. Song, L. Zhou & M. Wang*	219

Chapter 2: Multifunctional Materials Properties, Processing, and Manufactures

Study on the effect of hot press sintering temperature on the microstructure and properties of new Ti-Al-Nb alloy *S. Fang, M.C. Zhang, Q.Y. Yu & H.P. Xiong*	227
Antistatic finishing of polypropylene non-wovens with reduced graphene oxide *Y. Zhang & Q. Yu*	236
Construction and application of rare earth ibuprofen phenanthroline fluorescence system *L. Wang, W.B. Liu, D. Lu, C.F. Jin, N. Wen, Y.Q. Cheng & Y.G. Lv*	242
Microstructure and mechanical property of SiC/ZL114A composite fabricated by laser deposition manufacturing *C.S. Lv, L.Q. Wang & J.Z. Yi*	248
Study on the electromagnetic shielding efficiency of functional layer with reducing iron powder mortar *Y.J. Lu & Y. Gao*	257
High temperature oxidation characteristics of alloy steel in mixed C_2H_5OH-H_2O atmosphere *G.X. Cheng, C.M. Zheng & Q.C. Tian*	263
Design and experimental study of the metal rubber for elastic coupling *C.H. Lu, Y.Z. Zhang, L.R. Zhao & G. Tian*	271
Nonlinear dynamic characteristics of cylindrical dielectric elastomer actuator *L.J. Sun & J. Li*	276
Potential application of high-performance polymers in the oil and gas drilling engineering *G.F. Zhao & R.Y. Wang*	284

Chapter 3: Nanomaterials and Biomaterials

Electrospun the fifth generation solid dispersion of ibuprofen based on a hydrophilic polymeric excipient *T.B. Ning, S.R. Guo, W.L. He, W.H. Zhou, M.L. Wang & D.G. Yu*	293

Tunable band gap of graphene/graphene/arsenene van der waals heterostructures 299
D.Z. Zhou, J.C. Dong, X. Zhang, D.D. Han & Y. Wang

Improvement effect of coconut chaff on coastal salinity soil in the Hainan province 304
X. Deng, C.Y. Wu, G.S. Yang, Y. Li & Q.F. Li

Study on photodegradation of chlortetracycline hydrochloride by composite 310
N. Wen, S.X. Li, L. Wang & Y.G. Lv

Facile preparation of open-mouthed hollow SiO_2 microspheres 315
W.H. Zhang, J.S. Hu, A.J. Hou & Y.H. Cao

Surface modification and electrochemical properties of graphene cathode catalysts 319
X. Wang, X.D. Hou, Y. Ren, D.W. Deng & G.F. Zhang

In-situ generation and analysis of cellulose/silver nanoparticle composite filter paper with the assist of ultrasonic atomization 324
C.F. Shi, Z. Jin, H. Ma, M.W. Li, J.R. Ji & X.W. Zhang

Extraction optimization of antibacterial substances from *Cortex Phellodendri* against *Vibrio parahaemolyticus* 331
Y.A. Li, H.F. Zheng, S.Y. Guo, Z.Y. Zhang & L. Guo

Preparation of Au nanospheres/TiO_2 complexes and their photocatalytic performance of H_2 337
D.W. Zheng, Y.L. Zou, J.F. Liu, X. Zhang, S.Y. Wang & S.Z. Kang

One-pot synthesis of fulvic acid and cationic polyelectrolyte PDDA modified Fe_3O_4 nanoparticles for oil-water separation of hexadecane in water emulsions 343
Y.X. Cai, T.W. Mi, X.L. Ma & W. Wu

Preparation and properties of Eu^{2+} doped strontium phosphate blue phosphor by Sol-gel method 351
L.Q. Wu, Y. Wang, S.Y. Guan & J.R. Li

Multiple biotherapy effects of salidroside on tumors 361
X.P. Wang, D.Y. Yuan, W.H. Li & Y. Tian

Effects of spraying indonesian BioSilAc foliar fertilizer on growth and quality of vegetables in red soil regio 366
D.F. Huang, L.M. Wang & Y. Li

Chapter 4: Civil Materials and Sustainable Environment

Application of geological modeling constrained by body of volcano rock in Xushen gas field 373
Y. Xu

Research on government supervision mechanism for resource utilization of construction and demolition waste based on evolutionary game 377
D.D. Liu & L. Qiao

Experimental research on performance of interlayer bonding material for asphalt thin overlay 383
F.J. Bao

Effects on concrete mechanical performances by different replacement solutions of recycle coarse aggregates 389
L. Yu, S.Z. Lv & S.X. Zhang

Development of boron doped diamond sensor for heavy metal detection in water H.Z. Li, X. Yu, L. Pang, J.X. Pei & X.J. Liu	395
Application of orthogonal design method in analysis of injected water composition of whole formation model N. Jia	401
Velocity distribution in the annular gap between the walls of train and tunnel S.X. Sun, Y. Li, Y.X. Zhang & G.L. Li	406
Preparation of disodium terephthalate as intermediate of MOFs Z.W. Wang & L. Li	415
Microstructure degradation prompts durability reduction in a nickel-based single crystal superalloy C.G. Liu, Z.J. Sun, P.Y. Liu, H. Zhang, B. Shen, G. Yang, H.W. Wang, S. Zheng, Z.Y. Yang & Q.H. Wu	421
Straw bale buildings: Experiments, setbacks, and potentials in China M.X. Xie	428
Rock stability analysis for underground powerhouse of pumped storage power station based on block theory J.Y. Li, J.L. Guo, H.N. Liu & R.C. Xu	431
Analysis of influencing factors of Polymer organo chromium gel crosslinking density X. Zhao	437
Influence of SMA brace arrangement on seismic performance of frame structure Z.L. Yang, J.L. Liu & J.Q. Lin	443
Evaluating the spatial structure of a Pinus tabuliformis plantation using weighted Voronoi diagrams X. Zheng, J.Y. Li & F. Yan	449
Study on the technology of vacuum membrane distillation for desalination J. Huang, S.W. Cai, X.C. Zhang & H.L. Zhao	455
The management and control system's establishment of "three lines and one list" L.Y. Lv, C.P. Jin, Y. Xu, Y.F. Zhang & Y. Li	463
Research on steel structural detection based on ultrasonic phased array and infrared thermal imaging technology M.Y. Tian, Y. Ni, X.H. Chen, D.H. Hao, P.P. Zhu & Z.P. Wang	467
Study on the temperature-regulating effect of phase-change asphalt mixture J.X. Zhang, Y.G. Zhang, Y. Hou & Y.C. Zhuo	473
Study and application of water flooding development law in Daqing A oilfield B.F. Wu	481

Chapter 5: Electrochemical Valuation, Fracture Resistance, and Assessment

Structural dynamic analysis of a box-type launch system under different support modes X.X. Liu, B. Li, J.S. Liu & H. Liu	487
Servo feedforward compensation control of micro-texture machine tool based on ISOA Z.M. Wang, G.Q. Wu, J.F. Mao & K. Hu	493

An unbiased expression for calculating net power delivered by a directional coupler *J.Y. Li*	504
Simulation analysis on mechanical characteristics of rocker arm ground derrick joint *F.H. Meng, Y.J. Xia, Y. Ma, P. An & L.J. Sun*	509
Failure analysis and optimization of the light rail vehicle folding coupler in Hong Kong *T. Xu, J.G. Hu, W.J. Huang, W.B. Bao & D. Zhang*	518
Dynamic analysis of an impact-vibration system with multiple asymmetric rigid stops *Y. Yang*	525
Effect of cathode/anode area ratio on galvanic corrosion of TA1 titanium/316 stainless steel couples *S.Y. Guo, Z.D. Liu, B.K. Li, L. Chen, H.R. Ma & Y. Li*	532

Chapter 6: Designs Related to Materials Science and Engineering

Electromagnetic energy conversion technology and its application in modern engineering *H.L. Nie, W.F. Ma, Z.Q. Yu, J.J. Ren, K. Wang, J. Cao, W. Dang, T. Yao, X.B. Liang & K. Wang*	547
Stability analysis of high steep slope on the left side of the flood discharging tunnel exit of Wudongde hydropower station *H. Zhou, Z.J. Wang, D.B. Chen & Q.X. Cao*	554
Response characteristics of a sandwich plate with visco-elastomer core and periodically supported masses under random excitation *Z.G. Ruan & Z.G. Ying*	562
Application of rock slice material in early stage of geological experiment *Y. Liu*	570
Numerical investigation of superhydrophobic surface-induced drag reduction over NACA0012 *T. Kouser, Y.L. Xiong & D. Yang*	577
Evaluation of the model and index for durability of concrete during construction *Y. Yang, L. Yu, R.Y. Fang & Z.J. Zhao*	583
Research on equivalent construction technology of stealth coating in RCS measurement *Y.G. Xu, K. Fan, X.B. Wang, W. Gao & Y. Zhang*	589
Optimal control of vehicle semi-active suspension considering time delay stability *G.H. Yan & H.P. Shi*	595
Stability optimization of reticulated shells based on joint stiffness parameters *C.Y. Dun, B.X. Guan, Y. Guo, J.L. Wang & J.H. Sun*	602
Research on classification method of surface defects of hot rolled strip *K. Hu, G.Q. Wu, Z.H. Hu & Z.M. Wang*	610
Numerical study on multi-physics field in heating furnace with intermediate radiator *T.C. Jiang, W.J. Zhang & S. Liu*	616
Delamination failure analysis of aircraft composite material opening siding *J. He, F.L. Cong, L. Lin & B.T. Wang*	623
The super elastic strain sealing technology and material for tubing and casing connection in natural gas well *X.H. Wang, Y.P. Lv, J.D. Wang & B.C. Pan*	629

Study on structural design and parametric modeling of saddle of transmission line material ropeway *X.Y. Xin, Y.F. Wang, C. Liu, J. Qin, Z.S. Liu & F. Peng*	635
Analysis of influencing factors of water injection capacity in low permeability reservoir *M.G. Tang, Q.H. Liu, G.Q. Xue, P.R. Wang & H.M. Tang*	642
Low voltage multiloop series arc fault detection based on deep recurrent neural network *Q.F. Yu, W.H. Lu & Y. Yang*	647
Optimization strategy for power sellers to purchase and sell electricity considering use behavior *Z.K. Tan, Y.Q. Wang, J. Zhao, B. Liu & M Liu*	653
Theoretical studies on the electronic structure and optical properties of $NaTaO_3$ doped with different concentrations of cerium and nitrogen *G.W. Pang, D.Q. Pan, C.X. Liu, L.Q. Shi, X.D. Wang, L.Z. Liu, J.B. Liu, L. Ma, L.L. Zhang & B.C. Lei*	664
Research on shale gas fracturing technology *X.T. Gao*	675
Dynamic characteristics analysis and pressure regulating system research of a tire *B. Zheng, H. Li, Y.C. Jiang & X.C. Zhang*	680
Experimental analysis of energy absorbing effect of elastomeric energy absorbing device for suspension type crossing frame *F.H. Meng, Y.J. Xia & Y. Ma*	687
Effect of alloy elements on corrosion resistance of low alloy steel in marine atmosphere *G.P. Liu, L.C. Yan, K.W. Gao, C.Y. Lang & Z. Ji*	699
Game testing based on artificial intelligence emotional cognition *J.H. Cheng & J.Q. Li*	704
Nonlinear dynamic characteristics of mechanical vibration system with multi-gap asymmetric elastic constraint *F.W. Yin, G.W. Luo & D.Q. Liu*	710
Kinematic and dynamic simulation of landing gear retraction mechanism in aircraft *Z.J. Sun, J.T. Dai & L. Han*	718
Experimental research on the forming process of large-curvature composite honeycomb sandwich structure *K.X. He, Q.Y. Qiu & H.J. Ye*	725
Deformation behavior of strip steel matrix under shot blasting impact descaling *S. Wang, H.G. Liu, R.C. Hao, Z.X. Feng, X.C. Wang & Y.Z. Sun*	731
Author index	739

Preface

IWMSE 2021 [The 7th Annual International Workshop on Materials Science and Engineering] has been held on May 21st-23rd, 2021 (Virtual Conference) due to the COVID-19 situation and travel restriction. IWMSE 2021 has been converted into a virtual conference via the Tencent Meeting platform.

IWMSE 2021 hopes to provide an excellent international platform for all the invited speakers, authors, and participants. The conference enjoys a wide spread participation, and we sincerely wish that it would not only serve as an academic forum, but also a good opportunity to establish business cooperation. Any paper and topic around materials science and engineering would be warmly welcomed.

IWMSE 2021 proceeding tends to collect the most up-to-date, comprehensive, and worldwide state-of-art knowledge on materials science and engineering. All the accepted papers have been submitted to strict peer-review by 2-4 expert referees, and selected based on originality, significance and clarity for the purpose of the conference. The conference program is extremely rich, profound and featuring high-impact presentations of selected papers and additional late-breaking contributions. We sincerely hope that the conference would not only show the participants a broad overview of the latest research results on related fields, but also provide them with a significant platform for academic connection and exchange.

The Technical Program Committee members have been working very hard to meet the deadline of review. The final conference program consists of 108 papers divided into 6 sessions. The proceedings would be published in a volume by CRC Press/Balkema (Taylor & Francis Group).

We would like to express our sincere gratitude to all the TPC members and organizers for their hard work, precious time and endeavor preparing for the conference. Our deepest thanks also go to the volunteers and staffs for their long-hours work and generosity they've given to the conference. The last but not least, we would like to thank each and every of the authors, speakers and participants for their great contributions to the success of IWMSE 2021.

IWMSE 2021 Organizing Committee

Committees

Chair
Prof. Dr. Domenico Lombardo, Consiglio Nazionale delle Ricerche (CNR-IPCF) of the Department of Chemical Sciences and Materials Technologies, Italy
Prof. Ke Wang, East China University of Technology (ECUT), China

Editor
Prof. Dr. Domenico Lombardo, Consiglio Nazionale delle Ricerche (CNR-IPCF) of the Department of Chemical Sciences and Materials Technologies, Italy
Prof. Ke Wang, East China University of Technology (ECUT), China

Technical Program Committee
Prof. Engr. Shuaib Ahmad, NED University of Engineering and Technology, Pakistan
Prof. Osman Adiguzel, Firat University, Turkey
Assoc. Prof. Tehmina Ayub, NED University of Engineering & Technology, Pakistan
Dr. Gulsen AKDOGAN, Erciyes University, Turkey
Dr. Jeng, Shiang-Cheng, Taipei City University of Science and Technology, Taiwan
Assoc. Prof. Emre Burcu Özkaraova, Ondokuz Mayis University, Turkey
Prof. Kaouther Laabidi, University of Tunis Tunisia / University of Jeddah University, Saudi Arabia
Assoc. Prof. Gholamreza Khalaj, Islamic Azad University, Saveh Branch, Iran
Prof. Mahmoud Abdel-Shafy Elsheikh, Menoufia University, Egypt
Assoc. Prof. Zhanat Umarova, M.Auezov South Kazakhstan University, Kazakhstan
Dr. Woohyun Kim, LIG Nex1 Co. Ltd., Republic of Korea
Prof. Madalina Dumitriu, University Politehnica of Bucharest, Romania
Assoc. Prof. Mohamed M.S. Wahsh, National Research Centre (NRC), Egypt
Assoc. Prof. Benbahouche Lynda, EBT Department, Faculty of technology. UFAS Setif1, Algeria
Dr. Hedayat Omidvar, National Iranian Gas Company (NIGC), Iran
Prof. Chi-Wai Chow, National Chiao Tung University, Taiwan
Dr. Toufik Bentrcia, University of Batna 2, Algeria
Prof. Faheem Uddin, University of Karachi, Pakistan
Dr. Marcela-Elisabeta BARBINTA-PATRASCU , University of Bucharest , Romania
Prof. Nasr-Eddine CHAKRI, Badji-Mokhtar, Algeria
Prof. M.G.H.Zaidi, G. B. Pant University of Agriculture and Technology Pantnagar Uttarakhand, India
Prof. Chandraprabha M N, M S Ramaiah Institute of Technology, India
Prof. Angela Amphawan, Sunway University, Malaysia
Dr. Hao Yi, Chongqing University, China
Dr. Yasin POLAT, Nevsehir Haci Bektas Veli University, Turkey

Editors

Prof. Domenico Lombardo
Affiliation: Consiglio Nazionale delle Ricerche (CNR-IPCF) of the Department of Chemical Sciences and Materials Technologies, Italy

Editor biography:
Prof. Dr. Domenico Lombardo is a physicist and materials engineer in the Consiglio Nazionale delle Ricerche (CNR-IPCF) of the Department of Chemical Sciences and Materials Technologies (ITALY). He is performing both experimental and theoretical research in the soft matter, drug delivery and biotechnology fields, in collaboration with the industry, major Italian and European research institutions, and international partners. His research activity is focused on:

- Study of the self-assembly (and supramolecular) processes of nanostructured and hybrid (inorganic-inorganic) bio-materials.
- Study of the soft interaction between nanocarriers and biological (and model) membranes.

He is supervisor of a research unit at the CNR-IPCF (Italy), lecturer in several post-graduate courses at University. He has been leading several major research projects that are related to biotechnology, material science and engineering. He has been involved in instrumental development projects within various European laboratories and synchrotron radiation facility (such as Elettra in Trieste-Italy and LURE in Orsay-France).

He has been involved in the organization, in the scientific committees and as invited speaker of several workshops, seminars, and national and international conferences. He repeatedly participated at various expert committees at the national and international level. He regularly acts as a consultant, reviewer and editorial board member in various peer-reviewed international scientific journals. He is co-author of over 100 publications of which more than 70 on international peer-reviewed and indexed journals and a number of other publications (conference proceedings, book chapters etc.).

Prof. Ke Wang
Affiliation: East China University of Technology (ECUT), China

Editor biography:
Prof. Ke Wang obtained his Ph.D degree in microelectronics and solid state electronics from HUST. His expertise covers magnetic/optical/electronic thin films and devices. He had been working in the UK as research fellow for several years. Before that, he received Humboldt research award to carry out his work in RWTH-Aachen. Currently he is Head of the School of Mechanical and Electronic Engineering, ECUT. Prof. Wang serves as Editorial Board Member for over ten scientific journals and has published more than 70 peer reviewed papers.

Invited Talk

Invited Talk I: Prof. Rongguo Zhao
Affiliation: Professor, Xiangtan University, China

Biography: Prof. Rongguo Zhao, born in 1973, is a doctor, professor and doctoral supervisor. Deputy dean of College of Civil Engineering and Mechanics, Xiangtan University, deputy director of Institute of Rheological Mechanics, deputy director of Key Laboratory of Dynamics and Reliability of Engineering Structures of College of Hunan Province, young backbone teacher of Hunan Province, executive director of Applied Mechanics Society of Hunan Province, member of Mechanical Engineering Terminology Approval Committee of National Science and Technology Terminology Approval Committee, senior member of Chinese Mechanical Engineering Society, senior member of Chinese Materials Research Society. From 1994 to 1998, majored in Mechatronic Engineering, College of Mechanical Engineering, Xiangtan University, with a bachelor's degree in engineering. From 1998 to 2001, majored in general mechanics and fundamental mechanics in the Institute of Rheological Mechanics, Xiangtan University, specializing in the design and failure theory of viscoelastic materials, with a master's degree in engineering. From 2001 to 2004, he studied and worked in Engineering Mechanics major of China Academy of Engineering Physics, engaged in theoretical, experimental and numerical simulation research in engineering structure design, mechanical analysis and environmental assessment, and obtained a doctor's degree in engineering. From 2004 to now, he has been engaged in teaching and scientific research in College of Civil Engineering and Mechanics and Institute of Rheological Mechanics of Xiangtan University. Presided over or participated in 12 national and provincial scientific research projects, Presided over 4 enterprise research projects. He has edited two monographs as *Theoretical Mechanics, Rheological Theory of Porous Materials and Polymeric materials and Its Applications*, and participated in one monograph as *Mechanical Engineering Terms*. He has published more than 80 academic papers, including 43 papers indexed by SCI and EI. He won the National Xu Zhilun Excellent Teacher Award for Mechanics and the Third China Youth Award for Rheology.

Speech Title: Optimization design and fatigue life analysis of damping conical rubber spring for rail vehicle

Invited Talk II: Prof. Ping Wang
Affiliation: Associate Professor, Dalian University of Technology, China

Biography: Prof. Ping Wang, School of Energy and Power Engineering, Dalian University of Technology, PhD, associate professor, doctoral supervisor. She has published more than 50 academic papers and obtained 3 domestic invention patents.
Educational experience:

(1) From September 2000 to March 2008, Dalian University of technology, doctor
(2) From September 1986 to June 1989, Dalian University of technology, master
(3) From September 1982 to July 1986, Dalian University of technology, bachelor

Work experience:

(1) From May 2002 to now, associate professor, Dalian University of technology
(2) From July 1989 to August 2000, associate professor, Department of automotive engineering, Liaoning University of technology

Research field: seawater desalination, etc. Research direction: key technology of solution separation, mechanism of complex flow and enhanced heat and mass transfer in porous media. She has published more than 50 academic papers and obtained 3 domestic invention patents.

Speech Title: Effect of Parameters in Porous Medium Evaporator on the AGDD Performance

Keynote Speaker

Keynote Speech I: Prof. Jie Huang
Affiliation: Professor, Southwest University, Chongqing, China

Biography: Prof. Jie Huang, IEEE Member, male, is currently a professor and PhD Tutor in College of Engineering and Technology, Southwest University, Chongqing, China. He received the Ph.D degree in Microelectronics and Solid-State Electronics from Institute of Microelectronics of the Chinese Academy of Sciences in 2011. In March 2016, he conducted a one-year visiting study on compound semiconductors and monolithic THz circuits with Professor Jan Stake in the Terahertz and Millimeter Wave Laboratory of Chalmers University of Technology, Sweden. He has been mainly engaged in the research of compound semiconductor microwave devices and monolithic integrated circuits, radio frequency/microwave passive devices and modules, metamaterials, and microwave dielectric sensor. Recent research mainly focuses on measuring the complex permittivity and permeability of solid, liquid and powder materials by use of microwave resonant method, and sensing detection by using electromagnetic resonant absorption of meta-surface absorber, such as detecting water content of emulsified oil and identifying edible oil species. He has presided over 10 projects, including 2 projects from the National Natural Science Foundation of China. He currently have published more than 40 article in the microwave field, such as IEEE Trans. Microw. Theory Tech., IEEE Sensors J., Sensors, J. Electromagnet Wave, Microw. Op tic Tech. Lett., Prog. Electromagn. Res. C(M, Letters), Opt. Quant. Electron., Chin. Phys. B, and he is also a guest reviewer of these microwave journals.

Speech Title: Microwave Electromagnetic Induction Effect in the Near Field and Its Material Characterization

Keynote Speech II: Prof. Xizhong An
Affiliation: Professor, Northeastern University (NEU), China

Biography: Prof. Xizhong An received his Ph.D degree from University of Science and Technology Beijing in 2002 and worked as a postdoctoral fellow in the University of New South Wales, Australia from 2003-2005. Currently he is the full professor and director of Institute of Particle Technology in School of Metallurgy, Northeastern University. Prof. An's research interests include numerical and physical studies on particle packing, powder processing, granular matter, powder metallurgy, additive manufacturing, material design and optimization, etc. In recent years, he has undertaken more than 30 projects from overseas, Chinese government, local government, and enterprises. More than 140 papers were published, including one book chapter and >100 SCI journal papers. 24 patents were authorized. More than 20 plenary/keynote/invited presentations were given in international conferences. He has been the member of editorial board for many domestic and international journals and the director of many academic organizations.

Speech Title: Numerical Investigation on the Construction of Amorphous and Crystalline Granular Matter

Keynote Speech III: Prof. Domenico Lombardo

Affiliation: Professor, Consiglio Nazionale delle Ricerche (CNR-IPCF) of the Department of Chemical Sciences and Materials Technologies, Italy

Biography: Dr. Domenico Lombardo is a physicist and materials engineer in the Consiglio Nazionale delle Ricerche (CNR-IPCF) of the Department of Chemical Sciences and Materials Technologies (ITALY).
He is performing both experimental and theoretical research in the soft matter, drug delivery and biotechnology fields, in collaboration with the industry, major Italian and European research institutions, and international partners. His research activity is focused on:
- Study of the self-assembly (and supramolecular) processes of nanostructured and hybrid (inorganic-inorganic) bio-materials.
- Study of the soft interaction between nanocarriers and biological (and model) membranes.

He is supervisor of a research unit at the CNR-IPCF (Italy), lecturer in several post-graduate courses at University. He has been leading several major research projects that are related to biotechnology, material science and engineering. He has been involved in instrumental development projects within various European laboratories and synchrotron radiation facility (such as Elettra in Trieste-Italy and LURE in Orsay-France).
He has been involved in the organization, in the scientific committees and as invited speaker of several workshops, seminars, and national and international conferences. He repeatedly participated at various expert committees at the national and international level. He regularly acts as a consultant, reviewer and editorial board member in various peer-reviewed international scientific journals. He is co-author of over 100 publications of which more than 70 on international peer-reviewed and indexed journals and a number of other publications (conference proceedings, book chapters etc.).

Speech Title: Nanocarriers in Drug Delivery Processes: Interactions and Stability

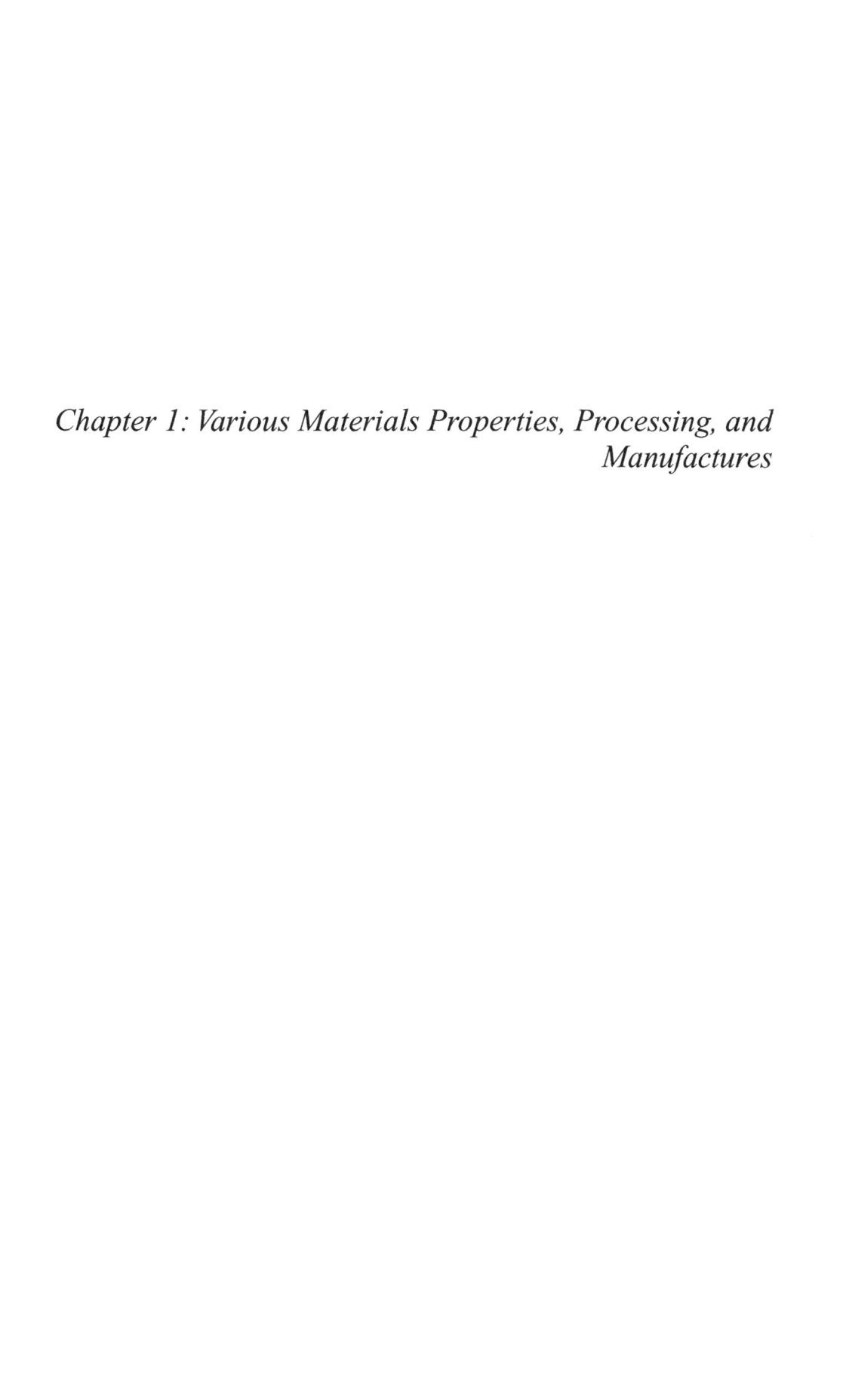

Chapter 1: Various Materials Properties, Processing, and Manufactures

Preparation and ferroelectric properties of strontium doped BaTiO3 nanotubes

Y.N. Sun, M.E. Ma, H. Li, T.Y. Yu & X.Y. Deng
College of Physics and Materials Science, Tianjin Normal University, Tianjin, China

ABSTRACT: In this paper, firstly, TiO_2 nanotubes were prepared by anodic oxidation method, and then $BaTiO_3$ nanotubes were prepared by hydrothermal synthesis with Ba $(OH)_2 8H_2O$. A series of strontium doped $BaTiO^3$ nanotubes with concentration gradient were prepared by adding different concentrations of $Sr(OH)_2 8H_2O$ in hydrothermal synthesis. The microstructure, crystal structure and ferroelectric properties of the samples were analyzed by using su8010 scanning electron microscope, D/max-2500 X-ray diffraction tester and TF-2000 analyzer. The results show that the morphology of the samples is nanotube structure and perovskite structure after annealing treatment. With the increase of strontium concentration, the remanent polarization intensity and coercivity field of the samples first increase and then decrease. When the concentration of strontium is 12wt%, the remanent polarization intensity and coercivity field reach the maximum value, reaching $0.0961 \mu C/cm^2$ and $24.93 kV/cm$, respectively.

1 INTRODUCTION

$BaTiO_3$ is a widely used ferroelectric material with typical perovskite structure [1]. It is widely used in advanced electronic equipment and power system due to its high storage density and high energy. Based on these characteristics, $BaTiO_3$ materials have been widely concerned [2–5]. Due to the rapid development of electronic components in the direction of high integration, high precision, high reliability, multi-function and miniaturization, higher requirements are put forward for the performance of $BaTiO_3$ materials [6–8]. The residual polarization strength of the pure phase $BaTiO_3$ nanotubes prepared by the hydrothermal method is low, and the ferroelectricity can be improved by doping other elements [9–11]. The concentration of strontium doped in this paper will have an important effect on the microstructure and ferroelectric properties of $BaTiO_3$ nanotube thin films. The doping concentration of strontium affects the Curie temperature, which in turn affects the synthesis of ferroelectric (tetragonal) phase or paraelectric (cubic) phase. The addition of strontium also affects the grain size of $BaTiO_3$, resulting in the change of grain boundary area and the ferroelectric properties of $BaTiO_3$ [12–13].

In this paper, strontium doped $BaTiO_3$ nanotubes were prepared by hydrothermal method. A series of strontium doped $BaTiO_3$ nanotubes with different strontium concentrations were prepared by changing the concentration of $Sr(OH)_2 \cdot 8H_2O$ in hydrothermal synthesis. The microstructure, crystal structure and ferroelectric properties of the composites were studied by using su8010 scanning electron microscope, D/max-2500 X-ray diffraction tester and TF-2000 analyzer.

2 EXPERIMENTAL SECTION

Preparation method of strontium doped $BaTiO_3$ nanotubes. The preparation process of $BaTiO_3$ nanotubes is shown in Figure 1.

2.1 Preparation of TiO₂ nanotube arrays (TiNTs)

Titanium (industrial grade, purity 99.4%), other reagents used are analytical pure. The pure titanium slices of 20 mm × 20 mm × 0.4 mm were ultrasonic cleaned in acetone cleaning agent, isopropanol, absolute ethanol and deionized water for 15 min, and then dried in an oven at 80° for standby. Prepare the electrolyte, dissolve 0.8g of ammonium fluoride in 15ml deionized water, and mix the ammonium fluoride solution with 150ml of ethylene glycol to prepare the required electrolyte. Titanium sheet was used as anode and platinum plate as cathode to prepare TiO_2 nanotube arrays in electrolyte. The results show that the distance between electrodes is 2cm, the oxidation voltage is 50V and the anodizing time is 4h. After the reaction, the titanium sheet was taken out and soaked in anhydrous ethanol and deionized water for 10 min respectively, and then the titanium sheet was dried in an oven at 80°C for standby.

2.2 Preparation of strontium doped BaTiO₃ nanotubes

A certain mass of Ba $(OH)_2 \cdot 8H_2O$ and $Sr(OH)_2 \cdot 8H_2O$ were weighed with an electronic balance and poured into a polytetrafluoroethylene (PTFE) hydrothermal reactor. A measuring cylinder was used to measure 70 ml of deionized water. The mixed solution of Ba $(OH)_2 \cdot 8H_2O$ and $Sr(OH)_2 \cdot 8H_2O$ with a concentration of 0.1 mol /L was prepared in the reactor (see Table 1 for specific parameters) The nanotubes were put into the groove of the scaffold, and together they were gently put into the polytetrafluoroethylene reactor containing hydrothermal reaction solution to ensure the full contact between the TiO_2 nanotubes and the reaction solution. Then put the sealed hydrothermal reactor into the high-temperature sintering furnace, raise the temperature by 180°C at the rate of 2°C / min, keep it for 2h, and take it out after the furnace cools down to room temperature. The samples were soaked in absolute ethanol and deionized water for 10 min, and then dried in an oven at 80°C. The treated samples were put into a crucible, annealed in muffle furnace, and heated to 450°C at a rate of 3°C/ min for 3h. After cooling down to room temperature, strontium doped $BaTiO_3$ nanotube samples were obtained and sealed.

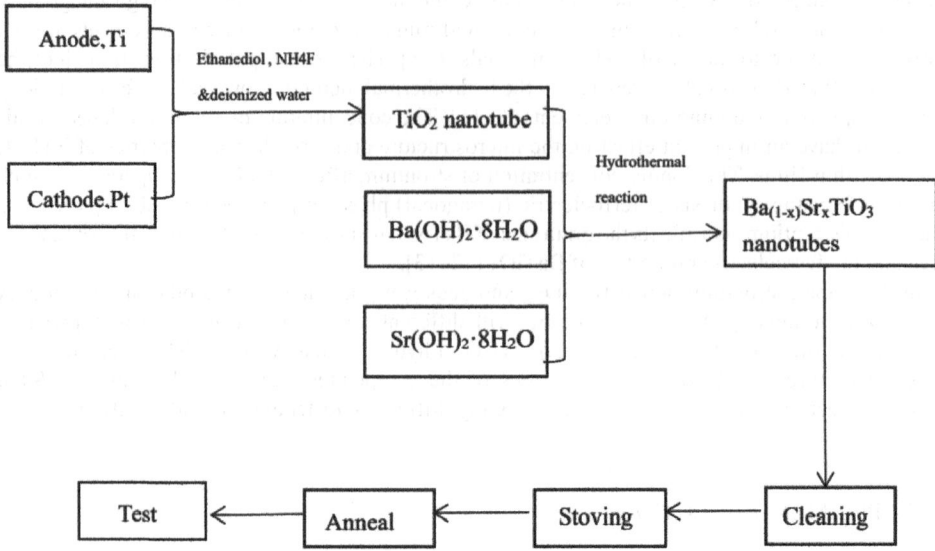

Figure 1. The strontium doped BaTiO₃ preparation process.

Table 1. Ba$_{(1-x)}$Sr$_x$TiO$_3$ nanotubes preparation process.

Number	X=?	Ba(OH)$_2$ concentration [g]	Sr(OH)$_2$ concentration [g]	Doping amount [wt%]
1	0	2.20820	0	0
2	0.08	2.03156	0.14883	8
3	0.10	1.98740	0.18603	10
4	0.12	1.94323	0.22324	12
5	0.14	1.89907	0.26044	14
6	0.16	1.85490	0.29765	16

3 RESULTS AND DISCUSSION

3.1 Surface morphology analysis of strontium doped BaTiO$_3$ nanotubes

Scanning electron microscopy (SEM) was used to observe the samples with different doping concentrations. It was observed that the samples were all nanotube structures after hydrothermal reaction. Compared with undope BaTiO$_3$ nanotubes, strontium doped BaTiO$_3$ nanotubes have thicker wall, smaller inner diameter and compact structure, but the surface becomes more uneven. With the increase of doping amount, the nanotubes are blocked and some of them become columnar. This is because with the increase of Sr(OH)$_2$·8H$_2$O doping amount, some of Sr(OH)$_2$·8H$_2$O can not participate in the reaction completely, which will form particles attached to the surface of nanotubes (Figure 2).

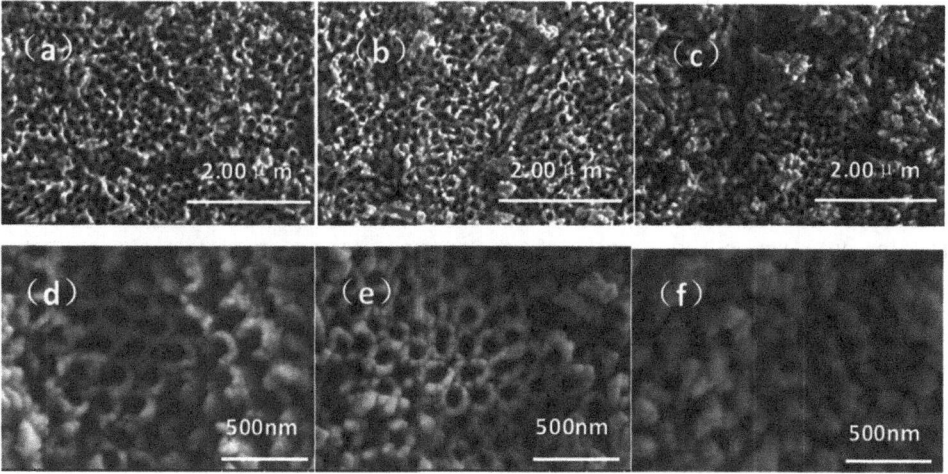

Figure 2. SEM images of strontium doped BaTiO$_3$ with different concentrations and the compositions are as follows: (a) undoped BaTiO$_3$; (b) BaTiO$_3$ with 12wt% Sr; (c) BaTiO$_3$ with 16wt% Sr; (d) undoped BaTiO$_3$; (E) BaTiO$_3$ with 12wt% Sr; (f) BaTiO$_3$ with 16wt% Sr.

3.2 The X-ray diffraction (XRD) analysis of strontium doped BaTiO$_3$ nanotubes is as follows

Figure 3 shows the X-ray diffraction pattern of 10° to 80° and the doping contents of Sr(OH)$_2$·8H$_2$O are 0wt%, 8wt%, 10wt%, 12wt%, 14wt% and 16wt%, respectively. The results show that the angles of diffraction peaks of six samples are basically the same under different conditions. It can be seen from the figure that all samples with different doping concentrations can form pure perovskite structure. This shows that all the doping amount of strontium is completely into the crystal of the ceramic.According to JSPDF standard card, the 48 degree diffraction peak belongs to TiO_2. It can be inferred that TiO_2 is completely transformed into $BaTiO_3$ with perovskite structure. The peak of 22 degree corresponds to (100) crystal face, 31 degree peak corresponds to (110) crystal face, 39 degree peak corresponds to (111) crystal face, 45 degree peak corresponds to (200) crystal face, 56 degree peak corresponds to (211) crystal face, peak near 66 degree and 71 degree corresponds to (220) and (212) crystal face. These peaks are basically consistent with the JSPDF standard card, indicating that the prepared sample is very pure.

Figure 4 shows an enlarged view of the (200) peak at 43–48 degrees. Because the radius of strontium ion is smaller than that of barium ion,it can be seen that according to the Bragg formula $d(hkl)=n\lambda/(2\sin\theta)$, the unit cell parameter $d(hkl)$ strain is small and $2\sin\theta$ value is correspondingly increased, so the peak direction is right. The success of Sr doping is further proved. When the doping amount of $Sr(OH)_2 \cdot 8H_2O$ was 12wt%, the peak of (200) shifted to the right. When the doping amount of $Sr(OH)_2 \cdot 8H_2O$ was further increased, the(200) peak did not move to the right, but began to move to the left. This is because with the increase of $Sr(OH)_2 \cdot 8H_2O$ doping amount, some $Sr(OH)_2 \cdot 8H_2O$ can not participate in the reaction completely, which will form particles attached to the surface of nanotubes.

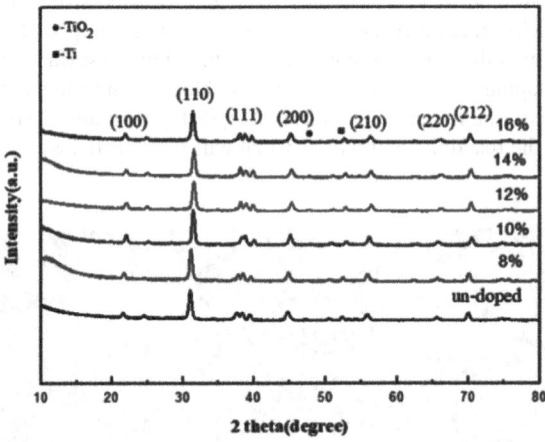

Figure 3. X-ray diffraction pattern of strontium doped $BaTiO_3$ *nanotubes*at 10–80 degrees.

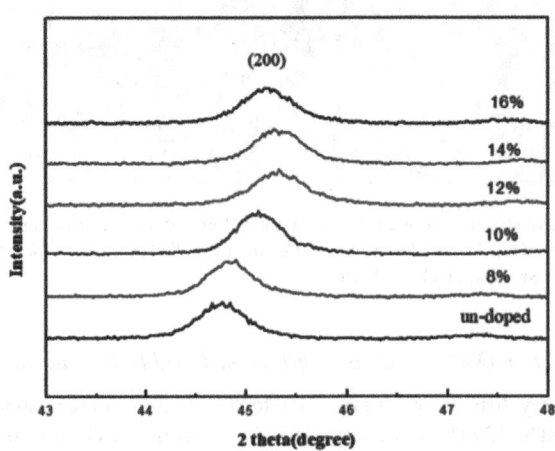

Figure 4. X-ray diffraction pattern of strontium doped $BaTiO_3$ *nanotubes* at 43–48 degrees.

3.3 Ferroelectric properties of strontium doped $BaTiO_3$ nanotubes

Figure 5 shows the measured hysteresis loops of strontium doped $BaTiO_3$ nanotubes at room temperature at a frequency of 1 Hz and an applied voltage of 25 v. P-E hysteresis loops confirm the ferroelectric properties of strontium doped BaTiO3 nanotubes. There are two current peaks in the first quadrant and the third quadrant of the curve, which are the characteristics of typical ferroelectrics, indicating that the samples have good ferroelectric properties. With the increase of doping concentration, the residual polarization first increases and then decreases. This may be because strontium ions will replace barium ions at the appropriate doping concentration, but if the doping concentration is too high, the substitution of titanium ions will lead to the decrease of ferroelectric properties.

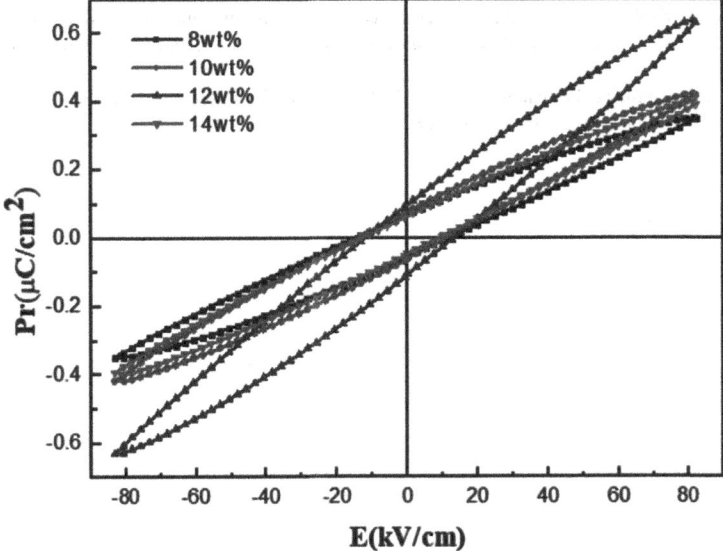

Figure 5. Hysteresis loops of strontium doped $BaTiO_3$ nanotubes.

ACKNOWLEDGEMENTS

This project was funded by Tianjin Normal University (Grant No.53H14049).

REFERENCES

[1] CHOI KJ, BIEGALSKI M, LI Y L, et al. Enhancement of ferroelectricity in strained BaTiO3 thin films [J]. *Science*, 2014, 306 5698 : 1005–1009.
[2] FU D, ITOH M, KOSHIHARA S. Crystal growth and piezoelectricity of BaTiO3–CaTiO3, solid solution [J].*Applied Physics Letters*, 2008, 93 (1) :1161.
[3] ZHANG D, BUTTON T W, SHERMAN VO, et al. Effects of glass additions on the microstructure and dielectric properties of barium strontium titanate (BST) ceramics [J]. *Journal of the European Ceramic Society*, 2010, 30 : 407–412.
[4] PARK SE, WADA S, CROSS LE, et al. Crystallographically engineered BaTiO3 single crystals for high-performance piezoelectrics [J]. *Journal of Applied Physics*, 1999, 865:2746–2750.
[5] AGNIHOTRI O P, MUSCA C A, FARAONE L. Topical review : Current status and issues in the surface passivation technology of mercury cadmium telluride infrared detectors [J]. *Semiconductor Science & Technology*, 1998, 13 :839.

[6] LIMMER S J, SERAJI S, WU Y, et al. Template -based growth of various oxide nanorods by sol -gel electrophoresis [J]. *Advanced Functional Materials*, 2002, 12(1):59–64.
[7] CREPALDI E L, SOLERILLIA G J DA A, GROSSO D, et al. Nanocrystallised titania and zirconia mesoporous thin films exhibiting enhanced thermal stability [J]. *New Journal of Chemistry*, 2003, 27: 9–13.
[8] JAIN K, LAKSHMIKUMAR S T. Porous alumina template based nanodevices [J]. *LETE Technical Review*, 2002, 195: 293–306.
[9] TING T H, WUK H. Synthesis and electromagnetic wave-absorbing properties of $BaTiO_3$/polyaniline structured composites in 2–40 GHz [J]. *Journal of Polymer Research*, 2013, 20(5): 127–132.
[10] HUO J, WANG L, YU H. Polymeric nanocomposites for electromagnetic wave absorption [J]. *Journal of Materials Science*, 2009, 44(15): 3917–3927.
[11] ZHU Y F, ZHANG L, NATSUKI T, et al. Facile synthesis of BaTiO3nanotubes and their microwave absorption properties [J].*ACS Applied Materials & Interfaces*, 2012, 4(4): 2101–2106.
[12] Mazon M A, Zaghete MA, Varela E L, et al. Barium Strontium Titanate Nanocrystalline Thin Films Prepared by Soft Chemical Method [J].*Journal of the European Ceramic Society*, 2007, 27(13–15): 3799–3802.
[13] Alaedin A S, Ramli N, Poopalan P, et al. AFM Studly of Multilayer $Ba_{(1-x)}$ Sr_x TiO3 thin films [J] .*Physical Review Letters*, 2010, 3(1):61–68.

Removal of wastewater-dissolved heavy metals by Na-carboxylate polyarylene ether sulfone

Y.N. Wang*, J.R. Zhang & Z.H. Lu
College of Food Science, Shenyang Agricultural University, Shenyang, Liaoning, China

Y. Fu*
Faculty of Light Industry and Chemical Engineering, Dalian Polytechnic University, Dalian, China

B.D. Wei
College of Food Science, Shenyang Agricultural University, Shenyang, Liaoning, China

ABSTRACT: In this paper, a type of polyarylene ether sulfone with sodium carboxylate groups (PAES-C-Na) was synthesized, and its dynamic adsorption capacity toward M^{2+} ions (Pb^{2+}, Cd^{2+}, and Cu^{2+} ions) was investigated. The experimental results showed that the carboxylate groups introduced on the polymer chain and the porous structure of PAES-C-Na make PAES-C-Na a suitable adsorbent for removal of M^{2+} ions in wastewater. PAES-C-Na exhibited the optimal adsorption capacity at pH=5.0. The obtained experimental results were further analyzed via Yoon-Nelson and Thomas models, which could be used to fit the breakthrough curves and thus predict the breakthrough time of the PAES-C-Na column.

1 INTRODUCTION

In recent years, with the rapid development of the modern industry, the enormous discharge of heavy metal ions has increased pollution to a great extent [1]. Hence, the elimination of the heavy metal ions has been of great interest because of their toxicity and non-biodegradability leading to serious environmental problems [2]. Up to date, various processes such as chemical precipitation, activated carbon adsorption, ion exchange, reverse osmosis technique membrane filtration and so forth have been developed to reduce the amount of heavy metal ions in sewage [3–6]. However, several drawbacks including high cost and secondary pollution retarded their pilot-scale application.

On the other hand, the chemical adsorption technique exhibited merits such as easy recovery and low cost and therefore has been regarded as one of the most efficient processes to remove heavy metal ions from wastewater [7,8]. Up to date, many types of adsorbents like activated carbon fibers [9] resins, and bio-sorbents [10] have been investigated. Among the various adsorbents, the preparation of polymer resins has attracted a lot of attention, considering its efficient interaction with heavy metal ions, designable structure and chemical stability even under serious conditions [11,12]. For example, Deng et al. prepared polyethersulfone/activated carbon (PES-AC)blended microspheres and depicted that increasing the ratio of activated carbon (AC) and propyleneglycol (PG) facilitated the adsorption capacity of polyethersulfone/activated carbon (PES-AC) microspheres. Yan et al selected a renewable micro-nanostructured polyethersulfone/polyethyleneimine (PES/PEI) nanofiber membrane as an adsorbent to remove anionic dyes or heavy metal ions in wastewater. The results indicated that PES/PEI nanofiber membranes possessed the attractive adsorption capacity due to the doped sulfone groups. It was also mentioned that the interaction between functional

---

*Y.N. Wang and Y. Fu contributed equally to this work and are co-first authors.

groups of the polymer (carboxyl, sulfonylimidazolyl etc.) and heavy metal ions played an important role in the M^{2+} adsorption [13]. In conclusion, these findings were favorable for the design and synthesis of novel polymer adsorbents.

In our previous study, PAES-C-Na containing both carboxyl and sulfonyl groups was successfully prepared. The porous structure of PAES-C-Na indicated that the as-prepared polymer might be a useful adsorbent for the efficient elimination of heavy metal ions in wastewater [14]. Further research studies on the dynamic adsorption of PAES-C-Na should be interesting. In this work, a deep endeavor was carried out to probe the dynamic adsorption performance of an adsorbent column filled with PAES-C-Na powder in order to determine its industrial application in removing the Pb^{2+}, Cd^{2+}, and Cu^{2+} metal ions from wastewater.

2 EXPERIMENTAL SECTION

2.1 Materials and instruments

4, 4'-bis(4-hydroxyphenyl)-pentanoic acid (DPA), anhydrous potassium carbonate (K_2CO_3), potassium hydroxide (KOH), N,N'-dimethylacetamide (DMAC), dimethyl sulfoxide (DMSO), 4,4'-dichlorodiphenylsulfone (DCDPS), tetrahydrofuran(THF), toluene, NaOH, lead nitrate ($Pb(NO_3)_2$), copper nitratetrihydrate ($Cu(NO_3)_2 \cdot 3H_2O$), cadmium nitratetetrahydrate ($Cd(NO_3)_2 \cdot 4H_2O$), and sodium hydroxide (NaOH) were purchased from the Sinopharm Chemical Reagent (China). Hydrochloric acid (HCl) was purchased from Tianjin Komiou Chemical Reagent Co., Ltd. All reagents are C.P. grade.

2.2 Preparation of PAES-C-Na

PAES-C-Na used as an adsorbent was prepared in the laboratory according to the previous literature [15]. 0.03 mol of K_2CO_3, 0.02 mol of DPA, and 0.01 mol of KOH were added into a three-necked flask which was preloaded with a mixture of toluene, DMAC and DMSO (totally 40 mL, volume ratio was set as 3:2:2). The mixture was mechanically stirred at room temperature for 2 h under the continuous nitrogen flow. Then it was slowly heated to 150°C and refluxed for 3 h in the condenser. The mixture was cooled down to room temperature. 0.02 mol of DCDPS was added into the flask. Subsequently, the mixture was slowly reheated to 150°C and further refluxed for 3 h. After complete separation of toluene and water in the trap, the residue in the flask was heated to 175°C and then it was maintained for 24 h. Upon the completion of the reaction, the mixture was cooled down to room temperature and poured into a 500 mL Erlenmeyer flask. THF and HCl (volume ratio 4:5) were added into the flask and acidified for 12 h. Then, the acidizing fluid was discarded and 400 mL of ionized water was added into flask. A white precipitate was obtained. The white precipitate was thoroughly washed with deionized water and ethanol and then dried at 100°C for 24 h in a vacuum drying oven to obtain PAES-C. PAES-C-Na could be obtained via a 4h alkalization procedure by using NaOH solution (0.4 mol/L) as the alkalization reagent.

2.3 Polymer characterization

The morphology structure of PAES-C-Na was checked by a JSM-7800F field emission scanning electron microscope (SEM). Atomic absorption spectroscopy (AAS) data were obtained from a flame atomic absorption spectrophotometer (Hitachi, model GH-9602).

2.4 Dynamic adsorption tests

All the dynamic adsorption experiments were operated in a series of abbreviated wet packing glass columns as shown in Figure 1. Typically, a tubular glass column with a 6 cm inner diameter and a 12 cm length was used as the adsorption column. The as-prepared PAES-C-Na powder was filled in the column and sandwiched by two layers of skim cotton (with a thickness ca. 2 cm). Both ends of the column were sealed up with a piston. The flow rate of simulation wastewater with a known M^{2+} concentration was precisely controlled by a P066–363T1 peristaltic pump. Samples were taken from the outlet at the fixed time intervals. Then, the concentration of the M^{2+} ions was determined by AAS. The adsorption quantity q_t (mg/g) of the PAES-C-Na can be calculated via the amount of M^{2+} ions adsorbed by the PAES-C-Na per unit mass at saturated adsorption equilibrium. With the fixed V_w and $C[M^{2+}]_0$, the adsorption quantity of the absorbent q_t could be expressed by the following formula:

$$q_t = (C_0 - C_t)\, V_s/m \qquad (1)$$

where m is the mass of the adsorbent filled in the column (g); V_s is the total volume of the wastewater flowing through the column at the saturated adsorption, equal to $V_w \times t$ (L); C_0 is the concentration of M^{2+} ions at the inlet end of the column (mg/L); and C_t is the concentration of M^{2+} ions at the outlet of the column (mg/L).

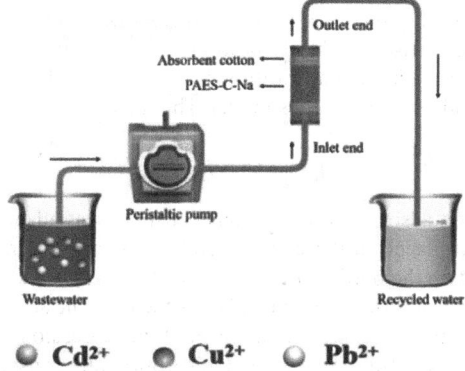

Figure 1. A schematic illustration of dynamic adsorption experiments.

3 RESULTS AND DISCUSSION

3.1 *Structures of polymers*

The SEM image of PAES-C-Na (Figure 2a) illustrated its porous structure, which provided favorable channels and a sufficient surface area for the adsorption of the M^{2+} ions. A SEM image of PAES-C-Na with Cu^{2+} ions after adsorption operation is shown in Figure 2b, demonstrating the stability of the PAES-C-Na porous structure before and after wastewater treatment.

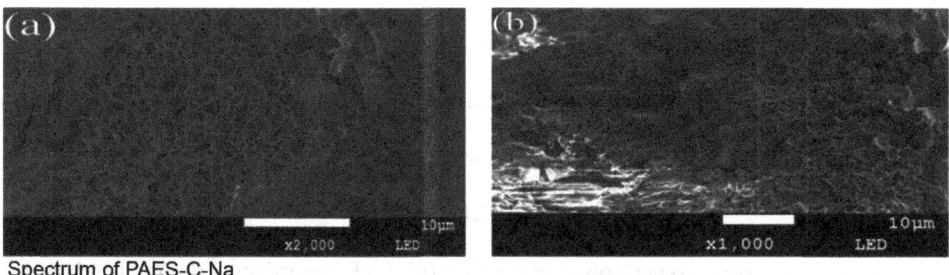

Figure 2. SEM images of PAES-C-Na: (a) before adsorption and (b) after adsorption of Cu^{2+} ions.

3.2 Breakthrough curve experiments

In our previous study, the as-prepared PAES-C-Na showed a good static adsorption capacity for M^{2+} ions. The relationship between the morphology structure and adsorption capacity of PAES-C-Na was well discussed. Generally, it was believed that the good adsorption capacity could be attributed to the porous density, the distribution of pore size and the incorporated functional groups of PAES-C-Na. In addition, a pair electron of O_{2p} in the sulfonyl group and delocalized π bond of three benzene rings could form a large delocalized π bond, arising from the special molecular structure of PAES-C-Na. This large π bond could also improve the adsorption capacity of PAES-C-Na due to its increased chelation probability with the d orbital of M^{2+} ions. Hence, PAES-C-Na could be used in water treatment because of its high mechanical strength, macroporous structures and oxygen-containing organic functional groups.

3.3 Breakthrough curves models analysis (Thomas model)

The models of Thomas and Yoon-Nelson were used to predict the performance and parameters of the fixed-bed column. The Thomas adsorption kinetics model is one of the dynamic adsorption models which could clearly describe the adsorption process of the adsorption column. Hence, the saturation adsorption capacity and adsorption rate constant of PAES-C-Na can be calculated according to the Thomas adsorption kinetics model. The linear form was as follows:

$$\ln(C_0/C_t - 1) = (K_{Th} q_{max} M)/V_w - C_0 t \qquad (2)$$

where q_{max} is the saturation adsorption of M^{2+} ions per unit mass adsorbent at equilibrium, M is the mass of the adsorbent placed in the column, and V_w is the flow rate of wastewater. Linear regression was performed on the outflow time t with $\ln(C_0/C_t-1)$. C_0 is the concentration of M^{2+} ions at the inlet end of the adsorption column and C_t is the concentration of M^{2+} ions at the outlet end of the adsorption column. The slope and intercept of the regression equation could be selected to calculate the saturated adsorption capacity and adsorption rate constants of PAES-C-Na under different conditions for Cu^{2+}, Pb^{2+} and Cd^{2+}.

Under the constant flow rate V_w, the values of the Thomas rate constant K_{Th} greatly decreased in all cases with increasing M^{2+} ion concentration C_0 from 20 to 40 mg/L. On the other hand, the value of q_{max} increased. These findings might be caused by the fact that the positive charge around the active sites of PAES-C-Na increased the repulsion force between M^{2+} ions, resulting in the decline of the adsorption rate. Obviously, more M^{2+} ions in the solution resulted in the increase of the concentration gradient which enhanced the collision probability between the M^{2+} ions and the active sites of PAES-C-Na. On the contrary, when C remained constant, the Thomas rate constant K_{Th} gradually increased with increasing V_w and the saturated adsorption capacity q_0 gradually declined.

3.4 Breakthrough curves model analysis (Yoon-Nelson model)

The Yoon-Nelson model assumed that the adsorption rates were directly related with the breakthrough probability and adsorption probability could be expressed as follows.

$$\ln(C_0/(C_0 - C_t)) = K_{YN} t - \tau K_{YN} \qquad (3)$$

According to the τ value, the equilibrium adsorption can be calculated as follows:

$$q_{max-YN} = C_0 V_w \tau / 1000 M \qquad (4)$$

Given a linear regression was performed on τ with $\ln(C_0/C_t-1)$, the values of the Yoon-Nelson rate constant K_{YN} and τ (the half time needed to absorb the adsorbent) could be calculated from the straight line and slope. Under the same flow rate V_w, the Yoon-Nelson rate constant K_{YN} of Cu^{2+},

Pb^{2+} and Cd^{2+} increased with increasing C_0. Additionally, the saturation adsorption capacity in the Yoon-Nelson model q$_{0-YN}$ also increased for all tested M^{2+} ions. The results revealed that the gradual increase of the C_0 improved the saturated adsorption capacity. When the C_0 remained unchanged, with the increase of V_w, the rate constant K$_{YN}$ of Yoon-Nelson gradually increased, and while a decline of the saturation adsorption q_{max} was noted.

No obvious fluctuation was observed for the (R^2) values in all cases. The experimental showed that either the Thomas model or the Yoon-Nelson model Pb^{2+} exhibited a (R^2) value ranging from 0.96 to 0.99, confirming the validity of the Thomas model and the Yoon-Nelson model for adsorption.

3.5 Competitive adsorption of Cu^{2+}, Pb^{2+} and Cd^{2+} ions by PAES-C-Na

The competitive adsorption tests for various M^{2+} ions (Cu^{2+}, Pb^{2+}, and Cd^{2+}) were conducted. Under the same experimental conditions, it was noted that the adsorption capacity for the mixed M^{2+} ions was less compared to the value for the single M^{2+} ions. For the saturation adsorption of single M^{2+} ions, the adsorption capacity of PAES-C-Na declined in the following order: Pb^{2+}> Cd^{2+}> Cu^{2+}. For the mixed system, the saturated adsorption order of PAES-C-Na was Pb^{2+}> Cu^{2+}> Cd^{2+}, and the order of M^{2+} ions adsorption in the mixed system was Cd^{2+}> Cu^{2+}> Pb^{2+}.

4 CONCLUSION

In the current work, the influence of different factors such as initial concentration of metal ions, flow rates, pH of solution etc. on the breakthrough curve of PAES-C-Na was deeply investigated. The polymer surface possessed a relatively uniform pore structure. The structure was relatively dense and tightly coupled, providing good adsorption channels and convenient porosity. In addition, the porous structure shown in the SEM image provided many favorable adsorption sites to improve the absorption capacity of PAES-C-Na. The breakthrough curve of the PAES-C-Na column was simulated by the Thomas model and the Yoon-Nelson model. Both the Thomas model and Yoon-Nelson model could clearly describe the column adsorption behavior with R^2> 0.95 in all cases.

REFERENCES

[1] Bai L, Hu H, Fu W, Wan J, Cheng X, Zhuge LandChen Q (2011) Synthesis of a novel silica-supported dithiocarbamate absorbent and its properties for the removal of heavy metal ions. J Hazard Mater 195: 261–275.
[2] Pellera FM, Giannis A, Kalderis D, Anastasiadou K, Stegmann R, Wang J YandGidarakos E (2012) Adsorption of Cu(II) ions from aqueous solutions on biochars prepared from agriculturalby-products. J Environ Manage 96:35–42.
[3] Sis HandUysal T (2014) Removal of heavy metal ions from aqueous medium using Kuluncak (Malatya) vermiculites and effect of precipitation on removal. Appl Clay Sci 95:1–8.
[4] Priest C, Zhou J, Klink S, Sedev Rand Ralston J (2012) Microfluidic solvent extraction of metal ions and complexes from leach solutions containing nanoparticles. Chem Eng Technol 35:1312–1319.
[5] Al-Rashdi BAM, Johnson DJandHilal N (2013) Removal of heavy metal ions by nanofiltration. Desalination 315:2–17.
[6] Noel Jacob K, Senthil Kumar S, Thanigaivelan A, Tarun Mand Mohan D (2013) Sulfonated polyethersulfone-based membranes for metal ion removal via a hybrid process. J Mater Sci 49:114–122.
[7] Kumar UandBandyopadhyay M (2006) Sorption of cadmium from aqueous solution using pretreated rice husk. Bioresour Technol 97:104–109.
[8] Zhao F, Tang WZ, Zhao D, Meng Y, Yin DandSillanpää M (2014) Adsorption kinetics, isotherms and mechanisms of Cd(II), Pb(II), Co(II) and Ni(II) by a modified magnetic polyacrylamide microcomposite adsorbent. Journal of Water Process Engineering 4:47–57.

[9] Xie R, Wang H, Chen Yand Jiang W (2013) Walnut shell-based activated carbon with excellent copper(II) adsorption and lower chromium(VI) removal prepared by acid-base modification. Environ Prog Sustainable Energy 32: 688–696.
[10] Amini Mand Younesi H (2009) Biosorption of Cd (II), Ni (II) and Pb (II) from aqueous solution by dried biomass of Aspergillus niger: Application of response surface methodology to the optimization of process parameters. CLEAN-Soil Air Water 37:776–786.
[11] Ahmed I, Ghonaim A, Abdel Hakim A, Moustafa M andKamalEl-Din A (2008) Synthesis and characterization of polymers for removing of some heavy metal ions of industrialwasteWater. J Appl Sci Res 4:1946–1958
[12] Türkmen D, Yılmaz E, Öztürk N, Akgöl Sand Denizli A (2009) Poly (hydroxyethyl methacrylate) nanobeads containing imidazole groups for removal of Cu (II) ions. Mater Sci Eng:C 29:2072–2078.
[13] Neagu V (2009) Removal of Cr (VI) onto functionalized pyridine copolymer with amide groups. J Hazard Mater 171:410–416.
[14] Wu X, Wang D, Zhang S, Cai Wand Yin Y (2015) Investigation of adsorption behaviors of Cu(II), Pb(II), and Cd(II) from water onto the high molecular weight poly (arylene ether sulfone) with pendant carboxyl groups. J Appl Polym Sci 132(20).
[15] Ma D, Wang D, Zhang Sand Cai W (2016) Cellular Poly (arylene ether sulfone) With Pendant Sodium Carboxylate Groups: A Floatable Absorbent for Heavy Metal Ion Removal fromWastewater. Cell Polym 35(4):209–216.

MgO-based desulfurizer for SO₂ removal of sintering flue gas: Performance and mechanism

W. Cao, W.J. Zhang & Q. Wu
School of Metallurgy, Northeastern University, Shenyang, China

ABSTRACT: Low-temperature desulfurization products of sintering flue gas are difficult to be treated, and the wastewater increases the cost of desulfurization; to solve these problems, the MgO-based dry desulfurization technology was studied. Application of sintering flue gas indicated that the desulfurization efficiency reached 97% steadily under the dry condition of 60°C. The absorption mechanism of SO_2 with the desulfurizer was investigated by the means of scanning electron microscopy and X-ray photoelectron spectroscopy; it was found that the absorption of SO_2 by a MgO-based desulfurizer belonged to chemical absorption, and the product was sulfate. The desulfurization process was as follows: first, SO_2 diffused to the MgO-based desulfurizer surface, then it was oxidized to sulfite by the oxygen defect site (O^{2-}), and finally, the sulfite was oxidized to the sulfate by the oxygen in the sintering flue gas.

1 INTRODUCTION

Steel industries release large amounts of SO_2, which is one of the causes of environmental problems. SO_2 is one of the main pollutants in the atmosphere. It can cause acid rain and human poisoning, which causes great harm to the environment and human beings. The sintering process is one of the important processes of steel production; it accounts for more than 60% of the total smoke emission [1, 2].

The wet limestone gypsum method (FGD) is the most widely used desulfurization technology at present, which has a high desulfurization efficiency. The method is often used in large power plants, in desulfurization of sintering flue gas, and in industrial boilers. However, this method can produce calcium sulfate, which blocks the equipment, and the chloride in the flue gas can also cause corrosion of the equipment [3–5]. In order to improve the deficiency, the semi-dry method and improved FGD methods have been applied [6, 7]. Among them, the most popular one is the wet magnesium oxide method [8]. This method is based on the reaction of magnesium hydroxide with water, and magnesium hydroxide can achieve a good effect for flue gas desulfurization. However, this method produces a large amount of wastewater with a complex composition during desulfurization [9]. On the one hand, the investment of desulfurization for some small iron and steel enterprises is high. On the other hand, most desulfurization methods produce wastewater which is difficult to be treated. In conclusion, the development of flue gas desulfurization should be efficient, the products must be recyclable, and there must be no secondary pollution.

In this paper, the MgO-based desulfurization technology was tested on sintering flue gas, and the absorption mechanism is discussed. The products of the technology can be used for resources, and the desulfurization process is environmentally friendly.

2 EXPERIMENTAL PREPARATION

2.1 *Materials*

Magnesium oxide was matured to form magnesium hydroxide slurry at 70° for 2h, and the dry magnesium oxide powder (90wt%) and the magnesium hydroxide slurry (10wt%) were put into a roller mixer for uniform mixing. At room temperature and pressure, the magnesium hydroxide mixture was made into spheres with a particle size of 3–5 mm by a disc pelletizing machine. The mixture was set aside for 2h.

2.2 *Method*

2.2.1 *Scanning electron microscope – energy-dispersive X-ray spectroscopy (SEM-EDS)*
The morphologies of the MgO-based desulfurizer were determined by ULTRA PLUS, Zeiss microscope (Germany) Co. Ltd. Energy-dispersive X-ray spectroscopy (EDS) (Mg, S, Ca, and Fe contents) was carried out using an elemental analyzer (Vario-EL).

2.2.2 *X-ray photoelectron spectroscopy (XPS)*
The X-ray photoelectron spectra were measured by using a Scienta SES 2002 spectrometer, which was operated at constant transmission energy (Ep = 30 eV) and used the Al K Alpha source.

2.3 *Engineering practice tests*

The engineering practice was arranged behind the sinter pellet production line. The research group placed a pipeline at the tail of the sinter to draw the industrial sinter flue gas into the dust collector, and the flue gas with a discharge of about 500 m3/h from the collector was used for the desulfurization test. The desulfurizer with a diameter of 3–5 mm was installed in the reactor in a multi-layer and piecewise manner (separated by orifice plates) and placed perpendicular to the smoke intake surface. The main characteristics of the sintering flue gas were high oxygen content and low temperature. The specific operating parameters of desulfurization process and flue gas composition parameters are shown in Table 1.

Table 1. Operating parameters and composition of sintering flue gas.

Operation parameters		Composition parameters	
Inlet flue gas temperature/ °C	60	Mass concentration of SO_2/ (mg· m^{-3})	140–200
Outlet flue gas temperature/ °C	33	Mass concentration of NO/ (mg· m^{-3})	120–180
Volume/ (m^3· h^{-1})	500	Mass concentration of O_2/ %	19.1–20.2
Inlet velocity/ (m· s^{-1})	16	Mass concentration of CO_2/ %	0–3.5
		Mass concentration of CO/ (mg· m^{-3})	0–20

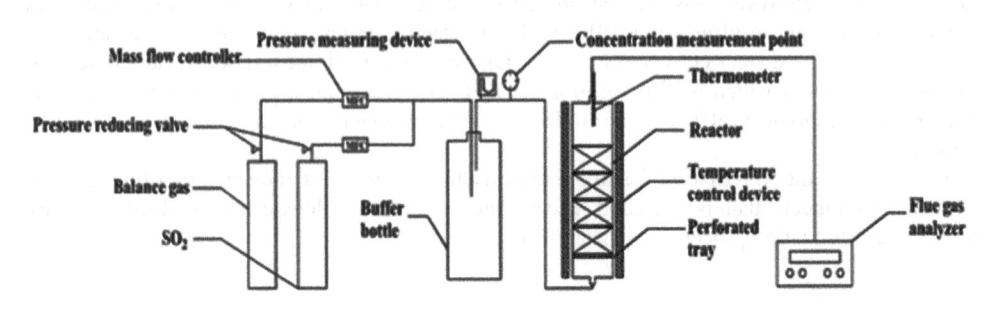

Figure 1. Experimental system diagram.

The reactor was 2.5 m in height with a diameter of 0.8 m. The reactor was divided into four compartments, each separated by perforated steel plates. The distribution of the desulfurizer was suitable for the reactor structure, which was four layers in total. Each layer was equipped with an inlet and outlet for manual feeding and unloading. The LAOying 3012H flue gas analyzer was used to monitor the concentration of SO_2 at the inlet and outlet of the reactor. The experimental system diagram of desulfurization is shown in Figure 1.

3 RESULTS AND ANALYSIS

3.1 *Stability test*

Desulfurization experiments with a cycle of 30 days were carried out, and the desulfurization efficiency is shown in Figure 2. The desulphurization efficiency of the MgO-based desulfurizer was stable and high after operation for 30 days, and the efficiency was stable at 97% in the whole process. On the 6th day, the efficiency dropped sharply, which was believed to be related to the operating conditions of the plant. The desulphurization efficiency was 95% on days 28–30; the slight decline might be due to the long operation time. Specifically, the magnesium-based desulfurizer led to the phenomenon of desulfurization saturation.

At present, the efficiency of the FGD desulfurization method used in industry is up to 90%, but its efficiency is greatly affected by the flue gas characteristics and the equipment operating conditions. Compared with FGD, the MgO-based dry desulfurization technology can ensure high efficiency and better stability.

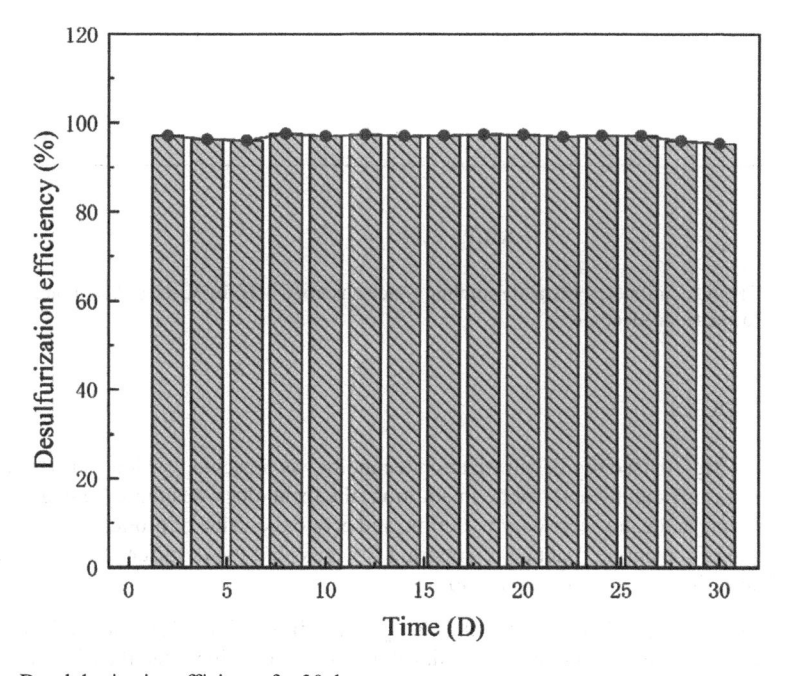

Figure 2. Desulphurization efficiency for 30 days.

3.2 *XPS studies*

The test results showed that SO_4^{2-} (170–169 eV) was obviously present, and SO_3^{2-} (168–167 eV) and S (163.5–161.5 eV) were absent, as presented in Figure 3 [10, 11] Although the results show no SO_3^{2-}, we consider that SO_2 was not directly oxidized to SO_4^{2-}. It is because the surface of the

MgOtype desulfurizer exhibits Schottky defect [11, 12]; SO_2 can react with O^{2-} (oxygen defect site) on the desulfurizer surface to produce SO_3^{2-}. In addition, the sintering flue gas had a high oxygen content, so SO_3^{2-} would be oxidized to SO_4^{2-}

The desulfurization process was as follows: SO_2 underwent redox reactions with the oxygen defect site on the surface of the MgO-based desulfurizer to produce SO_3^{2-}. Afterward O_2 in the flue gas further reacted with SO_3^{2-} to produce SO_4^{2-}. The mechanism of the reaction is as follows:

$$SO_2(g) + O^{2-} \xrightarrow{MgO} SO_3^{2-} \xrightarrow{O_2} SO_4^{2-}$$

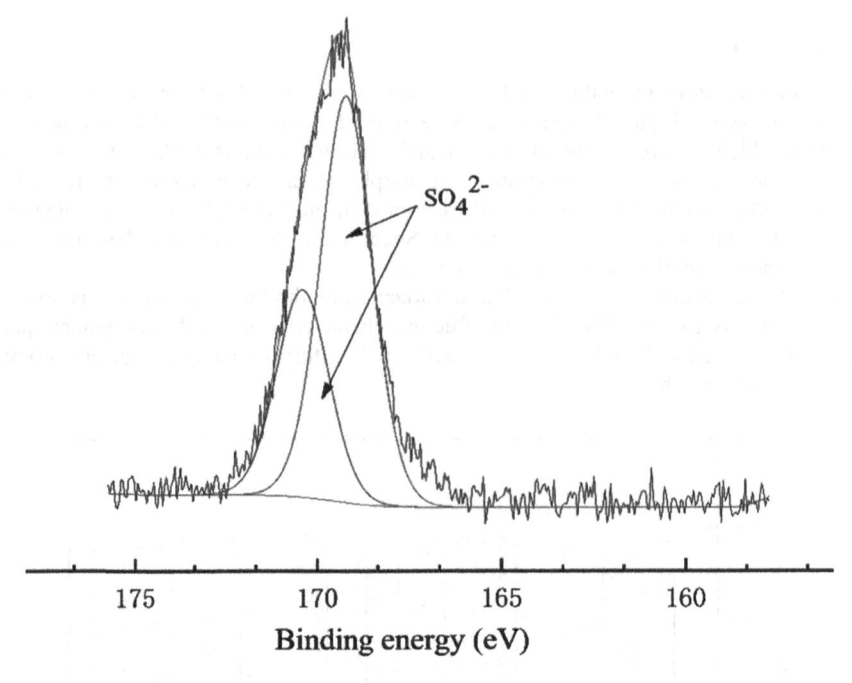

Figure 3. S 2P photoemission spectra of sampling which absorbed SO_2 for 30 days. A photon energy of 260 eV was used to excite the electrons.

4 CONCLUSIONS

In this paper, a MgO-based dry desulfurization technology is presented, which can remove SO_2 of sintering flue gas efficiently and stably. In addition, the technology has the advantages of simple equipment maintenance, no secondary pollution, and economic value of products. The technology is applied to the engineering practice of sintering flue gas, and the desulfurization mechanism is explored. The main conclusions are as follows:

(1) Under the drying condition of 60°, the flue gas desulfurization of the sintered flue gas was carried out by using the MgO-based dry desulfurization technology, and the efficiency reached 98.8%.
(2) Physical and chemical analyses showed that the technology used the chemical absorption of SO_2 by the desulfurizer to produce $MgSO_4$. The specific desulfurization process was as follows: SO_2 was oxidized to sulfite by the oxygen defect site (O^{2-}) on the surface of the MgO-based desulfurizer and then oxidized to sulfate by O_2 in the sintering flue gas.
(3) The technology has no wastewater, and its sulfate products can be made into fertilizers, building materials, sulfur, etc., which can realize recycling economy and resource recovery.

ACKNOWLEDGEMENT

This work was supported by the National Key Research and Development Program of China [project number 2017YFA0700300].

REFERENCES

[1] Yan Zhanhai, Shao Jiugang, Qi Chenglin and Jiang Xiaodong 2019 Selection of sintering flue gas purification process under the new environmental protection situation. Sintering and pelletizing. 44 73–77.
[2] Zhang Qi, Liu Shuai, Xu Hua Yian, Meng Zhiquan, Wang Gang and Xu Shi 2019 Development and practice of intelligent energy control system in iron and steel enterprises. Iron and Steel. 54 125–133.
[3] Yu Qingbo, Wang Xinhua and Yue Qiang 2002 Experimental study on desulfurization performance of self-excited dust removal and desulfurization unit. Journal of Northeastern University (natural science edition). 23 1101–03.
[4] Mingsheng D 2017 Pilot-plant studying on energy-saving turbulent pipe railings(TPR) desulfurization device. Environmental Science Technology. 40 136–142.
[5] Song Weiming, Zou Jianan, Li Shu and Yang Jan 2019 Test of coal coke reduction decomposition and sintering flue gas desulfurization gypsum. Iron and Steel. 54 110–115.
[6] Yue Yang, Yongtao Fan, Hongling Li and Yu Qi 2019 Study on thermal decomposition process of semidry flue gas desulfurization ash. Energy & Fuels. 33 9023–31.
[7] Shi Peiyang, Deng Zhiyin, Yuan Yiyi and Sun Jun 2010 Calcium sulfate whisker was synthesized by hydrothermal process of desulfurized gypsum. Journal of Northeastern University (natural science edition). 31 76–79.
[8] Li Juan, Qian Feng, Li Jianyu 2005 Study on the activity of desulphurization magnesium oxide in coal flue gas. Environmental Pollution and Prevention. 27 51–61.
[9] Lei Tao, Xueqian Wang, Ping Ning, Langlang Wang and Weijun Fan 2019 Removing sulfur dioxide from smelting flue and increasing resource utilization of copper tailing through the liquid catalytic oxidation. Fuel Processing Technology. 192 36–44.
[10] Rodriguez J A, Pérez M, Jirsak T, González L and Maiti A J S S 2001 Coadsorption of sodium and SO2 on MgO (100): alkali promoted S-O bond cleavage. Surf Sci. 477 279–288.
[11] Rodriguez J A, Jirsak T, Freitag A, Hanson J C, Larese J Z and Chaturvedi S J 1999. Interaction of SO2 with CeO2 and Cu/CeO2 catalysts: photoemission XANES and TPD studies. Catal Lett. 62 113–119.
[12] Antoshchenkova E, Hayoun M, Geneste G, Finocchi F 2010 Thermodynamics and kinetics of the schottky defect at terraces and steps on the MgO (001) surface. Phys Chem Chem Phys. 12 7251–57.

Effect of Mo and B on the structure and properties of high chromium cast iron surfacing metal

Q.Y. Zhao
Key Laboratory for Liquid-Solid Structural Evolution and Processing of Materials (Ministry of Education), Shandong University, Jinan, China

F.L. Tantai & H.F. Tian
Shandong Energy Heavy Equipment Group Dazu Remanufacturing Co. Ltd, Xintai, China

C. Wang
Key Laboratory for Liquid-Solid Structural Evolution and Processing of Materials (Ministry of Education), Shandong University, Jinan, China

W. Wang & Q.L. Hou
Shandong Energy Heavy Equipment Group Dazu Remanufacturing Co. Ltd, Xintai, China

J.S. Sun
Key Laboratory for Liquid-Solid Structural Evolution and Processing of Materials (Ministry of Education), Shandong University, Jinan, China

ABSTRACT: High chromium cast iron with Cr7C3 as the hard phase has excellent wear resistance, and is widely used as an anti-abrasive wear material in the industrial field. High-chromium cast iron surfacing materials are often used to improve the wear resistance of the workpiece surface. The surfacing metal has high hardness and excellent wear resistance, but it has high brittleness and poor crack resistance. In this paper, a flux-cored welding strip of high-chromium cast iron reinforced with boride is designed, using optical microscope (OM), scanning electron microscope (SEM), X-ray diffractometer, micro-hardness and other methods, the amount of Mo, B added to flux-cored welding strip TIG arc cladding surfacing metal was studied the influence of microstructure and performance. The results show that the main constituent phase of the surfacing metal with 43%Mo-7%B4C is Cr7C3, Mo2FeB2 composite strengthening phase, the number of hard phases in the surfacing metal is significantly increased, the size of the hard phase is also significantly reduced, and the distribution is more uniform. The surfacing metal with 34.4%Mo-5.6%B4C addition is mainly the constituent phase is Cr7C3 hard strengthening phase, and Mo2FeB2 is not formed.

1 INTRODUCTION

High chromium cast iron with Cr_7C_3 as the hard phase has excellent wear resistance, and is widely used as an anti-abrasive wear material in the industrial field [1].

The excellent wear resistance of hypereutectic high chromium cast iron surfacing metal is mainly due to the primary carbide Cr_7C_3 precipitated in the liquid phase solidification process [2]. The size, shape, volume fraction and distribution of Cr_7C_3 contribute to the structure and performance of the surfacing metal It has a great influence. The supporting austenite and ferrite around the primary Cr_7C_3 also affect the properties of hypereutectic high chromium cast iron [3, 4]. The high chromium cast iron surfacing metal is not suitable for wear conditions under impact load, and its application range is greatly restricted [5]. A large number of high hardness carbides in high chromium cast iron also greatly reduces its impact toughness and deformability [6]. Therefore, improving the impact resistance of the hypereutectic high chromium cast iron surfacing metal and reducing the tendency of cracking during the preparation of the surfacing metal have become an urgent problem to be solved [1].

In this paper, a flux-cored welding strip of high-chromium cast iron reinforced with ternary boride is designed to further improve the wear resistance and crack resistance of the surfacing metal. Using optical microscope (OM), scanning electron microscope (SEM), X-ray diffractometer, microhardness and wear test and other methods, the amount of Mo, B added to flux-cored welding strip TIG arc cladding surfacing metal was studied the influence of microstructure and performance. It provides technical data for the design of composite strengthened high-chromium cast iron surfacing materials.

2 MATERIALS AND METHODS

2.1 Materials

Commercial pure Mo powder, reduced Fe powder, TiFe powder, Cr_3C_2 powder, B_4C powder, and high-carbon ferrochrome powder were used as raw materials. Table 1 shows the technical parameters of each powder. The base metal employed in this study was Q235 steel, and its size is 10mm×50mm×200mm.

Table 1. Technical parameters of alloy powder.

Powder	Particle size(μm)	Chemical composition(wt%)
Mo	100–180	Fe<0.002,O<0.1,Si<0.001,Bal Fe
Fe	80–180	C<0.1,N<0.1,O<0.2,Bal Fe
Cr3C2	60	C=11.87,Cr=87.31,Si<0.16
B4C	60	B=76.53,C=20.85
High carbon ferrochrome	60	C=8.35,Si<0.5,Cr=69.8

2.2 Formula design of Mo and B reinforced high chromium cast iron flux-cored welding tape

A 304 stainless steel pipe with wall thickness of 0.2mm and diameter of 6mm is used to make flux-cored welding tape. The formula of flux-cored welding tape is shown in Table 2, where D0 does not add Mo powder and B_4C powder, and the formula with Mo_2FeB_2 content of 0 is used as a comparison standard for the influence of Mo_2FeB_2 on Cr_7C_3. The filling rate of flux-cored ribbon is 70%, and the content of Mo and B is changed to study the influence of the change on the morphology of Cr_7C_3

Table 2. Formula of flux-cored ribbon with different Mo and B content (Wt%).

Numbering	Mo	B4C	Fe	Cr3C2	High carbon ferrochrome
D0	0	0	50	40	10
D1	43	7	0	40	10
D2	38.7	6.3	5	40	10
D3	34.4	5.6	10	40	10
D4	30.1	4.9	15	40	10

The deposited metal of flux-cored strip is strengthened by the combination of Mo_2FeB_2 and Cr_7C_3. The appropriate amount of Mo and B can react with Fe to form Mo_2FeB_2, which has the effect of refining the Cr_7C_3 hard phase by hindering the free growth of the Cr_7C_3 hard phase formed by Cr and C Changing the form of hard phase of high chromium cast iron surfacing metal improving crack resistance and impact resistance.

2.3 *Method*

Use WSM-315 argon arc welding machine and use TIG arc cladding flux-cored welding tape. The cladding process parameters are: welding current 120–150A, arc voltage 20V, and argon flow 16L/min.

The XJP-6A metallurgical microscope was employed to analyze the microstructure of the surfacing metal, and the JSM-6600V scanning electron microscope was employed to analyze the fine microstructure and micro-area composition. The Rigiku D/MAX-RC XRD diffractometer was employed to analyze the phase of the surfacing metal. The DHV-1000 microhardness tester is employed to test the microhardness of the surfacing metal.

3 EFFECT OF MO AND B CONTENT ON MICROSTRUCTURE AND PROPERTIES OF SURFACING METAL

3.1 *Effect of Mo and B content on the microstructure of surfacing metal*

Mo, B, and Fe are the elements for in-situ synthesis of Mo_2FeB_2. Change the content of Mo and B to generate different content of Mo_2FeB_2 in the surfacing metal and study the influence of the content of Mo_2FeB_2 on the morphology and quantity of Cr_7C_3. Figure 1 shows the microstructure morphology of the middle part of the surfacing metal with 0%Mo-0%B4C, 43%Mo-7%B4C, 38.7%Mo-6.3%B4C, 34.4%Mo-5.6%B4C and 30.1%Mo-4.9%B4C. It can be seen that the surfacing metal is mainly composed of a large number of square, irregular polygon, long strip and block hard phases and matrix.Compared with figure 1a without Mo and B, the number of hard phases in the surfacing metal added with Mo and B is significantly increased, the size of the hard phase is also significantly reduced, and the distribution is more uniform.

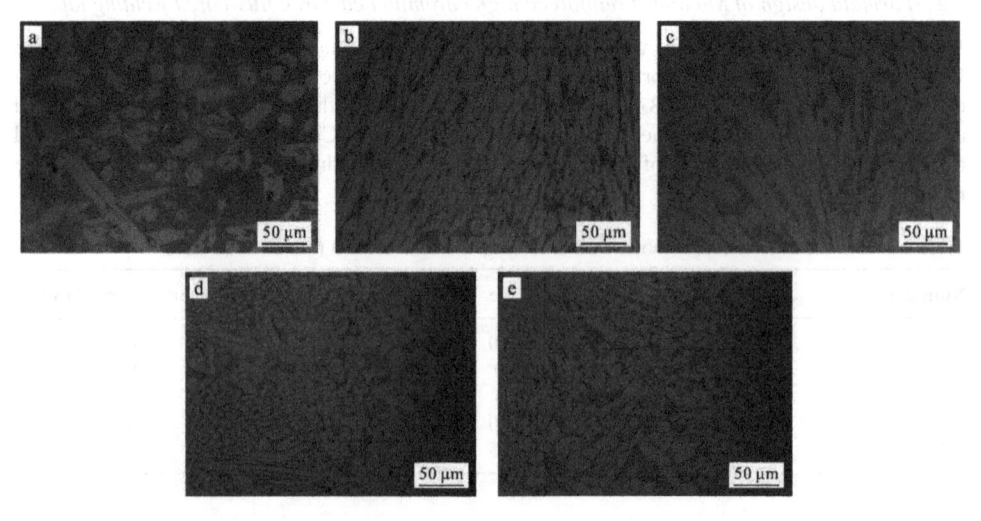

Figure 1. Metallographic structure of surfacing metal with different amounts of Mo and B added: (a) 0% Mo-0% B4C, (b) 43% Mo-7% B4C, (c) 38.7% Mo-6.3% B4C, (d) 34.4% Mo-5.6% B4C and (e) 30.1% Mo-4.9% B4C.

3.2 Effect of Mo and B content on phases of surfacing layer

Figure 2 is the X-ray diffraction pattern of the flux-cored tape overlay metal. It can be seen that the main composition phases of the surfacing metal with 43%Mo-7%B4C added in Figure 2a are Cr_7C_3, Mo_2FeB_2, NiCrFe and MoCrFe. Mo_2FeB_2 is a ternary boride formed by Mo, B, and Fe at high temperatures. Cr atoms will also replace the positions of Mo and Fe atoms in Mo_2FeB_2 to form a composite boride M_3B_2. NiCrFe is a solid solution formed by dissolving Cr and Ni in Fe, MoCrFe is a solid solution formed by dissolving Cr and Mo in Fe, Cr_7C_3 (long strips and massive white blocks in figure 1b), Mo_2FeB_2 (Square and irregular white blocks in figure 1b) mainly play a reinforcing role in the surfacing layer, forming a wear-resistant framework and improving wear resistance.

Figure 2b shows the main composition phases of the surfacing metal with 34.4%Mo-5.6%B_4C added are Cr_7C_3, α-FeCrNi and α-FeCrMo. There is no Mo_2FeB_2 diffraction peak in the X-ray diffraction pattern, which may be because the content of the added B element is too low to reach the level of forming Mo_2FeB_2. In Figure 1d, square and irregular block Mo_2FeB_2 hard phases are not observed either. Through XRD and phase analysis, it can be known that the long and massive hard phases are Cr_7C_3, and the matrix structure is α-FeCrNi and α-FeCrMo. From the X-ray diffraction results and microstructure analysis, the addition of Mo decreased from 38.7% to 34.4%, the addition of B_4C decreased from 7% to 5.6%, and the type of hard phase in the surfacing layer has changed significantly. The main phase of the surfacing alloy has also changed significantly. The hard phase of Mo_2FeB_2 is no longer formed in the surfacing layer, and all the Mo elements are dissolved into the matrix to play a solid solution strengthening effect.

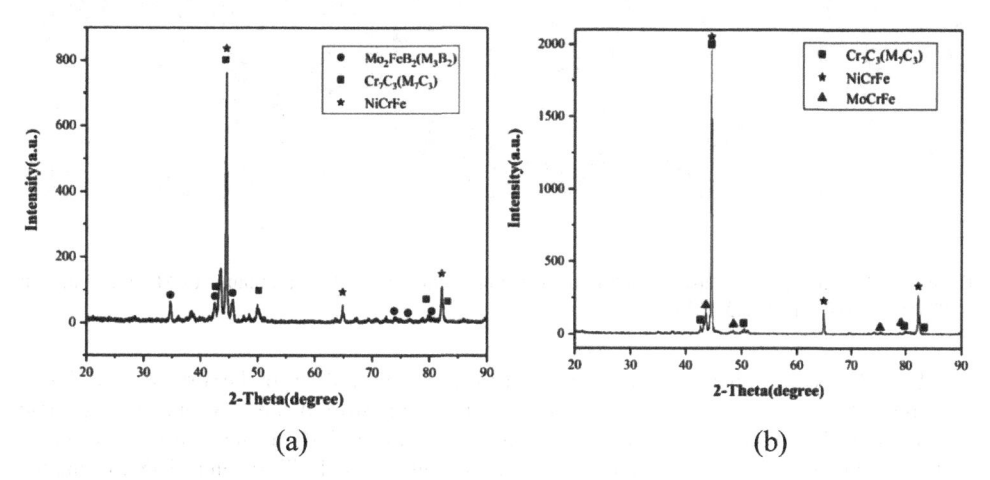

Figure 2. X-ray diffraction pattern of surfacing metal: (a) 43% Mo-7% B4C and (b) 34.4%Mo-5.6% B4C.

3.3 Effect of Mo and B content on size and volume fraction of hard phase of surfacing metal

Figure 3 is the SEM morphology of the middle part of the surfacing metal with 0%Mo-0%B_4C and 43%Mo-7%B_4C added to the flux-cored ribbon. Imagepro Plus software is used to process the backscattered electron image in the middle of the surfacing metal. The red represents the hard phase. As shown in Figure 4use Imagepro Plus software to calculate the size and distribution of the hard phase, and use ImageJ software to calculate the volume fraction of the hard phase.

(a) (b)

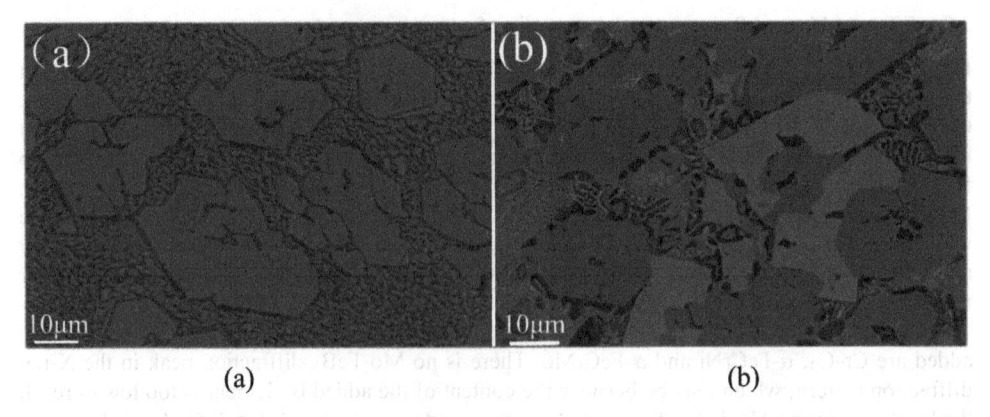

Figure 3. SEM morphology of surfacing metal with different amounts of Mo and B added: (a) 0%Mo-0%B4C and (b) 43%Mo-7%B4C.

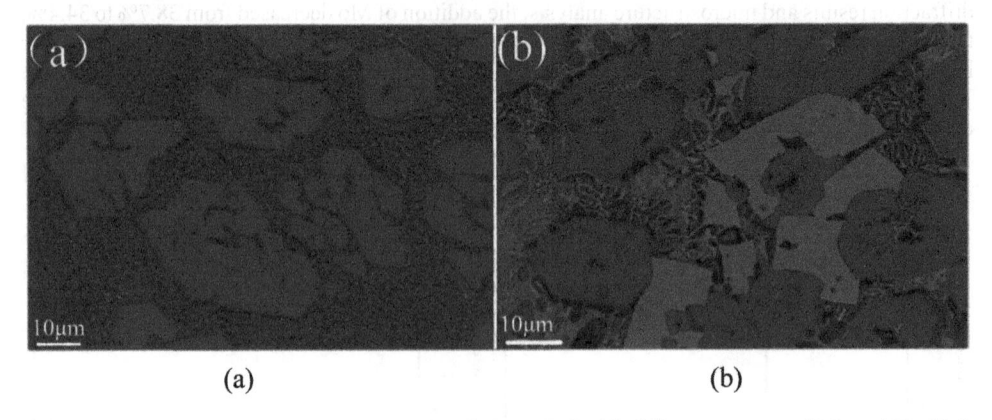

(a) (b)

Figure 4. Results of hard phase extraction of surfacing metal with different amounts of Mo and B added: (a)0%Mo-0%B4C and (b)43%Mo-7%B4C.

Table 3 shows the statistical results of the hard phase content and size of the surfacing metal with 0%Mo-0%B$_4$C and 43%Mo-7%B$_4$C added to the flux-cored strip. It can be seen that the volume fraction of the hard phase of the surfacing metal with 0%Mo-0%B$_4$C and 43%Mo-7%B$_4$C added to the flux-cored ribbon is 35.6% and 39%, and the average size is 15.79μm and 14.18μm, and the size variances are 33.13 and 26.22 respectively. Therefore, the addition of Mo and B increases the number of hard phases of the surfacing metal, refines the hard phase Cr_7C_3, and makes the size distribution of the hard phase of the surfacing metal more uniform.

Table 3. Statistical results of hard phase content and size of surfacing metal with different amounts of Mo and B.

The addition of Mo,B	Volume fraction of Hard phase	Average size of hard phase	Size variance of hard phase
0%Mo-0%B$_4$C	35.6%	15.79μm	33.13
43%Mo-7%B$_4$C	39%	14.18μm	26.22

3.4 Effect of Mo and B content on the hardness of surfacing layer

The average microhardness of the surfacing metal with 0%Mo-0%B_4C, 43%Mo-7%B_4C, 38.7%Mo-6.3%B_4C, 34.4%Mo-5.6%B_4C and 30.1%Mo-4.9%B_4C is 650.3HV0.5, 1105.4HV0.5, 1034.4HV0.5, 853.8HV0.5 and 807.9HV0.5. Respectively 4.2 times, 7.2 times, 6.7 times, 5.5 times and 5.2 times of the average microhardness of matrix Q235 153.9 HV0.5. Mo_2FeB_2 hard phase is formed in D1, D2 surfacing metal (Mo addition amount is 43% and 38.7% respectively), the average microhardness of these two groups is significantly higher than that of D3 (Mo addition amount is 34.4% B_4C addition amount is 5.6%) and D4 (Mo addition amoun is 30.1%, B4C addition amount is 4.9%) surfacing metals without Mo_2FeB_2 hard phase formation, which shows that the existence of Mo_2FeB_2 hard phase can significantly increase the microhardness of surfacing metal. This is related to the uniform distribution of the precipitated Mo_2FeB_2 hard phase in the surfacing metal, which hinders the growth of the Cr_7C_3 hard phase during the crystallization of the molten pool, refines the Cr_7C_3 hard phase, and prevents the coarsening of primary carbides. There is no Mo_2FeB_2 hard phase in D3 and D4, and the microhardness of the surfacing metal decreases with the decrease of Mo addition.

4 CONCLUSION

(1) The addition of Mo and B in the flux-cored welding strip affects the phase composition of the surfacing metal. The main constituent phase of the surfacing metal with 43%Mo7%B_4C is Cr_7C_3, Mo_2FeB_2, NiCrFe and MoCrFethe surfacing metal with 34.4%Mo5.6%B4C addition is mainly the constituent phases are Cr_7C_3, NiCrFe and FeCrMo, and Mo_2FeB_2 is not formed.

(2) Compared with those without Mo and B, the number of hard phases in the surfacing metal added with Mo and B is significantly increased, the size of the hard phase is also significantly reduced, and the distribution is more uniform.

(3) The average microhardness of surfacing metal with 0%Mo-0% B_4C, 43%Mo-7% B_4C, 38.7%Mo-6.3%B_4C, 34.4%Mo-5.6% B_4C and 30.1%Mo-4.9% B_4C is 650.3 HV0.5, 1105.4 HV0.5, 1034.4 HV0.5, 853.8 HV0.5 and 807.9 HV0.5. The surfacing metal with 43%Mo-7% B_4C added has the highest average microhardness. Compared with the absence of Mo and B, it has increased by 455.1 HV0.5.

ACKNOWLEDGEMENTS

The authors gratefully acknowledge the support provided by the Key Science and Technology Project of Shandong Provincial (2019JZZY020208).

REFERENCES

[1] Li L, Wang X, Li Z, Li Q, Zhao X. Research Progress of Grinding Roll Material and Surface Strengthening Technology of Ore Mill [J]. China Metal Bulletin, 2020,(06): 297–298.
[2] Jiang J, Li S, Hu S, Zhang J, Yao B, Lu X, Yu W, Zhou Y. Effects of Added TiC on Properties of High Chromium White Iron[J]. Rare Metal Materials and Engineering, 2020,49(02):701–705.
[3] Liu S, Jin Z, Wang Z, Shi Z, Zhou Y, Ren X, Yang Q. Refinement and homogenization of M_7C_3 carbide in hypereutectic Fe-Cr-C coating by Y_2O_3 and TiC[J]. Materials Characterization, 2017, 132: 41–45.
[4] Wu X, Xing J, Fu H, Zhi X. Effect of titanium on the morphology of primary M_7C_3 carbides in hypereutectic high chromium white iron[J]. Materials Science & Engineering A, 2006, 457(01): 180–185.
[5] Kou X. Effect of Modification Treatment on Microstructure and Properties of High Chromium Cast Iron[D]. Shaanxi University of Technology, 2015.
[6] Song L. Status and Development of Mine Wear Resistant Casting[J].Foundry Technology, 2009, 30(02): 295–298.

Effects of green and black tea on total phenolic content and textural properties of bread

Z.L. Ma, N. Duan, D.X. Zhang, P. Wu, Y.H. Guo & D.Y. Li
School of Laboratory Medicine, Hubei University of Chinese Medicine, Wuhan, China

ABSTRACT: In this study, the effects of adding green and black tea on the quality of bread were investigated. The green tea bread and black tea bread were made with the same concentration of tea water extract. The total phenolic content was detected by the Folin-phenol reagent method. The texture of bread was measured by a texture analyzer. Tea powder was extracted with boiling water at a 1:10 ratio (w/v) for producing the bread. The green tea bread (551.75μg/g) showed a higher level of total phenolic content than the black tea bread (269.96μg/g). The addition of both green tea and black tea could change the porosity of the bread. The hardness, gumminess, adhesiveness, and chewiness of the bread significantly increased after the addition of tea water extract. The texture of the green tea bread is better than that of the black tea bread. Our results suggest that green and black tea extract may act as functional materials to improve the nutrition of the bread.

1 INTRODUCTION

After water, tea is one of the most popular non-alcoholic drinks. Non-fermented (green tea), fully fermented (black tea), and semi-fermented (oolong tea) are the three forms commonly consumed tea [1]. There are many functional compounds in different teas, such as flavan-3-ols, theaflavins, thearubigins, etc. [2, 3]. Tea polyphenols, tea polysaccharide, tea pigment, and other components are the material basis of tea flavor quality and dietary healthcare function. There are some research studies focused on the effect of tea or components of tea on the flour product [4–8]. However, the comparison of green and black tea applied in the bread is rarely reported in the literature. In this work, water extracts of green and black tea were applied in bread processing. Total phenolic content and texture properties were compared between the two types of tea bread, which is of great importance for theoretical research and application of tea.

2 MATERIALS AND EXPERIMENTAL METHODS

2.1 *Materials and chemicals*

Green tea, black tea, high gluten wheat flour, sunflower seed oil, dried yeast, salt, and milk powder were purchased from the market. Gallic acid, Folin–Ciocalteu reagent, sodium carbonate, and methanol were all analytical reagents obtained from the China Sinopharm Group.

2.2 *Tea water extract processing*

The two kinds of tea were ground into powder, and tea powder was extracted with boiling water at 1:5, 1:10, 1:15 ratio (w/v) for 20 min. The extracted water solution was then filtered for bread preparation.

2.3 Preparation of tea bread

Different ratio of green and black tea aqueous extract [9] (w/v 1:10) were added instead of water in bread processing.

2.4 Sensory evaluation of tea bread

The participants, 10 men and 10 women, evaluated the bread from eight aspects including the appearance, internal structure, smell, taste, etc. and scored the overall acceptability. The sensory evaluation result is shown in Table 1.

2.5 Analysis of phenolic concentration in green and black tea bread

The bread was lyophilized and ground into powder. 50 mL methanol was added to the resulting powder (30 g) to extract phenolic compounds for 24 h. After centrifugation, the supernatant was used for the determination of phenolics in the two types of bread using our reported method with minor modification [10]. The bread extract or gallic acid standards (20 μL) was mixed with 100 μL of Folin-Ciocalteu reagent. Then, 80 μL of 75 mg/mL sodium carbonate was added and incubated for 30 min. At last, the absorbance at 760 nm was measured with a microplate reader (Thermo Fisher, USA). Values for the levels of phenolics were shown as μg gallic acid/g bread. All samples were triplicate determined in this study.

2.6 Texture profile analysis

The texture profile (hardness, elasticity, viscosity, chewability, cohesiveness, etc.) of green tea and black tea bread was measured by an FTC texture analyzer (FTC, USA). The surface and edge of the bread were removed, and the bread was cut into a cuboid with the length of 20 mm, width of 20 mm, and height of 10 mm. The test mode is TPA mode, and a cylindrical probe was used. The test parameters were as follows: maximum range was 25 N, the position of the upper surface of the sample from the probe was 30 mm, the deformation was 30%, the test speed was 30 mm/min, the trigger force was 0.1 N, and the number of cycles was 2. The texture profile of the bread is shown in Table 3.

3 RESULTS AND DISCUSSION

3.1 Determination of the optimum process of the tea bread

As shown in Table 1, various tea and water ratios influenced the appearance, texture, flavor, elasticity, crust, crumb, smell, and taste. At a ratio of 1:10, the bread was found to have the highest sensory score. Thus, the ratio of 1:10 was selected for further study.

Table 1. Sensory scores on the black tea bread.

Parameter	Appearance	Texture	Flavor	Impurity	Elasticity	Crust	Crumb	Smell and taste	Score
1:5	5.70	3.50	10.9	10.2	4.40	13.8	14.0	8.25	70.65
1:10	7.35	3.95	11.0	10.0	4.15	13.9	14.2	8.50	73.05
1:15	6.30	4.45	10.5	8.65	4.75	13.7	13.0	7.95	69.15

3.2 Comparison of the section of the control bread and tea bread

As shown in Figure 1, three types of bread had complete appearance. Green tea bread and black tea bread showed the color of green tea infusion and black tea infusion, respectively. It was reported that the color of cereal foods was very important for consumers; its change was an attractive character [7, 11,12]. In terms of internal structure, the porosity of the control bread was uniform and compact, which was the typical shape of the toast bread. The porosity of the green tea bread and black tea bread was coarse and uneven, and there were obvious holes in the top and bottom of the bread. As shown in the figure, the internal structure of the two types of tea bread was looser than the normal bread. Our results indicated that the addition of either green tea or black tea could increase the porosity of the bread. However, the addition of tea extract might result in the decrease in smoothness. Tea polyphenol may make the gluten network less well dispersed [13], contributing to the decreased smoothness of the tea bread.

Figure 1. Section of the control bread: (A), green tea bread (B), and black tea bread (C).

3.3 Comparison of the total phenolic content between green and black tea bread

The total phenolic content of green tea bread and black tea bread is exhibited in Table 2. The phenolic level in the green tea bread was 551.75 μg/g, whereas the phenolic level in the black tea bread was 269.96 μg/g, indicating that the phenols in tea were still retained during the bread processing. The green tea bread showed a higher level of phenolics than the black tea bread ($p < 0.05$), which was due to the different processing techniques of the two types of tea.

Table 2. Total phenolic content of green tea bread and black tea bread.

Sample	Total phenolic concentration
Green tea bread	551.75±37.49[a]
Black tea bread	269.96±37.41[b]

Values not sharing the same letter significantly differ ($p < 0.05$).

3.4 Texture analysis of the bread

The texture analysis including hardness, adhesiveness, springiness, gumminess, chewiness, and cohesiveness of three types of bread are shown in Table 3. Both black tea and green tea addition led to a marked change of hardness, gumminess, adhesiveness, and chewiness (p < 0.05), indicating that it might affect the feasibility of the bread added with the two kinds of tea at varied levels. However, it still depends on preferences of consumers. The addition of tea has no effect on springiness, which meant the changes of the gluten network were acceptable. The change of chewiness was correlated with the change of hardness and gumminess, which might be caused by the reaction of polyphenols with protein. Compared with other textural properties, the cohesiveness of the green tea bread was significantly decreased in response to green tea addition ($p < 0.05$). Additionally, it was lower than that of the black tea bread. Black tea is fully fermented, where the term 'fermented' refers to natural browning reactions induced by oxidative enzymes within the plant cell. It has been reported that the orange-red theaflavins and the red-brown thearubigins were the two main phenolic pigments in black tea [14]. According to the existing theory, green tea had the highest level of catechin. However, theafavins, thearubigins and other catechin polymeric pigments were the main polyphenols in black tea [15]. The components in black tea have higher molecular weights and are more complex than those in green tea. Therefore, the difference of the resulting polymeride might lead to the difference of cohesiveness.

Table 3. Textural properties of the bread.

Parameter	Hardness	Adhesiveness	Springiness	Gumminess	Chewiness	Cohesiveness
Green tea bread	2.79±0.57[a]	0.026±0.0029[a]	3.17±0.088[a]	1.87±0.38[a]	5.80±1.18[a]	0.52±0.046[a]
Black tea bread	2.34±0.43[a]	0.031±0.0013[b]	3.21±0.13[a]	1.57±0.27[a]	4.78±0.65[a]	0.60±0.032[b]
Control bread	1.73±0.31[b]	0.036±0.0036[c]	3.25±0.14[a]	1.11±0.20[b]	3.50±0.54[b]	0.64±0.017[b]

Values with the different letters in the same column are significant different ($p < 0.05$).

4 CONCLUSION

In the present study, green tea and black tea were extracted with boiling water at a 1:10 ratio (w/v) to process the bread. Compared with the control, the addition of tea aqueous extract caused the increase of porosity in the bread. In addition, the texture of the bread was changed by the addition of tea extract. We found that the green tea bread had a higher total phenolic content of 551.75 μg/g than the black tea bread. Polyphenols in green and black tea were reported to show many bioactivities. Thus, the nutritional and healthy characteristics of the tea bread would be studied in the future.

ACKNOWLEDGMENTS

The study was supported by the Scientific Research Plan of Hubei Provincial Department of Education. We thank the supportment from this funding (No. Q20182002).

REFERENCES

[1] Malongane F, McGaw L and Mudau F. The synergistic potential of various teas, herbs and therapeutic drugs in health improvement: a review. *J Sci Food Agric*. 2017, **97**(14): 4679–89.
[2] Hodges J, Zhu J, Yu Z, Vodovotz Y, Brock G, Sasaki G, Dey P and Bruno R. Intestinal-level anti-inflammatory bioactivities of catechin-rich green tea: Rationale, design, and methods of a double-blind, randomized, placebo-controlled crossover trial in metabolic syndrome and healthy adults. *Contemp Clin Trials Commun*. 2019, **17**: 100495.
[3] Khanum H, Faiza S, Sulochanamma G and Borse B. Quality, antioxidant activity and composition of Indian black teas. *J Food Sci Technol*, 2017, **54**(5):1266–72.
[4] Goh R, Gao J, Ananingsih V, Ranawana V, Henry C and Zhou W. Green tea catechins reduced the glycaemic potential of bread: An in vitro digestibility study. *Food Chem*, 2015, **180**: 203–10.
[5] Fu Z, Yoo M, Zhou W, Zhang L, Chen Y and Lu J. Effect of (-)-epigallocatechin gallate (EGCG) extracted from green tea in reducing the formation of acrylamide during the bread baking process. *Food Chem*, 2018, **242**: 162–8.
[6] Ning J, Hou G, Sun J, Wan X and Dubat A. Effect of green tea powder on the quality attributes and antioxidant activity of whole-wheat flour pan bread. *LWT - Food Sci Technol*, 2017, **79**: 342–8.

[7] Zhu F, Sakulnak R and Wang S. Effect of black tea on antioxidant, textural, and sensory properties of Chinese steamed bread. *Food Chem.*, 2016, **194**: 1217–23.
[8] Pasrija D, Ezhilarasi P, Indrani D and Anandharamakrishnan C. Microencapsulation of green tea polyphenols and its effect on incorporated bread quality. *LWT - Food Sci Technol*, 2015, **64**: 289–96.
[9] Wen H and Yang W. The development of the bread and tea flavor quality. *Food Nut China*, 2005(6): 40–2.
[10] Chen X, Ma Z and Kitts D. Effects of processing method and age of leaves on phytochemical profiles and bioactivity of coffee leaves. *Food Chem*, 2018, **249**: 143–53.
[11] Zhu F, Cai Y and Corke H. Evaluation of Asian salted noodles in the presence of Amaranthus betacyanin pigments. *Food Chem*, 2010, **118**: 663–9.
[12] Zhu F, Cai Y, Sun M and Corke H. Influence of Amaranthus betacyanin pigments on the physical properties and color of wheat flours. *J Agric Food Chem*, 2008, **56**(17): 8212–7.
[13] Ozdal T, Capanoglu E and Altay F. A review on protein–phenolic interactions and associated changes. *Food Res Int*, 2013, **51**: 954–70.
[14] Haslam E. Thoughts on thearubigins. *Phytochemistry*, 2003, **64**(1): 61–73.
[15] Li S, Lo C, Pan M, Lai C and Ho C. Black tea: chemical analysis and stability. *Food Funct*, 2013, **4**(1): 10–8.

Connecting technology of vertical components in fabricated concrete buildings: Current status and analysis

G.Z. Lu, Z.J. Wang, Z.L. Guo, Z.S. Zhang & Y. Zhang
Beijing Building Materials Academy of Sciences Research Co., Ltd., Shijingshan District, Beijing, China

ABSTRACT: Reinforced sleeve connection of precast vertical components is somewhat technically disadvantaged. Hence, non-reinforced sleeve connection is attracting increasing interest. In the present paper, two types of non-steel sleeve connecting structural systems are introduced, namely, the double-superimposed shear wall structural system and the prefabricated hollow slab shear wall structural system. Through detailed description of the technical parameters, production processes, connection nodes, construction procedures and BIM technique application, the merits and shortcomings of the two systems are analyzed and the development trend of the connection of precast vertical components is looked forward.

1 INTRODUCTION

Currently, vertical components in prefabricated concrete structural systems are mainly connected by means of reinforced sleeve grouting, which inserts a single ribbed steel bar into a metal sleeve and injects a grouting mixture into the sleeve to form a one-piece structure after the mixture hardens and to splice the force-transmitting steel bars [1]. Nevertheless, this technique is highly demanding for the in-depth design, production accuracy, and construction workmanship; the bar splicing and grouting saturation are not directly observable; there are concealed works and the detection method is unpopular. Hence there are some hidden safety hazards.

As such, "non-reinforced sleeve grouting connection", such as the double-superimposed shear wall system and the prefabricated hollow slab shear wall structure system, is attracting increasing interest.

2 DOUBLE-SUPERIMPOSED SHEAR WALL STRUCTURAL SYSTEM

2.1 *Brief introduction to the double-superimposed shear wall structural system*

Precast double-superimposed wall panels are shop fabricated with precast double-layer concrete slabs and steel trusses connecting the double-layer precast concrete slabs. This is referred to as a double-superimposed shear wall (in short, "double-skin wall"). After the wall panel is field mounted, concrete is poured to participate in the load-bearing of the structure to form a double-superimposed shear wall (Figure 1).

Figure 1. Component disconnection and sample room of double-superimposed shear wall.

2.2 *Main technical parameters of double-skin wall*

2.2.1 *Maximum applicable height*

The maximum applicable height of a building with a double-skin wall must be as specified in Table 1 [2].

Table 1. Maximum applicable height for buildings with double-superimposed shear wall.

Type of Structure	Seismic Fortification Intensity			
	6	7	8 (0.2g)	8 (0.3g)
Double-superimposed shear wall structure	90	80	60	50

Note: Height of a building is defined by the distance from the exterior ground to the main roof, excluding any part of the roof that is locally higher than the rest part.

The maximum applicable height of a building with a double-skin wall is smaller than the maximum height of reinforced sleeve connected shear wall. For instance, if the seismic intensity is 8 (0.2 g), the maximum applicable height of the reinforced sleeve connected shear wall is 90 m (the total shear force carried by precast shear wall is greater than 80% of the total shear force of the layer; the maximum applicable height is 80 m), which is greater than the maximum height 60 m of double-skin wall.

2.2.2 *Aseismic grade*

The aseismic design of a double-skin wall structural component must be determined for the aseismic intensity determined according to the category of fortification and the height of building, as well as in conformity with the applicable calculation and construction measures. The aseismic grade of a Class C double-superimposed shear wall structure must be determined according to Table 2 [3].

Table 2. Aseismic intensities of Class C double-superimposed shear wall structures.

Type of Structure		Aseismic Fortification Intensity				
		7			8	
Superimposed shear wall structure	Height (m)	≤ 24	>24 and ≤70	>70	≤ 24	>24 and ≤70
	Shear Wall	IV	III	II	III	II
Partial frame-supported superimposed shear wall structure	Height	≤ 24	>24 and ≤60		≤ 24	>24 and ≤50
	Field-fabricated support frame	II	II		I	I
	Shear wall at bottom-reinforced location	III	II		II	I
	Shear wall at other locations	IV	III		III	II

Note: The aseismic intensity of buildings close to or higher than the height specified for a grade must be determined according to the degree of irregularity, the site and foundation conditions of the individual buildings.

The aseismic grade of a Class B prefabricated integral superimposed shear wall structure must be 1 degree higher than the local aseismic fortification intensity [4].

The aseismic grade of double-skin wall structure components must be determined according to the same criteria for sleeve connected shear wall structures with no need of additional aseismic calculation or construction measures.

2.2.3 *Material requirements*

The strength grade of concrete of precast components in double-skin wall structures must preferably be a minimum of C30; the strength grade of post-cast concrete must preferably be not lower than that of the concrete of the precast part and in no case may be lower than C25.

The cavity of the double-skin wall is only 100 mm; there are also connecting bars in the cavity. Hence, after the post-cast concrete must preferably be self-compacting concrete conforming to the applicable performance requirements. Where common concrete is used, the maximum size of the concrete aggregate must not be greater than 20 mm. After the concrete is poured, it should be vibrated with a 30 mm diameter vibrating rod.

Where self-compacting concrete or fine aggregate concrete is used, the engineering cost will be increased accordingly.

2.2.4 *Internal force amplification of field fabricated wall limb*

For a fabricated integral shear wall structure having both field fabricated and precast wall limbs within the same layer, the horizontal internal seismic force of the field fabricated wall limb must preferably be multiplied with an enhancement coefficient not smaller than 2.1.

The enhancement coefficient of the internal force of field fabricated wall limb must be the same as that of sleeve connected shear wall. No additional enhancement will be needed.

2.2.5 *Provision of rough surface*

A rough surface must be provided for the cavity of a double-skin wall. The unevenness depth of the rough surface must be a minimum of 4 mm. During real production, panel A of the double-skin wall must preferably be rough surfaced (similar to compact panel). However, after panels A and B are clamped, the thickness of the cavity is only 100 mm; it also contains connecting bars. Hence it is not easy for panel B to be rough surfaced (see Figure 2). It is recommended to optimizing the production process and adding a rough surface.

Figure 2. No rough surface is provided on panel B of the double-skin wall.

2.3 Type and connection nodes of precast components

2.3.1 Type of precast components

Common precast components include double-superimposed shear walls, one-sided superimposed shear walls, and one-sided superimposed shear walls with sandwich insulation (see Figure 3).

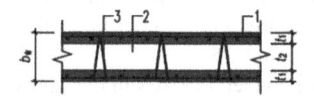

Double-superimposed shear wall construction

1-Precast part; 2-Post-cast part; 3-Steel truss; t_1-Thickness of precast panel; t_2. Thickness of the post-cast part; b_w. Calculated thickness of shear wall

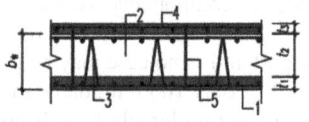

One-sided superimposed shear wall construction (without sandwich insulation)

1-Precast part; 2- Post-cast part; 3- Truss steel; 4- Steel mesh of the outer panel; 5-Connector; t_1--Thickness of precast panel; t_2-Thickness of the post-cast part; t_3-Thickness of outer panel; b_w- Calculated thickness of shear wall

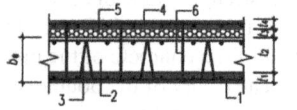

One-sided superimposed shear wall construction (with sandwich insulation)

1-Precast part; 2-Post-cast part; 3-Truss steel; 4-Steel mesh of outer panel; 5-Insulation; 6-Connector; t_1-Thickness of precast panel; t_2- Thickness of the post-cast part; t_3-Insulation thickness; t_4-Thickness of outer panel; b_w. Calculated thickness of shear wall

Figure 3. Common types of precast components.

2.3.2 Connection of the upper and lower precast panels

The height of the horizontal joint between the upper and lower precast wall panels must preferably be a minimum of 50 mm. The joint must be poured and compacted with field fabricated concrete.

Vertical connecting steel bars must be placed at the horizontal joint. The anchoring length of the vertical connecting steel bars must be a minimum of $1.2l_{aE}$; the spacing between the bars must not be greater than the spacing of bars in the precast wall panel and preferably not greater than 200 mm. The diameter of the bars must not be smaller than that of the bars in the precast wall panel (Figure 4).

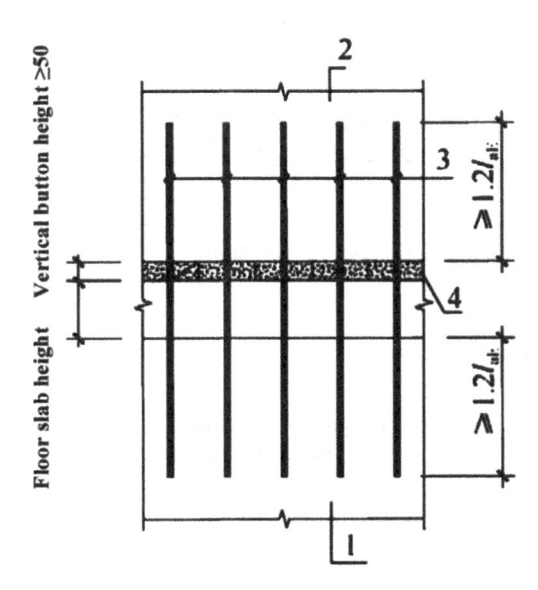

Figure 4. Connection of the upper and lower precast panels.

Wuhan University of Technology conducted a series of quasi-static seismic test on four double-superimposed shear walls and one fully field fabricated shear wall to compare the failure morphology, hysteresis loop, carrying capacity, displacement ductility, energy dissipation, and rigidity degeneration of the shear walls. The results show that the double-superimposed shear walls have roughly the same ultimate failure mode as the field fabricated shear wall [5]. According to these results, double-superimposed shear walls can be structurally designed as if they were field fabricated.

2.3.3 *Connection of precast wall panels within the same layer*
At the location of a non-edge component, the connection between the adjacent double-skin walls is as shown in Figure 5. The width of the post-cast section must not be smaller than the wall thickness and preferably a minimum of 200 mm. Inside the post-cast part, at least 4 vertical steel bars must be placed. The diameter of the steel bar must not be smaller than the diameter of the vertical bar in the precast wall panel and in no case may be smaller than 8 mm.

In the post-cast part, the height of the horizontal joint between the upper and lower precast wall panels must preferably be a minimum of 50 mm. The joint must be poured and compacted with field fabricated concrete.

The anchoring length of horizontal connecting bars in the double-skin wall must be a minimum of $1.2l_{aE}$. The spacing between the steel bars must preferable be the same as that of the steel bars in the precast wall and must be a maximum of 200 mm. The diameter of the bars must in no case be smaller than that of the bars in the precast wall panel.

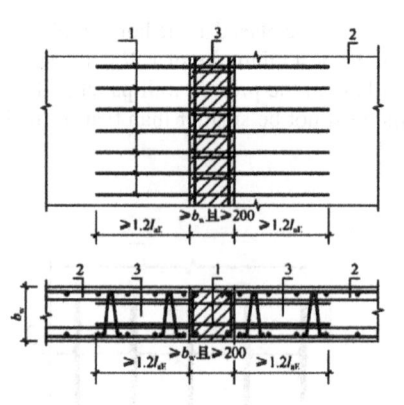

Figure 5. Connection of precast panels within the same layer.

2.4 Production of precast components

2.4.1 Production process

The production process of double-skin walls is marginally more complicated, predominantly in respect of the clamping of the panel A and panel B (see Figure 6).

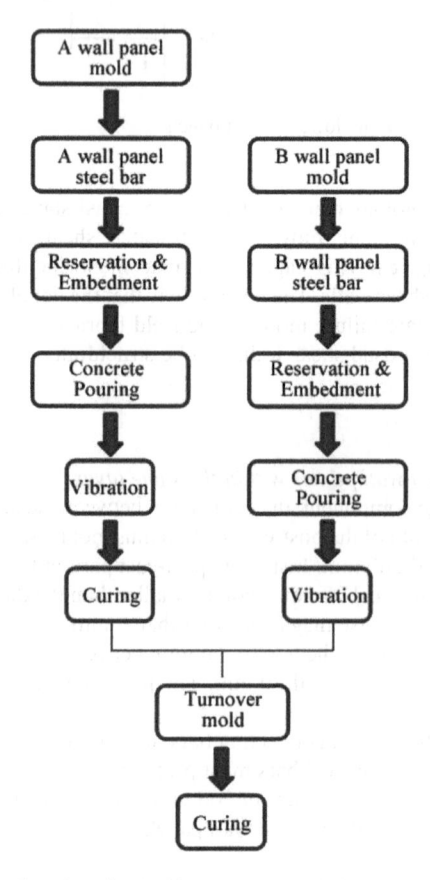

Figure 6. Production process flowchart for double-superimposed shear wall.

2.4.2 *Automated production*

When shop fabricating double-skin walls, automatic equipment, such as a truss machine, mesh machine, mold setting manipulator, torpedo ladle, spreader, vibrostand, turnover machine, and curing kiln, can be used to improve the production efficiency (Figure 7).

Figure 7. Automatic production equipment.

2.5 *Construction procedure*

Precast wall panels must be installed according to the construction drawing. Side stays must be placed according to the special construction plan. At the bottom of the double-skin wall, spacers for adjusting the joint thickness or bottom elevation may be mounted. Scaffolding for supporting the superimposed panel must be installed. Steel bars of fabricated parts such as horizontal joints, vertical joints, and edge components must be bound. Aluminum molds must be installed. Concrete must be poured and vibrated.

When installing a precast wall panel, check and correct reserved steel bars in the wall; check and adjust the reserved gaskets and screws. After installing the precast wall, check and adjust the installation position, installation elevation, perpendicularity, and the evenness between adjacent components; bind the steel bars of edge components; erect formwork at edge components and horizontal joints.

When pouring concrete, the floor slab and wall panel can be poured either at the same time or separately. When self-compacting concrete is used, it must be poured layer by layer. The thickness of each layer must be a maximum of 800 mm. After a layer of concrete is poured, the next layer must not be poured until after 1~1.5h. Vertical bars must be inserted before the concrete has initially set to ensure that the bars are continuously spliced.

When vibrating the concrete, a vibrating rod must be preferably used. Make sure that the reinforcement framework, facing brick and embedded parts are not touched. While vibrating, check from time to time whether there is slurry leakage or deformation or whether the embedded parts have been displaced.

2.6 Merits and shortcomings of double-superimposed shear wall structure system

The double-superimposed shear wall structure system involves indirect splicing with steel bars instead of sleeve connection with steel bars. It is less demanding for component fabrication and installation and therefore gained higher market recognition. No steel bar will come out from the end faces of the precast component. Theoretically, mold setting manipulators, turnover machines, torpedo ladles, spreaders and other automatic equipment can be used. Together with the plant MES system, information-based production can be realized to improve the production efficiency and release the production capacity of the plant. The precast components are with light dead weight and with less requirements for lifting weight of the tower crane. No formwork is needed during installation.

The maximum applicable height for buildings with a double-superimposed shear wall structure system is subject to limitations. It must be lower than that of buildings with a reinforced sleeve connected structure. These structures have higher requirements for the production equipment. Particularly, they need a mold platen turnover machine, which are not equipped in many component manufacturing factories. During construction, self-compacting concrete or fine aggregate concrete must be infilled, leading to a high production cost.

3 OTHER PREFABRICATED INTEGRAL STRUCTURE SYSTEMS

3.1 Prefabricated hollow slab shear wall structure system

The prefabricated hollow slab shear wall structure system is a prefabricated integral shear wall structure constructed from precast hollow shear wall, precast hollow edge components, precast hollow partition wall and precast connecting beams connected together at effective vertical and horizontal nodes and combined with post-cast concrete through the hollow holes in the wall panel (see Figure 8).

Figure 8. Prefabricated hollow slab shear wall structure system.

The prefabricated hollow slab shear wall structure system is indirectly spliced with steel bars instead of sleeve connection. It is less demanding for the processing and installation precision with higher market recognition. This system does not need grouting material and is therefore convenient for construction in winter. Precast components are highly standardized and modularized, making them easier for design, manufacturing and installation. Precast wall panels, precast under-window wall panels, precast infill panels, and precast laminated slab floor can all be molded in groups and manufactured on a production line. The production is advantageous for small land coverage and high level of automation. The manufacturing of the post-cast part at the construction site is highly standardized. The associated technical system used can well improve the construction efficiency at the site.

The disadvantage of this system is that small-sized panels are frequently used, therefore a lot of splices are needed. The installation speed is also compromised by the speed of the tower crane. Moreover, traditional external insulation is used. Hence there can be problems in terms of decoration, heat insulation and fireproofing performance.

3.2 *Prefabricated integral welded steel fabric composite concrete structure system*

All or part of the lateral resisting components are prefabricated integral concrete structures with welded steel fabric composite shear wall and composite columns, referred to as a composite structure [6]. The precast components of this structure are constructed with a formed steel bar cage and precast walls on its sides; the center is a cavity (Figure 9). It is similar to a double-superimposed shear wall.

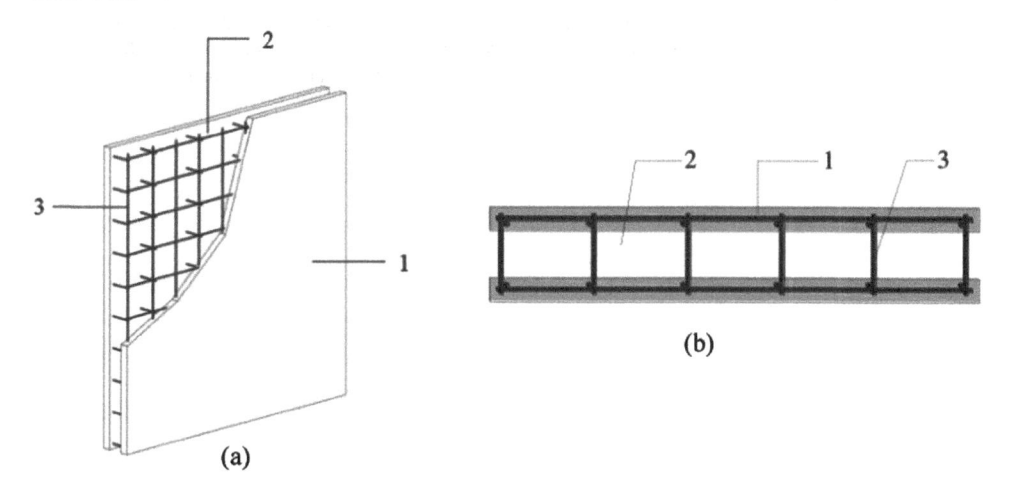

Figure 9. Prefabricated integral welded steel fabric composite concrete structure. Precast part; 2. Cavity part; 3. Formed steel bar cage: (a) 3D diagram of the precast hollow wall and (b) Section view of the precast hollow wall component.

The prefabricated integral welded steel fabric composite concrete structure system approximates a double-superimposed shear wall system. However, compared with a double-skin wall system, this system uses a formed steel bar cage with a higher level of automation. Besides, its maximum applicable height for buildings is higher than that of a double-skin wall system and equal to that of a reinforcement sleeve connected structure system.

4 CONCLUSIONS

The double-superimposed shear wall structure system and the prefabricated integral welded steel fabric composite concrete structure system are inherently light in dead weight. It requires no mold and has a high level of automation. In the meantime, it also has high requirements for the quality of the equipment and the infill of concrete.

The prefabricated hollow slab shear wall structure system has stimulated the research on non-sleeve structure systems in China. Instead of sleeve connection, it uses indirect splicing with steel bars in the cavity, answering better to the current situation of China. By emphasizing standardization and modularization, this system better conforms with the requirement of industrialization. On the other side, as building standardization in China is still at a low level, excessive emphasis on modularization can become too unrealistic.

The prefabricated integral structure systems available in the market so far all need to be further iterated and optimized based on their advantages of having a cavity to better suit to the market demand and speed up the assembly and industrialization process of China.

REFERENCES

[1] Sha An., Wang Xiaofeng. *Technical Specification for Grout Sleeve Splicing of Rebars* JGJ355–2015[S]. China Construction Industry Press, 2015.
[2] Liu Weidong, Yu Yinquan. *Technical Standard for Prefabricated Concrete Buildings* GB/T51231–2016 [S]. China Construction Industry Press, 2017.
[3] Wang Pingshan, Fan Hua. *Technical Specification for Superimposed Concrete Shear Wall Structure* DG/TJ 08–2266–2018 J 14282–2018 [S]. Tongji University Press, 2018.
[4] Gu Qian, Wen Siqing. *Technical Specification for Prefabricated Integral Superimposed Shear Wall Structures* DB42T1483–2018 [S]. Wuhan University of Technology Press, 2018.
[5] Dong Ge, Gu Qian, Tan Yuan. *Experimental Study on the Effect of Horizontal Joint Connection on the Seismic Behavior of Double-Superimposed Shear Wall* [J]. Vibration and Shock, 2020, 39(2):107–114.
[6] Ma Yunfei, Tian Chunyu, Zhang Meng. *Technical Specification for Prefabricated Integral Welded Steel Fabric Composite Concrete Structure* T/CECS 579–2019 [S]. China Planning Press, 2019.

Design of intermolecular charge-transfer fluorescent probes and determination of the content of drugs

C.F. Jin, L.S. Yu, Y Wang & Y.Q. Cheng
College of Pharmacy, Jiamusi University, Jiamusi, China

Y.G. Lv
College of Pharmacy, Jiamusi University, Jiamusi, China
College of Materials Science and Engineering, Jiamusi University, Jiamusi, China

ABSTRACT: Ofloxacin is capable of forming stable complexes with tricyanodihydrofuran derivatives. In this experiment, infrared, ultraviolet and fluorescence spectra will be analyzed to measure the pH, temperature and time of the most suitable reaction conditions. This will lead to the discovery of a method that can measure the content of ofloxacin with fast reaction speed, mild conditions and high sensitivity.

1 INTRODUCTION

Intramolecular charge-transfer compounds are extremely sensitive to their surroundings and are gaining importance in the field of bioanalysis and medicine as an effective fluorescent probe for chemical sensing and molecular recognition of fluorescence [1]. It is not only expected to be an effective fluorescent probe to probe the microenvironment, but also provides a very good research channel for chemical sensing and molecular recognition of fluorescence. Tricyanodihydrofuran, a D-π-A type molecule, has a P-πconjugation system. D-π-A complexes are usually composed of donor-conjugated bridge-acceptor. The electron donor and the electron acceptor are connected byπbridge to from a mutual transfer system [2–4]. The push-pull electrons of the electron donor and the electron acceptor cause great changes in the absorption and emission wavelengths of the structure, thus extending its application in the field of optics.

In this paper, we successfully constructed a fluorescent probe of tricyanodihydrofuran derivatives with ofloxacin, which was used for the determination of the drug content of of ofloxacin with high sensitivity and simple operation. By applying the intermolecular electron transfer theory to measure the drug content, this method provides an idea for the application of tricyanodihydrofuran electron absorbers, and also provides a new way for drug detection [5,6].

2 METHODS

2.1 *Preparation of solutions*

2.1.1 *Preparation of standard solution of tricyanodihydrofuran derivatives*
Weigh 3.5 mg of product III (tricyanodihydrofuran derivative) and dissolve in 46.67 mL of chloroform. Ultrasonically dissolve and dilute to a concentration of 1.5×10^{-4} mol/L.

2.1.2 *Preparation of standard solution of ofloxacin (OFX)*
Weigh 5 mg of Ofloxacin (AR≥99.0 %), dissolve in 50 mL of methanol and prepare 100 mg/L of Ofloxacin.

2.2 Determination of fluorescence spectra of tricyanodihydrofuran derivatives complexed with Ofloxacin

Three 5 mL stoppered tubes were washed with methanol and then dried. Take 2.0mL of the configured DCDHF-2-V-I solution and add it to one test tube, then take 2.0mL of the prepared Ofloxacin solution and add it to another test tube. Then 2.0mL of DCDHF-2-V-I solution and 2.0mL of ofloxacin solution were added into the third test tube and placed in a water bath at 40°C for reaction for 20 minutes. The fluorescence excitation and emission spectra of the three tubes were detected respectively. Firstly, its fluorescence emission spectrum is scanned at a certain excitation wavelength, and its fluorescence emission spectrum is obtained by taking the maximum emission wavelength as excitation.The results show that the maximum emission spectrum of the complex is 503nm with 359nm as the appropriate excitation wavelength. The fluorescence excitation and emission spectra of three kinds of branched tubes were obtained by this method.

Three 5 mL stoppered tubes were washed with methanol and then dried. The prepared solution of trisyandihydrofuran derivative 2.0mL was added into a tube. Then take the prepared ofloxacin solution standard solution and add 2.0ml into another test tube; then 2.0mL of the solution of trisyandihydrofuran derivative and 2.0mL of ofloxacin solution were added into the third test tube and placed in a water bath at 40°C for 20 minutes. After reaction, they were taken out of the water bath and left for a period of time to cool to room temperature. The fluorescence emission and excitation spectra were detected by measuring 2.0 mL of the standard solution ofloxacin, DCDHF-2-V-I and their complexes from three test tubes, adding them to the test dishes and then placing them in the fluorescence detection cell.

The factors affecting the fluorescence emission spectra of the complexes generated by ciprofloxacin and DCDHF-2-V-I were then investigated according to the above method, including temperature, pH, reaction time and dosagel[7–11].

3 RESULTS AND DISCUSSION

3.1 Results

3.1.1 Fluorescence excitation and emission spectroscopy

According to Figure 1, it is shown that both DCDHF-2-V-I and ofloxacin can emit fluorescence. When the two form a complex, intermolecular electron transfer occurs leading to luminescence changes, and the fluorescence intensity of their excitation and emission spectra is significantly enhanced after the formation of the ligand.

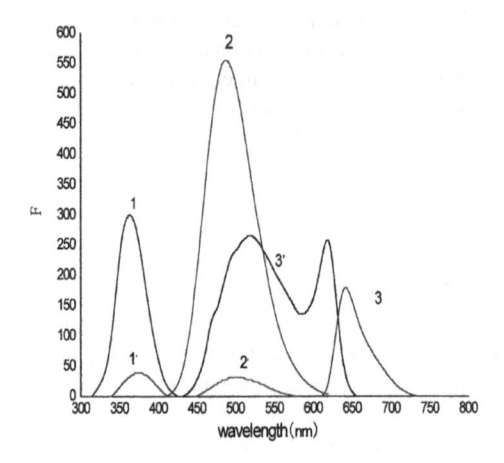

Figure 1. Fluorescence excitation and emission spectra of DCDHF-2-V-I and of ofloxacin:*1 ofloxacin emission spectrum .1'ofloxacin excitation spectrum.2 DCDHF-2-V-I ofloxacin emission spectrum.2'DCDHF-2-V-I ofloxacin excitation spectrum.3 DCDHF-2-V-I emission spectrum.3'DCDHF-2-V-I excitation spectrum.*

3.2 Discussion

3.2.1 Effect of pH on the reaction

Take 7 5 ml of tube, in turn, add 1 ml ofloxacin standard solution and DCDHF - 2 - V - I standard liquid, at 40°C water bath pot reaction for 20 minutes, add the pH buffer solution respectively in 4.5, 5.5, 5,6,6.5, 7,7.5 measuring the fluorescence emission spectrum, pH and fluorescence diagram as shown in figure 2, the fluorescence intensity of 550 strongest when pH = 5.5, show that the best luminous pH pH = 5.5 for.

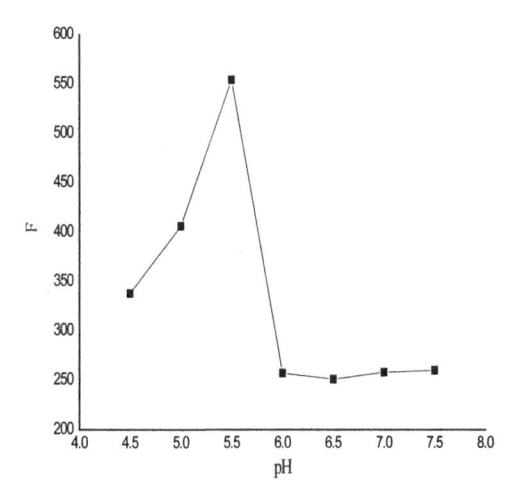

Figure 2. Effect of pH on the reaction.

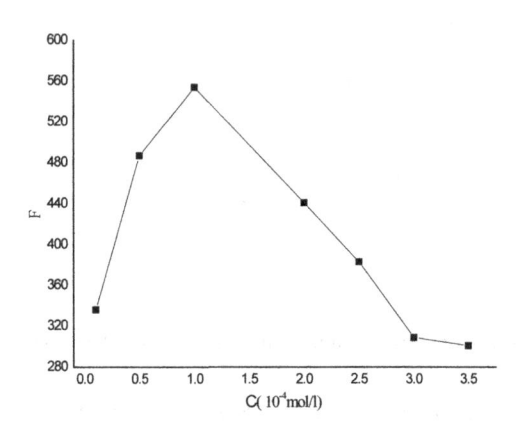

Figure 3. Effect of concentration of tricyanodihydrofuran derivatives on the reaction.

3.2.2 Effect of concentration of tricyanodihydrofuran derivatives on the reaction

Seven 5mL tubes with plug were taken, then 1mL ofloxacin standard solution and different volumes of DCDHF-2-V-I standard solution (diluted by 1mL) were added successively, and placed in a water bath at 40°C for reaction for 30 minutes. The tubes were scanned by fluorescence emission spectrum. The fluorescence spectra of the dihydrofuran derivatives showed the highest fluorescence at 1.25×10^{-4} mol/L. See Figure 3 below.

3.2.3 *Screening of conditions for the reaction of tricyanodihydrofuran derivatives with ofloxacin complexes*

3.2.3.1 *Effect of the amount of tricyanodihydrofuran derivatives* Ten 5mL tubes with plug were taken, and 1mL ofloxacin standard solution and DCDHF-2-V-I standard solution of different volumes were successively added into a water bath at 40°C with a gradient of 0.1-1.1mL for reaction for 30 minutes. The fluorescence emission spectrum of DCDHF-2-V-I 0.7mL showed the strongest fluorescence intensity.

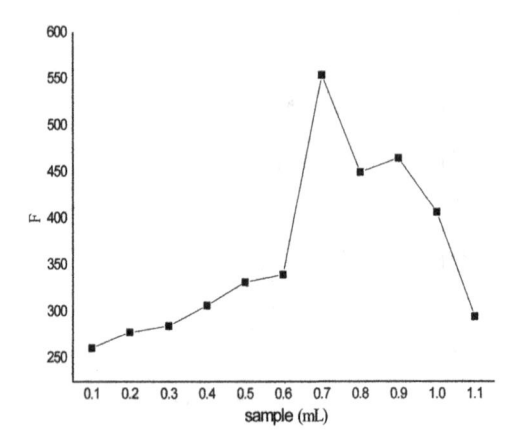

Figure 4. Effect of the amount.

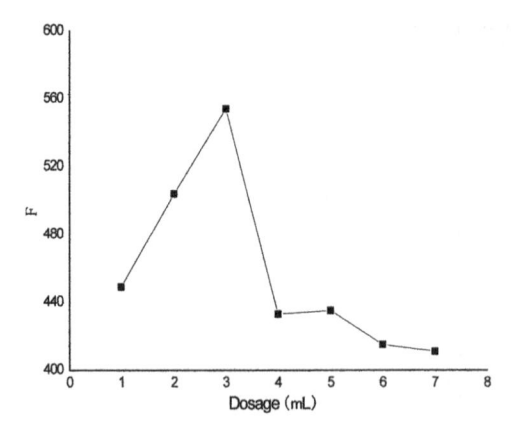

Figure 5. Effect of the dosage of of tricyanodihydrofuran derivativesofloxacin on the reaction.

3.2.3.2 *Effect of the dosage of ofloxacin on the reaction* Seven 10mL stopper tubes were taken, and 1mL DCDHF-2-V-I standard solution and 1mL ofloxacin standard solution were successively added and placed in a water bath at 40°C for 30 minutes to detect their fluorescence intensity. The results are shown in Figure 5: the fluorescence intensity of the charge-shift complex of of ofloxacin was maximum at 3.0 mL, when the conditions were optimal.

3.2.4 *Effect of temperature and time on the reaction*

3.2.4.1 *Effect of temperature on the reaction* Seven 5mL stoppered test tubes were taken with equal molar amounts of the two substances and placed in a water bath at an interval of 10°C for 50 minutes at 20 – 60°C respectively. The fluorescence intensity of the complexes was measured, and the results showed that the fluorescence intensity reached 560 at 30°C, and the fluorescence intensity was the strongest.

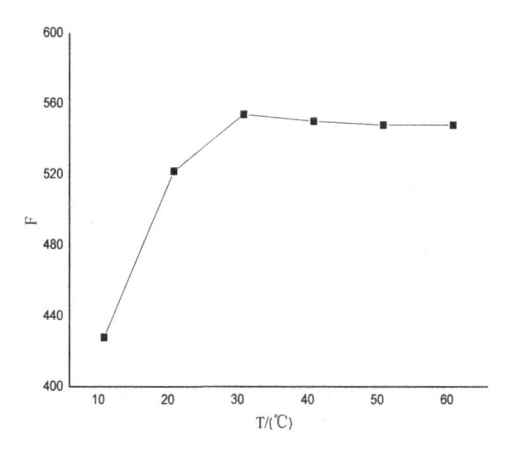

Figure 6. Effect of temperature on the reaction.

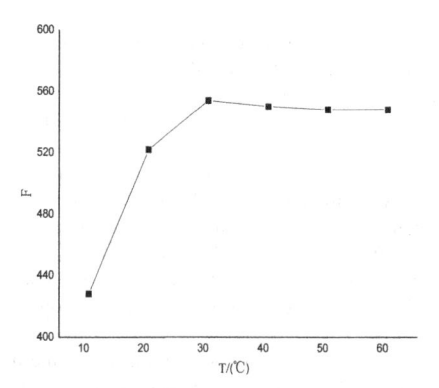

Figure 7. Effect of time on response.

3.2.4.2 *Effect of time on response.* Seven 5mL stoppered test tubes with equal molar amounts of the two substances were placed in a water bath at 30°C for 10–60 minutes (with an interval of 10 minutes), and the fluorescence intensity of the complex was strongest at 30 minutes.

3.3 *Application of tricyanodihydrofuran derivatives to the detection of of ofloxacin*

5 tablets ofloxacin (specification 100mg) were powdered in a mortar and pestle, and 1/5 of the total grinding amount (equivalent to 100mg of drug) was accurately weighed and dissolved with appropriate amount of anhydrous methanol and sonicated to aid dissolution. Allow to dissolve completely and fix the volume with anhydrous methanol (100 mL). Accurately pipette 10mL of the sample solution, dilute to 1×10^{-3} mol/L with anhydrous methanol and set aside. At pH 5.5, temperature 20°C, reaction time 40 min; take the above optimal conditions (where DCDHF-2-V-I dosage 0.7 mL), select the prepared standard solution ofloxacin, take 1–3mL 9 data (interval 0.3) for linear regression, with fluorescence intensity F as the vertical coordinate and concentration C as the horizontal coordinate, to obtain the linear equation $F = 442486C - 149.85$ with a correlation coefficient $r = 0.998$. The above optimal experimental conditions were selected for 11 parallel determinations of the standard solution of of ofloxacin and the RSD=1.13% was obtained.

The content of the prepared tablet solution of ofloxacin was measured by using the fluorescence intensity versus concentration to obtain 10.52; 10.49; 10.49; 10.50; 10.51, which is not much different from the 10 mg contained in the sample, and can be used to determine the content of ofloxacinr [12–14].

4 SUMMARY

Ofloxacin can form stable complexes with tricyanodihydrofuran derivatives. The optimum reaction conditions are pH 5.5, temperature 30°C and time 30 min. Therefore, this method can be used for the measurement of norfloxacin with fast reaction speed, mild conditions and high sensitivity.

ACKNOWLEDGMENTS

This work was financially supported by the National Science Foundation of China (No.213 46006), Department of scientific research project in Heilongjiang province (No.B2017015).

REFERENCES

[1] Hui Wang, Zhikuan Lu, Samuel J., et al. Modifications of DCDHF single molecule fluorophores to impart water solubility[J]. Tetrahedron Letters, 2007, 48: 3471–3474.
[2] Hui Wang, a Zhikuan Lu, a Samuel J, et al. Lord.The influence of tetrahydroquinoline rings in dicyanomethylenedihydrofuran (DCDHF) single-molecule fluorophores[J].Tetrahedron, 2007, 63(1):103–114.
[3] Mingqian He, Thomas M. Leslie, John A, et al. Sinicropi. r-Hydroxy Ketone Precursors Leading to a Novel Class of Electro-optic Acceptors[J]. Chem. Mater. 2002, 14(5):2393–2400.
[4] EnriqueFont-Sanchis, a Raquel E. Galian, b Francisco J, et al. Alkoxy-styryl DCDHF fluorophores[J]. Physical Chemistry Chemical Physics, 2010, 12(28):768–7771.
[5] Wissam Bentoumi. Concise Multigram-Scale Synthesis of Push–Pull Tricyanofuran-Based Hemicyanines with Giant Second-Order Nonlinearity: An Alternative for Electro-optic Materials [J]. Chem. Soc. Rev, 2014, 20: 8909–8913.
[6] Rurack. K., Genger, U.R. Rigidization, preorientation and electronic decoupling-the 'magic triangle' for the design of highly efficient fluorescent sensors and switches[J]. Chem. Soc. Rev., 2002, 33(22): 116–127.
[7] He L-F, Lin D-L, Li Y-Q, et al. Application of simultaneous fluorescence analysis and its new progress[J]. Advances in Chemistry, 2004, 16(6): 879–885.
[8] Pearce, D.A., Jotterand, N., Carrico, I.S., et al. Derivatives of 8-Hydroxy-2-methyl quinoline are powerful prototypes for zinc sensors in biological systems[J]. J. Am. Chem. Soc., 2001, 123(21): 5160–5161.
[9] Sigmund. H., Pfleiderer.W. A new type of labeling of nucleosides and nucleotides, [J]. Helv Chim Acta, 2003, 86(7):2299–2334.
[10] Adamczyk. M., Chan. C., Fino. J., et al. Synthesis of 5-and 6-hydroxymethyl fluoresce in phosphora inidites[J]. J. Org. Chem., 2000, 65(2): 596–601.
[11] Gong Wei. Design, synthesis and properties of novel second-order nonlinear optical chromophore molecules[D]. Master's thesis. Wuhan University, 2005.
[12] Seon-Yeong Gwona, Sung-Hoon Kim, Anion sensing and F3-induced reversible photoreaction of D-π-A type dye containing imidazole moiety as donor Spectrochimica Acta PartA[J]Molecular and Biomolecular Spectroscopy 2014,117: 810–813.
[13] J.W. Quilty D.G. Thomas, D.J. Clarke, R.D. Breukers. The effect of bulky groups on the electro-optic coefficient r 33 of a pyridine-donor nonlinear optical[J]. chromophore Optical Materials 2016,54:147–154.
[14] Venkatakrishnan Parthasarathy,Frederic Castet,Ravindra Pandey livier Mongin,PuspenduKumarDas, Mireille Blanchard-Desce. Unprecedented intramolecular cyclization in strongly dipolar extended merocyanine dyes: A route to novel dyes with improved transparency, nonlinear optical properties and thermal stability[J]. Dyes and Pigments 2016, 130:70–78.

Optimization design and fatigue life analysis of damping conical rubber spring for rail vehicle

R.G. Zhao
College of Civil Engineering and Mechanics, Xiangtan University, Xiangtan, China
Key Laboratory of Dynamics and Reliability of Engineering Structures of College of Hunan Province, Xiangtan University, Xiangtan, China

X.Q. Yang
College of Civil Engineering and Mechanics, Xiangtan University, Xiangtan, China

Y.J. Huang
Zhuzhou Time New Material Technology Co., Ltd., Zhuzhou, China

W.H. Liu, X. Zhou, Y.L. Liu & D.H. Ye
College of Civil Engineering and Mechanics, Xiangtan University, Xiangtan, China

ABSTRACT: The fatigue life prediction model for rubber materials is built by using the equivalent tearing energy method. The finite element model of a damping conical rubber spring for rail vehicles is established by using ABAQUS software. The finite element numerical simulation results show that the stiffness coefficient, and fatigue life as well, does not satisfy the structural design requirements. The influence of geometric parameters of the conical rubber spring on its fatigue life is analyzed by using the orthogonal numerical test method. The results show that the thickness and length of the first rubber layer significantly affect the fatigue life of the structure, and the thickness of the fourth rubber layer has the least effect. Taking the minimum strain as an optimization objective and the geometric parameters determined by the orthogonal numerical tests as the design variables, the second-order response surface model is established, and the Multi-Island Genetic Algorithm is used to carry out the single objective multi-factor optimization design for the rubber spring under the condition of stiffness constraint. The results show that the stiffness coefficient of the optimized structure is 0.630kN/mm, which falls within the range of the design stiffness coefficient, and the fatigue life is 4.963 million cycles, while the design fatigue life is 3.500 million cycles, so the fatigue life and stiffness all reaches the design requirement. The test verification indicates that the calculated fatigue life is well fit with the test one, and the simulated fatigue failure position is consistent with the test.

1 INTRODUCTION

The rubber material has many excellent properties, among them the most significant one is its visco-hyperelasticity, which makes it able to bear large strain, and plays an importance role in the fields of vibration isolation [1–3], noise reduction [4,5] and sealing [6–9]. In general, the rubber materials and their products are used under the conditions of quasi-static or alternating stress/strain environment. The fatigue life of a rubber product is one of the main indexes to check whether the product quality is qualified or not. The fatigue test of rubber products has the disadvantages of long period and high cost. Therefore, in the stage of design research and development for a rubber product, the finite element numerical simulation is applied to replace the traditional fatigue tests to predict the fatigue life of the rubber product, so as to effectively shorten the test time and reduce the product design cost [10–12].

Griffith [13] presented the energy criterion for crack propagation, which believed that there was a defect distribution in the material, and the failure started at the position of largest defect. Rivlin and Thomas [14] used the energy criterion for crack propagation to analyze the tearing behavior of rubber, and proposed the concepts of tearing energy and strain energy release rate. According to the energy criterion for crack propagation, Lake [15,16] found that the crack propagation rate is only related to the maximum strain energy release rate in the period, so the fatigue crack propagation rate of rubber material can be uniquely determined by the strain energy release rate. The rubber components in the actual service environment often need to bear loads in all directions, and the rubber materials are in a complex stress state. Mars [17], Verron [18] and Auyob [19] carried out a lot of experimental studies on the multiaxial fatigue of rubber materials, and proposed a various of damage parameters to describe the multiaxial fatigue behavior, so as to predict the fatigue life of rubber materials under the condition of complex stress state, in which, the tearing energy density proposed by Mars can be used to predict the cracking direction of rubber material.

Under the action of alternating load, the fatigue nucleation occurs at the positions of stress/strain concentration or the interface between particle and matrix of rubber material, and then fatigue cracks initiate and propagate [20]. For the rubber component without macro defects, the nucleation life and crack initiation life are the main factors controlling the fatigue life, while for the rubber component with macro cracks, the crack propagation life is almost the total fatigue life. In the process of fatigue crack propagation, the relation between crack propagation rate and tearing energy satisfies a power law. With the increase of tearing energy, the crack gradually propagates, and the ultimate failure occurs when the crack size reaches the critical length. The theory of cracking energy is a kind of theory of rubber fatigue analysis combining the fracture mechanics method and the S-N curve method, which can be used to predict fatigue life and crack initiation direction of rubber materials.

In this paper, in order to improve the axial fatigue life, based on the theory of cracking energy, the finite element modeling and optimization design of a damping conical rubber spring for rail vehicles are carried out by using ABAQUS, FE-safe and Isight software, so that the structural stiffness fits the design object, and the fatigue life satisfies the service requirements.

2 FATIGUE LIFE MODEL OF RUBBER MATERIALS

It is assumed that there is a microcrack with initial length c_0 in the rubber material before the load is applied. With the increase of the number of load cycles, the fatigue crack grows gradually, following with the strain energy releasing and the tearing energy producing. As the crack propagation reaches the critical length c_f, the fatigue failure occurs in the rubber material. The number of cycles that the crack undergoes from the initial length to the critical one is called as the fatigue life of rubber material, and its expression is written as

$$N_f = \int_{c_0}^{c_f} 1/f(T(c,t)) \mathrm{d}c \tag{1}$$

where, N_f is the fatigue life. $T(c, t)$ is the tearing energy, which is the function of the current crack length c and time t, and $f(T(c, t))$ is the expression of crack propagation model. In the stage of fatigue crack nucleation, the tearing energy can be expressed as the form of fracture energy, that is

$$T = 2\pi W_c c \tag{2}$$

where, T is the tearing energy, and W_c is the fracture energy. While in the stage of fatigue crack growth, the crack propagation rate of rubber is controlled by the equivalent tearing energy. Therefore, using the equivalent tearing energy instead of the stress intensity factor range in the Paris formula, the crack propagation rate equation can be obtained. The equivalent tearing energy is derived as [21]

$$T_{eq} = T_{max}^{\frac{F(R)}{F(0)}} T_c^{\frac{F(0)-F(R)}{F(0)}} \tag{3}$$

$$F(R) = F_0 e^{F_{exp}R} + F_1 R + F_2 R^2 + F_3 R^3 \tag{4}$$

where, T_{eq}, T_{max} and T_c are individually the equivalent, maximum and critical tearing energy, F is the function of the stress ratio R, and F_{exp}, F_0, F_1, F_2 and F_3 are the material parameters, respectively. When the stress ratio R is equal to zero, the fatigue load belongs to a pulse loading mode, in this case, the value of the function F is calculated as $F(0)=F_0$, while in the case of $R > 0$, the fatigue load is a tensile-tensile fatigue loading mode.

In order to calculate the equivalent tearing energy, it is necessary to input the fatigue parameters of rubber material into the material parameter table of FE-safe software. In this research, the fatigue parameters are obtained by the fatigue tests for rubber materials, and are given in Table 1.

Table 1. The fatigue parameters using for calculating the equivalent tearing energy of rubber.

Initial length c_0 (mm)	Critical length c_f (mm)	Critical tearing energy T_c (J/mm^2)	Critical growth rate r_c (mm/cycle)	F_0	F_1	F_2	F_3
0.178	1.000	33890	4.73625×10^{-3}	2.717	25.310	−102.700	164.250

3 FINITE ELEMENT MODEL AND FATIGUE LIFE ANALYSIS

3.1 *Constitutive model of rubber materials*

In the finite element fatigue analysis of rubber specimens or rubber products, it is necessary to select an appropriate hyperelastic constitutive model of rubber materials. In this research, the Mooney-Rivlin model is selected as the hyperelastic constitutive model of rubber materials, and the expression of the Mooney-Rivlin model in the form of strain energy density function is written as

$$U = C_{10}(\bar{I}_1 - 3) + C_{01}(\bar{I}_2 - 3) + \frac{1}{D_1}(J - 2)^2 \quad (5)$$

where, U is the strain energy density, $\bar{I}_1$ and $\bar{I}_2$ are individually the first- and second-order strain invariants, D_1 is the volume parameter, J is the volume deformation ratio, c_{10} and c_{01} are the material parameters. Using Equation (5), the experimental data obtained from the fatigue tests for the samples of rubber materials are fitted by using a nonlinear regression method, and the comparison between the theoretical curve and experimental data is shown in Figure 1.

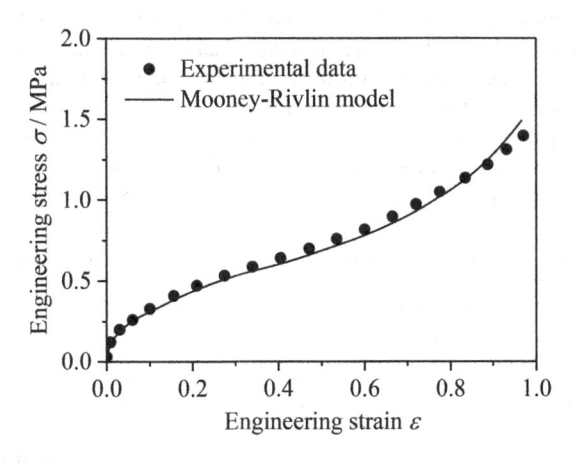

Figure 1. The engineering stress-strain curve.

It can be seen from Figure 1 that the theoretical curve is fitted well with the experimental data, the error between theoretical and experimental results is only 1.02%.

3.2 Finite element fatigue life analysis

In this paper, a damping conical rubber spring for rail vehicle is selected as a research object, and the finite element optimization design is carried out for such a structure. The conical rubber spring is composed of a shaft core, four-layer rubbers, three-layer spacer sleeves and an outer sleeve. In the process of finite element modeling, the structural symmetry of the conical rubber spring is considered, so a 1/2 model is used for the finite element numerical simulation. Firstly, the 3D model of the conical rubber spring is established by SolidWorks software, and then the model is inputted into HyperMesh for mesh generation. The C3D8H element is adopted for the rubber parts, and the C3D8R element is applied for the shaft core, spacer and outer sleeve. The finite element model is shown in Figure 2.

Figure 2. The finite element model of structure.

The finite element model of the conical rubber spring is inputted into ABAQUS software, and the material parameters in the model are shown in Table 2. According to the service environment of the product, the load is acted in the vertical direction of the conical rubber spring in a step-by-step loading mode. In the process of finite element numerical simulation, the eight-level loading mode is adopted. The vertical load is gradually acted on the structure from 0 to 35.1kN, and the numerical displacement results are calculated. The linear regression method is used to fit the numerical test data, and the stiffness curve of the structure is obtained, as shown in Figure 3. According to the linear regression analysis, the stiffness coefficient obtained by the numerical simulation analysis is about 0.724kN/mm, which is beyond the range of stiffness coefficient and does not satisfy the design requirement, since the test stiffness coefficient varies from 0.618kN/mm to 0.715kN/mm.

Table 2. The material parameters of finite element model of the conical rubber spring.

Component	Material	Elastic modulus E(GPa)	Poisson's ratio μ	Yield stress σ_Y (MPa)	c_{10}	c_{01}	D_1
Shaft core	Q235B	206	0.274	235	/	/	/
Spacer sleeve	Q235B	206	0.274	235	/	/	/
Outer sleeve	Q345	206	0.280	235	/	/	/
Rubber	NR	/	/	/	0.1969	0.0396	0.001

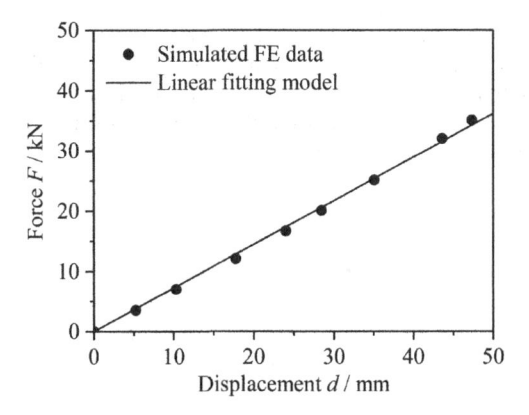

Figure 3. The stiffness curve obtained by FEM.

Under the condition of ultimate load, the stress distribution of the rubber lays are shown in Figure 4. It can be found from Figure 4 that when the load is 35.1kN, the metallic material is in the state of infinite life, while the maximum stress of the rubber material is 3.09MPa, which is located at the lower end of the first rubber layer near the shaft core, and the maximum stress exceeds the fatigue limit of rubber materials. Therefore, the nominal strain is selected to characterize the fatigue life of the structure, and the maximum strain on the first rubber layer is 0.7306. According to the rubber material parameters listing in Table 1, the fatigue life of the rubber materials is calculated by FE-safe software and the tearing energy method, and the fatigue life distribution of rubber lays is shown in Figure 5.

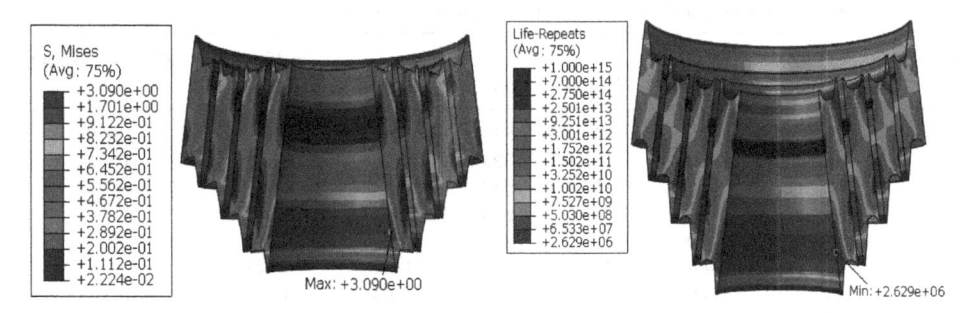

Figure 4. The stress distribution of rubber.

Figure 5. The fatigue life distribution of rubber.

It can be found from Figure 5 that the fatigue life of rubber material is 2.629×10^6 cycles, which is less than 3.5×10^6 cycles required by the design fatigue life. From the point of view of fatigue life, the fatigue life of the initial conical rubber spring does not fit the design requirements. Therefore, in order to better satisfy the fatigue life required in the engineering practices, the original structure needs to be optimized.

4 STRUCTURE OPTIMIZATION DESIGN

4.1 *Orthogonal numerical tests*

In order to satisfy the engineering requirements of the conical rubber spring, the geometric parameters of the overall shape of the structure can not be changed. Combining with the aforementioned finite element analysis of fatigue life, five geometric parameters are selected as the optimization parameters, including the thickness H_1 of the first layer of rubber, the thickness H_2 of the fourth layer of rubber, the length L of the first layer of rubber, the angle α between the spacer sleeve and the left horizontal line, and the distance R from the lower end of the first layer of rubber to the structural symmetry center, as shown in Figure 6.

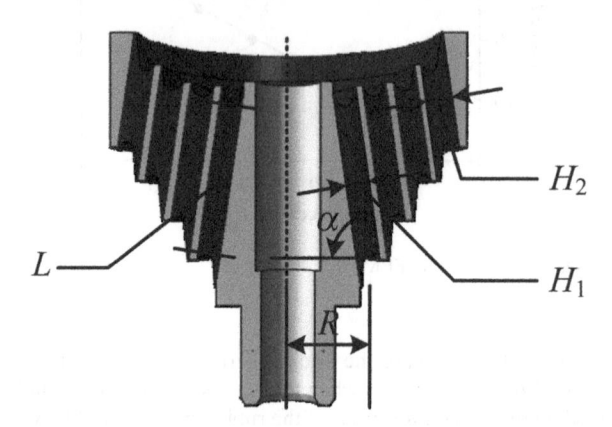

Figure 6. Geometric parameters for optimization design.

The influence of geometric parameters on fatigue life was analyzed by using the orthogonal test method. In the orthogonal tests of finite element simulation, each geometric parameter is selected for 4 levels. The orthogonal test is selected as a numerical test with 5 factors and 4 levels, and 16 sets of numerical tests are designed. The results of orthogonal tests are listed in Table 3.

Table 3. The results of orthogonal numerical tests for geometric parameters.

Sample No.	H_1 (mm)	H_2 (mm)	L (mm)	α (°)	R (mm)	Stiffness K (kN/mm)	Maximum strain ε_{max}
1	29.2	32.0	191.1	81.0	105.4	0.771	0.653
2	29.2	34.0	201.1	82.0	110.4	0.846	0.796
3	29.2	36.0	211.1	83.0	100.4	0.695	0.707
4	29.2	38.0	221.1	84.0	95.4	0.648	0.709
5	31.2	32.0	211.1	82.0	105.4	0.861	0.751
6	31.2	34.0	221.1	81.0	100.4	0.828	0.580
7	31.2	36.0	191.1	84.0	110.4	0.674	0.709
8	31.2	38.0	201.1	83.0	95.4	0.714	0.699
9	33.2	32.0	221.1	83.0	110.4	0.721	0.996
10	33.2	34.0	211.1	81.0	105.4	0.841	0.867
11	33.2	36.0	201.1	84.0	95.4	0.653	0.807
12	33.2	38.0	191.1	82.0	100.4	0.733	0.710
13	35.2	32.0	201.1	84.0	100.4	0.623	0.991
14	35.2	34.0	191.1	83.0	95.2	0.680	0.889
15	35.2	36.0	221.1	82.0	105.4	0.787	1.074
16	35.2	38.0	211.1	81.0	110.4	0.836	0.775

4.2 *Optimization design of geometric parameters*

According to the orthogonal numerical test results listed in Table 3, the effect of geometric parameters H_1, L, α and R on the fatigue life of the conical rubber spring is significant. Therefore, taking

H_1, L, α and R as the design variables, and taking the nominal strain ε_n as an optimization objective function, the mathematical model of structural optimization is established as follow

$$\min(\varepsilon_n) = f(H_1, L, \alpha, R) \qquad (6)$$

and the constraint conditions are that the stiffness coefficient K is limited from 0.618kN/mm to 0.715 kN/mm, the thickness H_1 is set from 29.2mm to 35.2mm, the length L is constrained from 191.1mm to 221.1mm, the angle α is selected from 81.0° to 84.0°, and the distance R is ranged from 95.4mm to 110.4mm.

Due to the complex relations among the four geometric parameters, an approximate model method is applied to construct the optimization objective function. The idea is to use DOE method to arrange the numerical test samples, and then get the response function values of the samples. According to the samples and the response values, the model is constructed by using a fitting and interpolation method. The objective function of the optimization mathematical model can be expressed as

$$\min(\varepsilon_n) = f(H_1, L, \alpha, R) = y(H_1, L, \alpha, R) = \tilde{y}(H_1, L, \alpha, R) + \varepsilon \qquad (7)$$

where, $\tilde{y}$ is the approximate value of the object function y, and ε is an infinitesimal value that is set in advance to ensure the convergence of the iterative calculation.

In order to ensure that the value of the established response surface model function is closer to that of the objective function, and also to reduce the amount of calculation and improve the optimization efficiency, the second-order response surface model is adopted. The designed sample points are inputted into Isight software, and the multi-island genetic algorithm (MIGA) in the global optimization is applied, the parameter factors are defined as the input variables, the stiffness coefficient is taken as a constraint condition, and the minimum strain is set as an output, the geometric parameters of the conical rubber spring are optimized. By using the MIGA optimization calculation, after 1001 iterations, the calculation just converges at the infinitesimal value of 1×10^{-6} set in advance, and the optimization results of geometric parameters of the conical rubber spring are listed in Table 4.

Table 4. The optimization design results of geometric parameters.

Optimization parameter	H_1 (mm)	L (mm)	α(°)	R (mm)	Nominal strain ε_n
Initial value	29.20	191.10	81.00	105.40	0.7306
Optimal value	31.04	195.10	83.20	95.70	0.6335

4.3 *Finite element simulation of optimized structure*

For the optimized conical rubber spring structure, a finite element modeling and meshing as well are carried out. The finite element model of the optimized structure is simulated under the same material parameters, fatigue test data and step-by-step loading conditions as the original structure. At various loading levels, the numerical calculation results of displacement of rubber material are obtained, and the displacement vs. load stiffness curve is plotted. The stiffness curve of the original structure model, and that of the optimized structure model are individually compared with the test stiffness curve, as shown in Figure 7.

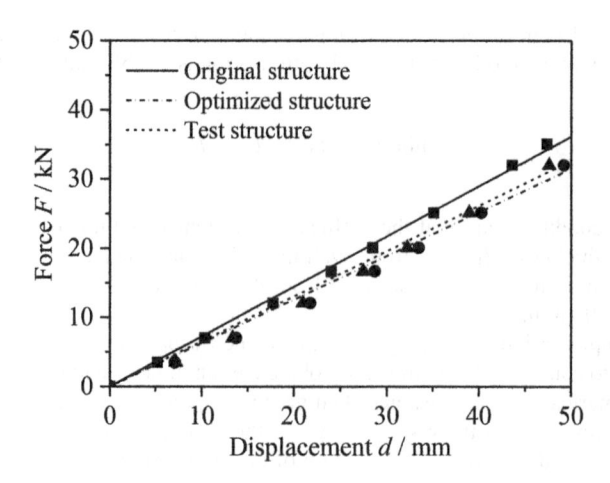

Figure 7. The comparisons of stiffness curves.

The finite element numerical calculation data and the experiment data are fitted by using the linear regression analysis method, the stiffness coefficients of the original structure and the optimized structure are individually 0.724kN/mm and 0.630kN/mm, and the stiffness coefficient obtained from the experiment data is 0.653kN/mm. The stiffness coefficient of the original structure is greater than that of the test structure, while the stiffness coefficient of the optimized structure is slightly less than that of the test one, and the value falls within the range of the stiffness coefficient constraint, which satisfies the requirement of the structural stiffness design.

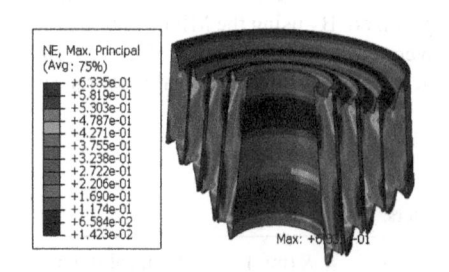

Figure 8. The strain distribution of rubber.

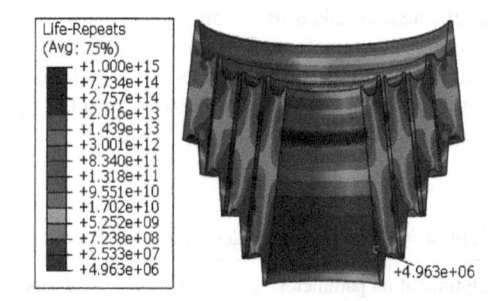

Figure 9. The fatigue life distribution of rubber.

Under the condition of ultimate load of 35.1kN, the strain distribution of the rubber lays of the optimized structure are shown in Figure 8. It can be found from Figure 8 that the maximum strain is 0.6835, which is also located at the lower end of the first rubber layer near the shaft core, but less than that of original structure. Simultaneously, according to the rubber material parameters listing in Table 1, the fatigue life of rubber is calculated by FE-safe software and the tearing energy method, and the fatigue life distribution of the optimized structure is shown in Figure 9. The fatigue life of the optimized structure is 4.963×10^6 cycles, while that of the original structure is only 2.629×10^6 cycles. The fatigue life of the optimized structure is 1.89 times of the original structure. The design fatigue life of the structure is 3.500×10^6 cycles, the fatigue life of the optimized structure is far more than the design fatigue life. Therefore, the fatigue life of the optimized conical rubber spring can also satisfy the requirement of the structural fatigue life design.

4.4 Test verification

According to the finite element numerical simulation of the optimized structure, the maximum strain amplitude of the rubber material in the damping conical rubber spring is 0.32, and the mean strain is 1.19. Under the condition of strain amplitude ε_a of 0.32, the Haigh curve of rubber material is shown in Figure 10. It can be seen from Figure 10 that as the mean strain ε_m is 1.19, the fatigue life of rubber material is 4.960×10^6 cycles, while the finite element simulation result is 4.963×10^6 cycles, and the relative error between the fatigue life of finite element calculation and that of test is only 0.12%.

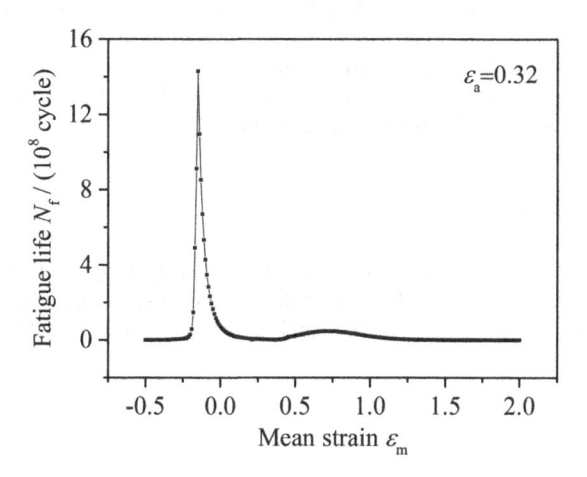

Figure 10. The fatigue life curve at a = 0.32.

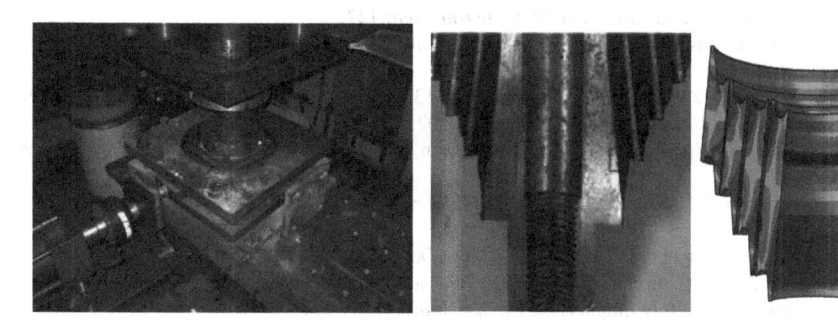

Figure 11. The fatigue test set.   Figure 12. The comparison of fatigue failure location.

The fatigue test for the optimized conical rubber spring is carried out in the laboratory, and the test set is shown in Figure 11. The lower end of the specimen is fixed on the test platform, and the upper end of the specimen is applied with alternating load in the vertical direction to test the fatigue life and failure behavior of the structure. The test results show that the fatigue life of the optimized structure is consistent with that of the finite element numerical simulation, and the service life of the optimized structure fully meets the fatigue life design requirements. Simultaneously, the fatigue failure mode of the conical rubber spring for rail vehicle is observed, and it is found that the lower end of the first layer of rubber appears some obvious fatigue failure morphology characteristics (as shown in the box in Figure 12), indicating that this location is the strain concentration position and the fatigue failure zone, which is completely consistent with the results of the finite element simulation.

5 CONCLUSIONS

The main conclusions are obtained as follows:

(a) The finite element model of a damping conical rubber spring for rail vehicles is constructed by using ABAQUS software, the finite element calculation results indicates that the fatigue life and the stiffness coefficient of the original structure do not satisfy the design requirements.
(b) The effect of geometric parameters on fatigue life of the structure is investigated, and it is found that the thickness and length of the first rubber layer significant effect on the fatigue life.
(c) Taking the minimum strain of rubber materials as an objective function, the optimization design for the conical rubber spring is carried out under the condition of stiffness constraint. The result shows that the fatigue life of optimized structure satisfy the design requirements.
(d) The test verification indicates that the calculated fatigue life is well fit with the test one, and the simulated fatigue failure position is consistent with the test.

ACKNOWLEDGEMENT

This work was supported by the High-level Talent Gathering Project of Hunan Province (no. 2019RS1059), the Key Project of the Reform of the Degree and Postgraduate Education in Hunan Province (no. 2019JGZD034), and the Research Project for Postgraduate Students of Hunan Province (no. CX20200648).

REFERENCES

[1] Sheng X W, Zheng W Q, Zhu Z H, Luo T J and Zheng Y H 2020 *Construction and Building Materials* **240** 117822
[2] Menga N, Bottiglione F and Carbone G 2019 *Journal of Sound and Vibration* **463** 114952
[3] Pan P, Shen S D, SHen Z Y and Gong R H 2018 *Measurement* **122** 554
[4] Dissanayake D G K, Weerasinghe D U, Thebuwanage L M and Bandara U A A N 2021 *Journal of Building Engineering* 33 101606
[5] Paje S E, Luong J, Vázquez V F, Bueno M and Miró R 2013 *Construction and Building Materials* **47** 789
[6] Jung J K, Kim I G, Kim K T, Ryu K S and Chung K S 2021, *Polymer Testing* **93** 107016
[7] He Y H, Fan Y, Wu P C, Xia C Y, Huang J, Wu J W, Qian L Q and Liu X M 2020 *Engineering Failure Analysis* **117** 104965
[8] Hu G, Zhang P, Wang G R, Zhang M and Li M 2017 *Journal of Natural Gas Science and Engineering* **38** 120
[9] Guan H Y, Yang F Y and Wang Q L 2011 *Materials and Design* **32** 2404
[10] Choi J, Quagliato L, Lee S, Shin J and Kim N 2021 *International Journal of Fatigue* **145** 106136
[11] Li Q, Zhao J C and Zhao B 2009 *Engineering Failure Analysis* **16** 2304.
[12] Zine A, Benseddiq N and Abdelaziz M N 2011 *International Journal of Fatigue* **33** 1360
[13] Griffith A A 1921 *Philosophical Transactions of the Royal Society of London* **A221**, 163
[14] Rivlin R S and Thomas A G 1953 *Journal of Polymer Science* **10** 291
[15] Lake G J, Lindley P B and Thomas A G 1965 *Rubber Chemistry and Technology* **38** 292
[16] Lake G J, Lindley P B and Thomas A G 1965 *Rubber Chemistry and Technology* **38** 301
[17] Mars W V and Fatemi A 2006 *International Journal of Fatigue* **28** 521
[18] Verron E and Andriyana A 2008 *Journal of Mechanics and Physics of Solids* **56** 417
[19] Auyob G, Nait-Abdelaziz M, Zaïri F, Gloaguen J M and Charrier P 2008 *Journal of Mechanics and Physics of Solids* **56** 417
[20] Loew P J, Peters B and Beex L A A 2020 *Mechanics of Materials* **142** 103238
[21] Mars W V 2001 *Tire Science and Technology* **29** 171

Experimental investigation on physical and mechanical properties of aluminium alloy 7050-T7451 after burning

B. Zhao, J.W. Wu, Q. Chen & B.W. Qi
Department of Engineering Mechanics, Northwestern Polytechnical, Xian, Shaan Xi, China

ABSTRACT: Fuel burn is one of the main forms of aircraft structural damage. A reliable evaluation of the residual performances of aircraft structures is needed to decide whether the structures should be repaired or replaced. At present, the burn simulation of aircraft structures is mainly achieved by heat treatment or thermal radiation, which is different from the real fuel burn damage. In this paper, a fuel burn test of aluminium alloy specimens was designed to get damaged specimens at 5 different burn temperatures, and then an experimental investigation was conducted to reveal the post-burning physical and mechanical properties of one widely used structural aluminium alloy, namely, 7050-T7451. The results of hardness tests, conductivity tests, tensile tests and fatigue tests of the damaged specimens present the changes of physical and mechanical properties of 7050-T7451 after burning, which may provide an accurate assessment of burning damage.

1 INTRODUCTION

Burn is one of the main forms of aircraft structural damage, which refers to the structural damage caused by the hit of weapons, or the leakage of oil resulted by the damage of aircraft fuel system meeting high temperature or fire source. At present, the structural materials mainly used in aircrafts are aluminium alloy and titanium alloy. In the process of thermal damage caused by burning, the microstructure and mechanical properties of alloys will change with temperature and cooling time, which is similar to the process of annealing, quenching and aging of alloys. Thermal damage which has a great influence on the strength of aircraft structures is difficult to detect, so simple and practical non-destructive testing method is always pursued by researchers to provide a reliable evaluation of the residual performances of aircraft structures [1].

Now the common methods of detection of thermal damage, prediction of damage and residual life are as follows [1–5]: (1) Colour test; (2) Hardness testing method; (3) Metallographic measurement; (4) Conductivity detection method; (5) Mechanical property test; (6) Finite element method [6,7]. Some results are achieved, like, literature [3] reviewed the modelling of residual mechanical behaviours of 5083-H116 and 6061-T651 after heat treatment, including empirical formula, and constitutive model based on physics and finite element calculation. In reference [9], a model for predicting the residual constitutive behaviour of heat-treated AA5083-H116 aluminium alloy was established based on microstructures. The model includes several sub models: (I) microstructure evolution model, (II) residual yield strength model, and (III) residual strain hardening model.

Although a lot of theoretical and experimental researches have been developed on the thermal damage of aluminium alloy materials, there are many contents still worthy of study on the damage and residual life prediction of aluminium alloy materials after burning:

(1) At present, the methods of heat treatment and thermal radiation are mostly used in burning simulation. For the sake of safety and convenience, few people use fuel burning to carry out

the test. Whether the damage caused by the heat treatment and thermal radiation is equivalent to the thermal damage of burning is worthy of study.

(2) Aircraft components usually work in fatigue state, so it is very important to study the fatigue properties of aluminium alloy after burning. But at present few researches we can reach.

In this paper, a fuel burn test of aluminium alloy material is designed, and hardness test, conductivity test, tensile and fatigue tests are carried out for the burned specimens, hoping to find some useful data to determine the residual life of the damaged aluminium alloy accurately.

2 BURN TEST

2.1 Test method

Before tests, brackets and thermocouples were arranged in a burning chamber like Figure 1 shows. The burning basin (D = 300 mm) was filled with No3 jet fuel (RP-3, GB6537-2006). By controlling the amount of fuel, the damaged specimens at different temperatures and with different burn time could be obtained. Figure 2 shows the burning field and the placement of test pieces, and the typical temperature curves of measuring points with time are shown in Figure 3.

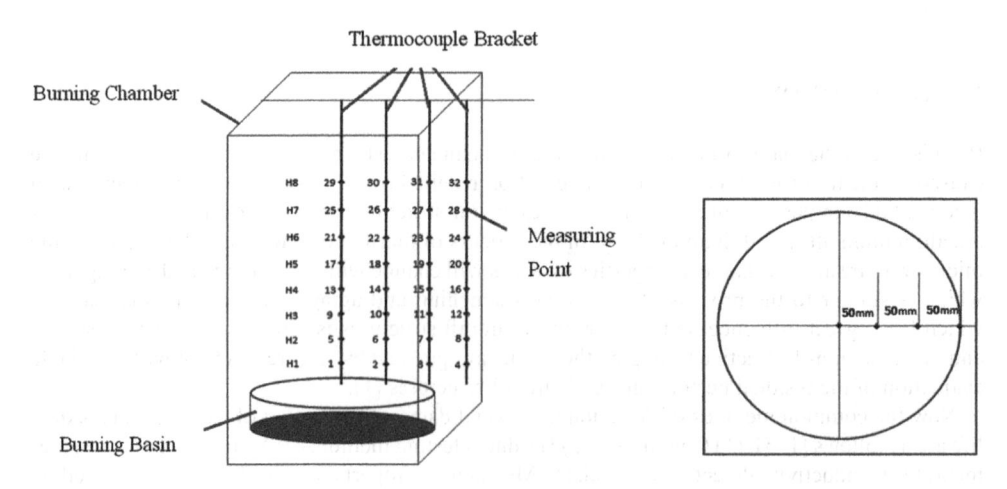

Figure 1. Thermocouple arrangement in burning field.

2.2 Fuel burn test

2.2.1 Effects of burning on physical properties of aluminium alloy

6 pieces of 7050-T7451 aluminium plates (350mm * 350mm * 2mm) were placed horizontally above the flame at H1 = 750mm, H2 = 800mm and H3 = 850mm (the sample is aligned with the center of the burning basin). After burning, the hardness and conductivity of 40 points were measured according to the positions shown in Figure 4.

(a)   (b)

Figure 2. Burning test device: (a) Burning field and (b) Placement of test pieces.

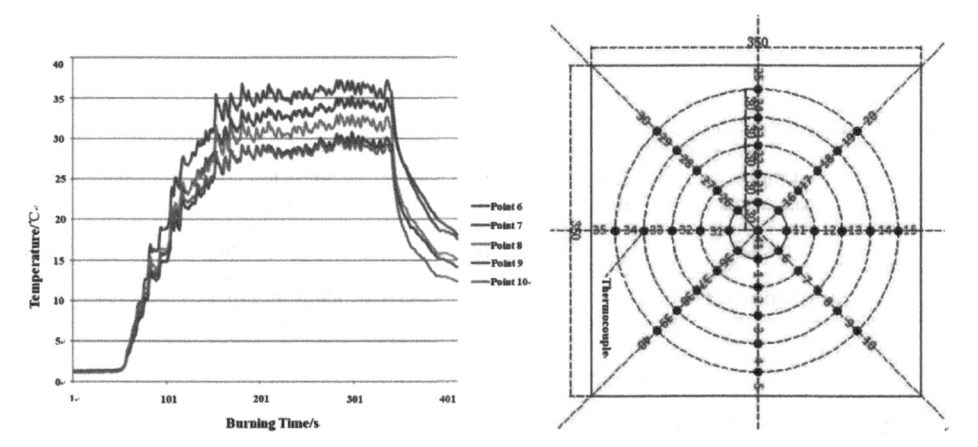

Figure 3. Curves of temperature versus time (Burning test of plate specimen, measuring points 6-10).

Figure 4. Measuring points of hardness and conductivity of aluminium plate after burning.

2.2.2 *Effects of burning on tensile properties*

Centers of tensile specimens were horizontally placed at positions with different temperatures in the flame. After burning, the burned specimens at different temperatures were obtained and cooled to room temperature. The tensile strength limit, yield limit and elastic modulus of the burned specimens were tested according to "HB 5143-1996 tensile test method for metals at room temperature". The shape and size of the specimen are shown in Figure 5. The test parameters and the number of test pieces are shown in Table 1.

2.2.3 *Effects of burning on fatigue properties*

Centers of the fatigue specimens were horizontally placed in the flame and burned at different temperatures, and then cooled to room temperature. The fatigue performances of the specimen were tested according to "HB 5287-1996 metallic materials axial constant amplitude low cycle fatigue test method". The shape and size of the test piece are shown in Figure 6. The test parameters and the number of test pieces are shown in Table 2.

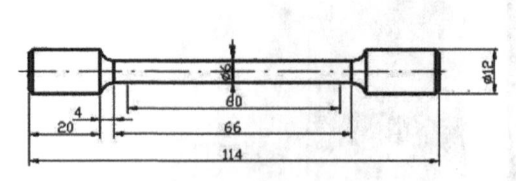

Figure 5. Tensile test specimen.

Figure 6. Fatigue test specimen.

Table 1. Summary of tensile test.

Materials	Quantities / piece	Burn temperature /°C	Burn time /min
7050-T7451	6	300	6
	6	350	6
	6	400	6
	6	450	6
	6	500	6

Table 2. Summary of fatigue test.

Materials	Fatigue parameters	Burn time/ min	Burn temperature /°C	Stress level /MPa	Quantities/ piece
7050-T7451	98Hz	6	300	460, 420, 380, 340,	3 for each group
	Symmetrical cycle		400	320, 300, 270, 240, 210	5 for each group

3 EXPERIMENTAL RESULTS

3.1 Physical properties

3.1.1 Hardness
The relationship between the burn temperature and hardness is shown in Figure 7. It can be seen that under the condition of burn time 6min, the hardness decreases sharply with the increase of burn temperature from 300°C to 350°C, and when the exposure temperature is 350°C it reaches the lowest value; then a rebound stage occurs thereafter up to 450°C; finally the hardness decreases with the increase of burn temperature up to 500°C.

3.1.2 Conductivity
The relationship between the burn temperature and conductivity is shown in Figure 8. It can be seen that under the condition of burn time 6 min, the conductivity do not change significantly at the burn temperature from 300°C to 350°C. The conductivity decreases with the increase of temperature up to 500°C.

3.2 Tensile properties

The relationships between burn temperature and yield limit, strength limit, elastic modulus are shown in Figure 9. From Figure 9 (a) and (b), it can be seen that a notable reduction in yield limit and strength limit occur from RT to 400°C, and when the exposure temperature is 400°C they reach the lowest values, then a rebound stage occurs thereafter up to 500°C. From Figure 9 (c), we can see that at 350°C the elastic modulus decreases to the lowest value, then increases at 400°C, finally decreases from 400°C to 500°C.

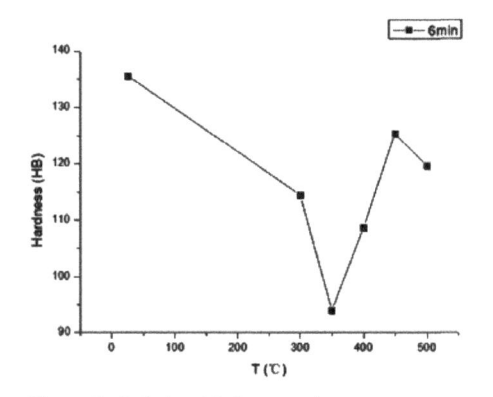

Figure 7. Relationship between burn temperature and hardness.

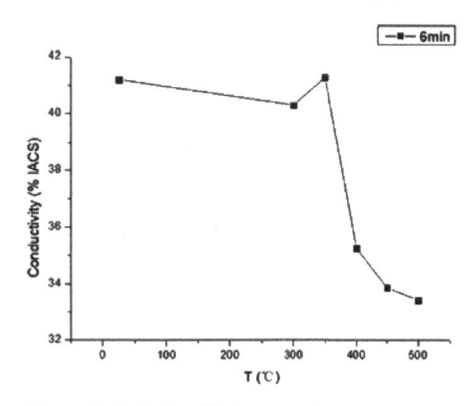

Figure 8. Relationship between burn temperature and conductivity.

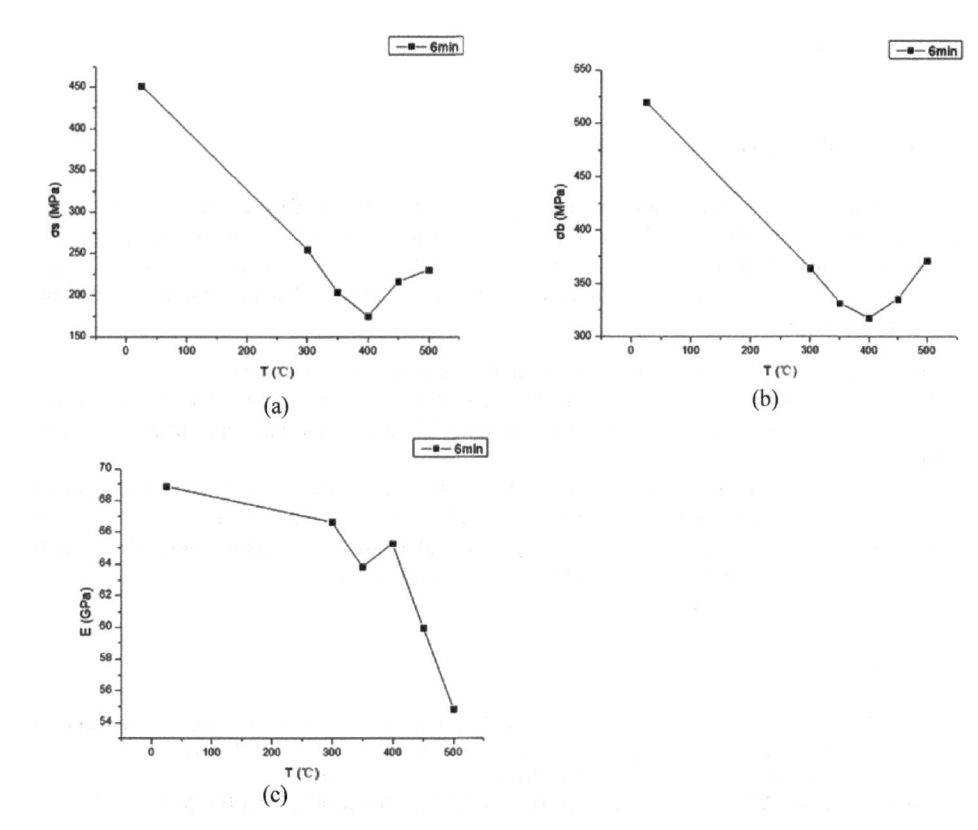

Figure 9. Effect of burn temperature and time on tensile properties of aluminium alloy: (a) Relationship between burn temperature and yield strength, (b) Relationship between burn temperature and strength limit, and (c) Relationship between burn temperature and elastic modulus.

3.3 *Fatigue properties*

The relationships between fatigue life and stress level are shown in Figure 10. It can be concluded that under the conditions of burn temperature 30°C and 400°C, burn time for 6 min, the overall fatigue lives decrease with the increase of burn temperature.

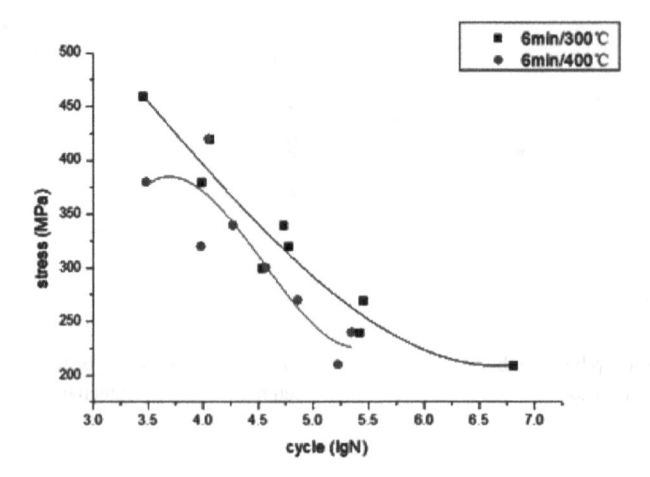

Figure 10. S-N curves.

4 DISCUSSION AND CONCLUSION

In this paper, a fuel burn test of aluminium alloy specimens was designed to get damaged specimens at different burn temperatures, and then a detailed experimental study on the post-burning physical and mechanical properties of aluminium alloy 7050-T7451, which is widely used in aircraft structures, were performed. Some useful conclusions are achieved and some influence factors on tests are discussed.

(1) After burning, for hardness, conductivity and elastic modulus the lowest values are at the exposure temperature 350°C; for yield limit and strength limit the lowest values are at the exposure temperature 400°C. The overall fatigue life decreases with the increase of burn temperature.
(2) Since the burning field is greatly affected by external environments, more measures need to be taken to obtain a more stable temperature field. And to get more accurate results a large number of specimens are required. It is suggested that a variety of methods should be used to determine the accurate damage of aluminium alloy structures.

REFERENCES

[1] Zhou ZP 2009 *Research and application of thermal damage detection method for titanium and aluminium alloy components* (Beijing: Beijing University of Aeronautics and Astronautics Press)
[2] Liu J, Jiang SM, Dong GK and Wang P 2106 *Avi. Maint. & Eng.* **8** 64-67
[3] Summers PT, Chen YY, Rippe CM, Allen B, Mouritz AP, Case SW and Lattimer BY 2015 *Fire Sci. Rev.* 4:3 1-36
[4] Liu Q, Baburamani P and Loader C Effect of High Temperature Exposure on the Mechanical Properties of Cold Expanded Open Holes in 7050-T7451 Aluminium Alloy 2009 *Air Vehicles Division Defence Science and Technology Organisation* Australia

[5] Calero J and Turk S Effects of Thermal Damage on the Strength 2008 *Air Vehicles Division Defence Science and Technology Organisation* Australia
[6] Afaghi Khatibi A, Kandare E, Feih S, Lattimer BY, Case SW and Mouritz AP 2014 *Comp. Mater. Sci.* **95** 242-249
[7] Liu Y, Liu H and Chen Z, 2019 *Const. & Building Mater.* **196** 256-266
[8] Rippe C, Case S, and Lattimer B 2017 *Fire Safety J.* **91** 561–567
[9] Summers PT, Mouritz AP, Case SW and Lattimer BY 2015 *Mater. Sci. & Eng.* A **632** 14–28

Dehydration kinetics of Mg(OH)$_2$ for thermochemical energy storage via model-free and model-fitting methods

D.Y. Wang, Q.P. Wu, M.T. Li, E.X. Ren, Z.Y. Chang & L. Zhu
School of Chemistry and Chemical Engineering, Beijing Institute of Technology, Beijing, P. R. China

ABSTRACT: Energy storage has attracted much attention due to the demand for renewable energy. Mg(OH)$_2$/MgO system is a promising thermochemical energy storage technology, basing on the reversible hydration and dehydration reaction. In this paper, the kinetic parameters of Mg(OH)$_2$ dehydration were derived through model-free and model-fitting methods. Contracting cylinder model is illustrated to be the mechanism function for Mg(OH)$_2$ dehydration. The kinetic control equation is established and clarified to be available for simulation.

1 INTRODUCTION

Thermal energy storage (TES) is a technology that stores thermal energy by a storage medium. There are several types of TES, i.e., sensible heat storage, latent heat storage, and thermochemical storage (TCS) [1]. TCS works on a reversible thermochemical reaction showing relatively high-storage density and long-term heat storage ability [2] Hydroxides are promising TCS materials via reversible hydration-dehydration reaction, such as magnesium hydroxide/magnesium oxide (Mg(OH)$_2$/MgO) system [3].

$$Mg(OH)_2\,(s) = MgO(s) + H_2O(g) \Delta H = 80.6 kJ \cdot mol^{-1} \qquad (1)$$

As a solid-gas reaction, the dehydration kinetics of Mg(OH)$_2$ can be expressed as equation (2). [4] The kinetics parameters can be derived via model-fitting methods and model-free methods. [5]

$$\frac{d\alpha}{dt} = A \exp\left(\frac{-E}{RT}\right) f(\alpha) \qquad (2)$$

where α is a conversion, β is the heating rate (K/min), A is the pre-exponential factor (s^{-1}), E is the activation energy (kJ/mol), R is the ideal gas molar constant (J/(mol·K)), T is the temperature (K), and $f(\alpha)$ is the kinetics mechanism function representing the specific reaction mode. In fact, the differential data, $d\alpha/dt$, are usually quite noisy, so the integral form of equation (2) is expressed as equation (3):

$$g(\alpha) = \int_0^\alpha \frac{d\alpha}{f(\alpha)} = A \exp\left(\frac{-E}{RT}\right) t \qquad (3)$$

where $g(\alpha)$ is the integral form of the reaction model.

Table 1. Kinetic models commonly used in gas–solid reaction.

Kinetic model	$f(\alpha)$	$g(\alpha)$
Avrami-Erofeev,n=2 (A2)	$2(1-\alpha)[-\ln(1-\alpha)]^{1/2}$	$[-\ln(1-\alpha)]^{1/2}$
Avrami-Erofeev,n=3 (A3)	$3(1-\alpha)[-\ln(1-\alpha)]^{2/3}$	$[-\ln(1-\alpha)]^{1/3}$
Avrami-Erofeev,n=3 (A4)	$4(1-\alpha)[-\ln(1-\alpha)]^{3/4}$	$[-\ln(1-\alpha)]^{1/4}$
1D Diffusion(D1)	$1/2\alpha$	α^2
2D Diffusion(D2)	$[-\ln(1-\alpha)]-1$	$(1-\alpha)\ln(1-\alpha)+\alpha$
3D Diffusion(D3)	$1.5(1-\alpha)^{1/3}[(1-\alpha)^{-1/3}-1]-1$	$[1-(1-\alpha^{1/3})]^2$
4D Diffusion(D4)	$1.5[(1-\alpha)^{1/3}-1]-1$	$1-2\alpha/3-(1-\alpha)^{2/3}$
First-order(F1)	$(-\alpha)$	$-\ln(1-\alpha)$
Second-order(F2)	$(1-\alpha)^2$	$(1-\alpha)^{-1}-1$
Contracting cylinder(R2)	$2(1-\alpha)^{1/2}$	$1-(1-\alpha)^{1/2}$
Contracting sphere(R3)	$3(1-\alpha)^{2/3}$	$1-(1-\alpha)^{1/3}$

The commonly applied kinetic models for gas–solid reactions in integral and differential forms are listed in Table 1. [6] However, model-free methods can accommodate more complex reactions. The model-free methods evaluate E without prior knowledge of the reaction model. The common model-free methods are OFW method, [7] KAS method, [8] and Starink method. [9]

OFW method:
$$\lg \beta = \ln\left(\frac{AE}{RG(\alpha)}\right) - 2.315 - \frac{0.4567E}{RT} \quad (4)$$

KAS method:
$$\ln\left(\frac{\beta}{T^2}\right) = \ln\left(\frac{AE}{RG(\alpha)}\right) - \frac{E}{RT} \quad (5)$$

Starink method:
$$\ln\left(\frac{\beta}{T^{1.92}}\right) = \ln\left(\frac{AE}{RG(\alpha)}\right) - 1.0008\frac{E}{RT} \quad (6)$$

Recently, Vyazovkin proposed a nonlinear model-free method based on a numerical algorithm, for n different heating rates, one can use the basic integral equation (3) to give the following equations: [10]

$$g(\alpha) = \frac{A}{\beta}\int_{T_{\alpha-\alpha}}^{T_\alpha} \exp\left(\frac{-E_\alpha}{RT}\right) dT \quad (7)$$

$$I(E_\alpha, T_\alpha) = \int_{T_{\alpha-\alpha}}^{T_\alpha} \exp\left(\frac{-E_\alpha}{RT}\right) dT \quad (8)$$

$$\frac{A_\alpha}{\beta_1}I(E_{\alpha,1}, T_{\alpha,1}) = \frac{A_\alpha}{\beta_2}I(E_{\alpha,2}, T_{\alpha,2}) = \cdots\cdots = \frac{A_\alpha}{\beta_n}I(E_{\alpha,n}, T_{\alpha,n}) \quad (9)$$

When the equality holds strictly, equation (9) is derived to (10)

$$\varphi(E_\alpha) = \sum_{i=1}^{n}\sum_{i\neq j}^{n}\frac{I(E_{\alpha,i}, T_{\alpha,i})\beta_j}{I(E_{\alpha,j}, T_{\alpha,j})\beta_i} = \min \quad (10)$$

In this paper, the dehydration kinetics of $Mg(OH)_2$ are investigated via the model-free and model-fitting methods.

2 EXPERIMENT SECTION

2.1 Materials and instruments

Commercial $Mg(OH)_2$ ($\geq$95%, Sinopharm chemical reagent Co. Ltd, Beijing). Thermogravimetric analysis (TGA) was performed on an SDT-Q600 thermal analyzer, high-purity nitrogen (99.99%) was used as the purge gas at a flow rate of 100 ml/min in the experiment. Non-isothermal dehydration kinetics was investigated from room temperature to 400°C at different heating rates of 3, 5, 8, and 10°C/min to record the weight loss curves. The conversion ratio of the dehydration process α is defined as:

$$\alpha = \frac{m_0 - m_t}{m_{H_2O}} \tag{11}$$

where m_t is the sample mass at time t/min, m_0 is the initial mass of the sample, and m_{H2O} is the mass of theoretical mass loss.

3 RESULT AND DISCUSSION

3.1 Dehydration of magnesium hydroxide

Figure 1(a) shows the TG-DSC curves of $Mg(OH)_2$ at a 10 K/min heating rate. onset-, peak-, and end-temperatures (Tonset, Tpeak, and Tend) for dehydration of $Mg(OH)_2$ are 610.5K, 652.K, and 716.5K, respectively. The enthalpy of reaction obtained by integrating the peak area is 1205J/g. The TG curves of $Mg(OH)_2$ dehydration at the various heating rates are shown in figure 1(b).

3.2 Activation energy from the model-free methods

According to equations (4), (5), and (6), the relationships between lg β, $\ln(\beta/T^2)$, $\ln(\beta/T^{1.92})$ and $1/T$ at a number of α are drawn in figure 2. The activation energy (E) is the interception of the fitting lines and listed in table 2. To further improve accuracy the E can be determined by minimizing the eq (1) The differences among all these E values are not large and the mean values are 153.96 ± 4.75, 150.99 ± 5.53, 151.93 ± 5.39, and 152.19 ± 5.49 kJ/mol for OFW, KAS, Starink, and Vyazovkin methods, respectively.

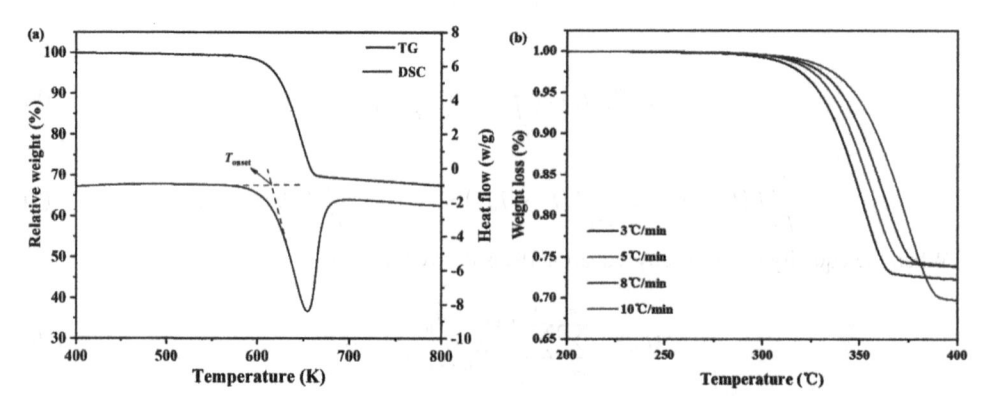

Figure 1. (a) TG-DSC curves of $Mg(OH)_2$ at a 10 K/min heating rate. (b) TG curves of $Mg(OH)_2$ at various heating rates.

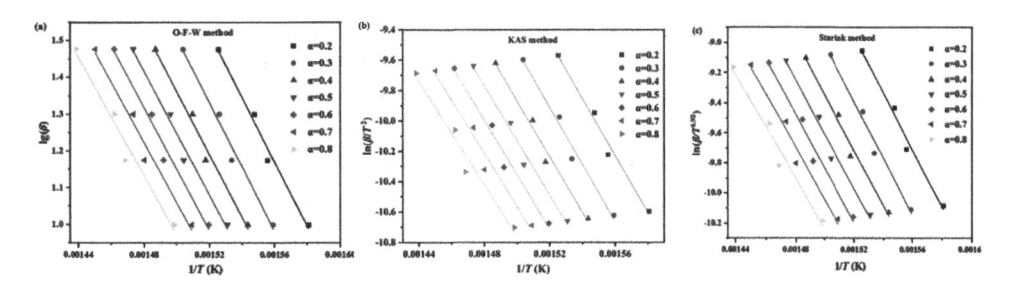

Figure 2. Linear fitting with different methods. (a) OFW method.(b) KAS method.(c) Starink method.

Table 2. The activation energy of Mg(OH)$_2$ calculated from model free methods.

α	OFW method		KAS method		Starink method		Vyazovkin method	
	E (kJ/mol)	Rs	E (kJ/mol)	Rs	E (kJ/mol)	Rs	E(kJ/mol)	Rs
0.2	159.31	0.9869	157.80	0.9851	157.91	0.9852	159.95	-
0.3	159.19	0.9838	156.70	0.9862	157.64	0.9817	155.80	-
0.4	156.53	0.9847	153.52	0.9825	154.57	0.9826	154.24	-
0.5	153.42	0.9841	150.68	0.9818	151.92	0.9819	153.51	-
0.6	152.96	0.9795	149.42	0.9765	150.88	0.9766	149.55	-
0.7	149.61	0.9754	146.61	0.9716	147.52	0.9717	149.45	-
0.8	146.71	0.9716	142.23	0.9671	143.07	0.9673	142.85	-
Mean	153.96 ± 4.75	0.9809	150.99± 5.53	0.9886	151.93 ± 5.39	0.9781	152.19± 5.49	-

3.3 Kinetics parameters and kinetics equation via model-fitting methods

Through Coats–Redfern integral method and ABSW differential method [1112], TGA data at a series of heating rates (3, 5, 8, and 10° C/min) were linearly fitted. E and $\ln A$ are listed in table 3. The results indicate that the largest Rs values accord to D3, D4, R2, and R3 models. The E values derived from these models are remarkably different. Compared with the E values listed in table 2, the closet number is calculated via model R2. Thereby, contracting cylinder (R2) model is the most fitting mechanism function for Mg(OH)$_2$ dehydration.

The CR integral equation can be expressed as:

$$\ln\left[\frac{G(\alpha)}{T^2}\right] = \ln\left(\frac{AR}{\beta E}\right) - \frac{E}{RT} \quad (12)$$

The ABSW differential equation can be expressed as:

$$\ln\left[\frac{d\alpha}{f(\alpha)\,dT}\right] = \ln\left(\frac{A}{\beta}\right) - \frac{E}{RT} \quad (13)$$

Based on R2 model, the mean E value is 166.69 ± 5.1 kJ/mol, and the average $\ln A$ is 25.45 ± 2.61. Thus, the kinetic control equation can be expressed as equation (14).

$$\frac{d\alpha}{dt} = 1.13 \times 10^{12} \times \exp\left(-\frac{1.67 \times 10^5}{RT}\right) \times 2 \times (1-\alpha)^{0.5} \quad (14)$$

To confirm the availability of this equation, the experimental TG data at the heating rate of 10°C/min were converted into α and the relation of $\alpha(t) - t$ is drawn in figure 3 as a black line.

The $\alpha(t)$s calculated via equation (14) are also drawn and linear fitted in figure 3.The largest deviation between these two lines is only 2.5%. This error is within an acceptable range. Thus, this kinetics control equation can be used for engineering design and heat storage system simulation.

Table 3. The Es and lnA for the dehydration of $Mg(OH)_2$ from various models.

Model	C-R method			ABSW method		
	E (kJ/mol)	lnA	Rs	E (kJ/mol)	lnA	Rs
A2	97.10	9.75	0.9988	124.87	21.94	0.9965
A3	61.00	6.81	0.9987	88.90	15.05	0.9948
A4	43.01	9.70	0.9986	70.91	11.52	0.9931
D1	305.95	46.10	0.9963	245.67	43.59	0.9745
D2	333.01	52.15	0.9990	307.24	54.46	0.9939
D3	375.68	58.78	0.9999	378.97	66.41	0.9999
D4	347.13	53.34	0.9996	331.84	57.57	0.9975
F1	205.22	29.99	0.9989	234.23	42.16	0.9979
F2	284.76	45.22	0.9875	374.22	68.61	0.9812
R2	**171.79**	**22.84**	**0.9998**	**161.58**	**28.05**	**0.9960**
R3	182.41	24.49	0.9999	185.52	32.12	0.9996

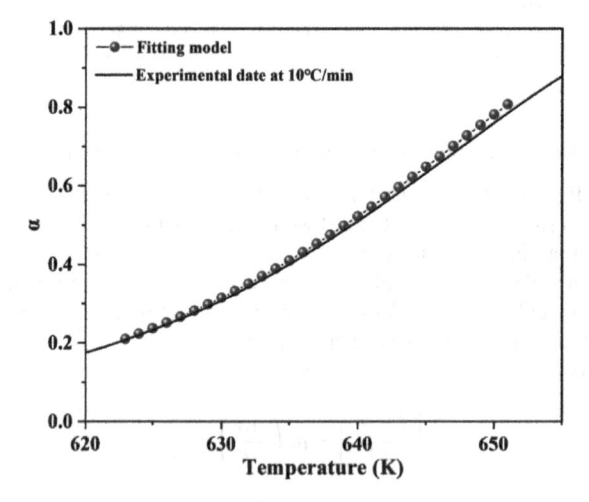

Figure 3. Experimental $\alpha(t)$ and calculated $\alpha(t)$ with kinetics equations.

4 CONCLUSION

The model-free methods are used to determine the kinetic parameters of $Mg(OH)_2$ dehydration, and the activation energy is derived from 150 to 154 kJ/mol. For the common model-fitting methods, the contracting cylinder model is found to describe well $Mg(OH)_2$ dehydration. The mean activation energy is derived to be 166.69 ± 5.1 kJ/mol close to the values calculated through model-free methods. The kinetic control equation is established and clarified to be available for engineering simulation

REFERENCES

[1] A.B. Awan, M. Zubair, R.P. Praveen, A.R. Bhatti, Design and comparative analysis of photovoltaic and parabolic trough based CSP plants 2019 *J. Sol. Energy* **183** 551–565.

[2] L. Sun, Q.P. Wu, L.L. Zhang, Y.T. Li, M.T. Li, T. Gao, Doping magnesium hydroxide with $Ce(NO_3)_3$: a promising candidate thermochemical energy storage materials 2019 IOP Conf. Ser.: Earth Environ. Sci. 295 032068.

[3] Y. Kato, N. Yamashita, K. Kobayashi, Y. Yoshizawa, Kinetic study of the hydration of magnesium oxide for a chemical heat pump 1996 J. Appl. Therm. Eng 16 853–862.

[4] M.T. Li, Y.T. Li, L. Sun , Z.B. Xu, Y. Zhao, Z.H. Meng, Q.P. Wu, Tremendous enhancement of heat storage efficiency for $Mg(OH)_2$-MgO-H2O thermochemical system with addition of $Ce(NO_3)_3$ and LiOH 2021 J. Nano Energy 81 105603.

[5] S. Vyazovkin, C.A. Wight, Kinetics in solids 1997 J.Ann Rev Phys Chem 48 125–149.

[6] S. Vyazovkin, A.K. Burnham, Criado JM, Pérez-Maqueda LA, Popescu C, Sbirrazzuoli N, ICTAC Kinetics committee recommendations for performing kinetic computations on thermal analysis data 2011 J. Thermochim Acta 520 1–19.

[7] J.H. Flynn, The 'Temperature Integral'—its use and abuse 1997 J. Thermochim Acta 300 83–92.

[8] S. Vyazovkin, Is the Kissinger equation applicable to the processes that occur on cooling 2002 J. Macromol. Rapid Commun 23 771–775.

[9] M.J. Starink, Activation energy determination for linear heating experiments: deviations due to neglecting the low temperature end of the temperature integral. 2007 J. Thermochimica Acta 42 483–489.

[10] S. Vyazovkin, Modification of the integral iso-conversional method to account for variation in the activation energy 2001 J. Comput. Chem 22 178–183.

[11] J. Yan, C.Y. Zhao, Thermodynamic and kinetic study of the dehydration process of $CaO/Ca(OH)_2$ thermochemical heat storage system with Li doping 2015 J. Chem. Eng. Sci 138 86–92.

[12] N. Sbirrazzuoli, Determination of pre-exponential factors and of the mathematical functions $f(\alpha)$ or $g(\alpha)$ that describe the reaction mechanism in a model-free way 2013 J. Thermochim. Acta 564 59–69.

Straw-based boards and panels: Experiment, setbacks, and underlying factors

M.X. Xie
Zhongnan University of Economics and Law, Wuhan, China

ABSTRACT: Straw of wheat, cotton, rice, and other crops poses serious environmental and agricultural problems. Prohibition on open field burning of crop straw on farmland helps control air pollution but makes it hard for farmers to handle crop straw. Manufacturing straw-based boards and panel has a bigger potential for using crop straw but so far has suffered setbacks. Straw-based boards and panels just account for a minimal percentage of the total sale of boards and panels, far from being accepted as a mainstream building material. The underlying factors include the lack of cost-effective technology, low market acceptance, and the insufficiency of consistent and concerted governmental policies. In the future, better technology as well as more consistent and concerted policies are needed for further development in the manufacturing of straw-based boards and panels.

1 INTRODUCTION

Straw of wheat, cotton, rice, and other crops poses serious environmental and agricultural problems in China as well as in other countries [1]. It is estimated that the amount of crop straw generated in China is more than 1040 million ton each year [2]. For various other reasons, crop straw is almost useless for household use in rural China now. Burning crop straw on farmland is prohibited because it causes serious air pollution, particularly in the harvest season. The government imposed a ban on open-field burning of crop straw more than a decade ago and imposed heavy punishment for violations. In sum, crop straw is a big problem for environment and agriculture, for farmers and law enforcement officers. Manufacturing boards, panels, and other building materials is a potential way to make use of crop straw and a potential way to solve the straw problem. This paper reviews China's exploration in this aspect in the past years, assesses the relatively small success and relatively large setbacks, reveals the underlying factors for this situation, and makes some suggestions on future development.

2 EXPLORATION OF STRAW-BASED BOARDS AND PANELS

2.1 *Production of straw-based boards and panels*

The exploration of manufacturing boards and panels with crop straw started in late 1970s in China, and modestly developed in the 1980s. Before 1990s, most of technology and production lines were imported from other countries. In the 1990s, some Chinese universities and research institutes joined the research effort, notably the China Academy of Forestry Sciences, Nanjing Forestry University, and Northeastern Forestry University [3–5]. After the turn of century, the development of straw-based building materials gained a new momentum [4]. For instance, a company of Sichuan Province established a production line with a capacity of 50000 m^3 in Southwestern China. Another company named Novofibre was established with foreign investment Shaanxi Province in 2007, which specialized in manufacturing boards and panels with wheat straw.

The processes vary in terms of straw particles and adhesives. The production processes generally include grinding crop straw into particles or powders, adding glue and other adhesives, and molding. Straw is first cut into short stems or ground into particles of different sizes, then mixed with different types of adhesives. Treatment of straw particles affects the property of particle and the property of final products. A research into enzymatically treated wheat straw revealed that the highest erosion of surface was reached via a combination of xylanases and pectinases [6]. Another research indicated that alkaline liquid treatment of wheat straw improved the surface set ability of straw and helped the infiltration of resin [7]. A research into rice straw indicated that fine grinding and removal of wax by extraction with hexane increased internal bonding [8]. Glue and molding are key factors for the properties of the final products. Urea-formaldehyde resin (UF resin), phenol-formaldehyde resin (PF resin), diphenyl-methane-diisocyanate (MDI) are some of the commonly used adhesives, particularly UF resin and PF resin [9]. Epoxidized sucrose soyate is also reported as a potential adhesive for straw-based low-density fiberboards [10]. MDI has the strongest adhesiveness with straw but is not commonly used because of its high price [6, 9, 11].

The properties of straw-based boards and panels mainly include water absorption, linear expansion, thickness swelling, internal bonding strength, static bending strength, modulus of rupture, modulus of elasticity, impact strength, and other aspects. The quality varies depending on the crop straw, size of particle, adhesives, processing procedures, and other factors [12]. Not all straw-based boards and panels have high quality but some do meet or exceed the standards for wood particleboards. For instance, it is reported that boards made with rice straw using 4% polymeric diphenylmethane diisocyanate (pMDI) and straw particles hammer milled with a 3.18 mm opening perforated plate exceeded the M-2 specification of American National Standard for Wood Particleboard [13]. Below is the key specification of the product of Novofibre which produces boards with wheat straw and MDI in Shaanxi province, the central part of China.

Table 1. Technical specifications of high-quality wheat-straw-based boards with MDI as adhesive.

Thickness	Unit (mm)	$\geq 5 \sim < 8$	$\geq 8 \sim < 15$	$\geq 15 \sim < 30$
Emission of formaldehyde	Mg/100g	≤ 3.0	≤ 3.0	≤ 3.0
TVOC of 72 h	Mg/m^2.h	≤ 0.5	≤ 0.5	≤ 0.5
Water content	%	≤ 2.5	≤ 2.5	≤ 2.5
Static bending strength	MPa	≥ 20	≥ 20	≥ 18
Modulus of elasticity	MPa	≥ 2400	≥ 2500	≥ 2600
Internal bonding strength	MPa	≥ 0.30	≥ 0.28	≥ 0.28

(Data source: Novofibre Company Ltd.)

The quality of Novofibre is relatively higher. Many other companies use other adhesives and have much lower qualities.

2.2 Market performance of straw-based boards and panels

Straw-based boards can be used for interior decoration. A comparable alternative is the board made of plaster or wood. Compared with wood-based or plaster-based boards and panels, the market performance of straw-based boards and panels are not quite good. As to the abovementioned Novofibre company in Shaanxi Province, the ex-works prices for boards of 2440 mm long and 1220 mm wide with thickness of 11.1, 15, or 18 mm range from around 20 US dollars to 40 US dollars. This price range is close to polywood boards. However, compared with polywood boards, such straw-based boards have inferior properties. They do not hold nails well and are dubious in water resistance. The adhesive used in the boards also make people concerned of possible health risks. For these reasons, this line of high-quality straw-based boards does not sell well. In contrast, boards of lower quality and lower price have a larger market. For boards of the same size, the ex-works price of boards of lower quality is just around 10 US dollars. This price range makes low quality boards better meet the demand of low-end market and for temporary uses. However, this reinforces the low image of straw-based building materials and makes straw-based boards of higher quality even harder to sell. Altogether, straw-based boards and panels just account for a minimal percentage of the total sale of boards and panels, far from being accepted as a mainstream building material.

2.3 Underlying factors of the setback

The setbacks of straw-based boards and panels also because of various reasons.

Firstly, current technology still falls short of expectation on high performance and cost-effectiveness. Morphological differences, wax, silica, and adhesive are key issues for the performance and cost-effectiveness of the processes. Trichome, protuberance, silica, and other features reduce the effect of adhesives. Fine grinding, extraction with hexane, fractionation and other technique may increase the self-bonding of boards. UF resin, PF resin, MDI, and other adhesives have been tested for their performance. Different combinations of the straw, the techniques of pretreating the straw, the adhesives, and the final processing technique leads to different properties of the final products. Besides, the performance of technology is different between tests in laboratories and mass production in factories. Some techniques used or developed in other countries have not been introduced into China for mass production. There is still a need to find a low-cost high-performance technology in laboratories as well as in factories.

Secondly, market acceptance is hard to foster. Not to mention the difficulty to change the low market image of straw-based boards and panels, the use of adhesive also deters a substantial percentage of consumers from buying straw-based boards. Consumers generally regard formaldehyde as risky for health and hold the belief that all adhesives have formaldehyde. Consumers may have environmental awareness and support environmental protection through comprehensive use of crop straw, but may not accept boards made with straw and glue for interior decoration of their homes. The environmental benefit of straw-based boards and panels is for the whole society, this advantage has not transformed into advantages to individual consumers. Besides, the high price of the current real estate market makes home-owners more flexible with the prices of building materials for interior decoration. After paying a high price for a condominium, a homeowner is less likely to care the marginal difference between the price of straw-based boards and panels and wood-based boards and panels. The marginal difference between high-quality straw-based boards and wood-based boards puts the former in a disadvantageous market position.

Thirdly, the governmental policies are not sufficiently consistent and concerted. The production chain of straw-based boards and panels from farmland to final production is subject to the regulation of various governmental agencies. The Ministry of Agriculture and Rural Affairs of the Central Government and the corresponding agencies of local governments have the responsibilities to supervise and regulate the handling of crop straw in rural areas. The National Development and Reform Commission of the Central Government and corresponding agencies of local governments have the responsibilities to formulate and implement industrial policies to encourage the comprehensive use of crop straw as raw materials. The Ministry of Housing and Rural-Urban Development of the Central Government and corresponding agencies of local governments have the responsibilities to regulate the construction industry, including formulating and enforce technical standards on building materials. Various other governmental agencies also have regulatory or law enforcement authorities over the use of crop straw as building materials. The Chinese government adopted a series of policies to encourage the environment-friendly handling and comprehensive use of crop straw, including promoting crop straw as building materials. However, these policies are not necessarily consistent and concerted, particularly when they are applied. One example is the lack of technical standards specifically made for straw-based building materials or inclusive enough to accommodate the special features of straw-based building materials. Currently the technical standards of the construction industry are mostly based on the cement, steel, plaster, and other tradition building materials [14]. This lack of technical standards makes developers reluctant to use straw-based building materials because of possible legal risks.

3 CONCLUSIONS AND SUGGESTIONS

Straw-based boards and panels may have a larger potential to make use of crop straw and save natural resources, but needs to overcome some major challenges, including technology, costs, and market acceptance. Further technical research is needed to improve the quality of straw-based building materials, to reduce secondary environmental pollution in the production processes, and to reduce the costs of production. There is a need to coordinated the policies of various governmental agencies to make the policies more concerted and have more synergy. The agencies in charge of the construction industry should take a more welcome attitude towards straw-based building materials. The governmental agency in charge of agriculture should take more policies to encourage the collection and sale of straw to enterprises of straw-based building materials. The environmental protection agencies need to strictly enforce environmental protection rules to force farmers to sell their straw to industrial users, including those manufacturing building materials with straw. Consistent and concerted governmental policies may give the market more incentive to invest in the development and use of new technology.

REFERENCES

[1] Bhattacharyya P, Bisen J, Bhaduri D, Priyadarsini S, Munda S, Chakraborti M, et al. Turn the wheel from waste to wealth: Economic and environmental gain of sustainable rice straw management practices over field burning in reference to India. *Sci Total Environ* [Internet]. 2021 Jun;775:145896. Available from: https://linkinghub.elsevier.com/retrieve/pii/S0048969721009633

[2] General Office of the National Development and Reform Commission of China, General Office of the Ministry of Agriculture of China. Guidelines on Drafting the Implementation Plan for the Comprehensive Use of Crop Straw in the 13th Five-Year Period. 2016.

[3] Liu H, Zhang L, Wang N. The Study of the Development of Straw-based Panels. *China New Technol Prod.* 2014;(3):131.

[4] Ren Y, Gu Y, Yang X, Chen J. Characteristics, Development Bottleneck and Countermeasures of Straw-based Building Materials. *Funct Mater.* 2016;47(6):06056–62.

[5] Chen Y. The Development of Straw-based Panel in China. *China For Prod Ind.* 2013;40(4):9–16.

[6] Hýsková P, Hýsek Š, Schönfelder O, Šedivka P, Lexa M, Jarský V. Utilization of agricultural rests: Straw-based composite panels made from enzymatic modified wheat and rapeseed straw. *Ind Crops Prod* [Internet]. 2020 Feb;144:112067. Available from: https://linkinghub.elsevier.com/retrieve/pii/S0926669019310775

[7] Zhu XD, Wang FH, Liu Y. Properties of Wheat-Straw Boards with Frw Based on Interface Treatment. *Phys Procedia* [Internet]. 2012;32:430–43. Available from: http://dx.doi.org/10.1016/j.phpro.2012.03.582

[8] Kurokochi Y, Sato M. Effect of surface structure, wax and silica on the properties of binder-less board made from rice straw. *Ind Crops Prod* [Internet]. 2015 Dec;77:949–53. Available from: https://linkinghub.elsevier.com/retrieve/pii/S0926669015304520

[9] Moslemi A, Zolfagharlou koohi M, Behzad T, Pizzi A. Addition of cellulose nanofibers extracted from rice straw to urea formaldehyde resin; effect on the adhesive characteristics and medium density fiberboard properties. *Int J Adhes Adhes* [Internet]. 2020 Jun;99:102582. Available from: https://linkinghub.elsevier.com/retrieve/pii/S0143749620300439

[10] Sitz ED, Bajwa DS, Webster DC, Monono EM, Wiesenborn DP, Bajwa SG. Epoxidized sucrose soyate—A novel green resin for crop straw based low density fiberboards. *Ind Crops Prod* [Internet]. 2017 Nov;107:400–8. Available from: https://linkinghub.elsevier.com/retrieve/pii/S092666901730287X

[11] Sun Y, Liu S, He J. Research Progress and Development Trend of Wood-based Boards Made of Crop Straws. For Mach Woodwork Equip. 2012;40(9):7–9.

[12] Koh CH (Alex), Kraniotis D. A review of material properties and performance of straw bale as building material. *Constr Build Mater* [Internet]. 2020 Oct;259:120385. Available from: https://linkinghub.elsevier.com/retrieve/pii/S0950061820323904

[13] Li X, Cai Z, Winandy JE, Basta AH. Selected properties of particleboard panels manufactured from rice straws of different geometries. *Bioresour Technol* [Internet]. 2010;101(12):4662–6. Available from: http://dx.doi.org/10.1016/j.biortech.2010.01.053

[14] Lv Z 2019 *New Horizons of Environmental Law* (Beijing: China University of Politics and Law Press) pp 249–255

Design of a distributed Bragg reflector of GaN-based light emitting diodes flip chips

J.W. Wang
School of Optoelectronics and Communication Engineering, Xiamen University of Technology, Xiamen, China

Y.C. Xu
School of Optoelectronics and Communication Engineering, Xiamen University of Technology, Xiamen, China
Fujian Key Laboratory of Optoelectronic Technology and Devices, Xiamen University of Technology, Xiamen, China

Q. Zhou, G.Y. Chen, M.M. Wang & P.P. Wu
School of Optoelectronics and Communication Engineering, Xiamen University of Technology, Xiamen, China

ABSTRACT: In order to improve the luminous efficiency of GaN-based light-emitting diodes flip chips and optimize the distributed Bragg reflector (DBR) and its preparation process, a high-reflectivity distributed Bragg reflector (DBR) was prepared by using high-low refractivity materials SiO_2/TiO_2. Firstly, the reflectivity characteristics of the distributed Bragg reflector is expounded and DBR reflectors with different cycles of alternating SiO_2/TiO_2 are designed and prepared for photoelectric performance tests and color rendering index tests at different color temperatures. By changing the lithography exposure to improve the coating of DBR reflectors, DBR with the size of 457μm*760μm is prepared, the lithography exposure was 60mJ, the undercut length is 1.46μm, and the slope angle is 30°. The experimental results show that the reflectivity of 18-cycle DBR reflectors with in the band of 300nm ~ 650nm is more than 98%.

1 INTRODUCTION

In recent years, the GaN-based light-emitting diodes (LED) have been comprehensively studied due to the advantages of wide emission spectral regime from ultraviolet to near infrared and high efficiency, especially for solid state lighting and full-color display [1]. The GaN-based light-emitting diodes have attracted great attention in many different solid state device applications because they have many advantages including high reliability, low power consumption and fast response speed [2]. As a new generation of high-efficiency solid-state light source, the luminous efficiency of GaN-based white LEDs has reached 130lm/W, but it is still far less than the theoretical maximum efficiency [3] The luminous efficiency of GaN LED is mainly determined by the internal quantum efficiency and light extraction efficiency of the device. The luminous efficiency of LED is determined by external quantum efficiency, which mainly depends on the extraction rate and internal quantum efficiency. There is a total reflection of light on the interface between GaN and air because that the refractivity of GaN is much higher than the refractivity of air. This phenomenon makes most of light reflect back to GaN and the light returned to GaN generates heat. Only a small amount of light can radiate into the air. If LEDs keep working at a high temperature for a long time, it will not only reduce the light extraction efficiency of LEDs, but also lead to the shortening of its life. Fortunately, on the one hand the enhancement of the light extraction efficiency can be achieved surface-textured and photonic crystal technology and some of them have been applied in practical production [4]. On the other hand, the main pattern to improve the internal quantum efficiency is precisely controlling the growth of epitaxy material and reducing lattice matching defect. While the internal quantum efficiency of GaN-based LEDs at room temperature is relatively low, the internal quantum efficiency is usually increased by improving the quality of GaN-based epitaxy material [5]. But the cost of such methods is relatively high Due to the limitation of materials absorption, electrode absorption and optical performance of reflectors, the light extraction efficiency of LED made by the mainstream process is comparatively low, which means it is practical to improve the light extraction efficiency by changing the LEDs package structure.

In order to improve the light extraction efficiency, flip chip structure is commonly used in GaN LED. The advantage of flip chip structure is that the light emitted can be directly radiated from the transparent substrate by the reflective layer on the surface of P-GaN, which avoids the blocking of the transmitted light by electrode on surface Therefore, choosing a reflector with high reflectivity is an effective method to further improve the light extraction efficiency of LED Highly reflective metals are often used as reflectors of traditional flip LED chips, such as aluminum (Al), silver (Ag) and so on. However, the reflectivity of these metals is different in the main band of the white light, which ranges from 400nm to 650nm. It is impossible to reflect all the photons, which is not conducive to improving the extraction efficiency of photons. In recent years, applications of distributed Bragg reflectors to improve the internal quantum efficiency and the light extraction efficiency of LEDs have attracted much attention. Jialiang Lu [6] combined the excellent heat dissipation materials with high reflection (such as Al and Ag) and the DBR structure with fewer cycles to form a metal enhanced DBR reflector to improve the reflection bandwidth and average reflectivity and reduce the DBR thickness, so as to significantly improve the heat dissipation performance of the chip and reduce the polarization effect. Yanfei Ma et al. [7] used the Ta_2O_5/SiO_2 hybrid distribution Bragg high output power emitter to design and manufacture AlGaN based near-ultraviolet LED with high output power at 368 nm. Compared with the LED without any DBR, the output power of LED with eight pairs of DBR and hybrid DBR configuration increased by 18% and 25% respectively. Seung-Min Lee [8] fabricated nanoporous GaN DBR by electrochemical etching, and the porosity of the nanoporous layer increased with the increase of etching voltage, resulting in the formation of a distributed Bragg reflector with high refractivity at high voltage. However, the preparation process was complicated comparatively. In order to optimize the gan-based flip LED distributed Bragg reflector and improve the light extraction efficiency, this paper will study a DBR reflector with high reflectivity formed by SiO2 and TiO_2 in the band of 300nm ~ 650nm.

2 THEORETICAL ANALYSIS AND DESIGN

DBR is a special all-dielectric reflective film composed of two alternating compounds of high and low refractivity, so that it can produce periodic modulation of refractivity in one dimension of space. Multilayers periodic nanostructures will produce strong interference phenomena and lead to the selective light reflection within a certain wavelength range. The extremely high reflectivity can be obtained by coating the multilayer quarter wavelength stacks with the alternating high and low refractivity on the optical glass substrate [9].

The reflectivity R is related to the film materials and the refractivity of its boundaries. Rs is the reflectivity of the substrate, n is single layer film refractivity, and ns is the substrate refractivity. If monolayer thickness is odd number of $1/4\lambda$ (central wavelength) times, R maxima appears. If n > ns, R is possible to increase; if n < ns, R is possible to reduce [10]. Let the quarter film system be S|(HL)PH| A, where S is the substrate; and ns is the refractivity of the substrate; A is air; P is an integer; H and L is relatively high refractivity and low refractivity film layers, which the corresponding refractivity are n_H and n_L respectively. And the thickness of the film layers is $1/4\lambda$.

There is a total of 2P+1 layers in the film system, and its equivalent admittance is:

$$y = \left(\frac{n_H}{n_L}\right)^{2P} \frac{n_H^2}{n_s} \tag{1}$$

Its reflectivity R is:

$$R = \left(\frac{1 - \left(\frac{n_H}{n_L}\right)^{2P} \frac{n_H^2}{n_s}}{1 + \left(\frac{n_H}{n_L}\right)^{2P} \frac{n_H^2}{n_s}}\right)^2 \tag{2}$$

As the number of film layers increases, the equivalent admittance of equation (1) increases. Then the reflectivity of formula (2) can be approximated by formula (3):

$$R \approx 1 - 4\left(\frac{n_L}{n_H}\right)^{2P} \frac{n_S}{n_H^2} \tag{3}$$

If there is no loss of absorption or scattering, the transmissivity will be:

$$T = 1 - R = 4\left(\frac{n_L}{n_H}\right)^{2P} \frac{n_S}{n_H^2} \tag{4}$$

It is concluded that the transmissivity decreases by the ratio of $\left(\frac{n_L}{n_H}\right)^2$ when a cycle of high and low refractivity film layer is increaseed. Therefore, the reflectivity is expected to keep increasing, which only is limited by the absorption and scattering of the interface between the film layers.

The oxidation films with high refractivity include $ZnO/TiO_2/Ta_2O_5/Nb_2O_5/ZrO_2/SiO_2/MgO$ etc. SiO_2 is one of the lowest refractivity materials with excellent film property (n≈1.45). SiO_2 is the best choice for film coating because of its smaller absorption and scattering and good transparency at 180nm to 8 μ m [11]. Besides, the film made up of TiO_2, combined with SiO_2 can reduce the absorption and scattering of light on account of the high refractivity of TiO_2 (n≈2.5) and high mechanical strength and all-band good transparency.

Hence, the high reflectivity SiO_2 and TiO_2 are selected DBR materials combination. According to formula (3) and (4), if a cycle of high and low refractivity film layer is increased, the transmissivity of film system will decrease and the reflectance will increase. In this experiment, a cycle is defined as a layers by alternating SiO_2/TiO_2 and the DBR reflector is consist of different cycles. Firstly, DBR reflectors with cycle numbers of 1/3/5/7 cycles were prepared and the reflectivity spectrum is shown in figure 1(a). The test data were shown as solid lines, and the theoretical values were depicted by dotted lines in gray. As shown in figure 1(a), the reflectivity in the blue light band can reach 50% with the increasing of cycle from 1 to 3, and the reflectivity can reach 75% and 90% with the increasing of cycle from 5 to 7. Moreover, the reflectivity has a promotion with the increasing of the DBR cycle in other wavelength.

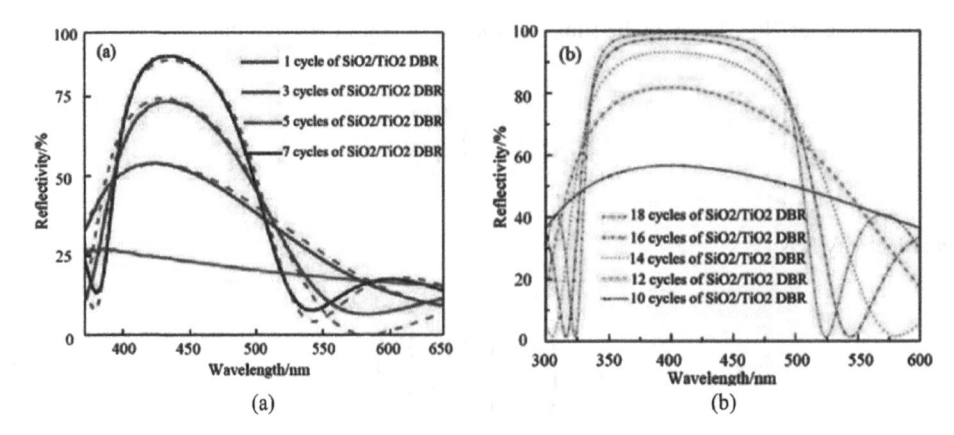

Figure 1. DBR reflectivity of different cycles: (a) Reflectivity of the 1/3/5/7 cycles DBR and (b) Reflectivity of the 10/12/14/18 cycles DBR.

In order to improve the reflectivity of DBR, the cycle of film continues being added. Therefore, alternate DBR structures of SiO_2/TiO_2 from 10 cycles to 18 cycles were prepared and the reflectivity is shown in figure 1(b). As can be seen from figure 1(b), the reflectivity of DBR increases in the blue light band and the spectrum becomes wider with the increase of the DBR cycles. However, the high reflectivity region is not significantly widened. The reflectivity will become saturated with the number of cycles always keeping increasing The main reason is that the absorptivity of DBR is increased as the thickness of material continues, so the reflectivity does not change much. The DBR spectrum range of 18 cycles can satisfy the reflection requirements of GaN-based blue LED from 300nm to 650nm.

As an flip chip reflector, DBR not only has high reflectivity, but also has sufficient reflection bandwidth. The peak of reflectivity increases gradually with the increase of the DBR cycles, which has exceeded 98%. The reflection bandwidth tends to be stable when the DBR cycles gradually increase to maximum. At the same time, the forward voltage of the chip will increase as the DBR cycles increase, which seriously reduces the heat dissipation of chips [12] So it's not the case that the more cycles the better. The choice of the cycle of the LED flip chip reflector should not only depend on high reflectivity of film system, but also hinge on account the forward voltage of the chip. Although more cycles make DBR satisfy the reflection requirements, from the perspective of both high reflectivity high bandwidth and heat dissipation, the DBR of 18 cycles is more practical.

3 EXPERIMENT

Before the preparation of the reflector, the metal electrode is deposited and keeps the current current. After the DBR is deposited on the metal surface, it made be insulated and etching.

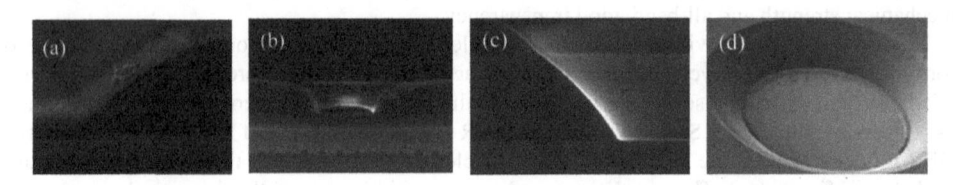

Figure 2. DBR Preparation: (a) Deposition, (b) Lithography, (c) Etching, and (d) Removing of photoresist.

The preparation of the DBR requires the photolithography process. Coating photoresist is the first step in the lithographic process of DBR. Photoresist is divided into positive photoresist and negative photoresist. The positive photoresist has effect for the protection of etching process, while the negative photoresist is often used for the lift-off due to the difference in the undercut length, as shown in figure 3.

Figure 3. Difference in the undercut length.

Overhung photoresist pattern is prepared by lift-off. Lift-off is widely used in the preparation of traditional DBR due to its easy preparation and low cost. The 457μm*760μm chip prepared in this paper was cut by FIB (Focused Ion beam), which made the DBR coating poor. The SEM of DBR fracture layer is shown in figure 4. It is significant to improve the performance of DBR coating to improve the chip passed yield.

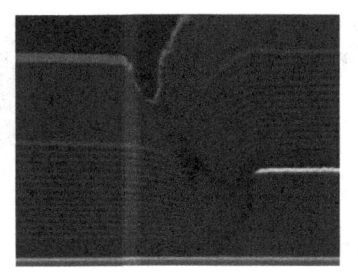

Figure 4. SEM image of DBR structure after lift-off.

As shown in figure 5, the fracture was found on the lateral wall after DBR lift-off. However, there was no fault in DBR on the side of Mesa. The measured angle of Mesa side slope is 35°, and that of DBR fault slope is 60°. Therefore, it can be refer that to improve DBR coating by reducing DBR slope angle is possible.

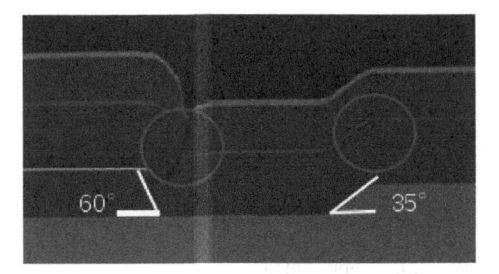

Figure 5. SEM of DBR sidewall fracture morphology after lift-off.

Since the exposure of the fracture layer in figure 6 is 50mJ, controlled experiments with exposure of 40mJ, 50mJ and 60mJ are designed. The corresponding length of undercut and DBR slope angle are shown in figure 6. The slope angle of DBR can be changed by changing the exposure. The undercut length decreases with the increase of the exposure, while the slope angle of DBR decreases with the decrease of the undercut length.

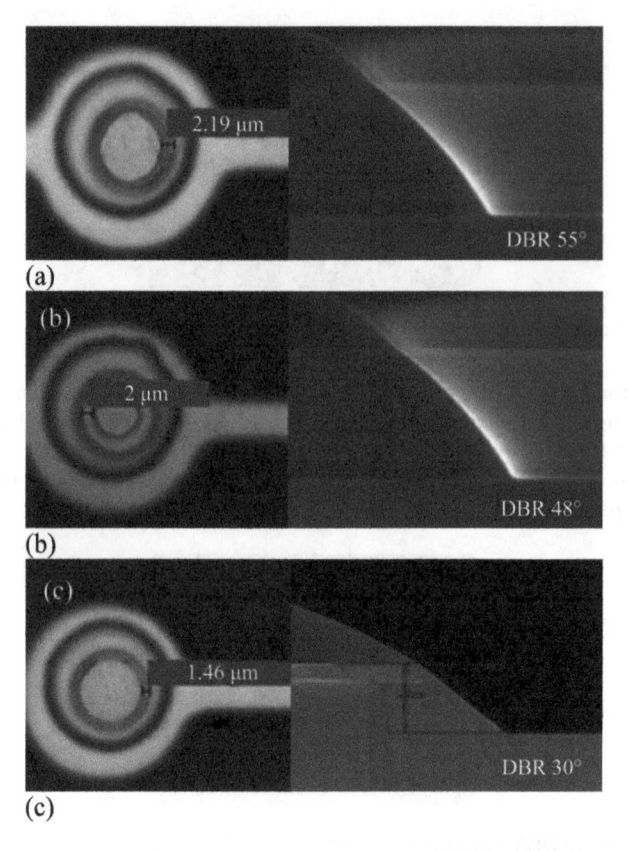

Figure 6. SEM image of undercut length and DBR slope angle at different exposures.

In order to improve DBR coating, the exposure of the lithography process was selected to be 60mJ. As shown in figure 6, the undercut length was 1.46m, the DBR slope angle was 30°. The DBR coating was improved obviously when exposure was 60mJ, as shown in figure 7.

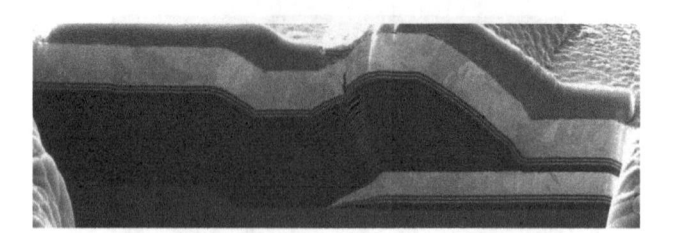

Figure 7. SEM image of DBR coating improvement.

In the experiment, the 10/14/18 cycles of SiO_2/TiO_2 alternating DBR structure were prepared and applied to the LED flip chip for photoelectric tests. According to figure 8, the output light power continues to rise with the increase of DBR cycles when the current changes from 0 to 100mA. Besides, the forward voltage gradually increases to saturation with the increase of DBR cycles.

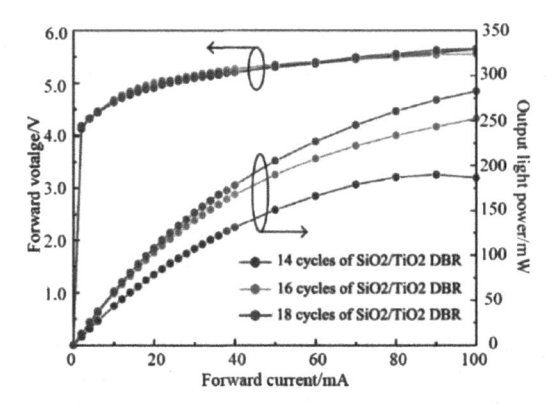

Figure 8. Photoelectricity curves of different DBR structure chips.

Color rendering index (CRI) is one of the most critical factors affecting package brightness and it is important to evaluate the ability of artificial light source to restore the color of objects [13]. The wider the spectrum is, the higher the CRI will be. The CRI of chip which Spectral Bandwidth is wider with the same bandwidth phosphor is relatively higher. During the packaging process, the different excitation intensity of phosphors in different bands results in different CRIs after packaging [14]. Therefore, 12 cycles and 18 cycles of DBR were selected for CRI test under the conditions of color temperature 3000K and 6500K. The difference of CRI between 12 cycles and 18 cycles of DBR excited by yellow phosphor was compared, as shown in figure 9 and figure 10.

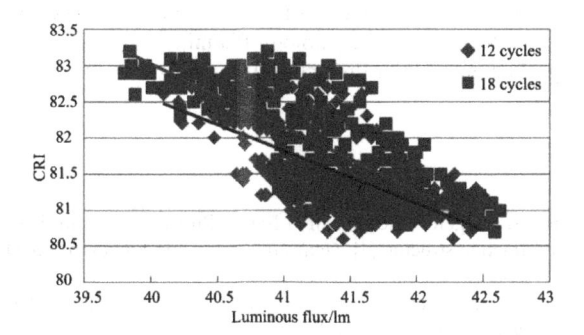

Figure 9. CRI of different cycles DBR at color temperature 3000K.

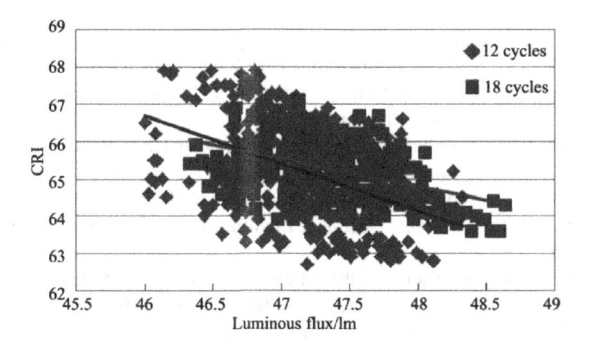

Figure 10. CRI of different cycles DBR at color temperature 6500K.

As can be seen from figure 9 and figure 1, the performance of 18 cycles DBR color rendering index is better than that of 12 cycles DBR under the conditions of color temperature 3000K and 6500K.

4 CONCLUSIONS

Distributed Bragg reflectors with high reflectivity were prepared by alternating high and low refractivity materials SiO_2/TiO_2 of different cycles. According to the results of photoelectric tests and color development index tests, 18-cycle DBR had high reflectivity. The 18-cycle DBR keeps chip voltage and heat dissipation in balance to make chip keep better performance and the preparation process was stable. By changing the lithographic exposure to change the slope angle, the coating property of DBR was further improved and the preparation process of DBR was optimized. The size of DBR is 457μm *760μm, the lithography exposure is 60mJ, the undercut length is 1.46μm, and the slope Angle is 30°. However, there are still some details that can be further optimized in this preparation process, such as the etching angle of DBR and the smoothness of the etching metal layer, which may be related to the etching power and the ratio of DBR metal in the preparation process and need to be further studied.

ACKNOWLEDGMENTS

Foundation: Project supported by general project of natural science foundation of Fujian province(2019J01876), the Education and Scientific Research Project for Young and Middle-aged Teachers of Fujian Provincial Education Department to Special Project of Provincial University (JK2017036), Xiamen Science and Technology Planning Project(3502Z20183060), Xiamen Science and Technology Planning Major Project(3502ZCQ20191002).

REFERENCES

[1] Zhao H F, Liu H, Sun Q,etc. Design of LED collimation system based on refractive/total reflection/reflection/refraction structure[J]. *Optics and Precision Engieering*,2011,19(7):1472–1479.(in Chinese)
[2] Zhu J Y, Ren J W, Li Y Y, etc. Spectral distribution synthesis of spectrally tunable light source based on LED [J]. *Journal of Luminescence*, 2010, 31(1): 882–887. (in Chinese)
[3] Zhong G G, Du X Q and Tian J. Simulation and analysis of light extraction efficiency of GaN-based flip-chip LED chips [J]. *Journal of Luminescence*, 2011, 32(8): 773–778.
[4] Wang Z B, Zhang W, Zhang J, etc. Research on improving the light extraction efficiency of LED based on surface plasmon bimetal grating structure [J]. *Journal of Luminescence*, 2013, 34(12): 1624–1630. (in Chinese)
[5] Li Z Q, Wang C, Li W C et al. Improving the LED light-emitting characteristics by using Ag/P-GaN double gratings[J].*Optical Precision Engineering*.2017,25(5):1185–1191. (in Chinese)
[6] Lu J J and Zheng C J.Optimization Design of LED Chip DBR Mirror[J].*Semiconductor Optoelectronics*,2018.12,39(6):798–801. (in Chinese)
[7] Ma Y , Fan B , Chen Z , et al. Enhanced Light Output of Near-Ultraviolet LEDs With TaO/SiO Hybrid DBR Reflector[J]. *IEEE Photonics Technology Letters*, 2017, 29(18):1564–1567.
[8] Seung-Min Lee, Jin-Ho Kang and June Key Lee,Fabrication of High Reflectivity Nanoporous Distributed Bragg Reflectors by Controlled Electrochemical Etching of GaN.Electron. Mater. Lett., Vol. 12, No. 5 (2016), pp. 673–678
[9] Dubey Raghvendra S and Ganesan V. Fabrication and Characterization of TiO2/SiO2 based Bragg Reflectors for Light Trapping Applications[J]. Results in Physics, 2017.
[10] Zhao B and Xu Y. Multilayer transparent medium thickness measuring device and method for confocal white light polarization interference, CN106546178A[P]. 2017. (in Chinese)

[11] SW Corzine, MRT Tan, CK Lin, et al. Distributed bragg reflector and method of fabrication: US, US7116483[P]. 2006.
[12] Li Z C. Research on SiO2/Si3N4 Distributed Bragg Reflector in Ultraviolet Band [D]. Nanjing: Nanjing University, 2012. (in Chinese)
[13] Hu Y B, Zhuang Q R, Liu S W, et al. Study on color synthesis index LED synthesis white light source[J]. Acta Optica Sinica, 2016, 36(03): 216–225.
[14] Jiang L, Ni K K and Liu M Q. Study on LED Light Efficiency and Color Rendering Index of White Light Using Blue Light Emitting Yellow Phosphor[J]. Journal of Lighting Engineering, 2014, 25(06): 16–19.

Application technology of thin layer repair materials for cement concrete pavement

Y. Liu
Research Institute of Highway Ministry of Transport, Beijing, China

X.J. Huang
Henan Huineng Road and Bridge Technology Co., Ltd, Zhengzhou, China

ABSTRACT: According to the damage characteristics of surface layer of cement concrete pavement, this paper adopts cement-based rapid repair materials for repairing. It analyzes the bonding mechanism of interface between old and new materials, researches the composition, performance index and mixing ratio design of the repair materials with respect to the rapid repair materials of BC type concrete, and finally puts forward the construction technology and quality control requirements of overlay method.

1 INTRODUCTION

It is widely known that the highway pavement can be divided into asphalt concrete pavement and cement concrete pavement according to material categories. Under the national condition and background that the asphalt resources are relatively scarce in China, by virtue of high strength, good stability, strong load diffusion capacity, long durability, strong wear resistance, low maintenance costs, high economic benefits and other advantages, cement concrete pavement is widely applied in highway pavements at all levels in China. With the popularization of cement concrete pavement, various distresses are exposed gradually. These distresses will seriously affect the comfort and safety of driving. Meanwhile, the heavy traffic loads aggravate the damage degree of cement concrete pavement. The damage forms of cement concrete pavement are divided into four categories: Joint class damage, which is reflected in that the vertical and horizontal joints are drawn apart relatively, and the joint sealant is seriously damaged; deformation class breakage, which is reflected in that there is obvious moving feeling when vehicle loads pass through cement concrete plates, and mud-pumping contamination appears at the joints; fracture class breakage, which is reflected in that there are cracks of plate corners, longitudinal cracks, transverse cracks and crossed cracks of plates, as well as large-area crushing of various plates; surface layer class breakage, which is reflected in the loss of fine aggregates and exposure of coarse aggregates on the surface of cement concrete pavement slabs [1]. Among them, the thin layer repair of surface layer damage and joint class damage generally adopts cement concrete as the repair materials. Due to traffic needs, it isn't allowed to close down traffic for a long time when repairing the pavement of roads in use, which is especially obvious in expressways and urban roads. Therefore, the performance requirements for rapid repair materials become higher and higher. With respect to the surface layer damage and joint class damage of cement concrete pavement, this paper adopts the BC type concrete rapid repair materials with excellent performance such as high strength in early stage, minimum inflation and strong bonding strength to conduct thin layer repair; meanwhile, it introduces the construction methods for such rapid repair material.

2 BONDING MECHANISM OF REPAIR MATERIALS

The most important aspect for repairing structures such as concrete buildings and concrete pavements is compatibility. Therefore, the research on the bonding mechanism between old and new materials is particularly important.

In Wall's viewpoint, the old concrete could be regarded as large aggregates, the interface between them was analogous to the relationship between cement and aggregates, thus simplifying the model research. In Park Choonkeun's viewpoint, macro-defect-free cement was the compound formed by the mutual crosslinking of positive ions in cement paste. If the positive ions were multivalent, the compound would be stable; if they were monovalent, the compound would be unstable, and the positive ions produced by the hydration reaction also could be converted into catalyst which made crosslinking reaction constantly occur. Xiao Liguang proposed that divalent calcium ions and divalent magnesium ions in cement could react with polymers to produce complexes, which lowered the concentration of divalent calcium ions, thus affecting the reaction speed and direction of hydration reaction. Moreover, polymers would adsorb hydrophilic group and impede the cohesion of the cement. It's not difficult to see that they sustained each other through the physical force dominated by the electrostatic repulsion between the van der Waals force and ions [2].

Among the viewpoints of numerous scholars, domestic experts are mainly inclined to the polymer-modified cement-based mechanism model proposed by Ohama from Japan. This model is mainly divided into three stages: Firstly, after polymer emulsions were mixed with cement particles, they were distributed in the paste uniformly, hydration reaction occurred gradually, the calcium hydroxide produced gradually reached saturation point, and the polymer particles adsorbed on the surface of cement particles, and this was the first stage; afterwards, due to the proceeding of reaction, pore water was consumed gradually, polymer particles started to flocculate and gather, which formed a layer of polymer sealing layer on the surface of cement particles; when it came to the third stage, the hydration reaction has reached a certain extent, the retiform conjugates formed by the hydration reaction between polymers and cement particles intertwined, which played the role of improving the cement structure [3].

3 DESIGN OF MIXING RATIO OF REPAIR MATERIALS

The rapid repair materials adopted in this paper are composed of cement, silica fume, fine aggregates, polymers and admixtures, etc.

3.1 Cementing materials

Cement: Adopt rapid hardening sulphoaluminate cement with the strength grade of Grade 42.5, referring to Table 1 for details of technical indexes.

Silica fume: Select high-quality silica fume with a density of $2.179 g/cm^3$, adding which with an appropriate amount can improve the impervious performance and compactness of mortar and contribute to guaranteeing the strength in later stage.

Table 1. Table of relevant parameters for thermodynamic analysis.

Initial Setting Time(min)	Final Setting Time(min)	Stability	3d Flexural Strength (MPa)	3d Compressive Strength (MPa)
25	39	Qualified	7.3	42.3

3.2 Fine aggregates

According to the requirements of specification, it is appropriate for the fineness modulus of natural sand used on the surface layer of cement concrete to be between 2.0 and 3.7. Meanwhile, the highway traffic load class shall satisfy corresponding technical requirements. This project adopts natural river sand with the fineness modulus of 2.5 obtained from the screening results, which belongs to medium sand.

3.3 Polymers

Adding an appropriate amount of polymers can effectively improve the bonding strength of rapid and hardening repair cement mortar. Due to the tendency that polymers transfer to the surface of fine aggregates, the polymers can permeate into the pores and cracks of mortar, and form the gelatinous film in the hardened mortar. The gelatinous film can connect the internal structure of cement concrete, playing a role of netting and filling. Meanwhile, filling the defects and micro-cracks at the old and new repair interface enhances the bonding performance of mortar. Re-dispersible emulsion powder is adopted as the polymer, referring to Table 2 for main technical data.

Table 2. Performance indexes of re-dispersible emulsion powder.

Nonvolatile Matter (%)	Ash Content (%)	Apparent Density (g/cm^3)	Minimum Film Formation Temperature (°C)
98	8–12	0.5–0.6	4

3.4 Admixtures

Water reducer. It is used to improve the workability of cement paste, enhance the early strength and reduce shrinkage cracks. Besides, the polycarboxylate superplasticizer adopted has the characteristics of high water reducing rate and long durability. The water reducing rate is 30%, and the water solubility is good.

Defoaming agent. It is used for reducing the bubbles produced in the mixing process, which further lowers the porosity of surface layer repaired effectively, and improves the impervious performance. Polysiloxane is adopted as the defoaming agent, and the mixed amount is 0.1% generally.

In addition, there is a small amount of early strength agent with the component of sodium nitrite ($NaNO_2$).

3.5 Water

Tap water is adopted for mixing and maintaining the repair materials, the mass ratio of materials and water is controlled at about 15%, and the water quality shall conform to the requirements of the current *Standards for Drinking Water Quality* (GB-5749).

3.6 Design of mixing ratio

In order to meet the requirement of opening to traffic within a relatively short period and ensure to satisfy the repair strength and performance requirements, the mixing ratio of the rapid repair materials shall meet the following requirements: the water-binder ratio is 0.14, the binder-sand ratio is 1:1.2, the content of silica fume in repair materials accounts for 6% of net weight of cement, and the polymer-cement ratio is 4%. The technical performance is tested through various tests such as fluidity test of cement mortar, compressive and flexural performance test and bonding strength test of old and new interface, and Table 3 is obtained.

Table 3. Technical performance indexes.

Test Item	Technical Requirement	Test Result	Remark
Initial Setting Time (min)	≥ 20	30	/
Final Setting Time (min)	≤ 90	40	/
	≥ 30	42	4h
Compressive Strength (MPa)	≥ 40	53	1d
	≥ 55	75	28d
Pulling Strength (MPa)	≥ 2.0	2.8	28d

4 CONSTRUCTION TECHNOLOGY AND QUALITY CONTROL

Overlay method is adopted for the thin layer repair of surface layer damage and joint class damage of cement concrete pavement.

4.1 *Construction technology of repair by overlay method*

The repair by overlay method is mainly used for repairs on surface cracks, crossed cracks, cracks near inspection well, dislocation, honeycomb and insufficient friction coefficient. This construction method is a repair method of chiseling away the damaged and loosened concrete on the surface of concrete pavement within a certain scope, and overlaying with rapid repair materials after being scrubbed clean. The specific repair technology is:

(1) Roughen the concrete with cracks or damaged parts;
(2) Drill holes at equal intervals along both sides of damaged cracks, and the bore diameter is required to be slightly greater than the diameter of casta hook;
(3) Fill the holes and slots with mortar, and install the casta hook;
(4) Remove the loosened surface exposed stones which haven't fallen off within the joint, and improve the bonding strength;
(5) After brushing a layer of interface agent on the repaired surface, pour the rapid repair materials to overlay and flatten the pavement.

4.2 *Construction machines & tools and personnel allocation*

Calculate the works quantities of pavement thin layer repair project, and select machines and tools and allocate operating personnel according to the magnitude of works quantity and characteristics of damaged pavement,.

With respect to the characteristics of smaller operating surface and scattered operating surface, it is relatively reasonable to adopt small artificial machines and tools for construction. The main construction equipment includes small-scale pavement cutting machine, generator unit, electric pneumatic pick, air blower, grinding machine, electric mixer and electric drill, as well as auxiliary tools such as steel wire brush, screwdriver, scraper knife and scale. There are 10 operating personnel equipped with dump truck and other transportation equipment, forming a construction squad to be stationed in site.

With respect to surface layer damage and joint class damage of large-area concentrated pavement, it is required to adopt large-scale paving machinery and conduct construction operation of continuous repair project. The main construction equipment includes generator unit, air compressor, cutting machine, pneumatic pick, electric drill, lorry mounted mixer, wet mortar spraying machine, slab vibrator and transport cart, and the specific number of operating personnel shall be determined according to the works quantities [4]

4.3 Construction quality control

Lineation: Delimit the peeling and surface damage areas of the cement pavement required to be repaired, and then operate within the designated area.

Pretreatment of operating surface: Use the special milling machine to clean the cement layer on the surface layer of peeling part, and the depth is required to be 2mm-3mm. In addition, use the high-pressure blower to blow away various dust, broken stones and concrete blocks within the delimited area. After using the high-pressure water gun to wash the delimited area thoroughly, keep the water within the operation area for a period of time to fully moisten the ground. Afterwards, use the high-pressure blower gun to blow the excessive water stored within the operating surface to the region outside the operating surface.

Mixing materials (Repair materials): Adopt the vertical mixer to evenly mix the repair materials and water of cement pavement according to the specified proportion, the mixture no longer has dry powder balls or bubbles, namely, the mixing is even, make sure to mix fully, and keep stirring once every 5 minutes, so as to prevent internal solidification.

Paving: Pour the well-stirred cement mixed repair materials onto the cleaned operating surface, flatten, press and polish. Ensure that the elevation of the repaired pit slot floor is consistent with the floor outside the delimited area. Besides, pressing and polishing for the second time isn't allowed after initial setting, and construction isn't allowed when the temperature is too low.

Maintenance: After the final setting, carry out the maintenance of spraying water and covering plastic film. As the repaired surface layer is relatively thin, the evaporation capacity is relatively significant. In order to guarantee the stable increase of strength of repair materials and facilitate the further setting and hardening of cementing materials, the humidity of the environment in which the repaired pavement is located shall be adjusted moderately. During the maintenance, vehicle traffic is strictly prohibited and pedestrians shall be prevented from treading on the repaired pavement. The traffic can be opened until the structural strength of pavement reaches the design value [5]

In addition, ensure that water and electricity are available on the operating site before construction, the operating personnel shall have corresponding occupation qualification and shall be employed with certificates, guarantee the operating proficiency, and organize the construction scientifically and effectively.

5 CONCLUSION

This paper adopts the rapid repair materials to carry out thin layer repairs for the surface layer damage and joint class damage of cement concrete pavement. BC type rapid repair materials have the advantages of short setting time, closer initial setting time and final setting time, high early strength, stable increase of later strength and strong bonding performance. The adoption of overlay method for construction by rapid repair materials has the advantages of good technical performance, good economic returns and short closure time of road. Through comparing with various technical parameters, it can be obtained that adopting BC type cement concrete pavement thin layer repair materials can resume the use of roads rapidly under the premise of guaranteeing the similarity to the original pavement color; after the paving is completed, it is required to maintain the pavement, ensure that the strength reaches the application standards, and guarantee the favorable repair effect.

REFERENCES

[1] He Jianjie. Research on the Rapid Repair Materials and Technologies of Cement Concrete Pavement [D]. Master's thesis. Chengdu: Southwest Jiaotong University, 2006:1
[2] Ye Yong. Research on the Rapid Repair Materials of Concrete Pavement [J]. Journal of Highway and Transportation Research and Development, 2014. 109(1):53–55

[3] Xie Zirong. Rapid Thin Layer Repair Design and Construct Method of Cement Concrete Pavement [J]. Shanxi Architecture, 2015, 41(23):130–132
[4] Huang Jinghai. The Application of Crack Repair Materials of Cement Concrete Pavement [J]. Technology & Economy in Areas of Communications, 2008, 46(2):20–22
[5] Xie Yongcheng. Rapid Repair Technology of Ultrathin Layer of Cement Concrete Pavement [J]. Highway, 2013(7):91–93

Study on microstructure of new nickel-based powder superalloy

G.H. Yang
School of Materials Science and Engineering, University of Science and Technology Beijing, Beijing, China

G.F. Tian
Science and Technology on Advanced High Temperature Structure Materials Laboratory, Beijing Institute of Aeronautical Materials, Beijing, China

X.L. Zhang, Z. Liu, Z. Ji*, X.Q. Yan & C.C. Jia
School of Materials Science and Engineering, University of Science and Technology Beijing, Beijing, China

ABSTRACT: Using metallographic optical microscope (OM), scanning electron microscope(SEM), electron backscatter diffraction(EBSD) and other methods to characterize the microstructure of the new nickel-based powder superalloy samples, the results show that there is no obvious PPB structure in the alloy; the size of the matrix phase γ grains is about 5μm, and the grain boundaries are clear. The precipitated phase γ' is distributed in the matrix phase crystal and accounts for more than 80% of the entire precipitated phase, and the crystal grain size is between 200–800nm. The precipitated phase γ' is less distributed in the grain boundaries of the matrix phase, and the size is between 0.8 – 1.8μm. The grain orientation of the alloy is mainly <111>, <101>, and <001>, among which the grains in the direction of <101> and <001> occupy most of the proportion. The alloy grains are mainly high-angle grain boundaries, and the low-angle grain boundary grains are the γ' phase located in the matrix and the secondary phase partially located on the grain boundaries. The grain size below 60μm occupies the majority, and the larger grain size only occupies a small part of the alloy. There are twins in the alloy, but the proportion of twin boundaries only accounts for a small part of the grain boundaries in the alloy.

1 INTRODUCTION

Compared with traditional casting and forging alloys, nickel-based powder superalloys have the advantages of uniform structure, no macro-segregation, high yield strength, and good fatigue performance. They can be used to manufacture high-performance aero-engine turbine disks [1, 2]. Nickel-based powder superalloys are mainly composed of γ phase, γ' phase, carbides, borides and so on. Because the amount of carbides and borides is particularly small, in theory, nickel-based powder superalloys can be regarded as two-phase alloys, namely the matrix γ phase and the strengthening phase γ' phase[3–5]. In the microstructure of the alloy, the size and distribution of the γ phase and γ' phase have a great influence on the mechanical properties of the alloy [6–8]. This paper uses metallographic optical microscope, scanning electron microscope, EBSD and other methods to study the microstructure of the new nickel-based powder alloy samples, and provides basic theoretical and practical guidance for improving the mechanical properties of nickel-based powder superalloys.

---

* Corresponding author

2 EXPERIMENTAL MATERIALS AND METHODS

2.1 *Materials*

The samples of new nickel-based superalloys used in this paper are prepared by powder metallurgy (PM) method.

2.2 *Microstructure characterization method*

Using metallurgical optical microscope (OM), scanning electron microscope (JSM-6480LV), in-situ scanning Auger probe-EBSD backscatter diffraction and other methods to study the microstructure of the new nickel-based powder superalloy.

3 RESULTS AND ANALYSIS

3.1 *Metallographic characterization*

Figure 1 is the metallographic structure of the new nickel-based superalloy. It can be seen that there is no obvious PPB structure, which is mainly composed of matrix γ phase and precipitated γ' phase.

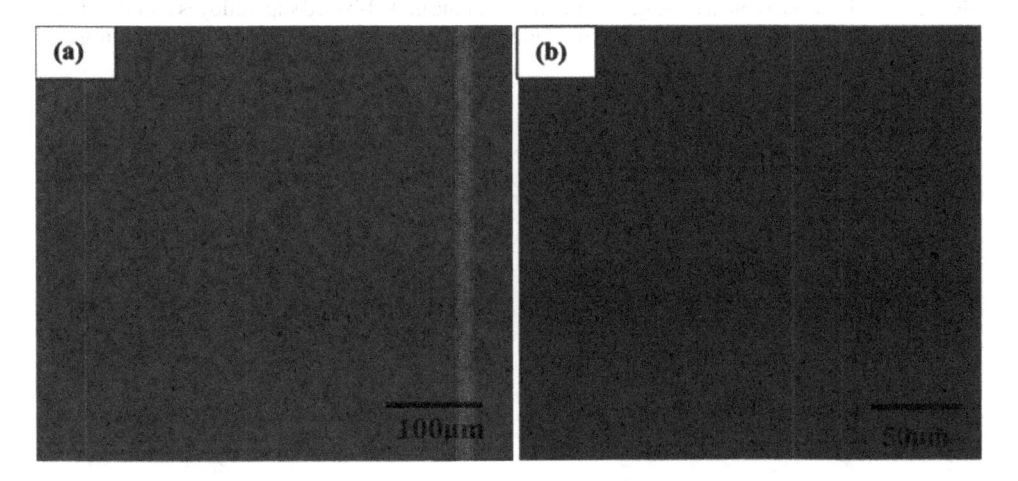

Figure 1. Photograph of nickel-based superalloy metallographic structure: (a) 100x; (b) 200x.

3.2 *SEM characterization*

The scanning electron microscope picture of the nickel-based superalloy in Figure 2 shows that the matrix γ phase is uniform in size, the γ grain size is about 5 μm, and the grain boundaries are clear. It can be seen from the metallographic diagram 1(b) that a part of the first γ' phase is distributed on the grain boundary of the matrix phase, and the size of the γ' phase is between 0.8 – 1.8μm. This is due to the energy on the grain boundary of the matrix γ phase High, more defects and vacancies are more favorable for the precipitation of secondary phases. More γ' phases are distributed in the crystals. The size of this γ' phase is smaller than that of the γ' phases distributed on the grain boundary. It is between 200–800 nm and occupies 80% of the total secondary phases. The above has played a major role in strengthening the alloy.

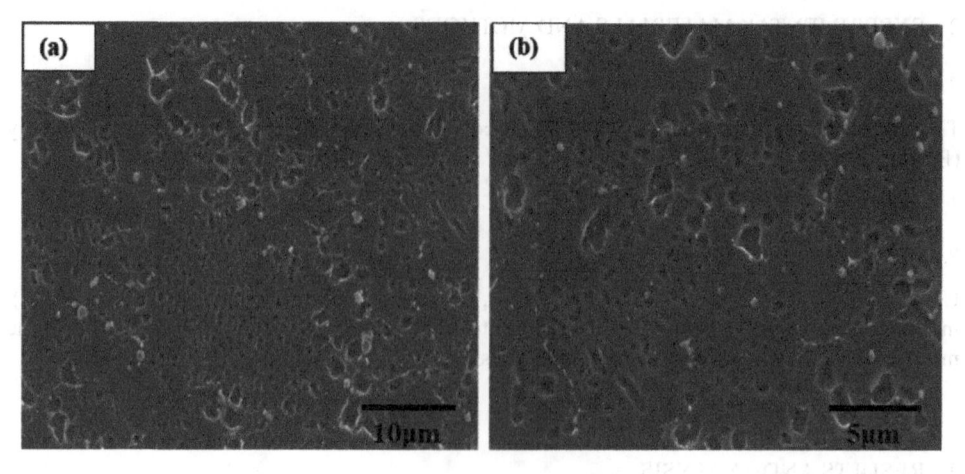

Figure 2. SEM photo of nickel-based superalloy :(a) 2000x; (b) 3000x.

3.3 *EBSD characterization*

Figure 3 shows the EBSD morphology and reverse pole diagram of the nickel-based powder superalloy sample. It can be seen that the grain orientation of the nickel-based superalloy is mainly <111>, <101>, and <001>, of which <101>, <001> Directional grains occupy most of the proportion.

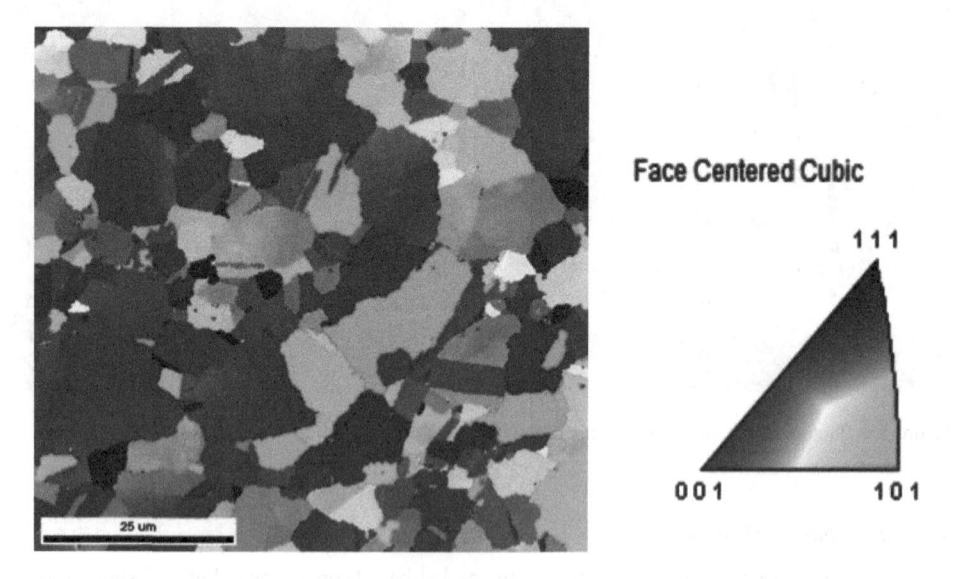

Figure 3. EBSD morphology and inverse pole diagram of nickel-based powder superalloy.

Figure 4 shows the EBSD grain boundary orientation of the nickel-based powder superalloy sample. It can be seen that the small-angle grain boundary grains with an orientation degree of 2° − 15° account for only 10%, and the orientation degree is a large angle of 15° − 70° The grain boundary grains account for nearly 90%, which indicates that the nickel-based superalloy grains are mainly high-angle grain boundaries, and the low-angle grain boundary grains are the γ' phase located on the matrix and the secondary phase partially located on the grain boundary.

Boundaries: Rotation Angle

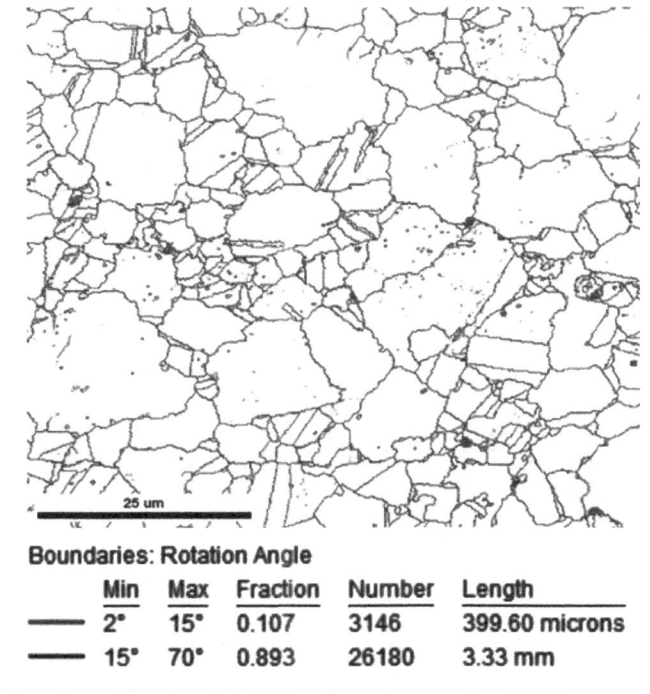

	Min	Max	Fraction	Number	Length
——	2°	15°	0.107	3146	399.60 microns
——	15°	70°	0.893	26180	3.33 mm

Figure 4. Grain boundary orientation of nickel-based powder superalloy EBSD.

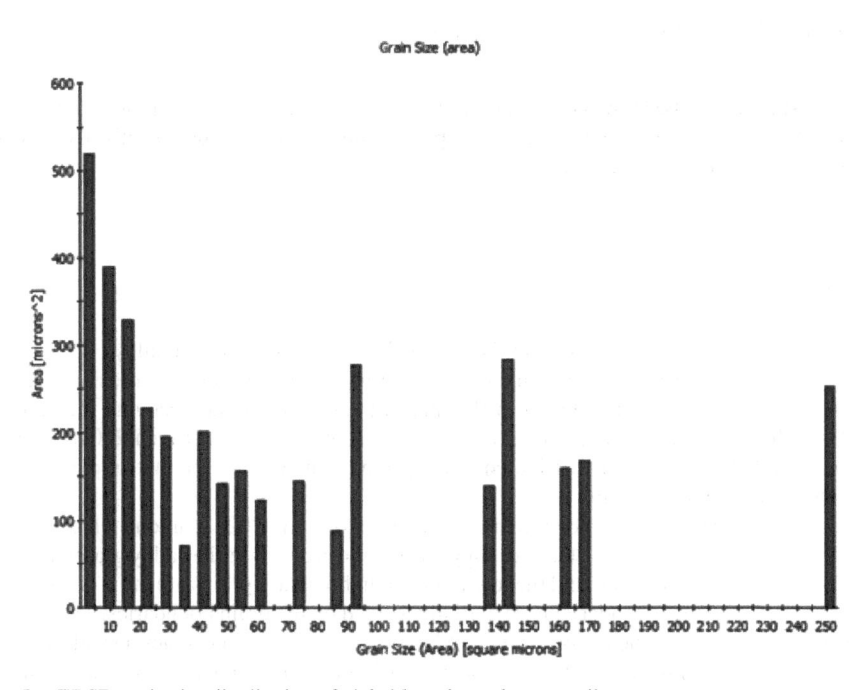

Figure 5. EBSD grain size distribution of nickel-based powder superalloy.

Figure 5 shows the EBSD grain size distribution of the nickel-based powder superalloy sample. It can be seen that the grain size below 60μm accounts for the majority, and the larger grain size only accounts for a small part of the nickel-based superalloy.

Figure 6. EBSD twin distribution of nickel-based powder superalloy.

Figure 6 shows the EBSD twin distribution of a nickel-based powder superalloy sample. It can be seen that there are twins in the alloy, but the proportion of twin boundaries only accounts for a small part of the grain boundaries in the alloy.

4 CONCLUSIONS

(1) There is no obvious PPB structure in the new nickel-based powder superalloy.
(2) The size of the matrix phase γ grains is about 5μm, and the grain boundaries are clear. The precipitated phase γ' is distributed in the matrix phase crystal and accounts for more than 80% of the entire precipitated phase, and the crystal grain size is between 200–800nm. The precipitated phase γ' is less distributed in the grain boundaries of the matrix phase, and the size is between 0.8 – 1.8μm.
(3) The grain orientation of the alloy is mainly <111>, <101>, and <001>, among which the grains in the direction of <101> and <001> occupy most of the proportion. The alloy grains are mainly high-angle grain boundaries, and the low-angle grain boundary grains are the γ' phase located in the matrix and the secondary phase partially located on the grain boundaries. The grain size below 60μm occupies the majority, and the larger grain size only occupies a small part of the alloy. There are twins in the alloy, but the proportion of twin boundaries only accounts for a small part of the grain boundaries in the alloy.

REFERENCES

[1] Hu Benfu, Liu Guoquan, Jia Chengchang et al. Materials Engineering. 2007, 8(2):49.
[2] Guo Weimin, Feng Di, Wu Jiantao et al. Materials Engineering. 2002, 3(l):44.
[3] Unocic R R, Viswanathan G B, Sarosi P M, et al. Mechanisms of creep deformation in polycrystalline Ni-base disk superalloys. Materials Science and Engineering. A, Structural Materials: Properties, Microstructure and Processing, 2008,(483–484):25–32.
[4] Konig W, Gerschwiler K. Machining nickel-based superalloys. Manufacturing Engineering, 1999, 122(3):102–108.
[5] Zhang Z, Yue Z. TCP phases growth and crack initiation and propagation in nickel-based single crystal superalloys containing Re. Journal of Alloys & Compounds, 2018, 746:84–92.
[6] Zhao Shuangqun, Xie Xishan, Smith Gaylord D et al. Materials Science and Engineering. 2003, A355:96.
[7] Xie Shishu, Pan Xianfeng, Yang Hongcai et al. Acta Metallurgica. 1999, 12(3):267.
[8] Guo Shiwen, Zhang Yusuo, Yang Hongcai et al. Journal of Northeastern University (Natural Science). 2003, 24(6):576.

Study on preparation and characteristics of Ni-Cr-W alloy laser cladding coating

L. Chen, Z.D. Liu, B.K. Li, S.Y. Guo, J.X. Li & Y. Li
North China Electric Power University, College of Energy Power and Mechanical Engineering, Key Laboratory of Power Transmission, Conversion and System of Power Station, Ministry of Education, Beijing, China

ABSTRACT: The Ni-Cr-W alloy coating with high wear resistance was prepared on the surface of the Q235 substrate by laser cladding technology. The phase composition and microstructure of laser coating were investigated by means of SEM, EDS, and XRD. The wear resistance of laser coating was analyzed using the hardness test and abrasive wear test. The results show that the laser cladding coating of Ni-Cr-W alloy was composed of solid solutions of Fe and Ni, low melting point eutectic phase Fe2B and Cr2B, and some carbides, and the laser coating is well combined with the substrate with a small heat-affected zone. Furthermore, the addition of a small amount of C on the Ni-Cr-W laser coating can significantly improve the hardness and wear resistance.

1 INTRODUCTION

At present, with the development of industrial modernization, the surface performance requirements of equipment parts are getting higher, especially in the field of high-end manufacturing. Laser cladding is a technology that originated in the 1960s. It can make the laser cladding coating and the substrate form a good metallurgical bond, while having a low dilution rate [1], small heat-affected zone, short cooling time, small deformation of the substrate, etc. The thickness of the laser cladding coating can be controlled according to requirements, and the ultra-high-speed laser cladding technology can be used to produce a laser cladding coating with a thickness of nanometers and good performance [2]. In many environments, wear resistance is always one of the most important properties. Nickel-based composite materials are widely developed and applied as wear-resistant, heat-resistant, and corrosion-resistant materials. WC particles [3] have a high melting point, high hardness, and good stability. Tungsten carbide particles have a zero wetting angle with iron-based metals and are easy to obtain compared with other cermet particles (such as titanium carbide, etc.); so WC particles are enhanced composite materials, and they have attracted more attention and have been widely used. Kaifang Dang et al. [4] applied a fiber laser to prepare a laser cladding Ni-based WC-reinforced composite coating on the substrate. The experimental results show that as the WC content increases, the dilution rate of the laser cladding coating first increases and then decreases. When the WC content is 20wt%, the dilution rate is the smallest. Tungsten carbide particles partially dissolve and interact with other elements to form block, strip, and spherical eutectic, which can increase the average hardness to 5 times that of the matrix, and the wear resistance is also significantly improved. Erfanmanesh, Mohammad et al. [5] studied the wear performance of WC-12Co and electroless Ni/WC-Co at high temperature, and the measurement found that the friction coefficient, wear rate, and microhardness of Ni/WC-Co were significantly lower than those of WC-Co. Farahmand, Parisa, et al. [6] added lanthanum oxide (La_2O_3) and molybdenum to the modified Ni-60% WC and replaced part of the ultrafine WC with nano-WC particles to enhance the chemical composition. The experimental results show that the optimal addition of nano-WC (5wt.%), La_2O_3 (1wt.%), and Mo (1wt.%) improves the grain size of the

Ni binder and increases the hardness, corrosion resistance, and abrasion resistance. Most of the abovementioned research studies added high content of WC to increase particles or trace elements to increase hardness and wear resistance, which requires advanced development technology and high cost. The main purpose of this article is to add low content of C and then use laser cladding technology to produce a WC-enhanced Ni-based laser cladding coating on the substrate, thereby improving the hardness and wear resistance of the equipment, extending the service life of the equipment, and reducing maintenance costs.

2 EXPERIMENTAL PROCEDURE

2.1 Sample preparation and analysis method

This sample is made by cladding WC-containing nickel-based powder on Q235 steel with a 3.3kW laser. A high-power laser beam with a size of 2 mm quickly heats and melts the powder blown out by the synchronous powder feeder and then interacts with the base material. Form a metallurgical bond laser cladding coating. The experiment uses a laser power of 2.5kW, a scanning speed of 6cm/s, and a laser step of 0.7mm. The cladding powder materials used in the experiment are W88-0C and W88-8C. The chemical composition of each layer is shown in Table 1. With a thickness of about 500μm, sample A is made by W88-0C with a total thickness of 1.5 mm, and sample B is made by W88-8C with a total thickness of 1.8 mm. Furthermore, the material of sample C is Q235.

Table 1. Chemical composition of material.

powder	Cr	W	B	Si	C	Fe	Ni
W88-0C	15	18	3.5	3.5	0.8	<1	Bal.
W88-8C	16±0.5	18±0.5	3.5	3.5	0.02	<1	Bal.

Wire cutting technology is used to cut the laser cladding sample into a wear sample with a size of 500×250mm and a test sample with a size of 10×10mm. When the sample is further tested, the multi-layer coating is wet-ground and polished [7], until a clean, shiny surface is obtained, and the polished surface is cleaned with distilled water in an ultrasonic bath. Energy spectrum analysis (EDS), scanning electron microscopy (SEM), and X-ray diffraction (XRD) phase composition analysis are carried out to check the microstructure. In this experiment, the SEM test uses a scanning electron microscope model S4800 and its attached EDS to observe and analyze the microstructure of the sample laser cladding coating. The XRD test uses an X-ray diffractometer model D/max to emit Cu-Kα ray. Phase analysis is conducted on the surface of the sample laser cladding coating.

2.2 Hardness test

A FM-300 microhardness tester with a set load of 500g was used to measure the hardness of the cross-section of the cladding sample along the thickness direction. The loading time is 15s, and the measurement is performed every 150μm along the depth direction to reduce the error and three times at the same depth from the surface of the laser cladding coating. Finally, the average of the 3 measurements is taken to draw a graph of the microhardness distribution curve of the laser cladding coating of the sample.

2.3 Wear resistance test

This test uses the MLS-225 wet sand rubber wheel type, semi-free type abrasion tester, and the bearing rubber wheel drives the sand, ore, and other abrasives mixed with water during the rotation to perform abrasive wear on the test materials. By calculating the weight loss of the material after a certain number of rotations of the rubber wheel, the final weight loss and the wear rate are obtained, the wear volume curve and the wear rate curve are drawn, and then the wear resistance of the sample is obtained. The specific main parameters are shown in Table 2.

Table 2. Experimental parameters of wear test.

Load/N	Rubber wheel diameter/mm	Rotation speed /r/min	Rubber wheel hardness	Mortar ratio
40	178	180	60	1.5:1

Before the start of the test, the sample was pre-grinded at a low speed to produce prefabricated wear marks. The purpose is to eliminate the experimental errors caused by the uneven surface quality and force of the sample. After pre-grinding, alcohol is used to clean and dry. A 1/10,000 analytical balance is used to weigh the sample mass. The result is the initial mass M_1 before abrasion, and the mass after each abrasion is recorded as Mi (i from one to five). The formal test includes a cycle of 2000r, a total of 5 cycles; after abrasion, the same method is used to dry and weigh, and the amount of wear G_i and the rate of wear g were calculated as follows:

$$G_i = M_1 - M_i \tag{1}$$

$$g = M_i - M_{i+1} \tag{2}$$

3 RESULTS AND DISCUSSION

3.1 *Material microstructure and phase analysis*

Figure 1 shows the morphology of the bottom, middle, and top of the cross-section of Ni-based cermet cladding sample A. The laser cladding coating is mainly composed of white ceramic phase and black phase. The total thickness of the laser cladding coating is about 1500μm. It can be seen from Figure 1(a) that the ceramic laser cladding coating close to the substrate is denser than the middle and top. The crystal grains are relatively small, and the shapes are mostly small irregular spots. As shown in Figure 1(b), a small amount of long strip and a small amount of scattered-like structure are distributed in the laser cladding coating [8]. The structure of the upper part of the laser cladding coating is similar to that of the middle part, but the degree of grain density is stronger than that of the middle part of the laser cladding coating, and the degree of mutual aggregation of crystal grains is greater than that of the middle part of the laser cladding coating. As shown in Figure 1(c), the upper part of the laser cladding coating tends to be smaller dense equiaxed crystals, and the morphology of equiaxed crystals will change with the growth conditions to form long strips, rhombuses, and polygons [9]. This is because there are overlaps in the laser cladding process. When the laser remelts the previous one, a temperature gradient is formed in the molten pool, which leads to directional solidification and epitaxial growth of part of the crystal grains, resulting in crystal grains of the same orientation and close size in the middle and the top. A point scan of the positions 1 and 2 is carried out in the laser cladding coating, as shown in Figure 1(b), and the element atoms at each position are shown in Table 3. It can be seen that the white dendritic phase is rich in Ni and Cr, and most of the tungsten atoms are dissolved. The black phase is rich in Ni and should be a nickel-based solid solution.

(a)　　　　　　　　　　(b)　　　　　　　　　　(c)

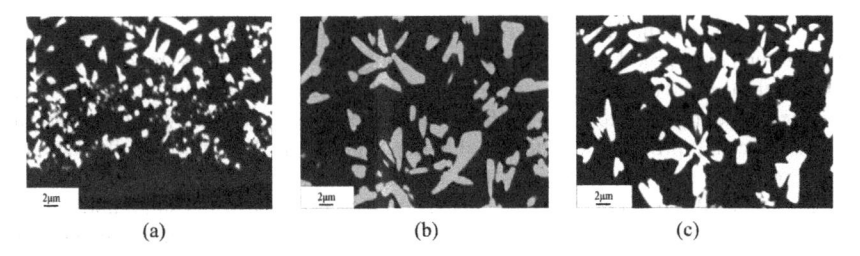

Figure 1. Microstructure of laser cladding coating W88-0C: (a) Bottom of the laser cladding coating, (b) Middle part of laser cladding coating and (c) Top of the laser cladding coating.

Figure 2 shows the morphology of the bottom, middle, and top of the cross-section of Ni-based cermet cladding sample B. The laser cladding coating is composed of two phases, white ceramic phase and black phase, with a thickness of about 1800μm. It can be seen from Figure 2(a) that under the same scale, the quantity in sample B is denser than that in sample A. The ceramic phase of the laser cladding coating is also tighter than the middle and top, and the shape of the crystal grains is still irregular. This is due to the increase in carbon content and the increase in carbides generated during the cooling process of the liquid metal. The nucleating particles reduce the nucleation work and further promote the formation of crystal grains. Therefore, sample B precipitated more ceramic particle phases. As shown in Figure 2(b), the morphology of the middle part of the laser cladding coating is not much different from that of the top part of the laser cladding coating. There are some fine dendrites in the middle part, and the overall distribution is uneven, but the grain density is stronger than that of the upper part of the laser cladding coating. The grain direction of the upper part of the laser cladding coating is stronger than that in the middle part, and the density is not as good as the bottom part of the laser cladding coating and the middle part of the laser cladding coating, and the shapes are different, as shown in Figure 2(c).

Table 3. Structure composition of different points of the laser cladding coating (atomic fraction, %).

Position	Ni	Cr	W	Fe	C
A-1	20.46	54.71	23.73	1.11	
A-2	48.31	43.37	7.78	0.05	
B-1	20.71	45.21	30.27	0.34	3.46
B-2	86.04	8.09	0.16	1.80	3.90

The dots are scanned at positions 1 and 2 in the laser cladding coating in Figure 2(b). The element atoms at each position are as shown in Table 3. The comparison shows that almost all tungsten atoms dissolve into the white dendritic phase, and the black phase is mainly Ni as the solvent. Dissolved in a small amount of Cr solid solution, the difference of C atom content is very small, indicating that the C distribution is relatively uniform.

(a)　　　　　　　　　　(b)　　　　　　　　　　(c)

Figure 2. Microstructure of laser cladding coating W88-8C: (a) Bottom of the laser cladding coating, (b) Middle part of laser cladding coating and (c) Top of the laser cladding coating.

Figures 3 and 4 show the XRD patterns of samples A and B of the nickel-based cermet laser cladding coating. For sample A, after calibration, its main phase components are solid solutions of Fe and Ni, low melting point eutectic phases Fe_2B and Cr_2B, and a small amount of carbide. The silicon in the sample has not been detected by XRD. It is inferred that silicon mainly exists as independent particles, which can improve the wettability of hard particles [11]. For sample B, the main phase is a solid solution of Fe and Ni, and there is also a low melting point eutectic phase Fe_2B. However, due to the increase of the C content, more WC is precipitated. Its size is small and is uniformly distributed in the laser cladding coating, which plays a role in dispersion enhancement [12].

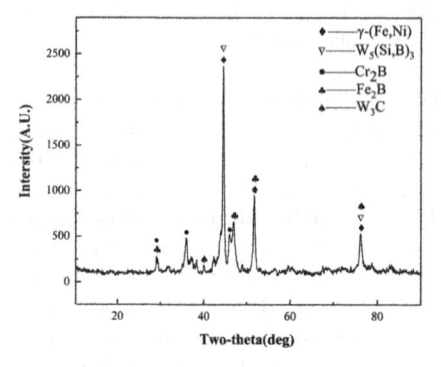

Figure 3. XRD pattern of sample A.

Figure 4. XRD pattern of sample B.

3.2 Microhardness

As shown in Figure 5, the average surface hardness of the laser cladding coating of sample A is 676.98HV; the average surface hardness of the laser cladding coating of sample B is 869.81HV; it can be clearly seen that the hardness of materials A and B is both stepped and distributed. The hardness of the cladding can reach 4–6 times that of the substrate. The three obvious steps correspond to the laser cladding coating, heat-affected zone [13], and the substrate of the material, and the hardness of material B is slightly higher than that of material A. This is because B contains higher C, so that more carbides can be precipitated, thereby increasing the hardness of the material, which can be reflected in the scanning electron microscope (Figure 2). In addition, the B material can play a role in dispersion strengthening due to the small volume and large number of carbides.

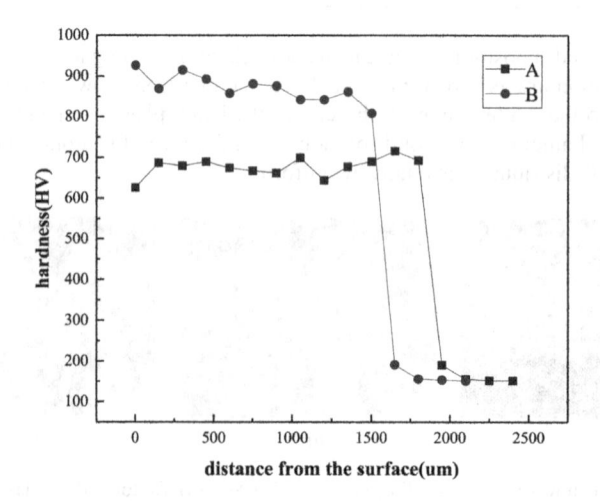

Figure 5. Microhardness of A and B.

3.3 Abrasion resistance study

The weight loss of abrasive wear of three materials is shown in Table 4:

Table 4. Sample wear (mg).

Turns	A	B	C
2000r	46.4	22.1	30.0
4000r	75.6	42.0	61.6
6000r	102.5	52.4	107.5
8000r	121.5	62.9	207.1
10000r	150.4	74.5	292.5

The wear of the three samples increases with the increase in the number of revolutions. After abrasion of 10000r, the sample B with a weight loss of 74.5mg has the smallest amount of wear, and the sample Q235 with a weight loss of 292.5mg has the largest amount of wear and the lowest hardness. The loss is much larger than that of samples A and B, indicating that the WC-containing nickel-based cladding coating prepared by laser cladding can indeed greatly improve the wear resistance of the substrate.

According to Figure 6 and Figure 7, the greater the hardness of the sample, the smaller the amount of wear. For sample A, it mainly forms Cr_3Si binary metal silicide as a hard reinforcing phase. In addition, the presence of Ni_3Si in the sample improves the brittleness of the material [14]. For sample B, the evenly distributed hard phase WC consumes more wear work during the wear process, so that the weight loss of the laser cladding coating is significantly reduced. In addition, the plastic deformation of the material is improved by increasing the particle phase and can hinder the wear scar. In development, this can be well reflected in the wear rate of the sample. The wear rate of sample C increased greatly with the increase of the number of turns, while the wear rate of samples A and B generally decreased with the increase of the number of revolutions. This is because the hard phase in the sample hindered the development of wear scars. The distribution of the hard phase in sample B is denser, so that it exhibits better performance in the wear test, and the wear rate curve of sample B is also relatively stable. The disadvantage of this experiment is that it is only carried out at room temperature. In actual situations, wear is often accompanied by high temperature, and the wear resistance decreases with the increase of temperature [15]. In short, an appropriate amount of C can improve the hardness and wear resistance of the laser cladding coating.

Figure 6. Wear curve of the cladding sample.

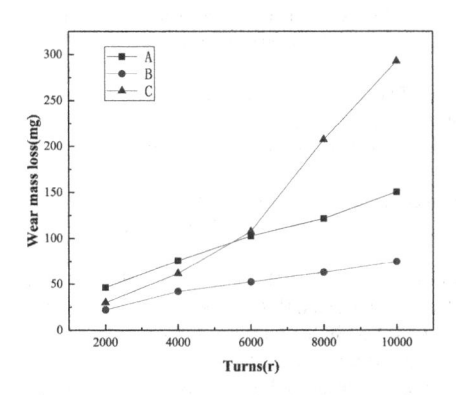

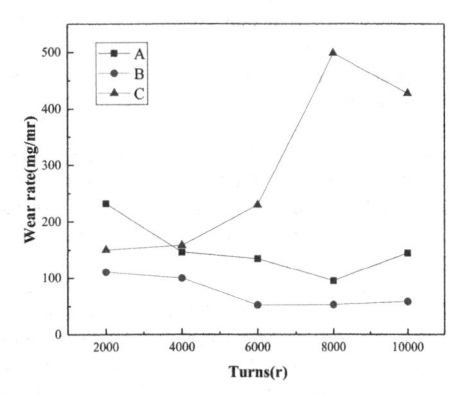

Figure 7. Wear rate curve of the cladding sample.

4 CONCLUSION

1) W88-0C and W88-8C alloy laser cladding coatings are prepared by laser cladding. The main phase components are Fe and Ni solid solutions, low melting point eutectic phases Fe_2B and Cr_2B, and some carbides. The alloy structure contains different sizes. With the increase of carbon content, more carbides are formed during the cladding cooling process, so the number of ceramic phase reinforcement phase grains in the structure of the W88-8C laser cladding coating is more.
2) The average surface hardness of the laser cladding coating of sample A is 4 times of 676.98HV; the average surface hardness of the laser cladding coating of sample B is 869.81HV, which can reach 4–6 times of the base Q235 steel, thereby significantly improving the hardness of the base material. The reason is that both samples A and B have hard phases, and sample B has more carbides.
3) After the substrate and the two laser cladding coatings are worn for 10000r, the wear amount of W88-8C is the smallest, and the total weight loss is 74.5mg. Q235 has the largest amount of wear, with a total weight loss of 292.5mg, and W88-8C has a medium amount of wear, with a total weight loss of 150.4mg, indicating that the WC-containing nickel-based cladding coating prepared by laser cladding has greatly improved the wear resistance of the substrate.

REFERENCES

[1] Lida Z, Pengsheng X, Qing L, et al. 2021 Recent research and development status of laser cladding: A review[J]. *Optics and Laser Technology*, 138.
[2] Habibi A H, Razavi R S, Borhani G H, et al. 2021 Effect of Argon Shroud Protection on the Laser Cladding of Nanostructured WC-12Co Powder[J]. *Journal of Materials Engineering and Performance*.
[3] Yefeng B, Linpo G, Chonghui Z, et al. 2021 Effects of WC on the cavitation erosion resistance of FeCoCrNiB0.2 high entropy alloy coating prepared by laser cladding[J]. *Materials Today Communications*, 26.
[4] Dang K, Jiang Z. 2020 Microstructure evolution and properties of a laser cladded Ni-Based WC reinforced composite coating[J]. *Materials Testing*, 62(11).
[5] Erfanmanesh M, Shoja-Razavi R, Abdollah-Pour H, et al. 2019 Friction and wear behavior of laser cladded WC-Co and Ni/WC-Co deposits at high temperature[J]. *International Journal of Refractory Metals and Hard Materials*, 81: 137–148.
[6] Farahmand P, Kovacevic R. 2015 Corrosion and wear behavior of laser cladded Ni–WC coatings[J]. *Surface and Coatings Technology*, 276: 121–135.
[7] Kumar V, Kumar S, Anand M, et al. 2020 Fiber Laser Cladding of WS2 + Cr on SS316 Substrate and its Characterization[J]. *Materials Today: Proceedings*, 22(Pt 4).
[8] Niranatlumpong P, Koiprasert H. 2011 Phase transformation of NiCrBSi–WC and NiBSi–WC arc sprayed coatings[J]. *Surface & Coatings Technology*, 206(2-3).
[9] Saravanakumar S, Gopalakrishnan S, Dinaharan I, et al. 2017 Assessment of microstructure and wear behavior of aluminum nitrate reinforced surface composite layers synthesized using friction stir processing on copper substrate[J]. *Surface & Coatings Technology*, 322.
[10] Dariusz B, Aneta B, Peter J. 2021 Laser cladding process of Fe/WC metal matrix composite coatings on low carbon steel using Yb: YAG disk laser[J]. *Optics & Laser Technology*, 136.
[11] Riquelme A, Rodrigo P, Escalera-Rodríguez M D, et al. 2020 Corrosion Resistance of Al/SiC Laser Cladding Coatings on AA6082[J]. *Coatings*, 10(7).
[12] Kartal M, Uysal M, Gul H, et al. 2015 Effect of surfactant concentration in the electrolyte on the tribological properties of nickel-tungsten carbide composite coatings produced by pulse electro co-deposition[J]. *Applied Surface Science*, 354.
[13] Xingchen Y, Cheng C, Zhaoyang D, et al. 2021 Microstructure, interface characteristics and tribological properties of laser cladded NiCrBSi-WC coatings on PH 13–8 Mo steel[J]. *Tribology International*, 157.
[14] Feng L-P, Fleury E, Zhang G-S. 2012 Preparation of Al-Cu-Fe-(Sn,Si) quasicrystalline bulks by laser multilayer cladding[J]. *International Journal of Minerals, Metallurgy, and Materials*, 19(5).
[15] Ye X, Ma M, Cao Y, et al. 2011 The Property Research on High-entropy Alloy AlxFeCoNiCuCr Coating by Laser Cladding[J]. *Physics Procedia*, 12.

Effects of nano-fillers on the surface structure and properties of expoxy resin modified by polyhedral oligomeric silsesquioxane

Y.H. Fang & P. Wang
State Key Laboratory of Power Transmission Equipment & System Security and New Technology, School of Electrical Engineering, Chonqing University, Chongqing, China

ABSTRACT: EP-POSS can be used as a protective coating in solar panels and other fields. ZnO and SiO_2 nanoparticles were used to improve the antibacterial and abrasion resistance of EP-POSS, and the change in the contact angle was determined to characterize the impact of nanoparticles on hydrophobicity. The microscopic morphology was analyzed by SEM, and the chemical structure of the surface was characterized by the FTIR spectrum. EDS and XPS were used to analyze the change of element content. The results showed that the addition of nanoparticles was not conducive to the hydrophobicity of the EP-POSS, but the introduction of the Si element had a promoting effect on maintaining its hydrophobicity.

1 INTRODUCTION

Polyhedral oligomeric silsesquioxane (POSS), suitable for preparing nano-level hybrid materials, is an organic–inorganic hybrid material with a small size effect, excellent mechanical strength, low dielectric constant, and good compatibility [1–3]. Epoxy resin modified by POSS (EP-POSS) has been studied in the previous research; it has the potential to be applied in the field of solar protective coatings with excellent hydrophobicity, heat resistance, and abrasion resistance [4]. However, it is also necessary to add antibacterial agents and other fillers into EP-POSS resin in the practical applications, such as ZnO nanoparticles and SiO_2 nanoparticles. Nano ZnO can be dissociated into zinc ions, which is not conducive to the growth of microorganisms and improves the antibacterial effect of the coating [5,6]. Nano SiO_2 is used as a nano-filler, which has the effect of improving material durability and resistance to oxidation [7]. In order to obtain EP-POSS that meets actual requirements, nano-fillers were used to modify EP-POSS, and the properties of the prepared EP-POSS nanocomposites were characterized.

2 RAW MATERIALS AND TEST METHODS

EP-POSS resin is self-made in the laboratory, and the preparation method refers to the previous research [4]; its molecular structure is shown in Figure 1. SiO_2 nanoparticles were purchased from Macleay Reagent Company, and its particle size is 20 nm. ZnO nanoparticles are a 30 wt% dispersion with a particle size of 30 nm and were purchased from Macleans Reagent Company. The thickness of graphene is 0.2 nm, its width is 2 μm, and its specific surface area is 800 mm^2/g; it was purchased from Macleans Reagent Company; all medicines are used directly.

The ZnO dispersion (1.67 g), SiO_2 (0.5 g and 0.25 g), and ZnO dispersion (0.84 g) were added into 10 g of EP-POSS. Then, the mixture was uniformly coated on the glass slide after 30 minutes of ultrasonic dispersion. After that, the films were dried at 150 ^oC in an oven for 15 minutes, The EP-POSS nanocomposite films with a thickness of 30 μm, (a) EP-POSS/ZnO, (b) EP-POSS/SiO_2, and (c) EP-POSS/SiO_2/ZnO, were obtained.

Figure 1. The molecular structure of EP-POSS.

The sample contact angle was recorded by an OCA 100 automatic video optical contact angle instrument of Data Physics. The FTIR test was conducted by the Bruker Vertex 80V spectrometer. XPS analysis was performed by the Thermo Scientific Escalab 250XI instrument. The microstructure was observed by a JSM-6060LV Jeol scanning electron microscope equipped with Oxford Instruments Inca 350 X-ACT energy spectrum analysis.

3 RESULTS AND DISCUSSION

The infrared spectrum is a common method for structural characterization, which is used to qualitatively analyze the changes in the surface structure of EP-POSS after adding nano-fillers; the infrared spectrum is shown in Figure 2. After adding ZnO and SiO_2 nanoparticles into EP-POSS, the three samples a, b, and c showed a broad peak at 3442 cm^{-1} and showed an obvious O-H bending vibration peak at 1642 cm^{-1}, which was due to the large specific surface area of the nanoparticle. The peaks at 1105cm^{-1} and 1040 cm^{-1} are attributed to Si-O-Si antisymmetric stretching vibration and C-O stretching vibration, respectively. These peaks become dull, indicating that the nanoparticles are complexed with the POSS cage structure in EP-POSS. In curve b, the Si-O symmetric stretching vibration peak intensity at 788 cm^{-1} is stronger than that at 719 cm^{-1}, indicating that the addition of SiO_2 increases the Si-O bond content.

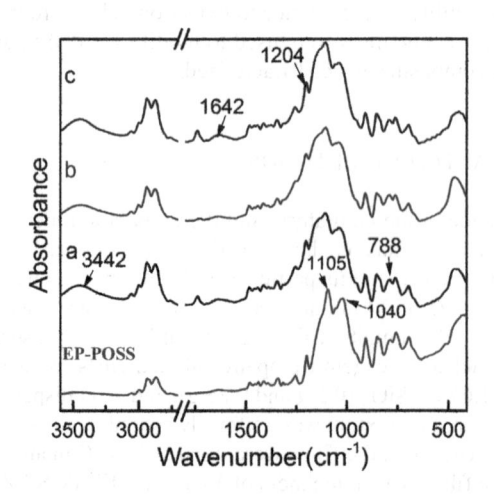

Figure 2. Infrared spectra of EP-POSS with inorganic nanoparticles, EP-POSS/ZnO (a), EP-POSS/SiO_2(b), and EP-POSS/ZnO/SiO_2(c).

EP-POSS has good hydrophobicity with a contact angle of 101°. The introduction of ZnO and SiO₂ nanoparticles into EP-POSS can improve its antibacterial properties and microstructure, but how does the introduction of inorganic nanoparticles affect the surface hydrophobicity of EP-POSS? The water contact angle test was carried out to characterize hydrophobicity, and the test result is shown in Figure 3. The contact angle of EP-POSS decreases after adding nanoparticles, and it drops sharply to 49.7° after adding ZnO. The contact angle of sample b decreases to 89.8°, which is slightly lower than EP-POSS. This shows that adding SiO_2 or ZnO nanoparticles will reduce the hydrophobic performance of EP-POSS, but ZnO has a bad impact on the hydrophobic performance, which is because the ZnO dispersion solution contains moisture and has good hydrophilicity. At the same time, the increased surface Si element content is beneficial to improve the hydrophobicity because the Si-C and Si-O bonds have a larger bond energy and a stronger shielding effect on the main chain.

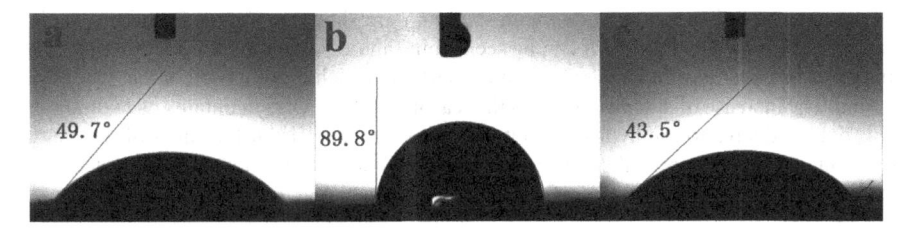

Figure 3. The contact angle test of EP-POSS/ZnO (a), EP-POSS/SiO₂(b), and EP-POSS/ZnO/SiO₂(c)

In order to study the influence of inorganic nanoparticles on the surface morphology, the films were tested by SEM, and the element content was measured by EDS. Figure 4 shows SEM images of EP-POSS and samples with different nano-fillers added, and Table 1 shows the corresponding element content. As can be seen from Figure 4d, the surface of EP-POSS is a smooth plane with an element content of 11.21 wt% Si on the surface. After adding ZnO, the surface presents randomly distributed tiny particles, which increases the surface roughness. The EDS test showed that the resin surface contained 1.75 wt% Zn element; ZnO nanoparticles contained partial hydrophilic $Zn(OH)_2$, which affected the hydrophilicity of the surface and resulted in a significant decrease in the contact angle. As shown in Figure 4b, the introduction of SiO₂ has little effect on the microstructure of the resin, because nano-SiO₂ particles can form complexes with -OH and POSS groups in the resin, showing good compatibility. However, the slight increase of surface roughness and the hydrophilicity of nano-SiO₂ particles make the contact angle decrease slightly. The test results showed that the addition of nanoparticles would increase the surface roughness and the hydrophilicity of EP-POSS. Although ZnO can improve antibacterial properties, it has an adverse effect on hydrophobicity. Compared with ZnO, SiO₂ has little effect on the hydrophobicity of EP-POSS.

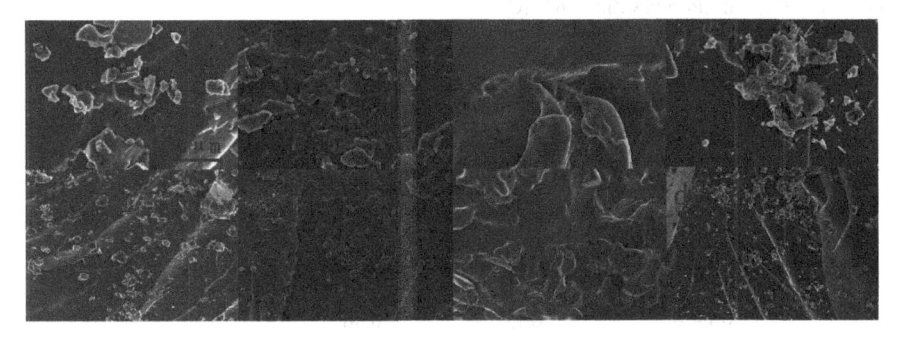

Figure 4. The SEM photos of EP-POSS/ZnO(a), EP-POSS/SiO₂(b), EP-POSS/ZnO/SiO₂(c), and EP-POSS (d).

Table 1. The content of surface elements.

Con/wt% Element	a	b	c	d
C	40.29	35.03	37.91	50.23
O	36.83	35.75	34.61	38.56
Si	21.13	29.71	25.65	11.21
Zn	1.75	0	1.83	0

In order to further understand the influence of the addition of inorganic nanoparticles such as ZnO and SiO_2 on the surface structure of EP-POSS, XPS was used to analyze the surface elements and structure. The full spectrum and silicon spectrum are shown in Figure 5, and the Si element content data are shown in Table 2. It can be seen from Figure 5 that the introduction of nano-SiO_2 can increase the content of the Si element on the surface, which is composed of the Si-C bond and Si-O bond. The increase of silicon content is the reason that EP-POSS maintains its hydrophobicity, but it is not conducive to the hydrophobicity by the large specific surface area of nanoparticles. These results provide theoretical support for the introduction of the silicon element structure into the molecular chain, thereby enhancing the hydrophobicity.

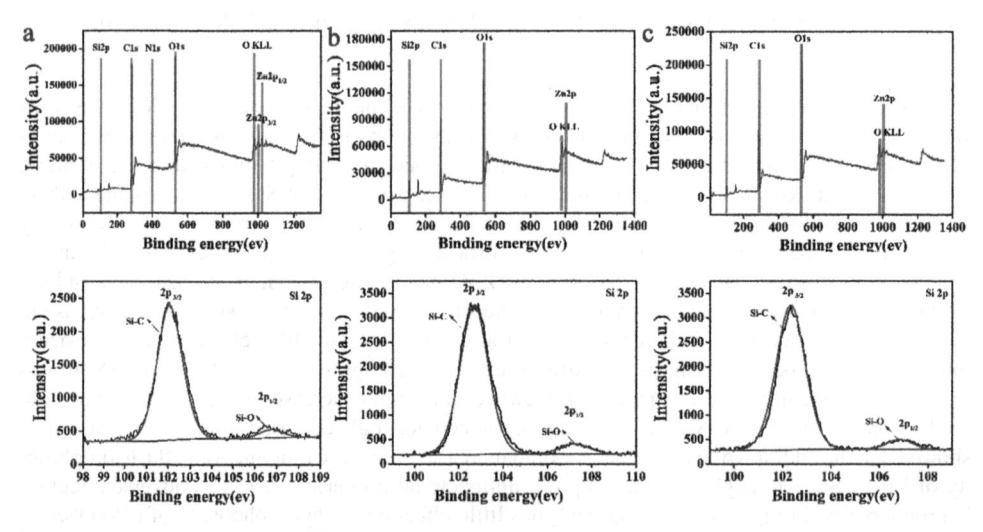

Figure 5. XPS spectra and silicon spectra of EP-POSS-added nanoparticles: EP-POSS/ZnO (a), EP-POSS/SiO_2 (b), and EP-POSS/ZnO/SiO_2 (c).

Table 2. Surface silicon structure and content of EP-POSS after the addition of inorganic nanoparticles.

Element	a	b	c
Si/%	2.03	5.44	2.99
Si-C/%	1.88	5.14	2.82
Si-O/%	0.15	0.30	0.17

4 CONCLUSION

Inorganic nanoparticles SiO_2 and ZnO, as nano-fillers, have good compatibility with EP-POSS, but have an adverse influence on hydrophobicity; this is due to the hydrophilicity of the nanoparticles. Nanoparticles easily interact with moisture and POSS groups to form a complex. Increasing the Si element content can increase the surface Si-C bond structure content, which is beneficial to the main chain shielding effect, thereby improving the hydrophobicity. This results provide the possibility that introducing Si-containing functional groups into the molecular structure can further improve the hydrophobicity of EP-POSS.

REFERENCES

[1] Jerman I; Šurca Vuk A; Kožcelj M; Švegl F and Orel B. Influence of amino functionalised POSS additive on the corrosion properties of (3-glycidoxypropyl)trimethoxysilane coatings on AA 2024 alloy. 2011 *Progress in Organic Coatings*, 72, 334–342

[2] Mishra K, Pandey G and Singh R.P. Enhancing the mechanical properties of an epoxy resin using polyhedral oligomeric silsesquioxane (POSS) as nano-reinforcement. 2017 *Polymer Testing*, 62, 210–218

[3] Laine R.M and Roll M.F. Polyhedral Phenylsilsesquioxanes. 2011 *Macromolecules*, 44, 1073–1109

[4] Yan F, Ping W, Li S and Lin W. Hydrophobic Epoxy Caged Silsesquioxane Film (EP-POSS): Synthesis and Performance Characterization. 2021 *Nanomaterials*, 11(2), 472

[5] Bai X, Li L, Liu, H, Tan, L, Liu, T and Meng X. Solvothermal synthesis of zno nanoparticles and anti-infection application in vivo. 2015 *Acs Applied Materials & Interfaces*, 7(2), 1308

[6] Xie Y, He Y, Irwin P L, Jin T and Shi X. Antibacterial activity and mechanism of action of zinc oxide nanoparticles against campylobacter jejuni. 2011 *Applied and Environmental Microbiology*, 77(7), 2325–2331

[7] Yi Z, Jun Z, Qing X and Yang H. Ultra-Stable and Durable Silicone Coating on Polycarbonate Surface Realized by Nanoscale Interfacial Engineering. 2020 *ACS applied materials & interfaces*, 12(11), 13296–13304

Effect of K$_2$CO$_3$ adding on properties of SiC porous ceramics

T.Y. Yu, H. Li, M.E. Ma, Y.N. Sun & X.Y. Deng
College of Physics and Materials Science, Tianjin Normal University, Tianjin, China

ABSTRACT: In this paper, 50-70 mesh SiC particles were used as the primary raw materials, K$_2$CO$_3$ was added into the binder which was composed of potash feldspar, kaolin and quartz, then graphite and activated carbon were used as pore-forming agent to prepare SiC porous ceramic supports. The effects of different contents of K$_2$CO$_3$ on the flexural strength, filtration pressure drop and porosity of porous SiC ceramics were investigated. The results indicated that the content of mullite phase in SiC porous ceramics increased by adding appropriate amount of K$_2$CO$_3$. When the content of K$_2$CO$_3$ in the binder was 15wt%, the samples prepared at 1300° for 90 minutes had the best comprehensive properties, at this point, the porosity was 24%, the flexural strength was 34.75MPa and the filtration pressure drop was comparatively low.

1 INTRODUCTION

SiC has uniform distribution of micropores and pores, high porosity, low volume density, developed specific surface as well as unique physical surface characteristics. In addition, it also has excellent properties such as high strength, high hardness, corrosion resistance, good high-temperature thermal stability and so on. So it has become one of the best high-temperature filter materials, and has great development potential in the field of high-temperature gas purification [1–5].

Mullite is a ceramic material with good high-temperature strength and thermal expansion coefficient similar to that of SiC porous ceramics. An appropriate amount of mullite can effectively improve the thermal shock resistance and creep resistance of SiC porous ceramics. The addition of potassium carbonate can promote the formation of mullite phase. Potassium carbonate decomposes into potassium oxide and carbon dioxide at high temperature. Carbon dioxide escapes in the form of gas. The change of alumina mole content in mullite is closely related to the content of potassium oxide in the sample. K atoms are conducive to the increase of liquid phase, and keep the amorphous state of liquid phase, as well as make the reaction of Si-O structure more active, thereby promoting the formation of mullite phase [6–10].

In this paper, 50-70 mesh SiC particles were used as the primary raw materials, sodium carboxymethyl cellulose (CMC) was used as dispersant, K$_2$CO$_3$ was added into the binder which was composed of potassium feldspar, kaolin and quartz, then graphite and activated carbon were used as pore-forming agents to prepare SiC porous ceramic supports. The effects of different contents of K$_2$CO$_3$ on the properties of SiC porous ceramics were studied by analyzing scanning electron microscopy (SEM), X-ray diffraction (XRD), flexural strength, filtration pressure drop and porosity.

2 EXPERIMENTAL SECTION

2.1 *Preparation of samples doped with different contents of K$_2$CO$_3$*.

50–70 mesh SiC particles were used as the primary raw materials, different contents of K$_2$CO$_3$ was added into the original binder (potash feldspar 64.53wt%, quartz 23.27wt% and kaolin 12.2wt%)

to prepare new binder (as shown in table 1). Sequentially adding sodium carboxymethyl cellulose (CMC, 2g/ml), binder (K_2CO_3, potassium feldspar, quartz and kaolin) and pore-forming agent (graphite and activated carbon) to the 50-70 mesh SiC particles by layer-by-layer coating method (the dosage of each substance is shown in table 2). In the meantime, stired evenly, then pressed and shaped the sample, finally, sintered it in a box-type high-temperature sintering furnace to 1300° and kept that temperature for 90 minutes to sinter (as shown in figure 1).

Table 1. The content of K_2CO_3 in the binder (%wt).

Number	1	2	3	4	5
K_2CO_3	0	5	10	15	20

Table 2. Experimental dosage of each substance (g).

Number	SiC	CMC	K_2CO_3	Potash feldspar	Kaolin	Quartz	Graphite	Activated carbon
1	16.8	0.2	0.000	1.355	0.256	0.489	1.05	1.05
2	16.8	0.2	0.105	1.287	0.243	0.464	1.05	1.05
3	16.8	0.2	0.210	1.220	0.231	0.440	1.05	1.05
4	16.8	0.2	0.315	1.152	0.218	0.415	1.05	1.05
5	16.8	0.2	0.420	1.084	0.205	0.391	1.05	1.05

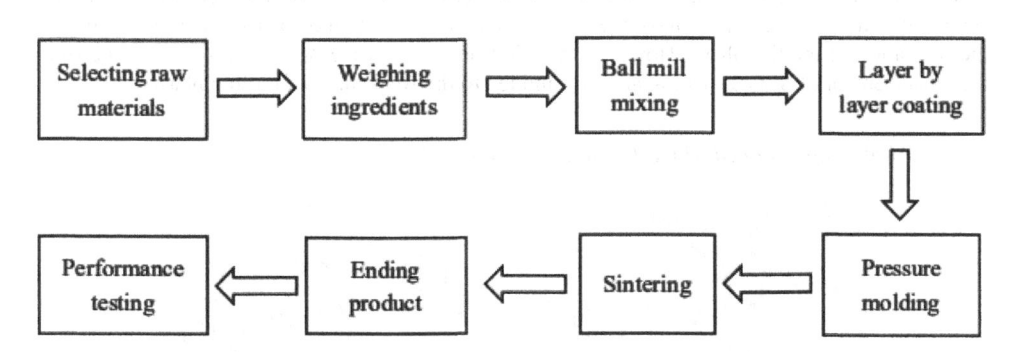

Figure 1. Experimental flow chart.

2.2 Characterization of SiC porous ceramic samples doped with different contents of K_2CO_3

The microstructure of the samples was observed by SEM S4800/TM3000. The diffraction pattern of the sample was obtained by X-ray diffractometer. The flexural strength of the sample was measured by YLN Electronic Press and three-point method. The porosity of the sample was obtained by Ultrasonic cleaning tank, Archimedes drainage method and Electronic balance analysis. The filtration pressure drop of the sample was obtained through the differential pressure meter, the glass flow meter and the multi-purpose vacuum pump for circulating water.

3 RESULTS AND DISCUSSION

3.1 XRD of samples doped with different contents of K_2CO_3

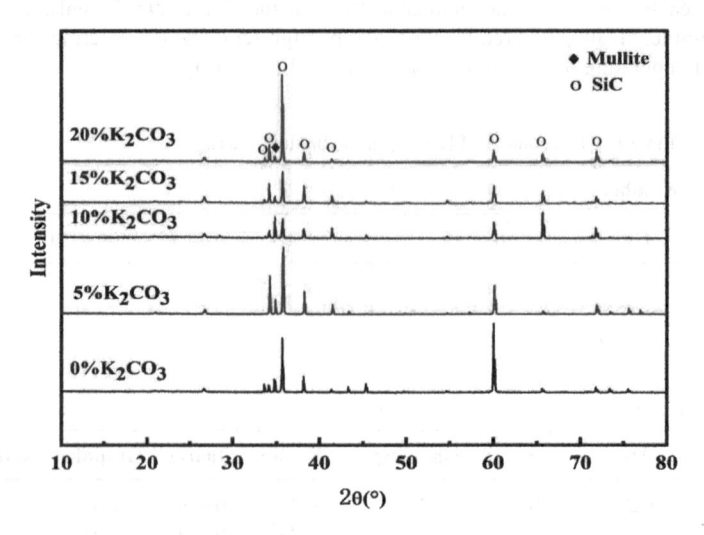

Figure 2. XRD of SiC porous ceramics doped with different contents of K_2CO_3.

It can be seen from the figure 2 that with the increasing of K_2CO_3 doping content, the mullite phase in the SiC porous ceramics sample increased first and then decreased. Because potassium carbonate decomposed into K_2O and CO_2 at high temperature. CO_2 escaped in the form of gas, and the change of alumina mole content in mullite was closely related to the content of potassium oxide in the sample. K atoms are conducive to the increase of liquid phase, and keep the amorphous state of liquid phase, as well as make the reaction of Si-O structure more active, thereby promoting the formation of mullite phase. However, the addition of excessive K_2CO_3 was not conducive to the enhancement of Si-O structural activity, and led to the weakening of mullite phase.

3.2 SEM of samples doped with different contents of K_2CO_3

Figure 3. SEM of SiC porous ceramics doped with different contents of K_2CO_3: (a) 0%wt K_2CO_3 in binder; (b) 5%wt K_2CO_3 in binder; (c) 10%wt K_2CO_3 in binder; (d) 15%wt K_2CO_3 in binder; (e) 20%wt K_2CO_3 in binder.

It can be seen from the figure 3 that when K_2CO_3 was not doped, the pore diameters formed by the accumulation of SiC particles were larger, and the surface of the particles was uneven. With the increase of K_2CO_3 content to a certain extent, the entire SiC porous ceramics gradually became dense. At the same time, the surface of SiC particles doped with K_2CO_3 was smoother than that of SiC particles without K_2CO_3. Because as the content of K_2CO_3 increased to a certain extent, the mullite phase increased. At the same time, the generated molten liquid phase covered the surface of the silicon carbide particles and filled the pores, so that the particles were more closely combined, and the microstructure of the ceramic matrix section looked more rounded. As the content of K_2CO_3 continued to increase, the mullite phase decreased and the pore size between particles increased.

3.3 *Flexural strength and porosity of samples doped with different contents of K_2CO_3*

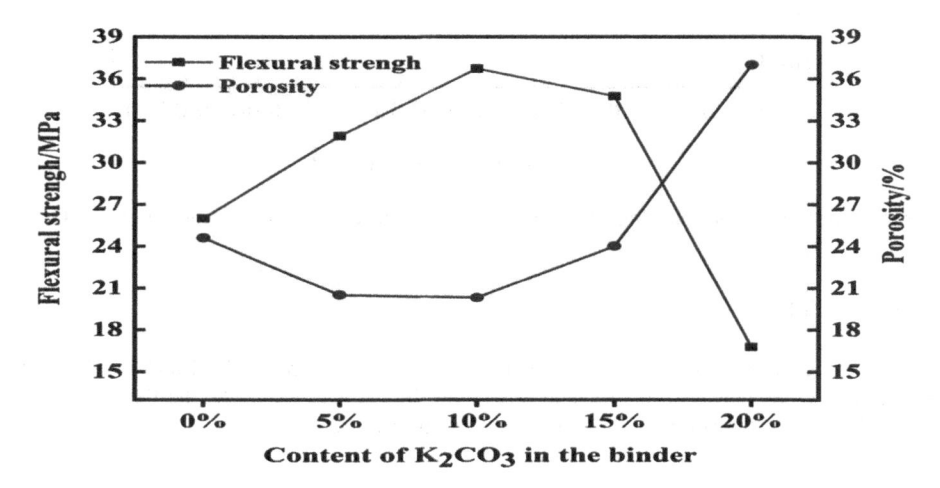

Figure 4. Flexural strength and porosity of SiC porous ceramics doped with different contents of K_2CO_3.

Flexural strength refers to the ability of materials to resist bending without breaking, and porosity is the characterization of the density of porous ceramic materials. It can be seen from the figure 4 that as the content of K_2CO_3 increased, the flexural strength of SiC porous ceramics first increased and then decreased, while the porosity decreased first and then increased. Because when the content of K_2CO_3 in the binder was 0%-10%wt, the content of mullite phase in the sample was more, and the mullite phase had the characteristic of high strength, so the whole SiC porous ceramic was gradually dense, the flexural strength of the sample enhanced, and the porosity reduced; when the content of K_2CO_3 in the binder was 15%-20%wt, the mullite phase in the sample decreased and the pore sizes between particles became larger, so the flexural strength of the sample decreased and the porosity increased.

3.4 *The filtration pressure drop of samples doped with different contents of K_2CO_3*

The filtration pressure drop is the resistance of gas flow. It can be seen from the figure 5 that when the content of K_2CO_3 in the binder was 10%wt, the porosity of SiC porous ceramic was the lowest, at the same time, the resistance to gas flow was the largest and the filtration pressure drop was the largest; when the content of K_2CO_3 in the binder was 20%wt, the porosity of SiC porous ceramic was the largest, in the meanwhile, the resistance to gas flow was the smallest and the filter pressure drop was the smallest.

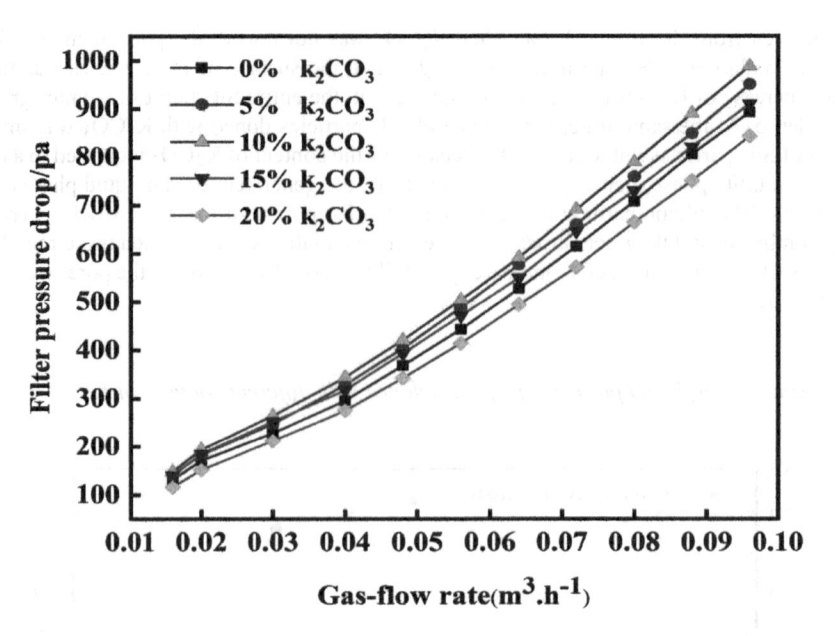

Figure 5. The filtration pressure drop of SiC porous ceramics doped with different contents of K_2CO_3.

4 CONCLUSION

When the content of K_2CO_3 in the binder was 15wt %, the samples prepared at 1300° for 90 minutes had the best comprehensive properties, at this point, the porosity was 24%, the flexural strength was 34.75MPa, and the filtration pressure drop was comparatively low.

ACKNOWLEDGMENTS

This project was funded by Tianjin Normal University (Grant No.53H14049)

REFERENCES

[1] Xiong H, Li H, Mao W, et al. *Materials Letters*, 2003, 57(22–23), 3417.
[2] Feng G J, Li Z R, Xu K, et al. *Transactions of The China Welding Institution*, 2014, 43(1).
[3] Li Z R, Xu X L, Liu W B, et al. *Transactions of the China Welding Institution*, 2012, 33(11), 9.
[4] Lv H, Chu J X, Kang Z J, et al. *Materials Review*, 2004, 18(1), 36.
[5] Xinming Ren, Beiyue Ma, Chang Su, et al. *In-situ synthesis of Fe x Si y phases and their effects on the properties of SiC porous ceramics*. Journal of Alloys and Compounds, 2019.
[6] Yu Li . *Preparation and properties of mullite-bonded silicon carbide porous ceramics for filtration* D. Hainan University, 2015.
[7] Jiadong Liu. *Preparation and performance of silicon carbide ceramic membrane for high-temperature filtration* D. China University of Geosciences (Beijing), 2019.
[8] Hong Jin. *Preparation and properties of mullite bonded silicon carbide porous ceramics* [D]. Hainan University, 2018.
[9] Liang Wang. *Preparation of mullite bonded silicon carbide porous ceramics* [D]. Wuhan University of Engineering, 2016.
[10] Junfeng Li. *Study on structural Design and Properties of Silicon Carbide porous Ceramics for High temperature Filtration* [D]. Tsinghua University, 2011.

Effects of TiC and rare earth on the microstructure and performance of LASER cladding Mo_2FeB_2

Z.X. Zhang
Key Laboratory for Liquid-Solid Structural Evolution and Processing of Materials (Ministry of Education), Shandong University, Jinan, China

H.F. Tian, F.L. Tantai, H. Liu & W. Wang
Shandong Energy Heavy Equipment Group Dazu Remanufacturing Co. Ltd, Xintai, China

J.S. Sun
Key Laboratory for Liquid-Solid Structural Evolution and Processing of Materials (Ministry of Education), Shandong University, Jinan, China

ABSTRACT: The hard phases of Mo_2FeB_2 in ternary boride cermet synthesized in situ by laser cladding are very coarse, which makes it difficult for Mo_2FeB_2 to withstand impact load and it has poor wear resistance. In this paper, with Mo powder, boron iron, iron powder, and Cr powder as the main raw materials, and by adding TiC, rare earth, ferrotitanium, and chromium carbide, the effects of TiC and rare earth on the microstructure and performance of the laser cladding layer and the amount and form of hard phase Mo_2FeB_2 were studied. Results revealed that the hard phase of Mo_2FeB_2 laser cladding metal is block or long strip, and the distribution is not uniform. Four parts of TiC were added to refine and distribute the hard phase evenly, forming a blocky or spherical shape. The number and size of hard phases with two rare earths added increased significantly, and they were cluster-like or large block-like, with smooth edges and corners, and some of them were connected to each other. The hard phase of in situ TiC synthesis by adding TiC was significantly increased and fine, and around the massive hard phase, obvious black sediments appeared, and the microhardness reached 1187.6 HV. Rare earth elements have a poor refining effect on the hard phase, but they can enhance the number and volume of hard phase in the matrix.

1 INTRODUCTION

Boride ceramic is a kind of cermet with good wear resistance, high melting point, and high hardness and is not easily corroded. Because of its excellent performance, it is widely used in mechanical processing, ore grinding, alloy smelting, and parts manufacturing [1]. Ternary boride cermet Mo_2FeB_2 has attracted much attention owing to its low cost, excellent performance, and perfect metallurgical bonding with steels [2–4]. At present, sintering methods have been primarily employed to prepare Mo_2FeB_2. But the sintering process is not easy to master and easy to bond; the metal reacts to form a hard and brittle third phase, which affects the performance [5].

In situ synthesis of Mo_2FeB_2 by laser cladding can not only overcome the disadvantages of high vacuum sintering process cost, long manufacturing cycle, and limited size and shape, but also be flexibly used to modify the surface of parts for heat resistance, wear resistance, and corrosion resistance. Laser cladding is due to the high temperature of the molten pool, and the in situ synthesized Mo_2FeB_2 cermet easily grows in the liquid molten pool metal, making the Mo_2FeB_2 coarse and difficult to withstand impact loads and exhibit poor wear resistance. Effective control of the size and shape of the laser cladding Mo_2FeB_2 has become the key to the preparation of Mo_2FeB_2 cermets by laser cladding.

The melting point of TiC is as high as 3140° C, and it can be precipitated in the molten pool as the non-spontaneous nucleation core of Mo_2FeB_2, so that the formed ternary boride Mo_2FeB_2 will be dispersed and distributed. At the same time, TiC and Mo_2FeB_2 phases formed by in situ reaction can form a pollution-free and weak interface bonding dual ceramic phase, which can effectively improve the strength and toughness of Mo_2FeB_2-based cermets [6]. The addition of rare earth to the cermet coating can improve the performance of the cermet coating by refining the crystal grains, purifying the molten pool, producing solid solution strengthening and dispersion strengthening, and reducing the dilution of the base material to the coating microstructure. The rare earth modification technology is used in the preparation process of cermet coatings such as laser surface cladding, which can significantly improve the strength and hardness of the coating, enhance the bonding strength of the coating and the substrate, improve the tribological properties of the coating, and improve the service performance of the cermet coating [7]. In this study, Mo powder, ferroboron, iron powder, and Cr powder are used as the main raw materials, and TiC, rare earth, ferro-titanium, and chromium carbide are added respectively to study the influence of TiC and rare earth on the structure and properties of the laser cladding layer. The influence of TiC and rare earth on the quantity and morphology of the hard phase Mo_2FeB_2 was also studied.

2 MATERIALS AND METHODS

Mo, FeB, Cr, Fe, titanium carbide, titanium iron, chromium carbide, rare earth magnesium silicon alloy, and other powders were used as raw materials; the laser cladding alloy powder is prepared by mechanical powder mixing. Mo, FeB, Cr, and Fe are used for in situ synthesis of Mo_2FeB_2 by laser cladding. TiC can be added in two ways: direct addition and in situ synthesis of TiC with ferro-titanium and chromium carbide. Table 1 shows the technical parameters of the test powder. The basic formula D of the laser cladding powder is designed as Fe-48Mo-6B-10Cr. In the basic formula D, 2 parts by mass of rare earth, 4 parts by mass of TiC, and parts by mass are added. 15 ferrotitanium + 14.6 chromium carbide is used to study the effects of rare earth, direct addition of TiC, and in situ synthesis of TiC on the structure and properties of the laser cladding layer and the number and morphology of the hard phase Mo_2FeB_2. The numbers are D+RE, D+TiC, and D+Ti-C.

Table 1. Technical parameters of the alloy powders.

Powder	Mean particle size (μm)	Chemical composition (wt.%)
Mo	100	Fe < 0.002, O < 0.1, Si < 0.001, Bal Mo
Fe	90	C < 0.1, N < 0.1, O < 0.2, Bal Fe
Cr	110	O < 0.2, Fe < 0.18, N < 0.045, Bal Cr
FeB	60	B = 22%, C < 0.27, Si < 0.71, Bal Fe
RE	100	Mg = 8%, RE = 8.1% (Y = 4.2%), Si = 41%, Bal Fe
Cr_3C_2	60	C=11.87, Cr=87.31, Si<0.16
TiFe	60	Ti=41, Si<2.8, Mn<1.25, C<0.06, Bal Fe
TiC	180	Fe<0.1, O<0.07, Si<0.004, Ca<0.004, Bal TiC

The test plate for laser cladding is Q235, and its size is 100mm × 50mm × 10mm. Using the preset powder method, the powder thickness is found to be 2 mm.

The laser used in the test is the FL-DLight3–6000 semiconductor laser produced by Xi'an Focuslight Technology Co., Ltd., with a nominal output power of 6000W. The process parameters of laser cladding are shown in Table 2.

Table 2. Process parameters of laser cladding.

Power	Scan speed	Spot size	Scanning channels	Focal length	Spacing	Powder thickness
3kW	400mm/min	15×2mm	2	285mm	6mm	2mm

A metallographic sample of 10mm×10mm×10mm on the laser cladding specimen was cut by a wire. Sand paper with different particle sizes (80/120/240/400/600/800/1000/1500) was used for grinding in turn. Then, the sample was polished with 1.5 μm diamond paste. Finally, the mixed acid solution (HF:20 vol.%, HCL:30 vol.%, HNO_3: 50 vol.%) was used for corrosion for 12–14s.

A JSM-6600V scanning electron microscope and its attached energy spectrum (EDS) for microstructure and phase composition analysis were used. The DHV-1000 micro Vickers hardness tester was used to test the microhardness, the load was 500 grams, and the load time was 10s. An XJP-6A optical microscope was used to observe the structure of the cladding layer.

3 RESULTS AND DISCUSSION

3.1 Laser cladding layer forming

Figure 1 shows the morphology of different powder laser cladding layers. It can be seen that the four powder cladding layers are continuous and smooth, and the forming effect is good. The surface finish of the cladding layer with TiC and RE has been improved. The surface of the four powder cladding layers has different degrees of spatter, especially the surface of the cladding layer with the addition of ferro-titanium and chromium carbide D+Ti-C formulations. This is due to the increase in the C content in the formulation. When the carbon monoxide and carbon dioxide gas produced by the metallurgical reaction escape from the molten pool, the molten metal splashes. The D+RE formula has less splashes, which may be due to the reaction of Mg, Si, and O due to the addition of rare earths to achieve the effect of inhibiting the generation of carbon monoxide and carbon dioxide, so the surface is the best.

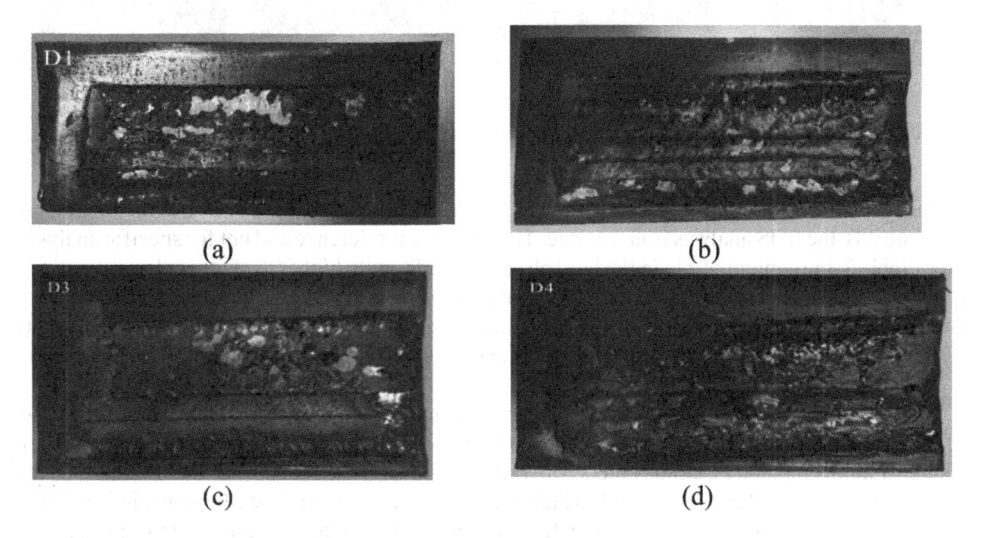

Figure 1. Surface forming of the laser cladding layer: (a) D, (b) D+TiC, (c) D+RE and (d) D+Ti−C.

3.2 The influence of TiC and rare earth on the microstructure of the cladding layer

Figure 2 shows the SEM microstructure of four powder laser cladding layers. The hard phase of the laser cladding layer of the basic formula D is lumpy or has long strips, with uneven distribution, as shown in Figure 2(a). Figure 2(b) shows the microstructure of the cladding layer of the D+TiC formulation with titanium carbide. Compared with Figure 2(a) of the basic formulation, it can be seen that the hard phase has become smaller and more uniform, and the shape of the hard phase is also changed. The lumps or long strips of formula D become smaller lumps or balls. Figure 2(c) shows the microstructure of the cladding layer with rare earth D+RE formulation. Compared with Figure 2(a) of the basic formulation, the number of hard phases is significantly increased and the size increases. At the same time, the shape of the hard phase also changes. It is in the shape of clusters or large clumps; some of the massive hard phases are interconnected, and the edges and corners of the hard phases are relatively smooth. Figure 2(d) shows the microstructure of the cladding layer with the addition of ferro-titanium and chromium carbide D+Ti-C. It can be seen that the hard phases are significantly increased and are finer, and there are obvious black deposits around the massive hard phases. Compared with Fig. 2(b) D+TiC, the width and content of the black deposit are higher.

For the D+Ti-C formula with ferro-titanium + chromium carbide, because the formula design idea is similar to the D+TiC formula, although the element content is different, the metallurgical reaction and phase change that occur are similar to the D+TiC with directly added TiC. Due to the higher Ti and C contents of the D+Ti-C formula, more TiC formed by the metallurgical reaction, the hard phase formed is denser, and the high-concentration Cr in the liquid phase region diffuses to Mo_2FeB_2, which improves the wettability between Mo_2FeB_2 and Fe, which makes the microstructure more uniform. However, due to the limited solid solubility of Cr in Mo_2FeB_2 and Fe, excessive addition of Cr will reduce the wettability. When the amount of Cr added is large, the Mo_2FeB_2 grains are easy to agglomerate. In addition, because Cr is supersaturated in the eutectic matrix, thick precipitates are formed around some hard phases; so there are obvious precipitates between the Mo_2FeB_2 hard phase and the matrix [8], as shown in Figure 2 (d).

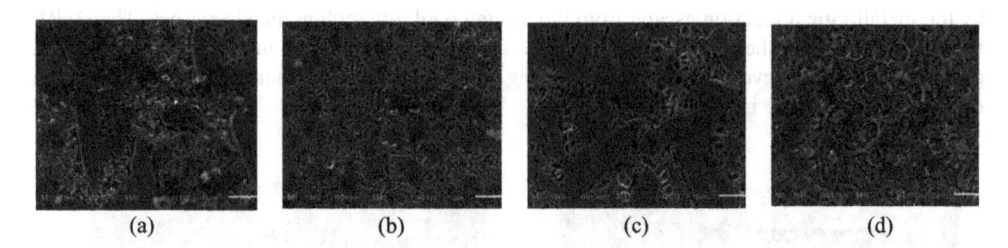

Figure 2. SEM microstructure of the laser cladding layer: (a) D, (b) D+TiC, (c) D+Re and (d) D+Ti−C.

The EDS analysis results of the hard phase in Figure 2 are shown in Table 3. Due to the small mass of element B, the EDS analysis is inaccurate. This is only for reference and not for specific analysis. From Table 3, the content of Mo in the hard phase is significantly higher than other elements, so Mo is the main element that composes the hard phase. According to the mass percentages of the three elements Mo, Fe, and Cr, based on the Cr element, the atomic percentages of the three elements can be calculated. The atomic ratios of Mo, Fe, and Cr in the hard phase of the cladding layer with D formula, D + TiC formula, D + RE formula, and D + Ti-C formula are 8.6:2.5:1, 7.29:6.4:1, 7.30:1.62:1, and 5.07:3.42:1, respectively. It can be seen from the atomic ratio that in the D formula and D+RE formula, the atomic ratio of Mo to Fe is greater than 2. This is because the position where Cr atoms replace Fe atoms enters Mo_2FeB_2; in the D + TiC and D + Ti-C formulations, the atomic ratio of Mo to Fe is less than 2, which is because Cr atoms replace Mo atoms in Mo_2FeB_2. Studies have shown that the Cr element will equivalently replace the positions of Mo and Fe atoms in Mo_2FeB_2 crystals to form composite boride M_3B_2.

Table 3. EDS analysis results of the hard phase of the cladding layer.

Sample	Element mass percentage (wt %)						
	Mo	Fe	Cr	B	Ti	C	Si
D	46.22	7.78	2.91	43.09			
D+TiC	45.44	23.23	3.37	24.98	0.95	2.02	
D+RE	48.31	6.28	3.60	41.67			0.14
D+Ti-C	43.98	17.20	4.71	28.59	0.23	5.29	

3.3 *The influence of TiC and rare earth on the hardness of the cladding layer*

Table 4 shows the microhardness of the four powder laser cladding layers. Compared with the matrix material and the basic formula D, the microhardness of the cladding layer with TiC, rare earth, and titanium iron + chromium carbide is significantly improved. Different additives have significantly different effects on the microhardness. Among them, the D+Ti-C formula with ferro-titanium + chromium carbide has the highest hardness.

Table 4. Average micro-Vickers hardness value of the laser cladding layer.

Powder formula	D	D+TiC	D+RE	D+Ti-C	Matrix
hardness/HV	434.5	913.9	879.5	1187.6	154.8

From Table 4, the hardness of the laser cladding layer of the basic formula D is about 3 times that of the base material, while the hardness of the laser cladding layer with the addition of TiC, rare earth elements, and titanium iron + chromium carbide formula is 6–8 times that of the base material. It can be seen from the SEM structure in Figure 2 that the number of hard phases in the laser cladding layer of the basic formula D is relatively small, and they are mainly distributed on the substrate in larger blocks. The hard phase in the cladding layer of the D+TiC formula is relatively small. It is small and uniform, and its quantity is more than that of the basic formula D, so its microhardness is higher, which is more than twice that of D, reaching 913.9HV. In the cladding layer with rare earth D+RE formulation, its hardness is significantly increased compared to the basic formulation D (see Figure 2), and the size increases, but the shape and distribution do not change much, and the microhardness is 879.5 HV, which is lower than those of D+TiC and D+Ti-C formulations, but it is also more than twice that of basic formulation D. In the laser cladding layer of the D+Ti-C formula with ferro-titanium + chromium carbide, the shape and distribution of the hard phase are similar to those of D+TiC, but the volume percentage increased, the number increased, and the microhardness reached the highest value of 1187.6 HV among the four powders.

4 CONCLUSIONS

(1) Using laser cladding technology, the cladding metal layer with ternary boride as the hard phase can be prepared on the surface of Q235 test board.
(2) The hard phase obtained by laser cladding is a massive or butterfly-shaped ternary boride, and the quantity and morphology of the hard phase will change greatly due to the difference of the reinforcing phase material.
(3) The effect of rare earth elements on the refinement of the hard phase is worse than that of titanium carbide, titanium iron, and chromium carbide. It cannot change the shape of the hard phase itself, but it can significantly enhance the hard phase in the matrix, quantity, and volume.

(4) Due to the refinement of the hard phase by TiC, the hard phase in the Ti element-added cladding layer is significantly denser and has a uniform shape distribution, and its microhardness can reach a value of 1187.6 HV.

ACKNOWLEDGEMENTS

The authors gratefully acknowledge the support provided by the Key Science and Technology Project of Shandong Provincial (2019JZZY020208).

REFERENCES

[1] Lingke Z and Hui W taoci cailiao biaomian gaixing jishu (Bei Jing: Chemical Industrial Press)
[2] Ivanov MB, V ershinina TN, Ivanisenko, VV 2019 The effect of composition and microstructure on hardness and toughness of Mo2FeB2 based cermets Materials Science & Engineering: A.763 138117
[3] Wu H, Zheng Y, Zhang JJ, et al.2019 Preparation of Mo2FeB2-based cermets with a core/rim structure by multi-step sintering approach Ceramics International.45(17) 22371–375
[4] Takagi, Ken-Ichi 2006 Development and application of high strength ternary boride base cermets J. Solid State Chem.179 2809–18
[5] Wenhu L and Futian L 2010 Sintering thermodynamic analysis of Mo2FeB2 cermet prepared by in-situ reaction [J] Powder Metallurgy Technology.28(03) 192–195
[6] Wenhu L 2019 Effect of Mo/TiC content on microstructure and properties of Mo2FeB2-TiC multiphase cermets [J]. Heat Treatment of Metals.44(08) 73–77.
[7] Keshan H,Xiyun C,Zhihua L 2009 Surface Modification on Cermet Coating Microstructure by Rare Earths and Its Application Development [J]. Lubrication Engineering. 34(03) 100–104+113
[8] Wu H, Zheng Y, Zhang J, et al.2020 Influence of Cr and W addition on microstructure and mechanical properties of multi-step sintered Mo2FeB2-based cermets [J]. Ceramics International.46 (8, Part A) 10963–970

The effects of carbon and nitrogen sources on biosurfactant fermentation by *Ochrobactrum intermedium* XY-1

Y.M. Sun, H. You, W.M. Si, L. Liu, X. Xu & X.Z. Li
School of Biological Engineering, Dalian Polytechnic University, Ganjingqu, Dalian, Liaoning, PR China

ABSTRACT: Carbon and nitrogen source is very important for biosurfactant production. In this paper, we investigated the influence of carbon sources and nitrogen sources on biosurfactant production by *Ochrobactrum intermedium* strain XY-1 by determining the cell density, as well as the surface tension and the emulsifying activity of culture broth. The results showed that *Ochrobactrum intermedium* strain XY-1 achieved higher biosurfactant yields with sucrose as a carbon source (10 g/L), sodium nitrate as a nitrogen source (4 g/L), and at a carbon to nitrogen ratio of 10:1, the highest emulsifying activity of fermentation supernatant to liquid paraffin was as high as 55%. The tested strain could grow and produce biosurfactants by utilizing potassium nitrate similar to sodium nitrate, maltose inferior to sucrose, and at a carbon to nitrogen ratio of 20:1 or 30:1 similar to 10:1, but could not utilize hydrocarbon carbon source and urea nitrogen source well.

1 INTRODUCTION

Compared with chemical surfactants, biosurfactants produced by microbial fermentation are low-toxic, highly biodegradable as well as ecological acceptable, therefore, are potential to be applied in the manufacture of medicine, cosmetics and food, the recovery of crude oil, and the remediation of environment. However, the yield is low and the cost is high, in biosurfactant production by fermentation. Some explorations to solve the economic problems in biosurfactants industrial manufacture have been done, such as isolating the microbes with high yield, optimizing the media compositions, and utilizing cheaper substrates [1].

Microbes producing biosurfactant include bacteria, fungi, and yeast. *Ochrobactrum* is an aerobic, Gram-negative bacterium of the family *Brucellaceae*. The researches on *Ochrobactrum* producing biosurfactant have been not so well as *Pseudomonas, Acinetobacter, Bacillus, Streptomyces, Rhodococcus, Achromobacter, Brevibacterium*, and *Arthobacter* [2]. Joy et al. [1] isolated a strain of *Ochrobactrum* producing lipopeptides, which could reduce the surface tension of water to 31.14 mN/m. Bezza et al. [2] isolated *Ochrobactrum intermedium* CN3 producing glycolipid biosurfactants, which showed high thermal stability and tolerance to extreme salinity and pH, as well as high emulsifying activity to cyclohexane, n-hexane and waste engine oil of 86%, 79% and 65%. Halophilic *Ochrobactrum* sp. strain BS-206 was able to produce Ochrosin, a biosurfactant with the structure of 4-dimethylaminobenzaldehyde, which could reduce the surface tension of water from 72 to 35 mN/m, and exhibited high emulsifying activity to p-toluene, n-hexane, tridecane, hexadecane [3]. Marine *Ochrobactrum intermedium* isolated from oily bilge wastes could converse hydrocarbon wastes to polyhydroxybutyrate (PHB) with high emulsifying activity [4].

In our previous study, *Ochrobactrum intermedium* strain XY-1 was screened as a biosurfactant producer. In this study, we investigated the influence of carbon sources and nitrogen sources on biosurfactant production by *Ochrobactrum intermedium* strain XY-1 by determining the cell density, as well as the surface tension and the emulsifying activity of culture broth.

2 MATERIALS AND METHODS

2.1 *Microorganism*

The strain XY-1 classified as *Ochrobactrum intermedium* was preserved in 20% glycerol at −80°C. The strain inoculated on the media slants was cultivated at 30°C for 24 h, and then preserved at −4°C for daily experiment. Before usage, the culture stored on slant was inoculated on fresh media slants and cultivated at 30°C for 24 h for the activation of the strain. The medium compositions included NaCl 15 g, beef extract powder 5 g, peptone 10 g, and agar 20g dissolved in 1 L of water.

2.2 *Biosurfactant fermentation*

The basic fermentation medium included KH_2PO_4 3.4 g, KH_2PO_4 1.5 g, $NaNO_3$ 4.0 g, $MgSO_4 \cdot 7H_2O$ 0.2 g, yeast extract powder 0.2 g, sucrose 10 g dissolved in 1 L of water, which was adjusted to pH 7 and sterilized in autoclave at 115°C for 20 min (for the carbohydrate carbon sources) or at 121°C for 20 min (for the hydrocarbon carbon sources). In order to obtain the fermentation starter, two loops of the activated strain was inoculated into 20 mL basic fermentation medium with 10 g/L sucrose and 4.25 g/L sodium nitrate in 50 mL Erlenmeyer flasks, and subsequently cultivated for 7 days in a shaker at 30°C and 160 rpm. The inoculation of fermentation was 8%. All the fermentation was carried out under the same condition above.

Sucrose, maltose, and starch (10 g/L) were selected as water-soluable carbohydrate carbon sources, and tridecane, liquid paraffin, pentacosane, and dotriacontane (20 g/L) were selected as water-immiscible hydrocarbon carbon sources to examine the effect of the carbon source on biosurfactant fermentation. Sodium nitrate, potassium nitrate, ammonium chloride, ammonium sulfate, ammonium nitrate, and urea at a nitrogen concentration of 0.05 mol/L were chosen as nitrogen sources to examine the effect of the nitrogen source on biosurfactant fermentation.

2.3 *Parameters measurement*

Throughout the fermentation process, we periodically sampled, and determined the cell density (OD_{600}) by the optical density of culture broth at 600 nm, and determined the surface tension (ST) by Du Nouy ring method, and determined the emulsifying activity (the E_{24} Index) by mixing sample with liquid paraffin. The diluted culture broth was directly utilized to determine OD_{600}, the supernatant obtained by centrifuging the culture broth at 4°C, 13300 g for 20 min was utilized to determine ST and the E_{24} Index. The E_{24} Index was expressed as the percentage of the emulsified layer height to the mixture total length of sample and liquid paraffin. The emulsified layer height was formed by intensely mixing the mixture of 1.5 mL of sample and 0.9 g of liquid paraffin for 2 min on a vortex and then leaved to set for 24 h or overnight.

3 EXPERIMENTAL RESULTS AND DISCUSSION

3.1 *The influence of carbon source on biosurfactant fermentation*

Both the cell density and the emulsifying activity of culture broth (figure 1) increased and achieved the maximum almost simultaneously, in which carbohydrate was used as carbon sources, revealing that the cell growth and the emulsifying activity of culture broth was positively correlated, that is to say that the biosurfactant production from the carbohydrate carbon sources was associated with the cell growth. Although the cell density and the emulsifying activity of culture broth initially increased faster, the maximum of the cell density and the emulsifying activity of culture broth was the lowest under the condition of starch carbon source, in tested carbohydrate carbon sources. Compared with the results with sucrose carbon source, the cell density exhibited lower and the emulsifying activity of culture broth achieved the maximum later using maltose as the carbon source, but tendencies in both the cell density and the emulsifying activity of culture broth throughout the fermentation process and the maximum of emulsifying activity of culture broth were similar to the situation using sucrose as the carbon source. Despite of larger decrease in the surface tension, the cell growth and the emulsifying activity of culture broth exhibited very poor with hydrocarbon as the carbon sources (figure 2).

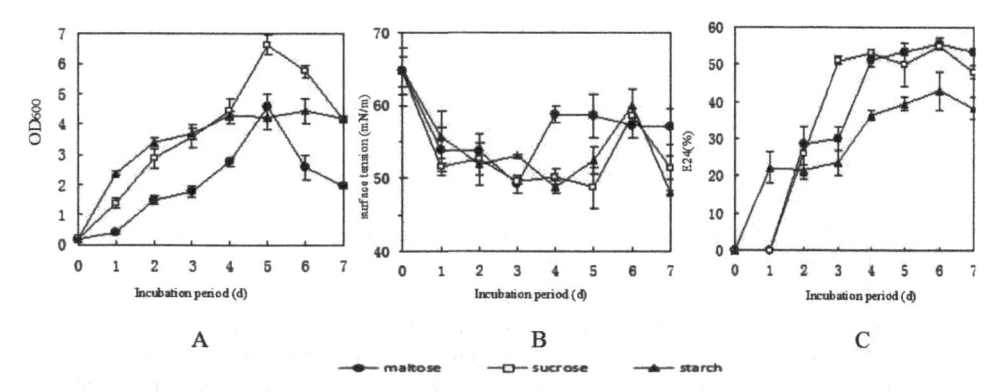

Figure 1. The performance of biosurfactant fermentation by using different carbohydrate carbon sources. A: cell density; B: surface tension; C: emulsifying activity.

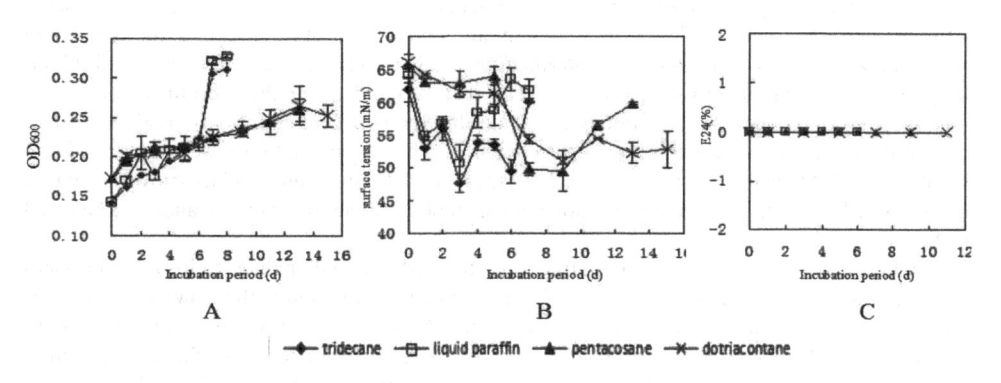

Figure 2. The performance of biosurfactant fermentation by using different hydrocarbon carbon sources. A: cell density; B: surface tension; C: emulsifying activity.

The surface tension of the culture broth with different carbon sources reduced slightly, demonstrating that the fermentation produced a small amount of biosurfactant reducing surface tension. The high emulsifying activity of the culture broth means that the fermentation produced a large amount of bioemulsifier. According to the results above, it could be concluded that the carbonhydrate carbon source was much better than the hydrocarbon carbon source for the cell growth and biosurfactant production of *Ochrobactrum intermedium* strain XY-1. By contrast, *Ochrobactrum intermedium* CN3 [2], *Ochrobactrum anthropi* strain RIPI5–1 [5] and *Ochrobactrum intermedium* sp. [4] could produce biosurfactant by using creosote, crude oil and hexadecane, respectively, starch carbon source could support biosurfactant production by halophilic *Ochrobactrum* sp. strain BS-206 [3]. Sucrose was selected for *Ochrobactrum intermedium* strain XY-1 to produce biosurfactants owing to the maximal bacterial growth and emulsifying activity (54%) on it.

3.2 The influence of nitrogen source on biosurfactant fermentation

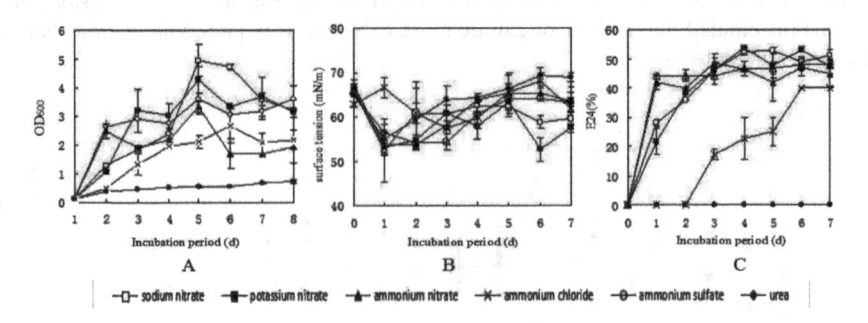

Figure 3. The performance of biosurfactant fermentation by using different nitrogen sources.
A: cell density; B: surface tension; C: emulsifying activity.

The cell density and the emulsifying activity of culture broth increased and achieved the maximum almost simultaneously in the fermentation with various nitrogen sources (figure 3), similar to that with various carbon sources. Although the cell density and the emulsifying activity initially of culture broth increased faster, the maximum of the cell density and the emulsifying activity of culture broth was not the highest in the medium with nitrogen sources containing ammonium. By contrast, the cell density and the emulsifying activity of culture broth initially increased slowly, the maximum of the cell density and the emulsifying activity of culture broth was the highest in the medium with sodium or potassium nitrate as nitrogen sources. The lowest, very small cell densities and almost no emulsifying activities of culture broth occurred throughout the fermentation process with urea nitrogen source, followed by that with ammounium chloride, revealing that urea could not be utilized well by *Ochrobactrum intermedium* strain XY-1, chloride element would inhibit the growth and metabolism of the producer. The surface tension of culture broth with different nitrogen sources reduced slightly, indicating again that the producer could only produce a small amount of biosurfactant reducing the surface tension. The above results showed that the nitrate was beneficial for *Ochrobactrum intermedium* strain XY-1 to grow and produce biosurfactant by metabolism. Therefore, sodium nitrate showing the stable fermentation performance was selected for further study.

It had been known that limiting the nitrogen in the medium could enhance biosurfactant yield [6]. Since nitrate would be assimilated after being reduced to ammounium, the slow assimilation of nitrate would simulate a condition of nitrogen limitation [7,8]. The results in this study were similar to those reports that sodium nitrate was the favorable nitrogen source for biosurfactant production by *Ochrobactrum anthropi* 2/3 [5] and halophilic *Ochrobactrum* sp. strain BS-206 [3], while urea could be beneficial to the growth and biosurfactant production of strain BS-206 *Ochrobactrum* sp. [3], unlike the experimental results in this study.

3.3 The influence of the ratio of carbon to nitrogen on biosurfactant fermentation

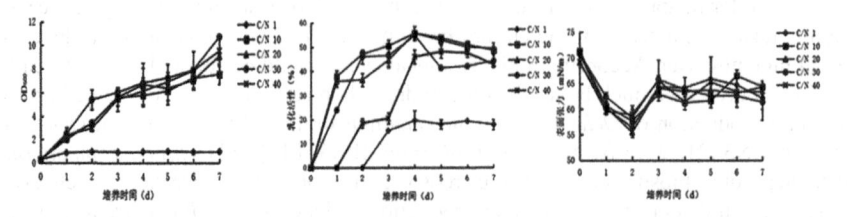

Figure 4. The performance of biosurfactant fermentation at different ratios of carbon to nitrogen.
A: cell density; B: surface tension; C: emulsifying activity.

Based on sodium nitrate concentration at 0.05 mol/L in medium, sucrose concentration was adjusted to achieve the ratio of C/N at 1:1, 10:1, 20:1, 30:1, and 40:1 to examine the effect of the C/N ratio on biosurfactant production. The poor cell density and the lower emulsifying activity of culture broth exhibited at the C/N ratio of 1, the lower cell density and the slowly increased emulsifying activity of culture broth exhibited at the C/N ratio of 40 (figure 4). It was indicated that the higher or lower C/N ratio was not beneficial to biosurfactants production, which was in accordance with the earlier conclusion [9]. It was found that there was no significant influence of the C/N ratios on the surface tension reduction, while a significant effect on the maximal emulsifying activity, the difference between the emulsifying activity at the C/N ratio of 10:1, 20:1, and 30:1 was not significant. Our finding is consistent with the result that nitrogen limitation can enhance biosurfactant production [6]. Therefore, the C/N ratio at 10:1 was economically viable for *Ochrobactrum intermedium* strain XY-1 to grow and produce biosurfactant.

The C/N ratio of 5:1 was optimal for halophilic *Ochrobactrum* sp. strain BS-206 to produce biosurfactant with sodium pyruvate as a carbon source [3]. *Ochrobactrum anthropi* 2/3 used palm oil decanter cake and CMSG as a carbon and nitrogen sources, obtained the higher yield of biosurfactant at the C/N ratio of 25:1 [5]. It is likely that there is difference between the nitrogen limitation required by strains, leading to the difference in the required C/N ratio.

In conclusion, *Ochrobactrum intermedium* strain XY-1 achieved higher biosurfactant yields with sucrose as the carbon source (10 g/L), and with sodium nitrate as the nitrogen source (4 g/L), and at the C/N ratio of 10:1, the highest emulsifying activity of fermentation supernatant to liquid paraffin was as high as 55%. The tested strain could grow and produce biosurfactants by utilizing potassium nitrate similar to sodium nitrate, maltose inferior to sucrose, and at the C/N ratio of 20:1 and 30:1 similar to 10:1, but could not utilize hydrocarbon and urea well. In further study, carbon and nitrogen sources of wide range should be detected, especially cost-free agro-industrial wastes.

ACKNOWLEDGMENTS

Authors wish to thank the financial supports received from the Natural and Scientific Funding of China (31371742) and the Science and Technology Department of Dalian (2013B11NC078).

REFERENCES

[1] Joy S, Rahman P K S M and Sharma S 2017 *Chem. Eng. J.* **317** 232–41
[2] Bezza F A, Beukes M, Chirwa E M N 2015 *Process Biochem.* **50** 1911–22
[3] Kumar C G, Sujitha P, Mamidyala S K, Usharani P, Das B and Reddy C R 2014 *Process Biochem.* **49** 1708–17
[4] Mahendhran K, Arthanari A, Dheenadayalan B and Ramanathan M 2018 *Bioresour. Technol. Rep.* **4** 66–73
[5] Holmes B, Popoff M Z, Kiredjian M and Kersters K 1998 *Int. J. Syst. Bacteriol.* **38** 406–16
[6] Guerra-Santos L H, Kappeli O and Fiechter A 1984 *Appl. Environ. Microbiol.* **48** 301–5.
[7] Guerra-Santos L H, Kappeli O and Fiechter A 1986 *Appl. Microbiol. Biotechnol.* **24** 443–48.
[8] Mulligan C N, and Gibbs B F 1989 *Appl. Environ. Microbiol.* **55** 3016–9.
[9] Cui X, Harling R, Mutch P, Darling D 2005 *Eur. J. Plant Pathol.* **111** 297–308.

Study on single-pass forming process parameters of laser direct deposition of Fe55 alloy

G.L. Yin
Key Laboratory for Anisotropy and Texture of Materials (Ministry of Education), Key Laboratory for Laser Application Technology and Equipment of Liaoning Province, School of Materials and Engineering, Northeastern University, Shenyang, China
School of Material Science and Engineering, Liaoning University of Technology, Jinzhou, China

S.Y. Chen, J. Liang, T. Cui & C.S. Liu
Key Laboratory for Anisotropy and Texture of Materials (Ministry of Education), Key Laboratory for Laser Application Technology and Equipment of Liaoning Province, School of Materials and Engineering, Northeastern University, Shenyang, China

M. Wang
Shenyang Dalu Laser Technology Co. Ltd. Shenyang, China

ABSTRACT: Using laser direct deposition, Fe55 alloy powder was deposited on the surface of the Q235 steel plate. The effects of scanning speed and laser power on the forming morphology, microstructure and hardness of single-pass deposition coating were analyzed. The results show that the height and width of the coating increase first and then decrease with the increase of laser power when the scanning speed is 5 mm/s. Compared with the 1700 W and 1900 W coatings, the microstructure in the middle of coating with 1800 W is the smallest, and its microhardness is the highest, with an average value of 892 HV. When laser power is 1800 W, the height and width of the coating decrease with the increase of scanning speed. Compared with the 4 mm/s and 6 mm/s coatings, the microstructure in the middle of coating with 5 mm/s is the smallest, and its microhardness is the highest, with an average value of 866.6 HV. The optimal process parameters such as laser power 1800 W, scanning speed 2 mm/s, defocus distance 304 mm, and powder feeding rate 0.14g/s are ascertained for singlepass deposition of Fe55 alloy powder

1 INTRODUCTION

As a rapid mature laser additive manufacturing technology, the laser direct deposition method has been widely used in the preparation of surface-modified coatings [1–4]. The coating quality of laser direct deposition is affected not only by the substrate material and deposition material, but also by the laser power, scanning speed, spot size, and powder feeding rate [5,6]. When there is a good match between the process parameters, the deposition coating with good performance and no defects can be obtained [7]. Moreover, the single-pass laser deposition is the basis of large area overlapping and high thickness multilayer deposition. Therefore, in this paper, the effects of laser power and scanning speed on the formability, microstructure, and hardness of single-pass coating are studied under constant other process parameters. This work provides a reasonable process specification and theoretical basis for the subsequent large area and high thickness multilayer deposition.

2 EXPERIMENTAL MATERIALS AND METHODS

In the test, the substrate material used was the Q235 steel plate with the size of 200 mm (length)×100 mm (width)×10 mm (thickness), and its chemical composition is 0.14~0.22 C, ≤0.3 Si, 0.30~0.65 Mn, ≤0.050 S, ≤0.045 P, and balance Fe (wt.%). Fe55 self-fluxing alloy powder which had spherical particles with sizes of -150~+400 mesh was used, and its chemical composition is 0.14~0.22 C, 16~18 Cr, 3.4~4.0 Si, 3.5~4.0 B, 1.0~2.0 Ni, and balance Fe (wt.%).

The single-pass laser deposition coatings were conducted on the Q235 substrate by an FL-Dlight 02-3000 W semiconductor laser. The process parameters of laser deposition are as follows: power 1500~2000 W, scanning speed 3~7 mm/s, defocus distance 304 mm, and powder feeding rate 0.14 g/s. The samples with the size of 10 mm×10 mm×10 mm were cut from each single coating, and cross sections of the samples were used for microstructure observation and hardness testing. The macroscopic morphology of the sample was observed by a macroscopic stereomicroscope (YMPUS-SZ61). The microstructure of single-pass coating was observed by an optical microscope (OM, OLYMPUS-GX71). The microstructural morphologies of single-pass coatings were observed by a scanning electron microscope (SEM, Shimadzu-SSX-550). The hardness of the cross section of the coating was measured using a digital hardness tester (MHVD-1000IS) with a load of 200 gf for 10 s.

3 RESULTS AND DISCUSSION

3.1 *The effect of laser process parameters on the cross-section morphology of singlepass coating*

3.1.1 *The effect of laser power on the cross-section morphology of singlepass coating*

Under the condition of scanning speed 5 mm/s and laser power 1500 W~2000 W, the cross-sectional morphology of the coating and the variation trend of height and width of the coating with laser power were determined and are shown in Figure 1 and Figure 2, respectively. It can be seen from Figure 1 that metallurgical bonding is achieved between the singlepass coating and the substrate under different powers. With the increase of laser power, the height and width of singlepass coating first increase and then decrease, as shown in Figure 2 because the substrate and deposited powder cannot absorb enough laser energy when the laser power is small and the molten pool area is insufficient. Meanwhile, part of Fe55 powder cannot melt and fly away from the molten pool during the deposition process, resulting in low width and height of singlepass coating (Figure 1 (a)). With the increase of laser power, enough energy can be obtained by deposition material in unit time, which increases the melting amount of powder in unit area, thus increasing the height and width of the coating (Figure 1 (b), (c)). However, when the laser power is too high, splashing and powder burning will occur in varying degrees during the deposition process, which will reduce the height and width of the coating to varying degrees (Figure 1 (d)–(f)). Therefore, laser powers 1700 W, 1800 W and 1900 W are considered as good power parameter ranges when the scanning speed is 5 mm/s.

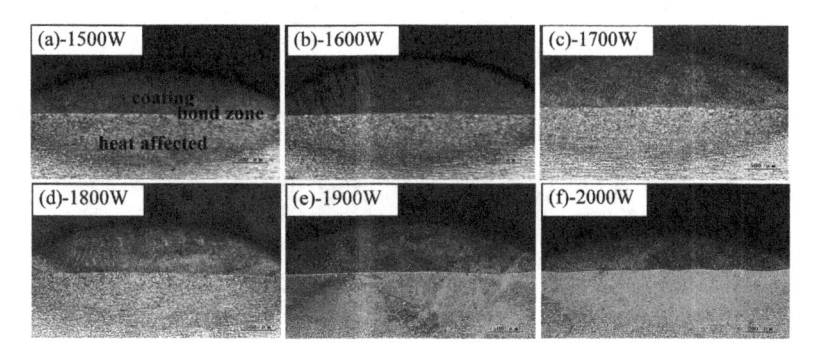

Figure 1. The cross-section morphology of the coating with different powers.

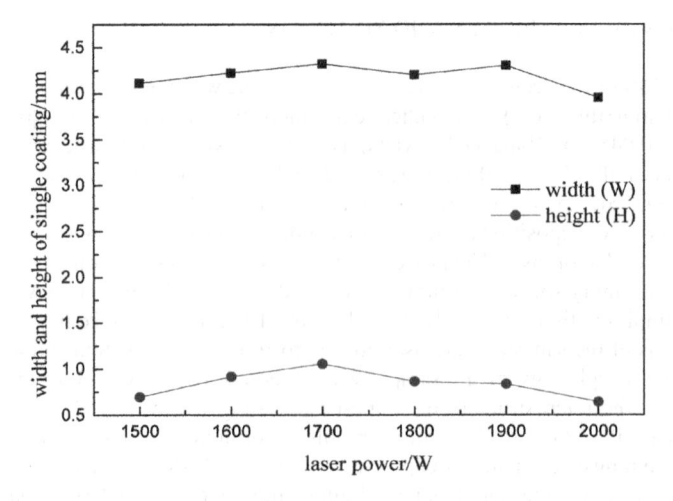

Figure 2. Variation trend of width and height of single-pass coating with laser power.

3.1.2 *The effect of scanning speed on the cross-section morphology of singlepass coating*

Under the condition of laser power 1800 W and scanning speed 3 mm/s~8 mm/s, the cross-sectional morphology of the coating and the variation trend of height and width of the coating with scanning speed are shown in Figure 3 and Figure 4, respectively. It can be seen that the influence of scanning speed on the formability of singlepass coating is very obvious. When the scanning speed is small, the coating height and width are larger (Figure 3 (a)). The height and width of the coating decrease with the increase of scanning speed (Figure 3 (b)–(f)). The reason can be as follows: With the increase of scanning speed, the laser residence time in the coating area is shorter, which shortens the contact time between the laser and the powder, so that the powder cannot have enough time to absorb heat and melt, resulting in the reduction of the height and width of the singlepass coating. Therefore, scanning speeds 4 mm/s, 5 mm/s and 6 mm/s are considered as good scanning speed parameter ranges when laser power is 1800 W.

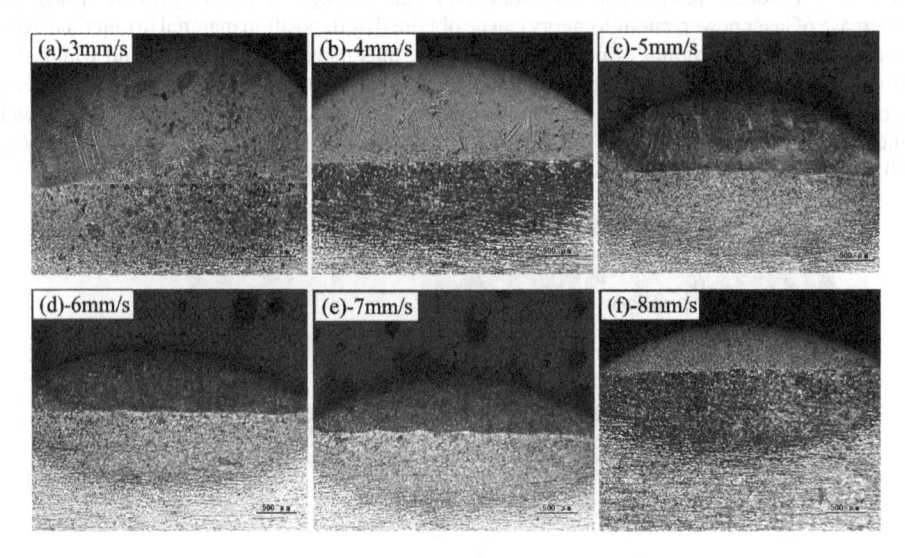

Figure 3. The cross-section morphology of the coating with different scanning speeds.

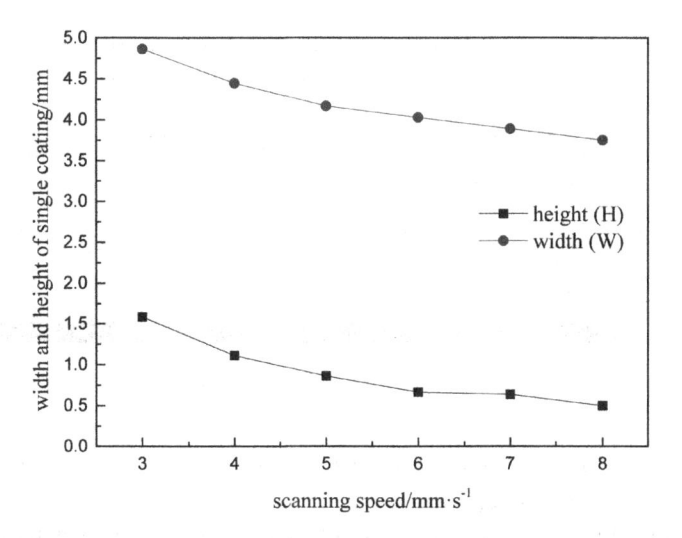

Figure 4. Variation trend of width and height of single-pass coating with scanning speeds.

3.2 *The effect of laser process parameters on the microstructure of singlepass coating*

3.2.1 *The effect of laser power on the microstructure of singlepass coating*

Figure 5 shows the SEM morphology of the middle part of the single-pass coating with laser powers of 1700 W, 1800 W, and 1900 W under a scanning speed of 5 mm/s. It can be seen that the microstructure of the coating is composed of dendrite or cellular matrix and eutectic between dendrites. When the power is 1700 W, the coating microstructure is mainly developed dendrite. With the increase of laser power, the microstructure is obviously refined, the long dendrites are significantly reduced, and the granular and short rod-shaped cellular crystals are significantly increased. When the power reaches 1900 W, the long dendrite disappears, but the structure becomes coarser, and the eutectic structure presents a network distribution. This is due to the increase of heat input and high temperature residence time of the coating with excessive power, which leads to the microstructure coarsening.

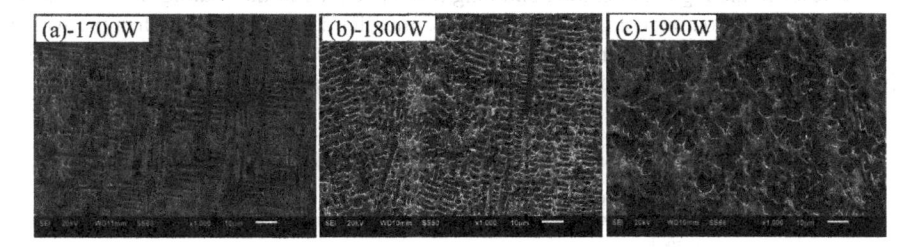

Figure 5. SEM images of the single-pass coating with different laser powers.

3.2.2 *The effect of scanning speed on the microstructure of singlepass coating*

Figure 6 shows the SEM morphology of the middle part of the singlepass coating with scanning speeds of 4 mm/s, 5 mm/s and 6 mm/s under a laser power of 1800 W. It can be seen that the coating microstructure is still composed of dendrite or cellular matrix and eutectic structure. With the increase of scanning speed, the developed dendrites gradually disappear and the microstructure of the coating is obviously refined. The reason is that the heat absorbed by the melt in the molten pool is relatively reduced with the increase of scanning speed, which makes the nucleation rate greater than the growth rate of the crystal, resulting in the inhibition of growth of dendrite and its gradual disappearance; thus the cellular crystal is formed. However, when the scanning speed is too fast or too slow, the better microstructure of the cladding layer cannot be obtained.

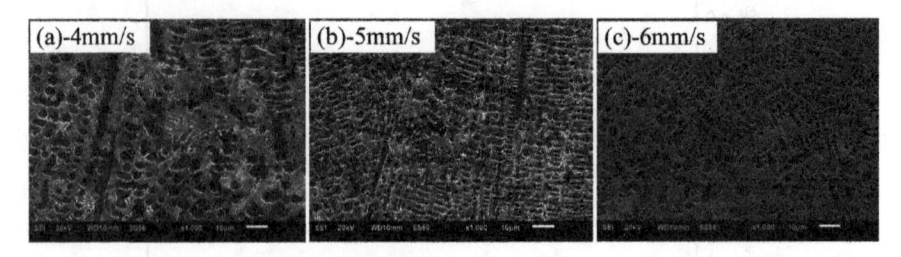

Figure 6. SEM images of the single-pass coating with different scanning speeds.

3.3 *The effect of laser process parameters on microhardness of singlepass coating*

Figure 7 shows the microhardness distribution of the single-pass coating with laser powers of 1700 W, 1800 W, and 1900 W under a scanning speed of 5 mm/s. It can be seen that the microhardness of the coating first increases and then decreases with the increase of laser power. The reason can be explained by two aspects. On the one hand, the dilution effect of the substrate on the coating increases, which leads to more Fe elements entering the coating and decreasing the proportion of the precipitated hard phase in the coating. On the other hand, the increase of energy density leads to coarsening of crystalline grains. As a result, the microhardness of the coating decrease. Moreover, it can also be seen from Figure 7 that the microhardness of single coating is above 650 HV, and the coating with 1800 W power gives the highest microhardness, with an average value of 892 HV. The hardness of the coating with 1700 W power is the lowest, and the average value is 775 HV. The difference of coating hardness is consistent with the size of the microstructure with different powers

Figure 8 shows the microhardness distribution of the singlepass coating with scanning speeds of 4 mm/s, 5 mm/s and 6 mm/s under a laser power of 1800 W. It can be seen that the microhardness of the coating is much higher than that of the substrate, and the microhardness of the coating with 5 mm/s is the highest, with an average value of 866.6 HV. This is mainly attributed to the fine structure of the coating. When the scanning speed is lower or higher, the reason for the low microhardness of the coating can be explained as follows: When the scanning speed is low, the high energy density of laser input leads to the high temperature and long residence time of the coating, which makes the microstructure of the coating coarsen and reduce its microhardness. However, when the scanning speed is high, the heat of the coating obtained per unit time is less, so it cannot form a good dense coating, which will inevitably reduce the hardness of the coating.

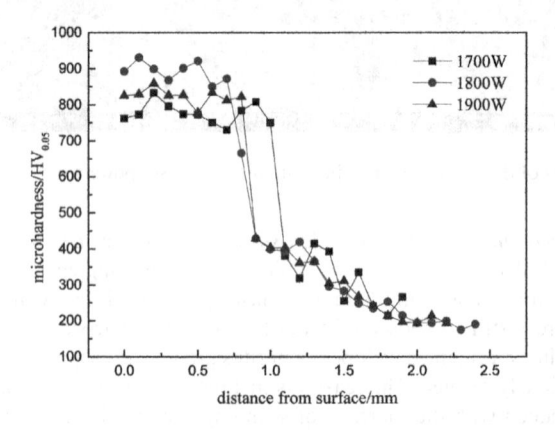

Figure 7. Microhardness of the coating with different powers.

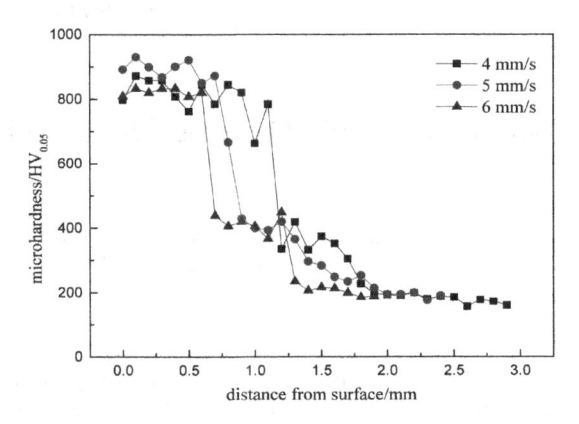

Figure 8. Microhardness of the coating with different scanning speeds.

Through the abovementioned analysis, a group of process parameters suitable for deposition of Fe55 alloy powder are optimized: laser power 1800 W, scanning speed 5 mm/s, defocus distance 304 mm, and powder feeding rate 0.14 g/s.

4 CONCLUSION

The height and width of the coating first increase and then decrease with the increase of laser power under 5 mm/s. The height and width of the coating decrease with the increase of scanning speed under 1800W.

The microstructure of the coating is composed of dendrite or cellular matrix and eutectic between dendrites. When the scanning speed is 5 mm/s, compared with the coatings of 1700 W and 1900 W, the middle part of 1800 W has the smallest microstructure and the highest microhardness, with an average value of 892 HV. When laser power is 5mm/s, compared with the coatings of 4mm/s and 5mm/s, the middle part of 5 mm/s has the smallest microstructure and the highest microhardness, with an average value of 866.6 HV.

The optimum process parameters of Fe55 alloy powder deposition were determined as follows: laser power 1800 W, scanning speed 5 mm/s, defocus distance 304 mm, and powder feeding rate 0.14 g/s.

ACKNOWLEDGMENTS

This work was financially supported by National Key R&D Program of China (2016YFB1100201), Green Manufacturing System Integration Project of the Industry and Information Ministry of China (2017), and Shenyang important scientific and technological achievements transformation project (20-203-5-6).

REFERENCES

[1] Chen X T, Chen S Y, Cui T, Liang J, Liu C S and Wang M 2020, *Optics and Laser Technology*, **126** 106080
[2] Liu K, Li Y J, Wang J and Ma Q S, 2015, *Journal of Alloys and Compounds*, **624** 234–240
[3] Zhu L N, Xu B S, Wang H D and Wang C B, 2012, *Journal of Materials Science*, **47** 2122–26
[4] Zhang H, Zou Y, Zou Z D and Wu D T. 2015, *Optics & Laser Technology*, **65** 119–125
[5] Sun K, Yao J W, Xu Z J and Lin S Z, 2007, *Agricultural Equipment & Vehicle Engineering*, 36–42
[6] Tan J H, Sun R L, Niu W, Liu Y N and Hao W J, 2020, *Materials Reports*, **34** 12094–100
[7] Yang N, Yang F,2010, *Journal of Henan Institute of Education (Natural Science Edition)*, **19** 15–17

Preparation and performances of SiO$_2$ aerogel heat-insulation coatings

G.Z. Lu, Z.J. Wang, Z.S. Zhang & Z.F. Yin
Beijing Building Materials Academy of Sciences Research Co., Ltd., Beijing, China
State Key Laboratory of Solid Waste Reuse for Building Materials, Beijing, China

ABSTRACT: In this paper, the SiO$_2$ aerogel heat-insulation coatings with high performance-price ratio are prepared by making use of SiO$_2$ aerogel prepared under the normal temperature and pressure drying process as the functional filler. The effects of such factors as SiO$_2$ aerogel amount, infrared powder, film thickness and emulsion varieties on coating performance are studied. As shown by the experimental results, SiO$_2$ aerogel significantly can improve the heat insulation properties of coatings, and can be widely applied in energy-saving of buildings, industrial pipelines and chemical storage tanks.

1 FOREWORD

In recent years, new coatings are used by various domestic industries in large quantities, including transportation, construction, petroleum and weaponry, to achieve the purpose of reducing the surface coating temperature and internal environmental temperature of equipment under solar radiation [1]. For example, in industrial buildings for storage of flammable materials warehouse, the temperature rise will bring great safety risks; in civil buildings, unreasonable indoor temperature will increase the use time and frequency of air conditioning, fans and other equipment, wasting electric power. Therefore, the use of various methods to reduce or prevent the temperature rise caused by intense solar radiation has become an important research topic. Among such methods, heat insulation coating is the most common and simplest one [2].

SiO$_2$ aerogel is a porous medium amorphous material with an irregular three-dimensional network frame structure. It has the following characteristics: low density, as low as 0.002g/cm3; low heat conductivity, vacuum conductivity at room temperature may reach 0.001W/(m·K); high specific surface area; large pore volume; low acoustic impedance [3] etc.; therefore, it is widely used in the field of high temperature insulation materials. In this study, SiO$_2$ aerogel is chosen as the main filler due to its unique nanostructure and solar spectrum-selectivity characteristic. During the preparation process, the effects of SiO$_2$ aerogel amount, infrared powder, film thickness, emulsion varieties, flame retardant and other factors on the performance of the coating are investigated, aiming at developing a SiO$_2$ aerogel heat-insulation coating with satisfactory heat insulation effect, dirt and weather resistance, and certain flame retardancy.

2 EXPERIMENT

2.1 *Experiment materials*

The materials used in the experiment are divided into four parts: base materials, functional materials, additives, pigments and fillers.

2.2 Instrument and equipment

Small high-speed disperser, Brookfield viscometer, electronic balance, infrared thermometer, heat conductivity tester and other instruments.

2.3 Preparation of SiO_2 aerogel

The SiO_2 aerogel is prepared in seven steps as follows with the drying process at room temperature and atmospheric pressure. (1) Sol preparation: mix tetraethyl orthosilicate, anhydrous ethanol and oxalic acid solution in a certain proportion, and stir in a magnetic stirrer for a certain time to obtain sol. (2) Gel preparation: add a certain amount of water, dimethylformamide and ammonia solution to the obtained sol, stir it evenly, pour it into a mold and let it stand for a certain time to obtain gel. (3) Preaging: keep the gel obtained for a certain time, so that the unfinished polycondensation reaction continues. (4) Aging: the two-step aging method is adopted, aging with water/anhydrous ethanol solution for 24h, and then aging with tetraethyl orthosilicate/anhydrous ethanol solution for 24h. (5) Solvent replacement: first, pure isopropanol is used for solvent replacement of the pore liquid in the gel, and then 75% isopropanol/n-hexane solution, 50% isopropanol/n-hexane solution, 25% isopropanol/n-hexane solution and pure n-hexane solution are used for solvent replacement respectively. (6) Surface modification: about 10vol% trimethylchlorosilane/n-hexane is used for surface modification until the gel is completely transparent. (7) Cleaning and drying: 10vol% isopropanol/n-hexane and pure hexane are used for surface cleaning in sequence. The drying is carried out at a certain temperature and normal pressure, thereby making the solvent evaporate slowly and obtaining SiO_2 aerogel eventually.

2.4 Preparation of SiO_2 aerogel heat-insulation coating

SiO_2 aerogel, water-based acrylic resin, dispersant and wetting dispersant are stirred uniformly in a high-speed disperser for 15min according to a certain ratio, and then defoaming agent, leveling agent, thickening agent, curing agent, dipropylene glycol methyl ether, dipropylene glycol butyl ether and deionized water are mixed uniformly in the disperser to prepare SiO_2 aerogel heat-insulation coating.

3 RESULT AND DISCUSSION

3.1 Analysis on performance of SiO_2 aerogel

The production of SiO_2 aerogel on the market takes expensive organic silicon as the raw material, and the supercritical technology with large investment is applied. The price of SiO_2 aerogel produced this way is as high as 2000~3000 yuan/kg, and the supercritical technology has the potential danger of high temperature and high pressure, leading to failure in continuous production and long production cycle (more than 100h), and limiting its application scope. Thus, how to improve the production process of SiO_2 aerogel and reduce the production cost has become the key to the successful wide application of SiO_2 aerogel in building materials.

In this paper, it is proposed that SiO_2 aerogel can be prepared with normal temperature and pressure drying technology. Sodium silicate is adopted as the raw material, and dehydration and drainage are conducted simultaneously in the process of drying. Benefiting from this, the reaction efficiency is increased, the production time is as short as 4~5h, the capacity of continuous mass production is realized, and the production cost is 30~50 yuan/kg, only 1/10~1/8 of the original cost.

Table 1. Comparison of technical indexes of aerogels prepared by two processes.

Item	Supercritical drying process	Normal temperature drying process
Pore diameter/nm	20	20
Pore volume/(cm^3/g)	3.27	2.63
Density/(kg/m^3)	40~100	120
Surface area/(m^2/g)	600~800	600
Heat conductivity at 25°C /(mW/m·k)	11~13	18

As shown in Table 1, the properties of SiO$_2$ aerogels prepared by the two processes are basically the same. Therefore, the normal temperature and pressure drying technology is innovative and the product price is reasonable, making the wide application of SiO$_2$ aerogel in the field of building materials possible.

3.2 *Determination of SiO$_2$ aerogel amount*

The optimal amount of SiO$_2$ aerogel is explored by single factor experiment method. Based on previous experiments, SiO$_2$ aerogel is added into the coating by 10%, 12%, 14%, 16%, 18% and 20% (mass fraction of the coating), respectively. The heat conductivity of the coating is determined by the heat conductivity tester according to the provisions of GB/T 10294, and the viscosity of the coating is measured by Brookfield viscometer to determine the optimal amount of SiO$_2$ aerogel. The test results are shown in Figure 1.

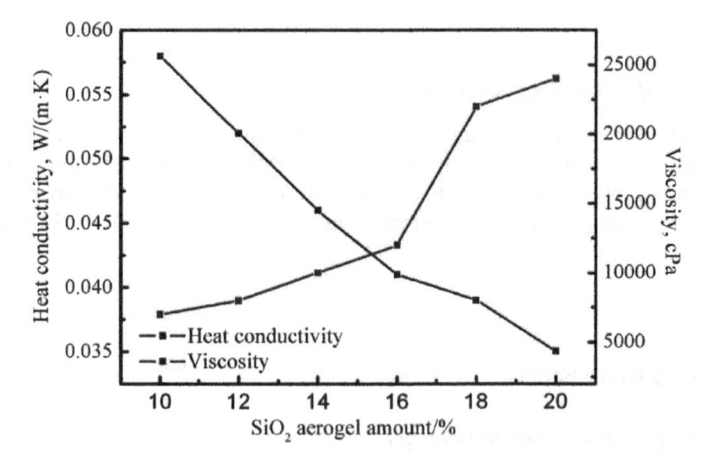

Figure 1. Variation of heat conductivity and viscosity of coating with the amount of SiO$_2$ aerogel.

As can be seen from the black curve in Figure 1, the heat conductivity of the coating gradually decreases with the increase of the amount of SiO$_2$ aerogel. The blue curve shows that the viscosity of the coating gradually increases with the increase of the amount of SiO$_2$ aerogel. When the amount of SiO$_2$ aerogel is greater than 16%, the viscosity of the coating increases rapidly. Too high viscosity will affect the construction performance of the coating. Considering such two factors of the coating, 16% SiO$_2$ aerogel is the optimal amount.

3.3 *Selection of emulsion varieties*

The performance of the coating is determined by its composition and structure. Since the SiO$_2$ aerogel heat-insulation coating is mainly used for the exterior decoration of buildings or oil tanks, outstanding weather resistance, color retention, adhesion and dirt resistance are required. 4 kinds of emul0sion products with high popularity in the market are selected for comparison. The basic properties of the SiO_2 aerogel heat-insulation coatings prepared with the same basic formula are as follows:

Table 2. Comparison of the main properties of 4 emulsion products.

Index	a	b	c	d
Dirt resistance/%	18	15	12	6
Weather resistance:	1	0	1	0
pulverizing (600h) discoloring	2	2	2	1

It can be seen from Table 2 that considering the conventional performance indexes of coatings, the comprehensive performance of the coating made of emulsion d are the best, so emulsion d is used in the preparation. Emulsion d is a silicon-acrylic emulsion. The introduction of inorganic silicon-oxygen structure into organic coatings greatly improves the performance of the coatings. The main chain of silicone resin is Si-O bond, and the bond energy is 452KJ/mol; the main chain of general latex paint molecule is C-C bond, and the bond energy is 356KJ/mol. Therefore, it can effectively resist the photooxidative degradation of the coating film by ultraviolet rays, and the silicone modified latex paint has satisfactory durability. The silicone resin molecule has organic groups, the molecular is highly symmetrical, the polarities cancel each other out, and the whole molecule is non-polar, thus achieving quite low surface tension and excellent water repellency [4,5]. Since water is one of the main elements causing building damage, it can effectively protect buildings. Meanwhile, it can improve the dirt resistance of the film, and the reason is that the dust dissolves in the rain to form a solution, or disperses in the rain to form a colloid, and then enters the pores of the film through the water absorption of the film, thus causing permanent contamination to the film. For the film with good water repellency, the water absorption is low, less dust enters the pores of the film, and the dirt resistance is improved. The silicone resin has a silica-oxygen skeleton structure, so the coating film boasts high air permeability. In addition, the main chain of silicone resin Si-O is helical, and its groups can be aligned directionally at the interface. It can also be cross-linked with the silicate materials of the base layer to form chemical bonds. Consequently, the adhesion between the coating film and the base layer is greatly enhanced, thus improving the adhesion of the coating film [5].

3.4 Influence of infrared powder on coating properties

Although the base material in the coating absorbs part of the energy in the sunlight, it eventually radiates the heat absorbed energy in the form of infrared radiation, thus playing a role of heat insulation. Adding an appropriate amount of infrared powder into the coating will improve the radiation performance of the coating and realize a high emissivity of the coating, $\varepsilon > 0.9$. From the perspective of the heat radiation mechanism of infrared powder, when it absorbs radiant heat energy, the internal movement of its molecules is changed and intensified, and the particles produce heat emission from high to low at the energy state level, thus reducing the temperature of the irradiated material. According to the results of spectroscopic analysis, the absorption of photons by molecules of matter will completely change the energy of photons into the energy of molecular vibration and rotation. In this way, the energy of molecular vibration and rotation is changed, the vibrations of lattice and bond group constantly collide with each other, and part of the absorbed energy returns to the external space in the form of infrared radiation.

Under the same experimental conditions, 1.0%, 2.0%, 3.0% and 4.0% infrared powder are added into the coating system, respectively, and the coating emissivity is measured to characterize the coating performance. The relationship between the amount of infrared powder and the coating emissivity is shown in Table 3.

Table 3. Relationship between the amount of infrared powder and the coating emissivity.

Amount/%	1.0	2.0	3.0	4.0
Emissivity	0.75	0.80	0.86	0.89

It can be seen from Table 3 that by adding 3%-4% of this infrared powder, the infrared radiation ability of the coating will be greatly improved, and the radiation coefficient of the coating will reach 0.88 or above.

3.5 Influence of additives on coating properties

Considering that the flame retardant must be compatible with the system, have good stability and meet the pH value requirements of the coating, magnesium hydroxide (MH) is selected as the flame retardant from halogen, antimony, inorganic phosphorus, boron (zinc borate), metal hydroxide and other flame retardants through many tests to achieve the goals of certain flame resistance, adjustable pH value and pH regulator saving. The flame-retardant mechanism is as follows:

1) Endothermic dehydration at high temperature. Endothermic dehydration removes heat from the combustion process, and the water vapor generated by temperature reduction dilutes oxygen and surround the flame, thus stopping the combustion. In fact, there is no water in the molecular structure of MH. The water they release thermally is generated by the decomposition of the hydroxyl group bonded to the metal. Therefore, heat absorption and water release is one of the flame-retardant modes of MH.
2) The oxides formed by dehydration of MH help to catalyze the formation of carbon, and the water-resistant inorganic layer is a heat barrier. The reason may be that they have poor heat conductivity and can reflect radiant heat. For example, MgO is a good insulator, and when MH loses water, it plays a flame-retardant function in this aspect.
3) The metal oxide layer formed after the dehydration of metal hydroxide is active, has a large surface area, and absorbs smoke and combustible material, endowing metal hydroxide with good smoke suppression effect.

It is found in the research that the amount of MH is related to the self-extinguishing property and storage stability of the coating. With the increase of MH amount, the self-extinguishing property of the coating enhances while the storage stability decreases. The relationship is shown in Table 4.

Table 4. Relationship between the amount of MH and self-extinguishing property and storage stability of the coating.

Amount/%	1.0	2.0	3.0	4.0	5.0
Self-extinguishing property/s	<5.0	<1.2	<0.8	<0.6	<0.5
Storage stability	Excellent	Excellent	Excellent	Good	Poor

By adding 2%-3% MH, the coating has excellent self-extinguishing property and storage stability.

3.6 Influence of film thickness on heat-insulation properties of coating

At present, the film of the coating in the market is generally 0.2–0.3mm thick for ideal heat-insulation effect. The thick coating material not only increases the required coating amount and the construction cost, but makes the dirt resistance worse. The comparison of the thickness and temperature difference between the SiO_2 aerogel heat-insulation coating we developed and other similar products is shown in Table 5.

Table 5. Comparison between the SiO$_2$ aerogel heat-insulation coating and other similar products.

Thickness, mm	0.075	0.08	0.1	0.15	0.2	0.25
Sample/°C	8.6	8.8	9.0	9.0	9.0	9.0
Contrast sample/°C	5.0	5.5	6.0	6.5	7.0	7.0

It can be seen from Table 5 that compared with similar products, the SiO$_2$ aerogel heat-insulation coating developed by us has a thinner coating film and better heat insulation effect. Moreover, the thickness of the film reaches a certain value, and its heat insulation effect is independent of the thickness of the film.

3.7 Characterization of heat insulation properties of coatings

The prepared SiO$_2$ aerogel heat-insulation coating is applied on the front and back of a 25cm×17cm ordinary household window glass with a wire rod coater, and then dried and cured at room temperature to obtain a transparent coating film sample. The heat insulation performance of the coating is tested with the heat insulation film performance tester. Two pieces of glass are placed at the left and right ends of the instrument. One piece is the glass coated with SiO$_2$ aerogel heat-insulation coating, and the other is the blank glass. When the temperature change is tested at the same temperature, the temperature difference and contrast ratio will be displayed on the instrument. The test results are shown in Table 6.

Table 6. Temperature variation of coating film.

	Item	\multicolumn{5}{c}{Irradiation time/min}				
		0	3	5	7	10
Irradiation on front side	Temperature of blank sample/°C	20	104	111	112.8	113
	Temperature of coating film sample/°C	19.8	83	84.2	85.6	86.4
Irradiation on back side	Temperature of blank sample/°C	19.8	84	89.4	93.6	92
	Temperature of coating film sample/°C	19.8	69	70	66.8	66.8

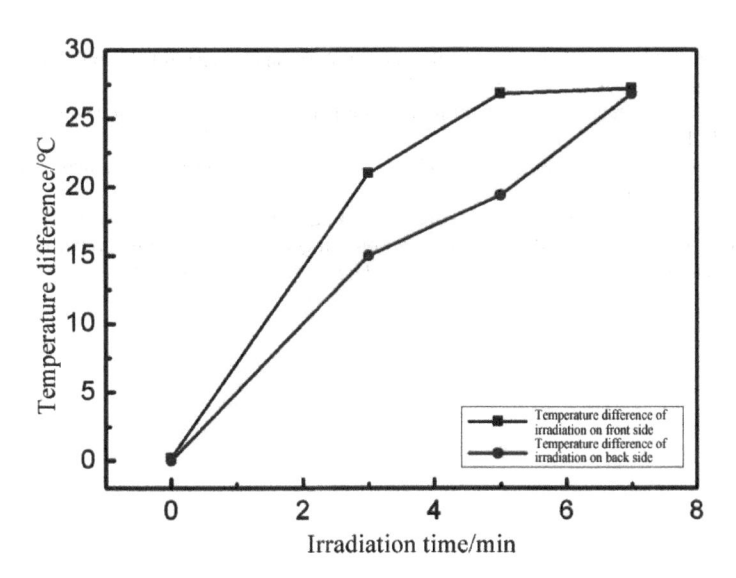

Figure 2. Temperature difference between coating film and blank glass.

It can be seen from Table 6 and Figure 2 that under the light of the tester, there is little difference in the temperature change between the front side and the back side of the irradiated glass. With the increase of the irradiation time, the temperature of the two pieces of glass is increasing. The temperature rise of the blank glass is greater than that of the coated glass. The temperature difference between the blank glass and the coated glass varies within the range of 0 ~ 25°C, and the maximum temperature difference reaches 26°C. SiO_2 aerogel is a nano-porous lightweight material with controllable structure. It has the characteristics of largest surface area up to 800~1000m2/g and a high porosity of 80%~98%. The heat conductivity of SiO_2 aerogel is below 0.013w/m·k, much lower than that of air at room temperature, 0.025w/m·k. These special properties are incomparable to other traditional materials; thus, it has become one of the best heat insulation properties of all solid materials, and known as a super heat insulation material. It is also confirmed by the test results that the heat insulation performance of SiO_2 aerogel is satisfactory.

4 CONCLUSION

SiO_2 aerogel heat-insulation coatings, with SiO_2 aerogel as filler, have a high reflection and dissipation effect on light and heat through the special material combination, greatly reducing the surface temperature of coated objects. The main conclusions are as follows:

1) SiO_2 aerogel prepared by normal pressure drying process is reasonable in cost and feasible in technology, and can be widely applied. The optimal amount of SiO_2 aerogel is 16%, and with such amount, the coating has good heat insulation performance.
2) SiO_2 aerogel heat-insulation coating has the following characteristics: (1) significant heat insulation effect: the maximum temperature difference of SiO_2 aerogel coating reaches 26°C through reasonable gradation. (2) high emissivity ε: by adding an appropriate amount of infrared powder, the emissivity of the coating is greatly improved. (3) high flame retardancy: the use of ultra-fin magnesium hydroxide as flame retardant reaches the goals of certain flame resistance, adjustable pH value and pH regulator saving. The most suitable amount of flame retardant is 2%-3%, and with such amount, the self-extinguishing time of the coating is less than 1s.

REFERENCES

[1] Gao, X.F., Liu, Z.H. et al. Energy saving effect and application prospect of far infrared radiation coatings. *Paint & Coatings Industry*, 2001, 31(8):30–33
[2] Fan, X.D., Cui, Z.Y. et al. Preparation of high-performance silicone acrylic emulsion. *Paint & Coatings Industry*, 2001, 31(12):11–14
[3] Yao, C., Zhao, S.L. et al. Characteristics and application of nano transparent coatings. *Paint & Coatings Industry*, 2007, 37(1):29–32
[4] Lu, G.Z. Development of reflective heat insulation coating for buildings. *China Coatings*, 2007.2(9):37–40
[5] Zhang, H.X., He, X.D. et al. Advances in studies on aerogel thermal insulation properties and composite aerogel insulation materials. *Journal of Materials Engineering*, 2007(9):94–96.

Optimization design for the free surface parameters of a spherical hinge suspender connecting locomotive structural frame and axle box

R.G. Zhao
College of Civil Engineering and Mechanics, Xiangtan University, Xiangtan, China
Key Laboratory of Dynamics and Reliability of Engineering Structures of College of Hunan Province, Xiangtan University, Xiangtan, China

W.H. Liu & N. Ji
College of Civil Engineering and Mechanics, Xiangtan University, Xiangtan, China

Y.J. Huang
Zhuzhou Time New Material Technology Co., Ltd., Zhuzhou, China

X.Q. Yang, X. Zhou & M. Zaheer
College of Civil Engineering and Mechanics, Xiangtan University, Xiangtan, China

ABSTRACT: The fatigue life of a spherical hinge suspender connecting the locomotive structural frame and axle box is the key factor affecting the reliability of locomotive body. Firstly, the ABAQUS software is used to establish the finite element model of a spherical hinge suspender, and the fatigue life of the spherical hinge suspender is predicted. The result shows that the fatigue life of the spherical hinge suspender does not satisfy with the expected life. Then, the SolidWorks is used to establish a 3D model of spherical hinge suspender, and the HyperMesh is adopted to mesh the rubber parts. The variables of support reaction force and displacement are derived by using ABAQUS to calculate the radial stiffness of spherical hinge suspender. The geometric parameters of free surface of spherical hinge are selected as the design variables, and the structural optimization with the fatigue life as an objective function is carried out under the condition of radial stiffness restraint. Under the Isight environment, the multi-island genetic algorithm is used to obtain the reasonable free surface geometric parameters of spherical hinge. The result shows that the fatigue life of the optimized spherical hinge suspender satisfies the design requirements. Finally, the stiffness tests for the original and optimized structures are individually carried out in a radial loading mode, the results show that the calculated stiffness coefficient of the optimized structure is consistent with the test one, and the simulation life is well fit with that obtained from the fatigue life curve.

1 INTRODUCTION

With the development of urbanization process, rail transit has become a main body of urban passenger transportation, but the vibration and noise problems caused by urban rail transit are not conducive to the implementation of the national standard *GB10070–1988: Standard of environmental vibration in urban area*. To improve the travelling comfortability and safety, it is necessary to add some vibration isolation elasticity supports on the rail vehicles. The spherical hinge rubber is a combination structure composing of a metal cylinder, a metal shaft and a rubber cylinder, and the rubber materials have the characteristics of reducing vibration and noise [1–3], slight wear [4,5] and long life [6,7]. Therefore, the spherical hinge rubbers are widely utilized in the rail transit vehicles.

As a rubber elastic shock absorber commonly used in rail vehicles, the fatigue life of spherical hinge suspender is one of the difficult problems restricting the development of rail transit [8].

Rong [9] investigated the influence of pre-compression on the radial stiffness and fatigue life of spherical hinge rubber based on the Ogden constitutive model, and found that a amount of proper circumferential pre-compression value could not only satisfy the matching relationship among the stiffness coefficients of spherical hinge in all directions, but also significantly improve the fatigue performance of spherical hinge, which indicated that the circumferential pre-compression value was a key factor in the spherical hinge rubber's optimization design. Xiang [10] gave a structural analysis and optimization for rubber mount of vehicle power train, and found that the structure shape, rubber cylinder geometry size and material properties of spherical hinge rubber have certain effects on the fatigue life of spherical hinge rubber. Based on the analysis of the ultimate load of V-type thrust rod, Shi [11] carried out a structural optimization design for spherical hinge rubber, and found that in order to delay the flow of rubber, the bulges or grooves could be manufactured on the metal-rubber contact surface, so as to decrease the stress concentration in the rubber cylinder, and thus improve the fatigue life of spherical hinge rubber.

In this paper, according to the fact that the fatigue life of a spherical hinge suspender in rail transit vehicles fails to satisfy with the actual application requirements, an optimization method based on the parameters of the rubber hyperelastic body free surface of the spherical hinge suspender is proposed. Selecting the free surface parameters of spherical hinge suspender as a design variable, the radial stiffness of spherical hinge suspender as a constraint condition, and the fatigue life of spherical hinge suspender as an optimization objective, the mathematical model is established and solved under Isight environment, so as to make the fatigue life of spherical hinge suspender connecting locomotive frame and axle box match the design requirements.

2 RADIAL STIFFNESS AND FATIGUE LIFE ANALYSIS OF THE INITIAL SPHERICAL HINGE

2.1 Materials and specimens

The spherical hinge suspender connecting the locomotive frame and the axle box body is composed of a metal cylinder, a metal shaft and a rubber cylinder. The front profile of the spherical hinge suspender is shown in Figure 1. Considering that the parts among the metal shaft, the rubber cylinder, and the metal cylinder are bonded together by using a vulcanization process, the bonding strength between the metal shaft and the rubber cylinder, and that between the rubber cylinder and the metal cylinder are large enough, so the contact surface between the metal shaft and the rubber cylinder, and that between rubber cylinder and metal cylinder are all set as a bonding constraint. Because a hexahedron element has obvious advantages over tetrahedron element in calculation accuracy, deformation characteristics, mesh number, degree of distortion resistance and number of meshing, in order to make the calculation results of spherical hinge suspender closer to the actual working conditions, the finite element model of spherical hinge suspender adopts the mesh model of hexahedron element, and the mesh density is set to 0.2 mm. In order to improve the computational efficiency of the finite element model, the chamfer of the metal shaft is simplified to a suitable angle. The metal parts of the model are simulated by the eight-node solid elements C3D8R, while the rubber part is simulated by the 3D eight-node hybrid elements C3D8H. The model is divided into 6979 elements, as shown in Figure 2.

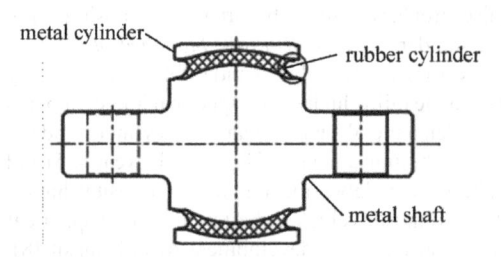

Figure 1. Front profile of the spherical hinge.

Figure 2. EM of spherical hinge.

According to the design requirements, the loading conditions and the design parameters of this type of spherical hinge suspender are shown in Table 1.

Table 1. The loading conditions and the design parameters for spherical hinge suspender.

Pre-compression value δ (mm)	Radial load F_r (kN)	Radial stiffness k_r (kN/mm)	Fatigue life N_f (cycle)
±1.5	±21.2	≥ 40	2000000

2.2 Rubber hyperelastic constitutive model

Rubber is a class of representative hyperelastic materials, which has a lots of constitutive models to describe its hyperelastic properties, such as the polynomial constitutive model based on the continuum mechanics theory and the constitutive model based on thermodynamic statistical theory. Considering the large deformation characteristics of rubber cylinder in the spherical hinge suspender, the modified Mooney-Rivlin model [6] is selected to describe the constitutive relationship of rubber material. The constitutive relation expressed in the form of strain energy density function W is written as

$$W = C_{10}(\bar{I}_1 - 3) + C_{01}(\bar{I}_2 - 3) + (J - 1)^2/D_1 \quad (1)$$

where, C_{10} and C_{01} are the material parameters, D_1 is the compressible parameter, $J=\det \boldsymbol{F}$, $\boldsymbol{F}$ is the deformation gradient tensor of rubber material, and

$$\bar{I}_1 = J^{-2/3} I_1, \quad \bar{I}_2 = J^{-4/3} I_2 \quad (2)$$

where, I_1 and I_2 are the first and second invariants of the strain tensor of rubber material, respectively. The material parameters of the rubber cylinder, the metal shaft, and the metal cylinder are individually listed in Table 2. For the metal shaft and the metal cylinder of the spherical hinge suspender, the linear elastic constitutive model is adopted in the finite element numerical simulations.

Table 2. The loading conditions and the design parameters of spherical hinge suspender.

Component	Element type	Constitutive model	Material parameters
Rubber cylinder	C3D8H	Mooney-Rivlin model	C_{10}=0.501 MPa C_{01}=0.123 MPa D_1=0.0001
Metal shaft	C3D8R	Linear elastic model	E=212 GPa μ=0.280
Metal cylinder	C3D8R	Linear elastic model	E=213 GPa μ=0.282

The strain amplitude has a great influence on the fatigue life of the spherical hinge suspender. With the increase of strain amplitude, it will accelerate the initiation of voids in the rubber materials and the peeling of rubber materials as well, resulting in a stress concentration, furthermore, it will increase the maximum principal strain of rubber materials, accelerate the oxidation reaction and ozone reaction of rubber materials, resulting in the reduction of fatigue life of rubber materials [12]. In order to evaluate the fatigue life of rubber products by Haigh curve, two indexes, strain amplitude ε_a and mean strain ε_m that can be determined by the maximum strain ε_{max} and the minimum strain ε_{min}, are needed. The strain amplitude ε_a is half of the difference between the maximum strain ε_{max} and the minimum strain ε_{min}, and the mean stress ε_m is half of the sum of the maximum strain ε_{max} and the minimum strain ε_{min}, i.e., $\varepsilon_a = (\varepsilon_{max} - \varepsilon_{min})/2$, and $\varepsilon_m = (\varepsilon_{max} + \varepsilon_{min})/2$.

The Haigh curve is fitted by the test data. As the strain amplitude is 0.35, the fatigue life vs. mean strain curve is obtained as shown in Figure 3. Considering the symmetry of spherical hinge suspender, the 1/4 structure model as shown in Figure 4 is used to study the distribution of strain amplitude and mean strain. The outer layer, middle layer and inner layer elements of rubber cylinder are analyzed, and 45 nodes are selected along the circumferential direction counterclockwise from 0° to 180°. The distribution of strain amplitude along the circumference direction of outer layer, middle layer and inner layer of rubber cylinder of spherical hinge suspender are individually shown in Figure 5, Figure 7 and Figure 9, the mean strains varying with the circumferential direction of outer layer, middle layer and inner layer of rubber cylinder are shown in Figure 6, Figure 8 and Figure 10, respectively.

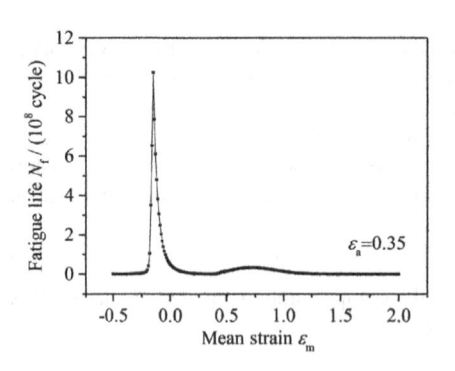

Figure 3. Haigh fatigue life curve at $\varepsilon_a = 0.35$.

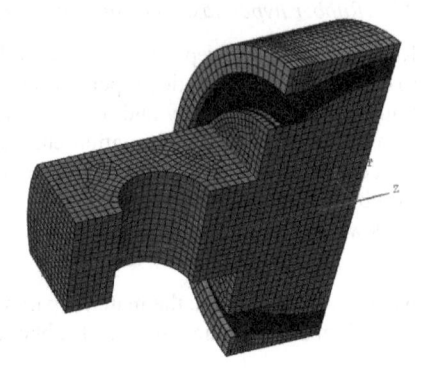

Figure 4. 1/4 model of spherical hinge.

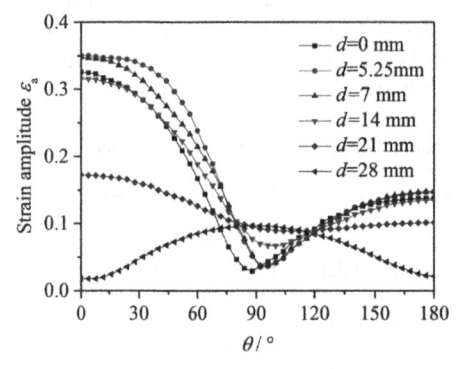

Figure 5. Strain amplitude on outer layer.

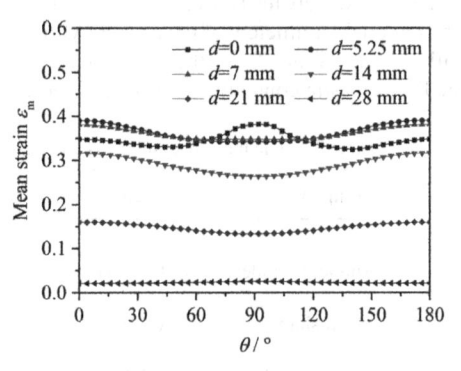

Figure 6. Mean strain on outer layer.

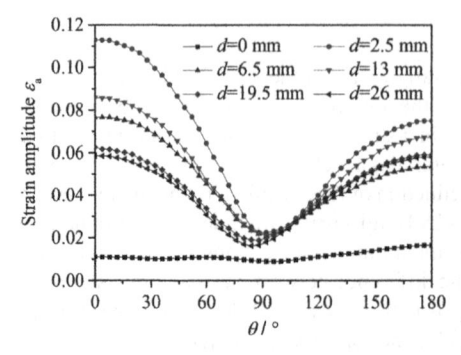

Figure 7. Strain amplitude in middle layer.

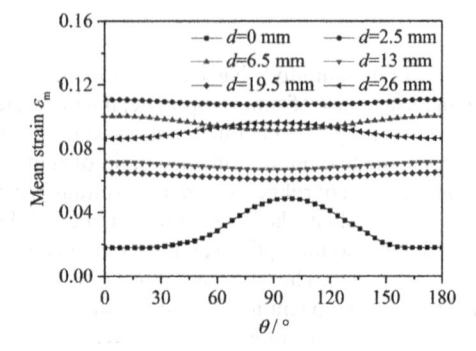

Figure 8. Mean strain in middle layer.

It is found from Figure 5 to Figure 10 that the strain amplitude of rubber cylinder is heterogeneous distributed along the circumference direction, and the mean strain of rubber cylinder is symmetrically distributed along 90 degrees. The maximum strain 0.35 and the mean strain 0.39 are found in the outer layer of the rubber cylinder at the distance 5.25mm away from the free end surface. It can be seen from Figure 7 and Figure 9 that the maximum strain amplitudes of the middle and inner layers of rubber cylinder are individually 0.11 and 0.33. Because rubber and metal are bonded together by using a vulcanization process, the deformations of metallic materials are relatively small compared with incompressible rubber body, which restricts the outward flowing of rubber body, resulting in the final dangerous point occurring in the metal-rubber contact area.

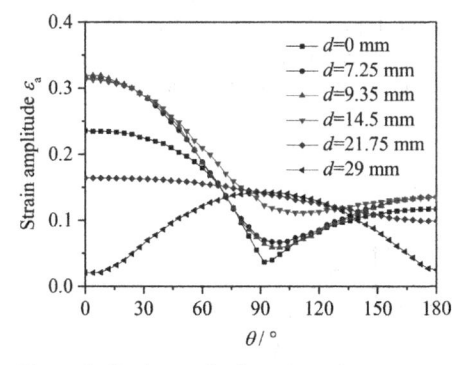

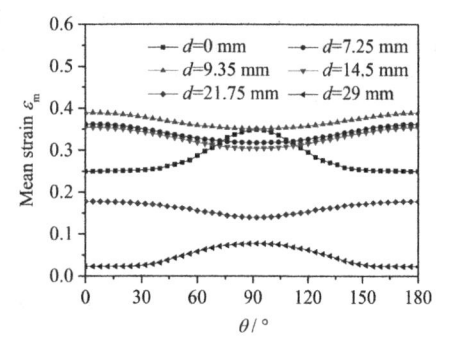

Figure 9. Strain amplitude on inner layer.

Figure 10. Mean strain on inner layer.

Simultaneously, it can be seen from Figure 9 and Figure 10 that at the distance 9.35 mm away from the inner layer of the rubber cylinder to the free end surface, the maximum strain and the mean strain are 0.33 and 0.39, respectively. So the fatigue failure location will not appear in the inner layer of the rubber cylinder. Furthermore, it can be seen from Figure 7 that the strain amplitude of the middle layer of rubber cylinder ranges from 0.01 to 0.11. Comparing with the strain amplitudes of the outer and inner layer of rubber cylinder, the strain amplitude is smaller. Based on the above analysis, the fatigue failure location appear in the outer layer of rubber cylinder at angle 0 and distance away from the free end surface 5.25 mm, and the corresponding strain amplitude and mean values are 0.35 and 0.39, respectively. According to the fatigue life curve described in Figure 4, the corresponding fatigue life at $\varepsilon_a = 0.35$ and at $\varepsilon_m = 0.39$ is about 1.91×10^6 cycles, while the design fatigue life of such a spherical hinge suspender is 2.00×10^6 cycles, which can not satisfy the design requirements.

3 PARAMETER OPTIMIZATION OF FREE SURFACE OF SPHERICAL HINGE SUSPENDER

3.1 *Optimization parameters and mathematical optimization model*

The Isight, SolidWorks, Hypermesh and ABAQUS software are used to optimize the free surface of spherical hinge. Firstly, a 3D model of the spherical hinge is constructed by using the SolidWorks software. Then, the 3D model is introduced into the ABAQUS software, the finite element analysis is carried out. The variables, such as support reactions and displacements, are derived, and the structural stiffness is calculated. Finally, the finite element analysis results are transmitted to the Isight software to operate the numerical tests. The five parameters controlling the free surface of rubber cylinder, as shown in Figure 11, are selected as the design variables.

The five free surface parameters affecting the fatigue life of spherical hinge suspender are set as the design variables, the radial stiffness K is determined as the constraint, and the fatigue life N_f is taken as an objective function. Thus, the following mathematical optimization models are established, i.e.,

$$\max (N_f) = f(H_1, H_2, H_3, L_1, L_2) \tag{3}$$

where, H_1 is the outer edge height of rubber bottom, whose value ranges from 0.2 mm to 0.8 mm. H_2 is the height of inflection point 2 to rubber bottom, whose value ranges from 1.4 mm to 2.0 mm. H_3 is the outer edge height of rubber top, whose value ranges from 0.2 mm to 0.8 mm. L_1 is the length from inflection point 1 to rubber outer edge, whose value ranges from 2.5 mm to 3.5 mm, L_2 is the length from inflection point 2 to rubber outer edge, whose value ranges from 1.2 mm to 1.8 mm, and the constraint condition is the radial stiffness $K \geq 40\text{kN/mm}$.

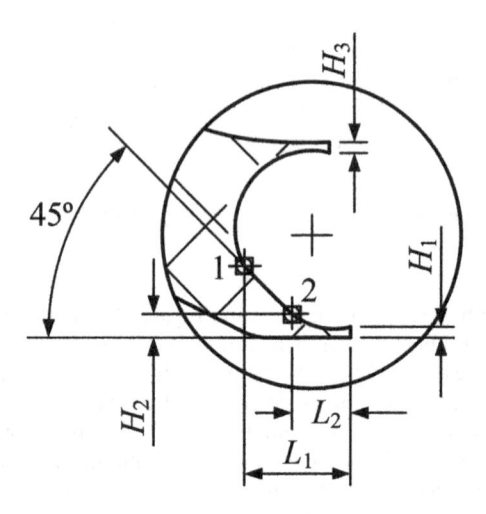

Figure 9. The parameters of free surface.

3.2 Optimization design under the Isight environment

The Optimized Latin Hypercube Sampling Method is selected as a numerical test design scheme, and 24 samples for the five design variables are tested in this paper. The approximate model is a second-order response surface model, the geometrical parameter values of test samples, and the corresponding calculated fatigue life values are listed in Table 3.

In this research, the Adaptive Multi-island Genetic Algorithm (AMGA) is adopted to carry out the optimization design of the spherical hinge suspender. After about 1000 times iteration, the initial and optimal geometrical parameter values of free surface of the rubber cylinder are listed in Table 4. The initial value of radial stiffness is set as 38kN/mm, the objective value is larger than 40 kN/mm, and the optimal value obtained by using Isight is 41kN/mm. The initial value of fatigue life is set as 1.91×10^6 cycles, the objective value is set as 2×10^6 cycles, and the optimal value obtained by using Isight is 2.34×10^6 cycles.

Table 3. The test samples and the calculated fatigue life values.

Sample No.	H_1 (mm)	H_2 (mm)	H_3 (mm)	L_1 (mm)	L_2 (mm)	K (kN/mm)	N_f (10^6 cycle)
1	0.226	1.765	0.252	2.848	1.435	45	1.31
2	0.670	1.817	0.722	3.022	1.278	46	1.72
3	0.330	1.687	0.617	2.630	1.252	38	1.30
4	0.357	1.791	0.461	2.500	1.774	41	1.46
5	0.696	1.609	0.487	2.978	1.800	37	1.50
6	0.200	1.661	0.513	3.152	1.722	47	1.26
7	0.513	1.400	0.748	3.022	1.330	32	1.80
8	0.722	1.974	0.565	2.717	1.617	49	1.89
9	0.643	1.530	0.330	2.935	1.226	37	1.67
10	0.252	1.478	0.435	3.239	1.304	42	1.38
11	0.565	1.948	0.357	2.804	1.200	40	1.48
12	0.461	1.922	0.278	3.109	1.748	34	1.75
13	0.748	1.557	0.513	3.500	1.409	47	1.90
14	0.591	1.896	0.670	3.370	1.670	36	1.67
15	0.800	1.843	0.304	3.196	1.487	45	1.39
16	0.774	1.583	0.591	2.543	1.461	43	2.05
17	0.409	1.870	0.383	3.457	1.357	46	1.78
18	0.487	1.452	0.696	3.413	1.696	42	1.58
19	0.383	1.426	0.409	2.674	1.565	37	2.01
20	0.435	1.635	0.800	2.761	1.643	38	1.78
21	0.513	1.504	0.226	3.283	1.591	39	1.37
22	0.304	2.000	0.643	2.891	1.513	41	1.48
23	0.278	1.713	0.774	3.326	1.383	43	1.51
24	0.617	1.713	0.200	2.587	1.513	48	1.35

Table 4. The comparison between initial and optimal geometrical parameters values.

Optimization parameter	H_1 (mm)	H_2 (mm)	H_3 (mm)	L_1 (mm)	L_2 (mm)
Initial value	0.500	1.700	0.500	3.000	1.500
Optimal value	0.719	1.402	0.711	2.508	1.457

3.3 Test verification

According to the geometrical dimensions before and after optimization, the spherical hinge rubber was processed, in which rubber cylinders were made of rubber materials with Shaw hardness of 65 degrees. In order to eliminate the influence of temperature on the mechanical properties of rubber, the spherical hinge rubber specimens were kept at constant temperature for 24 hours at 23 ± 2°C. Considering the test cost, the test conditions were that the pre-compression value of outer metal cylinder is of 1.5mm, and the radial load is of 30kN. The comparison between the radial stiffness curves of the spherical hinge rubber suspenders before and after optimization are shown in Figure 12.

It can be found from Figure 12 that as the load is 21.20kN, the test value of radial stiffness of the spherical hinge suspender is 43kN/mm. Under the Isight environment condition, the radial stiffness of the optimized spherical hinge suspender is 41kN/mm, and the relative error between the numerical results of radial stiffness and the test value is only 4.65%. The radial stiffness of the original spherical hinge suspender does not match the design requirements, while the radial stiffness of the optimized spherical hinge suspender fully satisfies the design requirement.

Figure 12. Load vs. displacement curve.

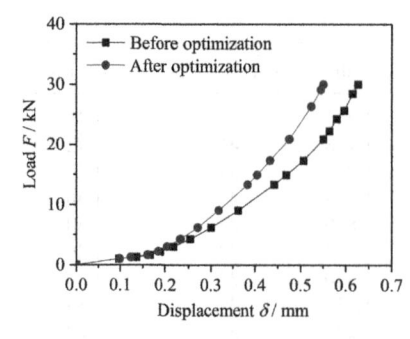

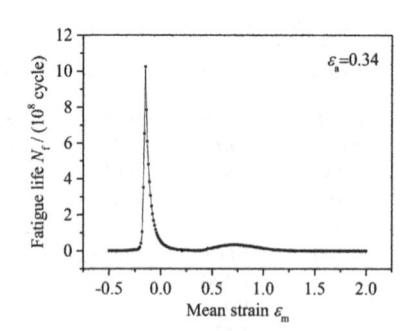

Figure 13. Haigh fatigue life curve at $\varepsilon_a = 0.34$.

According to the rubber fatigue test, as the strain amplitude is of 0.34, the Hiagh fatigue life curve of rubber material is shown in Figure 13. The strain amplitude and mean strain distribution of the optimized rubber cylinder of spherical hinge suspender are individually shown in Figure 14 and Figure 15. It can be seen from Figure 14 that the maximum strain amplitude of the optimized rubber cylinder is 0.34, and the corresponding average strain is 0.40 (as shown in Figure 15). According to the fatigue life vs. mean strain curve of rubber shown in Figure 13, the optimized fatigue life of spherical hinge suspender is 2.25×10^6 cycles, which is very close to the optimal value of fatigue life under the Isight environment, and its relative error is only 3.85%.

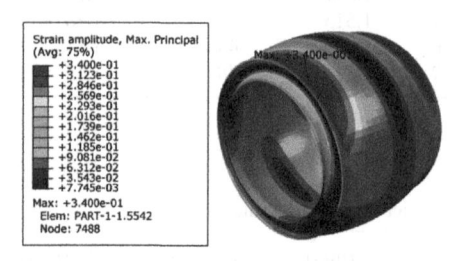

Figure 14. Distribution of strain amplitude.

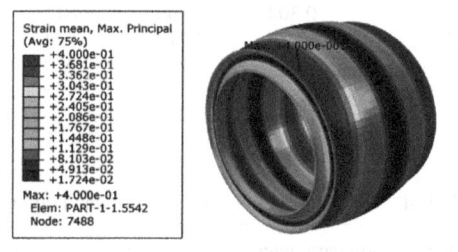

Figure 15. Distribution of mean strain.

4 CONCLUSIONS

The fatigue life of a spherical hinge suspender are studied by using the finite element numerical simulation method, and then the optimization design for the free surface parameters of this structure are carried out under the condition of radial stiffness constraint. The conclusions are given as follows:

(a) The distributions of strain amplitude and mean strain along the circumferential direction of the original spherical hinge suspender are studied. The results show that the strain amplitude and mean strain of the rubber cylinder of spherical hinge suspender are inhomogeneous along the circumferential direction. The maximum strain amplitude appears at 5.25 mm away from the outer layer of rubber cylinder to the free end surface. The corresponding strain amplitude and mean strain values are individually 0.35 and 0.39, and the fatigue life is 1.91×10^6 cycles, which fails to match the fatigue life requirement of the spherical hinge rubber suspender.

(b) The five free surface parameters of rubber cylinder affecting the fatigue life of spherical hinge suspender are set as the design variables. Under the Isight environment condition, the Adaptive Multi-island Genetic Algorithm is used to carry out the optimization design of the spherical hinge suspender. The calculated result show that the fatigue life of spherical hinge suspender designed according to optimal parameter is 2.34×10^6 cycles, which fully satisfies with the design fatigue life, and the optimal result is verified by the test stiffness curve of the spherical hinge suspender.

ACKNOWLEDGEMENT

This work was supported by the High-level Talent Gathering Project of Hunan Province (no. 2019RS1059), the Key Project of the Reform of the Degree and Postgraduate Education in Hunan Province (no. 2019JGZD034), and the Research Project for Postgraduate Students of Hunan Province (no. CX20200648).

REFERENCES

[1] Xia E L, Cao Z L, Zhu X W, Qiu S W, Xue Z G, He H and Li L X 2021 *Applied Acoustics* **175** 107780
[2] Gong D, Duan Y, Wang K and Zhou J S 2019 *Journal of Sound and Vibration* **449** 121
[3] Wang J, Peng L Q, Hou H B and Lin D W 2006 *World Rubber Industry* **33** 22
[4] Khafidh M, Setiyana B, Jamari J, Masen M A and Schipper D J 2018 *Wear* **412–413** 23
[5] Hakami F, Pramanik A, Ridgway N and Basak A K 2017 *Tribology International* **111** 148
[6] Xu X Q, Zhou X H and Liu Y Q 2021 *Engineering Structure* **227** 111449
[7] Xu Y J, Liu Y B, Kan C Z, Shen Z H and Shi Z M 2009 *Ocean Engineering* **36** 588
[8] Chang H, Cheng H T and Huang Y J 2015 *Rolling Stock* **53** 5
[9] Rong J G, Huang Y J, Tang X H and Yang J 2006 *Special Purpose Rubber Products* **27** 36
[10] Xiang Y X 2008 *Structure analysis and optimization of rubber mount for vehicle power train* (Master thesis: Chongqing University)
[11] Shi W K, Ke J, Wang Q, Teng T, Zhou Y F, Zhu Y, Liu T Y, Wu Z Y and Dong Y W 2013 *Journal of Xi'an Jiaotong University* **47** 132
[12] Mars W V 2001 *Tire Science and Technology* **29** 171

Using boron-doped diamond sensor to detect heavy metal in water

L. Pang
School of Economics and Management, China University of Geosciences (Beijing), Beijing, China

X. Yu & X. Qian
School of Materials Science and Technology, China University of Geosciences (Beijing), Beijing, China

T.Y. Yan & L.Q. Chen
School of Economics and Management, China University of Geosciences (Beijing), Beijing, China

H.Z. Li
School of Materials Science and Technology, China University of Geosciences (Beijing), Beijing, China

ABSTRACT: Detection of heavy metal pollutants in water has become a hot topic in the field of environmental protection. Rapid water detector is regarded as a unique measure to present the first-hand data for the detection and early warning. Available water detector has to use mercury film sensor as detection electrode, and subjects to low accuracy and secondary pollution. This work intends to use the BDD (Boron-doped diamond) film sensor, which is green and efficient, to replace the traditional mercury film sensors to explore the application of a new type of water quality detector in industry. An on-line quality detection system is constructed, and the detector system is connected with a data platform through network. The detected results are converted into digital signals, and are displayed on a mobile application software. With merits of fast detection and stable performance, it is rational to suppose that the designed system may provide data support and application reference for developing advanced devices for water heavy metal detection

1 INTRODUCTION

Heavy metal pollutants in water have a wide range of sources and cause serious harm. The pollutants may come from industrial production, mining metallurgy and domestic sewage, and are rather difficult to be degraded by microorganisms after entering the water. Such pollutants may endanger the aquatic ecological environment, and even harm human health through the food chain [1]. Online monitor of heavy metal pollutants is regarded as an important measure to prevent the pollutants. Rapid water detector is popular in water detection market due to its rapid detection and convenience.

Electrode is a core component of commercial detectors. Available detectors commonly use mercury film, and suffer from low accuracy and secondary pollution. BDD (Boron-doped diamond) film sensors have attracted much attention due to their high accuracy and environmental protection. Shi, D., et al. introduced the application of BDD electrode in seawater salinity detection [2]; Wang, P., et al. discussed the application of graphene modified BDD electrode in the detection of acetaminophen [3]; Ma, Z., et al. discussed the application of gold modified BDD electrode in the detection of clenbuterol [4]. Unfortunately, there are few studies on the application of BDD electrodes in the detection of heavy metals in water.

This works attempts to introduce BDD sensor to the field of water heavy metal detection. Using BDD to replace mercury film, a device for rapid detection of heavy metal pollutants in water is designed. With internet connecting device and data platform, the platform can convert the detected results into digital signals, and display on client in the user's application account. Output of this work may facilitate the development of BDD water detector, so as to meet the requirements of green and efficient detection in the field of online monitoring.

2 APPLICATION OF BBD SENSOR

ASV (Anodic stripping voltammetry) was used to detect heavy metal pollutants in water. The analysis process of heavy metal ions to be detected occurred at the electrode/solution interface. The electrode materials and their surface properties thus directly affect the selectivity and redox reactions of ions to be detected [5]. With a wide potential window and low background current, BDD sensor can simultaneously detect multiple heavy metal pollutants, and is expected to meet the requirements of efficient and accurate detections [6]. Graphene/BDD composite sensor was prepared by hot filament chemical vapor deposition and in-situ modification [7]. In the process of detecting ions by carrying BDD sensor in ASV method, the ions are enriched and dissolved on the surface of BDD, and then the dissolved signal is transmitted to the data platform to obtain detection information. The dissolution process ensures high electrochemical response and high reproducibility, which is the guarantee to obtain accurate detection information. For this reason, BDD electrochemical sensor is used to detect heavy metal pollutants in water.

With highly developed Internet, the core detection technology can be combined with the data platform. The data platform may make unified digital management, centralized storage of data resources and advanced analysis capabilities available. Its functions include data acquisition, data calculation, data analysis, data storage and data visualization. The new detector and data platform are connected by Internet, and the detection information is transmitted in real time. The results are displayed on the data platform, which is convenient for users to understand and judge water quality.

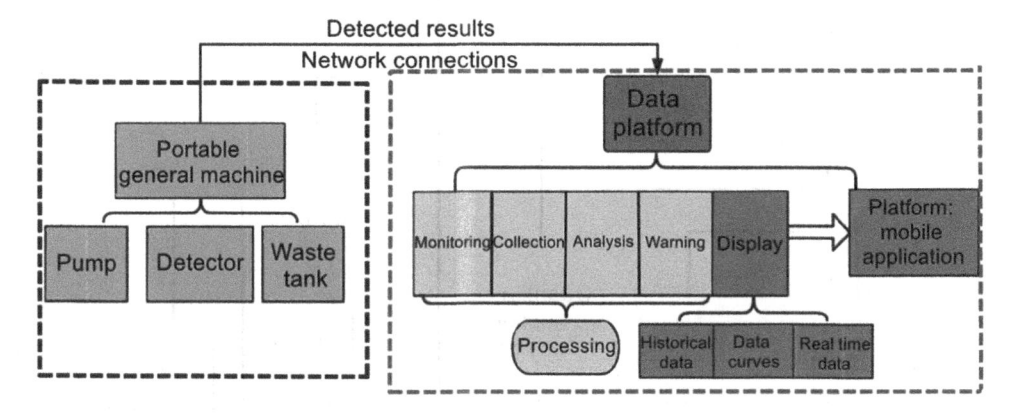

Figure 1. Structure diagram of detection system.

Figure 1 shows the schematic diagram of the detection system structure. As can be seen from Figure 1, the detection system consists of two parts: the first part is the portable general machine, and the second part is the data platform. The portable general machine and the data platform are connected through the network, and the detection information is transmitted from the machine to the platform and displayed in the form of data on the platform. The portable general machine consists of three parts: quantitative extraction device, water quality detection device and waste tank. A certain amount of water samples is extracted by the extraction device and pumped into the water quality detection device. The water quality detection device is composed of BDD film sensor as the electrode reactor, which can detect heavy metal ions in water. The waste tank is used for filling the waste liquid after detecting. Data platform includes two parts: data processing and data display. Data processing consists of real-time monitoring, data acquisition, data analysis and data early warning. The data display is composed of historical data, real-time data and data curve, which can be displayed on mobile application.

3 MAIN SYSTEM COMPONENTS

In the water detection system, the detection device is a critical part of the portable general machine, and whose two main components are quantitative extraction device and detector. In addition, the data platform contains data processing and data display.

3.1 *Water quality detection device*

Figure 2 shows the structure of the detection device. As is shown in Figure 2, the red device is extraction pump, which can extract the external water. The green device is quantitative extraction device, which can suction quantitative solution for subsequent detection. The blue device is detector (reactor), which can detect heavy metal ions. The yellow device 1–6 is working switch, which can pump in different liquids as required. Pumping water with the extraction pump, switch 2 controls the standard solution, switch 3 controls the de-ionized water, switch 4 connects to the air, and switch 6 collects wastes to the waste tank. The working principle of detection device is like this. Use deionized water to clean the detection device for detecting ions in water and in standard solution. Collect the waste liquid into the waste tank.

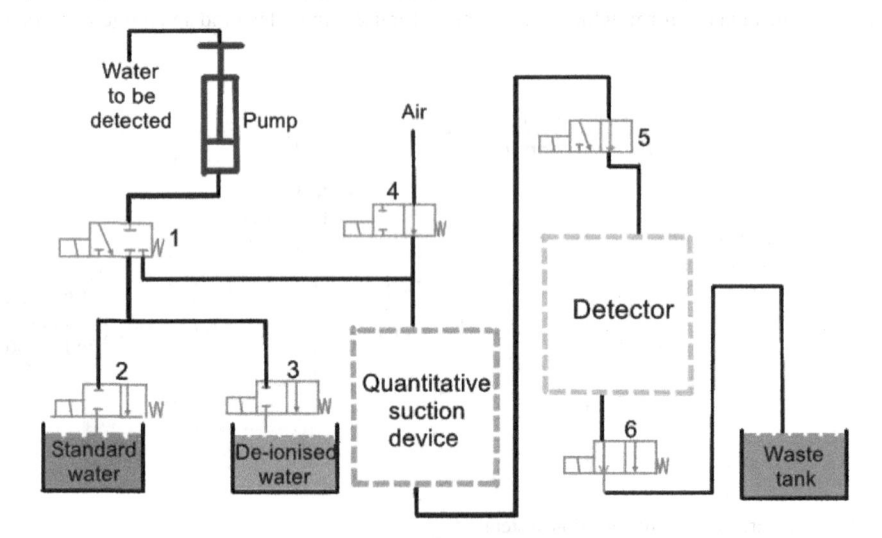

Figure 2. Structure drawing of detection device of water.

Using deionized water to clean detection device has four steps. (1) Turn on switch 1 and switch 3 to extract deionized water connected to switch 3. (2) Then, turn off switch 1 and switch 3, turn on switch 4 and pump deionized water into the extraction device driven by air. (3) After cleaning the extraction device, turn on switch 5 and pump the deionized water into the detector (reactor) from the device. (4) After cleaning the detector, turn on switch 6 and drain the deionized water into the waste tank. It should be noted that switches 4, 5, 6 need to be normally opened to drive liquid, and can be closed at the end of the work.

The detection of heavy metal ions in the water or standard solution includes three procedures. (1) Turn on switch 1 to connect the extraction pump to extract water, or turn on switch 2 to extract the standard solution. (2) Turn off the switches above, and turn on switch 4, to pump the liquid into the extraction device driven by air. (3) Turn on switch 5, use extraction device to pump a certain amount of liquid into the detector reactor for detecting heavy metal ions. After detecting, turn on switch 6, drain the liquid into the waste tank, complete the detection, and turn off the switch.

3.2 Quantitative extraction device and detector

Figure 3 shows the structure of extraction device. As shown in Figure 3, two pallets 2 are fixed on the backboard 8, the glass tube 1 is fixed between the two pallets, and the screw 4 and the support screw 5 are fixed between the two pallets by thread connection. Sensor frame 3 is fixed on screw 4 through two fastening nuts 9, and maintains balance by support screw 5. Illuminant 6 and photoelectric switch 7 are arranged in the corresponding holes of the sensor frame. Similarly three groups are settled on the frame. The whole quantitative system has three liquid labels, and the calibrated liquid volume is obtained by multiple injections of three determined volumes. The minimum injection is the volume difference between the left photoelectric switch and the middle photoelectric switch.

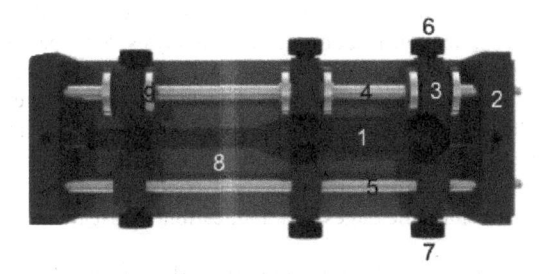

Figure 3. Structure diagram of quantitative extraction device 1. glass tube 2. pallet 3. sensor frame 4. screw 5. support screw 6. Illuminant 7. photoelectric switch 8. backboard 9. fastening nut.

Figure 4 shows the structure of the detector. In Figure 4, the stirring motor 1 is connected with the stirring rod 2, and the stirring rod 2 is inserted into the upper part of the reactor 4. The stirring rod 2 can stir deionized water and clean the reactor after detection. The reference electrode 3 is tilted on the upper side of the reactor, and the working electrode 5 is installed horizontally in order to replace easily. The auxiliary electrode 6 is placed at the bottom of the reactor. Reference electrode, working electrode and auxiliary electrode constitute three electrodes, and three electrodes and stirring rod 2 avoid each other. Once the detection is completed, the liquid is discharged from the outlet 7 into the waste tank.

Figure 4. Structure of extraction device 1. stirring motor 2. stirring rod 3. reference electrode 4. Reactor 5. working electrode 6. auxiliary electrode 7. Outlet

The three electrodes consist of two circuits. Reference electrode, controllable power supply and working electrode constitute the first circuit. Using the first circuit can measure/control the interface potential difference between working electrode and solution, and implement heavy metal ion enrichment and electrode scanning. Introducing an auxiliary electrode may detect the dissolution current of heavy metals in the circuit. The auxiliary electrode, polarization power supply and working electrode constitute the second circuit, and set a sensitive current meter in the circuit. The dissolution current may form the dissolution scanning spectrum, and the current magnitude corresponds to the working electrode scanning during detection reaction.

3.3 Data processing module

Data processing module includes four parts: data monitoring, data collection, data analysis and data early warning. (1) Data monitoring. Users can inspect and monitor the detecting data of the water by log in the application account in time. (2) Data collection. Data processing module can convert detect results into data, which is connected with the detector through network and data transmission device. Data processing module has the merits of collect and store data, and the users can check this data on the application accounts. (3) Data analysis. After passing the data processing module, the data is transmitted to the data display platform. The platform converts the collected data into digital signals. With support of digital simulation technology, the images can be automatically generated. Comparing historical data and real-time data, it is feasible to automatically draw the data curve. (4) Data early warning. The data platform analyses the data curve and compares them with the national/enterprise standards. Data warning will be carried out if the results exceed the standards. Warning ways include messages and application notifications to remind users timely and effectively.

The data platform and mobile application supporting the detector have many functions such as real-time monitoring, data acquisition and data analysis. In support of the data platform, the green and effective detection of various heavy metal pollutants in water can be realized, and it is convenient for users to check water quality anytime and anywhere. The information disclosure part of application enables people to understand the water quality in different places, and is conducive to mutual supervision.

3.4 Data display module

The data display module consists of the monitor of the data platform, and the detection device is connected by the network. The detection device is based on the principle of ASV method. The detected results are converted into digital signals through data processing, and specific parameters are generated on the monitor. Using digital simulation technology, images can be automatically generated on mobile application. Users log on accounts can check relevant data. The detection results of heavy metal ions in water can be seen like Figure 5.

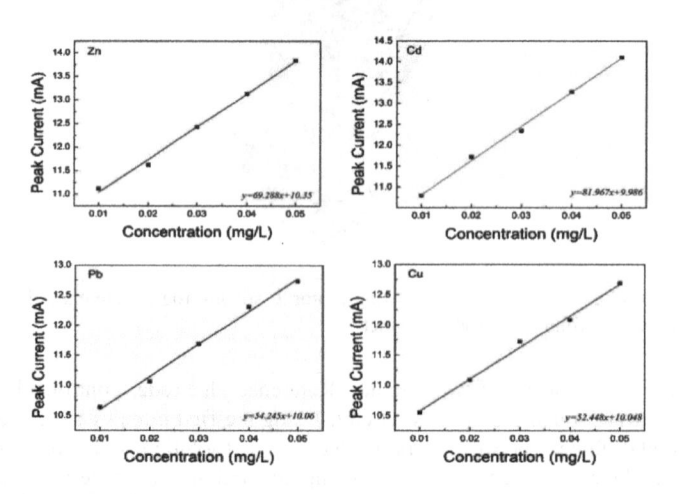

Figure 5. Detection results of heavy metal ions in water.

4 CONCLUSION

This work attempts to conduct an application research of BDD sensors for heavy metal detection in water. In cooperation with the water quality detector company, an advanced detection system was designed with BDD film as the electrode to overcome the shortcomings of the existing water quality detector. This work may provide theoretical guidance and data support for the application of heavy metal detection in water.

Based on the ASV, the feasibility of the detection system for heavy metal detection in water is discussed. The main work includes three parts: (1) BDD sensor is applied to the system of heavy metal detection in water. The BDD is combined with the Internet to facilitate people to observe and judge water quality. (2) Discuss the structure of water quality detection system. Water quality detection device is an important part of the portable general machine. Quantitative extraction device and detector are the key components of water quality detection device. Data processing and data display are two important components of the data platform. (3) The system makes online detection of heavy metal ions available. The detected results are converted into digital signals through data processing module, and specific parameters are generated on the display to facilitate the users to view the detection data.

ACKNOWLEDGMENTS

This work has been supported by funds from the Class D entrepreneurship practice project "LvDun company–Detection and early warning of heavy metal pollutants in water" of Innovation and entrepreneurship training project for College Students and the project "Practical Training Program "(289) for the cross-training of high-level talents in Beijing institutions of higher learning.

REFERENCES

[1] Borrill A, Reily N and Macpherson J 2019 *Analyst* **144** 6834
[2] Shi D, Huang N and Liu L 2020 *Appl. Surf. Sci.* **512** 145652
[3] Wang P, Yuan X and Cui Z 2021 *ACS Omega* **6** 6326
[4] Ma Z, Wang Q, Gao N and Li H. 2020 *Microchem. J.* **157** 104911
[5] Shah A, Sultan S and Zahid A. 2017 *Electrochim. Acta* **258** 1397
[6] Silwana B, Horst van der C and Iwuoha E 2016 *Electroanal.* **28** 1597
[7] Pei J X , Yu X, Zhang C, Liu X J 2020 *Appl. Surf. Sci.* **527** 146761

Pore defects and process control of pure molybdenum using wire arc additive manufacturing

Y.A. Qiao
School of Mechanical Engineering, Beijing Institute of Technology, Beijing, China

J.C. Wang
Institute of Advanced Structure Technology, Beijing Institute of Technology, Beijing, China

S.Y. Tang & C.M. Liu
School of Mechanical Engineering, Beijing Institute of Technology, Beijing, China

ABSTRACT: The deposits of pure molybdenum(Mo) by wire arc additive manufacturing (WAAM) were carried out with different arc parameters, and the microstructure and internal pore defects of the Mo deposit were analyzed. The results showed that there were pore defects in the pure molybdenum WAAM deposits and that the density of those deposits tended to increase with the increase of the heat input. However, having a huge background level could greatly reduce the density. Meanwhile, remelting could significantly reduce the porosity and increase the density of the pure Mo deposit.

1 INTRODUCTION

Molybdenum (Mo) is a refractory metal, and it is widely used in aerospace[1,2], nuclear[3–6], electronics industry[7], and other industry fields because of its high melting point, good mechanical properties, good thermal and electrical conductivity, and low coefficient of linear expansion[8,9]. However, Mo and its alloys have poor processing properties due to their high melting point and low plastic-brittle transition temperature, limiting the applications. Additive manufacturing (AM) provides a new method to fabricate molybdenum, but there are many defects and cracks in samples fabricated by AM[10]. This paper analyzes the mechanism of pore defects development in pure Mo fabricated by wire arc additive manufacturing (WAAM).

2 EXPERIMENTAL METHODS

The wire and substrate used in the experiment were Mo-1 prepared by powder metallurgy, with a wire diameter of 1.2 mm and a size of 75mm×75mm×5mm. The composition of Mo-1 is shown in Table 1. The substrate was polished and then cleaned with acetone before the experiment. The WAAM equipment was designed by the Beijing Institute of Technology (Figure 1) and was mainly composed of a computer numerical control (Sinunerik 828D of SIEMENS), a machine unit, a gas tungsten arc welding (GTAW) machine (Dynasty 700 of Miller), wire feeder control (Series 9700 of Jetline), and an argon protection. The experiment was protected by argon atmosphere. The purity of the argon was 99.999%, and the flow rate is 20 L/min. The processing parameters is shown in Table 2. Experiments 1 to 5 studied the influence that different background levels had on the pores in the deposit, while experiments 2 and 6 to 8 studied the influence that different peak times had on the pores in the deposit.

The density of the experimental samples was measured refer to the standard of GB/T1423–1996. The samples were take out of the top 15mm of the cylindrical deposit by Electrical Discharge Machining (EDM). The metallographic samples were polished with SiC papers and then corroded using the mixture solution of $NaOH:K_3[Fe(CN)_6]:H_2O$ with a ratio of 1:1:8. The microstructure was characterized through the optical microscope (Make: CEWEI, Shanghai, China, Model: LW600LJT).

Table 1. Mo-1 chemical composition (wt%).

Mo	Al	Ca	Fe	Mg	Ni	Si	N	C	O
Balance	<0.002	<0.002	<0.01	<0.002	<0.005	<0.01	<0.003	<0.01	<0.008

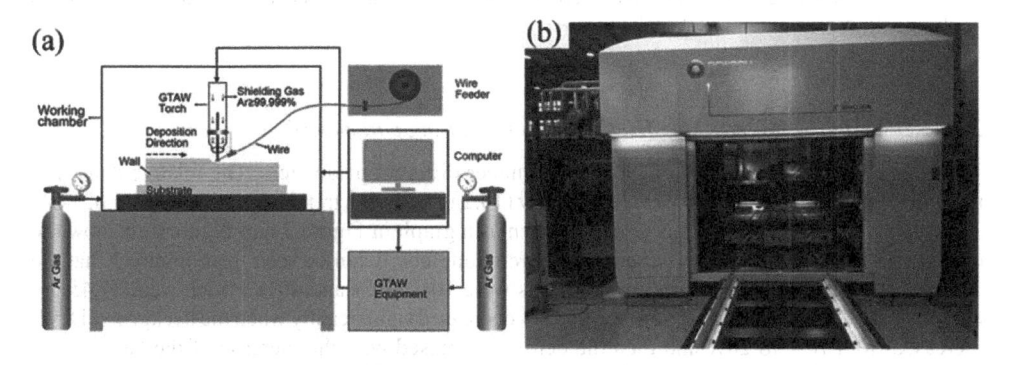

Figure 1. The WAAM equipment from the Beijing Institute of Technology.

Table 2. Parameter of the experiment.

number	Peak current(A)	Peak time(%)	Background level (A)	Wire feeding (cm/min)
1	300	10	3	30
2	300	10	6	30
3	300	10	9	30
4	300	10	12	30
5	300	10	15	30
6	300	20	6	30
7	300	30	6	30
8	300	50	6	30

3 RESULTS AND DISCUSSION

The samples of WAAM under different parameters are shown in Figure 2. Figure 2(a) shows the deposits under different background levels, with the experiments from left to right being experiments 1 to 5; Figure 2(b) shows the deposits under the action of different peak times and from left to right they are experiments 6–8. It can be found that the diameter of the sample increases with the increase of the background level and peak time. The reason for this is that as the heat input increases, the surface tension of the molten pool decreases, the melt metal will flow easily.

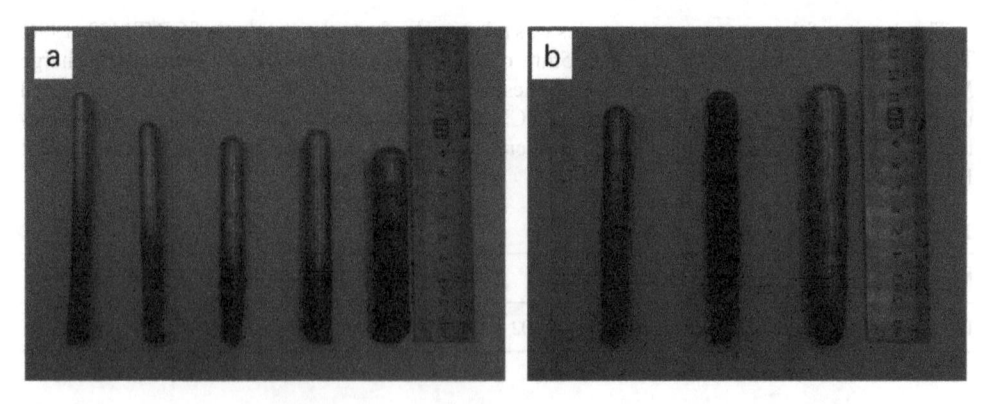

Figure 2. WAAM pure Mo samples under different parameters: (a)Deposits under different background levels; (b) Deposits under different peak times.

3.1 *Density of WAAM pure Mo under different processing parameters*

Table 3 and 4 show the density of the samples under different parameters. The relative density is the ratio of the actual density to pure Mo density ($10.2g/cm^3$). The trend in relative density of Table 3 and Table 4 with the arc parameters are shown as a graph in Figure 3 and Figure 4 to show the change of density with process parameters more intuitively. It can be seen from Figure 3 that the density was higher when the background levels were 10%, 30% and 40%, which were 87.451%, 87.745% and 88.235%, respectively. The density decreased significantly when the background level increased from 10% to 20% and then the density increased with the increase of the background level. But when the background level increased from 40% to 50%, the deposit density dropped to the bottom sharply. As can be seen from Figure 4, the density of the WAAM deposits increased with the increase of peak time, with the maximum density being 94.314% when the peak time was 50%.

Table 3. Density of deposits under different background levels.

Number	Actual density(g/cm^3)
1	8.92
2	8.56
3	8.95
4	9.00
5	8.43

Table 4. Density of deposits under different peak times.

Number	Actual density (g/cm^3)	Relative density (%)
2	8.56	83.922
6	8.86	86.863
7	9.01	88.333
8	9.62	94.314

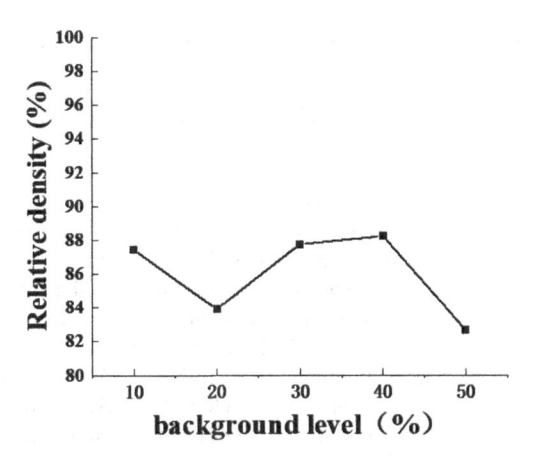

Figure 3. Relative density versus background level.

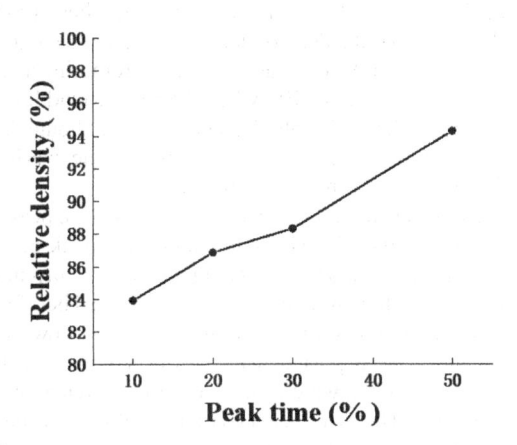

Figure 4. Relative density versus peak time.

In the WAAM process, holes are a relatively common defect and the main reason for the decrease of the density of the deposit. Holes defects in WAAM can be divided into two types. One type is caused by poor interlayer fusion, the shape and size of these holes are irregular, and they are mainly distributed at the overlap between layers. The second is the pores, which have a relatively regular shape, generally spherical or ellipsoidal, and a very smooth inner wall. During the pure Mo WAAM process, the low melting point composition (such as O, N, Ni, Al, Mg, Ga, Fe, etc.) inside the metal wire was volatilized under the action of arc temperature (the temperature of the molten pool can surpass 3000K). It will bubble and then gather and grow in the molten pool. When the bubbles reach to a certain size, they rise due to buoyancy until finally overflowing over the surface of the molten pool. The bubbles that did not overflow in time are the ones that form pores when the molten pool cools and solidifies. Whether the bubbles are able to overflow smoothly depends on the rising speed V_e of the bubbles in the molten pool and the solidification speed V_c of the molten metal in the molten pool. When $V_e > V_c$, the bubbles are able to smoothly overflow out of the molten pool. Conversely, when $V_e < V_c$, the bubbles will be left in the molten pool and form pores as the molten metal solidifies. Generally, V_e can be calculated by the Stocks formula, which is expressed as:

$$V_e = \frac{K(\rho_l - \rho_g)gR^2}{\eta}$$

where K is the coefficient; ρ_l is the density of the molten metal; ρ_g is the density of the bubble; g is the gravitational acceleration, R is the radius of the bubble, and η is the viscosity of the molten metal. According to this formula, the larger the bubble radius and the lower the viscosity of the molten pool, the faster the bubble rising speed. As the heat input of the arc increases, the viscosity of the molten metal in the molten pool decreases, and thus the bubbles are more likely to overflow. Therefore, the quantity of pores decreases as the heat input increases.

The reasons for the density change in Figure 3 are as follows: in WAAM process, the background level stage is the cooling stage of the molten pool. When the background level is too small (such as 10%), the molten pool cools quickly and the temperature of the molten pool is very low. The impurities in the raw materials don't have sufficient time to vaporize and form bubbles, so the deposits are denser. When the background level increases to 20%, the temperature of the molten pool rises; and the impurities begin to vaporize to form bubbles. However, as the viscosity of the molten pool is high at this time, according to the Stocks formula, the bubble rising speed V_e is slow. However, V_c is fast at this background level and so, a large number of bubbles become trapped in the deposit, meaning therefore that the density decreases. As the background level continues to increase, V_e increases accordingly, but V_c decreases, thereby promoting a large number of bubbles to overflow, and the density to thus increase. But when the background level continues to increase to 50%, the density of the deposit drops sharply. This is because although V_e is larger at this time, the current is so strong that the powerful arc force interferes with the overflow of the bubbles from the molten pool, which then results in a sharp drop in density.

The reasons for the change in density in Figure 4 are as follows: the peak time stage is WAAM's wire melting stage. At this stage, the Mo wire melts under the peak current's action to form a molten pool. The molten pool's temperature is at its highest in this current cycle, the viscosity of the molten metal is therefore at its lowest, and V_e is at its fastest. Therefore, the longer the peak time, the more conducive it is to impurity vaporisation and the overflow of bubbles, and thus, the more conducive it is to having a higher deposit density. However, having too long a peak time means having too short a background level time, which will lead to the molten pool not cooling sufficiently during the background level stage. This will then lead to the shaking or even collapse of the molten pool, which is of course not conducive to controlling the additive manufacturing morphology. As shown in Figure 2(b), the trembling of the molten pool leads to a lack of straightness in the Mo rod.

3.2 *Microstructure of WAAM pure molybdenum under different processing parameters*

The microstructure of the deposit's cross-section is as follows: Figure 5 shows the cross-section's microstructure (corresponding to experiments 1 to 5) under different background levels, while Figure 6 shows the cross-section (corresponding to experiments 6 to 8) at different peak times. It can be seen from Figure 5 and Figure 6 that the holes have smooth inner walls and a regular spherical shape, so it can be judged that the holes are pores. Furthermore, it can be seen from Figure 5 that the volume of a single pore tends to increase with the increase of the background level. The reason for this is that a larger background level means a higher molten pool temperature, which in turn leads to a greater molten pool fluidity, which then means a smaller surface tension, which finally promotes tiny bubbles to converge and merge into one big one. With the increase of the background level to 50% however, the arc force's powerful stirring effect in the molten pool greatly disturbs the movement track of the bubbles, and so large-volume bubbles are stirred and broken, which leads to a sharp increase in the number of smaller bubbles. Simultaneously, under the arc force's interference, the overflow of the bubbles is suppressed, the total volume of the pores increases and the density decreases, which thereby confirms the previous analysis.

Figure 5(b) and Figure 6 are the deposits at the peak time of 10%, 20%, 30% and 50% respectively. It can be seen that with the increase of peak time, the volume of a single pore has a tendency to be larger, which indicates that the increase of peak time is conducive to creating a confluence of bubbles from small to large. Leading on from this, large-volume bubbles mean a fast Ve, meaning that the density of the deposit increases with the increase of peak time.

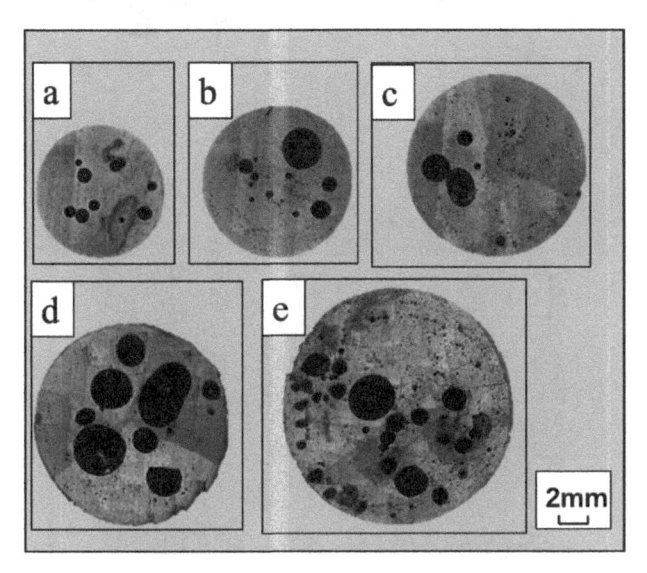

Figure 5. Microstructure of WAAM Mo deposit' cross-section under different background levels: Cross-section microstructure of (a) experiment 1; (b) experiment 2; (c) experiment 3; (d) experiment 4; (e) experiment 5.

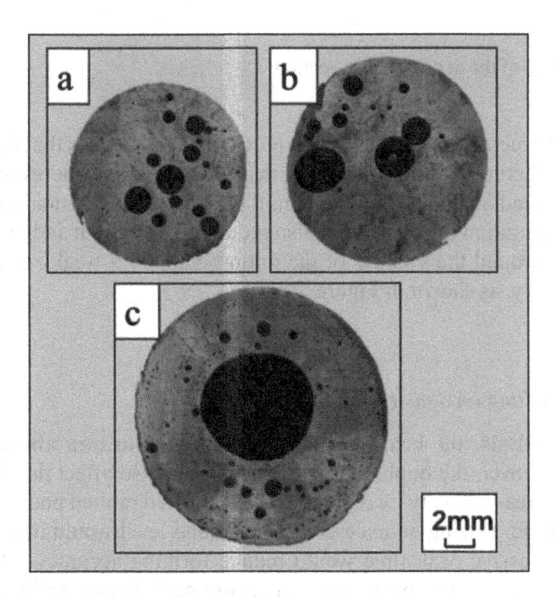

Figure 6. Microstructure of WAAM Mo deposit' cross-section under different peak times: Cross-section microstructure of (a) experiment 6; (b) experiment 7; (c) experiment 8.

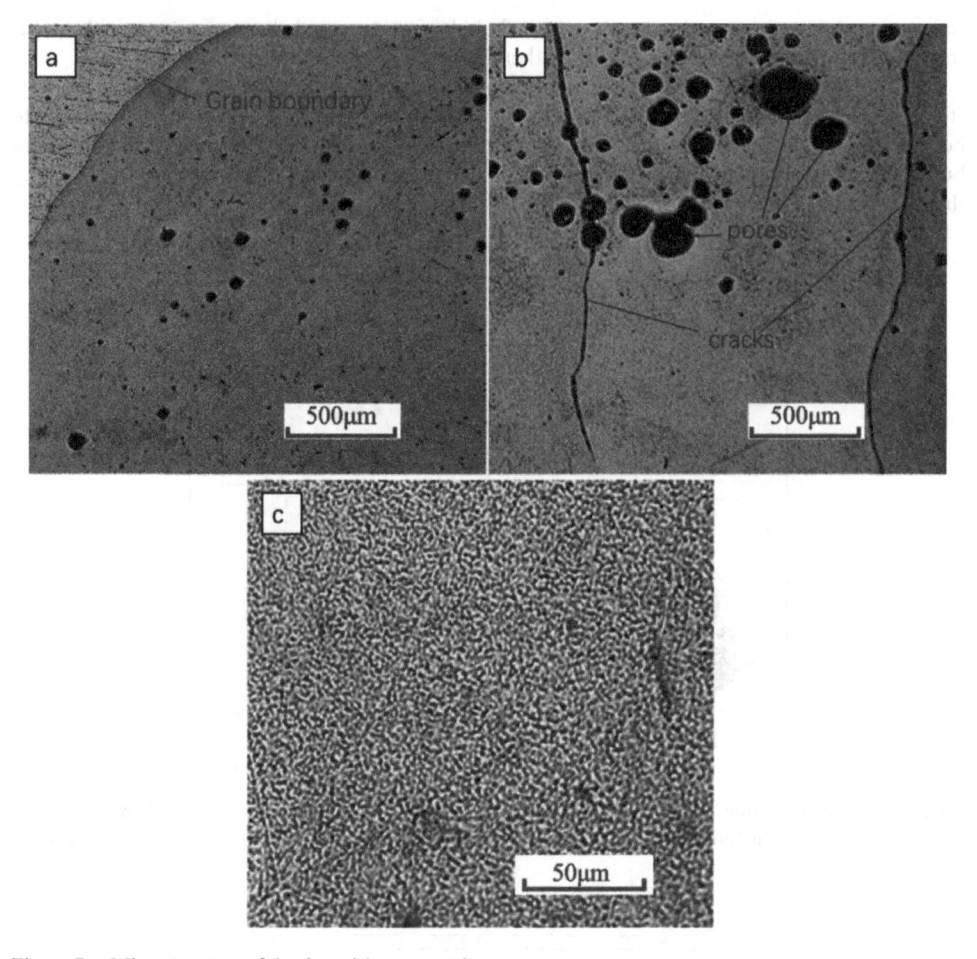

Figure 7. Microstructure of the deposit' cross-section.

Figure 7 is a microstructure graph of WAAM pure Mo which shows that the microstructure is a typical additive-manufactured large crystal, and that there are micro-pores on the surface—these micropores are distributed in chains along the grain boundary. The existence of these micro-pores not only cause a concentration of stress but it also reduces the strength and toughness of the metal, thus making the area around the vicinity of the grain boundary a weak zone and causing cracks along the grain boundary, as shown in Figure 7(b).

3.3 *The effect of remelting on density*

Through the above analysis, the heat input affects V_e which in turn affects the density of the WAAM pure Mo. Moreover, the bubble overflow time t can also affect the density, which means that it is possible to increase density by remelting the solidified molten pool as this would increase the bubble overflow time. Having an excessive background level would interfere with the bubble overflow, while an excessive peak time would reduce forming accuracy. Therefore, selecting a 30% background level and a 20% peak time can ensure both density and forming accuracy. The remelting strategy is to remelt every 2mm of deposited material, with remelting times of 5 or 10 arc cycles, as show in Table 5.

Table 5. WAAM parameters for Mo remelting.

Number	Peak current (A)	Peak time (%)	Background level (%)	Wire feeding (cm/min)	Remelted times
9	300	20	30	30	5
10	300	20	30	30	10

Table 6. Density of remelted deposits.

Number	Actual density (g/cm^3)	Relative density (%)
9	9.72	95.294
10	9.51	93.235

Table 6 shows the densities of the deposit after remelting (Experiments 9–10). Among them, the relative density in experiment 9 reached 95.294% after remelting five times, which is the highest value among the ten sets of experiments and is an increase of 12.353% compared with the deposit with the lowest relative density (Experiment 2). In addition, the forming accuracy is excellent, as is shown in Figure 8(a). The deposit's relative density after remelting ten times reached 93.235%, which is only lower than experiment 8, which had a huge peak time, and experiment 10, and is significantly higher than other experimental deposits. These experiments indicate that remelting can effectively increase the density of pure Mo deposits produced by WAAM.

The microstructures of experiments 9 and 10 are shown in Figure 8. The pores in experiments 9 and 10 appeared either near the grain boundaries (Figure 8(b)) or inside the grains (Figure 8(c)). The pores were low in number and volume and there were few micropores viewable under the ×500 lens. In addition, the overall organisation was relatively uniform and dense.

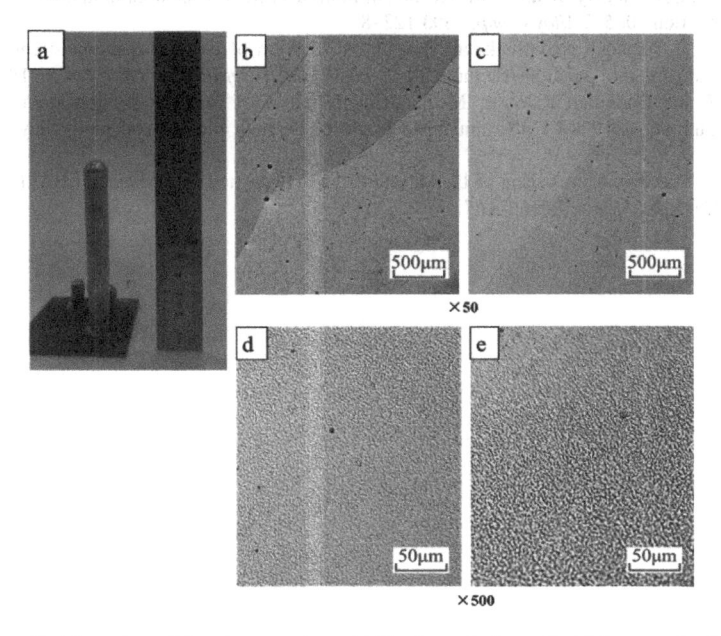

Figure 8. Remelted deposits and its microstructure: (a) The deposited remelted 5 times under the background level of 30%-peak time of 20%; (b) and (d) microstructure of experiment 9; (c) and (e) microstructure of experiment 10.

4 CONCLUSION

(1) The density of WAAM Mo deposits tends to increase with the increase of the background level. The density reaches its maximum at 40%, while when the base current exceeds 40%, the density is significantly reduced.
(2) The density of WAAM Mo deposits increases with the increase of peak time, but when the peak time exceeds 20%, the deposit's morphology quality is significantly reduced.
(3) Remelting can significantly reduce the internal porosity defects of pure Mo produced by WAAM. Using a process of 30% background level, - 20% peak time and five remelting times, a WAAM pure Mo deposit with a relative density of up to 95.294% with excellent morphology quality can be obtained

REFERENCES

[1] Tapia G and Elwany A. A review on process monitoring and control in metal-based additive manufacturing 2014 *J. Manuf Sci Eng-Trans ASME*. **136** 10
[2] Yang Z, Hu K, Hu DW, Han CL, Tong YG and Yang XY, et al. Diffusion bonding between TZM alloy and WRe alloy by spark plasma sintering 2018 *J. Alloy Compd.* **764** 582–90
[3] Gold RE and Harrod DL. Refractory metal alloys for fusion reactor applications. 1979 *J. Nucl Mater*. 85-6 805–15
[4] Miller MK, Kenik EA, Mousa MS, Russell KF and Bryhan AJ. Improvement in the ductility of molybdenum alloys due to grain boundary segregation 2002 *J. Scr Mater.* **46** 299–303.
[5] Xiao Y, Huang B, He B, Shi K, Lian YY and Liu X, et al. Effect of molybdenum doping on the microstructure, micro-hardness and thermal shock behavior of W-K-Mo-Ti-Y alloy 2016 *J. Alloy Compd.* **678** 533–40.
[6] Harimon MA, Hidayati NA, Miyashita Y, Otsuka Y, Mutoh Y and Yamamoto S, et al. High temperature fracture toughness of TZM alloys with different kinds of grain boundary particles 2017 *Int. J. Refract. Hard Met.* **66** 52–6.
[7] Sharma IG, Chakraborty SP and Suri AK. Preparation of TZM alloy by aluminothermic smelting and its characterization 2005 *J. Alloy Compd*. **393** 122–8.
[8] Mueller A J, Bianco R and Buckman R W Jr. Evaluation of oxide dispersion strengthened (ODS) molybdenum and molybdenum–rhenium alloys 2000 *J. Int J. Refract Met Hard Mater*. **18.4** 205–211
[9] Bianco R and Buckman R W Jr. Mechanical properties of oxide dispersion strengthened (ODS) molybdenum alloys 1998// TMS Annual Meeting Molybdenum and Molybdenum Alloys. San Antonio, 125
[10] Michael T. Stawovy. Comparison of LCAC and PM Mo deposited using Sciaky EBAM™2018 *J. Int J. Refract Met Hard Mater*. 73 162–167

Optimization and test evaluation of vibration control for an airborne radar frequency synthesizer

C. Peng & S.G. Yang
Research Institute of China Electronics Technology Corporation, Hefei, China

ABSTRACT: Optimized the design for the fatigue failure of the vibration isolator of an airborne radar frequency synthesizer in the test. Analyze the unfavorable factors in the original design scheme, and find out the main reason for the damage of the vibration isolator. The damping system of the frequency synthesizer is optimized from three aspects: the natural frequency of the system, the installation method of the vibration isolator and the limit design. Finally, the improved damping system was evaluated by dynamic phase noise and durability tests. The results showed that: the improved design not only meets the requirements of phase noise index, but also meets the requirements of durability, and solves the problem of the frequency synthesizer damping system. The problem of fatigue failure.

1 INTRODUCTION

The frequency synthesizer is used as the reference clock of the airborne radar, and its performance is related to the performance of the entire radar [1]. In order to improve the detection accuracy of the radar and enhance the ability of the radar to detect weak signals in a strong interference environment, the frequency synthesizer is required to have better frequency stability and fatigue durability.

Usually in a static environment, the frequency synthesizer has good performance, with low phase noise, low spurious and fast frequency conversion characteristics, but in a dynamic environment, these indicators will deteriorate sharply [2], which is a typical electromechanical coupling problem. On the one hand, the frequency synthesizer contains electronic components such as crystal oscillators and clock local oscillators, which are particularly sensitive to vibration, and small vibrations will lead to a decrease in performance indicators; on the other hand, in a dynamic environment, excessive vibration response will cause The strength damage and fatigue damage of the equipment.

However, for airborne radars, during take-off, landing and flight of the aircraft, they will be subjected to vibrations and shocks transmitted from the aircraft platform. Excessive vibrations or shocks will directly affect the performance of the frequency synthesizer, which will greatly reduce the radar detection. Distance and accuracy can't even work properly. In order to ensure the most reliable and full use of the design performance of the airborne radar in the various vibration environment tests experienced during the flight, it is necessary to design the vibration reduction of key core equipment such as frequency synthesizers. There are generally several methods for damping airborne equipment: optimizing the structure of the equipment for damping design [3], installing dampers; increasing system damping to reduce system vibration response [4], installing vibration absorbers to transfer the vibration response of electronic equipment to the vibrator of the vibration absorber [5], install a vibration isolator to isolate the vibration response of the electronic device to reduce the vibration transmitted to the electronic device [6, 7]. In view of the fact that the installation of vibration isolators for vibration reduction has the advantages of simple implementation and obvious vibration reduction effect, it has been favored by a large number of designers, especially in the field of airborne electronic equipment. There are many types of vibration isolators, but in the vibration damping design of airborne electronic equipment, in order to meet the requirements of durability, corrosivity and passiveness in the harsh environment of the airborne, metal rubber vibration isolators and steel wire rope vibration isolators are usually used.

For an airborne radar frequency synthesizer, the original damping design scheme has a better damping effect, and the telecommunications index meets the design requirements, but the problem of the vibration isolator wire rope breaking occurred during the test. Aiming at the question of whether the structural durability of the vibration isolator itself can meet the requirements of use, the article carried out the optimization design and improvement of the vibration isolation of an airborne radar frequency synthesizer to obtain an optimal vibration isolation design plan to meet the telecommunications performance at the same time Indexes and structural fatigue durability requirements.

2 AN AIRBORNE RADAR FREQUENCY SYNTHESIZER AND ORIGINAL VIBRATION REDUCTION DESIGN PLAN

2.1 Frequency synthesizer structure and vibration reduction scheme

As shown in Figure 1, an airborne frequency synthesizer is a cube structure composed of a clock local oscillator and a crystal oscillator. Its size is 200mm×125mm×58mm, and its total weight is 1.3kg. Due to the limitation of the installation space inside the antenna, the frequency synthesizer is installed by wall-mounted installation. In the previous plan, in order to reduce the vibration of the frequency synthesizer, two GG-18A wire rope vibration isolators (shown in Figure 1-b) produced by a certain manufacturer were used, and the installation method of the vibration isolators was parallel and horizontal installation. The sample parameters of the GG-18A vibration isolator are shown in Table 1.

Table 1. Sample parameters of wire rope vibration isolator produced by a certain manufacturer.

Type	Specified load /kg	Static displacement /mm	Natural frequency /mm	Resonance transmission rate	Size /mm
GG-18A	2.2	1–2.5 mm	10–16		
GG-18B	3.5	0.3–0.8 mm	17–28	≤ 1.5	80×25×18
GG-18C	4.2	0.2–0.4 mm	18–40		

(a) Frequency synthesizer and vibration isolation system

(b) Wire rope vibration isolator

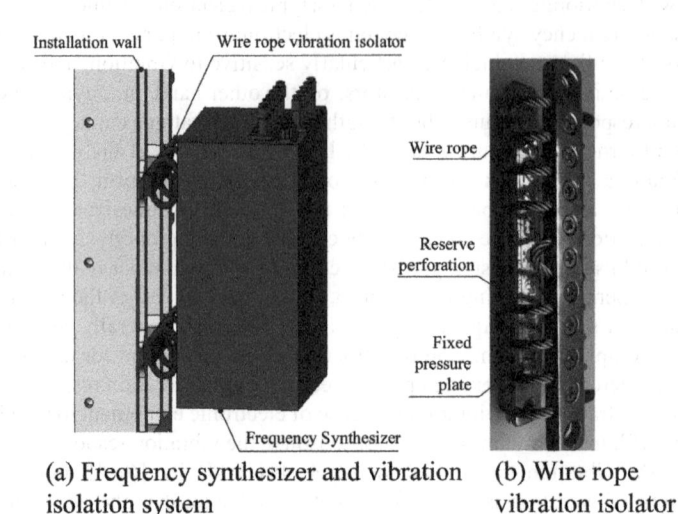

Figure 1. An airborne frequency synthesizer and its original vibration reduction design plan.

2.2 Test verification of the original damping plan

In order to verify the feasibility of the damping design of the airborne frequency synthesizer, a grounding test was carried out to verify it. The vibration level of the bottom test is the value of the functional vibration test, and the specific vibration input power spectrum curve is shown in Figure 2.

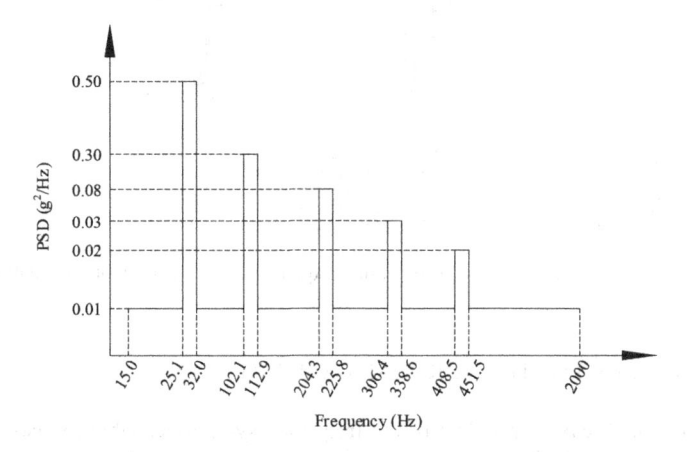

Figure 2. Vibration input power spectrum curve of an airborne frequency synthesizer function.

During the test, the phase noise index of the frequency synthesizer performed well. The phase noise characteristics of the synthesizer when vibrating in the Z direction (vertical direction) is worse than that of the other two directions. Without loss of generality, here is the test result of the phase noise index on a specific interface under Z-direction vibration. For details, see the test value of the original scheme in Table 2. The phase noise index test shows that the dynamic phase noise of the original scheme meets the index requirements.

Table 2. The dynamic phase noise index value and durability performance of the frequency synthesizer under several damping schemes.

Frequency point/Hz	10M	100M	960M	2.4G	8.1G	Durability
original scheme /dB	-153.4	-143.3	-123.2	-118.0	-109.3	not pass
alternative scheme/dB	-148.9	-140.6	-125.1	-115.2	-106.2	not pass
alternative scheme/dB	-142.0	-136.7	-116.3	-111.5	-99.3	pass
Phase noise index /dB	-145	-135	-120	-110	-100	pass

Although the phase noise index of the frequency synthesizer performed well during the test, the following problems occurred during the test: during the vibration process, the frequency synthesizer has a large low-frequency displacement response, and the vibration of the frequency synthesizer is very obvious; the upper and lower mounting plates of the vibration isolator Intermittent collisions occasionally occur; when the vibration direction is the Z direction, the frequency synthesizer vibrates the most severely, and during 24 minutes of vibration in this direction, the steel wire rope of the upper vibration isolator has a fatigue failure. It can be seen that the fatigue durability of the structure cannot meet the design requirements.

Figure 3. Fatigue failure of the steel wire rope of the frequency synthesizer vibration isolator.

3 DEFECT ANALYSIS OF THE ORIGINAL DAMPING PLAN

In order to solve the breaking problem of the frequency synthesizer vibration isolator, it is necessary to analyze the original damping design plan to find out its weak and unreasonable places. Analyze the breaking failure of the vibration isolator wire rope in the original design plan as follows:

(1) Low-frequency large displacement vibration caused by system resonance

The natural frequency of the vibration isolation system is similar to the first narrow-band peak frequency band of the input power spectrum curve of the vibration test. According to the estimation of the vibration isolator sample parameters, the first-order natural frequency of the vibration isolation system is about 18–29Hz, which is just close to the first narrow-band peak frequency band (25–32Hz) of the input power spectrum curve of the vibration test, which will lead to The system resonates. This is extremely detrimental to the wire rope vibration isolator, because resonance will cause large strains inside the vibration isolator structure. Fatigue damage with such large strain bands will consume more fatigue life. It can be seen that this low-frequency large displacement vibration causes fatigue damage. The main reason.

(2) Unlimited function

The wire rope vibration isolator itself does not have a limit function, and due to the limitation of the installation space inside the antenna, the frequency synthesizer is also designed with unlimited positions in the structure. In the process of random vibration of the frequency synthesizer, when the vibration displacement exceeds the distance between the upper and lower mounting plates of the vibration isolator at a certain time, a collision will occur.

(3) Improper installation of vibration isolators

The steel wire rope vibration isolator is made by spirally winding steel wire ropes, and is finally fixed and formed by two upper and lower pressure plates parallel to the axial direction. A round hole is reserved in the middle of the pressure plate to allow the wire rope to pass through. The specific structure is shown in Figure 1. Due to this configuration of the wire rope, its stiffness and damping in the three directions are different, and the degree of freedom around the axis of the circular hole will be released.

As shown in Figure 4, a simplified diagram of the force of the wire rope when the vibration isolator is installed in the horizontal direction and the axial direction in the vertical direction is given when the vibration isolator is vibrated in the three directions. It can be seen from the force diagram:

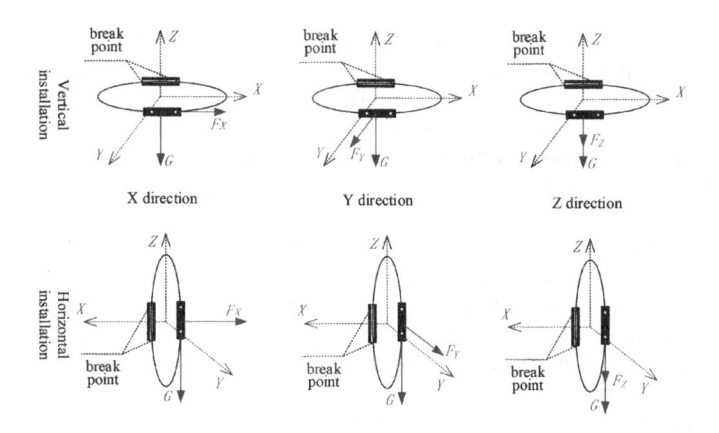

Figure 4. Diagram of the force of the vibration isolator under two different installation methods.

When the wire rope vibration isolator is installed in the vertical direction, the gravity G in the three directions and the vibration force F_Y in the Y direction are all along the axis of the vibration isolator. The displacement generated by these forces can be in the structural form. It has been released to a certain extent and will not bring too much burden to the breakpoint position. The breakpoint position only needs to bear the vibration force in the two directions of F_X and F_Z. When the wire rope vibration isolator is installed axially in the horizontal direction, only the vibration force F_X in the X direction is along the axial direction of the vibration isolator, that is, only F_X is released in the structural form. However, the gravity forces G, F_Y, and F_Z in the three directions are all perpendicular to the axis of the vibration isolator, which cannot be released from the structural form and needs to be directly borne by the position of the break point.

Based on the above analysis, for this wall-mounted installation method of the frequency synthesizer, installing the wire rope vibration isolator in the vertical direction is obviously better than installing it in the horizontal direction. In the original vibration reduction scheme, Choosing the latter is obviously not conducive to the fatigue resistance and durability of the vibration isolator.

4 OPTIMIZATION AND IMPROVEMENT OF DAMPING DESIGN

On the basis of the defect analysis of the original scheme, the frequency synthesizer vibration isolation scheme is optimized. During the optimization process, it is necessary to minimize the changes to the original interface and the space required for installation. The specific optimization and improvement plan are shown in Figure 5.

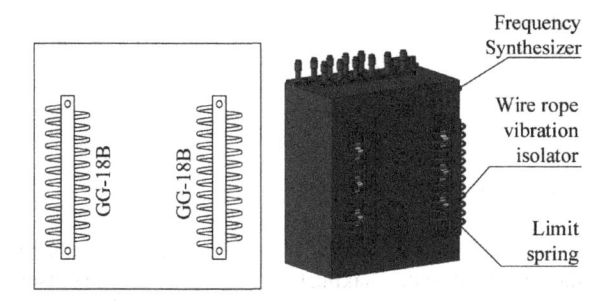

Figure 5. Improved vibration reduction design scheme.

In the selection of the vibration isolator model, due to the limitation of installation space, the steel wire rope is used for vibration isolation, and the model of the vibration isolator is adjusted to GG-18B. As shown in Table 1, this GG-18B vibration isolator has the same dimensions as GG-18A, and its stiffness is between GG-18A and GG-18C. Combined with the weight of the frequency synthesizer, the natural frequency of the system is approximately estimated to be between 40–65 Hz based on the sample parameters, which can effectively avoid the input power spectrum curve of the vibration test.

In terms of the arrangement of the vibration isolators, the original plan of the two vibration isolators installed horizontally up and down was changed to vertical installation on the left and right. In this way, the axial direction of the vibration isolator is changed from the axial direction along the horizontal direction to the axial direction along the vertical direction, which is beneficial to the fatigue durability of the vibration isolator.

Inside the wire rope vibration isolator, 3 sets of anti-collision pairs are installed on the upper and lower bottom plates of the wire rope. Under normal vibration conditions, collisions will not occur between the anti-collision pairs; when an impact or abnormal vibration occurs, the anti-collision pairs will absorb the energy of the impact and collision, thereby avoiding excessive impact of the frequency synthesizer..

5 TEST EVALUATION

In order to evaluate whether the optimized vibration isolation scheme can meet the design requirements immediately, it is tested and evaluated. In the test, under the two vibration isolation schemes of the improved scheme and the original scheme, the dynamic phase noise index, vibration response, and durability vibration test time of the frequency synthesizer were tested simultaneously.

5.1 Experimental test system

As shown in Figure 6, the experimental test system consists of two parts: phase noise test system and vibration test system. The phase noise test system is composed of a frequency synthesizer, a power supply and a phase noise tester; the vibration test system is composed of an acceleration sensor, a vibration data acquisition analyzer, a computer, a vibration table and its power amplifier.

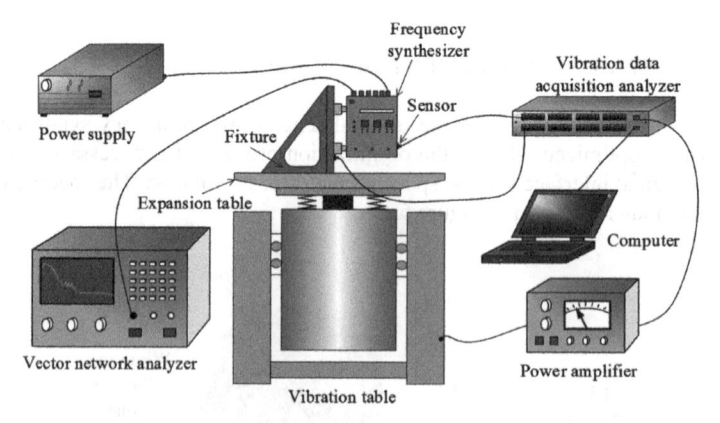

Figure 6. Vibration phase noise test evaluation system.

In the test, the frequency synthesizer is installed on the test fixture through the vibration isolator, and the test fixture is then installed on the shaking table. Two acceleration sensors, one as the control sensor C01 is installed on the fixture near the connection position of the vibration isolator and the test fixture; the other as the monitoring sensor M01 is installed on the frequency synthesizer.

5.2 Test and result analysis

Figure 7. Test and test site diagram.

Figure 7 shows the test site. The specific test results are shown in Table 3 and Figure 8. Without loss of generality, the test values of the phase noise value and vibration curve of the frequency synthesizer in the Z direction are also given here. Table 3 shows the two vibration isolation schemes. Synthesizer dynamic phase noise value; Figure 8 shows the vibration response curves of the monitoring points on the frequency synthesizer under two vibration isolation schemes.

Table 3. Comparison of dynamic phase noise value of frequency synthesizer before and after improvement.

Frequency point /Hz	10M	100M	960M	2.4G	8.1G	Durability
Original scheme /dB	-152.9	-141.1	-122.7	-118.2	-109.2	No Pass
Improved scheme /dB	-155.8	-143.9	-125.8	-122.6	-111.5	Pass
Phase noise index /dB	-145	-135	-120	-110	-100	Pass

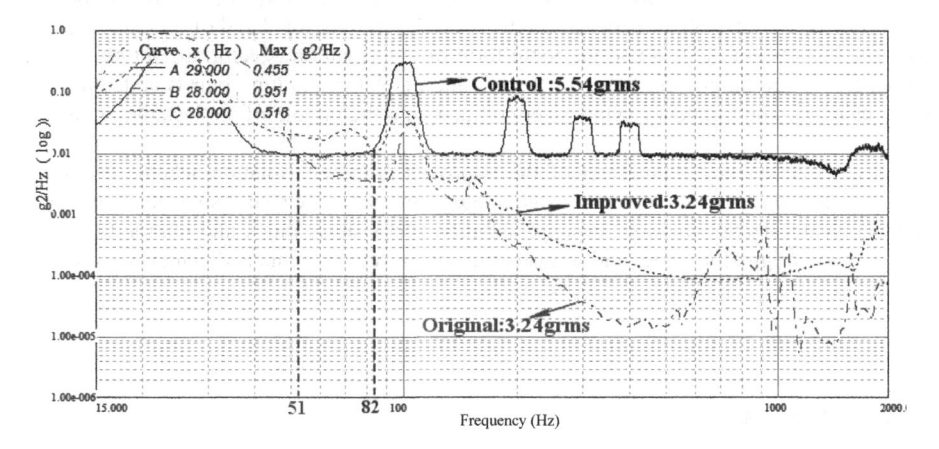

Figure 8. Comparison of vibration response of frequency synthesizer before and after improvement.

It can be seen from Figure 9 that the two solutions have a certain amplification effect on vibration in the low frequency band, and both have a significant attenuation effect on vibration in the high frequency band. Compared with the original scheme, the initial vibration isolation frequency of the improved scheme is increased from 51Hz to 82Hz. This is because the GG-18Y vibration isolator is used in the improved scheme, and its stiffness is greater than that of the GG-18 in the original scheme. Although the initial vibration isolation frequency has been increased, the improved scheme has significantly less amplifying effect on the vibration response in the low frequency band than the original scheme. This has a significant effect on the fatigue durability of the vibration isolator. The overall root-mean-square value of the vibration of the improved scheme is 3.56g. Compared with the 3.24g of the original scheme, an increase of 0.32g is not obvious, but the vibration response curve of the improved scheme is smoother than that of the original scheme, which shows that the vibration response of the new scheme is more smooth. It tends to a relatively stable state, and the sudden shock during the vibration process is small.

Tests show that the improved scheme has the most obvious improvement in fatigue durability compared to the original scheme. The original plan has a life of 33min, 29min and 14min in the three directions respectively; the improved plan only passed a 3-hour/direction durability vibration test with only one set of vibration isolator, which shows that its durability has been significantly improved.

Integrating the dynamic phase noise and durability test results, the improved scheme not only meets the requirements of telecommunications indicators, but also meets the requirements of fatigue durability.

6 CONCLUSION

In view of the problem of fatigue failure of the frequency synthesizer vibration isolation system of an airborne radar under the grounding functional vibration, the optimized design and experimental verification were carried out.

First of all, the structure of the frequency synthesizer and the original vibration isolation system and the grounding test conditions are described, and the grounding test verification conditions are explained. Secondly, it analyzes the shortcomings of the original damping design scheme, and points out that the low-frequency large displacement caused by system resonance and the improper installation of the vibration isolator are the main factors that cause the break of the frequency synthesizer vibration isolator. Thirdly, on the basis of absorbing the unfavorable factors of the original design, the vibration isolation system of the frequency synthesizer is optimized. The model of the vibration isolator is optimized to avoid the first narrow band of the power spectrum curve; the installation method and the force form of the vibration isolator are optimized, which is conducive to the fatigue durability of the vibration isolator; the limit pair is added to prevent sudden changes. The frequency synthesizer collided with the bottom of the installation under the impact.

Finally, the frequency synthesizer of the improved vibration isolation scheme is tested for dynamic phase noise, vibration response and fatigue durability. The results show that the improved frequency synthesizer's vibration response stability, phase noise index and durability index are better than the original scheme, and it meets the design requirements and achieves the expected purpose.

ACKNOWLEDGEMENT

Supported by Hefei Natural Science Foundation (2021).

REFERENCES

[1] Arnold E J, Yan J B, Hale R D, et al. Identifying and compensating for phase center errors in wing-mounted phased arrays for ice sheet sounding [J]. IEEE Transactions on Antennas and Propagation, 2014, 62(6): 3416–3421.
[2] Miller B, Arnold E J. Wing-integrated airborne antenna array beamforming sensitivity to wing deflections[C]//2018 IEEE Aerospace Conference. IEEE, 2018: 1–11.
[3] Baran İ. Optimization of vibration characteristics of a radar antenna structure[D]. , 2011.
[4] Kong Y, Huang H. Design and experiment of a passive damping device for the multi-panel solar array[J]. Advances in Mechanical Engineering, 2017, 9(2): 1–10.
[5] Huo G, Chen Z, Jiao Y, et al. Optimization of dynamic vibration absorbers for suppressing sound radiation of plate structures[J]. Journal of Vibration and Acoustics, 2021, 143(2): 021003.
[6] Veprik A M. Vibration protection of critical components of electronic equipment in harsh environmental conditions[J]. Journal of Sound and Vibration, 2003, 259(1): 161–175.
[7] Lu G, Zhou J, Cai G, et al. Active Vibration Control of a Large Space Antenna Structure Using Cable Actuator[J]. AIAA Journal, 2020: 1–12.

Catalytic hydroisomerization of n-Hexadecane over Pd/SAPO-41 bifunctional catalysts: Effect of molar ratio of ethylene glycol to water on their catalytic performance

J. Yang
Institute of Petrochemical, Heilongjiang Academy of Sciences, Harbin, China

X.M. Wei
School of Chemistry and Material Science, Heilongjiang University, Harbin, China

X.F. Bai
Institute of Petrochemical, Heilongjiang Academy of Sciences, Harbin, China
School of Chemistry and Material Science, Heilongjiang University, Harbin, China

ABSTRACT: Using di-n-butylamine (DBA) as template, SAPO-41 molecular sieves were hydrothermally synthesized in the ethylene glycol-water system. Pd/SAPO-41 bifunctional catalysts were prepared by incipient wetness impregnation. The prepared SAPO-41 and Pd/SAPO-41 were characterized by XRD, SEM, N2 physical adsorption, ^{29}Si MAS NMR and Py-IR. The effects of molar ratio of ethylene glycol to water in the initial gel on the structure, composition, morphology and acidity of (Pd/)SAPO-41 and the catalytic performance of n-hexadecane hydroisomerization over Pd/SAPO-41 were investigated. The crystallite size of SAPO-41 synthesized in the ethylene glycol-water system is smaller and the outer surface area is more abundant, which is more conducive to the formation and diffusion of the multi-branched isomer products on the outer surface of the molecular sieve. The 0.5Pd/S41(0.6)-Eg(1:4) catalyst exhibits the best catalytic performance with the n-hexadecane conversion rate of 87.3% and the *iso*-hexaoxane selectivity of 89.8%.

1 INTRODUCTION

Diesel hydroisomerization is the most important way to improve the low-temperature fluidity of diesel [1,2]. The hydroisomerization process is often accompanied by cracking reactions to produce light hydrocarbons, which reduces the conversion efficiency of diesel [3,4]. Therefore, on the basis of ensuring the conversion efficiency of diesel hydroisomerization, restraining the cracking reaction is the key to this process. The structure, acid properties, and morphology of molecular sieves used as acidic carriers for bifunctional catalysts are crucial to the hydroisomerization process [5,6].

Herein, a series of SAPO-41 molecular sieve samples were hydrothermally synthesized in the ethylene glycol-water system with different molar ratio of ethylene glycol to water in the initial gel. Pd/SAPO-41 bifunctional catalysts were prepared by incipient wetness impregnation. The hydroisomerization of n-hexadecane over Pd/SAPO-41 was carried out to investigate the effect of alcohol-water ratio on the catalytic performance.

2 EXPERIMENTAL SECTION

2.1 *Preparation of Pd/SAPO-41 bifunctional catalysts*

In the alcohol-water system, the traditional hydrothermal method was used to synthesize the SAPO-41 molecular sieves with the gel composition of 0.6 SiO_2:1.0 Al_2O_3:1.0 P_2O_5:3.0

DBA:55(H_2O+ ethylene glycol). The initial gel crystallized at 180°C for 24 hours and then cooled to room temperature. The products were centrifuged, washed, dried, and roasted to obtain SAPO-41 molecular sieve samples, which is denoted as S41(0.6)-Eg(y) (y=1:4, 2:3, 3:2), and y is the molar ratio of Eg to H_2O in the initial gel. Using Pd(NO_3)$_2$ as Pd precursor, the Pd/SAPO-41 bifunctional catalysts with the Pd loading amount of 0.5 wt.% (0.5Pd/ S41(0.6)-Eg(y))were prepared by incipient wetness impregnation.

2.2 *Characterization of 0.5Pd/ S41(0.6) and 0.5Pd/ S41(0.6)-Eg(y) catalysts*

See section 2.3 in reference [7].

2.3 *Catalytic hydroisomerization of n-hexadecane over 0.5Pd/ S41(0.6)-Eg(y) catalysts*

See section 2.4 in reference [7].

3 RESULT AND DISCUSSION

3.1 *Effect of glycol/water ratio on the structure, composition, morphology and acidity of S41(0.6)-Eg(y) molecular sieves*

XRD characterizations of S41(0.6)-Eg(y) molecular sieves were performed to determine the topological structure of each sample synthesized on the condition of different alcohol-water ratioes, and the results are shown in Figure 1.

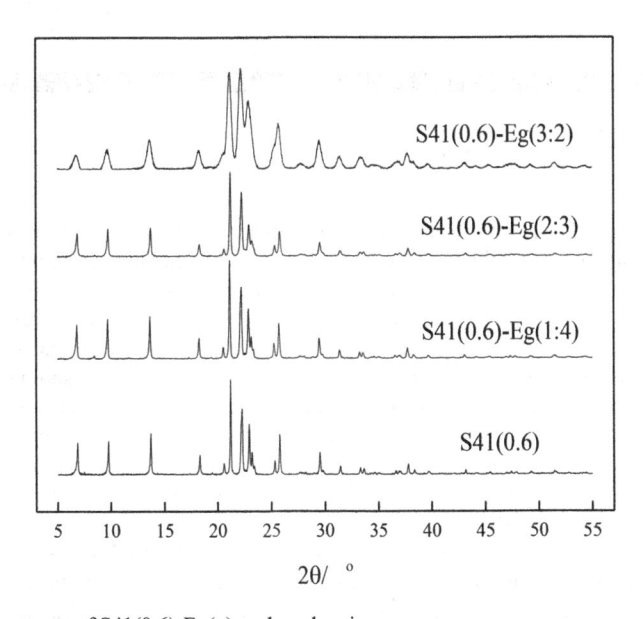

Figure 1. XRD patterns of S41(0.6)-Eg(y) molercular sieves.

It can be seen from Figure 1 that as the ratio of alcohol to water in the initial gel increases, the diffraction peak intensity of the sample gradually decreases, especially in the range of 2θ = 20.5 − 25.7°. When the alcohol-water ratio is 3:2, the diffraction peak at 2θ = 25.7° disappears, and the diffraction peak appears to be broadened, which may be due to a certain shrinkage of the crystal size in the sample.

In order to investigate the influence of the ratio of alcohol to water in the initial gel on the morphology of SAPO-41 molecular sieve samples, the S41(0.6)-Eg(y) molecular sieves were characterized by SEM, and the results are shown in Figure 2.

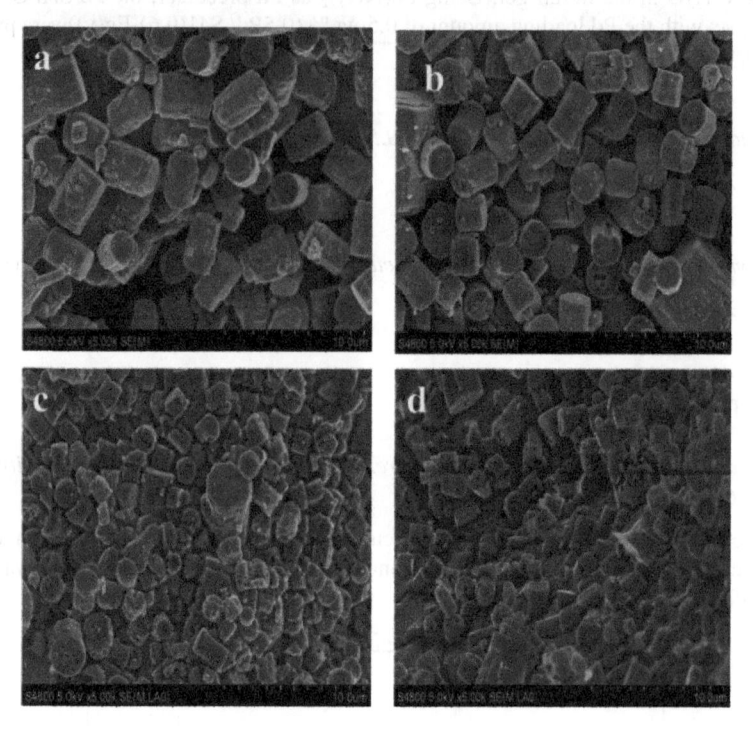

Figure 2. SEM images of S41(0.6)-Eg(y) molercular sieves.
a: S41(0.6); b: S41(0.6)-Eg(1:4); c: S41(0.6)-Eg(2:3); d: S41(0.6)-Eg(3:2)

As shown in Figure 2, each molecular sieve showed a typical morphology unique to SAPO-41 molecular sieve, indicating that the introduction of ethylene glycol did not affect the morphology of the sample. Compared with the synthetic molecular sieve samples under the pure water system, the aggregate size of the synthesized samples under the ethylene glycol-water system is significantly reduced, and the reduction trend of aggregate size is more obvious as the amount of ethylene glycol introduced in the gel increases. The length of aggregate in S41(0.6)-Eg(3:2) sample is reduced from 5 − 8μm to 2 − 4μm, and diameter is reduced from 4 − 6μm to 2 − 3μm. This may be because the introduction of ethylene glycol in the initial gel makes the depolymerization and polycondensation speed of the Si source and Al source in the system slower. Due to the hydrophobicity of the glycol molecule, it will inhibit the growth of crystal grains during the crystallization process, thereby making the aggregate size smaller [8].

N_2 physical adsorption of the S41(0.6)-Eg(y) molecular sieve samples was performed to investigate the influence of the ratio of alcohol to water in the initial gel on the pore characteristics of the synthesized molecular sieve, and the results are shown in Figure 2 and Table 1.

It can be seen from Figure 2 that the amount of N_2 adsorbed by each sample rises rapidly in the region with lower relative partial pressure, showing the characteristics of an I isotherm, indicating the existence of abundant microporous pore structure in the synthesized sample. Compared with synthetic molecular sieves in the water phase, S41(0.6)-Eg(y) molecular sieves have more obvious hysteresis loops at higher relative partial pressures, mainly due to the smaller crystallite size of the synthesized samples in the alcohol-water system. The accumulation formed more intercrystalline pores.

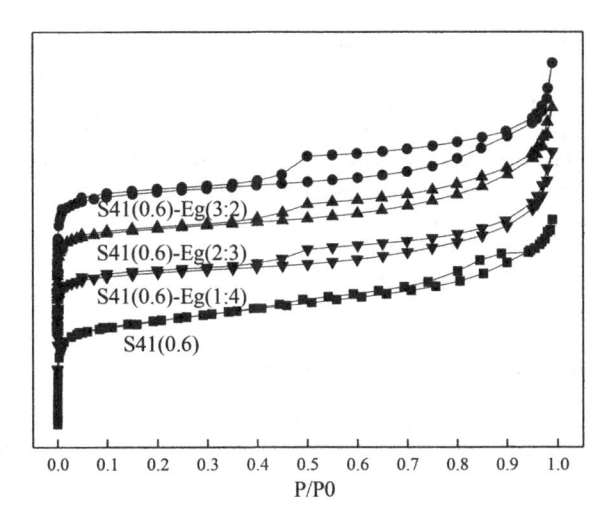

Figure 3. N$_2$ adsorption-desorption isotherm of S41(0.6)-Eg(y) molercular sieves.

Table 1. N$_2$ adsorption data of S41(0.6)-Eg(y) molecular sieves.

Sample	Surface area (m^2/g)			Pore volume (cm^3/g)		
	BET[a]	Micropore[b]	External	Total[c]	Micropore[b]	Mesopore
S41(0.6)	251	228	23	0.132	0.081	0.051
S41(0.6)-Eg(1:4)	254	226	28	0.138	0.078	0.060
S41(0.6)-Eg(2:3)	243	209	34	0.134	0.066	0.068
S41(0.6)-Eg(3:2)	243	194	49	0.130	0.054	0.076

[a] BET method, [b] t-plot method, [c] Volume adsorbed at $p/p_0 = 0.99$

As shown in Table 1, with the increase of the alcohol-water ratio in the initial gel, the micropore surface area and micropore volume of the sample decreased slightly, but the external surface area and mesopore volume gradually increased. The increase in mesoporous pore volume is mainly due to the accumulation of smaller grain sizes, which produces more intercrystalline pores and more defects on the surface of the molecular sieve.

In order to investigate the influence of the alcohol-water ratio in the initial gel on the substitution mode of Si atoms entering the molecular sieve framework during the crystallization process, ^{29}Si MAS NMR was used to analyze the synthesized S41(0.6)-Eg(y) molecular sieve samples, and the results are listed in Table 2.

Table 2. The Si species distribution of the S41(0.6)-Eg(y) molecular sieves.

Sample	δ /ppm				
	Si(4Al) -90~-92	Si(3Al) -95~-97	Si(2Al) -98~-102	Si(1Al) -105~-109	Si(0Al) -110~-115
S41(0.6)	32	11	13	12	32
S41(0.6)-Eg(1:4)	20	15	19	14	31
S41(0.6)-Eg(2:3)	21	18	14	17	30
S41(0.6)-Eg(3:2)	22	18	18	18	24

It can be seen from Table 2 that the distribution of Si atoms in the framework of each molecular sieve sample is similar, and the SAPO region and the silicon island region account for the largest proportion. For the S41(0.6)-Eg(y) sample, as the amount of ethylene glycol introduced in the initial gel increases, the proportion of Si(4Al) + Si(3Al) area gradually increases, and the proportion of silicon island area gradually decreases. It indicated that the distribution of Si atoms on the molecular sieve framework is more uniform, and more Si atoms enter the molecular sieve framework through the SMII substitution mechanism to form fewer silicon islands.

The 0.5Pd/S41(0.6)-Eg(y) bifunctional catalysts were characterized by pyridine adsorption infrared spectroscopy (Py-IR) to to compare the difference in the acidity of B acid with different strength, and the results are shown in Table 3.

Table 3. Py-IR datas of the 0.5Pd/S41(0.6)-Eg(y) catalysts.

Sample	Brønsted Acid sites (μmol/g)					
	150° C	250°C	350°C	Weak	Medium	Strong
0.5Pd/S41(0.6)	38	32	20	6	12	20
0.5Pd/S41(0.6)-Eg(1:4)	37	33	22	4	11	22
0.5Pd/S41(0.6)-Eg(2:3)	36	32	16	4	16	16
0.5Pd/S41(0.6)-Eg(3:2)	37	30	18	7	12	18

As shown in Table 3, the total acidity of B acids for each catalyst is not much different, about 36μmol/g, which is similar to the result that the proportion of Si (1~4Al) region in the framework of each sample obtained by ^{29}Si MAS NMR is consistent. But there is a big difference between the acidity of B acids with different strength. Among them, 0.5Pd/S41(0.6)-Eg(3:2) and 0.5Pd/S41(0.6)-Eg(2:3) catalysts have the largest acidity of weak acid and medium-strong acid. which may be because there are the most defective sites for the two catalysts mentioned above. The 0.5Pd/S41(0.6)-Eg(1:4) catalyst has the most strong B acid because of its more complete micropore structure.

The prepared 0.5Pd/S41(0.6)-Eg(y) bifunctional catalysts were characterized by H_2 chemisorption to compare the difference of metal Pd dispersion, and results are shown in Table 4.

Table 4. Metal Pd dispersion of 0.5Pd/S41(0.6)-Eg(y) catalysts.

Catalysts	Dispersion of Pd (%)	C_{Pd} (μmol/g)
0.5Pd/S41(0.6)	24.9	11.7
0.5Pd/S41(0.6)-Eg(1:4)	29.2	13.8
0.5Pd/S41(0.6)-Eg(2:3)	25.4	12.0
0.5Pd/S41(0.6)-Eg(3:2)	22.8	10.8

It can be seen from Table 4 that as the ratio of alcohol to water in the initial gel increases, the dispersion of Pd metal on the catalysts gradually decreases, which is consistent with the change trend of the amount of B acid in the catalysts. The main reason is that the interaction between the metal palladium particles and the acidic sites on the surface of SAPO-41 molecular sieve is weakened [9], and the degree of agglomeration between Pd particles increases.

3.2 Effect of the ratio of ethylene glycol to water on the reaction performance of Pd/SAPO-41 bifunctional catalysts

The performance of the hydroisomerization reaction of n-hexadecane on 0.5Pd/S41(0.6) and 0.5Pd/S41(0.6)-Eg(y) catalysts was compared, and the results are shown in Figure 5.

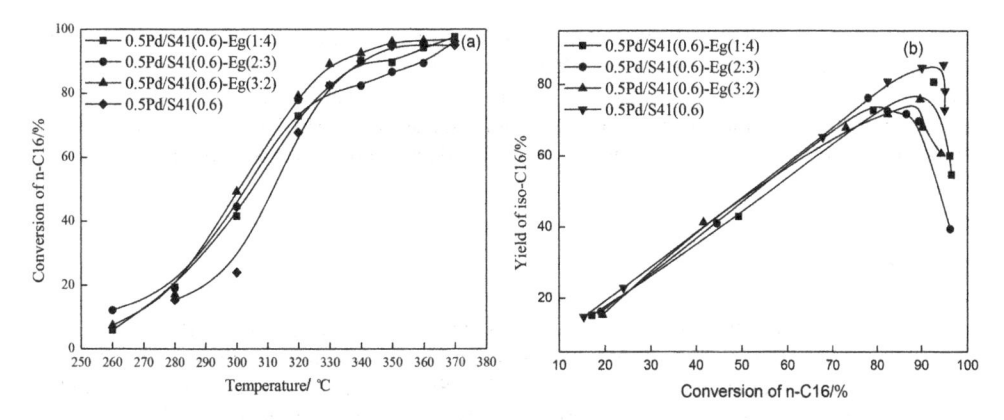

Figure 4. The conversion of *n*-hexadecane (a) and *iso*-hexadecane yield (b) in the hydroisomerization of *n*-hexadecane over 0.5Pd/S41(0.6)-Eg(y) catalysts.

It can be seen from Figure 4(a) that the conversion rate of n-hexadecane on the 0.5Pd/S41(0.6)-Eg(y) catalysts has a consistent trend with the reaction temperature. When the reaction temperature is lower than 320°C, with the increase of the reaction temperature, the conversion rate of n-hexadecane increases significantly, and there is little difference in the activity of each catalyst. When the reaction temperature is higher than 320°C, there is great difference in the activity of each catalyst at the same reaction temperature, where 0.5Pd/S41(0.6)-Eg(3:2)> 0.5Pd/S41(0.6)-Eg(1:4)> 0.5Pd/S41(0.6)-Eg(2:3)> 0.5Pd/S41(0.6)

As shown in Figure 4(b), the yield of isohexadecane on the 0.5Pd/S41(0.6) catalyst is higher than those of the 0.5Pd/S41(0.6)-Eg(y) catalysts. This may be because the 0.5Pd/S41(0.6) catalyst has more abundant microporous channels, which is conducive to the production of more single-branched isomer products. When the molecular sieve synthesized in an alcohol-water system is an acidic carrier, the 0.5Pd/S41(0.6)-Eg(1:4) catalyst exhibits the best catalytic performance. When the conversion rate of n-hexadecane is 87.3%, the selectivity of *iso*-hexaoxane reaches 89.8%. The main reason is that the metal dispersity of the catalyst is higher, and the matching degree between metal sites and acidity is stronger.

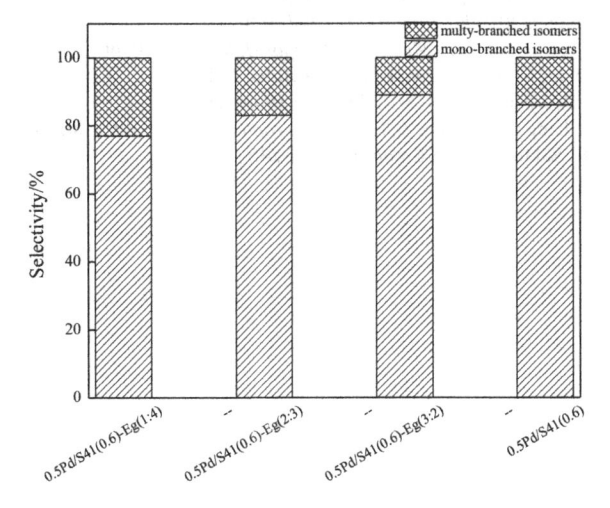

Figure 5. Distribution of mono-methy isomers and multy-branched isomers at *n*-C_{16} conversion ~70%.

It can be seen from Figure 5 that the 0.5Pd/S41(0.6)-Eg(1:4) catalyst has a higher selectivity for the multi-branched isomer product. When the conversion rate is 71.3%, the selectivity of the multi-branched isomer products is 22.9%. The main reason is that the smaller crystal grain size makes the catalyst have more pores and a richer outer surface area, which is more conducive to the formation and diffusion of multi-branched isomer products on the outer surface of the molecular sieve.

4 CONCLUSION

Using di-n-butylamine (DBA) as a template, a series of SAPO-41 molecular sieves (S41(0.6)-Eg(y)) were synthesized in the ethylene glycol-water system. The introduction of ethylene glycol makes the depolymerization and polycondensation rate of the Si source and Al source slower, and inhibits the aggregation and growth of crystal grains during the crystallization process, thereby making the crystal size of the molecular sieve smaller. However, when the amount of ethylene glycol added is too high (alcohol-water ratio is 2:3 and 3:2), it is not conducive to the depolymerization of Si source and Al source, and the microporous structure of the molecular sieve sample is incomplete. The 0.5Pd/S41(0.6)-Eg(1:4) catalyst exhibits the best catalytic performance with the n-hexadecane conversion rate of 87.3% and the iso-hexaoxane selectivity of 89.8%. The crystallite size of SAPO-41 synthesized in the ethylene glycol-water system is smaller and the outer surface area is more abundant, which is more conducive to the formation and diffusion of the multi-branched isomer products on the outer surface of the molecular sieve.

ACKNOWLEDGMENTS

This work was funded by the Science Foundation of Heilongjiang Academy of Sciences (WS2021SH01).

REFERENCES

[1] Wang W, Liu C and Wu W 2019 *Catal. Sci. Technol.* **9** 4162
[2] Zhang Y, Wang W, Jiang X 2018 *Catal. Sci. Technol.* **8** 817
[3] Lee S W, Ihm S K 2013 *Ind. Eng. Chem. Res.* **52** 15359
[4] Yang L, Wang W, Song X 2019 *Fuel ProcessTechnol.* **190** 13
[5] MonikaF,Andrzej ⁻,Karolina J 2020 *Micropor MesoporMater* **305** 110366
[6] Feng Z , Wang W, Wang Y 2019*Micropor MesoporMater* **274** 1
[7] Wei X, Kikhtyanin O V, Parmon V N 2018 *J Porous Mater* **25** 253
[8] Jegatheeswaran S, Cheng C M, Cheng C H 2015 *Micropor MesoporMater* **201** 24
[9] Yadav R, Singh A K, Sakthivel 2015 *Catalysis Today* **245** 155

Study on raw materials and structural performance of coral concrete composite structure

Y. Gao
Naval Logistics Academy, Tianjin, China

Y.J. Lu
Unit 91995 of PLA, Shanghai, China

ABSTRACT: Coral Reef Sand, as the concrete aggregate for the construction of marine island engineering, has the characteristics of light weight, environmental protection, energy saving and easy financing, and it has been used in the plain concrete engineering such as revetment and road. Considering the rich resources, combined with structural assembly and composite structure technology, a new type of composite structure with Coral Concrete-Filled Steel Tube(beam) with Wrapped FRP is proposed, it can be used in marine engineering. Through experimental research and analysis, the performance of raw materials and composite structure is obtained, which can provide the theoretical basis and design reference for application.

1 INTRODUCTION

Coral aggregate concrete has been used in road and revetment engineering. Due to the low strength and the defects of chloride ions, it is difficult to be used in structural load-carrying members, which limits the application. However, as a kind of energy-saving, environmental protection and sustainable marine aggregate, coral reef sand can be effectively used in the bearing components of marine structures, which has practical and long-term benefits for the construction of marine island.

The site conditions of marine engineering construction are mostly on the coast or island, where the resources of coral fragment and coral sand are rich. Considering the environmental protection and the rich resources, combined with the advantages of steel-concrete composite structure, using modular assembly technology of prefabricated structure, a composite structure called Coral Concrete-Filled Steel Tube/beam with Wrapped FRP(Fiber Reinforced Polymer) is proposed. It can be used as a structural bearing member, and its structural form is shown in Figure 1. This paper takes CFRP as research target. Through research and analysis, the basic properties of coral concrete composite structure materials and components are obtained.

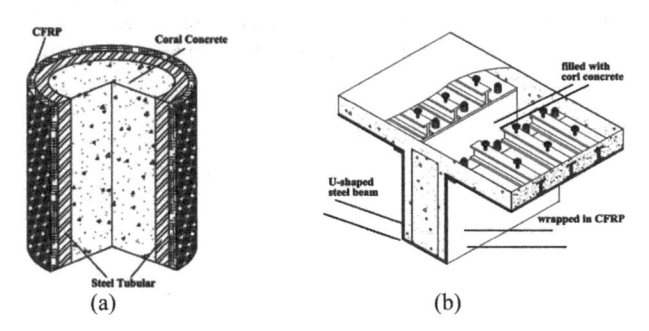

Figure 1. Composite structure construction of the steel tubes/beams filled with coral concrete and wrapped with CFRP: (a) composite column and (b) composite structure Beam-Slabs System.

2 STUDY ON BASIC PROPERTIES OF CORAL CONCRETE

Coral rubble is a natural lightweight aggregate with rough surface and high friction, which is light and porous. It has higher water absorption and lower strength than ordinary rubble, and its strength is close to the artificial lightweight aggregate. Coral reef sand is widely used in road, slope protection and breakwater due to its properties. In order to use it as coarse and fine aggregate for structural engineering, it is necessary to research the characteristics of raw materials.

2.1 Properties of coral aggregate

2.1.1 Gradation of aggregate

The coral reefs are taken from the far sea and porous. After crushing, the 5mm ~ 25mm ones are used as coarse aggregate, and the grading test is carried out on the 25mm ones. The results are shown in Table 1. The natural gradation of coral rubble meets the requirements, and it can be used as coarse aggregate. The fineness modulus is 2.726, which belongs to medium sand.

Table 1. Gradation of coral rubble.

Pore Size/mm	37.5	31.5	26.5	19	16	9.5	4.75
Reference Ranges/%	0	0–5	0–10	—	30–70	—	90–100
Coral Rubble/%	0	4.134	10.26	26.976	49.517	75.166	99.487

2.1.2 Strength of aggregate

The basic performance parameters of materials are tested according to the light aggregate and its test methods (GB/T 17431.1–2010). The cylinder compressive strength of aggregate is the basis. Through the test, the cylinder compressive strength is 6.23N/mm^2, which is higher than light coarse aggregate.

2.1.3 Other performance parameters

The physical properties of the coral aggregate were tested, including bulk density, apparent density and water absorption. The bulk density is 1040kg/m^3, the apparent density is 1860kg/m^3 and the water absorption is 10.32%.

2.2 Coral concrete

2.2.1 Preparation of coral concrete

42.5 ordinary portland cement is used for coral concrete preparation; Coarse aggregate is crushed and screened according to Table 1; Natural coral sand is used as fine aggregate; The mixture ratio of cement: Coral rubble : coral sand is 1:1.68:1.63 and the water cement ratio is 0.5 [1]. The test water is manually configured with reference to the South China Sea [2]. The proportion is shown in Table 2.

Table 2. Lon content of artificial seawater per 100L.

ionic composition	KCl	CaCl$_2$	NaCl	NaSO$_4$	NaHCO$_3$	MgCl
usage/g	74.5	108.2	2216	386.1	20.7	526.5

2.2.2 Strength of coral concrete

The standard cube and prism specimens were made [3]. The 28d cubic compressive strength was 35.4 MPa, the axial compressive strength was 33.96 MPa, and the elastic modulus was

30.96×10^3 N/mm². It reflects that when the coral reef sand is used, it can reach the strength of C30 grade concrete, which is higher than that of the concrete prepared with offshore coral sand [4,5]. Its strength index meets the requirements of the basic load-bearing component.

2.3 Pre-wetting effect and chloride ion permeability of coral concrete

2.3.1 Influence of pre-wetting degree on compressive strength

Due to the high porosity and salt content, the coral aggregate is not suitable for concrete. The reason is that the porosity of interface between mortar and coral aggregate is too large. In order to study its durability, it is necessary to test the influence of the pre-wetting of coral coarse aggregate on the strength. Through the analysis of the coral concrete with the same mix proportion in different pre-wetting degrees (30%, 50%, 70%) at different ages (7d, 28d, 56d), the variation is shown in Figure 2.

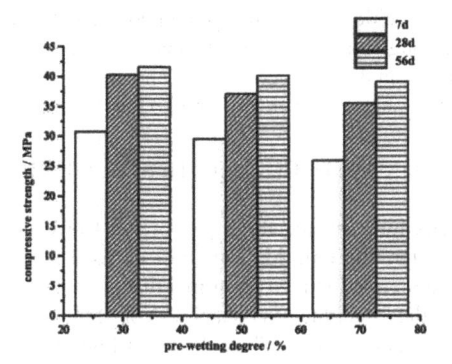

Figure 2. Effect of pre-wetting degree on compressive strength at different ages.

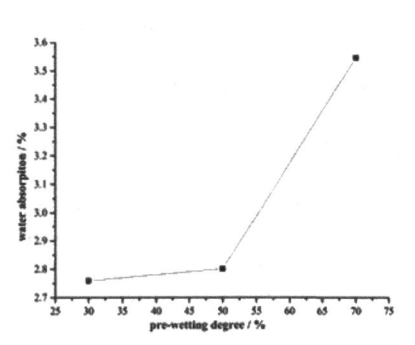

Figure 3. Effect of pre-wetting degree on water absorption.

The figure 2 shows that the strength growth lows down after the age of 28 days, and the hydration reaction is mainly at the age of 7–28 days, which is an important stage. The compressive strength of all ages decreased with the increase of pre-wetting degree. Therefore, the strength loss caused by the pre-wetting can be compensated by prolonging the curing time. In the early stage, the coral absorbs water from the cement base and reduces the local water cement ratio in the interfacial transition zone, which compacts the structure; Later, due to the decrease of water in the cement base, part of the water is exposed from the corals, which can maintain its hydration, and compact the pore structure [6]. Therefore, the advantage with higher degree of pre-wetting begins in the later stage, which can compensate for the early strength loss. When observing the influence of different pre-wetting degree (30%, 50%, 70%) on the water absorption at the age of 28 days, and the variation is shown in Figure 3. The figure shows that, with the increase of pre-wetting degree, the water absorption increases Especially when the pre-wetting degree exceeds 50%, the open macropores increase sharply, showing a significant increase in water absorption. Therefore, considering the durability, it is not suitable to use saturated or nearly saturated coral aggregate to prepare concrete.

2.3.2 Effect of seawater mixing on resistance to chloride ion penetration

Coral concrete was prepared from fresh water and sea water respectively. The electric flux and chloride diffusion coefficient at 28 d and 56 d age were measured. The variation is shown in Figure 4.

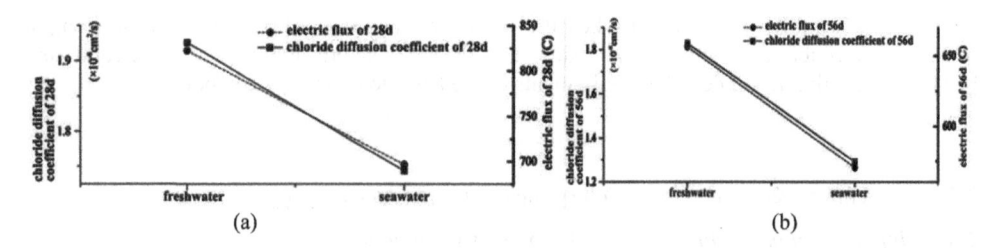

Figure 4. Effect of seawater on chloride diffusion coefficient and electric flux at different ages: (a) the age of 28d and (b) the age of 56d.

Figure 4 shows that the chloride diffusion coefficient and electric flux of coral concrete mixed with seawater are reduced. Taking fresh water mixing as reference, the chloride diffusion coefficient of coral concrete mixed with seawater is reduced by 9.5% in 28d, electric flux is reduced by 15.3% in 28d, chloride diffusion coefficient is reduced by 29.3% in 56d, and electric flux is reduced by 13.1% in 56d. Because of the increase of chemically combined chloride ion, the generated Friedel salt adheres to the pore wall and can fill the macropores. Although the pores are coarsened due to the heat of hydration, the porosity is reduced and reduce chloride ion diffusion coefficient and electric flux which indicates that seawater mixed plain concrete is not necessarily unfavorable to the durability.

3 STUDY ON THE PERFORMANCE OF CORAL CONCRETE COMPOSITE STRUCTURE

According to the basic properties of coral concrete, its strength can meet the requirements. However, due to the chloride content, and the mixing of seawater, the chloride ions will corrode steel. Therefore, when coral concrete is used in reinforced concrete load-carrying members, there is a problem of durability of steel corrosion.Based on this, the "steel tube / U-shaped steel beam coral concrete composite structure with outer FRP and inner coral concrete" as shown in Figure 1 is proposed. This paper researches the load-carrying capacity of coral concrete filled steel tube with FRP wrapped.

3.1 Structure

The structure of coral concrete filled steel tube wrapped with FRP is proposed as shown in Fig. 5.

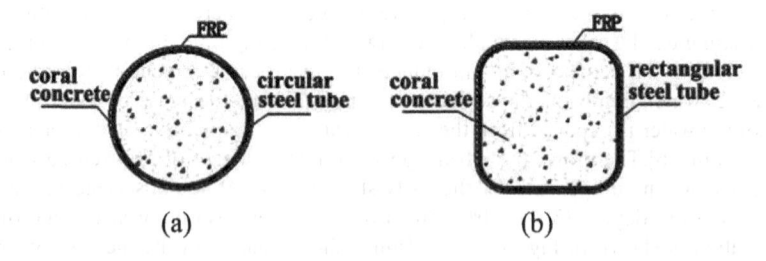

Figure 5. The cross section of coral concrete-filled steel tubular columns with wrapped FRP: (a) circular section and (b) rectangular section.

The composite member can make full use of the mechanical characteristics between the external casing pipe(tube) and the core concrete, thus improving the bearing capacity. In terms of structural function, the strength of coral concrete can be improved by the restraint of circular tube; Through the sealing of the outer pipe, the durability of coral concrete is enhanced; The corrosion resistance of steel pipe was strengthened by the external FRP; Through the inner lining of steel pipe, the brittleness of FRP can be made up. As shown in Figure 5, the different section has similar bearing capacity [7,8]. This paper describes the bearing capacity and calculation method of circular section.

3.2 Research on bearing capacity

3.2.1 Parameters of the specimen

Coral concrete was prepared with the mixture ratio before(measured cube compressive strength is 35.81 MPa, prism axial compressive strength is 33.92 MPa, elastic modulus is 30.98×10^3 N/mm^2). Steel pipes(the yield strength, ultimate strength and elastic modulus of the steel are 252.39 MPa, 319.40 MPa and 219.65×10^3 N/mm^2 respectively).The outer FRP is made of polymer carbon fiber cloth with a mass area ratio of 200 g/m^2, a theoretical thickness of 0.111 mm, a tensile strength of 3 245 MPa, a shear strength of 45.8 MPa and a tensile elastic modulus of 2.24 MPa$\times 10^5$ N/mm^2.Eighteen groups of 36 specimens were made to analyze the mechanical properties under axial compression. The specimen parameters are shown in Table 3.

Table 3. The basic parameters of specimens.

Group number	Inner diameter D/mm	Wall thickness t/mm	height L/mm	FRP Layer number	Group number	Inner diameter /mm	Wall thickness t/mm	height L/mm	FRP Layer number
SC1	110	1.5	300	0	FSC2	135	2	300	1
SC2	110	2	300	0	FSC3	135	2.5	300	1
SC3	135	2	300	0	FSC4	135	3	300	1
SC4	135	2	375	0	FSC5	135	2	375	1
SC5	135	2.5	300	0	FSC6	160	2	450	1
SC6	135	3	300	0	FSC1(2)	110	2	300	2
SC7	160	2	432	0	FSC2(2)	135	2	300	2
FSC0	110	1.5	300	1	FSC3(2)	135	2.5	300	2
FSC1	110	2	300	1	FSC4(2)	135	3	300	2

3.2.2 Load displacement variation

According to the characteristics of load displacement curve, the axial compression process can be divided into five stages: elastic stage, elastic-plastic stage, plastic stage, failure stage and restrengthening stage. The curve with different steel tube wall thickness under the full section loading conditions is shown in Figure 6, when the section is the same, with the increase of the wall thickness, the hoop coefficient increases.The smaller the elastic deformation process under axial pressure, the greater the vertical displacement when the specimen reaches the maximum bearing capacity, reflecting better ductility. In addition, the load displacement of short column specimens with different layers of FRP is different, as shown in Figure 7. The curve shows that: first, FRP has a certain improvement on the bearing capacity ; Secondly, the more FRP layers are, the greater the vertical displacement is. It is shown that the FRP cloth can not only improve the bearing capacity, but also improve the ductility.

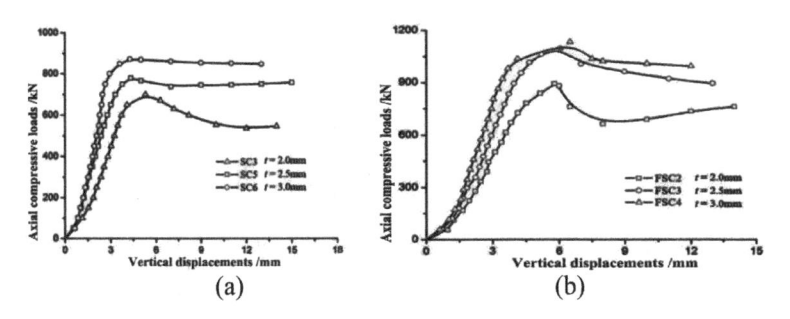

Figure 6. Load-displacement curve of different wall thickness specimens: (a) Without FRP, (b) With FRP.

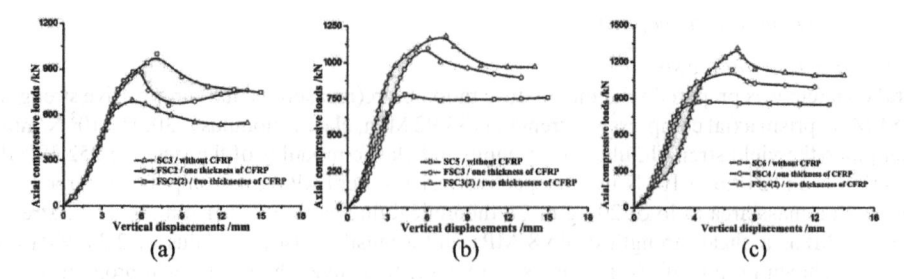

Figure 7. Load-displacement curve of same length-to-diameter ratio, different CFRP wrapped: (a) t = 2mm, (b) t = 2.5mm and (c) t = 3mm.

3.2.3 Load strain curve

In order to investigate the lateral constraint effect of CFRP cloth on steel tube, the relationship between circumferential strain and load at the same section of steel tube with or without CFRP wrapped is analyzed as shown in Figure 8, which reflects that under the same load condition, at the middle section, the hoop strain of steel tube without CFRP cloth is larger than that of steel tube with CFRP cloth, which shows the limiting effect of FRP cloth on the hoop strain of steel tube.

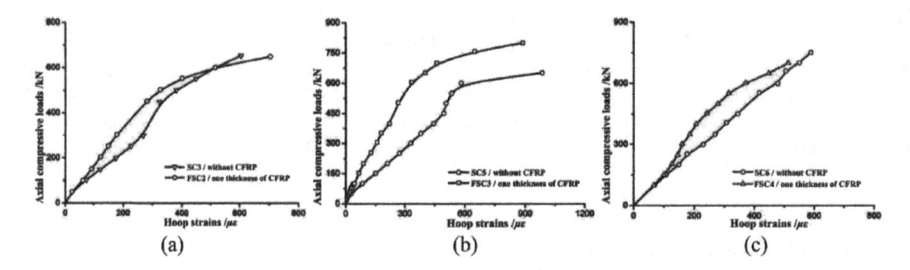

Figure 8. Load circumferential strain curve of same length-to-diameter ratio, different CFRP wrapped: (a) t = 2mm, (b) t = 2.5mm and (c)t = 3mm.

In addition, through the comparison of circumferential strain between steel pipe and FRP cloth, the deformation coordination between them can be further mastered. As shown in Figure 9, with the increase of axial compression load, the hoop strains all increase. Before the ultimate failure, there is no significant difference, and the growth path is basically the same. Therefore, it is considered that the steel pipe and FRP cloth can work together under the action of carbon fiber impregnant, and the bond slip is small, which has little effect on the mechanical properties. The results also provide experimental basis for the theoretical analysis and finite element numerical simulation of FRP coral concrete filled steel tube short columns.

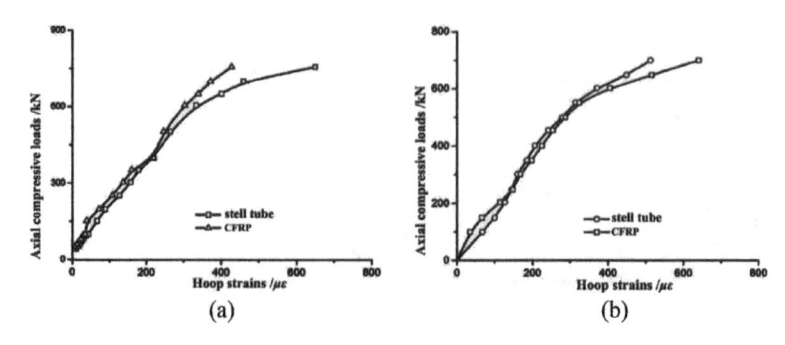

Figure 9. Load circumferential strain curves of specimens: (a) FSC3 and (b) FSC4.

3.3 Calculation of ultimate bearing capacity

Based on the research of the strengthening mechanism of coral concrete by steel tube hoop, the basic mechanical behavior and failure mode are obtained through the axial compression test ,the result shows that the factors influencing the stress mode include aspect ratio, material strength, steel content, section size, CFRP outsourcing conditions, etc. Starting from the stress conditions, the ultimate bearing capacity calculation method is analyzed by using the limit equilibrium theory, and a unified ultimate load-carrying calculation method is proposed.

3.3.1 Stress analysis

According to the experiment, the mechanical behavior of the material has the characteristics of three-way compression concrete. When the ultimate load is close, the volume deformation of concrete increases, the ultimate bearing capacity and volume compression deformation increase with the increase of the hoop coefficient. When the aspect ratio is large, the ultimate bearing capacity will decrease significantly. In the process of axial compression, the stress state and change process of FRP cloth, steel tube and core concrete are more complicated than that of ordinary concrete filled steel tube. With FRP cloth wrapped, the transverse deformation of steel pipe is also constrained under the bowstring effect. The steel pipe is subjected to radial force, axial pressure and circumferential force. FRP cloth is influenced by radial force and circumferential force of steel pipe.

The stress diagram of concrete, steel pipe and FRP cloth is shown in Figure 10. In the picture: $\sigma_{t,cf}$ is the hoop stress of FRP cloth; $\sigma_{r,cf}$ is the radial stress of FRP cloth; σ_1 is the longitudinal stress of steel pipe; σ_2 is the hoop stress of steel pipe; $\sigma_{c,c}$ is the longitudinal stress of core coral concrete; σ_r is the radial stress of steel pipe; p_c is the lateral stress (radial stress) of core coral concrete. In the limit state, both steel pipe and core coral concrete are subjected to radial pressure, and the bearing capacity can reach the peak.

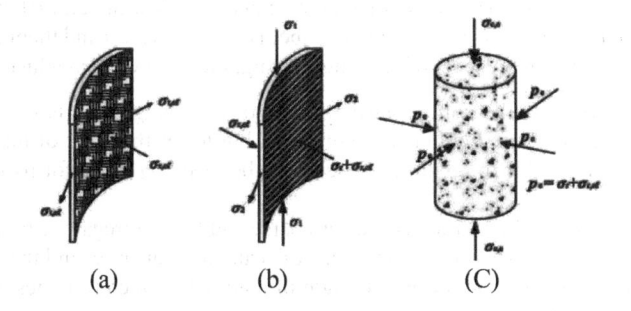

Figure 10. Force diagram of FRP coral concrete-filled steel tube: (a) FRP, (b) Steel tube and (c) core coral concrete.

3.3.2 Calculation formula and data check of ultimate bearing capacity

According to the limit equilibrium theory, under the limit state of axial compression bearing capacity, the calculation expression of bearing capacity N_u is obtained as shown in formula (1)

$$N_u = A_c f_{c,c}[1 + 0.5K_c(\xi_a + \xi_{cf})] \quad (1)$$

Where: A_c is the cross-sectional area of core coral concrete; $f_{c,c}$ is the axial compressive strength of core coral concrete; K_c is the lateral pressure coefficient (according to the analysis, the K_c can be calculated as 2.53); ξ_a is the hoop coefficient of steel tube to core concrete; ξ_{cf} is the hoop coefficient of core concrete under the condition of FRP cloth wrapped with steel pipe.

In order to verify the accuracy of formula (1), the measured value of ultimate strength of specimens in 3.2 bearing capacity test is compared with the theoretical analytical value of formula (1), as shown in Table 4. The ratio of the measured value N_u^t to the theoretical value N_u is between 0.868

and 1.073. It shows that the experimental results are in good agreement with the theoretical calculation results, and the analytical calculation method can provide a theoretical basis for engineering application.

Table 4. The measured and theoretical values comparison of FRP coral concrete-filled steel tube members ultimate axial compressive strength.

Specimen number	Member size $D \times t \times L/$ mm	hoop coefficient ξ_a	hoop coefficient ξ_{cf}	FRP Layers number	Slende-rness ratio λ	measured value $N_u^t/$ kN	calculated value $N_u/$ kN	N_u^t/N_u
FSC1	110×2.0×296	0.541	0.397	1	10.8	648.00	704.50	0.920
FSC2	135×2.0×300	0.441	0.323	1	8.8	893.00	954.47	0.936
FSC3	135×2.5×298	0.551	0.323	1	8.8	1 097.00	1 022.17	1.073
FSC4	135×3.0×300	0.662	0.323	1	8.8	1 134.00	1 089.86	1.041
FSC5	135×2.0×375	0.441	0.323	1	11.2	904.00	954.47	0.947
FSC6	160×2.0×450	0.372	0.273	1	11.2	1 102.00	1 237.73	0.890
FSC1(2)	110×2.0×296	0.541	0.794	2	10.8	800.00	866.20	0.924
FSC2(2)	135×2.0×300	0.441	0.647	2	8.8	1 001.00	1 152.93	0.868
FSC3(2)	135×2.5×298	0.551	0.647	2	8.8	1 174.00	1 220.62	0.962
FSC4(2)	135×3.0×300	0.662	0.647	2	8.8	1 310.00	1 288.32	1.017

4 CONCLUSION

The coral concrete-filled steel tube (bundle tube or steel beam) with wrapped FRP based on coral aggregate is proposed to be used as load-bearing member. Based on the test and theoretical analysis, combined with the application prospect of reef engineering, the following conclusions are drawn:

(1) The test results show that the compressive strength of coral aggregate is 6.23 MPa, which is slightly higher than that of light coarse aggregate, but lower than that of inland rocks and pebbles, so it is not suitable to be directly used as unbonded aggregate for roadbase backfill of road engineering;
(2) Coral Reef Sand with good gradation is used as coarse and fine aggregate to prepare concrete. Its mechanical properties meet the requirements of structural concrete, and the degree of pre-wetting has a great influence on the age strength of materials, which provides a reference for coral concrete construction;
(3) The results show that the chloride ion erosion has little effect on the durability of plain concrete, but it still has great erosion on the internal steel. Therefore, the composite structure of FRP coral concrete filled steel tube is proposed to be used as the bearing member;
(5) Through the experimental analysis on the bearing capacity of coral concrete-filled steel tube with wrapped FRP composite structure as pressure bearing member, it is considered that the new composite structure has large integral rigidity, high bearing strength and good ductility, which can meet the structural bearing requirements, and the analytical calculation formula can provide design basis for engineering application.

REFERENCES

[1] Gao Yi, Wei Zhuobin, Sun Xiao. Experimental research on basic mechanical properties of coral aggregate concrete[J]. Journal of Naval University of Engineering, 2017, 29(2):64–68.
[2] Sun Xiao, Wei Zhuobin, Gao Yi. Experimental research on the mix design of seawater coral aggregate concrete [J]. Sichuan Architecture, 2016, 36(1):204–206.

[3] GB/T 50081–2016, standard for test method of mechanical properties on ordinary concrete[S].
[4] Zhao Yanlin, Han Chao. Experimental study on the compression age strength of seawater coral concrete[J]. Concrete, 2011(2):43–45.
[5] Han Chao. Experimental research on the fundamental mechanical behavior of seawater coral concrete[D]. Nan Ning: Guangxi University, 2011
[6] Wei Zhuobin, Li Zhongxin, et al. Research on the influencing factors of performance of coral concrete and its early mechanical property[J]. Industrial Construction, 2017, 47(3):130–136.
[7] Zhao He, Gao Yi, Li Haoyang. Research of the compressive testes and bearing capacities of coral concrete filled square steel tube short columns [J]. China Concrete and Cement Products, 2019(3):43–47.
[8] Gao Yi, Li Haoyang, Zhao He. Experimental research on axial compression behavior of CFRP coral concrete filled square steel tube short columns[J]. Concrete, 2019(11):54–57.

Research on creep-fatigue lifetime of GH4133B superalloy at elevated temperature used in turbine disk of aero-engine

R.G. Zhao
College of Civil Engineering and Mechanics, Xiangtan University, Xiangtan, China
Key Laboratory of Dynamics and Reliability of Engineering Structures of College of Hunan Province, Xiangtan University, Xiangtan, China

X. Zhou
College of Civil Engineering and Mechanics, Xiangtan University, Xiangtan, China

Y.F. Liu
College of Aerospace Engineering, Chongqing University, Chongqing, China

X.Q. Yang, W.H. Liu, Y.L. Liu & D.H. Ye
College of Civil Engineering and Mechanics, Xiangtan University, Xiangtan, China

ABSTRACT: At high temperature of 650°C, the creep-fatigue lifetime tests for sheet specimens of GH4133B superalloy used in turbine disk of an aero-engine are conducted in two types of loading to study the creep-fatigue behaviour and to investigate the effect of loading parameters on fatigue lifetime. The results show that the creep displacement curve consists of three stages as initial creep, stable creep and accelerated creep, and the creep-fatigue interaction accelerates the accumulated fatigue damage, which leads to reducing fatigue lifetime. Under the condition of different stress ratios at a fixed maximum stress 700MPa, the creep-fatigue lifetime firstly decreases, and then increases with increasing stress ratio, there is a minimum creep-fatigue lifetime when the stress ratio is of 0.1. The creep-fatigue tests at various maximum stress levels show that there exists a mean stress threshold value about 330MPa for the superalloy, as the mean stress is above 330MPa, the influence of creep deformation on creep-fatigue damage is remarkable, resulting in the dramatic reduction of creep-fatigue lifetime.

1 INTRODUCTION

Nickel-based superalloys are widely used to manufacturing turbine disk of gas turbine aero-engine for their excellent high temperature performances, such as good fatigue resistance, adequate endurance strength and creep resistance, excellent corrosion resistance and oxidation resistance [1–3]. The turbine disk of aero-engine is suffering from severe operating conditions, such as repeated start-up and shut-down transient loading and sustained high stress at high temperature, in addition to highly corrosive hot exhaust gases, resulting in creep, fatigue, oxidation and corrosion damage as critical factors in predicting component lifetime, and a damage tolerance design criteria based on these damage variables has been introducing into the design of aeronautical materials and components [4–6].

GH4133B is a modified superalloy made by adding proper amount of Mg and Zr microalloying on the basis of nickel-based superalloy GH4133. GH4133B eliminates the notch sensitivity of GH4133 superalloy above 750°C, so that the service lifetime of material increases exponentially, and the rupture strength and plasticity are improved. Therefore, GH4133B is selected as the material of turbine disk of aero-engine in aviation material selection [7]. The hot compression and deformation behaviors of the FGH4096–GH4133B dual alloy were investigated, and a processing map approach was adopted to optimize hot forging process for thus dual superalloys [8,9]. Zhao [10] tested the accumulated fatigue damage for smooth specimens of GH4133B superalloy under a symmetrical fatigue cyclic loading mode at ambient temperature. A damage parameter of resistance change rate was selected to describe the variation of fatigue damage with cycle numbers, and the modified Chaboche model was adopted to derive the fatigue damage evolution equation, the results shown that the ratio of fatigue nucleation and small crack growth lifetime to the total fatigue lifetime increases with increasing stress amplitude. Hu [4] investigated the creep-fatigue interaction and the loading history effect on the creep-fatigue damage of GH4133B at 600°C in three types of loading as CF, F+C and C+F, a nonlinear damage evolution equation was proposed according to the fatigue accumulative damage theory, and found that the creep-fatigue damage in the cases of CF or F+C was larger than unity, while that in the case of C+F was smaller than unity.

Although many researches on creep-fatigue interaction of nickel-based superalloy has been achieved [12–15], there is no unified understanding for the micro-mechanism of creep-fatigue interaction. Commonly, under the condition of creep-fatigue interaction, the damage in the material increases with cyclic numbers. The dynamic load generates fatigue damage, and fatigue damage results in local fatigue crack initiation, while the static load in cyclic loading causes creep damage, and creep damage leads to voids accumulating at the grain boundary to form grain boundary cavitations. When the fatigue crack and grain boundary cavity meet inside the material, both of them mutually promote the formation of creep-fatigue interaction, which results in increasing the number of creep cavities, and accelerating the fatigue crack growth. Thus, the creep load shortens the fatigue crack initiation time, and the fatigue load accelerates the formation and growth of the creep cavity.

In this paper, the creep-fatigue tests for smooth sheet specimens of GH4133B superalloy are conducted at 650°C and standard atmospheric pressure. Adopting two types of stress cyclic loading, the creep-fatigue behaviors are investigated under a fixed maximum stress level at various stress ratios and under a fixed stress ratio at various maximum stress levels as well. The creep-fatigue lifetime is predicted, and the micro-mechanism of creep-fatigue interaction is discussed.

2 EXPERIMENTAL SECTION

2.1 *Materials and specimens*

The material used in this research is a powder metallurgy nickel-based superalloy with following chemical compositions in % weight: C 0.06, Cr 19 ~ 22, Al 0.75~1.15, Ti 2.5~3.0, Fe 1.5, Nb 1.3~1.7, Mg 0.001~0.01, Zr 0.01~0.1, B 0.01, Ce 0.01, Mn 0.35, Si 0.65, P 0.015, S 0.007, Cu 0.07, Bi 0.0001, Sn 0.0012, Sb 0.0025, Pb 0.001, As 0.0025, and the balance nickel. The specimens are machined from a turbine disk. The heat treatment process is as follows: $1080 \pm 10°C$, 8 hours, air cooling with $750 \pm 10°C$, 16 hours air cooling. The specimen is machined in the shape of sheet specimen, and the thickness of specimen is 3.0mm, as shown in Figure 1.

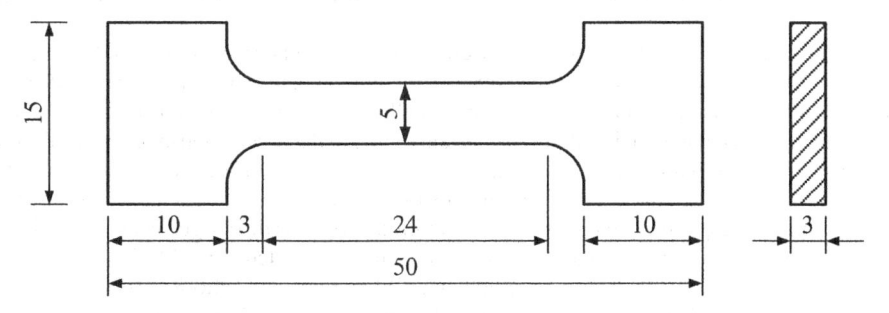

Figure 1. Dimensions of sheet specimen of GH4133B superalloy (unit: mm).

2.2 Creep-fatigue tests

Under the condition of atmospheric environment, the creep-fatigue tests are carried out at elevated temperature of 650°C on an MTS809 materials testing machine. The fatigue test waveform is sine wave with a frequency of 2Hz. The tests are divided into two groups. For the first group of creep-fatigue tests, the maximum stress is controlled as 700MPa, and the stress ratios are individually set as 0.01, 0.1, 0.2 and 0.4, and the specimens are numbered from S1 to S4. For the second group of creep-fatigue tests, the stress ratio is set as 0.1, and the maximum stress ranges from 900MPa to 550MPa, and the specimens are numbered from S5 to S8. A pair of self-made high temperature resistance fixtures is clamped at both ends of the MTS809 materials testing machine, and the two ends of the specimen are individually installed in the rabbets of the self-made fixtures. Adjusting the location of the high temperature furnace equipped on the testing machine, placing the specimen in the middle of the high temperature furnace, closing the high temperature furnace, turning on the power switch to raise the temperature to 650°C in the high temperature furnace, and then sustaining 10 minutes at 650°C, following that, the creep-fatigue test is conducted at 650°C, the test data of displacement of the specimen are collected, and the fatigue cycle numbers are recorded. The creep-fatigue lifetimes under different operated parameters are shown in Table 1.

Table 1. The creep-fatigue lifetimes of GH4133B superalloy at different operated parameters.

R	σ_{max} (MPa)	σ_{min} (MPa)	σ_m (MPa)	σ_a (MPa)	N_f	R	σ_{max} (MPa)	σ_{min} (MPa)	σ_m (MPa)	σ_a (MPa)	N_f
0.01	700.0	7.0	353.5	346.5	27903	0.1	900.0	90.0	495.0	405.0	3449
0.1	700.0	70.0	385.0	315.0	26823	0.1	800.0	80.0	440.0	360.0	10976
0.2	700.0	140.0	420.0	280.0	27928	0.1	600.0	60.0	330.0	270.0	54769
0.4	700.0	280.0	490.0	210.0	42659	0.1	550.0	55.0	302.5	247.5	178117

3 RESULTS AND DISCUSSION

3.1 Analysis of creep-fatigue test data

The maximum displacement and minimum one at both ends of specimen of GH4133B superalloy at every certain fatigue cycles are collected, and the corresponding fatigue cycles are recorded. The static component of creep deformation δ_s is calculated as a half of the sum of the maximum displacement d_{max} and the minimum one d_{min}, i.e., $\delta_s = (d_{max} + d_{min})/2$, and the dynamic component of creep deformation δ_a is calculated as a half of the difference of the maximum displacement d_{max} and the minimum one d_{min}, i.e., $\delta_a = (d_{max} - d_{min})/2$. At a fixed maximum stress as 700MPa, the creep-fatigue tests are conducted at various stress ratios at 650°C, and the creep-fatigue test data are recorded. The experimental curves of the static component of creep displacement versus fatigue cycle at a fixed maximum stress 700MPa under different stress ratios are shown in Figure 2.

It can be found from Figure 2 that the static creep displacement curve consists of three stages as initial creep, stable creep and accelerated creep. In the initial creep stage, the creep rate decreases gradually with increasing fatigue cycles, which is called as an unstable creep stage. In the stable creep stage, the creep rate remains almost a constant, while in the accelerated creep stage, the creep rate increases rapidly with increasing fatigue cycles, which leads to the ultimate fracture of the specimen. Fitting approximately the creep curve by using three segment broken line, and taking the abscissa values corresponding to the intersection of broken lines as the fatigue cycles of each creep stage, the initial creep lifetime, steady creep lifetime and accelerated creep lifetime of GH4113B superalloy under different stress ratios are obtained. When the stress ratio R is 0.01, 0.1, 0.2 and 0.4, the ratios of initial creep lifetime to total lifetime are 7.74%, 12.68%, 9.41% and 20.16% respectively, the ratios of steady creep lifetime to total lifetime are 87.59%, 78.66%, 86.15% and

75.58% respectively, and the ratios of accelerated creep lifetime to total lifetime of GH4133B alloy are 4.67%, 8.66%, 4.44% and 4.26% respectively. It can be seen that for GH4133B superalloy with different stress ratios, the ratio of steady creep lifetime to total lifetime is more than 75%, while the ratio of accelerated creep lifetime to total lifetime is less than 10%. When the stress ratio is 0.01, the ratio of initial creep lifetime to total lifetime is the smallest value, while the ratio of steady creep lifetime to total lifetime is the largest one. When the stress ratio is 0.4, the ratio of initial creep lifetime to total lifetime is the smallest value, The ratio of initial creep lifetime to total lifetime is the largest, while the ratio of steady creep lifetime to total lifetime is the smallest one.

Simultaneously, the curves of the dynamical components of creep displacement versus fatigue cycle at a fixed maximum stress as 700MPa under different stress ratios are shown in Figure 3. It can be seen from Figure 3 that the fluctuation value of creep displacement firstly decreases and then increases with increasing fatigue cycle. In the initial creep stage, the fluctuation value of creep displacement decreases with increasing fatigue cycle, while in the steady creep stage, the fluctuation value of creep displacement changes gently with increasing fatigue cycle, and until it reaches a minimum value. As the creep moves into the accelerated creep stage, the fluctuation value of creep displacement increases rapidly with the increase of fatigue cycle.

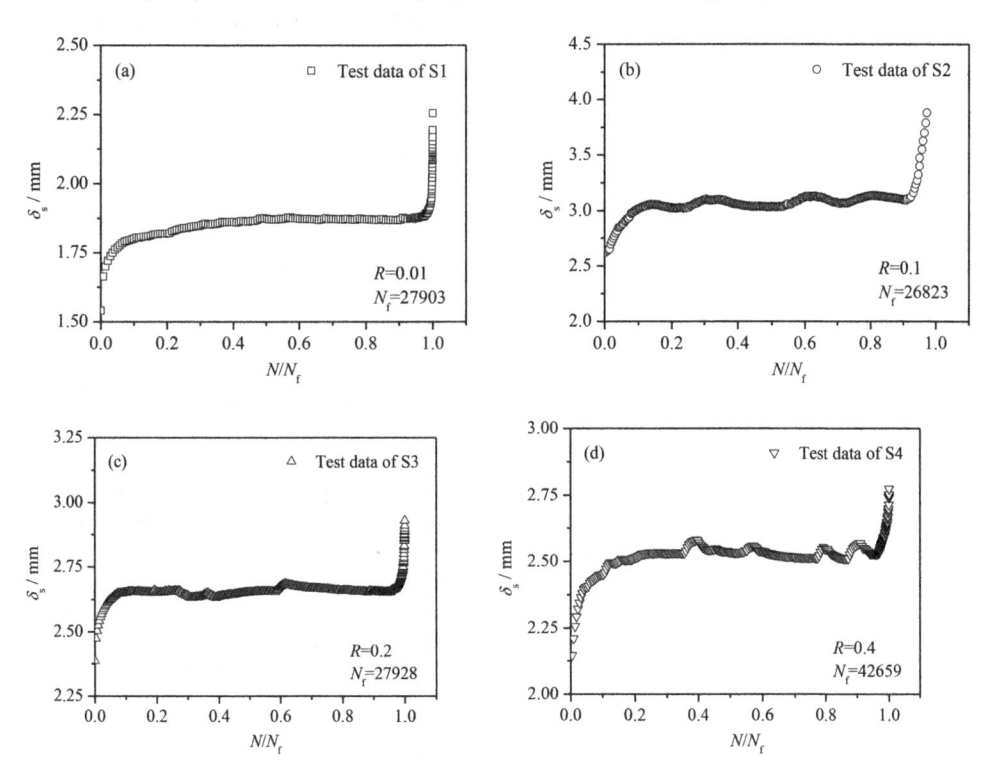

Figure 2. The static components of creep displacement versus fatigue cycle at $\sigma_{max} = 700$MPa.

3.2 Creep-fatigue lifetime analysis

As the maximum stress is set as 700MPa, the creep-fatigue lifetimes of GH4133B superalloy at various stress ratios are measured. According to the formulas that the mean stress σ_m is calculated as a half of the sum of the maximum stress σ_{max} and the minimum stress σ_{min}, i.e., $\sigma_s = (\sigma_{max} + \sigma_{min})/2$, and the stress amplitude σ_a is calculated as a half of the difference of the maximum stress σ_{max} and the minimum stress σ_{min}, i.e., $\sigma_a = (\sigma_{max} - \sigma_{min})/2$, as the stress ratio R ranges from 0.01 to 0.4, the mean stress σ_m increases from 353.5MPa to 490.0MPa, while the stress amplitude σ_a

decreases from 346.5MPa to 210.0MPa. It can be found from Table 1 that as the stress ratio is 0.4, the creep-fatigue lifetime is 42559 cycles. As the stress ratio decreases, the creep-fatigue lifetime significantly decreases until it reaches to a minimum value as 26828 cycles at stress ratio 0.1. While as the stress ratio decreases form 0.1 to 0.01, the creep-fatigue lifetime is reversal with decreasing stress ratio, as shown in Figure 4, which shows that the effect of creep-fatigue interaction on fatigue damage is the most significant at the stress ratio of 0.1, resulting in the shortening of creep-fatigue lifetime.

Under the condition of a fixed stress ratio as 0.1, the maximum stress σ_{max} ranges from 900MPa to 550MPa, following with which, the mean stress σ_m changes from 495.0MPa to 302.5MPa, and the stress amplitude σ_a decreases from 405.0MPa to 247.5MPa. The creep-fatigue lifetime of GH4133B superalloy decreases with increasing maximum stress, as shown in Table 1. Simultaneously, as the maximum stress σ_{max} increases from 550MPa to 600MPa, the creep-fatigue lifetime of the superalloy dramatically decreases, which suggests that in the case of maximum stress 550MPa, the mean stress is 302.5MPa, the effect of creep deformation on fatigue lifetime is not remarkable, while the maximum stress reaches 600MPa, and the corresponding mean stress is 330MPa, the effect of creep-fatigue interaction on cumulative fatigue damage is significant, resulting in that the creep-fatigue lifetime of GH4133B superalloy sharply decreases with increasing maximum stress or mean stress.

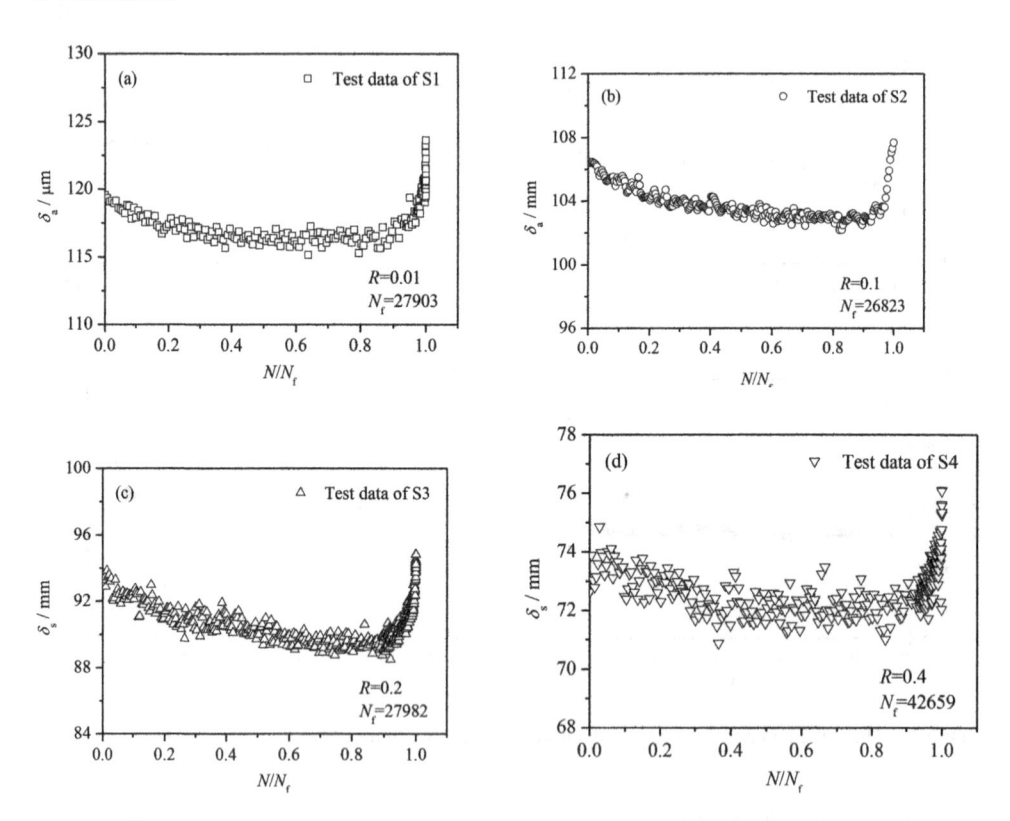

Figure 3. The dynamical components of creep displacement versus fatigue cycle at $\sigma_{max} = 700\text{MPa}$.

Supposing that there exists a function between the maximum stress σ_{max} and the creep-fatigue lifetime N_f for metallic alloy, the function can be expressed as

$$N_f = C\sigma_{max}^{-m} \qquad (1)$$

where, C and m are the material parameters. According to Eq. (1), the test data of creep-fatigue lifetime at various maximum stresses at a fixed stress ratio of 0.1 are fitted by using the nonlinear regression method. The material parameters C and m in Eq. (1) are individually identified as 7.46985×10^{26} and 7.88605. The comparison between the theoretical curve and experimental data are shown in Figure 5. It can be found from Figure 5 that the theoretical value of creep-fatigue lifetime is in good agreement with the experimental data, which shows that such function can be used to predict the creep-fatigue lifetime of GH4133B superalloy at various maximum stress levels.

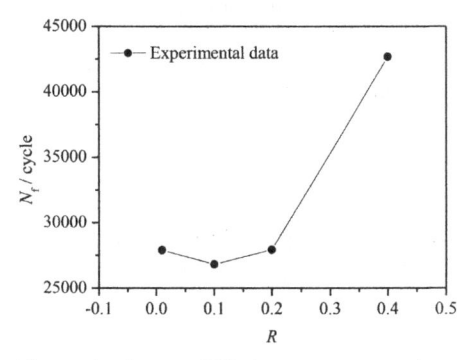

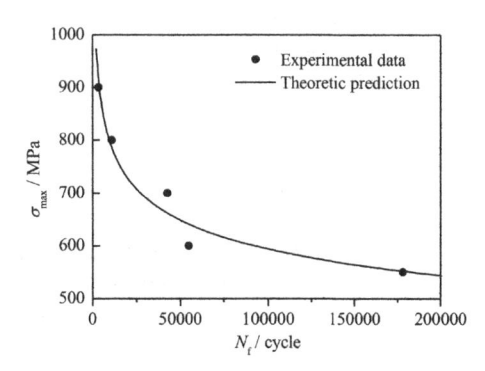

Figure 4. Curve of lifetime vs. stress ratio.

Figure 5. Curve of maximum stress vs. lifetime.

Simultaneously, the experimental curves of the static component and dynamic component of creep displacement versus the fatigue cycle at stress ratio 0.1 and at maximum stress level 550MPa are shown in Figure 6(a) and (b). It can be seen from Figure 6(a) that the static creep displacement curve is similar to those curves at a fixed maximum stress at various stress ratios (as shown in Figure 2). Whereas, the creep-fatigue test is interrupted due to the long fatigue lifetime at this operated fatigue parameters, so the creep recovery phenomena caused by the interruption test can be observed on the creep displacement curve. At a fixed stress ratio 0.1, the static creep displacement increases with increasing maximum stress level. When the maximum stress level drops to 550MPa, the static creep displacement is relatively minimal, which suggests that the effect of creep on fatigue lifetime is not significant at this case. Moreover, it can be seen from Figure 6(b) that the dynamic creep displacement curve is similar to those curves at a fixed maximum stress at various stress ratios (as shown in Figure 3). The fluctuation value of dynamic creep displacement decreases with increasing fatigue cycle in the initial creep stage, then slightly decreases with increasing fatigue cycle at the stable creep stage, and increases rapidly with increasing fatigue cycle at the accelerated creep stage.

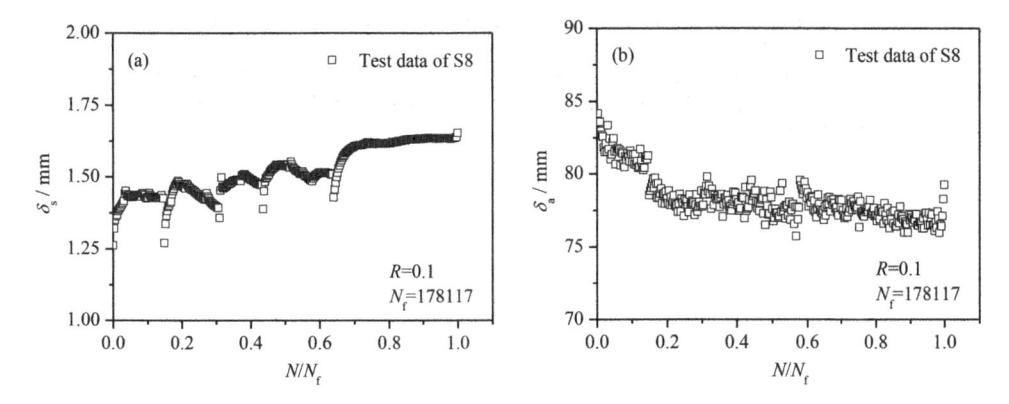

Figure 6. The Curves of creep displacement vs. fatigue cycle at $R = 0.1$ and $\sigma_{max} = 800$MPa.

4 CONCLUSIONS

The main conclusions are obtained as follows:

(a) The creep-fatigue interaction accelerates the creep deformation and accumulated fatigue damage of GH4133B superalloy operated at elevated temperature, resulting in dramatically reducing the service lifetime.
(b) As the stress ratio is close to 0.1, that is, the ratio of creep parameter and fatigue one is about 0.82, the effect of creep-fatigue interaction on fatigue lifetime is the most significant.
(c) There exists a mean stress threshold value about 330MPa for GH41133B superalloy, as the mean stress is above 330MPa, the effect of creep deformation on creep-fatigue damage is remarkable, resulting in the dramatic reduction of creep-fatigue lifetime.

ACKNOWLEDGEMENT

This work was supported by the High-level Talent Gathering Project of Hunan Province (no. 2019RS1059), the Key Project of the Reform of the Degree and Postgraduate Education in Hunan Province (no. 2019JGZD034), and the Research Project for Postgraduate Students of Hunan Province (no. CX20200648).

REFERENCES

[1] Pineau A and Antolovich S D 2009 *Engineering Fracture Analysis* **16**(8) 2668
[2] Jiang R, Song Y D and Reed P A 2020 *International Journal of Fatigue*, **141** 105887
[3] Pang H T and Reed P A S 2008 *International Journal of Fatigue* **30**(10–11) 2009
[4] Yu Q M, Yue Z F and Wen Z X 2008 *Materials Science and Engineering A* **477**(1,2) 319
[5] Pei C H, Zeng W and Yuan H 2020 *International Journal of Fatigue* **131** 105279
[6] Sulzer S, Li Z M, Zaefferer S, Haghighat S M H, Wilkinson A, Raabe D and Reed R 2020 *Acta Materialia* **185** 13
[7] Zhao R G, Li X J, Jiang Y Z, Luo X Y, Li J F, Li H C and Tan D H 2012 *Key Engineering Materials*, **512–515** 980
[8] Liu Y H, Yao Z K, Ning Y Q, Nan Y, Guo H Z, Qin C and Shi Z F 2014 *Materials and Design* **63** 829
[9] Liu Y H, Ning Y Q, Nan Y, Liang H Q, Li Y Z, Zhao Z L 2015 *Journal of Alloys and Compounds* **633** 505
[10] Zhao R G, Luo X Y, Jiang Y Z, Li H C, Li X J and Liu X H 2011 *Journal of Mechanical Engineering* **47**(6) 92 (in Chinese)
[11] Hu D Y and Wang R Q 2009 *Materials Science and Engineering A*, **515**(1,2) 183
[12] Wang R Z, Chen B, Zhang X C, Tu S T, Wang J and Zhang C C 2017 *International Journal of Fatigue* **97** 190
[13] Graverend J B, Cormier J, Jouiad M, Gallerneau F, Paulmier P, Hamon F 2010 *Materials Science and Engineering A* **527** 5295
[14] Wang R Z, Chen H, Zhang Y, Zhang X C, Tu S T 2020 *Materials and Design* **195** 108939
[15] Li K S, Wang R Z, Yuan G J, Zhu S P, Zhang X C, Tu S T and Miura H 2021 *International Journal of Fatigue* **143** 106031

Experimental research on mechanical performance of sulphoaluminate cement rapid repair material

Y. Liu
Research Institute of Highway Ministry of Transport, Beijing, China

X.J. Huang
Henan Huineng Road and Bridge Technology Co., Ltd, Zhengzhou, China

Z.Q. Hou
Research Institute of Highway Ministry of Transport, Beijing, China

ABSTRACT: In this paper, the mechanical properties of a new cement-based repair material, sulphoaluminate cement rapid repair material, are studied. The results show that the best retarding effect can be achieved when the gypsum content is 1%, and the retarder can be added appropriately to further achieve the retarding effect. The higher the temperature is, the faster the setting time is, and the basic strength requirements can be achieved within 4h, and the compressive strength can reach 40 MPa within 1 D, The bending strength is 7MPa, and there will be no shrinkage in the later stage. Moreover, if the interface smoothness is not good, it can provide a good help for the new and old concrete interface bonding. The old concrete surface is milled and washed to make the interface bonding force stronger. Through the experimental study, the new cement-based repair material sulphoaluminate cement has good market application value.

1 INTRODUCTION

Nowadays, highway pavements in many regions in China thoroughly show weakness and frequent defects under the alternate driving of overweight trucks. Moreover, the number of vehicles owned by residents in China keeps rising, transport vehicles are accelerated constantly, and these aspects cause the cement pavements to bear tougher pressure. Relevant departments and relevant industries take pavement repair as the objective, and expect to research and develop a sort of pavement repair material with a high setting and hardening rate, outstanding strength performance as well as stability and reliability. Cement concrete pavement has the advantages of favorable mechanical performance, excellent durability, good stability performance and low costs. However, as the setting time of concrete materials is long, the corresponding repair and maintenance time is also long, and the bonding performance of old and new concrete interfaces frequently has problems. Moreover, it often takes a period of time for maintenance to resume the normal traffic after the cement concrete is repaired. If the region repaired belongs to a traffic main artery or a business district, the influence will be greater [1, 2]

Sulphoaluminate cement rapid repair material is a new type of material based on sulphoaluminate cement-based material. It takes sand as the aggregate, can be added with gypsum, rubber powder, polypropylene fiber and additive, and is produced by mixing after optimizing the mix ratio. It has advantages of low price, strong toughness, long durability and impact resistance, even distribution of internal fibers, isotrocpic material, small discreteness, stable performance and good binding with other materials. The experimental material and mix ratio shall be selected reasonably, so as to guarantee the strength and durability of the material, and ensure to reach the expected effect.

2 EXPERIMENTAL SCHEME

The experiment adopts a cement-sand ratio of 1:1.4, the on-site experimental temperature is 19~21°C, the water-cement ratio W/C is 0.29, and standard curing box is adopted for curing. Refer to Table 1 for the details of materials and mixing amount used in the experiment, the sample numbers are 1, 2, 3, 4, 5 and 6 in turns, and the mechanical strength, setting time, bonding strength and dry-shrinkage performance of mortar are tested. The mix ratios of experiment are shown in Table 1.

Table 1. Table of mix ratios of experiment.

Serial No.	Cement	Coarse Sand	Fine Sand	Gypsum	Rubber Power	Water Reducer	Retarder	Defoaming Agent	Fiber	Water	Water-Cement Ratio
1	800	560	560	20	10	5	2.5	3	0.5	235	0.29
2	800	560	560	25	10	5	2.5	3	0.5	235	0.29
3	800	560	560	20	5	5	2.5	3	0.5	235	0.29
4	800	560	560	20	10	3	2.5	3	0.5	235	0.29
5	800	560	560	20	10	5	2.0	3	0.5	235	0.29
6	780	570	570	20	10	5	2.5	3	0.5	227	0.29

3 ANALYSES OF EXPERIMENTAL RESULTS

3.1 *Analysis of setting time experiment*

The setting time of cement-based rapid repair material determines the speed of road repair construction and the time of opening to traffic to a large extent. This experiment selects ingredients with various mixing amounts, estimates the specific time consumed by initial setting and final setting of cement, so as to further acquire the condition of mix ratio required by the construction. The experiment is carried out mainly according to Test Methods of Cement and Concrete for Highway Engineering. It adopts sulphoaluminate cement to research the initial setting time and final setting time of cement, meanwhile, it adds different ingredients, and research the influence of various ingredients on the setting time of cement [3] The experimental results are shown in Table 2.

Table 2. Table of determination results of setting time.

Serial No.	Initial Setting Time (min)	Final Setting Time (min)
1	34	52
2	20	35
3	52	83
4	18	30
5	40	58
6	29	48

The determination results of setting time indicate:

1. It can be known from Experiment 1, Experiment 3 and Experiment 6 that rubber powder has a delayed effect on the setting time of cement. It will restrict the hydration of C3A and the formation of ettringite, meanwhile, it makes the cement setting and hardening become slower, and the retarding effect is more obvious by addition of a small amount of rubber powder.
2. After adding the admixture, the addition of water reducer and retarder will also have a certain influence on the setting time of cement, under the premise of not affecting the mechanical property and workability of cement, adding an appropriate amount of admixture to adjust the setting time of cement is also a crucial choice.

3. Through the experiment, it can be clearly recognized that the temperature sensitivity of this material is relatively prominent, in short, the temperature continues to rise, and the time consumed for solidification will continue to reduce.

3.2 *Experimental analysis of mechanical performance of repair mortar*

The mortar sample is prepared according to the experimental regulations, and it is maintained in the standard curing box for 28 days. According to the specific experimental method for strength of cement mortar provided in Test Methods of Cement and Concrete for Highway Engineering, flexural strength experiment and compressive strength experiment are carried out for mortar sample. [4, 5] The experimental instruments are compression testing machine and electronic universal testing machine, the experimental results of flexural performance of repair mortar are shown in Table 3 and Table 4. The cross section of flexural test block after experiment is shown in Figure 1, the plane after compression resistance of experiment is shown in Figure 2.

Table 3. Mechanical performance data of repair mortar (flexural strength).

Serial No.	Age (MPa)			
	4h	1d	7d	28d
1	5.0	7.3	9.0	9.0
2	5.5	7.9	10.5	10.8
3	4.4	7.0	9.3	9.4
4	5.1	7.5	8.9	9.0
5	4.7	7.8	8.5	8.8
6	4.3	7.0	7.2	8.2

Table 4. Mechanical performance data of mortar (compressive strength).

No	Age (MPa)			
	4h	1d	7d	28d
1	21.1/21.7	43.4/43.3	48.2/48.1	48.9/49.1
2	24.1/24.0	45.0/44.3	50.3/50.5	50.5/50.5
3	26.5/26.3	40.2/40.0	48.5/48.3	48.7/48.8
4	22.3/22.4	44.3/44.4	47.3/47.5	48.3/48.2
5	25.3/25.4	44.4/44.7	47.2/47.2	47.7/47.5
6	24.3/24.2	40.6/40.4	44.4/44.3	45.8/45.9

Figure 1. Flexural surface of sample.

Figure 2. Compressive surface of sample.

The experimental results of strength experiment determination indicate:

1. It can be known from the experimental results that sulphoaluminate cement can meet the basic strength requirements within 4h, the compressive strength can reach 40MP and the flexural strength can reach 7MP within 1d, and the strength requirements conform to those of cement-based rapid repair material, but the strength in later stage increases gently, which basically reaches the maximum strength at 7d and completely conforms to the traffic requirements in a short term, the reduction phenomenon isn't shown in later stage, and the strength is stable.
2. The appropriate addition of gypsum and rubber powder can increase the strength of cement mortar, it can be known from the comparison of No. 1, No. 2, No. 3 and No. 6 that the blending ratio is optimal when the mixing amount of gypsum is 2.5%, which has a small influence on the strength and slightly increases the flexural strength. The mix ratio is optimal when the mixing amount of rubber power is 1.25%, the 1d compressive strength is 43.4/43.3MPa, and the flexural strength is 7.3MPa. According to the previous research on the reaction mechanism of sulphoaluminate cement, too much or too little addition of gypsum and rubber powder will influence the mortar strength.
3. The fluidity of sulphoaluminate cement-based repair material added with appropriate retarder and water reducer maintains well, which is still favorable after 25min. In order to meet the traffic requirements, the initial setting time must be controlled at about 30min and the final setting time must be controlled at about 45min. In order to guarantee the mechanical property of cement-based rapid repair material, 0.06% fibers are added to increase the flexural strength and the defoaming agent is added to reduce the bubbles on the surface layer, which is also to increase the strength of the repair material.

3.3 Experimental analysis of the bonding methods of repair mortar

Different material performance has different influences on the bonding strength, and the bonding performance is mainly related to the water-cement ratio, rubber powder and fiber, therefore, when selecting the mix ratio, for different water-cement ratios, different rubber powder outputs shall be compared with a certain of fiber output.

In the thin layer repair process of cement pavement, there are the following processing methods of bonding surface: (1) Shallow chiseling, short chiseling, and chiseling away some loose mortar and distress selectively; (2) When the interface hasn't been chiseled flat, the repair material shall be directly applied to the concrete surface, and dislocation repair on concrete surface with sufficient strength and settlement shall be carried out; (3) Deep chiseling, and the strength parts of the defective mortar and concrete shall be planed. [6, 7]

Table 5 is the interface processing methods designed. It is divided into Group a, Group b, Group c and Group d according to connection angle and connection shape, and experimental analyses are carried out.

Table 5. Processing methods of different interfaces.

		Horizontal	1	2	3	4
	a	Bonding Angle	30	45	60	90
Interface Processing	b	Rhombic Surface Bonding (Depth)/mm	1	2	3	4
Methods	c	Concave Curved Surface (Diameter of Curved Surface)/mm	1	2	3	4
	d	Convex Curved Surface (Diameter of Curved Surface)/mm	1	2	3	4

After multiple experiments, the integrated experimental results are shown in Table 6. There is data analysis obtaining that the weakest position of mortar as a whole is the position of interface connection, and the bonding strength of interfaces is the most important factor guaranteeing that the repaired surface course won't fall off and maintains the durability. Therefore, selecting the proper form of bonding surface is beneficial to the repair of pavement surface course [8]

Table 6. 4h Boding strength experimental results (MPa).

		Horizontal	1	2	3	4
Interface Processing Methods	a	Smooth Surface Bonding	4.5	3.3	2.5	2.0
	b	Rhombic Surface Bonding	1.9	2.5	4.1	4.8
	c	Concave Curved Surface	2.5	3.2	4.0	4.5
	d	Convex Curved Surface	2.3	3.0	3.6	4.0

By reading Table 6, it can be clearly recognized that when located in backgrounds of a1, b3, b4, c3, c4 and d4, all the strength of bonding surface is higher than 4MPa, which has significant similarity to the flexural strength of material. Through the experiment, it obtains that all the test blocks in the experiment have damage on the bonding surface during the experiment of bonding strength. According to research, when the shape of bonding surface is more complicated and the surface has no impurities, the bonding strength will be greater, the bonding depth will be bigger and the bonding force will increase correspondingly. The repair material directly bonds with the concrete slab, the test finds that the strength of directly vertical bonding is extremely low, while the bonding strength of a1 is high, which indicates that when there's no interface processing, it is necessary to make a small included angle to further guarantee the increase of bonding strength; for the concavo-convex edge bonding, if the depth maintains a higher trend, at this time, the bonding strength will be often in the lead. Finally, it should also be recognized that concrete corners after chiseling also can guarantee the smooth bonding of repair material and concrete [9, 10]

3.4 Experimental analysis of dry-shrinkage performance of repair mortar

The dry-shrinkage performance of mortar of concrete pavement repair material directly influences the bonding performance of old and new concrete interfaces. Therefore, the dry-shrinkage performance test on repair material is carried out. The experimental results are shown in Table 7.

Table 7. Dry-shrinkage performance data of repair mortar.

Experiment No.	Length of Mortar (mm)			Shrinkage Ratio (10^{-3})		
	3d	7d	28d	3d	7d	28d
1	15.987	15.985	15.982	0.688	0.938	1.13
2	15.957	15.953	15.948	2.69	2.94	3.25
3	15.966	15.961	15.959	2.13	2.43	2.56
4	15.979	15.975	15.973	1.31	1.56	1.68
5	15.968	15.960	15.957	2	2.5	2.69
6	15.989	15.978	15.977	0.813	1.38	1.44

It can be seen from Table 7 that the dry-shrinkage ratio of No. 1, No. 4 and No. 6 rapid repair mortar is smaller than that of No. 2, No. 3 and No. 5 rapid repair mortar. The shrinkage ratio of No. 1 mortar is the minimum and that of No. 2 mortar is the maximum. The appropriate addition of gypsum can lower the shrinkage ratio in later stage, thus it is beneficial to repair, moreover, with the extension of age, the difference of dry-shrinkage ratio between No. 1 mortar and mortar of other numbers will be bigger and bigger. On the one hand, as the functions of water reducer and retarder increasing gradually, under the same conditions, the fluidity of mortar shall be guaranteed, and the reduction of water consumption of mortar results in the reduction of dry shrinkage. On the other hand, the addition of rubber powder into cement mortar lowers the hydration heat release rate of mortar in early stage, the micro-crack and width of mortar in early stage reduce, the rubber powder after film formation forms the space grid structure in cement mortar, and the function of rubber powder closes the micro-pore structure of cement mortar, thus greatly improving the shrinkage performance of cement mortar.

4 CONCLUSION

This paper researches and analyzes five aspects of repair mortar, which are reaction mechanism, setting time, mechanical performance, bonding strength and dry-shrinkage strength. Starting from the mechanism of hydration reaction of cement, it analyzes and obtains that sulphoaluminate cement has the characteristics such as rapid hydration reaction, minimum inflation in early stage, small shrinkage in later stage and strong corrosion resistance. Therefore, it selects rapid hardening sulphoaluminate cement as the optimal choice for cement-based rapid repair mortar. Furthermore, it analyzes the influence of different admixtures on the setting time of rapid repair mortar, the optimal retarding function realizes when the mixing amount of gypsum is at 1%, an appropriate amount of retarder can also be added to realize the retarding function, and the setting time is rapider when the temperature is higher. In the experiment on the mechanical performance of cement-based rapid repair material, various mix ratios can reach the basic strength requirements within 4h, reach the compressive strength of 40MPa and flexural strength of 7MPa within 1d, and no reduction phenomenon occurs in later stage. Moreover, if the smoothness of interface is poor, it can further facilitate the bonding of old and new concrete interfaces on the contrary, and provide great reference for the interface processing of entity projects. The milling and rinsing of old concrete interface makes the bonding force of interface become stronger. Finally, it carries out the dry-shrinkage experiment, in which the addition of gypsum and polymer rubber powder to reduce dry shrinkage is beneficial to the bonding of old and new concrete in later stage.

REFERENCES

[1] Sun Jiaying, Wei Tao, Wang Xuewen. Influence of Polypropylene Fibermesh on the Properties of Road Concrete Materials [J]. Concrete, 2001(6):57–59.
[2] Strand D., Maedonld C.N., Emaktishnan V., Rajpathak V.N. Construction applications of polyolefin fiber reinforced concrete[J]. Proceedings of the Materials Engineering Conference, 1996: 103–112.
[3] Yu Zhenxing, Wang Lixin. Main Factors Affecting the Durability of Hydraulic Building and Preventive Measures [J]. Value Engineering, 2017, 36(19):191–192.
[4] Liu Zhiyong, Li Yantao, Liu Jinming. Research on the Efficiency of Rubber Powder and Mineral Powder in Cement-Based Materials of Polymers with a Low Polymer-Cement Ratio [J]. New Building Materials, 2001(7):16–18.
[5] Jiang Zhengping, Xu Huifen. A Study on the Cover Mortar for Repairing Surface Blemishes of Concrete Floor [J]. Journal of University of Science and Technology of Suzhou (Engineering and Technology), 2003, 16(4):49–55.
[6] Fowler D W. Polymer in concrete: a vision for the 21th century [C].Cement and Concrete Composites, 1999, 21(5/6):5.
[7] Ohama Y. Polymer-based admixtures [J]. Cement and Concrete, 1999, 28(2):4.
[8] Yang Q B, Wu X L. Factors influencing properties of phosphate cement-based binder for rapid repair of concrete [J]. Cement and Concrete Research, 1999, 29(3): 7.
[9] Zhang Xuehua, Ai Jun, Jiang Zhengping. Research and Application of Cement Concrete Pavement Rapid Repair Technology [J]. Forest Engineering, 2001, 6:2.
[10] Zhang Chunyu. The Research and Application of a Kind of New High-Performance Repair Material [J]. China Building Materials Science and Technology, 2003, 12(3):27–29.

In situ fatigue of new nickel-based powder superalloy at high temperature

X.L. Zhang
School of Materials Science and Engineering, University of Science and Technology Beijing, Beijing, China

G.F. Tian
Science and Technology on Advanced High Temperature Structure Materials Laboratory, Beijing Institute of Aeronautical, Beijing, China

G.H. Yang, Z. Liu, Z. Ji, X.Q. Yan & C.C. Jia
School of Materials Science and Engineering, University of Science and Technology Beijing, Beijing, China

ABSTRACT: In this paper, a low-cycle fatigue in situ observation was performed on a nickel-based powder alloy sample under a load stress of 650°C and 1050MPa using a scanning electron microscope to study the initiation, propagation and fracture of fatigue cracks. The results show that the fatigue cracks of the alloy mainly follow the direction perpendicular to the stress in the in situ fatigue test at 650°C. At 1528 Cycles, they sprout from the grain boundary, initially expand along the grain boundary, and then show a serpentine type cross grain Expansion, and finally the cycle breaks to 3214 Cycles. A slip zone is formed during the fatigue of the alloy, and the direction of the slip zone is 45° to the axis. Alloy fracture is a typical fatigue fracture, divided into fatigue source area, expansion area and instantaneous fracture area.

1 INTRODUCTION

The low-cycle fatigue of the belonging material refers to the fatigue failure process caused by the cyclic action of plastic strain under the action of alternating load, also known as plastic fatigue or strain fatigue [1–3]. Fatigue is a long-term damage accumulation process. In this process, the ultimate failure of materials lies in the local effects of certain materials. Due to the complex changes in the load, environment and other factors acting on components, so far, people still cannot fundamentally prevent accidents caused by low-cycle fatigue damage [4]

Turbine disks made of nickel-based powder superalloys are subjected to high centrifugal load, thermal load, aerodynamic load, vibration load and oxidation of environmental media during service. The working conditions are very harsh, which directly affects the safety, reliability and durability of the engine Sex. Among them, low-cycle fatigue damage is one of the important properties that affect nickel-based powder superalloys [5–7].In this paper, a low-cycle fatigue in situ observation was performed on a nickel-based powder alloy sample under a load stress of 650° and 1050MPa using a scanning electron microscope to study the initiation, propagation and fracture of fatigue cracks.

2 EXPERIMENTAL SECTION AND METHODS

2.1 *Materials*

The samples of new nickel-based superalloys used in this paper are prepared by powder metallurgy (PM) method.

2.2 In situ fatigue test

The nickel-based powder alloy sample was cut into the size of the sample shown in Figure 1. The thickness of the sample was 0.7cm, and the width of the narrowest part of the sample was 0.8cm. The specimen is subjected to fatigue test on a servo high temperature in situ fatigue testing machine. Test parameter selection: temperature 650°C, load stress ratio 0.1, loading frequency 10Hz, stress 1050MPa, loading load is 590N.

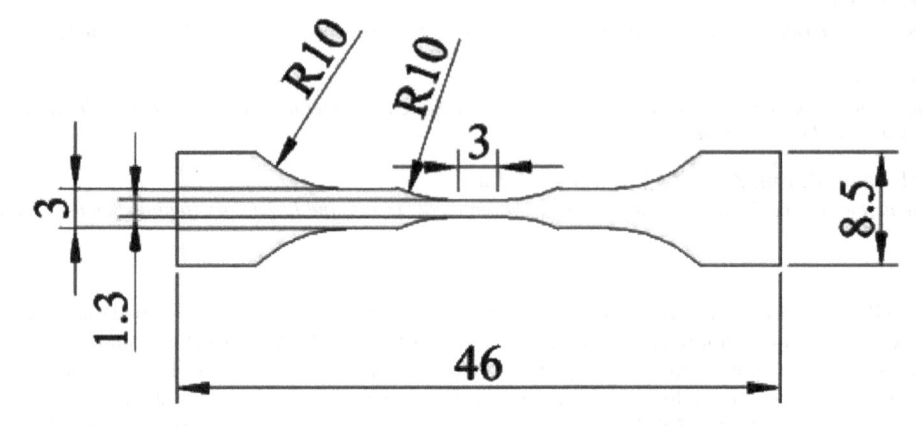

Figure 1. In situ fatigue test samples.

2.3 Characterization method

The in situ scanning Auger probe was used to observe the fatigue crack initiation, propagation and fracture of the new nickel-based powder alloy at high temperature.

3 RESULTS AND ANALYSIS

3.1 In situ observation of low cycle fatigue

Figure 2 shows the in situ fatigue morphology of the alloy at 650°C and 1050MPa load stress. Figure 2a is the SEM photograph of the sample before the fatigue test. When the sample is cycled to 1528 Cycles, as shown in Figure 2b, crack initiation is observed in the sample, and the crack initiates between the grain boundaries. This is because there are secondary phases on the grain boundaries. Strain occurs when the specimen is subjected to loading force, and the accumulation of cyclic strain will become the source of cracks, thereby forming cracks. When the sample is cycled to 2434 Cycles, it can be seen from Figure 2c that the initial crack propagation path is along the grain boundary. The reason is that there are defects such as vacancies and dislocations at the grain boundary. These defects can reduce the energy of crack growth. At this time, the energy required for the crack to expand along the grain boundary is much smaller and it is easier to expand. At the same time, the direction of crack propagation is perpendicular to the direction of loading force. When the cycle reaches 2852 Cycles, it can be seen from Figure 2d that the crack expands in a serpentine shape. In the later stage of the expansion, the crack expands from along the grain boundary to transgranular expansion. The expansion reduces the energy. In the later stage, due to the crack tip effect, the crack expansion has enough energy, no longer need to expand along the grain boundary, but to carry out transgranular expansion. The sample finally broke and failed after 3214 Cycles.

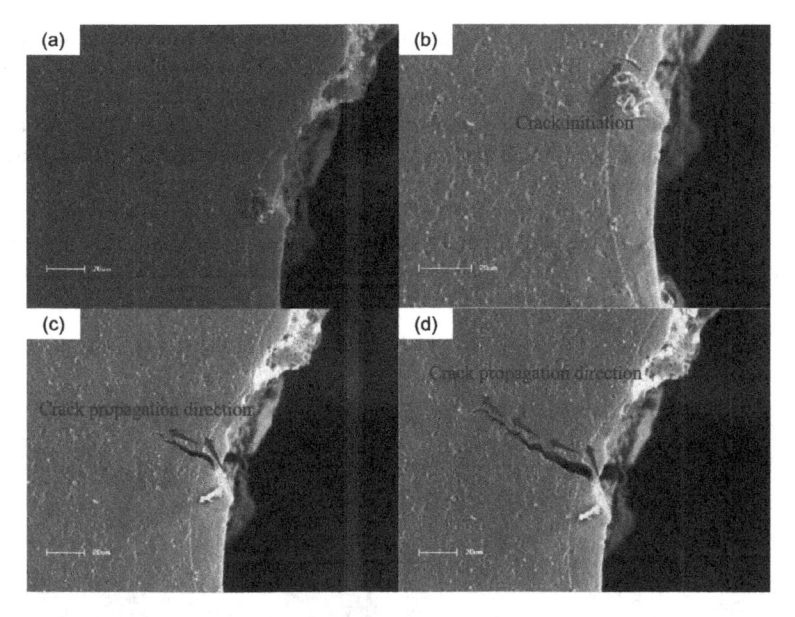

Figure 2. in situ fatigue morphology of the alloy at 650°C, 1050MPa load stress
(a) 0 cycles, (b) 1528 cycles, (c) 2434 cycles (d) 2852 cycles.

3.2 *Fatigue fracture analysis*

Figure 3 shows the SEM morphology of different areas on the fatigue fracture surface of a nickel-based superalloy. It can be seen from Figure 3 that there are a group of parallel lines in multiple areas. It can be inferred that a slip zone is formed during the fatigue process, and the direction of the slip zone is 45° to the axis. In the area slightly far away from the fracture, the slip zone gradually becomes less and sparse

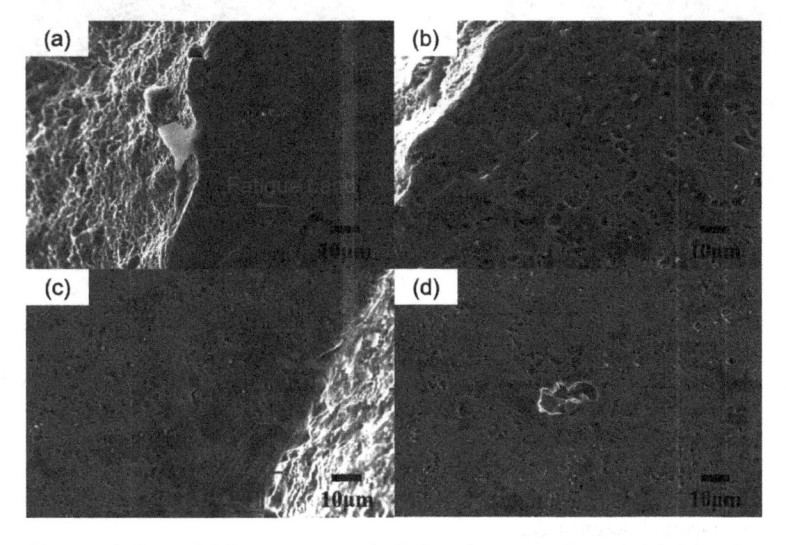

Figure 3. SEM morphology of different areas on the fatigue fracture surface of nickel-based superalloy
(a) 1000× (b) 1000× (c) 1000× (d) 1000×

Figure 4 shows the SEM pictures of different areas of the fatigue fracture of the nickel-based superalloy. It can be seen that the fracture is a typical fatigue fracture, which is divided into three areas: fatigue source area, expansion area and instantaneous fracture area. In Figure 4a and Figure 4b, we can clearly see the fatigue source zone, the expansion zone and the instantaneous fracture zone. The black-gray fan-shaped area is the fatigue source area and the expansion area. Macroscopic observation of the fracture shows that the fatigue source area is smoother, while the other two areas are rougher. In the first stage of fatigue, it can be seen that the fatigue expansion is insufficient. Near the fatigue source area, the fatigue band is fine, and the fatigue band can be seen in the area far from the fatigue source area, and the expansion stage near the fatigue source area It is a feature of cleavage plane and dimples. The farther from the fatigue source area, the more dimples, as shown in Figure 4c and Figure 4d.

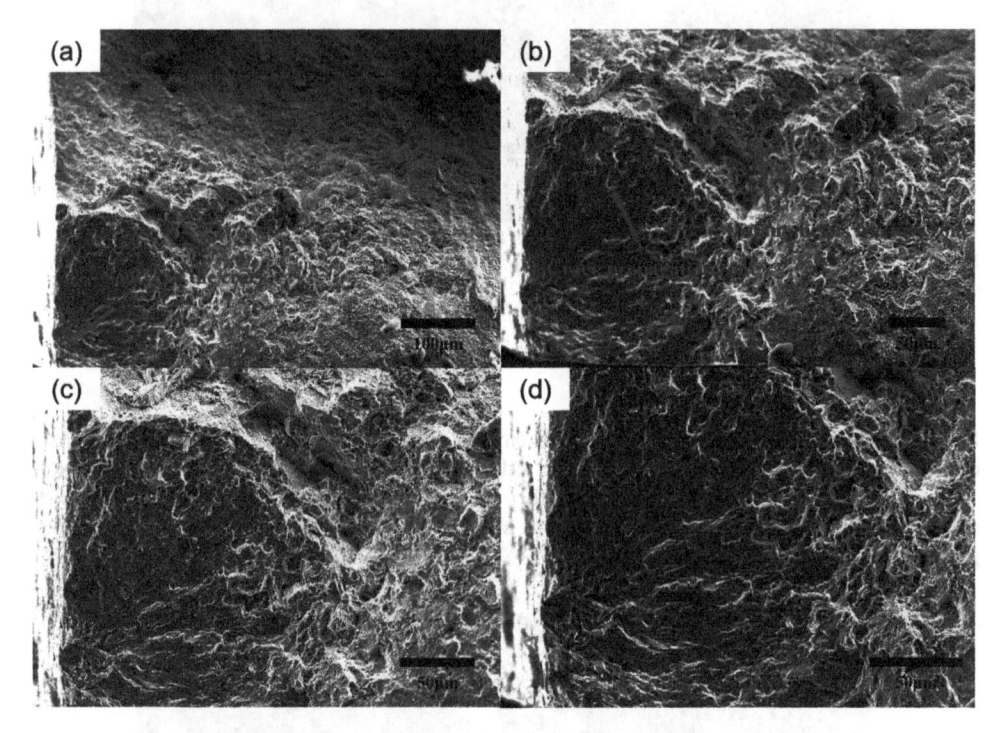

Figure 4. SEM photos of different areas of fatigue fracture of nickel-based superalloy
(a) 200× (b)300× (c)400× (d)500×

4 CONCLUSIONS

(1) In the in situ fatigue test at 650°C, the fatigue cracks of the alloy mainly oriented along the direction perpendicular to the stress, which sprang up from the grain boundary at 1528 Cycles, initially expanded along the grain boundary, and then showed a serpentine shape across the grain. Expansion, and finally the cycle breaks to 3214 Cycles.
(2) A slip zone is formed during alloy fatigue, and the direction of the slip zone is 45° to the axis.
(3) The alloy fracture is a typical fatigue fracture, divided into fatigue source area, expansion area and instantaneous fracture area.

REFERENCES

[1] He J, Sandstr M, Notargiacomo S. Low-Cycle Fatigue Properties of a Nickel-Based Superalloy Haynes 282 for Heavy Components J. *Journal of Materials Engineering & Performance*, 2017, 26(5):1–7.
[2] Veretimus D K, Veretimus N K. Low-cycle fatigue life prediction of a variable thickness disk J. *Journal of Physics: Conference Series*, 2019, 1348(1):012037 (7pp).
[3] Lin Y C, Deng J , Jiang Y Q , et al. Effects of initial δ phase on hot tensile deformation behaviors and fracture characteristics of a typical Ni-based superalloy J. *Materials & Design*, 2014, 598:251–262.
[4] Kashinga R J, Zhao L G, Silberschmidt V V, et al. Low cycle fatigue of a directionally solidified nickel-based superalloy: Testing, characterisation and modelling J. *MATERIALS SCIENCE AND ENGINEERING A-STRUCTURAL MATERIALS PROPERTIES MICROSTRUCTURE AND PROCESSING*, 2017,708: 503–513.
[5] Chen H M, Hu B F, Zhang Y W, et al. Research progress of nickel-based powder superalloys for aircraft turbine disks J. *Materials Review*, 2002, 16(011):17–19.
[6] Fukuda Y. Development of Advanced Ultra Supercritical Fossil Power Plants in Japan: Materials and High Temperature Corrosion Properties C. *Materials Science Forum*, 2011, 236–241.
[7] Xia Q X, Long J C, Zhu N Y, et al. Research on the microstructure evolution of Ni-based superalloy cylindrical parts during hot power spinning J. *Advance in advanced manufacturing*, 2019, 7(1):52–63.

Analysis of electrical contact failure of new beryllium bronze contact parts

Z.G. Kong & Y.C. Zhang
Research Laboratory of Reliability of Electrical Connection and Connector, Beijing University of Posts and Telecommunications, Haidian District, China

ABSTRACT: This paper introduces the failure cause analysis of a new type of beryllium bronze contact pair. A scanning electron microscope and an X-ray energy spectrometer are used to analyze and observe the failure samples. Combined with the actual failure phenomenon, the types of failure were determined and classified. On this basis, combined with the observation results of scanning electron microscopy and the results of static simulation analysis of ANSYS, the causes of different failures are analyzed, providing the corresponding solutions for different failure types.

1 INTRODUCTION

In traditional beryllium bronze contact, to maintain the stability of the pin and socket argued that if the contact by pin jack in the process of cooperation, beryllium bronze reed is being squeezed from pin: provided by the deformation as a result of the beryllium bronze reed due to its good elasticity, at the same time of produce deformation can produce a restoring force, the corresponding the restoring force is stable and reliable to ensure a contact of the contact force [1–3].

In the traditional beryllium bronze contact parts, the failure cases are also typical [4]. For example, in an electrical connector, the contact of pin and socket elastic parts processing and shaping is usually composed of beryllium bronze. As the current electrical connector-related technology has become increasingly mature, the industry also has a general classification of its failure mode:

a) due to the unreasonable size design, resulting in the male and female end of the overstress, which leads to the electrical connector pin fracture and failure.
b) due to the neglect of chamfering/fillet in the design and processing, or the plating of noble metal coating is too thin to protect the base metal, so that the copper alloy of the base is exposed and oxidized, resulting in excessive contact resistance and failure.
c) when designing the wiring between power supply and signal, there is no separate design of power supply pin, so that the power signal is transmitted through a thinner and longer signal pin, and the transmitted power exceeds the current-carrying capacity of the signal pin, resulting in the failure of the burning pin.
d) due to the poor isolation degree or the use of magnetic metal Ni as the intermediate coating in the design of a high-speed connector; the signal transmission on the connector is affected by large cross-talk or intermodulation interference, which affects the signal integrity, so that the connector is unable to achieve the designed indicators.

The failure case of a new type of beryllium bronze contact is analyzed in this paper. Different from the traditional beryllium bronze contact, the contact force of this new type of contact is not from the restorative force of the beryllium bronze reed, but from the applied pressure. The contact, which is made by applying applied pressure in different directions, enables the beryllium bronze reeds to come in contact with the fixed reeds in different positions, thus switching the corresponding working state. The new contact of beryllium bronze is shown in Figure 1. This new kind of beryllium bronze contact element is decided by a beryllium bronze reed and two different working conditions of beryllium bronze pieces; reeds of beryllium bronze with different contact can make contact work in different working conditions.

Figure 1. New beryllium bronze contact parts.

2 EXPERIMENT AND SIMULATION

2.1 *Failure analysis*

During the life test of this new type of beryllium bronze contact with a design contact life of 5 million cycles, two typical failure types were found: a) in the early stage of the life test, about thousands to tens of thousands of cycles, a relatively concentrated failure phenomenon occurred in individual batches of contacts. b) When the life test was carried out to about one million times, almost all batches of contact parts gradually failed.

Using visual inspection of failure of the contact, it was found that batches of beryllium bronze contact failure reeds and beryllium bronze pieces are intact, there are no visible scratches, base metal is not visible to the naked eye, reed and beryllium bronze pieces were not visible, fracture was observed, the reed thimble cooperate with applied pressure near the spindle hole, black pollutants can be observed to be attached on the surface of the beryllium bronze reed, and the closer the spindle hole, the more the number of black pollutants that can be observed.

Scanning electron microscopy (SEM) was used to observe and analyze the partially failed contact pairs in the life experiment [5,6]. Figure 2 shows beryllium bronze reeds with failed contact pairs. It can be seen that black contaminants can be observed on the reeds, and the amount of this black contaminant increases as the reed-thimble fit approaches spindle-shaped hole.

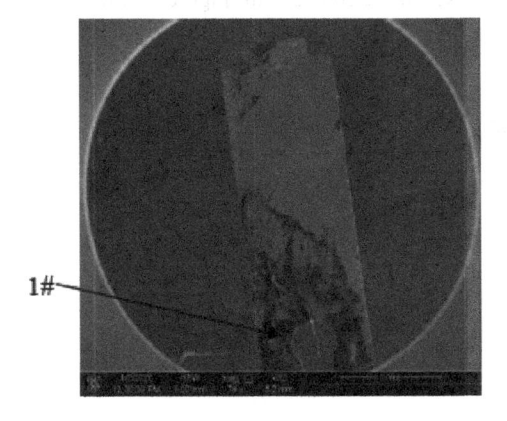

Figure 2. Failure sample of reed (1).

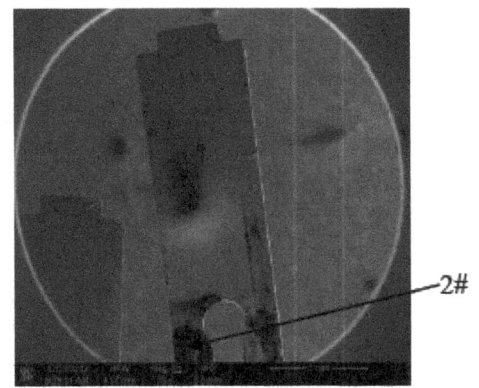

Figure 3. Failure sample of reed (2).

Figure 3 shows the beryllium bronze reed of another failed contact pair, very similar to that shown in Figure 2, where black contaminants are also observed near the spindle-shaped hole. Thus, can make a preliminary inference: in this new kind of beryllium bronze contact life experiments, as a result of beryllium bronze reed in thimble completed under the action of different working state of the switch, so the thimble when drive reed and reed in certain friction, so as to make the friction and wear debris accumulation near the thimble with reed spindle hole.

In order to further verify this hypothesis, the black contaminant part (the mark) of the failure contact pair shown in Figure 2 and Figure 3 was analyzed by an X-ray energy spectrometer. The spectral graph of each element at mark 1# in Figure 2 is shown in Figure 4, and the atomic ratio and mass ratio of each element content are shown in Table 1.

The spectrogram of each element marked 2# in Figure 3 is shown in Figure 5, and the atomic ratio and mass ratio of each element content are shown in Table 2.

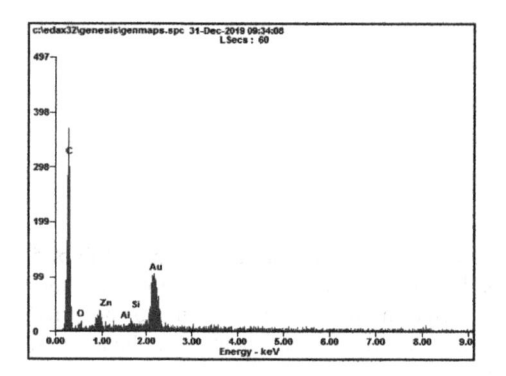

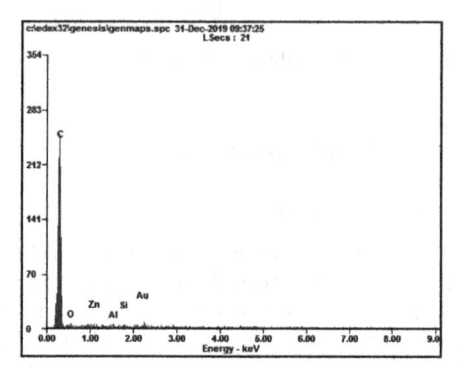

Figure 4. X-ray spectrometer graph at mark 1#.

Figure 5. X-ray spectrometer graph at mark 2#.

According to the analysis results of the X-ray energy spectrometer, the black contaminant covering the spindle-shaped hole on the surface of the beryllium bronze reed is a kind of carbon-based organic matter, while the thimble with the beryllium bronze reed is made of PAI, a kind of engineering plastic, and its main component is carbon. The black contaminants in the spindle-shaped hole were the product of friction and wear between the thimble and the beryllium bronze reed.

The beryllium bronze plates in the failure contact pair were observed and analyzed by scanning electron microscopy. The beryllium bronze plates not only played the role of current carriers in the new type of beryllium bronze contact pair, but also formed the corresponding working circuit on the beryllium bronze plates according to the different uses of the contact pair. Figure 6 shows one of the more typical observations.

Table 1. The atomic ratio and mass ratio of elements are marked 1#.

Element	Wt%	At%
CK	57.66	93.12
OK	02.11	02.56
ZnL	01.35	00.40
SiK	00.16	00.11
AlK	00.00	00.00
AuM	38.72	03.81
Matrix	Correction	ZAF

Table 2. The atomic ratio and mass ratio of elements are marked 2#.

Element	Wt%	At%
CK	93.93	96.87
OK	03.24	02.51
ZnL	00.75	00.14
SiK	00.41	00.18
AlK	00.48	00.22
AuM	01.20	00.08
Matrix	Correction	ZAF

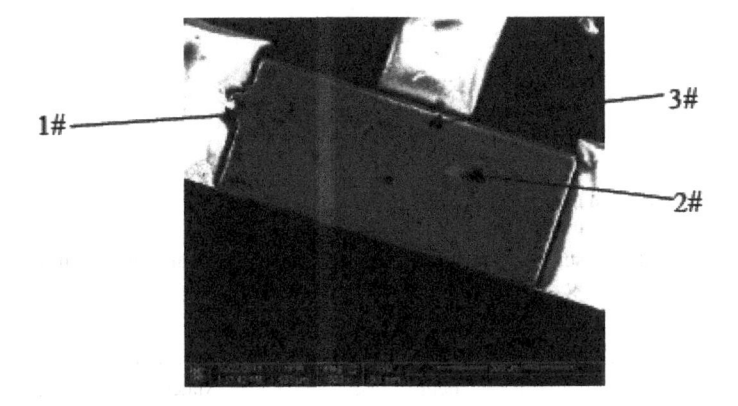

Figure 6. Scanning electron microscopy image of the beryllium bronze piece.

Scanning electron microscopy (SEM) results showed that there was an obvious contact mark with the reed on the coating of the beryllium bronze sheet, and the contact mark was not symmetrical along the centre line of the coating. At mark 1#, it can be seen from the contact trace that the reed and the beryllium bronze plate are in contact with a working circuit area outside the designed contact area. In the life experiment, the working circuit may be damaged by collision and may lead to failure. Bonding wear pits were observed at mark 2# due to bonding wear during contact, which proved that there was bonding wear between reed and beryllium bronze plate during contact. Bonding wear results in damage to the gilt layer and exposure of the base beryllium bronze, which increases the risk of electrochemical corrosion of beryllium bronze. At mark 3#, it can be found that the distance between the front end of the contact area and the working circuit is only about 100μm. The micron-level installation error and fretting can make the reed directly impact the working circuit when in contact, directly increasing the risk of contact pair failure.

2.2 *Simulation analysis*

In the life experiments of beryllium bronze contact pairs, the properties of the material itself will change with the increase of the number of life experiments. Because the essence of contact to contact is to apply external force to a point in the middle section of the cantilever beam, the free end of the cantilever beam produces a certain deflection and contacts the designed target and produces contact force on the contact surface. Therefore, the important index to determine whether the contact pair can maintain reliable contact is whether the contact pair can maintain its contact force for a long time. According to the relevant theoretical formulas for calculating the deflection angle of the cantilever beam in the mechanics of materials, the contact force of the contact pair will be affected by the elastic modulus of the reed, the deflection of the external force point, and the deflection of the contact point when the external size of the reed remains unchanged. Since the deflection of the force point is determined by the displacement of the thimble, which is driven by an electromagnetic electromechanical system, the displacement can be assumed to remain constant. The deflection of the contact point is determined by the spatial position of the reed relative to the beryllium bronze plate and therefore can be considered constant. Therefore, this paper uses the statics module in ANSYS Workbench to analyze the effect of the elastic modulus of beryllium bronze on the contact pair. The models involved in the simulation are shown in Figure 7.

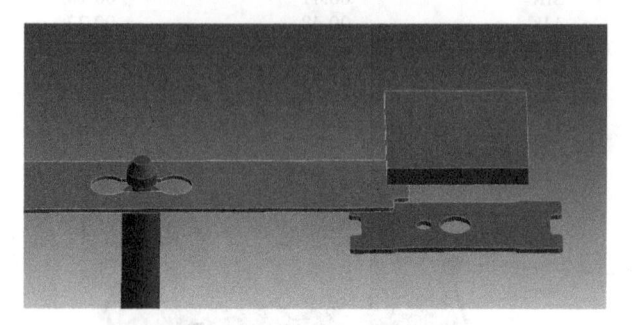

Figure 7. ANSYS static simulation model.

The simulated material settings and contact/constraint settings are shown in Tables 3 and 4, respectively.

Table 3. Simulation material settings.

Order	Item	Material
1	Reed	Beryllium bronze
2	Piece a	Beryllium bronze
3	Piece b	Beryllium bronze
4	Thimble	PAI

Table 4. Simulation contact/constraint settings.

Order	Contact/constraint	Settings	Comment
1	Fixed end of reed	Fixed support	(−)
2	Reed thimble	Frictional	Friction coefficient:0.15
3	Reed-piece a	Frictional	Friction coefficient:0.15
4	Reed-piece b	Frictional	Friction coefficient:0.15
5	Piece a upper surface	Fixed support	(−)
6	Piece b upper surface	Fixed support	(−)
7	Thimble displacement	Z-axis	Displacement:0.55 mm

In order to improve the simulation accuracy and ensure the quality of simulation mesh division, mesh subdivision was carried out at each contact side, as shown in Figure 8 and Figure 9, respectively. Considering that in the static simulation the large deflection is prone to the problem of cross-deflection between the models when a single load step is applied, the option of the large deflection is set to ON in the simulation setting, and the number of load steps is set to 100, with the displacement load being applied through 100 steps, ensuring the convergence of the simulation results.

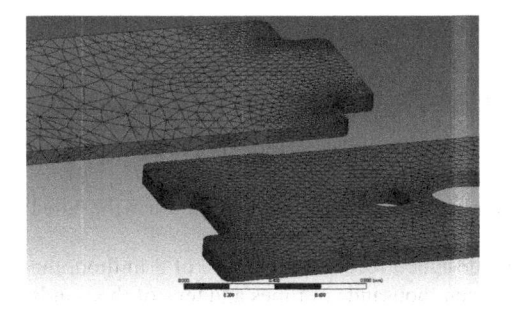

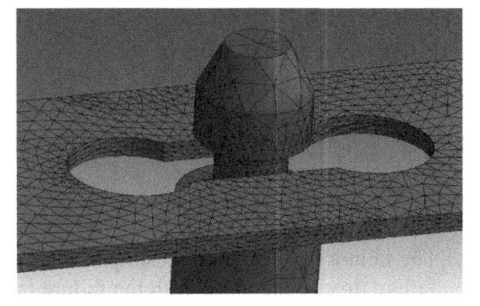

Figure 8. Finely divided meshes (a).

Figure 9. Finely divided meshes (b).

After running the simulation, the contact between the reed and the working piece 2 is shown in Figure 10. A probe was used to read the contact force on the contact surface between the reed and the working piece b, and the relationship between the contact force and the elastic modulus of beryllium bronze was obtained, as shown in Table 5.

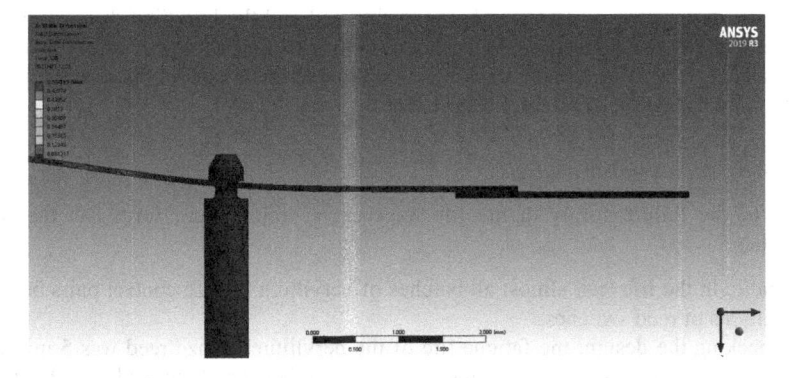

Figure 10. Contact between reed and working piece b.

Table 5. Simulation contact/constraint settings.

Elasticity Modulus E(GPa)	Contact Force (Simulation) F(g)	Contact Force (Calculating) F(g)
131	5.46	6.00
129	5.38	5.91
127	5.31	5.82
125	5.23	5.73
123	5.16	5.63
121	5.08	5.54
119	5.01	5.45

3 ANALYSIS AND DISCUSSION

3.1 Short-term failure cause analysis

From the expression form of short-term fault, this kind of fault has the following characteristics:

a) Contingency. Only a few batches of materials fail; the material performance of a majority of batches is normal.
b) Concentration. The beryllium bronze contact pair has a higher probability of failure in the failed material batch.
c) The failure occurred early. Compared with the designed reed switching life of 5 million, the failure time of the batch with the failure is between thousands of times and tens of thousands of times, and the time node of the failure is very early.

In accordance with the abovementioned three properties, and combined with scanning electron microscopy observation, presumably fault probably failed batches of beryllium bronze contact for assembly on production line produce a certain error, resulting in subsequent experiments, the life of a reed directly with the working circuit of beryllium bronze pieces produced contact collision, under the action of impact, caused the damage of the circuit work, leading to the failure to contact.

Therefore, this failure can be avoided in the following ways:

a) CCD test can be carried out on the beryllium bronze contact parts assembled on the production line to ensure the assembly accuracy [7].
b) During the design, the contact area between the reed and the beryllium bronze plate can be increased, so that the contact area is further away from the working circuit, so as to avoid the risk of damage and failure of the working circuit due to the impact of the reed on the working circuit under the action of external disturbance.

3.2 Long-term failure cause analysis

According to the failure forms in the life experiment, this failure form has the following characteristics:

a) Systematic. In the life test, almost all batches of beryllium bronze contact pairs began to fail after millions of reed switches.
b) After checking the design, the fatigue life of the beryllium bronze reed was 5 million times higher than the preset switching life, and no fracture or damage of the reed was observed in the failure samples.

From the mechanical point of view, combining the above two characteristics, it can be found that the failure mode of this new type of beryllium bronze contact pair conforms to the fatigue accumulation damage theorem, and this kind of failure is different from ordinary mechanical failure, which is a contact failure caused by fatigue accumulation damage under systematic stress. The time node of this failure is earlier than the fatigue strength failure. Decrease of contact force after the life test is the most likely cause of this failure. The decline of contact force leads to the decline of contact reliability, the degradation of contact resistance to external interference, and the increase of contact resistance, which leads to contact failure.

According to the scanning electron microscopy results, there are also reliability problems in the contact between the reed and the external pressure thimble. Due to the severe wear and tear of the thimble material, the core function of the reed is still current-carrying in the contact pair. When high-frequency signals are passed on the reed, the carbon-based wear products attached to the surface of the reed will lead to additional parasitic capacitance and parasitic inductance between the reed and its ground plane, which changes the impedance on the transmission line and leads to the emergence of signal integrity problems.

Therefore, the risk of failure of this new type of beryllium bronze contact can be reduced by the following two factors:

a) The contact force between the reed and the beryllium bronze plate should be appropriately increased during the design. In this way, sufficient contact force can be ensured to maintain stable contact between contactors after accumulated fatigue damage of the reed, and the influence of the nonlinear part of the transmission line can be suppressed when the contactors are used to transmit high-frequency signals, so that the transmitted signals can always maintain high quality.

b) More wear-resistant thimble materials should be chosen and the chamfer should be increased at the spindle hole with the thimble and reed during processing, so as to reduce the wear between the two under the condition of ensuring the stability of contact. When the amount of wear products is reduced, the risk of signal integrity problems is reduced.

4 CONCLUSION

In this paper, the failure cases of a new type of beryllium bronze contact parts during the life experiment are analyzed, and the following conclusions are drawn:

a) The failure types of the beryllium bronze contact parts during the life experiment can be divided into short term and long term. Short-term failures are sporadic and special, which only exist in some batches of materials, but the failure rate is higher in problem batches. The long-term failure is universal, which is caused by the system factors in the design of the contact parts.

b) Short-term failure is the main reason for the assembly error, and the design for the contact area reserved allowance is too small to reed and beryllium bronze piece work circuit; the risk of a collision impact is bigger, and playing in front of the CCD testing and design provides a sufficient margin for the contact area and can reduce the short-term failure of the fault probability.

c) There are two reasons for long-term failure. One is that the contact force is too small, which leads to the contact failure caused by insufficient contact force after fatigue damage of the reed. The reliable contact can be ensured by appropriately increasing the contact force during design. Second the wear-resisting performance of thimble is poor; after one million times of friction and wear, accumulation in the leaf surface of wear debris increases the risk of the signal integrity problems on a transmission line. More wear-resisting plunger materials that are appropriate when processing reed in addition on spindle hole chamfer can be chosen to reduce friction and wear.

ACKNOWLEDGEMENTS

The study is financially supported by National Key Research and Development Project of Ministry of Science and Technology (No.2018YFF01011604).

REFERENCES

[1] Liu Yonggang, Shen Xueliang, "Design outline of common contact parts for electrical connectors", *Electromechanical Components*, 2016,36(01):52–57.

[2] "Beryllium Bronze – New material, new technology and new equipment", *Technology Today*, 1999(02):15.

[3] "High strength beryllium bronze series material", *Aerospace Material & Technology*, 2003(02):61.

[4] Lin Jinbing, Wei Xiangyang, Li Ming, "Case study of connector failure engineering", *Electromechanical Components*, 2020,40(04):43–45.

[5] Han Zhengquan, Zhou Zuotao, Pan Wei, Tian Zhijing, "Application of modern analytical instruments in failure analysis of electronic components", *Electronic Instrumentation Customers*, 2020,27(06):103–106.

[6] Ling Yan, Zhong Jiaoli, Tang Xiaoshan, Li Dongyu, "The working principle and application of scanning electron microscope", *Shandong Chemical Industry*, 2018,47(09):78–79+83

[7] Li Hongda, Jiao Longfei, "Design and research of high precision non-contact laser caliper", *China Plant Engineering*, 2021(06):95–96.

Preparation and properties of laser cladding Cu-W alloy coating on pure copper electrical contacts

B.K. Li, Z.D. Liu, S.Y. Guo, L. Chen, J.X. Li & H.R. Ma
North China Electric Power University, College of Energy Power and Mechanical Engineering, Key Laboratory of Power Transmission, Conversion and System of Power Station, Ministry of Education, Beijing, China

ABSTRACT: In this study, laser cladding Cu-W alloy coating with different W content were prepared on the pure copper substrate by laser cladding. The phase composition and microstructure of Cu-W coating were analyzed by X-ray diffraction and scanning electron microscopy. The hardness and electrical conductivity of Cu-W coating were studied. The Cu-W coating consist of W and Cu and all coating are well bonded with the substrate. The results show that the hardness of coating increases with the increase of W content. The coating Cu-W40 shows the best conductivity. However, the coating of Cu-W has obvious segregation phenomenon, causing greatly change of the coating hardness value.

1 INTRODUCTION

Pure copper contact material has been widely used in the field of switching electrical appliances because of its high density, small expansion coefficient, good electrical and thermal conductivity. However, with the development of industry, the performance of pure copper can no longer meet some service conditions. The service environment has higher requirements on the performance of copper parts, such as high wear resistance, etc. [1–3] Therefore, this study adopted the method of adding W element to pure copper. Cu-W material is a composite material composed of Cu and W. Cu and W are not mutually soluble, nor do they form intermetallic compounds. Therefore, Cu-W material not only retains the good electrical and thermal conductivity of Cu, but also has the characteristics of W with high density, high strength, high melting point and low coefficient of expansion [4–5].

At the same time, the preparation process was improved. The research table at home and abroad showed that using advanced surface technology with excellent performance material as a protective layer could not only greatly reduce the process cost and technical difficulty, but also significantly prolong the service life of contact head parts [6].

There are many kinds of surface technology, including thermal spraying, plasma cladding and laser cladding coating, etc. The coating prepared by thermal spraying method has some limitations, such as low bonding strength (mechanical bonding), high porosity (1–12%), the existence of harmful phases in the coating, leading to the significant decrease of the corrosion resistance of the coating, and easy spalling and failure of the coating. These shortcomings make it difficult to ensure that the coating can work reliably for a long time under harsh conditions[7] Laser cladding technology, as an important means of surface modification, has the advantages of high flexibility, little thermal influence on the workpiece and high bonding strength between coating and substrate. These improve the surface hardness of the substrate and ensure the electrical conductivity of the pure copper substrate [8–10].

Although Laser cladding technology has made some achievements and economic benefits in strengthening steel surface. The special physical and chemical properties of pure copper and copper alloy, such as good thermal conductivity, low laser absorption rate and high reflectivity, are the major difficulties of laser cladding coating on copper surface. In this study, the cladding morphology was improved, and the Cu-W laser cladding coating was prepared on the pure copper substrate with better bonding strength by adjusting parameters such as laser power and laser scanning speed, as well as pretreating the substrate surface.

In this study, Laser cladding technology was used to prepare Cu-W20, Cu-W40 and Cu-W60 laser cladding coating on the surface of pure copper substrate. The surface and section of the laser cladding coating prepared by laser cladding technology were observed, and the section hardness and electrical conductivity of the laser cladding coating were detected and analyzed.

2 EXPERIMENTAL MATERIALS AND EQUIPMENT

2.1 *Experimental materials and equipment*

The substrates are pure copper with the dimension of $20 \times 20 \times 8$ mm. The surface of the substrate is polished with 200–400 mesh sandpaper, cleaned with acetone, and fixed on the workbench before laser cladding process.

Laser cladding technology of synchronous powder feeding is used in this experiment. The cladding material according to the composition in Table 1 above shall be dried for 1 hour at the drying temperature of 100°C, and then filled into the powder feeding equipment.

Table 1. The composition of powder laser cladding materials

	Composition/wt.%	
	W	Cu
a	20	80
b	40	60
c	60	40

All coating were prepare by a 3300 W fibber laser. The synchronous powder feeding method is used for single cladding. The total average thickness of the laser cladding coating is about 1400 μm after three cladding processes in the same way. The prepared laser cladding coating is polished. Some cladding technical parameters are shown in Table 2.

Table 2. Preparation parameters of laser cladding coating

Technical parameters	
Output power	3000 W
Relative scanning speed of laser beam and workpiece	6 cm/s
Spot diameter	2 mm
The gas that feeds the powder	N_2
Amount of powder feeding	3 kg/h
Shielding gas	Ar
The Shielding gas supply	18 ml/min
The lap rate of multi-channel cladding	60%
The thickness of single-layer laser cladding coating	400–600 μm

2.2 *Test method*

The phase analysis of the coating was carried out by an X-ray diffraction meter with a CuKα radiation source. The working voltage and current for the XRD were 40 kV and 100 mA with the incident radiation operating at a scan speed of 8/min and a 2θ scanning range of 20 to 80. The scanning electron microscopy operated at an accelerating voltage of 20 kV was used to observe the cross-sectional microstructure of coating A FM-300 hardness tester was used to measure the surface and section hardness of the coating The conductivity of Cu-W20, Cu-W40 and Cu-W60 laser cladding coating was measured by a digital portable fault current conductivity measuring instrument (FD102).

3 RESULTS AND ANALYSIS

3.1 *X-ray diffraction analysis*

Figure 1 shows the XRD patterns of the three laser cladding coating results. The results show that there are only two phases of Cu and W in the three materials. There are no other impurity phases on the surface, and no oxides are formed. The results show that Cu and W form pseudo alloys, which is consistent with the conclusion that Cu and W are not miscible and do not form intermetallic compounds.

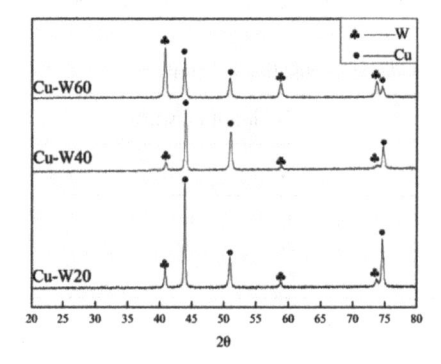

Figure 1. XRD results of Cu-W laser cladding coating.

3.2 *Structure analysis*

Firstly, the thickness of the laser cladding coating was measured,and the thickness was more than 1000 μm. The observation of the macroscopic surface of the laser cladding coating shows that the surface of the laser cladding coating is smooth, the overall thickness is uniform, and there are no large holes. The increase of copper also significantly reduces the porosity of the alloy.

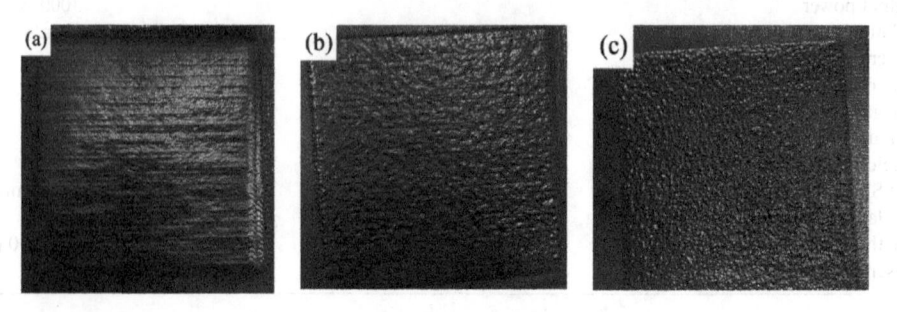

Figure 2. The surface of laser cladding coating (a) Cu-W20, (b) Cu-W40, (c) Cu-W60.

Figure 2 shows the surface morphology of the coating. By comparison, it can be found that with the increase of the content of W element in the laser cladding coating, the surface morphology gradually becomes worse, and a few holes begin to appear on the surface of the laser cladding coating By using nondestructive testing technology, no cracks can be found on the surface of the three laser cladding coating.

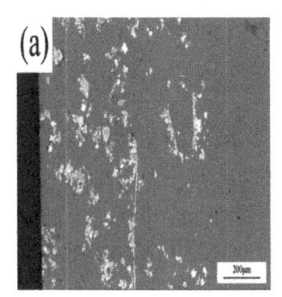

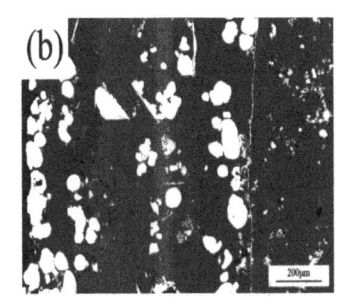

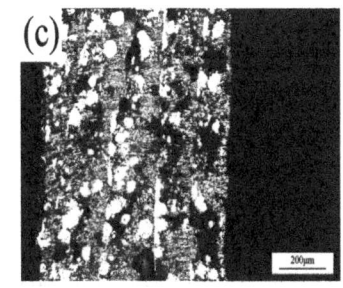

Figure 3. Scanning electron microscope image of laser cladding coating (a) Cu-W20, (b) Cu-W40, (c) Cu-W60.

Figure 3 shows the scanning electron microscopy (SEM) of the laser cladding coating of Cu-W20, Cu-W40 and Cu-W60, respectively. Cu-W material is a composite material composed of Cu and W, Cu and W are not mutually soluble, also do not form intermetallic compounds, so it has retained the good electrical and thermal conductivity of Cu, but also has the characteristics of W high density, high strength, high melting point and low coefficient of expansion. However, while retaining the excellent properties of the two materials, the agglomeration tendency of copper is more obvious, the distribution of tungsten particles in the copper is not uniform. Scanning electron microscopy (SEM) shows that, due to the high melting point of W, the laser cladding coating of the three kinds of components, to some extent, has the segregation of W element. The W particles in the laser cladding coating show the phenomenon of being coated by Cu element. The metal W in Cu-W20 and Cu-W40 is not uniformly distributed in copper, and the polarization is obvious. Due to the high content of W in Cu-W60, it has been able to fill the laser cladding coating. The phenomenon of W partial aggregation in Cu-W60 is not obvious. The three laser cladding coating are fused well with the substrate, showing metallurgical bonding.

Table 3. Composition of laser cladding coating.

	Composition/wt.%	
	Cu	W
Cu-W20	88.0	12.0
Cu-W40	67.3	32.7
Cu-W60	49.5	50.5

Table 3 shows the composition of each laser cladding coating It can be found that compared with before cladding, the content of W in the laser cladding coating decreases to a certain extent, which is related to the powder splashing during cladding.

3.3 *Hardness analysis*

The hardness of Cu-W20, Cu-W40 and Cu-W60 was measured using FM-300 hardness tester. The hardness test method is "S-type dot test method". First, start from the surface of the laser cladding coating, take three points in the horizontal direction of the laser cladding coating, the spacing of each point is 50 μm. Then dot to the substrate direction. The spacing of each point in the vertical direction of the laser cladding coating is 150 μm, and the load is 200 g/N. As shown in Figure 4, is the hardness test result 0 is the binding position between the substrate and the laser cladding coating.

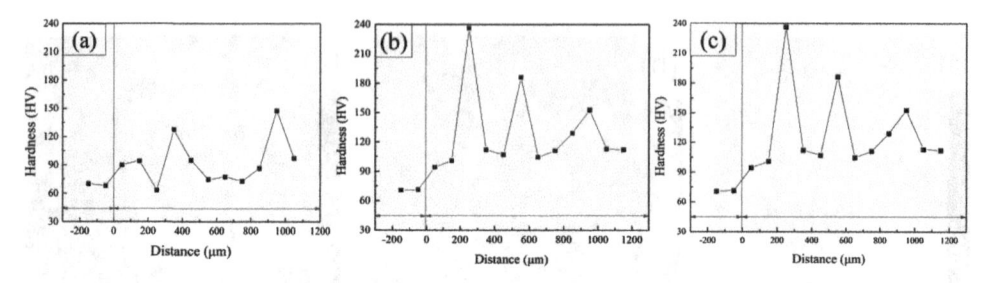

Figure 4. Hardness test results. (a) Cu-W20, (b) Cu-W40, (c) Cu-W60.

As can be seen from Figure 4, the Vickers hardness of substrate copper is about 60–70 HV and is not affected. In Figure 4(a), it can be seen that the hardness of Cu-W20 laser cladding coating is improved relative to that of the substrate due to the addition of 20% metal tungsten in the Cu laser cladding coating, with an average hardness of 101HV. But interlaced by measuring the hardness of the laser cladding coating value is large, from Cu – W20 laser cladding coating can be seen in the SEM photo, laser cladding coating W partial poly, and W particle size is bigger. Thus, in hardness tests, the hardness of the W offset area is significantly improved compared to that of the area with no W or less W content offset together. For example, when the distance between the laser cladding coating and the substrate is 950 μm, the hardness can reach more than 200HV. When no W was detected, the hardness even decreased to 63.4 HV.

The average cross-section hardness of Cu-W40 laser cladding coating is 134.2 HV, which is greatly improved compared with Cu-W20 laser cladding coating At the same time, the hardness of the laser cladding coating and SEM results show that the Cu-W40 laser cladding coating is similar to Cu-W20. Because the W in the laser cladding coating is segregated and the size of the W particles is large, the hardness values at different distances of the laser cladding coating are greatly different. At the same time, with the distance from the substrate increasing, the hardness of the laser cladding coating increases as a whole.

The average cross-section hardness of Cu-W60 laser cladding coating is 129.8 HV, which is not much different from that of Cu-W40. Compared with Cu-W20, the hardness is greatly improved. Table 4 shows the average hardness values of the surface and section of each laser cladding coating.

Table 4. The average hardness values of the surface and section of each laser cladding coating.

	Hardness/HV	
	Surface	Cross section
Cu-W20	98.7	101
Cu-W40	141.1	134.2
Cu-W60	190.3	129.8

The average surface hardness of Cu-W20 and Cu-W40 laser cladding coating is similar to the average section hardness of laser cladding coating. The average surface hardness of Cu-W60 laser cladding coating is larger than the average section hardness of laser cladding coating

On the whole, in the direction from the substrate to the laser cladding coating, there is a gradient of hardness growth in the transition region. In the laser cladding coating region, the hardness value of the whole is obviously higher than that of the substrate. And with the increase of the content of element W, the hardness value presents an upward trend. In addition, the hardness fluctuates greatly, which may be caused by the enrichment and uneven distribution of W element combined with the scanning electron microscope images.

3.4 *Conductivity analysis*

Before measurement, the Cu-W laser cladding coating sample was put on the metallographic pre-grinding machine. The surface was polished smooth and cleaned with 200#, 400#, 600#, 800#, 1000#, 1200# and 1500# waterproof sandpaper in turn to expose the metallic luster of the laser cladding coating without obvious scratches. The cross section and surface of the laser cladding coating were then polished with diamond grinding paste as a mirror, cleaned with alcohol ultrasonic oscillation. The sample was dried with a hair dryer. Then put it into the 20° constant temperature air conditioning room to stand for 24h, and then put the contact of the conductivity measuring instrument and the smooth surface of the material to measure the conductivity. Each laser cladding coating was measured for 5 times, and the average value was taken as the conductivity of the laser cladding coating The measurement results are shown in Table 5.

The electrical conductivity of pure copper substrates 98.15%IACS. Cu-W laser cladding

Table 5. Conductivity of each laser cladding coating.

	Cu-W20	Cu-W40	Cu-W60
Electrical conductivity/% IACS	32.40	56.60	30.50
	32.60	59.80	35.50
	31.80	58.70	35.35
	32.50	57.70	35.30
	31.70	57.10	35.50
Average conductivity/ %IACS	32.20	57.98	35.42

Coating is prepared on pure copper substrate by laser cladding technology, and the electrical conductivity of the laser cladding coating decreases obviously compared with that of pure copper substrate According to relevant literature, the continuous distribution of the tungsten phase will increase the resistivity of the material, resulting in a significant decrease in the electrical conductivity. The conductive properties of Cu-W40 laser cladding coating are similar to those of Cu-W20 and Cu-W60 laser cladding coating compared with those of traditional sintering and melting process. Material and process need further study.

4 CONCLUSIONS

Several main conclusions are drawn as follows:

(1) Cu-W20, Cu-W40 and Cu-W60 laser cladding coating were prepared by laser cladding coating method. XRD results show that the laser cladding coating is composed of Cu and W phases without impurity elements. Copper and tungsten do not dissolve into each other and do not form intermetallic compounds. SEM results show that the laser cladding coating fuses well with the substrate, showing metallurgical bonding. The composition distribution of laser cladding coating is uniform and W element segregation exists to a certain extent.

(2) Compared with the pure copper substrate, the hardness of Cu-W laser cladding coating is significantly increased, and the hardness value increases with the increase of the content of W element.
(3) While the surface hardness is increased, the electrical conductivity is not seriously decreased. The conductive properties of Cu-W40 laser cladding coating are similar to those of Cu-W monolithic contacts prepared by traditional sintering and melting process. However, on the whole, there is a W element rich area in the laser cladding coating, and the electrical conductivity under high W content still needs to be improved, so it is still necessary to further optimize the cladding process parameters to obtain the laser cladding coating with higher performance.
(4) In terms of comprehensive properties, among the three kinds of laser cladding coating, Cu-W40 laser coating has better surface morphology and microstructure, and has better electrical conductivity, hardness has also been greatly improved.

REFERENCES

[1] R. Lin, Wang L., Ma J., et al. 2018 Experiment investigation on vacuum arc of AMF contacts under different materials. [J] *AIP ADVANCES*, 2018, 8(9).
[2] L.L. Dong et al. 2018 Recent progress in development of tungsten-copper composites: Fabrication, modification and applications[J]. *International Journal of Refractory Metals and Hard Materials*, 2018, 75: 30–42.
[3] V.V. Bukhanovsky, N.I. Grechanyuk, R.V. Minakova, I. Mamuzich, V.V. Kharchenko, N.P. Rudnitsky. 2011 Production technology, structure and properties of Cu–W layered composite condensed materials for electrical contacts[J]. *International Journal of Refractory Metals and Hard Materials*, 2011, 29(5).
[4] Yang X., Liang S., Wang X., et al. 2010 Effect of WC and CeO 2 on microstructure and properties of W–Cu electrical contact material[J]. *International Journal of Refractory Metals and Hard Materials*, 2010, 28(2):305–311.
[5] Tetsuya Osaka, Yutaka Okinaka, Junji Sasano, Masaru Kato. 2006 Development of new electrolytic and electroless gold plating processes for electronics applications[J]. *Science and Technology of Advanced Materials*, 2006, 7(5).
[6] Pierre L, Fauchais, Joachim V.R. Heberlein, Maher I. Boulos. 2014 Thermal Spray Fundamentals[M]. *Springer*, Boston, MA:2014-01-01.
[7] Thomas Schopphoven, Johannes Henrich Schleifenbaum, Sadagopan Tharmakulasingam, Oliver Schulte. 2019 Setting Sights on a 3D Process[J]. *PhotonicsViews*, 2019, 16(5).
[8] Li, Zhang, Bultel, et al. 2019 Extreme High-Speed Laser Material Deposition (EHLA) of AISI 4340 Steel[J]. *coating*, 2019, 9(12):778.
[9] Boris Rottwinkel, Christian Nölke, Stefan Kaierle, Volker Wesling. 2014 Crack Repair of Single Crystal Turbine Blades Using Laser Cladding Technology[J]. *Procedia CIRP*, 2014, 22.
[10] Farahmand P, Kovacevic R. 2014 An experimental–numerical investigation of heat distribution and stress field in single- and multi-track laser cladding by a high-power direct diode laser[J]. *Optics & Laser Technology*, 2014, 63:154–168.
[11] Norhafzan B, Khairil C.M., Aqida S.N. 2021 Laser cladding process to enhanced surface properties of hot press forming die: A review[J]. *IOP Conference Series: Materials Science and Engineering*, 2021, 1078(1).
[12] Fang Liu, Changsheng Liu, Xingqi Tao, Suiyuan Chen. 2006 Laser cladding of Ni-based alloy on copper substrate[J]. *Journal of University of Science and Technology Beijing*, 2006, 13(4).
[13] Guojian Xu, Muneharu Kutsuna, Zhongjie Liu, Hong Zhang. 2005 Characteristics of Ni-based coating layer formed by laser and plasma cladding processes[J]. *Materials Science & Engineering A*, 2005, 417(1).
[14] Hongji Qi, Meipin Zhu, Ming Fang, Shuying Shao, Chaoyang Wei, Kui Yi, Jianda Shao. 2013 Development of high-power laser cladding coating[J]. *High Power Laser Science and Engineering*, 2013, 1(1).
[15] Mohit J., Ganesh S., Krista M. 2006 Microwave sintering : A new approach to fine-grain tungsten-I[J]. *International Journal of Powder Metallurgy*, 2006, 42(2).

Effect of laser energy density on microstructure and properties of SLM 24CrNiMoY alloy steel

M. Sun, S.Y. Chen, M.W. Wei, X.W. Song & L. Zhou
Key Laboratory for Anisotropy and Texture of Materials, Ministry of Education, Key Laboratory for Laser Application Technology and Equipment of Liaoning Province, School of Materials and Engineering, Northeastern University, Shenyang, Liaoning, China

M. Wang
Shenyang Dalu Laser Technology Co. Ltd. Shenyang, Liaoning, China

ABSTRACT: Energy density of selected laser melting technology (SLM) has an important effect on the microstructure and properties of alloy steel parts. In this paper the 24CrNiMoY alloy steel samples were prepared by SLM technology, the effects of different laser energy densities (102 J/mm^3, 120J/mm^3, 138J/mm^3, 156J/mm^3) on microstructure and mechanical properties of alloy steel samples were investigated. The results show: when the lap width was 0.07 mm, the scanning angle was 67°, the scanning line length was 10 mm, the scanning speed was 800 mm/s, while the laser power was changed to form different energy density, the internal defects of the sample first decrease and then increase with the increasing energy density, while the hardness and tensile properties also appeared similar trends as internal defects. The measurements when the energy density was 138 J/mm^3, the defects of the alloy steel sample were the least, the microstructure was mainly ultra-fine grain lath martensite, the mechanical properties achieved well. The average hardness was 386 HV0.2, the tensile strength was 1252 MPa, while the elongation was 12.00% after fracture. This 24CrNiMoY alloy steel sample has excellent matching properties between strength and toughness.

1 INTRODUCTION

The high-speed railway brake disc is one of the key parts to ensure the safe operation of high-speed railway. The traditional manufacturing method of brake disc is mainly casting heating treatment, which has a long process cycle and needs to improve the product performance [1]. With the rapid development of laser additive manufacturing technology, it has become one of the hot topics in recent years to study the preparation of large-scale and complex high-speed railway brake discs by selective laser melting (SLM) [2–5]. High speed railway brake discs are mainly made of 24CrNiMoY alloy steel. Based on the non-equilibrium solidification characteristics of SLM alloy steel, scholars around the world have studied the preparation of 24CrNiMoY alloy steel by SLM, the microstructure and properties of it have been deeply studied, while some important progress and achievements have been made [6–11]. These studies have established a good theoretical and technical basis for the microstructure evolution control properties of 24CrNiMoY alloy steel prepared by SLM. However, it is a complex non-equilibrium metallurgical process to produce alloy steel high-speed rail brake disc by SLM, while the influence of laser processing parameters on the microstructure, defects and properties of the formed alloy steel is deterministic. How to obtain 24CrNiMoY alloy steel samples with good properties (tensile strength > 1150 MPa, elongation of 10–12%, hardness of 340–350 HV$_{0.2}$) by SLM equipment is still one of the technical problems.

Therefore, on the basis of our previous research on the preparation of 24CrNiMoY alloy steel by SLM [1, 3, 4, 12], this article uses 1000W optical fiber SLM printing equipment to print 24CrNiMoY alloy steel samples under different energy density. The aim is to reveal the matching relationship between the structure and performance of 24CrNiMoY alloy steel samples prepared by the SLM equipment, and to find the most suitable SLM energy density for preparing 24CrNiMoY alloy steel samples with both high tensile strength and toughness. This paper can provide good technical references for SLM in printing large complex structure alloy steel brake disc samples.

2 EXPERIMENTAL MATERIALS AND METHOD

The experimental equipment for preparing alloy steel samples is a 1000 W optical fiber SLM 3D printer (260 × 300 × 300). The block sample of 10 mm × 10 mm × 10 mm was designed for the printed sample model, while the model slice forms the SLM device print file. After a series of optimization experiments, the scanning angle was 67°, the laser scanning speed V was 800 mm/s, the thickness of powder layer h was 0.04 mm, and the scanning distance was 0.07 mm. The parameters of the samples are shown in Table 1. Under the control of the forming software, the process of powder laying – melting – powder laying was circulated for obtaining 10 mm × 10 mm × 10 mm 24CrNiMoY alloy steel samples.

Table 1. Parameters of the SLM process.

Series	P/W	V/(mm/s)	h/mm	d_s/mm	EVD/(J/mm^3)
1#	230	800	0.04	0.07	102
2#	270	800	0.04	0.07	120
3#	310	800	0.04	0.07	138
4#	350	800	0.04	0.07	156

The samples were polished in sequence, then the samples were corroded. The corrosive liquid is nitric acid alcohol with a volume fraction of 4%. The inverted optical microscope (OLYMPUS-GX71) was used to observe the metallographic structure of the sample. X-ray analyzer (SmartLab-9000) was used for phase analysis, the parameter is Kα line of Cu target ($\lambda = 1.5406$ Å) with continuous scanning, the tube flow was 40 mA, the tube pressure was 40 kV, the sample scanning speed was 3°/min, the scanning angle range was 20°–100°. The field emission scanning electron microscope (FE-SEM, Shimadzu-SSX-550) was used to observe the microstructure and tensile fracture of the sample. The microscopic fine structure of the sample was analyzed using a transmission electron microscope (TEM, TECNAIG220), the resolution between the two points is 0.23 nm. A digital microhardness tester (WILSON-WOLPER-401MVD) was used to measure the microhardness of the cross section, its load was 200 g, load time 10 s, hit three rows side by side and take the average value. The universal testing machine (AG-X100kN) was used to test the tensile properties of the sample.

3 EXPERIMENTAL RESULTS AND ANALYSIS

3.1 *Analysis of porosity defects in molten pool*

Figure 1 shows the metallographic morphology of 24CrNiMoY alloy steel samples at different energy densities, there are two distinct areas in the graphic, namely the light gray laser melting zone (LMZ) and the gray-black heat affected zone (HAZ). The results show different amounts of porosity defects in the four samples, while the porosity defects of 138 J/mm^3 samples are less, it proves that the energy density has the solidification conditions to eliminate the porosity defects, while the energy density is one of the main factors affecting the formation of porosity defects [11]. The morphology and size of LMZ and HAZ in the sample molten pool vary with the energy density. In Figure 1(a) and (b), LMZ and HAZ are basically parallel and alternately distributed, it is compared with the sample prepared at 102 J/mm^3, the HAZ of the prepared sample at 120 J/mm^3 becomes wider and the boundary becomes more tortuous. With the increasing EVD, Figure 1(c) and (d) show that LMZ and HAZ are arranged in a cross pattern, while the depth of the molten pool increases significantly. It proves that high energy density lead to heat accumulated in the molten pool, then melting metal diffuses slowly and forms more obvious serrated morphology, finally the porosity defects increase significantly. Therefore, the laser energy density has a relatively obvious impact on the morphology and size of the LMZ and HAZ.

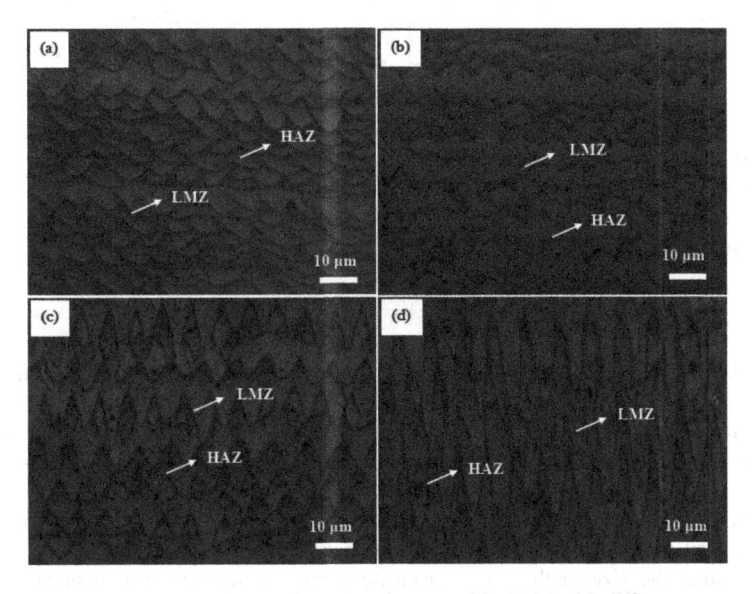

Figure 1. Metallographic images of 24CrNiMoY steel prepared by SLM with different energy densities (a) 102 J/mm^3, (b) 120 J/mm^3, (c) 138 J/mm^3, (d) 156 J/mm^3.

3.2 Structure morphology and phase analysis

Figure 2 is the SEM microstructure of 24CrNiMoY alloy steel samples prepared under different energy densities. The solidification structure is lath martensite, many parallel laths form a lath bundle, while lath martensite is composed of lath bundles in different orientations. The size of the lath martensite was formed by the alloy steel samples under different energy densities. The martensite lath width of the sample with 102 J/mm^3 (Figure 2(a)) is about 0.5–1 μm and the length. When the energy density was up to 156 J/mm^3, the martensite laths' width was 0.8–1.5 μm and 3–12 μm length. When the laser energy density was 120 J/mm^3 and at 138 J/mm^3, the martensite size of the prepared alloy steel samples was relatively smallest, while the fine martensite can improve toughness.

Figure 3 is the XRD patterns of 24CrNiMoY alloy steel samples prepared under different energy densities. The structure of samples is mainly composed of α-Fe(M) (M stands for C, Cr, Ni and other elements) solid solution, while martensite structure was added a small amount of γ-Fe retained austenite. When the energy density is too high (156 J/mm^3), the diffraction peak of retained austenite at 43.38°, it proves that high energy density lead to retained austenite decrease. While high energy density makes the molten pool form solidify slowly, it is more conducive to the complete transformation of austenite into martensite.

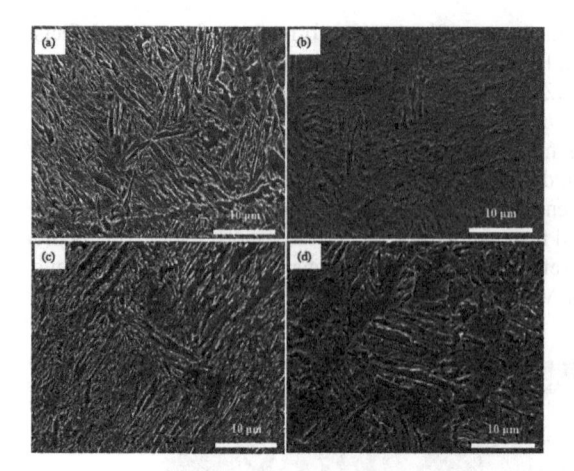

Figure 2. SLM images of 24CrNiMoY steel prepared by SLM with different energy densities.
(a) 102 J/mm^3, (b) 120 J/mm^3,
(c) 138 J/mm^3, (d) 156 J/mm^3

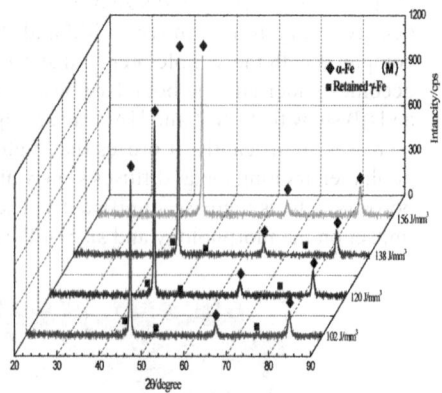

Figure 3. XRD of 24CrNiMoY steel prepared by SLM with different energy densities

3.3 Sample performance and analysis

Figure 4 shows the distribution of microhardness of 24CrNiMoY alloy steel under different energy densities. The average microhardness of the alloy steel samples is 334 HV$_{0.2}$, 365 HV$_{0.2}$, 386 HV$_{0.2}$ and 344 HV$_{0.2}$. When the energy density increases from 102 J/mm^3 to 156 J/mm^3, the microhardness of the deposited samples shows a trend of fluctuating. When the energy density increased to 156 J/mm^3, the average hardness of the sample decreased. Due to the laser energy density increases, the pore defects and martensite size in the prepared samples change. When the energy density is 138 J/mm^3, the sample preparation has the least defects in Figure 1(c), at the same time, the size of the lath martensite formed is the smallest, it makes the sample microhardness achieve the highest value. When the energy density increased to 156 J/mm3, obvious hole defects were observed in the sample, while the size of martensite increased relatively, it leads the microhardness decrease.

Figure 5 shows the stress-strain curve diagram of 24CrNiMoY alloy steel samples under different energy densities. When other conditions are fixed, the tensile strength and elongation after fracture of the deposited samples prepared under the conditions of energy density of 102 J/mm^3, 120 J/mm^3, 138 J/mm^3 and 156 J/mm^3 first increase and then decrease, among which the tensile strength is 1150 MPa, 1238 MPa, 1252 MPa and 1172 MPa respectively, while the elongation after fracture is 8.58%, 8.69%, 12.00% and 6.85% respectively. As the energy density increases, the tensile strength and yield strength increase, and when the energy density increases to 156 J/mm^3, the tensile strength and the yield strength reduced. While the energy density increases from 102 J/mm^3 to 138 J/mm^3, the pore defects of the sample gradually decrease, the size of the martensite lath is relatively refined, as a result, the mechanical properties of the prepared alloy steel sample are improved. When energy density is 138 J/mm^3, the maximum tensile strength of selected area laser melting 24CrNiMoY alloy steel is 1252 MPa, and the maximum elongation after fracture is 12.00%, which has a good matching relationship between strength and toughness. This result proves that the energy density has an important influence on the structure and properties of the formed alloy steel, different structures can be formed by adjusting different energy densities.

Figure 6 shows the tensile fracture morphology of 24CrNiMoY alloy steel samples under different energy densities. A large number of dimples and micropores can be observed in Figure 6(a) and (b). In Figure 6 (c), some equiaxed dimples and tearing edges and large hole defects are obviously reduced. While Figure 6(d) also shows some dimples and micropores, accompanied by microcracks,

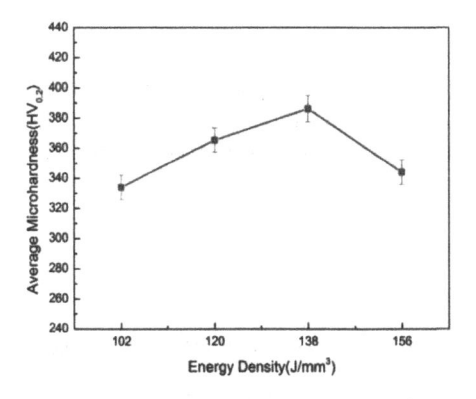

Figure 4. Microhardness of 24CrNiMoY steel prepared by SLM with different energy densities

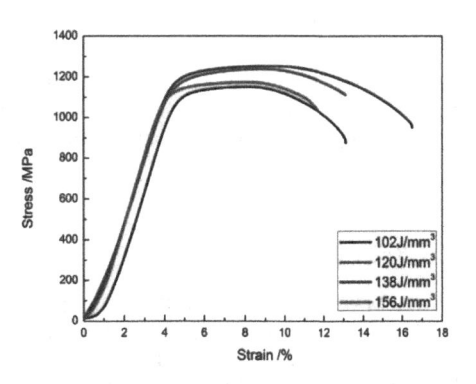

Figure 5. Stress-strain curve of alloy steel sample with different energy densities.

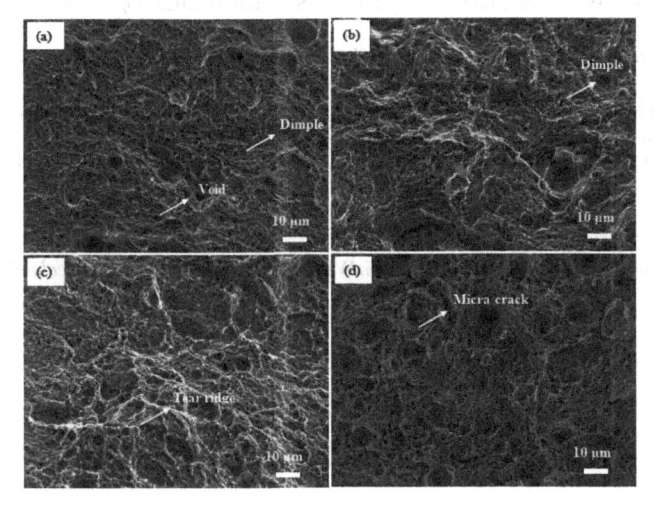

Figure 6. Fracture morphology of the alloy steel sample with different energy densities. (a) 102 J/mm^3, (b) 120 J/mm^3, (c) 138 J/mm^3, (d) 156 J/mm.

increasing defects, while tensile strength and elongation after the break is reduced. Therefore, in comparison, the sample with an energy density of 138 J/mm^3 has the highest tensile strength, it achieved the best toughness and the highest elongation after fracture.

4 CONCLUSIONS

The SLM 24CrNiMoY alloy steel samples without crack defects were prepared under different energy density, the microstructure was mainly composed of lath martensite and a small amount of retained austenite. The laser energy density has an important effect on the porosity defect of SLM 24 alloy steel samples, with the increase of energy density, the porosity defect in the samples decreases first and then increases. The energy density also affects the hardness and tensile properties of 24CrNiMoY alloy steel samples. When the energy density is 138 J/mm3, the obtained samples have a good matching relationship of strength and toughness. The average hardness is 386 HV0.2, the tensile strength is 1252 MPa, and the post-break elongation is 12.00%.

ACKNOWLEDGMENTS

This work was financially supported by National Key R&D Program of China (2016YFB1100201), Green Manufacturing System Integration Project of the Industry and Information Ministry of China (2017), Shenyang important scientific and technological achievements transformation project (20-203-5-6).

REFERENCES

[1] Wei M W, Chen S Y, Xi L Y, Jing Liang, Liu C S, 2018, *Optics & Laser Technology*, **107**, 99–109.
[2] Zhao X, Dong S Y, Yan S X, Liu X T, Liu Y X, Xia D, Lv Y H, He P, Xu B S, Han H S, 2020, *Materials Science & Engineering: A*, **771**, 138557.
[3] Wei M W, Chen S Y, Sun M, Liang J, Liu C S, Wang M, 2020, *Powder Technology*, **367** 724–739.
[4] Sun M, Chen S Y, Wei M W, Liang J, Liu C S, Wang M, 2021, *Powder Metallurgy*, **64(1)** 23–34.
[5] Zhou L, Chen S Y, Wei M W, Liang J, Liu C S, Wang M, 2019, *Materials Characterization*, **158**, 109931.
[6] Zuo P F, Chen S Y, Wei M W, Liang J, Liu C S, Wang M, 2019, *Optics & Laser Technology*, **119**, 105613.
[7] Cao L, Chen S Y, Wei M W, Guo Q, Liang J, Liu C S, Wang M, 2021, *Optics & Laser Technology*, **135**, 106661.
[8] Wang Q, Zhang Z H, Tong X, Dong S Y, Cui Z Q, Wang X, Ren L Q, 2020, *Optics & Laser Technology*, **128**, 106262.
[9] Zhang B Y, Zhang Z H, Zhou S H, Liu Q P, Zhang P 2021, *Materials Letters*, **282**, 128656
[10] Shi C F, Chen S Y, Xia Q, Li Z, 2018, *Powder Metallurgy*, **61(1)**, 73–80.
[11] Xi L Y, Chen S Y, Wei M W, Liang J, Liu C S, Wang M, 2019, *Journal of Materials Engineering and Performance*, **28**, 5521–2232.
[12] Guo Q, Chen S Y, Wei M W, Liang J, Liu C S Wang M, 2020, *Journal of Materials Engineering and Performance*, **29(10)**, 6439–6354.
[13] Zuo P F, Chen S Y, Wei M W, Liang J, Liu C S, Wang M, 2019, *Optics & Laser Technology*, **119**, 105613.

Chapter 2: Multifunctional Materials Properties, Processing, and Manufactures

Study on the effect of hot press sintering temperature on the microstructure and properties of new Ti-Al-Nb alloy

S. Fang, M.C. Zhang, Q.Y. Yu & H.P. Xiong
Beijing Institute of Aeronautical Materials, Beijing, China

ABSTRACT: In this paper, in order to explore the development of a titanium-based superalloy with a working temperature of 700°C to 750°C and good ductility at room temperature, using Ti+Al+Ti2AlNb ternary powder as the raw material, the temperature is 1200°C to 1400°C, 30MPa, 1.5h Directly mixed and sintered under hot pressing conditions, a new type of titanium-based superalloy (TiAl+Ti2AlNb) samples were prepared. The fracture morphology of hot-pressed sintered samples and tensile samples used scanning electron microscope (Scanning electron microscope, referred to as SEM) and X-ray diffractometer (XRD) for observation. In this paper, the effect of hot pressing sintering temperature on the microstructure and properties of the new Ti-Al-Nb alloy was studied accordingly.

1 INTRODUCTION

As a light metal structural material with good comprehensive performance, titanium alloy has been widely used in aviation, aerospace and other fields. However, as the temperature rises, the oxidation resistance and high temperature strength of titanium alloys are significantly reduced, which greatly limits the maximum use temperature. At present, the working temperature of advanced heat-resistant titanium alloys only reaches 550°C~600°C [1–3]. Therefore, the development of research on high temperature titanium alloys above 600°C has always been the development trend in the field of titanium alloy materials.

Ti2AlNb alloy is a Ti-Al intermetallic compound with orthogonal structure O phase as the matrix. Because it has good strength, plastic toughness and creep resistance at 650°C~700°C (typical room temperature performance data: tensile strength Around 1100MPa, the elongation can reach 8–14% [4]), and the density is low, so it has good application potential in the aviation and aerospace fields, but it is generally believed that the long-term stable working temperature of this material is difficult to reach 750°C. The γ-TiAl intermetallic compound has a series of advantages such as low density, high specific strength, high specific rigidity, and good high temperature performance. It is generally believed that its long-term use temperature can even reach 760°C~800°C (typical high temperature performance data: 800°C, tensile The tensile strength is about 500MPa [5, 6]).

On the one hand, Ti2AlNb alloys and γ-TiAl intermetallic compounds can replace conventional Ti-based alloys, thereby increasing the service temperature of materials; on the other hand, they can also replace Ni-based superalloys to achieve weight reduction. It can be seen that both Ti2AlNb alloy and γ-TiAl intermetallic compounds are new light-weight high-temperature structural materials with great application potential, which are of great significance for the further weight reduction of aerospace engines and the improvement of high-temperature service performance.

However, the biggest problem of γ-TiAl intermetallic compounds is poor plasticity at room temperature and high brittleness. In order to improve the room temperature plasticity of γ-TiAl materials, alloying is an important way to improve the room temperature plasticity and strength properties of TiAl-based alloys [7, 8]. A large number of experimental studies believe that alloying elements such as Al, V, Mn, Cr, Mo, B, and C can improve the plasticity of the alloy. In addition to the optimization of the composition of the material itself, a variety of hot working processes for the as-cast TiAl alloy ingot structure and billet have been developed, including isothermal forging, sheath forging, and sheath extrusion to refine the grain size and improve TiAl The room temperature plasticity. Nevertheless, the room temperature elongation of γ-TiAl material is difficult to exceed 3.0%. Considering that the TiAl intermetallic compound is very brittle as a single material and its forming process is extremely poor, the American GE Company, Oak Ridge National Laboratory and relevant European units proposed the research idea of micro-laminated titanium aluminum alloy and its composite materials [9–11], that is, through the Ti and Al element foil method (two types of foils are periodically stacked), through diffusion reaction, rolling and heat treatment, micro-laminated titanium aluminum alloy is obtained, and even SiC fiber is added to pass high temperature Sintering to obtain a reinforced TiAl intermetallic compound-based composite material. However, these researches are not only complicated in technology, but also are currently in the most basic research stage. Because the matrix material is still TiAl compound, the problem of room temperature brittleness is still difficult to solve. The brittleness of γ-TiAl seriously affects the safety and reliability of components. The use of γ-TiAl in the design of many high-temperature parts of aircraft and engines is greatly restricted [12, 13]. The high temperature stability of Ti2AlNb alloy, such as high temperature creep performance and oxidation resistance, needs to be further improved. Therefore, it should be said that there is still a lack of new titanium-based superalloys with long-term working temperature of 700°C to 750°C and good plasticity at room temperature.

2 EXPERIMENTAL SECTION

The raw materials used in the experiment are pure Ti powder, pure Al powder and Ti2AlNb alloy powder. The powder is produced by Xi'an Ouzhong Material Technology Co., Ltd. The chemical composition of the raw material powder used in the test is listed in Table 1.

Table 1. Chemical composition of pure Ti, Al and Ti2AlNb powder (wt%).

Element	Fe	C	O	N	H	Ti	/	Ti
Content, wt%	0.023	0.023	0.1661	0.0066	0.0023	Bal	/	
Element	Ti	O	Zn	Cu	Fe	Si	Al	Al
Content, wt%	<0.01	0.096	<0.01	<0.01	<0.01	<0.01	Bal	
Element	Al	Nb	C	N	O	H	/	Ti2AlNb
Content, wt%	10.25	42.90	0.008	0.002	0.068	0.002	/	

Through scanning electron microscopy secondary electron imaging, it is observed that the raw material powder used in the experiment is spherical particles of nearly equal diameter. As shown in Figure 1, The 400 mesh pure Ti powder has a relatively uniform particle size distribution. The 400 mesh pure Al powder has smaller spherical particles agglomerated near the 400 mesh powder particle size particles, and the 400 mesh Ti2AlNb alloy powder has smaller size spherical particles agglomerated near the 400 mesh powder particle size particles. Particles.

Figure 1. SEM photo of the raw material powder (400 mesh) used in the experiment (a) pure Ti powder; (b) pure Al powder; (c) Ti2AlNb alloy powder.

Weigh pure Ti powder and pure Al powder according to the volume ratio of 1:1, then weigh the Ti2AlNb powder according to the ratio of 10% by volume of Ti2AlNb powder, and put the weighed three raw material powders into a mortar to fully mix. Put the mixed raw materials into a graphite mold. The inner size of the graphite mold is Φ60 mm × 40 mm. The inner wall of the graphite mold, graphite paper and graphite sheet are uniformly coated with boron nitride in advance, and placed in an oven to dry to prevent sintering Carbon pick-up occurs during the process, resulting in sticky molds. Put the loaded graphite mold into a vacuum hot-pressing sintering furnace. First, pre-press the graphite mold to a pressure of 30MPa, and then release the pressure; secondly, vacuumize the vacuum hot-pressing sintering furnace and pass argon gas into it. Vacuum a second time, introduce argon gas, and then start to increase the temperature at a rate of 10°C/s. The pressure is maintained at 30 MPa from the temperature rise to the end of the sintering process, and wait until the temperature reaches the corresponding hot pressing sintering temperature (1200°C~1300°C), keep the corresponding time for 1.5h, stop heating and depressurize, and the alloy billet is cooled to room temperature with the furnace.

The hot-pressing sintering test was carried out at Beijing Jiaotong University. The hot-pressing sintering mold structure is shown in Figure 2. The mold material is high-strength graphite with a compressive strength of 35MPa, and the raw material powder (Ti powder + Al powder + Ti2AlNb powder) is placed in the sleeve , Relying on the upper indenter to transfer the sintering pressure. The sintering equipment is the ZT-40-21Y vacuum hot pressing sintering furnace developed by Shanghai Chenhua Technology Co., Ltd., as shown in Figure 2. The equipment adopts hydraulic pressure, the maximum pressure is 200kN, and its design vacuum degree can reach 1×10^{-3} Pa, temperature control accuracy is ±5°C.

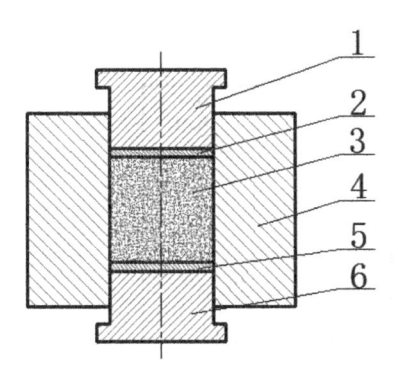

Figure 2. Hot-pressed sintering mold structure drawing and ZT-40–21Y vacuum furnace 1-Upper pressure head; 2-Upper gasket; 3-Powder; 4-Sleeve; 5-Lower gasket; 6-Lower pressure head.

X-ray diffractometer (X-ray diffractometer, referred to as XRD): The phase composition analysis of the hot-pressed sintered sample was completed on the D/max-rB 12kW rotating anode X-ray diffractometer. The diffractometer is a Cu target, the current and voltage are 40 mA, 40 kV, the X-ray wavelength is 0.15418 nm, the step length is 0.02°, and the scanning speed is 1°/min. The hot-pressed sintered material is cut by wire to obtain a sample with an external shape of 3 mm × 5 mm × 10 mm. The surface oxide scale of the sample is removed with sandpaper and polished to 1000 meshes, and then cleaned in an ultrasonic cleaning machine.

3 RESULTS AND DISCUSSION

Taking VTi2AlNb=10% as an example, using 400 mesh powder, the effect of hot pressing and sintering temperature on the microstructure was studied. As shown in Figure 3, it is a 400 mesh powder mixture (VTi2AlNb=10%) hot pressed at 1300°C The sintered microstructure morphology, it can be seen that not only a small amount of high Nb content structure (phase) is retained in the microstructure, but also a lath-like structure exists.

The phase of the hot-pressed sintered body was analyzed by XRD, and then a standard JCPDS card was used for calibration. As shown in Figure 4, it was found that the hot-pressed sintered body was mainly composed of Ti3Al, TiAl and Ti.

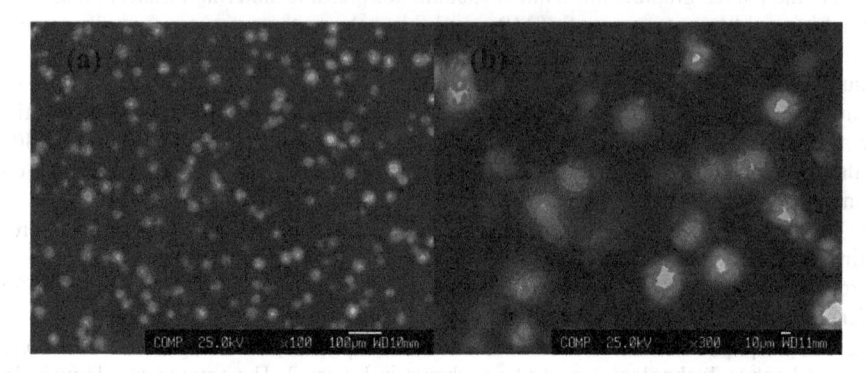

Figure 3. Microstructure morphology of 400 mesh powder mixture (VTi2AlNb=10%) after hot pressing and sintering at 1300°C (a) 100X; (b) 300X.

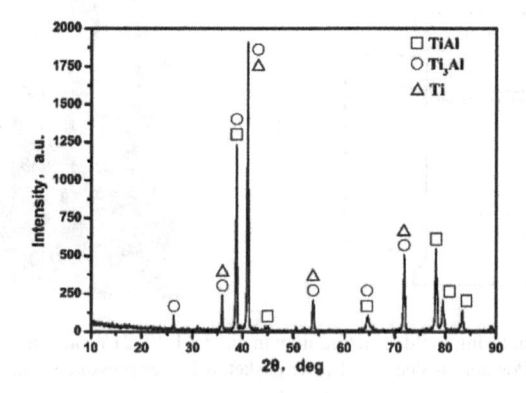

Figure 4. XRD pattern of 400 mesh powder mixture (VTi2AlNb=10%) after hot pressing and sintering at 1300°C.

Figure 5 shows the BEI structure and element line distribution of a 400 mesh powder mixture (VTi2AlNb=10%) after hot pressing and sintering at 1300°C. It can be seen that as the temperature rises to 1300°C, the Ti and Al elements in the lath structure do not change significantly, and basically do not contain Nb element, indicating that this is produced after the chemical reaction between Ti powder particles and Al powder particles , Has nothing to do with Ti2AlNb powder particles; Ti element, Al element and Nb element on both sides of the lath structure have certain fluctuations, which should be the product of the joint reaction of the three powder particles.

Combined with EDS, as shown in Figure 6, the EDS results of the average composition of precipitates 1 to 4 in the figure are shown in Table 2. Combined with the XRD analysis results, it can be determined that the precipitates 1 to 4 are TiAl(Nb)+Ti, TiAl(Nb), TiAl(Nb) and Ti3Al(Nb)+Ti.

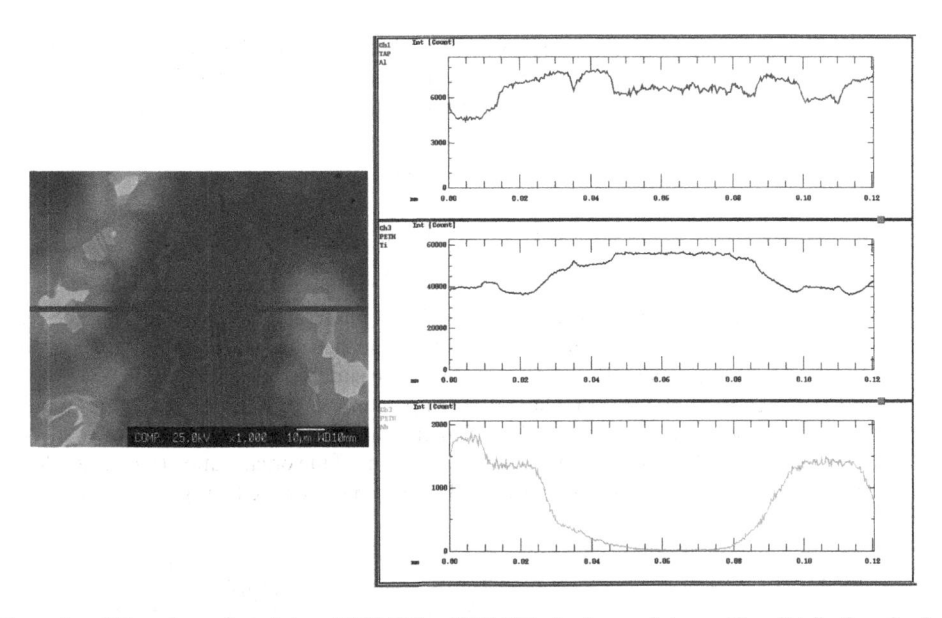

Figure 5. 400 mesh powder mixture (VTi2AlNb=10%) BEI structure and element line distribution after hot pressing and sintering at 1300°C.

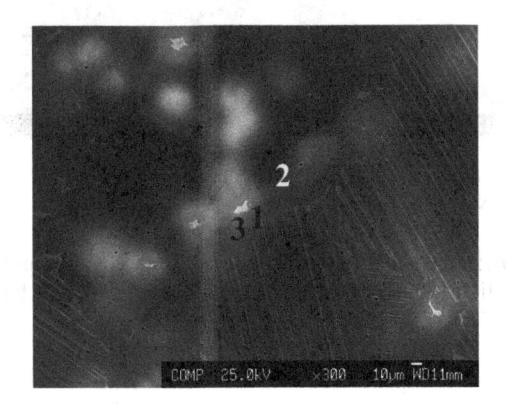

Figure 6. 400 mesh powder mixture (VTi2AlNb=10%) BEI structure after hot pressing and sintering at 1300°C.

Table 2. EDS analysis results of the components in each micro area in Figure 6 (atomic fraction/%).

Micro area	element Ti	Al	Nb	total	phase
1	55.4889	38.6287	5.8824	100.0	TiAl(Nb),Ti
2	49.2481	50.0483	0.7035	100.0	TiAl(Nb)
3	45.9479	49.0886	4.9636	100.0	TiAl(Nb)
4	78.2732	20.3769	1.3499	100.0	$Ti_3Al(Nb)$,Ti

Through the analysis of the type, morphology and distribution of the precipitated phases, it can be seen that when the sintering temperature is increased to 1300°C, the following chemical reactions mainly occur during the hot-pressing sintering process:

$$Al + Ti \rightarrow TiAl$$

$$TiAl + Ti \rightarrow Ti2Al$$

$$Ti2Al + Ti \rightarrow Ti3Al$$

Because the sintering temperature is too high, the Ti3Al phase generated by the interface reaction is dispersed and distributed in the microstructure and becomes lath. Therefore, a graded Ti/TiAl(Nb)/Ti3Al(Nb) microstructure is formed inside the sintered body.

Continue to increase the hot pressing sintering temperature to 1400°C. As shown in Figure 7, it is a 400-mesh powder mixture (VTi2AlNb=10%) after hot pressing and sintering at 1400°C. It can be seen that the microstructure is high in Nb. The content of the organization (phase) gradually decreases, the lath-like organization still exists, and the gap between the laths gradually increases.

Figure 7. Microstructure morphology of 400 mesh powder mixture (VTi2AlNb=10%) after hot pressing and sintering at 1400°C (a) 100X; (b) 300X.

XRD was used to analyze the phase of the hot-pressed sintered body, and then a standard JCPDS card was used for calibration. As shown in Figure 8, it was found that the hot-pressed sintered body was mainly composed of Ti3Al, TiAl and Ti.

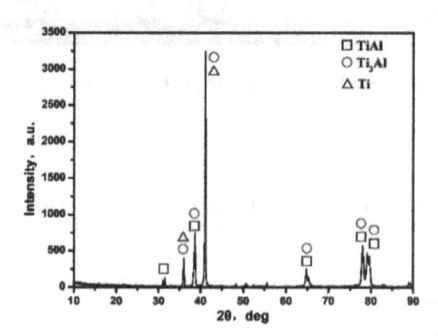

Figure 8. XRD pattern of 400 mesh powder mixture (VTi2AlNb=10%) after hot pressing and sintering at 1400°C.

Figure 9 shows the BEI structure and element line distribution of 400 mesh powder mixture (VTi2AlNb=10%) after hot pressing and sintering at 1400°C. Similar to Figure 5, it can be seen that as the temperature rises to 1400°C, the Ti and Al elements in the lath structure do not change significantly, and basically do not contain Nb. The Ti and Al elements on both sides of the lath structure There are certain fluctuations between the element and the Nb element.

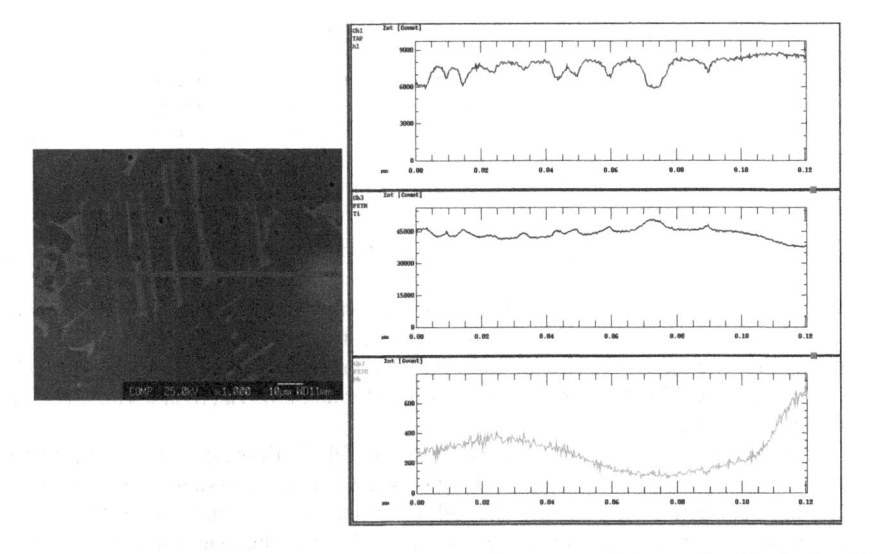

Figure 9. 400 mesh powder mixture (VTi2AlNb=10%) BEI structure and element line distribution after hot pressing and sintering at 1400°C.

Combined with EDS, as shown in Figure 10, the EDS results of the average composition of precipitates 1 to 4 in the figure are shown in Table 3. Combined with the XRD analysis results, it can be determined that the precipitates 1 to 4 are Ti3Al(Nb)+Ti, TiAl(Nb), TiAl(Nb) and TiAl(Nb)+Ti.

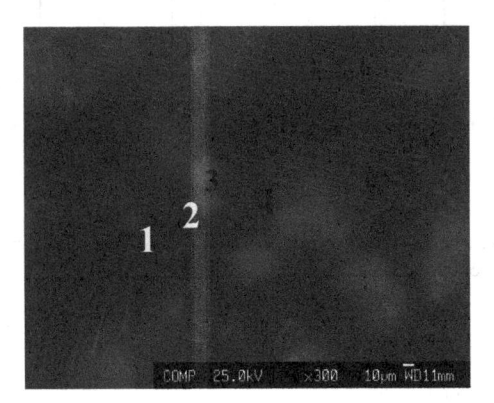

Figure 10. 400 mesh powder mixture (VTi2AlNb=10%) BEI structure after hot pressing and sintering at 1400°C.

Through the analysis of the type, morphology and distribution of the precipitated phases, it can be known that when the sintering temperature is increased to 1400°C, the following chemical reactions also mainly occur during the hot-pressing sintering process:

$$Al + Ti \rightarrow TiAl$$

$$TiAl + Ti \rightarrow Ti2Al$$
$$Ti2Al + Ti \rightarrow Ti3Al$$

Table 3. EDS analysis results of the components in each micro area in Figure 10 (atomic fraction/%)

Micro area	element Ti	Al	Nb	total	phase
1	76.3579	23.2528	0.3893	100.0	Ti$_3$Al(Nb),Ti
2	48.5074	49.6016	1.8910	100.0	TiAl(Nb)
3	43.3611	50.0001	6.6389	100.0	TiAl(Nb)
4	47.3704	50.2325	2.3971	100.0	TiAl(Nb),Ti

Because the sintering temperature is too high, the Ti3Al phase generated by the interface reaction is dispersed and distributed in the microstructure and becomes lath. Compared with sintering at 1200°C~1300°C, the gap between the lath structure becomes larger and the inside of the sintered body Formed a graded Ti/TiAl(Nb)/Ti3Al(Nb) microstructure.

It can be seen that with the increase of the sintering temperature, there is basically no Ti2AlNb phase, and the dispersed Ti2Al in the structure is transformed into Ti3Al, and Ti3Al is lath. The higher the temperature, the larger the lath gap.

The room temperature tensile properties and hardness of the hot-pressed sintered body after hot-pressing sintering are shown in Figure 11 and Figure 12. As the hot-pressing sintering temperature increases, the room-temperature tensile strength does not change significantly. When the hot-pressing sintering temperature is When the temperature is 1300°C, the room temperature elongation is the highest, and when the hot pressing and sintering temperature is 1200°C, the room temperature hardness is the highest.

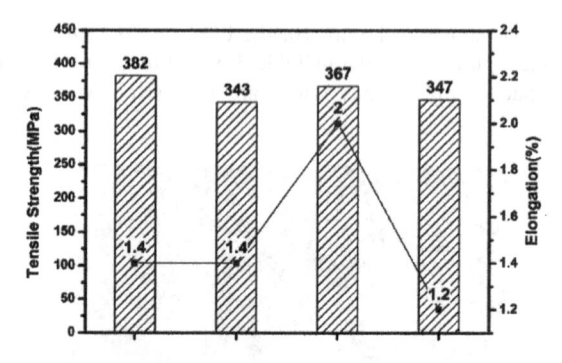

Figure 11. 400 mesh powder (Ti2AlNb composition: 10%) hot press sintering (1200°C~1400°C −30 MPa-1.5 h) room temperature tensile properties.

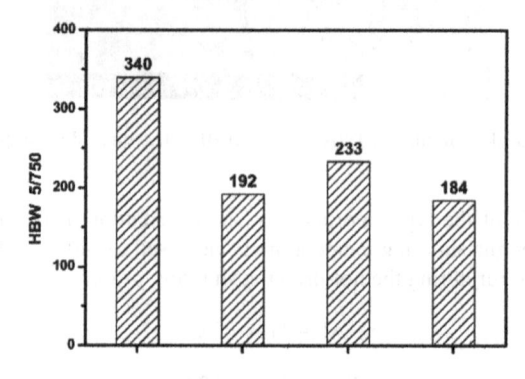

Figure 12. 400 mesh powder (Ti2AlNb component: 10%) hot-press sintering (1200°C~1400°C −30 MPa-1.5 h) body room temperature hardness.

4 CONCLUSIONS

(1) Under different hot pressing sintering temperature conditions, the hot pressing sintered body is mainly composed of Ti3Al, TiAl and Ti.
(2) With the increase of the sintering temperature, there is basically no Ti2AlNb phase. The dispersed Ti2Al in the structure is transformed into Ti3Al, and Ti3Al is lath-like. The higher the temperature, the larger the lath gap.
(3) With the increase of the hot pressing sintering temperature, the room temperature tensile strength does not change significantly. When the hot pressing sintering temperature is 1300°C, the room temperature elongation is the largest. When the hot pressing sintering temperature is 1200°C, the room temperature hardness highest.

REFERENCES

[1] China Aeronautical Materials Handbook Editorial Board.China Aeronautical Materials Handbook-Titanium Alloyand Copper Alloy[M]. Bei Jing: China Standard Press.2002: 140–142.
[2] LI Yujia, XUAN Fuzhen, TU Shandong. Influence ofstress ratios and residual stress on the fracture pattern ofTi-6Al-4V alloy in high cycle regime[J]. Journal ofMechanical Engineering. 2015, 51(6): 45–50.
[3] ZHANG Junhong, YANG Shuo, LIN Jiewei. Fatiguecrack growth rate of Ti-6Al-4V considering the effects offracture toughness and crack closure[J]. Chinese Journalof Mechanical Engineering. 2015, 28(2): 409–415.
[4] MAO Yong, LI Shiqiong, ZHANG Jianwei, et al. Studyon microstructure and mechanical properties ofTi-22Al-20Nb-7Ta Intermetallic Alloy[J]. ActaMetallurgica Sinica. 2000, 36(2): 135–140.
[5] YU Huichen, DONG Chengli, JIAO Zehui, et al. Hightemperature creep and fatigue behavior and lifeprediction method of a TiAl alloy[J]. Acta MetallurgicaSinica. 2013, 49(11): 1311–1317.
[6] TSIPAS S A, GORDO E. Molybdeno-aluminizing ofpowder metallurgy and wrought Ti and Ti-6Al-4V alloysby pack cementation process[J]. MaterialsCharacterization. 2016, 118: 494–504.
[7] NAZAROVAT I, IMAYEV V M, IMAYEV R M, et al.Study of microstructure and mechanical properties of Ti-45Al-(Fe, Nb) (at.%) alloy[J]. Intermetallics. 2017, 82: 26–31.
[8] RUDNEV V S, LUKIYANCHUK I V, VASILYEVA MS, et al. Aluminum- and titanium-supported plasmaelectrolytic multicomponent coatings with magnetic, catalytic, biocide or biocompatible properties[J]. Surface& Coatings Technology. 2016, 307: 1219–1235.
[9] SUN Yanbo, ZHAO Yeqing, ZHANG Di, et al. Multilayered Ti-Al intermetallic sheets fabricated by cold rolling and annealing of titanium and aluminum foils[J].Transactions Nonferrous Metals Society of China. 2011,21:1722–1727.
[10] J.G. Luo, V.L. Acoff. Processing gamma-based TiAl sheetmaterials by cyclic cold roll bonding and annealing ofelemental titanium and aluminum foils [J]. Mater Sci EngA, 2006, 433: 334–342.
[11] G.P. Chaudhari, V.L. Acoff .Titanium aluminide sheetsmade using roll bonding and reaction annealing [J].Intermetallics. 2010, 18: 472–478.
[12] WEI Dongbo, ZHANG Pingze, YAO Zunyao, et al.Preparation and high-temperature oxidation behavior ofplasma Cr-Ni alloying on Ti6Al4V alloy based on doubleglow plasma surface metallurgy[J]. Applied SurfaceScience. 2016, 388: 571–578.

Antistatic finishing of polypropylene non-wovens with reduced graphene oxide

Y. Zhang & Q. Yu
College of Textile and Fashion, Anhui Vocational and Technical College, Hefei, Anhui, China

ABSTRACT: With the extensive application of polypropylene non-woven in the field of medicine and health, it is particularly important to realize the antistatic property of polypropylene non-wovens. The antistatic property of finished polypropylene non-wovens mainly depends on the finishing process parameters. It is of guiding significance to study the effects of finishing process parameters such as $NaBH_4$ concentration, graphene oxide concentration, and impregnation time on antistatic properties of polypropylene non-woven. After a lot of experiments and tests, the influence of process parameters on antistatic properties of polypropylene non-woven was found out. It was found that when the concentration of $NaBH_4$ was 3.0 g/L, the concentration of graphene oxide was 3.5 g/L, the impregnation time was 60 min, and the antistatic property of polypropylene non-woven was the best. At the same time, it was found that the finished polypropylene non-woven can show a good antistatic effect after washing for less than 4 times.

1 INTRODUCTION

Polypropylene non-woven is a kind of material with a wide range of uses, especially in medical and health products, which has many applications, such as disposable protective clothing, surgical clothing, surgical cap, medical sheet, mask, etc. [1] Due to the non-polar structure of polypropylene fiber macromolecules, the normal moisture regain is zero. This makes polypropylene non-woven products to be used in the process; it is very easy to generate charge accumulation and static electricity and then absorb bacteria and dust. As medical and health products, there are potential bacterial infection, and the operation and use of medical equipment will also have adverse effects. This limits the application of polypropylene non-woven in the field of medical and health. Therefore, it is of great practical significance to study the antistatic property of polypropylene non-wovens.

The properties of reduced graphene oxide are similar to those of original graphene prepared from graphite directly. For example, they have excellent conductivity [2]. The application research of reduced graphene oxide as an antistatic finishing agent in the textile field is not deep enough [3–5], and the antistatic research of polypropylene non-wovens is rarer. In this study, graphene oxide was first prepared and then treated on polypropylene non-woven and then reduced to graphene by $NaBH_4$. The effect of processing parameters on the antistatic film of non-wovens was analyzed, and the process parameters were optimized. At the same time, the antistatic durability of finished polypropylene non-woven was discussed in order to play a guiding role in the production of medical polypropylene non-wovens.

2 EXPERIMENT

2.1 *Preparation of graphene oxide*

Graphene oxide was prepared by the modified Hummers method [6,7]. 69 ml of concentrated sulfuric acid and 1.5 g of sodium nitrate were added into a three mouth bottle, placed in an ice water bath, and cooled to 0°C, and then 3g of graphite powder was added. At the same time, stirring was continued. After mixing well, 9g of potassium permanganate was added slowly for several times and stirred for 60 min. During this process, the temperature should not exceed 20°C. Then, it was heated to 45°C and stirred for 60 min. 138 ml of distilled water was added for 15 min, heated to 98°C, and stirred for 60 min, and then hydrogen peroxide was added until there was no bubble. After standing for 12 h, using dilute hydrochloric acid and deionized water for multiple centrifugal washing and then drying in a drying oven at 60°C for 24 hours, graphene oxide was obtained.

2.2 Antistatic finishing of polypropylene non-wovens

(1) Polypropylene non-woven pretreatment. Before antistatic finishing, the non-woven should be pretreated. The non-woven was put into ethanol solution and then treated with ultrasonic for 30 minutes to remove the residual organic oil on the surface and then repeatedly washed with ionic water for 3 minutes before drying. (2) Graphene oxide finishing of non-woven. Graphene oxide solution was prepared by adding graphene oxide into ionic water and ultrasonic wave for 60 min. The pretreated nonwovens were soaked in graphene oxide solution for a certain time and then taken out. Then it was put into the drying oven at 60°C (3) Reduction of graphene oxide on non-woven. All cotton non-wovens finished with graphene oxide are soaked in $NaBH_4$ solution and then taken out for 60 min. After repeated cleaning with deionized water, it was dried in a 100°C drying oven.

2.3 Analysis and test methods

2.3.1 Scanning electron microscopy (SEM) [8]

The S-4800 scanning electron microscope was used to observe the surface morphology of polypropylene non-woven treated with reduced graphene oxide. The sample was fixed on the surface of the sample table with conductive adhesive, and the sample surface was sprayed with gold by an ion sputtering instrument and observed by a scanning electron microscope.

2.3.2 Antistatic property test of polypropylene nonwoven

A YG401 fabric inductive electrostatic tester is used for the half-life test. Referring to GB/T 12703.1–2010 "evaluation of electrostatic properties of textiles Part 1: static voltage and half-life" [9], the half-life test of non-woven samples is conducted under the humidity control and test conditions of 20°C and relative humidity of 35%. The antistatic grade of the sample after testing is determined in Table 1.

Table 1. Evaluation of electrostatic properties (halflife) of textiles

grade	requirements/S
A	≤ 2.0
B	≤ 5.0
C	≤ 15.0

The surface specific resistance of non-woven is characterized by a ZC36-type high resistance meter, which is tested according to GB/T 12703.4–2010 "textile electrostatic performance evaluation part 4: resistivity" [10]. The environmental conditions of humidity control and test atmosphere are as follows: temperature (20 ± 2)°C, relative humidity (35 ± 5)%, and ambient wind speed should be less than 0.1 m/s. The antistatic standard of fabric is determined in Table 2.

Table 2. Evaluation of electrostatic properties (resistivity) of textiles

grade	requirement/Ω
A	$\leq 1 \times 10^7$
B	$\geq 1 \times 10^7, <1 \times 10^{10}$
C	$\geq 1 \times 10^{10}, \leq 1 \times 10^{13}$

2.3.3 Antistatic durability test of polypropylene nonwoven

An Sw-8 color fastness tester was used for testing. According to GB/T 3921–2008 "textile color fastness test color fastness to soaping method" [11], the samples of non-woven were washed several times. The washing conditions were as follows: 5 g/L (40 ± 2)°C, 10 min, and a bath ratio of 1:50. After washing, half-life and surface specific resistance of the dried samples were determined

3 RESULTS AND DISCUSSION

3.1 Surface morphology of polypropylene non-wovens before and after finishing

Figure 1. Sample before finishing.

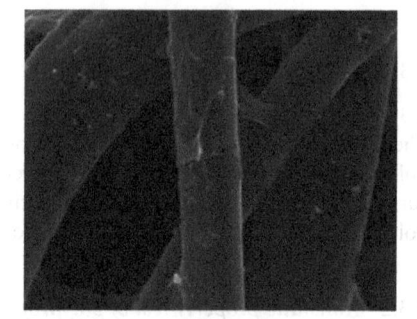

Figure 2. Sample after finishing.

It can be seen from Figure 1 that the surface of polypropylene fiber in the non-woven sample before finishing is smooth. It can be seen from Figure 2 that after antistatic finishing, reduced graphene oxide is adsorbed on the surface of the fiber, and reduced graphene oxide is combined with polypropylene fiber.

3.2 Factors influencing antistatic property of polypropylene non-woven

The concentration of NaBH4, aphene oxide and impregnation time were selected The effect of finishing parameters on antistatic property of PP nonwoven was discussed.

3.2.1 Effect of the $NaBH_4$ concentration on antistatic performance

Under the conditions of graphene oxide concentration of 4.0 g/L and impregnation time of 60 min, the effects of different concentrations of reducing agent $NaBH_4$ on antistatic properties of polypropylene non-woven are shown in Table 3.

It can be seen from Table 3 that the surface-specific resistance of the $1^{\#}$ non-woven sample without graphene oxide finishing is very large, with half-life > 180s, which is very easy to generate static electricity. The surface resistivity and half-life of the $2^{\#}$ non-woven treated with graphene oxide but not reduced by $NaBH_4$ reductant are significantly reduced compared with those of $1^{\#}$. It shows that graphene oxide with certain conductivity has been adsorbed on the non-woven sample.

Table 3. Effect of the NaBH$_4$ concentration on antistatic performance

sample	NaBH$_4$ concentration g/L	halflife S	surfacespecific resistance Ω
1#		>180	∞
2#	0.	3.2	2.44×10^7
3#	1.0	0.51	3.45×10^4
4#	2.0	0.26	9.86×10^3
5#	3.0	0.08	4.45×10^3
6#	4.0	0.08	5.05×10^3

It can also be seen from Table 3 that the surface-specific resistance and half-life of the non-woven sample decrease with the increase of the concentration of NaBH$_4$. This is because the reducing agent removes the oxygen-containing functional groups in graphene oxide adsorbed on the non-woven sample, strengthens the van der Waals force between layers, reconstructs the large π between layers, and increases the conductivity. However, when the concentration of NaBH$_4$ reaches 3.0 g/L, the surface-specific resistance and half-life of the samples do not decrease when the concentration of NaBH$_4$ is increased. This is because the graphene oxide adsorbed on the non-woven sample has been reduced by NaBH$_4$.

3.2.2 Effect of graphene oxide concentration on antistatic properties

The influence of graphene oxide concentration on antistatic property of polypropylene nonwoven is shown in Table 4 under the process conditions of NaBH$_4$ concentration of 4g/L and impregnation time of 60min.

Table 4. Effect of concentration of graphene oxide on antistatic properties.

sample	graphene oxide concentration g/L	halflife S	surface resistance Ω
1#	2.0	0.21	9.57×10^3
2#	2.5	0.16	8.34×10^3
3#	3.0	0.11	6.93×10^3
4#	3.5	0.07	4.14×10^3
5#	4.0	0.07	4.35×10^3
6#	4.5	0.07	4.09×10^3

The influence of NaBH$_4$ concentration on the antistatic property of non-woven was analyzed. In order to reduce the graphene oxide adsorbed on polypropylene non-woven to graphene, the concentration of NaBH$_4$ is 4 g/L. It can be seen from Table 4 that with the increase of the concentration of graphene oxide, the surface-specific resistance and half-life of the sample decrease, because the amount of graphene oxide adsorbed in polypropylene non-woven increases, and all of them are reduced to graphene. When the concentration of graphene oxide is 3.5 g/L, the surface-specific resistance and half-life of the sample reach the minimum, and the antistatic effect reaches the highest. When the concentration of graphene oxide increases again, the surface-specific resistance and half-life of the samples do not decrease, which is mainly because the adsorption of graphene oxide on the non-woven reaches saturation and no longer increases.

3.2.3 Effect of impregnation time on antistatic property

Under the process conditions of $NaBH_4$ concentration of 4 g/L and graphene oxide concentration of 3.5 g/L, the influence of impregnation time on antistatic property of polypropylene non-woven is shown in Table 5.

Table 5. Effect of immersion time on antistatic properties.

sample	immersion time min	halflife S	surface specific resistance Ω
1#	15	0.87	8.69×10^4
2#	30	0.39	1.07×10^4
3#	45	0.18	7.93×10^3
4#	60	0.08	4.21×10^3
5#	75	0.08	417×10^3
6#	90	0.08	414×10^3

It can be seen from Table 5 that the surface-specific resistance and half-life of the samples decrease with the increase of the immersion time of the non-woven in graphene oxide dispersion, which is due to the continuous increase of graphene oxide deposited on polypropylene non-woven. When the immersion time reaches 60 min, the surface-specific resistance and half-life of the sample reach the minimum and no longer decrease. The antistatic effect of the sample is the best. This is because the graphene oxide adsorbed on the non-woven is saturated and no longer increases.

Through the discussion and analysis of NaBH4 concentration, graphene oxide concentration, impregnation time, and other finishing process parameters, the best finishing process parameters can be obtained as reducing agent NaBH4 concentration of 3.0 g/L, graphene oxide concentration of 3.5 g/L, and impregnation time of 60 min.

3.3 Antistatic durability of polypropylene non-woven

The antistatic properties of polypropylene non-woven were tested after antistatic finishing with $NaBH_4$ concentration of 3.0 g/L, graphene oxide concentration of 3.5 g/L and immersion time of 60 min. The specific results are shown in Table 6.

Table 6. Influence of washing times on antistatic durability.

washing times times	halflife S	surface specific resistance Ω
0	0.07	4.82×10^3
2	0.47	1.29×10^4
4	1.9	1.71×10^7
6	4.4	4.51×10^8
8	>180	5.19×10^{12}

It can be seen from Table 6 that with the increase of washing times, the surface-specific resistance and half-life of the sample increase rapidly, and the antistatic effect decreases rapidly. After washing four times, the antistatic grade is lower than the standard. After washing eight times, the antistatic electricity basically disappeared. It shows that the reduced graphene oxide adsorbed on polypropylene non-woven has been washed off after washing eight times. The test results show that polypropylene non-woven should have a good antistatic effect, and washing should not be done for more than four times. This kind of non-woven can meet the use requirements in medical and health products with one-time or limited use times.

4 CONCLUSION

Reduced graphene oxide can be effectively adsorbed on polypropylene nonwovens. The optimum finishing parameters are $NaBH_4$ concentration of 3.0 g/L, graphene oxide concentration of 3.5 g/L, and impregnation time of 60 min. The finished PP non-woven can show a good antistatic effect after washing for less than four times. When washing more than eight times, there is no antistatic performance.

ACKNOWLEDGMENTS

The research funds were provided by the natural science research key project (KJ2017A478) and the quality engineering project (2016zy074). At the same time, I would like to thank the editors and reviewers for their review of this paper.

REFERENCES

[1] Z.X. Wang. Application and development of disposable medical nonwovens *[J.] Chemical fiber and textile technology*, 2020,49 (2): 25–28.
[2] G. Lu. Progress in preparation and application of graphene *[J]. Energy and environmental protection*, 2020,42 (5): 78–81,93.
[3] [M. Miao, D. Xu, N. Lu, et al. Study on antistatic modification of polypropylene nonwovens by graphene oxide *[J]. Industrial textiles*, 2017,35 (11): 39–43.
[4] M Miao, X.X. Wang, Y. Wang, et al. Preparation and antistatic properties of graphene oxide grafted polypropylene nonwovens *[J]. Acta textile Sinica*, 2019,40 (11): 125–130.
[5] X.Y. Gao, W.Q. Chen, G.Q. Wang, et al. Preparation of graphene modified polypropylene non-woven membrane *[J]. Applied Chemical Engineering*, 2018,47 (5): 999–1002.
[6] D.Y. Mao, D. Yang, J.P.Fan. Preparation and properties of graphene oxide hybrid molecularly imprinted composite membranes *[J]. Acta Chem Sinica*, 2020,71 (06): 2900–2911.
[7] D.D. Chen, W.L. Wang, Q. Wang, et al. Preparation of graphene oxide and antistatic finishing of polyester *[J]. Modern silk science and technology*, 2020,35 (01): 13–17.
[8] P.Yang, C.Q. Teng, F. Yu, et al. Interfacial properties of modified aramid fiber reinforced epoxy composites *[J]. Plastics industry*, 2015,43 (12): 23–27.
[9] R.M. Wu, J. Ang. discussion on evaluation method of antistatic property of textiles *[J]. China National Textile inspection*, 2019, (7): 59–61.
[10] W.L. Pan, X.W. Zhao. Test standard for antistatic properties of textiles *[J]. Printing and dyeing*, 2017, 43 (18): 43–47.
[11] C.M. Yao. Discussion on the test method of textile color fastness to soaping *[J]. China National Textile inspection*, 2018, (8): 82–85.

Construction and application of rare earth ibuprofen phenanthroline fluorescence system

L. Wang
College of Pharmacy, Jiamusi University, Jiamusi, China

W.B. Liu
School of Materials Science and Engineering, Jiamusi University, Jiamusi, China

D. Lu, C.F. Jin, N. Wen, Y.Q. Cheng & Y.G. Lv
College of Pharmacy, Jiamusi University, Jiamusi, China

ABSTRACT: The aim of this experiment was to find a new and reliable method for the determination of ibuprofen. The Tb^{3+}-Phen fluorescent probe was investigated and constructed for the determination of ibuprofen, and the fluorescence response of ibuprofen to different detection environments was investigated to find the best detection environment for the probe. The actual samples of ibuprofen were successfully assayed under optimal assay conditions, and the results were compared with the labeled levels, which proved that the levels determined in this study were in general agreement with the standard levels in the samples, and the reproducibility was good. Therefore, the method can be used for the determination of ibuprofen in drugs, and even for the determination of ibuprofen in different substances.

1 INTRODUCTION

Rare earth materials are new functional materials with various properties such as optical, electrical, magnetic and biological. Its wide range of applications can be used in various fields such as information, energy, biology and so on. At the same time, rare earth materials can also be applied to transform some traditional industries such as construction, chemical industry, agriculture, etc [1]. Nowadays, scientific researchers have extensive research on rare-earth organic complexes luminescent materials, which have great application value. Rare-earth ions can be combined with a variety of organic compounds through covalent bonding to form rare-earth complexes, and the 4f orbitals of rare-earth elements and organic complexes with the high absorption capacity of excitation light, so rare-earth complexes have special luminescence [2,3].

The second ligand used in this experiment is o-phenanthroline, which has good binding ability for transition metal and rare earth metal ions [2]. Ibuprofen has been widely used since its introduction. Ibuprofen is a chemically synthesized aromatic propionic acid non-steroidal anti-inflammatory drug that acts in vivo mainly as an inhibitor of COX-2 by inhibiting the action of COX-2 in order to reduce the synthesis and release of prostaglandins, acting almost identically to the rest of the aromatic propionic acid derivatives [4–8].This project focuses on the determination of ibuprofen tablets with the intention of designing a new simple and convenient fluorometric method for the quantitative determination of ibuprofen tablets.

2 METHODS

2.1 Preparation of ibuprofen standard stock solution

Ibuprofen standard 20 mg was accurately weighed and placed in a 100 mL flask. A small amount of anhydrous ethanol was first added for ultrasonic acceleration of dissolution, then anhydrous ethanol was added for constant volume to the scale to obtain 1.0×10^{-3} mol/L standard reserve solution, which was stored in a cool and dry place away from light, and diluted to the required concentration during use.

2.2 Preparation of ibuprofen sample solution

Buy a box of ibuprofen tablets (National Pharmaceutical Standard H53020571) manufactured by a company in Yunnan, take five tablets (500 mg) and crush them into powder, mix them evenly and weigh them precisely 1/10, dissolve them in anhydrous ethanol and fix the volume into a 100 mL volumetric flask. The sample solution of ibuprofen (10 mg/L) was obtained by pipetting 10 mL of the test solution, diluting 100 times with anhydrous ethanol, shaking well, and filtering the solution with 0.45 μm microporous membrane.

2.3 Determination of system fluorescence spectrometric method

1.1 mL of compound Tb^{3+} stock solution at a concentration of 1.0×10^{-4} mol/L, 2.2 mL of compound o-phenanthroline stock solution at a concentration of 1.0×10^{-3} mol/L, and 1.8 mL of potassium dihydrogen phosphate-disodium hydrogen phosphate (PBS) buffer solution (pH=7.0) were added to a series of clean, dry 10 mL colorimetric tubes. Take one as a blank solution, add ibuprofen standard solution of different diluents (1.0×10^{-3} mol/L) to other colorimetric tubes, then add anhydrous ethanol to the scale line, shake it to complete reaction. The fluorescence intensity (F) of the system at 552 nm was measured by irradiation with excitation light at 316 nm for 18 min at room temperature, using a 1 cm quartz cuvette.

2.4 Screening of optimal experimental conditions for the system

Optimization of system pH: Use potassium dihydrogen phosphate-disodium phosphate buffer solution (PBS) to adjust the pH of the system.

Selection of fluorescent probe concentration; Under the already screened conditions, different Tb^{3+} concentrationson the fluorescence intensity of the system to screen the optimal Tb^{3+} concentration of the system. Then the concentration of the selected Tb^{3+} solution was fixed to study the influence of different concentrations of o-phenanthroline on the system, so as to screen out the optimal concentration of o-phenanthroline in the system.

Selection of the amount of buffer solution; The single variable method was used to examine the changes in the response values of the system when the doses of buffer solutions were 1, 1.2, 1.4, 1.6, 1.8, 2.0, 2.2, and 2.4 mL, respectively, to screen the optimal buffer solution dosage.

Effect of reaction time; the changes in the fluorescence response values of the system were monitored at 5 min intervals from 5 to 40 min, and the optimal reaction time was screened.

2.5 Methodological examination

Specialized property examination: Under the present experimental conditions, the fluorescence profiles of the blank solvent, the fluorescent system containing IBP standard solution, and the fluorescent system containing the same concentration of IBP sample solution were examined.

Precision, Accuracy and Stability; The IBP stock solutions of 0.2, 1.0, 1.8 (10^{-6} mol/L) low, medium and high concentrations were aspirated precisely in 6 portions each, and the corresponding experimental results were recorded and the relative standard deviations (RSD %) were calculated. Take the same 6 copies of IBP stock solution. Determination was performed according to the method under "5.2.5", and the recovery was calculated according to the measured concentration and the actual concentration, the recovery (%) = measured concentration/actual concentration × 100%, and the relative standard deviation (RSD%) was calculated to examine the accuracy of the method. Take the same 6 copies of IBP stock solution. The stability of ibuprofen was investigated by placing it at room temperature for 0, 2, 4, 6, 8 and 10 h, respectively.

2.6 Determination of sample content and spiked recoveries

Under the optimized test conditions, the sample solutions were tested according to the standard curve method under "5.2.5.2", and each sample was measured four times in parallel, and the relative standard deviation (RSD %) was calculated when compared with the factory labeled content of ibuprofen tablets. Under the same conditions, 0.30 ml of the sample solution was taken for measurement, while 0.20 mL, 0.30 mL and 0.40 mL of the standard solution of ibuprofen (10 mg/L) were added in order of 80%, 100% and 120% of the sample solution taken. Each sample was measured six times in parallel, and the spiked recoveries and relative standard deviations (RSD%) were calculated.

3 RESULTS AND DISCUSSION

3.1 System fluorescence mapping

The maximum excitation wavelength (λex) and maximum emission wavelength (λem) of the system Tb^{3+}_Phen were 316 nm and 552 nm, respectively, under the condition of PBS buffer solution at pH = 7. The fluorescence intensity increased with the addition of a certain amount of ibuprofen to the system, and the finding could be used for the determination of ibuprofen content.

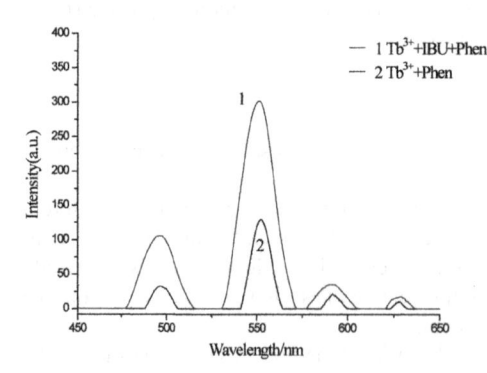

Figure 1. Fluorescence spectrum of system.
$C_{Tb}^{3+} = 1.2 \times 10^{-6}$ mol/L, pH = 7.0,
$C_{Phen} = 2.2 \times 10^{-5}$ mol/L,
$C_{IBU} = 5.0 \times 10^{-6}$ mol/L.

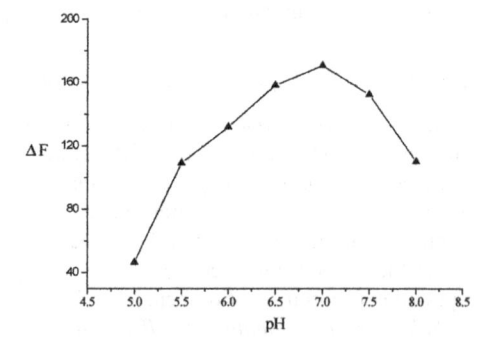

Figure 2. Effect of pH.
$C_{Tb}^{3+} = 1.2 \times 10^{-6}$ mol/L,
$C_{Phen} = 2.2 \times 10^{-5}$ mol/L,
$C_{IBU} = 5.0 \times 10^{-6}$ mol/L,
pH = 5.0; 5.5; 6.0; 6.5; 7.0; 7.5; 8.0.

3.2 Screening of optimal experimental conditions for the system

3.2.1 Optimization of system pH

Different pH buffer solution on system changes in fluorescence intensity value Δ F is shown in Figure 2. The results showed that the response value of the system reached the maximum when the pH of the buffer solution = 7. In summary, the PBS buffer solution with pH=7.0 was selected as the buffer condition for the best response of the system in this experiment.

3.2.2 Selection of fluorescent probe concentration

The system response reached a maximum when the probe concentration reached 1.2×10^{-6} mol/L. Therefore, the concentration of Tb^{3+} solution was fixed at 1.2×10^{-6} mol/L in this study to investigate the effect of different concentrations of o-phenanthroline on the system. It was found that the fluorescence response was more pronounced as the concentration of o-phenanthroline approached 2.2×10^{-5} mol/L, with no significant change as it continued to increase.

3.2.3 Selection of the amount of buffer solution

At 1.8 mL of buffer solution, the ΔF of the system reached a maximum. Therefore, in this experiment, the addition of 1.8 mL of potassium dihydrogen phosphate – disodium hydrogen phosphate (PBS) buffer solution (pH = 7.0) was chosen.

3.2.4 Effect of reaction time

The ΔF of the system reached its maximum value after 20 min of reaction, so the experiment was chosen to be conducted after 20 min of reaction. Meanwhile, to ensure that the time was kept within the maximum sensitivity, the reaction temperature was elevated to examine the effect of temperature on the system. After the measurement, it was found that the fluorescence was enhanced but not significantly after the temperature was higher than room temperature, and the sensitivity of the experiment could be reached at room temperature, so this study was still chosen to be conducted at room temperature 25°C.

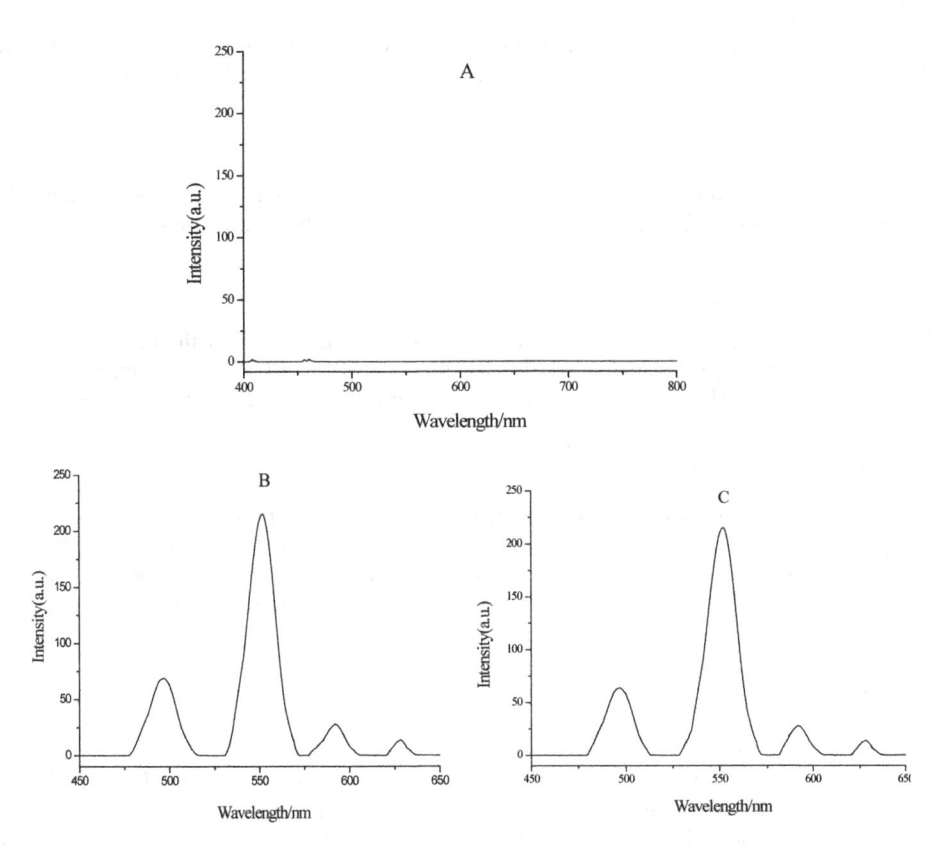

Figure 3. Blank solution (A), fluorescence system with standard IBP solution (B), fluorescence system with sample IBP solution (C) Fluorescence emission pattern.

3.3 Methodological examination

3.3.1 Specialized property examination
The fluorescence profiles of the blank solvent, the fluorescent system containing IBP standard solution, and the fluorescent system containing the same concentration of IBP sample solution are shown in Figure 3.

3.3.2 Precision
From the data in the table, it can be seen that the precision RSD of ibuprofen are less than 5.00%, and the method has good precision.

Table 1. Precision of ibuprofen (n = 6).

Concentration/(10^{-6}mol/L)	Average measured value mean ± SD /(10^{-6} mol/L)	RSD/%
0.2	0.20 ± 0.008	4.0
1.0	1.01 ± 0.049	4.7
1.8	1.79 ± 0.024	1.3

3.3.3 Accuracy
The recoveries of the low, medium and high concentrations of ibuprofen stock solution were (98.3 ± 4.08)%, (101.2 ± 4.71)% and (99.4 ± 2.34)%, respectively, with RSD values less than 5.00%.

3.3.4 Stability
Ibuprofen stock solution was placed at room temperature for 10 h and tested at two-hour intervals. TheRSD values of ibuprofen were all less than 5.00%, and its stability was good.

3.4 Determination of spiked recoveries

The average spiked recoveries of the method constructed in this study for the determination of ibuprofen content were in the range of 99.0%~105.0% with RSDs of 1.5%~3.1%, and the method recoveries were good. The method is reliable for the determination of actual samples of ibuprofen.

3.5 Sample content determination

There was no significant difference between the assay used in this study and the labeled content of ibuprofen tablets, while the relative standard deviations of the two samples were determined to be within the permissible limits.It shows that the novel method constructed in this study for the determination of ibuprofen content in ibuprofen tablets is practical and reliable.

4 SUMMARY

In this experiment, a Tb^{3+}-Phen fluorescent probe was investigated and constructed to detect ibuprofen content, and the probe was found to have a good fluorescence response to ibuprofen.By optimizing the probe detection conditions, the content of ibuprofen in actual samples was successfully detected under the optimal detection conditions. The results were compared with the labeled content, which proved that the content determined in this study was basically consistent with the standard content in samples, and the reproducibility was good.Thus, the method can be used to determine the amount of ibuprofen in drugs, even to the extent that it can be used to detect the amount of ibuprofen in different substances. It provides a new and reliable method for the determination of ibuprofen.

ACKNOWLEDGMENTS

This work was financially supported by the National Science Foundation of China (No.213 46006), Department of scientific research project in Heilongjiang province (No.B2017015).

REFERENCES

[1] K.D. Rainsford, M. Quadir, Gastrointestinal damage and bleeding from non-steroidal anti-inflammatory drugs.I.Clinicaland epidemiological aspects[J]. Inflammopharmacology, 1995, 3(2): 169–190.
[2] Zhang L. Market analysis of ibuprofen[J]. China Pharmacy, 2003, 14(01): 11–12.
[3] Peng M G, Li H J, Kang X, Du E D, Li D D. Photo-degradation ibuprofen by UV/H_2O_2 process: response surface analysis and degradation mechanism[J]. Water Science and Technology, 2017, 75(12): 2935–2951.
[4] Chen Biao, Wang Jianhua, Li Sakura, et al. Overview of research on ibuprofen extended-release formulations[J]. Journal of Pharmacy Practice, 2006, 24(2): 65–69.
[5] Su Huaide. China's first over-the-counter drug-ibuprofen[J]. China Pharmaceutical Journal, 2001, 3(3): 233–234.
[6] National Pharmacopoeia Committee. Pharmacopoeia of the People's Republic of China (1995 edition) [M]. Beijing: Chemical Industry Press, 1995: 115.
[7] Yang Baofeng. Pharmacology (6th ed.) [M]. Beijing: People's Health Publishing House, 2003: 160–179.
[8] Zhang Yitu, Chen Yao, Parliament Zhirong, et al. Human pharmacokinetics and bioequivalence of single-dose and multi-dose administration of ibuprofen extended-release capsules[J]. Chinese Clinical Pharmacology and Therapeutics, 2014, 19(3): 297–301.

Microstructure and mechanical property of SiC/ZL114A composite fabricated by laser deposition manufacturing

C.S. Lv & L.Q. Wang
School of Mechanical and Electrical Engineering, Harbin Engineering University, Harbin, Hei Longjiang, China

J.Z. Yi
School of Materials Science and Engineering, Shenyang Aerospace University, Shengyang, Liaoning, China

ABSTRACT: In this study, the SiC/ZL114A composite was fabricated by laser deposition manufacturing, and the microstructure, microhardness and tensile property were deeply studied by the optical microscope, microhardness tester and electronic universal testing machine. The results show that the deposited microstructure consists of ZL114A matrix structure and SiC particles. For single pass deposited sample, the SiC content in the middle of deposited layer increases with the laser power, but decreases in the layer surface. After multi-passes and layers deposition, SiC particles are redistributed and more evenly distributed in the deposited layer with the laser power increasing. The microhardness of deposited layer is higher in the layer surface and fluctuates greatly in the middle of deposition layer. Along with the increasing of laser power and SiC content, and the decreasing of scanning speed, the average microhardness of deposited layer increases with a higher fluctuation degree and a wider fluctuation range. When the SiC content is 10%, the strength of SiC/ZL114A composite is highest, and with the increasing of SiC content, the strength and plasticity decreases continuously.

1 INTRODUCTION

ZL114A (Al-Si-7Mg) alloy is the common Al-Si series hypoeutectic alloy with good cast property, welding property and comprehensive mechanical properties, which has been widely used in structural components under the action of high loading applying in the fields of aerospace and vehicle engineering, et al. [1]. Researchers attempt to improve the mechanical property of ZL114A alloy by the methods of changing alloy composition, adjusting heat treatment process and deformation process, but it appears to have little effect so far. Particle reinforced aluminum matrix composites well combine the metallic character of aluminum and ceramic character of reinforced particle showing superior wear-resisting property, mechanical property, high thermal expansivity and good dimensional stability [2]. It has the wide application space in the fields of aerospace, vehicle engineering and electronic engineering, which has attracted a lot of attention from researchers [3]. The common reinforcement particles in aluminum matrix composite mainly include the particles of SiC, Al_2O_3, SiO_2 et al. Among them, SiC particles have become the the most ideal, promising, and widely used reinforcement particle for aluminum matrix composites due to its high strength, hardness, modulus, and low expansion coefficient [4]. Generally, SiC particle well reinforced aluminum matrix composites exhibiting superior comprehensive property

The preparation technology is the key factor influencing the property and application for particle reinforced aluminum matrix composites. At present, the aluminum matrix composites were fabricated mainly by using the technologies of powder metallurgy, stir casting, squeeze casting and jet deposition, et al. [5]. However, as the aerospace component increasingly tends to the development with large-scale integrating, shape complicating and thinning-hollowing, the traditional preparation technology can be unable to meet the above demands limiting the development and application of aluminum matrix composites. Laser deposition manufacturing has become the mainstream development technology for manufacturing field due to the manufacturing advantage of lightweight, integration, flexibility and short production cycle [6], which has broad prospect in the preparation for aluminum matrix composites. Zhao et al. [7] studied the forming mechanism and structure characteristics of SiC particle reinforced aluminum matrix composite fabricated by laser selective melting (SLM) technology. Lian et al. [8] investigated the process of laser additive manufacturing for TiB_2 particle reinforced Al-Si composite and its property. Rao et al. [9] studied the forming mechanism and mechanical property of carbon nanotube (CNT) reinforced aluminum matrix composite, and the results showed that the density of composite markedly increases with the laser power density, and then improves the wear-resisting property and mechanical property. Simchi et al. [10] fabricated the SiC particle reinforced A356 composite and studied the effect of scanning speed, scanning interval, sintering atmosphere, particle size (7 and 17 μm) and particle content (up to 20 wt%) on density of composite.

The process of laser deposition manufacturing involves a series of physical and chemical reaction, such as mass transfer, heat transfer process and chemical reaction process, et al. Therefore, the key problem for the preparation of aluminum matrix composites with superior property depends on the laser deposition parameters including laser power, laser scanning speed, et al. and the morphology and content of reinforcement particles. In this study, the SiC particles are used as the reinforcement particles, and the SiC particle reinforced aluminum matrix composites were fabricated by laser deposition manufacturing. The effect of laser deposition parameters and particle content on the microstructure and mechanical property of aluminum matrix composites was deeply studied, which can provide experimental data and references improving the application of aluminum matrix composites in the complex field for aerospace.

2 MATERIALS AND METHODS

The ZL114A alloy after T6 heat treatment was selected as the substrate. The size of the substrate is 200 mm × 100 mm × 30 mm. Before test, the substrate surface was mechanical polished to remove the oxide layer, and then washed with acetone and dried. The preheating treatment was carried out to relieve the effect of temperature gradient on microstructure and mechanical property for the deposited composites. The ZL114A alloy powder was selected for deposition test, and its nominal chemical component was listed in Table 1. The alloy powder was fabricated by the vacuum plasma rotating electrode method, and the particle size is 40–160 μm as shown in Figure 1. The planetary ball mill (F-P8L) was employed to make the ZL114A alloy powders and SiC particles mix evenly. The ZrO_2 ceramic ball with the diameter of 5 mm was used as the milling ball and the ratio of ball to powder was set as 5:1. In order to avoid the oxidation of ZL114A alloy powder, the low speed ball mill was carried out with the rotate speed of 50 rpm, and the milling time was 6 h. The content of SiC particles was set as 0 wt%, 10 wt%, 20 wt% and 30 wt%, respectively.

The SiC/ZL114A composites was fabricated by a home-made LDM-800 laser deposition manufacturing system which mainly consists of a 6.0 kW fiber laser, three axis motion system, synchronous powder feeding system and argon protection box. The single pass test for laser deposition with multi-parameter combination was carried out to optimize the deposition parameters of laser power and laser scanning speed: the laser power is 1.2–2.0 kW, the scanning speed is 3–7 mm/s. The spot diameter of the laser is 3 mm. After the single scanning is done, the standing time of 15 s was set to alleviate the effect of heat accumulation during laser deposition on microstructure and mechanical properties for composites. In view of the lower density of ZL114A alloy powder, the powder feed rates was 2 g/min, the carrier gas pressure was 0.2 MPa and carrier gas flow was set as 2.2 L/min, which can efficiently deliver the ZL114A alloy powder to substrate surface.

Table 1. Composition of ZL114A aluminium alloy powder (mass fraction, %).

Element	Si	Mg	Ti	Al
Content	6.50–7.50	0.45–0.60	0.10–0.20	Bal.

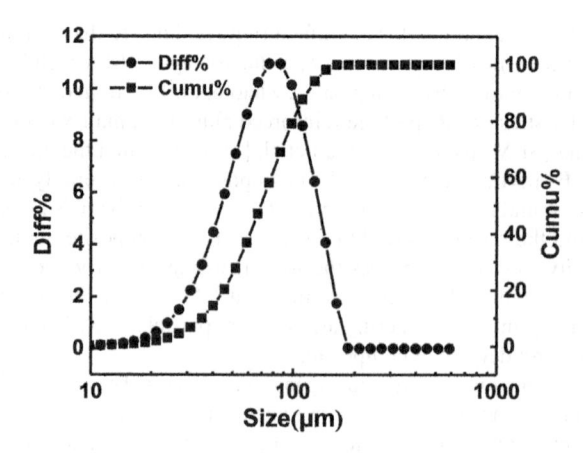

Figure 1. Size distribution of SiC particles.

The metallographic samples were prepared by cutting the deposition parts along the direction perpendicular to the laser scanning. These samples were inlayed, pre-grinded, polished in sequence, and then etched for 10–15 s by the corrosive liquid of 1 HF + 10 H_2O (volume ratio). The microstructure analysis for metallographic samples was carried out by the OLYMPUS-GX51 optical microscope. The microhardness was determined by the HVS-1000A microhardness tester with the loading of 100 g and the loading time of 10 s.

The tensile samples of SiC/ZL114A composites were fabricated by the local laser deposition methods, which can provide experimental data references for the application of SiC/ZL114A composites in laser deposition repair field. The detail of preparation and sampling process for tensile samples was shown in Figure 2. Firstly, an interpenetrated groove was machined on the ZL114A substrate by the mechanical processing as shown in Figure 2a. Secondly, the laser deposition preparation of SiC/ZL114A composites with different SiC contents was carried out on the groove region with the multi-passes and layers, and short side reciprocating scanning mode. The laser power was 2 kW and the laser scanning speed was 5 mm/s. Finally, the tensile samples were sampled as shown in Figure 2b, and the sample dimension schematic for tensile samples was shown in Figure 2c. The section height of laser deposition region in tensile samples is about 50% of the total height. The tension test at room temperature for SiC/ZL114A composites was measured by an electronic universal testing machine (INSTRON-5982) with the loading rate of 2 mm/min.

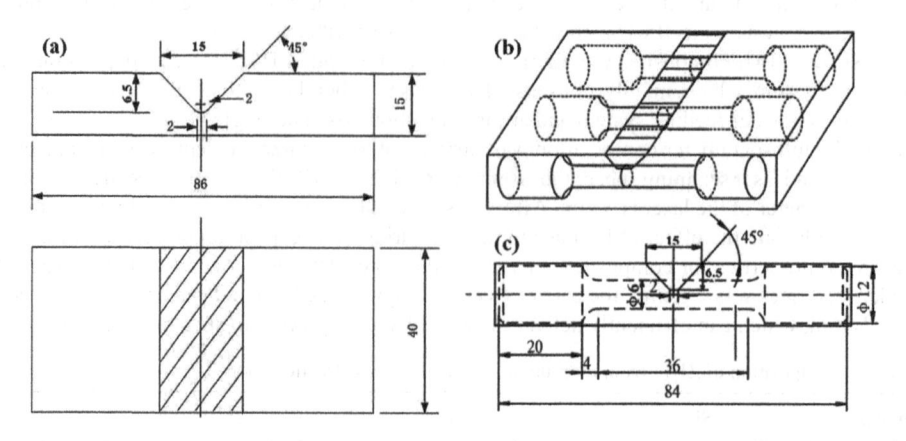

Figure 2. Schematic of preparation and sampling for tensile samples: (a) preparation schematic;(b) sampling schematic; (c) sample dimension schematic.

3 RESULTS AND DISCUSSION

Figure 3 shows the surface macro-morphology of laser deposited samples with single pass mode (Figure 3a) and multi-passes and layers mode (Figure 3b), and the corresponding deposition parameters are listed in Table 2. As shown in Figure 3 and Table 2, the forming size decreases with the laser power decreasing and the laser scanning speed increasing. It is the same for the variation of surface forming quality with laser deposition parameters. According to the laser energy density formula:

$$E = P/VD \quad (1)$$

Where E is the laser energy density, P is the laser power, V represents the laser scanning speed, and the D is on behalf of the spot diameter. It can be concluded that the laser energy density decreases with the laser power decreasing and the laser scanning speed increasing. At low laser energy density, the powder at unit mass absorbs less heat, which will cause the metal powder will not be completely melted reducing the size of the single-passes forming parts. In addition, the partial unmelted powder particles will be mixed inside the molten pool to form the defects of poor fusion, which will affect the surface forming quality, as shown in sample 1, 6, 11 and 12. However, if the laser energy density is too high, overburning and component ablation will occur to the deposited samples, which will easily cause bending and deformation of the formed parts affecting the machining accuracy. The multi-passes and layers (15 passes and 3 layers) samples were fabricated by the following parameters: the laser scanning speed is 5 mm/s, the laser power is 1.2 kW-2.0 kW. The sample size is 25 mm × 25 mm × 4 mm as shown in Figure 3b. At lower laser energy density (sample 19 and 20), poor fusion phenomena is obvious, which results in the significant decreases of the surface roughness for the deposited samples seriously affecting the appearance and dimensional accuracy of the formed parts.

Figure 3. Appearance of laser deposited samples fabricated by different process parameters:(a) single pass; (b) multi-passes and layers.

Figure 4 shows the macroscopic section morphology of 20%SiC/ZL114A composites at different laser powers. When the speed is fixed at 5 mm/s, with the laser power increasing, the laser energy density increases, which results in the full fusion for alloy powder leading to the increase of fusion depth and fusion width. In addition, in the deposited layers, the content of SiC particles increases with the laser power and the SiC particles mainly distribute in the surface and middle of the deposited layer. The content of SiC particles in the middle of the deposited layer increases with laser power, but surface SiC content differs, which mainly results from the convection behavior in the molten pool during laser deposition process [6]. At low laser power, the intensity of convection is poor resulting in the enrichment of SiC particles in the surface of the deposited layer due to the poor fluidity of SiC particles compared with the melted ZL114A alloy. With the laser power increasing, the intensity of convection increases driving the SiC particles to enrich in the middle of the deposited layer along with the convection of the melted metal in the molten pool. In addition, obvious pores can be observed inside the deposited layer as marked with circles. It is mainly because that there is no enough time for the gas in the molten pool to come up and overflow due to the fast cold characteristics for laser deposition process (the cooling rate is as high as 10^2–10^6 K/s), which leads to the formation of pore defects inside the deposited layers. It can be also observed that the

Table 2. Parameters of laser deposition test.

Sample number	Power (kW)	Scanning speed (mm/s)
1	1.2	3
2	1.4	3
3	1.6	3
4	1.8	3
5	2.0	3
6	1.2	5
7	1.4	5
8	1.6	5
9	1.8	5
10	2.0	5
11	1.2	7
12	1.4	7
13	1.6.	7
14	1.8	7
15	2.0	7
16	2.0	5
17	1.8	5
18	1.6	5
19	1.4	5
20	1.2	5

size of the pores is about 100 μm in Figure 4c and d, which is much higher than that in Figure 4a and b (about 40 μm). It is mainly because that the surface of SiC particle is easy to absorb gas, which results in the enrichment of gas around the SiC particles forming the pores with a bigger size [10].

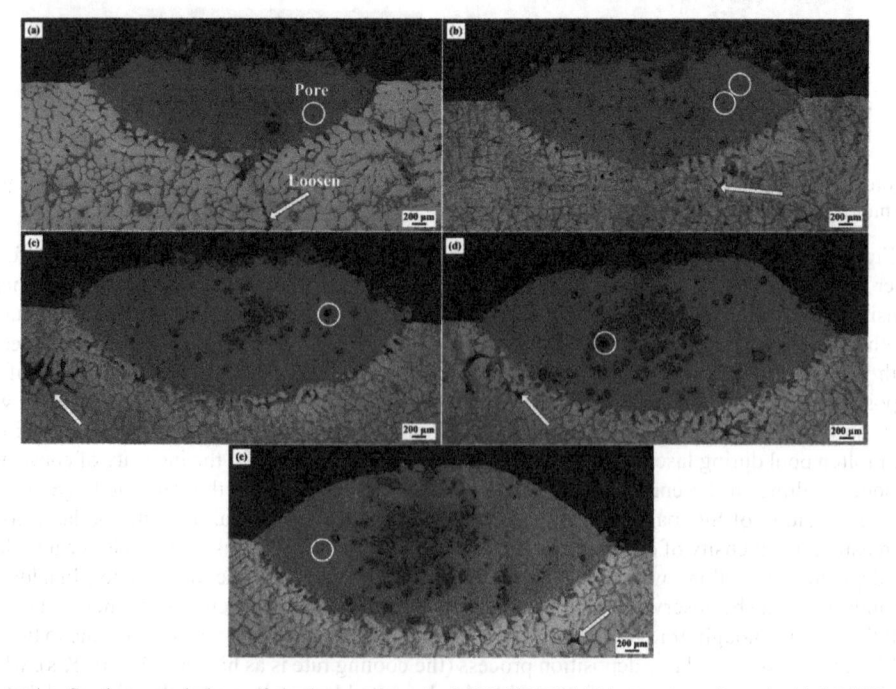

Figure 4. Section morphology of single pass deposited 20%SiC/ZL114A sample at 5 mm/s: (a) 1.2 kW; (b) 1.4 kW; (c) 1.6 kW; (d) 1.8 kW; (e) 2.0 kW.

Figure 5 shows surface micro-morphology of laser deposited single pass samples. The deposited layer is organized by the matrix (it is made of α-Al dendritic structure and the interdendritic Al-Si eutectic structure [1]) and SiC particles. The laser deposited microstructure is much finer compared with microstructure of ZL114A substrate, which results from the fast hot and fast cold characteristics for laser deposition process inhibiting grain growth. The top of laser deposited layer consists of equiaxial grain as shown in Figure 5a, and the microstructure coarses gradually and transforms into columnar dendrites as the distance from the surface increases as shown in Figure 5b and c. From the surface of the deposited layer to the bottom of the molten pool, the structural transformation depends on temperature gradient and solidification rate in the molten pool during the laser deposition process. For the surface of the deposited layer, there are many pathways for heat dissipation, such as the transfer of heat to a solidified deposit or the heat exchange with the outside environmental, which increases the component supercooling in the liquid/solid interface frontier. The surface layer is in a deep cold state and the solidification rate is fast. Under this condition, the grain growth rate is less than the nucleation rate resulting in the formation of fine equiaxial grains in the surface of the deposited layer. However, for the bottom of the deposited layer, the formation of nuclei occurs when the molten metal contacts with the cooler substrate, and then these crystal nuclei will gradually grow into columnar dendritic structure along the direction perpendicular to the pool. It is because that the internal heat of the molten pool cannot be dissipated in time and the heat exchange of the molten pool only occurs with the alloy substrate. In the area of SiC particle aggregation as shown in Figure 5d, the SiC particles and the substrate are closely coupled showing a good wetting behavior. Moreover, the bonding surface between the SiC particles and the substrate is relatively smooth and flat, and no interface products are observed.

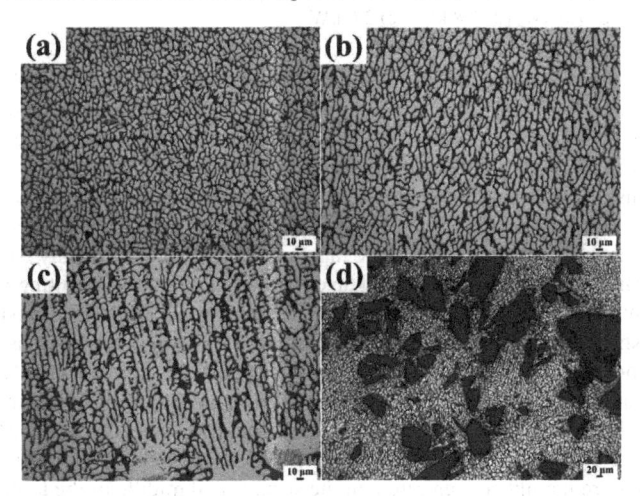

Figure 5. Microstructure of 20%SiC/ZL114A single pass deposited sample: (a) surface matrix microstructure; (b) central matrix microstructure; (c) bottom matrix microstructure; (d) SiC-accumulation region.

Figure 6 shows the macroscopic section morphology of multi-passes and layers deposited 20%SiC/ZL114A composites at different laser powers. As is mentioned above, the SiC particles aggregate at the surface of the deposited layer at low laser power as shown in Figure 4a and b. However, the surface and ends will be remelted during the multi-passes and layers lap deposition process. Due to the lower laser energy density (Figure 6a and b.), the fluidity of the liquid metal is poor resulting in aggregation of SiC particles. In addition, the combination effect between SiC particles and substrate is poor leads to the removal of partial SiC particles during the metallographic preparation process exhibiting many large size holes. With the increasing of laser power, the laser energy density increases, which will increase the convection intensity in the molten pool. In such a condition, the SiC particles evenly distribute in the middle of the single pass deposited sample as shown in Figure 4c, d and e. In the subsequent process of multi-passes and layers lap deposition, the SiC particles will be redistributed resulting in the even distribution of SiC particles inside the deposited layer as shown in Figure 6c, d and e. In addition, during lap deposition process, the introduced gas increases, which promotes the formation of pores. Due to the adsorption of SiC particles, these pores are much larger and mainly distribute around the SiC particles. And a few pores with small size are located in the regions without SiC particles.

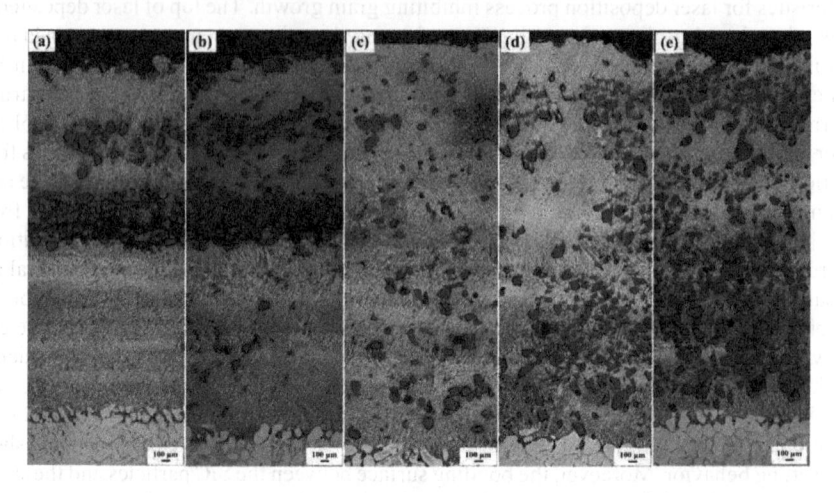

Figure 6. Section morphology of multi-passes and layers deposited 20% SiC/ZL114A sample at 5 mm/s: (a) 1.2 kW; (b) 1.4 kW; (c) 1.6 kW; (d) 1.8 kW; (e) 2.0 kW.

Figure 7 shows the microhardness distribution of laser deposited SiC/ZL114A composites. The microhardness of ZL114A alloy substrate is about 120 $HV_{0.1}$, which is similar to the microhardness of composites matrix. The addition of SiC particles increases the microhardness of composites, and the microhardness distribution depends on the distribution of SiC particles. Under different process conditions, the microhardness of single pass deposited layer all show similar distribution trend: the hardness for the surface of the deposited layer is high, and the hardness for the middle of deposited layer fluctuates greatly.It is because that the SiC particles are mainly concentrated on the surface and the middle of the laser deposited layer (shown in Figure 4) resulting in the increase of microhardness. With the increasing of laser power, the content of SiC particles inside the deposited layer increases (shown in Figure 4) resulting in the increase of microhardness, as shown in Figure 7a. At low laser powers (1.2 kW and 1.4 kW), most SiC particles are concentrated on the surface of the laser deposited layer, but a few SiC particles distribute dispersedly on the middle of the laser deposited layer, which results in the higher average microhardness of layer surface than that of layer middle. At high laser powers (1.6 kW, 1.8 kW and 2.0 kW), the content of SiC particles increases and these SiC particles are mainly concentrated on the surface and middle of the laser deposited layer, so the entire deposited layers exhibit the higher average microhardness with large fluctuations. It should be noted that due to the high microhardness of SiC particles (3020 $HV_{0.1}$), its microhardness value were not shown in the curve to avoid excessive microhardness fluctuations on the curves.

The effect of scanning speed on hardness of deposited layer was shown in Figure 7b. At constant power (2.0 kW), with the decreasing of scanning speed, the content of SiC particles increases inside the deposited layer resulting in the higher average hardness with a high fluctuation degree and a wide fluctuation range. This hardness distribution trend is similar to that under different laser powers.

The effect of SiC content on hardness of deposited layer was shown in Figure 7c. When the content of SiC particles is 10%, the SiC particles discretely distribute inside the deposited layer, so the region with high microhardness is small. With the increasing of SiC particles content, the SiC particles centrally distribute with a wide range inside the deposited layer, so the average microhardness of the deposited layer can be greatly improved.

The tension test at room temperature for SiC/ZL114A composites with different contents of SiC particles was carried out and the results were shown in Table 3. The average tensile strength and

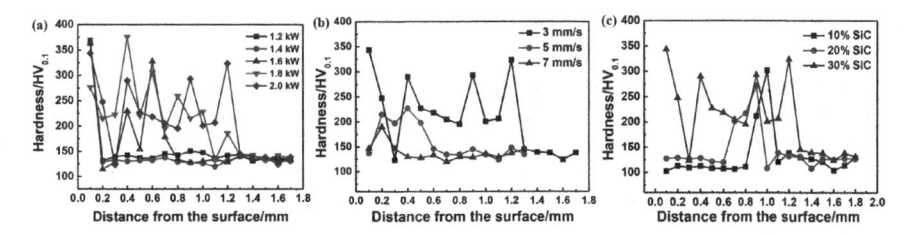

Figure 7. Microhardness distribution of laser deposited samples under different parameters: (a) laser power; (b) scanning speed; (c) SiC content.

average elongation of the ZL114A substrate are 238.2 Mpa and 4.5%, respectively, which are lower than the results obtained in reference [11]. It is maybe because that the solution treatment process is not reasonable or the ZL114A substrate so oversize that the solution treatment was incomplete for the whole ingot resulting in the decrease of tensile properties for ZL114A substrate. In addition, the large-sized (>200 μm) and irregular-shaped loose defects (shown by the arrows in Figure 4) widely distribute in the ZL114A substrate, which induces stress concentration decreasing the tensile properties for ZL114A substrate. When the content of SiC particles is 10%, the strength of laser deposited SiC/ZL114A parts is slightly higher than that of the ZL114A substrate, but the plasticity differs. On one hand, in SiC/ZL114A composite, the phase interfaces will be formed between the SiC particles and ZL114A substrate, through which the applied loading will be transformed from the substrate to the SiC particles. Generally, the loading capacity for applied loading will increase with the phase interfaces [12], so the tensile strength will be increased for the SiC/ZL114A composite. On the other hand, more lattice distortion and thermal mismatch will be generated in the ZL114A substrate due to the addition of SiC particles, which increases the dislocation density leading to a higher tensile strength of SiC/ZL114A composite [13]. Moreover, partial microstructure will undergo recrystallization transformation during the laser deposition process with multi-passes and layers. During this process, SiC particles can inhibit the movement of grain boundaries, which results in fine-grain strengthening improving the tensile strength [14]. However, the SiC particles tend to agglomerate when the content of SiC particles increases resulting in the formation of pores (shown in Figure 6), in which the severe stress concentration tends to occur and original crack will be formed under the tensile stress. Therefore, the tensile strength decreases with the SiC content. It is obvious that the elongation of SiC/ZL114A composite decreases with the SiC content due to the restriction of SiC particles on deformation. As concluded by Pang et al. [15], the tensile strength is only 139 Mpa for the laser deposition repaired (LDRed) ZL114A alloy with the repair proportion of 50%, but its elongation is much higher (up to 8.9%). It is because that the grain of substrate continues to grow resulting in a coarse structure due to the heat accumulation during the laser deposition repair process. By contrast, the addition of SiC particles in this work effectively inhibits the growth of grain improving the tensile strength. Therefore, the ZL114A aluminium alloy parts can be repaired by laser deposited SiC/ZL114A composites. In addition, the addition of SiC particles improves the microhardness of the laser deposited layer, which contributes to the improvement of wear-resisting property for the repaired region.

Table 3. Results of tensile testing at ambient temperature.

Sample	Tensile strength (MPa)	Yield strength (MPa)	Elongation (%)
ZL114A substrate	238.2 ± 5.1	210.4 ± 5.8	4.5 ± 0.8
10%SiC/ZL114A	239.8 ± 4.6	212.1 ± 4.3	4.0 ± 0.3
20%SiC/ZL114A	215.5 ± 3.2	206.5 ± 2.3	3.5 ± 0.4
30%SiC/ZL114A	201.3 ± 5.3	188.3 ± 7.6	1.6 ± 0.5
LDRed ZL114A[20]	139.0 ± 2.0	67.3 ± 1.7	8.9 ± 0.3

4 CONCLUSIONS

(1) The microstructure of laser deposited layer consists of the ZL114A matrix structure and SiC particles. The distribution and content of SiC particles in laser deposited layer closely relate to the laser energy density. The SiC content in the middle of deposited layer increases with the laser power, but decreases in the layer surface. After multi-passes and layers deposition, SiC particles are redistributed and more evenly distributed in the deposited layer with the laser power increasing.
(2) Under a temperature gradient, the microstructure of laser deposited layer transforms from the columnar crystal in the bottom of the molten pool to the fine equiaxial grain in the layer surface.
(3) The microhardness of the laser deposited layer is closely related to the distribution of SiC particles. With the increasing of laser power and SiC content, and the decreasing of scanning speed, the average microhardness of the deposited layer increases with a higher fluctuation degree and a wider fluctuation range.
(4) When the SiC content is 10%, the strength of SiC/ZL114A composite is highest, and with the increasing of SiC content, the strength and plasticity continuously decreases.

REFERENCES

[1] Wang C, Pang S, Yang G, et al. Microstructure properties and residual stress of casting ZL114A aluminum alloy repaired by laser deposition. Applied Laser, 2019, 39(4): 563– 568. (in Chinese)
[2] Zhu S Z, Ma G N, Wang D. Suppressed negative influence of natural aging in SiCp/6092Al composites. Materials Science and Engineering A, 2019, 767:138422.
[3] Li X P, Liu C Y, Ma M Z, et al. Microstructures and mechanical properties of AA6061-SiC composites prepared through spark plasma sintering and hot rolling. Materials Science and Engineering A, 2016, 650: 139–144.
[4] Song Y F, Ding X F, Zhao X J, et al. The effect of SiC addition on the dimensional stability of Al-Cu-Mg alloy. Journal of Alloys and Compounds, 2018, 750: 111–116.
[5] Chen B, Xi X, Tan C W, et al. Recent progress in laser additive manufacturing of aluminum matrix composites. Current Opinion in Chemical Engineering, 2020, 28: 28–35.
[6] Zhao T, Dahmen M, Cai W C, et al. Laser metal deposition for additive manufacturing of AA5024 and nanoparticulate TiC modified AA5024 alloy composites prepared with balling milling process. Optics and Laser Technology, 2020, 131: 106438.
[7] Zhao X, Gu D D, Xi L X, et al. Microstructure characteristics and its formation mechanism of selective laser melting SiC reinforced Al-based composites. Vacuum, 2019, 160:189–196.
[8] Lian Q, Wu Y, Wang H W, et al. Study on manufacturing processes and properties of TiB2 reinforced Al-Si composite by laser additive manufacturing. Hot Working Technology, 2017, 46(22): 113–117. (in Chinese)
[9] Rao X W, Gu D D, Xi L X. Forming mechanism and mechanical properties of carbon nanotube reinforced aluminum matrix composites by selective laser melting. Journal of mechanical engineering, 2019, 55(15): 1–9. (in Chinese)
[10] Simchi A, Godlinski D. Densification and microstructural evolution during laser sintering of A356/Si C composite powders. Journal of Materials Science, 2011, 46(5): 1446–1454.
[11] Li C D, Gu H M, Wang W, et al. Microstructure and properties of ZL114A aluminum alloy prepared by wire arc additive manufacturing. Rare Metal Materials and Engineering, 2019, 48(9): 2917–2922. (in Chinese)
[12] Wang Z W, Song M, Sun C, et al. Effects of particle size and distribution on the mechanical properties of SiC reinforced Al-Cu alloy composites. Materials Science and Engineering: A. 2011, 528(03): 1131–1137.
[13] Amirkhanlou S, Niroumand B. Effects of reinforcement distribution on low and high temperature tensile properties of Al356/SiCp cast composites produced by a novel reinforcement dispersion technique. Materials Science and Engineering: A. 2011, 528(24): 7186–7195.
[14] Tao Y Y, Ge X L, Xu X J, et al. Influences of SiC particle size and content on the mechanical properties and wear resistance of the composites with Al matrix. Key Engineering Materials. 2008, 375–376: 430–434.
[15] Pang S. Research on technology and performance of laser deposition repair casting ZL114A aluminium alloy. Master Dissertation. Shenyang: Shenyang Aerospace University, 2017. (in Chinese)

Study on the electromagnetic shielding efficiency of functional layer with reducing iron powder mortar

Y.J. Lu
Unit 91995 of PLA, Shanghai, China

Y. Gao*
Naval Logistics Academy, Tianjin, China

ABSTRACT: Based on the mortar mixed with reduced iron powder and high-performance copper nickel shielding fabric, the material was used in the construction of shielding functional layer of the model. The steel plate door and side wall of the model were selected as test positions, and the shielding effectiveness at test frequency point in 100kHz-3GHz frequency band was tested by the built-in radiation source method. The results show that the shielding effectiveness of the solid model can reach more than 20dB at the test frequency; The shielding effectiveness of the model is better at the test frequency point in the high frequency band, and the maximum shielding effectiveness is 80dB; The shielding effectiveness of the side wall is higher than that of the steel plate door, and it is more obvious at the high frequency test point.

1 INTRODUCTION

Cement-based materials have the advantages of low price and strong applicability, and they are widely used in various construction projects. Facing the increasingly complex space demand, a lot of research has been made on the functional materials prepared by adding special components to the common cement-based materials. In the field of electromagnetic shielding, some scholars have made cement-based materials by adding metal powder and fiber. The shielding efficiency of the materials in 100 KHz–1.5 GHz band has been studied [1]. Also, the shielding performance of the wire mesh reinforced cement-based materials has been studied. It is proved that the composite material composed of steel wire mesh and cement base has good shielding efficiency in 10 KHz–1.5 GHz frequency band [2]. Mortar is a kind of cement-based material, which is usually used for masonry and plastering. The plastering functional mortar prepared on its base can be used as the old project transformation. Without changing the main body of the project, the special needs of the original scene can be met by the form of additional functional layer.

At present, there are few studies on cement-based materials with reduced iron powder. Some studies have prepared concrete slabs with reduced iron powder as electromagnetic medium, and studied the electromagnetic wave reflectivity and transmittance in a certain frequency band. It is concluded that the amount of reduced iron powder has a greater impact on the electromagnetic wave transmittance [3]. As a kind of ultra-fine metal powder, reduced iron powder has good electromagnetic properties [4]. Compared with fiber material, its powder shape effect can make it more evenly distributed in cement-based materials. The plastering functional mortar prepared by adding reduced iron powder into cement mortar can provide a choice for the transformation of old projects with electromagnetic shielding needs. In this paper, the reduced iron powder mortar combined with high-performance copper nickel shielding fabric is used for the construction of the model shielding functional layer, and the shielding effectiveness in the frequency band of 100kHz-3GHz is studied.

---

* Corresponding author

2 STUDY ON THE SHIELDING FUNCTIONAL MATERIAL

2.1 Raw materials

Reduced iron powder: secondary reduced iron powder with particle size of 200 mesh; Cement: Ordinary portland cement of grade 42.5; Sand: medium river sand with fineness modulus of 2.5; Fly ash: first class fly ash with particle size of 1250 mesh; Superplasticizer: naphthalene superplasticizer. Shielding fabric: high performance copper nickel composite fabric, with shielding efficiency of 60 dB in 1 MHz–10 GHz band.

2.2 Model scheme design

According to the relevant requirements of the shielding effectiveness test of small shielding body, the design scheme and construction process of the model are determined.

2.2.1 Model size and wall thickness setting

The inner dimension of the model is set as 1m × 1m × 1m, outer dimension is 1.2m × 1.2m × 1.2m. In order to ensure a certain overall strength, the wall adopts C40 strength grade concrete, and the wall thickness is set as 10cm.

2.2.2 Setting of shielding door

Shielding door is an important part to ensure the overall shielding performance of the project, and it is necessary to reflect it in the model. Limited by the test conditions, 3 mm thick steel plate (40cm × 40cm) was used. The position of the door opening is set in the middle of the wall.

2.2.3 Grounding treatment

Good grounding treatment can effectively improve the shielding effectiveness of shielding body. When selecting grounding treatment, the following factors should be considered: conductivity, contact area and durability. Limited by the conditions, the conductive mortar grounding electrode is connected with the embedded flat iron in the model wall. The grounding treatment is shown in Figure 1.

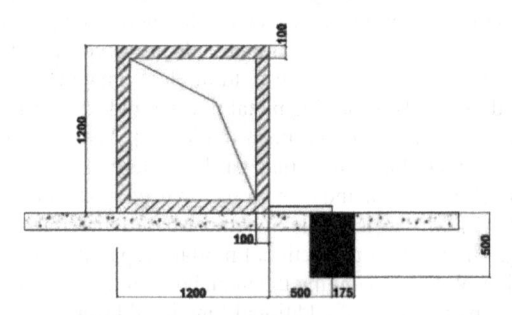

Figure 1. Structure diagram of grounding electrode.

2.2.4 Setting of shielding function layer

In order to simulate the adding position of shielding functional layer in practical engineering, the 10 mm thick reduced iron powder functional mortar layer was applied on the inner wall, ground and top surface. After the maintenance of the functional mortar layer, copper nickel shielding fabric was bonded on its surface. The proportion of reduced iron powder functional mortar is shown in Table 1.

Table 1. Mix proportion of functional mortar.

Reduced iron powder (kg/m^3)	cement (kg/m^3)	sand (kg/m^3)	Fly ash (kg/m^3)	Water (kg/m^3)
300	450	1423	50	300

2.3 Model construction

2.3.1 Construction of functional mortar layer

Select the raw materials of functional mortar according to the proportion. In order to prevent the reduced iron powder from settling at the bottom of the dry material due to its high density when mixing the dry material, first use the hand-held mortar mixer to fully mix the cement, fly ash and other dry materials, add water while mixing, and finally add the reduced iron powder to fully mix. In order to prevent the settlement or uneven mixing of reduced iron powder in the mortar caused by excessive mixing amount of mortar and affect the construction performance of functional mortar, the single mixing amount of mortar is controlled to a certain extent.

Remove the internal and external formwork, and spray water on the side wall for maintenance. After the maintenance of the side wall is completed, the construction of the functional mortar layer on the surface of the internal wall, ground and prefabricated roof is carried out. Before the construction of the functional mortar layer, the dust and stains on the surfac of the wall and prefabricated roof shall be removed. Apply 1–2mm thick interface mortar on the surface in advance, and plaster the functional mortar layer after the surface is slightly closed. When plastering, measures such as vertical lifting should be taken to ensure the surface flatness of the functional mortar layer,the method of layered plastering is adopted. The first layer is 5–6 mm thick. After 2 days of watering curing, the second layer is 5–6 mm thick for watering curing.

2.3.2 Construction of shielding fabric layer

Silver conductive adhesive with excellent properties is used as the adhesive of fabric lap joint. The fabric samples were tested in advance to test its bonding effect and conductivity. It is found that the bonding effect and surface smoothness of the samples are good, and there is no obvious impregnation phenomenon, and the conductivity of the samples before and after bonding has no obvious change.

Before bonding the shielding fabric, clean the surface of the functional mortar layer. In order to ensure the good adhesion between the fabric and the surface of the functional mortar layer, the surface of the mortar layer shall be filled with putty. After the putty is air dried, it shall be polished with sandpaper to ensure that there is no crack or pitting on the surface, and a layer of base film shall be evenly applied on the surface of the putty. Mix the wall adhesive and conductive adhesive, and evenly apply the adhesive on the wall surface. Paste the shielding fabric on the internal wall and ground. After preliminary pasting, use a brush to discharge bubbles and excess binder from the middle to the surrounding to make the shielding fabric stick to the wall. In order to ensure good conductive connection, the lap length between shielding fabrics shall not be less than 0.1 m, and the lap joint shall be bonded with conductive adhesive and avoid the corner position. After bonding, the overall bonding condition should be checked and the joint gap should be treated in time to avoid bulging, folding and edge warping of shielding fabric.

The construction of functional mortar is carried out on the precast concrete roof and the shielding fabric is bonded. After the construction of the internal functional layer is completed, the conductive adhesive is evenly applied around the fabric on the top surface of the roof and the top surface of the side wall, the prefabricated roof and the top surface of the side wall are accurately connected, and the joints are plastered and sealed to ensure the sealing connection between the main body of the solid model and the roof to maintain the overall internal good conductivity. Open the steel plate door to make the internal air circulation, so that the bonding effect of shielding fabric is better. The bonding of the inner shielding fabric is shown in figure 2. The final solid model is shown in Figure 3.

2.3.3 Construction of grounding electrode

A cylindrical pit with diameter of 0.35 m and depth of 0.5 m is excavated about 0.5 m away from the solid model wall, and conductive mortar is poured into the pit. The conductive mortar is made of 42.5 ordinary portland cement, medium river sand and 200 mesh ordinary flake graphite powder in the mass ratio of 1:1:0.5 [5]. The cylindrical pit is used as the grounding electrode. The stainless

Figure 2. Bonding of internal shielding fabric.

Figure 3. Solid model.

steel angle steel is buried 2m under the electrode center to ensure good conductive contact. The central stainless steel angle steel is connected with the flat iron inside the solid model wall by bolts.

2.3.4 *Installation of angle steel door frame*

Angle steel (⊥40 mm × 40 mm × 3 mm) is welded into a square door frame, one side is connected with the wall by bolts, and the bolt connection is sealed by conductive mortar; The other side is connected with the steel plate door by screws. In order to ensure good electrical connection, the screw hole distance is controlled at 5cm.

2.4 *Layout of physical model shielding effectiveness test*

The built-in radiation source method is used to test the shielding effectveness of the solid model. The frequency points include 100kHz, 300kHz, 30MHz, 100MHz, 450MHz, 1GHz and 3GHz. The test layout is shown in Figure 4.

When selecting the location of the measuring point, the paper focus on the situation that the gap or opening is easy to appear shielding defects. According to the setting of the model room, the steel plate door and side wall are selected as the test points of shielding effectiveness. The specific location of measuring points is shown in Figure 5.

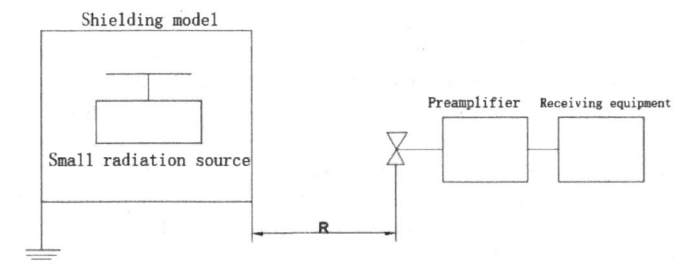

Figure 4. Arrangement of built-in radiation source method.

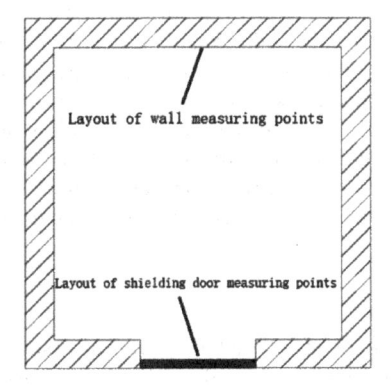

Figure 5. Arrangement of measuring points.

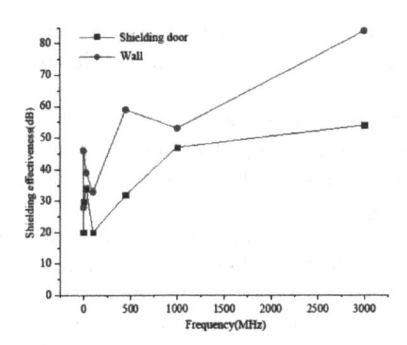

Figure 6. Shielding effectiveness of solid model.

3 RESULTS

Figure 6 is a line chart of shielding effectiveness of the solid model at different frequencies and locations. It can be seen that the shielding effectiveness increases with the increase of frequency, and the shielding effectiveness at the side wall is significantly better than that at the door.

The shielding effect at the low frequency point is poor, which mainly has two reasons: on the one hand, the low frequency electromagnetic wave has a long wavelength, which is easy to diffraction, and the shielding defect is easy to occur at the joint of angle steel door frame and the overlap of side wall fabric; On the other hand, when the shielding efficiency of low-frequency electromagnetic wave is tested, it is easy to be affected by external power supply and filter, which causes certain test errors. The two measuring points have big shielding defects at 100MHz frequency point, which is mainly because 100MHz is close to the receiving frequency of radio stations around the test site, which is easy to be affected, which causes the test results to be relatively small. With the increase of frequency, the shielding efficiency tends to increase, which is mainly because of the short wavelength, poor diffraction ability and strong directivity of high frequency electromagnetic wave, which is easy to reflect and absorb losses at the measuring point, so as to obtain more effective shielding.

The shielding efficiency of steel plate door is poor, which mainly has the following reasons: there are more gaps in the steel plate door relative to the side wall, such as the welding joint of angle steel door frame, the joint between steel plate door and angle steel door frame, the joint between angle steel door frame and wall and the gap of screw hole, which is more likely to cause shielding defects. In addition, the steel plate door is repeatedly opened and closed repeatedly, and the electrical sealing effect of conductive gasket is reduced, which leads to the decrease of shielding efficiency.

In the actual test, it is found that there is little difference between the shielding effectiveness without grounding treatment and that with grounding treatment. On the one hand, the built-in grounding flat iron is located in the concrete cushion, and the whole concrete cushion is higher than the ground, and the surrounding soil is relatively dry, which makes the connection effect between the shield and the earth not ideal; On the other hand, because the reinforcement is not arranged inside the wall, the electrical connection between the embedded flat iron and the concrete wall is poor, which leads to the poor electrical connectivity of the shield, so the improvement of the overall shielding efficiency of the solid model by grounding treatment is limited.

4 CONCLUSION

According to the relevant requirements of the built-in radiation source method in the small shield test specification, the dimension of the solid model is determined. In order to simulate the shielding treatment at the entrance of the project, the steel plate is used as the simple shielding door, and the electrical connection of the door is preliminarily set by combining the small spacing screw and the copper wire mesh conductive pad. The conductive mortar grounding pit combined with the central stainless steel strip is connected with the flat iron inside the solid model, thus to form the grounding setting;

According to the properties of plastering mortar and reduced iron powder, the plastering process and construction procedure of reduced iron powder functional mortar are determined. The fabric wall adhesive and the conductive adhesive at the lap joint of the fabric are selected to form the wall bonding process of the shielding fabric;

The shielding effectiveness of steel plate door and side wall was measured by internal radiation source method. The results show that the shielding effectiveness of the solid model can reach more than 20dB at the test frequency; The results show that the shielding effectiveness of the solid model is good at the test frequency point in the high frequency band, and the maximum shielding effectiveness is 80dB; The shielding effectiveness of the side wall is higher than that of the steel plate door, and it is more obvious at the high frequency test point.

REFERENCES

[1] Chen Yangru, Xiong Guoxuan, Zhang Zhibin. The application study of shielding media in cement-based materials, J. New Building Materials, 2010,37 (10) 80–82.
[2] Dai Yinsuo, Gong Lei, Yan Fengguo, et al. Research on electromagnetic protection performance of metal wire mesh reinforced cement-based materials J. China Concrete and Cement Products, 2015(12):6–9.
[3] Zheng Guozhi, Chen Bin, Zhang Zehai, et al. Reflection and transmission performances of concrete slabs mixed with reduced iron powder J. High Power Laser and Particle Beams, 2015, 27(4) 151–156.
[4] Xu Fangxing, zeng Guoxun, Zhang Haiyan, et al. Preparation and electromagnetic properties of flaky reduced iron powders J. Electronic components and materials, 2014, (12) 41–48.
[5] Shen ZhiQiang ,Kang qing,Wangzhengang.Experimental study on grounding performance of electromagnetic shielding solid model room J Journal Of Logistical Engineering University, 2012, 28(2): 82–85, 91.

High temperature oxidation characteristics of alloy steel in mixed C_2H_5OH-H_2O atmosphere

G.X. Cheng
Jiangxi Changhe Aircraft Industry Co., Ltd., Jingdezhen, China

C.M. Zheng & Q.C. Tian
State Key Laboratory of Advanced Special Steel, Shanghai University, Shanghai, China

ABSTRACT: Preparation of reasonable structured oxide scale on the surface of alloy steel piercing plugs is of great interest for achieving extended service life. In order to understand the effect of individual elements on the formation of oxide scale, high temperature oxidation of 10Mn, 15CrNi-MoW2 and 316L was conducted at 1020°C in drop-feeding mixed C_2H_5OH-H_2O atmosphere, and the oxidation characteristics were explored by using scanning electron spectroscope(SEM), X-ray diffraction(XRD), energy dispersive X-ray spectroscopy(EDS), metallographic microscope(OM) and microhardness tester. It is found that W promotes the growth of spinel phase of the oxide scale, elevates the spacing of crystal plane, and aggravates the inner oxidation of metal matrix. It is revealed that with the increase of Ni and Cr contents, the distribution of the elements exhibits different morphological features, and the formation of Ni-rich wall and Cr-rich oxide wall in 316L notably hinders the diffusion of O^{2-} from the outside to the inside oxide scale. It is reported that a double-walled structure with the Ni-rich wall and the Cr-rich oxide wall can cut off the inward invasion of O^{2-} and thus effectively improve the oxidation resistance.

1 INTRODUCTION

Piercing plug is a key deformation tool during manufacturing the seamless steel tubular product, while oxidation is the most economical and practical method for the surface treatment of the alloy steel piercing plug. The oxidation scale prepared on the surface of piercingplug plays an indispensable role of lubricating and heat-insulating between the plug and the billet during piercing process, and thus significantly prolongs the service life of plug [1–4]. Therefore, the constituents and mechanical properties of the oxide scales directly determine the plug service performance and attract much attention in the past researches. It was reported that the chromium-rich oxides (Cr_2O_3, $FeCr_2O_4$, $Cr_{1.5}Mn_{1.5}O_4$) enhanced the heat resistance to the steel matrix and hence demonstrated very good protective properties [5–7]. The Ni element could increase the bonding strength between the metal matrix and the oxide scale by altering their bonding ways [8]. Exactly, the alloying elements in the steel affect the formation of oxide scale through mass transfer at high temperatures [9].

Adjusting the content variables of such as Cr and Ni in the materials can tune up the oxidation procedure and thus regulate the quality of the oxide scale under drop-feeding mixed H_2O-C_2H_5OH atmosphere at present industrial production [10], which is of great significance in guiding the material selection and oxidation process design for preparation of piercing plug. Therefore, the oxidation experiments of piercing plug with varied contents of alloying elements such as C, Cr, Ni and Mn had been carried out to disclose how individual elements influencing the oxidation process, and some inspiring results have been found in our former research [4, 10]. It was confirmed that Ni did not possess the thermodynamic conditions of oxidation under drop-feeding atmosphere of

H_2O-C_2H_5OH at 1020°C, and the mass transfer of Ni occurred in the form of elementary substance [10]. However, it is still confusing that the "venation" pattern of Cr ions transport channels appeared in the inner oxide scale of H13 steel, while dot-like Ni elemental substance distributed dispersively in the scales of 20Cr2Ni3 and 30Cr3NiMo2V alloys [4]. Because interval sampling cannot be realized during the oxidation period under the industrial process of mass production, the evolution of oxidation cannot be observed in such experiment. Therefore, as an alternative in this paper, the content of alloying elements is extended lower to the carbon steel 10Mn and higher to the stainless steel 316L, while the steel 15CrNiMoW2isemployed as a comparative experimental material for better understanding the oxidation behavior of individual elements.

2 EXPERIMENTAL PROCEDURES

Table 1 lists the chemical composition of the materials marked as 10Mn, 15CrNiMoW2 and 316L, respectively. The three materials were prepared subsequently by vacuum melting, casting and forging. The cuboid samples with dimension of $8 \times 8 \times 16mm^3$ were spark cut and then polished with 600-grit SiC paper followed by ultrasonically cleaning in ethanol.

Table 1. The chemical composition of experimental materials (mass fraction/%).

Material	C	Mn	Cr	Ni	Mo	W	Fe
10Mn	0.10	1.20	0.30	–	–	–	Bal.
15CrNiMoW2	0.15	0.40	0.70	0.90	0.60	1.55	Bal.
316L	0.03	1.40	16.0	10.0	2.00	–	Bal.

The oxidation of the cuboid samples were carried out in the furnace under atmosphereof drop-feeding mixed H_2O-C_2H_5OH solution in mass production line. In the pre-oxidation period, the air in the furnace was removed by filling in nitrogen and alcohol with a flow of 2.5L/min and 50 drops/min, respectively. The oxidation treatment was conducted at 1020°C in the furnace for 4 hours by filling in H_2O-C_2H_5OH with a flow of 2.5L/h (The alcohol is ethanol with purity of 99%, water is deionized, and the volume ratio of alcohol to water is 1:4). After oxidation, the cuboid samples were cooled within furnace to 850°C, followed by air cooling to room temperature.

The cross-section samples of the oxidized cuboids were prepared and polished for observation of the scale structure using a DM6000M microscope (OM), and then the metallographic microstructures of 10Mn, 15CrNiMoW2and 316L were observed after etching with 4% nital or ferric chloride hydrochloric acid solution ($5gFeCl_3 + 50mLHCl + 100mLH_2O$), respectively. A Quanta450 scanning electron spectroscope(SEM)equipped with energy dispersive X-ray spectroscopy(EDS)was employed to characterize the distribution of the chemical composition of the oxide scale. The phase constituents of the oxide scale were identified by X-ray diffraction (XRD, D/MAX-rC X-ray diffractometer, operated at 40kV and 100mA), and all the diffraction profiles were obtained in continuous modes at a scan speed of 2°/min. AnEVERONE microhardness tester was used to measure the microhardness from the oxide scale to the metal matrix.

3 RESULTS AND DISCUSSION

Figure 1 presents the cross-sectional oxidation morphologies of the three materials under polished state, showing clearly two layers, which is in consistent with the previous reports [10–12]. The outer oxide scales crack to certain degrees, and the order of damage extent is as follows: 10Mn (relatively complete) <15CrNiMoW2 (partially damaged) < 316L (mostly shedding). Large and deep pores can be found in the inner oxide scale of 10Mn in Figure1a-I, while that of 15CrNiMoW2 are much shallow and small (Figure 1b-I). The oxide scales of both the 10Mn and 15CrNiMoW2 are wedge-shaped into the metal matrix, where the phenomenon of inner oxidation of 10Mn along grain boundary is obvious in Figure 1a-II, while the inner oxidationis more serious for 15CrNiMoW2 in Figure1b. As for 316L, the oxide scale penetrates the metal matrix showing a "hilly" shape. The inner oxide scale of 316L exhibits a large amount of gray-white worm-like substances among which a dense gray-white band is present near the boundary between the inner and outer oxide scales, as indicated by Band 1 in Figure1c, while the inner oxide scale of 15CrNiMoW2 exhibits some dot-like substances (Figure 1b-II), which has been proved to be the Ni elementary substance [4, 10]. There exists a dark-gray band at the boundary between the oxide scale and the metal matrix, as indicated by Band 3 in Figure 1c, where a layer of gray-white band sparsely covered in the upper part of the zone (Band 2 in Figure 1c). The total thickness of both 10Mn and 15CrNiMoW2 is close to 400 μm, but there are significant differences in the thickness of inner and outer oxide scales. The thickness of the outer oxide scale of 10Mn is about 260 μm, and the inner layer is about 140 μm, while the thickness of the inner oxide scale of 15CrNiMoW2 is almost equal to the outer counterpart, which is about 200 μm. The oxide scale of 316L is very thin (Figure 1c-I), and the thickness is only about 1/10 that of the other two materials. It is clear that the increase in the Cr and Ni contents can greatly enhance the high temperature oxidation resistance of the material.

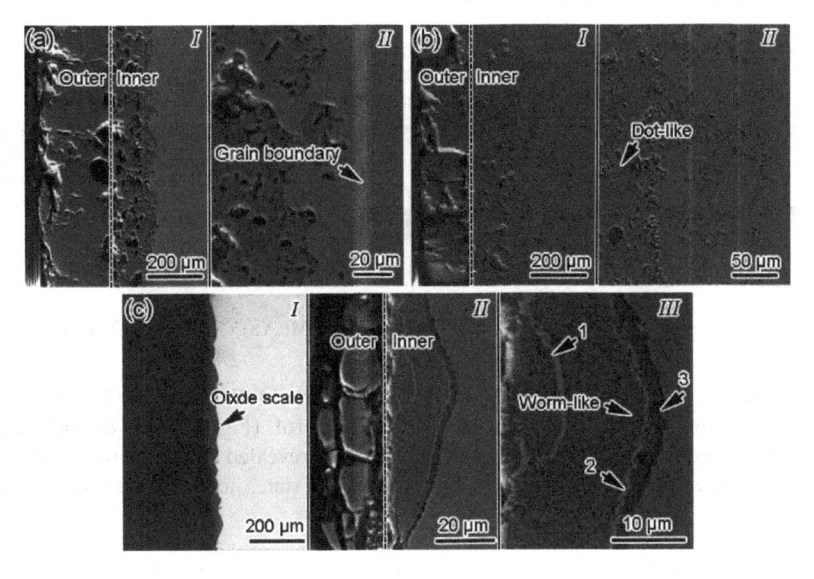

Figure 1. Cross-sectional oxidation morphologies of (a) 10Mn, (b) 15CrNiMoW2 and (c) 316L where a-II and b-II show the enlarged morphologies of a-I and b-I, respectively, and c-III shows the enlarged morphologies of c-II.

Figure 2 shows the elements distribution along the oxide scale of experimental materials. It can be seen that the oxide scale of 10Mn is dominated by Fe and O (Figure 2a), and there is a slight enrichment of Cr in the inner layer where Fe is relatively poor. Figure 2b shows the line scan of the inner oxide scale of 15CrNiMoW2. It can be seen that W, Mo, Ni and Cr distribute fluctuatedly and the O content level is lower in the oxide of 15CrNiMoW2 than that of 10Mn. This means that the impeding effect of alloying elements on ions transportation is significant. As for the two samples there is no obvious variation for Mn content. Mn generally does not exhibit enrichment in the process of oxidation at high temperatures [4,13]. Figure 2c shows that the outer oxide scale of 316L mainly consists of Fe and O while the inner layer is rich in Ni and Cr, and poor in Fe. Obviously, the worm-like substance is of Ni-rich, and it can be seen that both Band 1 and Band 2 is rich in Ni and Fe, and poor in Cr, while Band 3 is rich in Cr and Mn, and poor in Fe and Ni.

Figure 3 shows the distribution characteristics of the chemical compositionin the oxide scale. It can be seen that as for the 10Mn sample (Figure 3a), Cr enriches in the inner scale which is

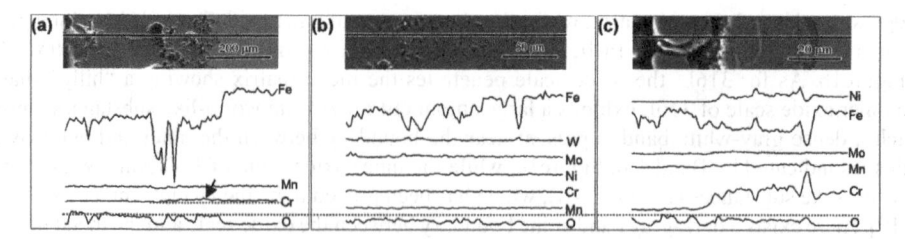

Figure 2. Line scan spectra of the oxide scale of (a) 10Mn, (b) 15CrNiMoW2 and (c) 316L.

consistent with Figure 2a. The same phenomenon is also observed for the 15CrNiMoW2 sample as that also have a small amount of Cr, and the distribution of W, Mo, Ni shows a similar characteristic as that of Cr (Figure 3b). Figure 3c shows the element mapping of 316L, it can be seen that Band 1 restraints the inward diffusion of O and the outward diffusion of Cr and Ni, effectively impedes the outward diffusion of Mn and Mo, while fails to take effect on the mass transfer of Fe. It is obvious that Band 2 and 3 cut off the inward diffusion of O.

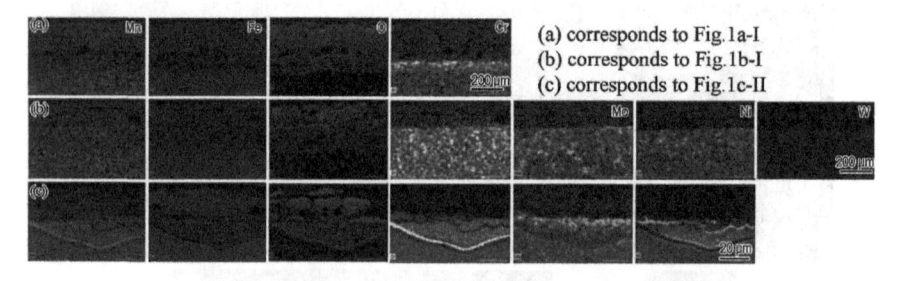

Figure 3. Chemical composition mapping of the scales of (a) 10Mn, (b) 15CrNiMoW2 and (c) 316L.

Figure 4 shows the microstructures of 10Mn, 15CrNiMoW2 and 316L. 15CrNiMoW2 (Figure 4a-2) is composed of ferrite doped with bainite, 316L (Figure 4b) is of austenite, while 10Mn (Figure 4c) consists of ferrite and some pearlite. It is revealed that oxidation is mainly along the grain boundaries for all the materials under the etched state, however, the inner oxidation of 15CrNiMoW2 is gravest.

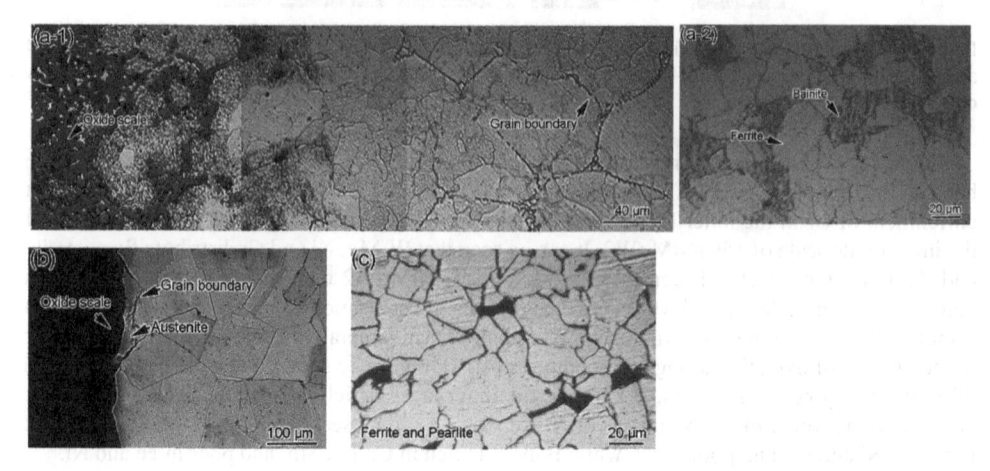

Figure 4. OM images of the metallurgical microstructure of (a) 15CrNiMoW2, (b) 316L and (c) 10Mn samples under etched state.

Figure 5 shows the microhardness variation of the three materials. The microhardness of the outer oxide scale is obviously higher than that of the inner layer. The microhardness of outer oxide scale for 10Mn is approximately identical to that for 15CrNiMoW2 (about 380HV), and the outer oxide scale of 316L is severely damaged and its microhardness cannot be measured. The microhardness of the inner layer of 10Mn is close to that of 15CrNiMoW2 (about 250HV), while 316L exhibits an obviously high value of about 310HV, for which thedenser oxide and the higher alloy contents in the scale may be accounted. The microhardness of the metal matrix increases gradually and then tends to be stable from the oxide/metal boundary to the inner matrix, and the value of three materials is as follows: 15CrNiMoW2 (160HV) > 316L (150HV) > 10Mn (125HV).

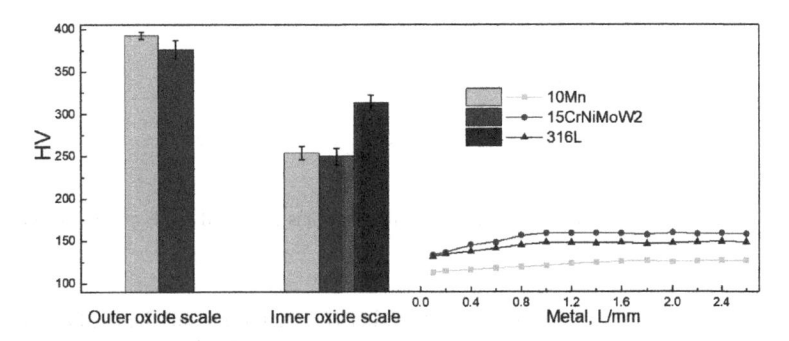

Figure 5. Microhardness distribution of the oxide scale and metal matrix (L—distance from oxide/metal boundary to matrix).

Figure 6 represents the phase constitutions of the outer oxide scale detected by XRD. It reveals that Me_3O_4 (Me_3O_4 stands for Fe_3O_4, $FeCr_2O_4$, $Mn_{1.5}Cr_{1.5}O_4$, $MnFe_2O_4$ and other spinel phases) and Fe_2O_3 are dominant in the outer oxide scale of the three materials, and a small amount of FeO, Cr_2O_3 and CrO also exist. The spinel peak of 15CrNiMoW2 is much stronger than that of other two materials.

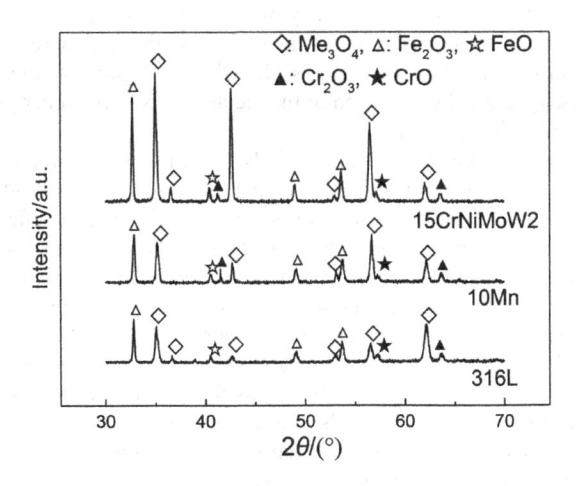

Figure 6. XRD profiles of the outer oxide scale of the three materialsoxidizedat 1020°Cin C_2H_5OH-H_2O atmosphere for 4 hours.

Since the discrepancy among the chemical composition of the three materials is most prominent in W content, it is inferred from the XRD profiles that the diffusion of W into the outer layer promotes the growth of spinel phase. The peak value of 15CrNiMoW2 is the highest atthe (311) of spinel phase. Figure 7 represents the profiles of (311) crystal plane of the three materials. The

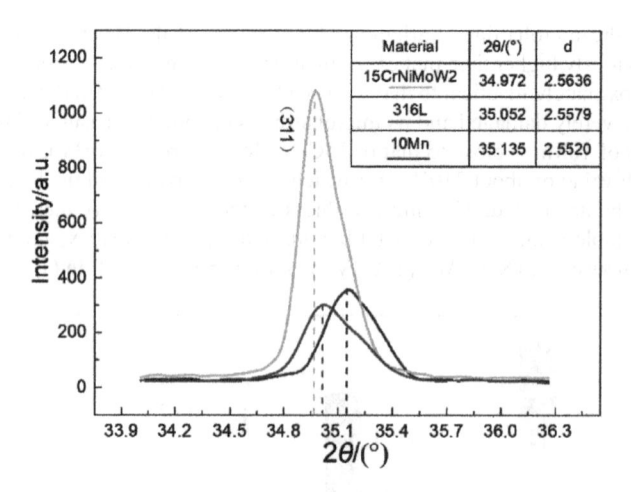

Figure 7. (311) crystal plane of the three materials and the values of 2θ and d(the spacing of crystal planes) are given in the upper right corner.

2θ of 15CrNiMoW2 is smallest than that of the other two materials and the (311) spacing of the corresponding material is determined as also shown in Figure 7. The spacing of crystal planes (d) of the three materials is 2.5636 for 15CrNiMoW2, 2.5579 for 316L and 2.5520 for 10Mn, respectively.

The diffusion coefficient of gap atoms is proportional to the square of the spacing of crystal planes [14], and has the form:

$$D = Pd^2\Gamma \qquad (1)$$

where D is the diffusion coefficient, P is the jumping probability in a given direction, d is the spacing of crystal planes, and Γ is the jumping frequency of interstitial atoms.

From this we can see that the diffusion coefficient of 15CrNiMoW2 increases, and O^{2-} is easier to enter the inner layer through the outer oxide scale, which will promote the growth of the inner oxide scale as well as aggravate the oxidation of the metal matrix to the inner grain along the grain boundary (Figure 1b and Figure 4a-1).

For Fe-Cr alloy, the difference of diffusion rate in Fe and Cr ions is an important factor affecting the phase composition and oxidation resistance of the oxide scale. The self-diffusion coefficient of FeO and Cr_2O_3 is 9×10^{-8} and 1×10^{-14} m^2/s, respectively [13]. It reflects the difficulty of alloying elements and oxygen passing through the oxide scale. Fe^{2+} is more easily diffused outward from the inner oxide scale, which leads to the phenomenon of Fe^{2+} depletion in the inner layer, while Cr^{3+} is not easy to trudge over a long distance and enriches in the inner layer (Figure 2 and Figure 3) Cr-rich oxide scale generally hinders ions diffusion [5–7]. The Cr content of 10Mn (0.3%) is lower than that of 15CrNiMoW2 (0.7%), therefore, a higher Cr^{3+} content is expected in the inner oxide scale, leading to about 43% increase in thickness for 15MnNiMoW (200μm) in comparison with 10Mn (140μm, Figures 1a-b). This is also responsible for the different morphology of pores in 10Mn and 15CrNiMoW2. It has been reported in the literature that many pores appear near the Cr-rich layer [15], because Fe^{2+} diffusion leaves a large number of vacancies, while the Cr^{3+} enrichment increases the opportunity of vacancy formation. Vacancies gather to form small pores [16], and then aggregate to form deep and large pores (Figure 1a-I).

According to the experimental results of Figures 2–3, the mass transfer model of 316L oxidation is given in Figure 8. In the initial stage of oxidation, O in the form of H_2O molecule reacts with the elements that diffuse outward from the metal matrix to form an initial oxide scale (Figure 8a). Once the oxide scale has covered the surface of the metal matrix, a selective permeability Cr-rich oxide layer (Band 3) has gradually formed at the boundary of the metal matrix, which cuts off the

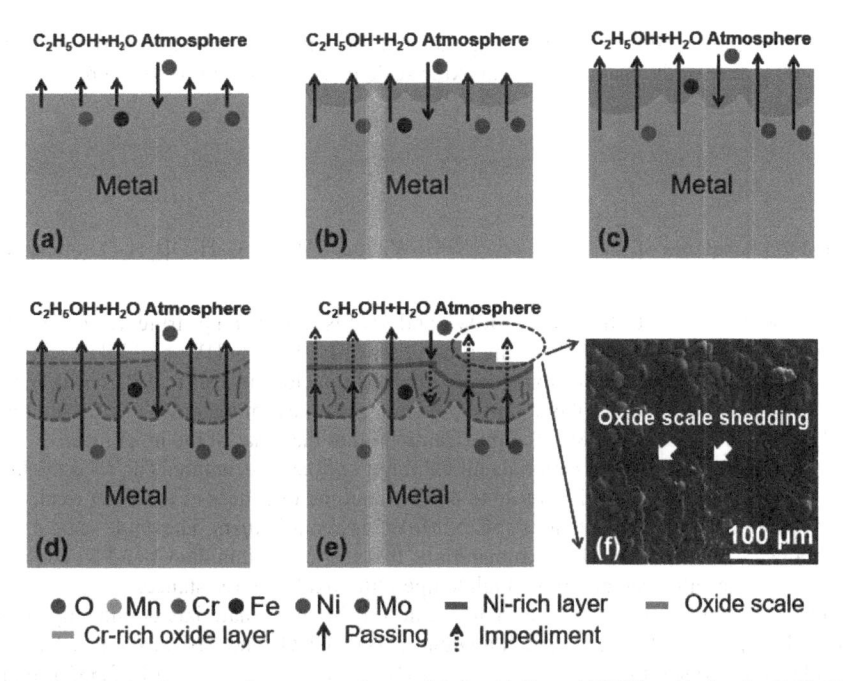

Figure 8. Schematic diagram of mass transfer model for 316L at 1020°C oxidation in $C_2H_5OH-H_2O$ atmosphere for 4 hours.

inward diffusion of O^{2-} (Figures 8b-c). In the later stage of oxidation, Ni gradually enriches at the boundary of the inner and outer oxide scales, forming a selective permeability Ni-rich layer (Band 1), which impedes the inward diffusion of O^{2-} and the outward diffusion of Cr^{3+}, Mn^{2+} and Mo. Moreover, a layer of non-compact Ni-rich layer (Band 2) is covered over Band 3, which is no-dense and has little effect on ions transport (Figures 8d-e). Fe^{2+} diffuses into the outer layer and reacts with O^{2-} to form the layered outer oxide scale (Figure 8f) which is easy to shed off.

Grain boundary is the main channel for outward diffusion of Cr [17], and it was found the "venation" pattern of Cr ions transport channels in the inner oxide scale of H13 (5.5% Cr) after oxidation at high temperatures, while that of 20Cr2Ni3 (2% Cr) and 30Cr3NiMo2V (3% Cr) did not exist [4]. The "venation" pattern of Cr ions transport channels is also not found in the oxide scale of the experimental material in this paper, but a Cr-rich oxide band (Band 3) exists in 316L (16% Cr). It can be inferred that with the increase of Cr content, Cr-rich oxide scale grows steadily and the "venation" pattern of Cr ions transport channels appear in the inner layer, when the content of Cr reaches a certain value, for example about 16%, a Cr-rich oxide wall like Band 3 will form at the boundary of the inner oxide scale and the metal matrix and then effectively hinder the ions transportation.

Ni-rich substances precipitate once the oxidation of Fe atoms leaves vacancies for the nearby Ni atoms to fill in to form a substitute Ni-rich crystal structure [18], and then the Ni-rich substances congregate and grow up to form walls as oxidation proceeds. Therefore, it is inferred that with the increase of Ni content, Ni distribution will go through the transition process from dot-like to warm-like and then to band-like. Band 1 and Band 2 are Ni-rich walls while Band 3 is a Cr-rich oxidewall. The coexistence of enriched Mn and Cr in Band 3 indicates the probability of complex reaction of MnO and Cr_2O_3 to form $(Mn,Cr)_3O_4$. The selective transmission mechanism of different walls results in a distinguished difference in the distribution of alloying elements. Band 1 notably hinders the diffusion of O^{2-} from the outside to the inside oxide scale while the double walls of Band 2 and 3 obviously cut off the inward invasion of O^{2-}, which causes the oxidation along the grain boundary inconspicuous and the oxide scale to bond with the metal matrix in a way that is no longer "wedge-shaped" but "hilly" along the metal matrix (Figure 1c-I), and thus enhance the oxidation resistance of the material with an oxide scale thickness of only 1/10 that of the other two materials.

4 CONCLUSIONS

The oxidation behaviors of the 10Mn, 15CrNiMoW2 and 316L in $C_2H_5OH-H_2O$ atmosphere are concluded as follows:

1) The microstructure of 10Mn, 15CrNiMoW2 and 316Lis ferrite and pearlite, ferrite and bainite, and austenite corresponding with the microhardness of 150HV, 160HV and 125HV, respectively. The microhardness of the inner oxide layer for 10Mn is close to that of 15CrNiMoW2, about 250HV, while that of 316L is about 310HV, the microhardness of the outer oxide scale for10Mn and 15CrNiMoW2 is about 380HV, significantly higher than that of the inner layer.
2) The oxide scales of the three materials all exhibit a two-layer structure. The total thickness of both 10Mn and 15CrNiMoW2 is close to $400\mu m$, but the thickness of the outer oxide scale of 10Mn is about $260\mu m$, while that of 15CrNiMoW2 is about $200\mu m$. The oxide scale of 316L is only about 1/10 that of the other two materials. It is the increase in the Cr and Ni contents that accounts for the enhancement of the high temperature oxidation resistance.
3) The phase constitutions of the outer oxide scale for the three materials dominantly consist of spinel Me_3O_4 and Fe_2O_3, and a small amount of FeO, Cr_2O_3 and CrO also exist. The spacing of (311) spinel crystal plane is 2.5636 for 15CrNiMoW2, 2.5579 for 316L, 2.5520 for 10Mn, respectively. W promotes the growth of spinel phase of the oxide scale, increases the thickness of the inner oxide scale and aggravates inner oxidation of the metal matrix.
4) The inner oxide scale of 316L penetrates the metal matrix showing a "hilly" shape, and exhibits a large amount of gray-white worm-like substances among which three bands are present. With the increase of Ni content, Ni distribution goes through the transformation process from dot-like to warm-like and then to band-like, while with the increase of Cr content, the oxide scale experiences a process from the "venation" pattern of Cr ions transport channels to aCr-rich oxide wall. A double-walled structure with a Ni-rich wall and Cr-rich oxide wall can cut off the inward invasion of O^{2-} and thus effectively enhance the oxidation resistance.

REFERENCES

[1] Ohnuki A, Hamazu S, Kawanami T et al. *Tetsu-to-Hagane*. 1986; 72(3): 450–457.
[2] Wang B, Yi DQ, Wu BT, et al. *Mater rev*. 2006; 20(6): 82–84.
[3] Zheng CM, Yuan HP, Tian QC. *Steel Pipe*. 2018; 47(6): 14–19.
[4] Zheng CM, Tian QC. *Acta Metall Sin*. 2019; 55(4): 427–435.
[5] Pettersson C, Jonsson T, Proff C, et al. *Oxid Met*. 2010; 74(1/2): 93–111.
[6] Smola1G,Gawel1R, Kyziol1K, et al. *Oxid Met*. 2019; 91(5/6): 625–640.
[7] DingR, TaylorMP, ChiuYL, et al. *Oxida Met*. 2019; 91(5/6): 589–607.
[8] Tian QC, Dong XP, Cui HC, et al. *Mater Sci Forum*. 2017; 896: 202–208.
[9] Saunders SRJ, Monteiro M, Rizzo F. *Prog MaterSci*. 2008; 53(5): 775–837.
[10] Zheng CM, Tian QC. 2019 *IOP Conf. Series: Mater Sci Eng*.490: 413–420.
[11] Yue ZW, Fu M, Li XG, et al. *Oxid Met*. 2010; 74(3/4): 157–165.
[12] Hayashi S, Narita T. *Oxid Met*. 2002; 58(3/4): 319–330.
[13] Wang C, Yu Y, Niu T, et al. *JIron Steel Res*. 2016; 28(5): 57–63.
[14] Hu GX, Cai X, Rong YH. 2000 *Fundamentals of Materials Science*. (Shanghai: Shanghai Jiao Tong University Press), p32.
[15] Viswanathan R, Sarver J, Tanzosh JM, *J Mater Eng Perform*. 2006; 15: 255–274.
[16] Quadakkers WJ, Ennis PJ, Zurek J, et al. *Mater High Temp*. 2005; 22(1/2): 47–60.
[17] Peng X, Yan J, Zhou Y, et al. *Acta Mater*. 2005; 53(19): 5079–5088.
[18] Zheng C, Zhao Y, Tian Q, 2020 *Journal of Physics: Conference Series 1637*, 011001: 557–563.

Design and experimental study of the metal rubber for elastic coupling

C.H. Lu, Y.Z. Zhang, L.R. Zhao & G. Tian
Department of Automotive Engineering, Hebei College of Industry and Technology, Shijiazhuang, China

ABSTRACT: In this paper, the cold stamping die and test fixture of the metal rubber used in elastic coupling were designed and manufactured, and the static and dynamic experimental research was carried out, which provided reference for the reasonable design of metal rubber components for elastic coupling.

1 INTRODUCTION

As shown in Figure 1, the rubber stopper type elastic coupling takes rubber as elastic element, and the elastic element is placed between the two coupling claws to realize the connection of the two coupling halves, which plays the role of transmitting motion and power, and has the functions of compensating the relative offset of two shafts, damping and buffering, so it is widely used in vehicles, ships and various machinery. However, due to the inherent characteristics of rubber, it is difficult to meet the operating requirements when working in high and low temperature, large temperature difference, corrosive media and other severe working conditions.

Figure 1. Rubber stopper type elastic coupling.

Metal rubber is a new type of nonlinear elastic and damping material, which is made of metal wire by spiral winding, fixed pitch drawing, laying blank, stamping forming and other processes, not only has excellent physical and mechanical properties of metal materials, but also has elasticity performance as rubber and stronger damping performance than rubber. It has excellent properties about absorbing vibration and shock. At the same time, it has other excellent properties, such as corrosion resistance, stable performance under high and low temperature conditions (20 ~ 537K), long fatigue life and can be preserved for a long time [1,2]. If the rubber components used in elastic coupling are replaced by the metal rubber, it will bring great benefits to the property, reliability and service life of elastic coupling.

In this paper, the cold stamping die and test fixture of the metal rubber used in elastic coupling were designed and manufactured, and the static and dynamic experimental research was carried out, which provided reference for the reasonable design of metal rubber components for elastic coupling and has certain engineering practical value.

2 PREPARATION OF METAL RUBBER ELASTIC COMPONENT

The metal rubber elastic components were formed by cold stamping, it's shape and size were pledged by the stamping die. The shape and size of metal rubber component for elastic coupling is shown in Figure 2. It can be seen that the cross-section shape of metal rubber component is relatively complex, which is composed of multiple arcs and straight lines. Among them, the upper and lower surfaces are curved surfaces, which is not easy to be formed by stamping, and the spatial expansion characteristics of metal rubber components after forming are relatively complex, which brings great difficulties to the design of stamping die.

By trial and error, the stamping die structure of metal rubber elastic component was determined as shown in Figure 3. The die is composed of punch, outer sleeve and cushion block. The end face of punch and cushion block has the same curved surface shape with metal rubber, and the machining precision of them should ensure the shape and size requirements of metal rubber elastic component. The metal rubber elastic components are shown in Figure 4.

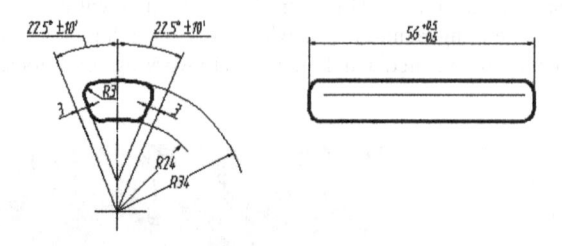

Figure 2. Schematic diagram of metal rubber elastic component.

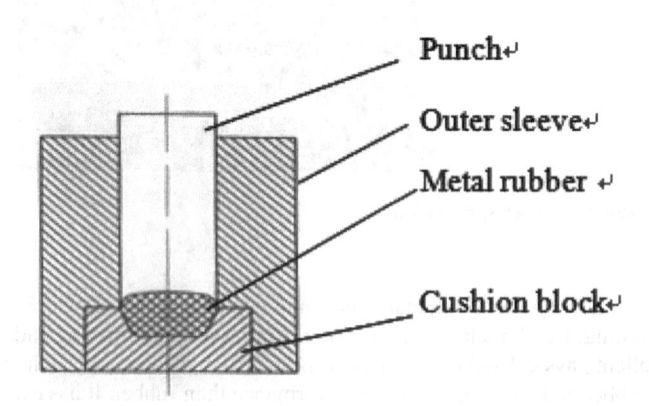

Figure 3. Cold stamping die.

Figure 4. Metal rubber elastic components.

3 STATIC AND DYNAMIC TEST

3.1 Test fixture

The torsional stiffness and damping loss factor of the metal rubber elastic components need to be tested through static and dynamic tests, which can provide reference for the reasonable design of metal rubber components for elastic coupling. Because it is difficult to test the torsional stiffness directly, the indirect measurement method based on compressive stiffness test was considered. According to the actual stress of the metal rubber elastic components in the elastic coupling, the test fixture of the metal rubber elastic components was designed as shown in Figure 5.

During the test, the test fixture was connected to the platform of the testing machine by bolts. The pressure head of the testing machine put stress on the metal rubber elastic component by the rotating arm. The torsional stiffness and damping loss factor of the metal rubber elastic element can be obtained by testing the displacement and force at the end of the rotating arm.

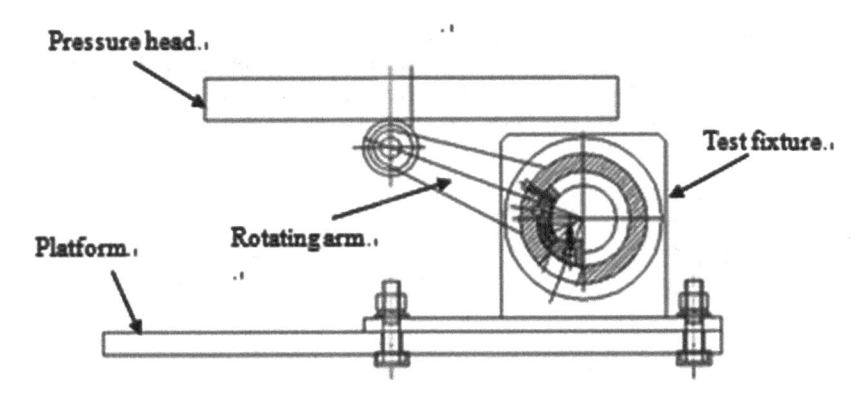

Figure 5. Schematic diagram of test tooling.

3.2 Static test

WDW-T200 microcomputer control electron universal testing machine was used for static test. The control mode is equal speed force control. The target force value is 800N and the speed is 0.01KN/s. The test curve was shown in Figure 6.

It can be seen from Figure 6 that the metal rubber elastic has the nonlinear characteristic of gradually hardening stiffness which can be explained by the microstructure of metal rubber. In the initial stage of deformation, all the micro-springs in metal rubber displayed linear elastic deformation. With the gradual compaction of the pores in metal rubber, the number of micro-springs participating in deformation on unit section increased, and the stiffness gradually increased. Soon after, the deformation continued to increase, the pores in metal rubber were fully compacted,

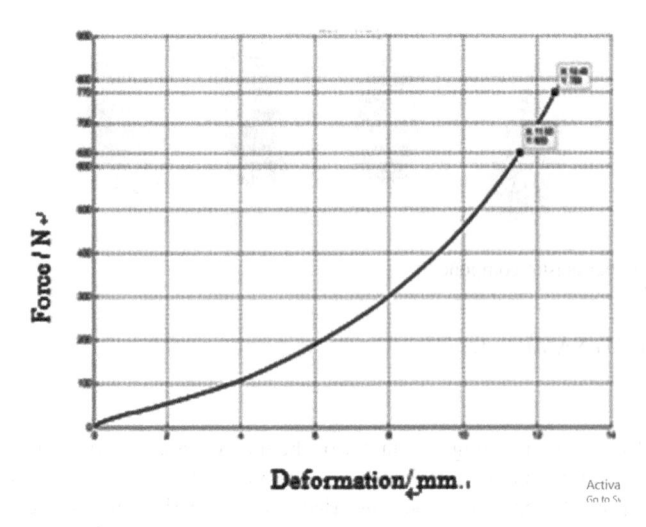

Figure 6. Load deformation test curve.

the micro-springs on some sections extruded each other, displayed plastic deformation, and the stiffness increased rapidly.

3.3 *Dynamic test*

The dynamic test system is mainly composed of PLS-20 electro-hydraulic servo fatigue testing machine and DH5936 vibration testing system. Figure 7 is the schematic diagram of the test system. The dynamic test adopts sinusoidal displacement control mode with amplitude of 1 mm and frequency of 1 Hz. The hysteresis loops was showed in Figure 8.

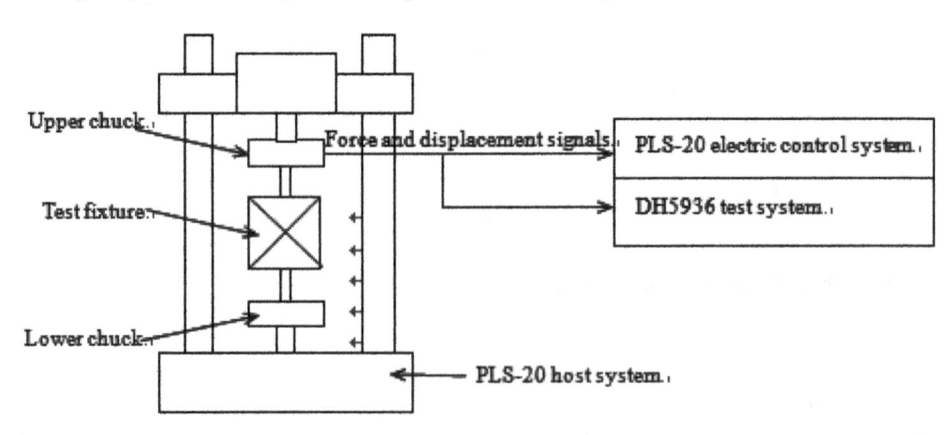

Figure 7. Schematic diagram of test system.

Damping loss factor η is an important index to measure the damping performance of damping materials[3], it can be represented as in equation (1).

$$\eta = \Delta W / 2\pi W = tg\varphi \tag{1}$$

In the equation, W is the maximum elastic energy storage of the material in a vibration period, ΔW is the energy consumption of the material in a vibration period and φ is the lag phase angle. If we can measure the energy consumption ΔW or lag phase angle φ, the damping loss factor η can be determined.

It can be seen from figure 8 that the metal rubber has nonlinear hysteresis characteristics, and its stress response contains high frequency harmonic components. Therefore, the lag phase angle φ can't be measured directly. However, the hysteresis loop formed by the restoring force and displacement reflects the fundamental characteristics of material energy consumption, and the area surrounded by the hysteresis loop is the energy consumed by the material in a vibration period. If the area of the hysteresis loop can be measured accurately, the damping energy consumption of the material can be quantitatively analyzed.

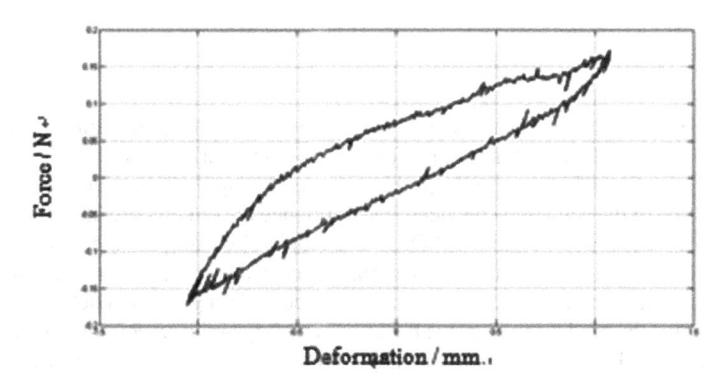

Figure 8. Hysteresis loop.

In the dynamic test, the simple harmonic displacement excitation ($x = x_0 \sin \omega t$) is applied, the restoring force is F_i, the sampling frequency is f, so the number of sampling points in a vibration period is $N = 2\pi f / \omega$. The area of the hysteresis loop can be calculated by the following equation (2).

$$\Delta W = \oint F_i dx = \frac{2\pi x_0}{N} \sum_{i=1}^{N} F_i \cos \frac{2\pi i}{N} \qquad (2)$$

By processing the dynamic test data and MATLAB programming, the damping loss factor η of metal rubber was 0.30, which was larger than that of ordinary rubber material.

4 CONCLUSIONS

The metal rubber components for elastic couplings have complex cross-section shapes and are not easy to be stamped. In this paper, the cold stamping die and test fixture of the metal rubber used in elastic coupling were designed and manufactured, the torsional stiffness and damping loss factors of metal rubber were measured by static test and dynamic test, which provides a reference basis for the reasonable design of metal rubber components for elastic coupling and has a certain engineering practical value.

REFERENCES

[1] Zhongying Li 2000 *Design of Metal Rubber Components*(Beijing: National Defense Industry Press).
[2] Huirong Hao 2019 *Research on elasticity, energy dissipation and fatigue characteristics of non-circular cross-section wire metal rubber*(Shijiazhuang: Ordnance Engineering College).
[3] Hongbai Bai, Chunhong Lu, Dongwei Li and Fengli Cao 2014 *Material and Engineering Application of Metal Rubber* (Beijing: Science Press).

Nonlinear dynamic characteristics of cylindrical dielectric elastomer actuator

L.J. Sun & J. Li
School of Mechanical and Power Engineering, East China University of Science and Technology, Shanghai, P.R. China

ABSTRACT: Dielectric elastomer actuators draw great interest in the emerging technology of soft actuations. This work studied the dynamic performance of a cylindrical dielectric elastomer actuator with coupled loads of tensile forces and variable voltages. Using the Euler-Lagrange equation, the governing equation of the actuator is deduced. The influence of film pre-stretch and spring stiffness on the resonance frequency of cylindrical DEA was studied by using the method of small perturbation of equilibrium position. When the damping viscosity is increased, the dynamic response stabilizes faster. The experimental results show that considering the viscoelastic properties of dielectric materials, the model can improve the accuracy of dynamic response prediction. The modelling and solving methods presented in this paper are helpful to analyse the viscoelastic dynamic response of DEAs, and provide theoretical support for the actuation and control methods of this kind of actuators in practical applications.

1 INTRODUCTION

Dielectric elastomer is a typical class of electro-active polymer [1,2] made from a soft membrane sandwiched between two compliant electrodes, and have garnered remarkable attention because of their large deformation, high energy density [3], fast response, low weight, variable stiffness, and low cost [4–6]. When subjected to voltage, the elastomer contracts along the electric field direction and expands in transverse directions [7–11]. Dielectric elastomer actuators (DEAs) can be made in many shapes and use a variety of actuation modes, such as axial elongation, radial bend, and in-plane or out-of-plane expansion. These properties of DEAs impart them with many applications such as artificial muscles, pumps, generators for harvesting energy, tactile sensors for Braille displays, and membrane resonators.

The working principle of the cylindrical DEAs is known that the electrostatic force generated by the electric field stimulation can transfer electrical energy into mechanical energy and make the cylindrical DEAs axial expansion. Although the working principle looks very simple, but the mechanical behavior of the cylindrical DEAs is not easy to modelling due to nonlinear properties of the material. When a voltage is applied to DE, the deformation of the material is time-dependent because of the viscoelasticity [12,13] of DE. Many previous works have demonstrated that the pre-stretch can improved the strain ability of the DEs [14,15]. Under different voltage amplitudes, the dielectric elastomer membrane is stretched to different amounts and thus changes its effective stiffness. Therefore, the resonant frequency of the actuator made of dielectric elastomer material will also change with the change of the excitation voltage [16–19]. Linear-elastic material model is not enough for the nonlinear dynamic response of the DEAs, many researchers have developed the viscoelastic model to demonstrated the time-dependent response of the DEAs. Li et al [10] presented the dynamic equations for the visco-hyperelastic DEs, and investigated how the viscoelasticity and strain-stiffening effect influence the resonance frequencies and oscillation amplitudes of the system. Zhang et al. [20] found that the spring stiffness of spring roll DEA has a more significant impact on dynamic performance compared to the effect of voltage. The above research results show that the viscoelasticity of material, pre-stretch of film, and spring stiffness have an impact on the dynamic output characteristics of DEAs.

In this paper, we investigate a viscoelastic model that incorporates the nonlinear effect of viscoelasticity and pre-stretch ratio, so as to study how these viscoelasticity and pre-stretch affect the performance and resonance frequency of DEAs. The content of this paper is organized as follow: first, we present a nonlinear rheological model to describe viscoelasticity, and propose a governing equation to describe the nonlinear dynamic response of the cylindrical DEAs. Next, based on the method of small-amplitude perturbation around the state of equilibrium, the influence of pre-stretch value of film and spring stiffness on the resonance frequency of actuator is analysed theoretically. Through simulation analysis, the function relationship between the resonant frequency of the actuator and the excitation voltage is studied. Last, the consistency of the strain time domain curves drawn by experimental data and simulation values verifies the correctness and effectiveness of the proposed model.

2 METHOD

To study the dynamic characteristics of cylindrical DEA undergoing axial deformation, we focus on a widely used configuration where a double layer dielectric elastomer is rolled around spring as shown in Figure 1.

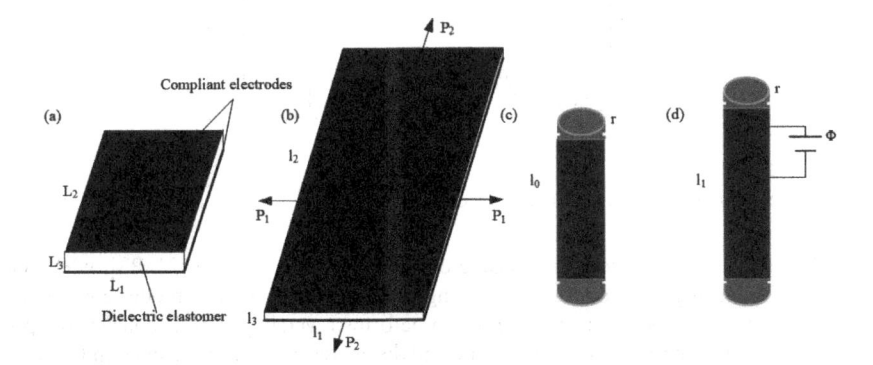

Figure 1. Diagram of cylindrical DEA structure.

Figure 1 illustrates the structure and production process of the cylindrical DEA. Figure 1(a) shows that the size of DE membrane in the reference state are length L_1, width L_2 and thickness L_3 in the Z, X and Y directions, respectively. Figure 1(b) illustrates that the DE membrane is pre-stretched by mechanical forces P_1 and P_2 in the L_1 and L_2 directions, respectively. Define the pre-stretches as $\lambda_{1p} = l_{1p}/L_1, \lambda_{2p} = l_{2p}/L_2, \lambda_{3p} = l_{3p}/L_3$. Two films of the same treatment are glued together. As shown in Figure 1(c), a spring with a stiffness of K, radius r, and free length of L_0 is axial pre-compressed to $\lambda_c = l_0/L_0$. A double layer of dielectric elastomer is rolled around the spring to form a cylindrical dielectric elastomer actuator. As shown in Figure 1(d), the actuator is axial deformed by applying voltage Φ.

We assume that the membrane and the spring are closely linked together, so the hoop stretch $\lambda_2 = \lambda_{2p}$ is prescribed. We define λ_1 as the axial stretch, the radial stretch $\lambda_3 = \lambda_1^{-1} * \lambda_2^{-1}$ is obtained because of the incompressibility in material. In the continuum mechanics, the materials coordinate (X,Y,Z) and the spatial coordinate (x,y,z) label a certain material point in the reference state and the deformable material point, respectively. The motion of the actuator can be written as:

$$x = \lambda_2 X \qquad (1)$$

$$y = \lambda_3 Y = \lambda_1^{-1} \lambda_2^{-1} Y \quad (2)$$

$$z = \lambda_1 Z \quad (3)$$

The dynamic governing equation of cylindrical DEA is obtained by the Euler-Lagrange equation:

$$\frac{\partial \ell}{\partial \lambda_i} - \frac{d}{dt}\left(\frac{\partial \ell}{\partial \dot\lambda_i}\right) = 0, \ell = T - U \quad (4)$$

Where $\dot\lambda_i$ denotes the rate of change λ_i. ℓ is the Lagrange, T is the kinetic energy, U is the potential of the conservative forces in the system.

The actuator contains a double layer membrane of DE, we assume that the spring used is massless, so the kinetic energy of the actuator takes the forms:

$$T = 2\int_\Omega \frac{1}{2}\rho\left(\dot{x}^2 + \dot{y}^2 + \dot{z}^2\right)d\Omega \quad (5)$$

Where ρ is the density of the membrane, Ω is the domain of membrane, $\dot{x}, \dot{y}$ and $\dot{z}$ are the deformation velocities of the membrane in three directions.

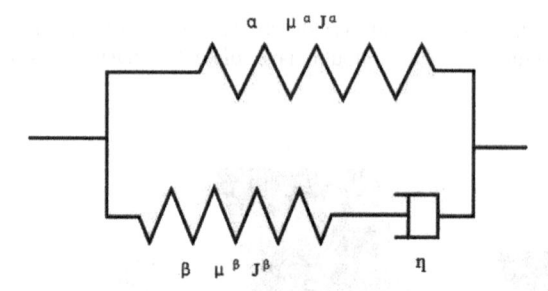

Figure 2. Viscoelastic model of dielectric elastomer.

The dielectric elastomer material is considered to be Newtonian Fluid [21–23], which can be represented by a rheological model of two springs and one dashpot. The model consists of two parallel units: unit A consists of a spring α which deforms reversibly; unit B consists of spring β with a series-wound dashpot that relaxes in time and dissipates energy, as sketched in Figure 2. We define that λ_i^e is the stretch of the spring β, and ξ_i is the stretch of the dashpot. By adopting the well-established multiplication rule, the stretch of spring β in the in-plane direction is $\lambda_i^e = \lambda_i / \xi_i, i = 1, 2$. Considering the strain-stiffening effect of the material, we choose the Gent model to describe the elastic energy density.

$$W(\lambda_1, \lambda_2, \xi_1, \xi_2) = -\frac{\mu^\alpha J^\alpha}{2}\ln\left(1 - \frac{\lambda_1^2 + \lambda_2^2 + \lambda_1^{-2}\lambda_2^{-2} - 3}{J^\alpha}\right)$$

$$-\frac{\mu^\beta J^\beta}{2}\ln\left(1 - \frac{\lambda_1^2 \xi_1^{-2} + \lambda_2^2 \xi_2^{-2} + \lambda_1^{-2}\lambda_2^{-2}\xi_1^2\xi_2^2 - 3}{J^\beta}\right) \quad (6)$$

where μ^α and μ^β are the shear moduli of spring α and β, J^α and J^β are the constant associated with the tensile limit of DE material.

Note that the potential of the actuator consists of the elastic energy of membrane, the potential energy induced by the mechanical forces, the elastic energy of spring, and the electrostatic energy. Based on the homogeneity assumption of dielectric elastomer materials, the elastic energy of membrane is obtained by multiplying the free energy density with the volume of the membrane. We get the potential of the system as follows:

$$U = 2L_1L_2L_3\left[-\frac{\mu^\alpha J^\alpha}{2}\ln\left(1-\frac{\lambda_1^2+\lambda_2^2+\lambda_1^{-2}\lambda_2^{-2}-3}{J^\alpha}\right) - \frac{P_1\lambda_1}{L_2L_3} - \frac{P_2\lambda_2}{L_1L_3}\right.$$
$$\left.-\frac{\mu^\beta J^\beta}{2}\ln\left(1-\frac{\lambda_1^2\xi_1^{-2}+\lambda_2^2\xi_2^{-2}+\lambda_1^{-2}\lambda_2^{-2}\xi_1^2\xi_2^2-3}{J^\beta}\right) - \frac{\varepsilon}{2}E_0^2\lambda_1^2\lambda_2^2\right] + \frac{K(L_0-L_1\lambda_1)^2}{2}$$
(7)

Where $E_0 = \Phi/L_3$ is the normal electric field. The rate of deformation in the dashpot (regarded as a Newtonian fluid) is described by $\xi_1^{-1} * d\xi_1/dt$ and $\xi_2^{-1} * d\xi_2/dt$, and the relationship between the rate of deformation of dashpot and the stretch can be described as [24,25]:

$$\frac{d\xi_1}{dt} = \frac{\xi_1}{3\eta}\left[\mu^\beta\left(\lambda_1^2\xi_1^{-2}-\xi_1^2\xi_2^2\lambda_1^{-2}\lambda_2^{-2}\right) - \frac{\mu^\beta}{2}\left(\lambda_2^2\xi_2^{-2}-\xi_1^2\xi_2^2\lambda_1^{-2}\lambda_2^{-2}\right)\right] \qquad (8)$$

$$\frac{d\xi_2}{dt} = \frac{\xi_2}{3\eta}\left[\mu^\beta\left(\lambda_2^2\xi_2^{-2}-\xi_1^2\xi_2^2\lambda_1^{-2}\lambda_2^{-2}\right) - \frac{\mu^\beta}{2}\left(\lambda_1^2\xi_1^{-2}-\xi_1^2\xi_2^2\lambda_1^{-2}\lambda_2^{-2}\right)\right] \qquad (9)$$

Where η is the viscosity of the dashpot.

Previous research data of our research group shows that that axial deformation of the actuator is influenced by the number of winding loops of the film N. when other conditions are determined, the largest axial deformation of the actuator is obtained when $N = 6$. Therefor, the number of winding loops of the actuator studied in this paper is $N = 6$. The length of the film in the first circle satisfies: $L_2, \lambda_{2p} = 2\pi r$; The thickness of the film after pre-stretching is $L_3\lambda_{1p}^{-1}\lambda_{2p}^{-1}$, and the double layer film is superposition during winding. The length of the film in the second round is $2\pi\left(r+2L_3\lambda_{1p}^{-1}\lambda_{2p}^{-1}\right) = L_{2,}\lambda_{2p}$, and so on, the circumferent length of the film after six rounds of winding is $L_2 = 3\times\left(4\pi r + 20\pi L_3\lambda_{1p}^{-1}\lambda_{2p}^{-1}\right)\lambda_{2p}^{-1}$. Because the film is tightly attached to the spring, within the safe deformation range of the film, the deformation of the film in the circumferential direction can be ignored, so that $\lambda_2 = \lambda_{2p} = \xi_2$. According to the incompressibility of the material, $\lambda_3 = \lambda_1^{-1}\lambda_2^{-1} = \lambda_1^{-1}\lambda_{2p}^{-1}$. Thus, the kinetic energy is derived from equation (5) as:

$$T = \frac{1}{3}\left(\rho L_1 L_2 L_3^3 \lambda_1^{-4}\dot\lambda_1^2\lambda_{2p}^{-2} + \rho L_1^3 L_2 L_3\dot\lambda_1^2\right) \qquad (10)$$

And the potential is obtained from equation (7) as:

$$U = 2L_1L_2L_3\left[-\frac{\mu^\alpha J^\alpha}{2}\ln\left(1-\frac{\lambda_1^2+\lambda_2^2+\lambda_1^{-2}\lambda_2^{-2}-3}{J^\alpha}\right) - \frac{\varepsilon}{2}E_0^2\lambda_1^2\lambda_{2p}^2\right.$$
$$\left.-\frac{\mu^\beta J^\beta}{2}\ln\left(1-\frac{\lambda_1^2\xi_1^{-2}+\lambda_1^{-2}\xi_1^2-2}{J^\beta}\right) - \frac{P_1}{L_2L_3}\lambda_1 - \frac{P_2}{L_1L_3}\lambda_{2p}\right] + \frac{1}{2}K(L_0-L_1\lambda_1)^2 \qquad (11)$$

Substituting equation (10) and equation (11) into the Euler-Lagrange equation (4), we obtain the governing equation:

$$\ddot\lambda_1 - \frac{2}{\lambda_1 + L_1^2L_3^{-2}\lambda_1^5\lambda_{2p}^{-2}}\dot\lambda_1^2 + \frac{3}{\rho\left(L_3^2\lambda_1^{-4}\lambda_{2p}^{-2}+L_1^2\right)}\left[\frac{\mu^\alpha\left(\lambda_1-\lambda_1^{-3}\lambda_{2p}^{-2}\right)}{1-\left(\lambda_1^2+\lambda_{2p}^2+\lambda_1^{-2}\lambda_{2p}^{-2}-3\right)/J^\alpha}\right.$$
$$\left.+\frac{\mu^\beta\left(\lambda_1\xi_1^{-2}-\lambda_1^{-3}\xi_1^2\right)}{1-\left(\lambda_1^2\xi_1^{-2}+\lambda_1^{-2}\xi_1^2-2\right)/J^\beta} - \frac{P_1}{L_2L_3} - \varepsilon E_0^2\lambda_1\lambda_{2p}^2\right] - \frac{3K(L_0-L_1\lambda)}{2\rho L_2L_3\left(L_3^2\lambda_1^{-4}\lambda_{2p}^{-2}+L_1^2\right)} = 0$$
(12)

For equation (12), the coefficient of damping nonlinear term is much smaller than other coefficients in the equation, so this term can be ignored in the governing equation:

$$\ddot{\lambda}_1 + g(\Phi, \lambda_1, \xi_1) = 0 \quad (13)$$

$$g(\Phi, \lambda_1, \xi_1) = \frac{3}{\rho \left(L_3^2 \lambda_1^{-4} \lambda_{2p}^{-2} + L_1^2\right)} \left[\frac{\mu^\alpha \left(\lambda_1 - \lambda_1^{-3} \lambda_{2p}^{-2}\right)}{1 - \left(\lambda_1^2 + \lambda_{2p}^2 + \lambda_1^{-2} \lambda_{2p}^{-2} - 3\right)/J^\alpha} \right.$$

$$\left. + \frac{\mu^\beta \left(\lambda_1 \xi_1^{-2} - \lambda_1^{-3} \xi_1^2\right)}{1 - \left(\lambda_1^2 \xi_1^{-2} + \lambda_1^{-2} \xi_1^2 - 2\right)/J^\beta} - \frac{P_1}{L_2 L_3} - \varepsilon E_0^2 \lambda_1 \lambda_{2p}^2 \right]$$

$$- \frac{3K(L_0 - L_1 \lambda_1)}{2\rho L_2 L_3 \left(L_3^2 \lambda_1^{-4} \lambda_{2p}^{-2} + L_1^2\right)} \quad (14)$$

The rate of deformation in the dashpot can be simplified by equation (8) and equation (9) as:

$$\frac{d\xi_1}{dt} = \frac{\xi_1}{3\eta} \left[\mu^\beta \left(\lambda_1^2 \xi_1^{-2} - \xi_1^2 \lambda_1^{-2}\right) - \frac{\mu^\beta}{2} \left(1 - \xi_1^2 \lambda_1^{-2}\right) \right] \quad (15)$$

$$\frac{d\xi_2}{dt} = 0 \quad (16)$$

3 RESULT

Note that when the forces P_1, P_2 and voltage Φ are static, the actuator may reach a state of equilibrium. In this case, since both the voltage and the forces are independent of the time t, in the state of equilibrium, the equation of motion, equation (13) and equation(15), can be safely reduced to the equilibrium:

$$\frac{3}{\rho \left(L_3^2 \lambda_{1eq}^{-4} \lambda_{2p}^{-2} + L_1^2\right)} \left[\frac{\mu^\alpha \left(\lambda_{1eq} - \lambda_{1eq}^{-3} \lambda_{2p}^{-2}\right)}{1 - \left(\lambda_{1eq}^2 + \lambda_{2p}^2 + \lambda_{1eq}^{-2} \lambda_{2p}^{-2} - 3\right)/J^\alpha} \right.$$

$$\left. - \frac{P_1}{L_2 L_3} - \varepsilon E_0^2 \lambda_{1eq} \lambda_{2p}^2 \right] - \frac{3K(L_0 - L_1 \lambda_{1eq})}{2\rho L_2 L_3 \left(L_3^2 \lambda_{1eq}^{-4} \lambda_{2p}^{-2} + L_1^2\right)} \quad (17)$$

$$\frac{d\xi_1}{dt} = 0, \xi_{1eq} = \lambda_{1eq} \quad (18)$$

Therefore, a state of equilibrium, λ_{1eq}, can be obtained by equation (17). As described by Zhu et al, when the DE membrane is perturbed from this equilibrium state, the time dependent stretch is expressed in the following form:

$$\lambda_1(t) = \lambda_{1eq} + \Delta(t) \quad (19)$$

Where $\Delta(t)$ is the amplitude of perturbation, and is taken to be small. Expanding the function $g(\Phi, \lambda_1, \xi_1)$ as a power series in Δ around the equilibrium strain λ_{1eq} and substituting the expressions $\lambda_1(t)$ of in equation (19) into equation (13), one obtains the governing equation of motion for Δ:

$$\ddot{\Delta} + \frac{\partial g(\Phi, \lambda_1, \xi_1)}{\partial \lambda_1}\bigg|_{\lambda_1 = \lambda_{1eq}} \Delta = 0 \quad (20)$$

Where:

$$\frac{\partial g(\Phi,\lambda_1,\xi_1)}{\partial \lambda_1}\bigg|_{\lambda_1=\lambda_{1eq}} = \frac{3}{\rho\left(L_3^2\lambda_{1eq}^{-4}\lambda_{2p}^{-2}+L_1^2\right)}\left[\frac{\mu^\alpha\left(1+3\lambda_{1eq}^{-4}\lambda_{2p}^{-2}\right)}{1-\left(\lambda_{1eq}^2+\lambda_{2p}^2+\lambda_{1eq}^{-2}\lambda_{2p}^{-2}-3\right)/J^\alpha}\right.$$

$$\left.+\frac{2\mu^\alpha\left(\lambda_{1eq}-\lambda_{1eq}^{-3}\lambda_{2p}^{-2}\right)^2}{J^\alpha\left[1-\left(\lambda_{1eq}^2+\lambda_{2p}^2+\lambda_{1eq}^{-2}\lambda_{2p}^{-2}-3\right)/J^\alpha\right]^2}+4\mu^\beta\lambda_{1eq}^{-2}-\varepsilon E_0^2\lambda_{2p}^2\right]+\frac{3KL_1}{2\rho L_2 L_3\left(L_3^2\lambda_{1eq}^{-4}\lambda_{2p}^{-2}+L_1^2\right)}$$

(21)

Based on equation (21), the resonance frequency of the cylindrical DEA is obtained as:

$$\omega^2 = \frac{\partial g(\Phi,\lambda_1,\xi_1)}{\partial \lambda_1}\bigg|_{\lambda_1=\lambda_{1eq}} \quad (22)$$

From equation (22), it is observed that the resonance frequency of the cylindrical DEA depends on the stretching state of the membrane.

4 DISCUSSION

4.1 *Resonant frequency analysis*

In the following three cases, the resonance frequency of the actuator is calculated as a function of voltage: different pre-stretch λ_{1p} under a fixed pre-stretch $\lambda_{2p}=5$ and a fixed spring stiffness $K=300N\cdot m^{-1}$ (Figure 3(a)); different pre-stretch λ_{2p} under a fixed pre-stretch $\lambda_{1p}=3$ and a fixed spring stiffness $K=300N\cdot m^{-1}$ ((Figure 3(b)); and different spring stiffness K under fixed pre-stretches $\lambda_{1p}=3, \lambda_{2p}=5$ (Figure 3(c)). In the calculation, the following parameters are used:

$r=6mm, L_0=70mm, K_1=200N\cdot m^{-1}, K_2=250N\cdot m^{-1}, K_3=300N\cdot m^{-1}, \lambda_c=0.6,$
$L_1=14mm, L_2=47.8mm, L_3=1mm, \rho=1.2\times 10^3 kg\cdot m^{-3}, \mu^\alpha=46000Pa,$
$J^\alpha=70, \mu^\beta=45000Pa, J^\beta=70, \tau_v=50s, \varepsilon=3.98\times 10^{-11}F\cdot m^{-1}$

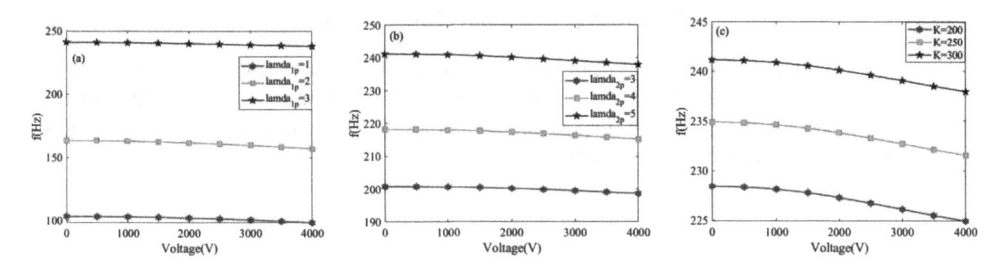

Figure 3. The resonant frequency of the actuator is plotted as a function of voltage (a) for different pre-stretches λ_{1p} at fixed value of pre-stretch $\lambda_{2p}=5$ and spring stiffness $K=300N\cdot m^{-1}$, (b) for different pre-stretches λ_{2p} at fixed value of pre-stretch $\lambda_{1p}=3$ and spring stiffness $K=300N\cdot m^{-1}$, and (c) for different spring stiffnesses K at fixed value of pre-stretch $\lambda_{1p}=3, \lambda_{2p}=5$.

An interesting observation is that in all three cases, the resonant frequency of the actuator decreases with the increase of voltage, but the decreasing rate is different. The actuator consists of a membrane and a spring in parallel. After the voltage is applied to the actuator, the aera of the film increases under the action of the electric field force, which makes the film in the original tensile state soften and causes the film stiffness to decrease. According to the relationship between the resonance frequency and the stiffness of the object, the reduction of the stiffness of the film leads to the reduction of the resonance frequency of the actuator.

4.2 Dynamic response analysis

We then investigate the effect of viscoelasticity on the dynamic responses of the actuator. For the dynamic analysis, we first investigate the (viscous) damping effects on free vibration. Under $\lambda_{1p} = 3, \lambda_{2p} = 5$ and DC voltage $\Phi = 2kV$, the equilibrium stretching ratio is $\lambda_{1eq} = 3.544$. From the relaxed state, we apply an initial condition of $\lambda_1(0) = \lambda_{1eq} = 3.544 (d\lambda_1/dt = 0, \xi_1(0) = \lambda_{1eq})$, and calculate the dynamic response of the actuator. As shown in Figure 4, the model captures the damping effect due to the viscoelasticity of the DE film that is characterized by the viscosity of the dashpot, η. We can see that the higher of viscosity, the more quickly of vibration attenuation. Experimental results also verify the correctness of the model.

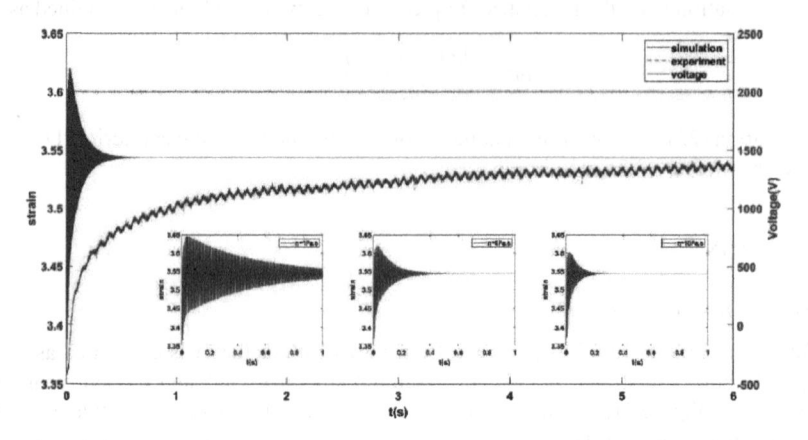

Figure 4. Dynamic responses of actuator with respect to different value of viscosity (a) $\eta = 1Ps$, (b) $\eta = 5Ps$, and (c) $\eta = 10Ps$.

Next, we investigate the vibration of the actuator under harmonic voltage actuations. In the following we keep the pre-stretch as $\lambda_{1p} = 3, \lambda_{2p} = 5$, and apply a harmonic voltage:

$$\Phi(t) = \Phi_{DC} + \Phi_{AC} \sin(2\pi f - \pi/2) \qquad (23)$$

The Φ_{DC} is the DC voltage, Φ_{AC} is the amplitude of the AC voltage and f is the frequency of the harmonic voltage. We set $\Phi_{DC} = \Phi_{AC} = 1kV$. The initial condition is set to be in fully relaxed state of the actuator without voltage, i.e., $\lambda_1(0) = \xi_1(0) = \lambda_{1eq}$. we study the dynamic responses of actuator under sinusoidal voltage signals.

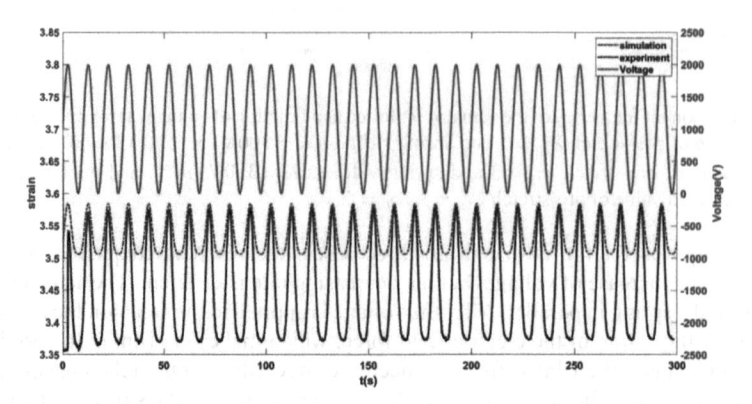

Figure 5. Time-domain responses of the actuator under frequency $f = 0.1Hz$.

As can be seen from the experimental data in Figure 5, under the excitation of sinusoidal voltage, the axial strain creep of the actuator was obvious in the first three periods. However, there is no creep phenomenon in the simulation, so there is room for improvement of the model. When the actuator reaches the strain stable state, the experimental data and the simulation values are in good agreement.

5 CONCLUSION

Using Euler-Lagrange equation, the dynamic equation of axial deformation of cylindrical actuator is established. Using the steady state micro-perturbation method, the influence of the pre-stretch values in two directions of the film and the spring stiffness on the resonance frequency of the actuator is analysed. Increasing the pre-stretch value of the film or the stiffness of the spring will lead to an increase in the overall stiffness of the actuator and an increase in the resonance frequency of the actuator. After the voltage is applied to the actuator, the film becomes soft due to the effect of electric field force, the overall stiffness of the actuator decreases, and the resonance frequency of the actuator decreases with the increase of the excitation voltage. Nonlinear dynamic analyses were carried out for the case of homogeneous deformation under DC and AC voltage. The results indicate that the proposed model can effectively capture the effect of viscoelastic damping on the oscillation amplitude and the experimental results are in good agreement with the simulation values.

ACKNOWLEDGMENTS

This work was supported in part by the National Nature Science Foundation of China (Grant No.52075172), and Natural Science Foundation of Shanghai (19ZR1413300).

REFERENCES

[1] Pelrine R, KornbluhR, PeiQ and Joseph J 2000 *Science* **287** 836–39
[2] BrochuP andPei Q 2010 *J, Macromol. Rapid Common.* **31** 10–36
[3] Wissler M and Mazza E 2005 *Smart Mater. Struct.* **14** 1396–1402
[4] KimS, LaschiC and TrimmerB 2013 *Trends Biotechnol.* **31** 287–94
[5] O'halloranA, O'malleyF and MchughP 2008 *J. Appl. Phys.* **104** 071101
[6] LiuL, ChenH, ShengJ, Zhang J, Wang Y and Jia S 2014 *Smart Mater. Struct.* **23** 025037
[7] BrochuP andPeiQ 2010 *Macromol. Rapid Commun.* **31** 10–36
[8] GuG, ZhuJ, Zhu L and Zhu X 2017*Bioinspir. Biomim.* **33** 1263–71
[9] Gu G, ZhuJ, ZhuL and Zhu X 2017 *Bioinspir.Biomim.* **12** 011003
[10] Li Y, Oh I, Chen J, Zhang H and Hu Y 2018 *Int. J. Solids. Struct.* **152–153** 28–38
[11] Quang T, Li J, Xuan F and Xiao T 2018 *Mater. Res. Express* **5** 065303
[12] Wang H and Qu S 2016 *J. Zhejiang Univ-Sc. A* **17** 22–36
[13] Guo J, Rui X, Park HS and Nguyen TD 2015 *ASME J. Appl. Mech.* **82** 091009
[14] Wang H, Li L, Zhu Y and Yang W 2016 *Smart Mater. Struct.* **25** 125008
[15] Wissler M and Mazza E 2007 *Sensor Actuat. A-Phys* **138** 384–393
[16] Sheng J, Chen H, Liu L, Zhang J, Wang Y and Jia S 2013 *J. Appl. Phys.* **114** 134101
[17] Zhu J, Cai S and Suo Z 2010 *Polym. Int.* **59** 378–83
[18] Zhu J, Cai S and Suo Z 2010 *Int. J. Solids Struct.* **47** 3254–62
[19] Feng C, Jiang L and Lau WM 2011 *J. Micromech. Microeng.* **21** 095002
[20] Zhang J, Chen H and Tang L 2015 *Appl. Phys. A* **119** 825–35
[21] Zhang J, Zhao J, Chen H and Li D 2017 *Smart Mater. Struct.* **26** 015010
[22] Bergstrom JS and Boyce MC1998 *Mech. Phys. Solids* **46** 931–54
[23] Lv X, Liu L, Liu Y and Leng J 2015 *Smart Matter. Struct.* **24** 115036
[24] Foo CC, Cai S, Koh SJA, Bauer S and Suo Z 2012 *J. Appl. Phys.* **111** 034102
[25] Foo CC, Koh SJA, Keplinger C, Kaltseis R, Bauer Sand Suo Z 2012 *J. Appl. Phys.* **111** 094107

Potential application of high-performance polymers in the oil and gas drilling engineering

G.F. Zhao & R.Y. Wang
SINOPEC Research Institute of Petroleum Engineering, Sinopec Science and Technology Research Center, Changping District, Beijing, China

ABSTRACT: The cross integration of information technology, biotechnology, new energy technology, and new material technology is triggering a new round of technological revolution and industrial transformation. Driven by new technologies, new materials, and new theories, drilling engineering technology integrates with nanomaterials, artificial intelligence and other new technologies. A number of key technologies, marked by intelligent well completion and nanorobots, have been continuously developed and improved, which has promoted the efficiency, automation and intelligence of drilling engineering. This article adopts qualitative analysis such as application environment comparison and quantitative analysis such as bibliometric analysis, to evaluate the application prospects of supermolecular hydrogel, polyimide aerogel, and 4D printed shape memory polymers in the field of drilling engineering. Specific transplantation suggestions have been put forward. It is recommended to continue to carry out the feasibility evaluation of the field application of high-performance polymers in drilling engineering, and it is expected to introduce new high-performance materials for downhole special tools, long-term downhole plugging materials, and new thermal insulation materials.

1 INTRODUCTION

The cross integration of information technology, biotechnology, new energy technology, and new material technology is triggering a new round of technological revolution and industrial transformation. Petroleum engineering technology, a multidisciplinary technical field, has benefited from the progress of materials, information, measurement and control and other related disciplines, and has maintained a good momentum of technological innovation. Petroleum engineering technology innovation has various forms, and cross-integration with related disciplines has become an important innovation method. Continuouslyexploring potential candidate technologies in non-oil and gas intersecting fields and strengthening interdisciplinary integration will surely bring new breakthroughs to the development of petroleum engineering technology and further accelerate the speed of field innovation.

The rapid development of high-tech technologies such as the Internet of Things, big data, cloud computing, virtual reality, quantum computing, 5G communications, high-performance materials, microelectronic components, and their integration with the oil and gas industry are expected to promote technological innovation in petroleum engineering, and produce a series of disruptive technology. The progress of key technologies in basic disciplines provides impetus for technological breakthroughs in the field of petroleum engineering, such as epoxy resin modification technology, high-end radio frequency chips, high-strength steel materials, high-precision algorithms and other fields that provide important prospects for technological innovation in petroleum engineering. At present, high-performance polymers have been applied in fields such as expandable screens, high-temperature-resistant sealing materials, and environmentally responsive plugging agents, and have achieved good economic benefits [1–3]. More and more attention has been paid to the application of high-performance polymers in thefield of petroleum engineering, which requires higher technical maturity and technical relevance of technologies in related disciplines. Therefore, it is necessary to conduct technology mining and application feasibility analysis of high-performance polymers, to provide new ideas and methods for the development of the next generation of petroleum engineering technology.

2 DEVELOPMENT TRENDS OF PETROLEUM ENGINEERING

Driven by new technologies, new materials, and new theories, downhole engineering technology is developing in the direction of high efficiency, automation, intelligence, low cost, and integration. More specifically, drilling equipment is combined with automatic control, information technology and artificial intelligence to develop towards high-power, modularity, automation and intelligence; downhole tools adopt new materials and new processes, and are towards high temperature resistance, long life, high reliability and automation; downhole measuring instruments integrate advanced detection, precise guidance and intelligent decision-making, and develop towards high performance, large capacity, precision and intelligence; wellbore working fluid integrates chemistry, geology and multidisciplinary results, and develops towards high efficiency, low cost, and environmental protection; the process technology focuses on the development of multiple resources, integrates the latest research results, and develops in the direction of efficient, economic, and engineering geological integration.

As the exploration and development of oil and gas turn to deep water, deep formations, unconventional formations, polar regions, and etc., the quality of oil and gas resources declines and the operating environment becomes more and more complexed.Oil and gas exploration is facing deeper, smaller, thinner, lower permeability and other complex geological conditions. The development of deep-water oil and gas, natural gas hydrates, and hot dry rocks has put forward new demands, driving the research and development of new technologies for deep-water engineering, environmental monitoring, and high-temperature drilling. Petroleum engineering technology is increasingly demanding new materials, especially high-performance polymers. The transplantation and application of new theories, new methods, new materials, and new equipment will further promote the development of petroleum engineering technology. BP spent 10 years investing in the establishment of the International Center for Advanced Materials (ICAM) to fund research on the basic understanding and use of advanced materials. The research of high-performance materials is also one of the latest research directions of Schlumberger Doll Research Center.

3 APPLICATION ANALYSIS OF HIGH-PERFORMANCE POLYMERS IN THE FIELD OF PETROLEUM ENGINEERING

Polymer materials are developed in pursuit of functionalization, lightweight, ecological, micromolding, and high performance. Polymer materials have obvious advantages and unique properties, such as high strength, high toughness, high (low) temperature resistance, wear resistance, corrosion resistance, thermal conductivity, high insulation and other characteristics. They also have the advantages ofsaving energy, protecting environment, enhancement and lightweight in the production process. Therefore,they arewidely used in aerospace, defence construction, automobile and other industries. They are indispensable in the field of new material revolution, and could meet the on-site needs of various oil and gas industries.

Equipment tools and fluids are developing in the direction of high performance, high reliability, high safety, functionalization and environmental protection. The reliability and temperature resistance of downhole rock breaking tools and measuring tools have been continuously improved, reducing complex downhole failures and realizing safe and efficient drilling. The downhole tools use new materials and new technology, and develop towards high temperature resistance, long life, high reliability and automation. The wellbore working fluid integrates with the achievements of chemistry, geology and other disciplines, and it is developing in the direction of high performance, low cost and environmental protection. Drilling and completion fluids,clean drilling fluids, completion fluids, fracturing fluids and other operating fluids are developing in the direction of high temperature resistance, multi-functionality, and recyclability, which not only meets operating requirements, protects oil and gas reservoirs, but also reduces environmental pollution. Obviously, high-performance polymers have broad application prospects in the field of petroleum engineering. The potential application of high-performance polymers in the field of petroleum engineering needs to overcome the high temperature and high pressure in the downhole operating environment and various corrosive components in the wellbore working fluid. International oil companies and oil service companies attach great importance to the application research of new materials such as nanomaterials and high-performance polymers, whichhas effectively promoted the progress of downhole nanofluids, tool surface treatment technology, intelligent cementing technology and so on. Even though most researchesare still in the laboratory stage, itis worthy of reference and learning by oil and gas companies in China.

3.1 *Supermolecular hydrogel*

In recent years, with the development of supramolecular science, non-covalent interactions with highly dynamic and reversible characteristics have shown significant advantages in preparing functional materials with regular structures and controllable properties [4]. A variety of reversible effects have been introduced into the field of shape memory polymer materials, and a series of new intelligent supramolecular hydrogels with shape memory functions at room temperature have been developed. Supramolecular hydrogel materials mainly refer to biological materials that are bound together through non-covalent interactions between small molecules. Since the non-covalent interactions in the supramolecular cross-linking network are usually reversible, many supramolecular materials exhibit self-healing properties.

Technical characteristics

Dual response to pH and sugar, supramolecular-polymer interpenetrating network (supramolecular shape memory, polymer covalent cross-linking), dynamic borate bond presents self-repairing properties. Its multiple shape memory functions at room temperature can be widelyapplied. It isself-healing and maintains shape memory function after repairing. The mechanical properties have been improved, and the strength has reached 0.12MPa@850%.

Recommendations for transplantation applications in the field of petroleum engineering

It is expected to use in drilling fluid leakage prevention and plugging while drilling, and cement ring crack repairing.Lost circulation is a common and complicated situation in the drilling process, which is extremely harmful. In drilling practice, we have gained experience on the causes and prevention of lost circulation, but there is a lack of broad-spectrum leak-proof and plugging technology. The main problems at present are: (1) Leakage causes huge losses, and it takes a lot of manpower, material and financial resources to plug the leakage; (2) The leakage plugging materials are not suitable for all leakage situations. Accordingly, related researchesthat need to be carried out include: (1) Application prospect and economic evaluation of supramolecular hydrogel technology in leak prevention and plugging; (2) Adaptability of supramolecular hydrogel in drilling fluid.

3.2 *Polyimide aerogel*

The porous material formed by the coalescence of polyimide molecules has a continuous three-dimensional framework network, and an open and interconnected nano-scale porous structure (the pore size is mainly in the range of tens of nanometers), with low density, high porosity and high specific surface area. Polyimide is a type of organic polymer with imide ring in the main chain. It has the advantages of high mechanical strength, good thermal stability and abrasion resistance. Whether it is used as a structural material or a functional material, its significant advantages have been fully recognized. Polyimide aerogel combines the high performance of the material itself with the characteristics of aerogel [5–7]. It is a new light, thin, clean and flexible thermal insulation material that combines the toughness, physical properties of plastic film and the thermal insulation performance of aerogel. It is suitable for a variety of industries, including aviation, oil and gas, radio frequency radar, automotive, construction, etc. The polyimide polymer structure avoids dust generation, making the product very clean and safe. Polyimide aerogel film material (50 microns thickness can be achieved) is expected to be used in deep-sea thermal insulation materials and dust-proof materials, and can overcome the shortcomings of silicon aerogels.

Technical characteristics

It can withstand extreme high and low temperatures, with aerogel-grade thermal conductivity and excellent sound insulation performance. Thermal insulation performance is only second to vacuum (thermal conductivity 0.02 W/mK). Its advantages include low density, soft and tough, excellent strength, clean and dustproof.

Recommendations for transplantation applications in the field of petroleum engineering

Technical requirements: In the process of geothermal drilling and natural gas hydrate coring, it is necessary to heat the geothermal pipeline and the natural gas hydrate coring tube. In addition, during the drilling process, various measurement-while-drilling electronic devices need to be insulated and protected. However, due to the limited working space, conventional thermal insulation materials are not thin enough and provide thermal insulation for downhole equipment.

The product can be easily incorporated into composite structures, just like the plastic films and blanks currently in use. It is expected to be applied in (1) Geothermal pipeline insulation, (2) Natural gas hydrate core tube insulation, (3) Thermal insulation protection of MWD electronic equipment, (4) Deep-sea thermal insulation and dust-proof materials.

Without the need of changing the original structure of the workpiece, it can be directly used for steam pipeline insulation and geothermal pipeline insulation through traditional processes; it can also be made into special shapes to protect downhole control devices from the high temperature of the formation.

Research content: selection of polyimide aerogel; structure design of polyimide aerogel; optimization of polyimide aerogel.

3.3 4D printed Shape Memory Polymers (4D-p-SMP)

Shape memory polymer material is a kind of smart material that can change shapes under external environmental stimuli. The deformable structure of this smart material has shown great application potential in many fields such as biomedicine, aerospace, etc., and it can be used as plugging device, bracket device, and releasing devices 4D printed shape memory polymers can form relatively complex, personalized, and high-precision structures, or structures that are difficult to achieve with traditional preparation techniques. It may provide new ideas for intelligent control and controllable plugging in the oil and gas field [8–11].

Technical characteristics

SMP can interact with stimulation conditions, so the structure printed by 4D printing SMP can produce corresponding shape changes after being stimulated by external conditions (such as temperature, humidity, electricity, pH, etc.). 4D printing SMP can not only perform simple shape changes, but also realize various functions such as self-deformation, self-assembly, and self-repair by pre-setting its deformation scheme (including target shape, properties, functions, etc.).

The flexible solar cell system based on 4D printing SMP realizes on-orbit controllable application. In addition, the expandable hinge has been successfully applied in the aerospace field. The hinge can complete the three processes of locking, unfolding and tempering after unfolding. At present, the intelligent structure can achieve more than 200 repeated tests on the ground, with high reliability; it can achieve light weight, self-locking, and rapid tempering after unfolding; the material can actively deform without the risk of jamming; and there is almost no impact.

Recommendations for transplantation applications in the field of petroleum engineering

Technical requirements: Many downhole operations need to be controlled by hydraulic pipelines or cables. The downhole space limits the number of these pipelines, thereby limiting the times of operations. Furthermore, manual risk operations cannot be avoided. Therefore, it is necessary to develop wireless control equipment to deal with the sudden change of environmental conditions. The 4D printed shape memory polymer can be deformed according to specific deformation conditions, providing a new type of control method that can be applied to downhole switch closing and the adjustmentof valves.It could save the effort in considering the space for hydraulic pipelines, power supply cables, transmission signal pipelines, etc., which optimizes and simplifies downhole operations and frees up part of the downhole space. Under special operating conditions, various intelligent structural switches can be set in advance foroperations such as shutting in wells, thereby avoiding HSE risks.

The main problems at present are: (1) The downhole control is through hydraulic or electric drive. Therefore, the number of hydraulic pipelines, cables, etc. is limited by the narrow downhole space; (2) Many operations need to be tripped and completed, which increases the time of non-drilling operations and reduces efficiency; (3) There are many manual and risky operations.

In order to achieve transplantation applications, relevant research needs to be carried out: (1) Demand analysis ofdownhole 4D printed shape memory polymer; (2) Optimization of downhole environment 4D printed shape memory polymer; (3) Application methods and application environmental analysis ofdownhole 4D printed shape memory polymer.

4 SUMMARY

High-performance polymers is one of the fastest-growing fields in materials in recent years, but related research is still dominated by basic research. Therefore, it is necessary to establish a database of candidate technologies for high-performance polymers, and to select suitable technologies based on the technical requirements of petroleum engineering. High-performance polymers have good application prospects in downhole fluids, downhole tools, etc. The introduction of high-performance polymer components can change the existing system, including temporary plugging agents for fracturing, precision controlled release materials for drilling fluids, and cementing sealing materials in cement, etc. These downhole fluids have a variety of components and complex mechanisms of action. High-performance polymers are required to improve the material performance without damaging the performance of the existing system or changing the existing construction conditions. High-performance polymers also need to improve phase change reliability, mechanical strength, and chemical durability. It is necessary to carry out research on the physical mechanism of shape memory, specific structure design, downhole mechanics evolution and prediction, and system parameter optimization to improve shape memory performance and comprehensive performance to meet the needs of oil and gas wellbore operations.

Interdisciplinarity is the source of innovative ideas. All disciplines have seen interdisciplinarity growth and there is no sign of slowing down. Interdisciplinary integration is an important driving force for technological progress. The accelerated innovation of petroleum engineering technology ushered in opportunities. The cross-integration of technology not only promotes the penetration and expansion of technology in various fields, but also drives better and faster development. The integration of new technologies will surely bring new breakthroughs to the development of petroleum engineering technology. Continuously exploring the frontiers and hotspots of high-performance polymer materials that are highly compatible with on-site needs, and carrying out targeted feasibility studies will promote the rapid and high-quality development of petroleum engineering technology.

REFERENCES

[1] Zhu S, Zhang J, Shi J, Zhao R, Liu M and Shu Y 2012 *Adv. Mater. Research* **490–495** 3802
[2] Lu B, Ding S, He L and Pang W 2019 *Petroleum Drilling Techniques* **47(1)** 1
[3] Santos L, Taleghani A and Li G 2016 *SPE* 181391

[4] Jian Y, Lu W, Zhang J and Chen T 2018 *ACTA POLYMERICA SINICA* **11** 1385
[5] Pei X, Ji P, Zhang W and He L 2016 *Polym. Bull.* **9** 262
[6] Husing N and Schubert U 1998 *Angew Chem Int Ed* **37(1/2)** 22
[7] Meador A, Malow J and Silva R 2012 *ACS Appl. Mater. Interfaces* **4** 536
[8] Li C, Zhang F, Wang Y, Zheng W, Liu Y and Leng J 2018 *SCIENTIA SINICA Technologica* **49** 13
[9] Zhang H, LanX and Leng J 2020 *Mater. Sci. and Tech.* **28** 157
[10] Wang L, Leng J and Du S 2020 *J. of Harbin Institute of Tech.* **52** 227
[11] Wang M, Guang X and Kong L 2018 *Petroleum Drilling Techniques*, **46(5)** 14

Chapter 3: Nanomaterials and Biomaterials

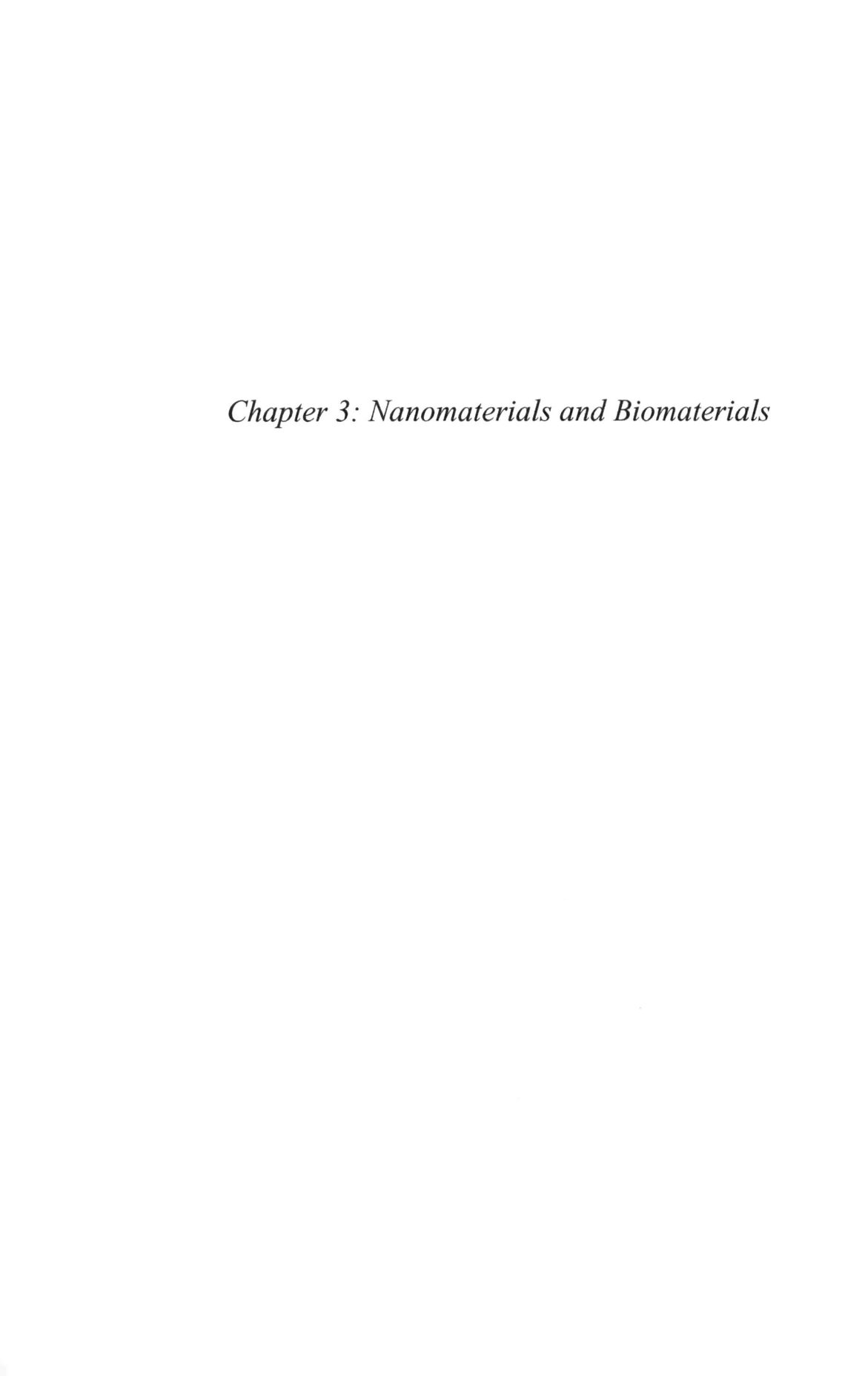

Electrospun the fifth generation solid dispersion of ibuprofen based on a hydrophilic polymeric excipient

T.B. Ning, S.R. Guo, W.L. He, W.H. Zhou, M.L. Wang & D.G. Yu
School of Materials Science & Engineering, University of Shanghai for Science & Technology, Shanghai, China

ABSTRACT: The dissolution of poorly water-soluble drug is always an important challenge in pharmaceutics during the past half century. Solid dispersion (SD) is viewed as a useful medicated material for resolving this issue. Today, SD is going into the fifth generation, in which complicated nanostructures are explored for promoting the fast dissolution and immediate release of poorly water-soluble from the loaded matrices. In this investigation, a modified coaxial electrospinning was exploited to prepare the fifth generation SD for the fast dissolution of ibuprofen (IBU), which was in the form of core-shell nanofibers. Different types of polyvinylpyrrolidone (PVP) were explored as the shell and core matrices. For easy electrospinning, the core fluid was pure solution of PVP K90 in ethanol (8% wt.). But for fast dissolution, the shell fluid was a blending solution of PVP K10 and drug IBU (5% wt. and 10% wt., respectively). The core-shell nanofibers had an average diameter of 780 ± 120 nm with a uniform structure and smooth surface. X-ray diffraction results demonstrated that IBU was distributed homogeneously and amorphously in the PVP K10 matrix due to favorable second-order interactions. In vitro dissolution tests demonstrated that the fifth generation SDs could rapidly release IBU within one minute, over 20 times faster than the raw IBU particles. The present investigation provides an example of the systematic design, preparation, characterization, and application of the advanced fifth generation SDs consisting of different types hydrophilic polymers and poorly water-soluble drug within the complicated nanostructures.

1 INTRODUCTION

The issue about the poorly water-soluble drugs' dissolution is always an important concern in the field of pharmaceutics during the past half century [1,2]. In Chinese Pharmacopoeia, the poorly water-soluble drugs refer to those drugs that are "slightly soluble, very slightly soluble, almost insoluble or insoluble". Traditional pharmaceutical industry, dosage forms including dispersible tablets and suspensions are frequently utilized for improving the therapeutic effects of them.

In pharmaceutical researches, both chemical methods and physical strategies are explored for providing useful solutions to elevate their solubilities, which including [3–6]: 1) chemical methods of making the insoluble active ingredients into salts, such as barbiturates, sulfonamides, aminosalicylic acid and other acidic drugs, alkali can be used to form salts with them to increase their solubility in water; 2) physical strategies of adding glycerol, ethanol, propylene glycol and other water-soluble organic solvents in water for increasing the solubility of some insoluble drugs, such as chloramphenicol; 3) physical strategies of solubility enhancers. Some drugs have strong lipophilicity and have strong affinity with the lipophilic group of a surfactant. Thus, surfactant can be utilized as a solubilizer for elevating the poorly water-soluble drugs' solubilities.

Among all types of solubility-enhanced methods, solid dispersion (SD) is a common one both in industry and also in scientific investigations. SD is thought to be a useful medicated material for increasing the drugs' soluble rates. During the past several decades, SD is developing from the first generation to now the fifth generation,which is characterized by the applications of complicated structures at nano scale for realizing the fast dissolution and immediate release of poorly water-soluble from the loaded matrices, often comprising pharmaceutical polymers [2]. Thus, the techniques that can be effectively exploited to convert the drug and polymer into a typical SD are highly desired.

Electrospinning [7–10], just as also its peer technique electrospraying [11, 12], is a hot method for preparing new types of SDs. Electrospinning has unique capability of creating both monolithic nanofibers and also complicated nanostructures through a single-step straight forward process [13, 14]. Electrospun nanofibers, regardless of simple or complex inner structures, have been broadly reported to be fine drug carries. Particularly, electrospinning is fast developing to all kinds of multiple-fluid electrospinning such as side-by-side, co-axial, tri-fluid coaxial, and other complicated ones [14–16]. These new electrospinning processes can provide a powerful tool for generating new sorts of fifth generation SDs.

In this investigation, a coaxial process was developed to prepare a new kind of fifth generation SD for the fast dissolution of ibuprofen (IBU). The prepared SDs were in the form of core-shell nanofibers. Different types of hydrophilic polyvinylpyrrolidone (PVP) were explored as the shell and core matrices. The prepared fifth generation SDs were subjected to a series of characterizations.

2 EXPERIMENTS

2.1 *Materials*

The hydrophilic polymeric matrices PVP K90 (Mw = 360,000, white powders) and PVP K10 (Mw = 10,000, white powders) were bought from the Sinopharm of Shanghai, China. The drug IBU was achieved from Shanghai Hua-shi large pharmacy, Ethanol (AR grade) was obtained from Shanghai No.1 Shiji factory. Water was doubly distilled just before usage.

2.2 *Preparation*

2.2.1 *Coaxial electrospinning system*
A home-made electrospinning system was utilized to conduct the experiments, which contained a power supply (ZGF 60 kV/2 mA, Wuhan Hua-Tian Corp., Wuhan, China) for providing the applied high voltage, two pumps (KDS100, Cole-Parmer, IL, USA) for driving the sheath and core working fluids, a home-made fiber collector, and a coaxial spinning head.

2.2.2 *Working fluids*
For easy electrospinning, the core fluid was pure solution of PVP K90 in ethanol (8% wt/v). But for fast dissolution, the shell fluid was a blending solution of PVP K10 and drug IBU (5% wt/v and 10% wt/v, respectively).

2.2.3 *Experimental conditions and observations*
After some pre-experiments, the optimized conditions were selected as follows: voltage = 12 kV; the distance of nozzle to collector = 20 cm; and the sheath and core fluid flow rates were 0.5 and 1.2 mL/h, respectively.

2.3 *Characterization*

2.3.1 *Morphology and inner structure*
The collected nanofibers were firstly gold sputter-coated for 60 seconds under a N2 atmosphere. Later, the gold-coated fibers were evaluated through a scanning electron microscopy (FESEM, Hitachi, Tokyo, Japan). For surface morphology, the fibers mats were directly sputtered with gold and observed under an applied electron voltage of 10 kV. For sampling the cross-sections, nanofiber strips were cut from the fiber mats and were immersed into the N2 (liquid), and then they were manually broken.

2.3.2 *Physical status*
The physical state of PVP, IBU and the nanofiber mats were assessed using the X-ray diffraction (XRD) analysis, which was conducted from a 2θ range from 5° to 60° using a X-ray diffractometer (Bruker-AXS with Cu K? radiation, Karlsruhu, Germany).

2.4 Materials functional performances

2.4.1 Quantitative measurement of IBU

A UV-2102PC UV-visible light spectroscopy (Shanghai Unico Instrument Co. Ltd., China) was utilized to detect the IBU concentration at a maximum absorbance at $\lambda max = 264$ nm. A series of standard solutions were prepared and were subjected to the measurement of absorbance. The absorbance values were regressed with the corresponding IBU concentration to achieve the standard equation for determining the IBU concentrations from unknown samples.

2.4.2 In vitro dissolution tests

The paddle method in the Chinese Pharmacopoeia (2015 Ed.) was exploited to carry out the in vitro dissolution experiments. About 50.0 mg nanofiber mats (or 5.0 mg) were weighted and were placed into the vessels (parallel six samples) loaded with 600 mL phosphate buffer saline (PBS, pH7.0, 0.1M). The PBS bulk solutions were kept at $37\pm0.5°C$ and the rotation rate of paddle was maintained at 50 rpm. More details can refer to reference [16].

3 RESULTS AND DISCUSSION

3.1 The working processes

A diagram showing the electrospinning system is given in Figure 1. Obviously, the modified coaxial system is similar with the traditional peer system and also the blending electrospinning system. In our home-made electrospinning system, a home-made concentric spinneret was exploited for leading the two fluids into the electric field in a coaxial manner. An aluminium-wrapped cardboard was explored as the fiber collector. Two pumps were utilized to send the sheath and core working fluids quantitatively. The high power supply can provide the high voltage.

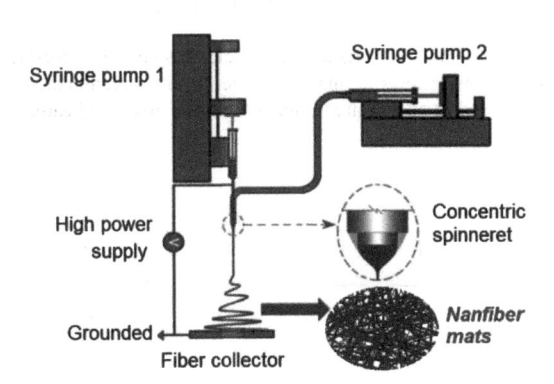

Figure 1. A diagram about the electrospinning system.

Under the predetermined conditions, the modified coaxial processes can be clearly recorded, which are shown in figure 2. Figure 2a. shows the connection of the home-made concentric spinneret with the power supply and the two syringe pumps. Under a magnifications of 4 times, the working processes can be discerned. A typical process is exhibited in Figure 2b, which is composed of a successive process from a Taylor cone, to the straight fluid jet, and to the the bending and whipping region. Under an even bigger magnification of 12 times, a typical compound Taylor cone was able to be captured (Figure 2c). By estimation, the core-shell Taylor cone is about 2.0 mm in height. The whole processes can be carried out smoothly and robustly.

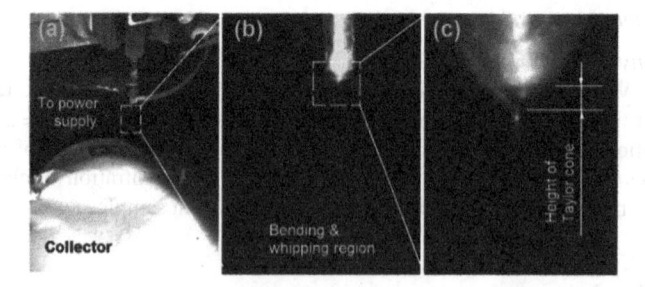

Figure 2. Observations of the modified coaxial processes: (a) The connection of spinneret; (b) The electrospinning process; (c)The Taylor cone.

3.2 *The as-prepared nanofibers*

Figure 3 shows the SEM morphology of nanofiber mats and also the inner structures of the electrospun nanofibers. As indicated by Figure 3.a, the nanofibers have a straight linear morphology. The up-right inset of Figure 3a shows an enlarged image of these randomly collected nanofibers. By estimation, these nanofibers had a diameter of about 780 ± 120 nm. Figure 3b provides a cross-section morphology of the electrospun nanofibers, its up-right inset shows and enlarged image. The arrow indicates a thin shell layer on the cross-section. electrospun nanofibers.

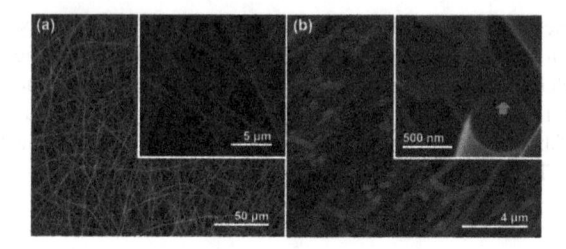

Figure 3. The SEM morphology and inner structures of the electrospun nanofibers: (a) surface morphology of the nanofiber mats, the up-right inset shows an enlarged image; (b) cross-section morphology of the electrospun nanofibers, the up-right inset shows and enlarged image with the arrow indicating a thin shell layer on the cross-section.

3.3 *Physical state*

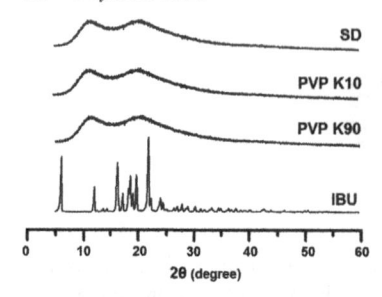

Figure 4. The XRD patterns of the raw IBU and PVP, and the electrospun SDs.

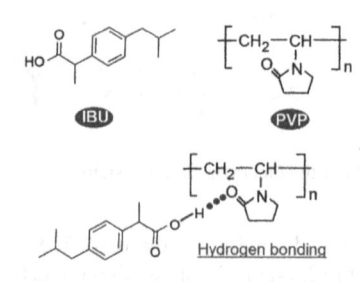

Figure 5. The molecular structures of IBU and PVP, and the hydrogen bonding between them.

XRD results in Figure 4 obviously demonstrates that IBU was distributed homogeneously and amorphously in the PVP K10 matrix. The raw IBU powders are crystalline materials, as suggested by the Brag sharp peaks from 5 to 30 degree of 2 theta. In sharp contrast, both PVP K10 and K90 have only two halos without any sharp peaks, which indicates that they are amorphous polymers. After electrospinning, the IBU molecules are successfully distributed within the PVP K10 shells uniformly. As indicated in Figure 5, each IBU molecule has a -OH group and one PVP molecule has numerous -C=O groups. This means that the IBU molecule can be hung on the PVP molecule through the secondary physical interaction, i.e. hydrogen bonding. Certainly the long carbon chain of PVP can interact with the benzene rings of IBU molecules through hydrophobic interactions. These secondary interactions do favor to the homogeneous distribution of IBU within the PVP K10 matrix and also the stability of the shell SDs.

3.4 In vitro dissolution

A broad scanning from 210 nm to 800 nm shows that PVP molecules has no ultra-violet and visible light absorbance. Thus, the drug dissolved in the bulk solutions can be determined through UV-vis spectroscopy. The UV-scanning curves of the series of standard IBU solutions are included in Figure 6. Clearly, IBU has a maximum absorbance at $\lambda max = 264$ nm.

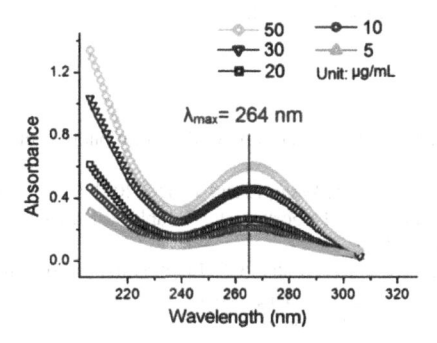

Figure 6. UV-scanning curves of the standard IBU solutions.

Based on the data about the absorbance values (A) of IBU at $\lambda max = 264$ nm, which indicates a highest accuracy and sensitivity, and the corresponding concentrations (C), a standard equation for determining IBU concentration in its aqueous solution can be achieved. The linear equation $A = 0.074 + 0.010C$ has a correlation efficiency of R=0.9998, whose regressed line is shown in figure 7.

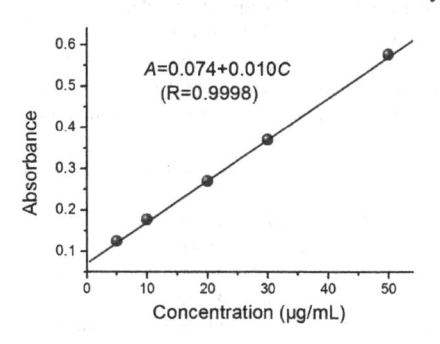

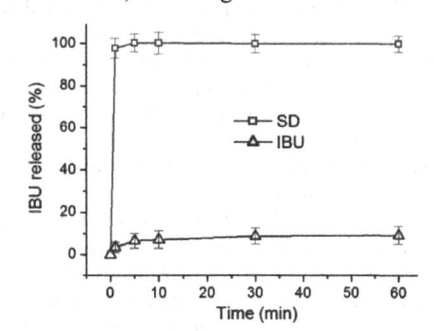

Figure 7. The standard equation for determining IBU concentration.

Figure 8. In vitro dissolution results.

In vitro dissolution results of the electrospun SDs and the raw IBU powders are concluded in figure 8. The electrospun core-shell SDs are able to release all the loaded IBU all at once, 97.7±4.8% at the first minute. In sharp contrast, the IBU powders dissolved only 3.4±1.5% at the first minute, and 13.4±3.7% in the first hour. Estimated by the first minute dissolution rate, the electrospun SDs have improved the dissolution rates of IBU 97.7%/3.4% = 28.7 times. The results demonstrated that the core-shell nanofiber-based SDs are able to greatly increase the drug dissolution rate. The reasons can be simply concluded as follows: 1) the small diameter of nanofiber; 2) the even smaller thickness of the shell layer of IBU-PVP K10; 3) the 3D web structure of the nanofiber mats; 4) the amorphous state of IBU presented in the shell layer of PVP K10. These favorable factors synergistically act to ensure a very fast dissolution of the insoluble IBU molecules.

4 CONCLUSIONS

The present investigation shows a new strategy for systematic designing and preparation of the advanced fifth generation SDs. The SDs were composed of electrospun core-shell nanofibers with the shell section being loaded with IBU. The prepared core-shell fibers were 780 ± 120 nm in diameter. XRD results demonstrated that IBU was distributed homogeneously and amorphously in the PVP K10 matrix. The dissolution tests showed that the fifth generation SDs could rapidly release IBU within one minute, over 20 times faster than the raw IBU particles. The present protocols can be similarly exploited for all types of insoluble drugs.

REFERENCES

[1] Yu DG 2021 Prefacebettering drug delivery knowledge from pharmaceutical techniques and excipients *Curr. Drug Deliv.* **18** 2–3
[2] Aidana Y, Wang YB, Li J, Chang SY, Wang K and Yu DG 2021 Fast dissolution electrospun medicated nanofifibers for effective delivery of poorly water soluble drugs *Curr. Drug Deliv.* **18** DOI:10.2174/1567201818666210215110359
[3] Wang Y Tian L Zhu T, Mei J Chen Z Yu DG 2021 Electrospun aspirin/Eudragit/lipid hybrid nanofibers for colon-targeted delivery using an energy-saving process. *Chem. Res. Chinese Universities* **37** https://doi.org/10.1007/s40242–021–1006–9
[4] Wang M, Li D, Li J, Li S, Chen Z, Yu DG, Liu Z and Guo JZ 2020 Electrospun Janus zein-PVP nanofibers provide a two-stage controlled release of poorly water-soluble drugs *Mater. Des.* **196** 109075.
[5] Al-Jbour ND, Beg MD, Gimbun J and Alam AMM 2019 An overview of chitosan nanofibers and their applications in the drug delivery process *Curr. Drug Deliv.* **16** 272
[6] Wang P, Wang M, Wan X, Zhou H, Zhang H and Yu DG 2020 Dual-stage release of ketoprofen from electrosprayed core-shell hybrid polyvinyl pyrrolidone/ethyl cellolose nanoparticles *Mater. Highlight.* **1–2** 14–21
[7] Kang S, Hou S, Chen X, Yu DG, Wang L, Li X and Williams GR 2020 Energy-saving electrospinning with a concentric Teflon-core rod spinneret to create medicated nanofibers *Polymers.* **12** 2421
[8] Ding Y, Dou C, Chang S, Xie Z, Yu DG, Liu Y and Shao J 2020 Core-shell Eudragit S100 nanofibers prepared via triaxial electrospinning to provide a colon-targeted extended drug release *Polymers.* **12** 2034
[9] Zhao K, Kang SX, Yang YY and Yu DG 2021 Electrospun functional nanofiber membrane for antibiotic removal in water: Review *Polymers* **13** 226
[10] Mofidfar M and Prausnitz MR 2019 Electrospun transdermal patch for contraceptive hormone delivery *Curr. Drug Deliv.* **16** 577–583
[11] Hou J, Yang Y, Yu DG Chen Z, Wang K, Liu Y and Williams GR 2021 Multifunctional fabrics finished using electrosprayed hybrid Janus particles containing nanocatalysts *Chem. Eng. J* **411** 128474
[12] Li D, Wang M, Song WL, Yu DG and Annie-Bligh SW 2021 Electrospun Janus beads-on-astring structures for different types of controlled release profiles of double drugs. *Biomolecules* **11** 635
[13] Wang M, Yu DG, Li X and Williams GR 2020 The development and bio-applications of multifluid electrospinning *Mater. Highlights* **1–2** 1–13
[14] Yang X, Chen S, Liu X, Yu M and Liu X 2019 Drug delivery based on nanotechnology for target bone disease *Curr. Drug Deliv.* **16** 782
[15] Wang M, Hou J, Yu DG Li S, Zhu J and Chen Z 2020 Electrospun tri-layer nanodepots for sustained release of acyclovir *J. Alloy. Compd.* **846** 156471
[16] Zheng X, Kang S, Wang K, Yang YY, Yu DG, Wan F, Williams GR and Annie Bligh S-W 2021 Combination of structure-performance and shape-performance relationships for better biphasic release in electrospun Janus fibers *Int. J. Pharm.* **596** 120203

Tunable band gap of graphene/graphene/arsenene van der waals heterostructures

D.Z. Zhou, J.C. Dong, X. Zhang, D.D. Han & Y. Wang
Institute of Microelectronics, Peking University, Beijing, China

ABSTRACT: Graphene is a promising two-dimensional (2D) material for electronic-device application, however, bandgap of graphene is too small to be used as semiconductor layers. In the present work, We perform first-principles investigation on the electronic properties of graphene/graphene/arsenene (G/G/A) heterostructures. We optimize band gap of the heterostructures by adjusting interlayer spacing d1 between the bilayer graphene and the arsenene and interlayer spacing d2 in the bilayer graphene. Remarkably, we found that bandgap of the G/G/A heterostructures achieves 127 meV when d1/d2 is 0.7, which is feasible for semiconductor applications. Our findings open up the possibilities for practical applications of graphene in nanoelectronics devices.

1 INTRODUCTION

Graphene has attracted intensive attentions because of the unique electronic and optical properties [1,2]. Numerous researchers are making effort to realize field-effect transistor (FET) based on graphene [3]. However, graphene has zero bandgap, making it hard for bandgap to serve as the semiconductor materials [4].

In recent years, arsenene has gained increasing attentions because it can be easily removed from gray-arsenic (called α-As). Remarkably, arsenene is wide-bandgap semiconductor and can be converted into direct band-gap semiconductor under biaxial strain [5]. Moreover, studies also show that arsenene tunneling field effect transistor (TFET) have high on-state current (I_{on}) of 101 μAcm^{-1} as well as a large on-off current ratio (I_{on}/I_{off}) over 10^{10} [4]. Thus, arsenene are recognized as a promising 2D material for the FET applications.

Therefore, efforts to make graphene and arsenene into multilayer stack via van der Waals forces have attracted tremendous attention. CongXin Xia et al. reported that the Schottky barrier of graphene/arsenene heterostructures can be tuned by interlayer distance [6]. And Zhen Luo et al. investigated the effect of interlayer orientation on the electronic properties of arsenene/graphene heterostructures [7].

In this work, we embed a monolayer graphene into the graphene/arsenene (G/A) bilayers to establish the graphene/graphene/arsenene (G/G/A) heterostructures. The effects of interlayer spacing on electronic properties of the G/G/A trilayer were explored.

2 COMPUTATIONAL DETAIL

We perform studies on G/G/A heterostructures using the first-principles calculations based on density functional theory (DFT). The exchange-correlation functional is based on generalized gradient approximation (GGA) expressed by Perdew-Burke-Ernzerhofer (PBE) functional. The cut-off energy is set to be 500 eV, and a $9 \times 9 \times 1$ k-mesh in the first Brillouin zone (FBZ) is employed.

The 2 × 2 arsenene supercell almost perfectly matches the 3 × 3 graphene supercell with a lattice mismatch of 1.89%. The optimized interlayer distance between arsenene and the top graphene layer (G_t) is $d_1 = 3.39$, and the distance between the top graphene and the bottom graphene layer (G_b) is $d_2 = 3.25$, shown in Figure 1(a–b). Therefore, the distance ratio of structure is d_1/d_2=1.04. The result of calculating binding energy is −3.69 eV. The evolution of the binding energy as a function of d_1/d_2 is given in Figure 2.

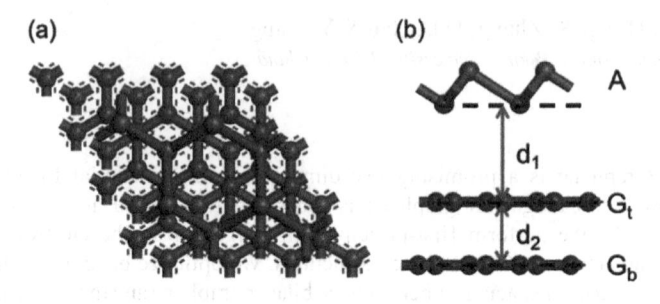

Figure 1. (a) Front view and (b) side view of G/G/A heterostructure (the middle graphene layer is G_t, and the bottom graphene layer is G_b).

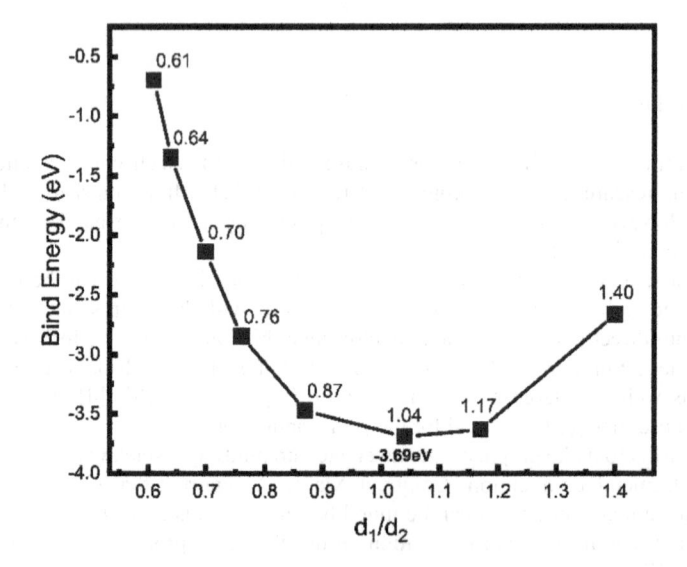

Figure 2. Binding energy of G/G/A heterostructures as a function of d_1/d_2. The lowest binding energy is obtained when d_1/d_2 equals to 1.04.

3 RESULTS AND DISCUSSION

Then, we made a investigation on band structure among monolayer arsenene, G/A bilayer, and G/G/A trilayer. For the G/G/A trilayer, d_1/d_2 of 1.04 was chosen as the reference. As shown in Figure 3(a), the arsenene exhibits an indirect bandgap of 1.58 eV, which is in good agreement with the previous study [8]. As shown in Figure 3(b), the G/A hybrid bilayer exhibits Dirac cone at G-point. As shown in Figure 3(c), the G/G/A trilayer exhibits a parabolic band structures at G-point with a bandgap of 44 meV.

Considering the vdW force is closely related to the interlayer spacing, we calculate the band structures of the G/G/A heterostructures with different d_1/d_2 ratio. As shown in Figure 4, the bandgap increases first and then decreases with d_1/d_2. Interestingly, the bandgap reaches its maximum of 127 meV when d_1/d_2 is 0.7. The results demonstrate that a relatively large bandgap of the G/G/A trilayer can be obtained by modulating d_1/d_2.

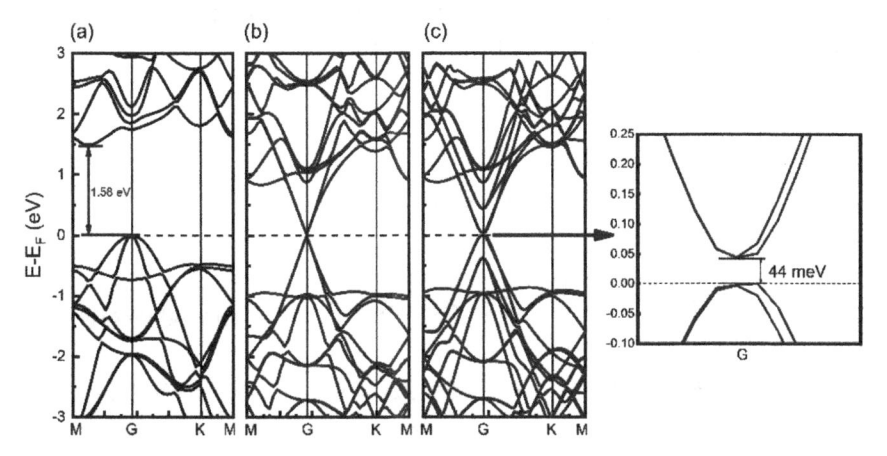

Figure 3. Band structures of (a) arsenene, (b) G/A bilayer, and (c) G/G/A trilayer. Arsenene shows an indirect bandgap of 1.58 eV. G/A bilayer shows a zero bandgap. G/G/A trilayer shows a bandgap of 44 meV.

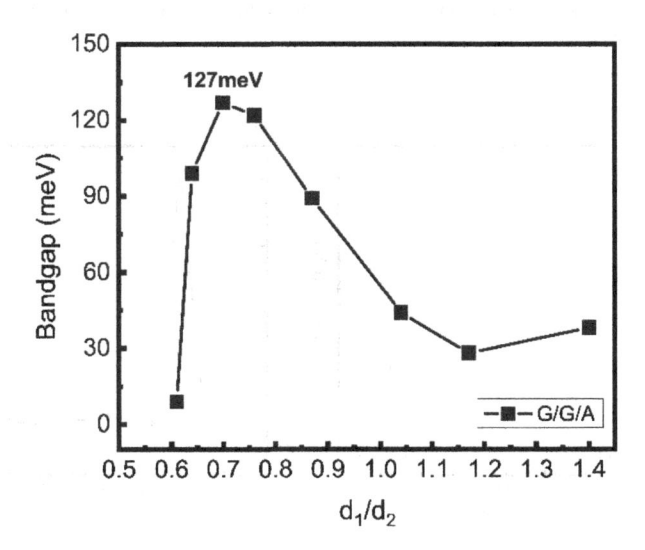

Figure 4. Bandgap of G/G/A trilayer with different d_1/d_2. The bandgap reaches its maximum of 127 meV when d_1/d_2 is 0.7.

In order to describe interlayer coupling, we calculated out charge density difference of the G/G/A trilayer with d_1/d_2 of 0.7, 1.04, and 1.40 as examples. As shown in Figure 5(a), when d_1/d_2 is 0.7, the charge accumulation regions (red regions) are near to arsenene, which indicates the intense charge transfer from G_t to arsenene. Thereby, there is a potential difference between G_t and G_b, which gives rise to a relatively large bandgap in G/G/A trilayer. When d_1/d_2 increases to 1.04, charge transfer and charge accumulation weakens (Figure 5(b)). When d_1/d_2 subsequently increases to 1.40, charge accumulation regions disappear, indicating that interlayer coupling between arsenene and G_t is very weak (Figure 5(c)).

We plotted electrostatic potential of the G/G/A trilayer to gain a further insight into the interlayer coupling, as shown in Figure 6(a). To quantify the results, we extracted the potential difference between G_t and G_b from electrostatic potential (Figure 6(b)). The potential difference defined as $V(G_b) - V(G_t)$. As shown in Figure 6(b), $V(G_b) - V(G_t)$ achieves its maximum (1.65 eV) as d_1/d_2 increases from 0.61 to 0.7. Then, $V(G_b) - V(G_t)$ decreases and maintains a value <0.2 eV as d_1/d_2 increases from 0.7 to 1.4. According to tight-binding model [9,10], bandgap of the G/G/A trilayer linearly depends on the interlayer electric field, hence we can comprehend the bandgap trend of G/G/A trilayer.

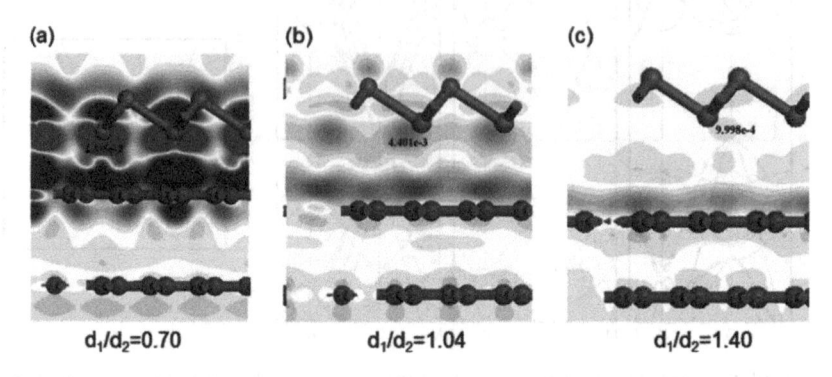

Figure 5. Charge density difference of G/G/A trilayer with d_1/d_2 equals to (a) 0.70, (b) 1.04 and (c) 1.40. Value and location of the largest charge density difference are shown in each panel. The red and blue regions indicate the increase and decrease of electrons, respectively. When d_1/d_2 is 0.7, intense charge transfer occurs between arsenene and G_t.

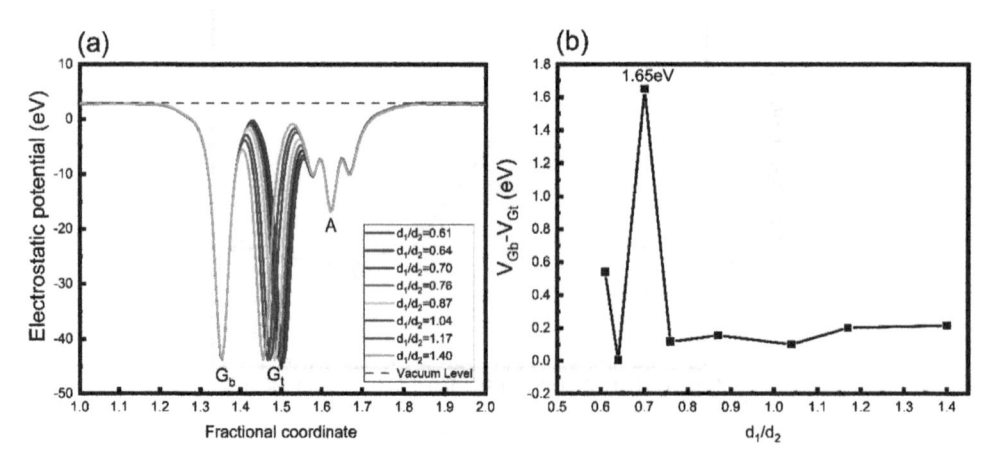

Figure 6. (a) Electrostatic potentials of the G/G/A trilayer with different d_1/d_2. (b) Potential difference between G_t and G_b with different d_1/d_2. When d_1/d_2 is 0.7, potential difference achieves its maximum (1.65 eV).

4 CONCLUSION

Herein, we research electronic properties of the G/G/A hybrid trilayer using first-principles calculations. Compared with the zero-bandgap G/A hybrid bilayer, the G/G/A trilayer ($d_1/d_2 = 1.04$) opens up a bandgap of 44 meV. Correlation between bandgap and d_1/d_2 is investigated in depth. We found that d_1/d_2 influences charge transfer and built-in electric field of the G/G/A trilayer, ultimately determining the bandgap. Significantly, the bandgap reaches its maximum value of 127 meV when d_1/d_2 is 0.70. Our research results accelerate the applications of graphene in electronic devices and widely broaden the categories of 2D semiconductor materials.

REFERENCES

[1] A. H. Castro Neto et al., Reviews of Modern Physics, pp. 109, 2009.
[2] Shen.W et al., Nanoscale, pp. 8329, 2018.
[3] J. Hicks et al., Nature Physics, pp. 49, 2010.
[4] H. Li et al., Semiconductor Science and Technology, pp. 085006, 2019.
[5] S. L. Zhang et al., Angewandte Chemie-International Edition, pp. 3112, 2015.
[6] C. X. Xia et al., Applied Physics Letters, pp.193107, 2015.
[7] Z. Luo et al., Japanese Journal of Applied Physics, 58(SB): SBBH01, 2019.
[8] W. Li et al., Physica E: Low-dimensional Systems and Nanostructures, pp. 6, 2017.
[9] E. McCann et al., Rep. Prog. Phys., pp.056503, 2013.
[10] A.H. Castro Neto et al., Rev. Mod. Phys., pp.109, 2009.

Improvement effect of coconut chaff on coastal salinity soil in the Hainan province

X. Deng & C.Y. Wu
Environment and Plant Protection Institute, Chinese Academy of Tropical Agricultural Sciences, Haikou, China
National Agricultural Experimental Station for Agricultural Environment, Danzhou, China

G.S. Yang
National Agricultural Experimental Station for Agricultural Environment, Danzhou, China

Y. Li
Environment and Plant Protection Institute, Chinese Academy of Tropical Agricultural Sciences, Haikou, China

Q.F. Li
Environment and Plant Protection Institute, Chinese Academy of Tropical Agricultural Sciences, Haikou, China
National Agricultural Experimental Station for Agricultural Environment, Danzhou, China
Hainan Key Laboratory of Tropical Eco-Circular Agriculture, Haikou, China

ABSTRACT: The coconut chaff and moderate salinity soil were used as soil improvement material and the improved object, respectively. The pot experiment of planting water spinach is used to study the improvement effect of the coconut chaff on moderate salinity soil under the conditions without chemical fertilizer utilization. Five treatments were set according to the volume ratios of the coconut chaff to moderate salinity soil. And the different volume ratios were 0:20, 1:20, 2:20, 3:20 and 4:20. The results showed that adding the coconut chaff could obviously improve the pH value of moderate salinity soil, and the soil pH value was changed from acidity to neutral. The contents of soil organic matter, total nitrogen, available phosphate and available potassium were increased by 11.1%~28.1%, 31.3%~35.6%, 13.0%~23.5% and 9.9%~44.8%, respectively. The richness index and diversity index of the bacterial community were significantly increased. However, the soil salinity was significantly decreased by 7.7%~24.8% than that of the control treatment without chemical fertilizer utilization. Especially, the treatments with the addition ratio of 2:20~4:20 could significantly improve the soil pH value, increase the soil nutrient contents, decrease soil salinity and change the bacterial community structure and enhance the cycling capacity of the carbon and nitrogen in coastal salinity soil. This demonstrates that the coconut chaff can improve the physicochemical properties and microbial diversity of the coastal salinity soil and provide a favorable soil environment for crop growth

1 INTRODUCTION

The salinity soil covers a large area and is widely distributed in China. However, the excess salt content in soil is a key factor for restricting land reclamation in coastal areas [1]. Soil salinization is a prominent problem that restricts agricultural development and improvement of the ecological environment, and soil improvement is an important way to efficiently utilize salinity soil resources [2]. At present, the improvement measures mainly include engineering [3,4], chemistry [5], physics [6], and biological measures [7–9]. However, the biological measures are generally believed to be the most effective ways to improve salinity soil among all kinds of technical measures. The coconut chaff with a variety of uses is mainly composed of cellulose, hemicellulose and other natural polymer substances, belonging to the pure natural biomass medium, which is a kind of renewable biomass resource containing potassium, calcium, nitrogen, phosphorus, magnesium, and other nutrient elements. However, research studies on the improvement of coastal saline soil by applying the coconut chaff have not been reported. The objective of this study is to screen the biomaterials that can improve coastal salinity soil and to determine the optimal application ratio. This study results will provide technical support and theoretical basis for the improvement of coastal salinity soil.

2 MATERIALS AND METHODS

The tested soil was collected from Nanxi Village (N 19°59.014′, E 110°37.453′), Wenchang City, Hainan Province. And the soil belongs to moderate salinity soil. Soil samples were taken from 0~20 cm tillage layer by applying the five-point sampling method. Soil samples were stored in a plastic bags and taken to the laboratory on the same day, the plant roots and stones were removed, and samples were passed through a 2 mm sieve after air drying. The tested coconut chaff was from Danzhou agricultural production base of Hainan University. The tested plant was water spinach. The pot experiment was conducted in the greenhouse of Haikou experimental base of Chinese Academy of Tropical Agricultural Sciences from May to July in 2019. Five treatments were set according to the volume ratios of the coconut chaff to moderate salinity soil. And the different volume ratios were 0:20, 1:20, 2:20, 3:20, and 4:20. Each treatment was repeated 3 times. The cultivation conditions and management methods of each pot were the same. All the coconut chaff and air drying soil were directly mixed for planting water spinach and watered one time a day. Ten plants with the same growth status were remained to observe in each pot after seeds sprouting. The potted soil samples were collected after the harvest of water spinach and passed through a 1 mm sieve after air drying to analyze the pH value, the contents of organic matter (OM), total nitrogen (TN), available phosphate (AP), available potassium (AK), and salinity in soil. All soil physicochemical indexes were determined by reference to the agricultural industry standards. The soil microbial diversity was sequenced by high-throughput sequencing, and the library construction and sequence determination were entrusted to Beijing Biomarker Biotechnology Co., Ltd. Estimation of means and standard deviations of samples using descriptive statistics was conducted using Excel 2010. Significant differences were analyzed by Duncan's method.

3 RESULTS

The soil salinity was obviously decreased after adding the coconut chaff under the conditions without chemical fertilizer utilization. The soil salinity was decreased by 7.7% ~ 24.8% after adding the coconut chaff (Figure 1). And it was gradually decreased with the increasing adding ratio of the coconut chaff in the treatments. The soil salinity of three treatments with volume ratios of 2:20, 3:20, and 4:20 were significantly lower than that of the control treatment ($P < 0.05$). However, the differences between the two treatments with adding the coconut chaff ratios of 2:20 and 3:20 were not significant ($p > 0.05$).

Note: Different lower case letters in each column indicate significant differences at $P < 0.05$. Error bars represent the standard deviations of the means (n = 3).

The pH value of coastal salinity soil was obviously improved after adding the coconut chaff (Figure 2). The pH value was 6.36 in coastal salinity soil without adding the coconut chaff, belonging to acid soil. However, the pH values were 7.13 ~ 7.41 in the treatments with adding the coconut chaff ratios of 1:20 ~ 4:20, belonging to neutral soil. The pH value of coastal salinity soil was increased by 0.77~1.05 units after adding the coconut chaff. And the soil pH value was decreased with the increasing adding ratio of the coconut chaff in the treatments with volume ratios of 1 : 20 ~ 4:20.

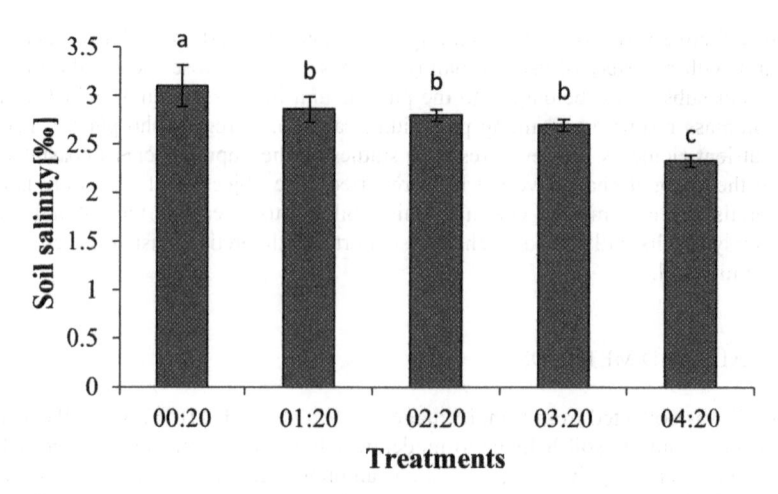

Figure 1. Effect of the coconut chaff on the salinity in coastal soil.

The pH value of four treatments with volume ratios of 1 : 20 ~ 4:20 was significantly higher than that of the control treatment ($P < 0.05$). However, the differences among the three treatments with adding the coconut chaff ratios of 1:20, 2:20, and 3:20 were not significant ($p > 0.05$).

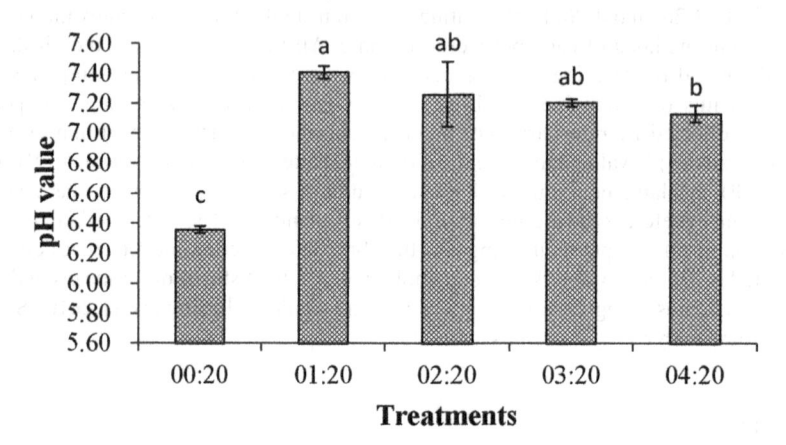

Figure 2. Effect of coconut chaff on the pH value in coastal soil.

The SOM was obviously increased after adding the coconut chaff in coastal salinity soil. The SOM contents were increased by 11.1% ~ 28.1% in the adding ratio range of 1:20 to 4:20 of the coconut chaff (Figure 3a). And it increased with the increasing adding ratios of the coconut chaff in the treatments with volume ratios of 1:20 ~ 3:20, while it slightly decreased in the treatment with adding the coconut chaff ratio of 4:20. The promoting effect of different treatments on the SOM was 3:20 > 4:20 > 2:20 > 1:20. The SOM contents of three treatments with volume ratios of 2:20 ~ 4:20 were significantly higher than that of the control treatment ($P < 0.05$). However, the differences among the three treatments with adding the coconut chaff ratios of 2:20, 3:20, and 4:20 were not significant ($p > 0.05$).

The contents of TN, AP, and AK were respectively increased by 31.3%~35.6%, 13.0% ~ 23.5%, and 9.9% ~ 44.8% after adding the coconut chaff in coastal salinity soil (Figure 3(b) ~ Figure 3(d)). The promoting effect of different treatments on the TN content in coastal salinity soil was 2:20 > 1:20 > 3:20 > 4:20. And the promoting effect of different treatments on the AP content in coastal

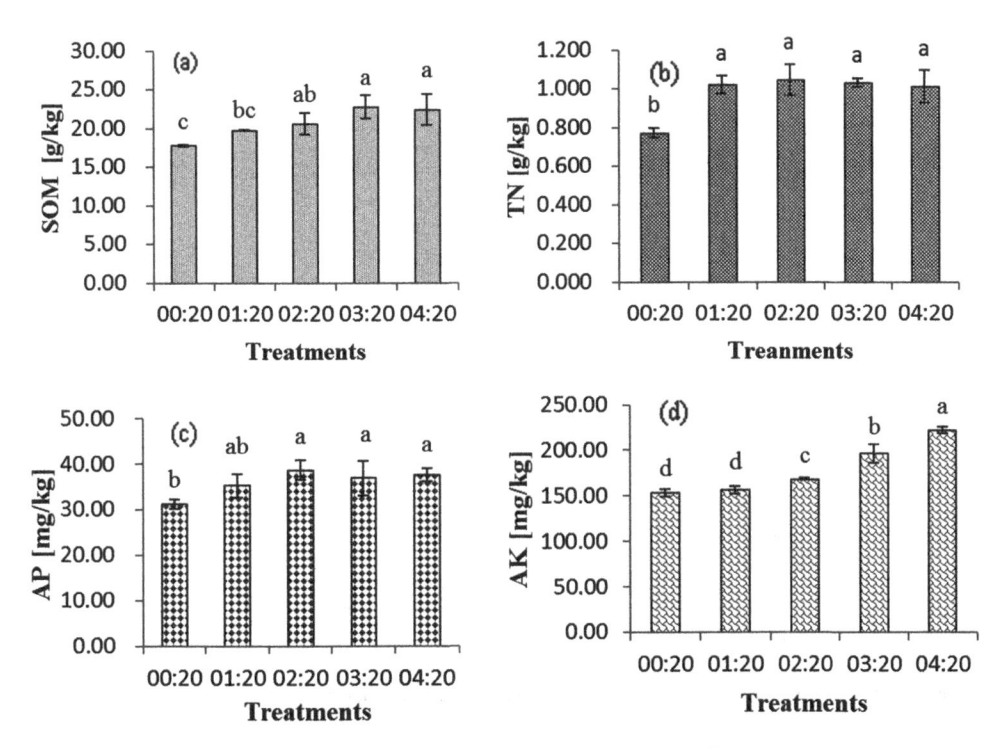

Figure 3. Effect of coconut chaff on SOM and content nutrient in coastal soil.

Figure 4. Clustering tree of bacterial relative abundance at the phylum level.

salinity soil was 2:20 > 4:20 > 3:20 > 1:20. The promoting effect of different treatments on the AK content in coastal salinity soil was 4:20 > 3:20 > 2:20 > 1:20. The TN contents in four treatments with volume ratios of 1:20 ~ 4:20 were significantly higher than that of the control treatment ($P < 0.05$). The contents of AP and AK in three treatments with volume ratios of 2:20 ~ 4:20 were significantly higher than that of the control treatment ($P < 0.05$).

The richness index (ACE index and Chao1 index) and diversity index (Shannon index) of bacteria were significantly increased in coastal saline soil after planting water spinach. The soil bacterial communities in the four treatments with volume ratios of 2:20 ~ 4:20 of the coconut chaff (B12, B22, B32 and B42) were clustered into one branch after harvesting the water spinach, while that of

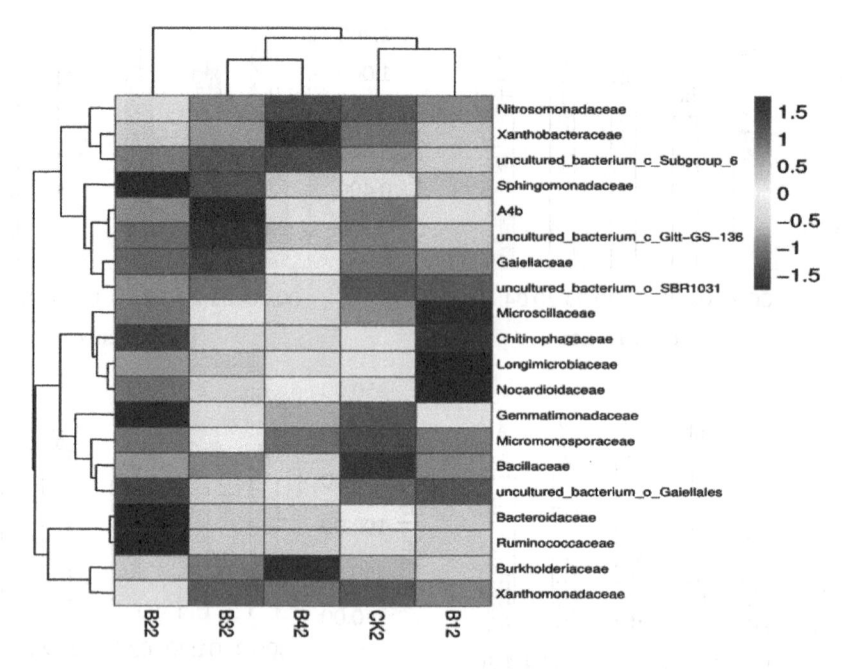

Figure 5. The heat map of bacterial species abundance at the family level.

the treatment without adding the coconut chaff (CK2) was clustered into another branch (Figure 4) which indicated that the bacterial community structure of coastal soil was significantly changed in four treatments with volume ratios of 1:20 ~ 4:20 of the coconut chaff in coastal salinity soil. The abundance of Nitrosomonadaceae was significantly increased in the three treatments with volume ratios of 2:20 ~ 4:20 of the coconut chaff (Figure 5), and the abundance was increased by 22.8%~193%, which indicated that the application of coconut chaff could enhance the cycling capacity of the carbon and nitrogen in coastal salinity soil and was more beneficial to the growth of crops.

4 CONCLUSIONS

In summary, the coconut chaff could obviously improve the pH value of moderately saline soil, and the pH value was changed from acidity to neutral. The contents of organic matter, total nitrogen, available phosphate, and available potassium of moderately saline soil were increased by 11.1% ~ 28.1%, 31.3% ~ 35.6%, 13.0% ~ 23.5%, and 9.9% ~ 44.8%, respectively. The richness index and diversity index of the bacterial community were significantly increased. However, the soil salinity was significantly decreased by 7.7% ~ 24.8% than that of the control treatment without chemical fertilizer utilization. Especially, the treatments with the addition ratio of 2:20 ~ 4:20 could significantly improve the soil pH value, increase the soil nutrient contents, decrease the soil salinity, change the bacterial community structure, and enhance the cycling capacity of the carbon and nitrogen in coastal salinity soil. This demonstrates that the coconut chaff can improve the physicochemical properties and microbial diversity of the coastal salinity soil and provide a favorable soil environment for crop growth.

ACKNOWLEDGEMENTS

We gratefully acknowledge the supported by Hainan Provincial Natural Science Foundation of China (320RC695) and Central Public-interest Scientific Institution Basal Research Fund for Chinese Academy of Tropical Agricultural Sciences (1630042017005, 1630042017006).

REFERENCES

[1] Wu Y Y, Liu R C and Zhao Y G. 2009, *Environmental Geology*, Vol.57, pp 1501–1508.
[2] Xu H G. 2005, *China Agricultural Science and Technology Press*, p5.
[3] Wang Y, Kang S and Li F. 2007, *Pedosphere*, Vol. 34, pp 303–317.
[4] Zhu J H. 2017, *Master thesis of Shandong Agricultural University*, pp 15–18.
[5] Huang H. 2015, *Master thesis of Dongbei Agricultural University*, pp 21–29.
[6] Song N. 2014, *Master thesis of Gansu Agricultural University*, pp 50–53.
[7] Li H Y. 2014, *Master thesis of Dongbei Agricultural University*, pp 31–38.
[8] Radic S, Stefanic P P and Lepedus H. 2013, *Environmental and Experimental Botany*, Vol.87, pp 39–48.
[9] Zhang Q, Chen FD, Feng G Y, Qi H, Wang S L, Liang Q L, Lei X P, Wang Y, Lin Y Z. 2016, *Tianjin Agricultural Sciences*, Vol.8, pp 35–39.

Study on photodegradation of chlortetracycline hydrochloride by composite

N. Wen, S.X. Li, L. Wang & Y.G. Lv
College of Pharmacy, Jiamusi University, Jiamusi, China

ABSTRACT: The prepared BiVO4 and its composite materials were subjected to photodegradation experiments on chlortetracycline-simulated target degradation products, and the catalytic activity of the photocatalyst was studied. The UV-visible spectrophotometer was used to detect the target degradation products under different photocatalytic reaction time. The absorbance is used to explore the degradation effect of the prepared photocatalyst on the target degradation product, and evaluate the photocatalytic performance. Under the simulated natural light irradiation, the degradation of chlortetracycline hydrochloride was carried out. After 160 min of photocatalytic reaction, the degradation rate of BiVO4 sample to chlortetracycline hydrochloride was only 32.4%; while the degradation rate of PGB sample reached 76.39%. The results of photodegradation reaction kinetics showed that the degradation rate of sample PGB reached 3.6 times of the degradation rate of pure BiVO.

1 INTRODUCTION

Antibiotics are metabolites of microorganisms, which can inhibit or kill a class of chemicals grown by other microorganisms at low concentrations, and can also be produced by artificial synthesis. As a major role in the emerging pollutant PPCPS, bacterial resistance caused by its extensive and extensive application and abuse has become a focus of threat to human health, and antibiotics and their derivatives are caused by trace pollutants to water. Pollution and the resulting environmental effects have also begun to attract attention in recent years [1,2].

Main source of antibiotic waste water Medical pollution sources: At present, China is one of the largest drug producing countries in the world. According to the statistics released by the World Health Organization, the use rate of iatrogenic antibiotics in China has exceeded 80%, and the average utilization rate of international iatrogenic antibiotics is 30%, it is obvious that the use rate of iatrogenic antibiotics in China is much higher than the normal standard; Industrial production: At present, China is in a stage of rapid development, and the industry is more developed. The industrial waste water produced by industrial antibiotics has high antibiotic concentration and contains many biologically toxic substances. The industrial production of antibiotics mainly includes: biological fermentation, drug synthesis, extraction of antibiotics and drug-forming stages; Source of life: The part of the antibiotic that is not metabolized in the human body will be directly excreted in the form of urine feces, and discharged into the sewage treatment plant. After being treated incompletely by the sewage treatment plant, it will be discharged into surface water and groundwater, causing pollution; Breeding: During the use of antibiotics in the livestock and poultry industry, most of the antibiotics in the animal body have a low absorption rate. Some of the residual antibiotics will be directly discharged in the form of feces, contaminating the soil and groundwater. Agricultural production: antibiotics are used in agricultural production. Most of them are accumulated in soil and sediments without treatment, and enter the groundwater and surface water through other means to cause antibiotic contamination [3–5].

At present, antibiotic pollution has became a sustainable issue in the world. Toxicity to aquatic organisms: During the use of antibiotics, not only can target disease be prevented, but also toxic effects on other organisms such as aquatic microorganisms, plants and animals. The emergence of induced bacterial resistance: In recent years, the problem of bacterial resistance caused by the large-scale production and consumption of antibiotics has become a research hot spot. Numerous studies have shown that the effective dose of antibiotics required to kill bacteria during the use of antibiotics continues to increase due to the continued development of resistance to pathogens. It is worth noting that even if the drug-resistant strain is not pathogenic, its drug resistance gene R-factor can be transferred to pathogenic bacteria. Therefore, residual antibiotics in the environment pose a great threat to human and animal and plant health.

The techniques for treating antibiotic waste water mainly include the following: biological treatment, physical treatment, chemical treatment, and combination treatment of various methods.

Biological treatment method: The most widely used biological treatment method in antibiotic waste water treatment technology. Biological treatment methods generally include: aerobic method, anaerobic method and anaerobic-aerobic combination method. (2) Physical treatment method: generally used to pretreat antibiotic waste water with high content or poor biochemicality. This method can also be used for subsequent deep treatment. Among them, coagulation, sedimentation, air flotation, adsorption and reverse osmosis have good treatment effects. Physical treatment does not completely remove antibiotics from the roots. It only transfers antibiotics from waste water, and the transferred antibiotics become a new pollutant.

Chemical treatment method: refers to the degradation of organic matter in waste water under the oxidation of oxidant. At present, oxidation and electrolysis methods are common as pretreatment and secondary treatment of antibiotic waste water [6–8].

Chemical, physical, biological and other methods for the routine treatment of antibiotic waste water can convert and decompose harmful substances and separate and dispose of pollutants, but some may cause secondary pollution, making it difficult to achieve relevant emissions of treated water. standard. Therefore, the pursuit of an economical and efficient method has important realization value and practical significance [9,10].

Among them, photo catalytic oxidation technology has been highly concerned by researchers. Many research experiments have confirmed that photo catalytic oxidation is a method for efficiently degrading antibiotic pollutants, and the method has a good effect on the final degradaocatalyst.

2 PREPARATION OF CHLORTETRACYCLINE HYDROCHLORIDE SOLUTION

Preparation of chlortetracycline hydrochloride solution: Dissolve 10 mg/L chlortetracycline hydrochloride solution, ultrasonically assisted dissolution, dilute to 500 ml volumetric flask with distilled water, and place in a refrigerator at 4°C for use.

2.1 *Photocatalysis research*

In this experiment, tetracycline hydrochloride was used as the target degradation agent. The photo catalytic activity of the photo catalyst was studied under visible light conditions. The absorbance of the target degradation product under different photo catalytic reaction time was detected by UV-Vis spectrophotometer. The degradation effect of the new ternary composite photocatalyst on the above target degradation products, and the evaluation of its photocatalytic performance.

In the photocatalytic performance evaluation of the sample, the degradation experiment uses a self-made photocatalytic reaction device. The photo catalytic device is composed of a light source, a stirrer, and a reactor. The light source is provided by a low-power halogen lamp, which effectively reduces energy compared with previous experiments. Consumption.

The forbidden band width of the ternary composite is relatively smallest, which indicates that the sample has relatively best absorption performance in the visible region, and therefore in photocatalytic degradation. It should exhibit the strongest photo catalytic activity.

3 ANALYSIS OF PHOTODEGRADATION EXPERIMENT RESULTS

3.1 *Analysis results of photodegradation of chlortetracycline hydrochloride*

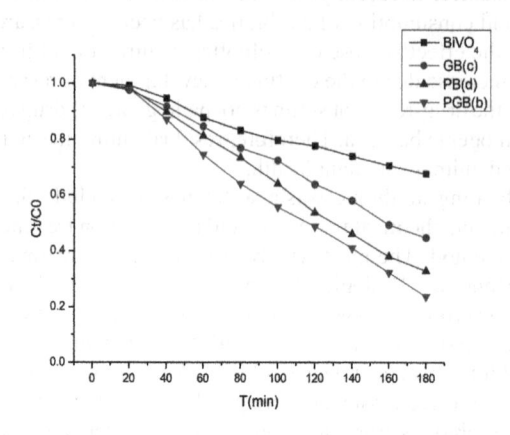

Figure 1. Photocatalytic degradation curve of Chlortetracycline Hydrochloride.

Under the simulated natural light irradiation, chlortetracycline hydrochloride was used as the simulated target degradant (Figure 1). The degradation rate of tetracycline hydrochloride aqueous solution was compared by the sample prepared in the comparison experiment. The photo degradation of tetracycline hydrochloride in different samples prepared by the experiment was investigated. The result is shown in the figure. After photocatalytic reaction for 160 min, the photocatalytic activity of BiVO4 sample was lower, the degradation rate of chlortetracycline hydrochloride was only 32.4%; the degradation rate of GB sample was 55.4%; the degradation rate of PB was 67.2%, while the degradation rate of PB sample was 67.2%. The degradation rate reached 76.39%. Since the concentration of chlortetracycline hydrochloride gradually decreases after the light is turned on, it can be said that the composite catalyst enhances the elimination of chlortetracycline hydrochloride in the solution by the photo catalyst.

3.2 *Analysis of kinetic Results of photodegradation reaction of chlortetracycline hydrochloride*

The experimental results were further studied (Figure 2). The sample was subjected to a dark reaction for 20 min in a solution of chlortetracycline hydrochloride to achieve an adsorption -desorption equilibrium. The time at which the reaction was completed and the start of illumination was set to time zero, and the photocatalytic reaction power of each sample was obtained. The analysis, that is, the relationship between $\ln(C_0/C_t)$ and time(t), as shown in the figure.

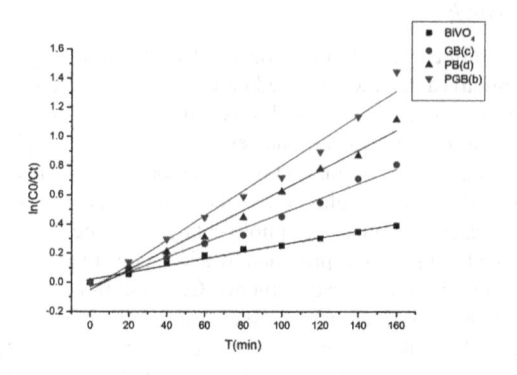

Figure 2. First order kinetics of Chlortetracycline Hydrochloride degradation.

The results show that the concentration of chlortetracycline decreases exponentially with the extension of illumination time, ie, $\ln(C_t/C_0)$ is linearly related to the degradation time, and accords with the first-order reaction kinetics equation. $\text{Ln}(C_0/C_t) = kt$, where C0 represents the original concentration of chlortetracycline hydrochloride, Ct represents the concentration after the reaction t, and k is the apparent reaction rate constant (min^{-1}). The correlation coefficient R^2 and the reaction rate constant k of each sample are shown in the table. It can be seen from the table that the reaction rate constants k of BiVO$_4$, GB(c), PB(d) and PGB(b) samples were calculated to be 0.00238 min^{-1}, 0.00503 min^{-1}, 0.00682 min^{-1} and 0.00852 min^{-1}, respectively. It was found by experiments that after 160 min of illumination, the degradation rate of sample PGB was the fastest, which was 3.6 times of the degradation rate of pure BiVO$_4$. This indicates that the composite photo catalyst PGB exhibits a faster photo catalytic reaction rate and a higher photo catalytic reactivity than pure BiVO$_4$. Therefore, the photo-generated carriers of the PGB sample have stronger redox ability, and the degradation and removal effect of chlortetracycline hydrochloride is better. In addition, the small size of the PGB sample is nanometer, which shortens the distance and time for the photo generated carriers to migrate from the inside of the photo catalyst to the surface, and reduces the recombination efficiency of the photogenerated carriers, which is consistent with the fluorescence emission spectrum. In addition, considering that each correlation coefficient R^2 in the table exceeds 0.9, it can be concluded that the degradation of the chlortetracycline hydrochloride solution of each sample accords with the first-order reaction kinetics equation (Table 1).

Table 1. Linear fitting data of photo catalytic reaction kinetics of each sample.

Sample name	R^2	k	intercept
BiVO$_4$	0.98614	0.00238	0.01969
GB(c)	0.98671	0.00503	0.02975
PB(d)	0.98484	0.00682	0.05076
PGB(b)	0.97746	0.00852	0.05338

3.3 Photodegradation of chlortetracycline by UV absorption spectroscopy

The picture shows the UV-visible absorption spectrum of degraded chlortetracycline hydrochloride. It can be seen that the intensity of the absorption peak at 366nm decreases with time, while the peak at 276 nm gradually decreases and a certain blue shift occurs. The peaks at 276 nm and 366 nm correspond to the E2 band and B band of the benzene ring, respectively, indicating that the sample prepared by us can degrade the benzene ring of chlortetracycline hydrochloride into a small molecular substance, and the catalyst has strong oxidation ability (Figure 3).

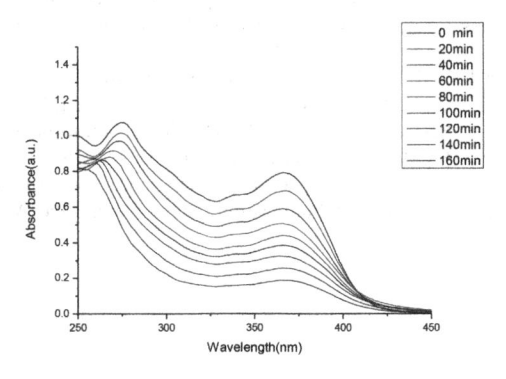

Figure 3. UV absorption spectra of Chlortetracycline Hydrochloride degraded by sample.

4 CONCLUSION

In this chapter, BiVO$_4$ and its composite materials were used to photo degradate the target degradation products of chlortetracycline hydrochloride to study the catalytic activity of photo catalysts. UV-visible spectrophotometer was used to detect target degradation under different photo catalytic reaction time. The absorbance of the material was used to explore the degradation effect of the prepared photo catalyst on the target degradation product, and to evaluate the photo catalytic performance. Under the simulated natural light irradiation, the degradation of chlortetracycline hydrochloride was carried out. After 160 min of photo catalytic reaction, the degradation rate of BiVO$_4$ sample to chlortetracycline hydrochloride was only 32.4%; while the degradation rate of PGB sample reached 76.39%. The results of photo degradation reaction kinetics showed that the degradation rate of sample PGB reached 3.6 times of the degradation rate of pure BiVO$_4$. The experimental results show that each sample prepared has a strong degradation effect on tetracycline hydrochloride, and the composite photo catalyst PGB exhibits a faster photo catalytic reaction rate and a higher photo catalytic reactivity than pure BiVO4. In addition, the degradation of tetracycline hydrochloride solution by each sample accords with the first-order reaction kinetics equation.

ACKNOWLEDGEMENTS

This work was financially supported by the National Science Foundation of China (No. 21346006), Department of scientific research project in Heilongjiang province (No. B2017015), Excellent discipline team project of Jiamusi University ((No.JDXKTD- 2019007).

REFERENCES

[1] Bian ZY, Zhu YQ, Zhang JX, Ding AZ, Wang H. Visible-light driven degradation of ibuprofen using abundant metal-loaded BiVO4 photo catalysts. Chemo sphere, 2014, 117: 527–531.
[2] Chen J, Li DZ, Wang JB, Wang P, Cao CS, Shao Y, Wang JX, Xian JJ. Morphological effect on photo catalytic degradation of Rhodamine B and conversion of active species over BaSb2O6. Appl Catal B-Environ, 2015, 163: 323–329.
[3] JingL, Wang M, Li X. et al. Covalently functionalized TiO$_2$ with ionic liquid: A high performance catalyst for photoelectrochemical water oxidation[J]. Appl.Catal B-Environ. 2015, 166–167:270–276.
[4] Zhao Y, He Y, He J.et al.Hierarchical porous TiO2templatedfromnaturalartemiacyst shellsfor photocatalysis applications[J]. RSC Adv. 2014, 4(39): 20393–20397.
[5] Schneider J, Matsuoka M, Takeuchi M.etal. Understanding TiO2Photocatalysis: Mechanisms and Materials [J].Chem. Rev. 2014, 114(19): 9919–9986.
[6] Bao N, Yin Z, Zhang Q. et al.Synthesis of flower-like monoclinicBiVO4 surface through TiO2 ceramic fiber with hetero structures and its photo-catalytic property[J]. Ceram.Int. 2016, 42(1):1791–1800.
[7] Wang Q, Jiang HY, Ding S T. et al. Butterfly-like BiVO4: synthesis and visible light photo-catalytic activity [J].Syn. React.Inorg.Met.2016, 46(4):483–488.
[8] Chen L, Wang J, Men D.et al. Enhanced photo-catalytic activity of hierarchically structured BiVO4 oriented along {040} facets with different morphologies[J]. Mater. Lett. 2015, 147:1–3.
[9] Zhang X, Ai Z,Jia F. et al. Selective synthesis and visible-light photo-catalytic activities of BiVO4 with different crystalline phases[J].Mater.Chem. Phys.2017, 103(1): 162–167
[10] Li JQ, Cui M , Liu ZX. et al. BiVO4 hollow spheres with hierarchical micro structures and enhanced photocatalytic performance under visible-light illumination[J].Phys. Status Solidi A, 2013, 210(9): 1881–1887.

Facile preparation of open-mouthed hollow SiO$_2$ microspheres

W.H. Zhang, J.S. Hu, A.J. Hou & Y.H. Cao
School of Chemistry and Chemical Engineering, Xianyang Normal University, Shaanxi Xianyang, China

ABSTRACT: Synthesis of open-mouthed hollow microspheres with simple and convenient method has been a challenging task in materials science. In this work, open-mouthed hollow SiO$_2$ microspheres were fabricated through a facile vacuum evaporation process. Firstly, taking spherical colloid aggregates of ammoniated PAA molecular in ethanol solution as soft templates, PAA/SiO$_2$ composite microspheres with core/shell structure were synthesized by a modified Stöber method. By adding appropriate volume of water, the mixture was then vacuum evaporated with a rotary evaporator. The formation of the open-mouthed hollow SiO$_2$ microspheres were induced by the cluster of polyelectrolyte molecules rush out from the thinnest shell of the microspheres under high negative pressure. This strategy offer a novel route for fabricating open-mouthed hollow microspheres and should also suitable to other soft templates derived from non-crosslinked molecular aggregates.

1 INTRODUCTION

In the recent over ten years, hollow microspheres with a surface hole, which are also referred to as open-mouthed hollow microspheres, have attracted significant attention due to their enhanced uptake capacity [1], diffusivity [2], catalytic performance [3] and other unique properties. Compared with conventional hollow spheres with closed shells, open-mouthed hollow microspheres are more suitable to be used to encapsulate large entities, such as protein, DNA, nanoparticles, etc. Moreover, this kind of hollow microspheres can encapsulate the target object more quickly and conveniently through their surface hole. So, open-mouthed hollow microspheres were considered as a new class of hollow structure [4]. Until now, various types of polymer-based or inorganic open-mouthed hollow microspheres have been synthesized by self-assembly [5], precipitation-phase separation method [6], Ostwald ripening [7], microfluidic fabrication technique [8], miniemulsion polymerization approach [9], co-electrospraying [10], template method [11], etc. Nevertheless, many of the current techniques used are complex, time-consuming, or require highly specialized equipment. And the challenges on design and preparation of open-mouthed hollow microspheres still remain.

Herein, we report a facile synthetic route for the preparation of open-mouthed hollow SiO$_2$microspheres. This method is based on an improved Stöber method while poly(acrylic acid) (PAA) is used as soft template. This strategy can be easily implemented in the laboratory for the whole process does not require the use of some specialized equipment, surfactants or other additives.

2 EXPERIMENTAL SECTION

2.1 *Materials*

Acrylic acid (AA) and tetraethoxysilane (TEOS) were obtained by Tianjin Kermel Chemical Reagent Co. Ltd. Ammonium persulfate, ammonium hydroxide(25–28%) and absolute ethyl alcohol were purchased from Sinopharm Chemical Reagent Co. Ltd. All chemicals were of analytical grade and used as received.

2.2 Synthesis of PAA

PAA was synthesized by solution polymerization. Briefly, a mixture of AA (5 mL), deionized water (100 mL) and initiator ammonium persulfate (0.3 g) were stirred at 75°C for 8 h under the protection of nitrogen. The obtained solution was evaporated and concentrated to about 20 mL. At last, PAA was obtained after the concentrated solution was freeze-dried.

2.3 Fabrication of open-mouthed hollow SiO_2 microspheres

PAA/SiO_2 microspheres with core/shell structure were prepared by the modified Stöber method [12]. Typically, 0.30 g of PAA dissolved in 4.5 mL of ammonium hydroxide was mixed with 90 mL of absolute ethanol. Then, 2.0 mL of TEOS was injected dropwise within 4 h by automatic sample injector under vigorous magnetic stirring at room temperature. After 8 h, PAA/SiO_2 composite microspheres at the nanometer scale were obtained.

Open-mouthed hollow SiO_2 microspheres were prepared by simple vacuum evaporation process. In brief, 100 mL of deionized water was poured into as-prepared PAA/SiO_2 dispersion, the mixture was vacuum concentrated to about 100 mL by rotary evaporation, open-mouthed hollow SiO_2 microspheres were obtained.

For comparison, hollow SiO_2 microspheres with intact shells were also prepared by dialyzing the PAA/SiO_2 colloids in deionized water for several days.

2.4 Characterization

The internal morphologies of hollow SiO_2 microspheres were observed by transmission electron microscope (TEM) on a Hitachi H-7650 transmission electron microscope operated at an acceleration voltage of 80 kV. The external morphologies of hollow SiO_2 microspheres were examined by scanning electron microscopy (SEM) on a Zeiss Sigma-500 field emission scanning electron microscope operated at 1.0 kV.

3 RESULTS AND DISCUSSION

Figure 1 illustrates the synthetic procedure used to synthesize open-mouthed hollow SiO_2 microspheres and corresponding hollow SiO_2 microspheres with intact shell. When PAA is dissolved in ammonia hydroxide, PAA chains undergo amination and become into polyelectrolyte. With the addition of a poor solvent (i.e., ethanol), the polyelectrolyte molecules tend to aggregate into numerous spherical colloid templates at nanoscale. The silica generated by TEOS hydrolysis will gradually wrap the colloid templates and form PAA/SiO_2 core/shell microspheres.

Due to the hydrophilic properties of polyelectrolyte, the ammoniated PAA molecular chains, which acted as core of PAA/SiO_2 composite microspheres, will escape slowly from the microspheres when the dispersion of core/shell microspheres are treated with mild dialysis operation, and then the hollow silica microspheres with complete shell will be obtained. However, when the dispersion of core/shell microspheres is treated by rotary evaporation, under the action of high negative pressure, the cluster of polyelectrolyte molecules instead of single polymer chain will rush out from the thinnest shell of the microspheres quickly, resulting in obvious cracks in the shell of the microspheres, and then the open-mouthed hollow SiO_2 microspheres are obtained.

We demonstrated the above synthesis concept by experimental examples, as shown in the TEM images of Figure 2a–b. The transmission electron microscopy (TEM) image of open-mouthed hollow SiO_2 microspheres (as shown in Figure 2a) revealed that the microsphere was about 50–100 nm in diameter and hollow in the interior. Meanwhile, one light spot with a clear edge can

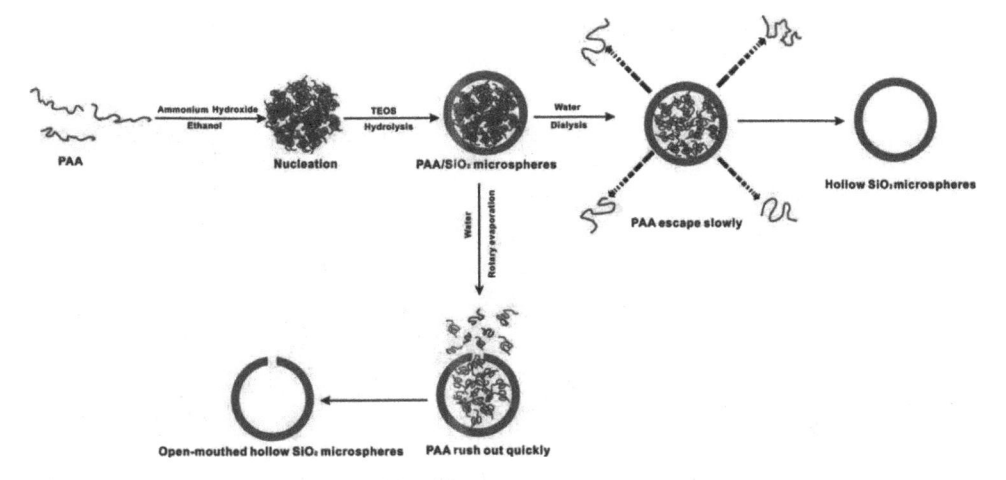

Figure 1. Schematic illustration of the formation of open-mouthed hollow SiO$_2$ microspheres and corresponding hollow SiO$_2$ microspheres with intact shell.

be seen on most of microsphere (as indicated by arrows), suggesting the presence of a single hole at the shell. As a comparison, we have also prepared hollow SiO$_2$ microspheres with intact shell through a conventional dialysis route. As shown in Figure 2b, all of the hollow microspheres have closed shells. It should be that PAA colloidal aggregates inside the microspheres are dissociated into single molecular chains and slowly escape from the shell when the dispersion of core/shell microspheres were dialyzed in the deionized water, thus the integrity of the shell was not destroyed.

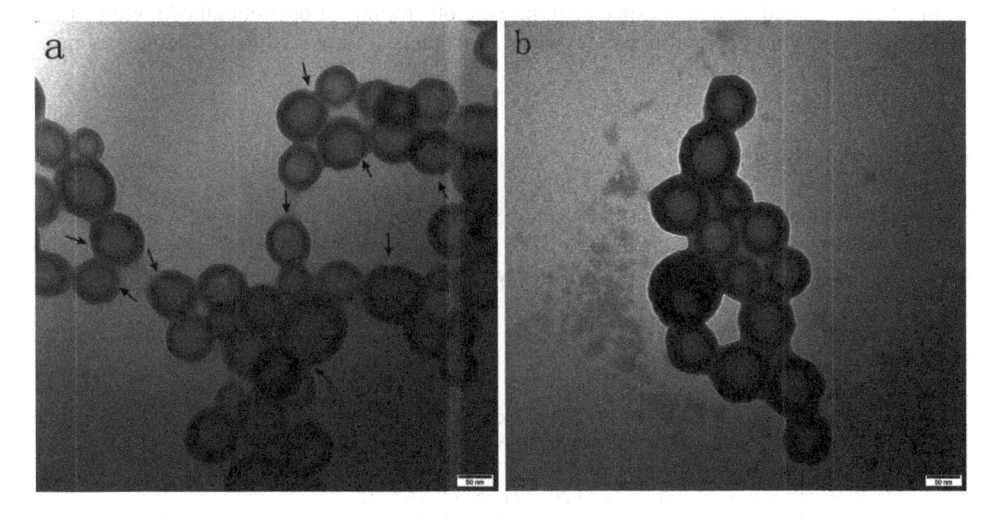

Figure 2. The TEM images of open-mouthed hollow SiO$_2$ microspheres (a) and corresponding hollow SiO$_2$ microspheres with intact shell (b).

The hole of open-mouthed hollow SiO$_2$ microspheres can clearly be seen by SEM observation (Figure 3a). Different from other methods for preparing open-mouthed hollow microspheres, the holes on the surface of the hollow microspheres prepared by this method are not single circular hole, but a flat and irregular crack, indicating that the PAA agglomerates have been disintegrated into some smaller clusters, rather than total PAA colloidal aggregates, before rushing out from the shell of the microsphere. The SEM image of hollow SiO$_2$ microspheres treated by dialysis

(Figure 3b) demonstrated that the microsphere shell was closed, while, a broken silica microsphere (as indicated by arrows) showed a hollow internal structure.

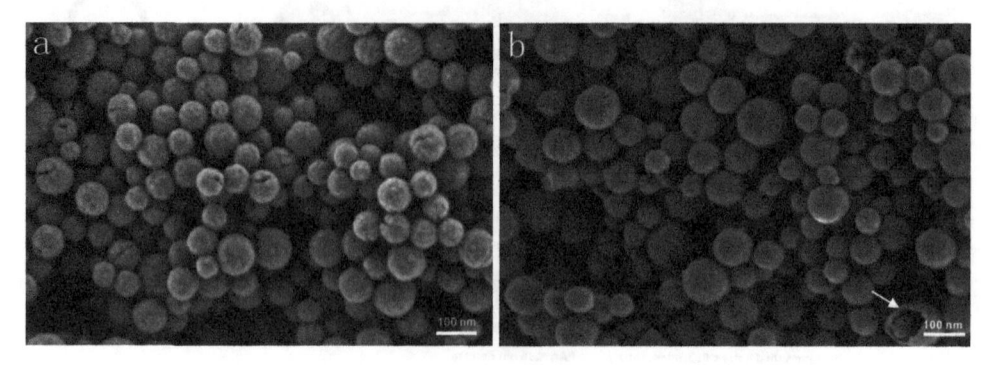

Figure 3. The SEM images of open-mouthed hollow SiO$_2$ microspheres (a) and corresponding hollow SiO$_2$ microspheres with intact shell (b).

4 CONCLUSION

In summary, open-mouthed hollow silica microspheres were fabricated by a simple vacuum evaporation process. This preparation method can be used to prepare various morphologies, internal structures and the particle sizes of the silica microspheres by changing the adding proportion of PAA, aqueous ammonia and TEOS as well as the the quantity of water before rotary evaporation. It can be easily scaled up and is expected to prepare other kinds of open-mouthed hollow microspheres as long as the uncrosslinked molecular aggregates are used as the soft templates in the synthetic process.

ACKNOWLEDGMENT

This work was supported by grants from Natural Science Basic Research Plan in Shaanxi Province (No. 2016JM5024), China Postdoctoral Science Foundation (No. 2014M560747), and Scientific Research Project of Xianyang Normal University (No. 13XSYK017).

REFERENCES

[1] Guan G, Zhang Z, Wang Z, Liu B, Gao D and Xie C 2007 *Adv. Mater.* **19** 2370
[2] Zhao J, Zeng M, Zheng K, He X, Xie M and Fu X 2017 *Materials* **10** 411
[3] Jia C, Yang P, Chen H-S and Wang J 2015 *CrystEngComm* **17** 2940
[4] Chang M-W, Stride E and Edirisinghe M 2010 *Langmuir* **26** 5115
[5] Xiong F, Han Y, Wang S, Li G, Qin T, Chen Y and Chu F 2017 *ACS Sustainable Chem. Eng.* **5** 2273
[6] Fu X, Liu J and He X 2014 *Colloids Surf., A* **453** 101
[7] Wang L, Xu Q, Liu L, Song Q, Lv H, Zhu G and Zhang D 2017 *J. Lumin.* **192** 1020
[8] Ju M, Ji X, Wang C, Shen R and Zhang L 2014 *Chem. Eng. J.* **250** 112
[9] Wang Z, Qiu T, Guo L, Ye J, He L and Li X 2019 *Chem. Eng. J.* **357** 348
[10] Zhou F-L, Chirazi A, Gough J E, Hubbard Cristinacce P L and Parker G M 2017 *Langmuir* **33** 13262
[11] Zhang Y, Chen J, Zhang G, Chen B and Yan H 2013 *Chin. J. Polym. Sci.* **31** 294
[12] Wan Y and Yu S-H 2008 *J. Phys. Chem. C* **112** 3641

Surface modification and electrochemical properties of graphene cathode catalysts

X. Wang & X.D. Hou
Key Laboratory of Materials Modification by Laser, Ion and Electron Beams (Ministry of Education), School of Materials Science and Engineering, Dalian University of Technology, Dalian, China

Y. Ren
School of Materials Science and Engineering, Henan University of Technology, Zhengzhou, China

D.W. Deng & G.F. Zhang
Key Laboratory of Materials Modification by Laser, Ion and Electron Beams (Ministry of Education), School of Materials Science and Engineering, Dalian University of Technology, Dalian, China

ABSTRACT: Development of low-cost cathode catalysts with both a high catalytic activity and durability is one of the great challenges for fuel cells. In this article, graphene cathode catalysts with a controllable platinum load between 4%–55% were obtained by pyrolyzing chloroplatinic acid on graphene at 480°C in argon atmosphere. The microstructure and morphology of the modified catalysts were characterized using scanning electron microscope and X-ray diffraction. Results showed that Pt particles were uniformly distributed on the layer of graphene. The electrochemical performance showed that with the increase of platinum content, oxygen reduction reaction (ORR) catalytic activity first increased and then decreased. Compared with the commercial Pt/C (20$wt.$% Pt) catalyst, the modified graphene catalyst (8$wt.$% Pt) in this paper has better ORR catalytic performance, durability and methanol resistance.

1 INTRODUCTION

Fuel cells have attracted worldwide attention due to advantages of high efficiency and low emission [1–3]. Catalytic materials are one of the important components of fuel cells. Because the reaction rate of the cathode oxygen reduction is six orders of magnitude slower than that of the anodic hydrogen oxidation, most of the studies have focused on the cathode catalysts [4]. The commercialization progress of fuel cells is seriously restricted by the complex preparation process, high cost and poor durability of the catalysts [5–7]. Researchers have developed many new materials for the cathode catalytic. However, the most effective ones are still platinum-based catalysts [8]. Thus, how to obtain high quality and low platinum-based oxygen reduction electrocatalysts is a technical challenge that people face at present. According to research, the selection of the carrier has a significant impact on the properties of the catalysts. It has been indicated that the support materials of catalysts should have the characteristics as follows: (1) good electrical conductivity, (2) large specific surface area, (3) high resistance, (4) appropriate porosity and porous structure and (5) interaction between the carrier and the catalytic material [9].

The honeycomb hexagonal lattice graphene with sp^2 hybridized orbitals is a two-dimensional carbon nanomaterial. The specific surface area of graphene is much higher than that of carbon nanotubes, Vulcan XC-72, and graphite [10]. Besides, graphene has a good electrical conductivity. It can be used not only as a catalyst carrier, but also as a catalyst for fuel cell. Usually, main preparation methods of graphene-supported metal catalysts include chemical reduction, electrochemical deposition, sol-gel, coprecipitation, hydrothermal, and solvent hot method. Among them, chemical reduction is the most common method to prepare graphene-based catalysts. However, the reducing agent and organic solvent will reduce the activity of bonding interface between graphene with nanoparticles, leading to poor electrochemical performance of the prepared composites [11].

In this paper, a low cost and simple operation method is adopted. Different from the existing reports, the platinum particles are loaded on graphene by the pyrolysis of $H_2PtCl_6 6H_2O$ without reducing agent and organic solvent. SEM and XRD were used to characterize the structure and morphology of the catalysts. The effects of different platinum loads on the catalytic activity of oxygen reduction were studied through peak potential, peak current density, initial potential and limit current density. The durability and methanol resistance of catalysts were evaluated by timing current method.

2 EXPERIMENTAL METHODS

Six 50mg of graphene were added to 10 mL $H_2PtCl_6H_2O$ solution with 0.625 g L^{-1}, 1.25 g L^{-1}, 2.5 g L^{-1}, 5 g L^{-1}, 8 g L^{-1} and 16 g L^{-1} respectively and ultrasonically vibrated for 2 h at 30°C to uniformly disperse graphene. The dispersed solution was lyophilized and dried into powder. The dried powder samples were placed in the crucibles and heated to 480°C at a rate of 5°C/min in a tubular furnace using inert argon as protective gas. After holding for 1 hour and cooling slowly to room temperature, the samples were taken out. The catalyst samples with theoretical platinum content of 4%, 8%, 16%, 27%, 38% and 55% were obtained. During the heat preservation process, pumping and injecting should be carried out every 20 min.

The surface morphology, elemental composition, and structure of the samples were observed by scanning electron microscope (SEM, Zeiss, SUPARR55), energy dispersive spectrometer (EDS), and X-ray diffractometer (XRD, PANalytical B.V., Empyrean) with a step of 0.02° width and a scanning range of 20~80°, respectively.

Electrochemical measurement was performed on electrochemical workstation (Shanghai Chenhua, CHI-760E) with rotating disk electrode (RDE) of 3 mm in diameter. A three-electrode system with platinum wire as the auxiliary electrode, Ag/AgCl as the reference electrode and modified glassy carbon electrode as the working electrode was used. Oxygen reduction catalytic activity was characterized by cyclic voltammetry (CV) and linear scanning voltammetry (LSV) curves. The scanning rate and range for the CV test was 50 mV/s and −0.8~0.4 V, respectively. In the LSV test, the scanning rate was 10 mV/s and the rotating speed was 400–2400 rpm. The stability and methanol resistance were obtained by means of chronoamperometry at an onset potential of −1 V and a rotational speed of 1600 rpm. All the above electrochemical measurements were carried out in O_2-saturated 0.1 M KOH electrolyte at room temperature (25°C).

3 RESULTS AND DISCUSSION

SEM images of graphene catalyst decorated with 8% platinum are shown in Figure 1. It can be seen from Figure 1a and b that the graphene has a lamellar folded structure, and platinum nanoparticles are uniformly attached on the graphene. Besides, spherical platinum particles deposited on graphene sheets have good dispersity and no obvious aggregation. The mass fraction of Pt in the sample obtained from EDS spectrum is 7.86%, which is close to the set value of 8%.

From XRD patterns (as shown in Figure 1c) of the catalyst samples, it can be clearly seen that the diffraction peaks of the Pt located at 39.8°, 46.3° and 67.7° are corresponding to (111), (200) and (220) crystal faces of face-centered cubic (FCC) Pt (PDF#04–0802) respectively. The diffraction peaks of platinum strengthen with the increase of platinum content and the peak position is basically unchanged. Consisting with SEM observation, the structure analysis indicates that Pt nanoparticles with a controllable content are successfully decorated on the graphene.

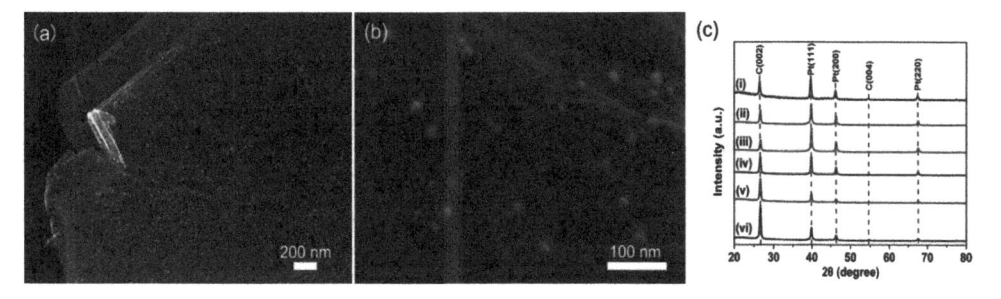

Figure 1. (a) SEM images of graphene decorated with 8wt.% Pt and (b) XRD patterns of catalyst samples with platinum content of (i) 55wt.%, (ii) 38wt.%, (iii) 27wt.%, (iv) 16wt.%, (v) 8wt.% and (vi) 4wt.%.

CV curves were measured by cyclic voltammetry to analyze the electrochemical properties of catalyst samples. As seen in Figure 2a and Table 1, the potential and current density of oxygen reduction peak change with the platinum content. The peak potential of oxygen reduction first shifts in a positive and then negative direction, and the peak current density first increases and then decreases with the increase of platinum content. Both the oxygen reduction peak potential and the peak current density reach a maximum value at 38%. The results demonstrate that the higher the platinum content, the better the oxygen reduction catalytic activity. However, it will lead to the aggregation of platinum particles and then decreases the catalytic activity if the platinum content up to 38%, as shown in Figure 2a.

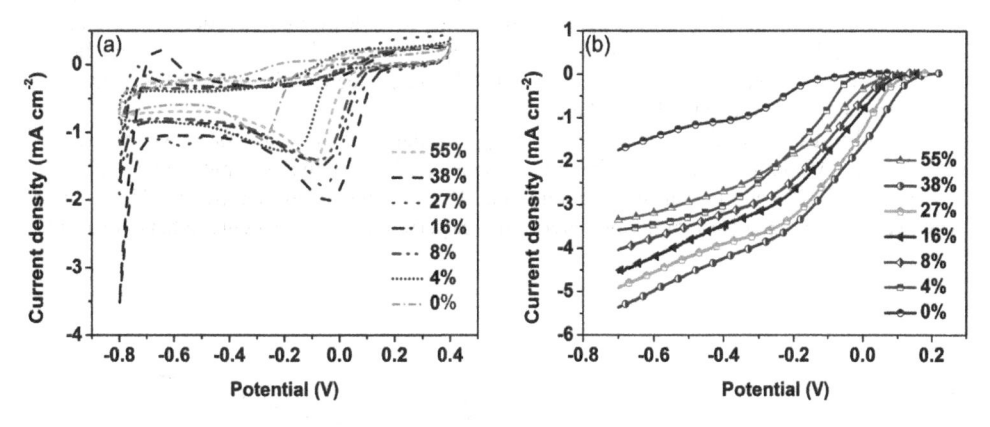

Figure 2. (a) CVs and (b) LSVs at a rotational speed of 1200 rpm of catalysts with different platinum contents in O_2-saturated 0.1 M KOH electrolyte.

The current between the working electrode and the auxiliary electrode in O_2-saturated 0.1 M KOH electrolyte was measured to obtain the LSV curves of the samples. It can be seen from Figure 2b that there are two current platforms for the graphene catalyst, and only one for the catalysts decorated with platinum. One current platform ensures that the oxygen reduction process is dominated by 4 electron pathway [12]. As it can be seen in Table 1, with the increase of platinum content, the onset potential first shifts in a positive then in a negative direction, and the current density first increases then decreases. Both of them reach an extreme value at 38%. As compared, the onset potential of commercial Pt/C (20wt.% Pt) measured under the same experimental condition is 0.048 V, less than that of the catalyst sample with 8% platinum content prepared in this paper, with a difference of 55 mV. It can be attributed to the good conductivity, the appropriate porosity of graphene, and the interaction between graphene and platinum. In conclusion, the catalytic activity of the oxygen

Table 1. Parameters of CVs and LSVs.

Pt content (%)	CV		LSV	
	E_{peak} (V vs. Ag/AgCl)	J_{peak} (mA cm^{-2})	E_{onset} (V vs. Ag/AgCl)	J (mA cm^{-2}) at −0.7 V
0	−0.270	1.10	−0.023	1.74
4	−0.188	1.28	0.015	3.57
8	−0.078	1.41	0.103	4.01
16	−0.058	1.48	0.125	4.54
27	−0.055	1.80	0.175	4.91
38	−0.040	2.00	0.189	5.36
55	−0.095	1.44	0.103	3.35

reduction becomes better with the increase of platinum content and will begin to decrease when the platinum content is over 38%, which is consistent with the result of the CV curves.

The commercialization of fuel cells was also hindered by a poor durability. A three-electrode system is used to test the durability of the catalyst samples in O_2-saturated 0.1 M KOH solution. The measurement results are given in Figure 3a. It is clear that the current decline ratios of the graphene catalysts decorated with Pt after the cycle of 15000 s are less than 50%. With the increase of platinum content, the relative current first increases and then decreases, and the graphene decorated 38wt.% Pt catalyst has the best durability due to its lowest current decline ratio (29%). Obviously, the durability of the graphene catalyst samples decorated with platinum is significantly higher than that of the commercial Pt/C catalyst (with a large decrease by 67%). This is due to the high resistance of graphene to electrochemical corrosion.

Methanol resistance is also a factor to evaluate performance of catalysts. As shown in Figure 3b, after adding 3 M methanol at 300 s, the current density of the catalyst sample with 8% platinum decreased by 5.32%, while that of the commercial Pt/C catalyst decreased by 39.43%. It can be clearly seen that the catalyst samples prepared in this experiment have excellent methanol resistance.

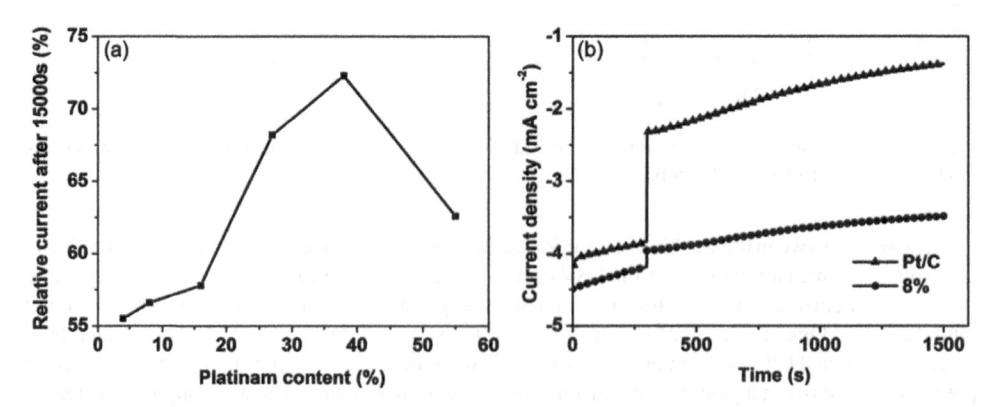

Figure 3. (a) Relationship between platinum content and relative current after 15000 s; (b) i-t curves of Pt/C and catalyst samples loaded with 8% Pt after adding 3 M methanol solution.

4 CONCLUSIONS

In this paper, platinum decorated graphene cathode catalysts were successfully prepared by pyrolysis of chloroplatinic acid. The electrochemical results show that the electrocatalytic activity, durability and methanol resistance of the prepared samples with 8% platinum loading are better than those of commercial Pt/C (20wt.%). The utilization rate of platinum is improved by the large specific surface area of graphene. The ORR catalytic activity of the catalysts is enhanced by the good electrical conductivity and porous structure. The durability of the catalysts is optimized by the high electrochemical corrosion resistance. Moreover, the preparation process is simple, almost pollution-free, low-cost, and controllable. It provides an idea for the commercialization of low platinum-based catalysts.

REFERENCES

[1] Wang Y, Diaz D F R, Chen K S, Wang Z and Adroher X C. Materials, technological status, and fundamentals of PEM fuel cells - A review. 2020 *J. Mater Today.* **32** 178–203
[2] Wu J F, Yuan X Z, Martin J J, Wang H J, Zhang J J, Shen J, Wu S H and Merida W. A review of PEM fuel cell durability: Degradation mechanisms and mitigation strategies. 2008 *J. J Power Sources.* **184** (1) 104–19
[3] Zhang S, Shao Y Y, Yin G P and Lin Y H. Recent progress in nanostructured electrocatalysts for PEM fuel cells. 2013 *J. J Mater Chem A* **1** (15) 4631–41
[4] Debe M K. Electrocatalyst approaches and challenges for automotive fuel cells. 2012 *J. Nature.* **486** (7401) 43–51
[5] Steele B C H and Heinzel A. Materials for fuel-cell technologies. 2001 *J. Nature.* **414** (6861) 345–52
[6] Nie Y, Li L and Wei Z D. Recent advancements in Pt and Pt-free catalysts for oxygen reduction reaction. 2015 *J. Chem Soc Rev.* **44** (8) 2168–201
[7] Tiwari J N, Tiwari R N, Singh G and Kim K S. Recent progress in the development of anode and cathode catalysts for direct methanol fuel cells. 2013 *J. Nano Energy.* **2** (5) 553–78
[8] Bing Y H, Liu H S, Zhang L, Ghosh D and Zhang J J. Nanostructured Pt-alloy electrocatalysts for PEM fuel cell oxygen reduction reaction. 2010 *J. Chem Soc Rev.* **39** (6) 2184–202
[9] Wang Y J, Fang B Z, Li H, Bi X T T and Wang H J. Progress in modified carbon support materials for Pt and Pt-alloy cathode catalysts in polymer electrolyte membrane fuel cells. 2016 *J. Prog Mater Sci.* **82** 445–98
[10] Vadahanambi S, Jung J H and Oh I K. Microwave syntheses of graphene and graphene decorated with metal nanoparticles. 2011 *J. Carbon.* **49** (13) 4449–57
[11] Zhou Y, Wei J, Wang X, Ni H and Zhu Y. Research progress in preparation methods of graphene supported metal catalysts. 2015 *J. Modern Chemical Industry.* **35** (7) 42–5
[12] Wu B, Li Q, Song K, Zhu X, Zhang B and Guo J. Preparation and Catalytic Characterization of Pt/SiC with Less Platinum for Oxygen Reduction Reaction. 2017 *J. Journal of Synthetic Crystals.* **46** (12) 2443–50.

In-situ generation and analysis of cellulose/silver nanoparticle composite filter paper with the assist of ultrasonic atomization

C.F. Shi, Z. Jin & H. Ma
State Key Laboratory of Silicon Materials, School of Materials Science and Engineering, Zhejiang University, Hangzhou, China

M.W. Li & J.R. Ji
State Key Laboratory for Diagnosis and Treatment of Infectious Diseases, National Clinical Research Center for Infectious Diseases, Collaborative Innovation Center for Diagnosis and Treatment of Infectious Diseases, The First Affiliated Hospital, College of Medicine, Zhejiang University, Hangzhou, China

X.W. Zhang
State Key Laboratory of Silicon Materials, School of Materials Science and Engineering, Zhejiang University, Hangzhou, China

ABSTRACT: The presence of active groups on the cellulose polymer chains make it a good carrier for the synthesis of silver nanoparticles. In this paper, pretreatment of cellulose filter paper with a low-concentration alkaline solution was followed by in-situ dispersion synthesis of silver nanoparticles on the cellulose filter paper by means of an ultrasonic atomizer at room temperature. The obtained cellulose/silver nanoparticle composites were characterized by scanning electron microscopy, Fourier transform infrared spectroscopy, X-ray diffraction, UV–visible absorption spectroscopy, X-ray photoelectron spectroscopy and transmission electron microscopy. The results show that the silver element in cellulose basically exists in the form of metallic state, the size of silver nanoparticles is between 5 nm and 10 nm, and the distribution of silver nanoparticles is uniform without obvious agglomeration. The inductively coupled plasma mass spectrometry test results show that the content of silver nanoparticles loaded on cellulose is 17.6–28.9 mg/kg, which meets the requirements of most Nano-silver materials. Antibacterial test shows that cellulose/silver nanoparticle composites have good antibacterial activity. This method can quickly synthesize silver nanoparticles on cellulose at room temperature.

1 INTRODUCTION

In recent years, with the development of nanotechnology, more and more nanomaterials have begun to be widely used. Silver is a special white precious metal, and its application history can be traced back to more than 4,000 years ago. Nano-silver materials have related applications in the fields of antibacterial, catalysis, sensors, data storage, and ultraviolet protection. For example, silver nanoparticles (Ag NPs) can be adsorbed on the surface of bacterial cell membrane, destroy the cell membrane, and penetrate into the bacterial cell, thereby affecting cell penetration and respiratory function, and achieving antibacterial effect. Due to its unique antibacterial mechanism, nano-silver material has stable broad-spectrum antibacterial properties. However, due to their small size and large surface energy, Ag NPs are prone to agglomeration, and difficult to separate and recycle from the system after use, which greatly limits the application of Ag NPs. In order to solve these problems, many studies are devoted to loading Ag NPs on specific carriers, such as metal oxides, carbon-based compounds, and polymers [1, 2]. Different carriers have different levels of difficulty and stability in loading Ag NPs. Therefore, it is necessary to find a carrier material suitable for stably loading Ag NPs.

Cellulose is the oldest and the most extensive natural polymer material in nature. There are many active groups including hydroxyl groups in cellulose, these groups have good chemical reactivity and adsorption [3], The hydroxyl and ether bonds on the cellulose polymer chain can form a strong electrostatic interaction with the metal ion, so that the metal ion can be firmly supported on the cellulose, and the hydroxyl group on the cellulose polymer chain has good reducibility, in addition, related research shows that the metal nanoparticles fixed on the surface of cellulose have a strong interaction with the cellulose material and can be considered harmless to humans and animals, thus these characteristics make the cellulose become a good carrier for metal nanoparticles [4, 5].

In this work, the reduction performance and template function of the cellulose will be used to reduce silver ions on the cellulose in situ to obtain Ag NPs with a particle size of about 10 nm with the help of an ultrasonic atomizer. This method does not use any organic solvents and toxic reagents, and the reaction is carried out at room temperature. Ag NPs are generated almost instantly, the generation of Ag NPs on the cellulose takes only a few minutes, and cellulose as a reducing agent and carrier is green, low cost and renewable. Therefore, this method has broad application prospects for the rapid room temperature preparation of cellulose/Ag NP composites. So far, similar methods have not been reported in other literatures.

2 EXPERIMENT

2.1 *Materials*

In this experiment, commercially available qualitative filter paper (area 2cm × 2cm) was used. $AgNO_3$ (AR) and sodium hydroxide (AR) were obtained from Sinopharm Chemical Reagent Co., Ltd. All the reagents were of analytical grade and used as received without additional purification.

2.2 *In situ synthesis of Ag NPs on filter paper cellulose*

The schematic diagram of in-situ loading of Ag NPs on the cellulose material of the filter paper is shown in Figure 1. First, soaking the filter paper cellulose in sodium hydroxide solutions of different concentrations (0.05 mol/L, 0.1 mol/L, 0.15 mol/L, 0.2 mol/L) for 1 minute, then rinse it with running deionized water and dry it at room temperature, the corresponding samples obtained were denoted by CE-1, CE-2, CE-3 and CE-4, and the sample without immersion treatment with alkali solution is denoted by CE-0. Then put silver nitrate solutions of different concentrations (0.2 wt%, 0.4 wt%, 0.6 wt%, 0.8 wt%, 1.0 wt%) into the ultrasonic atomizer for atomization, the model of ultrasonic atomizer is YUWELL 402B, and the atomized droplets enter the dark box through the pipeline. After the atomized droplets fill the dark box, place the treated filter paper in the dark box and ultrasonic atomization for 1 min, then rotate the filter paper vertically by 180° and continue ultrasonic atomization for 1 min, this process is hereinafter referred to simply as ultrasonic atomization. Then rinse the filter paper with running deionized water and dry it at room temperature. It can be found that due to the formation of Ag NPs, the filter paper changes from white to yellow. After the samples CE-0, CE-1, CE-2, CE-3 and CE-4 were treated by ultrasonic atomization with 0.2 wt% silver nitrate solution, the corresponding samples obtained were CE-Ag-0, CE-Ag -1, CE-Ag-2, CE-Ag-3 and CE-Ag-4. And the samples obtained after ultrasonic atomization treatment of CE-2 by silver nitrate solutions with concentrations of 0.2wt%, 0.4wt%, 0.6wt%, 0.8wt% and 1.0wt% are represented by Ag1, Ag2, Ag3, Ag4 and Ag5. The samples prepared above were carried out in the dark environment during the drying process and storage process to prevent sunlight from affecting the reaction process on the cellulose.

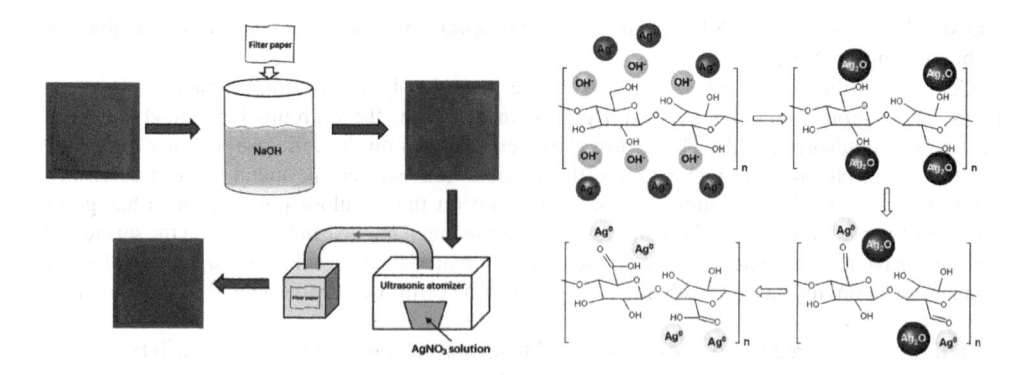

Figure 1. Schematic diagram of the synthesis of silver nanoparticles on cellulose.

Figure 2. Mechanism of the formation and loading of Ag NPs on cellulose.

2.3 Characterization

X-ray diffraction (XRD) patterns of the samples were obtained via a D/Max2550pc X-ray diffractometer with Cu Kα radiation (k = 1.54056 A°) at a scan rate of 4/min in the range 10°–80°. X-ray photoelectron spectroscopy (XPS) was used to investigate the chemical change of filter paper cellulose after loading Ag NPs and prove the existence of Ag NPs, using a VG ESCALAB MARK II spectrometer with an Mg Kα (1253.6 eV) X-ray source, and collected spectral data were calibrated by the C1s peak at 284.8 eV. Field emission scanning electron microscopy (FESEM, Hitachi SU-70) was used to characterize the microstructure of filter paper cellulose and cellulose/Ag NPs composite materials. The morphology of Ag NPs loaded on the cellulose was observed by transmission electron microscopy (TEM, JEOL-2000EX) and high-resolution transmission electron microscopy (HRTEM) at a 200 kV accelerating voltage. And the functional groups on the surface of cellulose were monitored by the Fourier transform infrared (FT-IR, Bruker Tensor 27) spectra, and the test range is 4000 to 500 cm^{-1}. UV–Visible absorption spectra of samples were obtained by UV-2600 UV–visible spectrophotometeran (Shimadzu Corporation) using a 350 nm excitation wavelength at room temperature. Furthermore, the amount of silver in samples were measured by Inductively Coupled Plasma Mass Spectrometer (ICPMS, Agilent 7700X).

2.4 Bacterial strain and antibacterial experiments

In this experiment, the antibacterial activity of cellulose filter paper with Ag NPs was tested against two bacterial strains viz. Escherichia coli (ATCC 11229) and Staphylococcus aureus (ATCC 6538). E. coli belongs to gram-negative bacteria, while S. aureus belongs to gram-positive bacteria. The strains were preserved in liquid nitrogen. The inhibition zone was used to characterize the antibacterial performance of the cellulose/Ag NPs composite material. Specifically, pick the test colony from the petri dish and place it in sterile physiological saline, adjust the bacterial solution to 0.5 McDonnell turbidity, apply the bacterial solution to the surface of the MH agar plate, and use sterile tweezers to stick the 6mm diameter cellulose/Ag NPs composites antibacterial sheet on the agar surface, place the plate at 37°C for 24 hours, and then use a ruler to measure the diameter of the inhibition zone around the sample paper sheet. Three replicate tests were carried out in the same conditions for each sample, and the average diameter of inhibition zone represents the antibacterial activity of the samples.

3 RESULTS AND DISCUSSION

3.1 Synthesis mechanism of Ag NPs on cellulose

Due to the existence of hydrogen bonds and other forces in the cellulose polymer chains, the cellulose polymer chains are tightly bonded to each other, so in the case of untreated, only part of the groups completely exposed on the surface of the filter paper cellulose can participate in the adsorption and the reduction of silver ions. Owing to the small number of these groups, the reducibility of untreated raw cellulose is relatively limited. In order to solve this problem, this experiment adopted the method of immersing in the alkaline solution to treat the filter paper cellulose, and the concentration of alkaline solution is much lower than the concentration that causes hydrolysis of cellulose, cellulose will exhibits higher reaction activation after treatment with alkaline solution. Soaking in the alkaline solution can fully swell the cellulose polymer and break the hydrogen bonds between the cellulose polymer chains, making more groups in the exposed or free state, thereby enhancing the adsorption and reduction ability of cellulose, in addition, in the alkaline environment, the reduction of hydroxyl groups will also be enhanced.

The reaction mechanism of cellulose to reduce silver ions under alkaline environment to obtain Ag NPs is shown in Figure 2. First, silver nitrate solution reacts with alkali solution to produce silver oxide. Due to the static electricity, silver oxide will be adsorbed by the hydroxyl groups on the cellulose polymer chains, and then under alkaline conditions, the silver oxide will react with the hydroxyl groups on the cellulose to form metallic silver and aldehyde groups, and the aldehyde groups will further reduce the silver oxide to generate metallic silver under alkaline conditions, and oxidized to carboxyl group eventually.

3.2 Characterization of cellulose loaded with Ag NPs

Figure 3 shows the SEM images of samples CE-Ag-0, CE-Ag-1, CE-Ag-2, CE-Ag-3 and CE-Ag-4, it can be seen from the figure that after immersing the cellulose in different concentrations of sodium hydroxide solution, the cellulose polymer chains will exhibit a certain swelling effect, this swelling effect will gradually enhance as the concentration of the alkaline solution increases, which is shown in the picture, the surface of the cellulose polymer become rougher. After the ultrasonic atomization process, it can be seen that the particle size of Ag NPs loaded on the cellulose is evenly distributed, and there is no obvious agglomeration, which is related to the characteristics of cellulose and ultrasonic atomization. The large number of active groups present on the surface of the cellulose polymer chains provide sufficient nucleation sites for the formation of Ag NPs, which effectively avoid the growth and reunion of Ag NPs. In addition, ultrasonic atomization produces mist droplets in the size of micrometers, these droplets start to form Ag NPs once the droplets contact with the filter paper, and the silver ions are sufficiently dispersed in the mist droplets, which make sure that the Ag NPs are loaded on the cellulose uniformly and quickly.

Figure 4 shows the SEM pictures of samples CE-Ag-0, Ag1, Ag2, Ag3, Ag4 and Ag5. For sample CE-Ag-0, the filter paper is directly subjected to ultrasonic atomization treatment without immersion in alkali solution, and Ag NPs can hardly be seen from the Figure 4a. It can be seen from Figure 4b-4f that as the concentration of silver nitrate increases, the particle size of the generated Ag NPs also increases accordingly, and the dispersibility gradually decreases. When the concentration of silver nitrate is 0.2wt% and 0.4wt%, the particle size distribution of Ag NPs is uniform, and there is no obvious agglomeration. When the concentration of silver nitrate continued to increase to 0.6 wt% and 0.8 wt%, some Ag NPs began to agglomerate. When the concentration of the silver nitrate increases to 1 wt%, there were basically only large-sized agglomerated silver particles in the field of view. This is because when the concentration of the silver nitrate increases, the Ag NPs generated per unit area increase, due to the large specific surface area and high surface energy of the Ag NPs, agglomeration is more likely to occur during the growth process of Ag NPs.

Figure 5 shows the UV-Visible absorption spectra of samples CE-0, Ag1, Ag2, Ag3, Ag4 and Ag5. From the absorption spectrum, it can be seen that the samples Ag1, Ag2, Ag3, Ag4 and

Ag5 which have undergone the alkali solution and ultrasonic atomization treatment have obvious absorption around 414nm. According to the literatures [6], this absorption peak corresponds to the ultraviolet absorption peak of the Ag NPs. These results can prove that the Ag NPs are successfully loaded on the filter paper cellulose.

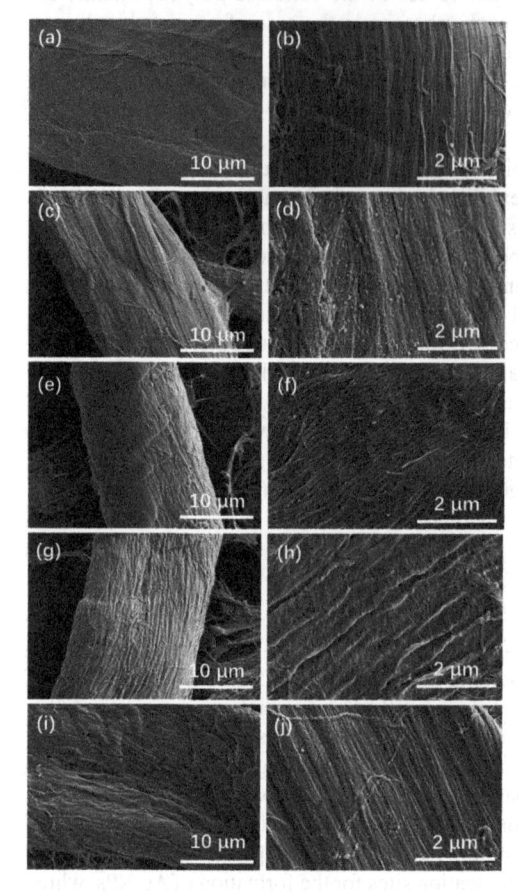

Figure 3. SEM images of samples (a), (b) CE-Ag-0, (c), (d) CE-Ag-1, (e), (f) CE-Ag-2, (g), (h) CE-Ag-3 and (i), (j) CE-Ag-4.

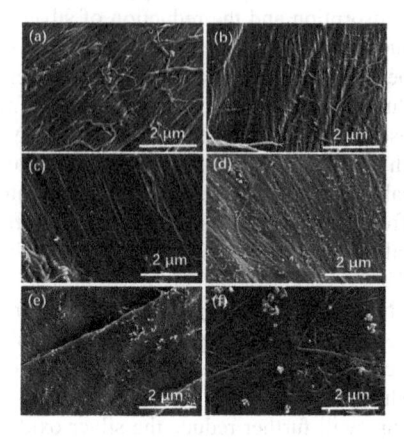

Figure 4. SEM images of samples (a) CE-Ag-0, (b) Ag1, (c) Ag2, (d) Ag3, (e) Ag4 and (f) Ag5.

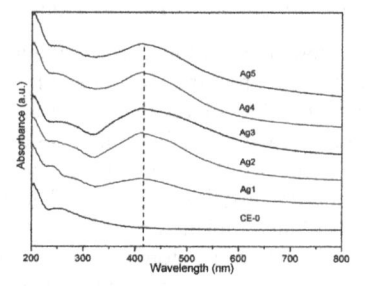

Figure 5. UV-Visible absorption spectra of samples CE-0, Ag1, Ag2, Ag3, Ag4 and Ag5.

The X-ray photoelectron spectroscopy (XPS) was used to characterize the chemical state of the surface elements of sample Ag1. The results are shown in Figure 6. Figure 6(a) clearly shows the Ag3d peak in cellulose, proving the presence of silver in cellulose. In order to further determine the state of the silver element, the high-resolution XPS pattern of silver was analyzed. As shown in Figure 6(b), the Ag 3d spectrum exhibited two contributions named 3d5/2 and 3d3/2 (resulting from the spin–orbit splitting), and the binding energy of Ag 3d5/2 and Ag 3d3/2 are located at 368.6 and 374.6 eV respectively, and the spin separation energy is 6ev. According to related literature, the orbital binding energy of Ag 3d5/2 and Ag 3d3/2 in Ag are about 368.6 eV and 374.6 eV. The high-resolution XPS pattern of C 1s was shown in Figure 6(c), C 1s was fitted with a maximum of four peaks centered at 284.8, 286.4, 287.7 and 288.6 eV, assigned to the sp3 carbon, C-O, O-C-O and O-C=O in cellulose polymer chains. The appearance of the O-C=O peak corresponds to the formation of carboxyl groups, indicating that redox reactions have occurred on the cellulose, further verifying that silver ions were reduced to metallic silver successfully. Thus, the XPS results show that most of the silver element in cellulose exists in the state of metallic.

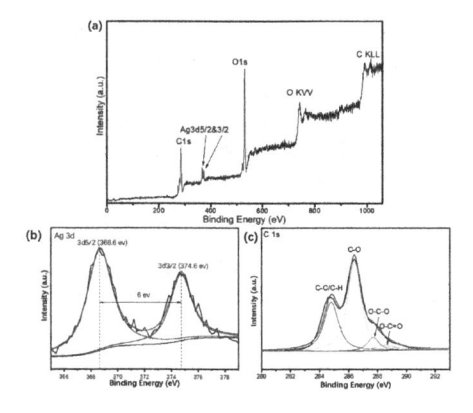

Figure 6. XPS survey spectra and high-resolution XPS pattern of sample Ag1.

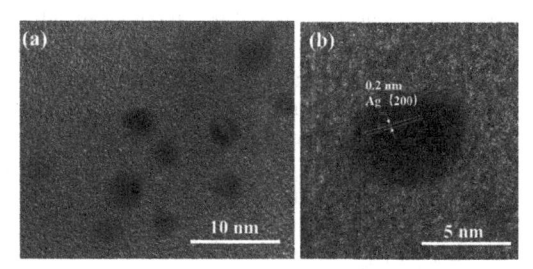

Figure 7. TEM and HRTEM images of Ag NPs collected from the sample Ag1.

The cellulose/silver nanoparticle composites were ultra-sonicated in an alcohol solution for 30 min, and then the Ag NPs were collected from the alcohol solution. Afterwards, the TEM and high-resolution TEM tests were carried out to investigate the morphology and size of Ag NPs collected from sample Ag1. As shown in Figure 7(a), it can be seen that the size of the Ag NPs is basically between 5 nm and 10 nm, and there is no agglomeration of Ag NPs. From the high-resolution TEM image Figure 7(b), the lattice fringes of Ag NP can be measured to be 0.2 nm, which corresponds to the (200) plane of the body-centered cubic structure of silver. The results show that Ag NPs were successfully loaded on cellulose.

Figure 8 shows the results of ICP-MS measurement of the silver content in samples Ag1, Ag2, Ag3, Ag4 and Ag5. According to the characterization results above, the silver content in the samples is also the content of Ag NPs loaded on cellulose. It can be seen that as the concentration of the silver nitrate increases in the ultrasonic atomization process, the content of Ag NPs loaded on the cellulose also increases. At present, there is no uniform specification for the content of Ag NPs in the material, and according to related research, when the concentration of Ag NPs reaches the ppb level, it can already play an antibacterial effect, and when the content of silver particles loaded on the cotton fabric reaches 18 ppm, the cotton fabric will exhibits good antibacterial properties [7].

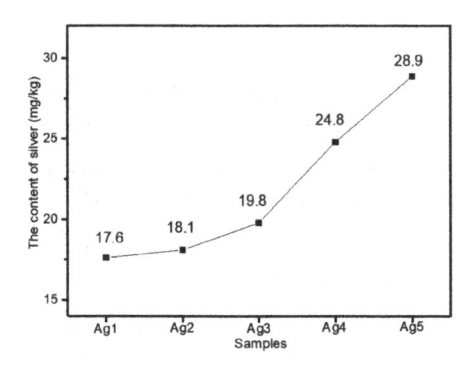

Figure 8. The content of silver in different samples.

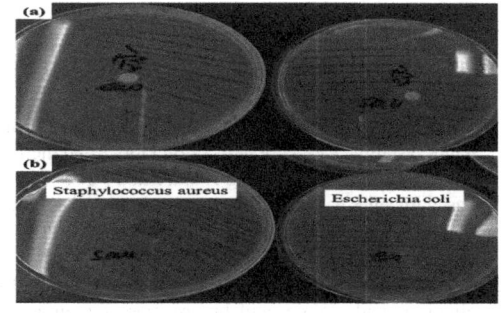

Figure 9. Antibacterial properties of samples (a) original cellulose filter paper, (b) cellulose/Ag NPs composite material Ag1 (left-against Staphylococcus aureus, right-against Escherichia coli).

It can be seen from Figure 9(a) that there is no inhibition zone around the original cellulose filter paper, indicating that the original cellulose has no inhibitory effect on Escherichia coli and Staphylococcus aureus. It can be seen from Figure 9(b) that the cellulose/Ag NPs composite material sample Ag1 has obvious inhibition zones in the MH agar plates of Escherichia coli and Staphylococcus aureus, and the diameter of inhibition zone in the petri dish of Escherichia coli is 1.1 mm, and the diameter of the inhibition zone in the petri dish of Staphylococcus aureus is 2.3 mm. Ag NPs can adsorb to the cell membrane surface of bacteria, destroy the cell membrane, and affect cell penetration and respiration. In addition, silver nanoparticles can also release silver ions, generate reactive oxygen species (ROS), and affect the function of DNA, thereby achieving antibacterial effects. The antibacterial test shows that the cellulose/Ag NPs composite material obtained in this experiment has good antibacterial activity.

4 CONCLUSION

The present research introduced a method for rapidly synthesis Ag NPs on cellulose filter paper at room temperature using the reducibility of cellulose itself. In this method, after treatment in alkaline solution, putting the cellulose in the mist droplets generated by ultrasonic atomization of 0.2 wt% silver nitrate solution for two minutes, then the content of Ag NPs loaded on the cellulose can reach 17.6 mg/kg. The study shows that as the concentration of the alkaline solution used to soak cellulose increases, the surface of the cellulose polymer will become rougher accordingly. With the increase of the concentration of silver nitrate for ultrasonic atomization, the dispersibility of Ag NPs on the cellulose will decrease and agglomeration will occur. And the results of the antibacterial test show that the cellulose/Ag NPs composites prepared by this method has good antibacterial activity. It is reasonable to believe that this method of rapidly loading Ag NPs on cellulose in situ at room temperature has promising applications in the preparation of cellulose/Ag NP related composite materials.

ACKNOWLEDGEMENTS

This work was supported by the opening foundation of the State Key Laboratory for Diagnosis and Treatment of Infectious Diseases and Collaborative Innovation Center for Diagnosis and Treatment of Infectious Diseases, The First Affiliated Hospital of Medical College, Zhejiang University, Grant No. 2017KF08, and the National Natural Science Foundation of China No. 50772098.

REFERENCES

[1] Phan D, Dorjjugder N, Khan M Q, Saito Y, Taguchi G, Lee H, Mukai Y and Kim I 2019 Synthesis and attachment of silver and copper nanoparticles on cellulose nanofibers and comparative antibacterial study *CELLULOSE* **26** 6629–40
[2] Jin Z, Wang S, Yang F, Dong P, Huang Z, Zhang X and Xiao Y 2019 A two-step preparation method for nanocrystalline Ag-decorated cotton fabrics and their antibacterial assessment *J MATER SCI* **54** 10447–56
[3] Hu H, Wu X, Wang H, Wang H and Zhou J 2019 Photo-reduction of Ag nanoparticles by using cellulose-based micelles as soft templates: Catalytic and antimicrobial activities *CARBOHYD POLYM* **213** 419–27
[4] Samyn P, Barhoum A, Ohlund T and Dufresne A 2018 Review: nanoparticles and nanostructured materials in papermaking *J MATER SCI* **53** 146–84
[5] Alavi M and Rai M 2019 Recent progress in nanoformulations of silver nanoparticles with cellulose, chitosan, and alginic acid biopolymers for antibacterial applications *APPL MICROBIOL BIOT* **103** 8669–76
[6] Khaksar M, Vasileiadis S, Sekine R, Brunetti G, Scheckel K G, Vasilev K, Lombi E and Donner E 2019 Chemical characterisation, antibacterial activity, and (nano)silver transformation of commercial personal care products exposed to household greywater *ENVIRON SCI-NANO* **6** 3027–38
[7] Mishra A and Butola B S 2017 Deposition of Ag doped TiO2 on cotton fabric for wash durable UV protective and antibacterial properties at very low silver concentration *CELLULOSE* **24** 3555–71.

Extraction optimization of antibacterial substances from *Cortex Phellodendri* against *Vibrio parahaemolyticus*

Y.A. Li, H.F. Zheng, S.Y. Guo, Z.Y. Zhang & L. Guo
Co-Innovation Center of Jiangsu Marine Bio-industry Technology, Jiangsu Ocean University, Lianyungang, China

ABSTRACT: This study aimed to optimize extraction process of antibacterial substances from *Cortex Phellodendri* against *Vibrio parahaemolyticus*. The extraction process was optimized by the single-factor experiment, Box-Behnken design and response surface methodology and determined as: anhydrous methanol, liquid-to-material ratio 20:1 mL/g, extraction temperature 91°C and extraction time 123 min. The factual diameters of the inhibition zone of the extracts of *C. Phellodendri* were 11.43 ± 0.29 mm, which was not significantly different from the predicted value (11.02 mm). The yield of antibacterial substances from *C. Phellodendri* was $15.80 \pm 0.49\%$ and the minimal inhibitory concentration is 1.0 mg/mL. The results provide a clue for further isolation of the antibacterial compounds of *C. Phellodendri*.

1 INTRODUCTION

At present, China's aquaculture output has exceeded 50 million tons, accounting for more than 78% of the world's total aquatic products, which is the only country where the total caught amount is less than the total farmed amount in the world. With the development of large-scale and intensive aquaculture modes, many problems related to aquatic animal diseases and water environmental pollution have been exposed, and these problems have severely restricted the stable development of aquaculture [1,2]. Bacterial diseases are the most vulnerable diseases to aquatic animals. Vibriosis caused by *Vibrio parahaemolyticus* and other bacteria of the genus *Vibrio* is one of the most serious bacterial diseases [3]. The improper treatment of vibriosis in aquatic animals not only brings huge economic losses to aquaculture industry, but also causes food safety issues [4]. In order to prevent the occurrence of diseases and reduce economic losses, a large number of antibiotics are used in the breeding process for a long time, which leads to the production of drug-resistant strains, ecological imbalance and weakening of the biological immune system, and even serious drug residues in aquatic products [5]. *V. parahaemolyticus* can cause seafood-borne diarrhea in people and infectious diarrhea related to aquatic products worldwide as one of the main reasons [3].

Chinese herbal medicine has received extensive attention in the development of antibiotic substitutes because of their wide sources, low price, rich active ingredients, low resistance, low toxicity, and low residue [6]. *Cortex Phellodendri* is a Chinese herbal medicine with anti-*V. parahaemolyticus* activity that we screened. The study aimed to optimize the extraction technology of the antibacterial components of *C. Phellodendri* against *V. parahaemolyticus*, obtain the antibacterial active substances, and provide a basis for further separation and identification of its antibacterial active compounds.

2 MATERIALS AND METHODS

2.1 Materials

C. Phellodendri was purchased from a local pharmacy in Lianyungang (Jiangsu, China); MH broth medium was purchased from Hangzhou Basebio Co., Ltd.

2.2 Single-factor trials

The crushed C. Phellodendri powder (1.0 g) was precise weighed and put into a 250 mL flat bottom flask for each reflux extraction. The extraction was implemented at different extraction conditions. The factors and levels of the single-factor trials are shown in Table 1. After each extraction is finished, the extract is obtained by vacuum filtration, and the antibacterial activity is determined by Oxford cup method after the volume is adjusted to 20 mL.

Table 1. Factors and levels in the single-factor trials.

Level	Solvent	Methanol concentration	Temperature (°C)	Time (min)	Liquid/material ratio (mL/g)
1	distilled water	60	60	60	10
2	50% ethanol	70	70	90	20
3	ethanol	80	80	120	30
4	50% methanol	90	90	150	40
5	methanol	100	100	180	50

2.3 Response surface methodology (RSM)

RSM was adopted to optimize the extraction conditions. Box-Behnken design (BBD) was used for experimental design, and Design Expert 7.0.0 was adopted to analyze the data and build the model [3]. BBD consists of twelve factor points and five center points. The response (Y, mm) is the inhibition zone diameter of the extract against $V.$ parahaemolyticus, temperature (X_1), time (X_2) and L/M ratio (X_3) are the independent variables. Table 2 showed the design and results of BBD. The experiment was carried out in three parallels, and the antibacterial activity is recorded as the average value of the diameter of the inhibition zone. The following equation (1) was applied to assess the model of this system:

$$Y = \beta_0 + \sum_{i=1}^{3} \beta_i X_i + \sum_{i=1}^{3} \beta_{ii} X_i^2 + \sum_{i=1}^{2}\sum_{j=i+1}^{3} \beta_{ij} X_i X_j \quad (1)$$

where Y is the response factor, X_i and X_j are the independent factors, β_0, β_i, β_{ii} and β_{ij} are the constants.

2.4 Agar diffusion method

The anti-$V.$ parahaemolyticus activity of the extract was determined by the agar diffusion method [7]. Twenty milliliters of sterilized MH broth agar medium (MH broth 21 g, agar 20 g, 1 L seawater) was poured into the plate. One hundred microliter of the $V.$ parahaemolyticus suspension with a concentration of 10^6 cells/mL was uniformly spread on the solidified medium. The Oxford cup was put in the plate, and two hundred microliters of the extract was put into the cup, then the plate was covered with a lid and cultured at 37°C for 24 h. The inhibition zone diameter was measured by the electronic digital vernier caliper.

2.5 Minimum inhibitory concentration (MIC) assay

The extract was diluted twice to prepare solutions of different concentrations, and the MIC value of the extract against *V. parahaemolyticus* was evaluated by the above-mentioned agar diffusion method. The MIC value was taken as the minimum concentration of the extract that can produce the inhibition zone, the test is carried out in parallel for 2 groups, and the result that meets the conditions at the same time is the positive result.

3 RESULTS AND DISCUSSION

3.1 Single factor investigations

Under the conditions of different extraction solvents, the inhibition zone diameters of the extract extracted with methanol was the biggest. The effect of different methanol concentrations on the inhibition zone of the extract was further investigated (Figure 1). With the increase of methanol concentration, the diameter of the inhibition zone gradually increased, and the effect of anhydrous methanol was the best, which indicated that the antibacterial substance was hydrophobic. Thus, anhydrous methanol was selected as the extraction solvent.

Figure 2 shows that as the temperature increases from 60°C to 90°C, the inhibition zone diameters gradually increase. Increasing temperature can improve extraction efficiency by reducing viscosity and increasing solubility [4]. Therefore, 90°C was adopted as the central point of temperature in the Box-Behnken design.

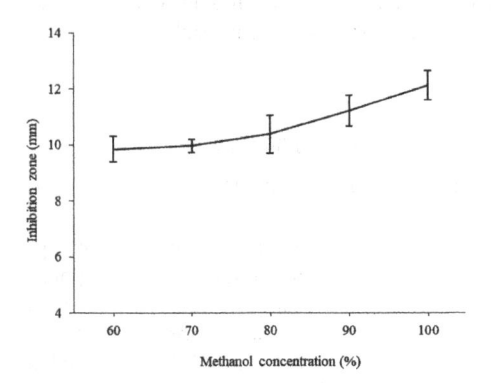

Figure 1. Effects of methanol concentration on the anti-*V. parahaemolyticus* activity of the extract from *C. Phellodendri*.

Figure 2. Effects of temperature on the anti-*V. parahaemolyticus* activity of the extract from *C. Phellodendri*.

Figure 3 shows that with the extraction time from 60 min to 120 min, the diameter of the inhibition zone gradually increases, and then does not increase. Thus, 120 min was chosen as the central point of temperature in the Box-Behnken design.

Figure 4 shows that when the L/M ratio changes from 10:1 mL/g to 20:1 mL/g, the diameter of the inhibition zone increases, but as the ratio of liquid to material continues to increase, the diameter of the inhibition zone does not increase and slightly reduce. Therefore, a L/M ratio of 20:1 mL/g was adopted as central point of temperature in the Box-Behnken design.

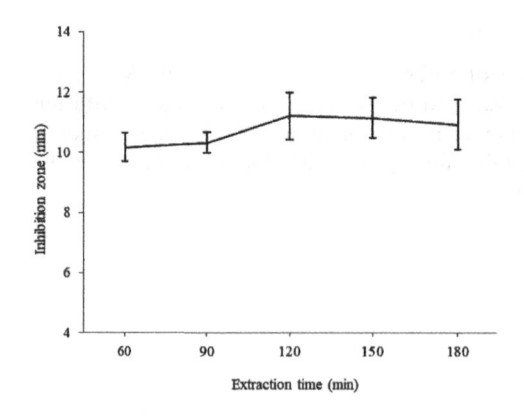

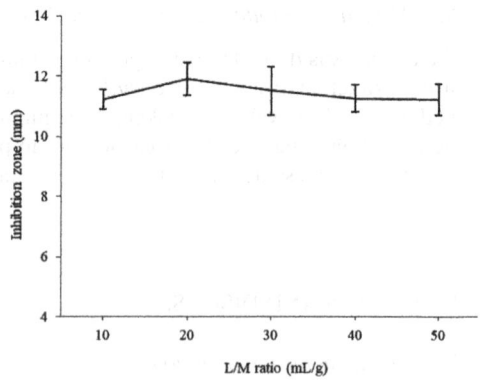

Figure 3. Effects of extraction time on the anti-*V. parahaemolyticus* activity of the extract from C. Phellodendri.

Figure 4. Effects of L/M ratio on the anti-*V. parahaemolyticus* activity of the extract from C. Phellodendri.

3.2 Optimization of extraction parameters

On the basis of the single-factor test results, extraction temperature (X_1), time (X_2) and L/M ratio (X_3) were used as 3 test variables for BBD optimization, and the inhibition zone diameter was used as the response variable (Y). The BBD matrix and results are shown in Table 2. RSM can provide an empirical relationship between the response and the test variable on the basis of parameter estimation. The equation of the diameter of the inhibition zone is established through the regression analysis of the data:

$$Y = 11.01 + 0.064X_1 + 0.085X_2 - 0.00375X_3 + 0.098X_1X_2 + 0.11X_1X_3 - 0.24X_2X_3 - 0.43X_1^2 - 0.45X_2^2 - 0.22X_3^2 \quad (2)$$

As can be seen from Table 3, P value < 0.05 indicates that the degree of model fit is significant, that is, the regression relationship between the independent variable and the response variable of the regression model is significant. In addition, the value of lack of fit is 0.3327, which means that there is no significant difference. The order of influence of each factor relative to the extraction efficiency of anti-*V. parahaemolyticus* activity of the extract from C. Phellodendri is time > extraction temperature > L/M ratio. The quadratic effects of temperature and time are also significant ($P < 0.05$).

The regression analysis was applied to the equation to obtain the optimal parameters predicted: $X_1 = 91°, X_2 = 123$ min, $X_3 = 20:1$ mL/g, and the corresponding predicted value is 11.02 mm. The actual value of the inhibition zone obtained by the verification experiment is 11.43 ± 0.29 mm (Figure 5), indicating that the BBD combined with RSM is feasible for the extraction optimization of the anti-*V. parahaemolyticus* components from C. Phellodendri.

3.3 MIC value of the antibacterial substances from C. Phellodendri

The extract extracted by the optimal extraction process was evaporated under reduced pressure and dried to obtain the solid antibacterial substances from C. Phellodendri against *V. parahaemolyticus* with the yield was $15.80 \pm 0.49\%$. The anti-*V. parahaemolyticus* active substance of C. Phellodendri was prepared by double dilution method to prepare a series of $0.125 \sim 2.0$ mg/mL solutions. The MIC value determined by the Oxford cup method was 1.0 mg/mL (Figure 6).

Table 2. BBD matrix and results.

No.	X_1	X_2	X_3	Y
1	−1	−1	0	10.25
2	+1	−1	0	10.11
3	−1	+1	0	9.95
4	+1	+1	0	10.21
5	−1	0	−1	10.26
6	+1	0	−1	10.23
7	−1	0	+1	10.27
8	+1	0	+1	10.69
9	0	−1	−1	10.01
10	0	+1	−1	10.92
11	0	−1	+1	10.23
12	0	+1	+1	10.20
13	0	0	0	10.66
14	0	0	0	11.00
15	0	0	0	11.20
16	0	0	0	11.27
17	0	0	0	10.92

Table 3. Analysis of variance (ANOVA) of the model.

Source	Sum of Squares	df	Mean Square	F Value	Prob > F
Model	2.42	9	0.27	3.7	0.0491
X_1	0.033	1	0.033	0.45	0.5225
X_2	0.057	1	0.057	0.79	0.4044
X_3	1.13E-04	1	1.13E-04	1.55E-03	0.9697
X_1X_2	0.039	1	0.039	0.53	0.4894
X_1X_3	0.053	1	0.053	0.73	0.4219
X_2X_3	0.22	1	0.22	3.06	0.1237
X_1^2	0.78	1	0.78	10.69	0.0137
X_2^2	0.85	1	0.85	11.66	0.0112
X_3^2	0.2	1	0.2	2.77	0.1399
Lack of Fit	0.27	3	0.091	1.55	0.3327

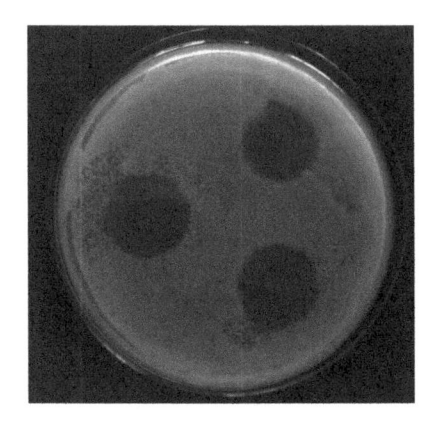

Figure 5. Validated inhibition zone diagram of the extract from C. Phellodendri.

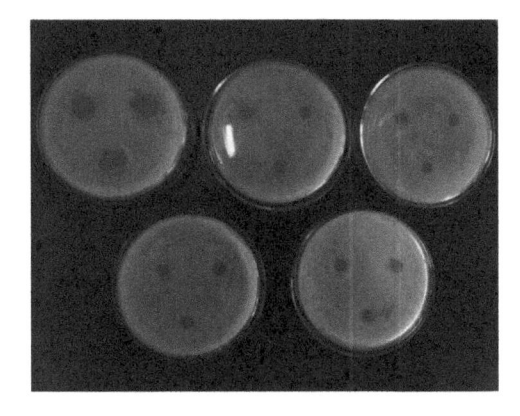

Figure 6. Minimum inhibitory concentration results of the antibacterial substances from C. Phellodendri against V. parahaemolyticus.

4 CONCLUSION

The extraction process of anti-*V. parahaemolyticus* components from *C. Phellodendri* was optimized by the single factor trial, BBD and RSM followed as: anhydrous methanol as extraction solvent, L/M ratio 20:1 mL/g, temperature at 91° and time for 123 min. Moreover, the yield of antibacterial substances from *C. Phellodendri* was 15.80 ± 0.49% and the MIC value is 1.0 mg/mL. The results provide a clue for further isolation of the antibacterial compounds of *C. Phellodendri*.

ACKNOWLEDGMENTS

The authors thank the funding of the Priority Academic Program Development of Jiangsu Higher Education Institutions (PAPD).

REFERENCES

[1] Wang XT, Lu YD, Wang LL, Li XY, Li SY, Che J, Xu YP 2015 *Feed and Livestock* **4** 18–22
[2] Wang YT 2012 *China Fisheries* **10** 7–19
[3] Wang X, Guo S, Zhang Z, Xu Q, Guo L 2020 *J. Phys.: Conf. Ser.* **1637** 012108
[4] Guo L, Wang X, Feng J, Xu X, Li X, Wang W, Sun Y, Xu F 2020 *Biotechnol Biotec Eq* **34** 1215–1223
[5] Martinez-Urtaza J, Lozano-Leon A, Varela-Pet J, Trinanes J, Pazos Y, Garcia-Martin O 2008 *Appl. Environ. Microbiol.* **74** 265–274
[6] Zhu F 2020 *Aquaculture* **526** 735422
[7] Xu X, Guo S, Chen H, Zhang Z, Li X, Wang W, Guo L 2021 *3 Biotech* **11** 193

Preparation of Au nanospheres/TiO₂ complexes and their photocatalytic performance of H₂

D.W. Zheng
New Energy Research Center, Research Institute of Petroleum Exploration & Development (RIPED), PetroChina, Beijing, China

Y.L. Zou & J.F. Liu
School of Chemical and Environmental Engineering, Shanghai Institute of Technology, Shanghai, China

X. Zhang & S.Y. Wang
New Energy Research Center, Research Institute of Petroleum Exploration & Development (RIPED), PetroChina, Beijing, China

S.Z. Kang
School of Chemical and Environmental Engineering, Shanghai Institute of Technology, Shanghai, China

ABSTRACT: Au nanospheres/TiO₂ composite was prepared by a sol-gel process. Meanwhile, its photocatslytic activity was investigated for H_2 evolution. The results indicate that it is an efficient photocatalyst with a H_2 evolution rate of approximate 1.18 mmolg^{-1}h^{-1} during first hour under visible irradiation. Additionally, the size of the Au nanospheres plays an importnt role in the hydrogen production. with increasing the size of Au nanospheres, the rate of H_2 evolution gradually increases, reaches a maximum and thereafter begins to decrease. It may be closely related to the change of plasmon resonance absorption induced by the size of Au nanospheres.

1 INTRODUCTION

With the deepening of TiO₂ research, many shortcomings of TiO₂ are exposed. For example, because it only absorbs ultraviolet light, the limitation of its application in photocatalysis is more obvious. Therefore, how to overcome these defects has been a research hotspot in the field of photocatalytic hydrogen production. The researchers have developed many effective methods for TiO₂ modification, such as doping [1], dye sensitization [2,3] and composite with other functional materials [4] etc.

Recently, it has been found that the surface plasmonic resonance absorption effect of gold and silver nanomaterials can make TiO₂ exhibit excellent catalytic performance for H_2 production under visible irradiation [5]. Silva et al. [6] prepared the Au/TiO₂ composite by deposition of gold nanoparticles on P25. The results showed that the composite can exhibit high photocatalytic activity for hydrogen production in both ultraviolet and visible light. The plasma resonance effect of gold nanoparticles is stronger than that of silver nanoparticles [7]. In addition, the higher work function of gold can better promote the charge separation in TiO₂. However, the relationship between the size and morphology of gold nanoparticles and the photocatalytic activity of H_2 production is still not clear. A deep understanding of the structure-effect relationship is of great significance for the utilization of the plasma resonance absorption effect of gold nanoparticles.

Based on the above considerations, we prepared four Au nanospheres with various sizes, and then Au nanospheres/TiO₂ composites were prepared by a simple sol-gel method. On this basis, the relationship between the size of the Au nanospheres and the photocatalytic activity of the composite was systematically discussed. Finally, the mechanism of photocatalytic hydrogen production of Au nanospheres/TiO$_2$ composite was preliminarily studied.

2 EXPERIMENTAL SECTION

2.1 Materials

TiO$_2$ nanoparticles (Degussa P25) were purchased from Shanghai Haiyi trade Co. Other agents were commercially available and used as received. All of reagents were used without further purification. Deionized water was used as the solvent.

2.2 Characterization

The powder X-ray diffraction (XRD) analysis was made on a PANalytical Xpert Pro MRD X-ray diffractometer (Netherlands). The UV-vis spectra were recorded with a Hitachi U-3900 (Japan) and a UV1000 (Shanghai Tianmei Science Instrument, China) UV-vis spectrophotometer. The High resolution transmission electron microscopy (HRTEM) pictures were taken on a JEOL JEM-2100F (Japan).

2.3 Preparation of Au nanospheres

2.3.1 Preparation of 12 nm Au nanospheres

The HAuCl$_4$ solution (50 mL 1 mM) was heated to boiling. Subsequently, 5 mL 1wt% of trisodium citrate solution was quickly added under stirring. Keep the solution boiling for 10 min, then continue to stir for another 15 min. After static and naturally cooled to room temperature, Au nanospheres dispersion was obtained. It is kept in dark at normal temperature.

2.3.2 Preparation of 30 nm Au nanospheres

The trisodium citrate solution (150 mL 2.2 mM) was heated to boiling. Under stirring, HAuCl$_4$ solution (1 mL 25 mM) was added. After stirred for 10 min, the Au seed solution was obtained. Next, the solution was naturally cooled to 90°C. Sodium citrate solution (1 mL 60 mM) and HAuCl$_4$ solution (1 mL 25 mM) were added in the solution above in 2 min, and the mixture was kept at 90°C for 30 min. The as-prepared Au nanoparticle dispersion was used as a new Au seed solution, and the above growth steps were repeated 14 times. After cooled to room temperature, the Au nanaspheres diepersion was kept in dark at normal temperature.

2.3.3 Preparation of 50 nm Au nanospheres

The HAuCl$_4$ solution (5.4 mL 25 mM) was diluted to 150 mL and heated to boiling. Under vigorous stirring, the trisodium citrate solution (18 mL 1 wt%) was dropped in and kept in boiling for 1 h. After naturally cooled to room temperature, the dispersion above was kept in dark at roon temperature.

2.4 Preparation of Au nanospheres/ TiO$_2$ complexes (Au/TiO$_2$)

Tetrabutyl titanate (3 mL) and ethanol (10 mL) were mixed at room temperature. The dispersion of Au nanospheres was added drop by drop and then stirred at room temperature for 12 h to form the sol. After the sol was transferred into a Teflon-lined stainless steel autoclave, the autoclave was heated at 120°C for 8 h. Next, the autoclave was cooled to room temperature; the samples were rinsed thoroughly with ethanol and double-distilled water, and dried in air at 50°C for 2 h. The finished product is obtained by grinding. The resulting products are marked as 12-Au/TiO$_2$, 30-Au/TiO$_2$ and 50-Au/TiO$_2$, and the numbers represent the size of the Au nanospheres. In addition, in order to investigate the effect of heat treatment on the properties of the composite, we also heat treatment some products, The Au/TiO$_2$ composites are calcined at 400° for 2 h in a muffle furnace.

2.5 *Electrochemical measurements*

The AC impedances of the samples were performed with a CHI660D electro-chemical system (Shanghai Chenhua Instruments, China) in a three-electrode system. The working electrode was Au/TiO$_2$ modified electrode. An Ag/AgCl electrode and a platinum electrode were used as the reference electrode and counter electrode, respectively. The supporting electrolyte solution was a mixture of KNO$_3$ (0.1mM), K$_3$Fe(CN)$_5$ (5 mM) and K$_3$Fe(CN)$_6$ (5 mM).

2.6 *Photocatalytic measurements*

Photocatalytic experiments were carried out with a CEL-SP2N water splitting system (Zhongjiao Jinyuan Instruments, Beijing, China). The photocatalytic reaction is carried out in a sealed quartz reactor. A 300 W xenon lamp with a filter ($\lambda > 420$ nm) was used as the light source. The light source is 1 cm away from the reactor. In a typical experiment, 60 mg Au/TiO$_2$ was added into 60 mL methanol solution (50 vol.%) under stirring. After dropping 390 μL 10 g L^{-1} chloroplatinic acid solution to the suspension, the system was pumped into vacuum state under stirring. Then, the suspension was reduced under UV light for 30 min to prepare Au/TiO$_2$ loaded with 1 mol% Pt (Au/TiO$_2$-Pt). Finally, the photocatalytic reaction device was vacuumed again, nitrogen was introduced, and the xenon lamp was turned on to start the photocatalytic reaction. The amount of hydrogen produced was measured with a gas chromatography.

3 RESULTS AND DISCUSSION

3.1 *Characterizations*

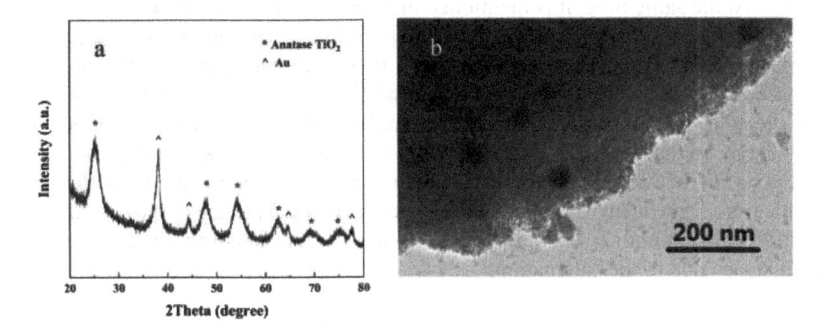

Figure 1. (a) XRD pattern and (b) TEM image of the as-prepared Au nanospheres/TiO$_2$ composite.

Figure. 1(a) shows the XRD pattern of the as-prepared Au nanospheres/TiO$_2$ composite. As can be seen from Figure 1(a), the XRD diffraction peak positions of the sample are well coincident with the standard XRD diffraction peaks of anatase TiO$_2$ (JCPDS No.21–1272) and Au (JCPDS No.04–0784), respectively. Therein, the diffraction peaks at 25.28°, 48.05°, 53.89°, 62.69°, 68.76° and 75.03° should correspond to the (101), (200), (105), (204), (116) and (215) crystal planes of anatase TiO$_2$, respectively. The peaks at 38.18°, 44.39°, 64.58° and 77.55° could be assigned to the diffraction peaks corresponding to (111), (200), (220) and (311) crystal planes of gold, respectively. This result indicates that the product is composed of Au and TiO$_2$ complexes.

Figuer 1(b) shows TEM image of as-prepared Au nanospheres/TiO$_2$ composite. From Figuer. 1(b),we can find that some dark nanoparticles are embedded in a lighter matrix. The lighter part consists of nanoparticles with 5 nm in size. Since Au atoms are heavier than Ti atoms, the contrast of the rich Au region is darker. Therefore, we refer to dark nanoparticles as Au nanospheres, while lighter nanoparticles are TiO$_2$. The above results further confirm that the sample is a composite composed of Au nanospheres and TiO$_2$ nanoparticles. And the Au nanospheres are embedded in a matrix composed of the TiO$_2$ nanoparticles.

3.2 Optical properties of Au nanosphere/TiO$_2$ composites

Figure 2 shows UV-vis spectra of the samples, which are recorded in the range of 250–800 nm. From Figure 2A, we can observe that Au nanospheres have strong absorption capacity in the visible light region, which should be attributed to its surface plasmon resonance effect. Moreover, as the size of the Au nanosphere increases, its absorption band gradually red-shifts from 520 nm to 550 nm. This phenomenon should be caused by the quantum size effect of the Au nanospheres [8]. These results indicate that we can modulate the surface plasmon resonance effect of Au nanospheres by controlling the size of the Au nanospheres.

From Figure 2B we can find that TiO$_2$ nanoparticles have no absorption in the visible region. In contrast, 30-Au/TiO$_2$ exhibits a very obvious absorption peak at 565 nm, which might be ascribed to the surface plasmon resonance effect of Au nanoshperes. Compared with the UV-vis spectrum of anatase TiO$_2$ nanoparticles, it can be observed that the introduction of Au nanospheres does not cause the absorption band of TiO$_2$ around 400 nm to move. This shows that the band structure of TiO$_2$ nanoparticles does not change due to the introduction of Au nanospheres. On the other hand, compared with the the UV-vis spectrum of 30 nm Au nanospheres, the plasmon resonance absorption of the Au nanospheres has a significant red shift in the composite. This shows that there exists some strong interaction between Au nanospheres and TiO$_2$ matrix.

Figure 2C shows the UV-Vis absorption spectra of 12-Au/TiO$_2$, 30-Au/TiO$_2$ and 50-Au/TiO$_2$. It can be seen from Figure 2C that when the size of the Au nanospheres becomes larger, the position of the absorption peak of the samples in the visible light region also undergoes a red shift. However, compared with the results shown in Figure 2A, the degree of red shift of the absorption peak is much smaller. At the same time, it is mentioned in Figure 2B that there is some strong interaction between Au nanospheres and TiO$_2$ matrix. Under its action, the plasmon resonance absorption of Au nanospheres in the matrix is red-shifted. From these experimental phenomena, we speculate that the interaction between the Au nanospheres and the TiO$_2$ matrix may be greatly affected by the size of the gold. The interaction between the small-sized Au nanospheres and the TiO$_2$ matrix is relatively large, so that the change in the position of the sample absorption peak caused by the change in the size of the Au nanospheres becomes weaker. After heat treatment, the absorption peak of the sample has a significant red shift. This may be due to: (1) The size of the Au nanospheres has increased after calcination; (2) The interaction between the Au nanospheres and the TiO$_2$ matrix has been enhanced due to the calcination.

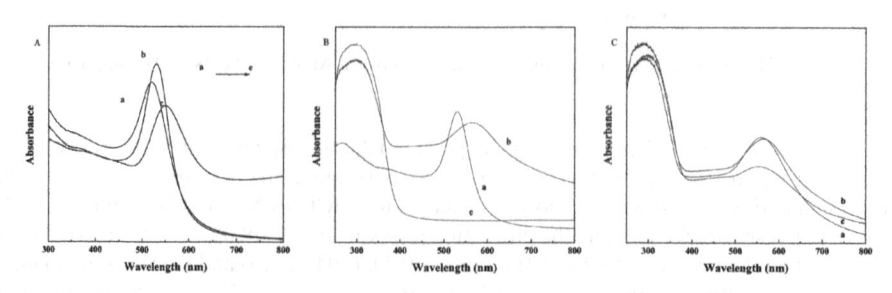

Figure 2. UV-vis spectra of A) Au nanospheres with varius sizes:(a) 12 nm, (b) 30 nm and (c) 50 nm; B) (a) Au nanospheres of 30 nm, (b) 30-Au/TiO$_2$ and (c) anatase TiO$_2$; C) Au nanospheres/TiO$_2$ composites: (a) 12-Au/TiO$_2$, (b) 30-Au/TiO$_2$ and (c) 50-Au/TiO$_2$.

3.3 Electrochemical properties of Au nanosphere/TiO$_2$ composite

Electrochemical impedance spectroscopy is a common method used to study the electron transfer properties of samples. Figure 3 shows the AC impedance diagrams of Au nanospheres/TiO$_2$ composites modified electrodes. The results show that the charge transfer resistance of the 30-Au/TiO$_2$ modified electrode is the smallest. The possible explanation is that as the size of the Au nanospheres increases, the effective contact between the Au nanospheres and TiO$_2$ decreases. On the other hand, as the size of the Au nanospheres decreases, the quantum size effect of the Au nanospheres gradually increases, so that the separation of Au energy levels gradually increases. The size is too large or too small is not favour to the electron transfer. As a result, 30-Au/TiO$_2$ exhibits the best electron transfer properties, implying that the photocatalystic activity of 30-Au/TiO$_2$ might be the highest.

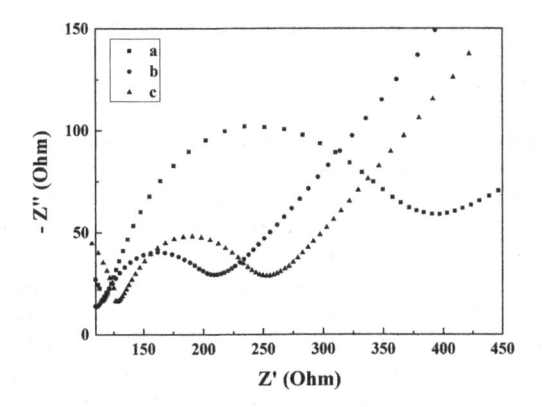

Figure 3. AC impedence of a) 12-Au/TiO$_2$, b) 30-Au/TiO$_2$, c) 50-Au/TiO$_2$.

3.4 Photocatalytic hydrogen production activity of Au nanosphere/TiO$_2$ composite

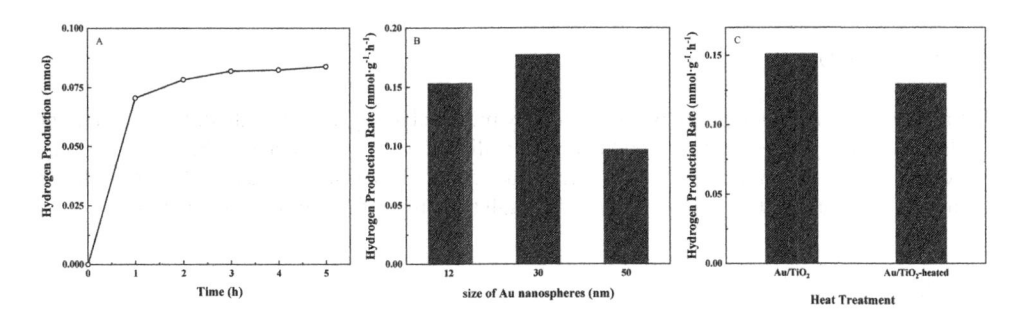

Figure 4. A) Irradiation time dependence of the amount of H$_2$ evolved from water over 30-Au/TiO$_2$; B) effect of the size of the Au nanospheres on the photocatalytic activity of Au/TiO$_2$; C) effect of heat treatment on the hydrogen production rate over 12-Au/TiO$_2$.

Figure 4A shows the kinetics curves of hydrogen production over 30-Au/TiO$_2$ loaded with Pt under visible irradiation. It can be clearly observed from Figure 4A that 30-Au/TiO$_2$ is an efficient visible photocatalyst for hydrogen production from water. Under its catalysis, hydrogen production rate is up to 1.18 mmol·g^{-1}·h^{-1}. Here, Au nanospheres play a very important role. Under visible light, the high photocatalytic activity of Au/TiO$_2$-Pt is attributed to the plasma resonance effect of Au nanospheres. The mechanism may be as follows: Au nanospheres absorb visible light due to plasma resonance effect. Then, the electrons are injected into the conduction band of TiO$_2$. Next, these excited electrons migrate to Pt under the action of Schottky energy barrier. Finally, the electrons react with water on Pt to produce hydrogen. In the photocatalytic process, Au nanospheres act as visible light capture units. At the same time, Pt acts as a charge separation center and a reactivity center. In addition, due to the interaction between Au nanospheres and TiO_2, the position of TiO_2 conduction band moves towards negative potential [6]. This may be another reason for the high photocatalytic hydrogen production activity of the system. Besides, we can also find that rapid hydrogen production can be achieved by using this catalyst. During the period of the first hours of irradiation, the hydrogen amount can reach 84% of the total hydrogen production. The photocatalytic hydrogen production reaction can reach equilibrium within 2 hours. This shows that the catalyst has a certain application potential in emergency hydrogen production in the future.

Figure 4B shows the effect of the size of Au nanospheres on the photocatalytic activity of Au/TiO_2-Pt. As can be seen from Figure 4B, when the size of Au nanospheres is 30 nm, the photocatalytic activity of Au/TiO_2-Pt is the highest. One possible explanation is that with the increase of the size of Au nanospheres, the interaction between Au nanospheres and TiO_2 tends to decrease, leading to a decrease in hydrogen production activity. Although the interaction between Au nanospheres and TiO_2 will gradually increase with the size of Au nanospheres decreasing, the number of Au nanospheres per unit TiO_2 increases, resulting in more defects in the TiO_2 matrix [9]. Too many defects become the recombination center of electron-hole pairs. Therefore, too small size of Au nanospheres may lead to a decrease in the photocatalytic hydrogen production activity of the system. Therefore, under the action of the above two factors, $30-Au/TiO_2$ shows the highest photocatalytic hydrogen production activity.

Figure 4C shows the effect of heat treatment on the efficiency of hydrogen production over Au/TiO_2-Pt. As can be seen from Figure 4C, after heat treatment, the hydrogen production rate of Au/TiO_2-Pt decreased significantly. One possible explanation is that after calcination, the crystallinity of the sample increases and the surface defect concentration decreases, which is very unfavorable to the Pt loadding on Au/TiO_2, which leads to the decrease of photocatalytic hydrogen production. In addition, after calcination, the size of Au nanospheres will increase, which makes the interaction between Au nanospheres and TiO_2 tend to decrease. This will also lead to a decrease in the photocatalytic hydrogen production activity. Therefore, heat treatment will have a negative effect on the photocatalytic activity of Au nanospheres /TiO_2 composite.

4 CONCLUSION

In this paper, Au/TiO_2 composite was successfully prepared using sol-gel method. It is an efficient visible photocatalyst for hydrogen production. In addition, the size of Au nanospheres plays a very important role in the photocatalytic hydrogen production. Only when the size of Au nanospheres is suitable, the composite can show high photocatalytic activity for hydrogen production.

REFERENCES

[1] Xiang Q, Yu J and Jaroniec M 2011 *Phys. Chem. Chem. Phys.* **13** 4853–4861
[2] Primo A, Corma A and Garcia H 2011 *Phys. Chem. Chem. Phys.* **13** 886–910
[3] Maitani M M, Zhan C, Mochizuki D, Suzuki E and Wada Y 2013 *Appl. Catal. B* **140–141**, 406–411
[4] Qi L, Yu J and Jaroniec M 2011 *Phys. Chem. Chem. Phys.* **13** 8915–8923
[5] Schneider J, Matsuoka M, Takeuchi M, Zhang J, Horiuchi Y, Anpo M and Bahnemann D W 2014 *Chem. Rev.* **114** 9919–9986
[6] Silva C G, Juarez R, Marino T, Molinari R and Garcia H 2011 *J. Am. Chem. Soc.* **133** 595–602
[7] Zhang Q, Lima D Q, Lee I, Zaera F, Chi M and Yin Y 2011 *Angew. Chem.* **123** 7226–7230
[8] Bastus N G, Comenge J and Puntes V 2011 *Langmuir* **27** 11098–11105
[9] Murdoch M, Waterhouse G I, Nadeem M A, Metson J B, Keane M A, Howe R F, Llorca J and Idriss H 2011 *Nat. Chem.* **3** 489–492

One-pot synthesis of fulvic acid and cationic polyelectrolyte PDDA modified Fe_3O_4 nanoparticles for oil-water separation of hexadecane in water emulsions

Y.X. Cai, T.W. Mi, X.L. Ma & W. Wu
College of Chemistry and Chemical Engineering, Xinjiang Normal University, Urumqi, China

ABSTRACT: Sewage from petrochemical industry usually contains micro/nano crude oil droplets, and forms stable oil in water emulsions. It is difficult to treat efficiently by conventional separation methods. In this paper, Fe_3O_4 nanoparticles modified with fulvic acid and poly (diallyldimethylammonium chloride) (Fe_3O_4/FA/PDDA MNPs) were synthesized by a green and economical one-pot co-precipitation method, and used for oil-water separation for hexadecane-water emulsion. Compared with positive charged Fe_3O_4/FA MNPs, negative charged Fe_3O_4/FA/PDDA MNPs exhibited better performance for oil-water separation. Combining with the influence of pH on the oil-water separation, the excellent performance of Fe_3O_4/FA/PDDA MNPs is attributed to its hydrophilicity and surface activity, and the electrostatic attraction between MNPs and the oil droplets simultaneously. Owing to their magnetic recovery performance and stability, Fe_3O_4/FA/PDDA MNPs maintained 99.8% separation efficiency after 7 cycles of oil-water separation experiments. This result should help to broaden application of MNPs in the treatment of emulsified oil-containing wastewater.

1 INTRODUCTION

Petroleum plays an important role in the development of social economy. With the increasing demand for it, oilfield development has gradually stepped into the aging stage, and tertiary recovery technology has been widely used to enhance oil recovery, which will lead to the discharge of a large amount of emulsified oil-containing wastewater. Due to its good stability and difficult treatment, it has caused a great negative impact on economy development and environment protection Traditional treatments, such as electrochemical method. [1], membrane separation. [2], gravity separation. [3] and flotation. [4] methods used in the treatment of it show various defects such as low efficiency, high energy consumption, high cost and easy to produce secondary pollution. Therefore, it is of great importance to develop efficient, green and low-cost treatment methods.

Magnetic nanoparticles (MNPs) have been widely concerned in oily sewage treatment due to their strong adsorption performance and easy recovery. [5,6] Because of the diffusional and solvent drag forces, the monodisperse MNPs below 50 nm are difficult to achieve magnetic separation at high gradient magnetic field. When MNPs form small clusters with a size of more than 50 nm, magnetic separation can be achieved, and at the same time, they have a large specific surface area and show strong adsorption performance. [7] In addition, small size monodisperse MNPs are prone to agglomeration and oxidation due to their large surface energy. Therefore, it is necessary to modify their surface to obtain suitable size and good stability of magnetic micro/nano heterostructures, and achieve excellent oil-water separation performance and reuse performance.

SiO_2 and carbon materials are usually used for coating the surface of MNPs. Peng et al. grafted triethylenetetramine onto the surface of Fe_3O_4 MNPs coated with SiO_2, which showed an excellent demulsification performance for the waste metalworking emulsions. [8] SiO_2 and

3-(trimethoxysilyl) propyl methacrylate modified magnetic nanoparticles (Fe_3O_4@SiO_2@MPS) were used to adsorb crude oil in water and the oil sorption capacity reached about 48.5 times of adsorbent weight. [9] Yang et al. prepared superhydrophobic magnetic hollow carbon microspheres with micro/nano hierarchical surface, which showed a superior separation efficiency (Es) for micro/nanoemulsions within an extremely short time. [10] Although the magnetic hybrid materials modified by inorganic materials possess good stability and excellent oil-water separation performance, the synthesis method is complex, rigorous demandant for synthesis conditions and high cost caused difficulty to achieve industrial production. [11]

Organic component, such as polymers biological macromolecules and micromolecules, with surface activity and functionality can be directly used to modify MNPs to form magnetic nanoclusters for oil-water separation. [11] Peng et al. reported oil-water separation of asphalt emulsion using MNPs grafted with surface-active ethyl cellulose. [12] Zhang et al. synthesized intelligent MNPs modified with cyclodextrin, which changed from hydrophilic to hydrophobic after adsorption on the water and oil interface, and could achieved rapid oil-water separation under the external magnetic field. [13] Palchoudhury et al. synthesized polyvinylpyrrolidone coated ferro oxide MNPs using a one-step process for oil-water separation of crude oilwater mixtures. [14] Although the surface of MNPs paved with organic components possess flexibility and creative functionality, it has various disadvantages, such as material stability, high cost of modification materials, and organic solvents using for the synthesis process. Therefore, it is still necessary to develop green, cost-effective, simple synthesis method and obtain stable magnetic micro/nano heteromaterials for oil-water separation.

Compared with humic acid, fulvic acid (FA) has lower molecular weight, and more oxygen-containing groups such as carboxyl group, hydroxyl group and ketone carbonyl group. Based on our previous research. [15], FA can be used to modify Fe_3O_4 nanoparticles (Fe_3O_4/FA MNPs) In consideration of its application in oil-water separation of emulsion with negative surface charge, the surface of the Fe_3O_4/FA MNPs needs to be modified into positive charge. Similar to SiO_2 coated on MNPs, FA acts as a soft-shell layer to protect and decorate Fe_3O_4 nanoparticles, on which can further connect positively charged materials. Poly (diallyldimethylammonium chloride) (PDDA), as a cationic polyelectrolyte, is usually used as a flocculant in wastewater treatment to remove anionic pollutants. [16] In addition, layer-by-layer self-assembly combines positively charged PDDAs with negatively charged materials to produce oil-water separation materials. For example, Xu et al. combined inorganic components of shell-like chitosan and montmorillonite clay with organic components of sodium polystyrene sulfonic acid and PDDA via typical layer-by-layer (LBL) method, the polyelectrolyte/clay hybrid film showed superhydrophobic properties under water corrosion resistance to sea water and low oil adhesion. [17] Hou et al. assembled the positively charged PDDA and negatively charged hydrophilic halloysite nanotubes layer by layer on the stainless-steel mesh. The as-prepared filter with underwater superhydrophobic property was used to separate various oil-water mixtures with a high separation efficiency above 97%. [18] Therefore, the negative charged Fe_3O_4/FA MNPs can be changed the surface charge by connection of PDDA.

In this paper, Fe_3O_4/FA MNPs were synthesized by co-precipitation method, then PDDA were connected by electrostatic interaction to Fe_3O_4/FA/PDDA MNPs with positive charged, stable properties and surface activity. They were used for oil-water separation of hexadecane/water emulsion. Through comparing the oil-water separation performance of Fe_3O_4/FA and Fe_3O_4/FA/PDDA MNPs, and pH of formed emulsions effecting on separation performance, the mechanism for oil-water separation by MNPs was detail discussed.

2 MATERIALS AND METHODS

2.1 *Materials*

Ferrous sulfate heptahydrate ($FeSO_4 \cdot 7H_2O$), ferric chloride hexahydrate ($FeCl_3 \cdot 6H_2O$) and ammonia solution ($NH_3 \cdot H_2O$ 25%) were purchased from Tianjin Zhiyuan Chemical Reagent Co., Ltd.

Fulvic acid and PDDA (Mw<10000) were purchased from Aladdin Reagent Co., Ltd. Hexadecane was purchased from Sigma-Aldrich All reagents were used as received without further purification.

2.2 Synthesis of Fe_3O_4 /FA and Fe_3O_4/FA/PDDA MNPs

Fe_3O_4/FA MNPs were synthesized by chemical co-precipitation method, as described in our previous work. [15]. According to the stoichiometric ratio (n(Fe^{3+}): n(Fe^{2+}) =2:1), 2.780 g $FeSO_4 \cdot 7H_2O$ and 5.405 g $FeCl_3 \cdot 6H_2O$ were dissolved in 100 mL of distilled water, and heated to 90°C. Subsequently, 10 mL of ammonium hydroxide (25%) solution and 0.770 g of FA were dissolved in 50 mL of water and quickly added to the reaction system. The mixture was stirred at 90°C for 30 minutes and then cooled to room temperature. The entire process was carried out under vacuum conditions. Finally, the black precipitate was collected by centrifugation, washed with distilled water, and dried in a vacuum oven at 65°C.

Fe_3O_4/FA/PDDA MNPs were synthesized by one-pot synthesis method. When Fe_3O_4/FA MNPs suspension cooled to room temperature, 50 mL of PDDA solution (10 wt.%) was added and stirred magnetically for 1.5 hours Fe_3O_4/FA/PDDA MNPs were obtained, washed with water and dried in a vacuum oven at 65°C for subsequent use

2.3 Preparation of emulsions

2 wt.% and 4 wt.% hexadecane-water emulsion obtained as follow. Hexadecane (400 μL or 800 μL) was mixed with water (20 mL) at a constant temperature of 25 ± 3°C for 15 minutes. Then the mixture sonicated (300 W, 40 kHz) for 10 minutes at 25 ± 3°C. [19]. As prepared emulsion was diluted to 100-fold and measured the size distribution and zeta potential. The results shown that the oil droplet size was approximately 200–300 nm, and the zeta potential was -20 to -30 mV. [20]

In order to investigate the effects of pH values ?on the oilwater separation. The pH values of the hexadecane-water emulsion were adjusted by NaOH and HCl solution (0.1 mol/L)

2.4 Oil-water separation test

A certain amount of Fe_3O_4/FA and Fe_3O_4/FA/PDDA MNPs were added to 20 mL of hexadecane-water emulsion (2 wt.% and 4 wt.%), and oscillated at room temperature with a speed of 240 rpm·min^{-1}. Then take 1 mL of sample from the emulsion every 30 minutes. After the MNPs were separated by a magnet, the absorbance (225 nm) of the emulsion was measured with an UV-visible spectrophotometer. The absorbance was converted to oil concentration of emulsion by the standard curve. [15]. The oil-water separation efficiency was calculated by the following formula 1:

$$Es = \frac{C_0 - C_e}{C_0} \times 100\% \qquad (1)$$

Where Es (%) is the oil-water separation efficiency, and C_0 (mg·L^{-1}) and C_e (mg·L^{-1}) represent the initial oil concentration of the emulsion and the oil concentration of the emulsion after adding MNPs, respectively.

2.5 Recycle tests

After the oil-water separation experiment, the MNPs were collected by the magnet, washed three times with water and ethanol. The regenerated MNPs were employed for the next oil-water separation experiments. Recycling procedure was totally performed for 12 rounds.

2.6 Characterization

Powder X-ray diffraction (XRD) analysis was carried out by using Bruker D2 with Cu Kα radiation ($\lambda = 1.5418$ Å, 50 KV). Fourier transform infrared (FT-IR) spectrum ranging from 400 to 4000 cm^{-1}

was recorded by Bruker Tensor 27 instrument. The magnetic property (VSM) was investigated by using Quantum Design PPMS-9 magnetometer at 300 K. Water contact angle (WCA) data was recorded on the KINO SL200KS Optical Contact Angle & Interface Tension Meter by pendant drop at room temperature. A droplet of deionized water (5 μL) was carefully dropped on the sample sheet using a microsyringe. The thermogravimetric analysis (TGA) was conducted on TA TGA 5500 in the temperature range from 30 to 900°C. UV-vis spectrophotometer was recorded on a Persee TU-1901 spectrometer. The zeta potentials and size distribution of the MNPs were determined by using a Malvern Zetasizer Nano Series Nano-ZS90.

3 RESULTS AND DISCUSSION

3.1 Characterization of MNPs

As shown in Figure 1A, XRD images of as synthesized sample at 2θ value of 29.8°, 35.2°, 43.2°, 57.0° and 62.8° were assigned to the (220), (311), (400), (511) and (440) lattice planes of the cubic spinel magnetite standard card (NO. 19–0629), respectively. The results suggested that the modification of FA and PDDA had no obvious influence on the crystal structure of Fe_3O_4 nanoparticles

FTIR spectra of Fe_3O_4/FA and Fe_3O_4/FA/PDDA MNPs were shown in Figure 1B. The absorption peak of 590 cm^{-1} was the stretching vibration of Fe-O. The absorption peaks at 1565 and 3410 cm^{-1} of curve (a) were C=O and -OH stretching vibration, respectively, indicating that the surface of Fe_3O_4 MNPs was successfully modified by FA. [21,22]. The absorption peaks at about 1460, 1639 and 2927 cm^{-1} of curve (b) were the bending vibration of $N-CH_3$, the stretching vibration of C=C and $-CH_2$.[23], which were the characteristic absorption peaks of PDDA. Absorption peaks of 3420 cm^{-1} and higher wave numbers were free water in the sample. The results proved that PDDA was modified on the surface of Fe_3O_4/FA MNPs.

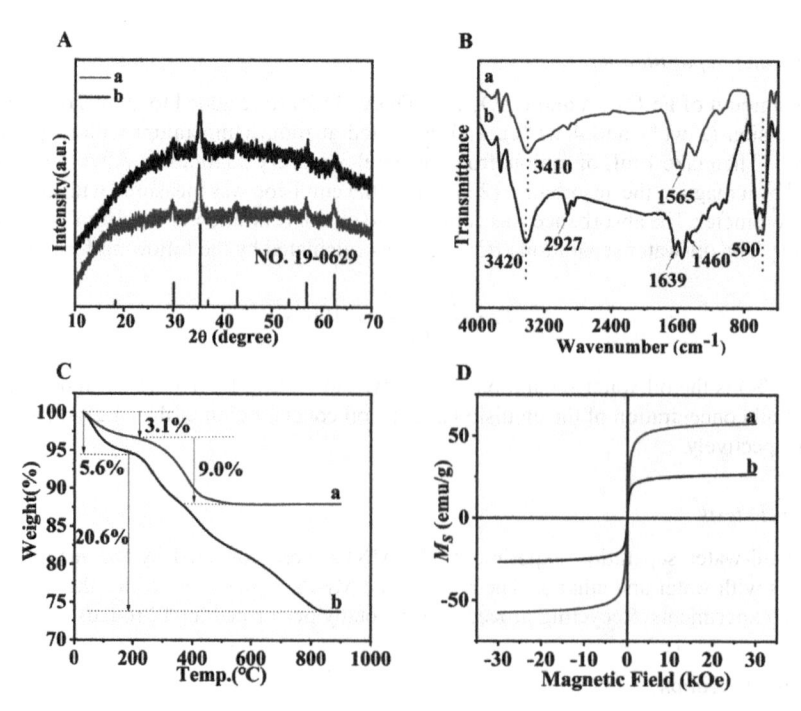

Figure 1. Characterization of MNPs (A) XRD analysis, (B) FT-IR spectra, (C) TGA curves, (D) Magnetization curves. Curve (a) and (b) are Fe_3O_4/FA and Fe_3O_4/FA/PDDA MNPs, respectively.

TGA results of Fe_3O_4/FA and Fe_3O_4/FA/PDDA MNPs were shown in Figure 1C for determining the concentration of organic groups coating on the MNPs. The weight loss of the two samples were 3.1% and 5.6% when the temperature reached 200°C, which resulted from the loss of water in the MNPs. The weight loss of 9.0% and 20.6% of them were observed from 200° to 900°, respectively. The results indicated the thermal decomposition of organic matters on the surface of MNPs. Compared with Fe_3O_4/FA MNPs, Fe_3O_4/FA/PDDA MNPs lost more weight, indicated more organic matters were modified on them. The above results also confirmed that FA and PDDA were successfully modified on the surface of magnetic materials.

Figure 1D showed the hysteresis loops of MNPs. No obvious remanence and coercivity were observed, indicating that the two MNPs were superparamagnetic. The saturation magnetization of Fe_3O_4/FA and Fe_3O_4/FA/PDDA MNPs is 57.5 and 26.6 emu·g^{-1}, respectively. The reduction of saturation magnetization was caused by the addition of non-magnetic PDDA. Although the saturation magnetization of Fe_3O_4/FA/PDDA MNPs reduced, it still showed strong magnetic response to the external magnetic field, and can be separated from emulsion quickly.

3.2 Oil-water separation performance

Oil-water separation performances of Fe_3O_4/FA and Fe_3O_4/FA/PDDA MNPs for hexadecane-water emulsion were compared as shown in Figure 2A. During the separation experiment, 2 mg of MNPs were added to 20 mL of emulsion (2 wt.%). After oscillating for 390 min, the Es of two MNPs reached 38.2% and 98.7%, respectively. Obviously, the oil-water separation rate and effect of Fe_3O_4/FA/PDDA MNPs was better than Fe_3O_4/FA MNPs

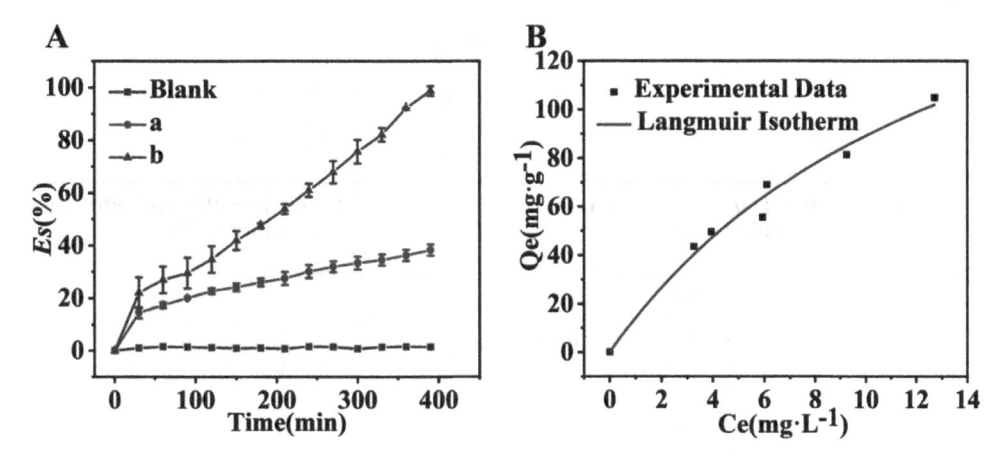

Figure 2. (A) Oil-water separation performance changed with oscillation time of emulsion for MNPs (B) Adsorption isotherm of oil to Fe_3O_4/FA/PDDA MNPs Curve (a) and (b) are Fe_3O_4/FA and Fe_3O_4/FA/PDDA MNPs, respectively. The insets show WCA results of (a) Fe_3O_4/FA and (b) Fe_3O_4/FA/PDDA MNPs

MNPs, using to the separation of oil and water for emulsion, firstly should be hydrophilic and evenly dispersed in the aqueous phase. [13] The water contact angles of Fe_3O_4/FA and Fe_3O_4/FA/PDDA MNPs were 28.7 ± 0.5° and 42.2 ± 0.4° (as shown in Figure 2B inset), respectively, and both of them exhibited hydrophilicity and could rapidly dispersed in the hexadecane-water emulsion. Then the aromatic ring and alkyl chain of FA and PDDA molecules connected with hydrophilic functional groups, which exhibited surface activity These properties could promote the migration of Fe_3O_4/FA and Fe_3O_4/FA/PDDA MNPs to the water-oil interface, thus reducing the interfacial tension to achieve oil-water separation In addition, the MNPs adhering oil droplets aggregated under external magnetic field, and the oil droplets also aggregated and demulsification accordingly. Therefore, both Fe_3O_4/FA and Fe_3O_4/FA/PDDA MNPs showed oilwater separation performance.

As mentioned above, though the WCA of Fe_3O_4/FA MNPs changed from $28.7 \pm 0.5°$ to $42.2 \pm 0.4°$ after dragged with PDDA, both of them were strong hydrophilicity. Hence it is assumed that the significant difference of oil-water separation performance between Fe_3O_4/FA and Fe_3O_4/FA/PDDA MNPs is not caused by the surface wettability. It was found that the zeta potential of Fe_3O_4/FA MNPs changed from -24.7 ± 2.4 mV to 26.4 ± 3.3 mV after PDDA modification. Both of oil droplets in emulsion and Fe_3O_4/FA MNPs were negative charged The electrostatic repulsion between them leaded to worse demulsification performance Positive charged Fe_3O_4/FA/PDDA MNPs showed electrostatic attraction with oil droplets Therefore, the oil-water separation performance of Fe_3O_4/FA/PDDA MNPs was obviously better than that of Fe_3O_4/FA MNPs.

To quantitatively analyze the oilwater separation performance as the function of the concentration of Fe_3O_4/FA/PDDA MNPs added in emulsion, Langmuir isotherm was used to fit the maximum oil adsorption capacity of MNPs. As shown in Figure 2B, the maximum adsorption capacity reached 215.58 mg·g^{-1}, which indicated that the positive charged Fe_3O_4/FA/PDDA MNPs showed good oil-water separation performance for emulsion [24,25].

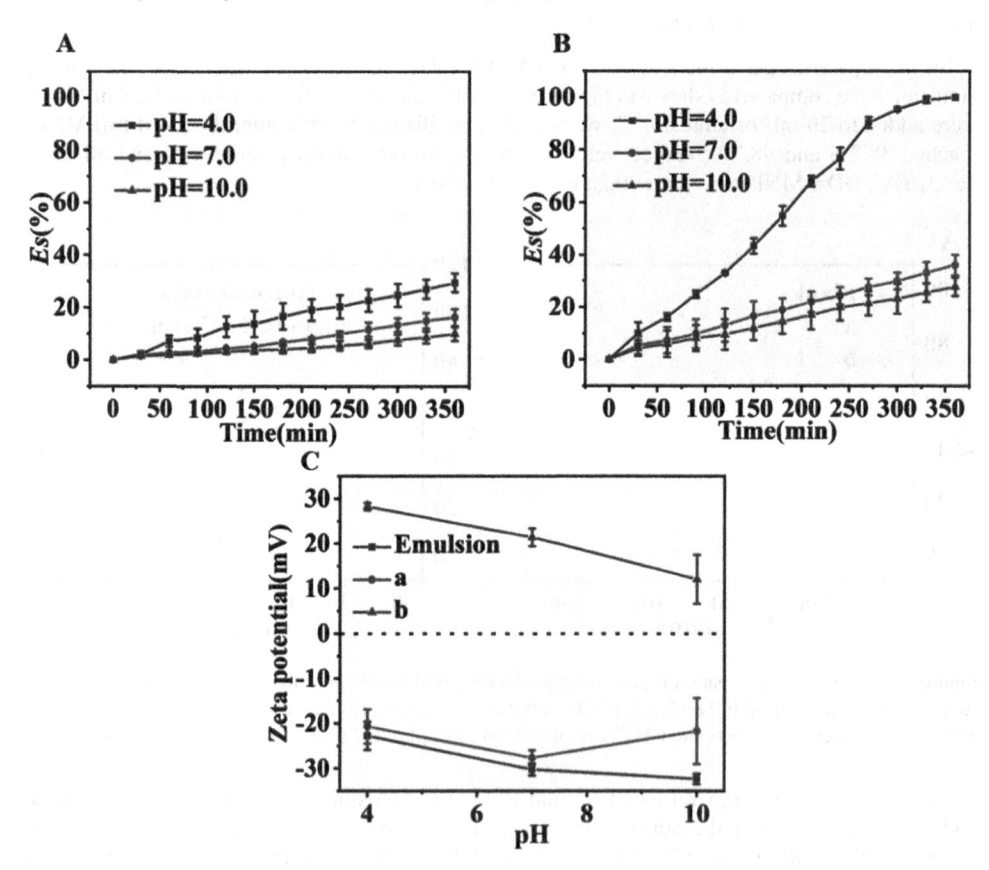

Figure 3. Variation curves of Es with the change of emulsion pH for (A) Fe_3O_4/FA and (B) Fe_3O_4/FA/PDDA MNPs. (C) Zeta potential of MNPs and emulsion changes with pH. Curve (a) and (b) are Fe_3O_4/FA and Fe_3O_4/FA/PDDA MNPs, respectively.

When adding 100 mg/L of MNPs in emulsions, the Es changed with pH was shown in Figure 3A and 3B The Es of Fe_3O_4/FA MNPs were 29.4%, 16.2% and 9.6% at pH 4.0, 7.0 and 10.0 after 3 hours, while Fe_3O_4/FA/PDDA MNPs were 100.0%, 36.0% and 27.6% Both of results showed that the Es of MNPs decreased with the increase of emulsion pH. Lü et al. prepared amino-functionalized

Fe_3O_4@SiO_2 nanoparticles with an isoelectric point about pH 6.8, and the zeta potential decreased gradually with increasing pH. In contrast, emulsified oil droplets were negatively charged over the whole pH range. The Es for the diesel-in-water emulsions decreased with increasing pH because of electrostatic interaction between MNPs and oil droplets.[26]. Zeta potential of Fe_3O_4/FA/PDDA MNPs decreased gradually with the increase of pH and exhibited positive charged over the pH ranging from 4 to 10 while zeta potentials of Fe_3O_4/FA MNPs and emulsions were negative charged in this pH ranges. (as shown in Figure 3C). Under acidic conditions, Fe_3O_4/FA/PDDA MNPs with high positive charged showed strong electrostatic attraction between Fe_3O_4/FA/PDDA MNPs and negative charged oil droplets. Thus, Fe_3O_4/FA/PDDA MNPs exhibited high Es with such small dosage in demulsification. Under neutral and alkaline conditions, the emulsion adsorbed hydroxyl group caused their more stability.[27], and therefore the Es decreased dramatically. For Fe_3O_4/FA MNPs, because of electrostatic repulsion between them, MNPs showed poor oil-water separation performance over the pH ranging from 4 to 10. Comparing the Es of Fe_3O_4/FA MNPs with Fe_3O_4/FA/PDDA MNPs at neutral and alkaline conditions, the latter were higher than the former, which attributed to the difference of electrostatic interaction The above results further illustrated the electrostatic interaction have significant influence on the oil-water separation performance of MNPs.

3.3 *Recyclability of Fe_3O_4/FA/PDDA MNPs*

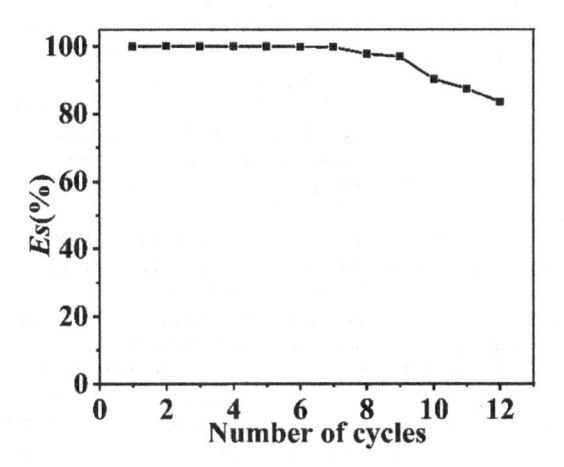

Figure 4. Recycling capacity of Fe_3O_4/FA/PDDA MNPs of oilwater separation for hexadecane-water emulsions.

The recycling performance of Fe_3O_4/FA/PDDA MNPs was investigated (the dosage of 500 mg·L^{-1}, the oscillation time of 180 min), as shown in Figure 4. The recyclability of Fe_3O_4/FA/PDDA MNPs nanoparticles was researched in 12 cycles. In the first 7 cycles, Es did not decrease and remained at 99.8%. The slight decrease of Es after 7 cycles was attributed to MNPs loss during the recycling process. The results showed that the Fe_3O_4/FA/PDDA MNPs possessed good recycling performance and strong stability.

4 CONCLUSION

In this paper, Fe_3O_4/FA/PDDA MNPs were synthesized by one-pot co-precipitation method. The samples were characterized by XRD, FT-IR, TGA and VSM. The Fe_3O_4/FA/PDDA MNPs and applied in oil-water separation of hexadecane-water emulsion. The results showed that

Fe$_3$O$_4$/FA/PDDA MNPs possessed excellent separation efficiency, and the maximum adsorption capacity for hexadecane was 215.58 mg·g^{-1}. The influence of emulsion pH on the oil-water separation performance were investigated. The results showed that surface activity and electrostatic interaction played an important role in oil-water separation. The oil-water separation cycle experiments conducted 12 times, and Es maintained at 99.8% after being reused for 7 times, which indicated that Fe$_3$O$_4$/FA/PDDA MNPs have good stability. In conclusion, the synthesis method of Fe$_3$O$_4$/FA/PDDA MNPs is simple and environmentally friendly, and these MNPs are expected to be applied in oil-containing wastewater treatment.

ACKNOWLEDGEMENTS

This work was supported by the National Natural Science Foundation of China (NNSFC) [grant number 21763025]; the University Research Program of Xinjiang Uygur Autonomous Region [grant numbers XJEDU2017S027]; Xinjiang key laboratory of energy storage and photoelectrocatalytic materials

REFERENCES

[1] Vigo C R and Ristenpart W D 2010 *Langmuir* **26** 10703–07
[2] Rajasekhar T, Trinadh M, Babu P V, Sainath A V S and Reddy A 2015 *J. Membr. Sci.* **481** 82–93
[3] Xue Z, Cao Y, Liu N, Feng L and Jiang L 2014 *J. Mater. Chem. A* **2** 2445–60
[4] Al-Shamrani A A, James A and Xiao H 2002 *Water Res.* **36** 1503–12
[5] Tempesti P, Bonini M, Ridi F and Baglioni P 2014 *J. Mater. Chem. A* **2** 1980–84
[6] Liu G, Cai M, Wang X, Zhou F and Liu W 2014 *ACS Appl. Mater. Interfaces* **6** 11625–32
[7] Ditsch A, Laibinis P E, Wang D I and Hatton T A 2005 *Langmuir* **21** 6006–18
[8] Peng K M, Xiong Y J, Lu L J, Liu J and Huang X F 2018 *ACS Sustain. Chem. Eng.* **6** 9682–90
[9] Kamgar A and Hassanajili S 2020 *J. Mol. Liq.* **315** 113709
[10] Yang N, Luo Z X, Chen S C, Wu G and Wang Y Z 2021 *Carbon* **174** 70–78
[11] Ali N, Bilal M, Khan A, Ali F, Yang Y, Khan M, Adil S F and Iqbal H M 2020 *J. Mol. Liq.* **312** 113434
[12] Peng J, Liu Q, Xu Z and Masliyah J 2012 *Energy Fuels* **26** 2705–10
[13] Zhang J R, Li Y M, Bao M T, Yang X L and Wang Z N 2016 *Environ. Sci. Technol.* **50** 8809–16
[14] Palchoudhury S and Lead J R 2014 *Environ. Sci. Technol.* **48** 14558–63
[15] Mi T W, Cai Y X, Wang Q, Habibul N, Ma X L, Su Z and Wu W 2020 *RSC Adv.* **10** 10309–14
[16] Ahmadiannamini P, Li X, Goyens W, Meesschaert B and Vankelecom I F 2010 *J. Membr. Sci.* **360** 250–58
[17] Xu L P, Peng J T, Liu Y B, Wen Y Q, Zhang X J, Jiang L and Wang S T 2013 *ACS nano* **7** 5077–83
[18] Hou K, Zeng Y C, Zhou C L, Chen J H, Wen X F, Xu S P, Cheng J, Lin Y G and Pi P H 2017 *Appl. Surf. Sci.* **416** 344–52
[19] Sang Y J, Yang F Y, Chen S L, Xu H B, Zhang S, Yuan Q H and Gan W 2015 *J. chem. Phys.* **142** 224704
[20] Wu W, Fang H, Yang F Y, Chen S L, Zhu X F, Yuan Q H and Gan W 2016 *J. Phys. Chem. C* **120** 6515–23
[21] Ghaeni N, Taleshi M S and Elmi F 2019 *Mar. Chem.* **213** 33–39
[22] Fu Q C, Zhou X D, Xu L H and Hu B W 2015 *J. Mol. Liq.* **208** 92–98
[23] Wang S Y, Yu D S and Dai L M 2011 *J. Am. Chem. Soc.* **133** 5182–85
[24] Javadian S, Khalilifard M and Sadrpoor S M 2019 *J. Water Process. Eng.* **32** 100961
[25] Elmobarak W F and Almomani F 2021 *Chemosphere* **265** 129054
[26] Lü T, Zhang S, Qi D M, Zhang D, Vance G F and Zhao H T 2017 *Appl. Surf. Sci.* **396** 1604–12
[27] Gan W, Wu W, Yang F Y, Hu D P, Fang H, Lan Z G and Yuan Q H 2017 *Soft Matter* **13** 7962–68

Preparation and properties of Eu^{2+} doped strontium phosphate blue phosphor by Sol-gel method

L.Q. Wu, Y. Wang, S.Y. Guan & J.R. Li
School of Applied Physics and Materials, Wuyi University, Jiangmen, China

ABSTRACT: By using strontium carbonate and diammonium hydrogen phosphate as raw materials, the rare earth ion Eu3+doped strontium phosphate blue phosphor was synthesized by sol-gel method. The Eu3+ of the intrinsic excitation of red light is reduced to Eu2+ of blue light by the redox method of activated carbon. Characterization and spectral testing of the sample by X-Ray diffraction and fluorescence spectrometer, respectively. The effects of different calcination temperature and rare earth ion concentration on the luminescence properties were studied. The spectral test of the sample shows that the Sr2P2O7: Eu2+ has the highest luminescence efficiency at annealing temperature of 800°C, and the luminescent material with doping concentration of 8.0%europium ions demonstrates the highest luminescence intensity. The spectrum of the sample has a strongest blue emission peak at 441.8 nm under 368nm UVA excitation, and the luminescence intensity has been increasing with the increase of Eu3+doping concentration. The concentration quenching effect was observed by the fluorescence spectra of the samples.

1 INTRODUCTION

In rare earth phosphate phosphors, rare earth ions can be used not only as matrix cations but also as activators. Due to the 4f layer of rare earth ion electrons, being provided with very rich electron transition energy levels, the spectrum from ultraviolet to infrared in visible area could be launched with stable physical performance, thus determine the spectral characteristics of rare earth ions doped phosphate phosphors in lighting source, and being widely used in the field of information display, etc.

For many kinds of phosphates, most of them have some special properties. Compared with silicate and aluminate, phosphates can be synthesized at a relatively low temperature. Due to its high luminous efficiency, good chemical stability, low synthesis temperature, low raw material price, and good absorption in both UV and N-UV regions, the phosphor system has great application potential in fluorescent matrix materials. In addition, the phosphates doped with rare earth ions have excellent thermal stability and charge stability, and rare earth ions-doped phosphate were regarded as one of the luminescent materials with practical value and prospect because of its simple synthesis method, high luminescence brightness and low annealing temperature [1–3].

Rare earth doped luminescent materials can be fabricated by many methods [4–7], such as high temperature solid phase method, sol-gel method, microemulsion method, thermal decomposition method, and co-precipitation method etc. Traditional high temperature solid phase method need very high calcination temperature, long sintering time, and difficult post-processing of ball grinding, that's the reason why luminescence brightness and luminous efficiency of the phosphor was comparatively lower. Sol-gel method mainly starts from solution, the synthesis temperature required is low, the particle size can be controlled to be nanometer scale with high purity, and the components of each raw material can be uniformly mixed at molecular level, which is conducive to the preparation of nanoscale phosphors that tend to the theoretical ratio level [8]. In this study, phosphate phosphors were synthesized by sol-gel method.

LED white light is generally obtained by the combination of inorganic phosphors and LED bule-light chips. There are many kinds of phosphors used for white LED. According to the different bands required, blue light excited phosphors, UV excited phosphors and UVA excited phosphors can be used for the excitation of phosphors. However, it is difficult to find strong blue phosphor for white LED applications, and the electro-optical conversion efficiency of UV LED chip is still relatively low. Therefore, people start to explore a new phosphor system, namely, blue phosphor for near ultraviolet (N-UV) chip excited white LED, which is a relatively active research field worldwide. As an example, Zhanchao Wu et al. reported a near-UV excited phosphor $SrMg_2(PO_4):Eu^{2+}$, which has a strongest excitation peak at 363nm and effectively emits blue light at the wavelength of 423nm, showing good thermal stability [8] Near UV chip has high energy, good excitation effect, it can improve the bule light luminous efficiency, so that the blue band can be compensated effectively in white LED spectrum, and widen the blue spectrum. RGB phosphor can be excited by near ultraviolet light to obtain better color rendering. A commercial blue phosphor $BaMgAl_{10}O_7(BAM)$ has been put into production, but it has some deficiencies such as weak blue light emission and low fluorescence efficiency [9]. In this study, by using strontium phosphate as matrix and Eu^{2+} as the activator, the effects of rare earth europium ion doping concentration and annealing temperature on the luminescence performance of strontium phosphate were studied in detail.

2 EXPERIMENT

2.1 *Principle analysis*

Rare earth luminescence is mainly the result of electron transition in the 4f and 5d orbitals of rare earth ions, and the cleavage degree of 5d energy level is also different in the heterogeneous matrix, and the transition from 4f to 5d is easily affected by environmental effects. Eu^{2+}, Dy^{3+}, Ce^{3+}, Tb^{3+} and other lanthanide rare earth ions can produce 4f→ 5d transition. In this experiment, Eu^{3+} ions are mainly doped, and Eu^{3+} is converted to Eu^{2+} by REDOX method.

The emission spectrum of Eu^{3+} ion is derived from the energy level transition emission from the excited state $^5D_0$ to the low energy level ^7F_J(J=0, 1, 2, 3, 4, 5, 6) in the $^4F_6$ electron configuration, which belongs to the typical linear spectrum. There are mainly $^5D_0$ magnetic dipole transition with a $^5D_0 \rightarrow {}^7F_1$ bimetallic grade transition, $^5D_1 \rightarrow {}^7F_2$ magnetic magnitude transition Eu^{3+} ions in the substrate cells for the center of symmetry, orange color, when the Eu^{3+} ions in the substrate cells for the asymmetric center, then will happen at the main level $^5D_0 \rightarrow {}^7F_2$ galvanic transition, in turn, produce the red light, but because ^5D level affected by any of the crystal field are extremely small, so can be ignored. On the contrary, the low level ^7F_J will be seriously affected by the crystal field and then split, resulting in the ultimate difference between the emission intensity of $^5D_0 \rightarrow {}^7F_1$ and the emission intensity ratio of $^5D_0 \rightarrow {}^7F_2$ of Eu^{3+} ion. So the color of Eu^{3+} will be different depending on the matrix materials.

Rare earth Eu ions have two valence states, Eu^{3+} and Eu^{2+}, and are easy to form luminous centers in crystal lattice, they are often used as phosphor activators. The 4f-4f transition forbearance of Eu^{3+} ion results in a narrow linear emission spectrum with low index and luminous efficiency and weak spectral line strength. The 5d → 4f energy level transition of Eu^{2+} makes its spectrum show wide peak emission, high luminous efficiency and strong spectral line strength, and its luminous performance is easy to be changed under the influence of coordination environment and crystal field environment. Therefore, Eu^{2+} has greater application prospect in the preparation of white LED light source [10].

If Eu^{3+} is converted to Eu^{2+} as much as possible in a single substrate, red light emission can be converted to blue light emission. By doping a small amount of alkaline earth metal ions (Mg^{2+}, Sr^{2+}, Ba^{2+}), the local environment of the cell where the activated ions exist is changed, so as to adjust the luminescence intensity of Eu^{2+}. Therefore, both the coordination environment and crystal size can affect the valence state and luminescence performance of europium divalent ion [11]. The linear spectral structure formed in Eu^{2+} ion transition and harmonic substructure, Eu^{2+} with 4F7 configuration shows broadband emission due to the parity of 5d-4f allowing electric dipole transition, which enables it to display higher color rendering effect and effectively improve the color rendering index of WLED devices [12]. Eu^{3+} ions, characterized by relatively simple energy level structure, high quantum efficiency and sensitivity to local environment, are often selected to study the basic physical properties of materials. Eu^{3+}, as a typical rare earth doped ion, has received extensive attention and research in both bulk materials and nanomaterials.

2.2 Experimental scheme

At present, the blue phosphors produced in the industry are generally deficient in low yield and weak luminescence intensity. There are a variety of matrix materials used, but few of them can really cooperate well with rare earth ions. Secondly, the luminescence properties of rare earth ion doped matrix materials can be influenced by the doping concentration of rare earth ion, annealing and sintering temperature, annealing time, water bath temperature and the amount of solvent added.

In this experiment, the self-designed sol-gel method was used to study the effects of europium ion doping concentration, annealing time and annealing temperature.

The main materials used in this experiment include europium oxide, strontium carbonate, diammonium hydrogen phosphate, concentrated nitric acid, methanol, acetic acid and so on. Strontium phosphate phosphors doped with europium were prepared by sol-gel method with strontium carbonate, diammonium phosphate and europium oxide as main raw materials. According to the experiment, $Sr_2P_2O_7$:Eu^{2+} stoicheometric ratio was determined, the raw materials were accurately weighed, various solutions were mixed in the beaker until they were clarified, and then poured into a three-neck flask for water bath heating at 80°C. Under the action of an electric agitator, the gel was formed. The doping amounts of Eu^{3+} ions were 2.0%, 3.5%, 6.0%, 8.0% and 10.0%, respectively. After gelation, the mixture was dried, ground into powder and put into crucible. The mixture was calcined at 800°C $\sim$ 900°C in muff furnace, and the heat treatment time was 4~5h. After cooling, the mixture was separated and ground to obtain nanometer blue phosphor.

The following key technological parameters were also considered in the experiment : (1) The color and condition of the solution when it was prepared; (2) Stirring time, stirring speed and water bath temperature of the mixed solution (sol); (3) The drying time and temperature of the gel;(4) Annealing temperature and annealing time; (5) Annealing process, etc.

3 SAMPLE CHARACTERIZATION AND RESULT ANALYSIS

3.1 XRD pattern

In this experiment, strontium phosphate phosphor doped with europium ion was prepared by sol-gel method, and REDOX reaction was carried out in Muff furnace, finally the phosphor with pure blue color was obtained. The luminescence properties of the samples are detected by fluorescence spectrometer, and then the structural properties are detected by XRD. XRD test was performed using X 'Pert PRO X-ray powder diffractometer for phase analysis of the sample (Cu target, tube voltage 40 kV, tube current 40 m, wavelength=0.154 06nm, the scanning range is 10° $\sim$ 90°).

On the basis of spectral testing, XRD test is carried out on the samples with the optimal doping concentration of Eu ions, the optimal temperature and annealing time. The sample annealing method is adopted in step annealing. The results are shown in Figure 1:

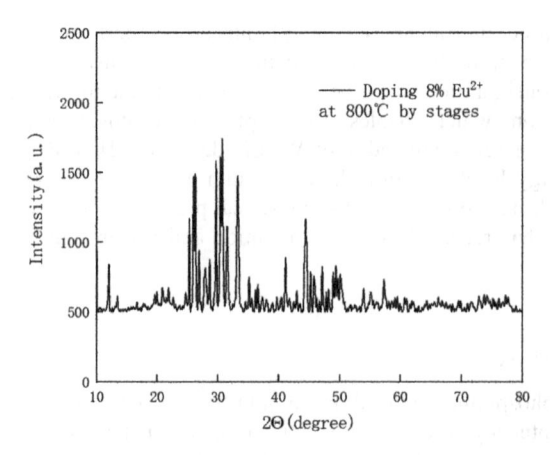

Figure 1. XRD pattern of 8% Eu^{3+} doped sample annealed at 800°C for 4 h.

The XRD pattern of the sample powder is the diffraction peak of the target product $Sr_2P_2O_7$. Compared with JCPDS standard card No.870561, the crystal structure belongs to the orthorhomytic system, and the orthogonal structure of phosphors mainly depends on the energy transfer between the luminescent center Eu^{2+}. This map shows that the solid phase reaction is basically complete and that doping Eu^{3+} ion does not change the crystal structure of the matrix material. XRD results show that the doped rare earth ions have no effect on the phase structure, and the obtained sample is $Sr_2P_2O_7:Eu^{2+}$ pure phase. On this basis, with the same doping concentration and the same annealing time, the influence of different temperatures on the luminescence intensity is explored. The XRD test of the samples is shown in Figure 2:

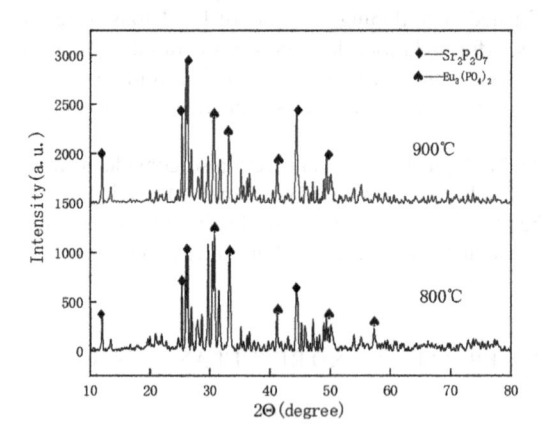

Figure 2. XRD pattern of 8% Eu^{3+} doped annealed at 800°C and 900°C for 4 h, respectively.

An interesting phenomenon can be seen from Figure 2, $Eu_3(PO_4)_2$ content is higher than that of $Sr_2P_2O_7$ at the reduction annealing and calcination temperature of 800°C. The characteristic peak value of Eu^{2+} is more prominent, and its luminescence performance is found to be better.

3.2 *Spectral analysis*

Hitachi F-4600 fluorescence spectrophotometer was used to test the spectra of the samples. In order to ensure that the testing conditions are in line with the actual production and testing situation of the enterprise, the voltage is set as 400V in the spectrometer, the width of both double slits is 2.5nm, the scanning speed is 2400nm/min, and the response time is 0.5s. Firstly, the emission spectrum of phosphor was measured with the intrinsic excitation wavelength of europium ion 394nm, and its peak wavelength was used as the monitoring wavelength to test the excitation spectrum of phosphor. Then the peak wavelength of the excitation spectrum of the phosphor was used as the excitation wavelength to accurately test the emission spectrum of the phosphor sample. Firstly, the emission spectra of samples with 8%Eu ion doping concentration were shown in figure 3 after annealing at 800°C by steps for 4 hours. The excitation wavelength of the spectrum is 394nm for europium ion.

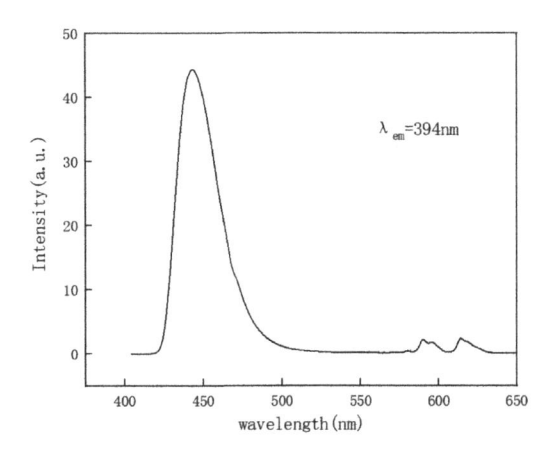

Figure 3. Emission spectra of annealed reduction at 800°C for 4 h at 8.0% Eu^{3+} ion doping concentration.

As seen from Figure 3, under the excitation of 394nm wavelength, the $Sr_2P_2O_7$: Eu^{2+} sample has a strong blue light emission spectrum with a peak wavelength of 441.8nm, which meets the requirements of commercial phosphor. At the same time, some orange-red emission peaks appeared in the band of 585 ~615nm, indicating that the sample reduction was not sufficient and a small part of the trivalent europium ions still existed. The excitation spectrum of the sample was tested with 441.8nm as the monitoring wavelength, as shown in Figure 4. It is not difficult to find that the excitation spectrum of the sample is relatively wide, and it can achieve strong excitation in the range of 300–400nm. The best excitation spectrum is in the range of 330 ~385nm. The peak excitation wavelength is 368nm near UV light. The preparation conditions of the sample were annealing temperature: 800°C, doping concentration of the sample: 8.0%, reduction annealing time: 4h. The annealing reduction method of heating up to 800°C is adopted in this use.

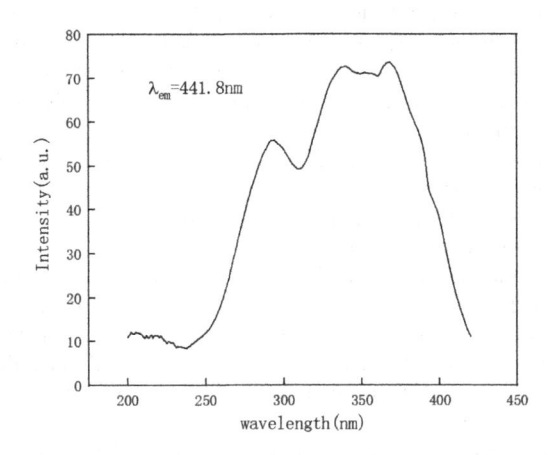

Figure 4. The excitation spectrum of $Sr_2P_2O_7$:Eu^{2+} under the monitor wavelength of 441.8nm.

According to the best effective excitation spectral wavelength, the luminescence spectrum of the sample was re-tested, as shown in Figure 5. It can be found that the most effective excitation wavelength of the phosphor sample is 368nm near-ULTRAVIOLET light, which is very suitable for the 365 ~370nm LED ULTRAVIOLET chip to obtain blue light in the range of 430–460nm. The color purity is very high. The orange red spectrum in the band of 585 ~615nm in Figure 3 disappeared, and the emission spectral intensity is 68% higher than that in 394nm excitation, fully meet the commercial requirements.

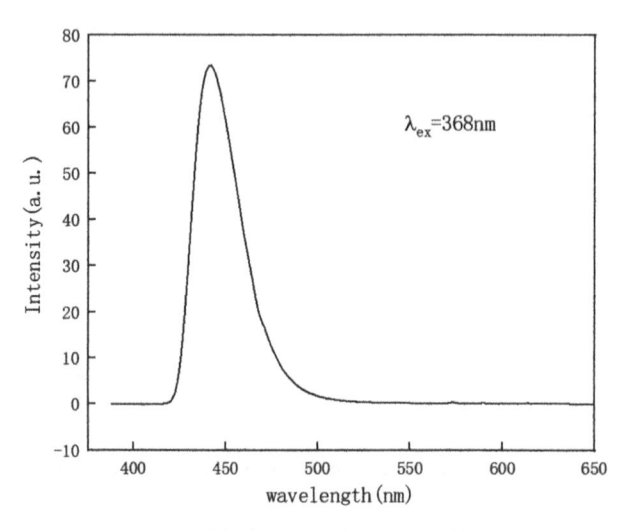

Figure 5. Emission spectra of $Sr_2P_2O_7:Eu^{2+}$ samples excited by 368 nm.

4 INFLUENCE OF DIFFERENT FACTORS ON LUMINESCENCE INTENSITY

4.1 Rare earth Eu^{3+} doping concentration

In the phosphors used in white leds, rare earth ions are the luminous centers. Under the irradiation of electron rays or ultraviolet light, a substance can transition from the ground state to the excited state, and then the transition from the excited state to the lower energy level state or the ground state can emit visible light of different wavelengths, which is the fluorescence phenomenon. Fluorescence is generated when the molecule is at the lowest vibrational level of the first electron excited singlet state and returns to the vibrational level of the ground state in the form of radiation transition. Electrons in the excited state usually return to the ground state in the form of radiative transition or non-radiative transition, and the radiative transition will produce fluorescence phenomenon. Non-radiative transition refers to radiating its excess energy in the form of heat. The possibility and degree of various transition modes are related to the structure of the fluorescent material itself and the physical and chemical environment when excited. In the process of fluorescence generation, due to the existence of various forms of non-radiation transition, energy is lost, so their maximum emission wavelength moves towards the direction of long wave.

Therefore, the concentration of rare earth ions will affect the luminescence intensity of phosphors. Because the Eu^{2+} ion has a very low 5d energy level, it gets excited and it gets to the 5d state.

The 5d orbital is exposed to the outer shell, so the 5d state is greatly affected by the external crystal field. When the substrate changes or the concentration of rare earth ions, the local environment in the crystal will change, making the local crystal field intensity change, affecting the emission intensity of the spectral properties.

The luminescence intensity of strontium phosphate is directly related to the doping concentration of rare earth ions. When the doping concentration of rare earth ions is too low, the luminescence intensity of the fluorite is very weak [2, 6], while when the doping concentration is too high, the luminescence intensity will be reduced because too many non-radiation luminescence centers will be formed [3, 13] The combination of matrix materials and rare earth elements to form the most effective luminescence mechanism is relatively complex, which will be affected by a variety of factors, even the co-mixing and proportion of various rare earth elements[1, 14–16]. In this paper, the effects of fixed Eu ion doping concentration on luminescent properties were studied. Figure 6 for phosphate doped Eu^{2+} strontium luminescent material, from figure 6 shows that when the doped europium ion, with the increase of europium ion concentration, the luminous intensity of phosphoric acid strontium in gradually increasing, until 8.0% when maximum, and then with the increase of rare earth ion concentration, the luminous performance is constantly reduced, this is the phenomenon of concentration quenching.

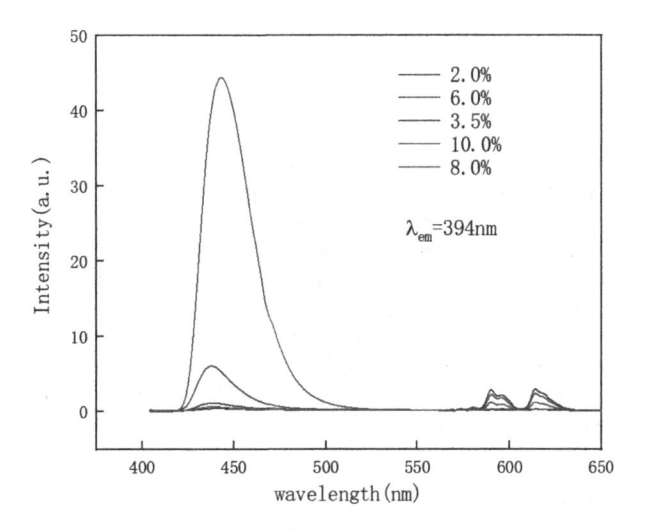

Figure 6. Luminescence spectra with different Eu^{3+} doping concentrations at annealing temperature of 800°c for 4 h.

Figure 6 demonstrates that, when the fluorescence powder was excited with the intrinsic excitation wavelength of europium ions at 394nm, all the spectra more or less had very weak emission spectra in the orange-red region, indicating that there were still very few Eu^{3+} ions that had not been reduced to the divalent europium ions. For different Eu^{2+} doping concentrations, the luminescence intensity of the sample increases with the increase of doping concentration before 8.0%, and the luminescence intensity almost reaches the best. However, when the doping concentration exceeds 8.0%, the luminescence intensity does not increase, but decreases. The luminescence intensity of the sample decreases when 10% Eu ion is doped, and we believe that concentration quenching phenomenon occurs here [5].

4.2 *Annealing temperature*

The preparation technology and conditions of sol-gel, such as stirring time, water bath temperature, baking temperature and time, annealing temperature and time, have great influence on the preparation of phosphors. The effects of different annealing and sintering temperatures on luminescence properties of luminescent materials are discussed. When the holding annealing time is 4 hours, the luminescence intensity of samples with 8.0% doping concentration at 800°C and 900°C is compared, as shown in Figure 7.

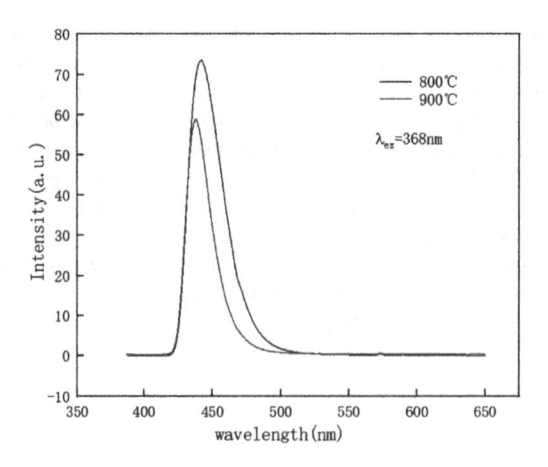

Figure 7. Influence of annealing temperature to luminescence intensity.

From Figure 7 we see, the luminescence intensity of the sample with annealing temperature of 900 °C is lower than that of the sample with annealing temperature of 800 °C. This result corresponds to the XRD test in Figure 2. With the increase of annealing temperature, the luminescence intensity of nano scale luminescent materials first increases, but after exceeding a certain temperature, the luminescence intensity begins to weaken, which is a temperature quenching effect in principle. This may be due to the fact that with the increase of temperature, the more sufficient the reaction of the sample is, the rare earth ions are completely mixed into strontium phosphate, which promotes the strong luminescence intensity. However, with the further increase of temperature, the internal structure of the material will change, and the chemical reaction of ion replacement will be reversed, resulting in the decrease of luminous intensity of the material.

4.3 Step annealing process

It is found that different reduction annealing schedules (heating forms) have great influence on luminescence intensity at the same doping concentration and temperature. A comparative experiment was designed for this experiment. Firstly, keep the same annealing temperature and the same doping concentration, and change the annealing schedule: one, directly raise the temperature, that is, directly raise the temperature from room temperature to 800°C for 3.5h; Two, it is divided into two sections for annealing treatment, that is, first heat up to 400°C for 1 hour and then heat up to 800°C for 4 hours annealing treatment. The results are shown in Figure 8:

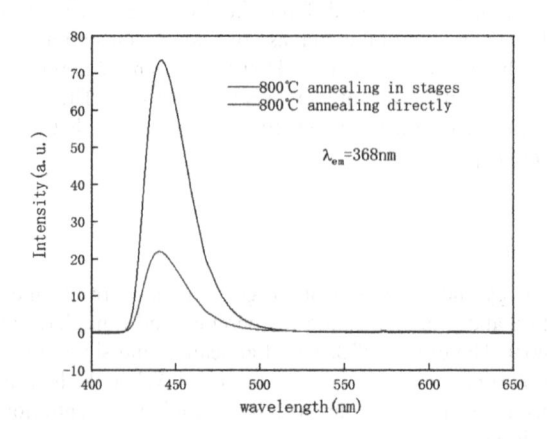

Figure 8. The influence to luminescence intensity of phosphor of two different temperature raise processes.

Figure 8 demonstrates that different annealing process have a significant impact on the luminescence performance of phosphors. Under the same doping concentration and annealing temperature, the luminescence intensity of the phosphor for 4 hours annealing treatment is more than three times that of 3.5 hours annealing at 800°C directly. The samples with annealing time of 4 hours, Eu^{3+} can be reduced to Eu^{2+} in a better way, and more luminous centers can be formed to obtain better luminous effect at one time.

5 CONCLUSION

In this paper, a strong blue phosphor $Sr_2P_2O_7:Eu^{2+}$ which can be effectively excited by near-UV LED chips was successfully synthesized by sol-gel method. A series of experimental studies were carried out on the effects of different rare earth ion doping concentrations, different annealing temperature and time, and different reduction annealing methods or processes on the luminescence properties of samples. The characterization and testing were conducted on the structure and luminescence properties of samples. Results can be summarized as follows:

(1) The luminescence properties of phosphor powder sample with rare earth ions doped concentration has a lot to do, with the increase of doping concentration, the luminous intensity of samples will increase accordingly, but when the concentration increased to a certain value and more than this value, with the increased of doping concentration of rare earth ions, the luminous intensity shown a decrease tendency, namely the concentration quenching effect.
(2) The annealing temperature required by sol-gel method is much lower than that of high temperature solid phase method, and the material purity is higher. The effect of annealing and reduction temperature is an important factor that must be considered in the preparation process.
(3) At a low annealing temperature and short annealing time, the reduction reaction is not sufficient, and the luminescent center formed by $Sr_2P_2O_7:Eu^{2+}$ ions is unsufficient. However, if the annealing temperature is too high and the annealing time is too long, the luminescence center of rare earth ions will decrease, the luminescence intensity of the material will be reduced, and temperature quenching effect will occur.
(4) There are many factors affecting the luminescent properties of phosphor materials, which are affected each other, thus increasing the complexity of the preparation of high-performance luminescent materials. This experiment found that the luminescent properties of 8.0%Eu^{2+} ion doped luminescent materials obtained by annealing for 4h at 800°C by stepwise annealing method is the best.
(5) The most effective excitation wavelength and best emission wavelength of $Sr_2P_2O_7:Eu^{2+}$ sample prepared by the experiment are 368nm and 441.8nm, respectively. The samples with high color purity and emission spectral intensity reaches the commercial standard, which is very suitable for UV LED chip excited blue emission phosphors in the range of 330nm-385nm [7].

ACKNOWLEDGEMENTS

Foundation: Key Laboratory of Optoelectronic materials and Applications in Guangdong Higher Education (2017KSYS011). National Natural Science Foundation of China (21975187).

REFERENCES

[1] Chen L, Zhang Y, Liu F, et al. A new green phosphor of $SrAl_2O_4;Eu^{2+}$, $Ce^{3+}Li^+$ for altemating current driven light-emitting diodes [J]. Materials Rasearch Bulletin, 2012, 47(12): 4071–4075
[2] Wang L L, Ni H Y, XIAO F M. Research on luminescence performance and electronic structure of SrAl2B2O7:Eu3+ [J].Chinese journal of rare earths, 2009, 27(03): 384–389.

[3] Botterman J, Smet P F. Persistent phosphor SrAl$_2$O$_4$:Eu,Dy in outdoor conditions: saved by the trap distribution [J]. Optics Express, 2015, 23(15): A868–A881.
[4] Chun Che Lin, Zhi Ren Xiao, Guang-Yu Guo, et al. Versatile Phosphate Phosphors ABPO4 in White Light-Emitting Diodes: Collocated Characteristic Analysis and Theoretical Calculations [J]. J. AM. CHEM. SOC. 2010, 132(9): 3020–3028.
[5] Li Y Q, Hirosaki N, Xie R J, et al. Crystal, electronic and luminescence properties of Eu^{2+}-doped Sr$_2$Al$_{2-x}$Si$_{1+x}$O$_{7-x}$N$_X$ [J]. Science and Technology of Advanced Materials, 2007 8(7): 607–616.
[6] Errui Yang, Guangshe Li, Jing Zheng, et al. Kinetic Control over YVO4:Eu3+Nanoparticles for Tailored Structure and Luminescence Properties [J]. J. Phys. Chem.C 2014, 118(7): 3820–3827.
[7] Zhou Qinqin, Wang Mengxia, Wang Yi, research progress of rare earth ion doped blue phosphors [J]. Guangzhou chemical industry, 2019, 47 (24) : 20–29.
[8] Wang G C. Preparation and property optimization of BaMgAl_(10)O_(17) : Eu~(2+) blue phosphor for white LED [D]. Hunan Agricultural University, 2012.
[9] Tana. Synthesis and Luminescence properties of blue phosphors made of strontium phosphate and calcium phosphate white LED [D].Central South University for Nationalities,2009.
[10] Shichun Mountain. On the F-F transition of Eu~(2+) [J]. Luminescence and Display, 1981(01): 39–46.
[11] Shi C S, Yecharen. Some criteria of Eu~(2+) ion F → F transition emission [J]. (in Chinese) Chinese Journal of Rare Earths, 1983(01): 84–99.
[12] Huang Xiaosong, Jiang Rongyun, Yang Jian, Liu Chunguang, Zhu Hancheng, Yan Duanting, Xu Changshan, Liu Yuxue. Preparation, luminescence properties and thermal stability of Green light emitting phosphors for white LED with Li~+/La~(3+) co-doping BaSi_2O_5: Eu~(2+) [J]. Acta luminescence,2020, 41(03): 271–280.
[13] Fan G D, Lin C. Luminescence of Eu2+ in -SR2P2O7 and energy transfer of Eu2+→Mn2+ [J].Journal of shaanxi university of science and technology (natural science edition), 2016, 34(02): 45–49+58.
[14] Lei Chen, Kuo-Ju Chen, Chun-Che Lin, et al. Combinatorial Approach to the Development of a Single Mass YVO$_4$:Bi^{3+},Eu^3+Phosphor with Red and Green Dual Colors for High Color Rendering White Light-Emitting Diodes [J]. J. Comb. Chem. 2010, 12(4): 587–594.
[15] Yizhu Lei, Qi Zou, Dehai Du. Energy transfer between Ce^{3+} and Eu^{2+} in doped KSrPO$_4$ [J]. Appl Phys A, 2009, 97(3): 635–639.
[16] Kshatri D S, Khare A. Characterization and optical properties of Dy^{3+} doped nanocrystalline SrAl$_2$O$_4$:Eu^{2+} phosphor [J]. Journal of Alloys and Compounds, 2014, 588: 488–495.
[17] Zhou Faguang, Preparation and luminescence properties investigation of several novel phosphate phosphors activated by Eu^{2+}/Eu^{2+}-Mn^{2+} Ions [D]. Bohai University, 2020.

Multiple biotherapy effects of salidroside on tumors

X.P. Wang, D.Y. Yuan, W.H. Li & Y. Tian
Department of Medicine, Xizang Minzu University, Xianyang, China

ABSTRACT: Salidroside, as a kind of natural herbal medicine, has the advantages of a wide range of treatment, low toxicity, and high efficiency. A large number of studies have shown that salidroside can inhibit the proliferation of tumorsin different ways, to achieve the purpose of treating tumors. Given the researches on the anti-tumor mechanisms of salidroside, it is concluded that salidroside can block the cancer cell cycle, promote cancer cell differentiation, and induce apoptosis or autophagy of cancer cells.

1 INTRODUCTION

Researches showed that the cancer incidence rate of Chinese people accounts for 22% of global cancer, while the mortality rate accounts for 27% of global cancer deaths [1]. In recent years, the pursuitof effective and low toxicity drugs has become a hot spot in basic and clinical research for cancer. Salidroside is the main derived effective component of the Chinese Tibetan herb, Rhodiola sachalinensis. It possesses the effects of invigorating health, promoting blood circulation, and protecting from blood stasis. The research shows that salidroside has the functions of immune regulation, inhibition of hypoxia, and treatment of tumors [2–4]. In recent years, with the development of molecular techniques and the continuous in-depth study of salidroside, it is found that salidroside indeed plays an active role in the treatment of tumors. Therefore, the recent progress in the mechanisms of anti-tumor effects of salidroside is generalized to provide a better basis for further clinical anti-tumor investigation.

2 ARRESTING CANCER CELL CYCLE

The growth of normal cells must have a complete cell cycle to proliferate, usually accompanying with the down-regulation of Cyclin-Dependent Kinase 4 (CDK4), cyclin D1, up-regulation of p27 (Kip1) and p21 (Cip1) expression to participatein the progression of G1-S-G2-M [5,6]. However, when some stimulating factors stimulate the cells, the cells often can not pass through the G1/S and G2/M detection points to inhibit the cell growth [7,8].

Cancer cell growth also depends on the normal replacement of the cell cycle in turn, so cell cycle arrest is an important way to inhibit cancer cells. In this process, cyclins and CDK are the most important key points [9]. The study found that salidroside inhibited the growth and proliferationof gastric cancer SGC-7901 in the G2/M phase by down-regulating the expression of cyclin B1 expression [10].

Similarly, salidroside could suppressthe growth of breast cancer cell lines MDA-MB-435 and MCF-7. The expression of cyclin B1 and Cdc2 was down-regulated in the G1 phase andthe cell cycle was arrested in the G2 phase, thus inhibiting the invasion and proliferation of breast cancer cells [11]. Besides, salidroside can block liver cancer cells in the G0/G1 phase, thus exerting the anti-cancer effect [12].

3 INDUCING APOPTOSIS OF CANCER CELLS

Tumor growth depends on the ratio between cell proliferation and cell death. In addition to the role of tumoroncogenes and suppressing genes, apoptosis-regulating genes also play important roles in the development of tumors. Therefore, imbalance of apoptosis is one of the mechanisms for tumor development [13–17]. The process of tumor cell apoptosis is the result of the activation and expression of a series of genes, which regulate and interact with each other. For example, caspases, an apoptotic protease, can cause cell division to form apoptotic bodies, which are recognized and phagocytized finally by adjacent phagocytes or other cells, and lead to the complete disintegration of abnormal cells, eliminating abnormal or damaged cells in the end.On the contrary, apoptosis-suppressing genes Bcl-2, p53, both can inhibit the apoptosis of tumor cells, promote the growth of tumor cells.Besides, the pro-apoptotic gene Bax participates in the process of tumor cell apoptosis. The research found that salidroside can induce apoptosis ofadenoma cells by up-regulating the expression of caspase-3 and caspase-8 [13]. The study also showed that salidroside can inhibit the expression of Bcl-2, promote the expression of caspase 3, caspase 8 and Bax, inducing apoptosis of hepatocellular cells, and suppressing the growth of cancer cells [18].

4 INDUCING DIFFERENTIATION OF CANCER CELLS

Inducing the differentiation of cancer cells does not directly kill cancer cells, but prevents cancer cells from progression and induces them to develop into normal cells under the function of some differentiation inducers. At present, retinoic acid (RA) is the most widely used differentiation inducer in clinical practice. RA is used in the treatment of a variety of malignant tumors, such as acute leukemia, thyroid cancer and breast cancer [19,20]. The initiation of cell differentiation is regulated by oncogenes such as c-myc and p53 [21,22]. As a proto-oncogene, c-myc participates in the process of cancer growth, proliferation, invasion and migration. In transformation cells, the expression of c-myc gene will increase, but with the differentiation of cells, the expression of c-myc gene will decrease, and the ability to promote cell proliferation will be lost. The research found that salidroside could cooperate with RA to downregulate c-myc and promote apoptosis of gastric cancer cells, to achieve the effect of treating gastric cancer [23]. When hepatocellular cells were treated with different concentrations of salidroside, the DNA content of the salidroside treatment group was lower than that of the control group, and the expression of c-myc decreased with the increase of the concentration of salidroside. Morphological observation also showed that the volume of the cells increased significantly in HCC cells with high expression of c-myc, especially the enlargement of nuclei and deep staining, however, when the expression of c-myc was low after treated with salidroside, the differentiation of HCC cells was better [24]. Therefore, salidroside can inhibit the expression of proto-oncogene c-myc, and promote cancer cells to develop towards normal cells, which can indeed induce differentiation of cancer cells.

5 AUTOPHAGY OF CANCER CELLS INDUCED BY SALIDROSIDE

Autophagy refers to the process in which cells phagocytize their own cytoplasm or organelles after receiving some stimulation, and form autophagy bodies through membrane encapsulation, and then autophagy bodiescombine with lysosomes to form autophagic lysosomes. Finally, autophagic lysosomes complete their degradation through internal proteolytic enzymes [25–27]. Autophagy activity changes in a variety of human tumor cells. In recent years, many studies have shown that autophagy plays an important role in the occurrence and development of tumors [25–27]. Inducing or inhibiting autophagy may become an effective anti-cancer method. Autophagy is regulated by a series of autophagyrelated genes (ATG) [28]. Becline1 and LC3 are involved in the formation of autophagy, and their expression level is closely related to autophagy ability [29]. LC3 exists in the cytoplasm as the inactive form, LC3-I. When autophagy occurs, LC3-I will be transformed into LC3-II and combine with the autophagy membrane to form an autophagic lysosome until it is degraded by lysosomal enzymes. Therefore, the content of LC3-II can reflect the level of autophagy in cells. The studyfound that salidroside could induce autophagy in SW1353 cells by upregulating LC3-II and down-regulating p62 expression [30]. Salidroside can induce autophagy in colon cancer cells, possibly by promoting the expression of LC3-II and becline 1 [31].

6 REGULATING CANCER CELLS THROUGH SIGNALING PATHWAYS

Salidroside can also inhibit cancer by regulating multiple signaling pathways [32–36]. The study showed that salidroside regulated EGFR/JAK2/STAT3 signal transduction through MMP2 and inhibits migration, invasion and angiogenesis of MDA-MB 231, breast cancer cells [37]. Salidroside induced apoptosis of human colon cancer cells by inhibiting PI3K/Akt/mTOR pathway and promoted the expression of LC3-II and becline 1 to induce autophagy [38]. The research found that salidroside could inhibit the proliferation, migration and invasion of colon cancer cell SW116 [39]. The mechanism is likely to be related to the inhibition of VEGF, VEGFR-2, MMP-2 and MMP-9 protein expression. Salidroside also inhibited the proliferation, migration and invasion of gastric cancer BGC-823 cells by down-regulating the activation of ROS mediated Src related signal pathway and inhibiting the expression of HSP70 [40].

7 SUMMARY

Salidroside, as a kind of plant-derived traditional Chinese herb, possesses the advantages of low toxicity and wide anti-cancer effects. Salidroside plays an important role through a variety of mechanisms, including inducing autophagy of cancer cells, inducing cancer cells to differentiate into normal cells, enhancing immune regulation to fight cancer, and blocking the cancer cell cycle to inhibit their proliferation. Recent studies proved that the anti-cancer mechanisms of salidroside involved the disturbing signal pathway and induced differentiation (Figure 1). Nevertheless, the detailed mechanisms for salidroside still needfurther investigation.

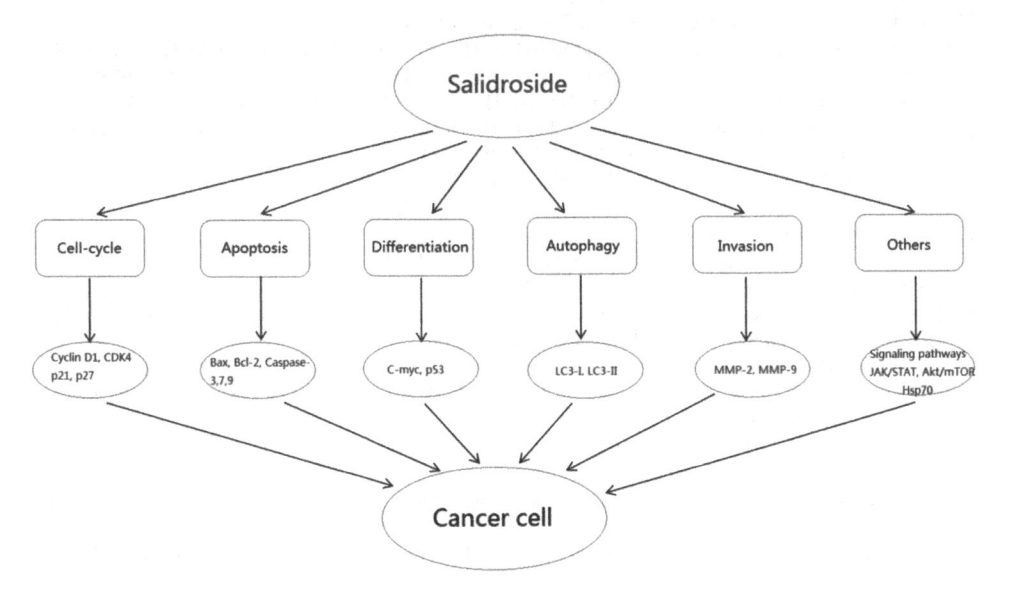

Figure 1. Mechanisms of salidroside against tumor.

ACKNOWLEDGEMENT

This work is supported by Scientific Research Program Funded by Shaanxi Provincial Education Department (No.18JS031),the Scientific Research Program of Shaanxi Administration of Traditional Chinese Medicine (No.15-SCJH001, JCPT001), the Natural Science Basic Research Plan in Shaanxi Province of China (No.2016JM8023, 2020JM-590)and the Key Scientific Research Program of Xizang Minzu University and Xizang Autonomous Region (No. 20MDT02, ZRKX2021000102).

REFERENCES

[1] Feng RM, Zong YN, Cao SM and Xu RH 2019 *Cancer Commun. (Lond).* **39** 22
[2] Chen M, Hui C, Chang Y, Wu P, Fu Y, Xu X, Fan R, Xu C, Chen Y, Wang L and Huang X 2016*Am. J. Transl. Res.***8**12
[3] Li Y, PhamV, BuiM, SongL, WuC, WaliaA, UchioE, Smith-LiuF and ZiX2017*Curr. Pharmacol. Rep.***3** 384
[4] Shi X, Zhao W, Yang Y, Wu S and Lv B 2018 *Mol. Med. Rep.* **17** 51
[5] Tachibana KE, Gonzalez MA and Coleman N 2005 *J. Pathol.***205**123
[6] Rastogi N and Mishra DP 2012*Cell Div.***7**26
[7] Abril-Rodriguez G and Ribas A 2017*Cancer Cell.***31**848
[8] Doğ an ŞÞiğva ZÖ, Balci Okcanoğlu T, Biray Avci Ç, Yilmaz Süslüer S, Kayabaşi Ç, Turna B, Dodurga Y, Nazli O and Gündüz C 2019*Gene.***687**261
[9] Yang P, Yin K, Zhong D, Liao Q and Li K 2015*Mol. Med. Rep.* **11**905
[10] Sun AQ and Ju XL 2020*Chin. J. Integr. Med.* **2020** doi: 10.1007/s11655-020-3190-8
[11] Hu X, Lin S, Yu D, Qiu S, Zhang X and MeiRuhuan 2010*Cell Biol. Toxicol.***26**499
[12] Ding SY, Wang MT, Dai DF, Peng JL and Wu WL 2020*Biochem. Biophys. Res. Commun.***527**1057
[13] TangC, ZhaoCC, YiH, GengZJ, WuXY, ZhangY, LiuY and FanG 2020*Front. Pharmacol.* **11** 976
[14] GuiD, CuiZ, ZhangL, YuC, YaoD, XuM, ChenM, WuP, LiG, WangL and HuangX 2017*BMC Pulm. Med.***17** 191
[15] YuG, LiN, ZhaoY, WangW and FengXL 2018 *Oncol. Lett.***15** 6513
[16] Pistritto G, Trisciuoglio D, Ceci C, Garufi A and D'Orazi G 2016*Aging (Albany NY).***8**603
[17] Mortezaee K, Salehi E, Mirtavoos-Mahyari H, Motevaseli E, Najafi M, Farhood B, Rosengren RJ and Sahebkar A. 2019 *J. Cell. Physiol.***234**12537
[18] Lu L, Liu S, Dong Q and Xin Y 2019*Mol. Med. Rep.* **19**4964
[19] Taibi G, Gueli MC, Nicotra CM, Cocciadiferro L and Carruba G 2014 *J. Enzyme Inhib. Med. Chem.***29**796
[20] Kogai T, Schultz JJ, Johnson LS, Huang M and Brent GA 2000 *Proc. Natl. Acad. Sci. USA.***97**8519
[21] Ouyang JF, Lou J, Yan C, Ren ZH, Qiao HX and Hong DS 2010*J. Pharm. Pharmacol.* **62**530
[22] Bertout JA, Majmundar AJ, Gordan JD, Lam JC, Ditsworth D, Keith B, Brown EJ, Nathanson KL and Simon MC 2009*Proc. Natl. Acad. Sci. USA.***106**14391
[23] Chen ZM, Wu Q, Chen YQ and Su WJ 1999 *Shi Yan Sheng Wu Xue Bao.***32**135
[24] Xie F, OUYANG X and Jiang M 2006 *Xi Nan Guo Fang Yi Yao.***12** 130
[25] LiT, XuK and LiuY 2018*Onco.l Lett.***16**3162
[26] LiuZ, LiX, Anne R. Simoneau, Mahtab Jafari and ZiX 2012*Mol. Carcinog.***51**257
[27] Klionsky DJ 2007*Nat. Rev. Mol. Cell Biol.***8**931
[28] Mujumdar N, Mackenzie TN, Dudeja V, Chugh R, Antonoff MB, Borja-Cacho D, Sangwan V, Dawra R, Vickers SM and Saluja AK 2010*Int. J. Cancer***139**598
[29] Noda T andYoshimori T 2009*Int. Immunol.***21**1199
[30] Zeng W, Xiao T, Cai A, Cai W, Liu H, Liu J, Li J, Tan M, Xie L, Liu Y, Yang X and Long Y 2017*Cell. Physiol. Biochem.***43**1487
[31] Li H and Chen C 2017 *BMC Complement. Altern. Med.***17**538
[32] HuangL, HuangZ, LinW, WangL, ZhuX, ChenX, YangS and LvC 2019*Int. J. Oncol.***54** 1969
[33] QinY, LiuH, LiM, ZhaiD, TangY, YangL, QiaoK, YangJ, ZhongW, ZhangQ, LiuY, YangG, SunT and YangC 2018*EBioMedicine.***38** 25
[34] ChenX, WuY, YangT, WeiM, WangY, DengX, ShenC, LiW, ZhangH, XuW, GouL, ZengY, ZhangY, WangZ and YangJ 2016*J. Cachexia Sarcopenia Muscle.***7**225

[35] Shang H, Wang S, Yao J, Guo C, Dong J and Liao L 2019 *Thorac. Cancer.* **10** 1469
[36] Chen X, Kou Y, Lu Y and Pu Y 2020 *J. Cell. Biochem.* **121** 165
[37] Kang DY, Sp N, Kim DH, Joung YH, Lee HG, Park YM and Yang YM 2018 *Int. J. Oncol.* **53** 877
[38] Fan XJ, Wang Y, Wang L, Zhu M. Oncol Rep, 2016, 36(6):3559–3567.
[39] He H, Chang X, Gao J, Zhu L, Miao M and Yan T 2015 *Inflammation.* **38** 2178
[40] Qi Z, Tang T, Sheng L, Ma Y, Liu Y, Yan L, Qi S, Ling L and Zhang Y 2018 *Mol. Med. Rep.* **18** 147

Effects of spraying Indonesian BioSilAc foliar fertilizer on growth and quality of vegetables in red soil region

D.F. Huang, L.M. Wang & Y. Li
Institute of Soil and Fertilizer, Fujian Academy of Agricultural Sciences, Fuzhou, China
Fujian Key Laboratory of Agro-products Qualitiy & Safety, Fuzhou, China

ABSTRACT: In order to explore the application effect of BioSilAc solid foliar fertilizer product produced by Indonesia Institute of biotechnology and Bio-industry on vegetables in red soil region, a field plot experiment was conducted to study the effects of different concentrations of BioSilAc foliar fertilizer (i.e., 200, 300, 400, 500, 600 g/hm 2times) on the agronomic traits, yield, nutrient uptake and nutritional quality of vegetables (pakchoi, Brassica chined L.) in red soil area. Results showed that spraying different concentrations of BioSilAc foliar fertilizer (L1~L5) could increase the plant height by 1.6%~11.3%, plant weight by 4.2% ~33.3%, leaf number by 8.0%~23.2%, yield by 1.7%~15.3% compared with CK (spraying the same amount of water), increase the uptake of phosphorus by 0.9%~16.7%, potassium by 7.3%~31.1%, reduce the nitrate content by 3.5%~61.7%, and increase the reduced Vc content by 0.1%~28.5%, and the content of total soluble sugar by 1.4%~56.2%. Among them, L3~L5 treatment (spraying BioSilAc foliar fertilizer 400~600g/hm 2times) had better effect on improving the yield and quality of pakchoi(Brassica chined L.) in red soil area.

1 INTRODUCTION

BioSilAc solid foliar fertilizer is developed by Indonesia Institute of Biotechnology and Bio-industry. It is a new type of water-soluble silicon dioxide formula fertilizer enriched with silicon soluble microorganisms. It contains 17.5%~20.0% of bioavailable silicon dioxide (SiO_2), more than 5% of available silicon dioxide in H_4SiO_4, $10^4 \sim 10^6$ propagule/tablet (4g) of non pathogenic silicon dioxide solubilizing microorganisms, and is mixed with humic acid coating [1]. It is developed by unique extraction and processing technology. The preparation of the extract and formula has a unique way in terms of composition and biological activity for plant growth and development, and has obtained the national invention patent of Indonesia [2,3].

In Indonesia, studies have shown that the mixed application of BioSilAc and 25%~50% reducing fertilizer could increase crop yield, and the yield increase rate was 12.1% for oil palm, 35% for rice, 20% for corn, 15% for sugarcane, 26% for black soybean and 50% for potato, which has achieved good crop yield increase effect[1].

In order to explore the application effect of BioSilAc solid foliar fertilizer product produced by Indonesia Institute of biotechnology and Bio-industry on vegetables in red soil region of Fujian province, southeast coast of China, we carried out the field fertilizer effect verification test of this foliar fertilizer product on the main vegetable crop variety (pakchoi, *Brassica chined* L.) in red soil region from January 2021. The purpose of this study was to provide a scientific basis for the introduction of this foliar fertilizer product to the popularization and application of vegetables in Fujian provinceand even in the vast red soil areas of southern China.

2 MATERIALS AND METHODS

The experimental site is located in the vegetable planting base of Honglai Town, Nan'an City, Fujian Province. The soil type is red soil, and the soil fertility is medium. Honglai town is located in the northeast of Nan'an City, adjacent to Luojiang District of Quanzhou City in the East, Kangmei town and Xuefeng Development Zone in the west, Fengzhou town in the south, Meishan town and Hongmei town in the north, with a land area of 87 km² and a population of more than 90100. The terrain belongs to a valley basin. Dongxi (Honglaixi), the main stream of Jinjiang River, flows through the whole town from north to south. The main peaks are Kuishan, Yujinshan, Shuangmao and Damaoshan, with an altitude of about 500 meters. The climate type belongs to south subtropical monsoon humid climate, with abundant sunshine and rain, pleasant climate, evergreen in four seasons, and unique advantages in physical geography and climate.

Six treatments were set up in the trial, which were based on Farmers' conventional fertilization, spraying the same amount of water as the control, and spraying different concentrations of BioSilAc foliar fertilizer products. The specific treatments were as follows: Treatment 1, control (spraying the same amount of water, 300 L/hm²·times); Treatment 2, low amount of BioSilAc solid flake products [that is, 50 tablets (200g)/hm²·times, completely dissolved in 300L water, foliar spray]; Treatment 3, low and medium amount of BioSilAc solid flake products [that is, 75 tablets (300g) hm²·times, completely dissolved in 300L water, foliar spray]; Treatment 4, medium amount of BioSilAc solid flake product [i.e., 100 tablets (400g) / hm²·times, completely dissolved in 300L water, foliar spray]; Treatment 5, medium and high amount of BioSilAc solid flake products [that is, 125 tablets (500g)/hm²·times, completely dissolved in 300L water, foliar spray]; Treatment 6, high volume of BioSilAc solid flake products [that is, 150 tablets (600g)/hm²·times, completely dissolved in 300L water, foliar spray]. The above six treatments were represented by CK, L1, L2, L3, L4 and L5 respectively. Each treatment was repeated for 3 times, and each plot area was 20 m², arranged randomly. The tested vegetable crop waspakchoi(*Brassica chined* L.), Huaguan 188 qinggencai F1 varieties, which were produced and provided by Shandong Hualiang Seed Industry Co., Ltd. Sowing date: January 18, 2021. During the growth period of vegetables, BioSilAc foliar fertilizer was sprayed twice on February 2 and February 23, respectively. On March 2, the main agronomic characters of vegetable plants were tested, sampled and analyzed.

The main agronomic traits of vegetable plant test method[4,5]: In each experimental plot, 10 normal growth vegetable plants were randomly selected, the roots were cut with a knife, and then the plant height, plant weight and leaf number of each plant were tested, and the average value of the observed value of 10 plants was taken. The yield of vegetables in the plot was the fresh weight of edible plants on the ground.

The nutrients and nutritional quality of vegetable plants were determined by conventional agro-chemical analysis methods[6~8]: The total nitrogen content was determined by Kjeldahl colorimetry, the total phosphorus content was determined by molybdenum antimony colorimetry, the total potassium content was determined by flame photometry, the nitrate content was determined by GB 5009.33–2016, the reduced Vc content was determined by 2,6-dichlorophenol indophenol titration, and the total soluble sugar content was determined by anthrone colorimetry.

Microsoft excel-2003 office software was used for data processing, and SPSS10.0 statistical software was used for variance analysis and significance test.

3 RESULTS AND ANALYSIS

3.1 *Effects of spraying BiosilAc foliar fertilizer on agronomic characters and yield of vegetable*

Results (Table 1) showed that: Compared with CK treatment, spraying different concentrations of BiosilAc foliar fertilizer (L1~L5) could improve the agronomic traits of plant growth and increase vegetable yield to a certain extent, in which the plant height increased by 1.6%~11.3%, the plant weight increased by 4.2%~33.3%, the number of leaves increased by 8.0% ~23.2%, and the yield increased by 1.7%~15.3%, Among them, L3~L5 treatment had better effect on agronomic traits and yield of vegetable plant. Analysis of variance showed that L4 treatment was significantly better than CK, L1, L2 and L3 treatment ($P < 0.05$), but there was no significant difference with L5 treatment ($P > 0.05$); L5 was significantly better than CK and L1 ($P < 0.01$), but there was no significant difference with L4, L3 and L2 ($P > 0.05$); The effect of L2~L5 treatment on the increase of leaf number of vegetable plant was significantly better than that of CK and L1 treatment ($P < 0.05$); The results showed that L3~L5 treatment was significantly better than CK and L1 treatment ($P < 0.05$), but there was no significant difference with L2 treatment ($P > 0.05$).

Table 1. Effects of spraying BiosilAc foliar fertilizer on agronomic characters and yield of vegetable.

Treatment	Plant height		Plant weight		Leaf bearing number		Yield	
	Value (cm)	Increasing rate (%)	Value (g/plant)	Increasing rate (%)	Value (piece / plant)	Increasing rate (%)	Value (kg/hm^2)	Increasing rate (%)
CK	12.4b	\	4.8Bb	\	11.2b	\	13098.6b	\
L1	12.6b	1.6	5.0Bb	4.2	12.1b	8.0	13320.8b	1.7
L2	13.1b	5.6	5.3ABab	10.4	12.8a	14.3	14078.6ab	7.5
L3	12.9b	4.0	5.7ABab	18.8	13.4a	19.6	14964.2a	14.2
L4	13.8a	11.3	5.9ABab	22.9	12.9a	15.2	15097.5a	15.3
L5	13.6ab	9.7	6.4Aa	33.3	13.8a	23.2	15008.5a	14.6

Note: Small and capital letters following the value indicate the difference is significant at 0.05 and 0.01 level, respectively (Inspect by LSD, following is the same).

3.2 Effect of spraying BiosilAc foliar fertilizer on the absorption of nitrogen, phosphorus and potassium by vegetable

Results (Table2) showed that: Compared with CK treatment, spraying different concentrations of BiosilAc foliar fertilizer (L1~L5) could promote the absorption of phosphorus and potassium by vegetable plants to a certain extent, with the increase rate range of 0.9% ~ 16.7% and 7.3% ~ 31.1% respectively, but it had little effect on the absorption of nitrogen by vegetable. The results of variance analysis showed that L5 was significantly better than CK ($P < 0.01$), and significantly better than L1, L2 and L3 ($P < 0.05$), but there was no significant difference between L5 and L4 ($P > 0.05$); However, there was no significant difference in the effects of different treatments on the uptake of nitrogen and phosphorus ($P > 0.05$).

Table 2. Effect of spraying BiosilAc foliar fertilizer on the absorption of nitrogen, phosphorus and potassium by vegetable.

Treatment	Nitrogen absorption		Phosphorus absorption		Potassium absorption	
	Value (kg/hm^2)	Increasing rate (%)	Value (kg/hm^2)	Increasing rate (%)	Value (kg/hm^2)	Increasing rate (%)
CK	30.7a	\	6.3a	\	50.8Bc	\
L1	29.0a	−5.5	6.4a	0.9	54.5ABbc	7.3
L2	28.9a	−5.6	6.5a	1.8	60.2 ABb	18.4
L3	33.6a	9.6	7.4a	16.7	59.1 ABb	16.3
L4	33.7a	10.1	7.0a	10.5	61.8 ABab	21.6
L5	33.2a	8.5	6.7a	6.3	66.6Aa	31.1

Note: The amount of N (P, K) absorbed by plant (kg / hm^2) = plant yield (kg/hm^2) × N (P, K) content of plant(%)

3.3 Effects of spraying BiosilAc foliar fertilizer on the nutritional quality of edible parts of vegetable

Results (Table 3) showed that: compared with CK treatment, spraying different concentrations of BioSilAc foliar fertilizer (L1~L5) could reduce the nitrate content of vegetables to a certain extent, with a reduction rate range of 3.5%~61.7%; The reduced Vc and water-soluble sugar content of vegetables were increased by 0.1%~28.5% and 1.4%~56.2% respectively. The results of variance analysis showed that L5 treatment was significantly better than CK, L1 and L2 treatment ($P< 0.01$), and significantly better than L3 and L4 treatment ($P< 0.05$); L3 and L4 treatments were significantly better than CK, L1 and L2 treatments ($P< 0.05$); L2 was also significantly better than CK and L1 ($P < 0.05$). The results showed that L4 treatment was significantly better than CK, L1 and L2 treatment ($P < 0.05$), but had no significant difference with L3 and L5 treatment ($P > 0.05$). L5 treatment was significantly better than CK, L1, L2 and L3 treatment ($P < 0.01$), and significantly better than L4 treatment ($P < 0.05$); L4 was significantly better than CK, L1, L2 and L3 ($P < 0.01$), but there was no significant difference among CK, L1, L2 and L3 ($P > 0.05$).

Table 3. Effects of spraying BiosilAc foliar fertilizer on the nutritional quality of edible parts of vegetable.

Treatment	Nitrate Content (mg/kg)	Increasing rate (%)	Reduced Vc Content (mg/kg)	Increasing rate (%)	Water soluble total sugar Content (mg/kg)	Increasing rate (%)
CK	1823Aa	\	8.48b	\	0.73Bc	\
L1	1759Aa	-3.5	8.63b	1.8	0.73Bc	0.0
L2	1554Ab	-14.8	8.49b	0.1	0.74Bc	1.4
L3	1322ABc	-27.5	9.55ab	12.6	0.80Bc	9.6
L4	1243ABc	-31.8	10.9a	28.5	1.00Ab	37.0
L5	699Bd	-61.7	9.3ab	9.7	1.14Aa	56.2

4 CONCLUSION

The results showed that, compared with CK (spraying the same amount of clean water) treatment, spraying different concentrations of BioSilAc foliar fertilizer (L1~L5) could increase the plant height by 1.6%~11.3%, plant weight by 4.2%~33.3%, leaf number by 8.0%~23.2%, vegetable yield by 1.7%~15.3%, increase the uptake of phosphorus by 0.9%~16.7%, potassium by 7.3%~31.1%, and reduce the nitrate content of vegetables by 3.5%~61.7%, the reduced Vc content of vegetables was increased by 0.1%~28.5%, and the water-soluble total sugar content was increased by 1.4%~56.2%. Among them, L3~L5 treatment (i.e., spraying BioSilAc foliar fertilizer 400~600 g/hm^2·times) had better effect on improving the yield and quality of pakchoi(*Brassica chined* L.) in red soil area.

ACKNOWLEDGMENTS

This work was supported by several projects, which included Fujian science and technology program spark project (2020S0050), Foreign cooperation project of Fujian Academy of Agricultural Sciences in 2020 "Introduction of Indonesian foliar fertilizer BioSilAc series products and its application in crops",Fujian Provincial Public Welfare Research Institute Basic Research Project(2020R1025007), Fuzhou Science and Technology Project(2018–G-65).

REFERENCES

[1] SantosoD, GunawanA, Budiani1A, Sari1D A and Priyono. 2018. Plant biostimulant to improve crops productivity and planters profit. *IOP Conf. Series: Earth and Environmental Science* 183 (2018) 012017.
[2] Santoso D and Priyono. 2014. Proses Produksi dan Formulasi Biostimulan dari Alga Coklat Sargassum sp. serta Penggunaannya untuk Pertumbuhan Tanaman. P-00201406718.(Indonesian Patent)
[3] Santoso D, Priyono and Budiani A. 2017. Formulasi Biostimulan Palmarin dan Penggunaannya untuk Peningkatan Rendemen Minyak dan Bobot Tandan Buah Segar (TBS) Kelapa Sawit. P-00201701786. (Indonesian Patent)
[4] YanC, ChanghaiY, JingliG. 2021. *Fertilizer and Health*, **48** (1): 10–12,26
[5] YanL, Shufen L, XihongW, Jigang F. 2019.*Shanghai Vegetable*, (4): 68–69
[6] QiY, xiaoyan L, KunZ,Wuning C. 2020. *Yangtze River Vegetable*, (8): 71–74
[7] RukunL. 2000. *Methods of Soil Agrochemical Analysis* (Beijing: China Agricultural Science and Technology Press)
[8] ChunshengL, Shouxiang Y.1996. *Agricultural Chemical Analysis* (Beijing: China Agricultural University Press)

Chapter 4: Civil Materials and Sustainable Environment

Application of geological modeling constrained by body of volcano rock in Xushen gas field

Y. Xu
Exploration and Development Research Institute, Daqing Oilfield Company Limited, Daqing, China

ABSTRACT: Volcanic reservoir belongs to the event deposits. So it is difficult to distinguish volcanic reservoir and sedimentary pyroclastic rock which is non-reservoir by conventional geological modeling methods. There is less wells in the study area, so the inter-well prediction accuracy is low. In respect of the issues above, the paper puts forward body of volcanic rock as constraint of neural network modeling. Firstly, based on the differences in seismic reflection characteristics between volcanic rocks and surrounding sedimentary pyroclastic rocks, volcano body are identified and characterized in 3D space which is also guided by the volcanic eruption model. Secondly, the most sensitive electrical parameters for volcanic reservoir prediction are optimized based on the reservoir evaluation results. The quantitative relationship between various seismic attributes and reservoir electrical parameters is established by neural network algorithm. Finally, constrained by the volcano body 3D geological model is established by using the neural network algorithm, which objectively reflects the distribution of volcanic mechanism and its internal attributes. A horizontal well in Xushen A block is optimized and designed by this method. The actual drilling reservoir is in good agreement with the prediction. For under-developed volcanic gas reservoirs, the model is established constrained by body of volcano rock to avoid the interference of sedimentary pyroclastic rock in the prediction. Deterministic method of neural network algorithm is applied to obtain the correlation between seismic attributes and reservoir parameters, thus reducing the uncertainty of inter-well prediction. Method of neural network geological modeling constrained by body of volcano rock begins with the reservoir sedimentary characteristics, and fully combines seismic data and drilling data by the neural network algorithm, which reduces the uncertainty of the model to the greatest extent. The paper provides methods and ideas to 3D geology modeling of volcanic rocks reservoir and other similar reservoir.

1 INTRODUCTION

As oil and gas reservoirs volcanic rock has been recognized for nearly a century. However, because its reserves and production are much lower than sedimentary reservoirs, the degree of research on volcanic reservoirs is relatively low. In recent years, large volcanic gas fields such as Xushen gas field in Songliao basin and Kelamili gas field in Junggar basin have been discovered and developed successively, which provides an important platform for geologists to study volcanic reservoirs. Compared with conventional sandstone and carbonate gas reservoir, the volcanic rock gas reservoir geological condition is more complex and the inside structure is complex. The development of volcanic reservoir is influenced by including eruption, tectonic action and other factors. Pore structure and its distribution are complex, which is difficult for quantitative description. Thus it is difficult to establish geological model which can reflect complex inside structure of volcanic rock.

In study of volcanic rock gas reservoir geology of recent years, a cooperative geological modeling method based on well, seismic and lithofacies is developed. It means that drilling, logging and mud logging data are used on well point and volcanic body, volcanic lithofacies and seismic prediction results are input as progressive constraints between wells. In this way the model can reflect the volcanic structure and the complex internal structure of reservoir in Xushen gas field.

2 VOLCANIC BODY DESCRIPTION

The distribution area of volcanic body has been the target of oil and gas exploration at home and abroad. It controls the distribution of volcanic rock facies, and is often a favorable reservoir area for volcanic rocks and a high oil-gas producing area. Therefore, the description of volcanic body is very important in the exploration and development of volcanic oil and gas reservoirs.

Volcanic eruptions are directly related to plate activities, and the eruption modes can be divided into three types according to the patterns of conduit: eruption, fissure eruption and composite central-fissure eruption. By comparing the seismic profile characteristics of volcanic rocks, it is concluded that the volcanic rocks in the Yingcheng formation in the study area are mainly eruptive with fissure type, and the craters along the controlling-depression faults in the plane are distributed in a beaded pattern and erupt for many times.

In Xushen A block, for example, based on the detailed geologic horizon calibration and seismic section features contrast, clear boundaries of volcanic body are recognized around the surrounding sedimentary rock, which is like a mushroom cloud inlaid in formation. Volcanic conduit can be recognized, the seismic event of volcano body internal presents the weak amplitude, intermittence, medium-low frequency. From seismic plane the size of volcano body is not big and the distribution area is about 2—5 square kilometers.

Powerful impetus during volcanic eruption produce the destruction of the volcanic crater formation, forming many fracture, micro fracture and fracture. The coherence algorithm based on eigenvalue can give good identification, so distribution of the crater can be clearly determined. From the picture volcanic crater are identified with coherent reflection(Figure1).

The volcanic body in the study area experienced multiple eruptions and tectonic transformation. The volcanic crater are identified by means of coherent and conventional horizontal slices of seismic bodies. Then volcanic body surface of each eruption cycle and the relationship between different volcanic bodies are depicted by fine comparison with combination of well and seismic.

Figure 1. Characteristics of volcanic edifice on seismic profile.

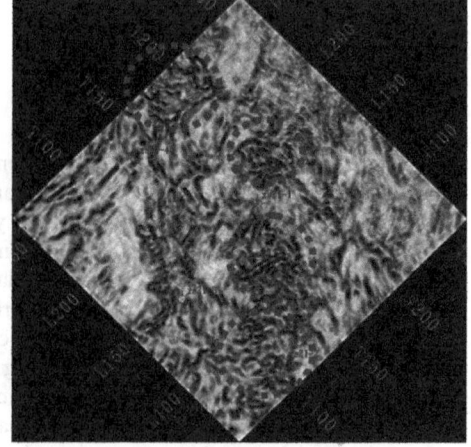

Figure 2. Characteristics of volcanic edifice on seismic coherent slice.

3 GEOLOGICAL MODELING CONSTRAINED BY VOLCANIC BODY

By well drilling in Xushen A block, pyroclastic rock deposits during each cycle of different volcanic eruption periods. It mainly consists of eruptive tuff. As hardly receives the weathering and leaching transformation function, its porosity and permeability are relatively poor. If the pyroclastic rock and volcanic rocks not distinguished during geology modeling, the accuracy will be reduced greatly. Therefore, in this paper a geology model is explored based on the results of volcano body characterization and then input as constrain condition during subsequent reservoir parameter modeling, so as to avoid the influence of pyroclastic rock formed between volcanic eruptions in reservoir prediction.

The logging response characteristics of volcanic reservoirs are influenced by physical properties: as the physical properties get better, the density value decreases, the acoustic time difference increases, and the depth and shallow lateral resistance increases with large difference, which performed a good correlation. According to the well logging parameters determined by the effective reservoir division of volcanic rock, density and deep lateral resistance are selected in the establishment of 3d geological attribute model. The volcanic reservoir is highly heterogeneous and varies greatly in thickness, and its internal structure is complex. According to the conventional properties of the study area, reservoir prediction and inversion body constraints are used to get a poor prediction effect. Therefore, from the perspective of multi-information blending modeling, the neural network method is tried to establish the relationship between density, resistivity and various seismic attributes. Finally, 7 highly correlated seismic attributes, including instantaneous frequency and elastic impedance, are selected to establish the neural network training model of density and resistivity.

Firstly, the seismic interpretation results of volcano body in Xushen block A are applied as constraints to establish a macroscopic lithological model, and the volcano rock is separated from surrounding sedimentary rocks to avoid the uncertainties involved in interpolation of sedimentary rocks and volcanic rocks. Then, density and resistivity curves are selected as the property objects to discretize the well data into the grid cells of the geological model, and the quality of the data before and after roughening is checked. By adjusting the height of the grid and the discretization algorithm, the discretization data keep a high degree of consistency with the original curve data. Then the rock density and resistivity models of reservoir are established by using neural network model.

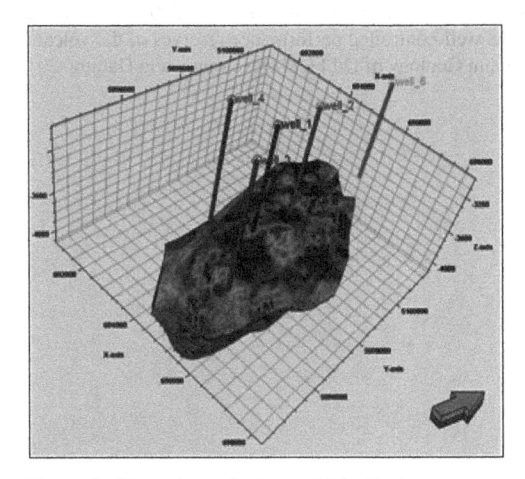

Figure 3. The volcano body model in Xushen A block.

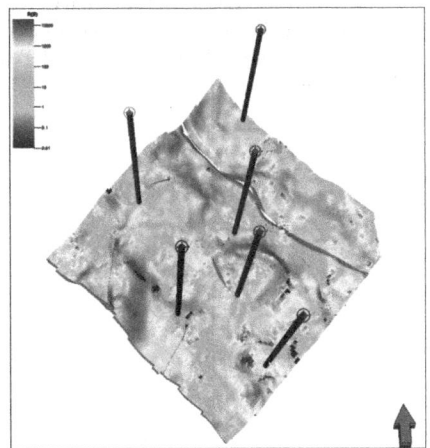

Figure 4. The rock density model in Xushen A block.

4 APPLICATION

In Xushen a block A horizontal well is designed based on the model above. According to the model prediction, the reservoir along horizontal well is made up of II and III type and the drilling rate of reservoirs with the prediction probability above 0.7 is 60.1%. Through the analysis of reservoir evaluation results after drilling, actual fact is in accordance with the model .The actual drilling rate is 64.2%, which is basically consistent with the prediction of uncertainty analysis.

5 CONCLUSION

Volcanic reservoir is controlled by volcano body with strong heterogeneity. In this article, constrained by volcano body the model is established by relationship between a variety of seismic attributes and reservoir parameters. Through the application of the model in the study area for horizontal well design, the results confirms that the geological modeling can objectively depict volcano body and its internal development and distribution of reservoir, reduce the uncertainty of prediction.

REFERENCES

[1] Xu Zhengshun, Pang Yanming, Wang Yuming, etal. Development technique of volcanic gas reservoir [M]. BeiJing: Petroleum Industry Press, 2009.
[2] Ran Qiquan Wang Yongjun Sun Huihui et al Volcanic gas reservoir characterization technique [M]. BeiJing: Science Press 2011.
[3] Xu ZhengshunPang YanmingWang Yuming,etalDevelopment of Volcanic Rock Reservoirs in the Xushen Gas Field in Daqing [J] Natural Gas Industry 2008, 28(12): 74–77.
[4] Pang Yanming, Chen Bingfeng, Yang Shuangling, Shu Ping, Xu Yan. 3D geologic modeling technology for the volcanic reservoir of Yingcheng Formation in the Xushen Gasfield, Songliao Basin [J]. Chinese Journal of Geology, 2009, 44(2): 759–768.
[5] Chen Bingfeng,Xu Yan,Yu Haisheng, Liu Chunsheng, Qu Licai. Studies on Uncertainty of 3D Geological Models of Volcanic Gas Reservoir [J]. Science Technology and Engineering, 2012, 12(19): 4602–4606.
[6] Wang Yuxue, Han Dakuang, Liu Wenling, Ran Qiquan ,Pang Yanming. The application of coherence technology in the volcanic reservoir prediction [J]. Geophysical Prospecting for Petroleum 2006, 45(2): 192–196.
[7] Liu Qi, Shu Ping, Li Songguang. Integrated description technique for deep volcanic gas reservoir in northern Songliao Basin [J]. Petroleum Geology of Oil Field Development in Daqing, 2005, 24(3): 21–23.
[8] Bi Xiaoming, Yang Fengrui. Characteristics of the well-controlled performance reserves of the volcanic-rock gas reservoirs in Xushen gasfield [J]. Petroleum Geology of Oil Field Development in Daqing. 2015, 34(4):60–64.

Research on government supervision mechanism for resource utilization of construction and demolition waste based on evolutionary game

D.D. Liu
School of Management Science and Engineering, Shandong University of Finance and Economics, Jinan, China

L. Qiao
School of Energy and Power Engineering, Shandong University, Jinan, China
Housing and Urban-Rural Development Institute of Shandong Province, Jinan, China

ABSTRACT: With the increasing of construction and demolition waste, the management of which is the foundation to ensure the normal operation and sustainable development of cities. Exploring the resource utilization of construction waste is of great significance to the development of circular economy. Construction and demolition waste management is a dynamic system involving environment, economy and society. In order to maximize their own interests, the relevant subjects of construction waste resource utilization will continue to make game choices. This paper introduces the economic game theory method, through the evolutionary game analysis of the participants under the market mechanism and government regulation, it shows that in the initial stage of construction and demolition waste resource utilization, the government needs to support enterprises through economic policies, and provide relevant preferential treatment, give certain punishment to related enterprises. It is necessary to choose a moderate degree of punishment. Under certain conditions, the greater the punishment, the better it is to promote the enterprises resource utilization of construction and demolition waste, but when the punishment is too strong, it will play a negative effect.

1 INTRODUCTION

With the acceleration of new urbanization process and the urban renewal in China, the construction and demolition waste is increasing, which has accounted for 30%–40% of the total urban waste [1].

According to statistics, the construction and demolition waste of 35 pilot cities is 1.315 billion tons in 2018, and is 1.369 billion tons in 2019. It is estimated that the annual total construction and demolition waste of China is no less than 3.5 billion tons. Construction waste management is a problem that must be solved at a certain stage of urbanization development, and it is the basis for ensuring the normal operation and sustainable development of the city.

The 13th Five-Year Plan points out: develop the circular economy, accelerate the utilization of construction and demolition waste and the establishment harmless treatment systems, and develop remanufacturing in line with standards. In 2020, the Ministry of industry and information technology's "standard conditions for construction waste recycling industry" (Revised Draft) requires for the first time that enterprises should fully accept the local construction and demolition waste that meets the requirements of relevant specifications, and the recycling rate of construction waste entering the plant should not be less than 95%.

The traditional extensive disposal methods such as open stacking and simple landfill can no longer meet the needs of the current social and economic development and ecological civilization construction. Utilization of Construction and Demolition Waste can save a lot of land, natural raw materials, coal and other resources and energy; reduce pollution, improve the urban environment, reduce the index of inhalable particles and fine particles; create new jobs, drive the development of equipment manufacturing industry and service industry, and create huge economic benefits [2]. Promoting the industrial development of comprehensive utilization of construction and demolition waste is of great significance to the development of circular economy, energy conservation and emission reduction, urbanization and the implementation of sustainable development strategy.

2 LITERATURE REVIEW

Construction waste management is a dynamic complex system which is composed of environmental, economic and social subsystems. It is a comprehensive problem involving sociology, economics, environmental science, management science, civil engineering and other disciplines.

Compared with the legal system and control measures of construction waste management in the United States, Germany, Japan and other developed countries, China's construction waste management is still in the primary stage. The relevant research is relatively less, and the relevant policies, regulations, standards and regulatory systems are not perfect.

The United States has passed legislation to enforce the utilization of construction waste, which has formed an effective theoretical basis and practical system. The United States enacted the "Solid Waste Disposal Act" in 1965 and the "Super Fund Act" in 1980. The law improved the standards for the reuse and recycling of construction waste, and provided detailed regulations on the waste treatment supported by state funds. Regulations and policies provide legal protection for the utilization of construction waste in the United States. Germany promulgated the "waste treatment law" in the 1870s, the "circular economy and waste removal law" in 1994, and successively formulated the "water and soil conservation and old waste act", "waste discharge law", and "sustainable promotion of ecological tax reform law". With the continuous improvement of the legal system, the recycling of construction waste in Germany has been effectively improved. Japan is an earlier country in the development of global resource conservation and utilization. Since the 1960s, it has started to study the effective management of construction waste. It has successively promulgated a number of relevant laws and regulations on resource conservation, established garbage collection and cleaning offices in many places throughout the country, continuously promoted new technologies, and improved the classification and reuse of waste.

In recent years, simulation experiment method has attracted the attention of scholars in the field of construction waste management. The construction waste management system involves various stakeholders, such as the construction party, the construction party, the design party, the transportation party, the recycling party, the landfill party, the government departments and so on. The research of construction waste management based on system dynamics includes policy, environment, economy, society and comprehensive impact.

Yuan Hongping [3] used system dynamics theory to study the construction waste on-site management problem, and established a construction waste on-site management model including three subsystems. Ding [4] built a subject-based simulation model to simulate construction waste management measures, revealing the relationship between stakeholders and environmental impact. Liu Tingting [5] applied system dynamics to model and simulate the environmental benefits of construction waste recycling, and conducted scenario analysis on the environmental benefit model. Chen Qijun [6] used system dynamics theory and method to build system simulation model to analyze the relationship between various factors in construction waste treatment industry. Jia Shuwei [7] introduced punishment and subsidy mechanism, and used the method of system dynamics and grey system theory to simulate the construction waste management model. Wang Dong [8] constructed a tripartite evolutionary game model of government, demolition enterprises and construction solid waste enterprises. Through the simulation of the evolutionary path of the game model, the change of the behavior strategy of the three parties under different incentive intensities was analyzed.

Construction waste management is mainly two related subjects of government and enterprise. In order to maximize their own interests, the related subjects will continue to make game choices. This paper introduces the game theory of economics and uses the theory of system dynamics to analyze the participants from the perspective of market mechanism and government regulation. The paper puts forward the policy suggestions for the utilization of construction waste suitable for the development stage of China, and explores the realization path of the construction waste resource utilization based on market government environment coordination. It provides theoretical reference and practical basis for formulating and issuing relevant policies of construction waste utilization, and improves the level of construction waste utilization in China.

3 STATIC GAME BETWEEN ENTERPRISES

Under the market mechanism, enterprises pursue the maximization of short-term interests in an unconstrained environment, without considering environmental pollution and sustainable development. If a few enterprises choose construction waste recycling products, but other enterprises do not, due to the lack of popularization of recycling products and poor social recognition, return of enterprises using recycled products is lower than that of non-renewable products, resulting in the extension of the payback period of investment funds. Therefore, in this mode, enterprises will not actively choose to use recycled products.

Suppose there are two identical enterprises A and B, they have two options for using recycled products : use and not use. If they don't use, their cost and their income is 0 ; If use, their cost is a and their profit is b, then the profit is $\varepsilon = b-a$.

Although the cost of recycled products is low, but the income is also low compared with non-renewable product, so suppose $0 < b < a$, that is to say, the profit of enterprises using recycled products is lower than that of enterprises without recycled products. The income matrix between the two owners is shown in Tables 1.

Table 1. Income moment of two enterprises.

		Enterprise A	
		use	not use
Enterprise B	use	E, ε	E, 0
	not use	0, ε	0, 0

Obviously, (not use, not use) is the only Nash equilibrium of this problem. Without the constraints of external forces, enterprises pursuing the maximization of short-term interests will not choose to use construction waste recycling products.

4 EVOLUTIONARY GAME BETWEEN GOVERNMENT AND ENTERPRISES

When the government incentives and supervision on whether enterprises use recycled products, enterprises and the government are faced with the game of use and supervision. The following is a further analysis of such problems.

Assuming that the subsidy under the government incentive policy is c, if the government checks that the enterprise does not use recycled products, it will impose a fine on enterprises, and the penalty is d ($d > c$). Assuming that the government needs to pay the cost of environmental governance for enterprises not using recycled products is m, and that the government obtains invisible economic and social benefits from the use of recycled products by enterprises is n. The relationship between the government and enterprises is shown in Tables 2.

Table 2. Income matrix between government and Enterprise.

		government	
		Subsidies, regulation	No subsidy, no regulation
enterprise	use	$\varepsilon + c$, n-c	E, n
	Not use	-d, d -m	0, -m

Obviously, there is no pure strategy Nash equilibrium in a single game. Because when the government chooses regulatory strategies, enterprises choose to use recycled products; When the government chooses not to supervise, enterprises will not use recycled products. In fact, this game behaviour between enterprises and the government is repeated and dynamic, and its decision-making behaviour is not based on complete rationality, but on bounded rationality. Therefore, the strategic choices are actually constantly adjusted, and they are constantly changing according to the changes of each other's strategies. To some extent, their choice is actually a trial and error process. Therefore, in this case, using evolutionary game tools to study the adjustment of both strategies is more in line with the reality.

Under the government regulation, assuming that enterprises can randomly and independently choose to use or not use renewable products, the government can also randomly and independently choose to supervise and not supervise, and repeat the game in multiple life cycles.

In the t-th life cycle, enterprises choose to use recycled products with probability p_t, and the government chooses to supervise with probability q_t, then the probability of enterprises choosing not to use is ($1-p_t$), and the probability of government choosing not to supervise is ($1-q_t$).

So in the life cycle t, the expected return when the enterprise chooses to use the recycled product strategy is

$$R_{e1} = q_t (\varepsilon + c) + (1 - q_t)\varepsilon$$

The expected return for selecting the strategy not to use recycled products is

$$R_{e2} = q_t (-d) + (1 - q_t) \cdot 0$$

The expected revenue when the government chooses the subsidy supervision strategy is

$$R_{g1} = p_t (n - c) + (1 - p_t)(d - m)$$

The expected return for choosing the non-subsidy and non-regulatory strategy is

$$R_{g2} = p_t n + (1 - p_t)(-m)$$

At this time, the expected return of enterprises is

$$R_e = p_t R_{e1} + (1 - p_t) R_{e2}$$

The expected return of the government is

$$R_g = q_t R_{g1} + (1 - q_t) R_{g2}$$

The probability change of enterprises choosing to use recycled products in the t life cycle can be described by the following discrete dynamical systems [9]

$$p_{t+1} = \frac{p_t R_{e1}}{p_t R_{e1} + (1 - p_t) R_{e2}} \tag{1}$$

The equation indicates that if the return obtained by enterprises when choosing to use the strategy of renewable products is greater than that obtained by the average social individual, then the probability of use will be increased. On the contrary, the probability of using renewable products in the next life cycle will be reduced.

The probability change of government choosing of subsidy regulation policy in t life cycle can be described by discrete dynamical system

$$q_{t+1} = \frac{q_t R_{e1}}{q_t R_{g1} + (1-q_t)R_{g2}} \tag{2}$$

It can be seen from the evolutionary phase diagram that system (1) (2) (see Figure 1) will evolve to equilibrium point 1 only when it is in region I. That is, the sustainable development situation of construction waste resource utilization will be formed. For the government, region III and region IV are invalid areas. In order to encourage enterprises to choose recycled products, the first thing is to make $q_t > \frac{a-b}{c+d}$

In the evolution phase diagram region I of the system, the evolution of system (1) is stable at equilibrium point 1, and the increase of the area of region I is conducive to the evolution of system(1). It can be seen from Figure 1 that the area of region I is

$$S_t = \left(1 - \frac{a-b}{c+d}\right)\left(1 - \frac{d}{c+d}\right)$$

Then

$$\frac{\partial S_t}{\partial B} = \frac{-c[(c+d)^2 - (a-b)(c+d) - (a-b)]}{(c+d)^2}$$

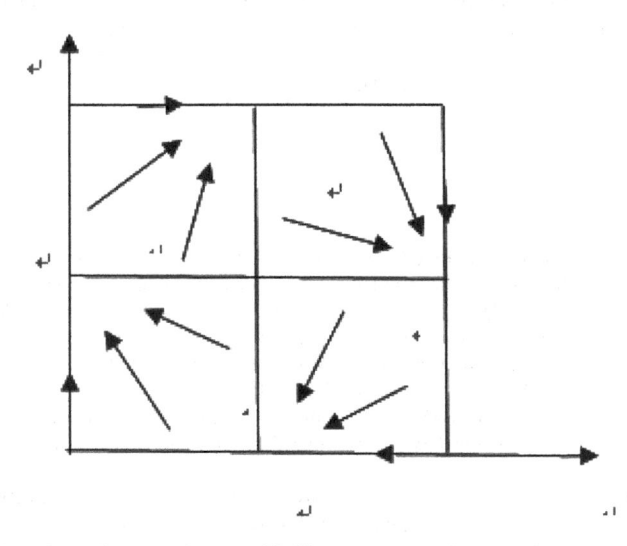

Figure 1. Evolution phase diagram of system (1) (2).

When $c < d < \frac{a-b}{2} - c + \frac{\sqrt{(a-b)^2 + 4(a+b)}}{2}$, $(c+d)^2 - (a-b)(c+d) - (a-b) < 0$, $\frac{\partial S_t}{\partial d} > 0$, S_t is about the increasing function of d, the penalty increases, the area of region I increases, which is beneficial to the evolution of system (1).

When $a > d > \frac{a-b}{2} - c + \frac{\sqrt{(a-b)^2 + 4(a+b)}}{2}$, $(c+d)^2 - (a-b)(c+d) - (a-b) > 0$, $\frac{\partial S_t}{\partial d} < 0$, S_t is about the decrease function of d, the penalty decreases, the area of region I decreases, which is not beneficial to the evolution of system (1).

Conclusion: in the initial stage of construction waste recycling, the social and consumer acceptance of construction waste recycled products is low, and the cost of construction enterprises includes positive externalities such as energy conservation and environmental protection. Enterprises lack the initiative to use recycled products. Therefore, it is necessary for the government to provide support through economic policies, provide certain benefits to enterprises that use renewable products, and give certain penalties to enterprises that do not use them. It is generally believed that the greater the punishment is, the more favourable it is to encourage enterprises to use recycled products. In fact, a moderate increase in punishment is conducive to the evolution of enterprises using recycled products, but when the punishment is too large, it will have a negative effect on the evolution of enterprises using recycled products. Therefore, choosing appropriate punishment is a key issue for the government to implement regulation.

5 CONCLUSIONS AND SUGGESTIONS

By the evolutionary game analysis of the participants of construction waste recycling, it can be seen that in the initial stage of construction waste utilization, the government needs to intervene and regulate the market. In the preliminary implementation of the strategy of resource utilization of construction waste, it is necessary to study and develop new materials and technologies and the investment amount is large The cost includes positive externalities such as energy conservation and environmental protection. The price is not superior to non-renewable products, and the choice of recycled products of construction waste is lack of positive initiative. So it is necessary for the government to support them through economic policies, provide certain benefits to enterprises that produce and use construction waste recycling products, and give certain penalties to enterprises that do not choose recycling products. In order to encourage enterprises to use construction waste and save the cost of subsidy supervision, firstly, the probability of subsidy supervision should be greater than $\frac{a-b}{c+d}$ but always tends to $\frac{a-b}{c+d}$. Secondly, appropriate punishment should be selected. Under certain conditions, the greater the punishment is, the more conducive it is to promote enterprises to choose renewable products. However, when the punishment is too large, it will play a negative role.

REFERENCES

[1] Zhang Q, Sun H. T. 2020. Comparative Analysis of Domestic and Foreign Construction Waste Treatment Related Policies and Regulations. Environmental Protection and Circular Economy, Vol 1: p85–87.
[2] Hu M. M, Zhang C. B, Zhang Q. 2016. Present Status of Construction and Demolition Waste Management and Recycling from the International Perspective. Journal of Engineering Studies, Vol 8: p365–373.
[3] Yuan H. P, Chen H, Liu Z. M. 2018. Construction Waste Management Model Based on System Dynamics. Science and Technology Management Research. Vol 4,210–216.
[4] Ding Z., Wang Y., Wu J. (2017) ABM Based Simulation Research on Construction Waste Management. Proceedings of the 20th International Symposium on Advancement of Construction Management and Real Estate. Springer, Singapore. p465–476.
[5] Liu T. T, Zhang J, Hu M. M. 2018. Analysis on environmental benefits of construction and demolition waste recycling: A case study in Chongqing. China Environmental Science. Vol 38: 3853–3867.
[6] Chen Q. J. etc. 2018. Analysis of China's Construction Waste Policy from the Perspective of Policy Tools: Based on the National Policy Texts from 2003 to 2018 [J]. Ecological Economy, Vol 36: p196–203.
[7] Jia S. W, Yan G. L. 2018. Dynamic Simulation of Construction Waste Management Model under Combination Policies [J]. *Systems Engineering-Theory & Practice.* Vol 38: p2966–2978.
[8] Wang D, Liu H. M, Hong W. M. 2020. Research on Classification Management on Site of C&D Solid Waste Based on Tripartite Evolutionary Game. *Engineering Economy.* Vol 30: p37–42.
[9] William H. Sandholm. Population Games and Evolutionary Dynamics[M]. Cambridge: *MIT Press*, 2011, 350–360.

Experimental research on performance of interlayer bonding material for asphalt thin overlay

F.J. Bao
*Ningbo Haishu District Transport Investment & Construction Co., Ltd,
Ningbo, China*

ABSTRACT: In the current situation of higher requirements for rapid maintenance, it has become a trend to maintain some roads with surface function attenuation in a convenient, economic and low disturbance way. Thin layer overlay is a reasonable and effective maintenance scheme. In order to improve the interlayer bonding performance of thin overlay and further improve the maintenance life of thin overlay, four different interlayer bonding materials were studied and tested, including SBS modified asphalt, SHELL70 # base asphalt, rubber asphalt and EVA modified asphalt. The test method is the lever pull-out method, which is a simple testing method of material bond strength. Through the pull-out force and bond strength of four different bonding materials with different spraying amount, the bond performance of them as interlayer materials was evaluated. The test results show that the four materials can meet the more stringent bond strength requirements, and the value is not less than 0.48mpa. The material with better bond performance is SBS modified asphalt, and the best recommended spraying amount is $0.6kg/m^2$.

1 INTRODUCTION

In recent years, with the rapid development of road traffic in China, the phenomenon of heavy traffic and heavy load are becoming more and more serious. Various surface functions of asphalt pavement are decaying rapidly, which directly affects the driving safety and comfort of cars [1]. Therefore, how to improve the use function of the built high-grade highway is one of the common concerns of highway maintenance.

Thin layer pavement is adopted to reconstruct the anti sliding performance of the pavement, improve the smoothness, prevent the aging of the original pavement when overlaid on the asphalt surface, restore the road surface function, and directly bear the load of driving vehicles and the influence of multiple external natural factors such as light, oxygen and water, so as to prolong the life of the pavement [2]. At the same time, the thin layer pavement has the characteristics of convenient construction, light weight, no increase of dead load of bridge deck, anti sliding, smoothness, noise reduction, wear resistance and environmental protection.

However, with the increase of traffic volume and vehicle load quality, shortly after putting into operation, diseases such as pushing, wrapping, cracking and "two-layer skin" appeared, which lost the purpose of preventive maintenance [3]. These diseases are usually caused by the poor adhesion between the thin overlay and the original pavement. In China's current asphalt pavement construction specifications, the treatment process of asphalt pavement interlayer and top surface of base course is relatively simple, and only the construction of prime coat and adhesive layer is generally described. For example, the selection of material and material consumption per unit area is completely determined by experience, and there is a lack of effective quality control index [4], which leads to insufficient shear resistance of interlayer materials Strength and bond strength.

Therefore, in order to effectively solve a series of thin overlay interlayer bonding problems, such as ultra-thin overlay, "white to black" pavement, and reduce the early disease of asphalt pavement [5,6], this paper studies the interlayer bonding effect evaluation of different asphalt materials through the appropriate evaluation method of interlayer material bonding effect.

2 MATERIALS AND EXPERIMENTAL SECTION

2.1 Materials

No matter what kind of tack coat material is used, asphalt is the main raw material of the tack coat. The quality of asphalt is directly related to the quality of the tack coat [7,8]. Through a large number of tests, it is found that the bonding effect of bonding materials and asphalt concrete surface is greatly affected by the types of asphalt, and the effect of using the same asphalt as the lower layer is obviously superior to that of different asphalt. The common interlayer bonding materials selected in this study include: SBS modified asphalt, shell 70 matrix asphalt, rubber asphalt, EVA modified asphalt.

The raw material shell70 # base asphalt is tested according to JTG E20–2011 test specification of China for asphalt and asphalt mixture of highway engineering. The main indexes are shown in Table 1. SBS modified asphalt is prepared by mixing shell 70 with 4.4% SBS at 185°C, and its main indexes are shown in Table 2. Rubber asphalt is prepared by mixing shell 70 with 18% 40 mesh waste radial tire rubber powder at 185°C. EVA modified asphalt is made by mixing 20% 40 mesh waste radial tire rubber powder and shearing repeatedly at 200°C.

Table 1. Main technical indexes of shell 70#.

Items			technical indicators	results
Penetration (25°C, 100g, 5S) (0.1mm)			60–80	64
Softening point (°)		≥	47	48
Ductility 15°C (cm)		≥	100	140
Viscosity at 135°C (Pa·s)		-	-	0.56
TFOT heating test 163°C (5h)	Mass loss (%)	≤	±0.8	0.08
	Residual penetration ratio (%)	≥	61	80
	Residual ductility 10°C (cm)	≥	6	7

Table 2. Main technical indexes of SBS modified asphalt.

Items	SBS modified asphalt
Penetration (25°C, 100g, 5S) (0.1mm)	51.2
Softening point (°C)	78.8
Ductility 15°C (cm)	35.8
Viscosity at 135°C (Pa·s)	2.31
Viscosity at 177°C (Pa·s)	0.75

2.2 Test method

(1) Drawing strength index

In the process of driving, due to the high-speed driving of the vehicle, the interaction between the wheel and the road surface is strong [9]. After the vehicle driving, the vacuum pumping action will be produced in contact with the ground, and the adsorption force is tensile stress. The adverse effect of tensile stress on the road surface is better than that of compressive stress on the road structure. When the overlay layer is thin, the effect is more obvious. It is believed that: once the upper structure layer and the lower structure layer are separated [10], sliding will occur between the layers, which will lead to fatigue cracking of the upper structure layer under the impact of vehicle load. Therefore, it is necessary to give a quantitative regulation on the vertical pull-out strength between layers.

According to the China Specification of Water borne Asphalt Based Waterproof Coating for Road and Bridge (JT/T535–2004), the bonding capacity of water borne asphalt based waterproof coating under standard test conditions shall not be less than 4 MPa, and according to the theory, the pull-out strength between layers should not be less than the bond strength of the bonding material itself, and the bonding layer will be adversely affected and the fatigue effect of load during the use of the pavement structure:

$$[\sigma] = 0.4 \times K(2-1)$$

Where: K-tensile strength coefficient

The pull-out strength coefficient needs to be obtained by a large number of pull-out tests in the actual pavement. In this paper, based on the selection of structural safety factor, the pull-out strength factor is taken as the more unfavorable condition, and the k-pull is 1.2,

$$[\sigma] = 0.4 \times 1.2 = 0.48 Mpa$$

According to the specification, to ensure the integrity and continuity of pavement structure layers, the interlayer bonding strength that can be improved by materials is more than 0.48MPa.

(2) Test method for bond strength of binder course materials

Because the asphalt pavement is paved in layers, or the surface of the old surface is smooth, the friction μ on the indirect contact surface of the layer will be greatly lower than the friction of the mixture itself, and the strength composition of the layer will turn to the dependence on the cohesive force C. In view of this, the research focus of the performance test of the interlayer material will turn to the research on its cohesive strength performance. At present, there is no standard method to evaluate the bond strength of materials [11]. Most scholars use MTs to carry out the pull-out test between mixtures to detect the bond strength of interlayer materials, or develop their own pull-out apparatus, such as some self-designed test methods proposed by Professor Shen Aiqin of Chang'an University. Therefore, this paper proposes an independent simple testing method of material bond strength: lever drawing method.

In the lever drawing instrument method, the model completely simulates the actual road surface. According to the commonly used grading of thin layer overlay, the thin layer asphalt concrete is formed in the laboratory after 1cm, and then the thin layer is spread on the old surface layer and rolled. Between the thin layer and the old surface layer, there are different amounts of bonding materials to be tested [12]. The waiting time is determined according to the type of tack coat oil. After the strength of tack coat oil is formed, the core is drilled until the tack coat is penetrated [13]. Finally, the bond strength of tack coat is measured by lever drawing instrument. The drawing diagram of thin layer overlay is shown in Figure 1.

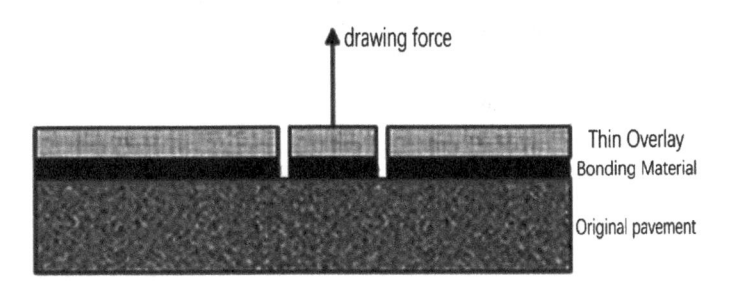

Figure 1. Drawing diagram of thin layer overlay.

The equipment used to measure the bond strength between layers by using the lever drawing instrument method are: self-made lever drawing instrument (lever arm length ratio is 1:6), weight (50kg in total), pulling head (specification d = 100mm), small roller (1t), core drilling sampler, asphalt mixture mixer, electronic scale, thermometer, oven, electric furnace, shovel, asphalt melting pot, etc.

3 TEST RESULTS AND ANALYSIS

3.1 *Bond strength test of bonding materials*

Clean up the old road surface, polish the surface, clean up the dust on the surface, and divide the old road surface. According to the common grading of thin layer overlay, the thin layer asphalt concrete with thickness of 1cm is formed in the laboratory, and then the thin layer is paved on the old surface layer and rolled. Put the adhesive layer oil in the same size cells according to different amount of drying cloth, and then pave the divided overlay. Use a small roller to roll back and forth for 6 times, so as to compact and promote the uniform distribution of viscous oil. Core drilling is carried out after the formation of the oil strength. The specific test is shown in Figure 2.

Figure 2. Drawing test chart of thin layer overlay.

The bond strength is tested by lever pull-out method, and the test data are shown in Table 3.

Table 3. Bond strength of asphalt on the bottom of asphalt concrete in 5°C water bath.

Asphalt type	Spraying amount(kg/m^2)	Average drawing force(KN)	Bond strength(MPa)
SHELL70#	0.3	1.01	1.15
	0.6	1.33	1.54
	0.8	1.15	1.30
	1.0	1.42	1.58
SBS Modified asphalt	0.3	1.71	1.90
	0.6	1.88	2.29
	0.8	2.31	2.57
	1.0	2.13	2.37
Asphalt Rubber	0.3	1.61	1.80
	0.6	1.83	2.03
	0.8	1.96	2.17
	1.0	1.89	2.10
EVA Modified asphalt	0.3	1.46	1.62
	0.6	1.78	1.95
	0.8	1.81	2.02
	1.0	1.77	1.94

3.2 Trend chart analysis

From table 3, we can draw the different changes of the bond strength of four kinds of asphalt in asphalt concrete with the amount of spraying. As shown in Figure 3–6:

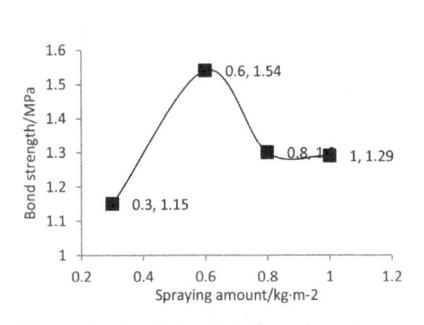

Figure 3. Bond strength of SHELL70# in asphalt concrete.

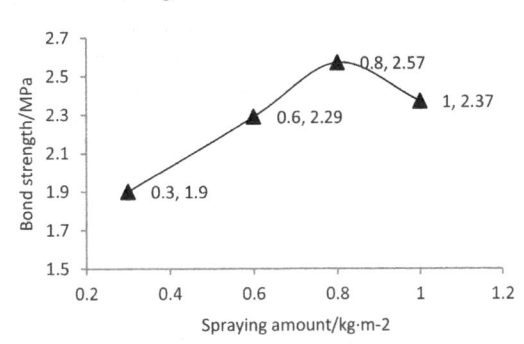

Figure 4. Bond strength of SBS Modified asphalt in asphalt concrete.

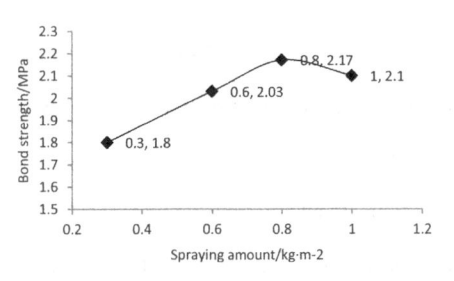

Figure 5. Bond strength of Asphalt Rubber in asphalt concrete.

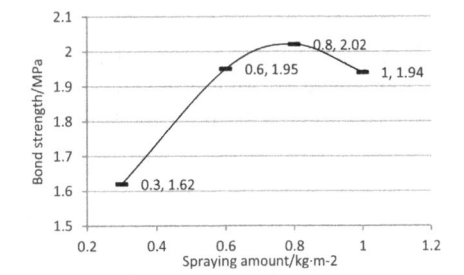

Figure 6. Bond strength of EVA modified asphalt in asphalt concrete.

It can be seen from Figure 3 that the maximum bond strength of shell70 # base asphalt is 1.58mpa. Since the bond strength has reached 1.54 MPa at the spraying amount of 0.6kg/m², and after the spraying amount of 0.6kg/m², the bond strength shows a downward trend and is in the fluctuation stage, so the spraying amount of the best bond strength is 0.55–0.65kg/m². It can be seen from the curve change in Figure 4 that the optimal amount of SBS modified asphalt is 0.8kg/m², and the maximum bond strength is 2.57mpa. It can also be seen from the table that the bonding strength of SBS modified asphalt is the largest among the four kinds of asphalt, that is, SBS modified asphalt is better as the bonding material. It can be seen from the curve change in Figure 5 that the change of adhesive strength of rubber asphalt is obvious, that is, the best spraying amount is 0.8kg/m², and the maximum adhesive strength is 2.17mpa. It can also be seen from the table that the bonding strength of rubber asphalt is slightly lower than that of SBS modified asphalt, indicating that rubber powder can also show greater strength at low temperature and can be used as a modifier for bonding materials. It can be seen from the curve change in Figure 6 that the spraying amount of EVA modified asphalt begins to decline after 0.7 kg / m², so the optimal spraying amount of EVA modified asphalt is in the range of 0.6–0.8 kg / m², and the maximum bonding strength is about 2.1 MPa.

4 CONCLUSIONS

Through the research, we can get the following conclusions:

(1) The four kinds of asphalt studied in this paper all have good bond strength, which can meet the more stringent bond strength requirements mentioned above, and the value is not less than 0.48mpa.
(2) In this study, the best application amount of the four interlayer bonding materials to achieve the maximum bond strength is concentrated in the range of $0.6kg/m^2$-$0.8kg/m^2$, the recommended amount of the specification is $0.3kg/m^2$-$0.6kg/m^2$, and the recommended amount of the adhesive layer material is $0.6kg/m^2$.
(3) The bond strength of four kinds of modified asphalt on the bottom of asphalt concrete from strong to weak is SBS modified asphalt > rubber asphalt > EVA modified asphalt > shell 70# Base asphalt.

ACKNOWLEDGMENTS

The authors would like to thank Sheng Wang(Ningbo Highway and Municipal Design Co., Ltd.) and Dr. Ming Huang,(Greenland Infrastructure Technology Research Institute) for helpful and constructive prophase studies.

REFERENCES

[1] Davide Ragni, Francesco Canestrari, Fatima Allou, et al. Shear-Torque Fatigue Performance of Geogrid-Reinforced Asphalt Interlayers. 2020, 12(11)
[2] Wang Wenwen. Research on binder and performance of asphalt pavement base and surface. Xi'an: Chang'an University, 2006
[3] Kai Yang, Rui Li, Yi Yu, et al. Evaluation of interlayer stability in asphalt pavements based on shear fatigue property. 2020, 258
[4] Ministry of communications. Technical code for construction of highway asphalt pavement (JTG F40–2004). Beijing: People's Communications Press, 2004
[5] Minh-Tu Le, Quang-Huy Nguyen, Mai Lan Nguyen. Numerical and Experimental Investigations of Asphalt Pavement Behaviour, Taking into Account Interface Bonding Conditions. 2020, 5(2)
[6] Cheng Chen, Yancai Yang, Yingwu Zhou et al. Comparative analysis of natural fiber reinforced polymer and carbon fiber reinforced polymer in strengthening of reinforced concrete beams. Journal of Cleaner Production, 2020, 263
[7] A. Bahgat Radwan, R.A. Shakoor Aluminum nitride (AlN) reinforced electrodeposited Ni–B nanocomposite coatings. Ceramics International, 2020, 46(7)
[8] A. Noory, F. Moghadas Nejad, A. Khodaii. Evaluation of geocomposite-reinforced bituminous pavements with Amirkabir University Shear Field Test. 2019, 20(2):259–279.
[9] Yue Xiao, Yefei Wang, Shaopeng Wu, et al. Assessment of bonding behaviours between ultrathin surface layer and asphalt mixture layer using modified pull test. 2015, 29(14):1508–1521.
[10] Quntao Tang, Honglie Shen, Hanyu Yao et al. Dopant-free random inverted nanopyramid ultrathin c -Si solar cell via low work function metal modified ITO and TiO_2 electron transporting layer. Journal of Alloys and Compounds, 2018, 769
[11] E. K. Tschegg, J. Macht, M. Jamek, et al. Mechanical and Fracture-Mechanical Properties of Asphalt-Concrete Interfaces. 2007, 104(5)
[12] Grilli, Andrea, Graziani, et al 2016 Volumetric properties and influence of water content on the compact ability of cold recycled mixtures Materials & Structures 49 1
[13] Choi Jaemin, Lee Seungyong, Sung Kitae, et al. A Fundamental Study on the Pullout Strength Equation of High Strength Anchor for Asphalt Pavement. 2017, 13(3):313–321.

Effects on concrete mechanical performances by different replacement solutions of recycle coarse aggregates

L. Yu
Research Institute of Highway Ministry of Transport, Beijing, China
Beijing Xinqiao Technology Development Co. Ltd, Beijing, China

S.Z. Lv
Shandong Highway Co. Ltd, Jinan, China

S.X. Zhang
Beijing Xinqiao Technology Development Co. Ltd, Beijing, China

ABSTRACT: In order to detect how the replacement solutions of the recycle coarse aggregate affect the mechanical performance of the concrete, all and part replacement solutions are made. The compressive strength and splitting tensile strength are determined. The results show that there is no obvious difference to the compressive strength between the two methods of equal ratio, replacing among 3 particle grade and single particle grade as long as the replacement amounts are same. Replacing all of the (20–31.5) mm and part of the (10–20) mm nature coarse aggregates can rarely influence the compressive strength and splitting tensile strength of the concrete. The replacing amount of (10–20) mm nature aggregates should not be greater than 50%, or else it may have an effect on the compressive strength.

1 INTRODUCTION

During the highway reconstruction and extension project, a large amount of old concrete will be produced [1]. To protect the environment and save resources, the concrete can be crushed into aggregates and then reused in the same construction [2]. The aggregate which is crushed by old concrete can be called recycle aggregate. Among these recycle aggregates, the recycle coarse ones can be used for producing new concrete instead of the nature aggregate [3,4]. The replacement solution needs to be studied. The solution mainly refers to two aspects as the replacing amount and the particle grade. The particle grade of the coarse aggregate in the bridge concrete structure can be divided into 3 grades. They are (5–10) mm, (0–20) mm, and (20–31.5) mm[5]. From the research before, it can be seen that replacing all particle grade by the same ratio are the most common method [6,7]. During the producing process, there are 3 stock bins to keep 3 particle grade nature aggregates. If the method of equal ratio replacing among 3 particle grade was used, 3 extra stock bins to keep 3 particle grade recycle aggregates are still needed. It is a very difficult thing for the field concrete producing plant because this can bring trouble management and added land occupation. But if all of the single particle grade aggregate is replaced, the stock bin used to store the single particle grade nature aggregate can be used to store the recycle ones with the same particle grade. So there is a new question. The single particle grade aggregate replacement can or cannot affect the mechanical properties of the concrete. That is the main research purpose.

2 EXPERIMENT DESIGN

2.1 Raw materials

The P.O 42.5 cement is used, and the performances are shown in Tble 1. Nature sand is used, and the accumulated retained percentage is shown in Table 2. The nature coarse aggregate is used. The particle grades are (5–10) mm, (10–20) mm, and (20–31.5) mm. The water absorption of the nature coarse aggregate is 0.45%. The crushing value is 10.5%. The apparent density is 2730kg/m^3. The recycle coarse aggregate is from a bridge concrete slab which is obtained from Shandong Ji Qing highway reconstruction and extension project. The properties of the recycle coarse aggregate are shown in Table 3. The polycarboxylic acid high performance water reducer is used to enhance the fluidity of the concrete. The rate of the water reducer is 30%. The recommended dosage is 0.5%–1.0% of the quality of the cement.

Table 1. Properties of cement.

Fineness (80μm square hole sieve) (%)	Standard consistency water consumption (%)	Setting time (h:min)		Compressive strength (MPa)		Splitting tensile strength (MPa)		Stability (boiling method)
		final	initial	3d	28d	3d	28d	
3.1	27.6	2:59	3:59	29.8	49.9	6.7	9.3	Qualified

Table 2. Accumulated retained percentage of nature sand.

Sieve size(mm)	4.75	2.36	1.18	0.6	0.3	0.15	0.075	<0.075
Accumulated retained percentage (%)	11.62	32.2	43.58	55.6	79	93.3	99.9	100

Table 3. Properties of recycle coarse aggregate.

Crushing value (%)	Water absorption (%)			Apparent density(g/cm^3)	Content of chloride iron (%)
	5–10	10–20	20–31.5		
29.3	6.7	5.9	4.3	2.645	0.0037

2.2 Mix proportion

In China, the common concrete used for highway is C30–C50 strength grade. So the experiments are carried out to make C30 and C50 concrete for the study. The polycarboxylic acid high performance water reducer is used to adjust the concrete slumps around (180±20) mm. The aggregate with particle grade (20–31.5) mm is recorded as 1#. The aggregate with particle grade (10–20) mm is recorded as 2#. The aggregate with particle grade (5–10) mm is recorded as 3#.

2.2.1 Equal ratio replacing

This method means the nature aggregates with 5–10, 10–20 and 20–31.5 particle grade are separately replaced with the same ratio. The mix proportions of C30 and C50 concrete are listed in Table 4 and Table 5.

Table 4. Mix proportion of C30 concrete (kg/m³).

Replacing ratio (%)	Cement	Water	Sand	1# Nature	1# Recycle	2# Nature	2# Recycle	3# Nature	3# Recycle
0	332	175	860	210	0	630	0	210	0
10	332	175	860	189	21	567	63	189	21
30	332	175	860	147	63	441	189	147	63
50	332	175	860	105	105	315	315	105	105
70	332	175	860	63	147	189	441	63	147
90	332	175	860	21	189	63	567	21	189
100	332	175	860	0	210	0	631	0	210

Table 5. Mix proportion of C50 concrete (kg/m³).

Replacing ratio (%)	Cement	Water	Sand	1# Nature	1# Recycle	2# Nature	2# Recycle	3# Nature	3# Recycle
0	472	165	763	210	0	630	0	210	0
10	472	165	763	189	21	567	63	189	21
30	472	165	763	147	63	441	189	147	63
50	472	165	763	105	105	315	315	105	105
70	472	165	763	63	147	189	441	63	147
90	472	165	763	21	189	63	567	21	189
100	472	165	763	0	210	0	630	0	210

2.2.2 Single particle grade replacment

This method means that one particle grade aggregate is first replaced. If it is insufficient, the second or the third aggregate is replaced. Only C30 concrete is planned to finish the single particle grade replacing experiments. The replacement ratios are 30% and 50%. The mix proportions of the concrete are listed in Table 6.

Table 6. Mix proportion of C30 concrete with single particle grade replacing method (kg/m³).

Replacing rate	Mark	Cement	Water	Sand	1# Nature	1# Recycle	2# Nature	2# Recycle	3# Nature	3# Recycle
30%	3+2	332	175	860	210	0	525	105	0	210
	3+1	332	175	860	105	105	631	0	0	210
	2	332	175	860	210	0	315	315	210	0
	2+1	332	175	860	0	210	526	105	210	0
50%	3+2	332	175	860	210	0	315	315	0	210
	3+2+1	332	175	860	0	210	525	105	0	210
	2	332	175	860	210	0	105	525	210	0
	2+1	332	175	860	0	210	315	315	210	0

2.3 Test method

The quality of all raw materials is determined. The cement, sand, aggregate are put into the mixing machine and stirred for 30 seconds. Water is poured with the water reducer into the mixing machine, and the fresh concrete is stirred for 90s. Enough specimens for the 7d, 28d, and 56d compressive strength and the splitting tensile strength tests are formed. The compressive strength test method of the concrete is ISO 4012. The splitting tensile strength test method of the concrete is ISO 4108.

3 RESULT ANALYSIS

3.1 Equal ratio replacing

As shown in Figure 1, the relationship between recycle coarse aggregate and compressive strength is illustrated on C30 and C50 concrete.

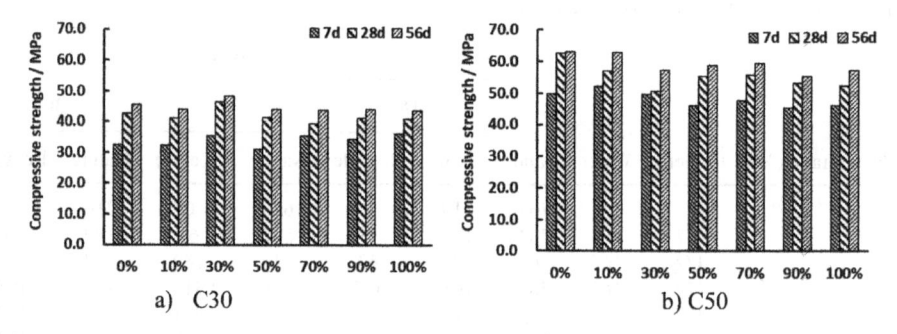

Figure 1. Compressive strength VS replacement ratio: a) C30 and b) C50.

For the C30 concrete, from Figure 2-a, it can be seen that the least compressive strength is 39.3MPa. The concrete with 30% recycle aggregate has the highest strength among these replacement ratios from 0% to 100%. The 28d compressive strength increased 1/3 compared to 7d compressive strength when the replacement ratio is below 50%, while above 50%, the increasing rate is only 1/6 when comparing 28d to 7d dates. From this, it can be concluded that the replacement ratio almost has no effect on the compressive strength, while when the ratio is above 50%, it is bad for the compressive strength increase during the later period.

C50 concrete is different with C30 ones; the 28d compressive strength is the highest one. There is almost no effect on the compressive strength when the replacement ratio is below 10%, while when the replacement ratio increases to 30% or above, the compressive strength of the C50 concrete decreases. It also can be seen that the 7d compressive strength is all below 50MPa. When the curing time is over 28d, the compressive strength can reach 50MPa line but is still lower than concrete without recycle coarse aggregate. These situations illustrate that any replacement ratio is bad for the C50 concrete.

The splitting tensile strength of the C30 and C50 concrete is shown in Figure 2.

From Figure 2-a, the splitting tensile strength of the C30 concretes without any recycle coarse aggregate can reach up to 2.5MPa, while the ones with recycle coarse aggregate are all as low as 2MPa. For the concretes with over 70% recycle coarse aggregate, the 28d splitting tensile strength is smaller than 7d ones. This phenomenon illustrates that it is bad for the later splitting tensile strength increasing. From Figure 2-b, for the C50 concretes, the 7d splitting tensile strength of concrete with recycle aggregate is more than the one without recycle aggregate. For any replacement ratio, the 28d splitting tensile strength is all lower than the 7d ones. This situation illustrates that the recycle coarse aggregate is bad for the increasing of later splitting tensile strength of concrete.

a) C30 b) C50

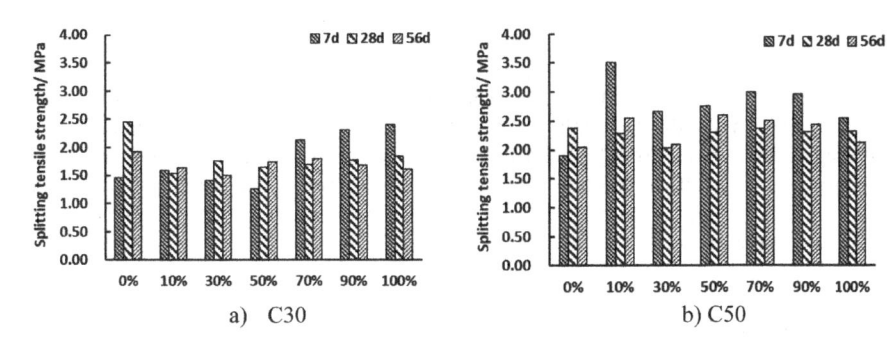

Figure 2. Splitting tensile strength VS replacement ratio: a) C30 and b) C50

3.2 *Single particle grade replacement*

As shown in Figure 3, the relationship between different single particle grade replacement solutions and compressive strength is shown. At the same time, the equal ratio replacement concretes are compared.

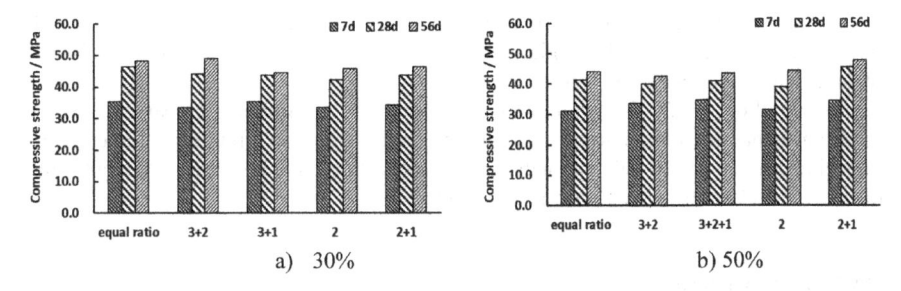

a) 30% b) 50%

Figure 3. Compressive strength VS replacement solution: a) 30% and b) 50%.

From Figure 3-a, when the replacement ratio is 30%, the 7d compressive strength of the concrete is 33–35 MPa. The 28d one is 42–44 MPa, and the 56d one is 45–49 MPa. The 2+1 replacement solution is similar to the equal ratio replacement solution. The Figure 3-b shows that there is no obvious difference between equal ratio replacement and 3.2 single particle grade replacement. The 2+1 replacement solution is better than the equal ratio replacement solution. So, from the abovementioned analysis, there is no obvious difference to the compressive strength between the two methods of equal ratio replacing among 3 particle grade and single particle grade replacement as long as the replacement amounts are same. There is one thing that needs attention, that is, the 2# aggregate should not be replaced very much more than 50%.

Figure 4 is about the splitting tensile strength of the concrete with single particle grade replacement solutions.

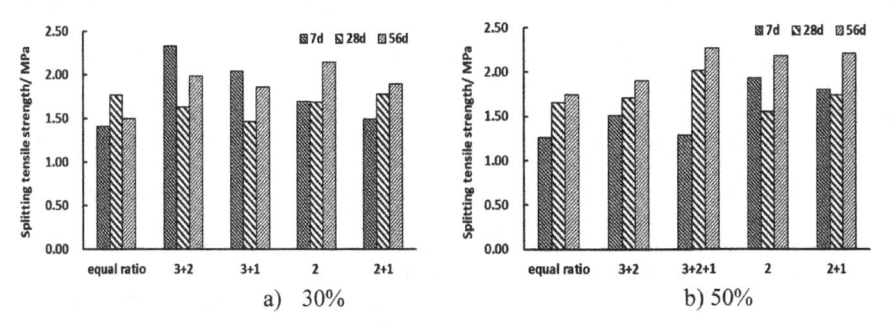

a) 30% b) 50%

Figure 4. Splitting tensile strength VS replacement solution: a) 30% and b) 50%.

Figure 4-a shows that the splitting tensile strength of the concretes with the single particle grade replacement solution is higher than the equal ratio replacement solutions. The single particle grade replacement solution is not good to the later splitting tensile strength increasing. From Figure 4-a, it also can been seen that the 2+1 replacement solution is as good as the equal ratio replacement at the 30% replacement ratio. From Figure 4-b, for the concrete with the 50% replacement ratio, only the 2# aggregate replacement solution is used, which is the worst. 3+2+1 replacement solution is the best. 2+1 replacement solution is the second best. Corresponding with the compressive strength date in Figure 3, the 2+1 replacement solution can be the best one.

4 CONCLUSION

To detect how the replacement solutions of the recycle coarse aggregate affect the mechanical performance of the concrete, equal ratio replacement among 3 particle grade and single particle grade replacement is compared. The compressive strength and splitting tensile strength are determined. The results are as follows.

(1) The 30% replacement ratio is the peak value to the C50 concrete, while the replacement ratio can be up to 100% for the C30 concrete.
(2) There is no obvious difference in the compressive strength between the two methods of equal ratio replacement, 3 particle grade and single particle grade replacement, as long as the replacement amounts are the same.
(3) Replacing all of the (20–31.5) mm and part of the (10–20) mm nature coarse aggregates can rarely influence the compressive strength and splitting tensile strength of the concrete. The replacing amount of (10–20) mm nature aggregates should not be greater than 50%, or else it may have an effect on the compressive strength.

ACKNOWLEDGMENTS

This work has been supported by 2019 technology innovation project of Research Institute of Highway Ministry of Transport. The item code is 2019–C534. This work also has been supported by grant from Shandong Highway Incorporated Company to Research Institute of Highway Ministry of Transport, China. The Technology Plan Item set up by Transportation Department of Shandong Province, China. The item code is 2018BZ3.

REFERENCES

[1] Ruoyu Jin, Bo Li, Ahmed Elamin, Shengqun Wang, Qurania Tsioulou, Dariusz Wanatowski. Experimental Investigation of Properties of Concrete containing recycled construction waste. International Journal of Civil Engineering. 2018, 4.10. (Published online)
[2] Hong-zhu Quan. Application overview and Technical Standards of Recycled Concrete in Other Countries. Journal of Qingdao Technological University. 2009, 30(4): 87–92, 126.
[3] WANG zhen shuang, WANG Lijiu, CUI Zhenglong, ZHOU Mei. Effect of recycled coarse aggregate on concrete compressive strength. Transactions of Tianjin University. 2011, 17(3): 229–234.
[4] Minkwan Ju, Kyoungsoo Park, Won-jun Park. Mechanical Behavior of Recycled Fine Aggregate Concrete with High Slump Property in Normal-and High-Strtength. International Journal of Concrete Structures and Materials. 2019, 13(61): 1–13.
[5] Rachit Sharma. Laboratory study on effect of construction wastes and admixtures on compressive strength of concrete. Arab J Sci Eng. Published online: 21 April 2017.
[6] C. Thomas, J. de Brito, A. Cimentada, J.A. Sainz-Aja. Macro- and micro- properties of multi-recycled aggregate concrete. Journal of Cleaner Production. 2019(10): 1–20.
[7] P. Recathi, R.S. Selvi, S. S. Velin. Investigation on Fresh and Hardened Properties of Recycled Aggregate Self Compacting Concrete. J. Inst. Eng. India Ser. A. 2013, 94(3): 179–185.

Development of boron doped diamond sensor for heavy metal detection in water

H.Z. Li, X. Yu, L. Pang & J.X. Pei
School of Materials Science and Technology, China University of Geosciences (Beijing), Beijing, China

X.J. Liu
Beijing Safeda Technology Limited Company, Beijing, China

ABSTRACT: The serious harm of heavy metal pollution in waters and the uncertainty of pollution sources have drawn an increasing concern. Online monitoring of heavy metal ions in water environments is a preferred measure to prevent the heavy metal pollution. Unfortunately, the mercury film electrode commonly used in available system cannot meet the field requirements of green and efficient detection. Based on the theory of ASV (anodic stripping voltammetry), a BDD (boron doped diamond) sensor was prepared to replace the mercury sensor, and the electrochemical response characters of the synthetic sensor to heavy metal ions were discussed. In cooperation with a professional water detector company, the developed sensor was installed on a commercial rapid detector to conduct its field tests in actual waters, so as to offer theoretical guidance and data support for heavy metal detection in waters.

1 INTRODUCTION

Heavy metal pollutants in waters may come from a wide range of sources and cause serious harm. The pollutant sources include industrial production, mining metallurgy, domestic sewage and etc. These pollutants are rather difficult to be degraded by microorganisms after entering water. As a result, the pollutants may not only damage the aqueous ecological environment, but also endanger the human health through the food chain [1]. Online monitoring of heavy metal content in waters is a unique measure to prevent the pollution. ASV (Anodic stripping voltammetry) has been successfully applied to various online monitoring detectors in domestic and foreign markets due to its high sensitivity and fast detection speed. However, available inspection electrode cannot meet the application requirements of green and efficient use for ASV. Mercury film sensors commonly used in ASV may volatilize mercury in the detection process, and result in secondary pollution of water [2]. Moreover, the heavy metal ions on the surface of the mercury film electrode may gather with the increase of detection times, and lead to the deterioration of the detection performance.

To overcome the drawbacks, this work attempts to develop a BDD (boron doped diamond) electrode to meet the application requirements. The detection performance of the electrode was improved by designing both the substrate and the sensing layer. Moreover, the interaction between the structure and performance of BDD electrode was investigated through a systematic characterization and detection test. Outcome of this work may facilitate the research and development of advanced detection products, and is conducive to the development and improvement of online monitoring system for the heavy metal detections in waters.

2 DETECTION METHOD OF HEAVY METALS IN WATER

Electrochemical analysis is regarded as a promising technique for rapid detection of heavy metals in water. The method can be used to conduct both qualitative and quantitative analyses by virtue of the electrochemical redox reaction of heavy metals to be detected, and can be divided into four types of anodic stripping voltammetry, potentiometric analysis, polarography and voltammetric titration [3]. ASV may be the first choice for the construction of online water monitoring system due to its simple structure, high sensitivity, low cost and convenience for miniaturization.

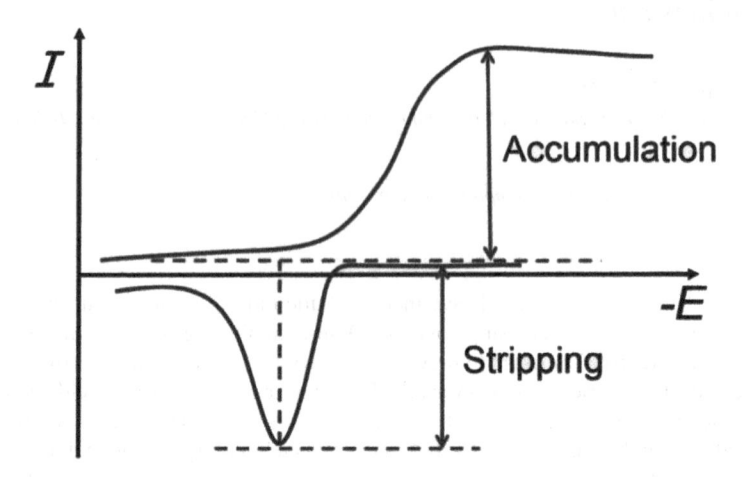

Figure 1. Working principle of ASV.

Being as an electrochemical analysis, ASV has attracted many attentions due to its wide applications in industry. Figure 1 illustrates the detection principles of ASV. As shown in Figure 1, the process of detecting heavy metal ions by ASV can be divided into two procedures. The first procedure is enrichment. Apply a constant potential negative to the ions on the detection electrode. This may reduce the ions into metals or compounds, and precipitate on the electrode. The second procedure is dissolution. Applying a reverse voltage on the electrode makes the metal re-oxidize into ions and dissolve into the water. The redox process of heavy metal ions may induce the change of voltage and current. Recording the current and voltage curves of the process, the dissolution peak potential value is used for qualitative analysis of the ions, and the dissolution peak current value is used for quantitative analysis of the ions [4].

Different with the detection methods like mass spectrometry and atomic spectrometry suffering from high cost and complex procedure, ASV may meet the requirements of on-site detection in waters due to its virtues [5]. In addition, the simple pre-treatment operation, short period, and convenient microminiaturization make this method have good industrialization prospective in the construction of online monitoring platform. Since the detection process occurs at the electrode interface, the electrode materials and its surface performances directly affect the selectivity and redox reaction of the ions [6]. But the electrode application still subjects to high cost, short life and low accuracy of fabricated electrode. The development of high-performance electrodes suitable for the detection of actual water environment has become an important topic to promote the application of ASV in the field of online monitoring of water.

3 DEVELOPMENT AND APPLICATIONS OF BORON DOPED DIAMOND SENSOR

At present, many professional companies such as Diamond detectors of UK, CSEM of Switzerland and Pressman of China have conducted the development and production for BDD electrodes. However, the market-oriented application of BDD electrodes in the field of heavy metal detection has not available. Main reason is that the developed electrode cannot meet the performance requirements, and does not meet the current market requirements for economic and efficient products.

Having a wide potential window and low background current, BDD sensor can simultaneously detect multiple heavy metal pollutants, and is expected to meet the requirements of efficient and accurate detection [7]. During using BDD sensor to detect heavy metal ions, the ions are enriched and dissolved on the surface of BDD, and the dissolution signal is transmitted to the data processing module to obtain the detection information. Two performances are necessary for obtaining detection information with accuracy. One is high electrochemical response, and includes both high intensity and high response speed of ion dissolution signal on the surface of BDD sensor. Another is high reproducibility, i.e., high surface stability of BDD sensor. For this purpose, BDD sensor was taken as the research object, and high-performance BDD electrochemical sensor was developed to meet the detection requirements of heavy metal pollution in water.

3.1 Boron doped diamond

Figure 2 shows the crystal structure diagram of boron-doped diamond. As shown in Figure 2, BDD is a p-type semiconductor formed by boron element doping into the diamond lattice and replacing part of the carbon atoms. Three main preparation methods for diamond include high temperature and high pressure, explosion, static synthesis and chemical vapor deposition. Among these, chemical vapor deposition is widely used to grow diamond film [8]. The basic principle of chemical vapor deposition is to activate and dissociate hydrocarbons such as methane in a specific way to form active groups. The active groups are gradually deposited on the surface of the substrate to form diamond or graphite phases. According to character of activation and dissociation, chemical vapor deposition (CVD) can be divided into hot-filament CVD, direct-current plasma CVD, and microwave plasma CVD [9].

Figure 2. Schematic of crystal structure of boron doped diamond.

The doping effect of boron element endows diamond with semiconductor properties, and different doping boron sources lead to different doping methods. At present, the boron sources has three options of gaseous (e.g. borane), liquid (e.g. trimethyl borate), and solid (e.g. boron trioxide). Among these, gaseous boron sources can be directly introduced into the reaction chamber for doping reaction, and enables a more precise control of the doping amount of boron elements. The problem is that the most of gaseous boron sources are corrosive and toxic, increasing the experimental risk [10]. Solid state boron sources and liquid state boron sources cannot be directly passed into the reaction chamber. This needs to be well mixed with the organic solution, and can be brought into the reaction chamber by hydrogen bubble to realize the doping.

3.2 Graphene modified self-supporting BDD

Current research on BDD sensors focuses on structural regulation and surface modification. Using structural regulation and surface modification may increase the electrochemical active area, and accelerate the ion mass transfer rate at the electrode interface. But such methods still cannot avoid early failure of the sensing layer. Whether the surface ions have high response speed and reliability is critical for evaluating the detection performance of the sensor. In this case, the self-supporting BDD electrochemical sensor was used as the research object, and the graphene modification was used to improve its electrochemical response for heavy metal pollutants.

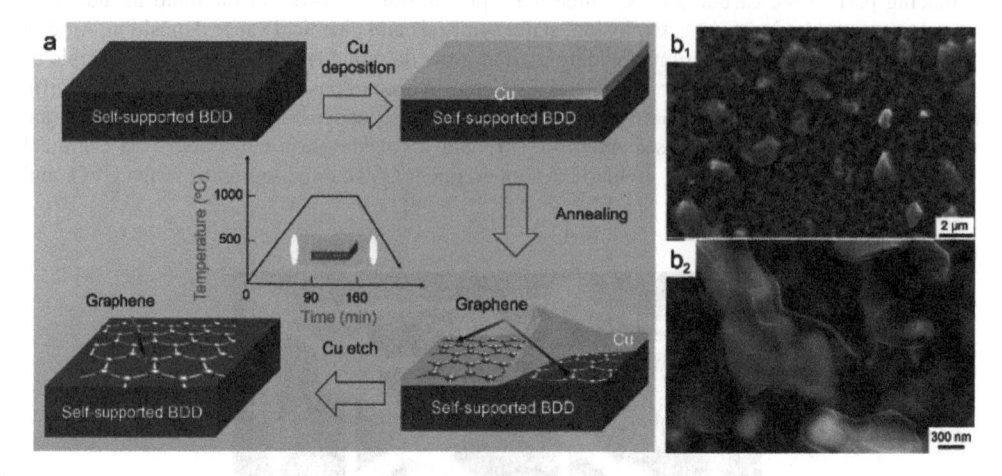

Figure 3. Preparation and characterization of in-situ graphene modified self-supporting BDD samples. (a) Schematic of sample preparation, (b) surface morphology.

Figure 3 shows the preparation and characterization of in-situ graphene modified self-supporting BDD samples [11]. As shown in Figure 3 (a), we constructed Cu-BDD structure by depositing copper film on the surface of self-supporting BDD via magnetron sputtering. Utilizing hot-filament CVD, the BDD was deposited on the surface of molybdenum for a long duration, and the BDD with a thickness of 200–500 μm was prepared and exfoliated into a self-supporting BDD. The pre-coating metal copper film was removed by etching to achieve in-situ modification of graphene on self-supporting BDD surface. Vacuum annealing treatment made the carbon element of BDD surface diffuse outward and it interacted with the catalytic metal copper to form graphene. The graphene/BDD composite sensor was thus prepared by in-situ modification. Compared with other modification methods, in-situ modification directly uses carbon atoms on the surface of BDD to generate graphene, and the resulting sensing platform can have a high uniformity and avoid creating an obvious interface with the substrate. In this process, the magnetron sputtering process parameters and annealing process parameters were adjusted to obtain graphene modified BDD with different properties. As shown in Figure 3 (b), the distribution of graphene film was uniform in a majority of BDD surface. The modified bilayer graphene had low defect content and high quality [12].

3.3 Electrochemical response of sensors to heavy metal pollutants

The electrochemical performance of the synthesized electrode was investigated by electrochemical workstation. The reaction kinetics, potential window, and specific surface area of electrochemical activity of the electrode were investigated by cyclic voltammetry test and impedance test. The optimal modification state and process parameters were determined by comparing the test results. Two ions of Cd and Pb was chosen for detection by three-electrode system. In that the two ions have serious harm and are rather difficult to be detected as compared with the other heavy metal ions. The detection performance of the sensor for Cd^{2+} and Pb^{2+} was explored under an optimized parameter set, containing standard curves, limits of detection, spiked recoveries, reproducibility, selectivity, and etc [13].

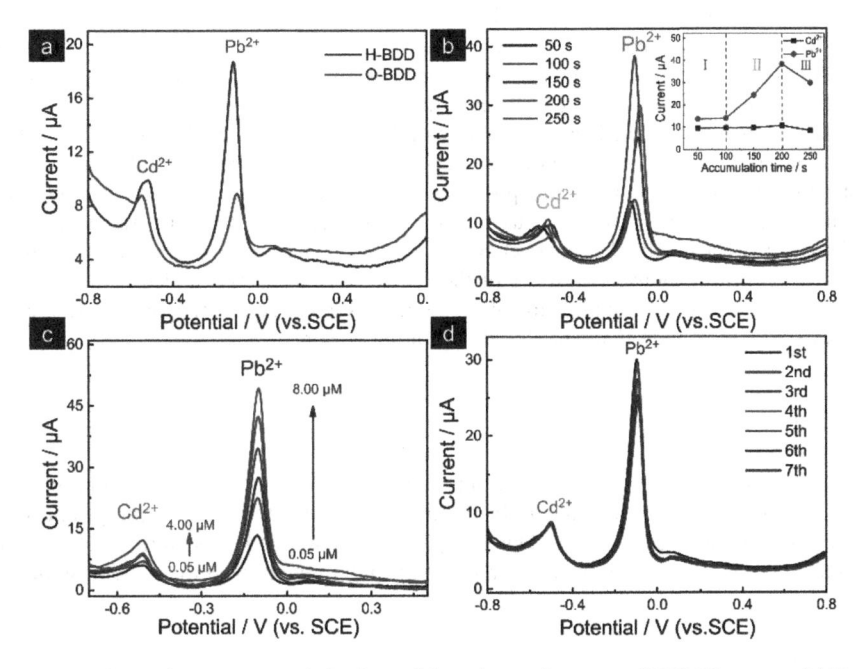

Figure 4. Experimental parameters optimization and detection performance of BDD film sensor. (a) Electrode terminals, (b) enrichment time, (c) standard curves, and (d) reproducibility.

Figure 4 shows the experimental parameters optimization and detection performance of BDD film sensor [13]. As shown in Figure 4, the synthesized electrode has good reproducibility and high accuracy. In the linear range of 1~100 ppb, the sensitivity of the sensor was 0.475 μA L μg-1 cm-2, and the detection limit was approximately 0.21 μg L-1. This means that the prepared film sensor has a wide linear working range, low detection limit, and good reproducibility.

3.4 Engineering applications of synthetic sensors for trail

Our group and a professional company are cooperated together to fabricate a commercial BDD sensor by means of our development of graphene modified BDD electrode in the laboratory and a rich production experience of Beijing Safeda Technology Co.,Ltd. in the field of heavy metal detectors in waters. The electrochemical response of synthetic sensors to heavy metal ions was discussed, and formed an interaction with on-line detection system for automatic treatment of water samples to provide a new idea and applicable solution for the industrial application of BDD sensor.

The integrated processing of the sensor was performed on the samples, and the engineering verification was carried out through real water sample detection. In order to evaluate the applicability of the sensor, the actual water samples were chosen for tests to evaluate the sensor performances. The evaluation of sensor reliability is divided into the following two aspects: (1) CRM (certified reference material) samples were taken as the research object, and Cd2+ and Pb2+ in the samples were simultaneously detected by sensors. The difference between the experimental results and the standard substance was compared; (2) Taking actual water samples as the research object, heavy metal pollutants in water were detected by sensors. The obtained test results were compared with that obtained by the other detection methods to measure the sensor feasibility.

4 CONCLUSION

The serious harm of heavy metal pollution in waters and the uncertainty of pollution sources have drawn an increasing concern. Online monitoring of heavy metal ions in water environments is a preferred measure to prevent the heavy metal pollution. Unfortunately, the mercury film electrode commonly used in available system cannot meet the requirements of green and efficient detection. This work conducted an investigation of development and application of a BDD sensor for detecting heavy metals in waters on basis of ASV. The main work includes three parts: (1) An integrated fabrication strategy was proposed for BDD electrochemical sensor modified with graphene. A self-supported BDD was used as the substrate, and a graphene was used to modify the substrate. (2) The electrochemical behavior was investigated for graphene modified BDD, and the electrochemical response characters of the sensor were discussed to detect the heavy metal pollutants. (3) The industrial application of synthetic sensor was conducted for trial. The real water samples were selected to evaluate the feasibility of the sensor.

ACKNOWLEDGMENT

This work has been supported by funds from the project "Practical Training Program "(289) for the cross-training of high-level talents in Beijing institutions of higher learning, and Class D entrepreneurship practice project "LvDun company–Detection and early warning of heavy metal pollutants in water" of Innovation and entrepreneurship training project for College Students.

REFERENCES

[1] Borrill A, Reily N and Macpherson J 2019 *Analyst* **144** 6834
[2] Economou A TrAC 2005 *TrAC-Trend. Anal. Chem.* **24** 334
[3] Mardegan A, Scopece P and Lamberti F 2012 *Electroanal.* **24** 798
[4] Manivannan A, Kawasaki R and Tryk D A 2004 Electrochim. Acta **49** 3313
[5] Zazoua A, Khedimallah N and Jaffrezic-Renault N 2018 *Anal. Lett.* **51** 336
[6] Shah A, Sultan S and Zahid A. 2017 *Electrochim. Acta* **258** 1397
[7] Silwana B, Horst van der C and Iwuoha E 2016 *Electroanal.* **28** 1597
[8] Núnez C, Arancibia V and Gómez M 2016 *Microchem. J.* **126** 70
[9] Lohani C R, Neupane L N and Kim J M 2012 *Sensor Actuat. B-Chem.* **161** 1088
[10] Barcelo C, Serrano N and Arino C 2016 *Electroanal.* **28** 640
[11] Pei J X , Yu X, Zhang C and Liu X J 2020 *Appl. Surf. Sci.* **527** 146761
[12] Yu X, Zhang Z, Liu F and Pei J X, Tian X Y 2019 *J. Alloy Compd.* **806** 1309
[13] Pei J X, Yu X , Zhang C and Liu X J 2019 *Int. J. Electrochem. Sc.* **14** 3393

Application of orthogonal design method in analysis of injected water composition of whole formation model

N. Jia

Exploration and Development Research Institute, Daqing Oilfield Company Limited, Daqing, China

ABSTRACT: Aiming at the phenomenon of water absorption in non oil reservoir, based on the dissection of typical blocks, an ideal model of the whole formation including effective sandstone, non effective sandstone and non reservoir is creatively established, and the regularity of influencing factors of water injection composition is studied. The orthogonal test is used to design the test schemes of three levels and six geological and development factors. The numerical simulation is used to calculate the injection water composition of each scheme, which makes the analysis of the influencing factors of the injection water composition of the whole formation model simpler and clearer, and clarifies the main controlling factors of the injection water composition.

1 INTRODUCTION

The peripheral oilfield of changyuan cover a wide range with various types, complicated geological conditions and various causes of high injection production ratio. Therefore, the present situatione of injection production ratio is analyzed by Oil production plant, oil field, oil layer and block, the variation law of injection production ratio in different types of blocks at different development stages is studied, and the influencing factors of injection production ratio are studied systematically. The existence of non oil layer water absorption in peripheral reservoirs of placanticline is proved theoretically, and the non oil layer water absorption is demonstrated from geological research, laboratory experiment, mechanism analysis and field test The objective existence of water. Aiming at the phenomenon of water absorption in non oil layer, the "whole formation" model is introduced Based on the dissection of typical blocks, the ideal model of the whole formation including effective sandstone, non effective sandstone and non reservoir is creatively established. The geological and development factors that have great influence on the injection production ratio are taken into account, and the numerical simulation method is applied to study the water injection composition and influencing factors under different factors and different levels. It provides a basis for genetic evaluation, reasonable injection production ratio optimization and effective management of high injection production ratio block in low and ultra-low permeability reservoir.

2 ANALYSIS OF INFLUENCING FACTORS OF INJECTION WATER COMPOSITION IN IDEAL MODEL OF WHOLE FORMATION

2.1 Orthogonal test design

Orthogonal test design is a kind of test design method to study multi factors and multi levels. According to the orthogonality, some representative points are selected from the comprehensive test. These representative points have the characteristics of uniform dispersion, neat and comparable.

Orthogonal table with equal levels of each factor, $L_n(r^m)$
L——Orthogonal design code;
n——Test number of orthogonal table;
r——Number of factor levels (number of index values);
m——Number of factors (number of indicators).
Design principle:
In any column of the table, the number of different numbers is the same; in any two columns of the table, the number of different peer pairs is the same.

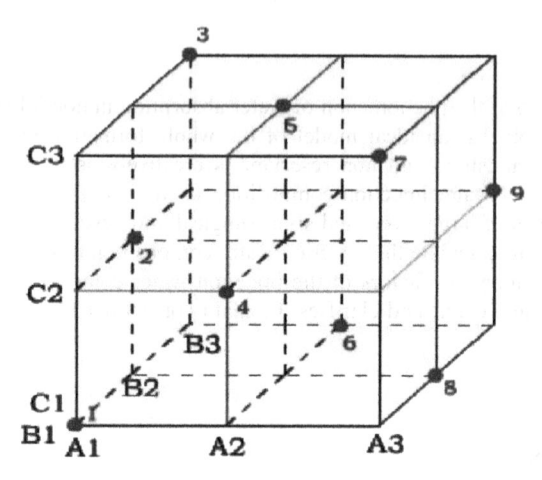

Figure 1. Stereogram of balanced dispersion of $L_9(3^3)$ test.

2.2 Analysis of main factors of injected water composition

According to the characteristics of waiwei reservoirs, the whole formation models of medium permeability, low permeability and ultra-low permeability reservoirs with 9 injection and 16 production are designed. Six geological and development factors and three horizontal test schemes were designed by orthogonal test, and the injection water composition of each scheme was calculated by numerical simulation (Table 1).

Table 1. Level setting of geological and development factors.

Level number	Reservoir type	Thickness ratio of non reservoir-non effective sandstone-effective sandstone	Viscosity (cp)	Non-effective sandstone permeability (mD)	Porosity (f)	Injection pressure (MPa)	Well spacing (m)
1	Medium permeability	5:0.8:1	5	0.6	0.06	16	180
2	Low permeability	10:1.7:1	8	0.8	0.08	20	300
3	Ultra-low permeability	20:2.5:1	12	1.0	0.10	24	420

The orthogonal test results show that the effective sandstone water storage is the main water storage in medium and low permeability reservoirs, and the non effective sandstone water storage proportion is higher in ultra-low permeability reservoirs, and there is also a certain proportion of non reservoir water storage. The extremely poor influence of various factors on water injection composition indicates that reservoir physical properties, thickness ratio, well spacing and other factors have a greater influence on water injection composition, especially reservoir physical properties, while crude oil viscosity, non reservoir physical properties and other factors have a relatively small influence within a certain level.

Comparing the influence of different factors and levels on the cumulative injection production ratio, we can see that the worse the physical properties of effective sandstone, the greater the thickness ratio of non effective sandstone and non reservoir, the higher the viscosity of crude oil, and the higher the cumulative injection production ratio (Figure 2).

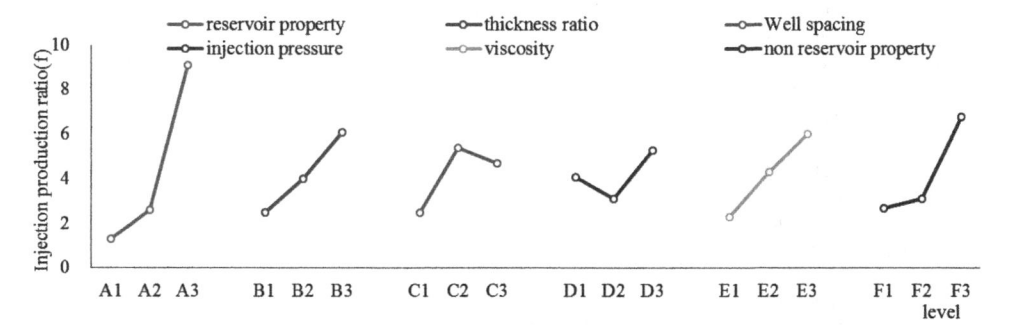

Figure 2. Influence of different factors and levels on cumulative injection production ratio.

Comparing the influence of different factors and different levels on the produced water, it can be seen that the worse the physical properties of effective sandstone, the larger the injection production well distance, the greater the thickness ratio of ineffective sandstone and non reservoir, the higher the proportion of unprocessed water (Figure 3).

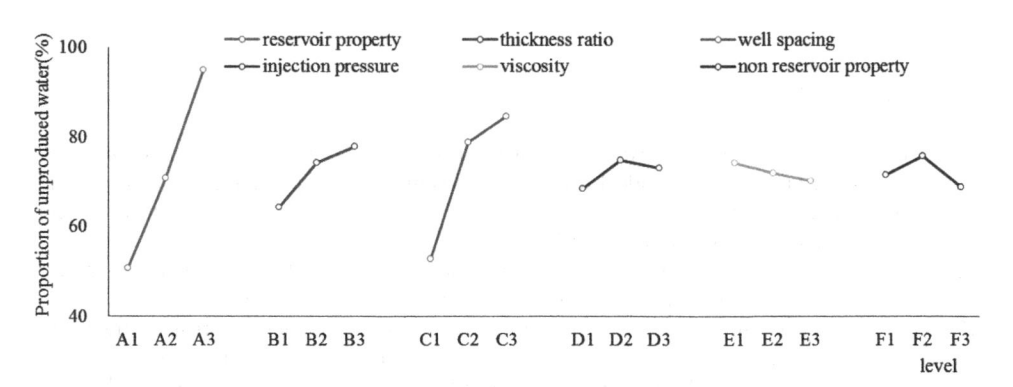

Figure 3. Influence relationship of different factors and levels on unprocessed water.

2.3 Composition of injected water in reservoirs with different permeability levels

In order to analyze the influence of physical properties on the composition of injected water, 16 sets of schemes are designed for the medium permeability, low permeability and ultra-low permeability reservoirs respectively. It can be seen from the results of orthogonal test that the composition of injected water in reservoirs with different physical properties is mainly affected by well spacing and thickness ratio; With the deterioration of reservoir physical properties, the influence of well spacing is gradually weakened, and the influence of thickness ratio is gradually enhanced (Table 2).

Table 2. Comparison of influence degree of different factors in different reservoirs on injection water composition.

Reservoir property	Target value	The range of different factors and the order of influence				
		Thickness ratio	Well spacing	Injection pressure	Viscosity	Physical property of undiscovered reservoir
Medium permeability	Effective sandstone water storage	13.9	41.4	0.8	0.9	1.8
	Ineffective sandstone water storage	0.05	0.1	0.02	0.01	0.02
	Non reservoir water storage	0.01	0.03	0.01	0.01	0
Low permeability	Effective sandstone water storage	14.1	21.4	2.7	6	2
	Ineffective sandstone water storage	3.7	8.5	0.7	7.7	2.4
	Non reservoir water storage	0.5	1.5	0.1	1.4	0.1
Ultra low permeability	Effective sandstone water storage	15	2.6	0.2	5.1	5.3
	Ineffective sandstone water storage	16.8	4.5	1.4	4.6	8.1
	Non reservoir water storage	2.6	0.6	0.1	0.3	1.3

3 CONCLUSION

(1) In view of the phenomenon of water absorption in non oil layers, the idea of "whole formation" can be applied to establish an ideal whole formation model including effective sandstone, non effective sandstone and non reservoir, and to carry out the regularity research on the influencing factors of water injection composition.
(2) By using the method of orthogonal test design, the analysis of influencing factors of injected water composition in the whole formation model becomes more simple.
(3) Reservoir physical properties, thickness ratio, well spacing and other factors have great influence on water injection composition, especially reservoir physical properties, while crude oil viscosity, non reservoir physical properties and other factors have relatively small influence in a certain level.
(4) The composition of injected water in reservoirs with different physical properties is mainly affected by well spacing and thickness ratio; With the deterioration of reservoir physical properties, the influence of well spacing is gradually weakened, and the influence of thickness ratio is gradually enhanced.
(5) The larger the thickness ratio of non-effective sandstone and non-reservoir is, the smaller the difference of physical properties between effective sandstone and non-effective sandstone and non-reservoir is. Therefore, with the deterioration of block physical properties, the proportion

of effective sandstone water storage decreases, while the proportion of non-effective sandstone and non-reservoir water storage increases.

REFERENCES

[1] WU Qiong. Study on the Mechanism of High Injection-Production Ratio(IPR) for Low-permeability Oil Reservoirs[J]. Special Oil and Gas Reserviors, 2012, 19(5): 82–85.
[2] WU Qiong, HAN Ying, WANG Yuying, et al. Study on the trend of injection-production ratio in the Xinli oilfield[J].Special Oil and Gas Reserviors, 2013, 20(3): 68–71.
[3] LI Yanjun. Strategy and effectiveness of development adjustment techniques for complex fault reservoirs—A case study on the glutenite reservoirs of Block Bei 301[J]. Special Oil and Gas Reserviors, 2013, 20(3):89–92.
[4] CANG Dan, ZHANG Jinyu. Cause analysis and Adjustment Countermeasures of injection production imbalance [J].China Petroleum and Chemical Standard and Quality, 2017, 23:89–90.
[5] XU Baoan.On How to Improve the Water Injection Development Results of Wang 78-2 Well Area in Wangchang Oil Field[J]. Journal of Jianghan Petroleum University of Staff and Workers:2013, 01:44–46.
[6] LI Hongchang, WU Yan, JIAN Jun.Making adjustment measures for water injection development in Wuqi by combining multiple means [J].China Petroleum and Chemical Standard and Quality, 2013, 05: 187–188.
[7] ZHANG Yuan, GAN Junqi, WANG Junwen et al.Effective accumulative injection-production ratio and its application during the effect evaluation of water injection development—taking ultra-low permeability reservoir in Da 45 Block of Honggang Oilfield in Jilin Province for example[J]. Oil Drilling & Production Technology, 2015, 37(6): 86–89.
[8] WANG Xuezhong. Computation Research of Useless Water Injection Rate[J]. Fault-Block Oil & Gas Field, 2007, 14(1):67–69.
[9] Zhang Q, Dong Y H,Tong S Q.Characterization of water imbition in sandstones studied using nuclear magnetic resonance[J].Journal of University of Chinese Academy of Sciences, 2017, 34(5): 610–617.
[10] ZHENG Junde, JIANG Hongfu, FENG Xiaoshu. Study on reasonable injection production ratio in Sazhong area[J]. Petroleum Geology and Recovery Efficiency, 2001(02): 55–57.
[11] LI Jieshi, GE Yunfeng, ZHANG Yangfan. A Study on Suitable Injection-production Ratio of Aobaota Oil[J]. Petroleum Geology & Oilfield Development in Daqing, 2002(02): 17–18.
[12] ANG Farong, ZUO Luo, HU Zhiming et al. Researching the Water Imbition Characteristic of Shale by Experiment[J]. Science Technology and Engineering, 2016, 16(25): 63–66+74.
[13] Yang Liu, Ge Hongkui, Cheng Yuanfang et al. Investigation on fracturing fluid imbition-ion diffusion and its influencing factors in shale reservoirs[J].China Offshore Oil and Gas, 2016, 28(04): 94–99.
[14] A Laboratory Investigation of The Effect of Rate on Recovery of Oil by Water Flooding[J]. J.G. Richardson, F.M. Perkins. Trans, AIME (1957) Vol. 210: 114.
[15] Waterflooding Heterogeneous Reservoirs-An Overview of Industry Experiences and Practices. Gulick K.E., et al. SPE40044, 1998 :1~7.

Velocity distribution in the annular gap between the walls of train and tunnel

S.X. Sun, Y. Li & Y.X. Zhang
School of Environmental Science and Municipal Engineering, Lanzhou Jiaotong University, Lanzhou, China
Engineering Research Center for Cold and Arid Regions Water Resource Comprehensive Utilization, Ministry of Education, Lanzhou, China

G.L. Li
China Railway First Survey & Design Institute Group Ltd, Xi'an, Shaanxi, China

ABSTRACT: To analyze the velocity distribution in annular gap between train and tunnel during the initial stage of train going through a tunnel when the piston wind was un-fully developed. The dynamic grid model using ANSYS/FLOTRAN was carried out. The standard $\kappa - \varepsilon$ turbulence model was used for numerical simulation analysis. The calculation model and method were verified by experimental results of full-scale train in reference. The results show that, during the initial stage of train going into a tunnel with constant speed, the air flow field is un-fully developed, the air velocity in annular gap flow downhill to the negative along front to back of the train, entrainment effect increased. The velocity distribution in annular gap at any specified moment got from numerical calculation is the reasonable supplement to results of the theory. during the initial stage of train going through a tunnel, the calculation method on the basis of theory of steady flow is not suitable to velocity distribution of piston wind in annular gap own to the un-fully developed character, the velocity distribution in annular gap only accords with type C proposed by Earashaw based on steady flow theory.

1 INTRODUCTION

When a train travels through a tunnel, it creates what is known as the piston effect. The piston wind is the important issue of aerodynamics, also the key to solve operation ventilation in tunnels. Along with the increasing of train lines and speed, it should not be surprising that the study on piston wind has become the main concern from experts and scholars both here and abroad. The study on unsteady flow field in annular gap between train and wall of tunnel is an important part of calculations about piston wind and air friction. In numerous studies on flow field of annular gap, incompressible and fully development steady flow was taken as the research presupposition, those are not suitable for the unsteady flow field in annular gap at initial stage of train into tunnels. In current *Code for Design on Operating Ventilation of Railway Tunnel*, the resistance coefficient for annular gap is simplified into the piston effect coefficient, which is too simple to study the flow field of annular gap. The front, rear and body of train are treated as a whole in the above method. The energy equation unites with continuity equation just in the front of headstock and after tailstock to calculate the piston wind speed. The coefficients such as local resistance of inlet and outlet with frictional resistance for annular gap are tied together to a simplified one known as 'piston effect coefficient'. The resistance characteristics of the still tunnel wall and moving train wall are out of consideration in last study on velocity distribution in the annular gap, also the interactions betweentrain and airflow. The calculation of piston wind in annular gap was overly simplified if just rely on pressure drop formulas of pipe flow and empirical coefficient moreover its application in calculation has heavy limitation.

There are mainly three following models for flow field [1].

Firstly, as proposed by Earnshaw, owing to the train is much longer than annular space, incompressible and fully development steady flow along the train in tunnels was taken as the research presupposition. The air flow in annular gap space can be approximated as axial flow in the concentric annular gap and divided into three conditions like as A, B, C in Figure 1 [1, 2].

Secondly, Couette flow model came forward. Reichardt put forward that flow in annular space can be regard as non-pressure Couette flow and the wall of train and tunnel both can be supposed to be smooth. Schlichting put forward that flow in annular space can be regard as pressure Couette flow and used pressure gradient to analyze velocity distribution in annular gap. The work shows the strong dependence of the annular velocity profile on the pressure gradient.

Thirdly, two-dimensional fully developed incompressible turbulence model proposed by Eder. Taking into account the influence of Reynolds stress transfer, velocity in annular gap sections is in accord with logarithmic distribution. Eder established $k - \varepsilon - uv$ equations with momentum, turbulent kinetic energy and dissipation rate and Reynolds stress transfer equation. The geometry size of annular gap, wall roughness of train and tunnel, speed of vehicle, average velocity in annular gap and adhesiveness of fluid motion were all reflected in above equations.

Those three methods described above mainly simulated the velocity distribution in annular gap with computer, verified with the results of physical model test based on parallel plate flow theories.

At present, the study and actual measurement about flow velocity distribution in annular gap of tunnel are little. Li Yan [3, 4] treated the velocity distribution in annular gap as combination of Poiseuille turbulence and Couetteturbulence, the velocity value was the sum of their respective result. Meanwhile, the theoretical formula was put forward. Based on the theoretical formula, the annular flow was Type B if tunnel length less than 5 km. It was Type A when longer. The compare results between calculated value based on theoretical formula and field measured value in Rilly-La-Montagne tunnel (3.440Km, <5Km) are shown in figure 2 [1, 2]. Figure 1 and figure 2 show that annular flow in Rilly-La-Montagne tunnel belong to Type C, not Type B derived by theory analysis. The deviation made by presupposition of incompressible and fully development steady flow in the annular space. The presupposition above has certain limitations to short tunnels or initial stage of train into tunnels when the flow is not fully development. So, the un-fully developed velocity distribution in annular gap will be necessary to be studied to analysis formation mechanism and compute value of the piston wind.

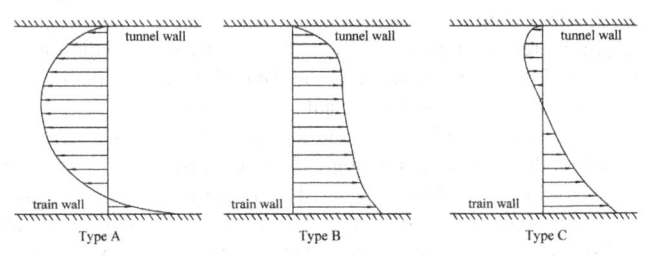

Figure 1. Velocity distribution in annular gap by Earashaw.

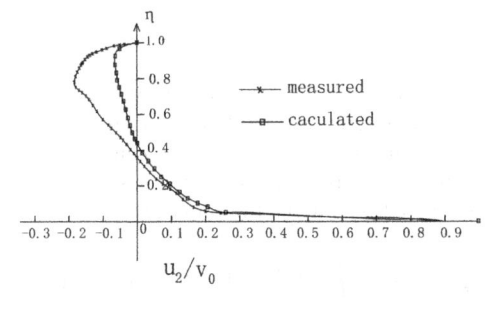

Figure 2. Compare between calculated value and measured value in references [3].

The computational models were constructed and solved using ANSYS/FLOTRAN, general purpose CFD software. By comparison from standard k-ε model, low-Reynolds number k-ε model, double k-ε model and RNG k-ε model, the standard k-ε model was selected to calculate the fluid velocity components in annular gap. The moving interfaces like train wall was involved in the fluid flow. So, the fluid domain with time and the finite element mesh must move to satisfy the boundary conditions at the moving interfaces. The Arbitrary Lagrangian-Eulerian (ALE) formulation was used to set up dynamic mesh.

The airflow characteristics during the initial stage of a single train going through a tunnel with constant speed were studied. The change trend of piston wind in annular gap was revealed and compared with existing research results. Those works contribute to further study formation mechanism of piston wind.

2 PHYSICAL MODEL AND KEY EQUATIONS

2.1 Physical model

The airflow is special in annular gap between the walls of train and tunnel, so the distribution of velocity. Considering the complexity of the actual tunnel and the feasibility of the calculation, the reasonable hypothesis was necessary for the model of tunnel and train to facilitate research of the piston wind. The air in tunnel is regarded as a continuum and no gaps between the air molecules. The air is homogeneous and impressible, satisfy the Stokes hypothesis. The equivalent diameter of domestic single-track tunnel in are about 6~7m. The piston wind speed are about 3~11m according to the general consideration. The Reynolds number (Re) for piston wind are about $1.2 \times 10^6 \sim 5.1 \times 10^6$. The coefficient of surface friction of tunnel (λ_t) is set to 0.02 generally for the tunnel at home and abroad. On the base of Moody chart widely used in engineering, under the condition of the above values of Re and λ_t, the airflow in tunnel is in the region of quadratic resistance law or near it. Ignoring the changes with temperature, the specific heat of air in tunnel is considered as a constant in the simulation.

A train passages through a tunnel can be divided into three distinct time periods for predicting the annulus airflow. These periods are, train entering the tunnel, train completely inside the tunnel and train exiting the tunnel.

The period 2 is the core concern in this paper.

The calculating parameters of tunnel and train for numerical simulation are shown in table 1. The speed of train is 12.17m/s with reference to the data of Qinghai-Tibet Line. The running time was set 2s, 10s and 20s separately. The reverse natural wind speed is 1.5m/s.

Tunnels and trains are all long and narrow and the train only has one main moving direction in tunnel. Owing to the length of tunnel is much greater than sectional area, two-dimensional model and Cartesian co-ordinates were used in this work. The numerical simulation assume the following, train fully enter into the tunnel with uniform speed, cross-sectional areas of tunnel and train are constant, both wall surfaces were smooth, no slope and curve. Seven sub-regions were divided along the direction of the length of tunnel (see figure 3). The sub-regions were meshed respectively and refined for region 1–3.

2.2 Key equations

Mobile grid and smooth handover interface technology was adopted to improve the unsteady numerical simulation. Unstructured grid was developed, meanwhile, wall functions was used near wall. Boundary of y^+ is between 30~35. The flow surrounding train was set two-dimensional model, viscous and incompressible unsteady turbulent for standard k-ε model. Φ was set as a parameter for flow field, for a control volume, flow control equations can be expressed as

$$\frac{d}{dt}\int_V \rho\Phi dV + \int_S [\rho\Phi U - \Gamma_\Phi grad\Phi]dS = \int_V S_\Phi dV \quad (1)$$

where $\Phi = 1, U, e, k, \varepsilon$, respectively, express continuity equation, momentum equation, energy equation, turbulent kinetic energy equation and turbulent kinetic energy dissipation rate equation. Γ_Φ is the generalized diffusion coefficient. S_Φ is the generalized source term.

For any control volume in the flow field, the general form of discrete equation for Eq. (1) can be written as

$$a_P \Phi_P = \sum_{j=1}^{n} a_j \Phi_{P_j} + s \tag{2}$$

where, a, s are the different parameters with different difference scheme of time and space.

Table 1. Parameters for numerical simulation.

Name	Value	Name	Value
Length of tunnel (m)	1000	Circumference of train (m)	14.3
Length of train (m)	140	Width of train (m)	6.0
Sectional area of tunnel (m^2)	32.29	Frictional resistance coefficient of annular air	0.02
Sectional area of train (m^2)	12.6	Frictional resistance coefficient of tunnel	0.02
Circumference of tunnel (m)	21.38	Local resistance coefficient of tunnel	0.5

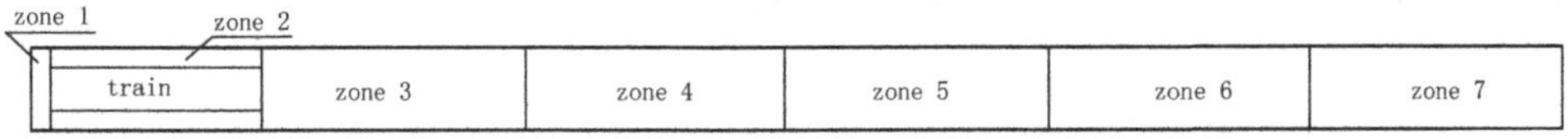

Figure 3. Division of the calculation area.

2.3 Arbitrary lagrangian–eulerian (ALE) formulation for moving domains

Base on the dynamic meshing theory foundation of ANSYS/FLOTRAN, for any control volume (V) of moving boundaries, the integral form of conservation equation of general scalar (ϕ) can be given by the following expressions:

$$\frac{d}{dt}\int_V \rho\phi dV + \int_{\partial V} \rho\phi (\vec{u} - \vec{u}_g) \cdot dA = \int_{\partial V} \Gamma \nabla \phi \cdot dA + \int_V S_\phi dV \tag{3}$$

where, ρ is the fluid density, $\vec{u}$ is the vector of airflow velocity, $\vec{u}_g$ is the velocity of moving grid, Γ is the diffusion coefficient, S_ϕ is the generalized source term. ∂V is the boundary of control body. By one order backward difference, the time derivative terms in eq. (3) can be equated to:

$$\frac{d}{dt}\int_V \rho\phi dV = \frac{(\rho\phi V)^{n+1} - (\rho\phi V)^n}{\Delta t} \tag{4}$$

where, n and n + 1 are the current time and next time respectively.
The V^{n+1} can be calculated by the following expression in the n + 1 time:

$$V^{n+1} = V^n + \frac{dV}{dt}\Delta t \tag{5}$$

where, $\frac{dV}{dt}$ is the time derivative of the control volume (v). For meeting the conservation rate of grid, the time derivative of the control volume (v) can be calculated with the following formula:

$$\frac{dV}{dt} = \int_{\partial V} \vec{u}_g \cdot \vec{dA} = \sum_j^{n_f} \vec{u}_{g,j} \cdot \vec{A}_j \tag{6}$$

where, n_f is the surface grid number for control volume (v), $\vec{A}_j$ is the vector of surface area. The dot product of vector $\vec{u}_{g,j} \cdot \vec{A}_j$ for each control volume can be calculated with the following formula:

$$\vec{u}_{g,j} \cdot \vec{A}_j = \frac{\delta V_j}{\Delta t} \tag{7}$$

where, δV_j is the changed volume caused by the expansion of surface j as the time step Δt.

2.4 Initial and boundary conditions

- Inlet boundary condition, velocity, $v_x = 0$ m/s, $v_y = 0$ m/s, displacement, $u_x = 0$ m, $u_y = 0$ m.
- Non-slip condition of fixed wall.
- Outlet boundary condition, relative pressure is 0 Pa, natural wind speed, $v_x = 1.5$ m/s, $v_y = 0$ m.

2.5 Calculation method

Continuity equation and standard k-ε equations were included in the key equations. The coupling equations of pressure and velocities were solved with the SIMPLE algorithm. The difference schemes of convective and diffusive terms were solved with TDMA method also known as triple diagonal matrix method. The algebraic equations were solved by the low relaxation iterative method. The calculation was considered to be convergent when each variable residual were less than 10^{-3} and computational domain energy were balanced [5–15].

3 RESULTS AND DISCUSSION

3.1 Theoretical calculation based on fully development steady flow theory

The annular gap flow under simulated condition was theoretically calculated and discussed first. The computation formulas are as follows

$$\eta = y/dh \tag{8}$$

where, η is the dimensionless distance from wall. y is the distance from train wall, m. d_h is the equivalent diameter of annular gap section, m. When $\eta = 0$, the velocity of annular gap flow equals train speed. When $\eta = 1$, the velocity of annular gap flow equals 0 m/s.

$$u_2 = -\left[u_{2pm} + \ln \eta^{2.5v_c^*}(1-|1-2\eta|)^{2.5v_p^*}\right] \tag{9}$$

$$v_c^* = \frac{v_0}{5.75 \log \frac{d_h}{2k_{so}} + 8.5} \tag{10}$$

$$u_{2pm} = \left(5.75 \log \frac{d_h}{4k_{st}} + 8.5\right) v_p^* \tag{11}$$

$$v_p^* = \frac{v_2}{5.75 \log \frac{d_h}{4k_{st}} + 4.75} \tag{12}$$

$$v_2 = \frac{\alpha v_0 - v_1}{1 - \alpha} \tag{13}$$

where, k_{so} is the surface equivalent roughness of train, 9.2mm. k_{st} is the surface equivalent roughness of tunnel, 5.0mm. u_2 is the velocity of annular gap flow. u_{2pm} is the max velocity of in annular gap center. v_0 is the speed of train, m/s. v_1 is the velocity of piston wind, m/s. a is the blockage ratio. v_2 is the mean velocity of annular gap flow.

The different points were selected on annular gap sections, and homologous η were got from Eq. (8). The theoretical calculation value of velocity in annular gap at $t = 20$s are shown in figure

4. If the piston wind is fully development at initial stage of train into tunnels, the computed results accords with the type B in Figure 1 which comes from theoretical calculation. But in fact, the length of tunnel under simulated condition is 1Km (<5Km), the annular flow should be similar to measured data of Rilly-La-Montagne tunnel, which is type C. So, steady flow theory is unsuitable for the initial stage of train into tunnels. There is much certain limitation in calculation of annular gap flow. The work of numerical simulation on unsteady flow field in annular gap is described below.

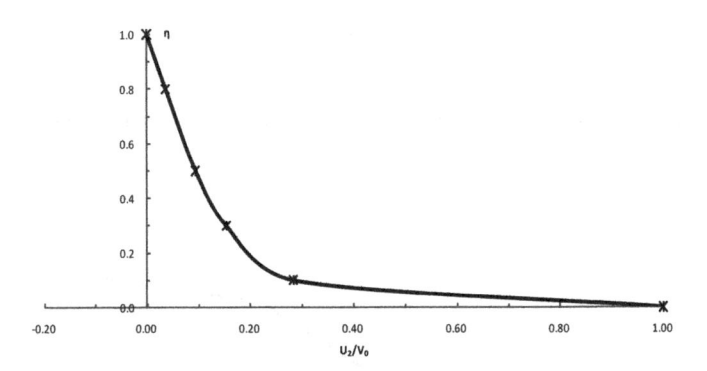

Figure 4. Theoretical calculation value of velocity in annular gap at $t = 20$s.

3.2 Numerical simulation based on unsteady flow theory

The annular space between train and tunnel wall is long and narrow when train running in tunnel. The surface of train is moving but the tunnel wall stand still. That adds complexity for calculation of piston wind. The velocity distribution of annular gap center line in the x direction is shown in figure 5. Where, the x axis is for the tunnel length, m. The y axis is for the speed of piston wind in annular, m/s, which is change along the tunnel. Given a time, the speed of piston wind in annular is different on each of points in tunnel. The impact of piston wind is gradually extending over time.

The velocity distribution of annular gap section ($t = 20$s, $x = 360$ m, front position) is shown in figure 6. The speed of piston wind near train surface is fast, slow near tunnel wall.

The velocity mean value of each annular gap section along the train at $t = 20$ s is shown in figure 7. The velocity in annular gap flow downhill to the negative along train front to back, entrainment effect is increasing.

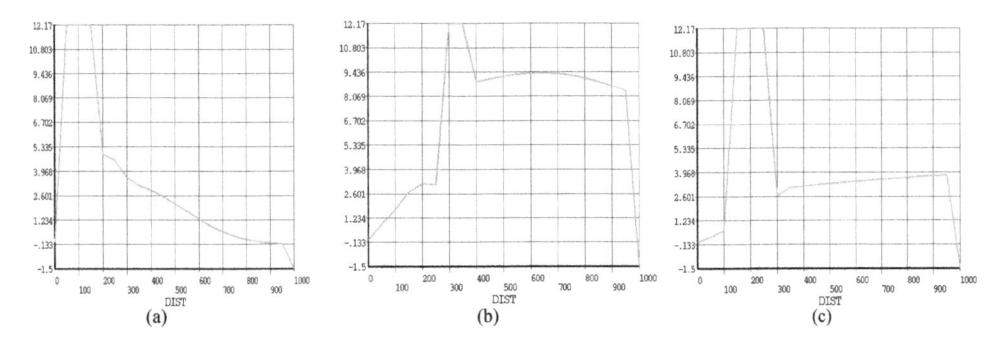

Figure 5. Velocity distribution of annular gap center in the x direction: (a) $t = 2s$, (b) $t = 10s$ and (c) $t = 20s$

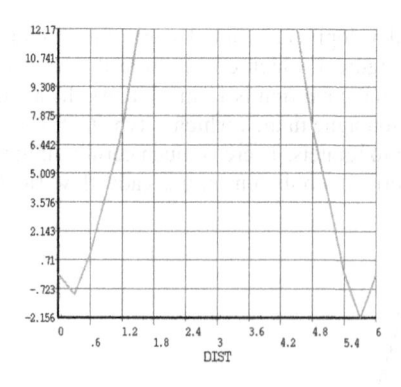

Figure 6. Velocity distribution of annular gap section ($t = 20$s, $x = 360$ m, front position).

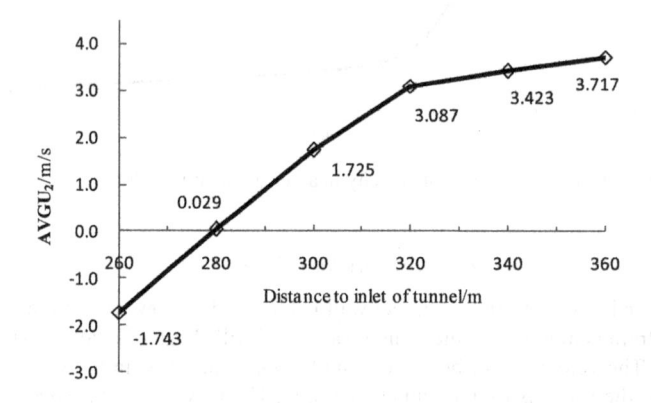

Figure 7. Mean value of each annular gap section along the train at $t = 20$s.

3.3 Results discussion of numerical simul-ation

Dimensionless is carried out for velocity of several annular gap sections along train. The u_2/v_0 and η are shown in figure 8. Where, u_2 is the velocity of annular gap flow come from numerical simulation. The mean values of u_2/v_0 with different η are shown in figure 9.

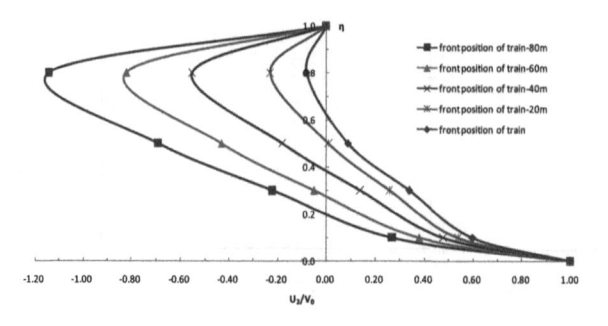

Figure 8. u_2/v_0 with different η of each annular gap section at $t = 20$s.

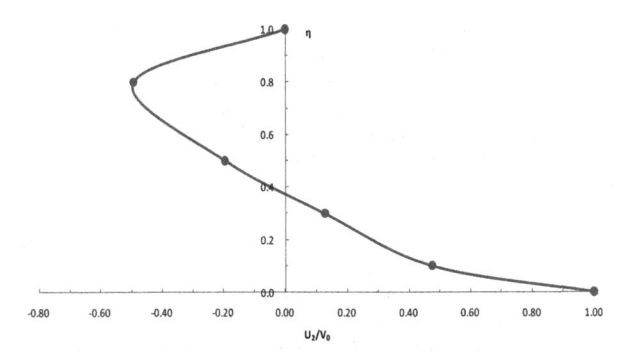

Figure 9. u_2/v_0 mean value with different η of each annular gap section at $t = 20$s.

As can be seen from the figure 8 and figure 9, the piston wind was un-fully developed at initial stage of train across tunnels. The air velocity in annular gap flow is changing along train front to back. The velocity distribution in annular gap only accords with type C proposed by Earashaw based on steady flow theory.

The data curve of simulation results is well similar to field measurement data in Rilly-La-Montagn tunnel (see figure 2). So, the results show that incompressible unsteady flow theory is more approach to practical situation.

4 CONCLUSION

The velocity distribution in annular gap between train and tunnel at initial stage of train across tunnels was studied numerically using ANSYS/FLOTRAN software. The standard k-ε model and dynamic mesh model were used for numerical simulation. Comparing with data from fully developed steady flow theory, the results of numerical simulation calculation based on incompressible unsteady flow is a very good agreement withrules in reality. During the initial stage of train into tunnel with constant speed, the air flow field is un-fully developed and the calculation method of constant flow theory is unfit. The velocity distribution in annular gap only accords with type C proposed by Earashaw based on steady flow theory. The air velocity in annular gap flow downhill to the negative along train front to back, entrainment effect increased. The impact of piston wind is gradually extending over time.

ACKNOWLEDGEMENT

This work was supported by the Technological Development Projects of Ministry of Railways (Project No. 20070301).

REFERENCES

[1] Li Y, Gao M L, Zhou M D and Li J X. Characteristics of flow field in annular gap space between train and tunnel [J]. Journal of the China Railway Society, 2009, 31(6): 117–120.
[2] Hite W R and Pope C W. The use of a hydraulic analogy for modeling the unsteady flows in railway tunnels [J]. 3th ISAVVT, Sheffield, UK, BHRA Fluid Engineering, Paper H1, 1979: 115–150.
[3] LI Yan, GAO Meng-li, ZHOU Ming-di, et al. theoretical study and calculation method of train piston wind in railway tunnels [J].Journal of the china railway society, 2010, 32(6):140–145. (in Chinese)
[4] LI Yan. Characteristic and theoretical study of train piston wind in railway tunnels [D]. Lanzhou, Lanzhou Jiaotong University, 2010. (in Chinese)

[5] Chow W K, K Y Wong and W Y Chung. Longitudinal ventilation for smoke control in a tilted tunnel by scale modeling [J].Tunnelling and Underground Space Technology, 2010, 25:122–128.
[6] J. Modic. Air velocity and concentration of noxious substances in a naturally ventilated tunnel [J]. Tunnelling and Underground Space Technology, 2003, 18:405–410.
[7] XIA Feng-yong. Numerical simulation research on twin tubes complementary ventilation system of extra-long highway tunnel [D].Xi'an: Chang'an University, 2012. (in Chinese)
[8] Jacques E. Numerical simulation of complex road tunnels [J]. Aerodynamics and ventilation of vehicle tunnels, 1991, 467–552.
[9] Cheng LH, Weng TH and Liu CW. Simulation of ventilation and fire in the underground facilities [J]. Fire Safety Journal, 2001, 36(6): 597–619.
[10] Tabarra M. Optimizing jet fan performance in longitudinally ventilated rectangular tunnels [J]. Separated and complex flows, ASMEFED, 1995, 217: 35–42.
[11] F. Colella, G. Rein, R. Carvel, P. Reszka and J L Torero. Analysis of the ventilation systems in the Dartford tunnels using a multi-scale modelling approach [J].Tunnelling and Underground Space Technology, 2010, 25:423–432.
[12] Wang Feng, Wang Mingnian, et al. Computational study of effects of traffic force on the ventilation in highway curved tunnels [J].Tunnelling and Underground Space Technology, 2011, 26(3): 481–489.
[13] Bogdan S, Birgmajer B and KOVACIC Z. Model predictive and fuzzy control of a road tunnel ventilation system. Transportation Research Part C: Emerging Technologies, 2008, 16(5): 574–592.
[14] D K Petersen, R C Arnold and W N Weins. Modelling of airflow and heat transfer in a train/tunnel annulus [J]. BHR Group Vehicle Tunnels, 2000: 829–845.
[15] Jojo S M Li and W K Chow. Numerical studies on performance evaluation of tunnel ventilation safety [J].Tunnelling and Underground Space Technology, 2003, 18: 435–452.
[16] Pope C W. The simulation of flows in railway tunnel using 1/25th scale moving model facilety [J]. In: Proc.7th ISAVVT, British Railways Board, 1991: 709–737.
[17] Yongye Li, Yuan Gao, Xihuan Sun and Xuelan Zhang. Study on flow velocity during wheeled capsule hydraulic transportation in a horizontal pipe [J]. Water, 2020, 12(4).
[18] Xiaomeng Jia, Xihuan Sun, Jiaorong Song and Michalis Xenos. Effect of Concentric Annular Gap Flow on Wall Shear Stress of Stationary Cylinder Pipe Vehicle under Different Reynolds Numbers [J]. Mathematical Problems in Engineering, 2020, 2020.
[19] Jianghong Zhao, Ching Wang and Xin Li. Gap flow with circumferential velocity in annular skirt of vortex gripper [J]. Precision Engineering, 2019, 57.
[20] Fahd M. Al-Oufi, Iain W. Cumming, Chris D. Rielly. Destabilisation of homogeneous bubbly flow in an annular gap bubble column [J]. The Canadian Journal of Chemical Engineering, 2010, 88(4).

Preparation of disodium terephthalate as intermediate of MOFs

Z.W. Wang
Polymer Material Engineering and Technology, Shenyang University of Technology, China

L. Li*
School of Petrochemical Technology, Shenyang University of Technology, China

ABSTRACT: The disodium terephthalate has been widely used in the preparation of organometallic frame compounds, battery materials and leather production AIDS. However, the increasing shortage of petroleum resources leads to the urgent need to develop new materials [1]. In contrast, in the field of plastic in China, white pollution is all over rivers, lakes and seas; At present, the utilization rate of plastic waste in China is less than 40%. However, general chemical recycling has not been widely used in China due to large investment scale, complex process technology, and incomplete recovery system [2]. This paper in the preparation of terephthalic acid disodium salt as the research object, and detailed the reaction system, reaction mechanism, such as material purity do the strict inspection, The reaction of 10 g polyester, 10 g sodium bicarbonate and 50 g ethylene glycol at 180°C was studied. 99% depolymerization rate and disodium terephthalate were obtained. aimed at improving chemical recycling of waste plastics, waste plastics, the preparation of new materials, make a contribution to human environmental protection career. The main results can be summarized as follows:

(1) Crude disodium terephthalate was prepared by depolymerization of PET.
(2) Refining of crude disodium terephthalate salt.
(3) Mechanism analysis of alcohol-alkali depolymerization.

1 EXPERIMENTAL METHOD

1.1 Reagents, materials and instruments

1.1.1 Reagent
The reagents and materials used in the experiment are as follows: the materials are 5-8mm fragments of waste polyester beverage bottles after crushing and cleaning;Reagents include sodium bicarbonate, zinc oxide, ethylene glycol, aluminum oxide, titanium dioxide, magnesium oxide, disodium terephthalate, all of which are analytical pure.

1.1.2 Instrument thermometer, condensing reflux, temperature controlled oil bath, mechanical stirrer, three-necked flask

1.2 The experimental method

1.2.1 Depolymerization method
Add PET, $NaHCO_3$, EG and ZnO into a three-necked flask with thermometer, condensing reflux device, temperature-controlled oil bath, mechanical stirring and temperature-controlled device in

---

* corresponding author

proportion, and add nitrogen into the flask to heat, stir and condensing reflux, and the reaction time is 1 hour. See Table 1 for the list of experimental ingredients.

1.2.2 Purification method

The depolymerized product was dissolved in deionized water, and then pumped and filtered. The obtained solid was undepolymerized polyester, which was weighed after drying. Calculate the depolymerization rate.

Table 1. Experimental ingredients table.

reagent	PET(Plastic bottle)	EG	NaHCO3	Catalyst
mass ratio	1	3-5	1	0.5%

$$\eta = \frac{m_1 - m_2 \times \%}{m_1} \tag{1}$$

In the equation: η—Disaggregation(%); $m1$— Polyester quality before depolymerization (g); $m2$—Quality of undissolved polyester(g). the obtained filtrate was recrystallized and repeated 3 times, and disodium terephthalate was obtained, and then qualitatively and quantitatively analyzed.

1.2.3 Analytical method

During the experiment, the infrared spectrum was used for qualitative analysis, the ultraviolet spectrum was used for quantitative analysis, and the particle size of the catalyst was measured by scanning electron microscope.

2 RESULTS AND DISCUSSION

2.1 Condition experiment

2.1.1 Comprehensive selection of catalysts.

PET 10g, NaHCO3 10g, The catalysts were zinc oxide, alumina, titanium dioxide and magnesium oxide with the mass ratio of 0.5%, and the depolymerization experiment was carried out according to the method of 1.2.1. The results are shown in Table 2. The effects of each group of catalysts are as follows:

Table 2. Catalyst selection results.

Catalyst and Manufacturer	ZnO	Al2O3	TiO2	MgO
Depolymerization rate(%)	95	70	83	99

In terms of depolymerization rate, magnesium oxide is the best catalyst, followed by zinc oxide and titanium dioxide. However, considering that titanium dioxide is a heavy metal, magnesium oxide is basic oxide, zinc oxide is amphoteric oxide; In the post-processing, such as acidification to make TPA zinc oxide can also be used as a catalyst, therefore, zinc oxide is selected as a catalyst in this experiment. Then, the influence of the amount of catalyst on the depolymerization rate was investigated, and the figure of the amount of catalyst on the depolymerization rate was drawn. The results are shown in Figure 1.

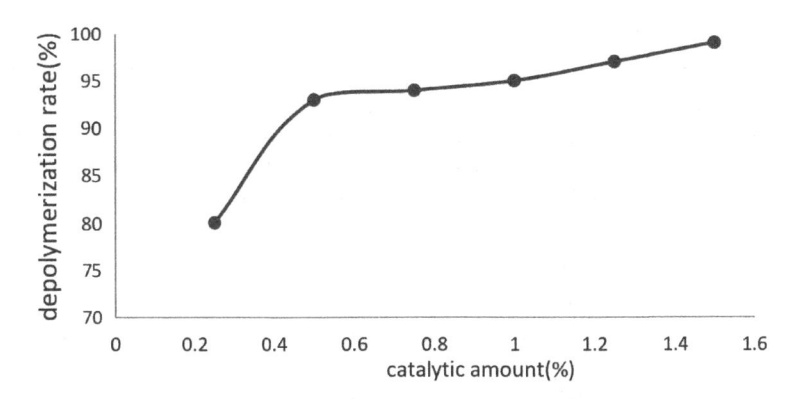

Figure 1. Selection of catalyst dosage.

2.1.2 Solvent dosage selection

The more glycol was added, the lighter color of the system was. According to the analysis, after the polyester swelling ethylene glycol can not polyester, isolated outside air, further reduce the occurrence of thermal oxygen reaction; And the reaction will produce ethylene glycol, further increase the volume, so take into account the volume of the reagent bottle; In addition, the reaction time caused by the concentration of catalyst and the recovery of catalyst by the concentration of catalyst should also be considered. So the solvent should be controlled in the polyester mass ratio of 3-5 times.

2.1.3 The choice of reaction temperature

According to 1.2.1. Design temperature condition experiment, other conditions remain unchanged, and investigate the depolymerization rate of polyester at 140°C, 150°C, 160°C, 170°C and 180°C, respectively. The results are shown in Figure 2. As the reaction system temperature reaches 140°C, the system begins to become cloudy. When the reaction reaches 180°C, the glycol boils and stops heating. As time goes on, PET is decomposed gradually, and the reaction stops after 40 minutes at 180°C. At this point, the decomposition rate of polyester reached 99%. After standing, the layer was layered, and the upper layer was ethylene glycol system, which was reddish brown. The lower layer is disodium terephthalate salt system, which is milky white. EG and NA2TP products can be obtained by vacuum filtration, distillation, ethanol washing, drying and recrystallization. After comprehensive consideration, the temperature should be controlled at about 180°C.

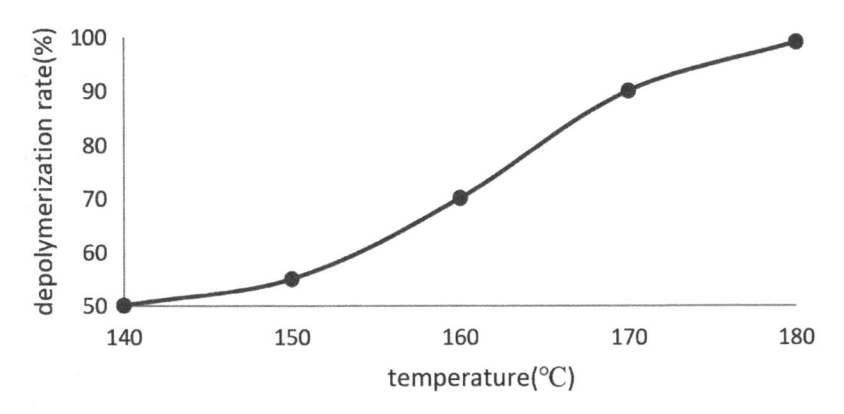

Figure 2. Influence of reaction temperature on polyester depolymerization.

2.2 Reproducibility experiment

Then the reaction time was extended, nitrogen protection was passed into the product for several times of recrystallization operation, repeated several experiments, and the above results were obtained. The experimental results of this experiment reproduce Yang Guang, Jiang Tao, Wu Limei.The mechanism analysis of alcohol-alkali combined depolymerization of waste polyester [4] experiment, in terms of depolymerization rate, product purity, system color effect is not ideal.

2.3 Catalyst and product characterization

2.3.1 Catalyst characterization
The catalyst was characterized in the experiment, and the results are shown in figure.3.

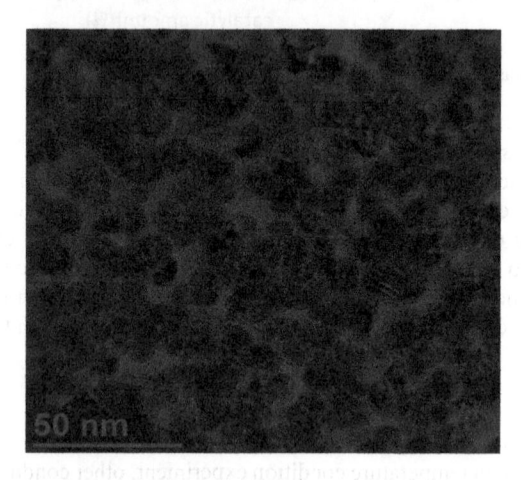

Figure 3. Scanning electron microscope photo of catalyst.

Micro-nano ZnO powder of the catalyst used in the experiment, with a particle size of 50nm, has high catalytic activity and selectivity.

2.3.2 Qualitative and quantitative analysis of products
The product was analyzed by infrared spectroscopy, and the results were shown in Figure 4.

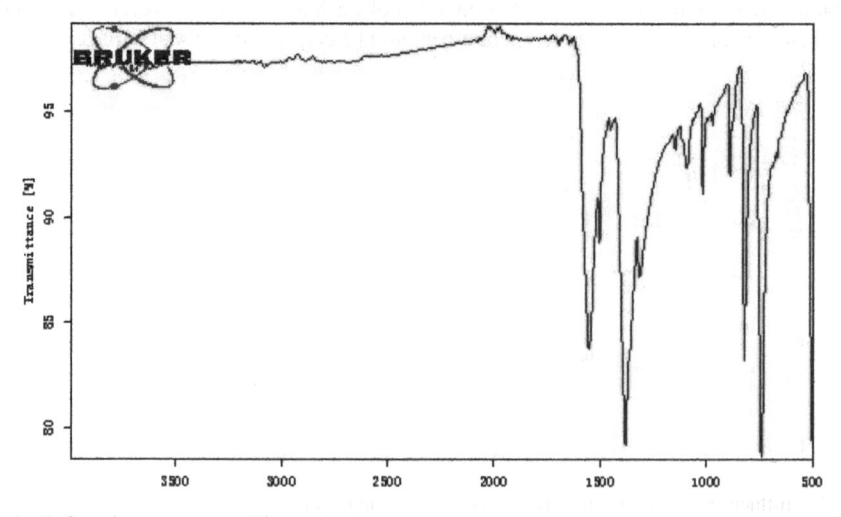

Figure 4. Infrared spectrogram of the product.

1403.29 and 1574.76 are characteristic peaks of carboxylic acid.830.97 May be substituted by benzene ring pairs.

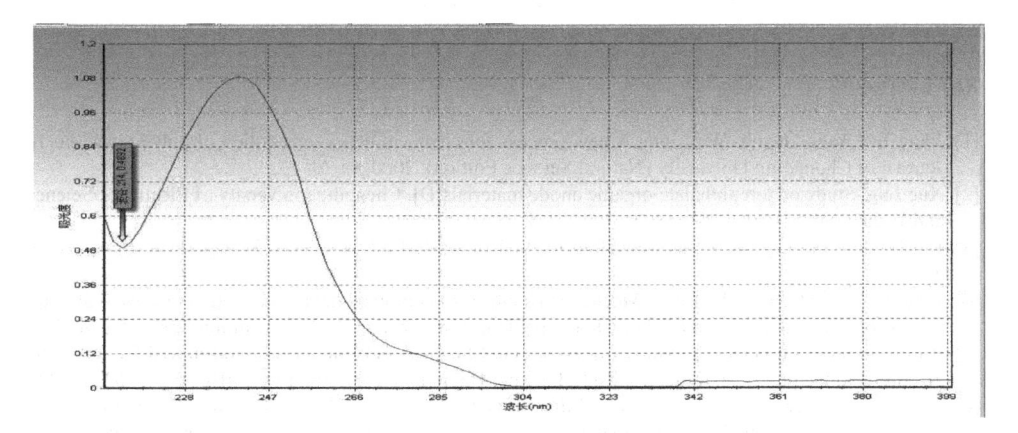

Figure 5. Ultraviolet spectra of the product The product purity is between 98.3 and 99.7%.

2.4 *Mechanism of depolymerization*

The experimental results show that the conversion rate of PET is consistent with the yield of monomer product. Since the TPA product from the alkaline hydrolysis of PET dissolves in solution and becomes its salt form, the terephthalate will be inactive for nucleophilic substitution in esterification. Therefore, the depolymerization of PET in sodium bicarbonate solution is an irreversible reaction. The oligomer after the reaction exists on the surface of solid PET, and the ester bond random fracture may occur on the PET surface. However, the reaction that occurs at the end of the polymer chain should be the dominant reaction. Then we found out that DMT was there; This further verified our conjecture that although the mechanism of alcohol-base joint depolymerization is more inclined to the end of alkalolysis to break one by one, alcoholysis mechanism also exists. We think this may be related to the amount of Lewis base added. Perhaps because the amount of sodium bicarbonate decreases with the prolongation of the reaction time, PET expands under EG and partial alcoholysis also occurs. At a certain temperature, the reactant even dissolves in EG, and the reaction volume increases rapidly. At this point, the presence of Lewis base and EG makes alkali hydrolysis and alcoholysis easier to proceed,promote each other and accelerate the reaction [4]. NaHCO3 and EG involve partial to complete depolymerization and the breakdown of terminal ester bonds one by one, in which a portion of the polymer is usually in contact with ethylene glycol. Glycol causes chain breaks by attacking ester bonds along the polymer skeleton.In this process, NaHCO3 acts as a catalyst to accelerate the reaction [5]. The intermolecular dehydration of side reaction EG may produce diethylene glycol, or may be caused by the ester bond between diethylene glycol and TPA in PET as an impurity and the decomposition of ethanol [6].When the reaction temperature is higher than 200°C, side reactions such as thermal decomposition may occur in the heating process. We need to control the temperature to reduce the impact of such side reactions [7].

3 CONCLUSION

In the process of preparing Na2TP, it is necessary to strictly control the temperature of 180°C and the rationality of the reaction process so as to reduce side reactions and problems existing in the operation process and impurities such as DMT and DEG. After the preparation of crude products, it is necessary to repeatedly recrystallization to remove $HCO3^{2-}$ impurities, so as to ensure the purity of Na2TP. Because in the application of metal-organic framework compounds, such as catalysts;

The purity of disodium terephthalate salts is particularly important. Therefore, improving the purity of the product is also the priority of the reaction. The key to the reaction is to strictly control the dose and the reaction system and improve the conversion rate.

REFERENCES

[1] Jiang Tao, Yang Guang, Wu Limei.A new process for chemical depolymerization of polyester waste[J]. Journal of Chongqing University (Natural Science Edition), 2000(5): 70–73.
[2] Xue Jing. Study on terephthalate organic anode materials[D]. Chengdu: University of Electronic Science and Technology of China 2016: 1–2.
[3] Yao Haibin, Yao Li, Liu Yayun.Recycling method of waste PET bottles[J]. Shanghai packaging 2016: 44–45.
[4] Jiang Tao, Yang Guang, Wu Limei.Mechanism analysis of combined depolymerization of waste polyester with alcohol and alkali[J]. Journal of Chongqing University (Natural Science Edition) 2000 (5): 74–77.
[5] R.López-Fonseca I. Duque-Ingunza B. de Rivas, ect.Chemical recycling of post-consumer PET wastes by glycolysis in the presence of metal salts[J]. Polymer Degradation and Stability, 2010: 95(6): 1022–1028.
[6] Genta, Minoru; Iwaya, Tomoko; Sasaki, Mitsuru; Goto, Motonobu; Hirose, Tsutomu. Depolymerization Mechanism of Poly(ethylene terephthalate) in Supercritical[J]. Methanol, Ind Eng Chem 2005: 44(11): 3894–3900.
[7] Jin-TaoDu, QianSuna, Xiao-FeiZeng, ect. ZnO nanodispersion as pseudohomogeneous catalyst for alcoholysis of polyethylene terephthalate[J]. Chemical Engineering Science, 2020(220): 4–6.

Microstructure degradation prompts durability reduction in a nickel-based single crystal superalloy

C.G. Liu
Science and Technology on Advanced High Temperature Structural Materials Laboratory, Beijing Institute of Aeronautical Materials, Beijing, China

Z.J. Sun
Science and Technology on Advanced High Temperature Structural Materials Laboratory, Beijing Institute of Aeronautical Materials, Beijing, China
School of Metallurgy and Ecological Engineering, University of Science and Technology Beijing, Beijing, China

P.Y. Liu & H. Zhang
Aviation Military Representative Office of Military Aviation Bureau of the Army Equipment Department in Beijing, Beijing, China

B. Shen, G. Yang, H.W. Wang, S. Zheng, Z.Y. Yang & Q.H. Wu
Science and Technology on Advanced High Temperature Structural Materials Laboratory, Beijing Institute of Aeronautical Materials, Beijing, China

ABSTRACT: Long-term serving nickel-based single crystal turbine blades tend to have structural degradation under a complex environment of high temperature and centrifugal force. Building the relationship between microstructural degradation and its mechanical properties is important to estimate the service life. The second-generation nickel-based single crystal alloy is still the most widely used commercial alloy, and its microstructure variation study during service is needed for the development of aeroplane materials. In this study, a second-generation nickel-based single crystal alloy was first stress aged at 1070°C/70 MPa to observe and study the law of structure degradation. Structure degeneration stages can be divided into two: the γ' phase size decreases first and then increases in the raft-shaped stage, while the γ phase size increases slowly in the coarsening stage. Stress rupture life under 1070°C/140 MPa was then determined; it exhibits a decreased life on long-time stress-aged samples from 370 h to 170 h. A fully raft-shaped structure decreases the rupture life and seriously coarsens the raft structure.

1 INTRODUCTION

Due to its superb high-temperature creep and fatigue properties, nickel-based single crystal superalloy is an irreplaceable material for advanced gas turbine blades [1]. The increase of the gas temperature at the front inlet of the turbine can effectively increase the thrust-to-weight ratio of the gas turbine [2,3]. Therefore, the structural stability of the nickel-based superalloys has a crucial impact on the performance and service reliability of the gas turbine [4, 5]. Nickel-based single crystal superalloys mainly improve its high-temperature performance through solid solution strengthening and precipitation phase strengthening [6, 7]. A high solid solution strengthening effect is realized by high melting point elements segregating in the γ phase, while the cubic γ' phase is embedded in the γ phase matrix to provide the effect of precipitated phase strengthening [8]. The morphology and volume content of γ' phase and γ phase play a significant role in the high temperature mechanical properties [9].

Nickel-based single crystal turbine blades rotate at high speed in a high-temperature environment during service. Under the action of high temperature and centrifugal force, the blades experience creep and deformation process, and the structure gradually becomes rafted, which thus affects its performance of the alloy obviously [10, 11]. Studies show that the raft structure can hinder the slip and climb of dislocations in the matrix in the creep process, thereby reducing the creep rate of the alloy, which is favorable to the enduring service of the alloy [12,13]. However, some studies have shown that the tensile properties and fatigue properties of nickel-based single crystal superalloys are reduced after rafting, which affects the safety and reliability of single crystal turbine blades in service [14,15,16]. For the purpose of ensuring the security and reliability of the service of nickel-based single crystal superalloy turbine blades, it is necessary to clarify the evolution of the single crystal alloy during service and its effect on the mechanical properties.

2 EXPERIMENTAL SECTION

The experimental material is a second-generation nickel-based single crystal superalloy, and the experimental content is listed in Table 1. The single crystal test bars with a size of $\Phi 21$ mm×210 mm was processed by the spiral crystal selection method in the HRS directional solidification furnace with the withdrawal rate of 3 mm/min. The single crystal test bars with crystal orientation less than 10° were selected as the experimental samples. The heat treatment system of the as-cast test bar is: 1290°C/1 h+ 1300°C/1 h+ 1310°C/2 h+1318°C/6 h (air cooling)+1130°C/4 h (air cooling) +870°C/32 h (air cooling). The heating process eliminates the eutectic structure, reduces the solidification segregation, and adjusts the size and morphology of the γ' phase during the aging heat treatment. A metallographic sample was taken along the (001) plane perpendicular to the stress axis and then was corroded after conventional grinding and polishing chemically. The erosion reagent was a 1% HF +33% HNO_3 +33% CH_3COOH +33% H_2O solution. A ZEISS SUPRA 55 field emission electron microscope was used to investigate the microstructure. In order to observe and analyze the degree of coarsening of the structure, the width of the γ phase and γ' phase was measured along the [001] direction parallel to the stress axis.

The stress aging samples have size of 12*190 mm and they are aged for 20–300 hs. After stress aging, the sample was taken out and the middle part was sampled for analysis, and two standard permanent samples were processed by sampling at both ends of the gauge length. The heat-treated test bar was processed into a standard tensile endurance specimen with a size of $\Phi 5 * 60$ mm. The durability performance was tested on a RD-100 endurance testing machine at 1100°C/70 MPa for 20 to 500 hs. At least two samples at each withdrawal rate were tested to obtain the average value.

Table 1. Chemical compositions of experimental alloy (wt.%).

Cr	Co	Mo	W	Re	Ta	Nb	Hf	Ni
4.0	8.0	2.0	7.0	3.0	7.0	0.5	0.2	Bal.

3 RESULTS AND DISCUSSIONS

Figure 1 shows the microstructure of the experimental Nibased single crystal superalloy after heat treatment. The cubic γ' phase is embedded in the γ phase matrix channels, and the average width of the γ phase matrix channel is about 40 nm, and the average size of the γ' phase is about 400 nm. The volume content of the γ' phase is about 68%, which predicts a good mechanical property according to its volume fraction ratio.

Figure 2. shows the SEM images on evolution of the γ/γ' phase structure of the experimental alloy during the stress aging process at 1070°C/70 MPa for different times. The γ' phase structure still maintains a cubic morphology when the stress aging is up to 20 h. The local γ' phase is connected

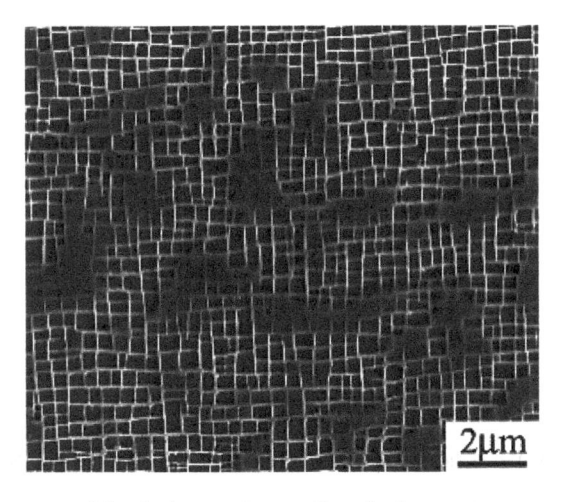

Figure 1. The microstructure of the single crystal superalloy after heat treatment.

together in the horizontal direction, while the γ phase vertical channels are refined in parallel to the stress direction with the disappearance of some vertical channels when the stress aging is up to 50 h. The raft structure has been formed, the cubic γ′ phase has completely disappeared, and the γ/γ′ phase is alternately distributed in layers when the stress aging is up to 100 h; the raft structure maintains a stable structure when the stress aging is up to 200 to 500 h. The evolution of the experimental alloy structure can be divided into three stages under the stress aging conditions at 1070°C/70 MPa: the initial stage before the raft is formed and the coarsening stage with a raft-shaped structure. In the first stage before rafting, only the γ/γ′ phase structure size increases (<100 h), while if the rafted γ/γ′ structure is formed, its size remains stable after 200 h.

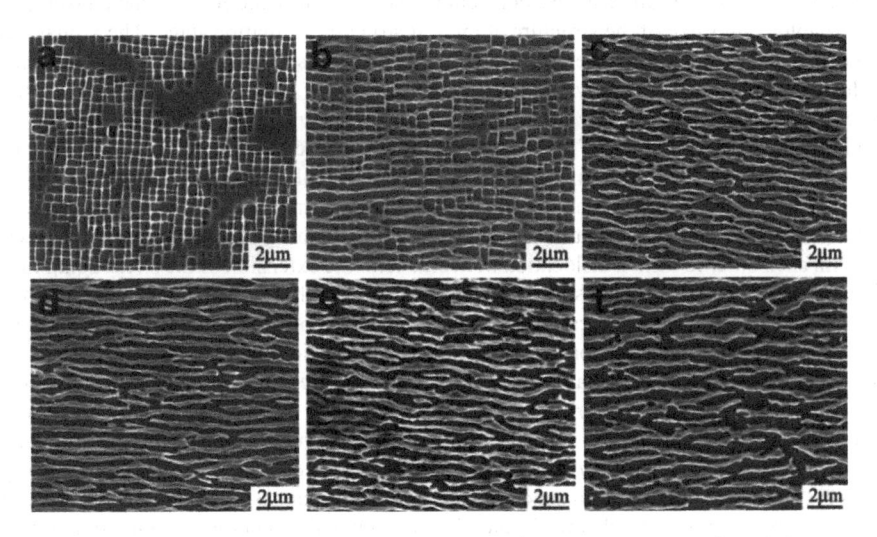

Figure 2. Microstructure evolution during stress aging under 1070°C/70 MPa. (a) 20 hours; (b) 50 hours; (c) 100 hours; (d) 200 hours; (e) 300 hours; and (f) 500 hours.

Figure 3 shows the evolution of the structure size of the experimental alloy during the stress aging process. The structural coarsening degree can be described by total width λ= T+S, which represents the coarsening of microstructure parallel to the stress axis, and T and S represent the

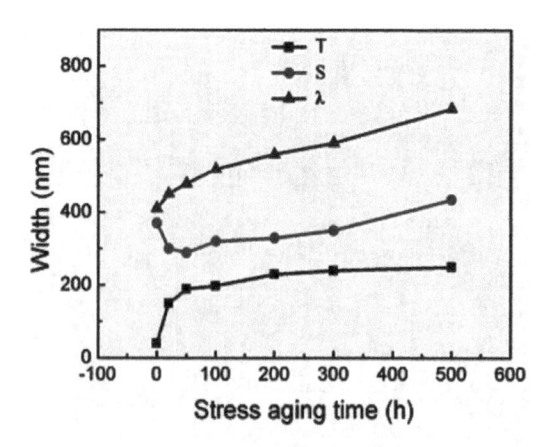

Figure 3. Size variation during the stress aging process where total width $\lambda = T + S$, T and S: average width of γ and γ' phase.

average width of γ and γ' phases, respectively. The coarsening degree of λ continues to increase in the stress aging process while it increases fast in the first 100 hs and it became slow after 100 hs. It indicates that the λ phase increases quickly before the formation of the raft and then increases gently with the increase of the stress aging time. The size of γ' phase, however, experience decrease in the first 50 hs and then increase after 50 hs. The coarsening of the structure was mainly attributed to the coarsening of the γ phase, which increased from tens of nanometers to nearly 200 nm in the first 50 hs stress aging.

The microstructure evolution of the single crystal alloy shown in Figure 2 and Figure 3 during the 1070°C/70 MPa stress aging process can be divided into two stages, namely, the raft structure formation stage and the raft structure coarsening stage. Significant changes have taken place in the morphology and size of the structure at the raft structure formation stage. The γ phase matrix channel rapidly coarsens, while the size of the γ' phase decreases in the beginning and then increases, and the total size coarsens rapidly. After stress aging for 50 h, the raft structure formed and total coarsening rate remained stable with the simultaneous γ and γ' phase increase.

Figure 4 shows the evolution of dislocations at the γ/γ' phase interface of the single crystal alloy during 1070°C /70 MPa stress aging. It can be seen that when the stress aging is up to 20 h, a small amount of a/2<011>{111} dislocations slip in the γ phase matrix channel. When the stress is aged for 50 h, the movement of dislocations at the γ/γ' phase interface is hindered, and the dislocations react with each other to form a dislocation network. With the extension of the stress aging time, the dislocation network still maintains a complete structure until the stress aging reaches 500 h. It can be seen that in the process of stress aging, the dislocation movement consists of two stages. The initial stage of stress aging is the formation stage of the dislocation network, and then the dislocation network enters the stabilization stage. In the initial stage of stress aging, under the combined effect of applied stress and mismatch stress between γ/γ' phase, the dislocations in the γ phase channel are activated, while a/2<011>{111} dislocations slip in the matrix. The number of dislocations in the channel increases, but they are blocked and accumulated when they slip to the two-phase interface. At the same time, the presence of external stress changes the internal stress field of the crystal, thereby changing the chemical potential gradient and promoting the directional diffusion of elements. The γ' phase forming elements such as Al and Ta diffuse in the opposite directions compared to the γ phase forming elements Cr and Mo. The directional diffusion of elements changes the morphology of the γ/γ' phase, so that the γ phase channel parallel to the stress axis gradually narrows and disappears. The γ phase channel perpendicular to the stress direction gradually widens. At the same time, the γ' phases are connected to each other in the horizontal direction to the raft completely in a short time. After the raft is formed, the γ phase vertical channel disappears, and

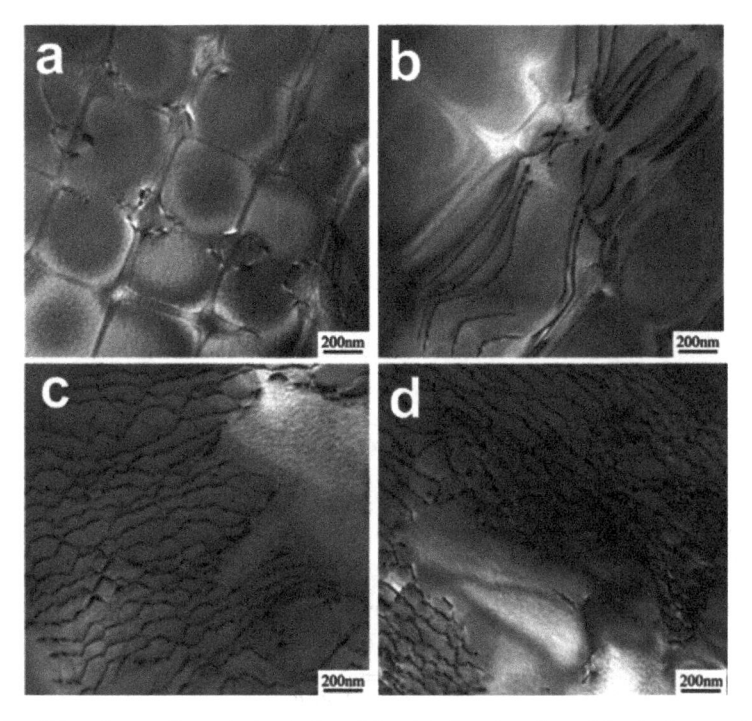

Figure 4. The dislocation configuration evolution of the γ/γ' phase interface of the single crystal alloy during the 1070°C/70 MPa stress aging process after (a) 20 hours, (b) 50 hours, (c) 200 hours, and (d) 500 hours B=[001], g=020.

the γ' and γ phases are reversedistributed, which indicates the end of the inter-diffusion process of the elements in the vertical channel. After that, the inter-diffusion of elements can only proceed at the γ/γ' phase interface parallel to the stress direction. In the coarsening stage of the raft, a dense and stable dislocation network formed at the γ/γ' phase interface prevents the dislocations in the matrix channel from slipping into the γ/γ' phase interface continuously. In the horizontal channel, the dislocation is blocked at the interface, when the dislocation slips and climbs along the γ/γ' phase interface, thereby reducing the dislocation movement rate, keeping the structure basically stable, and correspondingly reducing the coarsening rate.

Figure 5 shows the rupture life change curve of single crystal alloy after stress aging for different times. Obviously, the rupture life changes little compared with the initial rupture life until stress aging is up to 50 h. When the stress aging time reaches 200 h, the rupture life drops to about 320 h. With the extension of the stress aging time, the rupture life gradually decreases. When the stress aging reaches 500 h, the rupture life drops to about 200 h. It can be seen from Figure 5 that the rupture life of the alloy after stress aging decreases with the extension of the stress aging time. It can be seen that in the process of stress aging, the combined action of stress and temperature promotes the degradation and the coarsening of the structure and the rafting. The degradation of the structure will change the alloys' mechanical properties correspondingly. In the steady-state creep stage of nickel-based single crystal alloys under high temperature and low stress, the creep mechanism is the dislocation climbing mechanism, while the creep mechanism in the acceleration stage is the dislocation cut-in mechanism. It means a large number of dislocations in the matrix cut into the γ' phase which leads to creep rupture. During the stress aging process, the size of the γ phase channel increases with the extension of the stress aging time, and the channel size affects the movement of dislocations. The Orowan stress τ of the dislocation passing through the γ phase channel is closely

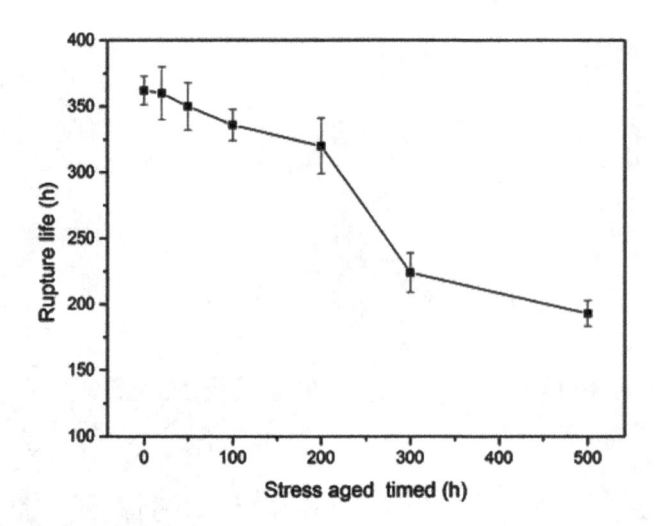

Figure 5. Rupture life change curve of single crystal alloy under 1070°C/140 MPa after pre-stress aging for different times under 1070°C/140 MPa.

related to the width of the γ channel, which is given by [17].

$$\tau = \frac{GB}{\lambda} = \sqrt{\frac{2}{3}} \frac{GB}{h} \qquad (1)$$

where τ is the critical shear stress, λ is the maximum channel size on the slip surface, G is the shear modulus, B is the Burgers vector, and h is the channel width. The γ phase channel is coarsened and the Orowan stress is reduced, which promotes the movement of dislocations in the matrix channel, so that the dislocations are more likely to overflow to form the dislocation network becoming a local stress concentration [18], which promotes the dislocation to cut into the γ' phase earlier, so that creep rupture occurs earlier. During the stress aging process, the cubic γ' phase gradually evolves into a raft shape, and a large number of dislocations are generated at the interface. The formation of dislocations effectively reduces the mismatch stress of the γ/γ' phase, and the rafting makes the partial coherent interface of the γ/γ' phase disappear at the same time. Therefore, it can be seen that the rafting reduces the precipitation strengthening ability of the γ' phase to a certain extent and reduces the strength of the alloy.

4 CONCLUSIONS

A second-generation nickel-based single crystal alloy was subjected to high temperature stress aging at 1070°C/70 MPa to observe and study the law of structure degradation. The rupture life under 1070°C/140 MPa was studied on alloys after stress aging for different times to build the relationship between structure degradation and rupture life.

1) The coarsening degree of λ continues to increase in the stress aging process while it increases fast in the first 100 hs and it became slow after 100 hs. The coarsening of the structure was mainly attributed to the coarsening of the γ phase, which increases from tens of nanometers to nearly 200 nm in the first 50 hs stress aging.
2) Microstructure evolution of single crystal alloy during the 1070°C/70 MPa stress aging process can be divided into two stages: the raft structure formation stage and the raft structure coarsening stage.
3) The rupture life of the stress aging structure in the raft-shaped stage is equivalent to the rupture life of the initial structure. The rupture life gradually decreases with the extension of the stress aging time during the coarsening stage.

4) The coarsening of the γ phase channel and the rafting of the γ/γ' phase organization after stress aging are the main factors for the decrease of the rupture life.

ACKNOWLEDGMENT

This work was supported by the Natural Science Foundation of China (No. 52001297).

CONFLICT OF INTEREST

The authors declare no conflict of interest.

REFERENCES

[1] R. C. Reed, The Superalloy Fundamentals and Applications, Cambridge University Press & Mechanical Industry Press, London, 2016.
[2] C.T. Sims, N.S. Stoloff, W.C. Hagel, Superalloys II: High Temperature Materials for Aerospace and Industrial Power, John and Wiley and Sons, New York, 1987.
[3] N.P. Padture, M. Gell, E.H. Jordan, Thermal Barrier Coatings for Gas-Turbine Engine Applications, Science, 296 (2002) 280–284.
[4] P. Caron, T. Khan, Evolution of Ni-based superalloys for single crystal gas turbine, Aerosp. Sci. Technol., 3 (1999) 513–523.
[5] T. Murakumo, T. Kobayashi, Y. Koizumi, H. Harada, Creep behaviour of Ni-base single-crystal superalloys with various γ' volume fraction, Acta Materialia, 52 (2004) 3737–3744.
[6] C. T. Sims, in The Superalloys, C. T. Sims, W. C. Hagel, Eds. (Wiley, New York, 1972), pp. 145–174.
[7] J. Sato, T. Omori, K. Oikawa, I. Ohnuma, R. Kainuma, K. Ishida, Cobalt-base high-temperature alloys, Science, 312 (2006) 90–91.
[8] M. Huang, J. Zhu, An overview of rhenium effect in single-crystal superalloys, Rare Metals, 35 (2016) 127–139.
[9] J. Chen, B. Zhao, Q. Feng, L. Cao, Z. Sun, Effects of Ru and Cr on γ/γ' microstructural evolution of Ni-based single crystal superalloys during heat treatment, Acta Metallurgica Sinica - ACTA METALL SIN, 46 (2010) 897–906.
[10] X. Guo, W. Zheng, C. Xiao, L. Li, S. Antonov, Y. Zheng, Q. Feng, Evaluation of microstructural degradation in a failed gas turbine blade due to overheating, Engineering Failure Analysis, 103 (2019) 308–318.
[11] A.M. Kolagar, N. Tabrizi, M. Cheraghzadeh, M.S. Shahriari, Failure analysis of gas turbine first stage blade made of nickel-based superalloy, Case Studies in Engineering Failure Analysis, 8 (2017) 61–68.
[12] T. Murakumo, T. Kobayashi, Y. Koizumi, H. Harada, Creep behaviour of Ni-base single-crystal superalloys with various γ' volume fraction, Acta Materialia, 52 (2004) 3737–3744.
[13] H. Gabrisch, D. Mukherji, Character of dislocations at the γ/γ' interfaces and internal stresses in nickel-base superalloys, Acta Materialia, 48 (2000) 3157–3167.
[14] Y. Zhao, J. Zhang, F. Song, Effect of trace boron on microstructural evolution and high temperature creep performance in Re-containing single crystal superalloys, Progress in Natural Science: Materials International, 30 (2020) 371–381.
[15] Z. Zhu, H. Basoalto, N. Warnken, R.C. Reed, A model for the creep deformation behaviour of nickel-based single crystal superalloys, Acta Materialia, 60 (2012) 4888–4900.
[16] D. Bürger, A. B. Parsa, M. Ramsperger, C. Körner, G. Eggeler, Creep properties of single crystal Ni-base superalloys (SX): A comparison between conventionally cast and additive manufactured CMSX-4 materials, Materials Science and Engineering: A, 2019, 762(5):138098.
[17] Pollock TM, Argon AS. Creep resistance of CMSX-3 nickel base superalloy single crystals. Acta Metall. Mater. 1992, 40(1):1–30.
[18] L., Agudo, Jácome, P., Nörtershäuser, C., et al. On the nature of γ' phase cutting and its effect on high temperature and low stress creep anisotropy of Ni-base single crystal superalloys. Acta Mater. 2014, 69:246–264.

Straw bale buildings: Experiments, setbacks, and potentials in China

M.X. Xie
Zhongnan University of Economics and Law, Wuhan, China

ABSTRACT: Straw of wheat, cotton, rice, and other crops poses serious environmental and agricultural problems. The technique of constructing straw bale buildings for residential use was experimented but suffered setbacks. The underlying factors include the drawbacks of straw bale buildings, unsuitability for urban areas, decreased housing demand in rural areas caused by the urbanization, costs, shortage of skilled personnel, and low image. However, straw bale buildings may have a big potential for tourism buildings. Improved straw bale building techniques can be used to restore buildings of ancient styles or exotic styles. Straw bale tourism buildings are more environment friendly and more likely to get governmental approval. In the future, straw stale buildings may be constructed for tourism instead of regular inhabitation.

1 INTRODUCTION

Straw, including crops straw, may be used as building materials. Until 1970s, straw was widely used as building materials in China. After being abandoned for some time, straw is reconsidered as building materials for better use of resources and environmental protection. This paper reviews China's exploration of using crop straw as building materials in the past years, assesses the relatively small success and relatively large setbacks, reveals the underlying factors for this situation, and makes some suggestions on future development.

2 EXPERIMENTS AND SETBACKS AS TO RESIDENTIAL BUILDINGS

2.1 Experiments of straw bale buildings

Straw, including crop straw such as rice straw, was directly used for building houses for thousands of years in China. Generally speaking, the building technology was rudimentary and the buildings were of low quality. With the economic development, such houses were gradually abandoned and replaced with brick buildings or concrete buildings. Since 1980s, straw was rarely used for new buildings. The limited number of straw buildings kept or restored are mainly used tourism. An example is the group of straw building of Wongding Village of Cangyuan County, Yunnan Province, a province in the southwestern part of China. Sadly, this group of straw buildings were burnt down in a fire accident on February 14, 2021 [1].

The sophisticated technique of constructing straw bale buildings was in fact introduced from outside. Adventist Development and Relief Agency (ADRA) played a major role in the introduction of this technique into China. At the turn of this century, ADRA initiated a project to construct new buildings with straw bales in cooperation with some local governments in North-eastern China. More than 600 straw bale buildings were constructed. Most of these buildings have brick-concrete frame construction with straw bale infill. One straw bale building was constructed with steel structure for experiment and exhibition. These building were constructed as pilot projects to demonstrate the feasibility and advantage of straw bale buildings [2].

2.2 Setbacks in promoting straw bale buildings as residential buildings

Although straw bale buildings yielded some success in some other countries [3], so far the exploration of straw bale buildings in China is generally a failure. The straw bale buildings are not well accepted and the building technique was not widely disseminated. Many buildings of the ADRA project are still there [2], but are not regularly inhabited. They are there mostly for demonstration rather than for inhabitation. After the ADRA project, few new straw bale buildings were constructed. This fact itself indicates that straw bales are not generally accepted as building materials.

2.3 Underlying factors of the setback

Various factors may account for the setback of promoting straw bale buildings. Some lie in straw bale buildings while some others economy or culture.

Firstly, straw bale buildings have some drawbacks. A recent research into the straw stale buildings of the ADRA project revealed the condensation issue and cracking issues. The design of these straw bale buildings failed to minimize thermal bridging through the straw bale wall [2].

Secondly, the straw bale buildings are only suitable for rural areas but the rapid urbanization in China significantly reduced the demand for new building in rural areas. Straw bale buildings are low and take much space. There are suitable for single family houses in rural areas but are not well suited for urbanized areas in China. Urbanized areas in China generally encourage high rises and discourage or even prohibit low-density low buildings. Most city dwellers live in multi-story condominium buildings instead of single-family houses. The rapid industrialization and urbanization draw a large percentage of young people from villages to cities [4]. As the rural population dwindles, the demand for new buildings in rural areas reduced significantly.

Thirdly, it is economically unworthy to construct straw bale buildings. Brick and concrete structure or steel structure is often necessary to make the straw bale building strong. It does not save much money to use straw bale as infill after paying for the cost of the brick and concrete structure or steel structure. Even worse, it may be even more expensive to build a high-quality straw bale building than a brick and concrete building.

Fourthly, there are very few specially trained engineers and construction personnel for high quality straw bale buildings. The ADRA project trained some personnel but the number is limited. Interview with the trainees of the ADRA project indicate that many of them could not find business with this training. Many of them returned or switched to the business of constructing building with bricks and concrete cement.

Last but not the least, buildings made with straw suffer a low image in China. No matter how much percentage of straw, buildings made with straw is generally regarded as a symbol of poverty in many parts of China. When farmers construct their new houses, they have no slightest interests in straw bale buildings.

3 POTENTIAL FOR TOURISM BUILDINGS

Straw bale buildings may have a big potential for tourism building which are not regularly used for residence, for the purpose of giving tourists a special experience and putting less pressure on the environment.

3.1 Restoration of ancient style buildings

The ancient China had a large number of straw buildings. Many of them enjoyed a high fame because a famous person lived in it or a famous story happened there. Even if there is no exact record of the ancient buildings, people may still construct ancient buildings based on the historical background, ancient building styles, the features of the scenery spots, and other considerations. The restored ancient buildings need also consider the current needs and requirements on accommodation, fire control, and other functions. These considerations in essence requires the construction of "modernized ancient buildings": on the one hand, they should have ancient style to give tourists an experience of ancient time; on the other hand, they should meet the basic needs of tourists and meet the basic requirements of modern building codes [5].

Straw bale buildings have a potential to meet these two-fold requirements. Straw bale buildings with straw roof have the appearance of ancient time. The reinforcement with iron mesh and the use of improved technology can make straw bale buildings stronger, safer, more durable, and more spacious. In sum, straw bale buildings have a big potential for the restoration of ancient style buildings. For instance, the restoration of Wongding Village should use improved straw bale building techniques while keeping the traditional style.

3.2 Construction of exotic buildings

Straw was used in the ancient time or is still used in the contemporary time by other cultures or countries. Different cultures have different style of straw buildings. Straw bale buildings of exotic style may give tourists an exotic experience. Therefore, it is advisable to construct straw bale buildings of exotic styles on beaches, theme parks, and other tourist resorts which features exotic styles.

3.3 Advantage in environmental protection

One advantage of straw bale buildings is that they are more environment friendly. Many tourist resorts are within or closed to protected areas. Even if they are not close or within protected areas, environmental protection is still needed to be considered. Straw bale buildings have much less environmental impacts in its construction process, in its use, and after its abandonment [6]. Therefore, they should be prioritized by tourism operators. In addition, the stringent environmental protection requirements also make government officials more inclined to prove tourist buildings of straw bale than brick and concrete [7].

4 CONCLUSIONS AND SUGGESTIONS

Constructing straw bale buildings for residential use generally has a deem prospect in China considering the disadvantages, needs, and costs. That being said, suggestions can still be made for its development in tourism. Well-designed straw bale building constructed with sophisticated technique may have a kind of occidental styles, ancient styles, or styles of minority nationalities. Therefore, it may be advisable to use straw bales to construct theme parks, restore ancient villages, or develop other types of tourist resorts. Besides, many tourist resorts are subject to restrictions on brick or concrete building for environmental considerations. The more environment-friendly straw bale buildings are more acceptable to the regulatory agencies.

REFERENCES

[1] Zhang Y 2021 *Yunnan Daily*. **2021**, February 15, p 1
[2] Yin X, Lawrence M, Maskell D 2018 *J. Build. Eng.* **18** 408–417, doi:10.1016/j.jobe.2018.04.009
[3] Zhu X.D, Wang F.H, Liu Y 2012 *Phys. Procedia* **32** 430–443, doi:10.1016/j.phpro.2012.03.582.
[4] You M 2015 *Hong Kong Law J.* **45** 621–650
[5] Wang M 2019 *Identification and Appreciation to Cultural Relics* **157** 082–083
[6] Zari M.P 2019 *Materials Today Sustainability* **3–4** 100010
[7] Lv Z 2019 *New Horizons of Environmental Law* (Beijing: China University Press) pp 249–255

Rock stability analysis for underground powerhouse of pumped storage power station based on block theory

J.Y. Li, J.L. Guo, H.N. Liu & R.C. Xu
College of Geosciences and Engineering, North China University of Water Resources and Electric Power, Zhengzhou, China

ABSTRACT: The stability of surrounding rock is one of the prominent problems in the excavation process of the underground powerhouse of pumped storage power station with large span and high side wall. Because it is hard to determine the location, shape, size and other characteristic parameters of the rock mass discontinuities, it is very difficult to find out all the possible unstable blocks in the surrounding rock. In this paper, according to the principles of block theory, using only the occurrence characteristics of the discontinuities, an abstract type analysis and an overall stability evaluation of the surrounding rock blocks of the cavern can be made. The calculation results of the engineering examples also show that the abstract type analysis results of the block theory have practical guiding significance for the selection of the tunnel routes and the support schemes.

1 INTRODUCTION

In order to adjust the peak stagger and improve the efficiency of the power grid, several pumped storage power stations have been built in China. However, the construction difficulty of the pumped storage power station is more and more complex, especially the underground powerhouse, which presents the trend of high side wall and long span. Therefore, during the excavation of the underground cavern, the stability of surrounding rock becomes one of the key problems while constructing pumped storage power station.

The block theory analysis method has great advantages in analysing the stability of surrounding rock of underground cavern. It not only requires less input parameters, but also the calculation results are very practical for support design, and the requirements for software and hardware resources for calculation are not high [1–3]. Block theory is put forward in the practice of hydropower engineering, and then further applied and developed in large hydropower projects [4–7]. In the analysis of block theory, the occurrence feature of discontinuities is an important input parameter. However, it is very difficult to obtain the accurate location, shape, size and other features of the discontinuities. Therefore, the mainstream method of identify the key block is generating the random discontinuities in terms of the principles of probability and statistics [8–10].

When the location of rock discontinuity is difficult to determine, the abstract type analysis is the greatest advantage of block theory according to the occurrence of discontinuities. The abstract type analysis is also of great significance for practical engineering, but it is a blind spot that has been ignored for a long time. Therefore, this paper analyses the surrounding rock stability of a underground powerhouse of a pumped storage station in Central China by using block theory. On the one hand, it has guiding role and reference value for the design and construction of the actual cavern project, on the other hand, it also clarifies some empirical problems in the process of applying block theory, and also extends the practical engineering applications.

2 ABSTRACT TYPE ANALYSIS WITH BLOCK THEORY

2.1 Project overview

The underground powerhouse of a pumped storage power station adopts the central development mode. The layout of the underground powerhouse avoids the relatively developed fault section, and is arranged in a strong mountain with a horizontal distance of 1450m from the water inlet to the outlet of the lower reservoir, with a buried depth of more than 550m. The longitudinal axis of the powerhouse is 110° and the excavation size of the main and the auxiliary powerhouse tunnels is 164.3m × 24.5m (26.0m) × 54.7m.

The surrounding rock of the cavern is late Yanshanian porphyry granite. The rock is compact and hard, and the rock mass integrity is good. It is mainly of class II ~ III, and the rock mass is generally slightly permeable. The geological structure in the area of the powerhouse is simple. There are mainly four groups of joints and fissures. The first two groups of joints are relatively developed. Most of the exposed length runs through the three walls. Most of the surfaces are covered with varying degrees of rust. Some of them are filled with altered fragments and debris. Water seepage is common along the joint surface. Six faults may be encountered in the powerhouse area, among which three obvious groups are mainly considered. The maximum width of the fault zone is 50cm ~ 70cm, all of which are fractured faults with strong alteration along the fault plane. The occurrence of joint faults and the calculated friction angle are shown in Table 1.

Table 1. Occurrence characteristics and mechanical index values of discontinuities.

Discontinuities	Dip Angle (°)	Dip Direction Angle (°)	Friction Angle (°)
J1	80	60	22
J2	85	45	22
J3	87	25	22
J4	75	343	22
F93	42	353	16
F97	76	50	16
F98	69	346	16

The in-situ stress field in the project area is mainly controlled by the gravity. According to the statistical results of joints and fissures in the powerhouse area, the representative discontinuities with large scale, long extension, poor properties, small intersection angle with the powerhouse axis are selected for mutual combination, and the stability of the surrounding rock block formed by cutting with the underground powerhouse is analysed.

2.2 Analysis of key blocks

Before the actual excavation of underground cavern, the block theory can analyse the abstract type of the overall project area according to the geological exploration data, especially the dip angle and the dip direction angle of discontinuities. The significance of abstract analysis is to understand all the block types, sliding mode, sliding direction and sliding force vector in the surrounding rock of underground cavern before excavation, which has an important reference for the selection of cavern axis and support measures.

With the excavation of underground cavern, the information of discontinuities revealed is more abundant. Especially when the position is determined, the positioning block analysis can be carried out in the local area. And the analysis results will be more specific, such as the actual shape and size of the block, the actual sliding direction of the block and the amount of the net sliding force can be determined. So more targeted measures for the excavation and support of the cavern can be taken to ensure the stability of the rock blocks.

Taking one of the discontinuities combinations (J2, J4, F97) as an example, the key block method is used to analyse the surrounding rock stability of underground powerhouse of pumped storage power station.

The discontinuities combination (J2, J4, F97) can cut the space into 8 parts at most, that is, corresponding to 8 types of joint pyramid (JP), namely (000, 001, 010, 011, 100, 101, 110, 111), as shown in Table 2. They correspond to 8 kinds of block movement modes, which are 1 wedge-shaped stable block, 1 shedding block, 3 single-sided sliding blocks and 3 double-sided sliding blocks. The combination of the joint pyramid and the free plane of the roof or side wall of the cavern forms a tetrahedral block. According to the calculated safety factor or net sliding force, whether the block is a key block can be determined.

Table 2. Discontinuities combination (J2, J4, F97) and its related block type.

Joint Pyramid	Sliding Mode	Sliding Force	Safety Factor	Angle Interval (°)	Maximum Sliding Area	Block Types
00	23	0.83	0.13	252.85–259.36	0.00	Real Key Blocks (very small)
001	2	0.86	0.11	259.36–305.26	171.09	Real Key Blocks (left wall)
010	3	0.90	0.07	–	–	Infinite Block
011	1	0.96	0.04	72.85–305.26	204.96	Real Key Blocks (roof)
100	-1	0.00	0.13	125.26–252.85	332.80	Stable Blocks (floor)
101	12	0.83	0.14	–	–	Infinite Block
110	13	-2.86	7.05	79.36–125.26	171.00	Potential Key Blocks (right wall)
111	0	1.00	0.00	72.85–79.36	0.00	Real Key Blocks (very small)

It can be seen from table 2 that the blocks formed by the combination of discontinuities (J2, J4, F97) have the following characteristics respectively.

The whole stereographic projection is used for analysis. The analysis results are shown in Figure 1 and Figure 2.

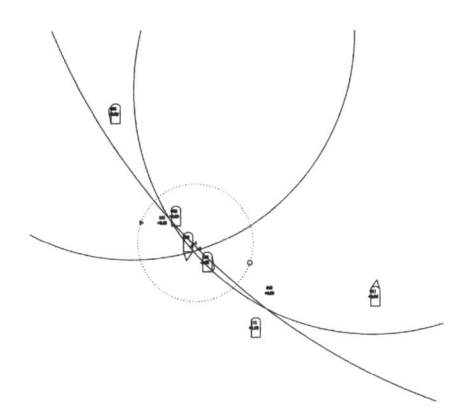

Figure 1. Joint pyramid and net sliding force by the whole stereographic projection.

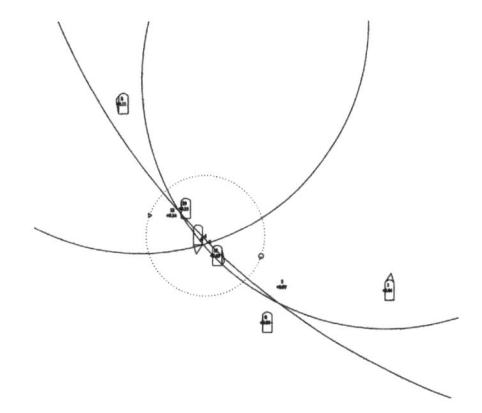

Figure 2. Sliding mode and safety factor by the whole stereographic projection.

The blocks corresponding to JP 010 and 101 are not completely cut on the surface of the cavern and are connected with the parent rock, so they belong to infinite blocks.

The block with JP of 011 is distributed in the roof, as shown in Figure 3. It slides along the single side of J2, with large net sliding force, low safety factor and large volume. It belongs to the most dangerous key block, so the support measures should be strengthened in the construction.

The block with JP of 100 is distributed in the floor of the cavern, as shown in Figure 4, which belongs to stable block.

The block with JP 001 is distributed on the left wall, as shown in Figure 5, sliding along J4 single side, with large anti sliding force and safety factor.

The block with JP 110 is distributed on the right wall, as shown in Figure 6, sliding along both sides of J2 and f97, the net sliding force is not large, and the volume is not large.

The blocks with JP of 000 and 111 have larger net sliding force and smaller safety factor, but their volume is very small and can be ignored.

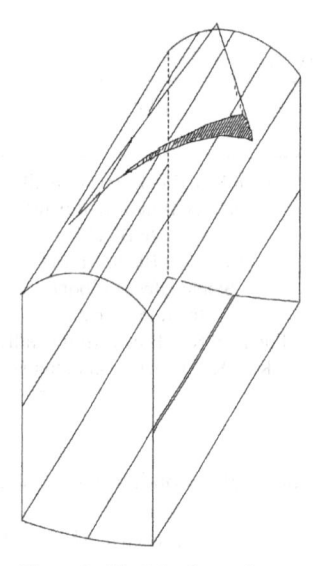

Figure 3. block in the roof.

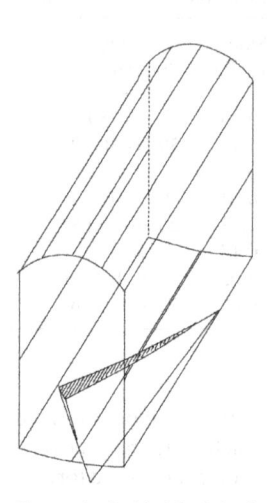

Figure 4. Stable block in the floor.

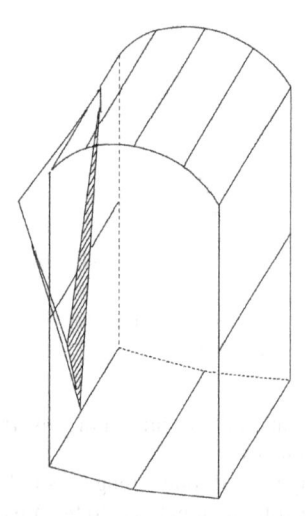

Figure 5. Key block in the left wall.

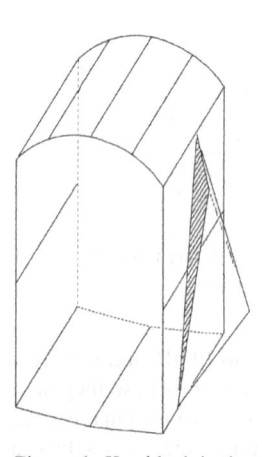

Figure 6. Key block in the right wall.

2.3 Distribution of key blocks

There are mainly 4 groups of joints and 3 faults in the area where the project is located, including 7 groups of discontinuities. In practical engineering rock mass, most of them are tetrahedral blocks cut by the combination of three groups of discontinuities and the free planes of underground cavern excavation. In order to study all the possible key blocks in the underground powerhouse area more comprehensively, the enumeration method is used to analyse the key blocks. There are 7 different types of planes, and every 3 planes can be combined. There are 35 combinations, and each combination of the planes can form 8 types of joint pyramids, so 280 block types can be obtained.

After excavation, local positioning block analysis is carried out after adding the location information of the planes, which is a derivative type of overall abstract type analysis and has all the characteristics of abstract block in overall type analysis.

The underground powerhouse belongs to high side wall and large-span underground cavern. According to the results of block theory analysis, all blocks with relatively large volume in surrounding rock and having potential threat to engineering safety can be found. According to the table 3, the key blocks are mainly distributed in the roof, floor, left wall and right wall of the cavern, and across the roof and left wall of the cavern, across the floor and right wall of the cavern. At the same time, it can be seen that there is no non convex block across the roof and right wall of the cavern, and across the floor and left wall of the cavern, so these two areas are relatively stable and safe.

Table 3. Number of blocks of different types distributed in different locations of caverns.

Location	Real Key Block	Potential Key Block	Stable Block
Roof	19	0	0
Across roof and left wall	14	1	0
Left Wall	8	1	0
Right Wall	5	1	4
Floor	0	0	20
Across floor and right wall	0	0	14

Note: (1) There are 2 small, 3 medium and 3 large blocks inside the left wall.
(2) The volume of blocks inside the right wall is small.

3 CONCLUSION

By using block theory method, through the combination of the discontinuities distributed in the area of the powerhouse, all the key blocks having potential threat to the project safety have been identified.

There are many key blocks distributed inside the roof of the underground powerhouse, and the volume of the key blocks is large, the net sliding force is large, and the safety factor is small. The roof is the most dangerous part in the underground cavern, which should be paid special attention to. There are several non-convex or curved key blocks across the roof and the left wall of the underground powerhouse, and the volume of the key blocks is large, the net sliding force is large, and the safety factor is small. There are some key blocks distributed inside the left wall of the underground powerhouse, but some of the real key blocks are small in volume. For some larger blocks, sliding may occur. In short, it is necessary to strengthen the support measures for the roof, left wall and upper left corner edge of the underground powerhouse to ensure the stability of surrounding rock.

There are some key blocks inside the right wall of underground powerhouse, but the volume of real key blocks is small, the anti-sliding force of potential key blocks is large, and there are some stable blocks scattered. The blocks distributed inside the floor of underground powerhouse are all stable blocks in the gravity stress field. The blocks distributed across the floor and right wall of underground powerhouse are also stable blocks in the same case.

The theoretical analysis of key blocks based on the occurrence of discontinuities belongs to abstract type analysis, that is, as long as one block type is composed of some types of planes, the kinematic and mechanical characteristics of corresponding actual blocks with detailed shape, size and position belong to the corresponding block type. The maximum sliding area on the section of underground cavern is the limit. When the position of the planes is determined, the projection of the key blocks cut by the planes must be a subset of the maximum sliding area. The net sliding force is the eigenvalue calculated according to the normalization of the active resultant force vector, which represents the direction the force.

The dip angle, dip direction angle and position of discontinuities are the key geometric factors affecting the stability of blocks. According to the measured data of the occurrence of joints and fissures, grouping depends on experience, and the selection of the combination of discontinuities cutting each other also depends on experience. The advantage of the exhaustive method used in this paper is that the analysis is complete, but it brings the problem of a large amount of calculation. If combined with the location and distribution characteristics of the discontinuities in the site, multiple discontinuities which are impossible to intersect can be removed, therefore, the computation amount can be reduced, and the efficiency is improved.

ACKNOWLEDGEMENTS

This work is sponsored by Natural Science Foundation of Henan (No.182300410109). We are grateful to the State Power Economic Research Institute for providing original data. We also greatly appreciate Prof. Genhua Shi's useful discussions.

REFERENCES

[1] Goodman R E and Shi G H 1985 *Block theory and its application to rock engineering* (New Jersey: Prentice-Hall)
[2] Liu Jinhua and Lv Zuheng 1988 *Application of block theory in stability analysis of engineering rock mass* (Beijing: Water Conservancy and Electric Power Press)
[3] Li Jianyong, Xiao Jun and Wang Ying 2010 Simulation method of rock stability analysis based on block theory *Computer Engineering and Applications* **46(21)** p 4
[4] Pei Juemin and Shi Genhua 1990 Analysis of key blocks in underground powerhouse cavern of hydropower station *Journal of rock mechanics and engineering* **9(1)** p 11
[5] I. M. Lee and J. K. Park 2000 Stability analysis of tunnel key block: a case study *Tunneling and Underground Space Technology*, **15(4)** p 453
[6] Huang Zhengjia, Wu Aiqing and Sheng Qian 2001 Usage of block theory in the Three Gorges Projects *Chinese Journal of Rock Mechanics and Engineering* **20(5)** p 648
[7] Liu Fuyang, Zhu Zhende and Sun Shaorui 2006 Block Theory and Its Application in Surrounding Rock Stability Analysis in Underground Opening *Chinese Journal of Underground Space and Engineering* **2(8)** p 1408
[8] Lu Bo, Chen Jianping and Wang liangkui 2002 Automatic search of complex finite block based on 3D network simulation and determination of its spatial geometry *Journal of rock mechanics and engineering* **21(8)** p 232
[9] Lu Bo, Ding Xiuli and Wu Aiqing 2007 Study on the classification method of occurrence data of random discontinuities in rock mass *Journal of rock mechanics and engineering* **26(9)** p 1809
[10] Wu Aiqing, Zhang Qihua 2005 Geometric identification of stochastic block in block theory *Journal of Hydraulic Engineering* **36(4)** p 426

Analysis of influencing factors of polymer organo chromium gel crosslinking density

X. Zhao
Exploration and Development Research Institute, Daqing Oilfield Company Limited, Daqing, China

ABSTRACT: At present, the polymer organochrome gels have been widely used in oil field EOR technology, and the gel crosslinking density is an important property of gel materials, which has an important influence on gel gelatinization performance. Therefore, the influence factors of gel crosslinking density have been studied. The effects of polymerization cross ratio, temperature, pH, Nacl concentration, Cacl2 concentration and other factors on crosslinking density were revealed by indoor chemical simulation method. The results show that the influence of poly-cross ratio and pH on crosslinking density is optimal. The crosslinking density increases with the increase of temperature. Nacl concentration increased and cross-linking density increased. Cacl2 concentration decreases and cross-linking density increases.

1 INTRODUCTION

The crosslinking density of the gel has an important influence on the mechanical properties of the gel material and the response of the gel to external stimuli. The cross-linking density reflects the number of cross-linking points in the microscopic view, and is closely related to the strength and dehydration degree of the gel in the macroscopic view. Therefore, the cross-linking density is an important parameter of the cross-linking system. The gel cross-linking density is affected by many factors. Through laboratory experiments, the influence of factors such as poly-to-cross ratio, temperature, pH, Nacl, $Cacl_2$, etc. on the gel cross-linking density is analyzed, and the optimal poly-to-cross ratio is determined. And pH, increase the cross-linking density of the gel, and provide a favorable reference for the formation of a uniform and stable gel.

2 EXPERIMENTAL PART

2.1 Laboratory equipment

Ultraviolet-visible spectrophotometric determination: Measured with UV-7211 ultraviolet-visible spectrometer from Shanghai Analytical Instrument Factory, and the measurement wavelength range is 180~860nm. Measure the light transmittance during the gel formation process to detect the ioncontent.

2.2 Detection method

Detection of weak gel crosslinking agent functional group: Weigh 0.2g sample and dissolve it in 30mL 1mol/L caustic soda solution in a 100mL volumetric flask, and use distilled water to bring the volume to the mark. Take out 25 mL of the resulting solution and place it in a 250 mL conical flask, add 50 mL of 0.1 mol/L potassium iodide solution, cover the flask, and place the mixture for 30 minutes. Then acidify with 10mL of 1mol/L hydrochloric acid solution, add 3~4 drops of 1% starch solution, and titrate the excess iodine with 0.1mol/L solution.

The percentage X of the sample is calculated according to the following formula:

Where: $\chi = \frac{(v_a-v_b) \times c \times 0.01501 \times 100}{25 \times m} \times 100\% = \frac{0.0606(v_a-v_b) \times c}{m}$

v_a—The number of milliliters of sodium thiosulfate consumed during the titration of the blank experiment;

v_b—The number of milliliters of sodium thiosulfate consumed during the titration of the test solution;

c—The equivalent concentration of sodium thiosulfate, mol/L;

3 RESULTS AND DISCUSSION

The effective content of the weak gel polymer was determined to be 91% by chemical recipitation and drying method, and the mass percentage of the functional group of the crosslinking agent was 13.7% by iodometric titration.

3.1 The effect of polymer/crosslinker functional group molar ratio on polymer gel crosslink ensity

Fix the polymer concentration to 3000mg/L, change the functional group concentration to determine the unreacted functional group concentration when different polymer/crosslinker functional group molar ratios are determined, and calculate the crosslinking density according to the formula 1 method, and the result is shown in Figure 1.

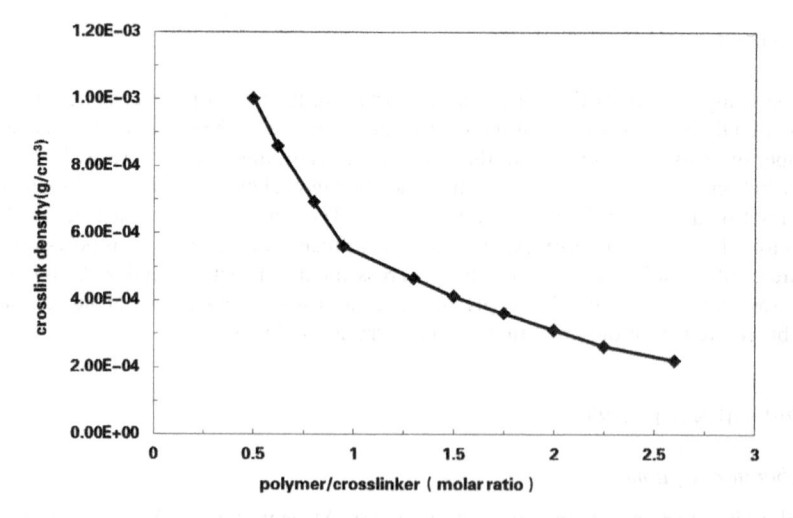

Figure 1. The effect of polymer/crosslinker functional group molar ratio on polymer gel crosslink density.

It can be seen from Figure 1 that when the polymer concentration is fixed, as the polymer/functional group molar ratio increases, that is, the crosslinker functional group concentration decreases, the gel crosslink density decreases. The mass ratio of polymer to crosslinking agent is about 1:3.66. The minimum polymer/crosslinking agent functional group molar ratio in this experiment is 1:1.87. The concentration of polymer reactive functional groups is greatly excessive, so as the crosslinking agent concentration decreases,Or the increase in the molar ratio of polymer/crosslinker functional groups, the decrease of crosslinking points and the decrease of crosslinking density.

In experiments, it was found that the two systems with polymer/crosslinker functional group molar ratios of 1:1.87 and 1:1.50 gelled faster during the gel preparation process, because the concentration of the crosslinker was large, indicating that local over-crosslinking occurred. And the polymer/crosslinker functional group molar ratio of 1.30:1 and 2.57:1 respectively formed a uniform and stable gel. The crosslinking density of the first two systems is relatively high, respectively 1.01×10^{-3} g/cm^3 and 8.69×10^{-4} g/cm^3, so that the local crosslinking density of the gel system is too large, and the measurement error is large at the same time. The cross-linking density of the latter two systems is relatively low, 4.79×10^{-4} g/cm^3 and 2.43×10^{-4} g/cm^3, respectively, indicating that under the conditions studied, when the cross-linking density exceeds 8.69×10^{-4} g After this point /cm^3, the gel system becomes unstable and dehydration occurs.

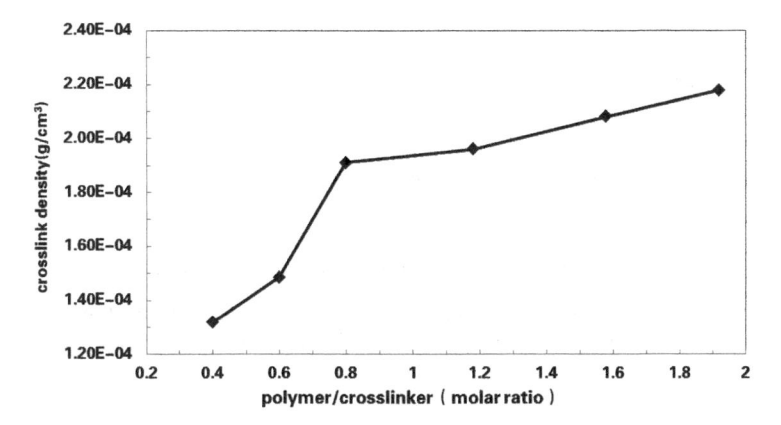

Figure 2. The effect of polymer/crosslinker functional group molar ratio on polymer gel crosslink density.

Figure 2 shows the effect of fixed crosslinking agent concentration at 1800mg/L, changing polymer concentration, polymer/crosslinking agent mass ratio and polymer/crosslinking agent functional group molar ratio on crosslinking density. It can be seen that as the polymer/crosslinker functional group molar ratio increases, that is, as the polymer concentration increases, the crosslink density first increases sharply and then tends to a constant value. The turning point is when the polymer/crosslinker functional group molar ratio is 1:0.8, this value is the cross-linking ratio. The amide group -CONH$_2$ in the polymer and the -CH$_2$OH in the crosslinking agent undergo a condensation reaction to crosslink into a gel. The polymer and the crosslinking agent react according to the following equation.

$$\sim\!\!CH_2CR\!\!\sim + R'CH_2OH \longrightarrow \sim\!\!CH_2CR\!\!\sim + H_2O$$
$$\quad\quad |\quad\quad\quad\quad\quad\quad\quad\quad\quad\quad\quad |$$
$$\quad CONH_2 \quad\quad\quad\quad\quad\quad\quad\quad CONHCH_2R'$$

3.2 The influence of pH value on the crosslinking density of polymer gel

The fixed polymer concentration is 800 mg/L, the cross-linking agent concentration is 1800 mg/L, and the effect of pH on the cross-linking density is shown in Figure 3. It can be seen from Figures 3 that in the studied pH range of 8–12, the crosslinking density first increases and then decreases with the increase of pH, which indicates that excessive pH is not conducive to the polymer and crosslinking agent. Cross-linked. The gel does not gel when the pH is lower than 7. The main reason is that increasing the pH value speeds up the cross-linking reaction of the cross-linking agent, thereby affecting the probability of collision and bonding between the polymer and the cross-linking agent, thereby increasing the cross-linking density and gel-forming properties. However, too high a pH value will cause the local cross-linking density of the three-dimensional network structure to be too large, thereby affecting the uniformity and stability of the gel structure. Under acidic conditions, the cross-linking agent will be inactivated due to side reactions such as hydrolysis and reduce the gel forming performance of the gel.

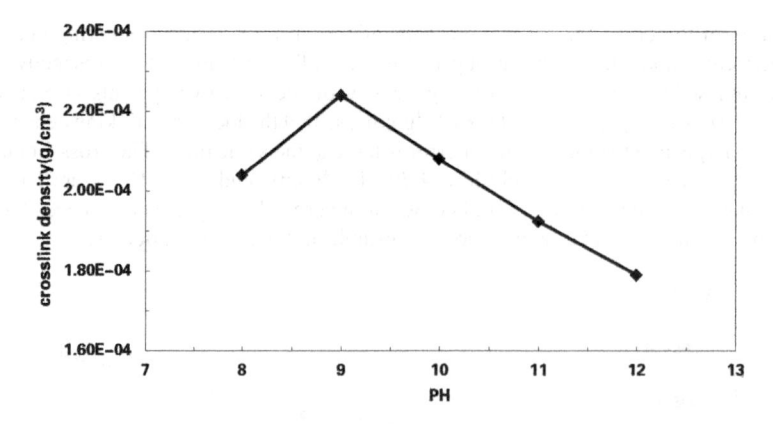

Figure 3. The effect of pH on the crosslinking density of polymer gels.

3.3 *The influence of temperature on the crosslinking density of polymer gel*

The fixed polymer concentration is 800 mg/L, the cross-linking agent concentration is 1800 mg/L, and the effect of cross-linking temperature on cross-linking density is shown in Figure 4.

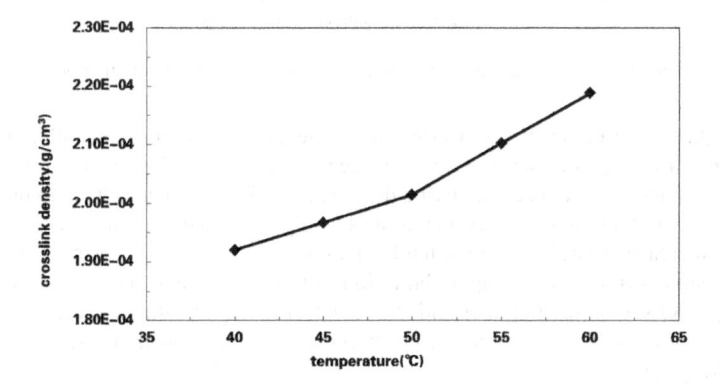

Figure 4. The effect of temperature on the crosslinking density of polymer gels.

It can be seen from Figure 4 that when the temperature increases from 40°C to 60°C, the value of the crosslink density gradually increases, which shows that the increase in temperature is beneficial to the crosslinking of the polymer and the crosslinking agent. Mainly because the temperature rises, the thermal movement of molecules increases, and the probability of collision and bonding between the polymer and the crosslinking agent increases. In the gel system, increasing the temperature is conducive to the formation of a uniform and stable gel.

3.4 *The effect of Nacl on the crosslinking density of polymer gel*

The fixed polymer concentration is 800 mg/L, the cross-linking agent concentration is 1800 mg/L, and the effect of Nacl concentration on the cross-linking density is shown in Figure 5. It can be seen from Fig. 5 that the increase of the Nacl concentration is beneficial to the cross-linking of the polymer and the cross-linking agent, so that the cross-linking density value increases. Therefore, in the gel system, containing a certain concentration of inorganic electrolyte is beneficial to the formation of a uniform and stable gel.

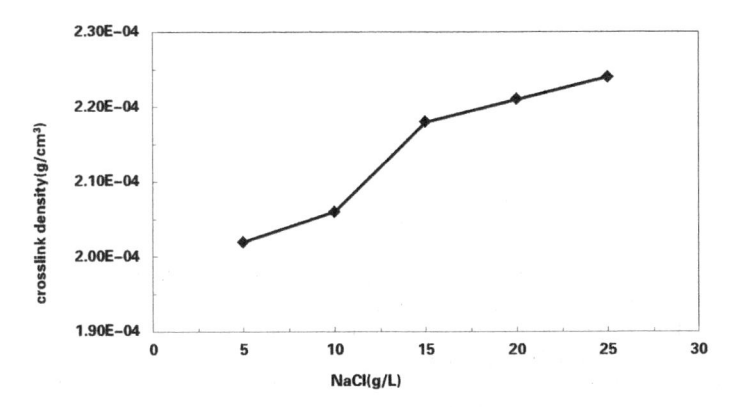

Figure 5. The effect of Nacl on the crosslinking density of polymer gels.

3.5 Effect of Cacl2 on the crosslinking density of polymer gel

The fixed polymer concentration is 800mg/L, the crosslinking agent concentration is 1800mg/L, and the influence of $CaCl_2$ concentration on the crosslinking density is shown in Figure 6.

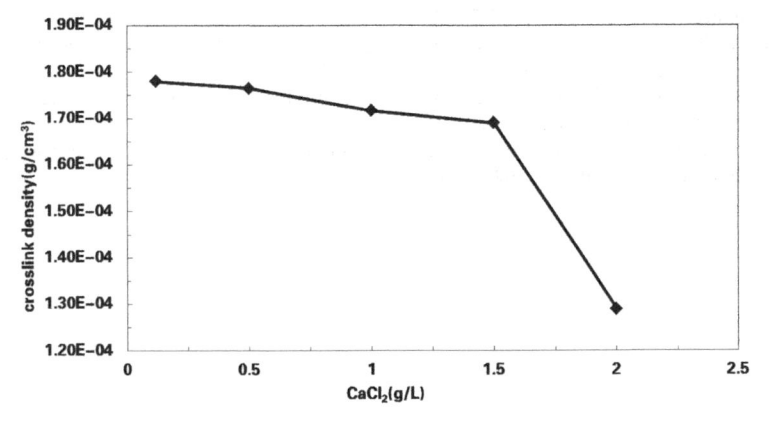

Figure 6. The effect of Cacl2 on the crosslinking density of polymer gels.

It can be seen from 7 that the increase in the concentration of $CaCl_2$ reduces the crosslinking density value, which is not conducive to the crosslinking of the polymer and the crosslinking agent. Therefore, in the gel. In the system, containing a certain concentration of inorganic electrolyte $CaCl_2$ is not conducive to the formation of a uniform and stable gel.

4 CONCLUSION

When the polymer/crosslinking agent functional group molar ratio is 1:0.8, it is the optimal poly-cross ratio. Gel will not form under the condition of pH value lower than 7, too high pH value will cause the local cross-linking density of the three-dimensional network structure to be too large. When the pH value is 9, the gel cross-linking density is the highest. Temperature has a certain effect on the cross-linking density of the gel, and increasing the temperature is beneficial to the formation of a uniform and stable gel. The increase in Nacl concentration increases the cross-linking density value, which is conducive to the cross-linking of the polymer and the cross-linking agent. As the concentration of $CaCl_2$ increases, the value of crosslinking density decreases. In the gel system, a certain concentration of inorganic electrolyte $Cacl_2$ is not conducive to the formation of a uniform and stable gel.

REFERENCES

[1] Li X J, Hou J R, Yue X A, et al. Effects of shear and absorption on in-depth profile control and oil displacement of weak gels[J]. Journal of China University of Petroleum(Edition of Natural Science), 2007, 31(6): 147–151.
[2] Xiong C M, Liu Y Z, Huang W, et al.Satus and solution of deep fluid diversion and profile control technique[J]. Oil Drilling & Prpduction Technology, 2016, 38(4): 504–509.
[3] Li C L, Zhao H L, Xin Q W, et al. Research and application on deep profile control technology in complex fault block reservoir[J]. Oil Drilling & Prpduction Technology, 2013, 35(5): 104–106.
[4] Yin L P,Xu F H,Ma D X,et al.Research and application of periodic injection of weak gel technology[J]. Petroleum Drilling Techniques, 2011, 39(4):95–98.
[5] Yang J Z. Xu G R, Liu G P, et al.Research and application of weak gel profile control technology in Bohai oilfield[J]. Well Testing, 2017, 26(3):58–59.
[6] Pan G M, Zhang C Q, Liu D, et al. Weak-gel profile-control flooding to enhance oil recovery in offshore heavy-oil reservoir[J]. Acta Petrolei Sinica, 2018, 25(3): 141–143.
[7] Pan W Y, Liu H, Tang X F, et al. Research on profile control charactedsfies of weak gel by multi-spots pressure measurement [J]. Petroleum Geology and Recovery Efficiency, 2009, 16(4): 83–85.
[8] Luo W L, He C Q, Feng L J, et al. Study and evaluation of high stability foamed gel[J]. Journal of Petrochemical Universities, 2018, 31(3): 28–34.
[9] Zhou D Y, Zhao J, Wang J, et al. Application of deep profile control technology in Lunnan oilfield, Tarim basin[J]. Xinjiang Petroleum Geology, 2014, 35(4): 457–460.
[10] Tang X F, Wu Q, Liu G F, et al. Field test on integral modifying and flooding project in regional reservoir with weak gel[J]. Acta Petrolei Sinica, 2003, 24(4): 58–61.
[11] Xiong C M, Tang X F. Technologies of water shut-off and profile control: An overview[J]. Petroleum Exploration and Development T, 2007, 34(1): 83–88.

Influence of SMA brace arrangement on seismic performance of frame structure

Z.L. Yang, J.L. Liu & J.Q. Lin
Key Laboratory of Earthquake Engineering and Engineering Vibration, Institute of Engineering Mechanics, China Earthquake Administration, Harbin, China

ABSTRACT: Shape memory alloy is an ideal energy dissipation material for engineering structures because of its good energy dissipation capacity. SMA braces can improve the seismic performance of frame structures, and the arrangement of braces has a great influence on the seismic performance of frame structures. In this paper, the interstorey drift ratio and the base shear force of the structure are selected as the evaluation criteria, and the dynamic time-history analysis method is used to study the influence of six SMA braces on the seismic performance of the structure. The results show that SMA brace arrangement 1 and arrangement 4 have obvious advantages.

1 INTRODUCTION

Steel frame structure is a structure form widely used at present. Traditional steel frame structure may produce large lateral deformation under earthquake action, which is easy to cause large residual displacement or damage [1]. SMA is a kind of good energy dissipating material, and SMA braces can effectively improve the seismic performance of steel frame structures, which has outstanding advantages over the buckling restrained braces and traditional structures [2]. Common since the restoration of frame structure way of supporting configuration with v-shaped, inverted V and diagonal bracing, etc., for the same steel frame structure, different ways of brace arrangement on seismic performance of structures, affect the brace force and the stress of the beam. As a result, this paper studies different SAM brace arrangement of structure seismic performance influence.

2 SMA CONSTITUTIVE MODEL

2.1 *SMA experiment*

The SMA wire with a diameter of 2mm is Ti-55.82at%Ni, purchased from Xi 'an Saite Siwei Metal Materials Co., Ltd., China. After cyclic training, the mechanical properties of SMA wires are stabilized. Then, amplitude variation experiments are carried out. The experimental loading frequency was 0.4Hz, and the amplitude was 1%, 2%, 3%, 4%, 5% and 6%, with each amplitude cycling for five times. The test was carried out at the Structural and Seismic Experimental Center of Harbin Institute of Technology.

Figure 1 shows the test results of mechanical properties of SMA under different amplitudes. As can be seen from Figure 1, under the same amplitude condition, the SMA curve barely changes, indicating that its mechanical properties are stable. This stable SMA can be directly used for structure controlled control to provide stable control for the structure [3].

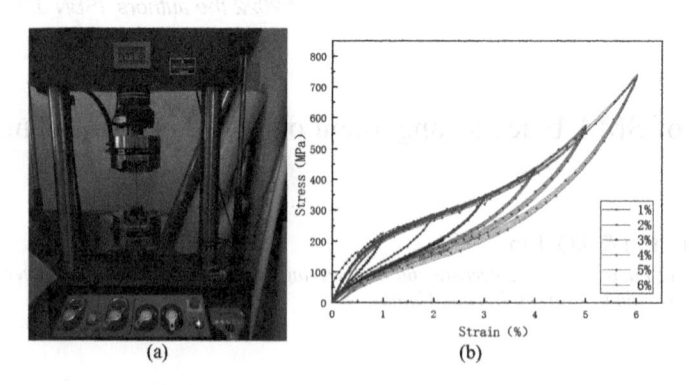

Figure 1. SMA performance test.

2.2 *SMA constitutive model establishment and parameter values*

As can be seen in Figure 1, the mechanical property parameters of the SMA experimental curve are shown in the following Table 1.

Table 1. SMA parameters.

	Parameters	Values
Stress	Start stress of martensitic transformation	212.38MPa
	Finish stress of martensitic transformation	427.89MPa
	Start stress of reverse martensite transformation	202.76MPa
	Finish stress of reverse martensite transformation	55.56MPa
	Maximum stress	737.44MPa
Stiffness	Elastic modulus of austenite	22262.05MPa
	Martensite modulus of elasticity	24470.65MPa
	Stiffness of martensite strengthening segment	14614.35MPa
Strain	Maximum residual strain	0.022
	Maximum strain	0.060

The VUMAT interface of ABAQUS platform is used to develop the subroutine, and PROTRAN language is used to write the material constitutive model subroutine, and the finite element simulation is obtained as shown in Figure 2. As can be seen in Figure 2, the constitutive model subroutine is in good agreement with the material experiment results, indicating that the material model subroutine written can correctly reflect the mechanical properties of material.

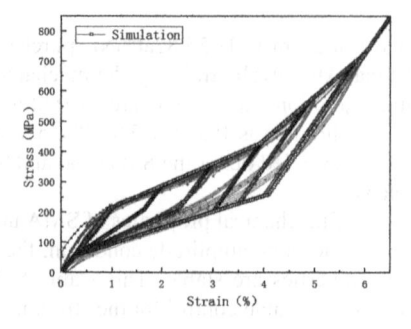

Figure 2. Finite element simulation of SMA performance.

3 OVERALL ARRANGEMENT OF STRUCTURE

The calculation example is a four-storey steel frame structure, the storey height is 3.9m, the span length is 7.8m, the plane size is 31.2m×31.2m, the dead load is $5kN/m^2$, the live load is $2kN/m^2$, the seismic fortification intensity of the structure is VIII degree, the surpassing probability is 10% within 50 years, the seismic design acceleration is 0.2g, and the site type is II. According to China's current seismic design code [4,5], each floor and each direction of the structure are equipped with four SMA braces. In order to compare the seismic performance of various brace arrangement forms, six brace arrangement forms are set, as shown in Figure 3.

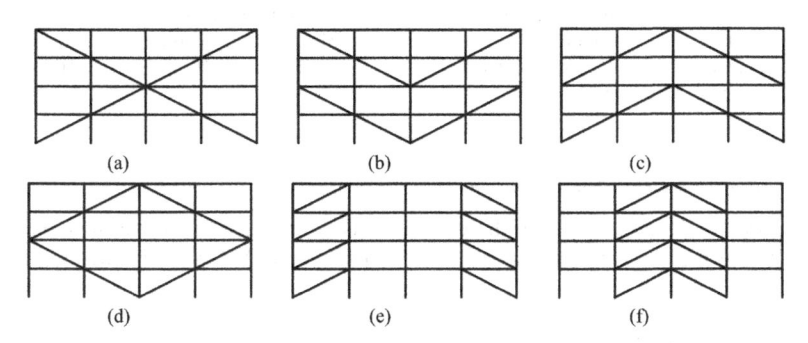

Figure 3. Arrangement of SMA braces: (a) BFS1, (b) BFS2, (c) BFS3, (d) BFS4, (e) BFS5, and (f) BFS6.

In order to compare the seismic performance of SMA self-centering braced structure, a pure frame structure was designed. The structural beams and columns are all made of Q235 steel, and the elastoplastic strengthening model is adopted. The elastic modulus of the steel is 206GPa, the yield strength is 235MPa, the hardening rate after yield is 2%, and the maximum ductility is 15. The beam and column sections are all H-shaped sections. The final beam-column section information and braces parameters of pure steel frame structures (SFS) and six SMA braced frame structures (BFS) are shown in Table 2.

In order to make the deformation of the braces concentrated in the SMA material, the other parts of the braces are rigid materials. The braces are only subjected to axial force, and their two ends are hinged with the beam-column joints [6].

Table 2. Beam-column section information and braces parameters.

	BFS			SFS	
Story	SMA length(m)	Area of SMA (mm^2)	Column (mm)	Column (mm)	Beam (mm)
1	0.78	1366.99	HW400×400×15×21	HW450×450×16×24	HN350×300×10×16
2	0.78	1097.38	HW400×400×15×21	HW450×450×16×24	HN350×300×10×16
3	0.78	738.28	HW350×350×12×21	HW400×400×15×21	HN350×300×10×16
4	0.78	290.12	HW350×350×12×21	HW400×400×15×21	HN350×300×10×16

4 DYNAMIC TIME-HISTORY ANALYSIS

4.1 Ground motion selection

The response spectrum of VIII degree rare occurrence level of Steel Frame structure was used as the target spectrum to match ground motion from the PEER database, and the corresponding PGA at this level was 0.4g. According to the research of Lv Hongshan et al. [7], the ground motion moment magnitude was required to be greater than 6.5, the site type was Chinese II type site, and the average shear wave velocity of 30m soil story was260m / s $\leq Vs\_30 \leq$ 510m/s. In dynamic analysis, the Rayliegh damping of the structure is 5%.

Finally, four ground motions are selected, and their information is shown in the Table 3.

Table 3. Ground motion data.

Ground motion	Time	Event	M_w	Station	Distance (Km)	Vs_30 (m/s)
EQ1	6/13/2008	Iwate	6.9	Kami Miyagi Miyazaki City	25.2	477.6
EQ2	2/9/1971	San Fernando	6.6	LA-Holloywood Stor FF	22.8	316.5
EQ3	6/28/1992	Landers	7.3	Mission Creek Fault	27.0	355.4
EQ4	9/3/2010	Darfield New Zealand	7.0	RKAC, N14E	16.5	295.7

4.2 Comparison of seismic performance

The interstorey drift ratio (IDR) and base shear force are selected as the evaluation criteria to compare the seismic performance of pure frame structure and SMA braced structure.

4.2.1 Interstorey drift ratio

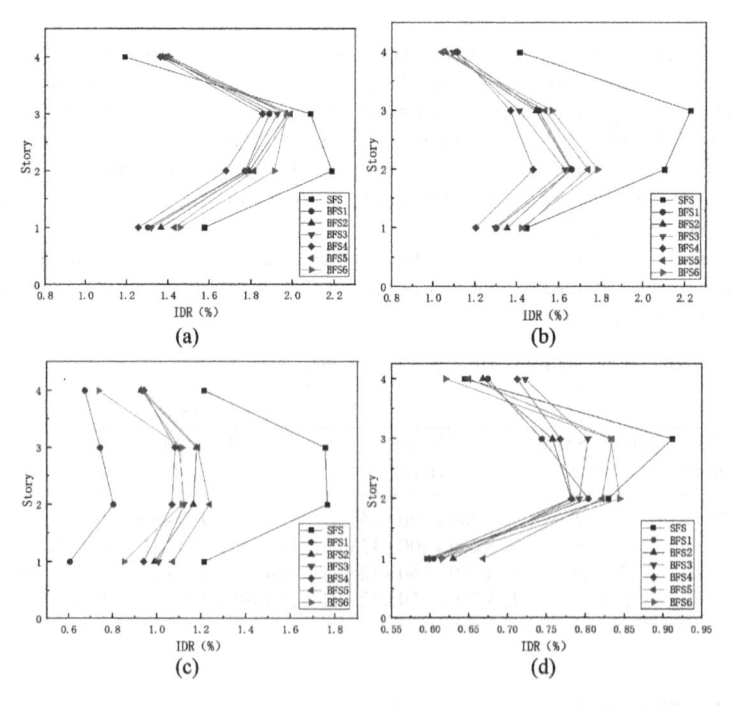

Figure 4. Comparison of structural displacements between storys: (a) EQ1, (b) EQ2, (c) EQ3, and (d) EQ4.

As can be seen from Figure 4, the IDR of SMA braced structure is smaller than that of steel frame structure, and this advantage is especially obvious under the action of ground motion EQ3. Under the action of ground motion EQ1, EQ2 and EQ4, the advantage of SMA braced structure is to reduce the maximum IDR.

4.2.2 Base shear force

The base shear force of the structure is shown in Figure 5. It can be seen from the Figure 5 that the base shear force of the SMA braced structure is significantly smaller than that of the steel frame structure, showing obvious advantages.

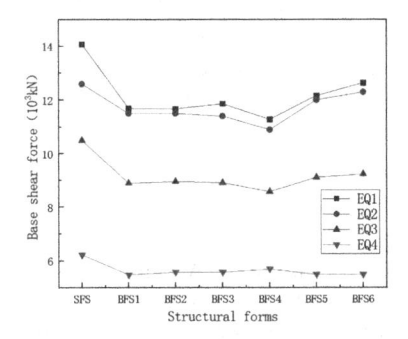

Figure 5. Base shear force comparison of different structural forms.

4.3 Improvement of seismic performance by different support arrangement forms

In order to quantitatively compare the improvement of seismic performance of SMA braced structure, steel frame structure was taken as reference to calculate the improvement of seismic performance of different brace arrangement:

$$\varepsilon = \frac{\gamma_{SFS} - \gamma_{BFS}}{\gamma_{BFS}} \times 100\% \qquad (1)$$

Where, is pure frame structure parameter, is the parameter of SMA support structure (Table 4).

Table 4. Improvement of seismic performance of the structure by SMA braced structure.

Evaluation criteria	Ground motion	Story	BFS1	BFS2	BFS3	BFS4	BFS5	BFS6
IDR	EQ1	1	21.27	15.72	19.78	**25.71**	10.51	8.41
		2	23.57	22.25	23.31	**30.34**	20.73	14.53
		3	10.70	5.95	8.51	**12.68**	5.05	6.02
		4	-14.36	-13.32	-13.89	**-12.63**	-14.67	-15.36
	EQ2	1	11.42	7.11	12.00	**20.44**	1.24	2.03
		2	26.52	26.87	28.98	**42.21**	20.95	17.84
		3	48.31	49.68	57.98	**62.87**	45.44	42.42
		4	27.38	33.42	28.98	26.34	**35.57**	33.76
	EQ3	1	**100.22**	22.26	20.14	28.65	12.90	41.90
		2	**120.32**	51.84	57.55	65.74	42.92	59.02
		3	**136.60**	49.10	59.88	62.72	48.50	58.11
		4	**79.45**	30.23	29.56	28.32	28.81	64.01
	EQ4	1	-0.72	-4.73	-2.30	**0.12**	-10.23	-2.50
		2	3.32	**6.20**	4.83	6.13	1.11	-1.65
		3	**22.34**	20.11	13.36	18.58	9.05	9.34
		4	-5.10	-4.22	-11.42	-10.17	-1.50	**3.24**
Base shear force	EQ1		20.4	20.7	18.7	**24.8**	15.8	11.4
	EQ2		9.6	9.6	10.5	**15.6**	5.0	2.4
	EQ3		18.1	17.2	17.8	**22.4**	15.3	13.8
	EQ4		**13.5**	11.5	11.6	9.2	13.2	13.2

Note: The parts in bold are the best performers.

Can be seen from the interstorey drift ratio, in part under the action of earthquake ground motion, the SMA first story and top story displacement interstorey drift ratio of the braced structure is larger than the pure frame structure, the interstorey drift ratio of the middle story control effect is obvious, this is because the SMA support structure to avoid large deformation is concentrated in one story, the lateral displacement of structure is more evenly distributed in each story.

By comparing the interstorey drift ratio and the base shear force, it is found that the seismic performance of the structure can be improved to different degrees by different SMA brace arrangements, among which SMA braced structures 1 and 4, namely BFS1 and BFS4, have advantages compared with other arrangement forms.

5 CONCLUSION

In this paper, the mechanical properties of SMA shape memory alloy are experimented, and a material constitutive model was compiled based on the VUMAT subroutine interface of ABAQUS software. A four-story frame structure was taken as the research object, and the influence of SMA brace arrangement on the seismic performance of the structure was compared. The results showed that:

(1) The dynamic time-history analysis results of four ground motions show that the seismic performance of SMA braced structure is significantly improved compared with that of pure frame structure;
(2) Taking interstorey drift ratio (IDR) and base shear force as evaluation criteria, the seismic performance of the six braced structures is compared. It is found that SMA brace arrangement 1 and 4 have advantages, so BFS1 and BFS4 are more recommended.

ACKNOWLEDGMENTS

This investigation is supported by the Project of Key Laboratory of Earthquake Engineering and Engineering Vibration, China Earthquake Administration (2019EEEVL0301).

REFERENCES

[1] Asgarian B , Moradi S . *Seismic response of steel braced frames with shape memory alloy braces.* Journal of Constructional Steel Research, 2011, 67(1): 65–74.
[2] Vafaei D , Eskandari R . *Seismic performance of steel mega braced frames equipped with shape-memory alloy braces under near-fault earthquakes.* The Structural Design of Tall and Special Buildings, 2016.
[3] Liu J L , Zhu S , Xu Y L , et al. *Displacement-based design approach for highway bridges with SMA isolators.* Smart Structures & Systems, 2011, 8(2): p.173–190.
[4] *Code for Seismic Design of Buildings.* China Building Industry Press,China, 2014:31–47.
[5] *Code for design of steel structures.* China Planning Press, 2017
[6] Auricchio F , Marfia S , Sacco E . *Modelling of SMA materials: Training and two way memory effects.* Computers & Structures, 2003, 81(24/25): 2301–2317.
[7] Hongshan L , Fengxin Z. *SITE COEFFICIENTS SUITABLE TO CHINA SITE CATEGORY.* Earthquake Science. 2007(01): 67–76.

Evaluating the spatial structure of a Pinus tabuliformis plantation using weighted Voronoi diagrams

X. Zheng
Planning and Design Institute of Forest Products Industry, National Forestry and Grassland Administration, Beijing, P.R. China

J.Y. Li & F. Yan
Precision Forestry Key Laboratory of Beijing, Beijing Forestry University, Beijing, P.R. China

ABSTRACT: The characteristics of the spatial layout of a Pinus tabuliformis plantation in Badaling, Beijing, were quantitatively analyzed with an aim to optimize its ecosystem structure. Using geographic information system (GIS) analysis, the number of adjacent trees of each focal tree was determined based on weighted Voronoi diagrams that were generated by improved Delaunay triangulation. Based on this analysis, spatial structure units in the forest were established, and the aggregation index, mingling index, neighborhood comparison, and degree of openness for each unit were calculated to describe the forest structure. The results indicated that the use of weighted Voronoi diagrams provides a reasonable approach to construct spatial structure units, which differs from the conventional method that is based on the nearest neighbor relationships between a focal tree and its four nearest neighbors. There were nine types of structural units in the study area, and the number of nearest neighbor trees ranged from 3–11, with the most frequent number being 5. The study findings provide support for the optimization of ecosystem structures of P. tabuliformis plantations and improve our conceptual understanding of forest community succession and restoration, in addition to the informatization and precision of forest spatial structure surveys.

1 INTRODUCTION

Forest stand spatial structure reveals the spatial features of all stands in a forest, and governs how different stands mutually influence and constrain each other [1, 2]. Forest stand spatial structure influences the competition among trees, spatial ecology, and stand growth stability [3, 4]. Therefore, research on the spatial layout of forest stands can provide important conceptual support for forest management and decision-making [5]. Spatial structure units are the fundamental units that constitute forest stand space, and each unit is a region comprising a focal tree and its nearest neighboring trees [3]. In studies of forest management, several parameters are commonly selected to analyze the characteristics of the forest spatial structure including aggregation index, mingling index, neighborhood comparison, and degree of openness. These parameters indicate the levels of competition, habitat space, tree species isolation, and other aspects in forest stands [6, 7]. Because the calculations of these parameters depend on the delineation of spatial structure units, the methodology of constructing optimal spatial structure units has been an area of active research. The conventional approach for constructing a forest spatial structure unit is selecting a focal tree and its four closest neighboring trees to form a quadrilateral, a method that may easily yield biased estimates of neighborhood-based parameters [8]. Studies have shown that Voronoi diagrams, as suggested by a Dutch climatologist, might express all lateral information of neighboring trees in forest stands [9, 10]. Voronoi diagrams are also referred to as Thiessen polygons [11], which have wide applications in computational geometry, urban planning, climatology, geology, geographic information system (GIS) analysis, image processing, and robot path planning, etc. [12]. At present, researchers in China and abroad are studying the characteristics of the spatial structure of different populations based on Voronoi diagrams [13]. Traditional uniform Voronoi diagrams are most commonly used, and their computation algorithms implicitly assume that neighboring trees are of equal competitiveness. However, in reality, these trees primarily engage in asymmetric competition. For example, trees with a larger diameter at breast height (DBH) and more developed root systems are considerably more competitive and consume more resources and living space. Therefore, this study introduced weighted Voronoi diagrams to evaluate a forest stand's spatial structure. The present study selected a P. tabuliformis plantation in Badaling as the study area and delineated spatial structure units of forest stands using weighted Voronoi diagrams to obtain neighborhood-based parameters. The study findings provide support for the optimization of ecosystem structures of P. tabuliformis plantations and improve our conceptual understanding of forest community succession and restoration [14], in addition to the informatization and precision of forest spatial structure surveys.

2 EXPERIMENTAL AREA AND SURVEY DATA

2.1 *Overview of the experimental area*

This study chose the P. tabuliformis plantation in Badaling National Park, located in Yanqing County in northwestern Beijing, as the study area. Badaling Forest Farm was established in 1958 (longitude: 115° 59'565"E; latitude: 40° 20'89"), and its highest elevation is 1,238 m above sea level. The percent green cover on the farm measures 96%, and the P. tabuliformis plantation has an average stand age of 43 years, and a plantation density of 800–1,200 trees·hm-2. Because the plantation comprised a single species, it had low biodiversity and poor ecosystem functioning given its improper forest structure. The forest had lagged stand tending and management, resulting in poor tree growth. The sampled trees in this study, located on a small slope within an area ranging from 690–725 m above sea level, were moderately healthy.

2.2 *Research area configuration and surveys*

In this study, a 90m × 90m research area was established in a representative region, which was further divided into four 45m × 45 m survey units. Trees in the survey units were tagged, and each tree was measured using a DBH tape and electronic total station (model: NTS-372R; South Surveying and Mapping Instrument Co., Ltd, Guangzhou, China). Basic data, including the identification number, species, tree height, DBH, crown width, and coordinates of relative locations in the survey unit of each tree, were recorded (as shown in Figure 1). A mixed forest containing both coniferous and broadleaf trees was present in the research area, in which there were 603 trees. The major tree species were P. tabuliformis, Armeniaca sibirica, Ulmus pumila, Pyrus betulifolia, and Celtis bungeana [14, 15]. In the sampling area, the average tree height was 8.7 m, and the average DBH was 14.6 cm (see details in Table 1).

Figure 1. Field measurement.

Table 1. Tree basic factor statistics.

	Tree height, m	Tree crown width (east-west), m	Tree crown width (south-north), m	DBH, cm
Maximum	13.2	8.02	8.00	36.9
Minimum	2.1	1.41	1.79	3.0
Mean	8.7	3.91	4.29	14.6
Stand deviation	1.8	1.42	1.59	4.5

3 METHODS

3.1 *Delaunay triangulation and the construction of weighted Voronoi diagrams*

3.1.1 *Generation of Delaunay triangulation*

Weighted Delaunay triangulation is actually a generalization of Delaunay triangulation, where each vertex P is denoted as (x, y, λ) and λ is the weight of P. In the context of this study, λ defines a tree's DBH. First, a weighted distance π is defined. Assuming a point a (x1, y1, λ1) and point b (x2, y2, λ2), the weighted distance π ab between points a and b is:

$$\pi_{ab} = \sqrt{(x_1 - x_2)^2 + (y_1 - y_2)^2 \cdot \left(\frac{1}{\lambda_1} - \frac{1}{\lambda_2}\right)^2} \qquad (1)$$

When λa = λb, this type of Delaunay triangulation may be regarded as a traditional Delaunay triangulation. Similarly, the center of the circumcircle C(σ) of each weighted Delaunay triangle is denoted as (xc, yc, λc), and λc is the average values of vertex weights of the triangle within that circle. Then, the circumcircle C is obtained with λc as the weight.

Provided the above conditions, the Bowyer-Watson algorithm may be used, which is an incremental algorithm that adds points to a Delaunay triangulation. In the first step, it was assumed that a 'super' triangle was added to a triangle list and that the triangle enclosed all sampling points. Second, discrete points, or the coordinates of each tree, were incrementally inserted, and whether these points were enclosed in the circumcircle was determined based on weighted calculations. In the triangle list, circumcircles enclosing the insertion points of triangles were determined. Then, the common sides affecting the triangles were deleted, and the insertion point was connected to all vertices of the influenced triangles, which completed the insertion of a point to the weighted Delaunay triangle list (Figure 2). In the third step, a local optimization procedure (LOP) was performed for the newly generated triangles, placing them in the triangle list. In the fourth step, the procedure was repeated until all points were inserted to complete the construction of the Delaunay triangulation [16].

3.1.2 *Construction of weighted Voronoi diagrams*

Based on the constructed weighted Delaunay triangulation, the center of each circumcircle of each Delaunay triangle was obtained. The centers of circumcircles of adjacent triangles were connected to form weighted Voronoi polygons generated by vertices of each triangle. This research used the ArcGIS Engine platform and C# language to develop a graphical software of the forest spatial structure, and the generated Voronoi diagrams are shown in the figure below.

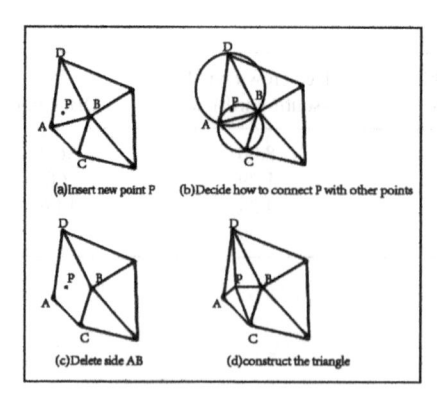

Figure 2. The plot of step 2.

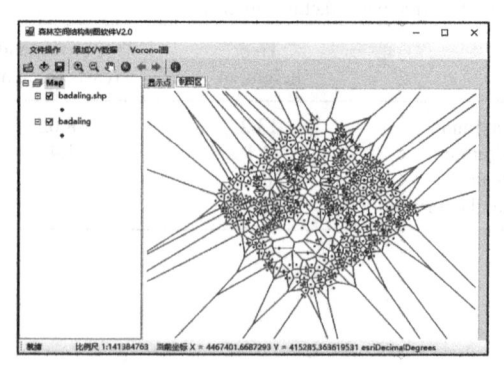

Figure 3. Voronoi diagrams generated by the forest space structure mapping software.

3.2 *Feasibility analysis of the research methodology and determination of the spatial structure parameters*

This research made use of frequency distribution histograms and a K-S test to evaluate whether the forest stand spatial structure parameters obtained based on traditional (n = 4) and weighted Voronoi diagrams followed a normal distribution. Correlation analysis and a variance analysis of the data obtained from the two methods were conducted to determine whether it was feasible to establish spatial structure units based on Voronoi diagrams. Additionally, the forest stand spatial structure parameters obtained based on weighted Voronoi diagrams were analyzed.

4 RESULTS AND ANALYSES

4.1 *Assessment of the normal distribution of the spatial structure parameters*

Inferring from the frequency distribution histograms of all parameters, data of the mingling index and the degree of openness obtained from the two types of Voronoi diagrams both followed a normal distribution. For the neighborhood comparison, the data obtained based on traditional (n = 4) Voronoi diagrams had a negatively skewed distribution, and data obtained based on weighted Voronoi diagrams had a positively skewed distribution.

The sample size in this study was 603 trees, which was low. To further verify the above results, K-S tests were performed. The significance of K-S tests for all parameters was P = 0.000, and therefore, null hypotheses were rejected, and the data of the parameters did not follow a normal distribution (results shown in Table 2). To further assess the relationships between the data obtained based on traditional (n = 4) and weighted Voronoi diagrams, correlation analysis and one-way analysis of variance (ANOVA) were conducted to analyze whether the data obtained from these two types of Voronoi diagrams were correlated and significantly differed from each other.

Table 2. The K-S test results of the structure indexes.

	K-S test results based on Voronoi		K-S test results based on n = 4 Voronoi	
	Test tatistics	Significance	Test statistics	Significance
Mingling index	0.143	0.000	0.183	0,000
Neighborhood comparison	0.100	0.000	0.116	0,000
Degree of openness	0.119	0.000	0.100	0,000

4.2 *Correlation analysis of spatial structure parameters*

Given the nonnormal data distributions of three spatial structure parameters in this study, Spearman's correlation coefficients (instead of Pearson's correlation coefficients) were calculated for the pairwise correlation of the mingling index, neighborhood comparison, and degree of openness among the data obtained based on traditional (n = 4) and weighted Voronoi diagrams. The results are shown in Table 3.

Table 3. The spatial structure index correlation analysis results.

spatial structure index	Correlation coefficient
Mingling index	0.905**
Neighborhood comparison	0.847**
Degree of openness	0.734**

NOTE: ** represents that the model has a significant difference at the 0.01 level.

From Table 2, the correlation of parameters was high for the data obtained based on the two types, as Spearman's correlation coefficients were greater than 0.7. This indicated that spatial structure parameters were highly correlated for data obtained from traditional (n = 4) and weighted Voronoi diagrams. Based on the above analyses, using weighted Voronoi diagrams to determine the spatial structure was a precise and effective method.

4.3 *Variance analysis of spatial structure parameters*

Before conducting one-way ANOVA, it was assumed that parameters determined from the two types were not significantly different. The ANOVA results revealed that the significance of the difference in spatial structure parameters obtained between the two types was $P = 0.000 < 0.05$. Therefore, null hypotheses were rejected, and there were significant differences between the determined spatial structure parameters based on the two types. Hence, this study establishes that using weighted Voronoi diagrams to determine forest spatial structure units is a new method that differs from the usage of traditional (n = 4) Voronoi diagrams.

5 CONCLUSION AND DISCUSSION

5.1 *Conclusion*

In this study, ArcGIS Engine was used as a platform to develop a graphical software for forest spatial structure. Through this software, weighted Delaunay triangulation was generated, and weighted Voronoi polygons were then obtained. The P. tabuliformis plantation in Badaling National Park, Yanqing County, Beijing, was selected as the research area. Tree individuals in the forest stand were considered discrete points on a plane to construct spatial structure units. Four spatial structure parameters – the aggregation index, mingling index, neighborhood comparison, and degree of openness – were calculated to quantitatively analyze the characteristics of the spatial structure of the plantation. The main conclusions of the research is that the use of weighted Voronoi diagrams to construct forest stand spatial structure units is a new method that specifically considers each focal tree, which is more reasonable and comprehensive than traditional (n = 4) Voronoi diagrams.

5.2 *Discussion*

In this study, some bias was associated with the determination of forest stand spatial structure units based on weighted Voronoi diagrams that used tree DBHs as weights. Future studies should integrate tree height, canopy breadth, and other factors for an overall analysis, in addition to generating comprehensive and reasonably weighted Voronoi diagrams that conform with forest stand conditions, to analyze the forest stand spatial structure.

ACKNOWLEDGEMENTS

Financial support for this study was supported by the Fundamental Research Funds for the Central Universities NO. 2017PT07 and National Natural Science Foundation of China (Project No. 31800468).

REFERENCES

[1] Hui, G.Y., Kiaus,V.G. Quantitative analysis method of forest spatial structure[M].Beijing: China Science and Technology Press, 2003:2–28.
[2] POMMERENING A. Evaluating Structural Indices by Reversing Forest Structural Analysis[J]. *Forest Ecology and Managenent*, 2006, 224(3): 266–277.
[3] ZHAO Chunyan, LI Jiping, LI Jianjun. Quantitative Analysis of Forest Stand Spatial Structure Based on Voronoi Diagram & Delaunay Triangulated Network [J]. *Scientia Silvae Sinicae*, 2010, 46(06): 78–84.
[4] LIU Shuai, WU Shuci, WANG Hong, et al. The stand spatial model and pattern based on voronoi diagram [J]. *Acta Ecologica Sinica*, 2014, 34 (06): 1436–1443.
[5] MASON W L, CONNOLLY T, POMMERENING A, et al. Spatial structure of semi-natural and plantation stands of Scots pine in northern Scotland [J]. *Forest*, 2007, 80(5): 567–586.
[6] FENG Yao. *Study on the Spatial Structure of Cunninghamia Lanceolata Non-commercial Forest Based on Voronoi Diagram* [D]. Changsha: Central South University of Forestry and Technology, 2014.
[7] TANG Mengping. *Study on Forest Spatial Structure Analysis and Optimal Management Model* [D]. Beijing: Beijing Forestry University, 2003.
[8] TANG Mengping,ZHOU Guomo,CHEN Yonggang, et al. Mingling of Evergreen Broad-Leaved Forests in Tianmu Mountain Based on Voronoi Diagram [J]. *Scientia Silvae Sinicae,* 2009, 45(06): 1–5.
[9] CHEN Jun, ZHAO Renliang, QIAO Chaofei. Voronoi diagram-based GIS spatial analysis [J]. *Geom Inf Sci Wuhan Univ*, 2003, 28(spec issue): 32–37.
[10] PENG Yipu,LIU Wenxi. Study on Delaunay Triangulation and Voronoi Diagram Application in GIS [J]. *Engineering of Surveying and Mapping*, 2002, 11(3): 39–41.
[11] WANG Xinsheng, LI Quan, GUO Qingsheng, et al. The generalization and construction of Voronoi diagram and its application on delimitating city's affected coverage [J]. *Journal of Central China Normal University(Nat.Sci.)*, 2002, 36(1): 107–111.
[12] QIN Xiwen, ZHANG Shuqing, LI Xiaofeng, et al. Spatial pattern of red-crowned crane nest-sites based on Voronoi diagram [J]. *Chinese Journal of Ecology*, 2008, 27(12): 2118–2122.
[13] TANG Mengping, CHEN Yonggang, SHI Yongjun, et al. Instraspecific and interspecific competition analy-sis of community dominant plant populations based on Voronoi diagram [J]. *Acta Ecologica Sinica*,2007,27(11):4707–4714.
[14] Zi Laibi·aibiMutiming. *Study on Modeling of Competition Indices of Pinus Tabulaeformis Individual Treesat Badaling Area* [D]. Beijing: Beijing Forestry University, 2012.
[15] WU Gang, FENG Zongwei. Study on the Social Characteristics and Biomass of the Pinus Tabulaeformis Forest Systems in China [J]. *Acta Ecologica Sinica*, 1994, 14 (4): 415–422.
[16] HESSE M, RODRIGUE J-P.The transport geography of logistics and freight distribution [J]. *J Transp Gergr*, 2004, 12(3): 171–184. doi: 10.1016/j.jtrangco.2003.12.004.

Study on the technology of vacuum membrane distillation for desalination

J. Huang
Department of Coast Defence and Engineering, Naval Logistics Academy of People's Liberation Army of China, Tianjin, China

S.W. Cai
Basic Department, Naval Logistics Academy of People's Liberation Army of China, Tianjin, China

X.C. Zhang & H.L. Zhao
The Institute of Seawater Desalination and Multipurpose Utilization, MNR, Tianjin, China

ABSTRACT: Harvesting clean water from seawater is an effective way to solve the water resources crisis problem. Membrane distillation has gradually attained much attention in seawater desalination area owning to high product quality, low operating pressure, high concentration ratio, and potential in using waste heat. Vacuum membrane distillation (VMD) is a kind of membrane distillation technology. In this article, the influence of main operating factors on the membrane performance and the optimal technological conditions of the VMD process had been systematically investigated. The results indicated that the permeation flux enhanced with the increase of vacuum pressure and feed temperature. Moreover, membrane fouling and membrane cleaning were also researched. The permeation flux returned to 3.45 $L \cdot m^{-2} \cdot h^{-1}$ after cleaning firstly, indicating the permeation flux recovery rate of 97.7%. The permeation flux decreases with the increase of cleaning times. The research could provide technology support and experimental evaluation for VMD desalination engineering.

1 INTRODUCTION

Water shortage has been regarded as a global challenge in recent years [1]. To solve the problem, it is necessary to desalinate seawater to meet the fresh water demand of people. Membrane distillation (MD) is widely regarded to possess the potential application in providing fresh water, which could effectively couple with solar energy and other renewable energy [2]. In addition, MD has several competitive advantages: such as high rejection rate, low operating temperature requirement, high efficiency and lower operating pressure [3–5]. The different pressure on permeate side and membrane surface is the driving force. Hollow fiber membrane play an important role of MD technology, it only allow water vapor through the membrane pores, but can not allow the other component pass the membrane [6]. Many polymers were used as hollow fiber, such as polytetrafluoroethylene (PTFE) [7], polyetherimide (PEI) [8], poly(vinylidene fluoride) (PVDF) [9].

The MD contains four types: sweeping gas membrane distillation (SGMD), air gap membrane distillation (AGMD), vacuum membrane distillation (VMD) and direct contact membrane distillation (DCMD) [10]. Nowadays, more and more attention are paid to the vacuum membrane distillation (VMD)

Owing to its higher permeate flux, lower conductive loss of heat, lower temperature and concentration polarization effects compared with other MD process [11]. Thus, it has been widely researched by scientists. There are several VMD studies in the article. For example, Xu et al. researched MD performance in the conditions of different operating factor and membrane materials [12]. The results showed that the membrane flux gradually increased with the increasing of feed temperature. However, the feed flow velocity has no obviously effect on permeate flux. Liu's group studied the VMD performances using PP modified by SiO_2 and PP nanoparticles as hollow fiber membranes, which indicating the membrane owing the potential applications in desalination [13]. Taylor's group designed a novel multi-effect hollow fiber membrane module using in VMD system for desalination [14]. Chang and his co-workers researched the simulated result and experiment result respectively for the effects of different operating conditions on VMD performances. The simulated result showed good agreement with the experiment result. At the same time, the author studied the energy efficiency of VMD system [15]. H. Kim and his co-workers considered PTFE owing a great prospect and market for vacuum-membrane distillation technology [16]. Chen's group analyzed thermo-economic and optimize operational parameters of multi-effect vacuum membrane distillation system [17]. In fact, membrane distillation technology has been widely applied in the industry field. N. Kjellander's group established two sets of pilot equipment on Hono island of the Atlantic coast as early as 1985, which initially confirming the operation and stability of water production device [18]. Singapore Memsys Cleanwater Company has developed a multi-effect plate-type vacuum membrane distillation enrichment process, which realized the energy recovery and cascade utilization in the membrane distillation process. The process can greatly reduce corrosion and scale, and the membrane flux at around 3 kg/ (m^2h). The research team at King Abdullah University of Technology in Saudi Arabia has developed a new device that combines two existing solar-driven technologies: photovoltaic and multi-stage membrane distillation, which can produce power and clean water. The entire unit is equivalent to that of a commercial solar cell, but the water purification output exceeds most of the existing units. The Institute of Seawater Desalination and Multipurpose Utilization in Tianjin has developed a new process of combining multi-effect membrane distillation and multi-stage flash and successfully prepared a multi-effect membrane distillation seawater desalination pilot device of 2 tons of water everyday.

At present, there is still a lack of systematic research for VMD process. This study systematically evaluated the effects of different operation conditions on the water production performance of membrane distillation seawater desalination, and discussed the methods to improve the system performance. Meanwhile, membrane fouling and cleaning test were also studied in this paper.

2 MATERIALS AND METHODS

2.1 *Membrane materials and membrane module*

Among the various membrane materials, PTFE has several advantages compared with other materials, such as strong hydrophobicity, excellent chemical stability, thermal stability and large loading density. Thus, in this study, as shown in Figure 1, the PTFE was used as hollow fiber membranes, which supplied by NanJing Cas-Bidun Newmem Technology Co. Ltd. The appearance and parameter of the PTFE membrane module were presented in Figure 2 and Table 1.

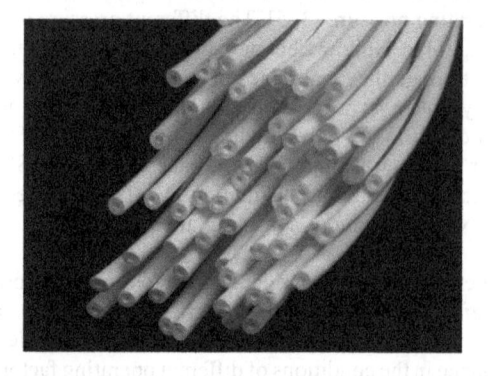

Figure 1. The PTFE hollow fiber membranes.

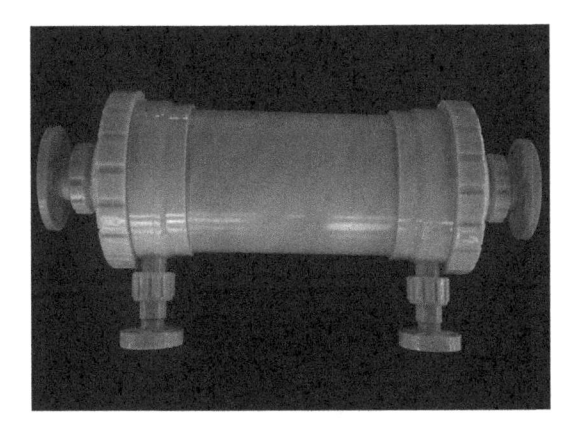

Figure 2. The appearance of PTFE membrane module.

Table 1. The parameter of the PTFE membrane module.

Content	Parameter
Length of membrane module(mm)	1110
Effective length of membranes(mm)	940
Effective area of membranes (m^2)	2

2.2 Feed solution

The sodium chloride (NaCl) solution was used as feed solution, which obtained by dissolving 200g NaCl in 100L distilled water at the room temperature. The NaCl was purchased from Heowns Company and used without further purification.

2.3 Experimental

Figure 3 showed the experimental procedure of the VMD. The experimental equipment mainly includes feed solution preheating system, membrane module, condensation recovery system and vacuum system. The feed solution was heated by the heat exchange module. When the feed solution reaches the preset temperature, it will be driven into the membrane module by magnetic pump and controlled by the flow control valve. In membrane module, the heated feed solution vapor contacted with the internal surface of the PTFE membrane filament directly enters the cold side through a hydrophobic PFTE membrane hole. A vacuum tank was installed to collected the fresh water and the steam was condensed in the condenser.

The permeate flux (J) is an important characteristic for the membrane distillation performance, which can be calculated using the equation (1).

$$J = W/S \bullet t \quad (1)$$

where, J is the membrane permeation flux (kg/m^2 h), W is the weight of the permeate (kg), S is the effective area of membranes(m^2) and t is the operation time in the VMD process (h).

The salt rejection rate (R) is used to evaluate the interception effect of the device on the target element, which can be calculated using the equation (2) given below:

$$R = \frac{D_f - D_p}{D_p} \times 100\% \quad (2)$$

where, R is the salt rejection rate(%) D_f is the feed conductivity (μS/cm) D_p is the distillate conductivity (μS/cm).

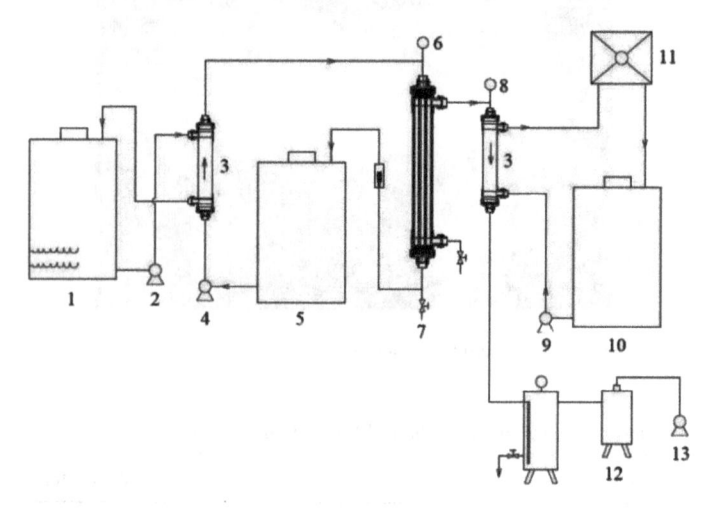

Figure 3. Vacuum membrane distillation for desalination equipment procedure: (1) Heating tank, (2) Hot water pump, (3) Heat exchanger, (4) Circulating pump, (5) Feed tank, (6) Pressure gauge, (7) Membrane module, (8) Vacuum gauge, (9) Cold water pump, (10) Cold water tank, (11) Cold cooler, (12) Separation of water gases, (13) Vacuum gauge.

3 RESULTS AND DISCUSSION

3.1 *Optimization of operating parameters on VMD procedure*

Much effort has been done to investigate the influence of different operating conditions on the VMD performance [19], but most of them were studied in the laboratory. In fact, there are a little difference between the experimental data and engineering data. Thus, it is necessary to research the system performance at different conditions in engineering.

3.1.1 *Effect of different vacuum pressure for system performance.*

Vacuum pressure influenced permeate flux remarkably in VMD process. The effect of different vacuum pressure on membrane flux and rejection rate is studied at the flow rate of 1.3 m³/h, feed temperature 75°C and the concentration of NaCl at 3.5%, as shown in Figure 4. When the vacuum pressure below 0.08 MPa, the permeate flux increased slightly with an increase in vacuum pressure. However, the permeate flux increased significantly after the vacuum above 0.08 MPa. The phenomenon is due to enhance the driving force across the membrane with improvement of vacuum pressure.

The conductivity of produced water decreased slightly with the increase of vacuum pressure, which is due to the feed solution vaporized easily at the condition of high vacuum. However, the conductivity of fresh water always keep below 10 μs/cm owing to the excellent water quality. The lower vacuum degree led to little membrane flux as a result of lower driving force of water vapor. The higher vacuum degree will result in more water vapor molecules penetrating the membrane, which improve the conductivity. In order to balance the water production and energy consumption, the vacuum pressure should between 0.08 MPa and 0.09 MPa.

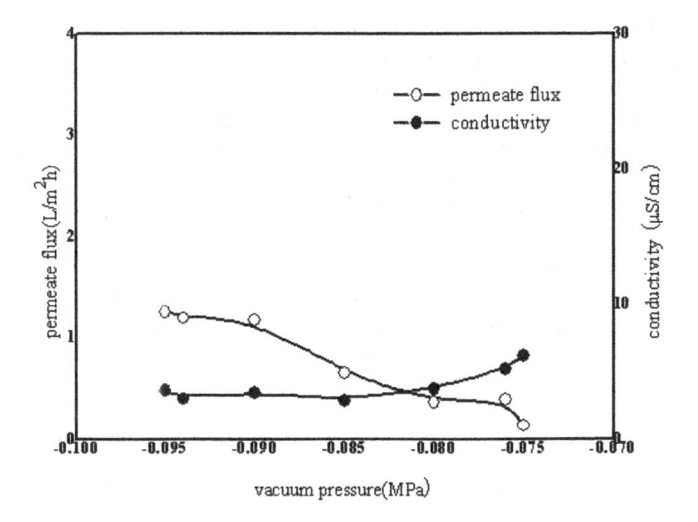

Figure 4. Variation law of membrane flux and water conductivity under different vacuum degree.

3.1.2 Effect of different feed circulation for system performance

The influence of different feed circulation on permeate flux and conductivity was researched at the vacuum pressure of 0.09 MPa, feed temperature 75°C and the concentration of NaCl at 3.5%. As seen in Figure 5, the permeate flux keep 1.3∼1.5 L/m²/h with the feed circulation from 0.8 m³/h to 1.2 m³/h, which indicating the feed circulation has no significant effect on permeate flux. Whereas, the permeate flux increased obviously when the feed circulation more than 1.2 m³/h. This might be because that the process is heat transfer limited within a laminar region for low circulation, turbulence in heat transfer boundary layer was not increased with the increase of feed circulation and temperature polarization was not decreased, thus permeate flux hardly improved [12]. However, the degree of turbulence improved and temperature polarization decreased with the increasing of feed circulation further. Thus, the lower feed circulation can reduce energy consumption and increase the life of the membrane module. As a result of the points discussed above, the optimized flow rate is 1.3 m³/h.

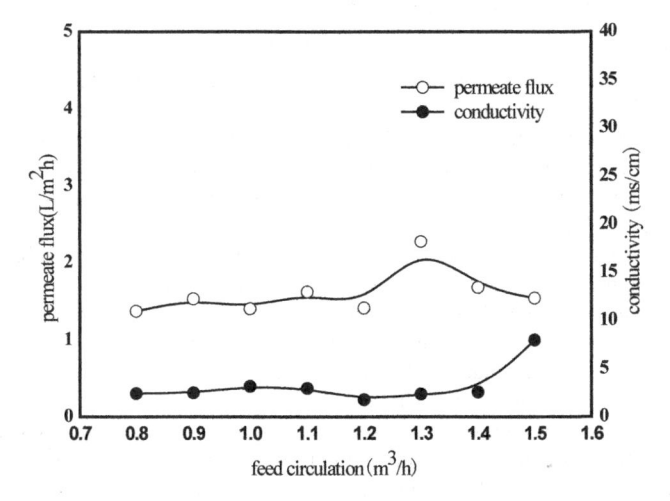

Figure 5. Variation law of membrane flux and water conductivity with different original water flow speed.

3.1.3 Effect of different feed temperature for system performance

As shown in Figure 6, the effect of different feed temperature on membrane flux and rejection rate is investigated at the flow rate of 1.3 m^3/h, vacuum 0.09MPa and the concentration of NaCl at 3.5%, the feed temperature was an important parameter in VMD performance. According to Antoine's equation, the water vapor pressure will be raise with the increasing feed temperature [20]. As we all know, VMD is a thermally driven process, when the water vapor pressure raised, the mass transfer force of water vapor through the membrane will be increase, which result a higher permeate flux [21]. However, when the temperature rises to a certain value, it gradually approaches the maximum flux of the membrane, so the increase of membrane flux slowed down and went into a plateau. At the same time, when the feed temperature become higher and higher, the energy consumption and thermal efficiency of the process should be considered, the increase of heat loss leads to the increase of energy consumption and the loss of membrane assembly. So, in order to save energy, it is suitable to chose 65 ~ 75°C as the optimized feed temperature.

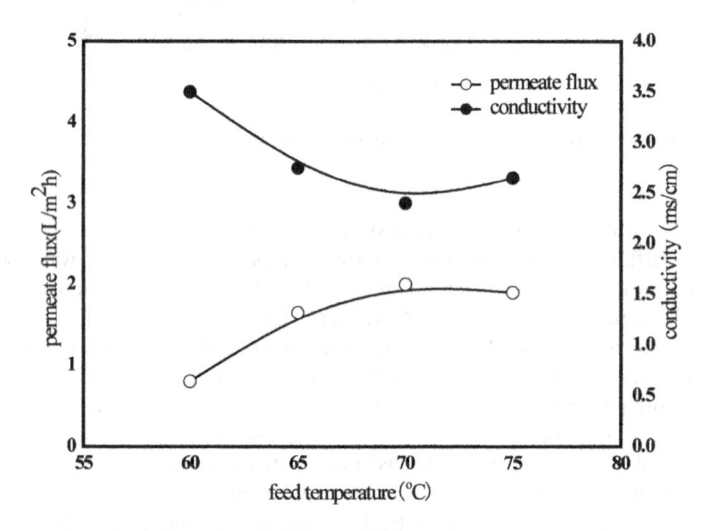

Figure 6. Variation law of membrane flux and water conductivity under different original water temperatures.

3.2 Membrane fouling and cleaning Test of PTFE

The Figure 7 showed that the VMD system ran for 200 h normally without failure, which indicated that PTFE had good stability. The permeate flux reduced 30.3% from 3.53 L/m^2/h to 2.4 L/m^2/h, which mainly influenced by concentration polarization. The accumulation of film surface pollutants in a long-term operation led to membrane hole blockage, which increased the temperature polarization coefficient (TPC) and concentration polarization coefficient (CPC), thus the permeate flux become lower and lower. In order to overcome this problem, it is needed to clean the pollutants regularly to ensure stable membrane flux.

In this study, we used deionized water to clean PTFE hollow fiber membrane surface and ultrasonic cleaning for 2 hours, then rinse with deionized water and put into a vacuum oven at 50°C for 24 h. After cleaning, the PTFE was reloaded in membrane module to check the recovery of the membrane flux. The experimental results are shown in Figure 8.

The film cleaning test shows that membrane cleaning can effectively alleviate membrane fouling and increase water production flux in VMD process. The produced flux of the original membrane was about 3.53 L/m^2/h, and the water flux decreased by 9.1% to 3.21 L/m^2/h after 100 hours of membrane distillation. After the first cleaning, the flux recovered to 3.45 L/m^2/h, and the recovery rate of water flux was 97.7%. PTFE water flux of the membrane decreases with the increase of running time and cleaning times.

In summary, membrane fouling leads to the decrease of water production flux. PTFE membrane can restore partial flux by reagent cleaning. For different treatment materials, membrane fouling components are different. In this study, the decline of permeate flux is mainly due to soluble salts. Deionized water can effectively remove these substances, thus restoring the surface morphology and microporous structure of the membrane and alleviating membrane fouling.

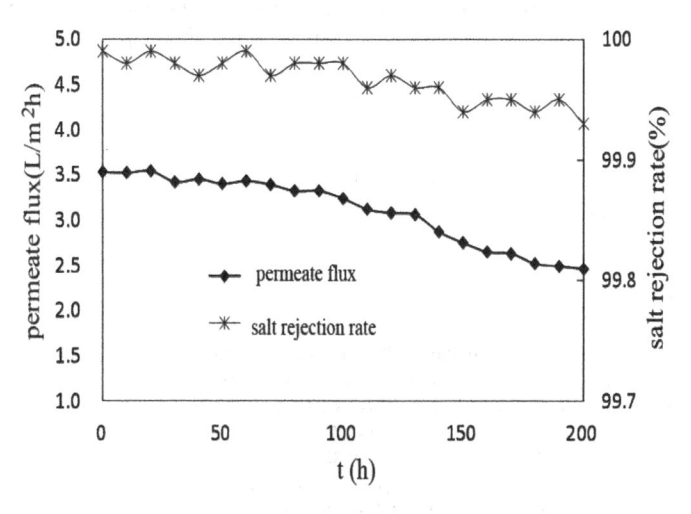

Figure 7. The stability test of membrane flux.

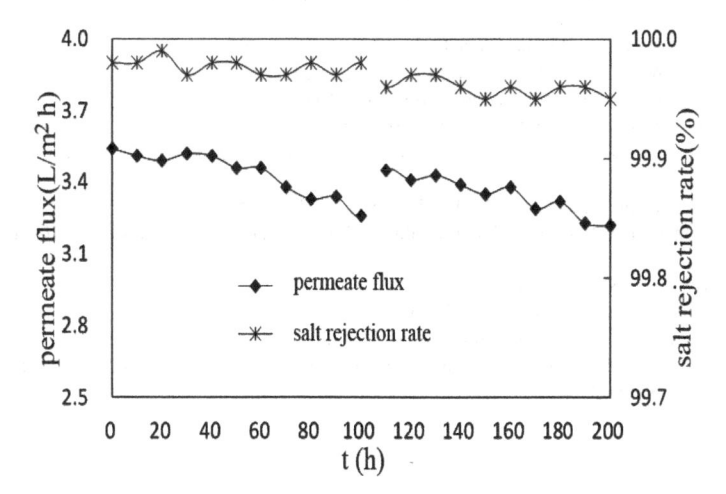

Figure 8. The influence of membrane cleaning on flux.

4 CONCLUSIONS

This article investigates the performance based on permeate flux and water productivity of VMD at different operating parameters and founds the optimized operation condition of the VMD. In general, the permeation flux gradually improved with the increase of vacuum pressure, feed circulation and feed temperature. However, this work founds that the feed circulation play an insignificant role on the permeation flux. The slight deviation may be owing to the influence of the engineering equipment, which not as sophisticated as an experimental instrument. Considering the temperature and energy consumption of membrane materials and cooling water consumption, the optimized operation condition of the VMD as following: 0.08 MPa ~ 0.09 MPa of vacuum pressure, 1.3 m^3/h of feed circulation, 65°C ~ 75 °C of feed temperature.

The pollutants are mainly soluble salts with concentrated brine as the feed solution for the VMD process, and the fouling can be effectively removed by deionized water, thus restoring the surface morphology and microporous structure of the membrane and alleviating the membrane fouling. This result can provide a guideline for the further study of membrane pollution in real sea water.

ACKNOWLEDGEMENT

This research was supported by the National Key Research and Development Project of China [grant numbers BHJ16C006].

REFERENCES

[1] Chennan L, Goswami Y and Stefanakos E 2013 *Renew. Sust. Energ. Rev.* 19 136–163.
[2] Zuo J, Chung T and Kosar W 2016 *J. Membr. Sci.* 523 103–110.
[3] Chiam C and Sarbatly R 2013 *Chem. Eng. Process. Process Intensif.* 74 27–54.
[4] Chang J, Zuo J and Lu K 2019 *Desalination* 449 16–25.
[5] Sun L, Wang L, Wang Z 2015 *J. Memb. Sci.* 488 30–39
[6] Zhou J, Zhang X, Sun B and Su W 2018 *Appl Therm Eng.* 144 571–582.
[7] Su C, Li Y, Cao H, Lu C, Li Y, Chang J and Duan F 2019 *J. Membr. Sci.* 583 200–208.
[8] Khayet M, Essalhia M, Qtaishatc M and Matsuurad T 2019 *Desalination* 466 107–117.
[9] Liu H, Chen Y, Zhang K, Wang C, Hu X, Cheng B and Zhang Y 2019 *J. Membr. Sci.* 578 43–52.
[10] Loulergue P 2018 *Sep. Purif. Technol.* 12 1–9.
[11] Lovineh S, Asghari M and Rajaei B 2013 *Desalination* 314 59–66.
[12] Xu J, Singh Y, Amy G and Ghaffour N 2016 *J. Membr. Sci.* 512 73–82.
[13] Liu Z, Pan Q and Xiao C 2019 *Desalination* 468 114060.
[14] Li Q, Omar A, Liu Q, Li X and Taylor R 2020 *Applied Energy* 276 115437.
[15] Chang Y, Ooin B, Ahmad A, Leo C and Lau W 2020 *Chemical Engineering Research and Design* 163 217–229.
[16] Kim H, Yun T, Hong S and Lee S 2021 *J. Membr. Sci.* 617 118524–30.
[17] Chen Q, Muhammad B, Akhtar F, Ybyraiymkul D, Muhammad W, Li Y and Ng K 2020 *Desalination* 483 114413–23.
[18] Nils K and Bo R 1988 EP19860904428.
[19] Zhang X, Guo Z, Zhang C and Luan J 2016 *Desalination* 385 117–125.
[20] Li B and Sirkar K 2005 *J. Membr. Sci.* 257 60–75.
[21] Guillén-Burrieza E, Blanco J, Zaragoza G, AlarcÓn D, Palenzuela P, Ibarra M and Gernjak W 2011 *J. Membr. Sci.* 379 386–396.

The management and control system's establishment of "three lines and one list"

L.Y. Lv, C.P. Jin, Y. Xu, Y.F. Zhang & Y. Li
Shandong Academy for Environmental Planning, Shandong, China

ABSTRACT: This paper discusses the technical routes and methods used in the compilation of "three lines and one list", and points out that "three lines and one list" is of great significance to strengthen the ecological environment protection, improve the land and space development and protection system, and promote high-quality development. Therefore, the implementation and application of "three lines and one list" achievements in serving the high-quality development of economy and society, boosting the high-level protection of ecological environment and promoting the digital supervision of ecological environment.

1 INTRODUCTION

In recent years, the Ministry of ecological environment has successively carried out the exploration of strategic environmental assessment in large regions, provinces and cities, and comprehensively promote the formation of "three lines and one list", which include red line of ecological protection, bottom line of environmental quality, upper limit line of resource utilization and the ecological environment access list [1]. At present, 31 provinces have completed the provincial achievement release of "three lines and one list", and some provinces have also completed the implementation and application of "three lines and one list". Now, effective working mode and complete technical methods have been basically formed, which laid a good foundation for the implementation of regional environmental assessment in China.

2 THE ESTABLISHMENT OF "THREE LINES AND ONE LIST" MANAGEMENT AND CONTROL SYSTEM

2.1 Collect and analyze data, establish working base map

The working base map is based on GIS software, using China Geodetic Coordinate System 2000. Build a basic database with high resolution, coordinate system and accurate positioning, integrating the basic geographic information data such as topography, river system, digital elevation and remote sensing image, and the multivariate data such as administrative division, land use, construction and development, economic and social, ecological basis (various protection areas, etc.), pollution sources (key industrial enterprises, sewage treatment plants, etc.), monitoring points (or sections) and the environmental quality data.

2.2 *Comprehensive analysis of economic and social environment, determine the evaluation framework and key point*

The comprehensive analysis conclusion of ecological environment foundation, situation and key problems is formed, combined with the ecological environment foundation, the resource and energy endowment, the ecological environment quality, the pollution emission characteristics, focusing on the urbanization situation, the key construction direction, the spatial layout, the industrial development layout, from the perspectives of improving environmental quality, optimizing environmental spatial pattern, controlling environmental pollution and innovating ecological civilization system.

2.3 Implement the ecological protection red line, identify and delimit the ecological space

Implement the ecological protection red line, and demarcate the ecological protection red line boundary further. Delimit the ecological space on the comprehensive consideration of integrity and stability of the regional ecosystem, fully connecting the achievements of urban development boundary and nature reserves, combined with the regional ecological security pattern, based on the important ecological function area, the protection area and other areas that need to be protected [2].

2.4 Determine the base line of environmental quality, build the environmental zoning management and control system

Following the principle of the environment quality "only optimization, not deterioration", connecting the relevant planning and policy's environmental quality objectives, standards and the requirements of reaching the standard within a time limit, scientifically evaluate the potential of improving environmental quality, and determining the objectives of air, water, coastal water and soil environmental quality in different region, basin and stage.

Specify atmospheric environment zone.Based on the air pollution source list, combined with the latest administrative divisions and the distribution of key industrial parks and industrial clusters, considering the key industries and future development space determined by current situation and relevant planning, we separately delimit the atmospheric priority protection area, high emission key control area, layout sensitive key control area, weak diffusion key control area, receptor sensitive key control area, and general control area [3].

Specify hydro-environment zone.Combined with the water function regionalization and the latest administrative regionalization, we refine the water control unit, reasonably determine the base line goal of each control unit's water environment quality in different stages. Combined with the current situation and improvement potential of the water environment, we determine the water environment capacity and allowable discharge of each control unit, and evaluate the impact of pollution source emissions on environmental quality. According to the results of water environment assessment and pollution source analysis, we separately delimit the hydro-environment priority protection area, the key hydro-environment control area and general hydro-environment control area, and propose clear space control requirement [4].

Specify soil risk zone. According to the results of soil environmental analysis, the main pollution causes, the distribution of pollution source and basic farm, connecting the management requirement of soil environmental quality standards, soil pollution prevention and control plans, focusing on the farm land and construction land, we delimit the soil pollution risk control zoning.The farm land is divided into priority protection area and key control area of agricultural land pollution risk, and the construction land is divided into key control area and general control area. Meanwhile, we put forward the space control requirements of each partition.

2.5 Formulate the resource utilization goal, clarify the resource management and control requirements

For the purpose of improving the environment quality and ensuring the ecological security, we determine the total amount and intensity requirement of the water resources, land resources and energy consumption.

2.5.1 *Water resources*

Based on the requirements of the water ecological function guarantee and water environment quality improvement, we calculate the ecological water demand and other indicators for the river sections involving important ecological service functions, water cut-off and heavy pollution, and clarify the control requirements of water surface area, ecological water level, river and lake shoreline. According to the calculation results of ecological water demand, we divide the relevant river sections into the ecological water supply areas, which are included in the key control areas of water resources to implement key control.

2.5.2 *Land resources*

Considering the ecological and environmental security, the areas with concentrated ecological protection red line, heavily polluted agricultural land or contaminated plots are identified as the key control areas of land resources, and the control requirements are put forward.

2.5.3 *Energy utilization*

With the improvement of atmospheric environment quality as the constraint, connecting the regional energy endowment and energy supply capacity, we calculate the future energy supply and demand, and determine the total coal consumption goal by using pollution emission contribution coefficient and other methods. Considering the requirements of atmospheric environment quality improvement, priority should be given to the high pollution fuel control areas in densely populated areas with high pollution emission intensity, and the control requirements should be put forward [5].

2.6 *Determine the environmental management and control unit*

Fully connecting the zoning results of ecological protection red line, general ecological space, water environment management and control, atmospheric environment management and control, soil pollution risk management and control, and resource utilization management and control, Integrating the boundary of ecological environment management and control, a fine environmental control zoning boundary is formed through matching and boundary checking with the data of regulatory detailed planning, land use patches, high-precision remote sensing image and topographic map [6].

Combined with the environmental control zoning boundary and the overall layout of regional ecological environment, the leading function of ecological environment and the main ecological environment problems and other factors, connecting the administrative boundary of township streets or key industrial parks (industrial gathering areas), we comprehensively delimit the control unit, the appropriate trade-offs should be made after the superposition and integration of various boundary types to avoid the unit delimitation fragmentation.

Unify the environmental control unit code, systematically analyze the regional functions and control requirements of ecological resources and environmental elements in each environmental control unit. By dividing the environmental control unit into three categories, namely priority protection, key control and general control, the environmental control unit is classified and managed, and a clear functional and clear boundary environmental control unit diagram is established.

2.7 *Implement the zoning control requirements of "three lines", and work out the list of ecological environment access*

Connecting the provisions of various industrial development and construction behaviors, coordinating the zoning control requirements of ecological protection red line, environmental quality base line and the upper limit of resource utilization, we establish the overall environmental access list covering the whole region [7].

3 STRENGTHEN THE IMPLEMENTATION AND APPLICATION OF ACHIEVEMENTS

3.1 Serve the economic and social development with high quality

The relevant departments at all levels should take the achievements as the prerequisite for comprehensive decision-making, and strengthen the analysis consistency and coordination in the process of local legislation, policy formulation, planning and law enforcement supervision, and should not make flexible breakthroughs or lower standards. In regional resource development, industrial layout and structural adjustment, urban construction, site selection and approval of major project, the ecological environment zoning control requirements should be taken as an important basis.

3.2 Promote the high level protection of ecological environment

The governments and departments at all levels should draw up the environmental protection plans and environmental quality standards to gradually achieve the objectives of regional ecological environment quality, on the basis of the regional and phased environmental quality baseline objective requirements determined by "Three Lines and One List" achievements. Priority should be given to ecological protection and restoration activities in the priority protection units to restore the ecosystem service functions. In the key control units, we will strengthen the pollutant emission control and environmental risk prevention and control, and solve the problems of ecological environment quality not up to the standard and the high ecological environment risk. The ecological and environmental departments should strengthen the application of "three lines and one single" achievements in the environmental management of ecological, water, atmosphere, soil and coastal waters, and make a deep fight against pollution.

3.3 Promote the digital supervision of ecological environment

The "three lines and one single" ecological environment zoning management and control requirements will be incorporated into the "three lines and one single" data application platform to promote the interconnection of the "three lines and one single" results with the ecological environment data systems such as environmental quality, pollutant discharge permit, monitoring and law enforcement, and other departments' business platforms such as the land and space basic information platform. On this basis, the "three lines and one single" achievements will be applied to the planning EIA review, the emission permit issuance and law enforcement supervision of construction projects, which will help improve the ecological environment quality continuously.

REFERENCES

[1] Hongdi L, Jun W, Changbo Q, Lei Y, Peipei Z, Nannan Z, Shangao X and Lu L, 2018 *Environmental Impact Assessment* **40** 2
[2] Zishu W, Wangfeng L and Yi L 2020 *Environmental Impact Assessment* **40** 2
[3] Nannan Z, Changbo Q, Qian W and Yuling T 2020 *Chinese Journal of Environmental Management* **10** 25
[4] Geng Z, Bin Z and Mengbo W 2020 *Journal of Green Science and Technology* **22** 129
[5] Maojie H and Ying W 2020 *Pollution Control Technology* **33** 52
[6] Shangao X, Jun W, Hongdi L, Lei Y, Nanan Z, Peipei Z and Changbo Qin 2019 *Pollution Control Technology* **33** 522
[7] Yang O, Xiaoli L and Yuanshi L 2018 *Environmental Impact Assessment* **40** 6
[8] Wenyan W, Yuanshi L, Yun J and Jingming R 2020 *Chinese Journal of Environmental Management* **10** 32

Research on steel structural detection based on ultrasonic phased array and infrared thermal imaging technology

M.Y. Tian, Y. Ni, X.H. Chen, D.H. Hao, P.P. Zhu & Z.P. Wang
Faculty of Civil Engineering and Mechanics, Jiangsu University, Zhenjiang, China

ABSTRACT: Screw fastener are a kind of widely used in steel structure connection basic components. Because of the special thread form, it causes great difficulty in detecting its internal failure. To evaluate the damage of steel structure screw fastener, this paper uses a phase-array ultrasonic detection method and infrared thermal image detection method to detect the damage of screw fastener. This paper mainly tests the overall steel structure through experimental methods. The results show that the combination of phase-array ultrasonic inspection technology and infrared imaging method can realize the damage detection of steel structure screw fastener, and provide a certain reference for the health monitoring and safety warning of steel structure screw fastener.

1 INTRODUCTION

At present, the non-destructive detection methods for fastener failure detection are as follows [1]: ultrasonic method, global/local vibration method, electro-mechanical impedance method, intelligent coating detection method, eddy current method, vision-based B/P neural network Monitoring method, X-ray method and acoustic emission method

Ultrasonic testing is a commonly used method to detect bolt fastener damage of the steel structure in civil engineering field In projects such as the construction of bridges and steel structure buildings, structural components are usually fastened with bolts and rivets to withstand higher structural loads. In the early stages of the fatigue process of screw fastener, the cracks are not visible because the bolt head and nut hide the crack initiation point. The information on crack initiation and propagation can usually be obtained by nondestructive testing such as ultrasound. The stress concentration caused by the geometric discontinuity at the bolt hole often leads to the initiation of fatigue cracks at the bolt and riveting point, which in turn leads to unexpected failure of the bolt fastener in the steel structures [2]. Due to the repeated relative motion, the joint interface of the joint part will also produce micro-motion [3]. The fretting zone is very sensitive to repetitive loads, and even under low stress can cause cracks. Many studies have studied fretting damage, but have not considered the impact of failure modes on the life cycle. Osegueda [4], Chang, Mal [5] and others use Lamb wave to detect cracks in rivet holes of steel plates. Although Lamb wave can detect tiny fatigue cracks, there are still limitations such as wave dispersion characteristics and mixed modes of different frequencies, which are difficult to identify single modes in reflected waves, and quantitative evaluation of cracks is not easy. In practical application, the oblique shooting method is often used to detect the cracks at the joints. In this method, the ultrasonic wave is incident on the surface of the material at an angle, propagates inside the material, reflects from the bottom of the material, and then radiates to the fatigue crack. Saka et al. [6] used different shear wave oblique incidencemethods to detect the vertical fatigue crack on the relative surface of the material. The limitation of oblique incidencemethod also includes dispersion and multiple reflections of mixed mode. When surface acoustic waves detect surface cracks and corrosion fatigue cracks, the path is simple and easy to identify. However, the study of using saw to detect and evaluate the fatigue crack of bolted connection in fastening state has not been reported. Therefore, in 2009, Wagle and

Kato [7] et al. Studied the fatigue crack of bolt specimen in fastening state by using the surface acoustic wave nondestructive testing technology as shown in Figure 1, and clarified the influence of tightening moment on the fatigue behavior of Al2024-T3 aluminum alloy plate bolt specimen such as fatigue life and failure mode under different stress amplitude. However, the ultrasonic detection method also has some limitations. It is only suitable for plane objects, and it is not suitable for small objects, and it can not get the visual image of defects.

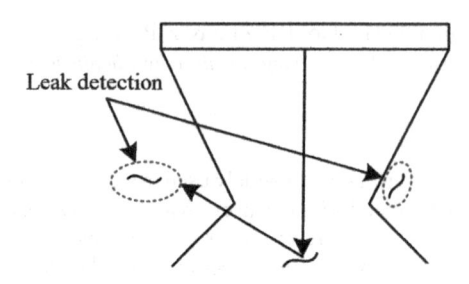

Figure 1. Ultrasonic testing device for acoustic surface waves.

At present, phase-array ultrasonic detection technology has been applied to many pipeline engineering detections in China and has formed the national standard of pipeline detection. Overseas R/D tech company of Canada has launched the phase-array ultrasonic scanning detector, which can realize the industrial application detection of more than 60 transducers, and can be applied to the off-line detection and imaging of larger thread damage. J. Ritter and others in the United States use a 16 element linear transducer composed of composite crystals to detect the surface wave and plate wave of the metal plate structure. The research group has studied the application of 16 elements phased array transducer in steel structure and concrete detection [8], but the phase-array ultrasonic detection technology still has a certain gap in the practical engineering application of screw fastener, many problems need to be solved. In order to improve the above shortcomings, the paper introduces the artificial crack between the threads and uses the combination of phased array technology and infrared thermal image detection technology to realize the detection of the crack between the threads.

2 THE BASIC PRINCIPLE OF STEEL STRUCTURAL DAMAGE INSPECTION

2.1 Basic principles of phase-array ultrasonic inspection

Ultrasonic phased array can quickly and multi-angle detect the damage of screw fastener and hole edges, and complete the rapid detection of damage of screw fastener and hole edges without moving or rarely moving the transducer. There are three main scanning methods for phase-array ultrasonic detection. As shown in Figure 2, (a) is linear scanning, (b) is fan-shaped scanning dynamic focusing, and (c) is dynamic focusing. In this experiment, the fan-shaped scan shown in Figure (b) is mainly used, and the incident angle of the ultrasonic wave can be freely changed by appropriately controlling the emission time in the emission wafer. The sound beam emitted by the same wafer in the array is moved within a scanning range for a certain depth of focus, and the depth of other different focal points can increase the scanning range.

The probe of the phase-array ultrasonic instrument is composed of a plurality of piezoelectric wafers that can be individually controlled in certain array order, instead of the probe of only one piezoelectric wafer in the traditional ultrasonic detector. The excitation (amplitude and delay) of the phased array probe wafer is controlled by the computer, so that the sound waves of each array element have the same phase when they reach the set focus, forming a specific transformable ultrasonic sound field to realize the detection of bolt defects.

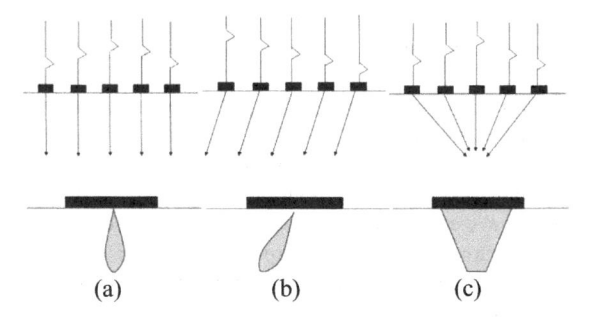

Figure 2. Ultrasonic phased array scanning mode: (a) linear scan, (b) Sector scan and (c) Dynamic focus.

2.2 *Basic principles of the infrared thermal image detection*

Infrared imaging technology can be divided into active infrared imaging technology and passive infrared imaging technology. The passive infrared imaging technology used in this article uses the infrared radiation emitted by the object itself to capture the image of the object. The infrared thermal imager is based on infrared imaging technology, using infrared detectors, photoelectric imaging objectives and photoelectric scanning systems. The working principle of infrared imaging is shown in Figure 3. The infrared signal radiated outward by the object is received by the photoelectric imaging objective. After passing through the grating and reaching the detector, the infrared number becomes an electrical signal, which is displayed on the screen and a real-time infrared thermal image can be obtained.

3 STEEL STRUCTURAL DAMAGE DETECTION EXPERIMENT

3.1 *Ultrasonic phased array damage detection experiment*

The instrument used in this experiment is TomoScan FOCUS LT, which can realize data compression and signal averaging in real time, and can use a variety of probes for detection. The experiment uses M42 high-strength bolts, bolt head height 26mm, inscribed circle diameter 64mm, plug body length 100mm, diameter 42mm. The size of the test sample of the bolt in civil engineering field is shown in Figure 4.

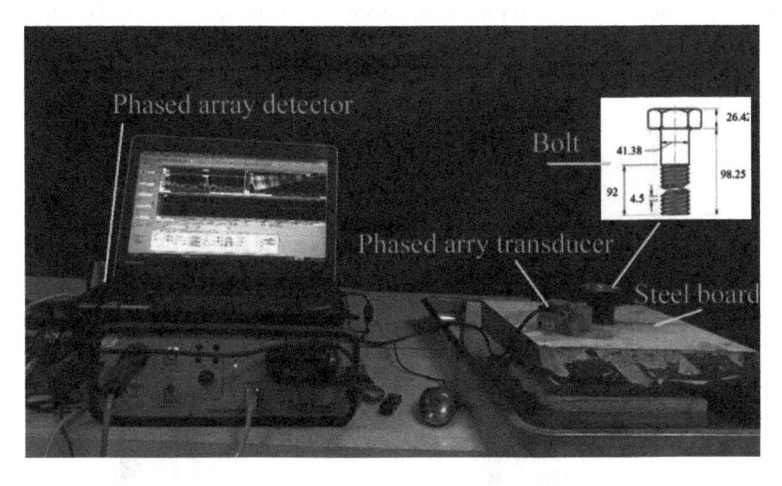

Figure 3. Ultrasonic phased instrument.

The first experiment is to use the inclined probe to detect the non-destructive surface of the bolt specimen. Figure 4 shows the A-scan inspection image and the fan-shaped scan diagram. From the figure, the waveform does not exceed the valve value, and there are many in the fan-shaped area. Oblique blue stripes are regularly distributed in the same direction, consistent with the thread distribution, and no damaged area is seen.

The second experiment is to use the inclined probe to detect the damaged surface of the bolt test piece. Figure 6 shows the A-scan inspection image and the fan-shaped scan diagram. From the figure, the waveforms have exceeded the valve value, and in the fan-shaped area, you can see To the damaged area, compared with the first experiment, it can be clearly seen that there are obvious defects, and the depth coordinate is about 18.1mm, which is also consistent with the actual prefabricated damage position.

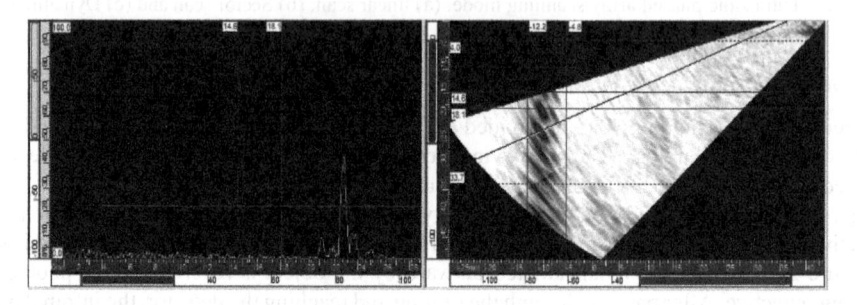

Figure 4. Image of ultrasonic phased array A scan and sector scan on damaged surface.

Therefore, the phase-array ultrasonic can be used to detect the damage of the bolt, and the position and size of the damage can be intuitively detected, and the detection result is consistent with the defect in the test piece.

3.2 *Infrared thermal image damage detection experiment*

The instrument used in this experiment is a German Infra Tec infrared thermal imager. The test instrument type is a short wave thermal imager with a spectral range of 1.5 μm–3.4 μm. It has a good radiation response to high-temperature targets and is more suitable for measuring high-temperature targets. The experimental instrument is shown in Figure 5, and the upper left corner is the bolt sample.

Place the experimental sample on the workbench, as shown in Figure 8. Due to the different specific heat capacity, the sample heats up slowly. Due to the air circulation in the middle-threaded hole, the temperature is higher than the sample but slightly lower than room temperature.

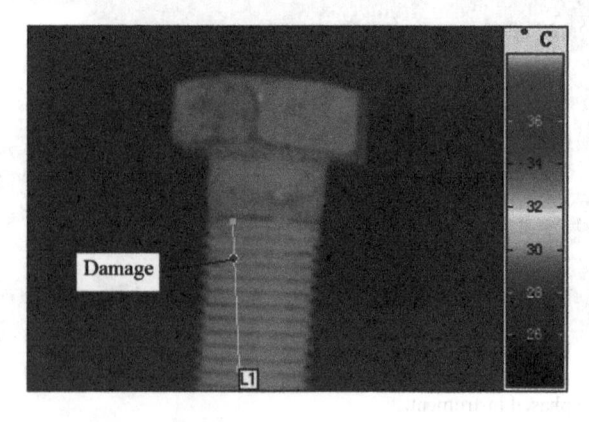

Figure 5. Thermal image of bolt specimen.

Take out the damaged bolts separately and use an infrared camera to collect data. In order to cause a significant temperature difference between the bolts and the environment, place the bolts in the sun in advance or use a hairdryer to increase the temperature. The thermal image is shown in Figure 6, the temperature difference between the bolt temperature and the room temperature is large, and the thread is clear, and a clear temperature gradient between the bolt and the air can be clearly seen.

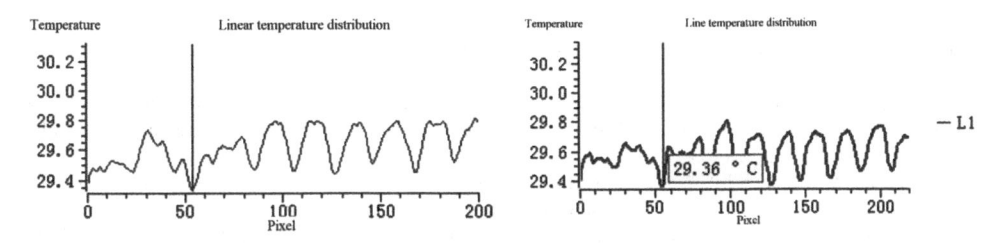

Figure 6. Line temperature distribution.

Set the sampling line from top to bottom, multiple sampling, and the line temperature distribution map is processed by the software. It can be seen from Figure 6 that from the starting point of the thread, an extremely low value appears at the 55th-pixel point, and the temperature is 29.36°C. Here the thread is deeper than the normal thread and is least affected by room temperature, so it is where the defect is. From the pixel point calculation, this point is about 17mm from the top of the thread, which is consistent with the premade damaged data. Then from the 85th-pixel point, the line temperature shows regularity. Due to the relationship of the thread, the outer end is greatly affected by the room temperature and the temperature is higher, and the bottom end of the thread is less affected by the room temperature and the temperature is lower, so it shows a wavy distribution.

4 CONCLUSION

Comparing the above two nondestructive testing methods, we can see that when using an phase-array ultrasonic instrument to detect the steel structure connection, the damage of the bolt in steel stuctures can be displayed with a clear image, and the shape and size of the damage in the deep thread can be seen. According to the image results,it can also accurately read the damage location depth of about 18.1mm. When using an infrared thermal imager to detect, when the bolt is connected to the experimental test combination, the temperature difference is not large, the thermal image cannot be displayed well, and the minor damage to the thread cannot be judged; when the bolt is unloaded, pass Equipment debugging can easily detect tiny temperature differences that are not visible to the naked eye, so as to determine whether there is any damage. Both methods can directly or indirectly obtain damage results, and the results are very similar to the preset damage data, but the image obtained by the phase-array ultrasonic method is clearer and has fewer restrictions. Therefore, it can be concluded that the phase-array ultrasonic is more suitable for detecting the damage of such steel structure screw fastener.

5 CONFLICT OF INTEREST

The authors declare no conflict of interest regarding the publication of this paper.

ACKNOWLEDGMENTS

This work was financially supported by the National Natural Science Foundation of China (11872191) and Postgraduate Research & Practice Innovation Program of Jiangsu Province (KYCX20_3073).

REFERENCES

[1] Chen Guoda, Xi Fengfei, Ji Shiming, Cao Huiqiang. Review of nondestructive testing methods for bolts [J]. Manufacturing technology and machine tools, 2017, (11): 22–28
[2] Shinde SR, Hoeppner DW. Fretting fatigue behavior in 7075-T6 aluminum alloy. Wear 2006; 261: 426–34.
[3] Chen YK, Han L, Chrysanthou A, O'Sullivan JM. Fretting wear in Self piercing riveted aluminum alloy sheet. Wear 2003; 255: 1463–70.
[4] Osegueda R, Kreinovich V, Nazarian S, Roldan E. Detection of cracks at rivet hole in thin plates using lamb wave scanning. El Paso, Texas: The University of Texas; 79968.
[5] Chang Z, Mal A. Scattering of Lamb waves from a rivet hole with edge cracks. Mech Mater 1999; 31:197–204.
[6] Akanda MAS, Saka M. Ultrasonic shear wave technique for sensitive detection and sizing of small closed cracks. JSME Int J, Ser A 2002; 45(2):252–61.
[7] Wagle S, Kato H. Ultrasonic detection of fretting fatigue damage at bolt joints of aluminum alloy plates[J]. International Journal of Fatigue, 2009, 31(8–9):1378–1385.
[8] Ziping Wang, Ying Luo, Guoqi Zhao, Fuh-Gwo Yuan. Design and optimization of an OPFC ultrasonic linear phased array transducer. International Journal of Mechanics and Materials in Design, 2017, 13(1):57–69.

Study on the temperature-regulating effect of phase-change asphalt mixture

J.X. Zhang, Y.G. Zhang, Y. Hou & Y.C. Zhuo
Beijing Key Laboratory of Traffic Engineering, Beijing University of Technology, Beijing, China

ABSTRACT: In order to investigate the influences of temperature on asphalt pavement, in this study, the paraffin wax-polyethylene and polyethylene wax phase-change materials were added into the asphalt mixture AC-16 and AC-20 based on the heat storage property of the phase-change materials, and the effects of Phase-change Material (PCM) on the temperature regulation of asphalt mixture was studied by indoor and outdoor temperature tests. Test results show that the addition of phase change materials can obviously reduce the pavement temperature, where the indoor maximum cooling change is about 4.3°C, and the biggest outdoor cooling change is about 5°C. It is discovered that, with the increase of dosage of phase change materials, the cooling rate gradually decreases. The cooling effect reaches the optimum state when the content is 0.25% (the mass ratio with asphalt mixture), and the gradation also has a certain influence on the cooling effect of phase change asphalt mixture.

1 INTRODUCTION

Temperature is one of the main environmental factors affecting the pavement service performance. Asphalt pavement is the main form of urban roads and expressways. In the high temperature environment in summer, asphalt pavement will absorb a lot of solar radiation heat, resulting in the sharp rise of the temperature of asphalt pavement. The temperature of the pavement can reach 60–70°C, which is far beyond the softening point of asphalt. Under the repeated loading from the vehicles, pavement diseases such as rutting, pit, etc. [1,2] easily occur, which will affect the driving safety, reduce the service performance of the road surface, and increase the maintenance cost.

In order to deal with the influence of high temperature on the road surface, there have been a lot of studies conducted by previous researchers, including preparation of thermal insulation coating, optimization of mineral grading, and use of modified asphalt to reduce road surface temperature [3]. Feng and Wang respectively added micropore ceramsite with particle size of 0~5mm and industrial waste ceramics into the ultra-thin wear layer. The study found that compared with the ordinary wear layer, the temperature difference between the upper and lower surfaces of the specimen increased from 3.5°C to 8.5°C. With the increase of the ceramic content, the thermal conductivity gradually decreased, and the temperature difference between the upper and lower surfaces increased [4]; Li et al. used the self-made unsaturated polyester resin as the coating substrate resin, and used the methyl ethyl ketone peroxide (MEKPO)/cobalt naphthenate system as the curing system of asphalt pavement thermal reflection coating. The test results showed that the coating had good water resistance, wear resistance and skid-resistance, and the pavement temperature could be reduced by nearly 10°C in the summer [5]; Zheng and Cheng et al. prepared the thermal reflective coatings by adding functional fillers into epoxy resin, and compared the surface temperatures of coated with the uncoated coatings. The temperature difference was 5.7~12.3°C, which showed a good cooling effect [6]. Most of the abovementioned methods of road cooling are passive to reduce the road temperature, which cannot fundamentally solve the problem of rise of road temperature. Phase change heat storage technology can realize the storage and release of heat energy by using the phase change of the material itself, so it can realize the autonomous cooling of asphalt mixture when used in the asphalt mixture. As a temperature self-regulating material, phase change material has attracted the pavement engineers' attention.

Phase change materials are materials that can absorb or release a large amount of energy when they undergo phase transformation. When the phase change material is added into the asphalt mixture, the excess heat will be absorbed when the phase change temperature of the phase change material is reached, so as to reduce the sensitivity of the pavement to the ambient temperature and reduce the pavement disease caused by the high temperature. Wang et al. added 1:1 solid-liquid paraffin as phase change material and expanded perlite as carrier into asphalt mixture to measure the temperature changes, and the test results show that adding PCM can reduce the pavement temperature by 4.4°C [7]; He and Yang et al. prepared the paraffin/expanded graphite composite phase change material by physical adsorption method, and added the phase change material into the cold-mix asphalt mixture. They found that the phase change material could reduce the road temperature by 5.2°C, and all the pavement performance indicators met the requirements of the specification [8]. In this study, paraffin-high density polyethylene composite phase change material was added into the asphalt mixture. Through indoor and outdoor temperature regulation tests, the temperature regulation effect of the phase change material was investigated.

2 TEST MATERIALS AND METHODS

2.1 Test materials

In this paper, two kinds of phase change materials (PCM) were selected to study the cooling effect of PCM, which were paraffin-high density polyethylene (PHDP)composite PCM and polyethylene wax (PW) PCM, as shown in Figure 1. 70# matrix asphalt was used in the experiment, and the basic indexes are shown in Table 1.

Figure 1. PCMs: (1) PHDP composite PCM and (2) PWPCM.

Table 1. Technical specifications for 70# matrix asphalt.

Types of asphalt	Penetration (25°C) (mm)	Softening point (°C)	Ductility (15°C)(mm)	Ductility (5°C)(mm)
70# matrix asphalt	70	42.80	134.59	—

The paraffin-high density polyethylene phase change material used in the test has good phase change temperature and latent heat, no supercooling phenomenon, good thermal stability and cyclic stability. In addition, the phase change material has good compatibility with asphalt material, and will not react with the asphalt material. The various properties of the phase change material are shown in Figure 2 to Figure 4.

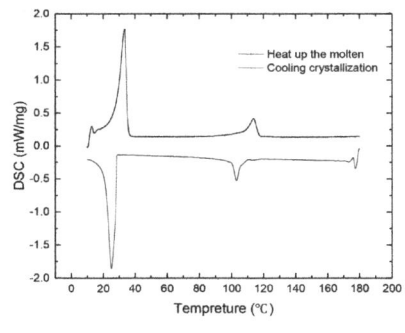

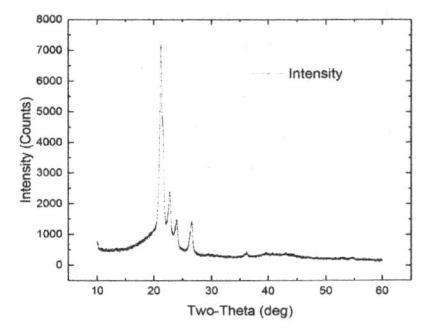

Figure 2. DSC curve of PHDP.

Figure 3. X-ray diffraction of PHDP.

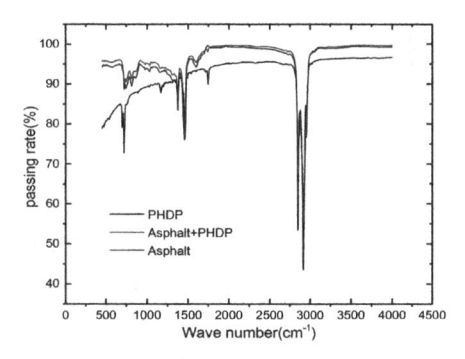

Figure 4. Infrared spectrum analysis.

Figure 2 shows the differential scanning calorimetry (DSC) curves of paraffin-HDP phase change materials. As we can see from the figure, in the process of heating melting, the initial phase transition temperature of paraffin-high density phase change material is 11.5°C, and the termination temperature is 37.9°C. The latent heat of phase transition is the area enclosed by the connection between DSC curve, phase transition starting point and phase transition ending point. It can be seen from the figure that the phase change material has a higher latent heat and thus can have a better cooling effect of phase change.

Figure 3 shows the X-ray diffraction (XRD) pattern of paraffin-high density polyethylene. It can be seen from the XRD pattern that it has a distinct characteristic diffraction peak with a strong diffraction Angle of 2θ of 21.279°. In addition, there are three relatively weak diffraction peaks, and the diffraction angles 2θ are 22.819°, 23.989° and 26.579°, respectively, which indicate that the PCM has good crystallization property.

Figure 4 is FT-IR diagram of PCM, matrix asphalt and mixture of PCM and asphalt. As can be seen from the figure, the curve of asphalt mixed with PCM is basically the same as that of matrix asphalt, indicating that the performance of asphalt material does not change after the addition of PCM, and the compatibility of PCM and asphalt is good, so it is possible to add paraffin-high density polyethylene (HDPE) to asphalt mixtures without affecting the properties of the asphalt itself. The temperature-regulating effect of phase change asphalt mixture is completely due to the phase change of phase-change material rather than the effect of other products produced by the reaction of phase-change material with asphalt.

2.2 Test methods

Two kinds of asphalt mixture, AC-16 and AC-20, were tested, where the grading curves of AC-16 and AC-20 are shown in Figure 5 and Figure 6, and the ratio of oil to stone was 4.59% and 4.32%, respectively. Common asphalt mixture, paraffin-high density polyethylene phase change asphalt mixture and polyethylene wax phase change asphalt mixture were prepared respectively according to the standard test procedures, where the content of paraffin-high density polyethylene phase change material is 0%, 0.25%, 0.35%, 0.5%, 0.75%, 1%, respectively. The content of polyethylene wax phase change material is 0.35% according to the user's specification, and the indoor and outdoor temperature tests were also carried out.

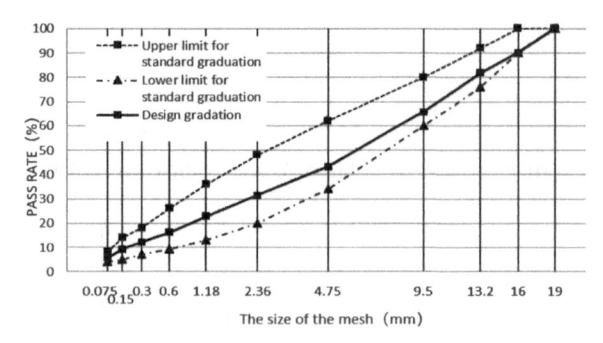

Figure 5. Grading curves of AC-16.

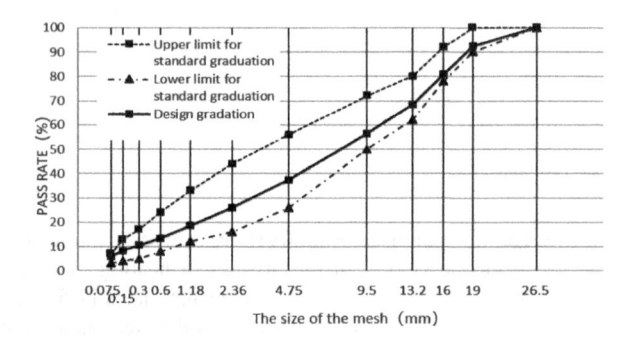

Figure 6. Grading curves of AC-20.

3 TEST PROGRAM

The temperature tests are divided into the indoor temperature test and outdoor temperature test, where the phase change-asphalt mixture mixed with phase change material and ordinary asphalt mixture were placed in outdoor natural conditions and indoor oven, respectively. The temperature change of specimen with temperature and time was recorded by a temperature recorder, and the temperature regulating performance of phase change asphalt mixture was investigated. At the beginning of the test, the molded phase-change asphalt mixture and common asphalt mixture rut board specimen were demolded. After the demodulation, a hole with 30mm deep was drilled at the center of the bottom of the rut specimen, then the temperature sensor was buried in the hole of the rut specimen and fixed with fine aggregate. The rutting specimen after treatment is shown in Figure 7.

3.1 Test methods

In the outdoor temperature test, asphalt mixture mixed with phase-change materials and ordinary asphalt mixture were placed in the real outdoor natural environment, so that the cooling effect of the addition of phase-change materials on asphalt mixture could be more truly reflected.

In the actual road traffic environment, the heat of asphalt pavement is mainly generated from the solar radiation. At the same time, it will also produce heat exchange with the atmosphere, roadbed and vehicles. According to the heat exchange law:

$$Q_{absorb} = Q_{release} \tag{1}$$

Where Q is heat.

When the heat absorbed from the external environment of the asphalt pavement is equal to the heat released by the asphalt pavement, the thermal equilibrium state is reached, and the temperature of the asphalt pavement is not changing. However, asphalt pavement is not an adiabatic system, so the heat flow equation of asphalt pavement is revised as:

$$q' = \alpha(G_s + G_{sky}) - G_{sur} - G_h \tag{2}$$

Where q' – Net heat flux of asphalt pavement surface, W/m²;
α – The absorption rate of asphalt against solar radiation;
G_s – The total radiation intensity of the sun, W/m²;
G_{sky} – Radiant heat of atmosphere, W/m²;
G_{sur} – Radiant heat of earth, W/m²;
G_h ——heat convection, W/m².
In Equation (2), when $q' = 0$,

$$\alpha(G_s + G_{sky}) - G_{sur} - G_h = 0 \tag{3}$$

Equation (3) is the heat balance equation of asphalt pavement surface.

The outdoor temperature test lasted for 8 hours, from 9:00 A.M. to 4:00 P.M., and the temperature was collected every 1 hour. After the temperature measurement, the test data was imported into the model and compared with the temperature data of rutting specimen of ordinary asphalt mixture to calculate the temperature decreasing range of phase-change asphalt mixture. The outdoor test process is shown in Figure 8.

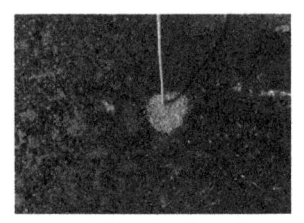

Figure 7. Setup of temperature sensor.

Figure 8. The outdoor test process.

3.2 *Indoor temperature test*

The outdoor test environment is closer to the real road condition, and it can reflect the cooling effect of phase-change asphalt mixture more truly and directly. However, the phase change cycle stability and long-term cooling performance of phase change asphalt mixture are greatly affected by the environment and weather, and the temperature collection efficiency is low. In order to eliminate the influence of external environment (wind and rainfall) on the temperature of phase change asphalt mixture, the indoor cooling test was adopted in this study. The treated phase change asphalt mixture and ordinary asphalt mixture were put into the incubator, and the temperature of the incubator was set at 40°C to minimize the ambient temperature in the high temperature environment in summer. The test time lasted for 8 hours, and the temperature data was collected every 15min.

4 RESULT AND ANALYSIS

4.1 Outdoor temperature test results

The temperature test was carried out according to the outdoor temperature test method, and the temperature changes of phase change asphalt mixture and ordinary asphalt mixture were tested respectively under the two grades. The temperature regulating performance of phase change asphalt mixture under different dosage was compared. The content of paraffin wax-high density polyethylene phase change material was AC-16:0.25% and AC - 20:0. 5% respectively. The test results are shown in Table 2.

Table 2. Outdoor temperature test results.

Time	The environment temperature/°C	Types of asphalt concrete					
		AC-16			AC-20		
		CAM	PHPCAM	PWPCAM	CAM	PHPCAM	PWPCAM
9:00	25	35	34.7	34.1	37.7	35.8	33.2
10:00	27	42.7	41.1	40.5	44.2	42.1	40.3
11:00	27	50.8	50.5	50.2	52.1	48.4	47.5
12:00	29	55.7	53.5	52.3	56.0	52.2	50.7
13:00	28	56.2	53.3	51.9	56.1	52.0	51.8
14:00	27	54.5	53.1	52.3	56.0	52.9	51.7
15:00	27	54	53.9	53.1	56.5	54.0	52.9
16:00	27	53.2	52.4	52.3	50.1	50.0	49.6

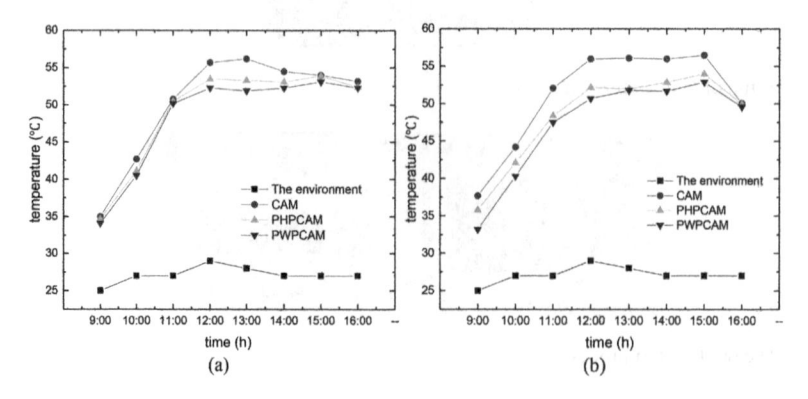

Figure 9. Outdoor cooling result curve.

Figure 9 shows the outdoor temperature test results. It can be seen from the figure that the temperature of asphalt mixture will rise greatly in the high temperature environment in summer, and the addition of phase change materials can reduce the temperature of asphalt pavement to a certain extent. As can be seen from the figure, as time goes by, the cooling range of asphalt mixture mixed with PCM increases first and then decreases, and the cooling range is the most obvious between 11:00 AM and 15:00 PM. The maximum cooling range of the two types of PCM asphalt mixtures can reach 4.3°C and 5.3°C respectively, and the cooling effect of AC-20 asphalt mixture is better than that of AC-16 asphalt mixture. The maximum cooling range of the asphalt mixture mixed with polyethylene wax is 5.3°C, and the maximum cooling range of the asphalt mixture mixed with paraffin-high density polyethylene is 4.1°C, indicating that the cooling effect of the asphalt mixture mixed with polyethylene wax is better than that of the asphalt mixture mixed with paraffin-high density polyethylene.

4.2 *Indoor temperature test results*

The indoor temperature test set the test temperature at 40°. The test results are shown in Figure 10.

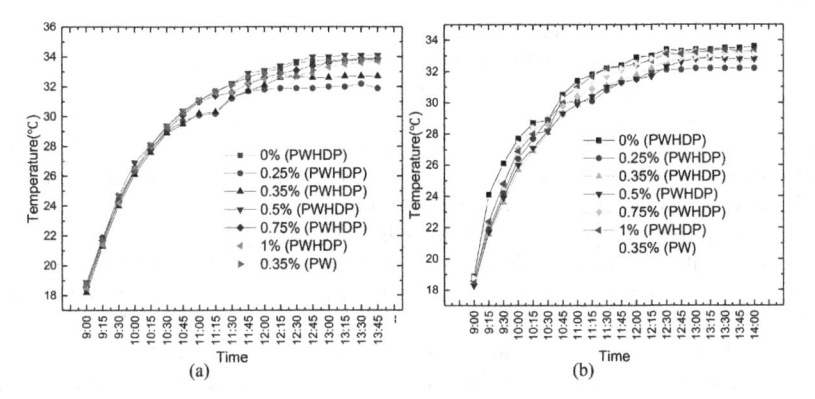

Figure 10. Indoor cooling result curve: (a) AC-16 and (b) AC-20.

Figure 10 shows the results of the indoor temperature control test. It can be seen from Figure 10 that the temperature of the indoor test is much lower than that of the outdoor cooling test. The reason is that the indoor test lacks direct sunlight irradiation and the asphalt mixture does not absorb the radiant heat of the sun, so the temperature does not rise above the pre-set test temperature. It can be seen from the figure that as time goes on, the temperature of asphalt mixture gradually increases. After about 3 hours of heating up, the temperature of asphalt mixture is basically stable without a significant rise or decrease. By comparing the temperature regulation under different dosage, it can be seen that with the increase of the dosage of phase-change material, the temperature of phase-change asphalt mixture is lower than that of ordinary asphalt mixture at the same time, and the maximum cooling range is 1.9°C for AC-16 asphalt mixture and 1.7°C for AC-20 asphalt mixture, respectively. With the increase of mixing amount, the decreasing range of temperature gradually decreases. In the heating process, when the temperature reaches the same temperature, the required time decreases with the increase of the content, which indicates that with the increase of the content of PCM, the wax content gradually increases, and the oiliness of asphalt mixture gradually increases, and the heat absorbed by the mixture spills out, which is more than the energy absorbed in the phase change process of PCM. Therefore, with the slight increase of temperature, the phase change cooling effect of PCM becomes less; When the temperature of asphalt mixture reaches 27.5°C and 30°C, and when the content of PCM is low, the temperature curve has obvious broken lines. The temperature of PCM does not rise for a period of time, indicating that in this temperature range, PCM absorbs part of the pavement heat through phase change, making the pavement temperature maintain at the same temperature for a period of time. All the conclusions show that with the increase of the content of phase change material, the effect of phase change heat absorption becomes less and less, and the temperature gradually increases.

5 RESULT AND ANALYSIS

The conclusions of this paper are as follows.

(1) The paraffin-high-density polyethylene phase change material and polyethylene wax phase change material can effectively reduce the pavement temperature when they are added into the asphalt mixture, and the polyethylene wax phase-change asphalt mixture has better cooling effect than paraffin-high-density polyethylene phase-change asphalt mixture, and the maximum temperature can be reduced by 4.3°C.
(2) Compared with the AC-16 asphalt mixture, the cooling rate of the AC-20 asphalt mixture is greater. Because the asphalt-aggregate ratio of the AC-20 asphalt mixture is lower than that of the AC-16 asphalt mixture, the content of coarse aggregate of the AC-20 asphalt mixture is more than that of the AC-16 asphalt mixture, and the content of fine aggregate is less, the AC-20 asphalt mixture absorbs less heat per unit volume, and the range of temperature decrease is larger.
(3) With the increase of phase change material content, the temperature regulating effect of phase-change asphalt mixture gradually becomes less. The reason is that with the increase of phase change materials, the influence on asphalt material gradually increases, the oil property of asphalt material increases, the penetration degree increases, the softening point decreases, and the viscosity increases, which makes the heat absorbed by the phase change asphalt mixture per unit volume exceeds the heat absorbed by the phase change material, resulting in the increase of the surface temperature of the asphalt mixture.
(4) It can be seen from the indoor test that there is an obvious inflection point in the temperature curve of asphalt mixture when the temperature reaches 27.5–30°C, and the temperature rises gently around the inflection point. The reason is that the phase change material absorbs part of the pavement heat at the inflection point, which makes the rising trend of road temperature slows down for a period of time, and increase the amount of energy needed to keep the pavement at a certain temperature. Thus it has the effect of slowing down the rise of pavement temperature and effectively preventing the occurrence of asphalt pavement high temperature disease.

REFERENCES

[1] LIN Fei-peng. Preparation and Properties of Mineral-based Composite Phase Change Materials for Asphalt Pavements[D]. Changsha University of Science & Technoogy, 2018.
[2] Stathopoulou M, Synnefa A, Cartalis C, et al. A Surface Heat Island Study of Athens Using High-Resolution Satellite Imagery and Measurements of the Optical and Thermal Properties of Commonly Used Building and Paving Materials[J]. International Journal of Sustainable Energy, 2009, 28(1–3):59–76. More references.
[3] MA Biao, WANG Xiao-man, LI Chao, LI Zhi-qiang, LIU Song-tao. Analysis of Application Prospect of Phase Change Materials in Asphalt Concrete Pavement[J]. Highway, 2009(12): 115–118.
[4] FENG De-cheng, WANG Guang-wei. Research on Application of Ultra-Thin Wearing Surface with Thermal Resistance Function[J]. Highway, 2009(3):62–65.
[5] LI Wen-zhen, LI Liang, SHI FeiWEI, He-guang, ZHONG Zhi-ming. Research and Development of Unsaturated Polyester Temperature Decreasing Coatings of Asphalt Pavement[J]. Journal of Chongqing Jiaotong University (natural sciences) (6):80–82+118.
[6] ZHENG Mu-lian, CHENG Cheng, WANG Yan-feng. HUANG Hu-jun. Experimental Study on Asphalt Pavement Cooling Technology Based on Improving Pavement Albedo[J]. Highway Traffic Technology (Application Technology Edition) (9):75–78.
[7] WANG Ruixin, JIANG Alan. The Preparation of Composite Phase Change Materials and Its Temperature Regulating Effect in Asphalt Concrete[J]. Low Temperature Architecture Technology, 2018, 40(4).
[8] HE Li-hong, YANG Fan, TONG Yu, ZHU Hong-zhou. Application of Phase Change Fine Aggregate in Cold Mix Asphalt Mixture[J]. Highway, 2016, 61(8) :181–185.

Advances in Materials Science and Engineering – Lombardo & Wang (Eds)
© 2022 the authors, ISBN 978-1-032-12707-1

Study and application of water flooding development law in Daqing A oilfield

B.F. Wu
Exploration and Development Research Institute, Daqing Oilfield Company Limited, Daqing, China

ABSTRACT: The A oilfield is a low-permeability reservoir with no fractures. The oil recovery rate of the oil layer with undeveloped fractures is low, the water flooding development effect is poor, the proportion of long shut-in wells and inefficient wells is large, and the economic benefits are poor. At this stage, the understanding of waterflooding development rules is not deep, and the applicability of the law research methods is poor. According to the core relative permeability curve data provided by Oilfield A, normalize it to obtain a normalized relative permeability curve. Non-dimensional oil and fluid index and water conteen thistory have been worked ot. The understanding of development laws and the application of research theories to practice provide a theoretical basis for subsequent adjustments and tapping of potential.

1 INTRODUCTION

The A oilfield is a low permeable reservoir with no fractures. It has the characteristics of low permeability, low production, deep buried reservoirs, and multiple and thin oil layers. Accurately calculating the fluid production and oil production indexes of low-permeability oilfields, and providing reliable basis for low-permeability oilfield productivity prediction and rational development of oilfields, has always been a concern of many scholars. At present, the oilfield is in medium-high water cut according to the indoor experiment and on-site production situation, it can be known that the low permeability oil field water breaks through, the fluid production index drops, which is unfavorable to the stable production of the oilfield. How to take measures to increase the problem of extracting liquid to ensure production has become an urgent problem to be solved. Accurate calculation the fluid production and oil production index of the permeable oilfield provides a reliable basis for the low permeable oil fields development.

2 RELATIONSHIP OF AIR PERMEABILITY AND EFFECTIVE PERMEABILITY OF ROCK SAMPLES IN A OILFIELD

The air permeability of the rock sample in the target reservoir of A oilfield is mainly distributed in 1–5md, and the effective permeability is mainly distributed in 0.5–2md. Effective permeability represents the ability of water and oil to flow through reservoir and is directly related to productivity. It can be concluded from the relation diagram between air permeability and effective permeability of A oilfield. Character K is used to present air permeability

$K > 50$md; Effective permeability is about 1/2 of air permeability.
$10 < K < 50$md: Effective permeability is between 1/4 and 1/3 of air permeability.
$2 < K < 10$md: Effective permeability is between 1/5 and 1/3 of air permeability.
$K < 2$md; Effective permeability varies by an order of magnitude from different reservoirs.

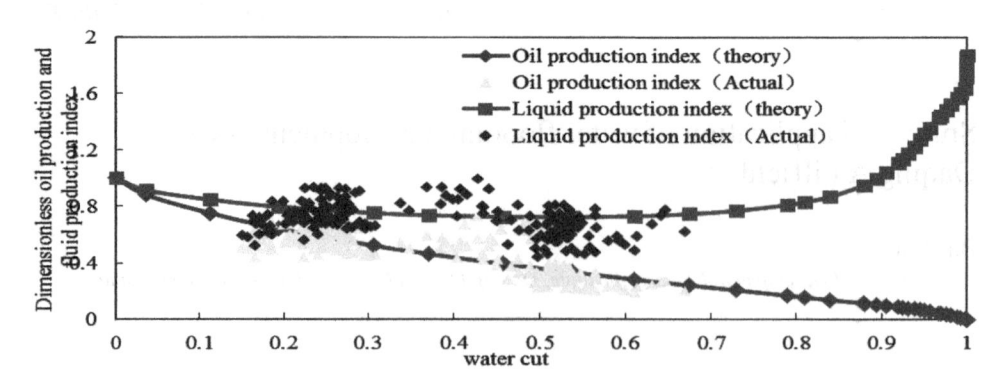

Figure 1. A comparison of the theory and practice of dimensionless oil production and fluid production index in A oilfield.

3 START VALUE STATISTICAL LAW OF RELATIVE PERMEABILITY CURVE

It is shown that the bound water saturation decreases with the increase of permeability. In the low and ultra-low target layers, the bound water saturation is generally in the range of 40% to 50%, and the peak is 44%.

The oil saturation of the rock sample increases with the increase of air permeability, but there is little relationship between the residual oil saturation and the air permeability. The residual oil saturation of the target reservoir in A oilfield is 28%–40%.

According to the statistics of the water permeability data of relative permeability rock samples with residual oil, it is concluded that the range of water permeability in the target reservoir is narrow, which is mainly distributed in 10% to 16%.

With the increase of permeability, the span of two-phase region increases. With the increase of bound water saturation, the range of two-phase region becomes smaller. The area of oil-water two-phase co-permeability in target reservoir is very narrow, mainly distributed in 24%-36%.

4 VARIATION RULE OF DIMENSIONLESS OIL RECOVERY AND PRODUCTION FLUID INDEX WITH WATER CUT

The oil production index decreases with the increase of water cut, and the liquid production index decreases with the increase of water cut, then stabilized, and finally increased in the high water cut period. The higher the permeability, the higher the extraction index rises. Water content of target reservoir is more than 80% and dimensionless production index increases gradually to the value greater than 1 (Figure 1).

The comparison between the actual non-dimensional production fluid index and the theoretical calculated value shows that the theoretical value is in good agreement with the actual value.

5 STUDY ON THE LAW OF RISING WATER CUT

In the process of non-piston displacement, the starting pressure gradient of low permeability reservoir and the flow velocity of fluid have certain influence on water cut. The fluid flow velocity under different reservoir conditions is calculated respectively. The influence of different flow velocity on water cut is analyzed. Only when the flow velocity is less than a certain value, it will exert a certain influence on water cut. The flow velocity of fluid in target reservoir is small, mainly distributed in the range of 1×10^{-5} m/s $\sim 1 \times 10^{-6}$ m/s, and its flow velocity has a certain influence on water cut.

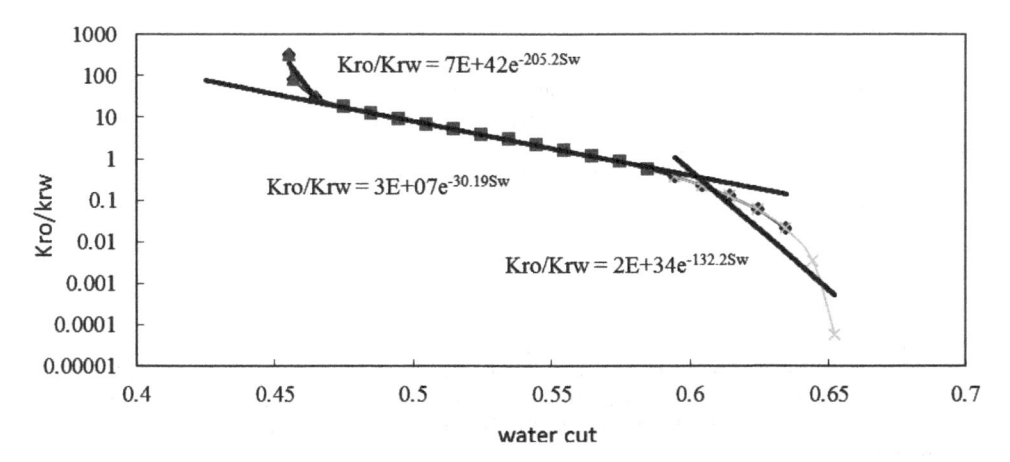

Figure 2. Logarithmic relation curve of relative permeability ratio and saturation Calculation method of water cut.

$$f_w = \cfrac{1}{1 + \cfrac{K_{ro}/\mu_o}{K_{rw}/\mu_w}} \left\{ 1 + \frac{KK_{ro}}{v(t)\mu_o} \left(\frac{p_c - p}{\sigma}\right)^{-s} [(\rho_o - \rho_w)g \sin\alpha + (G_o - G_w)] \right\} \quad (1)$$

6 DEVELOPMENT LAW ACCORDING TO DIFFERENT STAGES

By regression of the logarithmic relation curve of relative permeability ratio and saturation, the regression equations of different sections are given. Compared with a single regression, the regression equations of each section are more accurate, which can ensure that the calculated data are more in line with the reality. WOR calculation formula and the relationship between water cut and water oil ratio formula are used to calculate the water oil ratio of different sections, as well as water cut of different sections (Figure 2).

$$f_w = \frac{WOR}{1 + WOR} \quad (2)$$

$$\ln WOR = b(1 - S_{wi})R + bS_{wi} - \ln\left(a\frac{\mu_w}{\mu_o}\right) \quad (3)$$

Power, linear, and exponent functions have been applied to Regression of relative permeability curve according to different development stages, including oil phase relative permeability and water saturation curve and oil phase relative permeability and oil saturation curve. Then the relatively high precision function is given as the final regression function and finally through the regression function types decline types are received.

95% water content, the same degree of recovery in segmented and non-segmented.

98% water content, there is a difference in the degree of recovery, but the difference is very small.

Taking $5 < K < 10$ as an example, three functions are used to regress the relative permeability curve, and the function with relatively high accuracy is selected as the final regression function to determine the type of decline.

7 CONCLUSION

The lower the permeability, the faster the water content rises.

When the seepage velocity is relatively large, the starting pressure gradient has no obvious influence on the water cut curve.

When the seepage velocity is relatively small, the greater the starting pressure gradient, the faster the water cut curve will rise.

REFERENCES

[1] Gu Jianwei, Yu Hongjun, PengSongshui. Production characteristics of low-permeability oilfields under complex conditions. Journal of the University of Petroleum (Natural Science Edition), 2003;27(2):55–57.
[2] Gao Wenjun, LI Ning, Hou Chengcheng, Zhang Ningxian. Establishment and Optimization of Two Dimensionless Liquid Production Indexes and Water Cut Relationships. XinJiang Petroleum Geology, 2015.Vol. 36,No.1
[3] Lin Jiang, Li Zhifen, Zhang Qi. Study on liquid production index prediction for different water cut[J]. Petroleum Drilling Techniques, 2003, 31(4); 43–45.
[4] Yang Chao, Li Yanlan, Xu Bingxiang, et al. New nonlinear correction method of oilwater relative permeability curves and their application [J]. Oil & Gas Geology, 2013, 34(3); 394–398.
[5] Farquhar R A,Smar B G D. Stress sensitivity of low permeability sandstones from the rotliegendes sandstone[C].PE26501,1993:851–856.
[6] The Normalization Processing and Applicationof Dimensionless Fluid fluid production index in low permeability oilfield [J]. ScienceTechnologyandEngineering [J].
[7] Cui Yue, Shi Jingping. Experiments on the seepage features for the untabulated reservoirs in Daqing Oilfield [J].Petroleum Geology and Oilfield Development in Daqing,2017: Vol. 36 No. 2
[8] Gao Wenjun, Yao Jiangrong, Gong Xuecheng, et al. Study on oilwater relative permeability curves in water flooding oilfields [J].Xinjiang Petroleum Geology, 2014, 35(5); 100–105.
[9] Yu Qitai. Two types of changes of the rate of water cut with recovery factor and the corresponding relative oil -water permeabilitycurves for wate.

Chapter 5: Electrochemical Valuation, Fracture Resistance, and Assessment

Structural dynamic analysis of a box-type launch system under different support modes

X.X. Liu
Xi'an Institute of Modern Control Technology, Xi'an, China

B. Li
School of Mechanical Engineering, Nanjing University of Science and Technology, Nanjing China

J.S. Liu & H. Liu
Xi'an Institute of Modern Control Technology, Xi'an, China

ABSTRACT: The influence of support mode on the dynamic response of launch system is very obvious. In order to study the influence of support mode on the stability of a box launch system, the dynamic model of box-type launch system is established. The dynamic simulation of the launch system with rigid, semi-rigid and elastic support is carried out by using the explicit solver. the launch dynamic characteristics of the launch system under three support modes are simulated, the dynamic responses of key parts such as projectile table and instrument cabin under three support ways are obtained. The analysis results can provide a basis for the selection of support mode of a box-type launch system.

1 INTRODUCTION

The study of the dynamic response of the launch platform is of great significance for improving the launch attitude of the missile and revealing the general law of the launch response of the system. Yin Zengzhen et al [1] carried out the launch dynamics simulation of multi-rigid-body system and rigid-flexible coupling system. Xu Yue et al [2] established the flexible multi_body dynamics model of ship-borne missile vertical launch system (VLS), and carried out the dynamic simulation of the launch process. Sun Chuanbin et al [3] studied the vertical ejection response characteristics of the cold launch platform. Chen Yujun et al [4] analyzed the dynamic characteristics and feasibility of vehicle-borne missile launching at travel time. Shi Yichao et al [5] simulates the kinematics and dynamics of automata to verify the reliability of automata. The above literature shows that it is necessary to carry out the dynamic simulation of the weapon launch system in order to improve the reliability of the weapon launch system and optimize the structural parameters.

In this paper, on the basis of roadbed model parameters, hydraulic cylinder model and tire model, the dynamic analysis model of a box-type launch system is established, and the dynamic response of the launch system under three different support modes is taken as the research object. the dynamic simulation analysis of the launch system is completed.

2 RELATED BASIC MODE

2.1 *Model parameters of roadbed*

The roadbed is divided into four layers from top to bottom, and the parameters of each layer are shown in Table 1 [6].

Table 1. Parameters of roadbed model

Number of layers	d/(mm)	ρ/(g/cm³)	E	σ	α	β
1	20	2.4	1200	0.25	5	0
2	60	2.2	900	0.3	5	0
3	60	2.1	450	0.3	5	0
4	500	1.85	30	0.35	5	0

2.2 Model establishment of hydraulic cylinder

Hydraulic oil can actually be considered as the spring, its hydraulic stiffness:

$$C_T = \frac{P}{h} \quad (1)$$

where, P-external force; h-compression displacement.

2.3 Establishment of tire model of launch system

The equivalent simplified tire model is shown in figure 2, M is the load on the axle of the test device, C is the vertical dynamic damping of the tire, and K is the vertical dynamic stiffness coefficient of the tire.

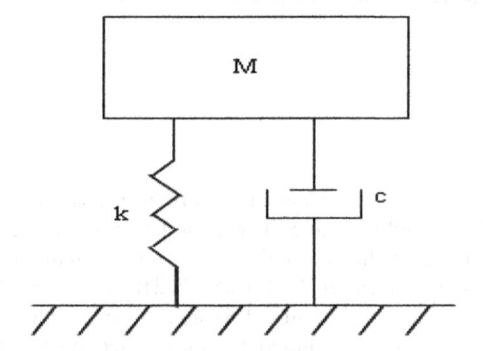

Figure 1. Vertical dynamic stiffness and damping model of tire.

3 LAUNCH SYSTEM DYNAMICS MODEL

The main components such as box body, projectile platform, Jack support leg, erection frame, support seat etc. are built, and the connection relationship between the parts is established. In the dynamic analysis of elastic support, it is necessary to establish a spring damper between the container and the foundation to simulate the effect of the tire, coupling and binding constraints are used between the other parts. The assembly and connection of the whole launch system is shown in figure 2. Before launch, the launch pad is subjected to the gravity of the projectile body, and during the launch process, the gravity gradually decreases to zero, and at the same time, the impact force of the gas flow begins to act on the support table. Before the dynamic calculation, the gas flow is simulated in FLUENT, and the relevant impact force of the gas flow is obtained and used as the load of the dynamic analysis. The load and boundary conditions are shown in figure 3.

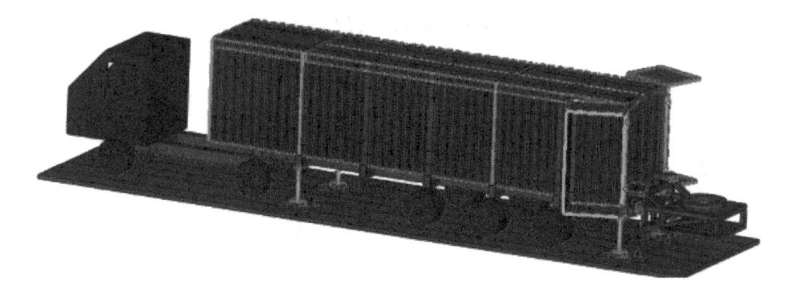

Figure 2. Schematic diagram of connection and assembly of a box launch system.

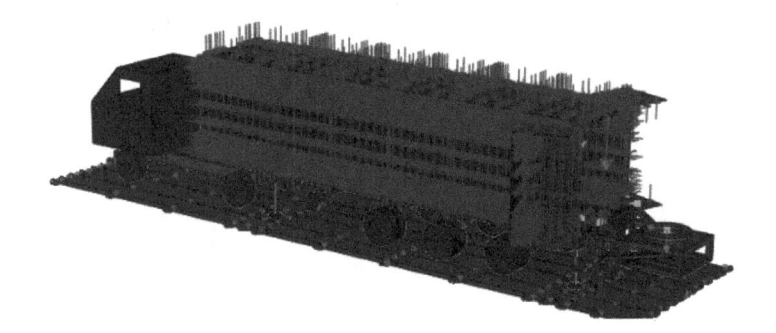

Figure 3. Schematic diagram of model constraint and load of a box launch system.

4 STRUCTURAL DYNAMIC ANALYSIS OF LAUNCH SYSTEM

4.1 *Analysis of launch system under rigid support*

In the launch vehicle model, the front and rear four Jack legs are used to land at the same time to establish a fully rigid support form.

4.1.1 *Vibration curve of projectile table*

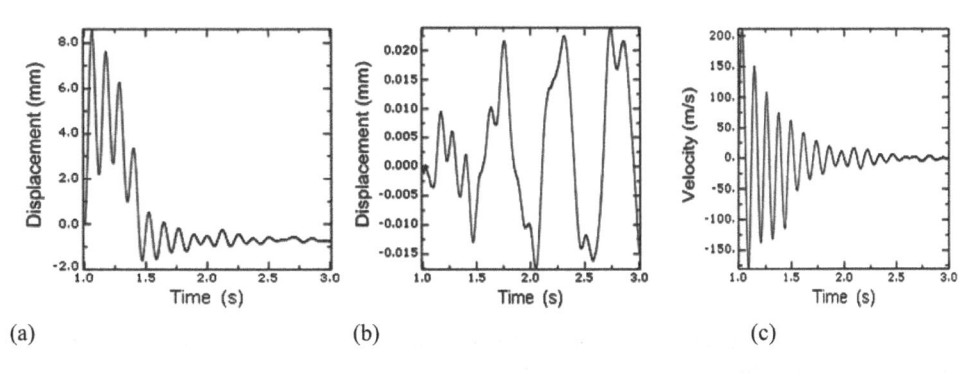

Figure 4. Vibration curve of projectile table under rigid support: (a) Vertical displacement, (b) Left and right lateral displacement and (c) Vertical velocity.

4.1.2 Vibration curve of instrument cabin under rigid support

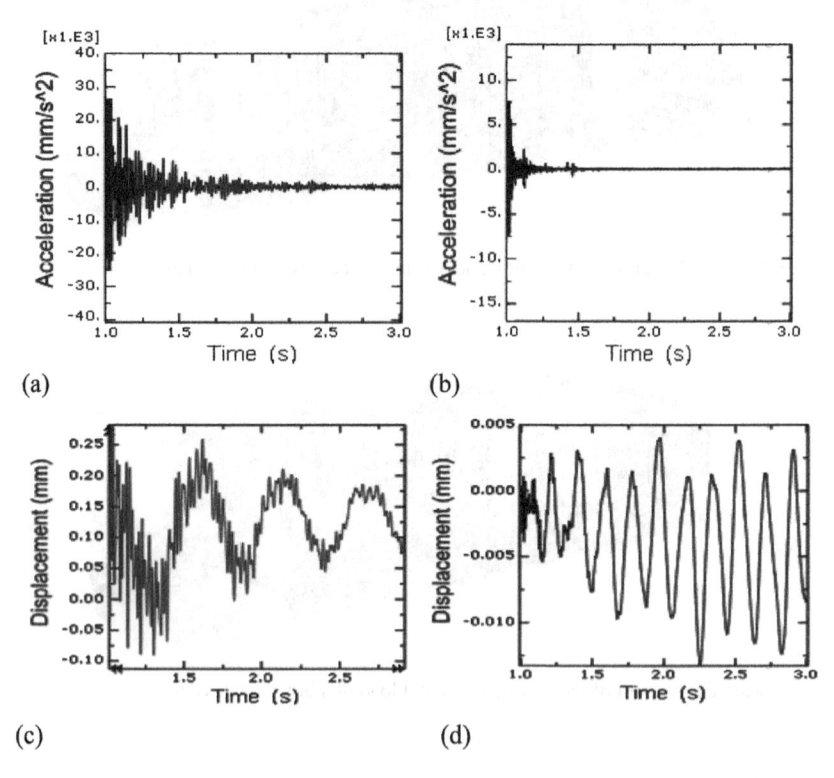

Figure 5. Vibration curve of instrument cabin under rigid support: (a) Vertical acceleration, (b) Left and right lateral acceleration, (c) Vertical displacement and (d) Left and right lateral displacement.

4.2 Analysis of launch system under elastic support

A pair of tires is established at the front, two pairs of tires are established at the connection between the front and the trailer, and three pairs of tires are established at the rear of the flat car.

4.2.1 Vibration curve of projectile table

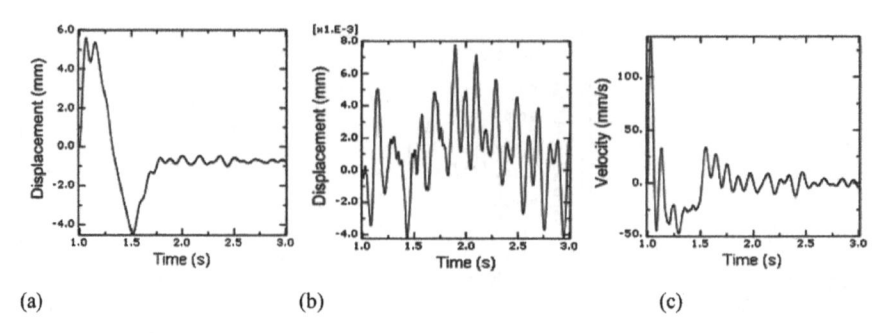

Figure 6. Vibration curve of projectile table under elastic support: (a) Vertical displacement, (b) Left and right lateral displacement and (c) Vertical velocity.

4.2.2 Instrument cabin vibration curve

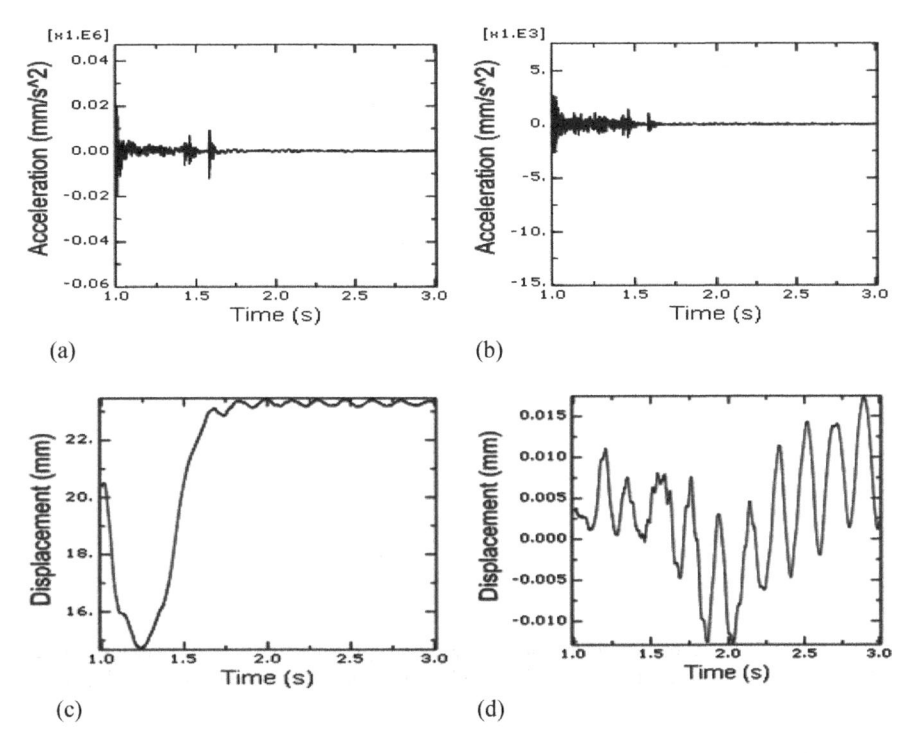

Figure 7. Vibration curve of instrument cabin under elastic support: (a) Vertical acceleration, (b) Left and right lateral acceleration, (c) Vertical displacement and (d) Left and right lateral displacement.

4.3 Analysis of launch system under semi-rigid support

The front wheel and rear Jack of the launch vehicle are supported on the ground, the front is elastic and the rear is rigid, and other conditions remain unchanged.

4.3.1 Vibration curve of projectile table

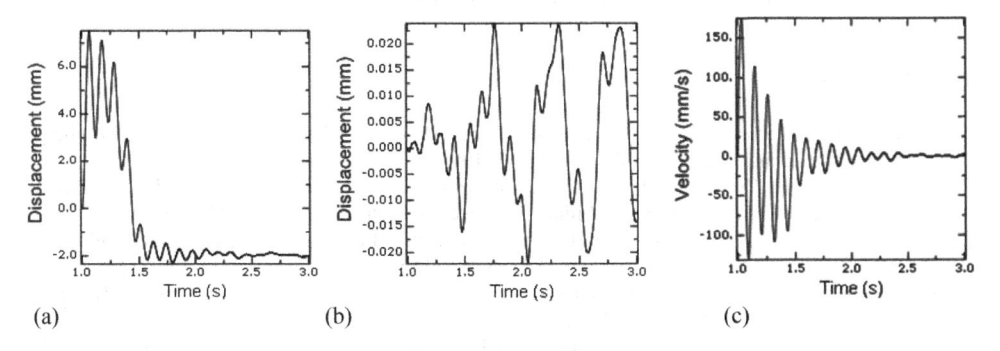

Figure 8. Vibration curve of projectile table under semi-rigid support mode: (a) Vertical displacement, (b) Left and right lateral displacement and (c) Vertical velocity.

4.3.2 Instrument cabin vibration curve

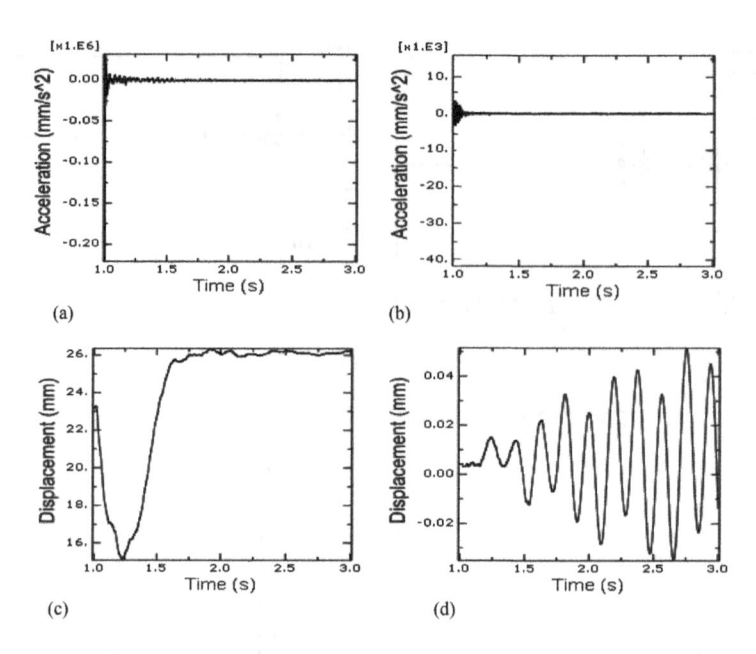

Figure 9. Vibration curve of instrument cabin under semi-rigid support mode: (a) Vertical acceleration, (b) Left and right lateral acceleration, (c) Vertical displacement and (d) Left and right lateral displacement.

5 CONCLUSION

1) Under the elastic support mode, the displacement of the launcher is small, and the damping effect is obvious, but it has a great influence on the vibration acceleration of the instrument cabin, which is easy to cause damage to important instruments and operators
2) Using rigid support and semi-rigid support, although the displacement of the launch pad is slightly larger, the acceleration of the left and right side and vertical direction of the instrument cabin is small, and the safety is good.
3) Compared with the rigid support mode, the semi-rigid support mode can reduce the acceleration amplitude of the instrument cabin and return to the stable state more quickly, so it is suggested that the semi-rigid support mode should be adopted in a certain box launch system.

REFERENCES

[1] Yin Zengzhen, Bi Shihua. Simulation Study on Launch Dynamics of Multi-flexible Vehicle Missile [J]. Journal of Missile and Guidance, 2009,29 (2): 183–190.
[2] Xu Yue, Tian Aimei, Zhang Zhenpeng. Flexible Multibody Dynamics Modeling and Simulation of Missile Vertical Launch System [J]. Journal of Military Science and Technology, 2008,29(9):1083–1087.
[3] Sun Shipin, Ma Dawei, Ren Jie, et al. Study on Vertical Ejection Response Characteristics of Cold Launch Platform [J]. Journal of Nanjing University of Science and Technology, 2015, 39 (5):516–522.
[4] Chen Yujun, Jiang Yi et al. Simulation analysis of launch dynamics of vehicle-borne missile [J]. Journal of Missile and guidance, 2011,31 (001): 55–58.
[5] Shi Yichao, Wang Yongjuan et al. Dynamic simulation study of a new automatic launch system [J]. Ordnance Automation, 2016, 35 (08): 25–32.
[6] Li Bo. Structural Analysis and dynamic Simulation of a Special Container launch system of a Rocket weapon [D]. Nanjing University of Science and Technology, 2008.

Servo feedforward compensation control of micro-texture machine tool based on ISOA

Z.M. Wang, G.Q. Wu, J.F. Mao & K. Hu
School of Electrical Engineering, Nantong University, Nantong, Jiangsu, China

ABSTRACT: To solve the problem of low speed servo tracking accuracy of micro-texture machine tools, based on the electromechanical coupling dynamic model, a control scheme combining improved seeker optimization algorithm (ISOA) and feedforward compensation PID is proposed, and the control structure of ISOA+ feedforward PID controller is designed, and the PID parameter optimization design is completed. In order to verify the actual effect of the control strategy, a simulation experiment platform is built by Matlab, and compared with the traditional intelligent optimization algorithm. The simulation results show that the feedforward compensation PID based on ISOA optimization has better optimization effect and greatly improves the speed tracking performance, which verifies its correctness and effectiveness.

1 INTRODUCTION

In order to further meet the technical requirements of modern machine tool processing industry, machine tool processing is bound to develop in the direction of high speed and precision. Micro-texture machine tools are widely used in the field of machine tools because they can accurately process complex micro-morphology and meet the technical requirements of rapid positioning system. Servo system and ball screw feed system constitute the servo feed system of micro-texture machine tool, and the performance of servo system affects the speed tracking accuracy [1] of micro-texture machine tool. In the process of designing servo system, PID controller is widely used in practical use [2] because of its strong applicability, simplicity and easy realization. However, the difficulty of PID parameter tuning has become a major difficulty in scientific research. With the deepening of intelligent algorithm research, people began to apply it to parameter optimization. In reference [3]. In order to solve the problem of overshoot and oscillation of system step response, SOA algorithm is proposed to adjust parameters, which improves the control accuracy of the system. In reference [4], aiming at the problem of low position tracking accuracy of micro-texture machine tools, the mathematical model of servo drive system of micro-texture machine tools is established, and the improved method of combining feedforward and feedback is adopted to improve the position tracking accuracy of machine tools. Literature [5] aims at the problem of low control precision of pump system, and uses seeker optimization algorithm to obtain good PID parameter self-tuning effect, which improves the operation stability of irrigation system.

To sum up, in order to further improve the speed servo tracking accuracy of micro-texture machine tools, based on the establishment of electromechanical coupling dynamic model, a feedforward PID control strategy based on ISOA is proposed to realize the PID parameter optimization design of micro-texture machine tool servo feed system. In order to verify the actual effect of the control strategy, a simulation experiment platform is built by Matlab, and compared with the traditional intelligent optimization algorithm. The simulation results show that the feedforward compensation PID based on ISOA optimization has better optimization effect and greatly improves the speed tracking performance, which verifies its correctness and effectiveness.

2 MODELING OF SERVO FEED SYSTEM OF MICRO TEXTURE MACHINE TOOL

2.1 Mathematical model of ac permanent magnet synchronous motor

Permanent magnet synchronous motor is assumed to be an ideal motor: eddy current loss is ignored, Permanent magnets also have no damping effect, Damping winding of motor is equivalent to d and q Two short-circuited damping windings on the shaft. The equivalent transformation of coordinates is established, and according to the principle of vector control, it is obtained $d - q$ Mathematical model of coordinate system, and its vector diagram model is as shown in Figure 1 Shown.

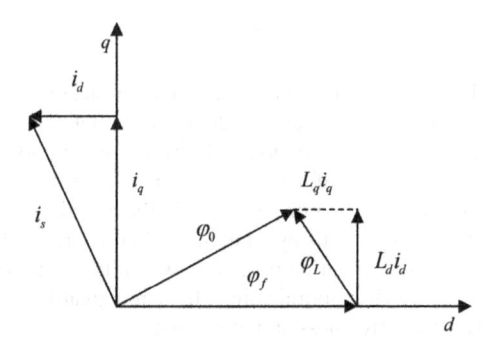

Figure 1. Vector of permanent magnet synchronous motor.

According to Figure 1, the current equation of PMSM can be deduced as follows

$$\begin{cases} \dot{i}_d = \dfrac{1}{L_d}(u_d - Ri_d + p_n\omega_r L_q i_q) \\ \dot{i}_q = \dfrac{1}{L_q}(u_q - Ri_q - p_n\omega_r L_d i_d - p_n\omega_r \varphi_f) \end{cases} \quad (1)$$

2.2 Modeling of ball screw feed system

The ball screw and servo motor are connected together by coupling, the ball screw is supported by bearings at both ends, the worktable is driven by the rotation of the ball screw nut, and makes linear reciprocating motion along the linear guide rail pair, and the grating ruler is used as a negative feedback device to form a complete closed loop of the system, and its model is shown in the figure below As shown in [6].

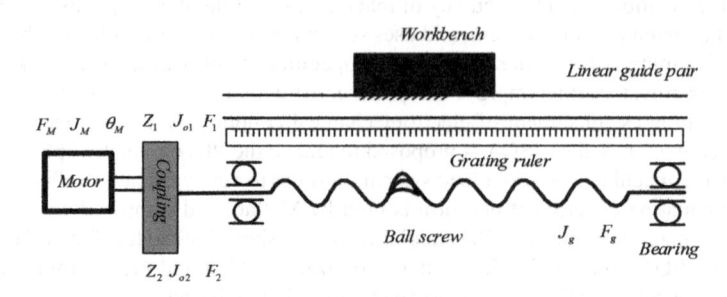

Figure 2. Model of ball screw feed mechanism.

$$\begin{cases} J = J_M + J_{o1} + \left(\dfrac{Z_1}{Z_2}\right)^2 (J_{o2} + J_g + J_M) \\ F = F_M + F_1 + \left(\dfrac{Z_1}{Z_2}\right)^2 (F_2 + F_g) \\ J\dfrac{d\omega_r}{dt} = T_e - F\omega_r - T_L \end{cases} \quad (2)$$

In which, T_L is disturbance, J is expressed as the sum of moments of inertia, it can be calculated by equivalent moments of inertia J_{o1}, J_{o2}, Rotational inertia of lead screw J_g And that rotational inertia of the motor shaft J_M Find out, F expressed as the total damping coefficient, which can be determined by the damping of the motor shaft F_M, Ball screw damping F_g And equivalent dam between lead screw and motor F_1, F_2 Find out, $\frac{Z_1}{Z_2}$ is the conversion ratio.

2.3 Electromechanical coupling dynamic model

Under the assumption of an ideal motor, make $L_d = L_q = L$, $F_M = 0$. In order to realize fast speed regulation, make $i_d = 0$. The state equation of the fully decoupled micro textured machine tool servo feed system is

$$\begin{bmatrix} \dot{i}_q \\ \dot{\omega}_r \end{bmatrix} = \begin{bmatrix} -\dfrac{R}{L} & -\dfrac{p_n \phi_f}{L} \\ \dfrac{3 p_n \phi_f}{2J} & F \end{bmatrix} \begin{bmatrix} i_q \\ \omega_r \end{bmatrix} + \begin{bmatrix} \dfrac{u_q}{L} \\ \dfrac{-T_L}{J} \end{bmatrix} \quad (3)$$

According to the formula (3), The transfer function of servo feed system of micro textured machine tool $G_M(s)$ is

$$G_M(s) = \dfrac{\dfrac{3}{2} p_n \varphi_f}{LJs^2 + (RJ + LF)s + FR + \dfrac{3}{2} p_n^2 \varphi_f^2} \quad (4)$$

3 DESIGN OF LINEAR CONTROLLER FOR FEED SYSTEM

3.1 PID control

The traditional PID controller used in the servo feed system of micro textured machine tool mainly has three parameters K_i, K_p and K_d. The PID controller block diagram is shown in the Figure 3

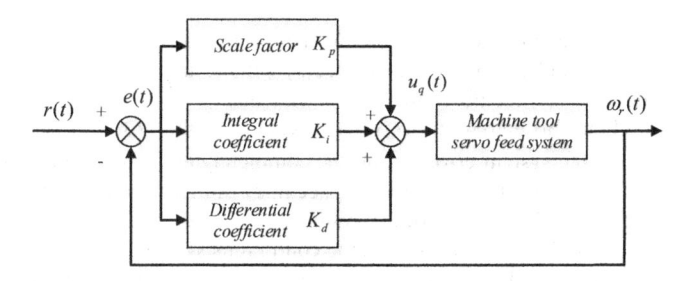

Figure 3. Block diagram of PID controller.

The control law is expressed as

$$u_q(t) = K_p \left(e(t) + \frac{1}{T_i} \int_0^t e(t)dt + T_d \dot{e}(t) \right) \tag{5}$$

In which, $e(t)$ is the velocity tracking error, T_i and T_d are integral time constant and differential time constant respectively.

3.2 Design of feedforward compensation controller

Considering that the feedback control link of traditional PID control has the problem of time delay, these problems will cause control delay or even failure, so the feedforward compensation PID control method is proposed. Its principle block diagram is shown in the Figure 4

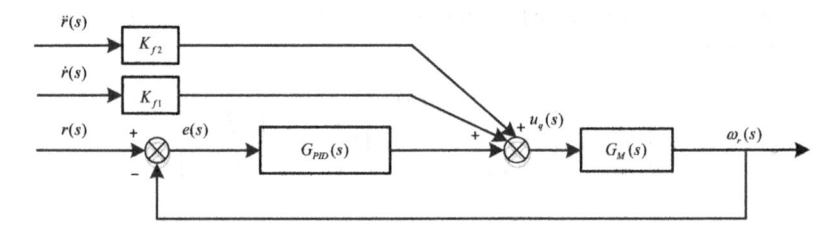

Figure 4. Feed forward PID controller block diagram.

System transfer function $\frac{\omega_r(s)}{r(s)}$ and the transfer function of input and systematic error $\frac{e(s)}{r(s)}$ for

$$\begin{cases} \dfrac{\omega_r(s)}{r(s)} = \dfrac{[G_{PID}(s) + K_{f1}\dot{r}(s) + K_{f2}\ddot{r}(s)]G_M(s)}{1 + G_{PID}(s)G_M(s)} \\ \dfrac{e(s)}{r(s)} = 1 - \dfrac{[G_{PID}(s) + K_{f1}\dot{r}(s) + K_{f2}\ddot{r}(s)]G_M(s)}{1 + G_{PID}(s)G_M(s)} \\ \qquad = \dfrac{1 - [K_{f1}\dot{r}(s) + K_{f2}\ddot{r}(s)]G_M(s)}{1 + G_{PID}(s)G_M(s)} \end{cases} \tag{6}$$

In which, $r(s)$ is the initial input signal, $u_q(s)$ is Voltage input under Laplace transform, $G_{PID}(s)$ is the PID transfer function under Laplace transform, $e(s)$ is the velocity tracking error under Laplace transform, K_{f1} and K_{f2} are all feedforward gain coefficients.

According to the above formula, when $f_1\dot{r}(s) + f_2\ddot{r}(s) = \frac{1}{G_M(s)}$, the system error can be eliminated theoretically. Although the system error can not be completely eliminated in practical use, it can be reduced to an acceptable level and the tracking accuracy of machine speed servo can be greatly improved without affecting the stability of the system.

For PID sampling period T_c, In a short time, the continuous system can be directly transformed into a discrete system through PID discretization, so the feedforward compensation PID discrete control law can be expressed as

$$u_q(k) = K_p \left(e(k) + \frac{T_c}{T_i} \sum_{j=0}^{k} e(j) + \frac{T_d}{T_i}(e(k) - e(k-1)) \right)$$

$$= K_p e(k) + K_i \sum_{\iota=0}^{k} e(\iota) + K_d (e(k) - e(k-1))$$

$$+ K_{f1}\dot{r}(k) + K_{f2}\ddot{r}(k) \qquad (7)$$

In which, $e(k)$ is the velocity tracking error.

For a servo feed system of micro textured machine tool with nonlinearity and hysteresis, it is difficult to adjust the parameters of PID controller, which is simple and easy to realize. Therefore, an improved seeker optimization algorithm is used to realize the global optimal control of the system.

4 SEEKER OPTIMIZATION ALGORITHM AND ITS IMPROVEMENT

4.1 SOA algorithm

Seeker optimization algorithmis a new intelligent algorithm of human population behavior. The objective function is constructed by using the absolute value of the time integral in order to obtain the appropriate dynamic iteration characteristics and construct the objective function f for

$$f = \begin{cases} \int_0^\infty [\eta_1|e(k)| + \eta_2 u^2(k)]dk, e(k) \geq 0 \\ \int_0^\infty [\eta_1|e(k)| + \eta_2 u^2(k) + \eta_3|e(k|]dk, e(k) < 0 \end{cases} \qquad (8)$$

In which, in order to obtain better control performance, η_1 takes 0.999, η_2 takes 0.001, η_3 takes 100.

Seeker optimization algorithm follows crowd search rules and uses *Fuzzy*. The approximation function of the system can obtain the search step size α_{fj}. The relationship is

$$\alpha_{fj} = \psi_{fj}\sqrt{-\ln(\varpi_{fj})} \qquad (9)$$

In which, ψ_{fj} is the parameter of membership function, ϖ_{fj} is the membership degree of the search space objective function.

Through the rational analysis of the pre-movement direction d_e, the altruistic direction d_a and the self-interested direction d_p, the search step size d_f is obtained as

$$d_f(t) = sign(\eta_0 d_p + m_1 d_e + m_2 d_a) \qquad (10)$$

In which, $sign()$ is a symbolic function, m_1 and m_2 for [0, 1] Interval random real numbers.

Get the search step size α_{fj} and search direction d_f after that, the individual position is updated to get the updated position $x_{fj}(t+1)$ for

$$x_{fj}(t+1) = x_{fj}(t) + \alpha_{fj}(t)d_i(t) \qquad (11)$$

4.2 Improved SOA algorithm

In order to solve the problem that seeker optimization algorithm is inefficient in the early stage and trapped in local extremum in the later stage, it can not find the global optimal solution *cauchy* Mutation operator, *Logistic* chaotic mapping function is iterated repeatedly [7]

$$x(\Phi + 1) = \mu x(\Phi)[1 - x(\Phi)] \tag{12}$$

In which, Φ is the number of iterations μ is Tuning parameters

Randomly generate an H-dimensional reference particle y_0 in the interval $(0,1)$, $y_0 = (y_{01}, y_{02}, y_{03} \ldots y_{0H})$, The chaotic population set is generated $y_{n+1,j}$ for

$$y_{n+1,j} = \mu y_{n,j}(1 - y_{n,j}) \tag{13}$$

Next, the formula will be Map to search space Middle, get Dimensional particle population for

$$x_{n+1,j} = \Gamma \times (2 \times y_{n+1,j} - 1) \tag{14}$$

In which, $n = 0, 1, 2, \ldots, N, j = 1, 2, \ldots, H$.

In order to solve the problem of falling into local extremum in the later stage of search, the method of introducing *cauchy*. The main performance of Cauchy distribution is that its distribution function has long two wings, and its generated random number has a wider mutation range. Therefore, the increased mutation range of Cauchy mutation operator makes the improved algorithm have a wider search range. The formula of Cauchy variation is [8].

$$x_f^k = \begin{cases} x_f^k \times cauchy(0,1), rand(0,1) \leq p \\ x_f^k, other \end{cases} \tag{15}$$

In which, p is random variation rate *cauchy()* is a quasi Cauchy distribution function. Then we use the Cauchy distribution pair G_{best} Perform mutation operation as

$$\begin{cases} \dot{G}_{bestj} = G_{bestj} + \Upsilon \times A(0,1) \\ \Upsilon = e^{-\frac{\lambda t}{T}} \end{cases} \tag{16}$$

In which, Υ is the variation weight, G_{bestj} is the global optimum, j is Dimension component, λ is a constant 10, $A()$ is the random number initially generated by Cauchy distribution.

5 DESIGN OF FEEDFORWARD CONTROLLER BASED ON ISOA ALGORITHM

5.1 Schematic diagram of isoa-pid controller

Using the feedforward compensation PID control strategy based on ISOA, the output speed tracking value and its error value are optimized respectively, and the three parameters K_p, K_i and K_d of the controller are adjusted. The principle of the control system is shown in the Figure 5.

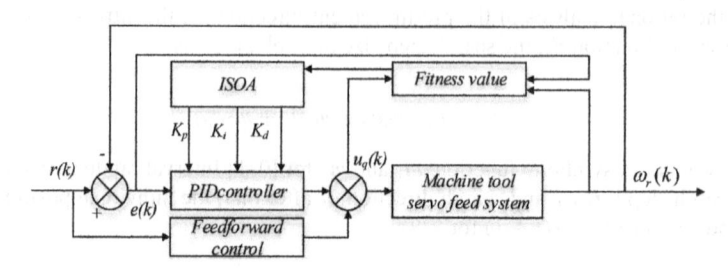

Figure 5. Schematic diagram of control system.

5.2 Algorithm flow of isoa-pid

The Figure 6 shows the ISOA optimization PID parameter flow chart

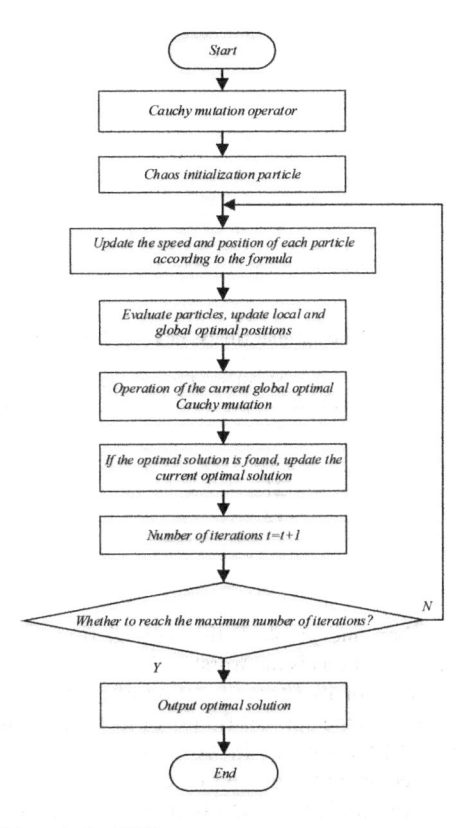

Figure 6. Flow chart of ISOA optimized PID parameters.

6 SIMULATION TEST AND RESULT ANALYSIS

In order to verify the correctness and effectiveness of the ISOA-feedforward PID algorithm, the coupling model parameters were designed, and the transfer function of servo feed system of micro textured machine tool was obtained. The results were compared with seeker optimization optimization (SOA), particle swarm optimization (PSO) and genetic optimization algorithm (GA) based on MATLAB simulation The transfer function of servo feed system of micro textured machine tool is

$$G(s) = \frac{1.05}{0.0000068s^2 + 0.00247s + 0.7925} \quad (17)$$

According to the mathematical model of micro textured machine tool servo feed system, the optimization algorithm programs based on ISOA, SOA, PSO and GA are written respectively, and the initial population is set as 30, dimension is 3, the sampling period is 1 ms the maximum number of iterations is 100, the minimum value of membership is 0.0111. simulation results show that the control fitness function of SOA, GA, ISOA and PSO is as follows Figure 7. table 1 shows the number of iterations and optimal fitness value of this algorithm.

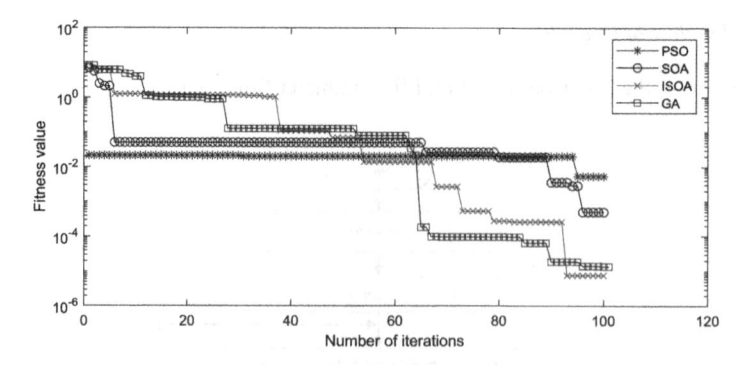

Figure 7. Control curve of fitness function.

Table 1. The number of iterations and the optimal fitness table of these four algorithms.

optimization algorithm	Number of iterations	Optimal fitness
PSO	3	5.3×10^{-3}
SOA	10	4.972×10^{-4}
GA	21	7.6817×10^{-5}
ISOA	12	1.3618×10^{-6}

As shown in the Table 1 Sum and graph 7, the optimal fitness value obtained by the PSO optimization algorithm after 3 iterations is the worst and is 5.3×10^{-3} the SOA optimization algorithm obtains the optimal fitness of its fitness function after 10 iterations is 4.9742×10^{-4} the GA optimization algorithm obtains the optimal fitness of its fitness function after 21 iterations is 7.6817×10^{-5} then the optimal fitness of the ISOA optimization algorithm is 1.3618×10^{-6} and only 12 Iterations. Therefore, compared with the four optimization algorithms, PSO optimization algorithm has the worst stability, while the ISOA optimization algorithm is the fastest and most stable in finding the optimal fitness value.

Using Equation(17) as the basis of the simulation, input the sinusoidal acceleration signal of $r = 0.5 \sin(2\pi t)$, and then use 6 algorithm simulations. Figure 8 is the sinusoidal response velocity tracking curve, and Figure 10 is the sinusoidal response velocity tracking error curve.

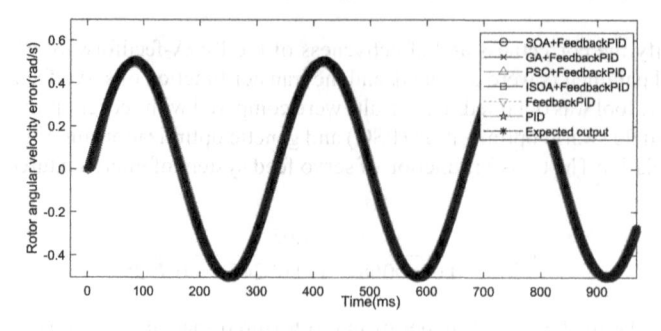

Figure 8. Sine response velocity tracking curve.

The sinusoidal response speed tracking curve and the sinusoidal response speed tracking error curve are respectively enlarged to make the comparison between the improved method and the traditional optimization algorithm more clear and intuitive. Figure 10 is an enlarged view of the

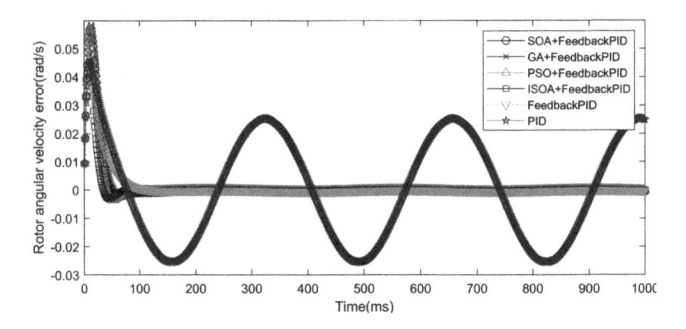

Figure 9. Velocity tracking error curve of sinusoidal response.

sinusoidal response speed tracking curve, Figure 11 and Figure 12 are a partial enlarged view of the sinusoidal response velocity tracking error curve.

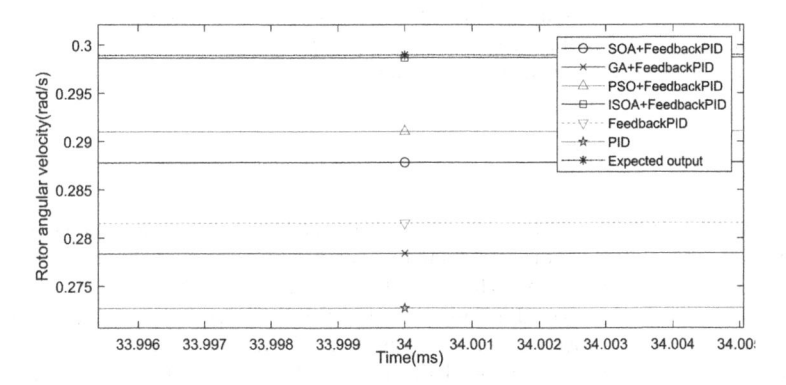

Figure 10. Enlarged part of sinusoidal response velocity tracking curve.

It can be seen from Figure 10 that Among these 6 algorithms, the ISOA+feedforward compensation PID speed servo tracking effect is the best, the tracking accuracy is higher, and the conventional PID tracking accuracy is the worst.

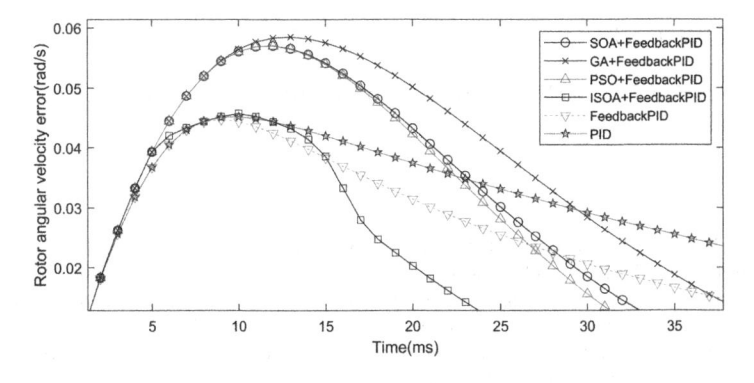

Figure 11. Enlarged part of velocity tracking error curve of sinusoidal response.

It can be seen from Figure 11 that Among these 6 algorithms, the peak error of GA+feedforward PID is the largest 0.058 rad/s, while the error peak of ISOA-feedforward PID is the smallest and is 0.046 rad/s.

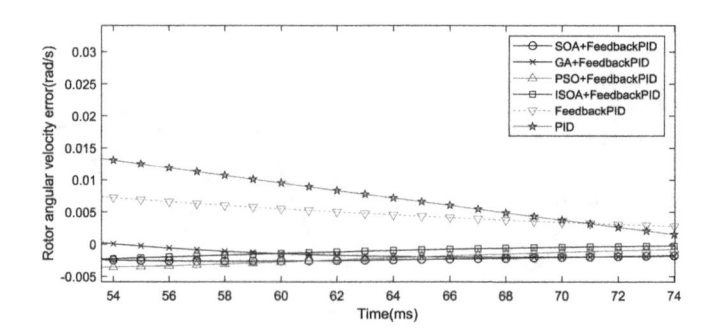

Figure 12. Partial enlarged view of sinusoidal response speed tracking error curve.

As shown in Figure 12, the ISOA+feedforward compensation PID method is the fastest to adjust and reach a steady state, followed by the PSO+feedforward compensation PID method to reach a steady state, and the speed tracking performance of other algorithms is slightly worse. In summary, for most nonlinear systems with hysteresis, the micro-texture machine tool servo motion control system is based on the ISOA optimized feedforward compensation PID control effect.

7 CONCLUSION

The feedforward compensation PID control optimized based on the ISOA algorithm has the shortest adjustment time in the micro-texture machine tool servo feed control, the speed servo tracking is accurate, and the dynamic response characteristics are better. The best control strategy. In the process of using the ISOA algorithm, adjusting the fitness function value and the size of the particle population can improve the optimization effect of PID parameters, and to a large extent eliminate the nonlinear and time-varying characteristics of the PID controller in the micro-texture machine tool servo feed system. And the problem of hysteresis, feedforward compensation is a commonly used model control method in time-delay systems. This method can obtain better dynamic characteristics and higher control accuracy without affecting system stability as much as possible.

ACKNOWLEDGMENTS

This work is supported by the Natural Science Foundation of China under No. 61273151.

REFERENCES

[1] Liu Zeyu, Wei Xin, Xie Xiaozhu, Hua Xiangang. Research on the application of micro-texture on the surface of laser machining tools [J]. Mechanical Design and Manufacturing, 2015(06):267–269+272.
[2] Yin chengqiang, Gao Jie, Sun Qun, et al. two-degree-of-freedom PID control based on improved Smith predictive control structure [J]. acta automatica sinica, 2020, 46(6):1274–1282.
[3] Liu Sheng, Liu Jianghua. Design of PID controller for servo system based on searcher first algorithm [J]. Control Engineering, 2017, 24(11):2189–2194.
[4] Wang Fan, Wu Guoqing, Zhang Xudong, Huang Zhenyi. Research on servo drive control system of laser micro-texture machining machine tool [J]. Modern Electronic Technology, 2020, 43(15):105–109.
[5] Xu Jinghui, Wang Lei, Tan Xiaoqiang, Wang Yichen, Zhao Zhongsheng, Shao Mingye. Research on intelligent irrigation control strategy based on SOA optimizing PID control parameters [J]. Journal of Agricultural Machinery, 2020, 51(04):261–267.

[6] Fu Zhenbiao, Wang Taiyong, Zhang lei, et al. Dynamic modeling and dynamic characteristics analysis of ball screw feed system [J]. Vibration and Shock, 2019, 38(16):56–63.
[7] Wang Rihong, Li Xiang, Li Na. Wolves algorithm based on Gaussian disturbance and chaos initialization [J]. Computer Engineering and Design, 2019, 40(10):2879–2884.
[8] Naseri H, Najafi S E, Saghaei A. Statistical process control (SPC) for short production run with Cauchy distribution, a case study with corrected numbers approach[J]. Communications in Statistics, 2020, 49(4/6):879–893.

An unbiased expression for calculating net power delivered by a directional coupler

J.Y. Li*
National Institute of Metrology (NIM), Beijing, China

ABSTRACT: A directional coupler is usually adopted for determining the net power delivered to the load in many applications, such as in a Power Flux Density (PFD) system. This paper proposes an unbiased expression for the net power, which requires to calibrate only one more S-parameter of the coupler as well as the reflection coefficient than the traditional method. The proposed method was evaluated against the traditional method and an improved traditional method in the literature by using a Monte Carlo analysis and with real measurements. The method outperforms the traditional methods under various coupler-directivity and load conditions. Its performance is consistent under all the conditions, suggesting its independency of the coupler directivity or the load reflection coefficient.

1 INTRODUCTION

A dual-directional coupler is usually adopted in a power delivery system. The coupler and two power sensors attached to its coupling ports are used to determine the net power delivered to the load. Figure 1 shows a schematic diagram of a Power Flux Density (PFD) system of such an application.

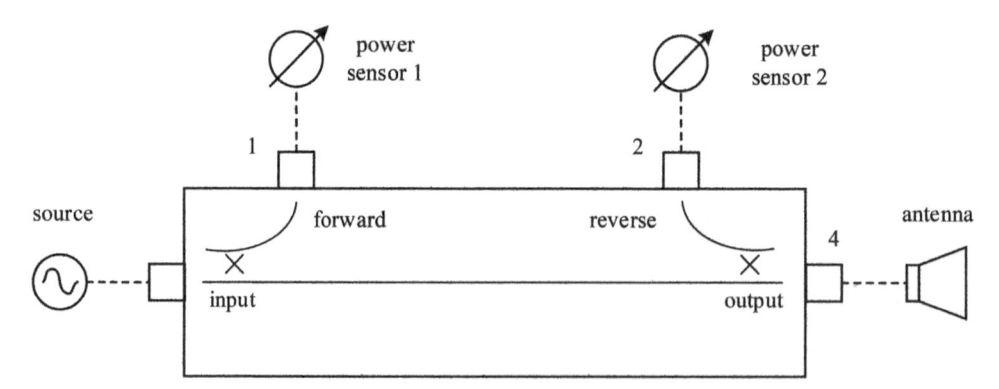

Figure 1. Power delivery by a dual-directional coupler.

Traditionally, the net power is calculated using the absolute values of three S-parameters of the coupler and the power measured at the coupling ports [1], [2], as shown in equation (1),

$$P_{net} = \frac{|S_{34}|^2}{|S_{13}|^2} \cdot P_1 - \frac{1}{|S_{24}|^2} \cdot P_2 \quad (1)$$

where P_1 and P_2 are the power emerging from Port 1 and Port 2 of the coupler, respectively. This expression is easy to use, involving no complex-number calculation. However, it is an approximate expression, only applicable to couplers with a high directivity, e.g. higher than 25 dB. An improved expression is proposed in [3] for couplers with lower directivities, which replaces the second term on the right hand side of equation(1) with a term involving the absolute reflection coefficient of the load. An exact expression for the net power using the full 4-port S-parameters of the coupler is proposed in [4]. The exact expression was used to evaluate the uncertainty introduced by the net power model [1], [3]. Nowadays, the widely used Vector Network Analyzers (VNAs) make it feasible to measure the full 4-port S-parameters of the coupler to calculate the net power with the exact expression. However, it is cumbersome to do so and it may introduce more uncertainty when using many measured S-parameters. A method for calculating the net power using scalar coupling coefficients is compared with a transfer method in [5]. The traditional method shown in equation (1) is modified by attaching a matched load on Port 2 in [6].

This paper proposes an unbiased expression for the net power, which outperforms the traditional expression in [2] and the improved traditional expression in [3]. The performance of the expression was evaluated using the improved Monte Carlo analysis based on the exact expression [3] and with real measurements.

2 METHODOLOGY

The traditional expression as shown in equation (1) fails when the directivity of the coupler is low and the absolute reflection coefficient of the load is high. Under such conditions, the power leaked to a coupling port from the nominally isolated main port is relatively high, resulting in an error in the power emerging from the coupling port, which is not accounted for in equation (1). This is especially true for Port 2 of the coupler. However, the reflected power from the load can be calculated more accurately using the incident power and the reflection coefficient of the load, as done in [3]. One only needs to improve the calculation of the incident power on the load, i.e. the emerging power from Port 4. The aim is to establish a relationship between the power emerging from Port 4 and that from Port 1.

The voltage waves emerging from Port 1 and Port 4 can be expressed using the full 4-port S-parameters of the coupler and the incident waves on the four ports. The power incident on Port 1 and Port 2 is much less than that incident on Port 4 as well as Port 3, because the former two are the multiplication of the power emerging from each coupling port and the squared absolute reflection coefficient of the corresponding power sensor, both of which are less than the counterparts at Port 4. Taking into consideration this fact and the reflected wave from the load, the waves emerging from Port 1, Port 2, and Port 4 can be approximated by equations (2), (3) and (4),

$$b_1 \approx a_3 \cdot S_{13} + b_4 \cdot S_{14}\Gamma_4 \tag{2}$$

$$b_2 \approx a_3 \cdot S_{23} + b_4 \cdot S_{24}\Gamma_4 \tag{3}$$

$$b_4 \approx a_3 \cdot S_{43} + b_4 \cdot S_{44}\Gamma_4 \tag{4}$$

where a_i is the incident voltage wave on Port i,
b_i is the emerging voltage wave from Port i
Γ_4 is the reflection coefficient of the load, and
S_{ij} is the S-parameter of the coupler.

By combining equations (2) and (4) and eliminating a_3, the relationship between b_1 and b_4 is obtained, from which the incident power on the load is obtained as shown in equation (5).

$$P_{\text{inc}} = \left| \frac{S_{43}}{S_{13}(1 - S_{44}\Gamma_4) + S_{43}S_{14}\Gamma_4} \right|^2 \cdot P_1 \tag{5}$$

Combining equations (3) and (4) and following a similar procedure and taking into account $a_4 = b_4 \Gamma_4$, the reflected power from the load can be obtained as shown in equation (6).

$$P_{\text{ref}} = \frac{1}{\left| S_{24} + \frac{S_{23}(1-S_{44})}{S_{43}\Gamma_4} \right|^2} \cdot P_2 \tag{6}$$

Then the net power can be calculated by subtracting P_{ref} from P_{inc} or using a simplified version that adopts the squared absolute reflection coefficient of the load to compute the reflected power as shown in equation (7).

$$P_{net} = \left| \frac{S_{43}}{S_{13}(1 - S_{44}\Gamma_4) + S_{43}S_{14}\Gamma_4} \right|^2 (1 - |\Gamma_4|^2) \cdot P_1 \qquad (7)$$

3 NUMERICAL RESULTS

The improved Monte Carlo analysis in [3] was adopted to evaluate the performance of the improved method for calculating the net power. Two case studies are presented in this section. Firstly, simulation under the conditions in Case 2 in [2] was carried out to compare the proposed method with the traditional and the improved traditional methods for various coupler directivities and various load reflection coefficients. Then, real measurements of the S-parameters of a coupler and of the reflection coefficient of an antenna were used to calculate the net power to compare the three methods.

3.1 Case 1

The difference in decibel between the net power obtained from the proposed expression in equation (7) and the net power from the exact expression using the full 4-port S-parameters of the coupler was calculated under the conditions in Case 2 in [2]. The statistical results are summarized in Table 1. The mean values and standard deviation from the traditional method and the improved traditional method in [3] are listed for comparison.

Table 1. Comparison between the proposed method and the traditional method as well as the improved traditional method.

$|\Gamma_4|$	Directivity [dB]	$|S_{14}|$ = $|S_{23}|$ [dB]	Proposed method (Based on equation (7))				Traditional method (Based on equation(1))		Improved traditional method (In reference [3])	
			Mean [dB]	Standard deviation [dB]	Max/min [dB]	Kurtosis	Mean [dB]	Standard deviation [dB]	Mean [dB]	Standard deviation [dB]
0.05	30	0.0032	0.0000	0.02	0.02/−0.02	1.49	−0.005	0.03	0.000	0.02
0.05	20	0.0100	−0.0001	0.02	0.02/−0.02	1.50	−0.049	0.05	0.000	0.04
0.05	10	0.0316	−0.0001	0.02	0.03/−0.03	1.52	−0.512	0.16	0.000	0.09
0.1	30	0.0032	0.0000	0.02	0.02/−0.02	1.50	−0.005	0.04	0.000	0.04
0.1	20	0.0100	0.0000	0.02	0.02/−0.02	1.50	−0.050	0.10	0.000	0.07
0.1	10	0.0316	0.0000	0.03	0.03/−0.03	1.53	−0.522	0.32	0.000	0.19
0.15	30	0.0032	0.0000	0.02	0.02/−0.02	1.50	−0.005	0.07	0.000	0.06
0.15	20	0.0100	0.0000	0.02	0.02/−0.02	1.50	−0.051	0.14	0.000	0.10
0.15	10	0.0316	0.0000	0.03	0.03/−0.03	1.53	−0.538	0.48	−0.001	0.28
0.224	30	0.0032	0.0000	0.02	0.02/−0.02	1.50	−0.006	0.10	0.000	0.08
0.224	20	0.0100	0.0000	0.02	0.02/−0.02	1.51	−0.054	0.22	0.000	0.15
0.224	10	0.0316	0.0000	0.03	0.03/−0.03	1.53	−0.580	0.76	0.000	0.42
0.333	30	0.0032	0.0000	0.02	0.02/−0.02	1.52	−0.006	0.16	0.000	0.12
0.333	20	0.0100	0.0001	0.02	0.02/−0.02	1.51	−0.061	0.35	0.000	0.22
0.333	10	0.0316	0.0001	0.03	0.03/−0.03	1.54	−0.689	1.25	−0.003	0.62
0.5	30	0.0032	−0.0001	0.02	0.02/−0.02	1.53	−0.011	0.28	0.000	0.18
0.5	20	0.0100	0.0001	0.02	0.03/−0.03	1.53	−0.090	0.63	0.000	0.33
0.5	10	0.0316	0.0000	0.03	0.03/−0.03	1.56	invalid net power		0.003	0.94

It is obvious that the proposed method is superior to the traditional method in terms of the mean value and the standard deviation under all the directivity and load conditions. It is superior to or as equally good as the improved traditional method under all the conditions. The almost-zero mean values and the symmetric maximum and minimum values show that no bias is introduced by the method. The good consistency in the mean values, the standard deviation, the max/min values, and the kurtosis values under all the conditions shows that the method is independent of the coupler directivity and the load conditions.

3.2 Case 2

Real measurements of the S-parameters of a dual-directional coupler and of the reflection coefficient of a broadband antenna over the frequency range 300 MHz to 1 GHz were used to calculate the net power in this case study.

The VSWR (Voltage Standing Wave Ratio) of the antenna is shown in Figure 2. It can be seen that most of the VSWR values over 750 MHz are greater than 1.5, which corresponds to an absolute reflection coefficient of 0.2 and is not a very small value.

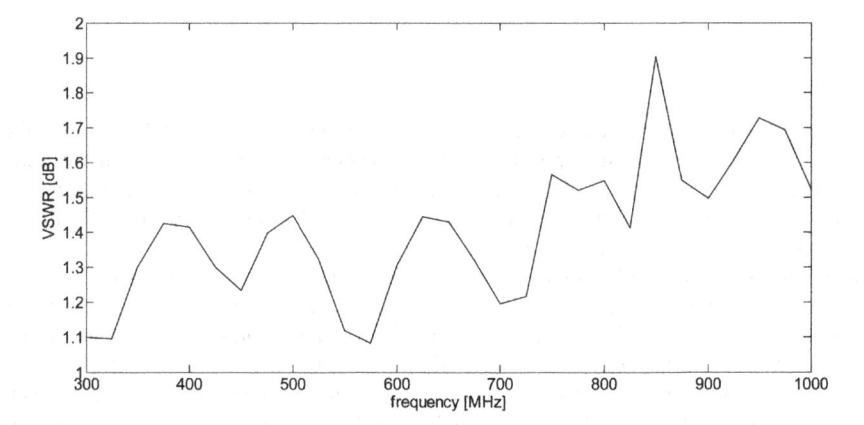

Figure 2. The VSWR of the antenna.

The directivity values for the two coupling ports of the coupler over the frequency range are shown in Figure 3. It is obvious that the performance of the two coupling lines is not similar. It is also noted that the directivity with Port 4 as the input port is less than 25 dB at 1 GHz, which is not a high directivity value.

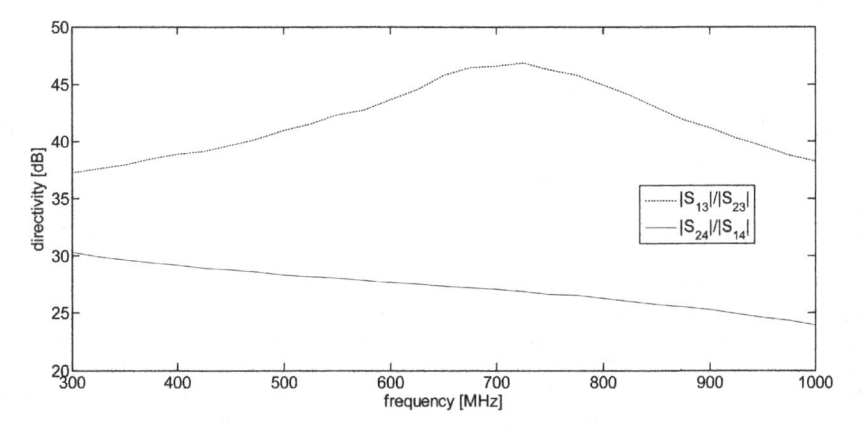

Figure 3. The directivity of the coupler.

The error in the net power introduced by the model for the three methods is compared in Figure 4. It is obvious that the proposed method outperforms the traditional and the improved traditional methods. The error for the proposed method is mainly within ±0.02 dB, which is consistent with the results in Table 1.

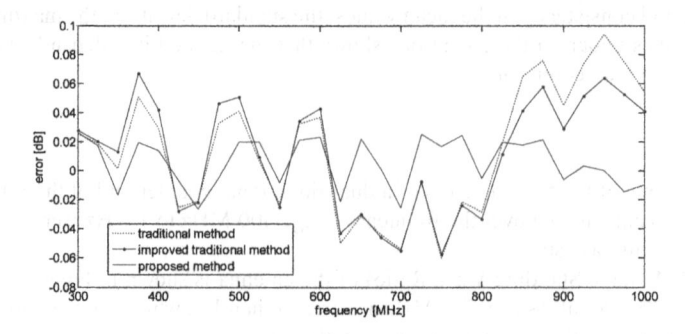

Figure 4. Comparison between the methods based on real measurements.

4 CONCLUSION

This paper proposes an improved expression for calculating the net power delivered to the load by a directional coupler. It uses only one more S-parameter of the coupler than the traditional method. It involves computation of complex numbers, which is readily obtained with a computer. The complex S-parameters of the coupler and the complex reflection coefficient of the load can be easily measured with a widely used VNA.

The method was evaluated against the traditional and the improved traditional methods in the literature using the Monte Carlo analysis. Under various coupler-directivity and load conditions, the proposed method outperforms the other methods in terms of bias and the model uncertainty component. Moreover, its performance under all the conditions is rather consistent, suggesting its independency of the coupler directivity or the load reflection coefficient. With only the power at the forward coupling port required, the method can be used for uni-directional couplers. Nevertheless, when using equation (7) with a dual-directional coupler, equation (6) can be used for monitoring the intact operation of the power delivery system.

ACKNOWLEDGMENTS

This work is supported by the National Key Research and Development Program of China. (No. 2017YFF0206203).

REFERENCES

[1] Chen Z and Lewis D 2005 Evaluating Uncertainties in Net Power Delivery using Dual Directional Couplers *Proc. IEEE Int. Symp. on Electromagnetic Compatibility* pp 782–786
[2] IEEE Std. 1309–2013 *IEEE Standard for Calibration of Electromagnetic Field Sensors and Probes, Excluding Antennas, from 9 kHz to 40 GHz*
[3] Li J 2021 Improved Net Power Delivery using a Directional Coupler *Proc. 6th Annual Int. Conf. on Information System and Artificial Intelligence* (Accepted)
[4] Kanda M and Orr R D 1985 A radio-frequency power delivery system: Procedures for error analysis and self-calibration *National Bureau of Standards Technical Note 1083*
[5] Li D, Song Z and Meng D 2017 Comparison of two measurement methods on net power delivery with dual directional couplers *Proc. IEEE Conf. on Antenna Measurements & Applications (CAMA)* pp 374–376
[6] Xiong Y, Wan S and Chen Z 2019 Experimental Comparison Between Two Different Radio-Frequency Delivery Methods *Proc. Joint Int. Symp. on Electromagnetic Compatibility, Sapporo and Asia-Pacific Int. Symp. on Electromagnetic Compatibility (EMC Sapporo/APEMC)* pp 637–640

Simulation analysis on mechanical characteristics of rocker arm ground derrick joint

F.H. Meng, Y.J. Xia, Y. Ma, P. An & L.J. Sun
China Electric Power Research Institute, Beijing, China

ABSTRACT: In order to obtain the mechanical characteristics of the derrick joint, this paper first establishes the finite element model of the derrick, calculates and analyses the overall force of the derrick under different working conditions, and then extracts the real force information of the joint to determine the force form and law of the derrick joint. The simulation results show that the double rocker arm ground derrick joint is subjected to axial force, shear force and bending moment at the same time, in which the axial force of the member is far greater than the maximum shear-force of the member, and the axial stress of the member is basically greater than the maximum bending stress of the member. When the main chord is under tension, the axial force distribution of welded joints is 30: − 1:1: − 1: − 1; when the main chord is under pressure, the axial force distribution of welded joints is − 40: − 2:1: − 2:1. Finally, the finite element model of the joint is established. According to the mechanical characteristics of the joint, the ultimate bearing capacity of the welded joint is analysed, which provides the basis for the subsequent test of the mechanical characteristics of the joint.

1 INTRODUCTION

In order to promote the optimization and adjustment of energy structure and solve the problem of reverse distribution of energy consumption and energy distribution in China, China's power grid construction has developed rapidly and still has a large space to meet the energy demand of rapid economic development [1–3]. Derrick is one of the important tower construction equipment for extra-high voltage line (UHV) and various voltage level power grid construction, which has the advantages of simple structure, convenient transportation and strong terrain adaptability [4]. With the improvement of the voltage level of the transmission line, the size of the tower increases significantly, the complexity of the tower material lifting process and the one-time lifting weight become larger, the section, lifting height and rated lifting weight of the derrick increase, and the large double rocker arm ground derrick is popularized, as shown in Figure 1(a).

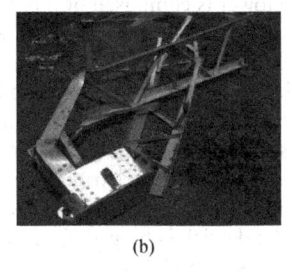

(a)   (b)

Figure 1. Large derrick and typical accident cases: (a) Double rocker arm ground derrick and (b) Failure accident of derrick joint.

Derrick is a kind of lattice light lifting equipment, which belongs to high-rise structure [6,7]. Its failure is generally controlled by buckling failure [8]. As a statically indeterminate space system, there are a lot of members in the derrick structure, and the restraint way is special. Under the load, its stress is more complex [9,10]. Finally, they all converge to the node, which requires the node to have a higher bearing capacity. At present, safety accidents caused by improper design or insufficient strength of derrick joints occur frequently, as shown in Figure 1(b). Therefore, the design and research of derrick joint considering the ultimate bearing capacity has important engineering significance to improve the construction safety [11,12].

In order to obtain the mechanical characteristics of the derrick joint, this paper first establishes the finite element model of the derrick, calculates and analyses the overall force of the derrick under different working conditions, and then extracts the real force information of the joint to determine the force form and law of the derrick joint. Finally, the finite element model of the joint is established. According to the mechanical characteristics of the joint, the ultimate bearing capacity of the welded joint is analysed, which provides the basis for the subsequent test of the mechanical characteristics of the joint.

2 FINITE ELEMENT ANALYSIS OF DOUBLE ROCKER ARM GROUND DERRICK

Taking the full steel double rocker arm ground derrick (ZB-DYG-12/12 × 700 × (2 × 40)) with the maximum independent height of 24m as the research object, the whole structure of the derrick was analysed by numerical simulation. According to the design parameters of the derrick, the finite element model of the derrick is established. The finite element model of the whole structure of the derrick is shown in Figure 2(a).

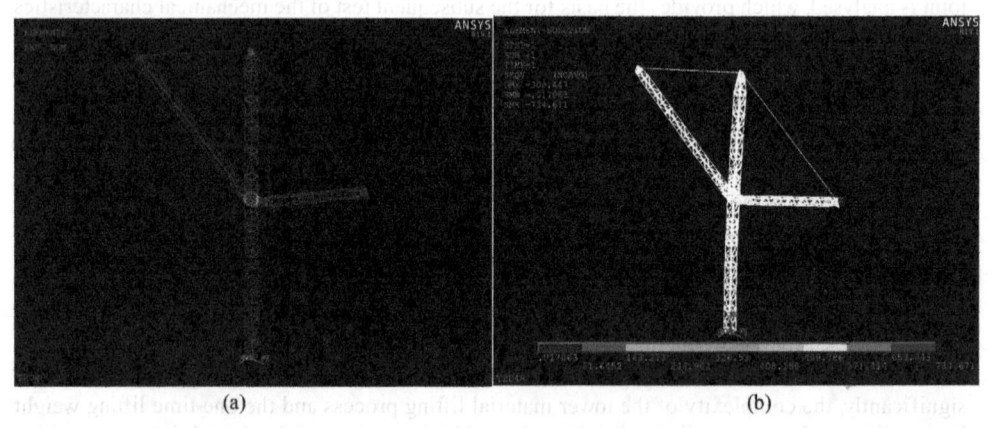

Figure 2. FE model of double rocker arm ground derrick: (a) Double rocker arm ground derrick model and (b) Mises stress nephogram of derrick.

The body of the tower is composed of three 2.0m standard sections, one 2.0m transition section and two 2.0m tower top sections. The cross section of the standard section is 700mm×700mm square. The main chord of derrick is ⊥80×6 angle steel, the straight web bar is ⊥65×5 angle steel, the inclined web bar is ⊥63×5 angle steel, and the straight web bar at the end of standard section is ⊥80×6 angle steel, all made of Q355. Beam 188 element (three-dimensional linear finite strain beam) is used to simulate the main chord, straight web and diagonal web of the derrick, and the connection between the members is rigid. The luffing wire rope is simulated by link180 element (three-dimensional tension or compression bar only), and the connection between the cable and the main body of the derrick is hinged. Q355 elastic modulus $E = 2.06 \times 10^5$ Mpa, poisson's ratio $v = 0.3$, wire rope elastic modulus $E = 1.20 \times 10^5$ Mpa, Poisson's ratio $v = 0.3$. In order to obtain the force law of the rocker arm ground derrick joint, several typical working conditions are selected for analysis.

2.1 Case 1

The ultimate working condition of the derrick is selected as the simulation analysis object. The right boom inclination angle of the derrick is 30°, the load is 4T, the left boom inclination angle of the derrick is 30°, the load is 2T, the maximum torque difference between the two sides is 16/T·m, and the wind load is 3690N. Figure 2(b) is the general Mises stress nephogram of the derrick, Figure 3(a) is the Mises stress nephogram of the derrick tower body, and Figure 3(b) is the axial force nephogram of the derrick body. It can be seen from Figure 3(b) that the tension force of the main chord near the standard section of the tower body is the largest, with the maximum pressure of 113kN and the maximum tension of 54kN. The nodes near these two positions are the most disadvantageous nodes on the main chord of the tower.

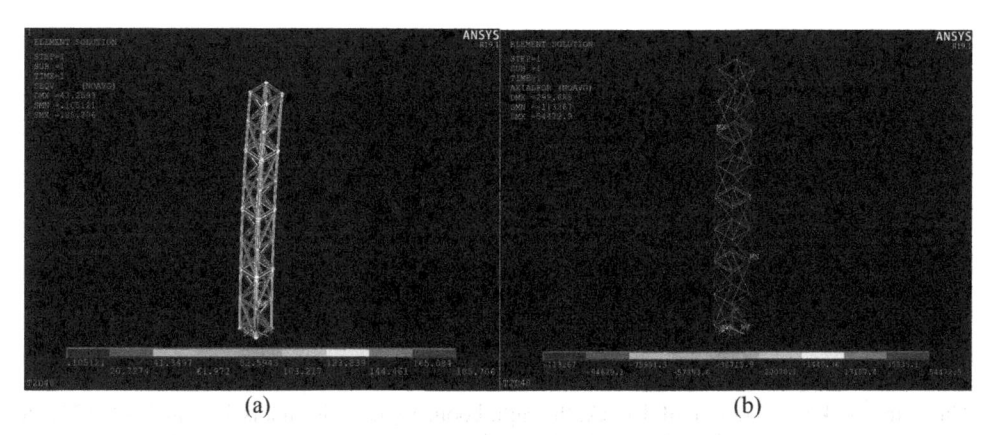

(a)   (b)

Figure 3. Results of FE analysis of double rocker arm ground derrick: (a) Mises stress nephogram of tower body and (b) Cloud chart of axial force of tower body.

The joints in the standard section of the rocker arm derrick are mainly welded joints, which bear tension and pressure through the main chord. Therefore, according to the results of finite element analysis, the axial force, the maximum shear force, the axial stress and the maximum bending stress of the welded joint member near the maximum axial tension and the maximum axial pressure on the main chord of the derrick are extracted, as shown in Table 1 and Table 2 respectively, and the number of the welded joint member is shown in Figure 4.

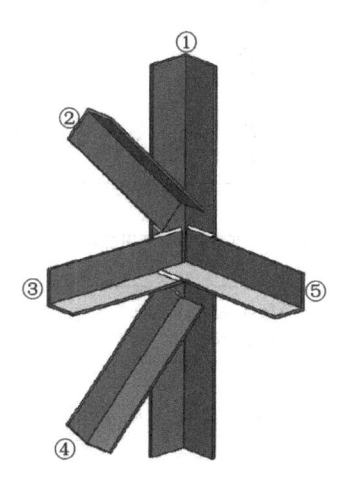

Figure 4. Welded joint model and member number.

Table 1. Stress analysis of welded joint members at the maximum tension of main chord. (30° − 4T-30° − 2T, 16T · m)

Member number	Axial force (N)	Maximum shear force (N)	Axial stress (MPa)	Maximum bending stress (MPa)
①	52201	421	56.5	−16
②	−1940	129	−3.2	3
③	1882	−157	3.1	2.1
④	−1918	−190	−3.1	2.1
⑤	−1887	−97	−3	0.5

Table 2. Stress analysis of welded joint members at the maximum pressure of main chord (30° − 4T-30° − 2T, 16T · m).

Member number	Axial force (N)	Maximum shear force (N)	Axial stress (MPa)	Maximum bending stress (MPa)
①	−108670	1202	−118	−67
②	−5603	338	−9.2	−10
③	2831	58	4.7	1.3
④	−5633	322	−9.3	−9
⑤	2882	−157	4.7	2.1

2.2 Case 2

Under the working condition of derrick, the right boom angle of derrick is 30°, the load is 2T, the left boom angle of derrick is 30°, the load is 0, the maximum torque difference between the two sides is 16/T·m, and the wind load is 3690N. According to the results of finite element analysis, the axial force, the maximum shear force, the axial stress and the maximum bending stress of the welded joint members near the maximum axial tension and the maximum axial pressure on the main chord of the derrick are extracted, as shown in Table 3 and Table 4 respectively.

Table 3. Stress analysis of welded joint members at the maximum tension of main chord (30° − 2T-30° − 0T, 16T · m).

Member number	Axial force (N)	Maximum shear force (N)	Axial stress (MPa)	Maximum bending stress (MPa)
①	63536	586	69	−19
②	−1968.0	149	−3.2	2.5
③	2092.6	−128	3.5	3
④	−1847.8	−177	−2.9	1
⑤	−1977.3	−68	−2.9	1.8

Table 4. Stress analysis of welded joint members at the maximum pressure of main chord (30° − 2T-30° − 0T, 16T · m).

Member number	Axial force (N)	Maximum shear force (N)	Axial stress (MPa)	Maximum bending stress (MPa)
①	−97339	1024	−105	−59
②	−4831.1	303	−8	−3.7
③	2441.5	49	4	0.9
④	−4863.4	293	−8	−5
⑤	2492.6	−128	4.2	3

2.3 Case 3

Under the working condition of derrick, the right boom angle of derrick is 3°, the load is 4T, the left boom angle of derrick is 3°, the load is 4T, the maximum torque difference between the two sides is 0/T·m, and the wind load is 3690N. Under this working condition, the main chord only bears pressure. According to the results of finite element analysis, the axial force, maximum shear force, axial stress and maximum bending stress of the welded joint members near the maximum axial pressure on the main chord are extracted, as shown in Table 5.

Table 5. Stress analysis of welded joint members at the maximum pressure of main chord (3° – 4T-3° – 4T, 0T · m).

Member number	Axial force (N)	Maximum shear force (N)	Axial stress (MPa)	Maximum bending stress (MPa)
①	−34013	530.15	−36.8	−9.1
②	−2308.5	117.91	−3.8	−1.4
③	1170.6	−88	1.9	1.1
④	−2365.2	100	−3.9	1.6
⑤	1170.6	−88	1.9	1.1

2.4 Case 4

Under the working condition of derrick, the right boom angle of derrick is 45°, the load is 4T, the left boom angle of derrick is 45°, the load is 4T, the maximum torque difference between the two sides is 0/T·m, and the wind load is 3690N. Under this working condition, the main chord only bears pressure. According to the results of finite element analysis, the axial force, maximum shear force, axial stress and maximum bending stress of the welded joint members near the maximum axial pressure on the main chord are extracted, as shown in Table 6.

Table 6. Stress analysis of welded joint members at the maximum pressure of main chord (45° − 4T − 45° − 4T, 0T · m).

Member number	Axial force (N)	Maximum shear force (N)	Axial stress (MPa)	Maximum bending stress (MPa)
①	−38525	571	−42	−28
②	−2517.6	131	−4.1	−1.9
③	1272.8	−86	2.1	1.1
④	−2559.8	114	−4.2	3.0
⑤	1217.0	−92	2.0	1.2

2.5 Result analysis

It can be seen from Table 1 to Table 6 that the double rocker arm ground derrick joint is subjected to axial force, shear force and bending moment at the same time, in which the axial force of the member is far greater than the maximum shear-force of the member, and the axial stress of the member is basically greater than the maximum bending stress of the member. Only axial loading is considered in the subsequent simulation of the mechanical characteristics of the derrick joints.

In addition, according to Table 1 and Table 3, the stress forms of double rocker arm ground derrick welded joints under different working conditions of maximum axial tension position are basically the same, and the distribution of axial force value bears certain regularity, basically according to the axial force ratio of 30: − 1:1: − 1: − 1; according to Table 2, Table 4, Table 5 and Table 6, the stress forms of welded joints under different working conditions of maximum axial compression position are basically the same. The axial force distribution of the bar has a certain regularity, basically according to the axial force ratio of − 40: − 2:1: − 2:1.

3 FINITE ELEMENT ANALYSIS OF ULTIMATE BEARING CAPACITY OF WELDED JOINTS

3.1 Simulated conditions

According to the finite element analysis results of the whole derrick, the main force transmission path of the derrick standard section is the axial force transmission, and the shear stress and bending stress of the bar are negligible. According to the analysis results in Section 2, there are mainly two ultimate load-bearing conditions. The simulated case 1 carries out multi axial loading according to the axial force ratio of 30: – 1:1: – 1: – 1 at the maximum tension of the main chord, as shown in Figure 5 (a), and the simulated case 2 carries out multi axial loading according to the axial force ratio of – 40: – 2:1: – 2:1 at the maximum pressure of the main chord, as shown in Figure 5 (b).

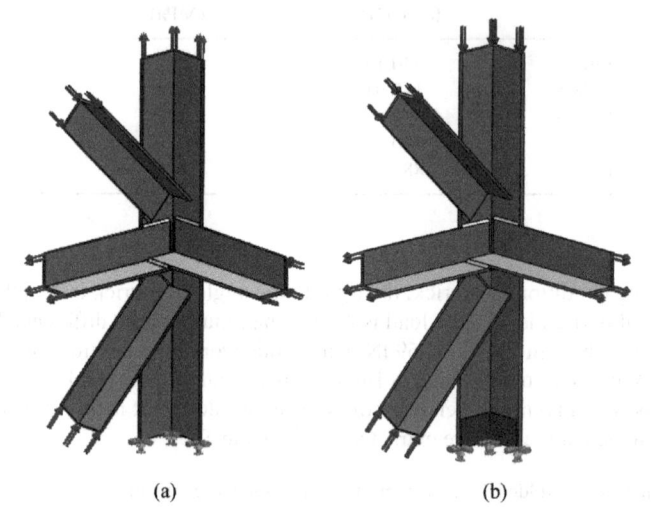

Figure 5. Simulation condition of ultimate bearing capacity of welded joints: (a) Simulated case1 and (b) Simulated case2.

3.2 Finite element analysis results of welded joints

In the finite element analysis, the main chord, diagonal web and straight web are Q355, and the fixed point is located at the end of the main chord, as shown in Figure 5. In simulated case 1, when the multi axial load reaches the stress value shown in Table 7, it is found that the maximum stress of right-angle weld is 409MPa, reaching the yield critical value, as shown in Figure 6. At the same time, the maximum stress of the main chord is 395 MPa, reaching the yield limit of 345 MPa, as shown in Figure 7. Therefore, the end of the main chord needs to be reinforced by adding a 4mm thick angle iron block. The maximum stress of the main chord is 290 MPa, which does not reach the yield limit, as shown in Figure 8(a), and the maximum displacement is 3.4mm, as shown in Figure 8(b).

Table 7. Simulated case 1 – ultimate bearing capacity of welded joint.

Member number	1	①	②	③	④	⑤
Ultimate bearing capacity (N)	One side fixed	120000	−4000	4000	−4000	−4000

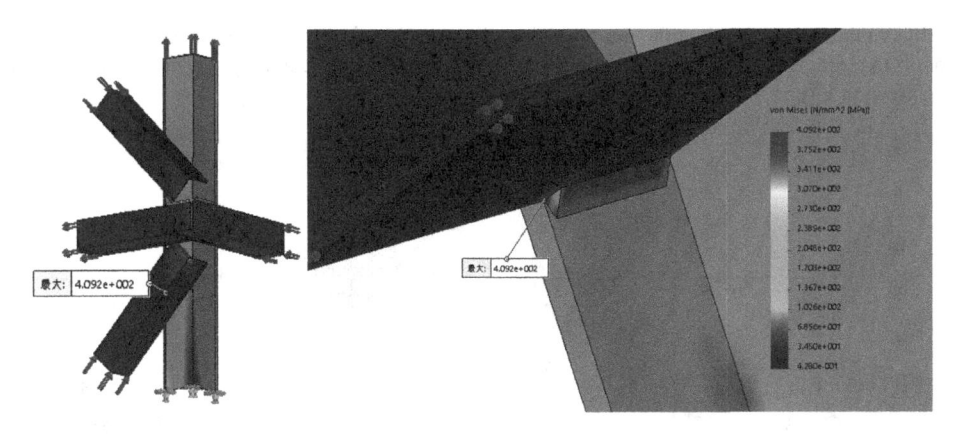

Figure 6. Simulation case 1: simulation results of welding joint – stress nephogram of weld line.

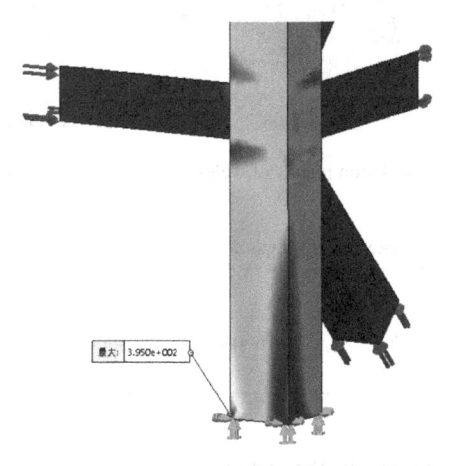

Figure 7. Simulation case 1: simulation results of welding joint – stress nephogram of link.

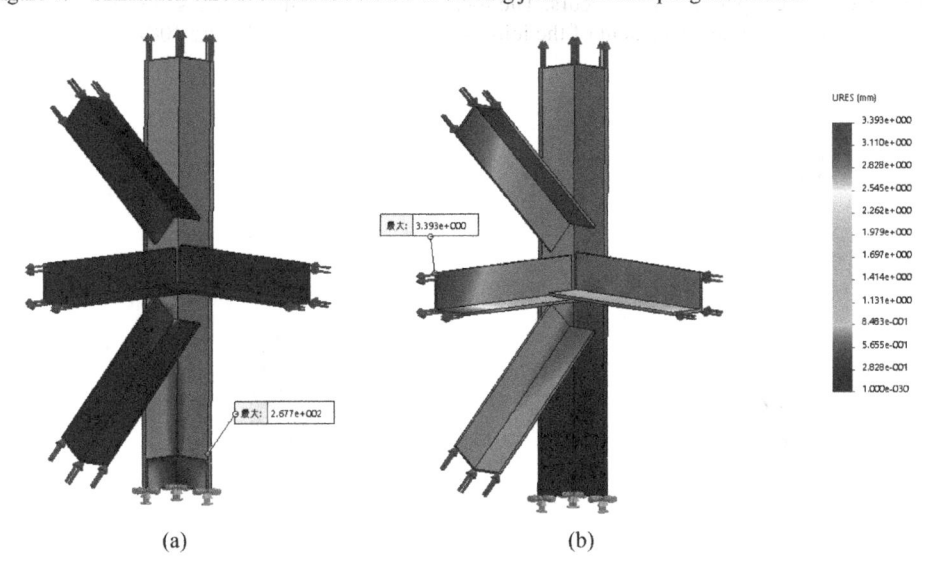

(a)　　　　　　　　　　　(b)

Figure 8. Simulation case 1: simulation results of enhanced welding joint: (a) Stress nephogram of link and (b) Displacement nephogram of welding joint

In simulated case 2, when the multi axial loading reaches the stress value shown in Table 8, it is found that the maximum stress of right-angle weld is 436 MPa, reaching the yield critical value, as shown in Figure 9.

Figure 9. Simulation case 2: simulation results of welding joint – stress nephogram of weld line.

Table 8. Simulated case 2 – ultimate bearing capacity of welded joint.

Member number		①	②	③	④	⑤
Ultimate bearing capacity (N)	One side fixed	−160000	−8000	4000	−8000	4000

When the right-angle weld yields, the maximum stress on the main chord of the joint is 262 MPa, and no yield phenomenon occurs. The location of the maximum stress is shown in Figure 10(a). The maximum displacement of the joint is 1.3mm, as shown in Figure 10(b).

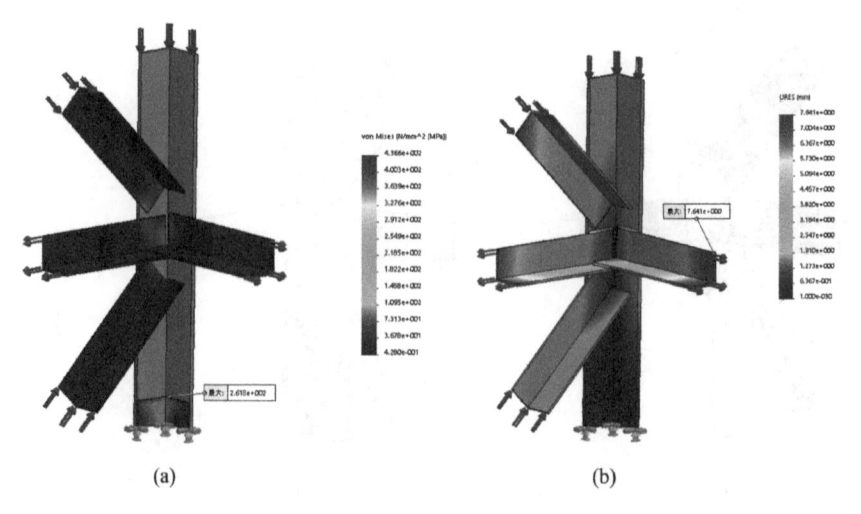

(a)   (b)

Figure 10. Simulation case 2: simulation results of enhanced welding joint: (a) Stress nephogram of link and (b) Displacement nephogram of welding joint

4 CONCLUSION

The joints in the derrick plays the role of connecting the converging members and transferring the load, the safety of the joint becomes very important.

In this paper, the finite element simulation method is used to extract the stress form of the joint in the whole finite element analysis of the derrick, and the method of increasing the proportion of multi axial load step by step until the joint failure is adopted to simulate the ultimate bearing capacity of the joint. Through the finite element analysis, the ultimate bearing capacity of each joint is obtained, which can be used to guide the ultimate bearing capacity test of the derrick joint. Based on the obtained results and their interpretations, the main findings are presented below:

The double rocker arm ground derrick joint is subjected to axial force, shear force and bending moment at the same time, in which the axial force of the member is far greater than the maximum shear-force of the member, and the axial stress of the member is basically greater than the maximum bending stress of the member.

When the main chord is under tension, the axial force distribution of welded joints is 30: – 1:1: – 1: – 1; when the main chord is under pressure, the axial force distribution of welded joints is – 40: – 2:1: – 2:1.

Finally, the findings of this study must be extended and validated to field tests on a real and full-scale double rocker arm ground derrick.

ACKNOWLEDGMENTS

Authors gratefully acknowledge the financial support of Science and Technology Research Project of State Grid (Research on intelligent generation technology of design and construction scheme of derrick for tower assembly based on limit state method, Grant No. 5442GC200005).

REFERENCES

[1] Ma H C, Cheng J, Li Q 2018 Application of floor type double rocker arm holding pole in construction of UHV steel pipe tower.*Shandong Electric Power*.**45**47–50

[2] Ni D, Xia Y J, Zhang Y D 2017 Development and engineering application of composite holding pole.*J.Mechanical Design*.**34** 95–98

[3] Shi F, Shi Y, Huang C Y 2016 Mechanical performance analysis of coupling structure of double flat arm holding pole and UHV transmission line tower *Steel Construction*.**31** 59–61

[4] Wang Y H,Li H,Dong J 2016 Analysis and Research on overall stability bearing capacity of derrick structure *Steel Construction*.**31** 73–85

[5] LalikK, Dominik I, CwiakalaP 2017 Integrated stress measurement system in tower crane mast *Measurement*.**102** 47–56

[6] Chwastek S 2020 Optimization of crane mechanisms to reduce vibration *Automation in Construction*. **119**103–335

[7] Chi P, Dong J, Xia Z Y 2016 Analysis and experimental study on stress state of rocker arm holding pole of overhead transmission line.*Steel Construction*.**31** 38–41

[8] Huang Y H, Wang R H, Zou J H 2010 Finite element analysis and experimental study on high strength bolted friction grip connections in steel bridges *J. Constructional Steel Research*.**66**803–815

[9] YangJ, Dewolf J T 1999 Mathematical Model for Relaxation in High-Strength Bolted Connections *J. Structural Engineering*.**125**803–809

[10] Mcevily A J 2007 Analysis of the effect of cold-working of rivet holes on the fatigue life of an aluminum alloy *International Journal of Fatigue*.**29**575–586

[11] Ho K C, Chau K T 1997 An infinite plane loaded by a rivet of adifferent material.*International Journal of Solids & Structures*. **34**2477–2496

[12] Borba N Z, Afonso C R 2017 On the Process-Related Rivet Microstructural Evolution,Material Flow and Mechanical Properties of Ti-6Al-4V/GFRP FrictionRiveted Joints.*Materials*. **10**1–20

Failure analysis and optimization of the light rail vehicle folding coupler in Hong Kong

T. Xu, J.G. Hu, W.J. Huang, W.B. Bao & D. Zhang
D&R Department, CRRC Nanjing Puzhen Rolling Stock Co., Ltd., Nanjing, China

ABSTRACT: A number of fracture failure of the light rail vehicles folding coupler hooks occurred during service in Hong Kong. It has threatened the operation of the Mass Transit Railway. Two modifications of the hook were proposed in the new vehicles design. One proposed modification was extending the hook and welded to the coupler head. The other proposed modification was using an adapter which connects hook and coupler head. By macroscopic analysis of the fracture surface, it was judged as a fatigue fracture because of cold work or welding. The modification of using an adapter was adopted. And the simulation analysis of fatigue showed that the utilization of endurance limit is lower than 100% when loaded with the cyclic loads which was required by EN12663 *'Railway Applications—Structural Requirements of Railway Vehicle Bodies'*. The modification of using an adapter can fulfil the requirement of the standard EN12663.

1 INTRODUCTION

The Mass Transit Railway Corporation Limited (MTR) contract K1846-15E was awarded to CRRC Nanjing Puzhen Co., Ltd. (CRRC PZ) for the supply of forty new light rail vehicles (LRVs). The new LRVs are fully compatible for coupling with the existing Phase III and Phase IV fleets achieving a standard of smoothness in coupling operation.

The folding couplers are installed at the front and the rear of the vehicles and they can be folded during the single car operation. This design can prevent accidents because of the coupler protruding injury to pedestrians and cars.

2 INTRODUCTION TO FOLDING COUPLER

The automatic Scharfenberg 330 coupler was installed at the front and the rear of the LRVs. The detailed technological parameters of the coupler are listed in Table 1, e.g., the vehicle starting acceleration, coupler installing height and folding angle.

Table 1. Technological parameters of vehicle and the coupler.

Parameters	Installing height (mm)	Coupler free angle		Weight (kg)	Folding angle (°)	Acceleration and deceleration (m/s²)	
		Yaw (°)	Pitch (°)			Starting	Braking
Value	500	±60	±8	265	97	1.3	

Every coupler has a hook and a catch lock. When it is not in use, the coupler should be folded under the underframe, and the coupler head is fixed by the catch lock, as shown in the following Figure 1 which used in Phase III and Phase IV.

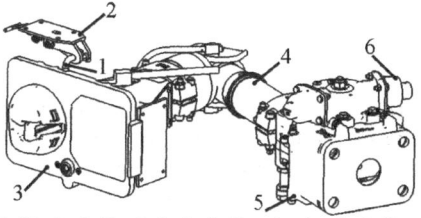

1 Hook, 2 Catch lock, 3 Coupler head, 4 Coupler Shank, 5 Elastomeric Draft Gear, 6 Mechanically Operated Re-Centering Feature

Figure 1. Coupler folding interface.

3 FAILURE AND OPTIMIZATION

3.1 *Fracture failure*

During the operation of Phase III and Phase IV LRVs in Hong Kong, there were a number of fracture failure of hooks, as shown in the following Figure 2. The fracture failure occurred very abruptly with no warning, which caused accidents and threatened the safety of LRVs operation.

Figure 2. The fracture failure of hooks.

3.2 *Design proposals of hook*

There were two design proposals of the hook proposed by VOITH, the supplier of coupler, in the new LRVs. As shown in the following Figure 3, one proposed modification was extending the hook length of the side which was freely suspended and welded to the coupler head directly. The other proposed modification was using an adapter which connects the hook and coupler head by the threaded connect.

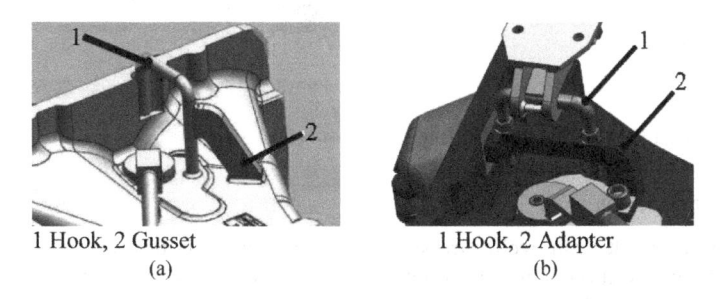

1 Hook, 2 Gusset 1 Hook, 2 Adapter
(a) (b)

Figure 3. Hook modification: (a) modification by welding and (b) modification of using an adapter.

In order to shorten design cycle, CRRC PZ decided to analyze the fracture failure first and selected one proposal to carry out simulation analysis. A comprehensive analysis of the hook fracture cause was carried out.

4 FRACTURE ANALYSIS

4.1 Macroscopic analysis of fracture surface

The macroscopic analysis of metal fracture is the basis of fracture failure. Through the macroscopic fracture analysis, the macroscopic exhibition and properties of fracture can be determined directly, as well as the position, quantity and crack propagation direction of fracture source area [1].

The macroscopic analysis of metal fracture is mainly based on the pattern of fracture, roughness, edge and position [2]. The ductile fracture is usually composed of three elements. They are fiber part, radiation zone and cutting zone. Ductile fractures are characterized by extensive plastic deformation prior to and during crack propagation as shown in the Figure 4. On the other hand, fatigue fracture takes place at stresses below the net section yield strength, with very little observable plastic deformation and a minimal absorption of energy. Such fracture occurs very abruptly with little or no warning and can take place in all classes of materials. The most characteristic feature usually found on fatigue fracture surfaces is beach marks, which are centered around a common point that corresponds to the fatigue crack origin [3], as shown in Figure 4.

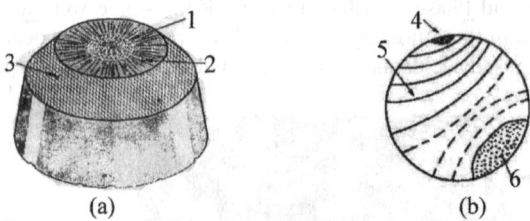

(a)   (b)
1 Fiber part, 2 Radiation zone, 3 Cutting zone, 4 Origin, 5 Beach mark, 6 Rupture region

Figure 4. Typical fracture surface: (a) Tensile damage fracture and (b) Fatigue fracture.

From the fracture failure of hooks, the fracture occurs very abruptly with no warning and with low elonggation, low reduction in area, no necking, as shown in Figure 5. Although there are no clear photos of the beach mark found on fatigue fracture surfaces. And there is no need to do the metallographical inspection. It can be judged as a fatigue fracture from the macroscopic properties and the stress analysis of the hook.

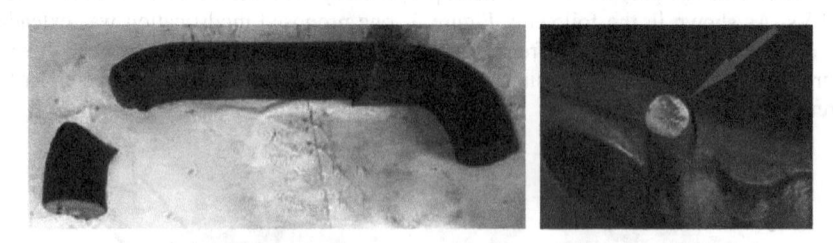

Figure 5. The macroscopic properties of fracture.

4.2 Fatigue fracture analysis

Fatigue fractures result from cyclic stressing, which progressively propagates a crack or cracks until the remaining section is no longer able to support the applied load. The process of fatigue consists of three stages [4]:

- Stage I: Initial fatigue damage leading to crack nucleation and crack initiation
- Stage II: Progressive cyclic growth of a crack until the remaining uncracked cross section of a part becomes too weak to sustain the loads imposed
- Stage III: Final, sudden fracture of the remaining cross section.

As shown in Figure 6, the fracture occured in the cold formed area and the weld heat-affected zone (HAZ) of the hook. Due to cold work and welding, there may be nonuniformity and residual stress that affect the properties of the material.

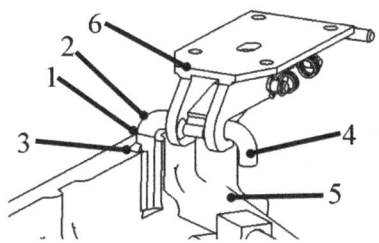

1 Fracture surface, 2 Cold formed area, 3 Weld, 4 Hook, 5 Coupler head, 6 Catch lock

Figure 6. The analysis of position of fracture.

Also in welding, hydrogen from the arc atmosphere can dissolve into the liquid weld pool, and if the weld cools rapidly, a significant quantity may be retained in the weldment at low temperatures [5]. Sufficient hydrogen levels will cause cracking in the weld metal and the HAZ [6].

And when the coupler is folded and fixed, the hook subjects to cyclic stress produced by the vibration and shock of the coupler [7]. Under the action of cyclic loads, the HAZ of the hook becomes an initiation site for a fatigue crack caused by cold work or welding. The fatigue crack propagates through the material until complete fracture results [8].

4.3 Optimization design

By comparing two design proposals of the hook proposed by VOITH, there still may be residual stress and hydrogen embrittlement in the modification by welding. Even with the tempering, there would still be a risk of fracture under the action of cyclic stress. And it is difficult to be adjusted and replaced comparing with the modification of using an adapter. Therefore CRRC PZ and MTR determine that the modification of using an adapter will be adopted in the new LRVs of Phase V. The simulation analysis of the structure will carry out.

5 SIMULATION ANALYSIS

5.1 FEM model

All parts are meshed with 2nd order elements. The edge length of the elements was chosen sufficiently small depending on the accuracy of the calculation. For solids where possible hexahedral elements are used, all other solids are meshed with tetrahedral elements [9]. The meshing is visualized in Figure 7. The entire model contains 460578 nodes and 224392 elements.

5.2 Calculation loads and material property

According to EN 12663–1 [10], the system must withstand the cyclic loads listed in Table 2 in a way that it 'sustained for the specified life without detriment to the structural safety'.

According to EN 10083–3 [11], the parts and material properties relevant for strength assessment are listed in Table 3.

Figure 7. Meshing.

Table 2. Cyclic loads.

Load	Name	Value[g]		
Cyclic loads	combined positive acceleration	0.15gx	0.15gy	1.15gz
	combined negative acceleration	–0.15gx	–0.15gy	+0.85gz

5.3 Boundary conditions

The loads and constraints are applied as Figure 8. The constraints are also set in a manner that the calculated system behaviour is not better than in reality. Loads include acceleration load and pretension load of bolts according to VDI 2230 [12].

Table 3. Material property.

No.	Part	Re [MPa]	Rm [MPa]
1	Hook	700	900
2	Adapter	400	650
3	Catch lock	355	400

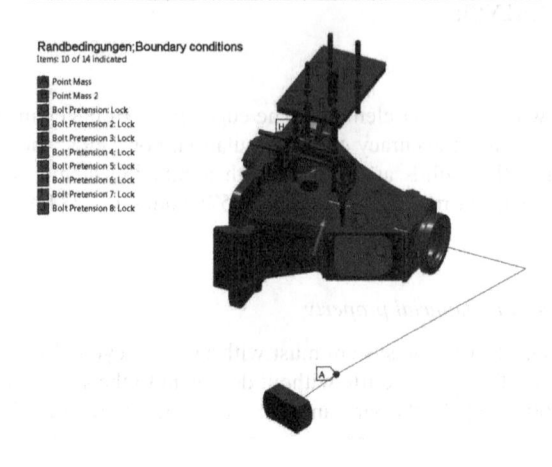

Figure 8. Boundary conditions.

5.4 Evaluation criterion

The fatigue strength is evaluated against cyclic loads. According to EN12663–1, all dynamic stress cycles remain below the material endurance limit, no additional safety factors are necessary in these calculations. The utilization of the material endurance limit should below 100%.

5.5 Calculation results

The global utilization of the material endurance limit is below 100%. The detailed results are listed in Table 4. The modification of using an adapter can fulfil the requirement of the standard EN12663.

Table 4. Calculation result of endurance limit Utilization.

Part	Utilization [%]	Figure	Annotation
Hook	32.32	Figure 9	Lower than 100%
Adapter	8.62	Figure 10	Lower than 100%

The maximum endurance limit utilization of hook is 32.32%, which is located in the fixed position of the catch lock as shown in Figure 9.

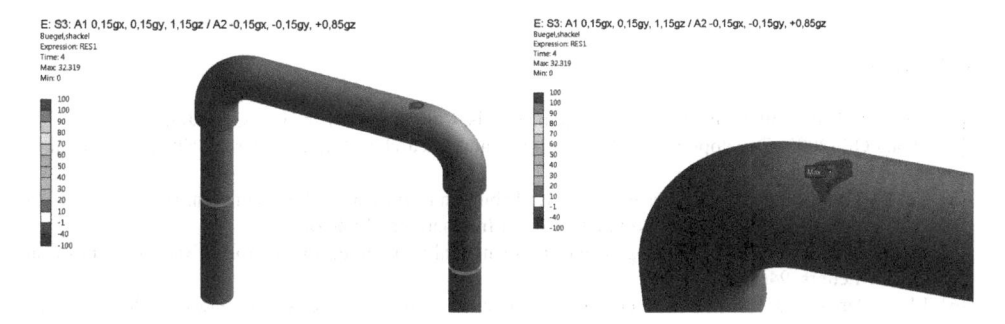

Figure 9. Maximum endurance limit utilization of hook.

As shown in Figure 10, the maximum endurance limit utilization of adapter is 8.62%, which is located at the connection between the adapter and the coupler head.

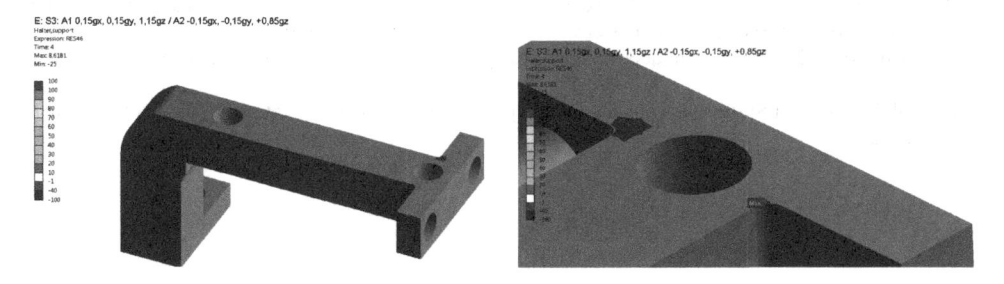

Figure 10. Maximum endurance limit utilization of adapter.

6 CONCLUSIONS

The reason of fracture failure of the LRVs coupler in Hong Kong was found by an analysis of the macroscopic properties and the stress of the hook. The conclusions from this failure analysis and optimization are as follows:

1. There is residual stress or hydrogen embrittlement caused by cold work and welding in the hook of Phase III and IV LRVs folding coupler in Hong Kong. It leads to crack nucleation and crack initiation.
2. The couplers are installed at the end of the vehicles where subjected to the great shock and random vibration. Under the action of cyclic loads produced by the shock and vibration, the crack propagates through the hooks until complete fracture results.
3. By comparing two design proposals of the hook, the modification by welding needs to remove residual stress by temper aging. And there still be a risk of fracture. Therefore the modification of using an adapter was adopted in the new LRVs of Phase V.
4. The simulation analysis of fatigue shows that the maximum utilization of endurance limit is 32.32% lower than 100%. The modification of using an adapter can fulfil the requirement of the standard EN12663.
5. With the modification of using an adapter, the height of hook can be adjusted easily by the nuts. They are easy to be replaced by the spares. This design can be widely used in the other urban rail transit vehicles.

Because of the cyclic loads, it should be considered at design time how the structure of the folding coupler prevent crack in the weld and HAZ. Also the looseproof technique of the threaded connection should be used to avoid the folding coupler detaching from vehicles.

REFERENCES

[1] Yang XJ (2019) Failure analysis of metallic materials. Chemical Industry Press, Bei Jing, 54–61
[2] Zhong QP (2005) Development of 'Fractography' and research of fracture micromechanism. Journal of mechanical strength.
[3] Cui YX (1998) Fracture Analysis of Metal. Ha Erbing industry college book concern, Ha Erbing ,21–28
[4] F.C. Campbell. (2012) Fatigue and Fracture ASM International, State of Ohio.
[5] LU G, KAXIRAS E (2005) Hydrogen embrittlement of aluminum: the crucial role of vacancies. Physical Review Letters. 94(15): 155501.
[6] T.L. Anderson (2011) Fracture Mechanics Fundamentals and Applications. 3th Edition, Taylor & Francis Group, New York.
[7] Wang QS (2017) Study on vibration behavior of carbody underframe suspended systems for high speed trains. Southwest Jiaotong University Doctor Degree Dissertation, 105–135.
[8] Gustavo M, Castelluccio, Juan E(2013) Fracture testing of the heat affected zone from welded steel pipes using an in situ stage. Engineering Fracture Mechanics, 98:52–63
[9] Forschungskuratorium Maschinenbau. (2012). FKM Guideline. Analytical Strength Assessment of Components in Mechanical Engineering (6th, revised edition, 2012). (V. Verlag, Ed.)
[10] EN 12663–1 (2010) Railway applications - Structural requirements of railway vehicle bodies - Part 1: Locomotives and passenger rolling stock (and alternative method for freight wagons).
[11] EN 10083–3 (2007) Steels for quenching and tempering - Part 3: Technical delivery conditions for alloy steels.
[12] VDI 2230 (2003) Systematic calculation of high duty bolted joints - Joints with one cylindrical bolt, German.

Dynamic analysis of an impact-vibration system with multiple asymmetric rigid stops

Y. Yang
School of Mechatronic Engineering, Lanzhou Jiaotong University, Lanzhou, China
Key Laboratory of System Dynamics and Reliability of Rail Transport Equipment of Gansu Province, Lanzhou, China

ABSTRACT: The dynamic behavior of an impact-vibration system with clearances represented by multiple asymmetric rigid constraints is studied. Three Poincaré sections are defined to describe the periodic impact-vibration characteristics on different constraints. Local bifurcation and global bifurcation of the system are investigated by numerical simulation. Pattern types, bifurcation forms, transition law of periodic-impact vibration groups are analysed. Periodic-impact motions are defined by the periodically exciting force and differ by the numbers of impact occurring at the asymmetric rigid stops of the clearances. Due to the existence of asymmetric multiple clearances, the system exhibits complex dynamic behavior.

1 INTRODUCTION

The existence of the constraints and clearances among many mechanical systems, bring about the impact vibration and noise, which affect the service life of the equipment. Therefore, the study on dynamic behavior of mechanical systems with clearances and constraints has important application value. Many scholars and engineers have conducted theoretical derivation, numerical simulation, experimental verification and other researches for this purpose as partly seen in refs [1–4]. Stanton [5] discussed the influence of parameters such as dimensionless frequency and damping on the dynamic performance of the model by using the theory of electrical dynamics, and studied the theory in predicting chaos. Thota, Gritli [6, 7] studied the nonlinear dynamic behavior such as bifurcation characteristics, chattering and chaotic of the single-degree-of-freedom oscillators through bifurcation diagrams, time-traces, phase diagrams and Poincaré sections. Gatti, Cirillo and their co-workers [8, 9] carried out researches on the nonlinear dynamics of the two-degree-of-freedom system from the motion response of the parameters and the invariant manifolds in phase space. Luo, Wagg, Li, et al. [10–12] discussed the dynamic behavior of different types of elastic and rigid constraint impact vibration systems with clearances at low frequency, such as chattering, sticking, sliding and periodic motions. Oruganti [13] demonstrated the vibro-impact phenomena of a torsional system with multiple clearances from experimental perspective. Tiwari, Walha, Al-Solihat, et al. [14–16] applied nonlinear dynamics theory on mechanical systems such as bearing rotors, gear transmissions, discussed the establishment of dynamic models, parameter correlation, and dynamic response.

The dynamic characteristics and parameter matching of collision systems with clearances have always been paid attention to by the engineering community. However, there are few studies on the dynamics analysis of impact-vibration systems with multiple clearances, especially the influences of dynamic behavior caused by clearances asymmetry. A model of periodically-forced system with clearances represented by multiple asymmetric rigid constraints is established, the existence domains of periodic motion and its transition law are analyzed by means of numerical simulation method, and focus on the periodic-impact motion characteristics on the bifurcation boundary affected by the clearances asymmetry.

2 MECHANICAL MODEL

The mechanical model of a two-degree-of-freedom system with clearances represented by three asymmetric rigid constraints is shown in Figure.1. X_1 and X_2 represent the vibration displacements of two mass blocks M_1 and M_2. The studied mechanical model is: two mass blocks are connected by a linear spring with stiffness K_0 and a viscous dashpot with damping coefficient C_0, and the mass M_1 is attached to the right supporting base by a linear spring with stiffness K_1 and a viscous dashpot with damping coefficient C_1, the mass M_2 is attached to the left supporting base by a linear spring with stiffness K_2 and a viscous dashpot with damping coefficient C_2. There are periodic forces $P_i \sin(\Omega T + \tau)$ $(i = 1, 2)$ acting on the mass $M_i (i = 1, 2)$, where the amplitude of the periodic force is P_i $(i = 1, 2)$. The clearances $B_1, -B_1$ and B_2 restrict the vibration amplitude of the system. For small periodic forcing amplitude the system exhibits linear forced vibration. With the amplitude of the periodic force increases, the mass block M_1 impacts mutually with the mass block M_2 at the rigid stops A_1 or $\bar{A}_1$, when the relative displacement $X_1 - X_2$ equals to the clearance B_1 or $-B_1$. Similarly, the mass block M_2 hit the rigid stop A_2 when the displacement X_2 equals to the clearance B_2. After the impacts, the kinetic energy of the system will be partially dissipated, which is represented by the coefficient of restitution $0 < R < 1$.

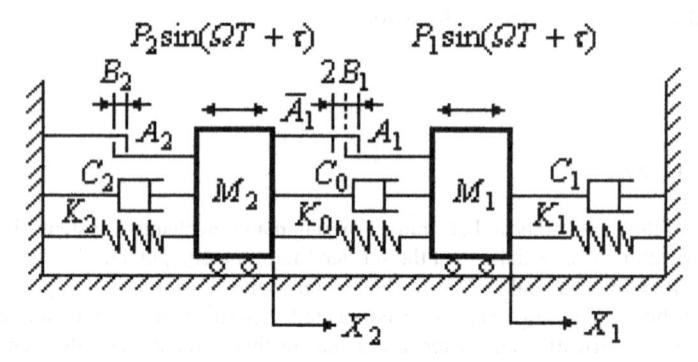

Figure 1. Mechanical model of vibration system with asymmetrical multiple clearances

When $|X_1 - X_2| < B_1$ and $|X_2| < B_2$, the differential equations of system motion between two adjacent impacts are:

$$\begin{cases} M_1\ddot{X}_1 + C_1\dot{X}_1 + C_0(\dot{X}_1 - \dot{X}_2) + K_1X_1 + K_0(X_1 - X_2) = P_1 \sin(\Omega T + \tau) \\ M_2\ddot{X}_2 + C_2\dot{X}_2 + C_0(\dot{X}_2 - \dot{X}_1) + K_2X_2 + K_0(X_2 - X_1) = P_2 \sin(\Omega T + \tau) \end{cases} \quad (1)$$

In order to make the analysis general, the dynamic of the system with series non-smooth impacts is transformed into a dimensionless form, and all dimensionless variables and parameters are given by:

$$\mu_m = \frac{M_2}{M_1 + M_2}, \mu_c = \frac{C_2}{C_1 + C_2}, \mu_k = \frac{K_2}{K_1 + K_2}, \mu_{c0} = \frac{C_0}{C_1 + C_0}, \mu_{k0} = \frac{K_0}{K_1 + K_0}, \omega_n = \sqrt{\frac{K_1}{M_1}},$$

$$t = \omega_n T, \omega = \frac{\Omega}{\omega_n}, x_i = \frac{X_i K_1}{P_1 + P_2} (i = 1, 2), \zeta = \frac{C_1}{2\sqrt{K_1 M_1}}, f_{20} = \frac{P_2}{P_1 + P_2},$$

$$\delta_2 = \frac{B_2 K_1}{P_1 + P_2}, \delta_1 = \frac{B_1 K_1}{P_1 + P_2}. \quad (2)$$

The definition range of some dimensionless parameters are: $\mu_m \in (0,1)$, $\mu_c \in (0,1)$, $\mu_k \in (0,1)$, $\mu_{c0} \in (0,1)$, $\mu_{k0} \in (0,1)$, $f_{20} \in [0,1]$, $R \in (0,1)$.

Therefore, the dimensionless equations of the system are:

$$\begin{cases} \ddot{x}_1 + 2\zeta\dot{x}_1 + 2\zeta\dfrac{\mu_{c0}}{1-\mu_{c0}}(\dot{x}_1 - \dot{x}_2) + x_1 + \dfrac{\mu_{k0}}{1-\mu_{k0}}(x_1 - x_2) = (1-f_{20})\sin(\omega t + \tau) \\ \dfrac{\mu_m}{1-\mu_m}\ddot{x}_2 + 2\zeta\dfrac{\mu_c}{1-\mu_c}\dot{x}_2 + 2\zeta\dfrac{\mu_{c0}}{1-\mu_{c0}}(\dot{x}_2 - \dot{x}_1) + \dfrac{\mu_k}{1-\mu_k}x_2 + \dfrac{\mu_{k0}}{1-\mu_{k0}}(x_2 - x_1) \\ \qquad = f_{20}\sin(\omega t + \tau) \end{cases} \quad (3)$$

The impact equations at the rigid constraints A_1 and $\bar{A}_1$ with $|x_1 - x_2| = \delta_1$ are described by:

$$\begin{cases} \dfrac{1-\mu_m}{\mu_m}\dot{x}_{1+} + \dot{x}_{2+} = \dfrac{1-\mu_m}{\mu_m}\dot{x}_{1-} + \dot{x}_{2-} \\ R = (\dot{x}_{2+} - \dot{x}_{1+})/(\dot{x}_{1-} - \dot{x}_{2-}) \end{cases} \quad (4)$$

The impact equation at the rigid constraint A_2 with $x_2 = \delta_2$ is described by:

$$\dot{x}_{2+} = -R\dot{x}_{2-} \quad (5)$$

Where, $\dot{x}_{1-}, \dot{x}_{1+}$ denote the velocity of the mass block M_1 immediately before and after impact, $\dot{x}_{2-}, \dot{x}_{2+}$ denote the velocity of the mass block M_2 immediately before and after impact, respectively.

3 DYNAMICAL CHARACTERISTICS OF THE SYSTEM WITH MULTIPLE CLEARANCES

Numerical simulation of collision system with multiple clearances can intuitively provides information about pattern type, transition law and bifurcation forms of the periodic motions in the parameter domain, systematically reveal the correlation between system dynamics and parameters. Based on the definition ranges of model dimensionless parameters, taking $\mu_m = 0.5$, $\mu_c = 0.5$, $\mu_k = 0.5$, $\mu_{c0} = 0.5$, $\mu_{k0} = 0.5$, $\zeta = 0.1$, $R = 0.8$, $f_{20} = 1.0$ to study the dynamical characteristics of the system. Define three Poincaré sections:

Impact mapping section:

$$\sigma_{p1} = \{(x_1, \dot{x}_1, x_2, \dot{x}_2, t) \in R^4 \times T |\; |x_1 - x_2| = \delta_1, \dot{x}_1 - \dot{x}_2 = \dot{x}_{1+} - \dot{x}_{2+}\};$$

$$\sigma_{p2} = \{(x_1, \dot{x}_1, x_2, \dot{x}_2, t) \in R^4 \times T |\; x_2 = \delta_2, \dot{x}_2 = \dot{x}_{2+}\}.$$

Time mapping section: $\sigma_n = \{(x_1, \dot{x}_1, x_2, \dot{x}_2, t) \in R^4 \times T | (x_1 - x_2) = (x_1 - x_2)_{\min}, \mod(t = 2\pi/\omega)\}$.

From the impact mapping section σ_{p1}, we can confirm the number of left impacts p_1 ($p_1 = 0, 1, 2, 3, \ldots$) at rigid constraint $\bar{A}_1$ and the number of right impacts q_1 ($q_1 = 0, 1, 2, 3, \ldots$) at rigid constraint A_1; from the impact mapping section σ_{p2}, we can confirm the number of impacts p_2 ($p_2 = 0, 1, 2, 3, \ldots$) at rigid constraint A_2. The number of the forcing period in the motion period $T_n = 2n\pi/\omega (n = 1, 2, 3, \ldots)$ can be identified from the time mapping section σ_n. Combining these above Poincaré sections, it can be identified that the pattern type $n - p_1 - q_1$ of the system at rigid constraints A_1 and $\bar{A}_1$, and the pattern type $n - p_2$ of the system at rigid constraint A_2 under the certain parameter conditions.

It can be found from ref [17] that when the clearance of the vibro-impact system is small, its dynamic performance is more complicated, so take a smaller clearance $\delta_i = 0.18$ ($i = 1, 2$), and let the rigid constraints $A_1, \bar{A}_1$ and A_2 equal gap changes. The global bifurcation diagrams shown in Figure.2 that the system mainly presents $1 - p_1 - q_1, 1 - p_2$ fundamental periodic-impact motions in the mapping section σ_{p1} and σ_{p2}, in the part without marked with symbols in the figures, the system mainly presents long-periodic multi-impact motion or chaotic motion. Figure.2(a) is the bifurcation diagram of the instantaneous relative velocity before impact $\dot{x}_{1-} - \dot{x}_{2-}$ of the masses M_1 and M_2. at the constraints A_1 and $\bar{A}_1$ change with the excitation frequency ω, corresponding to Poincaré mapping σ_{p1}, Figure.2(b) is the bifurcation diagram of the instantaneous velocity before impact $\dot{x}_{2-}$ of the mass M_2 change with the excitation frequency ω, corresponding to Poincaré mapping σ_{p2}. The relative displacement $X_1 - X_2$ of the two mass blocks change with the excitation frequency ω on the Poincaré mapping σ_n is shown in the bifurcation diagram of Figure.3(c), which can determine the number n ($n = 1, 2, 3, \ldots$) of the forcing period in every period $T_n = 2n\pi/\omega$ of the impact motions. From Figure.2(a), we can observe that in the mapping section σ_{p1}, the number of mutual impacts p of mass blocks M_1 and M_2 on the left constraint $\bar{A}_1$ is obviously greater than the number of mutual impacts q on the right constraint A_1. In the high frequency domain of $\omega > 1.827$, the system exhibits stable 1-1-1 fundamental periodic-impact motion, which means, the number of impacts of two masses on both sides of the middle rigid constraints A_1 and $\bar{A}_1$ is equal to one. With the excitation frequency ω decreases, the system undergoes a narrow sub-harmonic motion and chaotic motion window, then transitions to periodic-impact motions dominated by 1-1-2, 1-1-3, 1-2-4, etc. When ω is small, the system presents 1-0-q unilateral impact motion under low-frequency vibration, mass blocks M_1 and M_2 only impact with the right side A_1 of the middle constraint. As ω continues to decrease, the number of right impacts q of the two masses on the constraint A_1 continues to increase. When ω decreases sufficiently, the system exhibits $1 - 0 - \bar{q}$ chattering impacts with sticking characteristics. In the mapping section σ_{p2} shown in Figure. 2(b), the periodic-impact motions is similar to the above-mentioned, except that the system presents 1-0 impact less motion in the high-frequency domain. In order to deeply study the pattern type, transition law and bifurcation characteristics of the periodic-impact motions of the system, the numerical simulation results need to be further refined locally.

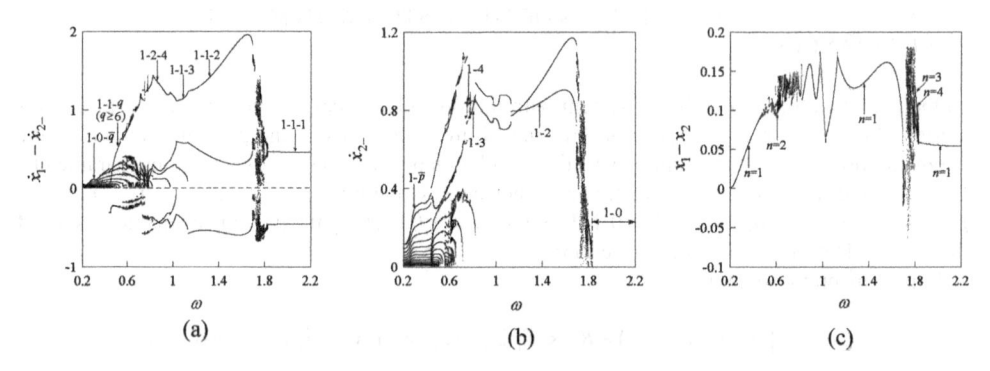

Figure 2. Global bifurcation diagrams for the system for $\delta_1 = \delta_2 = 0.18$: (a) bifurcation diagram for relative impact velocity $\dot{x}_{1-} - \dot{x}_{2-}$ (δ_1), (b) bifurcation diagram for impact velocity $\dot{x}_{2-}$ (δ_2) and (c) bifurcation diagram for the relative displacement $x_1 x_2$ ($\delta_1 = \delta_2$).

4 MOTION ANALYSIS ON THE BOUNDARY OF GRAZING BIFURCATION

Figure 3 (a)–(c) is the local amplification and details of Figure. 2(a)–(c) when the excitation frequency $\omega \in [0.665, 0.687]$, it exactly passes through the existence domain surrounded by the bare-grazing bifurcation boundary of 1-2-4 (1-4) motion and the period doubling bifurcation boundary of the sub-harmonic motions. Figure. 3(a) can determine the number of left impact number $p_1 = 2n$ at the rigid constraint A_1 and the right impact number $q_1 = 4n + 1, 4n + 2$ or $5n$ at the rigid constraint $\bar{A}_1$. Combining Figure. 3 (a) and (c), we can observe the process of bare-grazing bifurcation of the 1-2-4 fundamental periodic-impact motion with increasing excitation frequency ω, the sub-harmonic impact motion windows about 5-10-21, 4-8-17, 3-6-13, 2-4-9,..., n-$2n$-$(4n+1)$($n = 2, 3, 4, 5, \ldots$) and other sub-harmonic impact motions such as 3-6-14 are described representatively. Nearby $\omega=0.683546$, the system transitions from chaotic motion to n-$2n$-$5n$ sub-harmonic impact motion, finally, 1-2-5 motion is generated with increasing ω by inverse period doubling bifurcations of 2-4-9, 4-8-20 sub-harmonic impact motion. Similarly, the impact number $p_2=4n$ of the mass M_2 at the constraint A_2 within an excitation forcing period can be determined from Figure. 3(b). Combining Figure. 3 (b) and (c), we can observe the process of bare-grazing bifurcation of the 1–4 fundamental periodic-impact motion with increasing excitation frequency ω, the sub-harmonic impact motion windows of 5-20,4-16, 3-12,2-8,..., $n - 4n$ $(n = 2, 3, 4, 5, ...)$ are described. These sub-harmonic motions continuously undergo inverse period doubling bifurcations with increasing frequency ω, 4–16 motion transitions to 2-8,1-4 sub-harmonic motion.

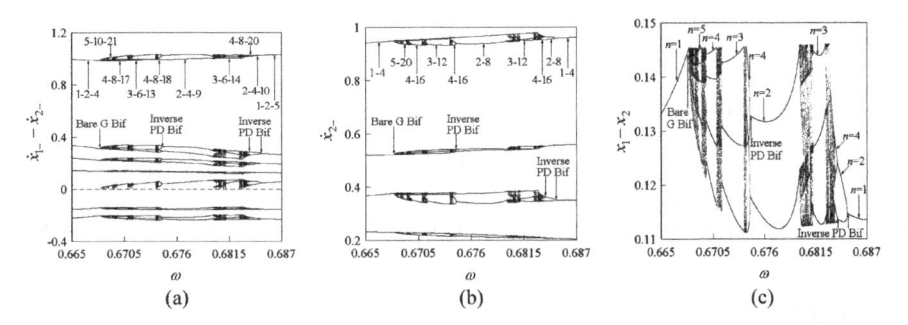

Figure 3. The local amplification and details of Figure. 2, $\omega \in [0.665, 0.687]$: (a) bifurcation diagram on Poincaré mapping σ_{p1}, (b) bifurcation diagram on Poincaré mapping σ_{p2} and (c) bifurcation diagram on Poincaré mapping σ_n.

Figure.4 (a)–(c) are the motion phase diagrams near the critical value of the 1-2-4 periodic motion bare-grazing contact instability for the masses M_1 and M_2 on Poincaré mapping σ_{p1}, which describe the transition process when the excitation frequency ω gradually increases and exceeds $\omega = 0.668254$, 1-2-4 motion degenerates to chaotic motion via the bare-grazing bifurcation. The evolution process of the motions shown in Figure. 3 and Figure. 4 is consistent. With ω increasing, the main branch of transition, for certain fixed value of clearance $\delta_1 = \delta_2 = 0.18$ can be expressed as follows:

$\omega \uparrow: 1 - 2 - 4 \to$ Bare G Bif $\to \ldots \to n - 2n - (2n + 1) \to$ Bare G Bif $\to \ldots \to 5 - 10 - 21 \to$ Bare G Bif $\to \ldots \to 4 - 8 - 17 \to$ Bare G Bif $\to \ldots \to 3 - 6 - 13 \to$ Bare G Bif $\to \ldots \to 4 - 8 - 18 \to$ Inverse PD Bif $\to 2 - 4 - 9$

(6)

Where the symbol Bare G Bif denotes bare-grazing bifurcation; Inverse PD Bif denotes inverse period doubling bifurcation.

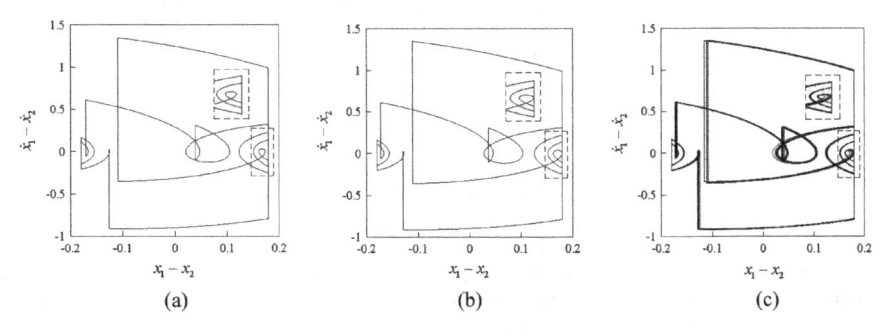

Figure 4. Phase plane portraits of the impacts masses M_1 and M_2: (a) 1-2-4 motion, $\omega = 0.667$, (b) 1-2-4 motion with grazing contact, $\omega = 0.668254$ and (c) chaotic motion, $\omega = 0.668255$.

5 CONCLUSIONS

In this paper we consider a vibration system with clearances represented by three asymmetrical rigid stops. The influence of system dynamics behavior is based on two important parameters: the dimensionless excitation frequency ω and clearances δ_i. As the frequency ω decreases, the system generates grazing bifurcation, the number of impacts at the asymmetrical rigid constraints increase one by one, when the number p is large enough, the system presents the dynamic characteristics of chattering-impact. Certain fundamental periodic-impact vibration groups with grazing bifurcation sequences is interrupted by bare-grazing bifurcation, inverse period doubling bifurcation, sub-harmonic motions and chaos, the chief transition law of the periodic motions on the boundary of grazing bifurcation is summarized.

ACKNOWLEDGMENTS

The authors gratefully acknowledge the support by National Natural Science Foundation of China (11862011), Gansu Science and Technology Planning Project (18YF1WA059, 20JR10RA232) and Innovation and Entrepreneurship Talents Training Project of Lanzhou City of China(2014-RC-33).

REFERENCES

[1] Shaw, S.W. and Holmes, P.J. 1983. A periodically forced piecewise linear oscillator. *Journal of Sound and Vibration*, 90(1), 129–155.
[2] Feldman, M., Zimmerman, Y., Sheer, S., and Bucher, I. 2017. Decomposition of stiffness and friction tangential contact forces during periodic motion. *Mechanical Systems and Signal Processing*, 94, 400–414.
[3] Galdi, Giovanni, P. 2016. On bifurcating time-periodic flow of a navier-stokes liquid past a cylinder. *Archive for Rational Mechanics and Analysis*, 222(1), 285–315.
[4] Zhang, Z., Liu, Y., Sieber, J. 2020. Calculating the Lyapunov exponents of a piecewise smooth soft impacting system with a time-delayed feedback controller. *Communications in Nonlinear Science and Numerical Simulation*, 10545, 1–28.
[5] Stanton, S.C., Mann, B.P., Owens, B.A.M. 2012. Melnikov theoretic methods for characterizing the dynamics of the bistable piezoelectric inertial generator in complex spectral environments. *Physica D Nonlinear Phenomena*, 241(6), 711–720.
[6] Thota, P., Zhao, X., Dankowicz, H. 2006. Co-dimension-two grazing bifurcations in single-degree-of-freedom impact oscillators. *Journal of Computational and Nonlinear Dynamics*, 2(4), 328–335.
[7] Gritli, Hassène, Belghith, S. 2018. Diversity in the nonlinear dynamic behavior of a one-degree-of-freedom impact mechanical oscillator under ogy-based state-feedback control law: order, chaos and exhibition of the border-collision bifurcation. *Mechanism and Machine Theory*, 124, 1–41.
[8] Gatti, G., Kovacic, I., Brennan, M.J. 2010. On the response of a harmonically excited two degree-of-freedom system consisting of a linear and a nonlinear quasi-zero stiffness oscillator. *Journal of Sound and Vibration*, 329(10), 1823–1835.
[9] Cirillo, G.I., Mauroy, A., Renson,L., Kerschen, G., Sepulchre, R. 2016. A spectral characterization of nonlinear normal modes. *Journal of Sound and Vibration*, 377, 284–301.
[10] Lyu, X.H., Gao, Q.F., Luo, G.W. 2020. Dynamic characteristics of a mechanical impact oscillator with a clearance. *International Journal of Mechanical Sciences*, 178: 105605.
[11] Wagg, D.J. 2005. Periodic sticking motion in a two-degree-of-freedom impact oscillator. *International Journal of Non-Linear Mechanics*, 40(8), 1076–1087.
[12] Li, G.F., Ding, W.C. 2018. Global behavior of a vibro-impact system with asymmetric clearances. *Journal of Sound and Vibration*, 423, 180–194.
[13] Oruganti, P.S., Krak, M.D., Singh, R. 2018. Step responses of a torsional system with multiple clearances: study of vibro-impact phenomenon using experimental and computational methods. *Mechanical Systems and Signal Processing*, 99(jan.15), 83–106.
[14] Tiwari, S., Bhaduri, S. 2017. Dynamic analysis of rotor-bearing system for flexible bearing support condition. *International Journal of Mechanical Engineering and Technology*, 8(7), 1785–92.

[15] Walha, L., Driss, Y., Khabou, M.T., Fakhfakh, T., Haddar, M. 2011. Effects of eccentricity defect on the nonlinear dynamic behavior of the mechanism clutch-helical two stage gear. *Mechanism and Machine Theory*, 46(7), 986–997.

[16] Al-Solihat, M.K., Behdinan, K. 2019. Nonlinear dynamic response and transmissibility of a flexible rotor system mounted on viscoelastic elements. *Nonlinear Dynamics*, 97(2), 1581–1600.

[17] Luo, G.W., Lv, X.H., Shi, Y.Q. 2014. Vibro-impact dynamics of a two-degree-of freedom periodically-forced system with a clearance: diversity and parameter matching of periodic-impact motions. *International Journal of Non-Linear Mechanics*, 65(oct.), 173–195.

Effect of cathode/anode area ratio on galvanic corrosion of TA1 titanium/316 stainless steel couples

S.Y. Guo, Z.D. Liu, B.K. Li, L. Chen, H.R. Ma & Y. Li
North China Electric Power University, College of Energy Power and Mechanical Engineering, Key Laboratory of Power Transmission, Conversion and System of Power Station, Ministry of Education, Beijing, China

ABSTRACT: This paper focuses on the relationship between the corrosion law of TA1 titanium and 316 stainless steel when coupled in 3.5% NaCl solution and the area ratio of cathode to anode. 316 stainless steel and TA1 titanium were processed into 316 /TA1 couple pairs with area ratios of 1:1, 1:2, 1:4, immersed in 3.5% NaCl solution, each sample's open circuit potential and polarization curve were measured every day by electrochemical workstation. And the test period is twenty days. After the end of the corrosion, the sample was subjected to microstructure analysis and energy spectrum analysis (EDS) using a scanning electron microscope (SEM). According to the experimental data analysis, a dense passivation film is formed on the surface of TA1 titanium and 316 stainless steel to increase the corrosion potential. After pitting corrosion on the surface of 316 stainless steel, polarity reversal occurs. TA1 titanium is used as cathode and 316 stainless steel is used as anode to accelerated its corrosion. In the 316 /TA1 galvanic couple, in the range of small area ratio, the galvanic current density increases with the area ratio of cathode to anode.

1 INTRODUCTION

Titanium is widely used in important fields such as national defense, biomedicine and petrochemical due to its high strength, good corrosion resistance, high heat resistance, and strong weldability. Its application as a corrosion-resistant structural material in corrosive environments is more and more widely [1, 2].

316 stainless steel is often used to make marine corrosion-resistant equipment, such as corrosion-resistant containers, equipment linings, pipelines and acid-resistant equipment parts. It has better resistance to chloride corrosion than 304 stainless steel, and is often used as "marine steel" [3].

This paper works on the corrosion characteristics of galvanic corrosion between TA1 titanium-316 stainless steel. Compared with conventional metal materials, these two metals have higher density, denser structure and better corrosion resistance. The pure titanium oxide film is easily formed on the surfaces which can reduced the corrosion rate. Study the corrosion characteristics of TA1 titanium and stainless steel in the marine environment is better for using these materials in marine equipment, effective protection of metals that may be corroded in some cases. It is a great strategic significance to effectively avoid the dangers caused by galvanic corrosion, prevent problems before they occur, increase the life of marine equipment, reduce the loss of national public property, and speed up the construction of national defense [4–6].

The uniform corrosion behavior of titanium and titanium alloys in seawater and the galvanic corrosion behavior of steel have an important impact on the surface service life and structural safety of marine equipment [7, 8]. The research on galvanic corrosion of titanium and titanium alloys in seawater roughly began in the 1980s [9].

Since the 21st century, in order to study the corrosive behavior of titanium and titanium alloys, scholars have done a lot of research on different TA1 titanium with different experimental methods.

Zhang Yifei et al. [10] studied the void of TC4 in LiBr solution by full immersion method. It is found that the cumulative mass loss of TC4 in distilled water during cavitation is lower than that in LiBr solution. Compared with static conditions, the self-corrosion potential of TC4 is positively shifted. Under cavitation conditions, the conductivity increases and electrochemical The corrosion rate is increased; Yang Fan et al. [11] conducted a total immersion corrosion test on Ti35 alloy annealed for 8 hours in nitric acid, according to the polarization curve and electrochemical impedance spectroscopy of the alloy, and found that the Ti35 alloy the grain size is significantly increased after long-term annealing, the corrosion current density is reduced, and the corrosion resistance in concentrated nitric acid is greatly improved. In 2013, Si Weihua et al. [12] studied the titanium alloy TA28 and Ti6321 and the ship hull 907A by electrochemical methods. According to the results, when using TA28 or Ti6321, there will be slight electrochemical corrosion on the hull, and proper insulation can effectively control the electrochemical corrosion between the titanium alloy and the hull steel.

In the application of the galvanic corrosion test, Hao Xiaobo et al. [13] found that the self-corrosion potential of titanium alloy is higher than that of steel, and the steel acts as an anode when the two contacts are accelerated. As the area ratio of the TA1 titanium steel pair increases, the quality loss of steel, the galvanic corrosion current and the galvanic corrosion coefficient gradually increase. When the area ratio of the two pairs is 1:1, the galvanic corrosion sensitivity coefficient reaches level D. Therefore, cathodic protection of steel is required when they are in contact; Hu Yaojun et al. [14] studied the corrosion resistance of a new titanium alloy Ti-3AR (Ti-Al-Mo-Ni) at 60°C, and the mechanism of resistance corrosion is analyzed by electrochemical method. The results show that the corrosion potential of Ti in the corrosion solution is increased by adding Mo and Ni to Ti-3AR, so Ti-3AR has the same corrosion resistance as Ti-12.

In April 2019, Chen Yunfei, Li Zhengxian, et al. [15] conducted a galvanic corrosion experiment of the T2/TC4 galvanic pair in static artificial seawater. Through the analysis of electrochemical workstation, SEM, EDS, XRD, etc. they found that T2/TC4 reacts strongly compared with T2 self-corrosion, the copper ion release rate of T2/TC4 galvanic corrosion is increased by dozens of times, which can meet the marine antifouling requirements that inhibit the adhesion of most marine organisms and achieve a better marine anti-fouling effect.

It can be seen that the application types and application ranges of titanium and titanium alloys are constantly expanding, the development and application of new surface treatment technologies are also booming. Although the research on galvanic corrosion in seawater is ongoing, there is still a lack of more systematic and comprehensive research. The discovery and exploration, and the current research on this aspect is not sufficient, and the understanding of the relevant mechanism is not clear enough. Therefore, the research on the galvanic corrosion performance between TA1 titanium and stainless steel has important academic significance and engineering value.

2 EXPERIMENT

The materials selected for this experiment are TA1 titanium and 316 stainless steel, and the specific components are shown in Table 1.

Table 1. Chemical compositions of TA1 titanium and 316 stainless steel.

	Element	Ti	Al	Mn	C
TA1	Content(wt.%)	Margin	0.60~0.79	0.09~0.15	0.13

	Element	Fe	Cr	Ni	Mo	Mn	Si
316	Content(wt.%)	Margin	16.0~18.5	10.0~14.0	2.0~3.0	≤2.00	≤1.00

The TA1 titanium plate is processed to obtain each TA1 titanium sample of 10mm×10mm×2mm, 10mm×20mm×2mm, 20mm×20mm×2mm, and three 316 stainless steel samples of 10mm×10mm×4mm are obtained by processing the 316 stainless steel plate. Divided into three groups, the working area ratios of 316 stainless steel and TA1 titanium are 1:1, 1:2, and 1:4 respectively. They are welded with copper wires and sealed with epoxy resin. After solidification, the working surface is water-ground to 1200 with sandpaper. Mesh, clean, dry, and set aside. Connect each group of TA1 titanium and stainless steel samples with wires, immerse them in a 0.6mol/L NaCl solution, the distance between the opposite sides of the galvanic couple is 30mm, the temperature is set to 25°, and the corrosion period is 20 days.

Monitor the galvanic current of the sample every day, and measure the weak polarization curves of the two electrodes in 0.6mol/L NaCl solution when the corrosion period is 0, 4, 10, and 20 days. The scan interval is -0.03V~ +0.03V, the scan rate is 0.001V/s.

The electrochemical workstation(CHI660E) was used to test the galvanic current, galvanic potential, open circuit potential and polarization curve of the galvanic pair.

When the experiment lasted for 20 days, the sample was taken out to observe the surface morphology, and the surface of the sample was analyzed with the scanning electron microscope (FEI Quanta200F).

3 EXPERIMENTAL RESULTS, ANALYSIS AND DISCUSSION

3.1 Self-corrosion potential analysis

Figure 1 shows the change curve of self-corrosion potential of 316 stainless steel and TA1 titanium 30 minutes before the start of corrosion. It can be seen that the overall self-corrosion potential of 316 stainless steel and TA1 titanium has little change. This shows that 316 stainless steel and TA1 titanium have relatively stable chemical properties in 3.5% NaCl solution, and both have a certain self-passivation ability, that is, the surface can generate a dense and stable passivation film. The self-corrosion potential of 316 stainless steel is -0.1394V, which is higher than the self-corrosion potential of TA1 titanium -0.3382V. The difference between the two (198.8mV) is greater than the threshold potential difference (50mV) for galvanic corrosion. Therefore, 316 stainless steel and TA1 galvanic corrosion occurs when titanium is coupled.

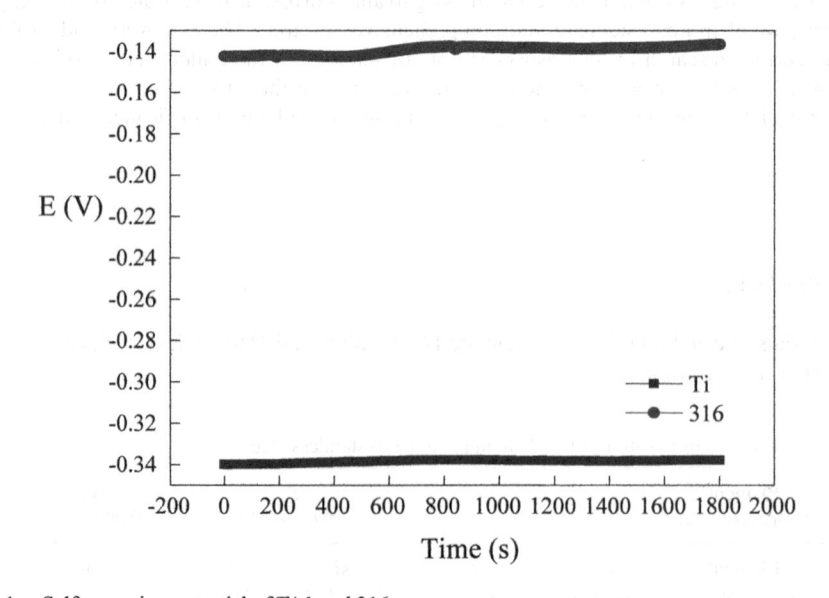

Figure 1. Self-corrosion potential of TA1 and 316.

After connection, in the 316/TA1 galvanic couple, TA1 titanium with a more negative potential was used as the anode in the early stage of the experiment, and electron loss accelerated corrosion, and 316 stainless steel with a more positive potential was used as the cathode, which was protected and the corrosion rate was slowed down.

3.2 Galvanic current analysis

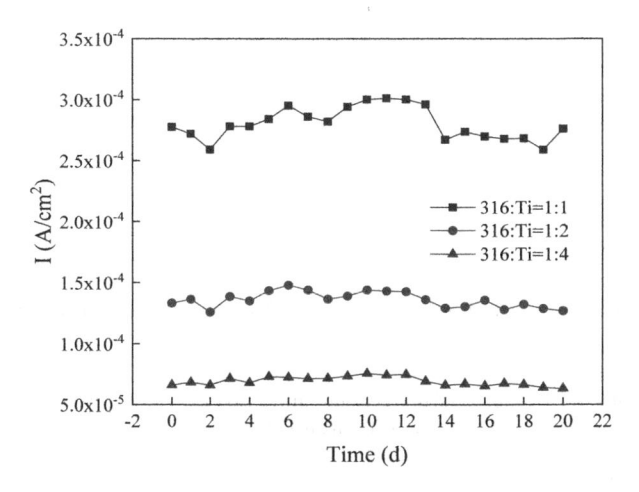

Figure 2. Galvanic current density.

As shown in Figure 2, after TA1 titanium is coupled with 316 stainless steel, since the self-corrosion potential of TA1 titanium is more negative, TA1 titanium is used as anode and 316 stainless steel is used as cathode. On the whole, the current density of the three galvanic couples with area ratios of cathode and anode of 1:1, 1:2, and 1:4 are relatively stable, and there are no obvious trend of change. Among them, the 316/TA1 galvanic couple with an area ratio of yin and yang of 1:1 has the largest average galvanic current density, which is 0.280mA/cm2, and the change range is less than 0.042mA/cm^2 within a 20-day period. The area ratio of cathode and anode is 1:2. The current density of the 316/TA1 galvanic couple is 0.136mA/cm^2, and the range of change does not exceed 0.021mA/cm^2. T current density of the 316/TA1 galvanic couple with an area ratio of 1:4 is the smallest, which is 0.069mA/cm^2, The amplitude of change is less than 0.012mA/cm^2. It can be seen that in the range of relatively small area, the increase of the galvanic current density basically increases linearly with the increase of the area ratio of the cathode and anode, which conforms to the "oxygen-enriched area principle".

3.3 Weak polarization curve analysis

Observe the polarization curve change trend of TA1 titanium and 316 stainless steel in the first group of samples from Figure 3. From the start day to the 4th day, the corrosion potential of TA1 titanium moves positively, and the corrosion potential of 316 stainless steel moves negatively, because a dense and stable oxide film layer is formed on the surface of TA1 titanium, which has the effect of hindering and slowing down the corrosion rate. The corrosion resistance in the corrosive solution has been improved, and the corrosion has been slowed down. Although a dense oxide film can be formed on the surface of stainless steel, pitting corrosion has appeared on the surface of 316 stainless steel, and the corrosion has accelerated. From the 4th to the 10th day, due to the process of dissolved oxygen's consumption and replenishment, the corrosion potential of TA1 titanium and 316 stainless steel has a slight negative move, which is relatively stable. From the 10th day to the end of the test period, the polarization curve is basically stable, and the corrosion potential of TA1 titanium moves positively and gradually stabilizes at around -0.17V. At this time, TA1 titanium is protected as the cathode and the corrosion is slowed; 316 stainless steel is used as the anode, and the corrosion is accelerated.

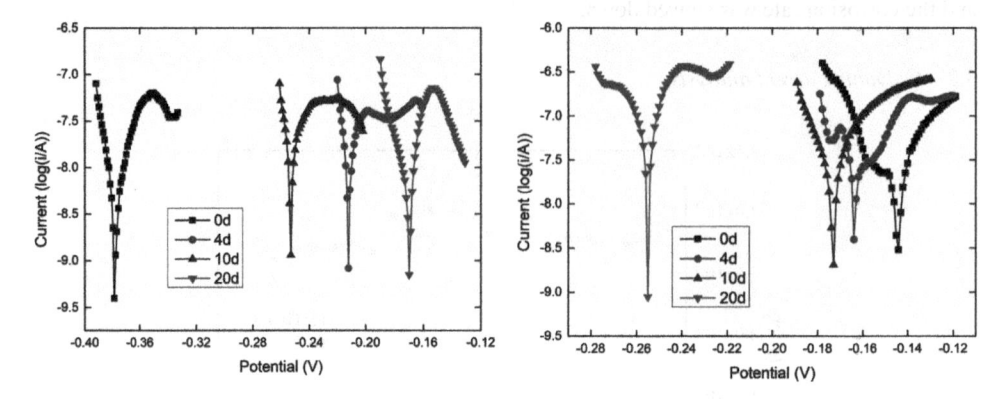

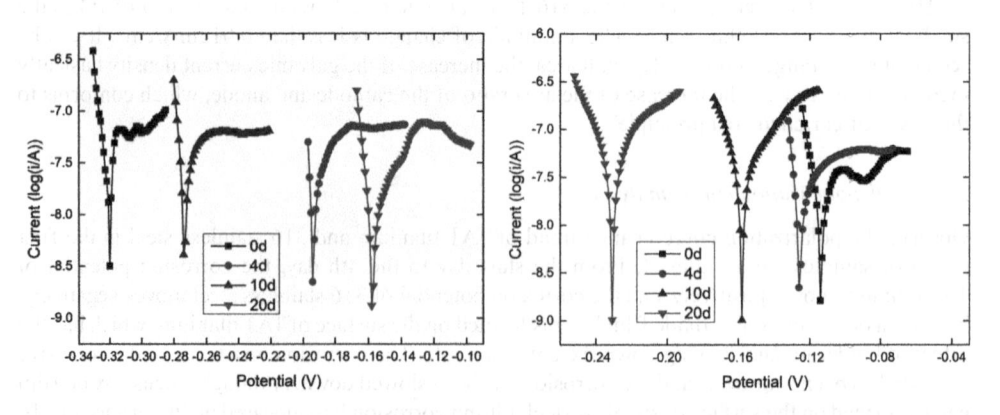

Figure 3. Polarization curve of TA1 titanium sample and 316 stainless steel sample with area ratio of 1:1.

Table 2. Change trend of corrosion current ($\log(i/A)$) of TA1 titanium and 316 stainless steel samples with an area of 1:1.

Time(d)		4	10	20
TA1	−8.34	7.55	−7.58	−7.56
316	−7.51	−7.47	−7.32	−7.40

Table 2 shows the changes in the surface corrosion current of TA1 titanium and 316 stainless steel obtained by fitting the polarization curve during the period. It can be seen that after the 4th day, the logarithmic value of the surface corrosion current of TA1 titanium is basically stable around 7.56; Throughout the experiment, the logarithmic value of the 316 stainless steel corrosion current was relatively stable, and the fluctuation range did not exceed 0.20.

Figure 4. Polarization curve of TA1 titanium sample and 316 stainless steel sample with area ratio of 2:1.

Observe the polarization curve trend of TA1 titanium and 316 stainless steel in the second group of samples from Figure 4. From the start day to the 4th day, the corrosion potential of TA1 titanium is in a positive move. This is due to the formation of an oxide film layer on the surface of TA1 titanium that hinders and slows down the corrosion rate, so that the corrosion resistance of TA1 titanium in the corrosion solution has been improved. At the same time, due to the occurrence of pitting corrosion, the potential of 316 stainless steel becomes more negative, and the corrosion resistance of stainless steel is affected and corrosion is accelerated. From the 4th to the 10th day, due to the process of dissolved oxygen's consumption and replenishment, the corrosion potential of TA1 titanium and 316 stainless steel has a slight negative move, and the fluctuation does not exceed 0.08V. From the 10th day to the end of the test period, TA1 titanium was protected as a cathode, the corrosion was slowed down, the polarization curve was basically stable, the corrosion potential became more positively, and gradually stabilized near -0.16V; 316 stainless steel was corroded as the anode, the corrosion was accelerated, and the potential became more negative, and finally stabilized at about -0.23V.

Table 3 shows the change of the corrosion current on the surface of TA1 titanium and 316 stainless steel obtained by fitting the polarization curve during the period. During the zeroth day to the fourth day, due to the production of the oxide film on the 316 stainless steel surface, the corrosion current decreases. From the fourth day to the twentieth day, the logarithmic value of the corrosion current stabilizes at around -7.64, and the fluctuation range does not exceed 0.04. The logarithmic value of the corrosion current of TA1 titanium gradually stabilized at about -7.45 after the tenth day, and the range of corrosion current change was extremely small.

Table 3. Change trend of corrosion current ($\log(i/A)$) of TA1 titanium and 316 stainless steel samples with an area of 2:1.

Time(d)	0	4	10	20
TA1	−7.94	−7.62	−7.48	−7.43
316	−7.17	−7.62	−7.66	−7.64

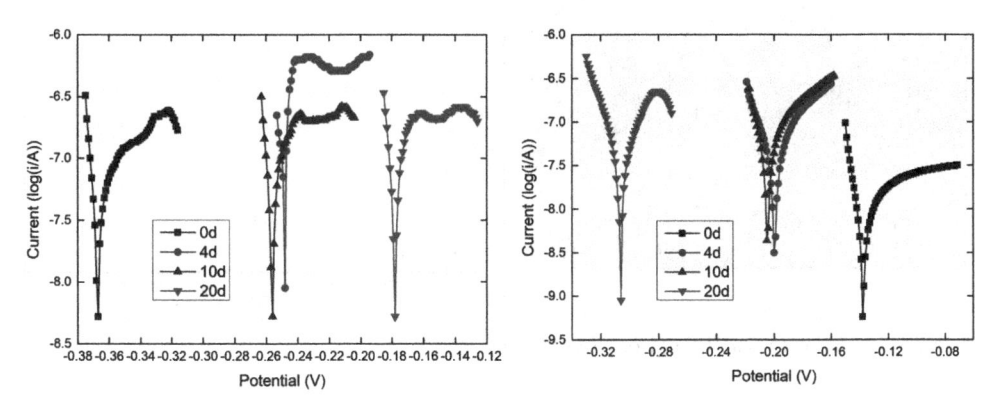

Figure 5. Polarization curve of TA1 titanium sample and 316 stainless steel sample with area ratio of 4:1.

First observe the polarization curve change trend of TA1 titanium and 316 stainless steel samples in the third group of samples from Figure 5. From the 0th day to the 4th day, the positive shift of the corrosion potential of TA1 titanium was observed. A passive film was formed on the surface of TA1 titanium which can hinder and slow down the corrosion rate. The corrosion resistance of TA1 titanium in solution was improved; 316 stainless steel corrosion potential shifts negatively; pitting corrosion on the surface of stainless steel was appeared, the passive film of stainless steel was destroyed, the corrosion resistance reduced. From the 4th to the 10th day, due to the process of dissolved oxygen consumption and replenishment, the corrosion potential of TA1 titanium moved in the negative direction, and the corrosion potential of 316 stainless steel showed small fluctuations. From the 10th day to the end of the test period, the corrosion potential of TA1 titanium moved positively and gradually stabilized. The potential of 316 stainless steel was lower than that of TA1 titanium. TA1 titanium was protected as a cathode and the corrosion slowed down.

Table 4. Change trend of corrosion current (log(i/A)) of TA1 titanium and 316 stainless steel samples with an area of 4:1.

Time(d)	0	4	10	20
TA1	−7.94	−7.21	−7.12	−7.17
316	−7.49	−7.99	−7.41	−7.48

Table 4 shows the changes in the surface corrosion current of TA1 titanium and 316 stainless steel with an area ratio of 4:1 obtained by fitting the polarization curve during the period. From the 0th day to the 4th day, a passive film was formed on the surface of 316 stainless steel, and the corrosion current decreased; from the 4th to the 10th day, pitting corrosion occurred and the corrosion current increased; after the 10th day, the logarithm of the corrosion current is relatively stable which is about -7.44. The logarithmic value of the corrosion current of TA1 titanium was stable within a range from the fourth day to the twentieth day, and the fluctuation range did not exceed 0.09.

3.4 *Corrosion products, SEM And EDS*

3.4.1 *Analysis of TA1 titanium and 316 stainless steel samples with an area ratio of 1:1*

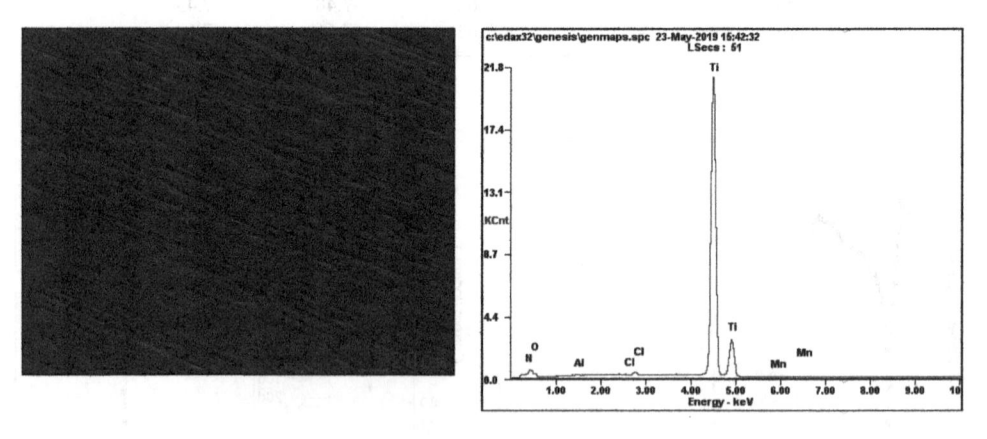

Figure 6. SEM image and EDS of No. 1 TA1 titanium.

Table 5. Surface element content of No. 1 TA1 titanium.

Element	Ti	O	Al	Mn	Cl
Content(wt.%)	83.69	9.26	0.66	0.09	0.13

Observing Figure 6, it can be seen that there is no obvious corrosion trace on the overall surface of TA1 titanium. The selected point of the energy spectrum is the gray film-like structure with the most similar morphologies on the surface, and the similar morphology is in the center of TA1 titanium. And the edge is evenly distributed without segregation. According to the energy spectrum elements in Table 5, compared with the initial element content of the material, the oxygen element content has been greatly improved, so it can be judged that a large amount of dense titanium oxide passivation film is formed on the surface of TA1 titanium in the solution. It can effectively avoid the corrosion of the surface of TA1 titanium in the 3.5% NaCl solution. The content of Al, Mn and Cl elements is low, and there is basically no change.

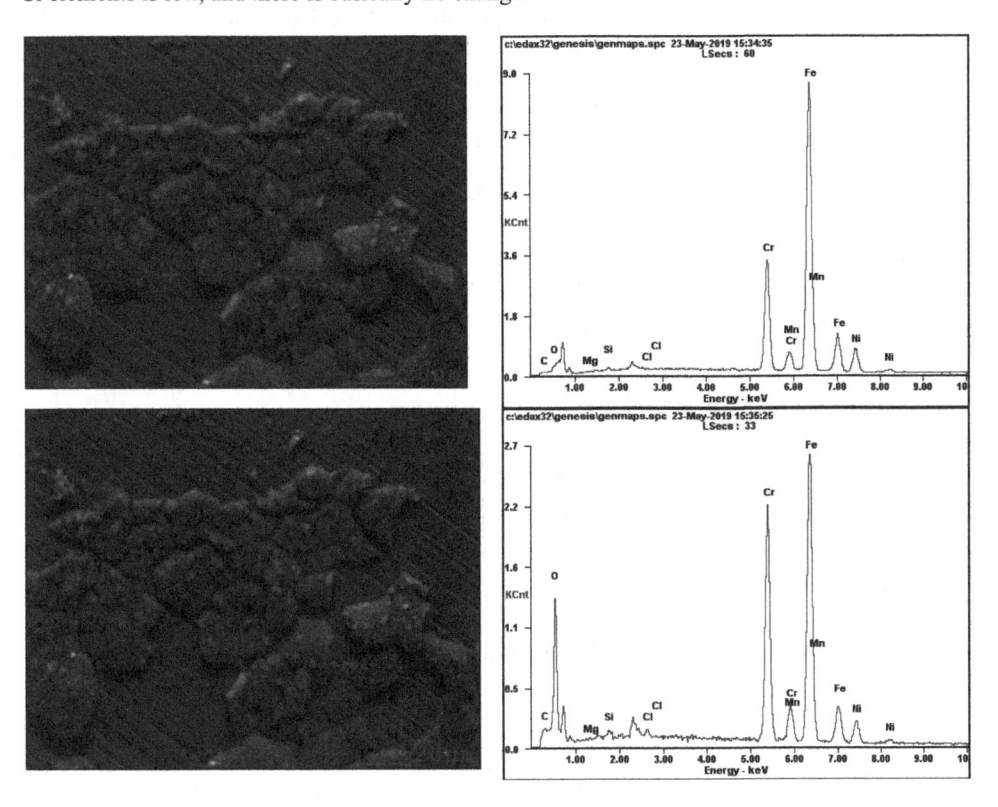

Figure 7. SEM images and EDS of 316 stainless steel non-corroded spots and corroded spots.

Observing Figure 7, you can see a large number of gray lumps and white granular attachments, dense gray patches appear, and white particles are attached to the gray patches and the surface of the sample, and the patches are only distributed on the upper edge of the entire sample. The Table 6 shows the element changes. First of all, the iron element has dropped from 67.76% to 38.99%, and the oxygen element has risen from 1.91% to 20.69%, which has risen sharply. It can be seen that in the uncorroded area, the elements are mostly mixed with metals in the alloy. That is to say (Fe, Cr, Ni) solid solution and the oxygen content is low; in the corrosion area, the corrosion product is mainly iron oxide and ferrous oxide, and the surface of the sample is macroscopically reddish brown. The nickel element is reduced from 9.08% to 4.61%. It can be seen that some Ni element forms nickel oxide and exists on the surface of the stainless steel in the form of a nickel oxide film. Other metal elements still exist in the sample in the form of solid solution.

Table 6. Element content (wt.%) of 316 stainless steel in non-corroded area and corroded area.

Element	Fe	Cr	Ni	O	Mo	Mn	Si
non-corroded area	67.76	17.18	9.08	1.91	1.44	1.22	0.64
corroded area	38.99	20.62	4.61	20.69	2.17	0.73	0.84

3.4.2 *Analysis of TA1 titanium and 316 stainless steel samples with an area ratio of 2:1*

Observing Figure 8, you can see that the surface of the TA1 titanium sample is uneven, there are a lot of scratches, there are a lot of scratches, gray areas and some fine structures, and the gray areas are suspected to be the oxide film formed on the surface of TA1 titanium. The uneven areas are all over the surface of the sample, and the surface has more traces caused by manual polishing. According to the change of the self-corrosion potential of No.2 TA1 titanium in Figure 4, it is found that the self-corrosion potential of the sample rises slowly, and compared with the surface element content in Table 7 and the initial element content of TA1 titanium, it can be found that the Ti element has decreased by a small part, and the O element has risen to 8.28%. It can be seen that the 316 stainless steel/TA1 titanium galvanic couple is in the corrosion solution. A large amount of titanium oxide film is formed on the surface of TA1 titanium, which hinders the progress of corrosion, as a result, the self-corrosion potential is on the rise.

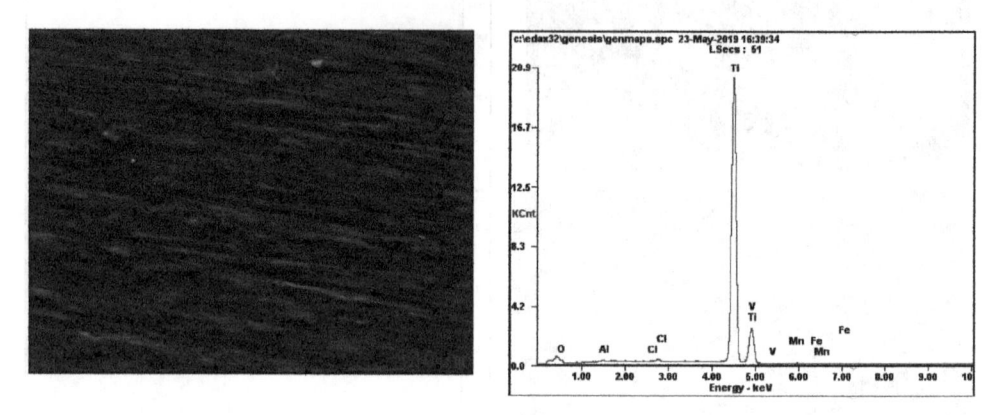

Figure 8.   SEM image and EDS of No. 2 TA1 titanium.

Table 7.   Surface element content of No. 2 TA1 titanium.

Element	Ti	O	Al	Cl
Content(wt.%)	91.27	8.28	0.38	0.07

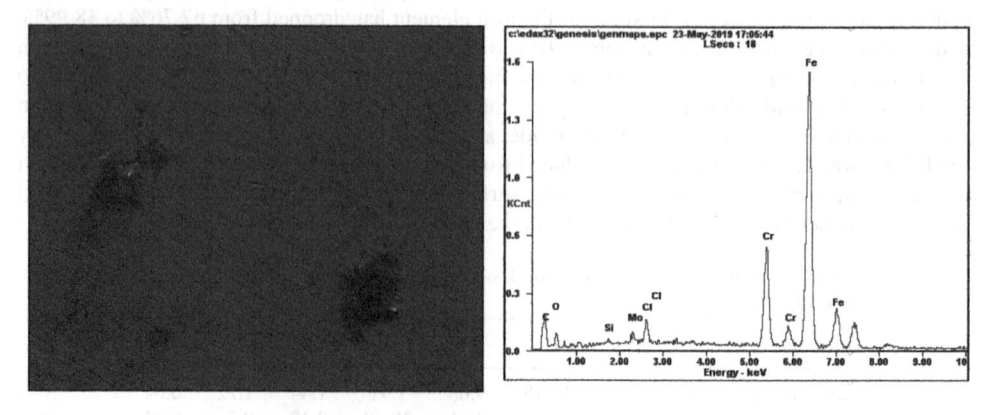

Figure 9.   SEM image and EDS of No. 2 316 stainless steel.

Observing Figure 9, the gray dots are present on the surface of the stainless steel, and the distribution is less. According to Table 8, the change of element content in the gray area can be observed. First, iron element decreased from 67.76% to 52.91%, the reduction was 14.85%, and oxygen element increased from 1.91% to 4.30%, and C element change from a small amount to 13.07%. No obvious traces of iron oxide were found when observing the macroscopic surface. It can be judged that the main components of the gray area are the precipitated phase of iron carbides and the oxide of Cr that protects stainless steel. Other metal elements exist in the sample in the form of solid solution.

Table 8. Surface element content of No. 2 316 stainless steel.

Element	Fe	Cr	C	O	Cl	Mo	Si
Content(wt.%)	52.91	24.97	13.07	4.30	2.13	2.06	0.57

3.4.3 *Analysis of TA1 titanium and 316 stainless steel samples with an area ratio of 4:1*

Observing the SEM surface image in Figure 10 and the content of each element in TA1 Titanium No. 3, which are basically consistent with the image and element changes of TA1 titanium in Figure 6, and no obvious corrosion has occurred, so do not make concrete analysis.

Figure 10. SEM image of No. 3 TA1 titanium.

Observing Figure 11, you can see a large number of gray granular and white granular attachments, and the attachment surface of the gray particulates is wider, and the white particulates are less. And macroscopically, the attachment appears at the edge of the 316 stainless steel sample surface, and the iconic color of iron oxide appears. The white particles are attached to the gray patches and the surface of the sample. According to the energy spectrum, the element changes can be observed.

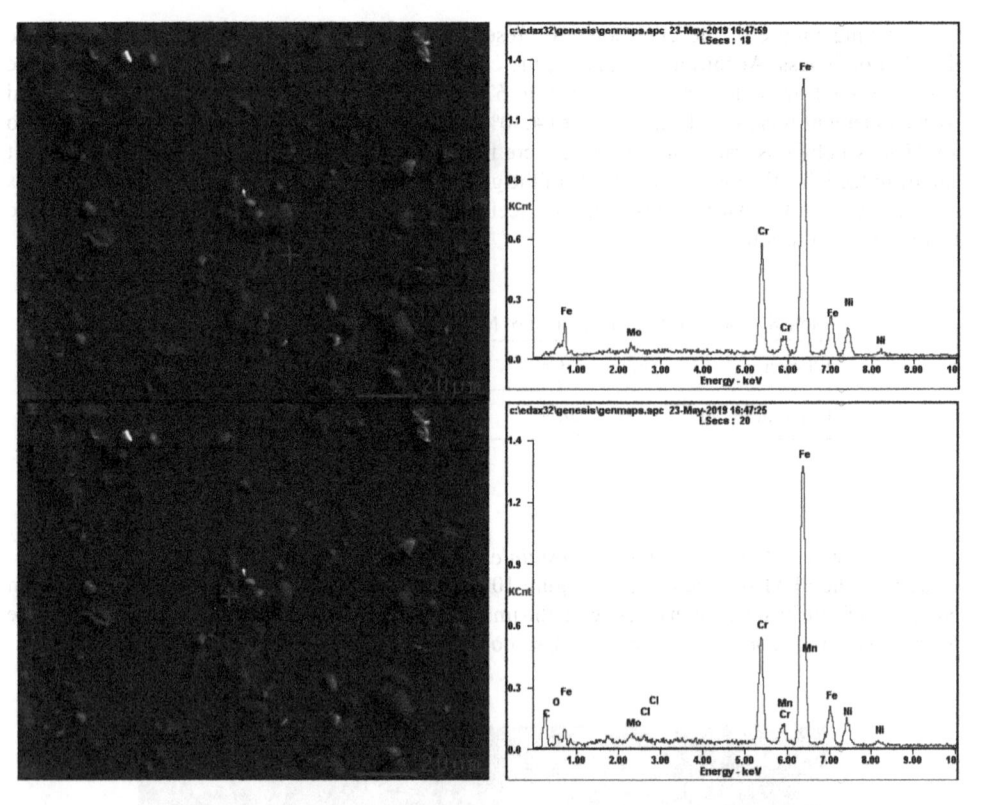

Figure 11. SEM images and EDS of 316 stainless steel non-corroded spots and corroded spots.

Table 9. Element content (wt.%) of 316 stainless steel in non-corroded area and corroded area.

Element	Fe	Cr	Ni	Mo	O	C
non-corroded area	70.38	16.8	10.5	2.32	0	0
corroded area	54.01	20.03	13.11	2.02	3.62	7.21

Compared with Table 9, iron element decreased from 70.38% to 54.01%, carbon elements increased to 7.21% and oxygen-based elements increased to 3.62%. Cr and Ni contents increased slightly compared with the original sample. It can be seen that in this region, a corrosion product with a ferocarbon compound is a main material, and a small amount of oxide is contained.

4 CONCLUSION

(1) When the TA1 titanium and 316 stainless steel were coupled in a 3.5% NaCl solution, TA1 titanium was used as an anode, and 316 stainless steel was used as a cathode. After a period of time, TA1 titanium and 316 stainless steel surface generate a layer of passivation film that can slowly reduce corrosion, and can improve the corrosion resistance of alloy surfaces. When a 316 stainless steel surface oxide film occurs, the surface corrosion speed is accelerated, and the corrosion resistance is lowered. The self-corrosion potential is lower than TA1 titanium, at that time TA1 titanium is used as a cathode, and the polarity is reversed.

(2) According to experimental data, the inclusion of 316 / TA1 electrical ratio of 1: 1, 1: 2, 1: 4, respectively. The variation of the galvanic current density accords with "the principle of oxygen-rich area". In other words, when the area ratio is small, the galvanic current density increases linearly with the increase of the area ratio of anode and cathode.

REFERENCES

[1] Filip R, and, Kubiak K, Ziaja W, Sieniawski J. 2003 The effect of microstructure on the mechanical properties of two-phase titanium alloys[J]. *Journal of Materials Processing Technology*.
[2] Dong Y, Guo J. 2011 Corrosion Mechanism of Titanium Alloys and Development of Corrosion-Resistance Titanium Alloys[J]. *Titanium Industry Progress*.
[3] Ge H H, Zhou G D, Wu W Q. 2003 Passivation model of 316 stainless steel in simulated cooling water and the effect of sulfide on the passive film[J]. *Applied Surface Science*, 211(1–4):321–334.
[4] Wood R J K. 2006 Erosion–corrosion interactions and their effect on marine and offshore materials[J]. *Wear*, 261(9):1012–1023.
[5] Meng H, Hu X, Neville A.2007 A systematic erosion–corrosion study of two stainless steels in marine conditions via experimental design[J]. *Wear*, 263(1):355–362.
[6] Huang G. 2000 STUDY OF THE CORROSION POTENTIAL OF METALS IN SEAWATER[J]. *Corrosion & Protection*.
[7] Soares C G, Garbatov Y, Zayed A, et al. 2009 Influence of Environmental Factors on Corrosion of Ship Structures in Marine Atmosphere[J]. *Corrosion Science*, 51: 2014–2026.
[8] Moshrefi R , Mahjani M G , Ehsani A , et al. 2011 A study of the galvanic corrosion of titanium/L 316 stainless steel in artificial seawater using electrochemical noise (EN) measurements and electrochemical impedance spectroscopy (EIS)[J]. *Anti Corrosion Methods & Materials*, 58(5):250–257.
[9] Doig P. Flewitt P.E.J. 1978 An analysis of galvanic corrosion: Coplanar electrodes with one electrode infinitely large[J]. *Philosophical Magazine B Physics of Condensed Matter*, 38(1):27–40.
[10] Zhang Y F, Lin C, Du N, etal. 2015 Cavitation corrosion behavior of TC4 titanium alloy in lithium bromide solution [J]. *Corros. Prot.* 36: 522.
[11] Yang F, Ju H J, Mao X N, etal. 2016 Corrosion behavior of annealed Ti35 alloy in concentrated nitric acid [J]. *Rare Met. Cem. Carbides*. 44(1): 47.
[12] Si W H. 2013 An investigation on of galvanic corrosion between exhaust system of titanium alloy and hull steel [J]. *Dev. Appl. Mater.* 28(2): 34.
[13] Hao X , Liu Y , Li B , et al. 2019 Corrosion behavior of marine Ti70 alloy in artificial seawater[J]. *Corrosion Science and Protection Technology*. 31(1):27–32.
[14] Hu Y. 2000 Study on Corrosion-Resistance of a New Titanium Alloy Ti-3AR. *Rare Metal Materials and Engineering*.
[15] Chen Y , Li Z , Liu L , et al. 2019 Galvanic Corrosion Behavior of T2/TC4 Galvanic Couple in Static Artificial Seawater.

Chapter 6: Designs Related to Materials Science and Engineering

Electromagnetic energy conversion technology and its application in modern engineering

H.L. Nie
Institute of Safety Assessment and Integrity, State Key Laboratory for Performance and Structure Safety of Petroleum Tubular Goods and Equipment Materials, Tubular Goods Research Center of CNPC, Xi'an, China
Northwestern Polytechnical University, Xi'an, Shaanxi, China

W.F. Ma
Institute of Safety Assessment and Integrity, State Key Laboratory for Performance and Structure Safety of Petroleum Tubular Goods and Equipment Materials, Tubular Goods Research Center of CNPC, Xi'an, China

Z.Q. Yu
China Classification Society Quality Certification Company Sichuan Branch, Chengdu, China

J.J. Ren, K. Wang, J. Cao, W. Dang, T. Yao, X.B. Liang & K. Wang
Institute of Safety Assessment and Integrity, State Key Laboratory for Performance and Structure Safety of Petroleum Tubular Goods and Equipment Materials, Tubular Goods Research Center of CNPC, Xi'an, China

ABSTRACT: Electromagnetic energy is a kind of energy closely related to our life. It can be transformed into other forms of energy in an instant, which promotes the development of modern industry. With the establishment of Maxwell's electromagnetic equations, the understanding of electromagnetic energy is more profound and perfect, and the application of electromagnetic energy in modern society is more and more extensive, which promotes the development of some traditional industries. This paper reviews the main historical events in the development of electromagnetic energy, and summarizes and analyses the innovative applications of electromagnetic energy conversion technology in modern industry.

1 INTRODUCTION

Magnetic phenomenon is a basic phenomenon in nature. The ancient Chinese people discovered the magnet as early as the Spring and Autumn Period and the Warring States Period, and invented the oldest Guide-Si Nan. Around the thirteenth century, people found that the magnet has two poles, and named the N pole and S pole, and through experiments confirmed that the opposite magnetic poles attract, same magnetic poles repel each other [1].

In 1600, British doctor Gilbert published a book "Magnetism, Magnets and the Earth as a Giant Magnet", which summarized the previous studies on magnetism, and carefully discussed the nature of geomagnetism, a large number of experiments were also recorded. This was a systematic research work on magnetism [2].

In 1785, Coulomb published the inverse square law of electric power obtained by torsion scale experiment, namely Coulomb's law, which was recognized by the world and brought electricity and magnetism into the stage of quantitative research [3].

It was only after Oster discovered the effect of electromagnetism that scientists really connected electricity and magnetism. After that, French scientist Ampere carried on the research to the electromagnetic force, put forward the famous right hand rule, and successfully explained the cause of geomagnetism. Then Ampere established the interaction rule between current elements - Ampere's law, and put forward the hypothesis that the origin of magnetism is electric current - Ampere's molecular current hypothesis. At the same time, the French scientist Biot Savard and Laplace analysed and summarized the law of magnetic field generated by current element-Biot-Savard law based on the experiment. The basic theoretical system of electromagnetism was established [4].

In 1831, the British Physicist Faraday discovered the phenomenon of electromagnetic induction, which further confirmed the unity of electrical and magnetic phenomena [5]. In 1865, Maxwell [6] generalized the law of electromagnetic field with a set of equations combining Faraday's idea of electromagnetic near action with the law of electrodynamics initiated by Ampere, and established the theory of electromagnetic field. Maxwell's predicted the electromagnetic properties of light according, which finally realized the second great synthesis in the history of physics.

Nowadays, electromagnetism is applied in various fields of modern society. The trend of multidisciplinary integration also urges scientists and engineers to continuously explore more application space of electromagnetic energy technology. Researchers use the mutual conversion of electromagnetic energy and other energies to achieve the basic functions of various modern devices. This paper summarizes the application of electromagnetic energy conversion technology in modern production, and puts forward the future development trend of electromagnetic application field.

2 ELECTROMAGNETIC FORMING TECHNOLOGY

Electromagnetic forming technology is a new metal processing technology proposed in the 20th century. It has been highly valued by various countries, and has been used in many cutting-edge technology fields, such as aviation, atomic energy, instrument, automobile, etc.

Electromagnetic pulse forming is a successful case of the principle of electromagnetic energy conversion. In the forming process, the electromagnetic energy, the deformation energy of the structure and the heat energy of the equipment are transformed with each other, which involves a complex process of multi-physical field coupling.

The principle of electromagnetic forming technology is shown in Figure. 1. The capacitor is used to store electric energy in advance. At the moment when the discharge switch is closed, a typical RLC oscillation circuit is formed between the capacitor and the coil through the cable. A strong alternating current will be generated in the working coil, and then a strong pulsed magnetic field will be formed around it and pass through the workpiece. According to the electromagnetic induction law, an induced current will be generated on the surface of the metal workpiece. The direction of the induced current is opposite to that of the current of the main coil. The induced current is subjected to the Lorentz magnetic field force of the magnetic field of the main coil, and the workpiece will be deformed under the action of the magnetic field force [7]. The whole deformation process involves subjects such as electricity, electrodynamics, electromagnetism, plastic dynamics and thermodynamics [8].

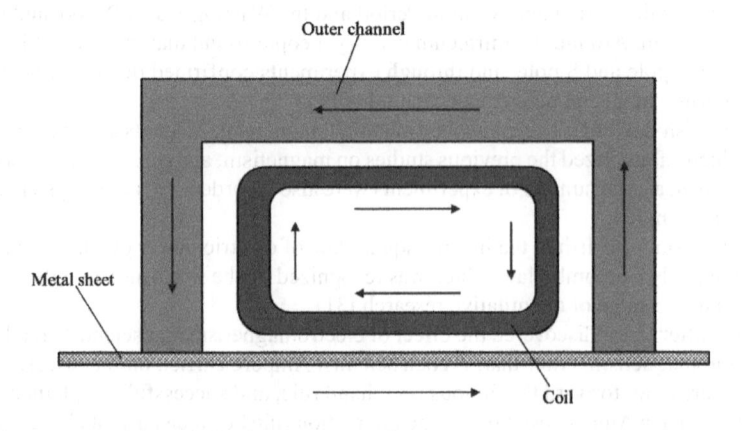

Figure 1. The diagram of the split Hopkinson pressure bar.

The first practical and industrial Electromagnetic forming device was invented by Brower and Harvey [9] in 1958, which was a milestone achievement. In China, there are also research institutions such as Harbin Institute of Technology and highly engaged in the research of electromagnetic pulse forming technology, but generally speaking, it lags behind western countries. In the future, electromagnetic forming technology will be an important application of electromagnetic energy conversion technology.

3 ELECTROMAGNETIC DRIVING TECHNOLOGY

3.1 *Electromagnetic Launch*

Electromagnetic Launch (EML) is also a specific application of the principle of electromagnetic energy conversion. It is an important research direction in the military field, involving Launch devices of artillery, rockets and other chemical weapons.

The principle of electromagnetic launch technology is similar to that of a linear motor: a changing current creates a changing magnetic field in a coil, which is also wrapped around the object being driven to produce an opposing magnetic field. The two magnetic fields repel each other, and the electromagnetic repulsion is used to accelerate the object being driven. The driven object may also be a magnetic material whose magnetic field direction is opposite to that of the fixed coil during the driving process, and the magnetic repulsion between the two causes the driven object to accelerate [10].

Therefore, electromagnetic launch technology is a new linear propulsion technology, generally suitable for short range, large load launch conditions, has a wide range of application prospects in military, industrial and civil fields.

The principle of electromagnetic launch was proposed in the early 19th century. After decades of research, it developed rapidly in the 1970s, which benefited from the great progress in ultra-high power pulse technology and electronic technology. In an epoch-making event, in 1978, Marshall et al. accelerated a 3g-weight polycarbonate projectile to an initial speed of 5.9 km/s, demonstrating that it was feasible to propel units to high velocities by electromagnetism [11].

3.2 *Electromagnetic orbital launch technology*

Electromagnetic orbital launch is another kind of electromagnetic drive technology, its main principle is to produce Lorentz force on the target object through the change of magnetic flux in the loop, so as to push the target object to move.

This technology uses electromagnetic driving force to generate loads, so it converts electromagnetic energy into mechanical energy of the driven object, and can accelerate a variety of objects, such as projectile, satellite, rocket, aircraft, etc. [12]. Whether in military or civil fields, it has considerable development prospects and application value.

The earliest railgun was developed in 1958 by Los Alamos National Laboratory in the United States, which was the first to demonstrate the feasibility of orbital launch using plasma armatures [13]. Since then, many scholars have stolen the armature rail gun research, including Marshall, Brast, Sitzman and so on [14–16]. China's electromagnetic rail gun started late, and there are still many technical problems, which need to be further studied and solved by many scholars.

3.3 *Electromagnetic driven Hopkinson bar*

Electromagnetic driven technology is also used in the field of dynamic impact testing of materials. In 2010, Guo Weiguo et al. developed a kind of miniature Hopkinson bar [17] using electromagnetic driving technology. The principle of generating stress wave is still impacting method, but the striking bar is driven by the electromagnetic force.

In 2014, Liu Zhanwei et al. also developed a multi-stage driven miniature Hopkinson bar with the same principle and carried out some material tests. At the same time, theoretical and experimental analysis was conducted on the influencing factors in the experiment [18].

This electromagnetic driven method is still in the trial stage, and compared with the more mature gas driven method, there are more influencing factors, such as the influence of the coil surrounding on the bullet on the generated waveform, the electromagnetic interference problem in the experiment, and so on.

More importantly, this method only changes the driving method of the striking bar, while the basic principle of generating stress wave remains unchanged. Therefore, the problems existing in the traditional Hopkinson bar still cannot be solved in this method [19, 20].

4 ELECTROMAGNETIC BRAKING TECHNOLOGY

In the early 1990s, Japan took the lead in the study of a new non-contact braking method – permanent magnet eddy current braking [21]. The principle is to cut the magnetic induction line through the moving conductor, and produce the magnetic field force that hinders the movement, so as to realize the non-contact braking, and avoid the disadvantages of the traditional mechanical braking mode, such as large friction force and large heat.

The earliest application of the technology is on vehicle braking [22]. According to the installation position of braking device, eddy current braking can be divided into linear and rotary type. From the view of energy conversion, the principle of two permanent magnet eddy current brake is the same, its essence is using Faraday electromagnetic induction principle to convert the kinetic energy of an object into its internal electromagnetic energy, and at the same time the electromagnetic energy is converted into heat energy by the resistance of the object.

5 TRANSIENT MAGNETIC TECHNOLOGY

5.1 *Electromagnetic riveting technology*

The research on the principle and application of transient magnetic technology first appeared in the assembly of aviation structures. Its application is not the dynamic mechanical properties of materials, but the riveting of aviation structures. In the 1960s, in order to solve the problems existing in ordinary riveting, Huber A Schmitt et al. in Boeing Company took the lead in researching electromagnetic riveting technology, and applied for the patent of strong impact electromagnetic riveting device in 1968 [23].

In 1986, Zieve Peter [24] successfully developed low-voltage electromagnetic riveting, which solved the problems existing in the quality and application of high-voltage riveting, thus enabling the rapid development of electromagnetic riveting technology. Electromagnetic riveting technology has been applied in the manufacturing of Boeing and Airbus series aircraft. Nowadays, low-voltage electromagnetic riveting technology has been developed, and the size and duration of riveting force can be controlled more accurately.

The technical principle of the electromagnetic riveter is that a coil and stress wave amplifier are added between the discharge coil and the workpiece. At the moment when the discharge switch is closed, a strong magnetic field is generated around the primary coil through the rapidly changing impulse current. The secondary coil coupled with the primary coil generates an induced current under the action of a strong magnetic field, and then generates an eddy current magnetic field. The interaction between the two magnetic fields generates eddy current repulsion force, which is transmitted to the rivet through the amplifier, forming the rivet. Electromagnetic riveting is also called stress wave riveting because of the high frequency of eddy currents, which propagate in the form of stress waves in amplifiers and rivets [25].

There are many scholars on the parameters of electromagnetic riveting process and simulation studies. At the same time, the circuit control system of electromagnetic riveting is constantly optimized, which makes the manufacturing technology and theory of electromagnetic riveting equipment very mature. There are also many investigations on the riveting quality, which confirmed the stability and efficiency of electromagnetic riveting [26,27].

Nowadays, electromagnetic riveting technology has been very mature both in theory and in technology.

5.2 Electromagnetic ramp wave loading technology

Ramp wave loading refers to the quasi-isentropic loading process in which compression wave rather than shock wave enters the sample [28]. It is a new loading technology developed and improved in recent ten years in the field of impact dynamics and high energy density physics. The typical representatives are magnetic driven isentropic loading technology, laser isentropic loading technology and Pillow fly-plate loading technology.

High-voltage ramp wave loading device is also proposed by using the electromagnetic energy conversion technology [29]. The device adopted the principle of electromagnetic pulse timing discharge to reassemble multiple discharge pulses in time sequence, so as to obtain the required rising edge and realize the ramp loading of the fly-plate [30]. In addition, the device can also realize the generation and acceleration of plasma, which is of great help to the study of space mechanics.

5.3 Electromagnetic Hopkinson bar

In fact, the author of this paper has also done some research on the electromagnetic energy shock technology and made some progress. In 2018, we applied the principle of the stress pulse generated in the electromagnetic riveter to the split Hopkinson pressure bar, replacing the air gun in the traditional split Hopkinson pressure bar with the electromagnetic energy gun [31].

In the process of stress wave produced, the electromagnetic energy in the loading gun is directly converted into a stress wave inside the coil material. The whole process does not require any macroscopic displacement of components, so the time error caused by macroscopic mechanical movement is reduced, which make the dynamic biaxial loading technology possible. In addition, some traditional Hopkinson bar cannot achieve low strain rate, while one can obtain a long enough stress pulse to meet the requirements of low strain rate loading using the electromagnetic Hopkinson bar [32].

However, the problem of electromagnetic interference has not been fundamentally solved, which makes the data acquisition in the biaxial loading technology become the biggest problem in the technology.

6 SUMMARY AND PROSPECTS

This paper reviews the development history of electromagnetic theory, and summarizes the application status of electromagnetic energy conversion technology in the current key technical fields. The electromagnetic forming technology, electromagnetic driven technology, electromagnetic braking technology and electromagnetic transient technology are mainly introduced.

Obviously, the introduction of electromagnetic energy conversion technology has solved the problem that the traditional energy method cannot solve, promoting the development of modern industry to a higher level.

Although electromagnetic energy conversion technology has many advantages, it also has some serious defects. Among them, electromagnetic interference is a very serious problem. For many fields with higher requirements for data acquisition accuracy, electromagnetic interference has become one of the problems that need to be solved urgently.

For some high frequency, high energy electromagnetic energy application field, there are still no effective solution. In order to make the interference to a minimum, a lot of laboratory built a shielding room alone to shield the electromagnetic interference, this makes the equipment occupies a large area and the cost is higher, which is one of the reasons why the electromagnetic energy conversion technology cannot be used in more fields at present.

ACKNOWLEDGMENTS

This work was supported by the China Postdoctoral Science Fund (2019M653785) and the Basic Research and Strategic Reserve Technology Research Fund of China National Petroleum Corporation (2019D-5008 (2019Z-01)). The author Jun Cao is also very grateful for the support received from the Young Scientists Fund of the National Natural Science Foundation of China [grant number 51904332] and Natural Science Fundation of Shannxi Province, China [grant number 2020JQ-934.

REFERENCES

[1] Zwanziger D. Dirac magnetic poles forbidden in s-matrix theory. *Phys Rev*, 1965, 137(3B):647–648.
[2] Gilbert NE. Electricity and magnetism. *the Macmillan*, 1956
[3] Desheng S. Coulomb's Contribution to Electricity and Magnetism – The Bicentennial of the Discovery of Coulomb's Law. *physical*, 1985(12):0-0. (in Chinese)
[4] Razuvaev OI. Method for calculating equivalent ionospheric currents according to meridional-chain magnetic data. *Geomagn Aeronomy*, 1991, 31:75–80.
[5] Ning YN, Jackson DA. Faraday effect optical current clamp using a bulk-glass sensing element. *Optics Letters*, 1993, 18(10):835–837.
[6] Maxwell JC. On physical lines of force. *Philos Mag*, 2010, 90(St):11–23.
[7] Deng J, Jiang H, Li C, et al. The Electromagnetic-Structural Couple Analysis in the Electromagnetic Bulging of Tube with Die. International Conference on Physical & Numerical Simulation of Materials Processing, 2007.
[8] Seth M, Vohnout VJ, Daehn GS. Formability of steel sheet in high velocity impact. *J Mater Process Tech*, 2005, 168(3):390–400.
[9] Brower DF. Electromagnetic forming apparatus: US patent, 1972.
[10] Wang DM, Liu P, Liu HQ, et al. The design and structural analysis of a coilgun for low acceleration of heavy loads. IEEE T Magn, 1999, 35(1):160–165.
[11] Oberly CE, Kozlowski, et al. Principles of application of high temperature superconductors to electromagnetic launch technology. *Magnetics, IEEE Transactions on*, 1991, 27(1):509–514.
[12] LI J, Yan P, Yuan W. Electromagnetic gun technology and its development. High Voltage Engineering, 2014, 4(014): 1052–1064. (in Chinese)
[13] Bostick WH. Propulsion of plasma by magnetic means. *Journal of Nuclear Energy*, 1958, 7(3): 278–279.
[14] Marshall RA. High Current and High Current Density Pulse Tests of Brushes and Collectors for Homopolar Energy Stores. *IEEE Transactions on Components Hybrids & Manufacturing Technology*, 2003, 4(1):127–131.
[15] Brast DE, Saule DR. Feasibility study for development of a hypervelocity gun, MB-R-65/40. San Ramon, USA: MB Associates, 1965.
[16] Sitzman A, Surls D, Mallick J. Design, construction, and testing of an inductive pulsed-power supply for a small railgun. *IEEE Transactions on Magnetics*, 2007, 43(1): 270–274.
[17] Weiguo G, Rong Z, Tengfei W, et al. Electromagnetic Driving Technique Applied to Split-Hopkinson Pressure Bar Device. *Experimental Mechanics*, 2010, 25(6):682–689. (in Chinese)
[18] Liu Z, Chen X, Lv X, et al. A mini desktop impact test system using multistage electromagnetic launch. *Measurement*, 2014, 49:68–76.
[19] Hopkinson B. A method of measuring the pressure produced in the detonation of high explosives or by the impact of bullets. *Philosophical Transactions of the Royal Society of London Series A*. 1914:437–56.
[20] Baker WE, Yew C. Strain-rate effects in the propagation of torsional plastic waves. *Journal of Applied Mechanics*. 1966;33(4):917–23.
[21] Kuwahara T, Araki K. Development of permanent magnet type eddy current retarder. *JSAE Review (Japan)*.1992, 13(1):92–96.

[22] Collan HK. Rapid Optimization of a Magnetic Induction Brake. *IEEE Transactions On Magnetics*, 1996. 32(4):3040–3044.
[23] Schmitt HA, Sekhon JS. High-impact portable riveting apparatus. US Patent, 3559269, 1968.
[24] Zieve PB. Low voltage electromagnetic riveter. Ph. D. Dissertation of University of Washington. 1986: 1~140.
[25] Deng J, Tang C, Zheng Y, Zhan Y. Effect of coil parameters on rivet deformation in low voltage electromagnetic riveting. *Advanced Materials Research*. 2013(602–604):1887–90.
[26] Deng JH, Tang C, Fu MW, Zhan YR. Effect of discharge voltage on the deformation of Ti Grade 1 rivet in electromagnetic riveting. *Mat Sci Eng A-Struct*. 2014(591):26–32.
[27] Zhang X, Yu HP, Li CF. Multi-filed coupling numerical simulation and experimental investigation in electromagnetic riveting. *Int J Adv Manuf Tech*. 2014 (73):1751–1763.
[28] Hare DE, Forbes J W, Reisman DB, et al. Isentropic compression loading of octahydro-1, 3, 5, 7-tetranitro-1, 3, 5, 7-tetrazocine (HMX) and the pressure-induced phase transition at 27 GPa. *Appl Phys Lett*, 2004, 85(6): 949–951.
[29] Hall CA, Asay JR, Knudson MD, et al. Experimental configuration for isentropic compression of solids using pulsed magnetic loading. *Rev Sci Instrum*, 2001, 72(9): 3587–3595
[30] Wang GJ, Luo BQ, Zhang XP, et al. A 4 MA, 500 ns pulsed power generator CQ-4 for characterization of material behaviours under ramp wave loading. *Rev Sci Instrum*, 2013, 84(1): 151–170.
[31] Nie H, Suo T, Wu B, Li Y, Zhao H. A versatile split Hopkinson pressure bar using electromagnetic loading. *Int J of Impact Eng*. 2018,116: 94–104.
[32] Nie H, Suo T, Shi X, Liu H, Li Y, Zhao H. Symmetric split Hopkinson compression and tension tests using synchronized electromagnetic stress pulse generators. *Int J of Impact Eng*. 2018,122, 73–82.

Stability analysis of high steep slope on the left side of the flood discharging tunnel exit of Wudongde hydropower station

H. Zhou, Z.J. Wang, D.B. Chen & Q.X. Cao
Changjiang Institute of Survey, Planning, Design and Research, Wuhan, China

ABSTRACT: The left side slope of the flood discharging tunnel exit of Wudongde Hydropower Station is high and steep, the geological conditions are complex, the rock mass quality is poor and the deformation of the slope is large. Based on the engineering geological conditions revealed by excavation of left side slope, the overall three-dimensional geological generalization model of slope is established according to the terrain and geological data. Combined with the results of safety monitoring and geophysical detection of the slope, the mechanical parameters of rock mass are analyzed by $FLAC^{3D}3D$ numerical model. On this basis, the distribution of displacement field, stress field and plastic area of slope under excavation unloading condition is studied, and the overall stability of slope is analyzed in depth.

1 INTRODUCTION

The stability of high rock slope is the most important engineering geology and geotechnical engineering problem in the construction of large-scale water conservancy and hydropower projects. Its stability control has become one of the key technologies for the success or failure of water conservancy and hydropower projects, which affects and restricts the development of hydraulic resources and hydropower projects [1, 2]. At present, there have been many research results on high rock slope. However, because of the complex composition of lithology and geological structure of rock high slope and many factors affecting the stability of slope, the engineering design of high slope is still prior to theory. Therefore, it is necessary to study this kind of rock slope in depth. Wang H T et al. [3] used the strength reduction method to simulate and analyze the high and steep slope project reinforced by prestressed anchor cable, and studied the influence of different anchor cables on the slope stability. Bai J L et al. [4] combined with the mechanics theory of unloading rock mass, $FLAC^{3D}$ software is used to simulate the slope with different excavation rate, and the stability of slope under different unloading conditions is compared and analyzed. Yang Y et al. [5] studied the distribution of stress field and displacement field in the excavation and unloading process of left bank slope by $RFPA^{3D}$ software, and revealed the failure and instability mechanism of rock mass. Xu P H et al. [6] studied the stability of the left bank slope of Jinping First Hydropower Station under the condition of excavation and flood discharge and fog by $FLAC^{3D}$ software.

In this paper, the left side slope of the flood discharging tunnel exit of Wudongde Hydropower Station is taken as the research object. Based on the engineering geological conditions revealed by slope excavation, a three-dimensional geological generalization model of the slope is established according to the terrain and geological data. Combined with the results of slope safety monitoring and geophysical detection, the $FLAC^{3D}3D$ numerical model is used to carry out the back analysis of the mechanical parameters of the slope rock mass The distribution of displacement field, stress field and plastic zone of the slope under the condition of excavation unloading is studied, and the overall stability of the slope is analyzed.

2 OVERVIEW OF THE LEFT SIDE SLOPE OF THE FLOOD DISCHARGING TUNNEL

The left side slope of the flood discharging tunnel exit of Wudongde Hydropower Station is located on the left side of artificial water cushion pond and downstream of Huashan debris flow ditch outlet. The position and construction image of slope project are shown in Figure 1. The exposed strata of the slope are quaternary overburden (Q^{col}), fold basement falling snow formation (Pt_{2l}), sedimentary cover guanyinya formation (Z_{2g}) and dengying formation (Z_{2d}). The engineering slope with an elevation of 925m and its natural slope above it is composed of Z_{2g}, Z_{2d}, P_{2y}, P_{3em}, T_{3bg}, J_{1y}, J_{2x} and other sedimentary cover strata, which tend to be a gentle reverse slope, which is inclined to the upstream and the inclined slope. The slope of the project below 925m is composed of the Pt_{2l} fold basement stratum, which is distributed near the crossing river. The slope is steep and lateral slope, which is inclined to the right bank downstream. F_6, F_7, F_9 and other faults are developed on the slope, among which the larger one is Huashan fault F_6.

Figure 1. Construction image of the exit slope of the flood discharging tunnel of Wudongde Hydropower Station.

The left side slope is constructed in two phases: the first phase mainly excavates all the overburden, and the artificial slope is excavated to the elevation of 850m-895m; the second phase mainly includes the rest of the water cushion pond, the tailrace below 850m and the downstream section of the drainage channel. The left side slope is constructed by first excavating the upper cover layer and then excavating the artificial slope. The single excavation slope ratio is 1:0.2-1:0.3. The artificial slope structure is a transverse slope, with the tailridge as the boundary divided into two sections, the upstream section is the left side slope of the water cushion pond, with the maximum height of 183.5m; the downstream section of the tailridge is the left side slope of the tailrace, with the maximum height of about 125m. During the excavation of slope, the slope is supported by system shotcrete anchor, and the upper part is set with system anchor cable. The exposed block of slope is also supported randomly by anchor cable, anchor pile and anchor bolt.

3 STABILITY ANALYSIS OF LEFT SIDE SLOPE

3.1 *Establishment of finite element calculation model*

Combined with the geological conditions exposed by excavation of left side slope at the outlet of the flood discharging tunnel, the three-dimensional numerical analysis model of the left side slope (cushion pond, tailcanal) and the front side slope of the exit is established after the lithology and geological defects of the slope are properly simplified. The calculation range is 1650m × 1000m × 1550m ($x \times y \times z$), the highest is 1760m, in which the X axis is positive with the vertical slope pointing to the inside of the mountain, the Y axis points to the upstream side with the parallel slope and the Z axis is vertical upward (according to the right-hand coordinate system).

The calculation area simulates strata Pt_{21}^{10}, Pt_{21}, Z_{2d}, Z_{2g}, Huashan fault F_6, fault F_9, P_{2y}, P_{3em}, T_{3bg}, J_{1y} and J_{2x}, and considers the class IV_2, IV_1 and III_2 rock mass in the main rock strata Pt_{21} and Pt_{21}^{10}, and considers the disturbance unloading relaxation zone of rock mass excavation, and the relaxation depth is determined according to the measured data of slope geophysical prospecting and unloading relaxation area in geological data. The calculation domain is divided into 753839 units and 147449 nodes. The numerical model is shown in Figure 2.

According to the actual excavation and support sequence, the slope excavation at all levels and the support of anchor, anchor pile and prestressed anchor cable are simulated. Layered rock mass is based on the elastic-plastic layer model with joints. The constitutive model can simulate shear failure and tensile failure along the plane and rock mass, considering the anisotropy of the strength of rock mass in two directions, parallel and vertical.

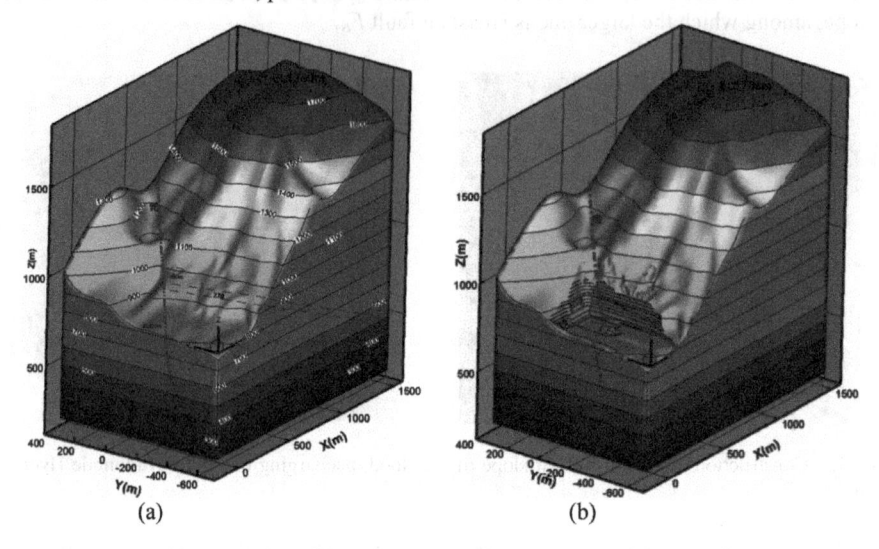

Figure 2. Numerical calculation model of left side slope before and after excavation: (a) Before excavation and (b) After excavation.

3.2 Calculation conditions and mechanical parameters

Based on the analysis of the monitoring data of the internal and external deformation of the left side slope at the exit of the flood discharging tunnel, the mechanical parameters of the rock mass and the structural surface of the slope are obtained by using the incremental displacement intelligent inversion method. Considering the unloading relaxation effect of rock mass with poor quality caused by excavation and unloading of high steep slope, the rock mass quality is reduced obviously. When the mechanical parameters are inversion, the cohesive force of unloading relaxation rock mass in the larger deformation part is reduced by 20%-50% of the non unloading rock mass and 50%-60% of the friction coefficient of the non unloading rock. The model mainly inverts the deformation modulus, friction coefficient and cohesion of grade IV_2, IV_1 and III_2 rock mass in Pt_{21}^{10} and Pt_{21}. The mechanical parameters of slope rock mass obtained by inversion are shown in Table 1.

3.3 Numerical analysis of slope stability

3.3.1 Displacement field distribution

The overall deformation law of the slope conforms to the general law. With the continuous downward excavation of the slope, the rock mass outside the slope is gradually excavated, and the height of the artificial slope increases gradually. Especially with the excavation of the middle and lower rock mass, the excavation unloading relaxation of the middle and lower slope weakens the supporting effect on the upper rock mass, and the displacement value of the slope body gradually increases. The deformation vector of slope rock mass is shown as the tendency of deformation outside the slope and downward deformation. The horizontal displacement outside the slope is greater than the

Table 1. Mechanical parameters of slope rock mass.

lithology	unloading relaxation	bulk density	deformation modulus E (GPa)		poisson's ratio μ	shear strength			
						f''		c'(MPa)	
			geological	inversion		geological	inversion	geological	inversion
$Pt_{21}^{10}(IV_2)$	non nloading	26.8	1~2	2	0.33	0.5~0.7	0.7	0.2~0.4	0.4
	unloading	26.6	/	1	0.35	/	0.42	/	0.2
$Pt_{21}^{10}(IV_1)$	non nloading	26.8	3~5	5	0.3	0.7~0.8	0.7	0.4~0.7	0.7
	unloading	26.7	/	2.5	0.33	/	0.42	/	0.35
$Pt_{21}^{10}(III_2)$	non nloading	26.8	5~7	7	0.26	0.40.9	1	0.7~0.9	1
	unloading	26.7	/	4	0.3	/	0.8	/	0.7
$Pt_{21}(IV_2)$	non nloading	26.8	1~2	2	0.33	0.5~0.7	0.7	0.2~0.4	0.4
	unloading	26.6	/	1	0.35	/	0.42	/	0.2
$Pt_{21}(IV_1)$	non nloading	26.8	3~5	5	0.3	0.7~0.8	0.7	0.4~0.7	0.7
	unloading	26.7	/	2.5	0.33	/	0.42	/	0.35
$Pt_{21}(III_2)$	non nloading	26.8	5~7	7	0.26	0.8~0.9	1	0.7~0.9	1
	unloading	26.7	/	4	0.3	/	0.8	/	0.7
$Z_{2g}(IV_2)$	non nloading	26.8	1~2	2	0.33	0.5~0.7	0.7	0.2~0.4	0.4
	unloading	26.6	/	1	0.35	/	0.42	/	0.2
F_6 cataclastic rock ~ broken powder	non nloading	26.6	1	1	0.33	0.5	0.5	0.2	0.2
	unloading	26.6	/	0.6	0.35	/	0.25	/	0.1
F_6 mud with rock debris		20	0.15	0.15	0.36	0.25~0.35	0.25	0.01~0.05	0.01
Z_{2d}		27.3	5~7	10	0.3	0.8~0.9	1	0.7~0.9	1
P_{3em}, P_{2y}		27	14~18	18	0.25	1.0~1.2	1.2	1.2~1.4	1.4
J_{2x}, J_l, T_{3bg}		27	5~10	10	0.28	0.8~1	1	0.5~0.7	1

vertical displacement, and is larger than the displacement towards the upstream and downstream direction. The deformation characteristics of the slope after excavation are shown in Figure 3. The displacement of rock mass in this area is larger than that of the deep part, which is decreasing gradually in the slope due to the relaxation of excavation and unloading. The deformation of slope is controlled by the poor lithology, upper hard and lower soft rock structure, F_6 fault cutting, excavation slope and front slope restraint, and other factors such as slope blasting excavation and flood season. The large deformation of slope is located in grade IV_2 rock mass with poor rock mass quality in upper and lower part of F_6 fault F_6 and class IV_2 rock mass with elevation below 830m. Among them, fault F_6 is steep inclined to mountain, and upper plate deformation is larger than lower plate, and the fault and lower rock mass are squeezed, and the lower slope body plays a reverse impedance body for use; however, when the middle and lower slope toe of slope is weakened, poor rock mass excavation is poor, and the unloading and relaxation of slope body, the rock mass quality is loose under the condition of reducing the mechanical parameters, the upper part of the slope is pulled and the upper part is outward subsidence deformation.

The deformation value of grade IV_2 rock mass of the left side slope fault F_6 is generally 200mm-350mm, and the deformation of partial F_6 fault and the affected zone elevation is 820m-880m, 350mm-380mm; the deformation in horizontal direction (X direction) is the main deformation

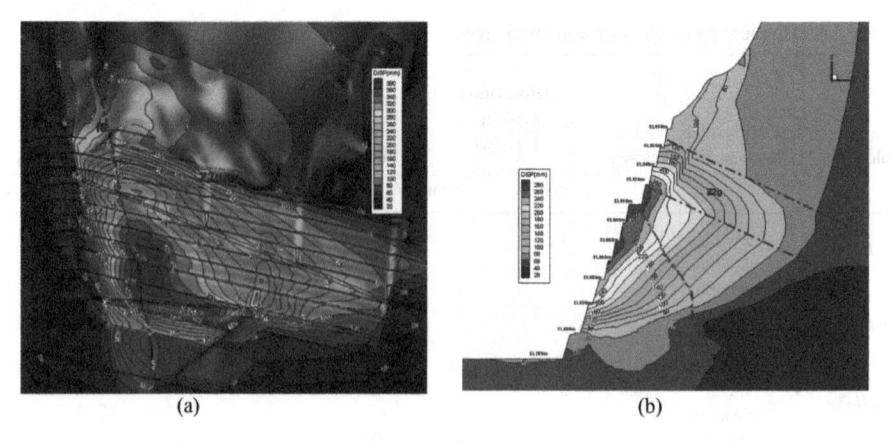

Figure 3. Deformation feature cloud map of left side slope after excavation: (a) Integral displacement field of slope and (b) Displacement field of typical section.

on the whole of all grades, and the deformation outside the horizontal direction is about 250mm-350mm. The settlement deformation is mainly under the vertical direction (Z direction) of the middle and upper part of the slope, and the settlement is mainly under the vertical direction of the slope below the elevation of 910m, with the amount of about 40mm-100mm, and the settlement decreases gradually from the top to the bottom; the middle and lower part is characterized by upward deformation, with the measurement value of about 50mm-170mm, and the part of F_6 fault reaches 240mm, which is located at the exposed part of the slope fault with an elevation of 813m. The deformation of parallel slope (Y direction) is about -40mm-68mm, which is shown as the deformation of the upper plate rock body in F_6, with the maximum deformation of 60mm-68mm (located at the elevation of 850m-865m), while the lower the lower the elevation, the greater the deformation of the lower the rock mass in the lower part of the lower part of the rock mass, the maximum 42mm (located at the elevation of 806m).

The deformation of grade IV_2 rock mass is obvious under excavation and unloading, and the parts near the front slope and the downstream side elevation 850m concrete mixing system are constrained by the outer mountain. The deformation outside the slope is obviously larger than that of other parts in the elevation direction from 925m to 830m. The 7-meter wide debris flow ditch discharge channel plays a role of spreading and transferring stress on a wide platform. In a certain extent, the slope is divided into two sections, the lower slope is equivalent to the base of the upper slope; when it intersects with F_6 or is a rock mass with poor quality, the rock mass will deform outwards under the unloading relaxation of the slope, and the foundation support is weakened, and the side slope is under the continuous action of the self weight of the mountain body. The slope rock, especially the upper rock, deforms outward and cracks outward; but because of the F_6 reverse slope, the bottom of the upper slope is in a reverse structure, the sliding force of the upper slope is obtuse angle with the direction of the bottom sliding surface, and can only deform outward along the face of the slope and continue to squeeze the fault and the lower rock mass; when the lower rock body has good properties, the overall stability of the slope is guaranteed. But when the lower rock mass is deformed or damaged due to unloading relaxation, there is a possibility of continuous deformation of the upper slope.

3.3.2 Stress field distribution
The stress field of natural state of slope is mainly controlled by the stress field of self weight of slope, slope shape and rock structure of slope. The stress relaxation of rock mass is obvious in the weathering unloading zone. The maximum and minimum main stress distribution of slope is shown in Figure 4.

After all the excavation of the slope is completed, the unloading of the slope is relaxed, and the relaxation range is distributed along the slope, and the general stress disturbance depth is 30m-60m. After the excavation of the slope, there is a certain degree of stress concentration at the toe and junction of the slope, with the stress level of about 2MPa-5MPa, and the local two slope joints reach 12MPa (Figure 5, 6). After excavation, there are large-scale tensile stress areas in the slope, mainly on slopes at all levels; the tensile stress of the slope is generally 0MPa-0.4MPa, the maximum tensile stress is 1MPa, and the places with large tensile force appear near the slope opening line and the junction area with large lithologic changes in the slope.

3.3.3 Distribution of plastic zone

After the excavation of the left side slope is completed, the scope and depth of the slope plastic area are larger. The plastic zone of rock mass increases with the excavation. The slope surface is mainly distributed on all levels of slope. The plastic zone is mainly located in fault F_6 and structural impact zone, Pt_{21}^{10}, Z_{2g} in IV_2 and IV_1. The rock mass basically enters into plastic state, mainly due to shear failure, and there is tension on the fault, slope surface and local rock mass Shear failure. The distribution of plastic zone after the excavation of left side slope is shown in Figure 7.

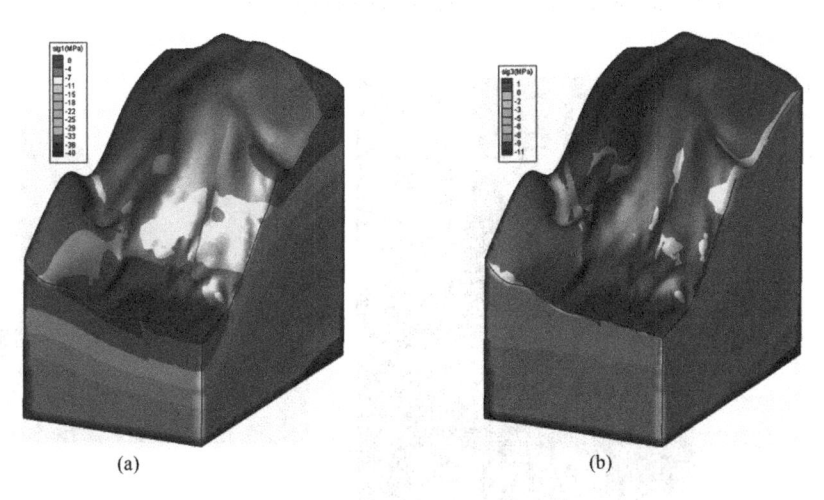

Figure 4. Cloud chart of the maximum and minimum main stress of the left side slope in natural state: (a) Maximum principal stress and (b) Minimum principal stress.

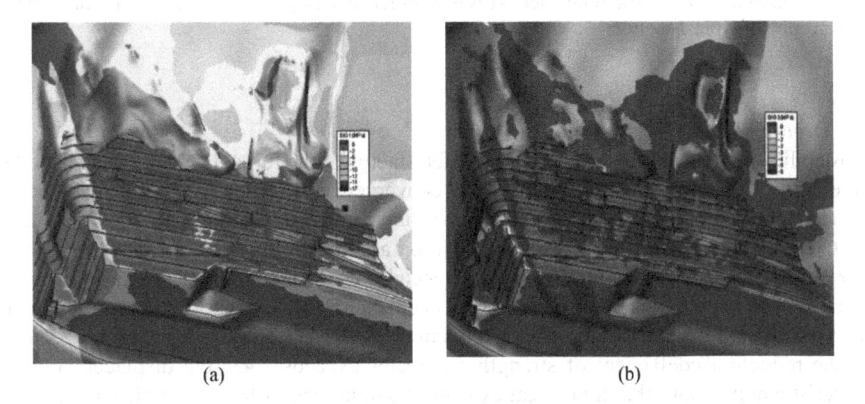

Figure 5. Cloud chart of the maximum and minimum main stress of the left side slope after excavation: (a) Maximum principal stress and (b) Minimum principal stress.

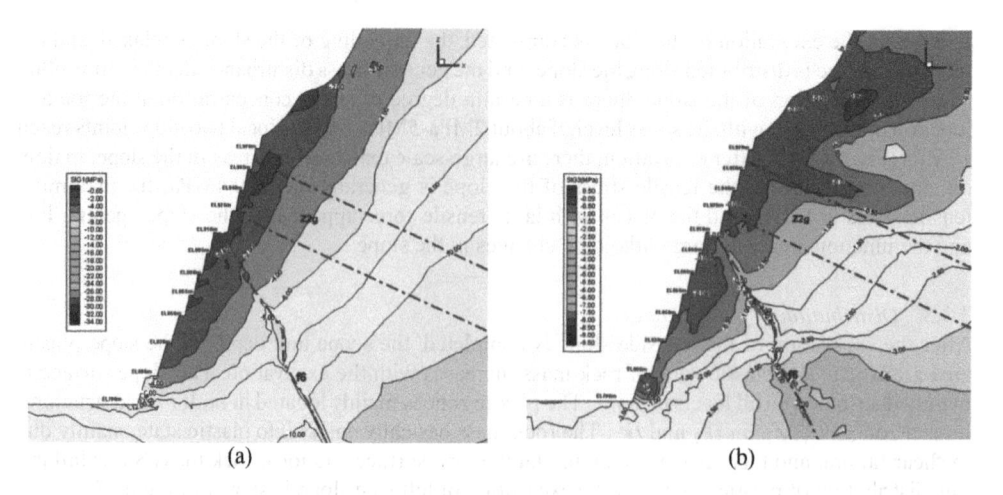

Figure 6. Cloud chart of maximum main stress and minimum main stress of typical section of left side slope after excavation: (a) Maximum principal stress and (b) Minimum principal stress.

Figure 7. Distribution of plastic zone after excavation of left side slope: (a) Integral plastic zone of slope and (b) Plastic zone of typical section.

3.3.4 Overall stability

The instability mode and the corresponding safety factor of the left side slope are analyzed by the finite element strength reduction method, the convergence criterion and the displacement trend of the characteristic points. The instability mode and failure area of the left side slope under the limit state are shown in Figure 8. The deformation and instability are concentrated in the upper and lower part of F_6 upper and lower slope, especially near the unloading relaxation area. Under the condition of systematic anchoring and new deep anchor cable support and grouting measures, the curve of displacement variation with reduction coefficient of strength parameter shows that when the reduction coefficient of strength parameter exceeds 1.48, the displacement curve of characteristic point above the slope shear exit appears nonlinear inflection point, and the rock mass on the upper part of the slope has the trend of overall instability. The safety coefficient of the left side slope can be judged as 1.48.

Figure 8. Unstable area of left side slope in critical state: (a) Instability area of slope in critical state and (b) Instability area of typical section.

4 MAIN CONCLUSIONS OF STABILITY ANALYSIS

The geological conditions of the left side slope of Wudongdeflood discharging tunnel exit are complex and the rock mass quality is poor. Under the condition of quick excavation and unloading, the relaxation deformation depth of rock mass is obviously larger than that of the general rock slope, and the deformation of the slope is large, which has adverse effect on the slope stability. The three-dimensional stability of slope shows that the stability safety coefficient of slope is 1.48 under the condition of systematic anchorage and new deep anchor cable support and grouting measures, and the overall stability of the slope is guaranteed.

ACKNOWLEDGMENTS

The research of this paper is supported by the youth talent promotion project of China Association for Science and Technology and the independent innovation project of Changjiang Institute of Survey, Planning, Design and Research (CX2020Z47).

REFERENCES

[1] Huang R Q, 2005. Main characteristics of high rock slopes in southwestern China and their dynamic evolution [J]. Advances inEarth Science, 20(3): 292–297.
[2] Song S W, Feng X M, Xiang B Y, et al. , 2011. Research on key technologies for high and steep rock slopes of hydropower engineering in southwest China [J]. Chinese Journal of Rock Mechanics and Engineering, 30(1): 1–22.
[3] Wang H T, Zhang X H, Song C, et al. , 2018. Effect of Bolt Parameters on the Stability of Prestressed Anchor Slope [J]. Highway, 63(04):18–24.
[4] Bai J L, Wang L H, Tang K Y, et al. , 2014.Influence of Excavation Unloading Rate on the Stress and Strain of Rock Slope [J]. Journal of Yangtze River Scientific Research Institute, 31(06): 60–64+68.
[5] Yang Y, Xu N W, Li T, et al. , 2018.Stability analysis of left bank rock slope at Baihetan hydropower station based on $RFPA^{3D}$ software and microseismic monitoring [J]. Rock and Soil Mechanics, 39(06): 2193–2202.
[6] Xu P H, Huang R Q, Chen J P, et al. , 2009.Study of the stability analysis and three dimensional numerical simulation of $IV^{\#} - VI^{\#}$ ridges in left bank at Jinping First Hydropower Station [J]. Rock and Soil Mechanics, 30(04): 1023–1028.

Response characteristics of a sandwich plate with visco-elastomer core and periodically supported masses under random excitation

Z.G. Ruan & Z.G. Ying
Department of Mechanics, School of Aeronautics and Astronautics, Zhejiang University, Hangzhou, PR China

ABSTRACT: Vibration control of structures with supported masses under random loading is an important problem. Adjustability of dynamic characteristics such as frequency response and response spectrum is important properties for structural vibration control. In the present paper, stochastic response adjustability of VE sandwich plates having periodically distributional supported masses under random base motion loading is studied. The VEC has dynamic properties changeable. Partial differential equations of coupling motions of the sandwich plate with supporting masses are derived and converted into ordinary differential equations for coupling mode motion using Galerkin method. Then FR and RS expressions of the plate system are obtained based on random dynamics theory. Numerical results are given to show the effects of sandwich plate parameters and periodic masses on stochastic response characteristics of the plate.

1 INTRODUCTION

Structural vibration control, especially stochastic vibration control is an important research subject in engineering, e.g., in aerospace, vehicle and civil engineering. To attenuate vibration, sandwich structures are designed which consist of two stiff skins and one soft core. Sandwich structures have been increasingly studied, and composite beams and plates with supported masses were presented to model vibration-sensitive instruments supported on structures under random loading. With the development of new materials, using smart material, e.g., magneto-rheological liquid (MRL) and magneto-rheological visco-elastomer (MRVE) as cores of composite structures is a potential method to mitigate the excessive vibration and consequently to improve the dynamic performance of structures, which can rapidly change in rheological properties under applied magnetic fields in millisecond with changeless structural design.

In past decades, many researches were reported on engineering application of smart materials such as MRL [1–8] and MRVE [9–14]. Sandwich structures composed of MR materials gained increasing attention, especially MRVE, which avoid the problem of magnetic particle settlement in MRL. The sandwich beams with MRVE as core layer have been investigated on static behaviour [15], harmonic vibration and changeable dynamic rigidity [16, 17], frequency-response characteristics [18], torsional vibration characteristics [19], stochastic vibration response [20, 21], and vibration stability [22, 23]. Experiment researches on MRVE sandwich beams were presented [24, 25]. At the same time, some researchers put their effort into sandwich plates. The effects of MRVE type [26], magnetic field and thickness [27, 28], temperature [29], and temperature-dependent material properties and boundary conditions [30] on dynamics characteristics of MRVE sandwich plates were studied. Ying et al. [31] investigated random vibration mitigation of MRVE sandwich plates by applying different localized magnetic field distributions. However, stochastic vibration properties of VE sandwich plates having periodically supported masses are not studied.

The present paper studies on a VE sandwich plate with supported masses, which is used for modelling vibration-sensitive instruments supported on a structure. Stochastic response variation of the sandwich plate having supporting masses subjected to base loading is explored, in which supported masses are distributed periodically. First, partial differential equations for transverse and longitudinal vibrations of the plate system are derived. Second, they are simplified to coupling modes vibration equations. Third, frequency response (FR) expression and response spectrum (RS) expression of the plate system are obtained. Finally, numerical results are given to show effects of sandwich plate parameters and distributed masses on stochastic response characteristics of the plate.

2 VIBRATION FORMULAS OF PLATE SYSTEM

2.1 Description of sandwich plate

A sandwich plate having visco-elastomer core (VEC) and distributed supporting masses is as Figure 1, where the plate is excited by transverse base motion. The plate length is a and plate width is b. Upper and lower face layers have equal Young's modulus of E_1, Poisson's ratio of μ, density of ρ_1 and thickness of h_1. Middle layer has density of ρ_2 and thickness of h_2. The supporting masses are fixed on the plate. The ith concentrated mass has mass of $m_i \times ab$. The size of mass can be disregarded by comparing it with the plate. The base has vertical motion w0 which is a random loading.

The VEC has adjustable dynamic properties through diffident external magnetic fields. The Young's modulus of the VEC is much smaller than that of the face layers and neglected, while its shearing deformation is larger than that of the face layers and needs to consider. Linear dynamic stress-stress relation is used and thus shear stresses τ_{2xz} and τ_{2yz} of the core are expressed by the corresponding shear strain γ_{2xz} and γ_{2yz} as

$$\tau_{2xz} = G_{21}\gamma_{2xz} + G_{c1}\dot{\gamma}_{2xz}, \qquad \tau_{2yz} = G_{21}\gamma_{2yz} + G_{c1}\dot{\gamma}_{2yz} \qquad (1)$$

where G_{21} and G_{c1} are constant coefficients (adjustable by external magnetic fields).
The basic assumptions of the sandwich plate are as described in references [14] and [31].

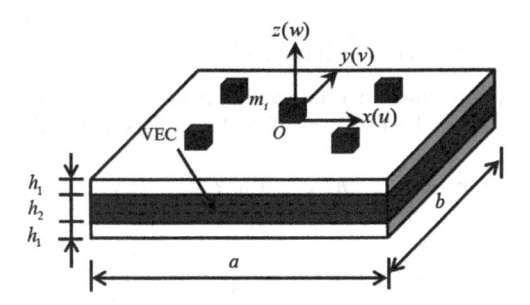

Figure 1. VEC sandwich plate with distributed supporting masses.

2.2 Vibration equations of sandwich plate system

Based on the assumptions, transverse motion displacement of the plate is $w = w(x, y, t)$, where t denotes time. Longitudinal displacements (in x-axis and y-axis directions) of upper and lower face layers can be expressed respectively as

$$u_1(x,y,z_1,t) = u_{10}(x,y,t) - z_1\frac{\partial w}{\partial x}, \qquad v_1(x,y,z_1,t) = v_{10}(x,y,t) - z_1\frac{\partial w}{\partial y} \qquad (2)$$

$$u_3(x,y,z_3,t) = u_{30}(x,y,t) - z_3\frac{\partial w}{\partial x}, \qquad v_3(x,y,z_3,t) = v_{30}(x,y,t) - z_3\frac{\partial w}{\partial y} \qquad (3)$$

where u_{10}, v_{10}, u_{30} and v_{30} are respectively mid-layer displacements of the two face layers, and z_1 and z_3 are coordinates of the face layers. Utilizing equations (2)–(3), shear strains of core layer of the sandwich plate can be determined. So shear stresses of the core layer are expressed as

$$\tau_{2xz} = G_{21}\gamma_{xz} + G_{c1}\dot{\gamma}_{xz} = G_{21}\left(\frac{h_a}{h_2}\frac{\partial w}{\partial x} + \frac{u_{10} - u_{30}}{h_2}\right) + G_{c1}\left(\frac{h_a}{h_2}\frac{\partial \dot{w}}{\partial x} + \frac{\dot{u}_{10} - \dot{u}_{30}}{h_2}\right) \quad (4)$$

$$\tau_{2yz} = G_{21}\gamma_{yz} + G_{c1}\dot{\gamma}_{yz} = G_{21}\left(\frac{h_a}{h_2}\frac{\partial w}{\partial y} + \frac{v_{10} - v_{30}}{h_2}\right) + G_{c1}\left(\frac{h_a}{h_2}\frac{\partial \dot{w}}{\partial y} + \frac{\dot{v}_{10} - \dot{v}_{30}}{h_2}\right) \quad (5)$$

where $h_a = h_1 + h_2$. Longitudinal normal strains and shear strains of the two face layers can be determined based on geometrical relationship and equations (2)–(3). Then normal stresses and shear stresses of the face layers are given by

$$\sigma_{1x} = \frac{E_1}{1-\mu^2}(\varepsilon_{1x} + \mu\varepsilon_{1y}) = \frac{E_1}{1-\mu^2}\left[\left(\frac{\partial u_{10}}{\partial x} - z_1\frac{\partial^2 w}{\partial x^2}\right) + \mu\left(\frac{\partial v_{10}}{\partial y} - z_1\frac{\partial^2 w}{\partial y^2}\right)\right] \quad (6)$$

$$\sigma_{1y} = \frac{E_1}{1-\mu^2}(\varepsilon_{1y} + \mu\varepsilon_{1x}) = \frac{E_1}{1-\mu^2}\left[\left(\frac{\partial v_{10}}{\partial y} - z_1\frac{\partial^2 w}{\partial y^2}\right) + \mu\left(\frac{\partial u_{10}}{\partial x} - z_1\frac{\partial^2 w}{\partial x^2}\right)\right] \quad (7)$$

$$\tau_{1xy} = \frac{E_1}{2(1+\mu)}\gamma_{xy} = \frac{E_1}{2(1+\mu)}\left(\frac{\partial u_{10}}{\partial y} + \frac{\partial v_{10}}{\partial x} - 2z_1\frac{\partial^2 w}{\partial x \partial y}\right) \quad (8)$$

$$\sigma_{3x} = \frac{E_1}{1-\mu^2}(\varepsilon_{3x} + \mu\varepsilon_{3y}) = \frac{E_1}{1-\mu^2}\left[\left(\frac{\partial u_{30}}{\partial x} - z_3\frac{\partial^2 w}{\partial x^2}\right) + \mu\left(\frac{\partial v_{30}}{\partial y} - z_3\frac{\partial^2 w}{\partial y^2}\right)\right] \quad (9)$$

$$\sigma_{3y} = \frac{E_1}{1-\mu^2}(\varepsilon_{3y} + \mu\varepsilon_{3x}) = \frac{E_1}{1-\mu^2}\left[\left(\frac{\partial v_{30}}{\partial y} - z_3\frac{\partial^2 w}{\partial y^2}\right) + \mu\left(\frac{\partial u_{30}}{\partial x} - z_3\frac{\partial^2 w}{\partial x^2}\right)\right] \quad (10)$$

$$\tau_{3xy} = \frac{E_1}{2(1+\mu)}\gamma_{xy} = \frac{E_1}{2(1+\mu)}\left(\frac{\partial u_{30}}{\partial y} + \frac{\partial v_{30}}{\partial x} - 2z_3\frac{\partial^2 w}{\partial x \partial y}\right) \quad (11)$$

By using equilibrium conditions in x-axis and y-axis directions with equations (6)–(11) and boundary conditions, the other shear stresses of the face layers obtained are

$$\tau_{1xz} = -\frac{E_1}{1-\mu^2}\left\{\frac{\partial^2 u_{10}}{\partial x^2}\left(z_1 - \frac{h_1}{2}\right) + \frac{\partial^3 w}{\partial x^3}\left(\frac{h_1^2}{8} - \frac{z_1^2}{2}\right)\right.$$
$$+ \mu\left[\frac{\partial^2 v_{10}}{\partial y \partial x}\left(z_1 - \frac{h_1}{2}\right) + \frac{\partial^3 w}{\partial y^2 \partial x}\left(\frac{h_1^2}{8} - \frac{z_1^2}{2}\right)\right]\right\}$$
$$- \frac{E_1}{2(1+\mu)}\left[\frac{\partial^2 u_{10}}{\partial y^2}\left(z_1 - \frac{h_1}{2}\right) + \frac{\partial^2 v_{10}}{\partial y \partial x}\left(z_1 - \frac{h_1}{2}\right) + \frac{\partial^3 w}{\partial y^2 \partial x}\left(\frac{h_1^2}{4} - z_1^2\right)\right] \quad (12)$$

$$\tau_{1yz} = -\frac{E_1}{1-\mu^2}\left\{\frac{\partial^2 v_{10}}{\partial y^2}\left(z_1 - \frac{h_1}{2}\right) + \frac{\partial^3 w}{\partial y^3}\left(\frac{h_1^2}{8} - \frac{z_1^2}{2}\right)\right.$$
$$+ \mu\left[\frac{\partial^2 u_{10}}{\partial y \partial x}\left(z_1 - \frac{h_1}{2}\right) + \frac{\partial^3 w}{\partial x^2 \partial y}\left(\frac{h_1^2}{8} - \frac{z_1^2}{2}\right)\right]\right\}$$
$$- \frac{E_1}{2(1+\mu)}\left[\frac{\partial^2 v_{10}}{\partial x^2}\left(z_1 - \frac{h_1}{2}\right) + \frac{\partial^2 u_{10}}{\partial y \partial x}\left(z_1 - \frac{h_1}{2}\right) + \frac{\partial^3 w}{\partial x^2 \partial y}\left(\frac{h_1^2}{4} - z_1^2\right)\right] \quad (13)$$

$$\tau_{3xz} = -\frac{E_1}{1-\mu^2}\left\{\frac{\partial^2 u_{30}}{\partial x^2}\left(z_3+\frac{h_1}{2}\right)+\frac{\partial^3 w}{\partial x^3}\left(\frac{h_1^2}{8}-\frac{z_3^2}{2}\right)\right.$$
$$\left.+\mu\left[\frac{\partial^2 v_{30}}{\partial y \partial x}\left(z_3+\frac{h_1}{2}\right)+\frac{\partial^3 w}{\partial y^2 \partial x}\left(\frac{h_1^2}{8}-\frac{z_3^2}{2}\right)\right]\right\}$$
$$-\frac{E_1}{2(1+\mu)}\left[\frac{\partial^2 u_{30}}{\partial y^2}\left(z_3+\frac{h_1}{2}\right)+\frac{\partial^2 v_{30}}{\partial y \partial x}\left(z_3+\frac{h_1}{2}\right)+\frac{\partial^3 w}{\partial y^2 \partial x}\left(\frac{h_1^2}{4}-z_3^2\right)\right] \quad (14)$$

$$\tau_{3yz} = -\frac{E_1}{1-\mu^2}\left\{\frac{\partial^2 v_{30}}{\partial y^2}\left(z_3+\frac{h_1}{2}\right)+\frac{\partial^3 w}{\partial y^3}\left(\frac{h_1^2}{8}-\frac{z_3^2}{2}\right)\right.$$
$$\left.+\mu\left[\frac{\partial^2 u_{30}}{\partial y \partial x}\left(z_3+\frac{h_1}{2}\right)+\frac{\partial^3 w}{\partial x^2 \partial y}\left(\frac{h_1^2}{8}-\frac{z_3^2}{2}\right)\right]\right\}$$
$$-\frac{E_1}{2(1+\mu)}\left[\frac{\partial^2 v_{30}}{\partial x^2}\left(z_3+\frac{h_1}{2}\right)+\frac{\partial^2 u_{10}}{\partial y \partial x}\left(z_3+\frac{h_1}{2}\right)+\frac{\partial^3 w}{\partial x^2 \partial y}\left(\frac{h_1^2}{4}-z_3^2\right)\right] \quad (15)$$

Based on continuity relations of the shear stresses between face and core layers of the sandwich plate, equations governing longitudinal displacements are derived and given by

$$\frac{E_1 h_1}{1-\mu^2}\left(\frac{\partial^2 u}{\partial x^2}+\mu\frac{\partial^2 v}{\partial y \partial x}\right)+\frac{E_1 h_1}{2(1+\mu)}\left(\frac{\partial^2 u}{\partial y^2}+\frac{\partial^2 v}{\partial y \partial x}\right) = G_{21}\left(\frac{h_a}{h_2}\frac{\partial w}{\partial x}+\frac{2}{h_2}u\right)$$
$$+G_{c1}\left(\frac{h_a}{h_2}\frac{\partial \dot{w}}{\partial x}+\frac{2}{h_2}\dot{u}\right) \quad (16)$$

$$\frac{E_1 h_1}{1-\mu^2}\left(\frac{\partial^2 v}{\partial y^2}+\mu\frac{\partial^2 u}{\partial y \partial x}\right)+\frac{E_1 h_1}{2(1+\mu)}\left(\frac{\partial^2 v}{\partial x^2}+\frac{\partial^2 u}{\partial y \partial x}\right) = G_{21}\left(\frac{h_a}{h_2}\frac{\partial w}{\partial y}+\frac{2}{h_2}v\right)$$
$$+G_{c1}\left(\frac{h_a}{h_2}\frac{\partial \dot{w}}{\partial y}+\frac{2}{h_2}\dot{v}\right) \quad (17)$$

Where $u=u_{10}=-u_{30}$ and $v=v_{10}=-v_{30}$. Based on dynamic relation of an element of the sandwich plate with supported masses (in z-axis direction) and using shear stresses (4), (5) and (12)–(15), equation governing transverse displacement of the sandwich plate system is obtained as

$$\left[\rho h_t + \sum_{k=1}^{n_a} m_k ab\delta(x-x_k)\delta(y-y_k)\right]\ddot{w}+D_1 h_1^3\left(\frac{\partial^4 w}{\partial x^4}+2\frac{\partial^4 w}{\partial y^2 \partial x^2}+\frac{\partial w^4}{\partial y^4}\right)$$
$$-\left[G_{21}\frac{h_a}{h_2}\left(h_a\frac{\partial^2 w}{\partial x^2}+2\frac{\partial u}{\partial x}\right)+G_{c1}\frac{h_a}{h_2}\left(h_a\frac{\partial^2 \dot{w}}{\partial x^2}+2\frac{\partial \dot{u}}{\partial x}\right)\right]-\left[G_{21}\frac{h_a}{h_2}\left(h_a\frac{\partial^2 w}{\partial y^2}+2\frac{\partial v}{\partial y}\right)\right.$$
$$\left.+G_{c1}\frac{h_a}{h_2}\left(h_a\frac{\partial^2 \dot{w}}{\partial y^2}+\frac{\partial \dot{v}}{\partial y}\right)\right] = -\left[\rho h_t+\sum_{k=1}^{n_a} m_k ab\delta(x-x_k)\delta(y-y_k)\right]\ddot{w}_0 \quad (18)$$

where $D_1 = E_1/6(1-\mu^2)$, $\delta(\cdot)$ is the Dirac delta function, n_a is total number of masses, (x_k, y_k) are the coordinates of the k-th mass, and $\rho h_t = 2\rho_1 h_1 + \rho_2 h_2$. The partial differential equations (16)–(18) describe coupled transverse and longitudinal motions of the sandwich plate system.

Consider a simply supported rectangular plate. It has boundary constraint conditions as

$$w\bigg|_{x=\pm a/2} = w\bigg|_{y=\pm b/2} = 0, \quad \frac{\partial^2 w}{\partial x^2}\bigg|_{x=\pm a/2} = \frac{\partial^2 w}{\partial y^2}\bigg|_{y=\pm b/2} = 0, \left(\frac{\partial u}{\partial x} + \mu \frac{\partial v}{\partial y}\right)\bigg|_{x=\pm a/2}$$

$$= \left(\frac{\partial v}{\partial y} + \mu \frac{\partial u}{\partial x}\right)\bigg|_{y=\pm b/2} = 0 \tag{19}$$

The differential equations and boundary conditions can be transformed in the non-dimensional (ND) form by using ND coordinates and displacements (w_a is the amplitude of base motion w_0)

$$\bar{x}_k = \frac{x_k}{a}, \quad \bar{y}_k = \frac{y_k}{b}, \quad \bar{x} = \frac{x}{a}, \quad \bar{y} = \frac{y}{b}, \quad \bar{u} = \frac{u}{w_a}, \quad \bar{v} = \frac{v}{w_a}, \quad \bar{w} = \frac{w}{w_a}, \quad \bar{w}_0 = \frac{w_0}{w_a} \tag{20}$$

3 STOCHASTIC RESPONSE CHARACTERISTICS OF SANDWICH PLATE

Based on boundary conditions (19), the ND vibration displacements of the sandwich plate can be expanded as

$$\bar{u} = \sum_{i=1}^{N_1} \sum_{j=1}^{N_2} r_{ij}(t) \sin\left[(2i-1)\pi\bar{x}\right] \cos\left[(2j-1)\pi\bar{y}\right],$$

$$\bar{v} = \sum_{i=1}^{N_1} \sum_{j=1}^{N_2} s_{ij}(t) \cos\left[(2i-1)\pi\bar{x}\right] \sin\left[(2j-1)\pi\bar{y}\right]$$

$$\bar{w} = \sum_{i=1}^{N_1} \sum_{j=1}^{N_2} q_{ij}(t) \cos\left[(2i-1)\pi\bar{x}\right] \cos\left[(2j-1)\pi\bar{y}\right] \tag{21}$$

where $r_{ij}(t)$, $s_{ij}(t)$ and $q_{ij}(t)$ are generalized displacement components varying with time, N_1 and N_2 are term numbers. Utilizing Galerkin method and expansion (21), motion equations (16)–(18) can be converted into vibration equations for q_{ij}, r_{ij} and s_{ij} (in ordinary differential form). After getting rid of r_{ij} and s_{ij}, the vibration equations for q_{ij} can be obtained and rewritten in the matrix form

$$\mathbf{M}\ddot{\mathbf{Q}} + \mathbf{C}\dot{\mathbf{Q}} + \mathbf{KQ} = \mathbf{F}(t) \tag{22}$$

where generalized displacement vector $\mathbf{Q} = [\mathbf{Q}_1^T \ \mathbf{Q}_2^T \ \cdots \ \mathbf{Q}_{N_2}^T]^T$ and $\mathbf{Q}_j = [q_{1j} \ q_{2j} \ \cdots \ q_{N_1 j}]^T$, $\mathbf{M}$, $\mathbf{C}$ and $\mathbf{K}$ are generalized mass, damping and stiffness matrices, respectively, and $\mathbf{F} = -\ddot{w}_0 \mathbf{F}_C$ is generalized excitation vector.

Equation (22) describes a coupling multiple degree-of-freedom vibration system which represents dynamics of the VE sandwich plate with supporting masses subjected to random base loading. The stochastic dynamic response of the plate system is evaluated generally utilizing RS expression. For system (22), the FR and RS matrices are expressed as

$$\mathbf{H}(\omega) = (\mathbf{K} + j\omega\mathbf{C} - \omega^2\mathbf{M})^{-1}, \quad \mathbf{S_Q}(\omega) = \mathbf{H}(\omega)\mathbf{F}_C\mathbf{F}_C^T\mathbf{H}^{*T}(\omega)S_{\ddot{w}_0} \tag{23}$$

where $j = \sqrt{-1}$, ω is vibration frequency, superscript "*" denotes complex conjugate and $S_{\ddot{w}_0}(\omega)$ is the loading PSD. Then the RS expression for the ND transverse displacement of the plate is

$$S_{\bar{w}}(\omega, \bar{x}, \bar{y}) = \mathbf{\Phi}^T(\bar{x}, \bar{y}) \mathbf{S_Q}(\omega) \mathbf{\Phi}(\bar{x}, \bar{y}), \quad \mathbf{\Phi}(\bar{x}, \bar{y}) = [\mathbf{\Phi}_1^T \ \mathbf{\Phi}_2^T \ \cdots \ \mathbf{\Phi}_{N_2}^T]^T$$

$$\mathbf{\Phi}_j = [\phi_{1j} \ \phi_{2j} \ \cdots \ \phi_{N_1 j}]^T, \quad \phi_{ij} = \cos(2i-1)\pi\bar{x} \cos(2j-1)\pi\bar{y} \tag{24}$$

The response statistics of the plate system under random loading can be estimated using the RS expression. For instance, the MS displacement response of the sandwich plate is

$$E\left[\bar{w}^2(\bar{x},\bar{y})\right] = 2\int_0^{+\infty} S_{\bar{w}}(\omega,\bar{x},\bar{y})\,d\omega \tag{25}$$

4 NUMERICAL RESULTS

To show further stochastic response adjustability, a VE sandwich plate having periodically supported masses subjected to base motion loading is considered. Its basic variables are specified as follows: $a=4$m, $b=2$m, $\rho_1=3000$kg/m³, $\rho2=1200$kg/m³, $\mu=0.3$, $E_1=10$GPa, $w_a=1$, $G_{21}=2$MPa, $G_{c1}=0.003$MPa·s, $h_1=0.05$m, $h2=0.2$m, $x_1=y_1=0$, $m_1=240$kg/m². The loading is considered as white noise with PSD of 1.0×10^6. The numbers N_1 and N_2 in expansion (33) are determined based on the convergence of displacement responses. Numerical results on stochastic responses and RS densities on the plate mid-point are given in Figures 2–7.

Figure 2 shows the RMS ND displacement ($\bar{w}$) of the VE sandwich plate varying with the ND PSD of base loading. The RMS displacement of the plate obtained by numerical simulation is also given, which validates the result obtained by the proposed analysis method.

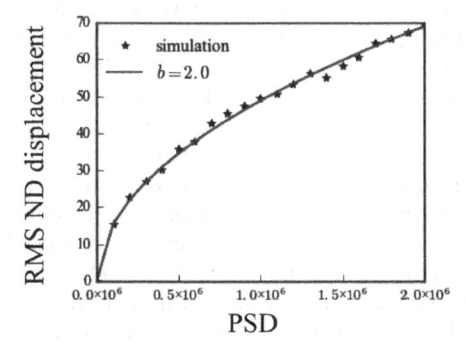

Figure 2. RMS ND displacement of plate as function of PSD of base loading.

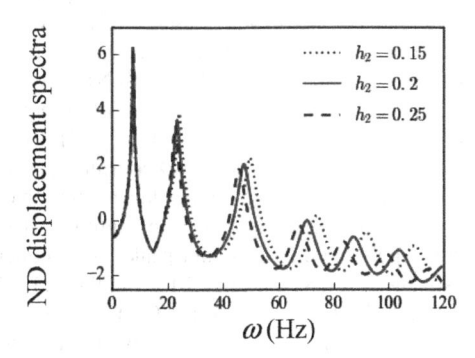

Figure 3. LND RS of plate displacement under different middle layer thicknesses h_2.

4.1 Effect of core modulus and layer thickness

Effect of core layer thickness and modulus (stiffness coefficient G_{21} and damping coefficient G_{c1}) on the RS of the plate is considered which is illustrated by Figures 3–4. The logarithmic ND (LND) RS of the plate displacement under different middle (core) layer thicknesses h_2 is shown in Figure 3. It is seen that the core layer thickness has more effects on the high-order resonant responses and frequencies (e.g., the spectral amplitudes of the third resonance are 9.60, 7.77, 6.42, and the third resonant frequencies are 49.76, 47.57, 45.78 Hz for h_2=0.15, 0.2, 2.5m, respectively).

The LND RS of the plate displacement under different damping coefficients G_{c1} is shown in Figure 4. It is seen that the resonant peaks decrease as the core layer damping coefficient increases, while the damping coefficient has little effect on resonant frequencies. Large core layer stiffness coefficient G_{21} has also certain effect on the resonant peaks and frequencies. Thus, the plate response as dynamic characteristics can be improved through changing damping coefficients G_{c1} and stiffness coefficients G_{21}.

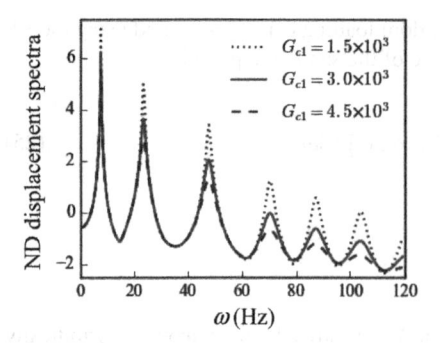

Figure 4. LND RS of plate displacement under different damping coefficients G_{c1}.

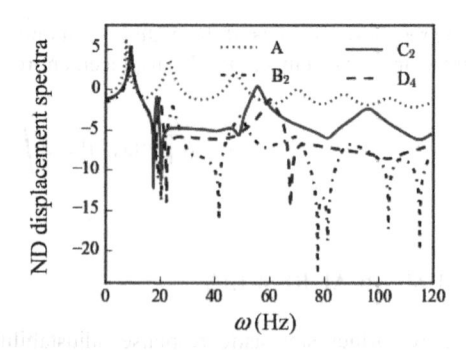

Figure 5. LND RS of plate displacement with distributed masses for cases A, B_2, C_2, D_4.

4.2 *Effect of periodically supported masses*

The distribution of masses on the sandwich plate is also considered in several cases. Case A is only one mass (240kg/m^2) on the plate with ND coordinates (0, 0). Case B_1 is two equal masses (120kg/m^2) on the plate with ND coordinates (-0.2, 0), (0.2, 0), respectively. Case B_2 is two equal masses (120kg/m^2) on the plate with ND coordinates (-0.25, 0), (0.25, 0), respectively. Case B_3 is two equal masses (120kg/m^2) on the plate with ND coordinates (–0.3, 0), (0.3, 0), respectively. Case C_2 is three equal masses (80kg/m^2) on the plate with ND coordinates (–0.25, 0), (0, 0), (0.25, 0), respectively. Case D_1 is four equal masses (60kg/m^2) on the plate with ND coordinates (–0.2, –0.2), (–0.2, 0.2), (0.2, –0.2), (0.2, 0.2), respectively. Case D_2 is four equal masses (60kg/m^2) on the plate with ND coordinates (–0.25, –0.25), (–0.25, 0.25), (0.25, –0.25), (0.25, 0.25), respectively. Case D_3 is four equal masses (60kg/m^2) on the plate with ND coordinates (–0.3, –0.3), (–0.3, 0.3), (0.3, –0.3), (0.3, 0.3), respectively. Case D_4 is four equal masses (60kg/m^2) on the plate with ND coordinates (–0.3, 0), (–0.1, 0), (0.1, 0), (0.3, 0), respectively.

The LND RS of the plate displacement with distributed masses for cases A, B_2, C_2, D_4 (different numbers of masses) is shown in Figure 5. It is seen that the mass distribution has large effect on the RS including resonant peaks and frequencies (e.g., the spectral amplitudes of the second resonance are 38.27, 6.16×10^{-5}, 0.394, 0.3744, and the resonant frequencies are 23.49, 19.90, 18.91, 20.10Hz for cases A, B_2, C_2, D_2, respectively). Figure 6 illustrates that the mass distribution for cases B_1, B_2, B_3 (identical numbers of masses) has much effect on the RS including resonant peaks and frequencies. The influence of small perturbation to the mass distribution on the RS is slight. Figure 7 shows the effect of the mass distribution for cases D_1, D_2, D_3 and D_4 (three masses) on the RS. Thus, the plate response as dynamic characteristics can be further improved through suitable mass distribution.

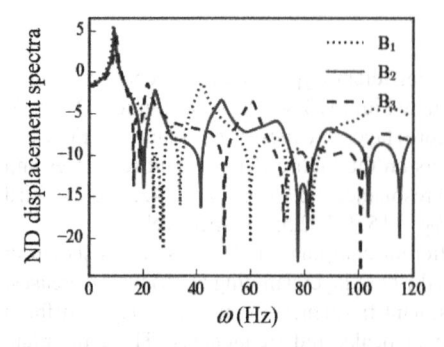

Figure 6. LND RS of plate displacement with distributed masses for cases B_1, B_2, B_3.

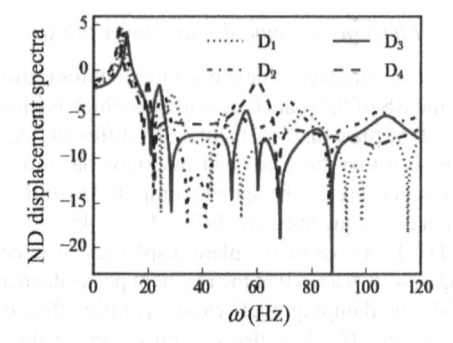

Figure 7. LND RS of plate displacement with distributed masses for cases D_1, D_2, D_3, D_4.

5 CONCLUSIONS

Stochastic response adjustability characteristics of the VE sandwich plate having periodically distributional supported masses under random base motion loading are studied. The equations governing transverse and longitudinal coupling vibrations of the plate system subjected to base loading are obtained. They are further converted into ordinary differential equations for multiple modes coupling motion. The FR expression and RS expression of the plate system are derived which can be used for dynamic optimization analysis of composite structures.

Numerical results have illustrated that: (1) the plate response as dynamic characteristics is adjustable by core layer thickness, especially for high-order resonant amplitudes and frequencies; (2) the response characteristics of the plate vibration are improvable by suitably choosing stiffness coefficient and damping coefficient; (3) the plate dynamic characteristics including resonant amplitudes and frequencies can be improved by the periodic mass distribution. These inferences will be potential for stochastic dynamic optimization of smart composite structures with distributed supporting masses.

ACKNOWLEDGMENTS

The study was supported by the National Natural Science Foundation of China under grant nos. 12072312 and 11572279. The support is gratefully acknowledged.

REFERENCES

[1] Dyke S J, Spencer B F, Sain M K and Carlson J D 1996 Smart Materials & Structures 5 565
[2] Spencer B F and Nagarajaiah S 2003 J. Engineering Mechanics 129 845
[3] Wang D H and Liao W H 2011 Smart Materials & Structures 20 023001
[4] Casciati F, Rodellar J and Yildirim U 2012 J. Intelligent Material System & Structures 23 1181
[5] Ghobadi E, Khajehsaeid H, Asiaban R, et al. 2020 Polymer Testing 87 106512
[6] Zhu X, Jing X and Cheng L 2012 J. Intelligent Material Systems & Structures 28 839
[7] Kumbhar B K, Patil S R and Sawant S M 2015 Engineering Science & Technology 18 432
[8] Jiang W, Zhang Y and Xuan S, et al. 2011 J. Magnetism & Magnetic Materials 323 3246
[9] Kallio M, Lindroos T, Aalto S, et al. 2007 Smart Materials & Structures 16 506
[10] Koo J H, Khan F, Jang D D and Jung H J 2010 Smart Materials & Structures 19 117002
[11] Ying Z G, Ni, Y Q and Sajjadi M 2013 Science China Technological Sciences 56 878
[12] York D, Wang X J and Gordaninejad F 2007 J. Intelligent Material Systems & Structures 18 1221
[13] Hu W and Wereley N M 2008 Smart Materials & Structures 17 045021
[14] Ni Y Q, Ying Z G and Chen Z H 2011 J. Sound & Vibration 330 4369
[15] Aguib S, Nour A, Benkoussas B, et al. 2016 Composite Structures 139 111
[16] Zhou G Y and Wang Q 2005 Smart Materials & Structures 14 1001
[17] Zhou G Y and Wang Q 2006 Smart Materials & Structures 15 59
[18] Choi W J, Xiong Y P and Shenoi R A 2010 Advances in Structural Engineering 26 837
[19] Bornassi S and Navazi H M 2018 J. Intelligent Material Systems & Structures 29 2406
[20] Ying Z G and Ni Y Q 2017 Int. J. Structural Stability & Dynamics 17 1750075
[21] Ying Z G and Ni Y Q 2009 Smart Materials & Structures 18 095005
[22] Dwivedy S K, Mahendra N and Sahu K C 2009 J. Sound & Vibration 325 686
[23] Nayak B, Dwivedy S K and Murthy K S R K 2011 J. Sound & Vibration 330 1837
[24] Hu G, Guo M, Li W, et al. 2011 Smart Materials & Structures 20 127001
[25] Chikh N, Nour A, Aguib S, et al. 2016 Acta Mechanica Solida Sinica 29 271
[26] Mikhasev G I, Eremeyev V A, Wilde K, et al. 2019 J. Intelligent Material Systems & Structures 30 2748
[27] Yeh, J Y 2013 Smart Materials & Structures 22 035010
[28] Asgari M, Rayyat R A M, Yousefi M, et al. 2019 J. Intelligent Material Systems & Structures 30 140
[29] Jeyaraj P, Padmanabhan C and Ganesan N 2011 J. Sandwich Structures & Materials 13 509
[30] Mohammadimehr M, Akhavan A S, Okhravi S, et al. 2017 J. Intelligent Material Systems & Structures 29 863
[31] Ying Z G, Ni Y Q and Ye S Q 2014 Smart Materials & Structures 23 025019

Application of rock slice material in early stage of geological experiment

Y. Liu
Key Laboratory of Dense Oil and Mudstone Reservoir Formation in Heilongjiang Province, Daqing, China

ABSTRACT: In the study of rock microstructure, it is an important basic work to make good use of materials to make thin sections. Due to the different materials and different technologies, the transparency and reflectivity of the thin slices made of colloidal materials are different. With the increase of the use degree of the materials, the transparency of the chemical components and stable components in the materials will gradually decrease, until the soft minerals in the thin sections are opaque. Therefore, different materials and processes were used to grind thin slices for identification and analysis.

1 INTRODUCTION

Abrasives refer to the materials used for grinding, grinding and polishing rock slices. They are mainly granular or micro powder minerals or mineral complexes with certain hardness, cutting and self sharpening. The abrasives can be directly used to grind the thickness of rocks, or they can be made into abrasives for grinding thin slices.

2 BASIC PROPERTIES OF ABRASIVES

Most of the materials used as abrasives are natural or artificial mineral synthetic materials, which generally have the following basic characteristics.

① The hardness is higher than that of the processed material. The greater the hardness difference is, the higher the grinding efficiency is
② The abrasive grains are not easy to be deformed and broken under pressure during grinding
③ The self sharpening property is good, even if the abrasive grinding process is broken, there will still be new sharp edges and corners of particles, which is not easy to wear.
④ During the grinding process, the hardness and strength of the inherent particles can be maintained when the particles are grinded with the grinding disc at high temperature. The heat generated by high-speed friction will not soften or melt the sharp edges and corners of the abrasive.
⑤ It has stable chemical properties and does not react with processed materials.
⑥ The particle size and shape are uniform, and the particle size of each type of abrasive is limited in a certain range within the grinding thickness [1, 2].

2.1 *Size classification and application scope of abrasives*

The size of abrasive refers to the size of abrasive particle. For irregular abrasive particles, the maximum diameter is taken as single micron (m). The grading standards of abrasive particle size in the world are not consistent. The standard in China is to divide the abrasive degree into two categories and 29 grades. The abrasive with particle diameter greater than 63m is called abrasive. The particle size number is represented by the symbol, which is divided into 17 levels. The number of abrasive particles on the area of one square inch is represented by "- the larger the surface value (value), the smaller the abrasive particle size; The particles with a diameter less than 63m are called micro powder abrasives. The particle size number is represented by the symbol "W", which is divided into 12 grades. The subscript value (W value) is the upper limit value of abrasive particle size. The smaller the value, the smaller the abrasive particle size [3, 4].

The 29 grades of abrasive particle size from large to small are 12, 14, 16, 20, 24, 30, 36, 46, 60, 70, 80, 100, 120, 150, 180, 240, 280, W40, W28, W20, W14, W10, W7, W5, w3.5, w2.5, w1.5, W1, w0.5.

At present, carborundum abrasives are widely used in rock slice processing. See Table 1 for the comparison of abrasive particle size and particle size.

Table 1. Comparison of grain size and particle size of common emery abrasives.

Material specification	Particle size / μ M	Particle size of micro powder abrasive		
		Material parameters	Parameter value	Particle size / μ M
12#	2000~1600	W40	320#	40~28
14#	1600~1250	W28	400#	28~20
16#	1250~1000	W20	500#	20~14
20#	1000~800	W14	600#	14~10
24#	800~630	W10	800#	10~7
30#	630~500	W7	1000#	7~5
36#	500~400	W5	1200#	5~3.5
46#	400~315	W3.5	1500#	3.5~2.5
60#	315~250	W2.5	1800#	2.5~1.5
70#	250~200	W1.5	2000#	1.5~1
80#	200~160	W1	2400#	1~0.5
100#	160~125	W0.5	3200#	≤ 0.5
120#	125~100			
150#	100~80			
180#	80~63			
240#	63~50			
280#	50~40			

At present, in the process of rock slice processing in Daqing oilfield exploration and Development Research Institute, the range of application of different particle size abrasives is different. Therefore, it is necessary to select and use the abrasive materials with corresponding particle size according to the types of rock slice production (Table 2).

Table 2. Application scope of common particle size processing.

Abrasive material size number	Scope of application
24#~36#	Surface treatment of specimens
40#~90#	Plastic treatment of specimen
100#~120#	Rough grinding after cutting
150#~240#	Shaping after glue penetration
280#~W40	Post cutting treatment of specimens
W28~W10	Final fine grinding thickness
W7~W0.5	Final thickness of polishing

2.2 Types and characteristics of common abrasives

According to the cause of formation, abrasives can be divided into natural abrasives and man-made materials. Natural abrasives mainly include diamond and corundum, and artificial abrasives can be divided into three series of diamond series, carbide series and corundum series, with more than ten kinds of abrasive materials [5].

The following introduces the main common grinding materials for sheet materials

2.2.1 *Silicon carbide (SIC)*
Silicon carbide (SIC) was synthesized in 1891, also known as "artificial emery". Silicon carbide can be divided into black silicon carbide (low temperature synthesis) and green silicon carbide (high temperature synthesis) according to the color. The Mohs hardness of the former is 3100–3280kgf / mm, and the latter is 3200–3400kgf / mm 2. Both of them are brittle and have the characteristics of high temperature resistance and corrosion resistance. Green silicon carbide has less impurities, higher hardness and brittleness than black silicon carbide, stronger grinding force and better self sharpening property. Silicon carbide is often used to treat the surface of rock samples, reshape, grind and so on. It can be used to grind all kinds of concave convex and hard materials except for a variety of sedimentary rocks. Because of its low price, it is widely used in grinding.

2.2.2 *Boron carbide (BC)*
Boron carbide (BC) artificial abrasive is dark gray to black, its hardness is higher than that of silicon carbide, its microhardness is 400000 / mm2, its wear resistance is good, and its cutting ability is 40% - 50% of that of diamond. It is a kind of advanced abrasives. It is often used instead of diamond to grind and polish thin rock slices.

Figure 1. Emery Micropowder.

3 ADHESIVE COLLOID

3.1 *Epoxy resin a B mixed type*

Epoxy resin is a kind of high molecular polymer containing epoxy group, which is a kind of adhesive with strong bonding and bonding ability.

3.1.1 *Strong adhesion*
The molecular structure of epoxy resin has hydroxyl group, ether group and very active epoxy group, which makes epoxy resin have high bonding strength. Hydroxyl group and ether group have high polarity, which makes the epoxy resin molecules and the adjacent interface produce electromagnetic attraction; while the epoxy group can form chemical composition with the medium surface, which makes the epoxy resin adhesive force especially strong. In addition to a small number of substances (such as polyethylene, polytetrafluoroethylene and polyvinyl chloride plastic), epoxy resin has great adhesion to many other substances[7, 8].

3.1.2 *Small shrinkage*

Epoxy resin reacts with hardener directly, so there will not be a lot of bubbles in the curing process, so the shrinkage rate of thermosetting resin is small. In addition, its coefficient of thermal expansion is also small.

3.1.3 *High stability*

The epoxy resin without hardener will not be hardened by heating, so it has high stability and can be stored for a long time without deterioration. The epoxy resin with hardener has high stability and strong mechanical impact resistance.

3.2 Cyanoacrylate adhesive

Cyanoacrylate adhesive has strong ability to penetrate into samples, especially for loose and broken samples. It has good transparency, homogeneous optical property and refractive index of 1.46–1.50, which is better than ordinary fir adhesive.

4 PRODUCTION METHOD, PROCESS AND PRECAUTIONS

The production process is divided into six steps: cutting sample, cementation of sample, surface treatment of rock, sticking, slicing after sticking and fine grinding. The specific production process, materials and precautions required by the process are as follows:

Table 3. Materials and precautions.

Production process	Materials required	Matters needing attention
cutting	Cutting machine, rock sample, oven	① The samples were labeled and some sections were selected; ② Cut about $10 \times 34mm2$; ③ Dry the sample with non fluorescent paper and dry the pore water in the container
Sample cementation	Ethyl cyanoacrylate rubber, rubber gloves	① Low molecular weight colloids were chosen to be non fluorescent; ② The samples were cemented with ethyl cyanoacrylate to cement the pores of rocks;
Plane treatment	Emery	① The particle size of SiC powder should be fine, and the particle diameter is $7 \sim 14\,\mu M$; ② When rough grinding, slightly remove the surface colloid, remember to use too much force;
Adhesive sheet	Adhesive and slide	① It is necessary to check whether the flatness of the slide is the same before sticking the slide, and the optical slide is selected; ② Choose fluorescent or fluorescent glue
Slice after sticking	Rock slice cutting machine	① The slice should be stable and the cutting thickness should be $0.09 \sim 0.8mm$; ② After cutting, clean the sample with air gun;
Fine grinding	Silicon carbide, particle size	① SiC particles $\leq 0.3\,\mu M$; ② Slight fine grinding can be done, do not use too much force; ③ The final thickness of grinding inclusion sheet should be about $0.06 \sim 0.03mm$

Firstly, low molecular weight, non fluorescent colloid was selected for sample preparation, and non fluorescent gel was used to invade the pores of rock, and the low temperature treatment method was used to ensure that the organic matter content of the sample did not volatilize. In the production process, the cooling fluid and grinding fluid are cut, and the mixed grinding method is used to bond the rock slices. The rock slices are not cut to 0.08mm in the pore area and polished to 0.03mm in thickness. Finally, the complete microscopic geological information such as organic matter, mineral type, structure distribution, geometric morphology and sedimentary structure on the thin section of the rock are completely preserved.

1. According to the characteristics that potassium ion is more active than calcium ion, potassium ion can replace calcium ion in pore fracture, and bring water molecules to reduce interlayer cracks and expansion coefficient.
2. Using ethyl cyanoacrylate adhesive has the advantages of low cost, less pollution and good permeability. It is very convenient for cementing particles. It has short time point in curing process and good mechanical impact resistance. The refractive index of adhesive is close to that of optical resin adhesive (1.46–1.50). Another epoxy resin AB adhesive has low cost, less pollution and good permeability, which is very convenient for bonding rubber core (rock cuttings), The refractive index of cured adhesive is 1.542 ~ 1.548 (close to 1.53 ~ 1.54) of optical resin adhesive; the curing time is 1:20 hours when heating 60 ~ +90°C. When the temperature is adjusted to 70°C, it takes only 2 hours to reach the saturation state. The curing process is characterized by short time point, controllable, tasteless, pollution-free, strong adhesion, strong waterproof performance and mechanical impact resistance (Figure 2).

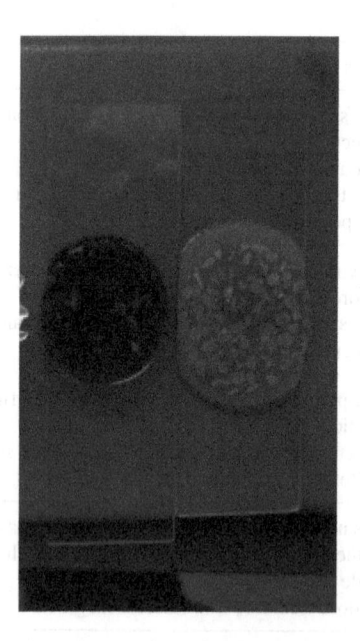

Figure 2. Ethyl cyanoacrylate adhesive, Epoxy resin AB adhesive.

3. Research on the fusibility of grinding fluid and mineral. According to the comparison of material standards used in thin rock slices, a practical separation grid is established, and the experimental data of various materials are analyzed. The area of the thin section after grinding can retain the mineral composition of the rock sample to the maximum extent. After grinding the suspension, the grinding liquid can make the flake meet the identification standard under the microscope (Figure 3).

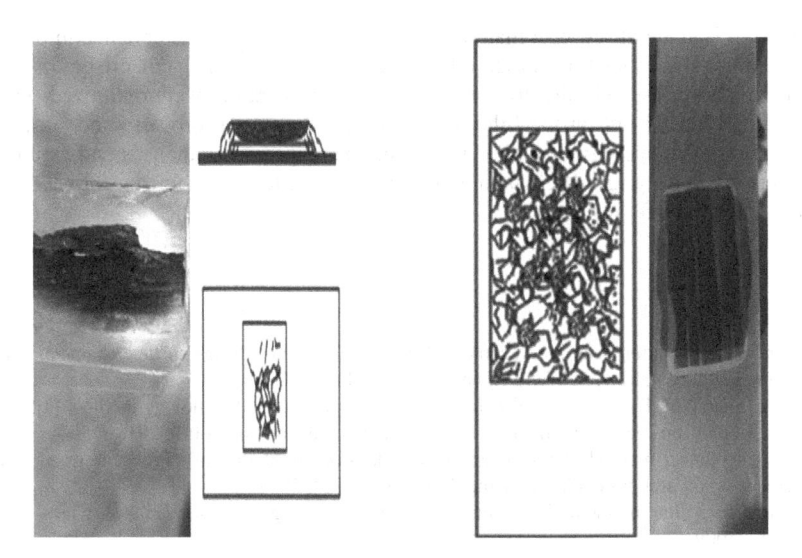

Figure 3. Thin section of rock after grinding with emery, Thin section of rock after micro powder grinding.

4. The mixed grinding fluid has fine particles, no agglomeration, uniform particle size of 0.06mm, stable chemical properties, and no chemical effect on the surface of rocks, minerals and glass; the optical specimen after direct grinding is smooth and bright; the surface is smooth as a mirror under the microscope, without pits, fine cracks and scratch marks; the polishing degree of the same mineral in the central part and edge part of the optical plate is also fine. When grinding about 0.04 ~ 0.03mm, put the sample into the microscope for observation, with less bubbles and high transmittance, until all the interference colors of quartz, feldspar and other minerals under the microscope are grade I gray white (0.03mm); the milled thin section area can retain the mineral composition of the rock sample to the maximum extent (Figure 4).

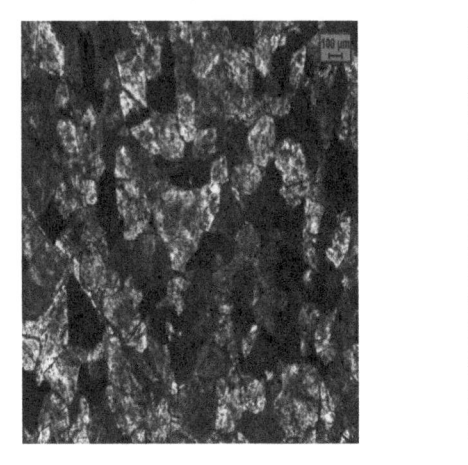

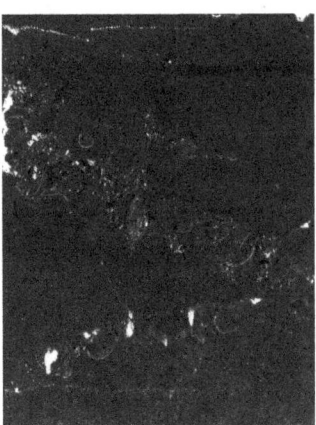

Figure 4. photos of emery and micro powder under microscope.

5 CONCLUSION

Focusing on thin section, the fine subdivision process is studied, and the research framework is determined for the purpose of "material, method, production and process". Through repeated experiments, it is made of low cost materials. Ensure that the ground sample maximizes and truly reflects the mineral composition, structural characteristics and local characteristics of the sample. Based on the above research, the process and material preparation were optimized. The process optimization is carried out on the lithology, and the thin section process mode is established. A new production technology, combining conventional exploration analysis and exploration and development test in Daqing Oilfield laboratory, can play a better role in unconventional thin section production and scientific research.

REFERENCES

[1] Yang Zhenjie shale composition and shale wellbore surface and cuttings surface characteristics have an impact on wellbore stability. Oilfield chemistry. 2000 (1) 7. –77. 96
[2] Liu Yang, Zhao Xin, Han Bo, et al. Casting thin section production technology method. [a]; proceedings of International Conference on oil and gas exploration and development [C]; 2018
[3] Liu Yang, technical method of unconventional rock slice production [a], Proceedings of International Conference on oil and gas exploration and development [C], 2018
[4] Discussion on evaluation method of shale wellbore stability logging in shihuanhuan, China University of petroleum, 2009
[5] Xu Tongtai, a brief introduction to the classification of unstable shale in foreign countries. 1989 (2). 32–34
[6] Meng F W, Galamay A R, Ni P, et al. The major composition of a middle-late Eocene salt lake in the Yunying Depression of Jianghan Basin of Middle China based on analyses of fluid inclusions in halite [J]. Journal of Asian Earth Sciences, 2014, 85: 97–105.
[7] Vreeland R H, Jones J, Monson A, et al. Isolation of live Cretaceous(121–112 million years old) halophilic Archaea from primary salt crystals [J]. Geomicrobiology Journal, 2007. 24: 275–282.
[8] Schubert B A, Lowenstein T K, Timofeeff M N. Microscopic identification of prokaryotes in modern and ancient halite, Saline Valley and Death Valley, California [J]. Astrobiology, 2009, 9: 467–482.

Advances in Materials Science and Engineering – Lombardo & Wang (Eds)
© 2022 the authors, ISBN 978-1-032-12707-1

Numerical investigation of superhydrophobic surface-induced drag reduction over NACA0012

T. Kouser & Y.L. Xiong
Department of Mechanics, Huazhong University of Science and Technology, Wuhan, China
Hubei Key Laboratory of Engineering Structural Analysis and Safety Assessment, Wuhan, China

D. Yang
School of Naval Architecture and Ocean Engineering, Huazhong University of Science and Technology (HUST), Wuhan, China

ABSTRACT: In the present study, the effect of superhydrophobic surfaces (SHSs) is elucidated numerically for a fixed Reynolds number (Re) 1000. Flow over NACA0012 for a range of angles of attack 0° – 10° with 5° variance in angles is considered in this context. SHS takes into account by imposing alternating slip and no-slip boundary conditions. Gas fraction (G.F) and solid interface fraction are kept at the same ratio of 0.5. Appreciable drag reduction (DR) and increment in lift force are observed as well as modification in the wake is captured in comparison to the three-dimensional (3D) hydrofoil with no-slip surface. A prominent difference in wake flow is observed for SHS versus the smooth surface of the hydrofoil. As the angle of attack increases, the superhydrophobic surface causes to delay the separation bubble and the vortex shedding frequency. Thus, flow remains a two-dimensional (2D) laminar pattern and alters the critical angle of attack to an extent.

1 INTRODUCTION

Reduction in frictional force is a major problem in the area of fluid mechanics from a practical perspective. Among DR technologies, a surface with random roughness features (superhydrophobic surface) is a rapidly growing drag reduction (DR) method. Air is trapped in between the micro/nano ridges (meniscus) present on such surfaces. These surfaces are intensively found in nature [1, 2]. The repellency mechanism of liquid in different directions depends on their hierarchical features. Mechanism of superhydrophobic surfaces induced drag reduction is based on the theory of slip length (δ) defined as $u = \delta(\partial u)(\partial y)^{-1}$. Gas-fraction (meniscus ratio) is an important characteristic as it provides effective slip to fluids.

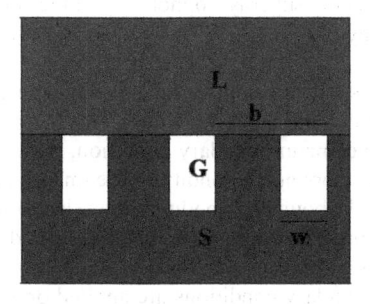

Figure 1. Schematic representation of Superhydrophobic surfaces.

DOI 10.1201/9781003225850-86

A detailed microscopic study of lotus leaf is conducted by Bhartlott and Neinhuis in 1997 as a pioneering work on superhydrophobic surfaces [3]. Therefore, the superhydrophobic surface is also known as the lotus leaf effect. Numerous materials with random or patterned artificial superhydrophobic surface characteristics are fabricated at the industrial level by utilizing various chemical and mechanical techniques providing promising DR results [4, 5].

In addition to the drag reduction effect, several studies are conducted to control the flow around blunt bodies utilizing superhydrophobic surfaces. Literature has revealed that such surfaces can delay the onset of vortex shedding and critical Reynolds number. Macroscale changes in the flow pattern are observed as the result of microscopic modification of surfaces [6–8]. Few studies are available in an attempt to control flow around streamlined bodies as compare to bluff bodies. 18% DR is reported via experimental investigation for flow over Jurkowski hydrofoil with SHS [9]. However, only zero angle of attack was considered and a detailed study is not conducted to investigate the flow field modification. Effect of SHS are found complex related to angles of attack $0°-20°$ while flow over NACA0012 for chord Reynolds number 2000–10000 [10].

It is well known that for streamlined bodies, flow dramatically changes as a function of angle of attack. Thus, it is purposeful to investigate the flow around the streamlined body. One can expect that usage of superhydrophobic surfaces can be beneficial at a commercial scale such as anti-fouling of ship hull, propeller blade, enhance the performance of underwater lifting surfaces and many other flow control-strategies. Thus, in the present study, flow over NACA0012 is studied numerically at a fixed Reynolds number of 1000 for angles of attack $0°-10°$. Superhydrophobic surface is modelled by means of slip and no-slip boundary conditions [11, 12].

2 MATHEMATICAL AND PHYSICAL MODEL WITH BOUNDARY CONDITIONS

A symmetric hydrofoil NACA0012 of sharp trailing edge is used for computational purposes, defined by the equation combined with incompressible Navier-Stokes equations [13].

$$y(x) = b_1(b_2 x^{1/2} + b_3 x + b_4 x^2 + b_5 x^3 + b_6 x^4) \tag{1}$$

$b_1 = 0.594689180$, $b_2 = 0.298222773$, $b_3 = -0.127125232$,
$b_4 = -0.357907906$, $b_5 = 0.291984971$, $b_6 = -0.105174606$.

$$\frac{\partial u_i}{\partial x_i} = 0 \tag{2}$$

$$\frac{\partial u_i}{\partial t} + u_i \frac{\partial u_i}{\partial u_j} = -\frac{\partial p}{\partial x_i} + \frac{1}{Re}\frac{\partial^2 u_i}{\partial x_j^2} \tag{3}$$

The Reynolds number is defined as $Re = (U_{inf} c) v^{-1}$. Alternating no-slip and slip boundary conditions are applied to model the superhydrophobicity of the surface. Gas fraction for superhydrophobic surface is defined by $G.F = (b-w) b^{-1} = 0.5$, where b is the pitch length and w is air interface between micro-grates.

(24c, 12c, 12c) being dimensions of computational domain along upstream (x), cross-stream (y) and downstream (z) are adopted. Uniform flow velocity U_{inf} is applied at the inlet, in a streamwise direction. At the outlet, the Neumann boundary condition, as well as pressure is specified as a reference value of zero. Periodic boundary conditions are imposed at lateral boundaries normal to the span of the hydrofoil. Periodic boundary conditions have been usually employed to approximate the flow past bluff bodies of an infinite spanwise length. Top and bottom boundaries are considered as symmetric boundaries to prevent the thick layer of fluid due to the no-slip wall. While periodic shear-free and slip boundary conditions are applied on the hydrofoil surface to achieve superhydrophobic characteristics.

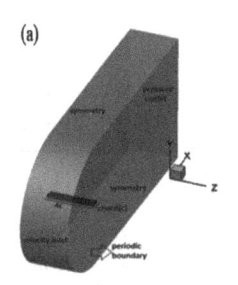

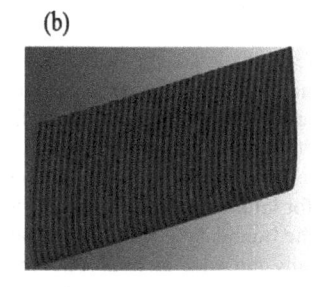

Figure 2. (a) Schematic 3D model of computational domain (b) Periodic shear-free and slip boundary conditions over surface of NACA0012.

Convergence of hydrodynamic forces for coarser and finer grid points at angle of attack 5° and G.F= 0.5 are analyzed to verify methodology and alternating periodic slip and no-slip boundaries selection. $\overline{Cl} = F_d(0.5\rho U_{inf} A)^{-1}$ and $\overline{Cd} = F_d(0.5\rho U_{inf} A)^{-1}$ becomes convergent as the number of alternating parts are increased. Difference between values of $\overline{Cl}$ becomes less than 2% for finer mesh points and solution remains trivial in case of δc^{-1} for both finer and coarser grid points. Mesh with 16102400 hexahedra cells and 80 alternating shear-free and no-slip boundary parts are adopted for simulation purposes and no-slip boundary conditions.

3 RESULTS AND ANALYSIS

3.1 *Hydrodynamic coefficients*

Low viscous friction is an interesting characteristic of superhydrophobic surfaces where liquid barely touches the surface. Therefore, SHSs are vigorously studied to deduce drag reduction over bluff bodies. Recently, interesting results for drag reduction are observed for streamlined bodies. It is concluded that drag reduction and lift growth are a function of gas-fraction and angles of attack.

Comparative results for mean lift and drag coefficients are listed in Table 1. With the introduction of superhydrophobicity, lift and drag forces increase as the angle of attack grows. However, drag reduction is observed for each angle of attack. Constantini [7] explained the mechanism of drag reduction over slip and no-slip boundary conditions in detail. Texture of superhydrophobic surface traps air to form meniscus and reduces frictional force experienced by fluid while flowing present between solid surface and liquid.

DR rate of 19% (compared to smooth surface hydrofoil (G.F = 0)) for $\alpha = 0°$ and $\alpha = 5°$ remains the same, while there is no lift at angle of attack 0° in both cases. A maximum lift of 33% is computed at $\alpha = 5°$ from the current simulation and a maximum DR of 25% is calculated at $\alpha = 10°$. 18% DR was found at $\alpha = 0°$ [9], which validates current findings.

Table 1. Results of coefficient of drag and lift for flow over three-dimensional NACA0012 with varying gas-fraction at Re=10^3.

	$\alpha = 0°$	$\alpha = 5°$	$\alpha = 10°$
$\overline{Cl}$ (G.F = 0)	0	0.2458	0.4153
$\overline{Cl}$ (G.F = 0.5)	0	0.33	0.5090
$\overline{Cd}$ (G.F = 0)	0.1203	0.1287	0.1657
$\overline{Cd}$ (G.F = 0.5)	0.101	0.108	0.132

3.2 Effective slip length and velocity

Table 2 enlists the normalized slip velocity and the effective slip length over the NACA0012 surface for gas-fraction 0.5 respectively. Surface heterogeneities are responsible for the deduced effective slip. Effective slip length is calculated from the computed mean velocity over the surface of hydrofoil constructed through alternating micro-grates and then applied the slip length definition. From the table, it is evident that effective slip length depends on G.F which is also observed from literature existed for laminar and fully developed turbulent flows [14].

In our model, the boundary is assumed as heterogeneous in the slip region otherwise smooth. Slip length is usually regarded as hydrophobic performance with angle of attack. However, it is not necessary that in presence of angle of attack, slip length follow the same rule. From figure 3, it is clear that with the increase in the angle of incidence normalized slip length does not increase. The same behavior is reported by Voronov [15]. Furthermore, the presence of meniscus enhances the slip velocity but normalized slip velocity decays with regard to the increase in angle of attack.

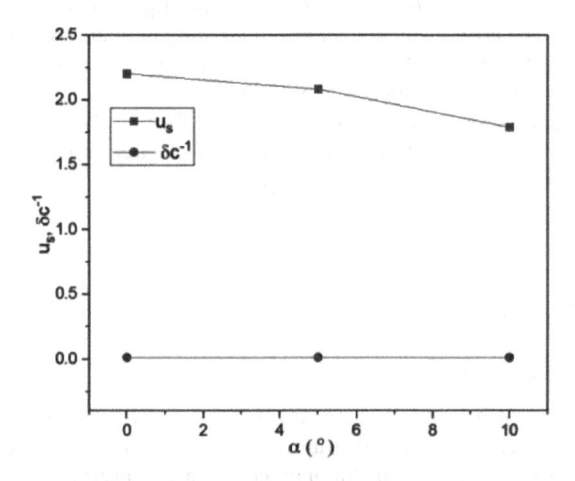

Figure 3. Information about normalized slip velocity and slip length over the surface of hydrofoil.

3.3 Wake behavior

Superhydrophobicity is well known for the delay of the separation and the intensity of vortex shedding frequency to a large extent inflow over circular cylinders. Figure 4 shows that the flow field is a function of the angle of attack and strongly depends on the heterogeneity of the surface. Comparative wake flow for 5° and 10° demonstrates the effect of superhydrophobicity.

In the case of no-slip surface hydrofoil, flow is laminar and a separation bubble grows near the trailing edge. At the angle of attack 10°, flow becomes fully developed two-dimensional. Von-Karman vortices are shed in wake. Strouhal number is calculated as $St = (fc)U^{-1}$. However, in presence of a heterogeneous surface, the vortex street diminishes and the wake becomes laminar attached to the trailing edge of the hydrofoil $\alpha = 10°$, thus alters the critical angle of attack to higher for Re= 1000. An explanation for this flow behavior is surface heterogeneity traps the air layer and thickens the boundary layer on the surface. Due to meniscus liquid-solid contact area becomes lower and decreases velocity gradient. As a result, an increase in wall slip and a decrease in shear stress is observed.

Figure 5 represents the behavior of drag and lift forces for the angle of attack 10°. From the figure it is clear that for no-slip surface fluctuation is symmetric. Therefore, the wake is two-dimensional. For slip/no-slip hydrofoil surface figure indicates the steady laminar attached flow.

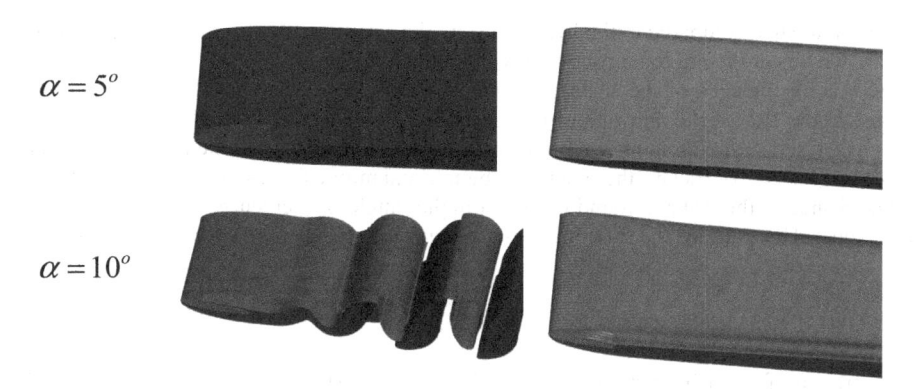

Figure 4. wake pattern for threedimensional flow over a hydrofoil with smooth surface versus heterogeneous surface at $\alpha = 5°, 10°$. Blue colour exhibits for positive vortex while grey colour indicates the negative vertex. Left: G.F=0, right: G.F=0.5.

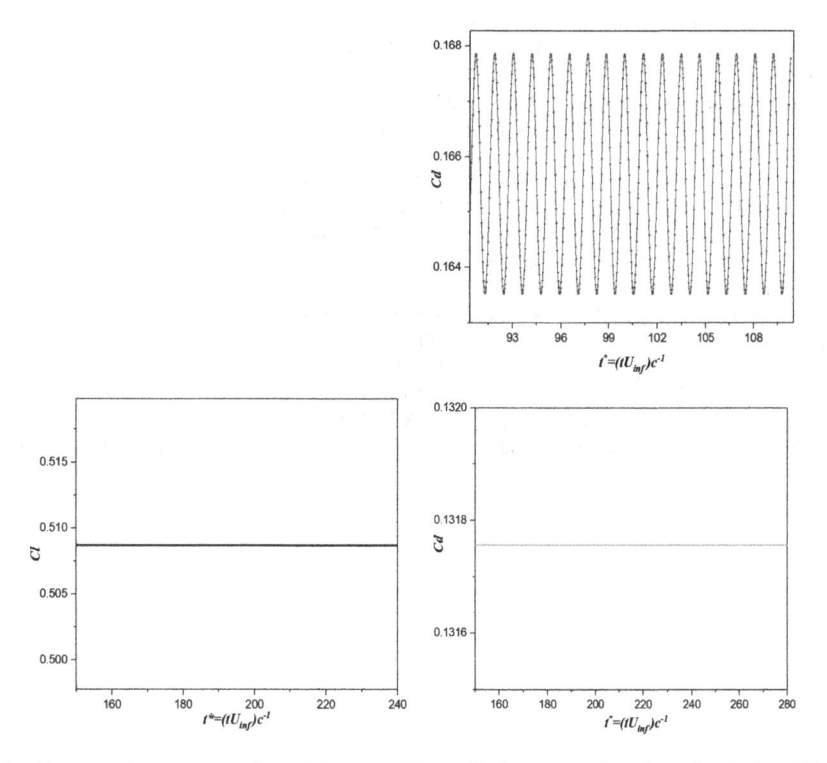

Figure 5. Comparative representation of drag and lift coefficients at angle of attack 10° for G.F=0(Top) versus G.F=0.5(Bottom).

4 CONCLUDING REMARKS

In the present work, the flow field computations over a hydrofoil with a superhydrophobic surface are performed. Numerical results demonstrate the SHs implementation as a promising DR technology. Appreciable drag reduction, lift increment, and corresponding changes in velocity profile are observed when compared to the hydrofoil with the smooth surface of the same computational domain. Results are found to be consistent with the literature.

Microscopic changes of the surface directly affect the boundary layer leads to macroscopic changes in the flow field. Flow field modifications are functions of angles of attack and gas-fractions at fixed Reynolds number. Although, the number of grates are limited due to computational resource and cost. SHS delays the separation bubble and mimic the instability of wake. Therefore, no disturbance in the wake is found even at a higher angle of incidence, and flow remains two-dimensional. Thus, reduce the frictional force and enhance the performance of hydrofoil.

ACKNOWLEDGEMENT

This work was funded by National Natural Science Foundation of China (Nos 11872187 and 51779097). The authors thanks to SCTS/CGCL HPCC of HUST for providing computing resource.

REFERENCES

[1] Yunqing G, Tao L, Jiegang M, Zhengzan S and Peijian Z 2017 *Applied bionics and biomechanics*.
[2] Darmanin T and Guittard F 2015 *Materials Today* **18(5)** 273–85.
[3] Barthlott W and Neinhuis C 1997 *Planta* **202(1)** 1–8.
[4] Zhang S, Ouyang X, Li J, Gao S, Han S, Liu L and Wei H 2015 *Langmuir* **31(1)** 587–93.
[5] Li Z, Marlena J, Pranantyo D, Nguyen BL and Yap CH 2019 *Journal of Materials Chemistry A* **7(27)** 16387–96.
[6] Legendre D, Lauga E and Magnaudet J 2009 *J. Fluid Mech.* **633** 437–447.
[7] Costantini R, Mollicone JP and Battista F 2018 *Physics of Fluids* **30(2)** 025102.
[8] Xiang Y, Huang S, Lv P, Xue Y, Su Q and Duan H 2017 *Physical review letters* **119(13)** 134501.
[9] Gogte S, Mammoli A and Vorobieff P 2016 *International Journal of Computational Methods and Experimental Measurements* **4(4)** 493–501.
[10] Lee J, Kim H and Park H 2018 *Experiments in Fluids* **59(7)** 111.
[11] You D and Moin P 2007 *Physics of Fluids* **19(8)** 081701.
[12] Ren Q, Xiong YL, Yang D and Duan J 2018 *Acta Mechanica* **229(9)** 3613–27.
[13] Balakumar P 2017 *In47th AIAA Fluid Dynamics Conference* 3978.
[14] Park H, Park H and Kim J 2013 *Physics of Fluids* **25(11)** 110815.
[15] Voronov RS and Papavassiliou DV and Lee LL 2007 *Chemical physics letters* **441(4–6)** 273–6.

Evaluation of the model and index for durability of concrete during construction

Y. Yang
Sichuan Jiu Ma Highway Co. Ltd, Chengdu, China

L. Yu
Beijing Xinqiao Technology Development Co. Ltd, Beijing, China

R.Y. Fang
Sichuan Jiu Ma Highway Co. Ltd, Chengdu, China

Z.J. Zhao
Research Institute of Highway Ministry of Transport, Beijing, China

ABSTRACT: In order to distinguish the durability of concrete during construction and control it timely, the mixing amount of fly ash, air content, ratio of water and bind, and water in unit are selected as construction control indexes. By changing these indexes, concrete was made for determining the diffusion coefficient of chloride ion, carbonation depth, and dry shrinkage value. Based on a fuzzy theory, the weight from construction control indexes to durability of concrete is calculated. A new index called Synthesize Evaluation Index from Construction to Durability is proposed. Based on the AHP method, the evaluation model is established between this index and construction control indexes. The range value of the synthesize index is given. And the level standard and evaluation steps are put forward. During real construction, by calculating the synthesize index, the durability of concrete can be obtained. And the durability of concrete during construction can be controlled by adjusting the construction control indexes so that the good durability concrete can be obtained

1 INTRODUCTION

During construction, the working performance and compressive strength are two priority parameters. The durability of the concrete has not been paid enough attention. Results of the study by Chen Can [1] show that only considering the strength as the acceptance check is unreasonable. The permeability index should be considered timely. What is more, when some cracks appear in concrete during the service period, the common handling methods include pouring some mending materials. The study by Haochuan Xu [2] shows that the epoxy resin always used for mending crack. There are a number of studies that have exploited mending materials. Yangjie Chen and Xiongfei Zhang [3] studied the polyurethane modification epoxy resin as mending materials and exploited the quick mending materials under the indoor temperature. But these mending methods all belong to controlling after crack. If the lack of durability can be found during construction, the useful methods such as adjusting the index of the concrete mix proportion can be used to avoid cracking after the short period of construction. Control can be changed after the event to control the event. Zhang Duo [4] optimized the mix proportion of concrete by taking advantage of the neural network method to predict the strength and durability. But he did not get the quantitative relation between mix proportion indexes and the durability indexes. Xia Liang [5] used a concrete quality control system including controlling before construction, during construction, and after construction. But he used only a qualitative method and there is no quantitative method to evaluate the performance of concrete during construction. By now, there is no good method to distinguish the durability of the concrete during construction.

This study aims to establish the quantitative relationship between parameters of construction and indexes of concrete durability performance. There are some construction parameters that can affect the performance of concrete durability. The mainly parameters include raw materials (cement grade, the sand fineness modulus, and mud content), mix proportion indexes (water paste ratio, fly ash content, aggregate grading, unit water mass, and sand rate), working performance (slump and air content), the density of vibration, the homogeneity of pave, floating situation, curing indexes (watering height), environment factors (temperature and humidity), and so on. This paper selected only four factors, water paste ratio, fly ash content, unit water content, and air content, to establish the model between construction indexes and durability ones. This model is to support kind of evaluation method by theory. This model is only appropriate for a common mountainous area environment. If used in other environments, the range of the index needs to be standardized anew.

2 EXPERIMENT DESIGN

2.1 Raw materials

P.O 42.5 cement has been used. The properties of the cement are listed in Table 1. The properties of fly ash are shown in Table 2. The crushed aggregate with continuous grading is used. The particle size range is from 5mm to 31.5mm. The nature sand is used with a fineness modulus of 2.85. The tap water is used. The polycarboxylate superplasticizer is used. The water reducer rate is 30%, and the solid content is 40%.

Table 1. Properties of cement.

Setting time (h:min)		Standard consistency water consumption (%)	Splitting tensile strength (MPa)		Compressive strength (MPa)		Fineness (80μm square hole sieve) (%)	Stability (boiling method)
Initial	Final		3d	28	3d	28d		
3:10	5:15	30.1	5.1	8.1	26.0	47.7	1.25	Qualified

Table 2. Properties of fly ash.

Fineness (45μm square hole sieve) (%)	Water demand ratio (%)	Loss on ignition (%)	Water content (%)	Stability (boiling method)
4.4	92	1.9	0.6	Qualified

2.2 Mix proportion and serial number

Table 1 is the mix proportion of the concrete. The groups from 1 to 6 have different water paste ratios. The groups from 7 to 10 have different fly ash contents. The sand rates of all of these groups are all 38%. The amount of water reducer is 1% of the paste mass. In the following text, the code of the fly ash is F. The code of the air content is A. The code of the water paste ratio is W/B. The code of the unit water amount is U. The mix proportions of the 10 group concrete are listed in Table 3.

Table 3. Mix proportions of concrete.

ID	Paste (kg/m³)	W/B	Water (kg/m³)	Cement (%)	Fly ash (%)
1	460	0.30	138	80	20
2	440	0.32	141	80	20
3	420	0.34	143	80	20
4	400	0.36	144	80	20
5	380	0.38	144	80	20
6	360	0.40	144	80	20
7	420	0.34	143	60	40
8	420	0.34	143	70	30
9	420	0.34	143	90	10
10	420	0.34	143	100	0

2.3 Experiment methods

The chloridion diffusion coefficient (RCM), the depth of carbonization (C), and dry shrinkage value (D) are the three main indexes considered. The three indexes are tested according to the standard of GB/T 50082. The experiment is similar to the RCM method. The air content index is tested according to the standard of GB/T 50080.

3 EXPERIMENT DESIGN

3.1 Test results

The corresponding relationship between the construction indexes of water and paste ratio (W/U, U) and the durability indexes (RCM, C, and D) can be obtained from group 1 to group 6. The corresponding relationship between the fly ash content and the air content can be obtained from group 7 to 10. The detail data are listed in Table 4.

Table 4. Construction indexes and durability indexes list.

ID	W/B	U (kg/m³)	RCM ($\times 10^{-12}$ m²/s)	C(mm)	D (10 µε)
1	0.30	138	3.00	5.0	28
2	0.32	141	3.60	10.0	27
3	0.34	143	3.80	16.0	28.5
4	0.36	144	4.40	20.0	29.6
5	0.38	144	5.20	28.0	33.5
6	0.40	144	6.00	28.0	28.0

ID	F (%)	A (%)	RCM ($\times 10^{-12}$ m²/s)	C (mm)	D (10 µε)
7	40.0	2.10	2.90	24.0	26.9
8	30.0	2.10	3.20	18.0	27.5
9	10.0	1.90	4.20	8.0	30.0
10	0.0	2.10	4.60	7.0	31.5

3.2 Association analysis

The Gray Relative Analysis method [6] is used to calculate the gray relational grade between the construction indexes and the durability indexes. The data are listed in Table 5.

Table 5. The gray relational grade between the construction indexes and durability indexes.

Indexes	RCM ($\times 10^{-12}m^2/s$)	C (mm)	D ($\mu\varepsilon$)
U	0.638	0.706	0.817
W/B	0.538	0.658	0.597
A	0.955	0.525	0.964
F	0.525	0.693	0.466

The abovementioned data are normalized. The weights from the construction indexes to durability indexes are obtained and shown in table 6.

Table 6. The weight from the construction indexes to durability indexes.

Indexes	RCM ($\times 10^{-12}m^2/s$)	C (mm)	D ($\mu\varepsilon$)
U	0.241	0.273	0.287
W/B	0.203	0.255	0.210
A	0.359	0.203	0.339
F	0.199	0.269	0.164

3.3 The evaluating model

A new concept is reported in this paper, that is, the Synthesize Evaluation Index from Construction to Durability (SEICD). The independent variables include RCM, C, and D.

3.3.1 The index system established

Based on the Analytic Hierarchy Process method [7], the index system is established. The SEICD is the top layer and numbered layer O. Under the top layer, there are three indexes. They are RCM, C, and D. This layer is numbered A. The bottom layer includes F, A, W/B, and U and is numbered B. The weight from B to A is shown in Table 6.

3.3.2 Weight calculation from A to O.

Based on the scaling theory [8], 30 people who are designers, constructors, maintainers, and researchers are investigated. Table 7 is a sample of the marking table. Table 8 is the result of their marking.

Table 7. Sample of the marking table.

Marking factors:
The factors of durability of concrete: A_1—Chlorine iron content; A_2—Carbonation depth; A_3—Dry shrinking value.
Marking rules:
1 score— A_i and A_j have the same important level 3分 —A_i is a little more important than A_j; 5分—A_i is more important than A_j; 7分—A_i is much more important than A_j; 9分— A_i is absolutely more important than A_j. the reciprocal is the result comparing A_j to A_i
Marking result:
Please mark o
$A_1 \rightarrow A_2$: _____ score;
$A_1 \rightarrow A_3$: _____ score;
$A_2 \rightarrow A_3$: _____ score.

Table 8. The result of marking.

Factor	1	2	3	4	5	6	7	8	9	10	11	12	13	14	15
$A_1 \to A_2$	1	5	3	5	1	3	1	1	3	1	1	1	5	3	1
$A_1 \to A_3$	5	1	1	3	1	3	3	5	1	3	3	3	5	3	5
$A_2 \to A_3$	1	1	1/3	1	1	1/3	3	1	1/3	1	1/3	1	1/3	1	1/3
Factor	16	17	18	19	20	21	22	23	24	25	26	27	28	29	30
$A_1 \to A_2$	1	5	7	3	5	1	1	5	1	3	5	1	3	1	5
$A_1 \to A_3$	1	5	1	3	3	3	5	1	5	3	3	5	3	1	3
$A_2 \to A_3$	1/5	1	1	1/3	1	3	1	1/3	1	5	1	1	1/3	1	1/5

The important degree is calculated. The important degree from A1 to A2 is 2.73. The important degree from A1 to A3 is 3.00. The important degree from A2 to A3 is 1.01. By calculating the judgment matrix from Ai to Aj, the weighting values from the A layer to O layer are 0.59, 0.21, and 0.20.

3.3.3 Weight calculation from B to O

Based on the weight from B to A (Table 6) and the weights from A to O, the weight from B (F,A,W/B,U) to O can be obtained. They are 0.206, 0.322, 0.215, and 0.257. So, the model between the SEICD and the construction indexes (F,A,W/B,U) can be put forward as formula 1.

$$SEICD = \omega_1 F + \omega_2 A + \omega_3 W/B + \omega_4 U' \quad (1)$$

In this formula, F is the fly ash content. A is the air content of the fresh concrete. W/B is the water and binder ratio. U' is the relative error between the real water content and the design water content. The units of these four factors are all %. In the formula, $\omega_i = (0.206, 0.322, 0.215, 0.257)$, and $i = 1, 2, 3, 4$.

In the formula, U' is used instead of the former U. The reason is the uniformity of these four independent variables. The weights of U and U' are the same. What is more, according to the mix proportion of the concrete, the value range of the F is between 20% and 30%. A is below 4%. W/B is between 0.30 and 0.40. U' is between -5% and 5%. So, the range of the SEICD can be calculated from 0.097 to 0.285.

3.4 The level divided and evaluation procedure of the SEICD

3.4.1 Level divided

In order to use the SEICD value to evaluate the durability of the concrete during construction, the levels are divided into three. The range of every level is shown in Table 9.

Table 9. Range of the SEICD level.

Level	Good	Middle	Bad
Range	0.144–0.097	0.191–0.144	0.285–0.191

3.4.2 Evaluation procedure of the SEICD

In order to use this evaluation method, the operating steps can be listed as follows:

Step 1. During construction, record the values of F, A, W/B, and U'. If they are in the value range of the independent variables, substitute them in formula 1.

Step 2. Compare the calculated value from formula 1 and the range of the SEICD level in Table 9; the level of the concrete can be assured.

Step 3. If the level is good or middle, the SEICD is reasonable and the mix proportion of the concrete can be used. If the level is bad, go back to the step 1. Adjust the values of F, A, W/B, and U and recalculate the value of the SEICD until it is good or middle.

4 EXPERIMENT DESIGN

In order to distinguish and control the durability of the concrete during construction, four construction indexes are selected. By changing the value of these four construction indexes, three durability indexes of the concrete have been tested. Based on the Gray Relative Analysis method, the weights from four construction indexes to three durability indexes have been calculated. The new concept of Synthesize Evaluation Index from Construction to Durability (SEICD) is put forward. Based on the Analytic Hierarchy Process, the evaluation model from construction indexes to SEICD has been established. The level range of the DEICD has been divided. The evaluation steps have been put forward to indicate how to use this evaluation method. In the practice construction, the calculated SEICD can be used to evaluate the durability of the concrete in the service period. The most important thing is that the durability can be adjusted timely to satisfy the need of the concrete during the construction period. Control can be changed after the event to control the event.

ACKNOWLEDGMENTS

This work has been supported by the grant from Sichuan Jiu Ma Highway Incorporated Company to Beijing Xinqiao Technology Development Co.LTD, China. The Technology Plan Item set up by Transportation Department of Sichuan Province, China. The title of the item is Concrete Materials and Construction Quality Controlling Technology in the Environment of Supreme Temperature and Altitude.

DECLARATION OF COMPETING INTEREST

The authors declare that they have no known competing financial interest or personal relationships that could have appeared to influence the word reported in this paper.

REFERENCES

[1] Can Chen. Study on the relationship between the material properties in construction period and the concrete durability in the later period. Thesis of master degree of Zhejiang University. 2013. China.
[2] Haochuan Xu. The solution of concrete cracking. Commercial concrete. 2019, (8), pp4–6.
[3] Yangjie Chen, Xiongfei Zhang. Study on the crack quick mending materials of epoxy rein modified with polyurethane. The Chinese and foreign road.2019, 39(4), pp229–233.
[4] Duo Zhang. Study on the technology of bridge standard construction based on the durability of concrete. Thesis of master degree of Zhengzhou University. 2017. China
[5] Liang Xia. High performance concrete and construction controlling technology. Thesis of master degree of Anhui Jianzhu University. 2015. China.
[6] Jijian Xie, Chengping Liu. Fuzzy mathematics and application. Huazhong University of Science and Technology Press.2013.China.
[7] Bokun Zhang. Comprehensive evaluation of the situation of extradosed cable-stayed bridge based on AHP and Fuzzy mathematics theory. Thesis of master degree of Jilin University. 2015. China.
[8] Lei Yu. Durability of concrete and optimization method. China Building Industry Press. 2018. China.

Research on equivalent construction technology of stealth coating in RCS measurement

Y.G. Xu
Science and Technology on Electromagnetic Scattering Laboratory, Shanghai, China

K. Fan
Shanghai Radio Equipment Laboratory, Shanghai, China

X.B. Wang, W. Gao & Y. Zhang
Science and Technology on Electromagnetic Scattering Laboratory, Shanghai, China

ABSTRACT: The acquisition of the target characteristics of stealth aircraft plays an important role in the development of anti-stealth weapons and equipment in various countries. For non-cooperative stealth targets, the stealth coating is widely used on the stealth aircraft surface, and it is difficult to construct the electromagnetic target model in the absence of unknown electromagnetic properties of the stealth material. In this work, based on the coated plate specimens, the reflectance approximation method is used to construct the scaled model. The RCS value of the scaled model can maintain good consistency with the original sample. Based on the diversity of the scaled model, the scattering characteristics of each scaled model is nearly the same, the concept of the equivalent structure of the stealth coating is proposed, and the corresponding materials are extended to several other typical samples (coated slit plate, coated dihedral plate) and verified respectively. The research method in this work has a driving role for the accurate acquisition of the characteristics of non-cooperative stealth targets.

1 INTRODUCTION

For the stealth aircraft, the early use of electromagnetic scaled measurement is an effective method to obtain the target electromagnetic scattering property, and it is of low cost and high efficiency [1–4]. Previous studies have solved the scaled ratio of all-metal targets. The scaled structure of dielectric loss materials without frequency dispersion effects has also been explored. However, the magnetic stealth coating with frequency dispersion effects are still under study. For the uniform stealth coating, the reflectivity approximation method was used in the previous period to construct the scaled stealth material [5]. The results showed that it had good applicability for the flat-plate target and was optimized in the design process [6–8]. The design method can supply a series of approximate scaled stealth materials. At this time, considering the following two situations: (1) If the scale ratio is further performed, the scale ratio is taken as the reciprocal of the aforementioned scale factor, that is, the amplification measurement, the stealth coating is constructed according to the strict scaled measurement theory, and the material structure size is enlarged while the dielectric constant and magnetic permeability remain unchanged. At this time, the model must satisfy the similarity criterion. (2) Two optimized scale stealth materials by reflectivity approximation method are selected, and the two models still maintain similarity. As a result, the same phenomenon can occur, and the two materials at the same measurement frequency can maintain the same reflectivity, and at the same time they can bring about a substantially uniform scattering characteristic (mainly RCS) after being applied to the model. For the two stealth materials involved, we define them as approximate equivalent stealth materials. In particular, there are cases where the material thicknesses are equal in the equivalent design.

In order to verify the above mentioned construction material, based on the theoretical calculation of the coating plate, the material electromagnetic parameter library is established, and the composition of the equivalent coating is obtained through interpolation calculation. Then the simulation calculation and verification of various samples are carried out. The RCS variation law of the sample before and after the equivalent design is compared to provide a new idea for the acquisition of electromagnetic scattering characteristics of the non-cooperative stealth aircraft.

2 RESEARCH METHODS

2.1 Sample model and RCS calculation

A coated metal plate was selected. Figure 1 shows a rectangular metal plate with absorbing material of 2a in length and 2b in length. The coating thickness is d, the relative complex permittivity and permeability of the absorbing material are uniform, and the materials can be seen as isotropic uniform material. For the convenience of analysis, the coordinate system is established. The flat plate is located in the XOZ plane. The incident wave of the radar is a horizontally polarized uniform plane wave, the incident surface is the XOY plane, and the incident angle is θ0.

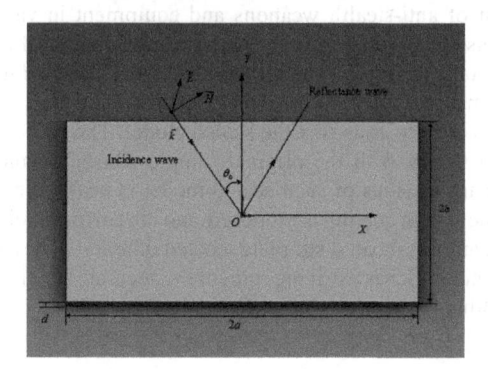

Figure 1. Metal plate model and calculation of absorbing material.

Since $0 < \theta_0 < \pi/2$, the backscattering field is mainly the microwave diffraction of the two edges. Based on the uniform diffraction formula, the two-dimensional scattering field of the impedance mode, the three-dimensional scattering of the rectangular plate and the backscattering RCS are defined. The relationship between the far-field backscattering RCS for a rectangle is as follows [1]:

$$RCS = \frac{b^2}{4\pi} \left| e^{-j2ka\sin\theta_0} \left\{ \Gamma^h (1 - ctg\frac{\theta_0}{2}) - (1 - ctg\frac{\pi - \theta_0}{2}) \right\} \right.$$

$$\left. + e^{j2ka\sin\theta_0} \left\{ \Gamma^h \left(1 + ctg\frac{\theta_0}{2}\right) - \left(1 + ctg\frac{\pi - \theta_0}{2}\right) \right\} \right|^2 \quad (1)$$

where Γ^h is the reflection coefficient of the oblique wave,

$$\Gamma^h = \frac{(\sqrt{\mu_r}\cos\theta_0 - \sqrt{\varepsilon_r}\cos\theta)e^{j2k_yd} - (\sqrt{\mu_r}\cos\theta_0 + \sqrt{\varepsilon_r}\cos\theta)}{(\sqrt{\mu_r}\cos\theta_0 + \sqrt{\varepsilon_r}\cos\theta)e^{j2k_yd} - (\sqrt{\mu_r}\cos\theta_0 - \sqrt{\varepsilon_r}\cos\theta)} \quad (2)$$

in which, $k_y = k\sqrt{\varepsilon_r\mu_r}\cos\theta$, $\sin\theta = \sin\theta_0/\sqrt{\varepsilon_r\mu_r}$, $\varepsilon_r = \varepsilon' - j\varepsilon''$ and $\mu_r = \mu' - j\mu''$ are the complex permittivity and permeability respectively.

According to the theoretical formula, Γ^h is kept basically unchanged, the RCS of the flat sample can be basically unchanged, so the electromagnetic parameters and the thickness of the material can be adjusted as long as the parameters are adjusted. The RCS of the plate sample could be also adjusted. When electromagnetic waves are obliquely incident on the surface of the absorbing material, there are mainly two types, namely TE and TM modes. When the electromagnetic wave is of the TE type, the magnetic field vector is parallel to the incident plane at this time. The reflection loss could be interpreted as Γ^h, the reflection loss (RL) of the coating could be described as $RL = 20\lg|\Gamma^h|$.

2.2 Equivalent electromagnetic parameter construction process

The design of the electromagnetic parameters (including εr and μr) corresponding to the scaled material is based on the effective medium theory, and the electromagnetic parameter library of the corresponding material is established by using the results of the electromagnetic parameters of the previous test. The optimized electromagnetic parameters can be obtained through the optimization process. The optimization algorithm used is the genetic algorithm. The objective function of the genetic algorithm is the reflectivity of the material under vertical and horizontal polarization. The optimization variables are material thickness and material addition ratio. After the optimization results are obtained, the scale material can be constructed. In particular, if the reduction ratio coefficient is 1, that is, there is no change in the size of the metal sample, only the material thickness and the electromagnetic parameters are changed, so the equivalent structural material of the prototype material can be obtained, and the construction flow chart is shown in Figure 2.

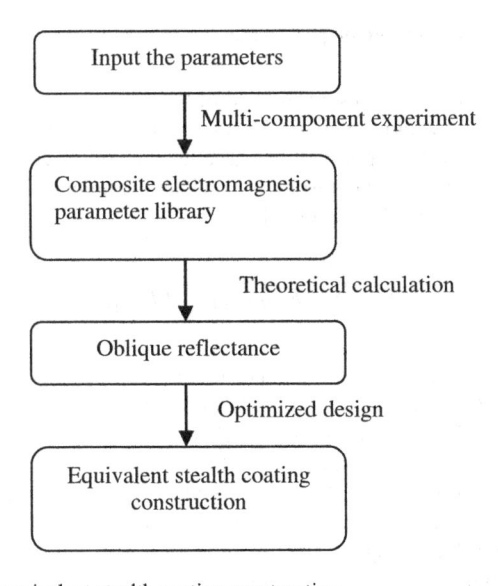

Figure 2. Flow chart of equivalent stealth coating construction.

3 RESULTS AND ANALYSIS

3.1 Electromagnetic parameter construction of equivalent stealth material

Step 1: Assuming that the thickness of the absorbing coating is 0.5 mm, the selected prototype frequency is 8 GHz, with $\varepsilon_{full-size} = 21.99 - j2.24$, $\mu_{full-size} = 2.10 - j2.35$.

Step 2: Testing the absorbing materials of various components. The additive of the material still is the carbonyl iron particle. The main variable is the volume content of the particles.

According to the experiment, the electromagnetic parameter library of the material can be obtained [5].

Step 3: Calculating the oblique reflectivity of the absorbing material, and using the formulas to obtain the oblique reflectivity of the material, the results are as shown in Figure 3. It can be seen that the oblique reflectivity of the absorbing material under the VV polarization gradually increases as the angle increases to 90°, and the change trend of the curve is basically the negative cosine function curve. The reflectivity of the absorbing material under HH polarization will decrease first and then increase, reach a minimum around 80°, and then increase sharply to 0dB. Therefore, it can be seen that in the equivalent design process, the difficulty lies in the oblique reflectance design of HH polarization.

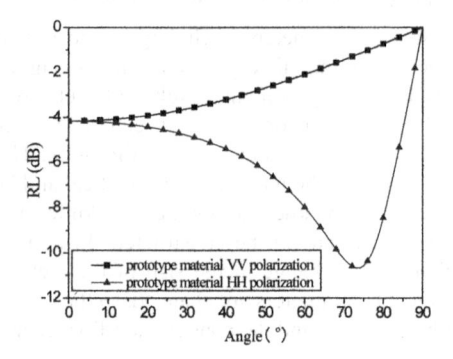

Figure 3. Oblique reflectance of the prototype material.

Step 4: The structural optimization design method of the equivalent material is the genetic algorithm, and the optimization objective function is the sum of the reflectance differences at equal spacing angles. The optimization variable is the thickness of the absorbing material and the volume content of the absorbent, such as the angular spacing selected as 5°, the optimization function can be expressed as:

$$F = \sum_{i=1}^{18} \left| RL_{HH}\big|_{f=f_i} - RL_{opt}\big|_{f=f_i} \right| \tag{3}$$

Where, f is the frequency, fi is the i-th frequency point, $i = 1, 2, \ldots, 18$, the corresponding angle is 5i degree, RLHH and RLopt are the oblique reflectance of underlying prototype material and the optimized material under HH polarization. The thickness and the absorbent are added to establish an initial population than the two variables, and then the thickness and the volume content are decoded to obtain a thickness value and volume content. Furthermore, the Hermit interpolation calculation is performed according to each proportional electromagnetic parameter in the material parameter library, and the electromagnetic parameters at the ratio are obtained, and then RLopt can be obtained according to the reflectance formula.

After the optimization design, the electromagnetic parameters and formula of the material can be obtained as follows: the volume content is 0.4, equivalent material thickness deff=0.6mm, electromagnetic parameters of the equivalent material are εeff=16.99−j0.38, μ scale=1.78−j1.95. The oblique reflectivity of the equivalent stealth materials is shown in Figure 4. It can be seen that for the oblique reflectivity of the material at 8 GHz, the reflectivity of the material under VV polarization increases with the increase of the angle. The difference is very small, the maximum occurs at 0° position is 0.03dB. While for HH polarization, the angle of reflection decreases first and then increases to 0, and the maximum value is 0.40dB around 74°. The overall material the oblique reflectance is basically consistent, and the maximum reflectance deviation cumulative value is 4.28 dB.

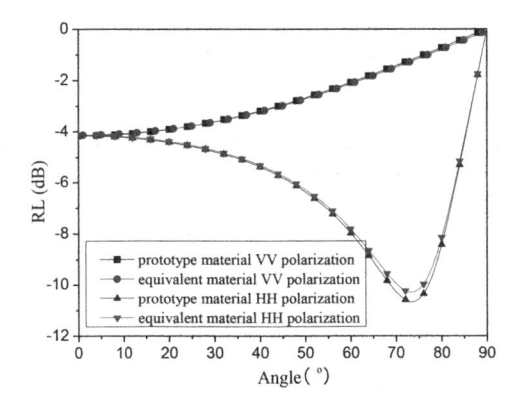

Figure 4. Oblique RL of the equivalent materials.

Selecting the full-size absorbing plate sample with side length 400mm, the frequency is 8GHz, the two electromagnetic parameters of the sample are used in the theoretical formula to calculate the flat sample RCS. Figure 5 is RCS curve of the equivalent absorbing plate samples. It can be seen that the RCS of the equivalent sample is very close to that of the prototype plate sample, and the mean difference is within 0.05 dB. Although there are many peaks in the calculated RCS curve, from the overall results, the RCS values of the equivalent samples are in good agreement.

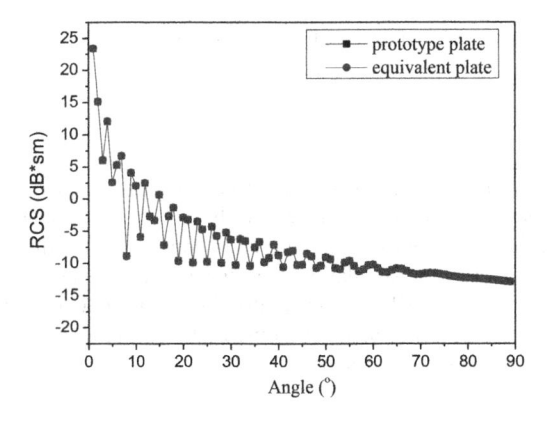

Figure 5. RCS of the prototype and equivalent plate.

3.2 RCS simulation calculation of typical structural equivalent samples

In order to verify the design method on the equivalent material, the sample with the coated gap and the dihedral angle were verified. The width of the flat slit sample is 300mm×300mm, the slit is diagonal, the length is 300mm, the gap width is 2mm, the incident angle of electromagnetic wave is $0° \sim 90°$. While the size of the single flat plate in the dihedral angle is 300mm×300mm, the dihedral angle is 90°, the incident angle of electromagnetic waves is $0° \sim 45°$, and the polarization modes are vertical polarization and horizontal polarization. The target RCS value simulation is completed by numerical simulation, and the simulation software FEKO software was selected. The calculation method is the moment method combined with the fast multi-level sub-algorithm.

The calculation results are shown in Figure 6. It can be seen that the RCS of the equivalent samples still maintains a good consistency. For the dihedral angle, the maximum RCS deviation appears at the peak-to-valley position. The RCS mean deviation is less than 0.2dB, and the error is higher than the flat sample, which may be caused by the cumulative deviation of the reflectance under multiple reflections of the dihedral angle. For the coated plate with the coating, the electromagnetic wave is vertically polarized, the geometric mean is 0.16dB, and the RCS deviation near 22° had the maximum value. The angle is at the RCS curve valley position, and RCS is below -30dBm2. The value can be ignored. When the electromagnetic wave is horizontally polarized, the deviation of the RCS of the equivalent samples between 55° and 60° is obvious. The electromagnetic scattering characteristics are more complicated at these angles, and the RCS value when the electromagnetic wave is incident at both ends is basically below -30 dBm2. The effects of this type of electromagnetic scattering characteristics are temporarily negligible. However, comparing the results of the two samples, it can be seen that the method has good applicability to the target model.

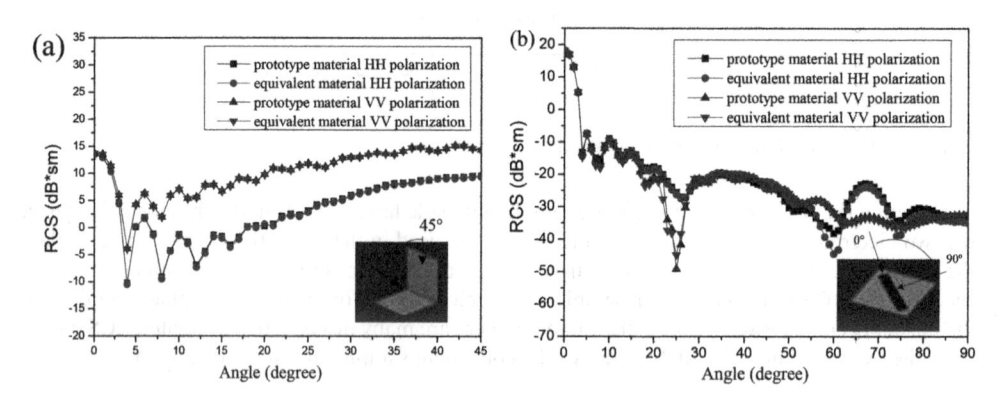

Figure 6. RCS of different models before and after equivalent, (a) dihedral plate, (b) coated slit plate.

4 CONCLUSION

In this paper, the equivalent construction of stealth coating plate is carried out. In order to sovle the problem that the electromagnetic parameters in the non-cooperative target are unknown and difficult to construct, the equivalent stealth material is constructed by the engineering method of reflectance approximation, and the simulation is performed for the flat specimen. On the basis of this, the other two samples are verified. The results show that the constructed equivalent stealth material has good applicability, and the adopted method can promote the accurate acquisition of non-cooperative stealth target characteristics.

REFERENCES

[1] Liu Tiejun and Zhang Xiangyang 1992 Acta. Electronica. Sinica. **20** 12
[2] Shi Zhendong, Yang Shiwen and Ding Chunsheng 1993 J. Appl. Sci. **11** 109
[3] Sinclair G 1948 Proc. IRE. **36** 1364
[4] Sang Jianhua 2013 *Stealth Technology of Aircraft* (Beijing: Aviation Industry Press)
[5] Xu YG, Yuan LM, Gao W, Wang XB, Liang ZC and Liao Y 2018 J. Phys. D Appl. Phys. **51** 065004
[6] Liming Yuan, Yonggang Xu, Wei Gao, Fei Dai, Qilin Wu. Design of scale model of plate-shaped absorber in wide frequency range. 2018. Chinese Physics B 27(4) 303–308.
[7] Liming Yuan, Zhijie Xie, Fei Dai, Yonggang Xu, Yuan Zhang. An effective methodology to6 design scale model of non-metallic structural entity, MATEC Web of Conferences, 2018.
[8] Feiming Wei, Yonggang Xu, Wei Gao, Dianliang Zheng, Ting Liu. Designing on scaled absorbing materials at THz band using Ni@mica composite. 2019, Applied Physics A, 125(9) 655.

Optimal control of vehicle semi-active suspension considering time delay stability

G.H. Yan & H.P. Shi
College of Automobile and Transportation, Tianjin University of Technology and Education, Tianjin, P.R.China

ABSTRACT: With the building of the dynamics' differential equation model of 1/4 vehicle semi-active suspension with time-delay, the stability conditions of the model were deduced by the PID control strategy and the theory of linear ordinary differential equation with time-delay, the stability of model was analyzed by the Routh-Hurwitz stability criterion and the critical instability delay-time was discussed and calculated. By the example simulation, the results show that when the critical delay-time is 0.132s, comparing with PID control method of without time-delay, the system was being on the critical stability. When the critical delay-time is 0.16s, the system was being on the instability and chaos vibration station. The calculation and simulation results proved that the theory of Routh-Hurwitz stability criterion lay a foundation for the design and instability mechanism of active suspension.

1 INTRODUCTION

Automotive semi-active suspension uses multiple sensors to monitor vehicle vibration in real time, the control unit calculates the optimal control signal by the control algorithm established in ECU, and the control instructions are sent to the variable damper to generate variable damping forces, By the variable damping force to ease the impact and vibration from the road, improve the comfort of the vehicle ride comfort and stability. The magnetorheological fluid semi-active suspension shock absorber developed by LORD and Delphi has been successfully used in commercial applications.

There will be a series of time delay factors in semi-active suspension system, including: information measurement and transmission time delay control calculation time delay and variable damper response time delay and so on. These time delays are inevitable. The existence of time delay has great influence on the control performance of semi-active suspension. With the increasing of the sensors, the complexity of the algorithm, and the multi-objective control, these factors will lead to the increase of the time-delay. Finally, the semi-active-suspension will become an unstable system, and even generates serious instability of the control system and negative effects.

A group of international and domestic academics has studied the time-delay systems in theory in the 1930s. For instance, references [1]: The gradient method is applied to reduce the cost via explicit algebraic formula with respect to the hidden final state/costate and to the switching times. The reference [2]: Two different air spring models are put forwarded. The reference [3]: The vertical dynamics characters of the automobile are represented through a qLPV model.

The above research results have done theoretical research from different aspects. However, these research papers are mainly aimed at the control ways. The reference [4] proposed a quasi-linear-parameter-varying algorithm. In reference [5]: an adaptive optimal control constraining semi-active vehicle suspension system is presented. The reference [6] researched on performance of semi-active suspension which ued magnetorheological damper based on LQG algorithm strategy. In this paper, the application of LQG algorithm in semi-active suspension control is emphatically studied.

Above research results have done theoretical research from different aspects. However, these research contents rarely consider the real delay-time of active control system and were mostly in the hypothesis of no delay exist in the system. But the time delay inevitably exists in the semi-active suspension system. These time delay factors include information measurement and transmission time delay, control calculation time delay and variable damper response time delay. The existence of time delay has great influence on the control performance of semi-active suspension. With the complexity of the control algorithm, and the multi-objective control, these factors will lead to the increase of the total time delay.

Based on the above reasons, this paper established a two-degrees of freedom model for semi-active suspension containing delay and analysed the influence of delay on the control effect. In this paper, the factors to be considered in the optimal control of semi-active suspension with delay are obtained.

2 MODELING AND CONTROL ALGORITHM OF SEMI-ACTIVE SUSPENSION WITH TIME DELAY

2.1 Establishment of model

The 1/4 vehicle semi-active suspension dynamic model established by this paper is shown in Figure 1. This model has two-degrees of freedom: vertical movement of sprung mass and upsprung mass. In order to simplify the research problem, the following assumptions are made for the vehicle vibration system: the vehicle body is rigid and the tire damping is ignored, the tire is simplified as a spring suspension with equivalent stiffness, the spring stiffness is linear and the vehicle travels on the road at uniform speed.

By Newton's laws of motion, the dynamic equation at any time has been acquired:

$$\begin{cases} m_b\ddot{Z}_b(t) + C_0[\dot{Z}_b(t) - \dot{Z}_w(t)] + C_{S(t-\tau)}[\dot{Z}_b(t) - \dot{Z}_w(t)] + K_s[Z_b(t) - Z_w(t)] = 0 \\ m_w\ddot{Z}_w(t) + K_t[Z_w(t) - Z_r(t)] - C_0[\dot{Z}_b(t) - \dot{Z}_w(t)] - C_{S(t-\tau)}[\dot{Z}_b(t) - \dot{Z}_w(t)] \\ \quad - K_s[Z_b(t) - Z_w(t)] = 0 \end{cases} \quad (1)$$

In equation (1), m_b is sprung mass, m_w is unsprung mass, K_s is stiffness of suspension spring, C_0 is Zero control damping coefficient, $C_{S(t-\tau)}$ is controlled damping coefficient at time t, K_t is tyre equivalent stiffness, Z_b is sprung mass vertical displacement, Z_w is wheel vertical displacement, Z_r is pavement disturbance, τ is time-delay.

Select $X(t) = [\dot{Z}_b(t), \dot{Z}_w(t) \, Z_b(t), Z_w(t), Z_r(t)]$ as variable, then the state space equation is:

$$\dot{X}(t) = AX(t) + Ew(t) \quad (2)$$

In the equation (2): A is system matrix, E is the disturbance matrix of road. Thereinto:

$$A = \begin{bmatrix} -\frac{C_0+C_{s(t-\tau)}}{m_b} & \frac{C_0+C_{s(t-\tau)}}{m_b} & -\frac{K_s}{m_b} & \frac{K_s}{m_b} & 0 \\ \frac{C_0+C_{s(t-\tau)}}{m_w} & -\frac{C_0+C_{s(t-\tau)}}{m_w} & \frac{K_s}{m_w} & -\frac{K_s+K_t}{m_b} & \frac{K_t}{m_w} \\ 1 & 0 & 0 & 0 & 0 \\ 0 & 1 & 0 & 0 & 0 \\ 0 & 0 & 0 & 0 & -2\pi f_0 \end{bmatrix}, E = \begin{bmatrix} 0 & 0 & 0 & 0 & 2\pi\sqrt{G_0 V} \end{bmatrix}^T.$$

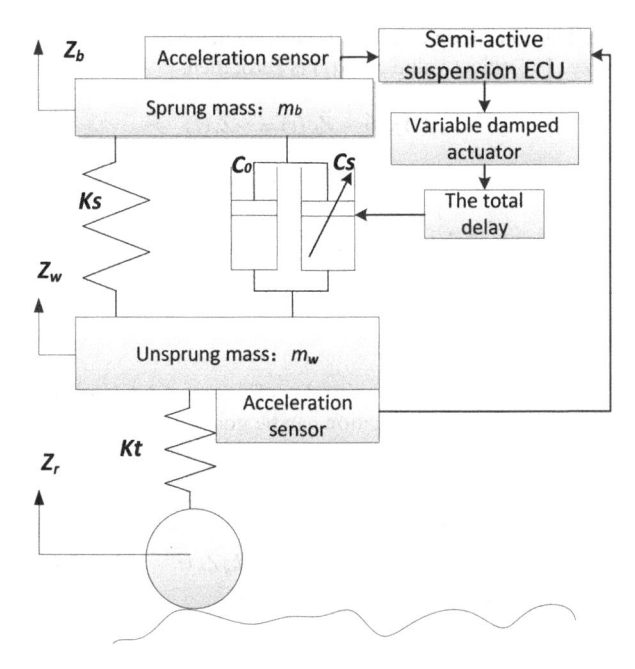

Figure 1. Semi-active suspension dynamic model.

The white gaussian noise is adopted as input variable of road in this model which is:

$$\dot{Z}_r(t) = -2\pi f_0 Z_r(t) + 2\pi \sqrt{G_0 V} w(t) \quad (3)$$

In the equation (3): f_0 is cutoff frequency in low frequency, which is 0.01 Hz, G_0 is road roughness coefficient, V is vehicle speed, $w(t)$ is the gaussian white noise.

Select $Y(t) = [\ddot{Z}_b(t)\ Z_b(t) - Z_w(t)\ Z_w(t) - Z_r(t)]$ as output variable, which expresses the output value of the acceleration, the suspension dynamic deflection and the tire dynamic transformation. Then the output state space equation is:

$$Y(t) = CX(t) \quad (4)$$

In the above equation: The output matrix, $C = \begin{bmatrix} -\frac{C_0+C_{s(t-\tau)}}{m_b} & \frac{C_0+C_{s(t-\tau)}}{m_b} & -\frac{K_s}{m_b} & \frac{K_s}{m_b} & 0 \\ 0 & 0 & 1 & -1 & 0 \\ 0 & 0 & 0 & 1 & -1 \end{bmatrix}$.

2.2 The Semi-active suspension PID control

PID controller is the most commonly used in controlling system, PID controller is the linear controller whose control accord to the deviation between control object value $obj(t)$ and the real output value out(t) [7].

$$e(t) = obj(t) - out(t) \quad (5)$$

The optimal control objective of active suspension is to obtain the good vehicle ride comfort and handling stability, and to reduce sprung mass vertical acceleration $\ddot{Z}_b(t)$ and dynamic deformation of tire $Z_w(t) - Z_r(t)$ as far as possible, at the same time restrict the vertical dynamic deflection $Z_b(t) - Z_w(t)$ so as to prevent suspension impact bumper block.

The $\ddot{Z}_b(t)$ is selected as control objective. $\ddot{Z}_b(t)$ can be made as little as possible by modulating the active control force. The object value of $\ddot{Z}_b(t)$ is zero, then:

$$e(t) = 0 - \ddot{Z}_b(t) = -\ddot{Z}_b(t) \qquad (6)$$

Therefore, the output of the controlled damping coefficient C_S with time delay τ is:

$$C_{S(t-\tau)} = -k_p \ddot{Z}_b(t-\tau) - k_i \int \ddot{Z}_b(t-\tau)dt - k_d \frac{d\ddot{Z}_b(t-\tau)}{dt} = -k_p \ddot{Z}_b(t-\tau) - k_i \dot{Z}_b(t-\tau) - k_d \dddot{Z}_b(t-\tau) \qquad (7)$$

In the above equation: k_p, k_i, k_d are PID control coefficients. From equation (7) to equation (1), the semi-active-suspension differential equation can be got:

$$\begin{cases} \ddot{Z}_b(t) = \frac{1}{m_b}[C_0 - k_p \ddot{Z}_b(t-\tau) - k_i \dot{Z}_b(t-\tau) - k_d \dddot{Z}_b(t-\tau)][\dot{Z}_w(t) - \dot{Z}_b(t)] \\ \quad - \frac{K_s}{m_b}[Z_b(t) - Z_w(t)] \\ \ddot{Z}_w(t) = [C_0 - k_p \ddot{Z}_b(t-\tau) - k_i \dot{Z}_b(t-\tau) - k_d \dddot{Z}_b(t-\tau)][\dot{Z}_b(t) - \dot{Z}_w(t)] \\ \quad - \frac{K_t}{m_w}[Z_w(t) - Z_r(t)] + \frac{K_s}{m_w}[Z_b(t) - Z_w(t)] \end{cases} \qquad (8)$$

3 STABILITY ANALYSIS OF SEMI-ACTIVE-SUSPENSION WITH DELAY

3.1 Stability analysis of differential equation

Let the solution of the equation (8) as follows:

$$Z_j(t) = Z_j e^{st}, j = b, w \qquad (9)$$

In the above equation: Z_b, Z_w is the Laplace transform variable, s is the system variable.

According to the nonzero solution condition and bring (9) into (8), then equation (10) could be got:

$$\begin{vmatrix} s^2 + \frac{C_0}{m_w}s + \frac{K_s}{m_w} + \frac{K_t}{m_w} - \frac{Q}{m_w}e^{-\tau s} & -\frac{C_0}{m_w}s - \frac{K_s}{m_w} \\ -\frac{C_0}{m_b}s - \frac{K_s}{m_b} & s^2 + \frac{C_0}{m_b}s + \frac{K_s}{m_b} + \frac{Q}{m_b}e^{-\tau s} \end{vmatrix} = 0 \qquad (10)$$

In the above equation: $Q = k_d s^3 + k_p s^2 + k_i s$.

According Routh-Hurwitz stability criterion, the sufficient and necessary condition of equation (8) is all the root of (10) has negative real part, if equation (10) has imaginary root $s = i\omega$ the system is under the critical unstable state. Bring $s = i\omega$ into (10) and the real and imaginary part were (11) and (12):

$$m_b m_w \omega^4 - (K_s m_w + K_s m_b + K_t m_b)\omega^2 + K_s K_t = 0 \qquad (11)$$

$$-C_0 \omega^3(m_b + m_w) + K_t C_0 \omega + (K_t - \omega^2 m_w)(k_i \omega - k_p \omega^2 - k_d \omega^3)e^{-\tau s} = 0 \qquad (12)$$

Accord to the numerical solution of steady equation, the real root of equation (11) is ω. Bring the root ω into equation (12), the critical time-delay of equation (8) which can be expressed as $\tau = f(K_s, K_t, C_0, m_w, m_b)$. If the equation (11) has no real root then the semi-active-suspension will be stable at random time-delay.

So, by equation (11) we can learn that if the parameters K_s, K_t, C_0, m_w, m_b were designed suitably then the ω will have no real root. Then, by equation (12), the critical time-delay τ can be guaranteed in a certain range. Through the designing of the value of $K_s, K_t, C_0, m_w, m_b, k_p, k_i, k_d$, we can ensure the stability of the semi-active suspension [8].

3.2 Parameters design

Select the appropriate parameters can not only let the time-delay be made in the permission range of actuator, but also can obtain the more ideal control effect, so as to reduce the vertical vibration acceleration and improve riding comfort of vehicle. The determination methods of PID parameters include: trial-and-error method, the critical ratio method, the decay curve method, etc. In this paper the k_p, k_i, k_d were set by the trial-and-error method. In the meantime, the critical delay time can satisfy the whole time delay demand of the semi-active suspension by adjusting the parameters according to Equation (12).

Following the above steps and according to the data in Table.1, the PID parameters can be determined as: $k_p = 176$, $k_i = 1260$, $k_d = 0.79$. By computer aided calculation, it can be calculated that the system critical time $\tau = 0.132$s.

4 TEST ANALYSIS

4.1 Simulation parameters

This paper adopted Matlab7.0/Simulink to do simulation, the suspension parameters values of a vehicle are shown in Table.1.

Table 1. Semi-active suspension parameters.

parameters	value
m_b / kg	345
m_w / kg	42
Ks / N·M^{-1}	18000
C_0/N·(ms^{-1})$^{-1}$	1200
Kt/N·M^{-1}	198000

Set the simulation time to 8s, the vehicle speed to 55 km·h^{-1}. This paper has designed two simulation schemes:

Scheme 1: Set the B grade road surface Gaussian white noise as input signal, its roughness coefficient $G_0 = 64 \times 10^{-6}$ m^3.

Scheme 2: Use a 0.06m deep pit square wave as input signal, thereinto:

$$Z_r(t) = \begin{cases} 0(0 \leq t \leq 1; 5 \leq t \leq 8) \\ 0.05(1 < t < 5) \end{cases}$$

4.2 Simulation test analyses

This paper had the simulation tests in the two road surface input schemes in three kind of time-delays $\tau = 0$s, $\tau = 0.132$s, $\tau = 0.16$s, the simulation results were shown in Figue 2 and Figue 3, The quantitative contrast were listed in Table 2.

1) When the system has no time-delay ($\tau = 0$), compared with the passive suspension PID control strategy can make the sprung mass acceleration root-mean-square of active suspension reduce 60% or so, which greatly improved the vehicle ride performance and ride comfort.

2) As the time-delay increases the amplitude range and root mean square value of the sprung mass are gradually increasing.

3) As the critical time-delay of semi-active-suspension $\tau = 0.132$s, the system amplitude increased 1.2 times to 1.5 times, the amplitude of the acceleration was reduced by 0.7 to 1.5 times or so compared with the system which has no time-delay, the semi-active-suspension is still in a critical stability state.

4) Take the critical time-delay $\tau = 0.18$s, the amplitude of sprung mass vertical acceleration increased 2.6 times dramatically. Compared with the passive suspension, the amplitude range and

root mean square value have no obvious change. It shows that the semi-active-suspension has no control effect when the time delay exceeds the critical time delay.

5) The total time-delay value is the key factors which affect the integral control effect. If the time-delay of hardware system can be reduced and take time-delay compensation strategy in control

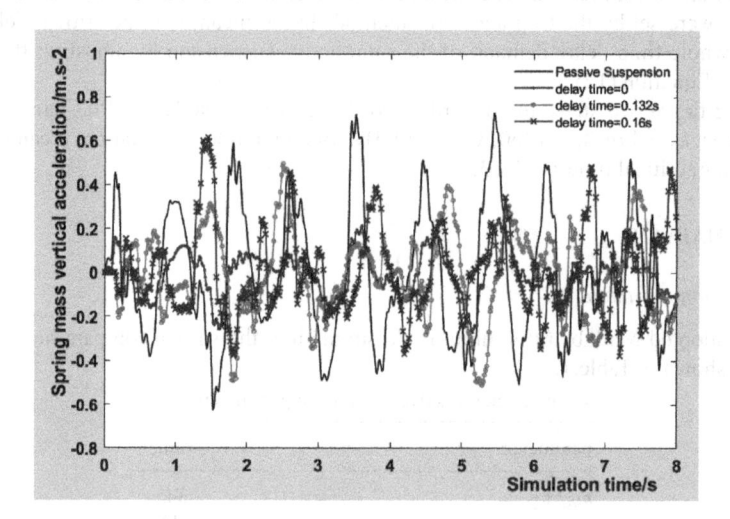

Figure 2. Spring load quality vertical acceleration under input scheme 1.

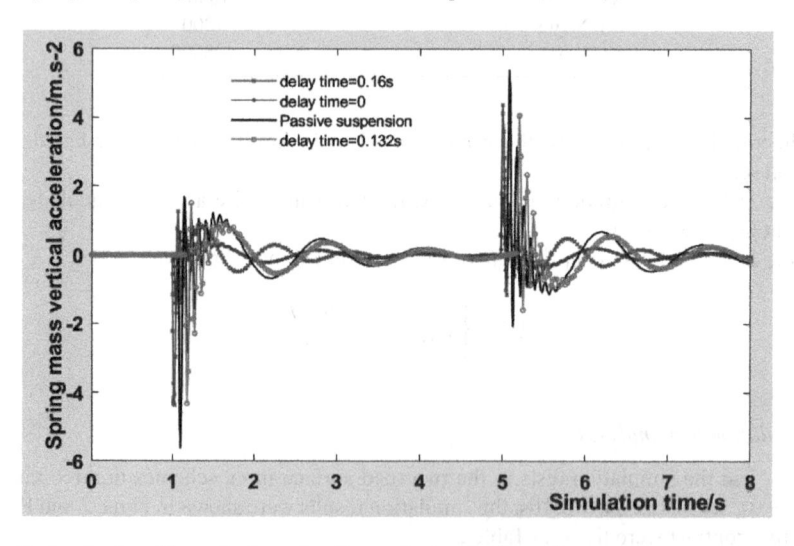

Figure 3. Spring load quality vertical acceleration under input scheme 2.

Table 2. The simulation results of spring load quality vertical acceleration(Units:m·s^{-2}).

τ/s	Amplitude range /m/s^{-2}		Root mean square value	
	B Grade Road Surface	Square wave pit	B Grade Road Surface	Square wave pit
	(−0.22,0.21)	(−1.89,1.78)	0.0913	0.2869
0.132	(−048,049)	(−4.76,3.51)	0.2021	0.4842
0.16	(−041,064)	(−4.13,4.25)	0.4635	1.2317
Passive suspension	(−0.63,068)	(−5.21,5.32)	0.3213	1.2179

software thus the system stability can be improved greatly, and the vehicle ride performance and ride comfort can be improved accordingly.

5 CONCLUSION

1) According to the theory of linear ordinary differential equation which contains time-delay the stable condition of dynamic model containing time-delay is deduced, according to the Routh-Hurwitz stability criterion that the system stable conditions directly related to the parameters K_s, K_t, C_0, m_w, m_b and k_p, k_i, k_d. These parameters should be considered comprehensively in the design of semi-active suspension system.

2) Simulation results show that the increase of time-delay can cause the serious deterioration of the control effect of the semi-active suspension, and which will not be improved by the advanced and complicated control algorithm or by the expensive hardware equipment. So, the time-delay and control strategies of the semi-active suspension should be considered comprehensively.

3) If the PID control algorithm is adopted in the design of semi-active suspension vehicle, firstly, the approaching delay time of the system can be calculated according to the time-delay stability calculation method proposed in (Equation (6)–(12)) in this paper. Then the total delay time of the system must be measured by practical experiments. Finally, the system parameters (K_s, K_t, C_0, m_w, m_b and k_p, k_i, k_d) should be comprehensively designed. In this way, the total time delay of the system can be controlled within the critical time delay and the stability of the system can be ensured, and the vibration comfort of semi-active suspension can be improved.

ACKNOWLEDGEMENTS

This work has been supported by Tianjin natural science foundation (Grant No: 18JCQNJC01500) and Scientific research and develop fund of Tianjin University of Technology and Education (Grant No: KJ15–12).

REFERENCES

[1] Vicente Costanza, Pablo S. Rivadeneira, John A. Gómez Múnera. *An efficient cost reduction procedure for bounded-control LQR problems*. 2018, 37(2) p 1175–1196.

[2] M.M. Moheyeldein, Ali M. Abd-El-Tawwab, K.A. Abd El-gwwad, et al. *An analytical study of the performance indices of air spring suspensions over the passive suspension*. Beni-Suef University Journal of Basic and Applied. 2018

[3] Morato Marcelo Menezes, Normey Rico Julio Elias, Sename Olivier. *Sub-optimal recursively feasible Linear Parameter-Varying predictive algorithm for semi-active suspension control*. 2020, 14(18) p 2764–2775. Morato Marcelo Menezes, Normey Rico Julio Elias, Sename Olivier. *Sub-optimal recursively feasible Linear Parameter-Varying predictive algorithm for semi-active suspension control*. 2020, 14(18) p 2764–2775.

[4] Wang X. *Semi-active adaptive optimal control of vehicle suspension with a magnetorheological damper based on policy iteration*. Journal of intelligent material systems and structures, 2018, 29(2) p255–264.

[5] Sha S, Wang Z, Du H. *Research on performance of vehicle semi-active suspension applied magnetorheological damper based on linear quadratic Gaussian control*. Noise & Vibration Worldwide, Vol 51, Issue 7–9, 2020 p 119–126

[6] Guang Hui Yan, Shuo Zhang. Research on Modeling and Optimization Control of Heavy Truck Cab Active Suspension System, Applied Mechanics and Materials, 2014

[7] Guang Hui Yan, Shao Hua Wang, Zhi Wei Guan, Chen Fu Liu. *PID Control Strategy of Vehicle Active Suspension Based on Considering Time-Delay and Stability*. Advanced Materials Research, 2013

Stability optimization of reticulated shells based on joint stiffness parameters

C.Y. Dun
College of Urban and Rural Construction, Hebei Agricultural University, Baoding, China

B.X. Guan
Chengde College of applied technology, Chengde, China

Y. Guo, J.L. Wang & J.H. Sun
College of Urban and Rural Construction, Hebei Agricultural University, Baoding, China

ABSTRACT: Instability failure is one of the main failure modes of reticulated shell structures. In this paper, the concept and calculation formula of joint stiffness parameters of reticulated shell structure are established based on the mechanical characteristics of the bar system structure before instability. Through the analysis of a K6 single-layer reticulated shells, it is found that the instability point and instability area appear in the nodes and areas with the minimum joint stiffness parameters after considering the initial geometric imperfection of the reticulated shells; strengthening the member sections in the area with the minimum joint stiffness parameters can effectively improve the stability ultimate bearing capacity of the reticulated shells with less steel consumption.

1 INTRODUCTION

Reticulated shell structure has been widely used in gymnasiums, cinemas, airports, railway stations and other large public buildings, many of which have become urban landmarks. Instability failure is the main failure form of single-layer reticulated shell. Through a lot of research [1–4], the nonlinear stability analysis method of reticulated shell has been well solved, and the general software has been applied in engineering. In the aspect of optimal design of reticulated shells, Wang Fawu [5] proposed an optimal design method based on the optimal criterion method considering the conditions of displacement, stress, member stability and overall structural stability. Chen Shiying et al [6–8] studied the structural optimization of reticulated shells through the change of member grouping and mesh density. Liu Wenzheng et al [9] proposed a criterion for determining the stiffness uniformity of single-layer spherical reticulated shells based on the joint configuration. Lu Mingfei et al [10] defined the relative change gradient of the node configuration degree reflecting the static stability characteristics of the structure, and optimized the stability of the cylindrical reticulated shell with the objective of maximizing the minimum value of the change gradient. Yu Jinkun et al. [11] optimized the circumferential grid, radial grid and ratio of height to span for reticulated shells under the general contracting mode.

In this paper, based on the analysis of the load carrying state of the frame structures before instability, the concept of the node stiffness of the reticulated shell structures is proposed and the calculation formula is given. On this basis, the instability nodes and regions of a Kewitt spherical reticulated shell is analyzed, and the stability ultimate bearing capacity of the reticulated shell is optimized by using the node stiffness.

2 JOINT STIFFNESS PARAMETERS

The stability loads of the bar system structures are not only related to the tensile and compressive stiffness EA of the bars, but also have great relationship with the angle between the bar and the tangent plane of the joint. The following two examples are used to illustrate this concept.

Example 1, Two bar truss structure subjected to a vertical load shown as in Figure 1. The initial length of two bars $L_0 = 1000mm$, the initial angle is $\alpha_0 = 20°$, the cross-sectional area of each bar is $A = 50mm^2$, and the elastic modulus of the material is $E = 210GPa$ The nonlinear stability analysis of this structure is carried out. Figure 2 shows the change of stability load with the increase of α value.

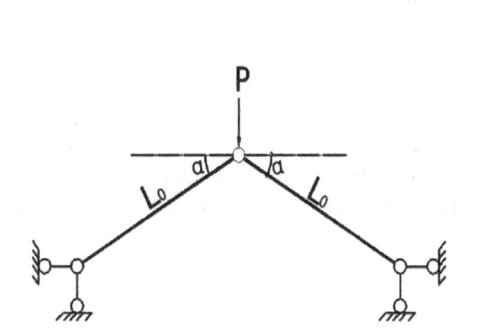

Figure 1. Two bar truss structure.

Figure 2. Load-displacement of truss structure.

Example 2, Figure 3 shows a spatial hexagon frame system, with six side joints of sliding hinge supports. The initial $h = 44.45mm$, $L = 609.6mm$, and Table 1 shows the specific parameters of the frame. With the increase of h, the angle between each member and the vertex tangent plane increases. Figure 4 shows the nonlinear stability analysis results with the change of α.

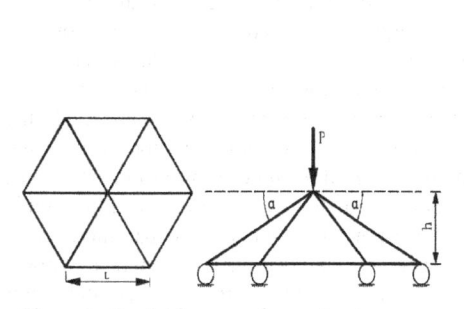

Figure 3. Spatial hexagon frame structure.

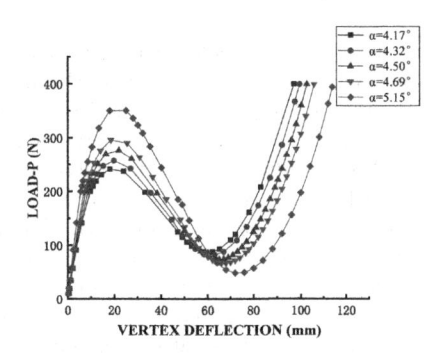

Figure 4. Load-displacement of frame structure.

Table 1. Parameters of hexagon frame structure.

Bar parameters	Quantitative value
Elastic modulus E	3092MPa
Shear modulus G	1096MPa
Cross section A	318.7mm^2
X Moment of inertia	13777.3mm^4
Y Moment of inertia	8449.5mm^4
Z Moment of inertia	8449.5mm^4

It can be clearly seen from Figure 2 and Figure 4 that the ultimate load of stability increases with the increase of angle α, regardless of the two bars truss structure or the spatial hexagon frame structure, under the condition of constant member section. Before the instability of single-layer reticulated shell structures, the membrane forces of reticulated shell make the members mainly bear the axial force while the bending moments and torsion moment are very small. Therefor the tensile and compressive stiffness EA of the connecting members and the angle between each member and the tangent plane of the joint constitute the main part of the joint stiffness. Based on the above analysis, taking the tension and compression stiffness of the members and the angle between the tangent plane of the joint and each member as variables, the joint stiffness parameters are constructed as follows.

Joint stiffness parameters: $$S_n = \sum_{i=1}^{m} \alpha_i EA_i$$

where E - elastic modulus of member material.

A - the cross-sectional area of the members connected to the node.

α - the angle between the member connected with the joint and the tangent plane of the joint.

n - number of members connected to nodes.

It can be clearly seen from the formula that the joint stiffness S_n mainly reflects the resistance of the joint to the loads. The instability of reticulated shell structure generally appears from the node or local area, then gradually develops to nearby area and finally leads to the collapse of the whole structure. Therefore, the stiffness of weak joints and weak areas is the key to control the stable bearing capacity of reticulated shells.

3 THE PREDICTION OF INSTABILITY NODES AND REGIONS OF SPHERICAL RETICULATED SHELLS

As shown in Figure 8, the K6 reticulated shell has a span of 40m, a rise to span ratio of 1/3, 8 radial grids and 6 circumferential grids. The support condition is fixed hinge support. For type cross sections (ϕ 60 × 3, ϕ 70 × 3, ϕ 76 × 4 and ϕ 89 × 4) for members of the structure are designed Due to the symmetry of the structure, the configuration of one sector is shown as in Figure 5. ANSYS software is used to analyze the whole process of stability of the model. In the numerical analysis, the BEAM188 elements are used to consider the geometric nonlinearity and material nonlinearity. The uniform imperfection method is used to impose initial imperfections, and the distribution of initial geometric imperfections is calculated by the lowest buckling mode The maximum value of initial imperfections is 1/300 of span. The joint stiffness parameters of a sector are shown in Table 2, and the circumferential average joint stiffness parameters are shown in Figure 6 The whole process curve of nonlinear stability and the deformation configuration after instability of the structure are shown in Figure 7 and Figure 8 respectively.

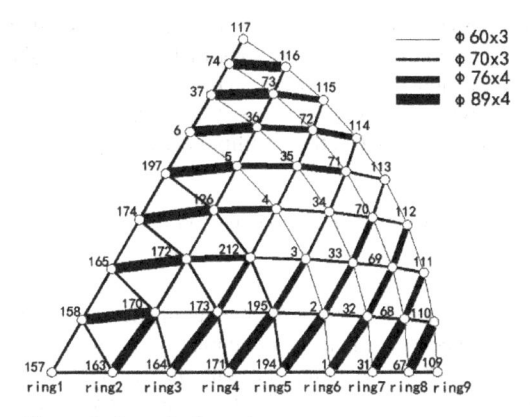

Figure 5. Layout of members.

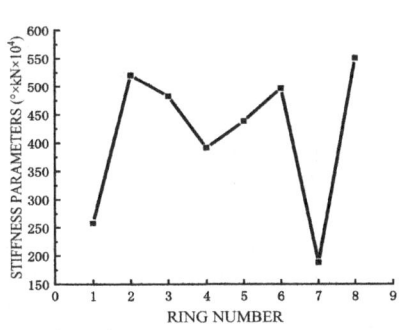

Figure 6. Average joint stiffness parameters of circumferential joints.

Table 2. Joint stiffness parameters.

Joint number	157	163	158	164	170	165	175	171	173
$\sum \alpha_j EA_j (\times 10^4)$	264.2	441.3	559.7	504.8	460.9	504.8	460.9	402.5	386.2
Joint number	172	174	177	194	195	212	196	197	198
$\sum \alpha_j EA_j (\times 10^4)$	386.1	402.6	386.2	428.7	446.1	425.4	446.1	428.7	457.2
Joint number	213	1	2	3	4	5	6	7	8
$\sum \alpha_j EA_j (\times 10^4)$	425.3	607.2	498.9	437.6	437.606	498.9	607.0	498.9	437.5
Joint number	31	32	33	34	35	36	37	38	39
$\sum \alpha_j EA_j (\times 10^4)$	225.6	225.8	148.1	124.2	147.9	225.9	255.6	226.1	148.0
Joint number	40	67	68	69	70	71	72	73	74
$\sum \alpha_j EA_j (\times 10^4)$	124.3	490.3	536.5	578.6	566.2	566.2	578.6	536.5	490.3
Joint number	75	76	77						
$\sum \alpha_j EA_j (\times 10^4)$	536.4	578.4	504.2						

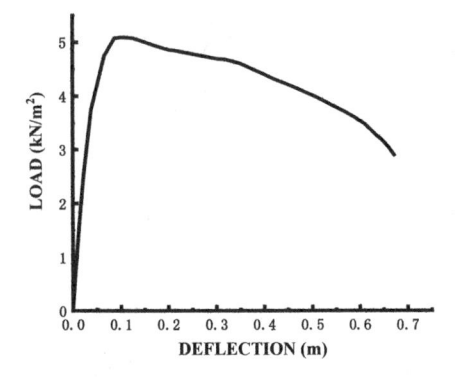

Figure 7. Load-displacement (node 34).

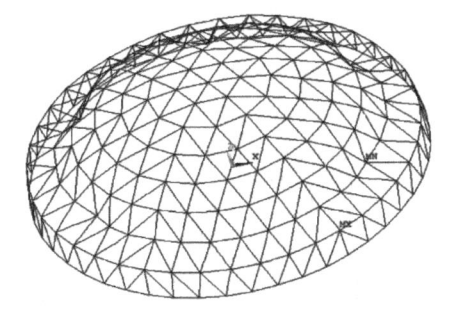

Figure 8. Deformation configuration.

It can be found from Figure 7 and Figure 8 that the instability point of the structure appears on the node 34 shown in Figure 5 and the nodes corresponding to the other five sectors, and the stability limit load is 5.09kN/m². Table 2 shows that the stiffness parameters of node 34 and its corresponding nodes are the smallest in the whole reticulated shell structure. At the same time, Figure 6 shows that compared with other rings, the average stiffness parameters of ring 7 are significantly smaller than those of other rings Considering the initial geometric imperfections of the structure, the buckling points of the reticulated shells occur at the nodes and regions with the minimum stiffness parameters. This is because the nodes and areas with the smallest joint stiffness are the weakest part of the reticulated shell structure, which is the key to control the stability bearing capacity.

4 OPTIMIZATION OF STABILITY BEARING CAPACITY OF SPHERICAL RETICULATED SHELLS

As the ultimate bearing capacity of single-layer reticulated shell structure is generally determined by its stable bearing capacity, how to improve its stable bearing capacity economically and effectively is a very important problem in the design process without changing the grid division and rise to span ratio of reticulated shell. Generally, the stable bearing capacity of structure is improved by increasing the section of members. Because there are many members in reticulated shell structure, to strengthen members in which location and how many members to strength can economically and effectively improve the stability bearing capacity of the structure This is an urgent problem to study. In this paper, taking the above-mentioned Kewitte reticulated shell with rise to span ratio of 1/3 as an example, the stability ultimate bearing capacity of the reticulated shell is optimized based on the joint stiffness parameters. In the research process, in order to measure the optimization effect, the parameter $\Delta P_{CR}/\Delta W$ is introduced, where ΔP_{CR} is the increased stability limit load value and ΔW is the increased steel consumption per unit area.

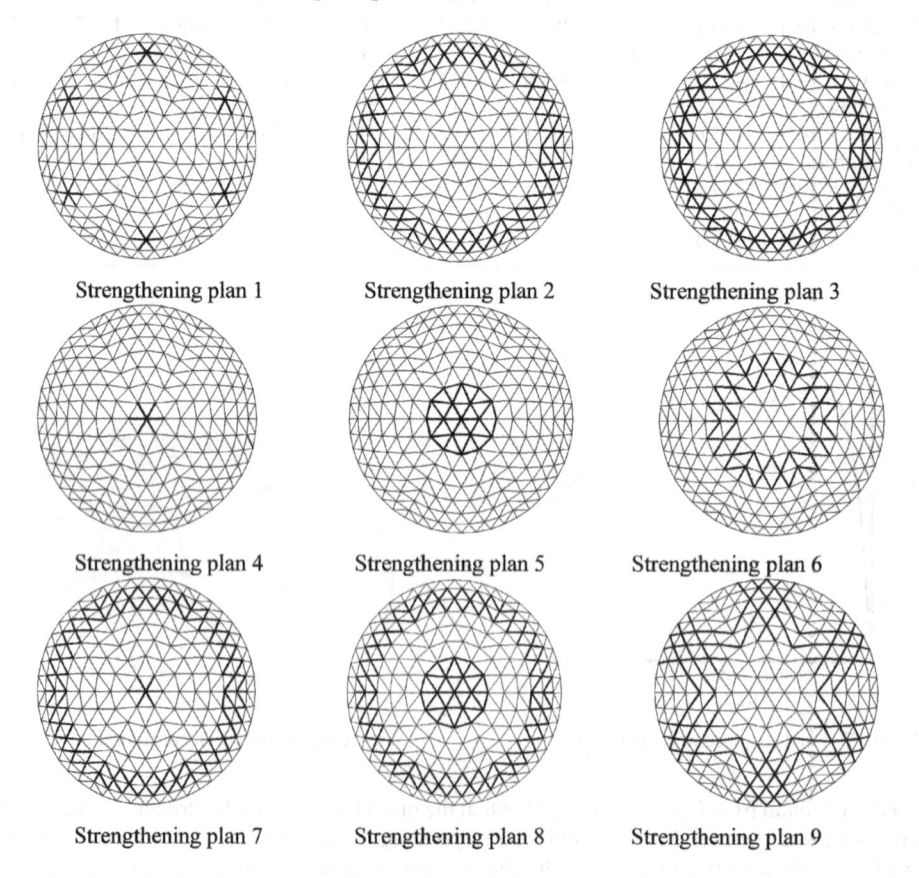

Figure 9. Strengthening plan of structure.

Figure 9 shows the optimization plans of increasing the cross section of some members, in which the thick solid line represents the strengthened members. Plan 1 to plan 3 show the strengthening members located in the seventh ring with the smallest stiffness. Among them, plan 1 is to strengthen only the members connected with six instability nodes; plan 2 is to strengthen all members of the seventh ring except the circumferential members; and plan 3 is to strengthen all members of the seventh ring including the circumferential members. Plan 4–6 is to strengthen the member near the top node which has the second smallest joint stiffness. Plan 7 is the combination of plan 2 and plan 4; and plan 8 is the combination of plan 2 and plan 5. Plan 9 is to strengthen the diagonal members near the six unstable nodes. In the above strengthening plans, the original section of the member is increased to $\Phi 89 \times 4$, and the different analysis results are shown in Table 3.

Table 3. Analysis results of optimization plan.

Strengthening plan	Replacing number of members	Replacing cross section	P_{cr} (kN/m^2)	$\Delta P_{cr}/\Delta W$(kN/kg)	Steel consumption (kg/m^2)	Instability nodes
Initial structure			5.09		11.96	34
Plan 1	36	$\Phi 89 \times 4$	5.18	0.234	12.30	34
Plan 2	108	$\Phi 89 \times 4$	5.60	0.59	12.81	34
Plan 3	144	$\Phi 89 \times 4$	5.60	0.42	13.15	34
Plan 4	6	$\Phi 89 \times 4$	5.09	0	12.01	34
Plan 5	30	$\Phi 89 \times 4$	5.09	0	12.22	34
Plan 6	114	$\Phi 89 \times 4$	5.09	0	12.43	34
Plan 7	114	$\Phi 89 \times 4$	5.60	0.55	12.86	34
Plan 8	138	$\Phi 89 \times 4$	5.62	0.47	13.07	34
Plan 9	112	$\Phi 89 \times 4$	5.71	0.54	13.06	34

Table 4. Further optimization analysis results of plan 2.

Strengthening plan	Replacing number of members	Replacing section	P_{cr} (kN/m^2)	$\Delta P_{cr}/\Delta W$(kN/kg)	Steel consumption (kg/m^2)	Instability Instability
Plan 2.1	144	$\Phi 102 \times 4$	5.86	0.314	14.41	34
Plan 2.2	144	$\Phi 114 \times 4$	6.29	0.414	14.86	157
Plan 2.3	144	$\Phi 121 \times 4$	6.47	0.440	15.12	157

From the optimization analysis results in Table 3, it can be seen that plan 1, which only strengthens the members connected with the instability node, the stability ultimate bearing capacity increases from 5.09kN/m^2 to 5.18kN/m^2, and the increase range is very small. This is because there are too few strengthened members. The stability ultimate bearing capacity of plan 2 is increased to 5.60kN/m^2, and is increased by 10.02% when the steel consumption is only increased by 0.85kg/m^2. Its $\Delta P_{cr}/\Delta W$ is the highest among all plans and is the better plan. On the basis of plan 2, plan 3 strengthens the circumferential members of ring 7. Table 3 indicates that the steel consumption increases, but the stability ultimate bearing capacity does not increase. This is because the circumferential members mainly bear tensile force. The analysis results in Table 3 show that plan 4–6, which strengthens some members near the top node of the reticulated shell, does not improve the stability ultimate bearing capacity of the reticulated shell, and is the worst strengthening plan. Plan 7 and plan 9 all contain members to strengthen the minimum area of average joint stiffness, so the ultimate bearing capacity of the reticulated shell is improved in some degrees, but they are not as good as plan 2.

In order to study the influence of different member section size on the strengthening effect, plan 2 is further investigated. At this time, the number and position of strengthening members remain unchanged, but the member section size is increased. The analysis results are shown in Table 4. It can be seen from the table that with the increase of the cross section, the ultimate bearing capacity of stability increases continuously, and $\Delta P_{cr}/\Delta W$ remains at a high level although it decreases. Figure 10 shows the load displacement curve of plan 2 after optimization.

The analysis results of the above different optimization plans show that strengthening the members in the minimum joint stiffness area can quickly improve the stability ultimate bearing capacity of the reticulated shell, and save the steel consumption. The joint stiffness parameters proposed in this paper can be used to guide the optimization of the stability bearing capacity of spherical reticulated shells. It can be found from the calculation formula that the concept of joint stiffness parameter is clear, the calculation is simple, and it is convenient for engineering application.

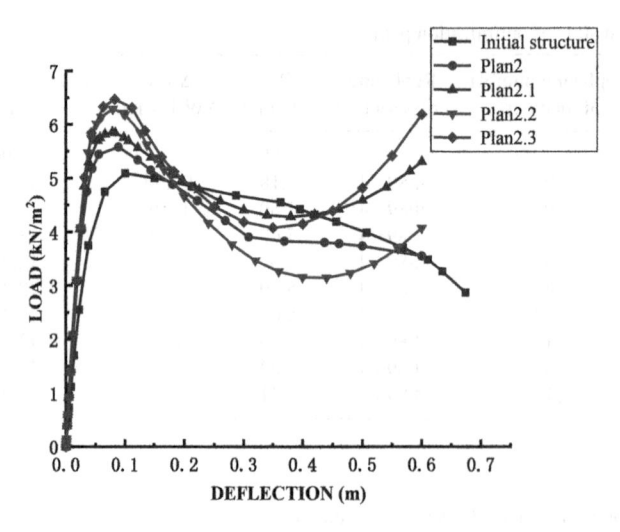

Figure 10. Load-displacement of reticulated shell after optimization.

5 CONCLUSION

In this paper, the concept of joint stiffness parameters of reticulated shell is proposed and its calculation formula is constructed. The instability joint of K6 single-layer spherical reticulated shell is predicted by using the joint stiffness parameters, and its stability bearing capacity optimization is analyzed.

(1) Considering the initial imperfections, the instability point of the spherical reticulated shell structure occurs in the nodes and regions with the smallest node stiffness, and the node stiffness constructed can be used to predict the instability nodes and regions of the spherical reticulated shell.
(2) Strengthening the member section in the areas of minimum joint stiffness can effectively improve the stability bearing capacity of the reticulated shell, and the steel consumption is more economical.
(3) The joint stiffness parameters and the index $\Delta P_{cr}/\Delta W$ are simple and easy calculated. This makes the theory of this paper can be easily applied to the optimal design of reticulated shells.

REFERENCES

[1] Shen S Z, Chen X. Stability of reticulated shells [M]. Beijing: Science Press, 1999.
[2] Fan F, Cao Z G, Ma H H, Yan J C. Elastoplastic stability of reticulated shells [M]. Beijing: Science Press

[3] Shiro K, Tetsuo Y. Evaluation of Elasto-plastic Buckling Strength of Two-way Grid Shells using Continuum Analogy [J]. International Journal of Space Structures, 2009, 17(4):249–261.
[4] He Y J Zhou X H, Liu D Research on stability of single-layer inverted catenary cylindrical reticulated shells[J], Thin-Walled Structures 82 (2014) 233–244
[5] Wang F W, Tang G. Sectional optimum design of single-layer lattice shells considering structural stability[J] Spatial Structures,2006(03):31–34+6
[6] Chen S Y LU X Y Zhu H Y. The effect of bar group and mesh density on shell optimization design for cross section [J]. Mechanics in Engineering, 2011, 33(02):71–74.
[7] Chen S Y LU X Y Wang H L, Zhao X Y. Influence of section optimization on stress, displacement and global stability of latticed shells [J] Spatial Structures, 2014, 20(01):45–52.
[8] Chen S Y LI W G LU X Y. Section optimization design of reticulated dome to ensure constant stability[J]. Journal of CHNA university of petroleum, 2018, 42(03):147–153.
[9] Liu W Z Luo Y F. Uniform stiffness criterion of single-layer spherical shells based on nodal well-formedness [J] Journal of Building Structure, 2015, 36(11):38–45
[10] Lu M F Ye J H. Stability optimization design for single-layer cylindrical domes based on joint well-formedness [J]. Journal of Vibration and Shock, 2018, 37(09):74–79.
[11] Yu J K Xiao J C Wu X Y, Luo J, Ma K J Multi-objective optimization of latticed shell under engineering procurement construction[J], Journal of Guizhou University (Natural Science Edition) 2017,34(01) :99–102+135

Research on classification method of surface defects of hot rolled strip

K. Hu & G.Q. Wu
School of Electrical Engineering, Nantong University, Nantong, China

Z.H. Hu
School of Information Science and Technology, Nantong University, Nantong, China

Z.M. Wang
School of Electrical Engineering, Nantong University, Nantong, China

ABSTRACT: The classification of metal surface defects is an important content of defect detection. Aiming at the surface defects of hot-rolled strip steel, an improved VGG16 network structure classification method is proposed. Based on the VGG16 network, the attention mechanism CBAM is introduced to enhance the defect feature learning ability, and data enhancement processing is performed on the defect image to solve the scarcity of data. After experimental verification, on the data set NEU-CLS, the classification accuracy of 6 types of defects reaches 98.15%.

1 INTRODUCTION

China is a large country in the production of the steel industry. Strip steel and sheet steel are widely used in the automotive industry, defense industry, aerospace and other fields due to their good surface quality and mechanical properties. However, with the continuous improvement of product quality, how Detecting and controlling surface defects of steel plates and improving the surface quality of steel plates have become the main tasks of quality improvement of metallurgical enterprises.

In the production process of strip steel, due to many factors such as raw materials, processing technology, equipment, environment aging, etc., many defects such as cracks, scars, holes, and stains will appear on the surface. These defects, affect the wear resistance and corrosion resistance of the steel plate. And electromagnetic characteristics. Traditional defect detection mainly relies on manual visual inspection, which is labor intensive and easy to miss and misdetect. Therefore, it is inevitable that machine vision inspection will replace manual inspection, and it is now also one of the research hotspots. Pernkopf et al. Proposed a method for detecting defects on the surface of steel blocks with three-digit features, using Bayesian network classifiers to extract defect features for classification [1]. Aiming at the problem of classification of surface defects of hot-rolled steel strip, the literature [2] uses multi-scale geometric analysis methods to extract features of images, reduce dimensionality through graph embedding algorithms, and finally use support vector machine (SVM) to classify defects, and finally achieve good classification results. Yi et al. Proposed a method of symmetrical surround saliency mapping and deep convolutional neural network, which achieved good results in the classification of 7 types of strip defects [3]. The literature [4] proposes a residual neural network (ResNet50) method to classify three types of metal surface defects with an accuracy of 96.91%.

Based on the VGG16 network, this paper carries out classification experiments on the surface defect data of hot-rolled strip steel, introduces the attention mechanism CBAM [5], enhances the weight coefficient of defects, and makes it better to extract defect features. The accuracy rate of the

NEU-CLS steel surface defect dataset released by Northeastern University on the final improved VGG16 network reached 98.15%.

2 HOT ROLLED STRIP STEEL SURFACE DEFECT DATA SET

Hot-rolled steel strip surface defect data set NEU-CLS is a defect data set publicly released by Northeastern University [6]. The image acquisition is obtained by shooting the surface of the steel strip with four area scanning CCD cameras. The initial gray image size is 1024×1024 pixels. In order to save calculation time, the original image is down-sampled and the sampled image is 200×200 pixels. The data set includes 6 typical defect categories, namely roll-in scale (RS), patches (Pa), crazing (Cr), pitted surface (PS), inclusion (In) and scratches (Sc). The defect diagram is shown in the Figure 1. Each type of defect contains 300 images for a total of 1800 images.

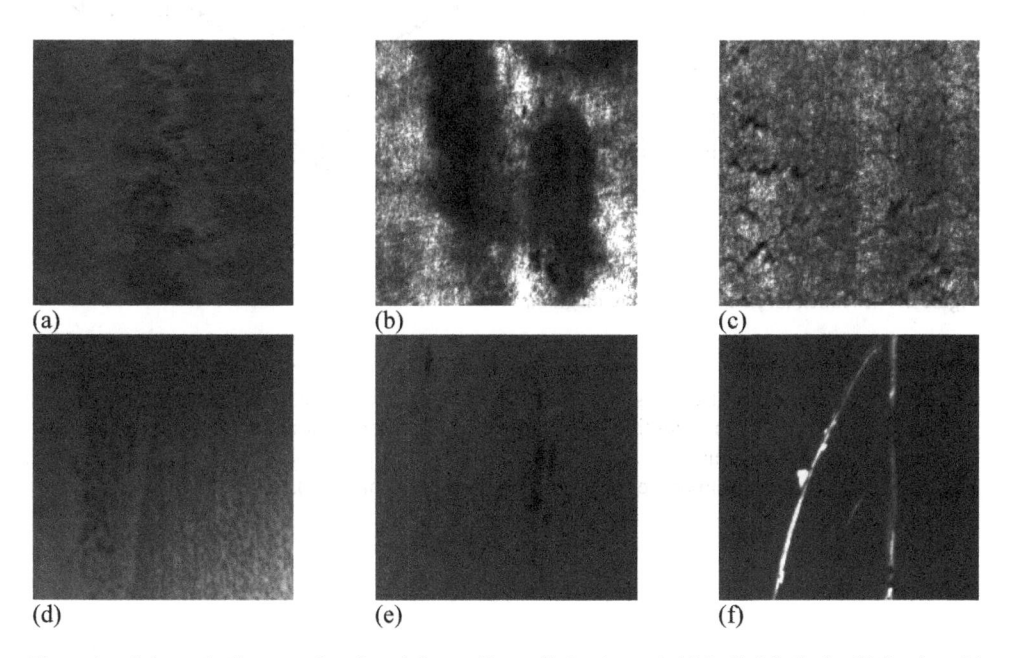

Figure 1. Schematic diagram of surface defects of hot rolled strip steel: (a) Rolled-in Scale, (b) Patches, (c) Crazing, (d) Pitted Surface, (e) Inclusion and (f) Scratches.

3 DEFECT CLASSIFICATION METHOD

3.1 *Attention mechanism*

For the convolutional neural network model, there are three main factors that affect its performance, including: depth, width, and cardinality; the literature proposes a network performance influencing factor that affects feature attention, namely the attention mechanism. The attention mechanism includes two aspects: one is to obtain channels with the obvious characteristics by judging the image characteristics of different channels, and the other is to obtain areas with obvious characteristics by judging the advantages and disadvantages of image features in different regions. By introducing the attention mechanism, the feature learning ability of the network can be strengthened, thereby improving the accuracy of network learning. The structure of the attention mechanism is shown in the Figure 2.

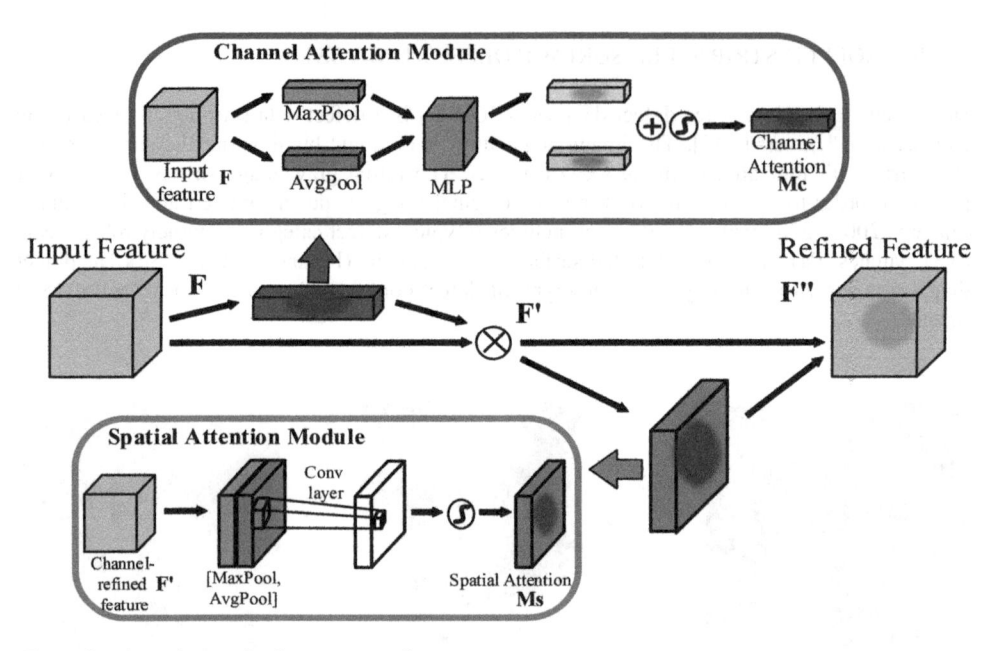

Figure 2. Attention mechanism structure diagram.

The feature map in the figure is $F \in R^{C \times H \times W}$, F' is the feature map of the execution channel attention, $M_c \in R^{C \times 1 \times 1}$ is the attention map of the execution channel attention, F'' is the feature map of executive spatial attention, $M_s \in R^{1 \times H \times W}$ is the attention map of the execution channel attention, and the attention mechanism process formula is shown in formula (1).

$$F' = M_c(F) \otimes F$$
$$F'' = M_s(F') \otimes F' \qquad (1)$$

In the channel attention process, first use the average pooling method to compress the spatial dimension of the feature map to obtain F_{avg}^c, use the maximum pooling method to gather the important features of the feature map to obtain F_{max}^c, and then pass F_{avg}^c and F_{max}^c to the multi-layer perceptron (MLP) to obtain Attention map $M_c \in R^{C \times 1 \times 1}$, the output feature vector is obtained by summing the elements. The calculation formula is as shown in formula (2), σ represents the sigmoid function, and $W_0 \in R^{C/r \times C}$ and $W_1 \in R^{C \times C/r}$ are MLP weights.

$$M_c(F) = \sigma(MLP(AvgPool(F))) + MLP(MaxPool(F)))$$
$$= \sigma(W_1(W_0(F_{avg}^c)) + W_1(W_0(F_{max}^c))) \qquad (2)$$

In the process of spatial attention, the average pooling method and the maximum pooling method are also used to obtain F_{avg}^s and F_{max}^s and connect them, and then the spatial attention map $M_s \in R^{H \times W}$ is obtained through a layer of convoluted layer for the feature area. The calculation formula is shown in formula (3), where $f^{7\times 7}$ is a 7×7 convolution operation.

$$\begin{aligned}M_s(F) &= \sigma(f^{7\times 7}([AvgPool(F); MaxPool(F)])) \\ &= \sigma(f^{7\times 7}([F_{avg}^s; F_{max}^s]))\end{aligned} \quad (3)$$

3.2 Improved VGG16 network structure

VGG is a convolutional neural network with simple structure, strong scalability and good mobility for classification. Its unique 3×3 convolution kernel structure enhances the nonlinear ability of the network. In this paper, VGG is selected as the hot rolled strip steel The defect classification network model, and the introduction of an attention mechanism to strengthen the learning of defect features, thereby improving the learning ability of the network, the improved VGG network structure is shown in the Figure 3.

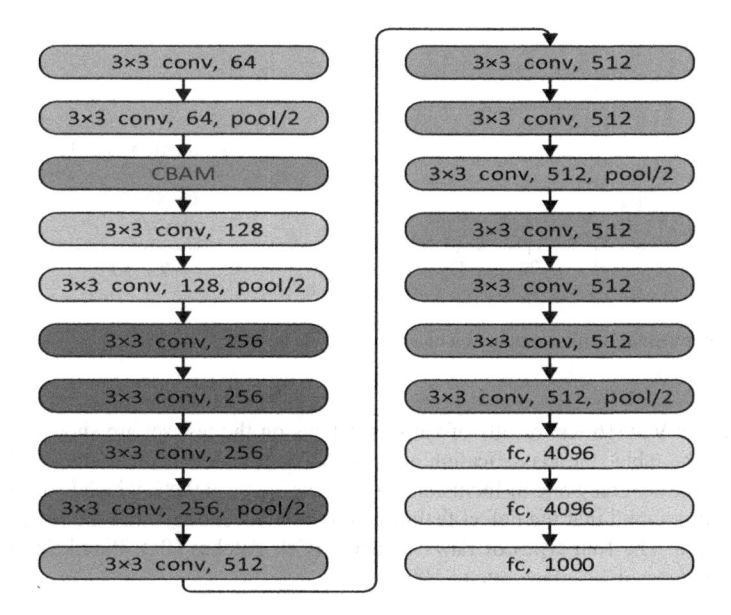

Figure 3. Improved VGG16 network structure diagram.

In view of the diversity of surface defects on hot-rolled steel strip and the limited number of images, the introduction of attention mechanism can learn defect features better and more accurately, thereby improving the accuracy of defect recognition.

4 ANALYSIS OF CLASSIFICATION EXPERIMENT RESULTS

The deep neural network in this experiment is built on the Tensorflow2.2 deep learning framework based on the Ubuntu18.04.4LTS system and implemented by Python programming. The computer processor is Intel○RCoreTMi7–7700K CPU@4.20GHz X 8, and the GPU is GeForce GTX1060, Cuda version is 10.1. In this experiment, the data set is divided into training set and testing set according to 7:3. The number of iterations of the training set is 150. The batch size of the sample training is 32. The learning rate uses the Adam optimizer. The initial learning rate is 0.00001; to prevent over-simulation to solve the problem, the dropout ratio is 0.2. Due to the limited amount of data, data enhancement techniques such as random image rotation, image scaling, and image translation are used to enhance the training data set to prevent over-fitting in learning.

In order to further illustrate the superiority of the improved network model, and make parallel comparisons with AlexNet, VGG16Net, and ResNet50, the experimental results are shown in the figure. From the marks in the figure, we can see that the improved VGG16Net has faster convergence speed and higher accuracy, with an accuracy rate of 98.15% higher than other basic network models, and an improvement of 1% compared to VGG16Net.

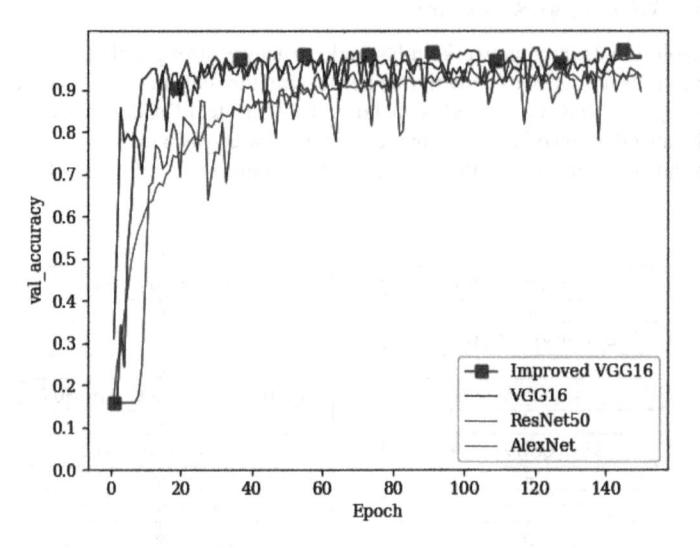

Figure 4. Experimental results comparison chart.

The improved VGG16 test results of various defects on the test set are shown in the Table 1. According to the table, the classification accuracy of various defect features is relatively high. Among them, the accuracy rate of Inclusion (In) defects is low at 93%, which is observed through the data set. It is found that the defect feature is not obvious enough, and the model is relatively difficult to learn. The four types of flaws: Crazing (Cr), Patches (Pa), Pitted Surface (PS), and Scratches (Sc), have an accuracy rate of 100%.

Table 1. Various defect test results.

	Precision	Recall	F1-score	Support
Cr	1.00	0.97	0.98	92
In	0.93	1.00	0.96	89
Pa	1.00	1.00	1.00	85
Ps	1.00	0.97	0.98	101
Rs	0.97	1.00	0.98	87
Sc	1.00	0.95	0.98	86

As shown in the Figure 5, it is a visual heat map of the confusion matrix of the test set. From the figure, you can see the test situation of various defects more intuitively. The abscissa is the test result, and the ordinate is the real result. The graph shows that various test results are extremely consistent with the real results. The data is concentrated on the diagonal, and the color depth indicates the degree of concentration.

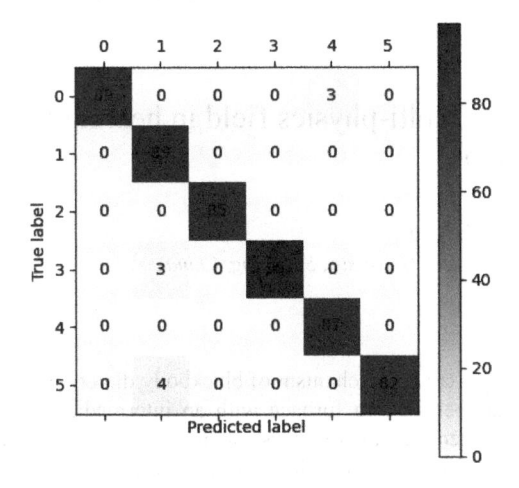

Figure 5. Visualization heat map of test set confusion matrix.

5 CONCLUSION

Aiming at the classification of surface defects of hot-rolled strip steel, an improved VGG16 defect classification method is proposed. The attention mechanism CBAM is introduced into the basic VGG16 network model, which improves the learning performance of diverse defect features and improves the classification effect of the model. The experimental results show that the improved VGG16 network model has certain practical application significance for the classification of surface defects of hot-rolled strip steel.

ACKNOWLEDGMENTS

This work is supported by the Natural Science Foundation of China under No 61273151.

REFERENCES

[1] Pernkopf F. Detection of surface defects on raw s-teel blocks using Bayesian network classifiers[J]. Pattern Analysis & Applications, 2004, 7(3):333–342.
[2] Xu K, Ai Y H, Wu X Y. Application of multi-scale feature extraction to surface defect classification of hot-rolled steels[J]. Journal of Mineral Metallurgy and Materials (English Edition), 2013, 020(001):37–41.
[3] Yi L, Li G, Jiang M. An End-to-End Steel Strip Surface Defects Recognition System Based on Convolutional Neural Networks[J]. steel research international, 2016.
[4] Konovalenko I, Maruschak P, Brezinová J, Viňáš J, Brezina J. Steel Surface Defect Classification Using Deep Residual Neural Network. Metals. 2020; 10(6):846.
[5] Woo S, Park J, Lee J Y, et al. CBAM: Convolutional Bl-ock Attention Module[J]. 2018.
[6] Song K, Yan Y. A noise robust method based on completed local binary patterns for hot-rolled steel strip surface defects[J]. Applied Surface Science, 2013, 285(Pt.B):858–864.

Numerical study on multi-physics field in heating furnace with intermediate radiator

T.C. Jiang, W.J. Zhang & S. Liu
School of Metallurgy, Northeastern University, Shenyang, China

ABSTRACT: The energy-saving mechanism of blackbody directional radiation technology was studied. Based on a chamber heating furnace with an intermediate radiator in a steel plant a physical model with intermediate radiators was established. In this study, a transient 3-D mathematical combustion model has been developed to simulate the temperature distribution in the furnace. The research results show that the addition of the directional radiator changes the flow in the combustion chamber of the furnace body and the angle coefficient of the furnace wall to the slab solid radiation, and enhances the radiant heat transfer intensity of the slab surface and the flue gas combustion temperature from the angle of gas radiation and solid radiation. The flue gas combustion temperature is reduced by about 21 K, the CO mass fraction reduced about 85%.

1 INTRODUCTION

Heating furnace is an important part of the steel smelting process. There are currently about 120,000 heating furnaces in China, with a total energy consumption of 2.5tce. The average energy efficiency is about 18.6% lower than the international advanced level [1–3]. Based on the current situation of industrial heating furnaces, researchers have developed regenerative combustion technology, high and low temperature flue gas recovery technology and other methods to improve the heating furnace [4–6], but the technology above cannot investigate the mechanism of heating enhancement. In industrial production, it is always necessary to transform the structure of the heating furnace that has been used for several years. Therefore, the structural improvement of the furnace body based on the heating mechanism has become an important research direction.

2 PHYSICAL MODEL AND BASIC HYPOTHESIS

2.1 *Physical model*

The process of heating ingot is a complex process. The structure is simplified, and the position of the burner, the size of the furnace body and other indispensable parts in the combustion process are retained, aiming to directly seek the influence of the intermediate radiator on the heating effect. Based on the structure of the traditional chamber furnace, a calculation model of the chamber furnace with intermediate radiators is established shown as Figure 1.

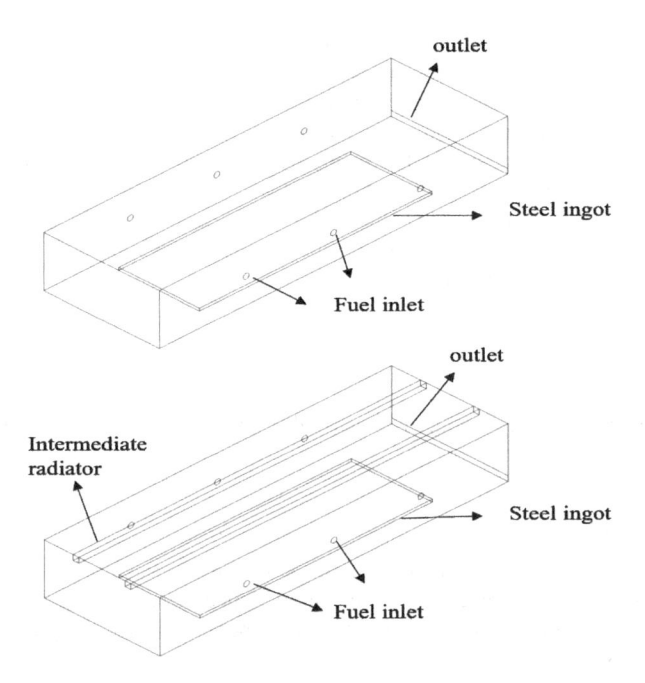

Figure 1. The model of the chamber furnace.

2.2 Boundary conditions and basic assumptions

The size of the chamber furnace is 6m*1m*2m, the size of the intermediate radiators is 4m*0.1m*0.1m, hexahedral mesh is used to discretized furnace model, residual value is limited under $1*10^{-5}$. The physical parameters and boundary conditions are shown in Table 1.

Table 1. Furnace dimensions and operating parameters.

Furnace parameters	Value
Furnace structure	—
Length	6m
Width	2m
Height	1m
Number of burners	6
Burners diameter	0.1
Burn condition	—
Gas flow velocity	3.5m/s
Excess air coefficient	1.15
Gas flow temperature	350K
Gas flow components	—
Methane	0.85
Ethane	0.09
Propane	0.06

Taking into account the economy of computing resources and the accuracy of calculation results, the following assumptions are made: (1) Turbulent mixing controls the chemical reaction speed, ignoring the influence of time scale, thus avoiding the calculation of Arrhenius chemical kinetics. (2) The methane-air one-step reaction mechanism was used to model the combustion, and add ethane and propane to the reactants, ignored the CO production in the intermediate reaction, and used the eddy model to consider the turbulence-chemical interaction. (3) The wall reflection is assumed to be diffuse reflection. (4) The combustion gas and the inner wall of the furnace are assumed to be ash. (5) Ignore the oxidation process on the surface of the oxidized steel billet. (6) The temperature changes in the furnace due to the lack of sealing and the temperature difference between the inside and outside of the furnace are not considered. (7) Since the mass of the gas itself is small, the gas gravity is ignored.

3 MATHEMATICAL MODEL

In the study, the mathematical model is consisted of three parts, fluid flow model, combustion model, and radiation model.

3.1 Fluid flow model

The flow in the heating furnace involves complex situations. In this paper, the Reynolds Average Method (RANS) is used to solve the N-S equation. Since the continuity equation, momentum equation, energy equation, etc. have similarities in numerical formats, they are expressed by the following formulas (1)–(4):

$$\frac{\partial}{\partial t}(\rho\phi) + \frac{\partial}{\partial x_i}(\rho u_i \phi) = \frac{\partial}{\partial x_i}\left(\Gamma_\phi \frac{\partial \phi}{\partial x_i}\right) + S_\phi \tag{1}$$

$$\frac{\partial}{\partial t}(\rho k) + \frac{\partial}{\partial x_i}(\rho k u_i) = \frac{\partial}{\partial x_j}\left[\left(\mu + \frac{\mu_t}{\sigma_k}\right)\frac{\partial k}{\partial x_j}\right] + G_k + G_b - \rho\varepsilon - Y_M + S_k \tag{2}$$

$$\frac{\partial}{\partial t}(\rho\varepsilon) + \frac{\partial}{\partial x_i}(\rho\varepsilon u_i) = \frac{\partial}{\partial x_j}\left[\left(\mu + \frac{\mu_t}{\sigma_\varepsilon}\right)\frac{\partial k}{\partial x_j}\right] + C_{1\varepsilon}\frac{\varepsilon}{k}(G_k + C_{3\varepsilon}G_b) - C_{2\varepsilon}\rho\frac{\varepsilon^2}{k} + S_\varepsilon \tag{3}$$

$$\mu_t = \rho C_\mu \frac{k^2}{\varepsilon} \tag{4}$$

In this formula, ρ was the density of incompressible fluid, the above formula is continuity equation when $\phi = 1$; u_i is the components in three dimensions x, y, z; Γ_ϕ is generalized diffusion coefficient; G_k is the turbulent kinetic energy caused by velocity gradient; G_b is turbulent kinetic energy caused by buoyancy; Y_M is the influence of fluid wave expansion on dissipation rate; $C_{1\varepsilon} = 1.44$, $C_{2\varepsilon} = 1.92$, C_μ, $\sigma_k = 1.0$, $\sigma_\varepsilon = 1.3$; S_ϕ is generalized source term; S_k, S_ε is source term for turbulent kinetic energy and dissipation rate considering other influencing factors.

3.2 Combustion model

The chemical reaction part of the combustion model is determined by the component transport model. The component transport model is mainly based on the convection, diffusion and reaction of each component to determine the conservation equation. The convection-diffusion conservation equation is expressed as follows:

$$\frac{\partial}{\partial t}(\rho Y_i) + \nabla \cdot (\rho \vec{v} Y_i) = -\nabla \cdot \vec{J}_i + R_i + S_i \tag{5}$$

In the equation, Y_i is mass fraction of reaction species, R_i is generation rate of production, S_i is generation rate of production in user defined process. The chemical reaction product model based on turbulent flow is as follows [5]:

$$R_{i,r} = v'_{i,r} M_{w,i} A \rho \frac{\varepsilon}{k} \min_{\Re} \left(\frac{Y_\Re}{v'_{\Re,r} M_{w,\Re}} \right) \tag{6}$$

$$R_{i,r} = v'_{i,r} M_{w,i} A B \rho \frac{\varepsilon}{k} \frac{\sum_P Y_P}{\sum_j^N v''_{i,r} M_{w,j}} \tag{7}$$

$Y_\Re$ is mass fraction of reaction species, Y_p is mass fraction of production species, $B = 5.0$ is constant.

3.3 Radiation model

Considering the calculation economy, the P1 radiation model is used for radiation calculation. The basic heat transfer equation and the expression of radiant heat flow are as follows:

$$\frac{dI(\vec{r}, \vec{s})}{ds} + (a + \sigma_s) I(\vec{r}, \vec{s}) = an^2 \frac{\sigma_s}{\pi} \int_0^{4\pi} I(\vec{r}, \vec{s}') \Phi(\vec{s} \cdot \vec{s}') d\Omega' \tag{8}$$

$$q_r = -\frac{1}{3(\alpha + \sigma_s) - C\sigma_s} \nabla G \tag{9}$$

$\vec{r}$ is position vector, $\vec{s}$ is direction vector, $\vec{s}'$ is scattering direction vector, s is ray travel length, a is absorption coefficient, n is Refractive index, σ_s is Scattering coefficient, σ is Stefan-Boltzmann constant ($5.669 \times 10^{-8} W/m^2 \cdot K^4$), I is Radiation intensity, Φ is phase angle function, Ω represents the solid angle of space. The gas radiation is calculated using weighted sum of gray gases model (WSGGM), and the total emissivity over distance is:

$$\varepsilon = \sum_{i=0}^{I} a_{\varepsilon,i}(T) \left(1 - e^{-\kappa_i ps}\right) \tag{10}$$

$a_{\varepsilon,i}$ is absorption coefficient of specific species, κ_i is absorption coefficient of graybody, p is absorbent gas partial pressure.

4 CALCULATION RESULTS

The heating furnace mainly involves flow, heat transfer, combustion and other field distributions. The addition of the intermediate radiator changes the multi-physical field distribution, which will be explained below.

4.1 Flow field distribution

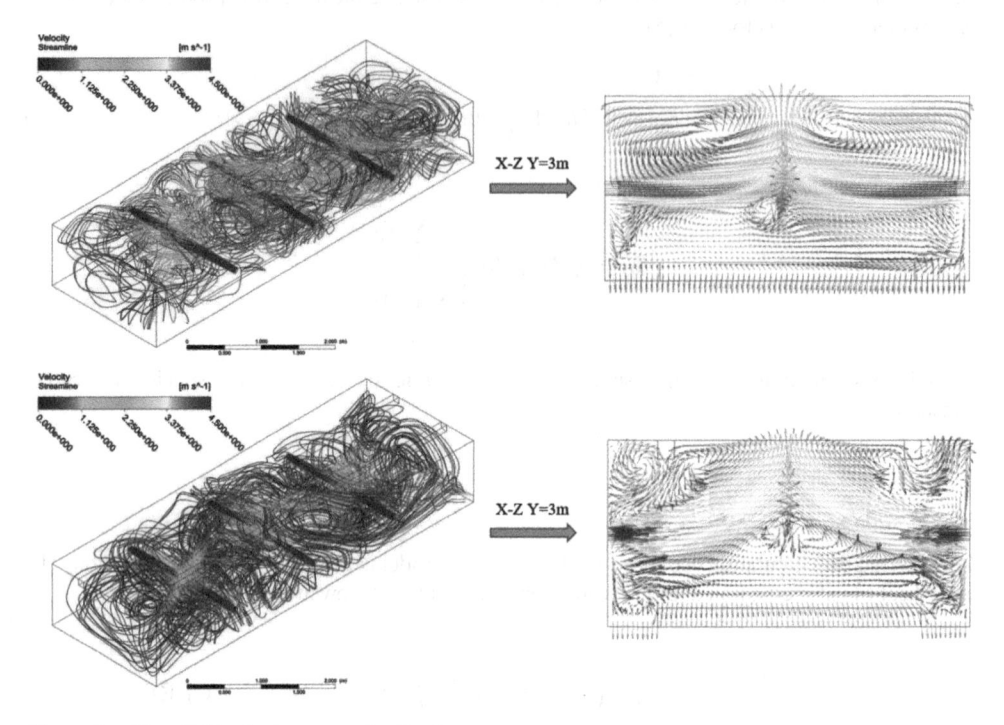

Figure 2. Flow field velocity vector distribution.

Figure 2 shows the flue gas flow in the combustion chamber after adding an intermediate radiator. It can be clearly seen that the intermediate radiator changes the gas flow distribution in the combustion chamber. Generally, high-speed gas enters the combustion chamber and mixes in the middle part of the combustion chamber before burning, but the flow velocity is slower at the corners of the combustion chamber and there is less disturbance in the furnace. After adding the intermediate radiator, the flow boundary layer is destroyed to form more vortices, which promotes the mixed combustion of the fuel [2].

4.2 Temperature and burning field distribution

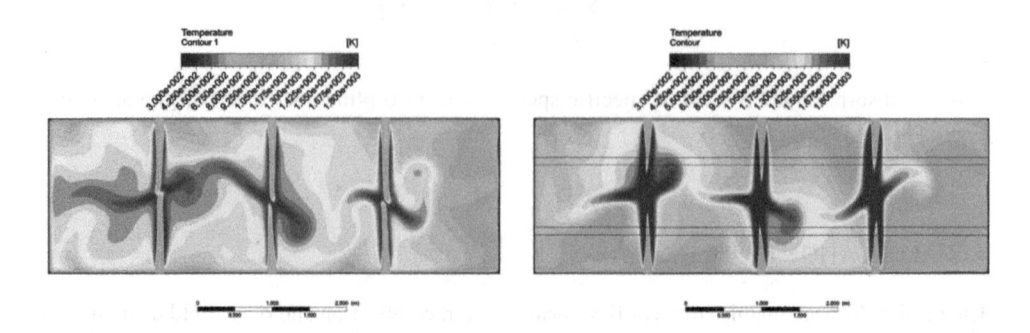

Figure 3. Temperature distribution.

Figure 3 shows the combustion state after the fuel enters the combustion chamber. In a conventional chamber heating furnace, the maximum combustion temperature is 1985K. After adding the intermediate radiator, the maximum flame temperature in the furnace is reduced by 21K, but the flame distribution area is significantly increased.

4.3 Species fraction distribution

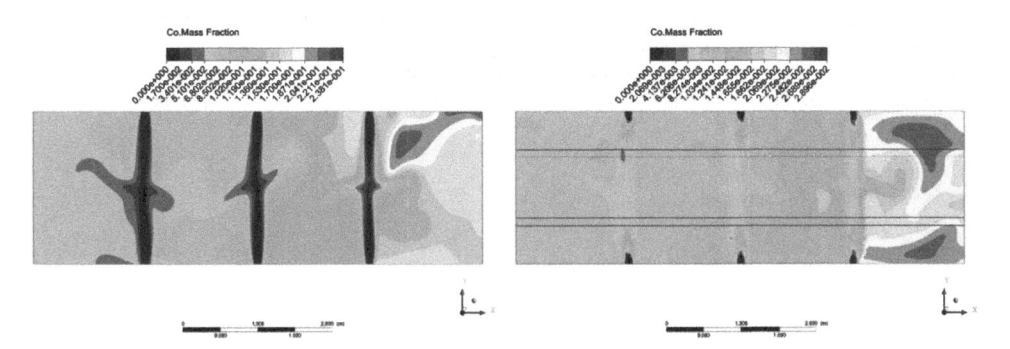

Figure 4. CO fraction distribution.

Figure 4 shows the CO product mass fraction calculated based on the methane-air two-step combustion reaction mechanism under the two working conditions. It is obvious that due to the addition of the intermediate radiator, the gas mixture is more fully mixed and the CO production is reduced. Thereby promoting the process of combustion reaction.

5 CONCLUSION

In this paper, the physical state of the chamber heating furnace with the added intermediate radiator is numerically simulated. Mathematical models are established from the perspectives of flow, combustion and combustion products to calculate, which provides a direction for the structural improvement of the chamber heating furnace.

In terms of flow distribution, the intermediate radiator has a positive effect on the flow distribution in the combustion chamber, promotes gas mixing, and forms more vortexes to make combustion more complete.

In terms of combustion, due to the positive influence of the intermediate radiator on the flow, the distribution volume of the high-temperature zone in the furnace is larger, and the intermediate radiator has a higher emissivity, which promotes the temperature change of the flue gas in the furnace from both gas radiation and solid radiation. However, due to the reduced volume of the furnace, the maximum combustion temperature has been reduced by 21K.

Regarding the concentration of combustibles, since this paper adopts the methane-air two-step reaction model, the intermediate product CO is used to characterize the adequacy of combustion. After the combustion enters the stable phase, the CO mass fraction in the improved chamber furnace model is reduced by about 8.6 And the distribution is more uniform, proving the positive effect of the intermediate radiator in promoting the combustion process in the furnace.

REFERENCES

[1] Cai Jiu-ju, Ye zhu, Sun Wen-qiang. Analysis of Influencing Factors of Energy Consumption in China's Iron and Steel Industry from 1995 to 2010[J]. *Journal of Northeastern University (Natural Science Edition)*, 2013(10); 1438–1441.

[2] Zhang Wei-jun, Wang Fang, Qi Feng-sheng, et al. Simulation study on flow field, temperature field and concentration field of push-steel slab heating furnace[J]. *Journal of Northeastern University (Natural Science Edition)*, 2011, 32(002); 266–269.

[3] Li Zhi-min, Wei Yu-wen. A new energy-saving approach for industrial heating furnaces-the mechanism of black body enhanced radiation heat transfer and energy-saving [J]. *Heat Treatment Technology and Equipment*, 2008, 029(002); 36–40.

[4] Qi Feng-sheng, Wang Zi-song, Li Bao-kuan, et al. Numerical study on characteristics of combustion and pollutant formation in a reheating furnace[J]. *Thermal Science*, 2018; 2103–2112.

[5] Elmabrouk E M. Enhancing the heat transfer in a heat treatment furnace through improving the combustion process in the radiation tubes[J]. *Ima Journal of Applied Mathematics*, 2012, 77(1); 59–71.

[6] Yi Z, Su Z, Li G ,et al. Development of a double model slab tracking control system for the continuous reheating furnace[J]. *International Journal of Heat and Mass Transfer*, 2017, 113(oct.); 861–874.

Delamination failure analysis of aircraft composite material opening siding

J. He
College of Aerospace Engineering, Shenyang Aerospace University, Shenyang, China

F.L. Cong
College of Civil Aviation, Shenyang Aerospace University, Shenyang, China

L. Lin & B.T. Wang
College of Aerospace Engineering, Shenyang Aerospace University, Shenyang, China

ABSTRACT: Fiber-reinforced composite materials have the characteristics of high specific strength, high specific stiffness, good fatigue resistance and good design ability, and are widely used in aviation, aerospace and other fields. The introduction of fiber-reinforced composite materials has greatly enhanced the integrity of the aircraft, but due to the needs of inspection, disassembly, maintenance, and process limitations, there are often some holes in fiber-reinforced composite structures, such as those used for Bolt holes for bolt connections, through-holes for inspection, and assembly holes, etc. There will be stress concentration around the opening during loading, and the laminate is prone to delamination and degumming under the action of shearing force. This subject mainly analyzes the problem of delamination and degumming of aircraft composite laminates with holes under shear load.

This paper mainly uses ABAQUS finite element analysis software to establish a finite element analysis model for composite laminates based on the Hashin failure criterion, analyzes the evolution process of layering, and determines the location of the layer and the degree of damage based on the results of the finite element analysis.

1 INTRODUCTION

A composite material is a new type of material with excellent specific strength and specific modulus [1–3]. In recent years, advanced composite materials with high-performance fibers as reinforcing materials have been widely used in military and civil aircraft. Carbon fiber laminate has become a typical representative of aerospace composite materials due to its high strength and lightweight [3, 4]. Carbon fiber siding is an important part of the main load-bearing structure such as wings and fuselage. Due to assembly and maintenance, system panels, etc., openings in carbon fiber wall panels are common operating conditions. An excellent opening structure design can reflect the advanced nature of the aircraft to a certain extent [5–7].

During actual load-bearing, stress concentration often occurs around these openings, and laminates are prone to delamination and degumming, which reduces the fatigue strength of the structure and affects the fatigue performance of the structure. Because composite materials have the structural characteristics of multiple layers, multiple components, and complex manufacturing processes, and the performance of composite materials is easily affected by factors such as temperature and environment, the failure modes of their microstructures are also different, sometimes independently and sometimes interacting occur. The research on composite laminates with openings is of great significance in practical engineering [7, 8]. When studying composite laminates, the cumulative damage of single-layer boards should be considered. Due to the inhomogeneity and anisotropy of composite materials, the damage mechanism is quite complicated under the action of load stress. Composite material damage is a gradual accumulation process. Damages that are not easily visible in aviation composite material components will be destroyed when they accumulate to a certain degree. Such sudden damage is often the main cause of air crashes [2–7].

The damage process of composite laminates is simulated by the finite element method and progressive damage is used to analyze the redistribution state of load and the development after the damage of laminates. This software analysis of the damage process of composite laminates can well predict the failure state and failure load. To prevent the damage of laminates, it is of great significance, and the prediction of damage in aviation composite materials can provide great help [5, 9].

In this paper, the finite element analysis software ABAQUS is used to analyze the delamination degumming problem of a three-dimensional composite laminate with holes under shear load. This paper chooses the connection between the aircraft composite wing skin and the aluminum alloy frame plate. Through simulation calculation, the material failure process of the composite material is determined, the layer evolution process is analyzed, and the layer position and damage degree are determined according to the finite element analysis results.

2 METHODS

2.1 *Modeling of open siding*

The component is divided into three parts, namely alloy plate, nut bolt and carbon fiber plate, as shown in Figure 1.

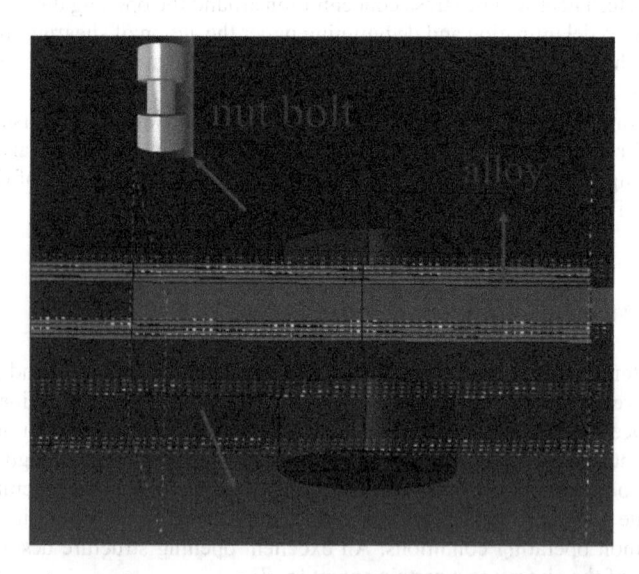

Figure 1. A perforated composite structure, including carbon fiber composite, aluminum alloy and cap bolts.

2.2 *Hashin failure criterion*

The simplified three-dimensional expression of the Hashin failure criterion is as follows:
Fiber drawing mode ($\sigma_{11} \geq 0$):

$$\frac{\sigma_{11}^2}{X_T^2} + \alpha \left[\frac{\sigma_{12}^2 + \sigma_{13}^2}{S_{12}^2} \right] = 1 \qquad (1)$$

Fiber compression mode ($\sigma_{11} < 0$):

$$\left(\frac{\sigma_{11}}{X_C}\right)^2 = 1 \tag{2}$$

Matrix stretching mode ($\sigma_{22} + \sigma_{33} > 0$):

$$\frac{(\sigma_{22}+\sigma_{33})^2}{Y_T^2} + \frac{\sigma_{12}^2+\sigma_{13}^2}{S_{12}^2} + \frac{\sigma_{23}^2 - \sigma_{22}\sigma_{23}}{S_{23}^2} = 1 \tag{3}$$

Matrix compression mode ($\sigma_{22} + \sigma_{33} < 0$):

$$\left[\left(\frac{Y_C}{2S_{23}}\right)^2 - 1\right]\frac{\sigma_{22}+\sigma_{33}}{Y_C} + \left(\frac{\sigma_{22}+\sigma_{33}}{2S_{23}}\right)^2 + \frac{\sigma_{23}^2 - \sigma_{22}\sigma_{33}}{S_{23}^2} + \frac{\sigma_{12}^2+\sigma_{13}^2}{S_{12}^2} = 1 \tag{4}$$

The formula, X_T, X_C indicates the longitudinal tensile and compressive strength of the single-layer board; Y_T, Y_C indicates the transverse tensile and compressive strength of the single-layer board; S_{12} indicates the shear strength of the single-layer board in the 1–2 direction; σ indicates the normal stress of the material.

It can be seen that the Hashin three-dimensional failure criterion can predict a total of four failure modes, namely: fiber tensile failure, fiber compression failure, matrix tensile failure and matrix compression failure. And in the fiber tensile failure, the influence brought by the shear effect is taken into account. In the simulated cloud diagram, I found that there are only three of the seven modes of Hashin failure damage mentioned in the previous section of the carbon fiberboard. They are matrix compression damage, matrix tensile damage and shear damage.

3 RESULTS AND DISCUSSION

The damaged cloud image also started from 0.05 seconds at the earliest. By 0.07 seconds, the carbon fiber board had been severely damaged. At this time, the red area on the cloud image was the most. After 0.07 seconds, the carbon fiber board has completely failed, which is of no research significance. Figure 2 shows the three failure modes of carbon fiberboard.

When loaded to 0.05s, the carbon fiber board only bears the matrix compression damage and shear damage, but not the matrix tensile damage. The compression damage and shear damage of the matrix are exactly the same at this time. When the load is loaded to 0.06s, the carbon fiber board still only bears the matrix compression damage and shear damage and does not bear the matrix tensile damage. The matrix compression damage and shear damage are still completely the same, but the damage has been further expanded when compared to 0.05s. The most obvious difference when loading to 0.07s is that the carbon fiber board itself has already undergone deformation damage. At this time, the carbon fiberboard is damaged by the compression of the matrix, the tensile damage of the matrix and the shearing damage at the same time. Among them, the compression damage of the matrix is the smallest, followed by the tensile damage of the matrix, and the sheer damage is the largest. After 0.07s, the material is completely damaged and failed, which is of no research significance, and the damaged cloud image will not continue to change.

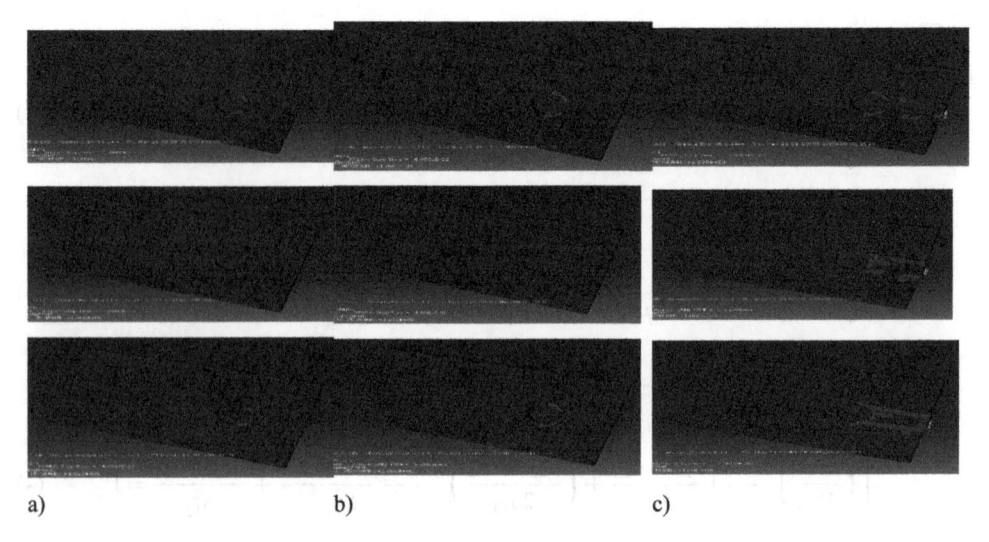

Figure 2. The matrix compression damage, matrix tensile damage and shear damage of the carbon fiber laminate are at a) 0.05, b) 0.06 and c)0.07s.

It can be seen from Figure 3 that as time increases, the component stress value initially increases, and the displacement (strain) progresses to about 0.2mm at about 0.042s, at which time the stress value reaches the first peak on the curve. At the 0.2mm stage before reaching the first peak, which is the elastic stage of the alloy plate, as the load increases, the stress increases almost proportionally. If the load is removed at this time, the alloy plate parts can be restored to their original state, showing elastic deformation. At this stage, the elastic modulus E of the material can be measured. Continue to load from the first peak to 0.055s when the displacement (strain) progresses to about 0.3mm. At this time, the curve reaches the highest peak, which means that the whole has reached the maximum tensile strength. When the load continues to about 0.067s, the displacement (strain) is about 0.38mm, and then the curve suddenly slides down cliff and the displacement drops to 0 at about 0.42mm at about 0.068s. At this time, the whole component has been completely unloaded and completely damaged and invalidated. For aluminum alloy plates, from the first peak point of the curve to the point where the cliff suddenly occurs, we can determine that this is the plasticity stage of the aluminum alloy plate. The "basic judgment" is used here because the material is a combination of aluminum alloy plate and composite carbon fiber plate. We cannot only discuss its elasticity and plasticity, which also involves composite material mechanics. So we have no way to determine the specific elastic stage and plastic stage, only a rough range can be given.

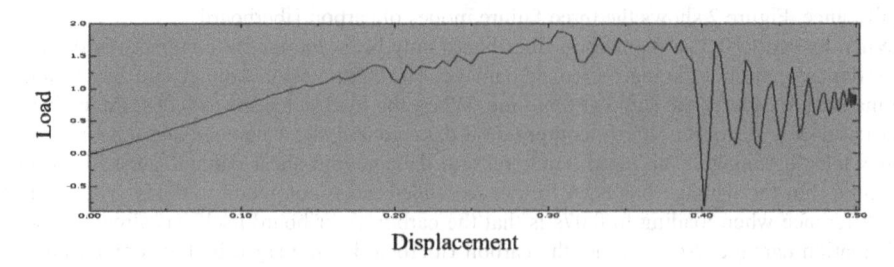

Figure 3. The load of the composite structure is the curve of the displacement change.

The stress in the stress cloud diagram is Mises stress, that is, Mises stress. It is a yield criterion proposed by von Mises in 1913.

Figure 4 shows the stress contours of different components. From the stress cloud diagram, we can know: When the whole component is loaded to 0.055s and the displacement reaches 0.31mm, the maximum stress is tolerated, and the maximum stress value is 738.6MPa. The stress value begins to decrease when the load continues, and it can be concluded that this is the maximum tensile strength of the overall material.

When the aluminum alloy plate is loaded to 0.055s and the displacement reaches 0.31mm, it bears the maximum stress, and the maximum stress value is 144.00MPa. The stress value begins to decrease when the load continues, and it can be concluded that this is the maximum tensile strength of the aluminum alloy plate. The largest damage is on the side away from the load direction around the hole, and the rest of the damage shows a V-shaped spread on the cloud image.

When the composite carbon fiberboard is loaded to a displacement of 0.060s to 0.35mm, it bears the maximum stress, and the maximum stress value is 548.3MPa. The stress value begins to decrease when the load continues, and it can be concluded that this is the maximum tensile strength of the composite carbon fiberboard. The largest damage is on the side close to the load direction around the hole, and the rest of the damage shows a V-shaped spread on the cloud image.

We can see that under the action of shear load, the tensile strength of the composite carbon fiberboard is much higher than that of the aluminum alloy plate, and when the two materials are combined, the overall tensile strength is improved to a certain extent.

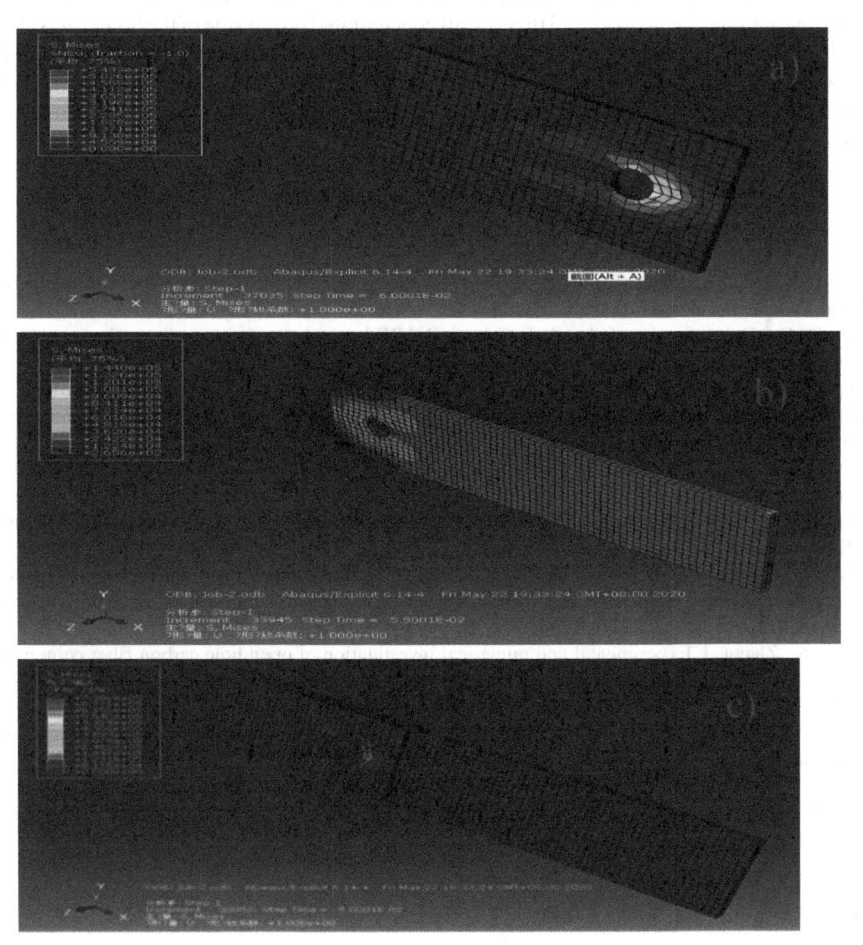

Figure 4. Stress cloud of a)Stress cloud diagram of composite carbon fiberboard, b)Aluminum alloy plate and c)the whole frame.

4 CONCLUSIONS

In this paper, the delamination failure of composite open wall panels is studied, and a constitutive relationship model is established based on the mechanical properties of each constituent material. The Cohesive unit is added to the carbon fiberboard, and the Hashin failure criterion is selected as the judgment basis. Based on ABAQUS finite element analysis, the finite element modeling of the porous carbon fiber resin matrix composite laminate is carried out. And through simulation calculation, the damage and failure process of each material under shear load is obtained, and the following conclusions are drawn:

1) The Hashin failure is considered for the composite carbon fiberboard. According to the Hashin failure criterion, it can be known that the composite carbon fiber board suffers from matrix compression damage, matrix tensile damage and shear damage. Among them, the compression damage of the matrix is the smallest, followed by the tensile damage of the matrix, and the sheer damage is the largest.
2) Under the condition that the total simulation time is 0.1s and the total displacement is 0.5mm, the overall member reaches the maximum tensile strength at 0.055s and the displacement reaches 0.31mm, and then begins to fail.
3) The aluminum alloy plate and the composite carbon fiber plate have different degrees of damage under the action of shear load, and they are all concentrated on the side of the hole that is stretched by the shear load. The farther away from the hole, the smaller the degree of damage. The order of failure is that the Cohesive unit fails first, followed by the aluminum alloy plate, and finally the composite carbon fiber plate. In the actual simulation, it can also be known that the tensile strength of the composite carbon fiberboard is much better than that of the alloy board.

REFERENCES

[1] Lin, Y., Liu, C. Interlaminar Failure Behavior of Glare Laminates Under Double Beam Five-Point-Bending Load. *Composite Structures*. 2018, 201, 79–85.
[2] Liu, C., Du, D. Interlaminar failure behavior of GLARE laminates under short-beam three-point-bending load. *Composites Part B: Engineering*. 2016, 97, 361–367.
[3] Han, S., Meng, Q. Mechanical and electrical properties of graphene and carbon nanotube reinforced epoxy adhesives: Experimental and numerical analysis. *Composites Part A: Applied Science and Manufacturing*. 2019, 120, 116–126.
[4] Hu, Y., Zhang, Y. Mechanical properties of Ti/CF/PMR polyimide fiber metal laminates with various layup configurations. *Composite Structures*. 2019, 229, 111408.
[5] Du, D., Hu, Y. Open-hole tensile progressive damage and failure prediction of carbon fiber-reinforced PEEK–titanium laminates. *Composites Part B: Engineering*. 2016, 91, 65–74.
[6] Zhou, S., Zhang, J. Experimental and numerical investigation of open hole carbon fiber composite laminates under compression with three different stacking sequences. *Journal of Materials Research and Technology*. 2019, 8, (3), 2957–2968.
[7] Zhou, S., Zhang, J. Experimental and numerical investigation of open hole carbon fiber composite laminates under compression with three different stacking sequences. *Journal of Materials Research and Technology*. 2019, 8, (3), 2957–2968.
[8] Lau, W. L., Reizes, J. Heat and mass transfer model to predict the operational performance of a steam sterilisation autoclave including products. *International journal of heat and mass transfer*. 2015, 90, 800–811.
[9] Pott, P. C., Schmitz-Watjen, H. Influence of the material for preformed moulds on the polymerization temperature of resin materials for temporary FPDs. *J Adv Prosthodont*. 2017, 9, (4), 294–301.

The super elastic strain sealing technology and material for tubing and casing connection in natural gas well

X.H. Wang
State Key Laboratory for Performance and Structural Safety of Petroleum Tubular Goods and Equipment Materials, Tubular Goods Research Institute of China National Petroleum Corporation (CNPC), Xi'an, P.R. China

Y.P. Lv
Xi'an Shiyou University, Xi'an, P. R. China
Baoji Petroleum Steel Pipe Co., Ltd. Baoji, P. R. China

J.D. Wang
State Key Laboratory for Performance and Structural Safety of Petroleum Tubular Goods and Equipment Materials, Tubular Goods Research Institute of China National Petroleum Corporation (CNPC), Xi'an, P.R. China

B.C. Pan
Baoji Petroleum Steel Pipe Co., Ltd., Baoji, P. R. China

ABSTRACT: In view of the leakage of the oil tubing and casing connections in high temperature and high pressure natural gas well, this paper first summarized the important results of the research on the sealing method and leakage mechanism. Based on this, a new sealing method and manufacturing method of the tubing and casing connections were designed, the sealing material to implement this method was selected. The sealing test were carried out to verify the validity of this method. It was proposed to increase the elastic strain design on the basis of the sealing surface pressure stress design, so that the sealing surface produces a large enough elastic strain to block the leakage channel produced on the sealing surface of oil casing and tubing. The elastic materials can be used to achieve elastic strain seal design, and a supper elastic alloy Ti-Ni-X was selected. Several elastic strain sealing structures were designed, and a kind of the sealing ring supper-elastic strain sealing casing was produced. In accordance with the standard API RP 5C5 regulations, the casing specimens successfully passed CAL- IV class B series sealing detection.

1 INTRODUCTION

At least one-third of China's high-temperature and high-pressure natural gas wells operate at a high pressure in the well bore annulus. Tubing/casing connection leakage is one of the main reasons, such as in the Tarim oilfield, a variety of premium tubing from domestic and abroad had been replaced many times, so far the well bore sealing is still unable to guarantee, the current premium connection cannot completely ensure the seal integrity of the well bore [1, 2]. Therefore, it is important to significantly improve the sealing performance of the tubing/casing connection by strengthening the research on the sealing mechanism and innovating the method of sealing.

2 THE REVIEW OF SEALING METHOD AND LEAKAGE MECHANISM

Oil tubing and casing seal system are often the thread seals, the torque shoulder seals, the resilient seals, radial metal contact seals [3].

Thread seals are not reliable for sealing high pressure gases. Gaps exist in the threads providing a leak path through the connector. These gaps must be plugged by the thread compound to effect a seal [3]. In order to improve leak tightness, thick tin plating was applied for API Buttress Casings [4].

Torque shoulder seal ability is very high, However, if a high axial compression load causes a partial plastic deformation of the torque shoulder, the seal ability may be affected. Similarly, if the high axial stretch load reduces the pressure of the shoulder pressure, it will also affect the sealing.

Most premium gas tight connection utilize radial metal seal, the seal contact pressure in excess of the expected well pressure determine the sealing performance of the connection. the actual effective contact pressure is a function of the mechanical interference fit of the two seal surfaces and additional energy from the internal and external applied pressure [5]. A properly designed metal-to-metal radial seal exhibits a pressure energization effect, This effect causes the bearing pressure at the seal to increase at a faster rate than the pressure being sealed [3].

One problem with premium connection is that the negative torque shoulder advantage disappears when the stretch load separates the torque shoulder or compress load to deform the torque shoulder. An innovative feature can amplify the seal contact force using the stiffness the pin extension, instead of using the reaction force provided by the negative shoulder angle [6]. Some connection incorporate resilient seals as a back up to the metal seals, but the resilient seal may create too much radial stress [5], this may cause the couplings to crack. Proper installation of the sealing ring is difficult, if not installed properly, the sealing ring will come off.

A simplified model [7] to determine the connection leak probability of a string of pipe showed that single metal gas seal leak probability was 10^{-6}, then the gas leakage probability of 500 tube-connected string is 10^{-2}. Even though the sealing contact pressure is significantly higher than the fluid pressure, there is still a small leakage channel on the sealing surface [8]. The distribution of the contact stress on seal surface in axial and circumference was detected by ultrasonic waves [9], even if there is a perfect makeup torque curve, but the sealing surface is still intermittent, indicating that the sealing surface is damaged, forming a leak channel [9].

3 THE ELASTIC STRAIN DESIGN THEORY OF SEALING STRUCTURE

The seal design method of the current premium connection is the contact pressure stress design method, that is, the shape and size of the sealing surface are optimized by using the finite element simulation calculation method, and the pressure stress value and reasonable pressure stress distribution are formed on the sealing surface to achieve the gas seal. The criterion does not take into account the roughness of the sealing surface and the structural accuracy. In fact, the sealing surface must be smooth enough to ensure sealing. Otherwise, even if the design stress is exactly the same, the sealing may not be the same due to the different manufacturing accuracy.

When running into the wellbore, the casing or tubing may encounter resistance and need to be lifted or even rotated for many times. In the oil and gas production process, high-pressure air flow causes the oil tubing string to vibrate. So the thread connection may be loose, Sealing surface pressure stress drop or even disappear, resulting in the thread connection leakage.

This paper proposes that the elastic strain design should be used as a complement to the pressure stress design method, that is, the elastic strain of the sealing surface should be greater than the existing or potential leakage channel, which can be expressed as:

$$k\varepsilon = \frac{\varepsilon_e}{\delta} \geq k_{es} \qquad (1)$$

In the equation (1), $k\varepsilon$ is the elastic strain design factor, ε_e is the sealing surface elastic strain variable, the unit is % or mm; δ is the amount of leakage channel on the sealing surface caused by the roughness of the sealing surface, defects, damage, as well as the seal circumference diameter deviation, the loose thread during service, the unit is % or mm; $k\varepsilon s$ is a safety margin, obviously leakage does not occur when $k_{es} \geq 1$.

whether the sealing structure of a premium casing meets the above design requirements is estimated below. The casing steel grade is P110, the outer diameter is 140mm.

There may many kinds of leakage channel, one is the machining deviation of the sealing circumference diameter, or connection loose. The design value of sealing circumference diameter deviation is less than 0.04%, it can be considered that the leakage channel of the sealing surface is $\delta \leq 0.04\%$. The design value of sealing surface pressure stress is $\sigma_e = 600 \sim 700$MPa, and according to the stress strain curve of steel, the elastic strain of sealing surface is $\varepsilon_e = 0.3\%$. Then the elastic strain design factor $k\varepsilon = \frac{\varepsilon_e}{\delta} \geq 7.5$.

The results show that if the machining quality meets the design requirements, the casing connector is well sealed. However, if there is external force damage, the sealing surface appears defects deeper than 12μm, or the connector make up torque is insufficient, or the pipe string vibration causes the connection to loosen, the sealing surface appear loose more than 0.3% in the diameter direction, the casing connection will lose the sealing, Then the elastic strain design factor $k\varepsilon = \frac{\varepsilon_e}{\delta} \leq 1$.

In order to satisfy equation (1), either the depth of the seal surface defect is controlled to within 12 m, or it is ensure that the connection does not loosen by more than 0.3%, however this is difficult. Either the sealing surface elasticity ε_e should be greater than below at least:

$$\varepsilon_e \geq \delta k_{es} = \frac{12}{4000} k_{es} = 0.3\% k_{es} \qquad (2)$$

If the safety margin is valued at $k_{es} = 1$, according to equation (2), the elastic strain of the sealing surface material should reach to $\varepsilon_e = 0.3\%$. In order to ensure the sealing structure sealing, the safety margin $k\varepsilon$ should be large enough. For example, if the safety margin is $k_{es} = 7$, the elastic strain of the sealing surface material should reach to $\varepsilon_e = 2.1\%$. This obviously exceeds the elasticity of steel materials, but many alloy materials are highly elastic or even ultra-elastic and can be used to manufacture sealing structures.

4 SUPPER-ELASTIC STRAIN SEALING TECHNOLOGY

There are many super-elastic materials, including metal and non-metallic materials. Due to the various disadvantages of non-metallic materials, the super-elastic shape memory alloy are preferred to achieve super-elastic strain seal design in this paper. The super-elastic shape memory alloy can be applied to the sealing structure surface of the casing or tubing connections, by the additive manufacturing technology such as welding, spraying, electroplating, 3D printing, etc., or the seal ring made of super-elastic shape memory alloy can be placed at the sealing area of the connections.

During the connection makeup, with the increase of torque, the contact pressure stress of the sealing surface increases gradually from 0 and produces elastic strain (see the straight line section $0M_{S2}$ in Figure 1(a)). With the increase of the makeup torque, when the contact stress reaches the critical value, the martensite phase change occurs in the sealing surface alloy, resulting in a pseudo-elastic strain (see the curve segment $M_{s2}M_{f2}$ in Figure 1(a)). If the contact pressure stress exceeds a certain value, the martensite transformation is completed and enters the martensite loading stage, the alloy has an elastic strain (see the straight line section $M_{f2}\sigma_y$ in Figure 1(a)), until the contact pressure stress reaches the yield strength σ_y of the alloy, plastic deformation occurs. Therefore, the contact pressure stress should be controlled below the alloy yield strength. During the tubing or casing connection breakout, with the decrease of torque, the contact pressure stress of the sealing surface decreases gradually, the alloy material begins to exhibit the elastic unloading characteristic, and when the contact stress is reduced to critical value, the austenite transformation occurs in the alloy, resulting in a nonlinear strain response (see the curve segment $A_{s2}A_{f2}$ in Figure 1(a)).When the alloy is fully converted to the austenite, it is also represented by the austenite elastic unloading performance (see the linear segment σ_e in Figure 1(a)), until the contact pressure stress reaches zero, the strain also returns to zero, and the residual strain is also zero. This nonlinear elasticity of alloy is phase-change pseudo-elasticity, which significantly exceeds the linear elasticity, so it is called super-elastic. The corresponding stress strain curve is the lag ring curve of the "Little Banner" shape shown in Figure 1(a). The super-elastic strain of the alloy can reach to 8%.

After the tubing or casing is brought down into the oil well, the temperature increases, and the sealing surface material will also have a martensite inverse phase change, i.e. austenite phase change, resulting in a strain response, the seal surface contact will be more closely. During the tubing service process in the well, due to vibration and other factors, sealing surface pressure stress could reduce, than the material will also have a martensite reverse phase change, resulting in a strain response, thus the sealing surface is always in close contact. If a well needs to be repaired, after the breakout of the threaded connection, the seal surface pressure stress will drop to zero, then the strain is almost completely restored to its original state, the tubing can still be put into use again.

This paper selected a Ti-Ni-X alloy, which stress-strain cycle curve is showed in Figure 1(b)(c). the compression stress is 600MPa at the strain compression 5%, the residual strain was 0.162% after unloading, almost restored to the pre-deformation amount. Tensile strain / %

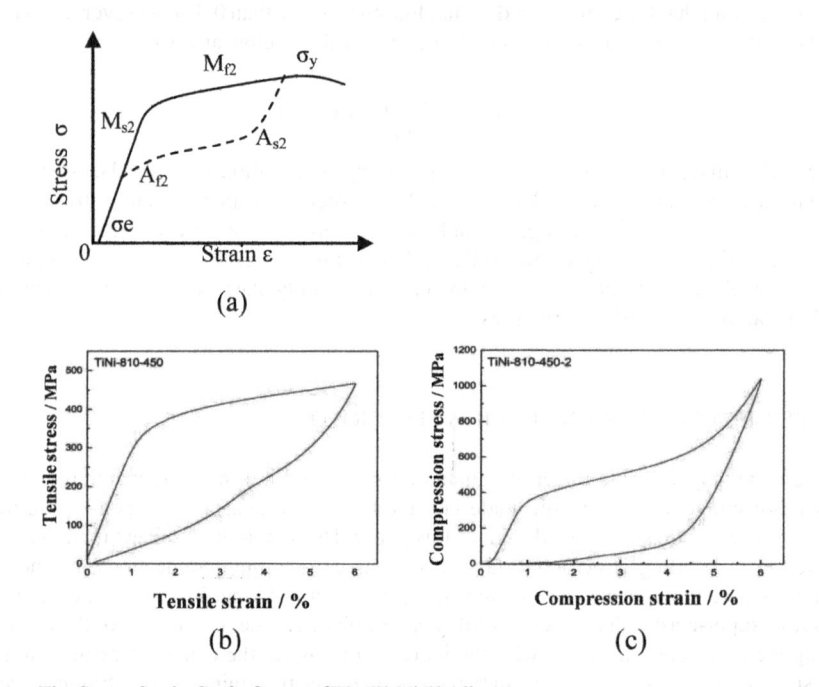

Figure 1. The Stress-Strain Cycle Curve of The Ti-Ni-X Alloy.

In the process of loading and unloading, the super-elastic shape memory alloy material can absorb a lot of energy, showing damping characteristics. The stress-strain relationship has a hysteresis effect, which can hinder the vibration loose buckle of the tubing or casing connection, thus further improve the structural integrity and sealing integrity of the well pipe.

5 SUPER ELASTIC STRAIN SEAL RING AND SEAL TEST RESULTS

According to the theory and method described above, this paper designs a sealing ring-form sealing structure of tubing or casing connection. The sealing ring is manufactured with superelastic material, as shown in Figure 2, the nose end of the pin connection is designed as a shoulder form, the sealing surface is a conical surface. Before makeup, the super-elastic seal ring package onto the nose end of pin connection, or placed in the inner the box connection. When the pin and box connections was assembled together, the seal ring is elastically compressed in the direction of the wall thickness, the inner and outer surface of the seal ring is in excessive contact with the main seal surface of the pin and box connections, forming the main seal.

The main seal surface can be designed as a cylindrical surface, conical surface, step cylindrical face, or other surface, and various shape seals of the pin and the box connection can be freely paired. The metal seal ring is cylindrical, conical, or otherwise shaped.

In order to verify the technology, as shown in Figure 2, a kind of super-elastic strain seal casing with the above alloy seal ring was manufactured, the casing specification was $\Phi 139.7 \times 9.17$mm, and the steel grade was P110. The sealing ring was manufactured from Ti-Ni-X alloy.

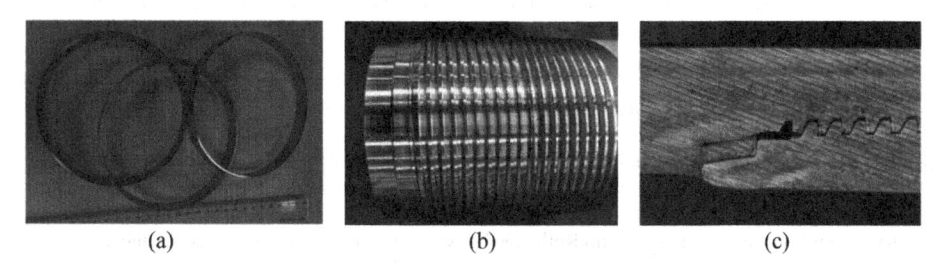

Figure 2. Super-elastic Alloy Seal Ring Sealed Structure: (a) Alloy Seal Ring, (b) Pin Connection and (c) Assembled connection.

The makeup and breakout testing was carried for the casing, and the thread compound was shell III type, the dosage was 25 to 30 g, the makeup speed was 5 to 10 rpm, and the condition of the thread and the metal seal ring was checked after each breakout. A total of three times makeup and two times breakout tests were carried, the maximum value of the makeup torque was 13020 N.m, and the minimum was 11050 N.m. After each breakout, it could be seen that the sealing ring fit tightly on the end seal cone of the pin connection, the ring was in good shape and without damage.

In accordance with API RP 5C5 (2017) regulations, the casing has undergone a CAL III and IV Class IV B-series sealing test. According to API specified minimum yield strength, the specified outer diameter, the specified wall thickness, the testing parameters of the casing specimens were calculated, the high temperature yield strength was calculated according to the proportional coefficient $k_{temp} = 0.875$. The tensile and compression efficiency of the connection is 100%. The maximum test load reached to 95% of the pipe body reference curves, the maximum test temperature was 180°, and the maximum internal pressure reached to 90MPa. The test results were good and there was no leakage.

6 CONCLUSION

Although the design method of premium connection of oil well casing and tubing has been improved, there is still leakage problem. In order to block the leakage channel of sealing structure, the elastic strain design should be increased on the basis of stress design, and thus the super-elastic strain sealing technology were developed. The super-elastic material can be used to achieve super-elastic strain sealing technology. the effectiveness of the super-elastic strain sealing technology was verified, and this technology can be used to develop a new generation of premium casing and tubing connection.

REFERENCES

[1] Yang Xiangtong, lushuanlu, Xie junfeng, et al. 2018, Cause analysis on leakage of special thread joint oil tube for high pressure gas well[J]. *Physical Testing and Chemical Analysis (Part A:Physical Testing)* **54(7)** p519–522.

[2] Ding Liangliang, Yang Xiangtong, Zhang Hong, et al. 2017, Design and application of annular pressure management charts for high pressure gas wells[J]. *Natural Gas Industry* **37(3)** p83–85

[3] Michael J, Jellison, Manuel Alfredo Davila, et al 1996, How to evaluate and select premium casing connectors[M]. lADC/SPE 35037 *The 1996 IADC/SPE drilling conference held at New Orleans, New Orleans, Louisiana* p51–60

[4] Kazuhiko Itoh, Hajime Watanab and Yohichi Yazaki. 1983, Development of Tin Plated High Seal Buttress Casing Joint[M]. SPE 51402. *The 106th ISIJ Meeting at Akita University in Akita.* p224–227.

[5] A.B.Bradley, S.Nagasaku and E. Vergy, 2005, Premium connection design, testing, and installation for HPHT sour wells[M]. SPE 97585 SPE high pressure-high temperature sour well design a*pplied technology workshop held in the woodlands, texas, USA* p1–8

[6] M.Sugino, K.Nakamura S.Yamaguchi, et al. 2010, Development of an innovative high-performance premium threaded connection for OCTG[M], *2010 offshore technology conference held in houston, texas, USA.* p1–6.

[7] Gloria A. Valigura, Andrew Tallin, 2005, Connection for HPHT well applications and connection leak probability[M]. SPE97588. *2005 SPE high pressure/high temperature sour well design workshop held in the woodlands, TX, USA* p1–4

[8] Jueren Xie, Cam Matthews and Andrew Hamilton, 2016, A study of sealability evaluation criteria for casing connections in thermal wells[M]. SPE180720. *SPE Canada heavy oil technical conference held in Calgary, Alberta, Canada* p1–10

[9] Kirk Hamilton, Brian Wagg and Tim Roth, 2007, Using ultrasonic techniques to accurately examine seal-surface-contact stress in premium connections[M]. SPE110675. *the SPE annual technical conference and exhibition, Anaheim, California.* p696–704

[10] He Zhirong, 2015, Superelasticity of Ti-Ni-V Shape Memory Alloys[J]. *Rare Metal Materials And Engineering* **44(7)** p1639–1642

[11] Zhuang Peng, Xue Suduo and Wei Jieliang, 2015, Research on mechanical performance of superelastic NiTi Shapememory alloy bars[J]. *Journal of architecture and civil engineering* **32(1)** p96–102

Study on structural design and parametric modeling of saddle of transmission line material ropeway

X.Y. Xin & Y.F. Wang
State Grid Shaanxi Electric Power Company Construction Branch, Xian, China

C. Liu & J. Qin
China Electric Power Research Institute, Beijing, China

Z.S. Liu
State Grid Shaanxi Electric Power Company Construction Branch, Xian, China

F. Peng
China Electric Power Research Institute, Beijing, China

ABSTRACT: Aiming at the saddle of transmission line material ropeway, the structural design of components was carried out with the combination of 3D design software. Parametric modeling method was adopted to carry out the modeling of components with different structural parameters. The stress status of components of saddle was analyzed, the design load of components was proposed, and the strength of the designed components was checked through finite element simulation analysis. The results show that the components can meet the requirements of structural strength. The proposed method can realize parametric modeling of solid model and effectively improve the design efficiency of material ropeway.

1 INTRODUCTION

Nowadays, 3D design software is widely used in electric power equipment design, such as CATIA, Solid works, etc. Structural design is often checked by finite element simulation analysis. Through simulation, optimized and improved design schemes need to establish finite element simulation model based on geometric data, which will bring huge workload and seriously affect the development schedule cost [1, 2]. Through parametric modeling method, according to the geometric relationship between structures, the model can be changed in time with the modification of geometric parameters, which greatly improves the efficiency of structural design [3–5].

Transmission line material ropeway is composed of several components. At present, the components have fewer specifications, the standardization of the components is low, which is not convenient for transportation in mountainous terrain [6, 9].

Based on the 3D design software, this paper realizes the parametric modeling of the solid model of the components. Combined with the finite element software, the strength checking calculation is carried out to improve the efficiency of the material ropeway design.

2 DESCRIPTION OF COMPONENTS OF MATERIAL ROPEWAY

Material ropeway mainly includes trestle (crossbeam, leg), saddle, bracket, running car, steering pulley, in this paper, the saddle is taken as the research object to carry out component structural design and parametric modeling.

According to the number of carrying ropes, the types of saddle include saddle of one carrying rope, saddle of double carrying rope, saddle of four carrying rope.

2.1 Saddle of one carrying rope

The saddle of one carrying rope is mainly composed of frame, saddle, saddle shaft and support roller.

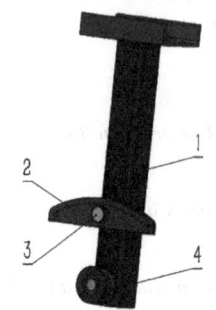

1-frame; 2-saddle; 3-saddle shaft; 4-support roller

Figure 1. Schematic diagram of saddle of one carrying rope.

2.2 Saddle of double carrying rope

The saddle of double carrying rope is mainly composed of frame, saddle, saddle shaft, support roller, connecting beam.

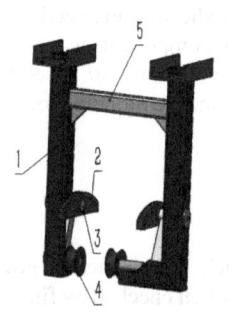

1-frame; 2-saddle; 3-saddle shaft; 4-support roller; 5-connecting beam

Figure 2. Schematic diagram of saddle of double carrying rope.

2.3 Saddle of four carrying rope

The saddle of four carrying rope is mainly composed of frame, saddle, saddle shaft, support roller, connecting beam.

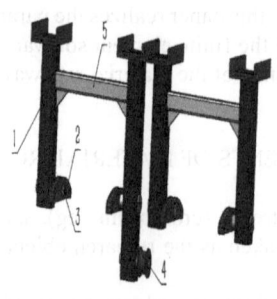

1-frame; 2-saddle; 3-saddle shaft; 4-support roller; 5-connecting beam

Figure 3. Schematic diagram of saddle of four carrying rope.

3 STRUCTURAL DESIGN PARAMETERS OF SADDLE

3.1 Frame

The structural design parameters of frame are shown in the Table 1.

Table 1. Structural design parameters of frame.

Serial number	Structural design parameter	Value range of parameter (mm)
1	Height of frame SF-H	500~1500
2	Section length of frame SF-L	50~200
3	Section width of frame SF-W	50~200
4	Section thickness of frame SF-T	5~15

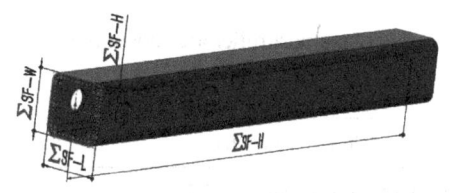

Figure 4. Schematic diagram of structural design parameters of frame.

3.2 Saddle

The structural design parameters of saddle are shown in the Table 2.

Table 2. Structural design parameters of saddle.

Serial number	Structural design parameter	Value range of parameter (mm)
1	Length of saddle SS-L	200~600
2	Height of saddle SS-H	20~50
3	Thickness of saddle SS-T	20~60
4	Diameter of saddle SS-D	30~60

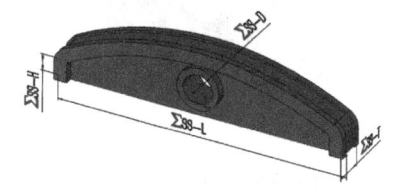

Figure 5. Schematic diagram of structural design parameters of saddle.

3.3 Saddle shaft

The structural design parameters of saddle shaft are shown in the Table 3.

Table 3. Structural design parameters of saddle shaft.

Serial number	Structural design parameter	Value range of parameter (mm)
1	Diameter of saddle shaft SP-D	It can be calculated according to the strength calculation formula of the pin
2	Length of saddle shaft SP-L	Determined by the width of frame

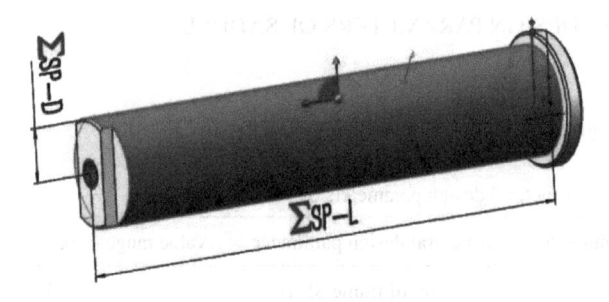

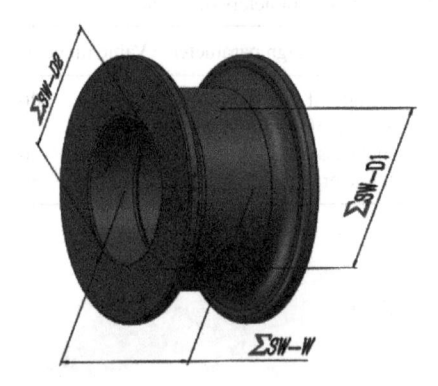

Figure 6. Schematic diagram of structural design parameters of saddle shaft.

3.4 *Support roller*

The structural design parameters of support roller are shown in the Table 4.

Table 4. Structural design parameters of support roller.

Serial number	Structural design parameter	Value range of parameter (mm)
1	Diameter of support roller SW-D1	60~140
2	Width of support roller SW-W	60~120
3	Shaft diameter of support roller SW-D2	It can be calculated according to the strength calculation formula of the pin

Figure 7. Schematic diagram of structural design parameters of support roller.

3.5 *Connecting beam*

The structural design parameters of connecting beam are shown in the Table 5.

Table 5. Structural design parameters of connecting beam.

Serial number	Structural design parameter	Value range of parameter (mm)
1	Height of connecting beam SB-H	300~1000
2	Section length of connecting beam SB-L	50~200
3	Section width of connecting beam SB-W	50~200
4	Section thickness of connecting beam SB-T	5~15

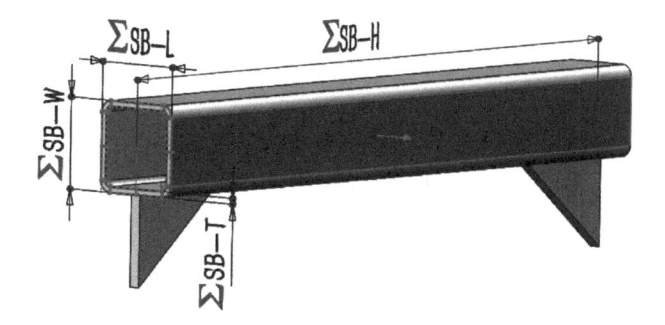

Figure 8. Schematic diagram of structural design parameters of connecting beam.

4 EXAMPLE OF PARAMETRIC MODELING OF COMPONENTS OF MATERIAL ROPEWAY

The main parametric structural design parameters of saddle are shown in the Table 6. The material of frame, saddle, support roller and connecting beam is Q355, and the material of saddle shaft is 40Cr.

Table 6. Example of parametric modeling of saddle.

Serial number	Structural design parameters	Structural parameters 1 (mm)	Structural parameters 2 (mm)
1	Height of frame SF-H	700	980
2	Section length of frame SF-L	90	120
3	Section width of frame SF-W	90	120
4	Section thickness of frame SF-T	8	8
5	Length of saddle SS-L	360	440
6	Thickness of saddle SS-T	45	45
7	Diameter of saddle SS-D	35	45
8	Diameter of saddle shaft SP-D	35	45
9	Height of connecting beam SB-H	620	620
10	Section length of connecting beam SB-L	50	80
11	Section width of connecting beam SB-W	50	80
12	Section thickness of connecting beam SB-T	5	5

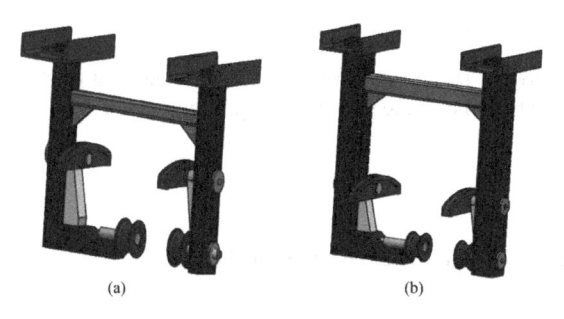

Figure 9. Example of parametric modeling of saddle: (a) Structural parameters 1 and (b) Structural parameters 2.

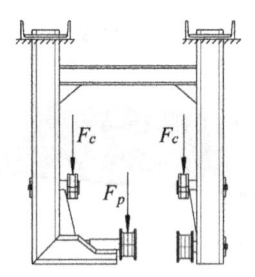

Figure 10. Schematic diagram of force analysis of saddle.

The schematic diagram of force analysis of saddle is as shown in the Figure 10, the saddle bears the load of carrying rope F_c and the load of return rope F_p.

4.1 *Structural parameters 1*

The maximum applied load of the saddle is set as 50kN, and the maximum stress of the saddle is 478MPa through finite element simulation, which occurs at the joint of the saddle shaft and frame. The overall stress of the saddle can meet the design requirements of the ultimate strength (considering the ultimate strength of the material as 507MPa).

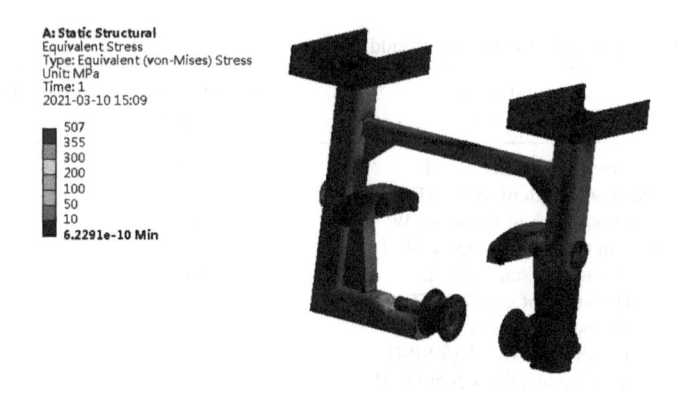

Figure 11. Stress cloud of saddle.

According to the safety factor equal to 2.5, the rated load of the saddle with structural parameter 1 is 20kN.

4.2 *Structural parameters 2*

Referring to the method of structural parameter 1, the rated load of the saddle with structural parameter 2 is 30kN, which meets the strength design requirements.

5 CONCLUSION

The structural design and parametric modeling method of saddle of transmission line material ropeway are proposed.

(1) According to the type of saddle, the structure composition of parametric components is determined, and the structural design parameters of each component are put forward.

(2) The component modeling with different structural parameters is carried out, the stress condition of the components is analyzed, and the finite element analysis is carried out. The results show that the structural design with parametric modeling meets the use requirements.

The proposed method of structural design and parametric modeling can realize parametric modeling of solid model and effectively improve the efficiency of material ropeway design.

ACKNOWLEDGMENTS

This work was financially supported by Science and Technology Project of State Grid Shaanxi Electric Power Company (Research on path planning and component selection system of material ropeway based on oblique photography technology). The number of the project is SGSNJS00GGJS2000108.

REFERENCES

[1] Zongbo Hu, Xiaofeng Tang, Xuncheng Wu and Xin Jiang 2013 Application of implicit parametric modeling technology in early CAE analysis *Journal of Shanghai University of Engineering Science* **27(04)** pp 328–332
[2] Yanan Li, Caifu Qian 2019 Stress analysis and strength assessment of high pressure heater based on VB and ANSYS parametric modeling *Pressure Vessel Technology* **36(1)** pp 48–53
[3] Kaiming Zheng 2019 Lightweight design of aluminum frame body structure for small electric vehicles based on parametric model *Jilin University*
[4] Libin Duan, Ningcong Xiao, Zhaohui Hu, Guangyao Li and Aiguo Cheng 2017 An efficient lightweight design strategy for body-in-white based on implicit parameterization technique *Structural & Multidisciplinary Optimization* **55(5)** pp 1927–1943
[5] Shuaishuai Qin 2020 3D parametric modeling and stability calculation software development of ship hull *Dalian University of Technology*
[6] Q/GDW 11189—2018 Special aerial material ropeway with transit materiel of over head transmission line engineering
[7] Infrastructure Department of State Grid Corporation of China 2010 Standardization manual for material ropeway transportation of over head transmission line engineering of State Grid Corporation of China *China Electric Power Press*
[8] Zhusen Sun, Qian Miao, Ming Jiang 2011 Standardized construction scheme for aerial ropeway of electric transmission line engineering *Electric Power Construction* **32(3)** pp 117–120
[9] Qian Miao, Xuesong Bai 2009 Study on transportation technology and equipment in freight cableway system *Electric Power Construction* **30(12)** pp 93–96

# Analysis of influencing factors of water injection capacity in low permeability reservoir

M.G. Tang, Q.H. Liu, G.Q. Xue, P.R. Wang & H.M. Tang
Yufu International Building, Research Institute of South China Sea West of CNOOC, Haikou City, China

ABSTRACT: Low-permeability oil reservoirs account have a large proportion in oil reservoir resources of various oil-producing countries. The improvement of the recovery rate of low-permeability oil reservoirs is one of the focuses of scholars from various countries. Now, as a secondary oil recovery technology, oilfield water injection development is the main development method used in low permeability reservoirs. A systematic summary of many factors that affect the water injection area of an oil field can provide a certain reference basis for solving the problems of oil field water injection well pressure increase and water injection decrease.

1 GEOLOGICAL FACTORS

1.1 *Pore throat*

The throats of low-permeability and ultra-low-permeability reservoirs are mainly of medium-pore, medium-fine throat and small-pore, fine-throat type. The injected fluid mainly enters the reservoir along the fine throats in the middle pores and a few large pore throats, the affected area is relatively large, but fluid circulation is more difficult. The throat with small pores and fine throat is a poor type of reservoir, during the water injection development period, the start-up pressure of this type of throat reservoir is generally relatively high, the injected fluid pushes forward very slowly. The oil in this type of throat is difficult to be displaced, resulting in poor water injection effect and difficulty in improving reservoir production.

1.2 *Interlayer interference*

If a water injection well in an oil field needs to inject water into multiple oil layers, and each reservoir permeability contrast larger, there will be very small water injection in the low-to-medium permeability layer, and water channeling in the high-permeability layer, the efficiency of injection wells will decrease. If the injected water from the high-permeability layer enters the oil well in advance, it will cause the oil well to see water in advance or even violent flooding.

1.3 *Clay mineral sensitivity*

1.3.1 Water sensitivity
Many oil layers have clay interlayers, some rocks contain clay, therefore, during the water injection process of the oil layer, the clay tends to swell with water, which causes blockage and reduces the water injection rate. In low-permeability reservoirs, as water injection progresses, some mineral particles will migrate, when they accumulate in the pore throat, it is easy to cause blockage, resulting in a decrease in water injection, and ultimately a decrease in oil production [1].

1.3.2 Speed sensitivity

When the water injection rate exceeds a certain value, Speed-sensitive minerals (illite, kaolinite, etc.) will begin to fall off as the flow rate of the injected water increases. the shed particles flow in the formation with the injected water, and finally settle in tiny pores or cracks, causing the pores to be blocked [2].

1.3.3 Salinity sensitivity

The damage mechanism of salinity sensitivity is similar with water sensitivity, because the illite/montmorillonite-mixed layer in rock minerals is usually attached to the surface of rock clastics in the form of flaky film. When the mineralization of the injected water is too low, the clay minerals in the formation will swell, fall, and migrate, blocking the pores and causing the permeability of the formation to decrease.

1.3.4 Acid sensitivity

Acid sensitivity refers to the phenomenon that acid-sensitive minerals in the formation react with acidic fluids after being injected into the formation and produce sediments or release tiny particles, which reduces the permeability of the formation. The acid-sensitive minerals in the formation include chlorite, calcite, iron argillaceous, stucco and so on. Sensitive minerals will react to form $Fe(OH)_3$, $Al(OH)_3$ and other precipitates and muddy particles after encountering acidic liquids, these products will cause the blockage of the pore throat of the reservoir, reduce the permeability of the reservoir, and affect the water injection work [3].

1.3.5 Alkali sensitivity

In an alkaline environment with a high pH value, the negative charge content on the mineral surface increases, resulting in a strong rejection reaction between the crystal layers in the reservoir, causing some colloidal particles on the mineral surface to fall off, migrate and block the pore throat. Silicate minerals and silica minerals in the reservoir may generate sediments to block pore throats in an alkaline environment. In a strong alkaline environment, clay minerals will produce new silicate and silica gel deposits to block the pore throat [4].

1.3.6 Stress sensitivity

Formation clay minerals are easily deformed or broken under the pressure of the overlying rock, producing tiny particles, which will block the throat after being transported by water. Moreover, as the pressure of the overlying rock increases, the microscopic pores of the reservoir rock are compressed, and the average pore throat radius of the rock decreases, resulting in a decrease in rock permeability. Combining the influence of the above factors will result in a significant decrease in the formation permeability [5].

1.4 Poor stratum connectivity

If there are faults or fan-stacked zones with poor physical properties in the reservoir, low-permeability barrier zones similar to separators will be formed on the main flow lines of water injection wells and oil production wells, it takes a lot of pressure for the leading edge of water injection to cross the barrier zone, so the injected water will flow to both sides of the barrier zone, at the same time, the injection pressure of the water injection well will increase and the water injection rate will decrease. In addition, due to the insufficient injection-production well pattern or the heterogeneity of the formation, the water absorption capacity of the injection layer will be poor.

1.5 Unbalanced injection-production

If high-intensity water injection is performed for a short period of time, the diffusion rate of the water injected into the formation is slower than the injection rate, which will cause the injection pressure of the injection well to rise rapidly, and the water injection volume begins to decrease.

At the same time, if the intensity of the water injection operation is too strong, the formation rock structure will destroyed, cause sand blocking and the injection pressure to rise.

2 DEVELOPMENT FACTORS

2.1 *Water quality*

The solid particles injected into the water in the oil field and the insoluble precipitation formed in the formation will cause the blockage of the reservoir, make the injection well high-pressure under-injection [6]. It is mainly manifested in the following aspects:

2.1.1 *Solid suspended particles*
When the suspended solid particles in the injected water enter the formation, they will block the throat of the reservoir, causing the injection pressure to rise.

2.1.2 *Corrosion of the pipeline*
The corrosiveness of the injected water is mainly due to the corrosive gases such as carbon dioxide and hydrogen sulfide in the injected water, these substances will produce iron sulfide and iron oxide after corrosion of the inner wall. The chloride and sulfate contained in the injected water will also corrode the water injection pipeline and produce solid materials that block the bottom layer.

2.1.3 *Scaling*
If the injected water does not match the formation water, insoluble precipitation and scaling will occur, in addition, the mixing of water from two or more different ground sources may also cause scaling. When the injected water is not matched, it will block the pores of the formation and cause difficulty in water injection.

2.1.4 *The growth of microorganisms*
Bacteria (such as SRB and TGB) contained in the injected water are easy to breed on pipe walls and water storage tanks, and will produce hydrogen sulfide gas and organic acids, which will increase the corrosiveness of the water, t will also corrode iron and produce iron sulfide precipitation, block the water flow channel and affect the water injection work.

2.1.5 *The water contains oil*
If excessive water is injected oil content, water is injected into the formation, the water droplets will be clogged in the reservoir at shouted impede the flow of injected water, resulting in water pressure.

2.2 *Management operational issues*

During drilling operations, if the mud enters the formation due to operator errors, a filter cake or serious blockage will be formed in the formation, which will result in the failure of normal water injection, it's phenomenon is that both forward water injection and reverse water injection are invalid [7]. In oilfield development, if workers do not regularly clean the wells as required, the pollutants at the bottom of the well will be carried deep into the formation by the injected water, causing more serious blockage problems and higher costs for clearing blockages. When the staff is performing acidizing operations, if the acidizing operations are not correct, the rock skeleton of the reservoir will be destroyed, resulting in a decrease in reservoir permeability and a decrease in water injection.

2.3 Reservoir pressure rise

After the water injection development of the oil field, as the water injection progresses, the reservoir pressure will slowly rise. At the same time, in order to achieve the daily injection rate, the injection pressure of the water injection well needs to be continuously increased, which interacts with the rise in formation pressure [8, 9].

2.4 Wettability transformation

The long-term mechanical scouring of the high-speed fluid will cause the adsorbed oil film on the rock surface to fall off, which will cause the wettability of the rock to change from neutral or partial lipophilic to hydrophilic. In addition, the long-term erosion of the rock by the alkaline injected water will also cause the wettability of the oil layer to change to the hydrophilic direction. After the wettability changes to the hydrophilic direction, the water film adsorbed on the pore wall becomes thicker, resulting in a reduction in the effective throat flow radius, the Jia Min effect will appear at the throat, causing the injection pressure of the water injection well to increase continuously, making it difficult to meet the injection requirements[10].

2.5 Water injection profile control measures

Although the water flooding area has been increased by profile control and water shutoff measures, the water flooding recovery rate has been improved, however, profile control and water shutoff also have a downside. It changes the main flow channel of injected water and forces the injected water to flow from the original high-permeability oil layer and micro-fractures to the medium-low permeability oil layer [11]. The medium and low permeability reservoirs have poor physical properties, complex pore structure, mostly small and medium pores and throats, the propagation speed of water injection pressure in the formation is significantly reduced. The difficulty of water injection increases rapidly, which affects the efficient progress of water injection.

REFERENCES

[1] Dejin Han, Pingchuan Dong, Na Shi. Research on reservoir water sensitivity experiment and formation mechanism[J]. Daqing Petroleum Geology and Development, 2008(05):14–17.

[2] Dongbo Shao, Jianwen Chen. Sensitivity characteristics and controlling factors of tight sandstone reservoirs in the Ordos Basin: Taking Chang 6 reservoir of Yanchang Formation in Xin'an bian area as an example[J]. Journal of Xi'an Shiyou University (Natural Science Edition), 2017, 32(03):55–60.

[3] Yuxia Wang, Lifa Zhou, Zunsheng Jiao, etc. Sensitivity Evaluation of Tight Sandstone Reservoir of Yanchang Formation in Shanbei Area, Ordos Basin[J]. Journal of Jilin University (Earth Science Edition), 2018, 48(04):981–990.

[4] Chunjie Duan, Xuguang Wei. Research on Reservoir Sensitivity of Deep, High Pressure and Low Permeability Sandstone Reservoir[J]. Geological Science and Technology Information, 2013.32(03): 94–99.

[5] Jijia Liao, Hongming Tang, Xiaomin Zhu, etc. Study on rock stress sensitivity of ultra-low permeability sandstone reservoirs in Chang 8 reservoir of Xifeng Oilfield[J]. Journal of China University of Petroleum (Natural Science Edition), 2012, 36(02):27–33.

[6] Weiliang Chen. Discussion on Countermeasures of High Pressure Under-injection for Water Injection Wells in Low Permeability Oilfields[J]. China Petroleum and Chemical Standards and Quality, 2018, 38(15):68–69.

[7] Guolong Chen, Research and Application of Deep Profile Control Technology for Water Injection Wells in Low Permeability Oilfields[J]. Chemical Engineering and Equipment, 2018(01):162–163.

[8] Chao Wang, Junfeng Wu. Research on Technical Countermeasures for Low Permeability Oilfield Development and Adjustment[J]. Chemical Engineering and Equipment, 2017(07):237–239.

[9] Jing Zhang, Jun Xiong, Zhenjun Wu. Research on Decreasing Factors of Water Absorption Capacity of Water Injection Well in Low Permeability Oilfield[J]. Chemical Management, 2016(33):86.

[10] Haitao Li, Qirui Ma, Donghao Li. Microscopic mechanism of low salinity water injection to increase the recovery of sandstone reservoirs[J]. Oil drilling and oil production technology, 2017, 39(02):151–157.

[11] Yong Lin. Feasibility Analysis of Application of Profile Control and Water Plugging Technology in Xinzhan Oilfield[J]. China Petroleum a, 2016(S1):222–223.

Low voltage multiloop series arc fault detection based on deep recurrent neural network

Q.F. Yu
School of Electrical Engineering and Automation, Henan Polytechnic University, Jiaozuo, China
Postdoctoral Programme of Beijing Research Institute, Dalian University of Technology, Beijing, China

W.H. Lu & Y. Yang
School of Electrical Engineering and Automation, Henan Polytechnic University, Jiaozuo, China

ABSTRACT: In the low-voltage residential power distribution system, due to the diversification and variability of the load series and parallel connection forms of the load end in the actual distribution network, it is of great significance to detect the branch fault through the waveform change of the trunk line. Based on the periodicity and timing characteristics of the current signal, in this paper, a method of multi branch series arc fault detection based on deep recurrent neural network is proposed. The experimental platform is built to collect the trunk current signals under different series faults of branches. The experimental results show that the final detection result reaches 95.67%, which confirms the use of depth. The feasibility of using recurrent neural network to identify multi branch series arc fault in low voltage system is discussed, and provides a new idea and beneficial exploration for the accurate detection of low-voltage series arc fault.

1 INTRODUCTION

In the low-voltage power distribution system, when a fault arc occurs, the luminous discharge between the electrodes can generate a lot of heat. The heat generated during arc burning can make the temperature reach 20000K [1], which can ignite the combustible materials around the line, causing fire and even explosion. For AC system, arc faults can be divided into series arc fault, parallel arc fault, and grounding arc fault, the common circuit breaker can not detect the series fault arc in time when the series fault arc occurs [2]. Therefore, it is of great practical significance to effectively identify the series fault arcs in transmission lines.

Scholars have carried out a lot of research work in the field of series arc faults detection, and achieved certain research results. Some scholars analyzed the series arc faults from the aspects of arc physical phenomenon [3] and arc mathematical model [4]. Mathematical model detection method has many detection parameters and complex algorithm, and it is still only in the research stage. The detection method based on the physical phenomenon of arc is only applied to arc detection of fixed equipment such as switch cabinet because of the randomness of fault points in practical application. Therefore, neural network algorithm has been gradually applied to the detection of series arc faults, and has achieved certain research results.

In the low-voltage distribution system, due to the diversity and complexity of the load series and parallel connection in the actual distribution network, in the multi branch load, the main circuit current is affected by the branch load, the waveform is complex, and the fault current wave characteristic is hidden in the normal current, so it is difficult to extract the arc characteristics, which increases the difficulty of analyzing the series arc faults of the branch from the change of the trunk current waveform. With the development of artificial intelligence technology, LSTM network is gradually favored by scholars. Therefore, this paper extracts current characteristics from the trunk current, determines the series fault arc in the line through the change of the trunk current waveform, cuts off the connection of the line, and detects the series fault arc by using the deep LSTM network Measurement.

2 EXPERIMENTAL PLATFORM CONSTRUCTION AND SAMPLES ANALYSIS

In order to simulate the series arc fault generator with multi branches on the low voltage electrical side, it is necessary to build an experimental platform according to GB14287.4–2014 electrical fire monitoring system Part 4: arc fault detector [5]. The experimental platform of arc fault is shown in Figure.1. With 220 V, 50 Hz AC as the input voltage, the disconnector is set as the main circuit switch and the air circuit breaker is used as the circuit protection device. When the current is too large, the air circuit breaker is disconnected, and 100w1p cement is set on the trunk road The resistance is used as sampling resistance, 4mm copper wire is used as circuit connecting line, tiepiescope HS801 five in one virtual comprehensive tester is used for oscilloscope.

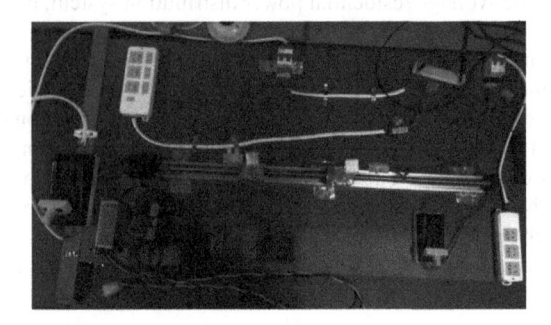

Figure 1. Experimental platform of arc fault.

In order to better simulate the series parallel connection form of the load end line in the actual power grid, the electrical equipment in the common low-voltage distribution network is classified according to the load characteristics, and several typical electrical equipment are selected as the experimental objects. According to the load characteristics, the experimental equipment is divided into three types of loads: the kettle, the heater belong to the resistive load, the air conditioner and fan belong to the inductive load, and the computer and TV belong to the nonlinear load. Resistive load only works through resistance type elements, inductive load through the motor with coil electrified to generate magnetic field as the core component, and nonlinear load contains rectifier equipment. The experimental current signal data are collected by the virtual tester HS801, the data sampling frequency is set to 50kHz, a total of 10,000 points are collected for each sampled data, the sampling time is 0.2S, and a total of 10 current cycles. And filter the collected current data. In the 10 cycle waveforms collected at a time, there are 8 or more fault arc half-wave waveforms (or 4 full-wave fault cycle waveforms appear), can judge this the group current data is arc fault data. The data collection scheme is shown in Table 1.

The current waveform of the trunk line is affected by the branch load. When the branch load changes, the trunk current will change. As shown in Figure 2, the heater is a resistive load. When the kettle works alone, the normal current waveform is sine wave. When the kettle has series arc fault, there is a "flat shoulder" phenomenon at the zero crossing point of the current waveform, which is the temporary load break caused by the extinction of the arc. When the arc reburning, the current waveform increases instantaneously and there are a lot of "spikes" in the waveform, but the waveform is complete Because the difference between the fan current and the kettle current is too big, the difference between the main circuit current and the single circuit kettle is not big. When the kettle fails, the main circuit current waveform appears "spike" compared with the normal, and there is a slight "flat shoulder" phenomenon. Compared with the single branch heater fault current, the "spike" is reduced, and the "flat shoulder" phenomenon is less obvious.

Table 1. Proposed experimental scheme and load parameters.

Experimental Category	Branch 1 Loads	Branch 2 Loads	Data Collection Normal	Fault
1	Kettle(1500W)	Television(120W)	3000	3000
2	Kettle(1500W)	Electric Fan(60W)	3000	3000
3	Kettle(1500W)	Heater(400W)	3000	3000
4	Kettle(1500W)	Break Off	3000	3000
5	Air Conditioner(2000W)	Television(120W)	3000	3000
6	Air Conditioner(2000W)	Electric Fan(60W)	3000	3000
7	Air Conditioner(2000W)	Heater(400W)	3000	3000
8	Air Conditioner(2000W)	Break Off	3000	3000
9	Television(120W)	Computer(350W)	3000	3000
10	Television(120W)	Electric fan(60W)	3000	3000
11	Television(120W)	Heater(400W)	3000	3000
12	Television(120W)	Break Off	3000	3000

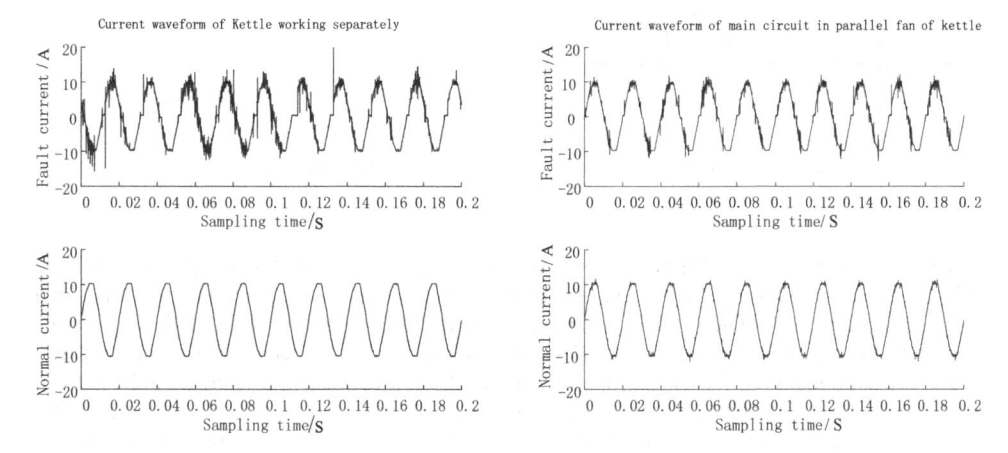

Figure 2. Current waveform of trunk road when Kettle branch is fault branch.

3 IMPROVED LSTM NETWORK

German scholar s. Hochreiter Added a gating structure to RNN network, and obtained an improved cyclic neural network model LSTM network. Compared with the traditional RNN network, LSTM network adds three gate units to control the transmission of information, which are input gate, output gate and forgetting gate. Gate unit mechanism can effectively forget useless information and retain effective input information, so as to reduce errors and effectively process long-term sequences. The core of LSTM network is unit state C, which keeps the memory of unit state at time t, remembers or forgets the previous state through the forgetting gate, updates the unit state through the input gate, and controls the output signal to be transmitted to the next cell through the output gate. The unit structure of LSTM network is shown in Figure 3. Its mathematical descriptions are presented as follows.

$$f_t = \text{Sigmoid}(W_f[h_{t-1}, x_t] + b_f) \tag{1}$$

$$i_t = \text{Sigmoid}(W_i[h_{t-1}, x_t] + b_i) \quad (2)$$

$$\tilde{C}_t = \text{Tanh}(W_c[h_{t-1}, x_t] + b_c) \quad (3)$$

$$C_t = f_t * C_{t-1} + i_t * \tilde{C}_t \quad (4)$$

$$o_t = \text{Sigmoid}(W_o[h_{t-1}, x_t] + b_o) \quad (5)$$

$$h_t = o_t * \text{Tanh}(C_t) \quad (6)$$

In the above formula, x_t represents the input at time t, h_{t-1} represents the output at time t-1, h_t represents the output at time t, $[h_{,t-1}, x_t]$ represents the connection of the current time x_t and the previous time output $h_{,t-1}$ into a new vector as the input, + represents the addition by elements, * represents the multiplication by elements, W_f, W_i, W_c, W_o is the weight matrix, b_f, b_i, b_c, b_o is the offset value, Sigmoid and Tanh are the internal activation functions.

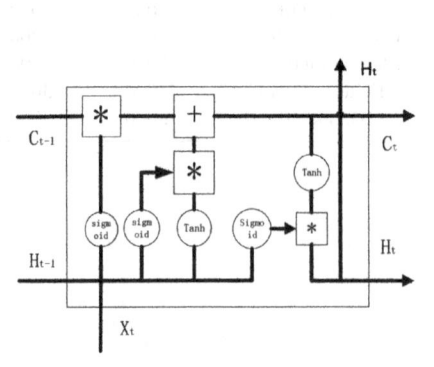

Figure 3. Schematic diagram of LSTM network.

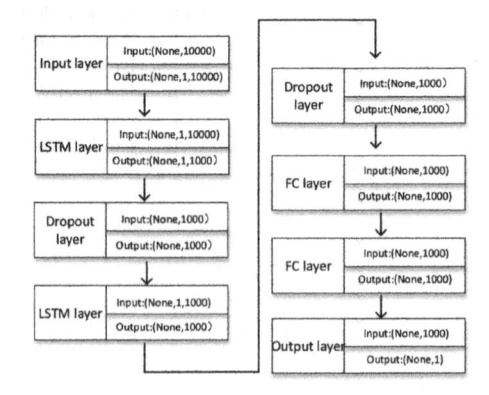

Figure 4. LSTM network model.

LSTM network has good processing results in timing problem. In this paper, the sensitivity of LSTM network to sequential current signal is used to detect multi branch fault current signal through improved LSTM network. In order to enhance the learning ability of the model, the original LSTM network is improved by adding LSTM network layer, dropout layer and full connection layer. If there are too many LSTM layers, the training results are difficult to converge. In the network model constructed in this paper, the LSTM layer is two layers, the output of the upper LSTM layer is taken as the input of the next LSTM layer, and the dropout layer is added after the second LSTM layer to reduce over fitting and improve the generalization energy Finally, softmax is made in the output layer, and the results are classified. The network model results are shown in Figure 4. The Adam optimizer, which uses momentum to improve traditional gradient descent and promote hyperparametric dynamic adjustment, is selected in this model. After several training experiments, we can see that when the learning rate is set to 0.001 and each batch of 500 samples, the model training accuracy and test accuracy are higher. In this paper, the model is used to detect and classify the current data, and the test set is sent to the trained model for binary classification. The binary classification confusion matrix of classification results is shown in Table 2.

Table 2. Binary confusion matrix.

Practical markup	Predictive markup		In Total
	Predictive Positive	Predictive Negative	
Practical Positive	TP	FN	TP+FN
Practical Negative	FP	TN	FP+TN
In Total	TP+FP	FN+TN	TP+FP+FN+TN

According to the confusion matrix, the precision and recall of the model are determined. The performance of network model needs to be judged according to different standards. Accuracy is an important index to judge whether the model is good or not. Accuracy refers to the probability of correctly classified samples in the total samples. The calculation of accuracy is shown in equation 7. Due to the particularity of low-voltage distribution system, line fault is easy to cause fire and other hazards. In low-voltage distribution system, it is more necessary to detect the fault and cut off the line when the line fault occurs. Therefore, the model needs to ensure the accuracy rate of the actual fault arc identification, that is, the recall rate of the model, which refers to the actual current signal of the model The recall rate is shown in equation 8.

$$\text{Accuracy} = \frac{TP + TN}{TP + FN + FP + EN} \tag{7}$$

$$\text{Recall} = \frac{TP}{TP + FN} \tag{8}$$

4 EXPERIMENTAL RESULTS AND ANALYSIS

The operating system of constructing deep LSTM network is Ubuntu 16.04 system in this study, and the programming environment is pychar community 2018.2 software. The deep LSTM network model is constructed by tensorflow GPU of version 1.10.0.

4.1 *Training results and test results*

The experimental data were classified according to 5:1 training set and test set. A total of 72000 data samples were collected, 60000 samples were used as training sets, the remaining 12000 samples were used as test sets, and the training samples are sent to the network model, after 120 iterations, the training process is over. To present the change of training accuracy and loss function intuitively, the two curves are plotted. Figure 5 describe accuracy curve, Figure 6 describe loss curve, with the increase of the number of iterations, the accuracy rate gradually increases. When the iteration reaches 60 times, the accuracy rate converges, which is basically stable at 97%. During the training process, the loss value shows a downward trend, tends to be stable when iterating to about 60 times, and finally converges to about 0.09.

Test samples were randomly scrambled and sent into the trained deep LSTM network model to judged and evaluated the accuracy of the network model. The detection results are shown in Figure 7, the average accuracy rate is 95.67%, and the average recall rate is 98.17%. The experimental results show that the depth LSTM network has a good detection effect in multi branch series arc fault detection.

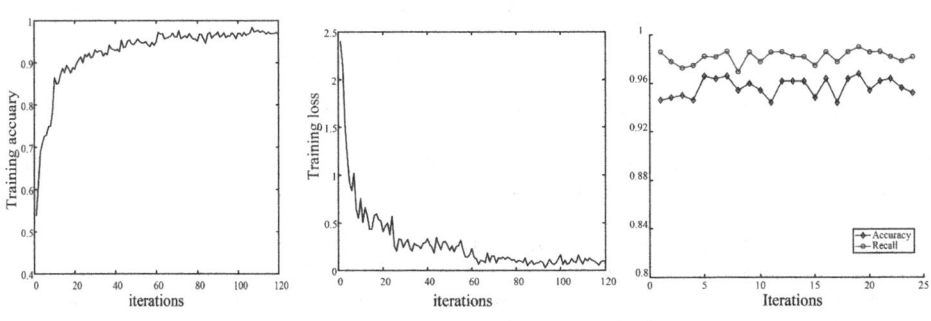

Figure 5. Model training accuracy curve.

Figure 6. Model training loss curve.

Figure 7. Accuracy curve of training set.

4.2 Compared with other experimental phenomena

In order to verify the recognition and classification effect of the proposed algorithm for series arc fault, the load circuit, application method, application range, sample number and detection accuracy are compared with some methods. The results are shown in Table 3. Siegle [6] used Fourier coefficients, Mel frequency cepstrum data and wavelet features as feature vectors to input deep neural network; Yu [7] used convolution neural network to detect series fault arc in single circuit. The application scope included resistive, inductive and nonlinear loads, but the number of data samples was different In addition, the design of the number of load loops is also different. The scheme designed in this paper can not only detect different loads, but also detect multi branch loads. When the number of samples exceeds other methods, the accuracy still reaches 95.67%. The experimental results show that our method has high accuracy.

Table 3. Compared with other experimental phenomena.

Literature sources	Methods	Number loop	Application Range	Sample size	Detection Accuracy
Siegle [6]	Deep neural network algorithm for multi-feature fusion	Single loop	resistive, inductive, Nonlinear loads	30600	95.61%
Yu [7]	parallel Alexnet	Single loop	resistive, inductive, nonlinear loads	7200	92%
Our	Deep LSTM	Multiloop	resistive, inductive, nonlinear loads	72000	95.67%

5 CONCLUSION

The actual power line wiring is complex, the current signal is hidden, and it is difficult to detect the fault arc on the trunk road side. Based on the recursive network, the signal with time sequence characteristics can be effectively processed. In this paper, a multi branch series fault arc detection method based on improved deep LSTM network is proposed. The experimental results show that the method can effectively identify the series fault arc under multiple branches, provides an effective theoretical basis for electrical fire prevention.

REFERENCES

[1] Yin Z, Wang L, Zhang Y, et al. A novel arc fault detection method integrated random forest, improved multi-scale permutation entropy and wavelet packet transform[J]. Electronics, 2019, 8(4): 396.
[2] Lippert, K.J.; Domitrovich, T.A. AFCIs—From a standards perspective. IEEE Trans. Ind. Appl. 2014, 50, 1478–1482.
[3] Pulkkinen J P. Commercial arc fault detection devices in military electromagnetic environment [J].IEEE Electromagneti Compatibility Magazine,2018,7 (4) :49–52.
[4] Su J J, Xu Zh H. Multivariable fault arc diagnosis method based on EMD and PNN [J]. Power automation equipment, 2019,39(4): 106–113.
[5] Fire detection and alarm sub committee of national fire protection standardization technical committee.gb14287.4–2014, electrical fire monitoring system Part 4: fault arc detector [s]. Beijing: China Standard Press, 2014.
[6] Siegel J E, Pratt S, Sun Y, et al. Real-time Deep Neural Networks for internet-enabled arc-fault detection[J]. Engineering Applications of Artificial Intelligence, 2018, 74(SEP.):35–42.
[7] Yu Q Huang G, Yang Y. Low voltage AC series arc fault detection method based on parallel deep convolutional neural network[J]. IOP Conference Series Material science and Engineering, 2019, 490(7):072020.

Optimization strategy for power sellers to purchase and sell electricity considering use behavior

Z.K. Tan
Electric Power Research Institute of Guizhou Power Grid Co. Ltd., Guiyang, China

Y.Q. Wang & J. Zhao
College of Electrical Engineering, Guizhou University, Guiyang, China

B. Liu
Electric Power Research Institute of Guizhou Power Grid Co. Ltd., Guiyang, China

M. Liu
College of Electrical Engineering, Guizhou University, Guiyang, China

ABSTRACT: As one of the response strategies of demand response, real-time electricity price can provide sufficient and effective price signals for power users, and then effectively guide users to participate in demand response so as to improve their own power consumption curve. Under this background, considering the degree of user participation, the degree of regret matching mechanism is introduced to describe the probability that users choose to participate in a certain demand response, and a power seller model considering user choice behavior and user demand response is established in order to solve the problems of users' enthusiasm to participate in the operation of power market and the maximization of users' income in the development of smart grid. The electricity purchase and sale strategies of power sellers with and without user choice behavior are compared and analyzed, and the feasibility of the model is verified through the simulation of an example.

1 INTRODUCTION

The opening of electricity sales market is an important part of the new round of power system reform In fact, the key is to establish the operation mechanism of electricity sales market and cultivate the main body of market competition on the side of electricity sales. The main body of the electricity seller is mainly composed of independent electricity sellers, who make their own decisions to purchase electricity. In addition, power users can make a power purchase plan through negotiation with the electricity sellers [1–4].

A large number of applications of intelligent control and two-way communication technology, as well as distributed power generation and electric vehicles on the user side, have greatly improved the ability of users to actively respond and participate in power grid regulation with the development of smart grid. As a consequence, the power purchase and sale decision of the power sales company with reasonable consideration of user demand response has become a new topic that urgently needs to be studied [5]. The demand response is considered in the current research, which is usually real-time electricity price, time-of-use electricity price and interruptible load in the optimal decision-making of purchasing and selling electricity [6]. The dynamic optimization scheme of demand response based on reinforcement learning is proposed in reference [7]. The dynamic optimization income function of the electricity seller is established, and the response load is determined by the user demand response benefit function through the establishment of the demand response model of the electricity seller and the user, and considering the relationship between the user comfort cost function. In addition, the initial demand response income function of the seller is transformed into an immediate response function. Literature [8] considers the complementarity and diversity of power load so as to integrate and improve user resources, and then analyzes the effect of load rate on power purchase cost by reducing the cost of electricity purchase and promoting the income of power sellers. Taking into account the comprehensive load characteristics of users, the combination optimization model of electricity sellers is established. Literature [9] mainly constructs a load combination optimization model aiming at maximizing the profit of electricity purchase and sale for power sellers in the medium and long-term market, and then uses demand response technology to improve the comprehensive load characteristics of users. It maximizes the profit of electricity sellers in the market environment. Literature [10] analyzes the operation mode of e-commerce sellers and divides different types of users; furthermore, the demand response behavior of users is modeled according to the development of demand response under the environment of the Internet of things and the elasticity coefficient of electricity price; based on the peak-valley time-of-use electricity price, a multi-type load coordination operation optimization model is established with the goal of maximizing the profit of e-commerce sellers. Very few literatures consider the impact of user demand response on power sellers' purchasing and selling strategies under real-time electricity prices. As a consequence, it will be studied in this paper.

Whether users choose to participate in demand response should follow the wishes of users and their right of choice in the study of power purchase and sale model. As a result, the benefit settlement of power sellers' optimization strategy for electricity purchase and sale is closer to the actual market trading results [11,12]. They have independent optimization power and choice power when users choose whether or not to participate in demand response. Retailers should take into account user self-optimization and their choice behavior when making decisions on power purchase and sale strategies. A regret matching mechanism is introduced to describe the probability that users choose to participate in a demand response. In addition, a power seller model considering user choice behavior is established. The strategies of power sellers considering and not considering users' choice behavior are compared and analyzed.

2 USER BEHAVIOR MODEL

2.1 User behavior analysis

Generally speaking, choosing to participate in demand response or not and the degree of participation response after they choose to participate are mainly affected by the pre-achieved benefits after participating demand response and the impact of the user's own electricity behavior on the income loss caused by the impact of the participation demand response. The profit from participating in demand response is the direct reason that determines the degree of user participation in demand response. It is assumed that they have to reflect their own degree of participation demand response to the system operator after each user response. A user behavior model is constructed according to the "regret matching mechanism" according to the above description. The "regret matching mechanism" describes how to obtain an "average degree of regret" based on user reporting and past participation, and then decide whether to participate in the next strategy and the degree of participation [13,14]. For the "average regret degree", it describes the difference between the average profit of the user choosing a new demand response strategy to replace the old demand response strategy. As a consequence, the higher the value is, the greater the probability that the user will re-choose the demand response strategy.

2.2 Regret degree matching mechanism model

The "regret matching mechanism" describes that the participants of an event make a decision on whether to participate in the next related event and the degree of participation based on the "regret measure" of the results of all previous related events. The "regret measure" describes the real benefits gained by the participants in the event based on past decisions and participation in the decision-making. If other factors that affect the choice of events are not changed, we replace all the past and real participation in the event into a new decision a, and finally calculate the increase of the benefits. The probability of the event participant to re-select the demand response strategy is greater if the calculated "regret measure" value is higher. In addition, its occurrence probability is positively correlated with its calculated value according to the principles described above.

Its own benefit is the difference between the benefit of reducing electricity demand during the peak period and the loss caused by the transfer of load during the trough period when the user participates in the demand response:

$$W_i = D_{i0}p_{i0} - D_i p_i = D_{i0}p_{i0} - (D_{i0} + \frac{D_{i0}\alpha_{ii}(p_i - p_{i0})}{p_{i0}} + \sum_{j=1}^{24} \alpha_{ij} \frac{D_{i0}}{p_{j0}}(p_j - p_{j0}))p_i \quad (1)$$

$$= D_{i0}\left(p_{i0} - p_i - \frac{\alpha_{ii}(p_i - p_{i0})}{p_{i0}} + \sum_{j=1}^{24} \frac{\alpha_{ij}(p_j - p_{j0})}{p_{j0}}\right)$$

In the formula, D_{i0} indicates the initial electricity load of the user; x_i, x_j, x_{i0}, x_{j0} is the catalogue electricity price of the power grid merchant and the sales price of the electricity seller, respectively. α_{ii} denotes its own coefficient of price elasticity; In addition, α_{ij} represents cross-coefficient of price elasticity.

According to the historical participation degree reported by the user, the demand response strategy set of user participation is discretized, that is, $\prod_{ui} = \{u_i | u_i = 0, 0.1, 0.2 \cdots 1; \forall_i\}$. The user can decide the degree of participation in this event before the demand response is really involved after receiving the demand response signal sent by the system. For a user, the average degree of regret for not choosing a new policy u_i^k, but still choosing my old strategy u_i^j in the nth event can be expressed by the formula [14]:

$$R_n^i(j,k) = max(D_n^i(j,k), 0) \quad (2)$$

In the formula: $R_n^i(j,k)$ choose to replace the average regret measure for users to participate in decision-making. n is the number of times users participate in the decision, and h is the number of responses to the current demand response. $x_h^i(\beta_{i,h}^j)$ and $x_h^i(\beta_{i,h}^k)$ are the participation rates for participation policies u_i^j and u_i^k for the user in the h th event, respectively.

According to the above formula, if $D_n^i(j,k) > 0$, the user will have a greater bias to replace the old policy and choose the new one. If in the nth event, the user selects decision j, then in the n + 1 event, the probability that the user chooses to replace to decision k and still selects decision j is:

$$x_{n+1}^i(\beta_{i,h+1}^k) = (1 - \sigma) \, min\left(\frac{1}{\lambda}R_n^i(j,k), \frac{1}{S-1}\right) + \sigma\frac{1}{S} \quad (3)$$

$$x_{n+1}^i(\beta_{i,n+1}^j) = 1 - \sum_{\beta_i^k \in \Pi_\beta, \beta_i^j \neq \beta_i^k} x_{n+1}^i(\beta_{i,n+1}^k) \quad (4)$$

In the formula: σ is the weight coefficient, and the range of values is (0,1); S is the number of elements in the user's u_i, π is the inertia coefficient that the user still chooses the original strategy, and needs to meet $\Pi > 2W_i^{max}(S - 1)$. Among them, W_i^{max} is the maximum income value for users to participate in the demand response project. In addition, when there is a difference in the type of users, the value of π is also different.

According to the above described model, first, randomly generate users to participate in a policy transfer distribution x_1^i, and ensure $x_1^i \geq \sigma/S$, and then calculate the x_n^i, of the user policy transfer distribution according to the regret matching mechanism. It can match the user participation under the current policy according to the distribution sampling. In formula (3), σ/S represents a random fluctuation value.Characterizing the subjectivity and randomness of users in the behavior choice of demand response strategy in order to ensure that each demand response strategy has the lowest selected record.

Based on the above description, users will determine the participation degree and choice of demand response in the next stage according to their historical participation and their actual income value. Users can obtain a series of participation degree and probability of user participation in demand response in the next stage. For simple expression, the j participation degree and probability of user i are assumed to be $\beta_{i,j}$ and $p_{i,j}$, respectively. When the user is rational and conservative, the user reports the willingness β_i of the seller to participate:

$$\beta_i = \beta_{i,j} p_{i,j} \tag{5}$$

3 ELECTRICITY SELLER'S ELECTRICITY PURCHASE AND SALE MODEL CONSIDERING USER BEHAVIOR

3.1 Profit model of electricity purchase and sale by electricity sellers

The difference between users' contracted purchases and actual purchases will be settled on an hourly basis in the spot market. As a consequence, users' electricity consumption needs to be broken down to hourly consumption. The contract electricity quantity and its price are usually determined in advance between the user and the seller in order to restrict and regulate the interests of both parties more effectively. As a consequence, the purchase cost of the bilateral contract between the consumer and the seller is F_b:.

$$F_b = \sum_{t=1}^{T} (\frac{x_{bt}^n + x_{rt}}{2}) Q_{bt}^n \tag{6}$$

In the formula, t is the total time to make the purchase and sale strategy; x_{bt}^n is the contract electricity price signed by the bilateral contract n in the period t. x_{rt} is the real-time electricity price in the t period; Q_{bt}^n is the agreed transaction quantiThe demand response model is introduced into the demand response model under real-time electricity price in order to cut the peak and fill the valley for the user load and maximize the interests of the users participating in the load regulation [15]. The load adjustment or load transfer discussed in this paper is carried out for one day, that is, 24 hours. Available, that is, the adjusted electricity consumption of the user is Q_{ri}.

$$Q_{ri} = Q_{i0} + \frac{Q_{i0}\alpha_{ii}(x_i - x_{i0})}{x_{i0}} + \sum_{j=1}^{24} \alpha_{ij} \frac{Q_{i0}}{p_{j0}} (x_j - x_{j0}) \tag{7}$$

In the formula, Q_{i0} is the user load under the grid merchant catalogue price; x_i, x_j, x_{i0}, x_{j0}, is the i, j time grid merchant catalogue price and the seller's sales price, respectively.

The regret matching mechanism is introduced to describe the user's willingness to choose to participate in the demand response. Assuming that the user i chooses to participate in the demand response, the afterload of the demand response is:

$$Q'_{ri} = \beta_i Q_{i0} + \frac{\beta_i Q_{i0} \alpha_{ii}(x_i - x_{i0})}{x_{i0}} + \sum_{j=1}^{24} \alpha_{ij} \frac{\beta_i Q_{i0}}{p_{j0}} (x_j - x_{j0}) \tag{8}$$

The real-time balance transaction is used to realize the real-time balance of the electricity difference. We can get the user's actual electricity consumption and the income of demand response after optimizing the user's power consumption curve:

$$F_{RT} = \sum_{t \in T} (Q'_{ri} - \beta_i \sum_{b=1}^{N} Q_{b,t}) \cdot x_{r,t} \tag{9}$$

The daily income of e-commerce sellers in the retail market is as follows:

$$F = \sum_{t=1} Q_t^s \cdot p_t \tag{10}$$

Q_t^s is the electricity consumption after the user participates in the demand response adjustment at t time. In addition, p_t is the real-time electricity price implemented by the electricity seller to the user.

This study considers purchasing electricity from bilateral contracts in the medium-and long-term market and the spot market, taking the maximum profit of the sellers as the optimization goal, and taking the electricity price p_t and the medium-and long-term purchase ratio ϕ as the optimization variables. In addition, the objective function is that the profits of the electricity sellers are as follows:

$$\text{Max} K = F - F_b - F_{RT} = \sum_{t=1} Q'_{ri} \cdot p_t + \sum_{t=1}(1-\beta_i)Q_{t0} \cdot p_{t0}$$
$$- \sum_{b=1}^{N}\sum_{t=1}^{T}(\frac{x_{b,t}+x_{rt}}{2})Q_{b,t} - \sum_{t \in T}(Q'_{ri} - \beta_i \sum_{b=1}^{N} Q_{b,t}) \cdot x_{r,t} \tag{11}$$

In the formula, T = 24, N is the number of bilateral contracts signed by the power sellers (N = 3).

The scientific basis of setting electricity price based on marginal cost is that the selling company can determine the price according to the increase or decrease of enterprise cost caused by the increase or decrease of users' electricity demand. In addition, users can adjust their own power demand according to the fluctuation of price in order to achieve the purpose of reasonable consumption of electricity [16]. The marginal cost is expressed as the derivative of the total cost in mathematical form, that is,

$$p_{t0} = C'(Q''_{ri}) \tag{12}$$

$$C = F_b^k + F_{Rt} = \sum_{b=1}^{N}\sum_{t=1}^{T}(\frac{x_{b,t}+x_{rt}}{2})Q_{b,t} + \sum_{t \in T}(Q'_{ri} - \beta_i \sum_{b=1}^{N}Q_{b,t}) \cdot x_{r,t}$$
$$= \sum_{b=1}^{N}\sum_{t=1}^{T}(\frac{x_{b,t}+x_{rt}}{2})\alpha_b \phi(Q'_{ri} + (1-\beta_i)Q_{t0}) + \sum_{t \in T}(Q'_{ri} - \beta_i \phi(Q'_{ri} + (1-\beta_i)Q_{t0}))) \cdot x_{r,t}$$
$$= \sum_{b=1}^{N}\sum_{t=1}^{T}(\frac{x_{b,t}+x_{rt}}{2})\alpha_b \phi Q''_{ri} + \sum_{t \in T}(Q''_{ri} - (1-\beta_i)Q_{t0}) - \beta_i \phi Q''_{ri})) \cdot x_{r,t}$$
$$= \sum_{t=1}^{T} Q''_{ri}(\sum_{b=1}^{N}(\frac{x_{b,t}+x_{rt}}{2})\alpha_b \phi + (1-\beta_i \phi)x_{r,t}) - (1-\beta_i)Q_{t0}x_{r,t} \tag{13}$$

$$Q''_{ri} = Q'_{ri} + (1-\beta_i)Q_{t0} \tag{14}$$

In the formula, C' is the marginal cost, p_{t0} is the initial electricity price, Q''_{ri} is the user's electricity demand, Q_{t0} is the user's initial electricity demand, C is the total cost of electricity purchased by the electricity seller, and α_b is the electricity purchase ratio of the b contract signed between the power seller and the power generator. For the above optimization problems, the selling company first sets the price to the marginal cost of the electricity demand provided by the selling company and the fluctuation of user demand according to the results of load forecasting. Based on this initial value, the selling company and users iteratively calculate the real-time electricity price p_t and Q''_{ri} demand for each period of time in advance, and then implement them according to the corresponding plan. It can be solved by distributed iteration for each user because each power user is independent of each other, and the solution steps are expressed as

$$p_t^k = C'(Q''^{(k)}_{ri}) \tag{15}$$

$$Q_{ri}^{\prime\prime(k+1)} = Q_{ri}^{\prime\prime(k)} + \chi\left(\frac{\partial W(Q_{ri}^{\prime\prime(k)})}{\partial Q_{ri}^{\prime\prime(k)}} - p_t^k\right) \qquad (16)$$

In the formula, χ represents the step of iteration and k represents the number of iterations. Figure 1 shows the dynamic coupling relationship of the solution of this variable.

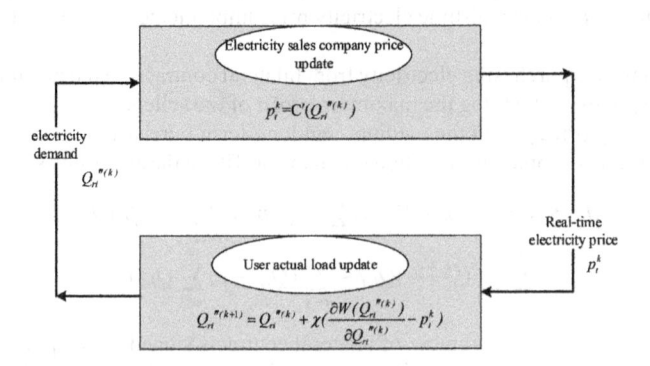

Figure 1. Dynamic Coupling relationship between load and electricity Price.

3.2 *Solving process*

The model solving process is shown in the Figure 2.

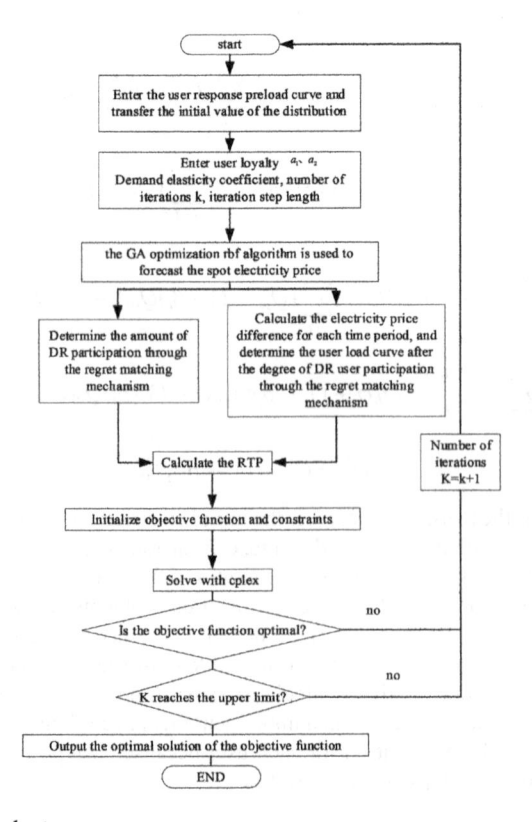

Figure 2. Solving flow chart.

First of all, the forecast power load value and demand response coefficient are inputted(Reference [13]), user loyalty a1 = 0.4, a2 = 0.4, iteration times k = 20. According to the historical data of pjm, the GA optimization rbf algorithm is used to forecast the spot electricity price. Under the above conditions, according to the regret matching mechanism, the actual participation of users in the demand response to the release of electricity sellers is calculated. In addition, the real-time electricity price and the medium-and long-term market purchase ratio are taken as variables. Moreover, the cplex toolbox in matlab is selected to optimize the objective function. Finally, the optimal electricity price which meets the constraints and maximizes the profit of electricity sellers is obtained.

4 EXAMPLE SIMULATION

According to the historical data of pjm, the spot price forecasting curve 3 is obtained by using GA optimized rbf algorithm (Figure 3).

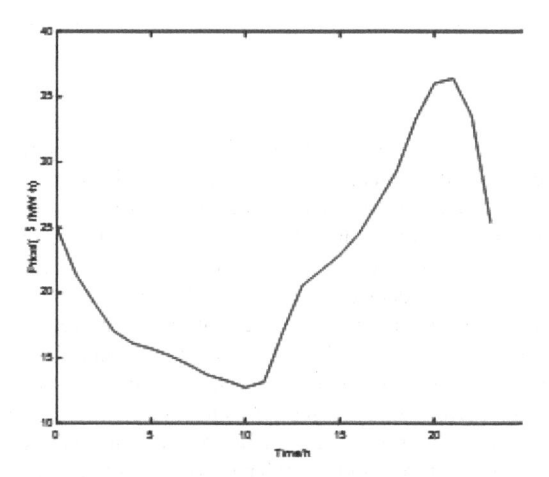

Figure 3. Spot electricity price forecasting curve.

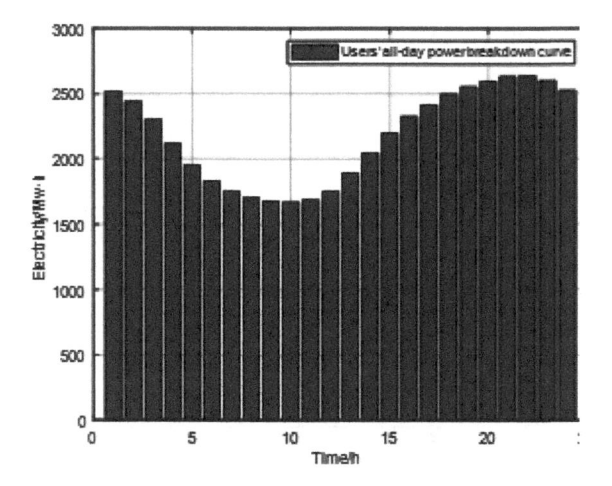

Figure 4. Load decomposition curve.

Assuming that the electricity seller decides the medium-and long-term market purchase volume and publishes the electricity price to the users one month in advance, this study takes the typical summer load of a certain area in the PJM electricity market as an example. It forecasts the total amount of the agent user of the electricity seller in this area every day in the next month according to its historical data. In addition, takeing 24 hours as the optimization cycle, it calculates the medium- and long-term market purchase volume and electricity price of each day in the next month. The load curve of the decision day is obtained by decomposing the forecast load as shown in Figure 4. The volume and price information of bilateral contracts that e-sellers can choose in the medium-and long-term market is shown in Table 1.

Table 1. Bilateral contracts available in the medium-and long-term market.

No.	Contract electricity price ($/MW·h)	Minimum load/ MW·h	Maximum load/ MW·h
1	10	8000	15000
2	15	10000	17000
3	12	14000	21000

According to the electricity price optimization process in Figure 1, the electricity price and user load are continuously optimized by distributed iteration. In addition, the real-time electricity price which maximizes the profit of purchasing and selling electricity is finally output, as shown in Figure 5. Curve 1 is the electricity price curve after considering the user behavior (that is, the degree of user participation). Besides, curve 2 is the user curve when the user fully participates in the demand response (that is, the situation considered in the third chapter). The electricity price obtained by considering the user behavior is lower than that obtained without considering the user behavior during the peak period. However, the electricity valley period is on the high side, which is due to the fact that the user does not fully participate in the demand response. Only some users participate in the demand response to realize the optimization of their own load, which leads to the increase of the actual transaction load of users. Besides, the electricity sellers need to buy more electricity in the spot market. In addition, the influence of user load leads to the change of electricity price. As a consequence, if the impact of user behavior is not taken into account, it will lead to the optimization that the sales electricity price is too high or too low in the actual transaction. It will lead to an increase in the electricity cost of users or an increase in the loss of electricity sellers.

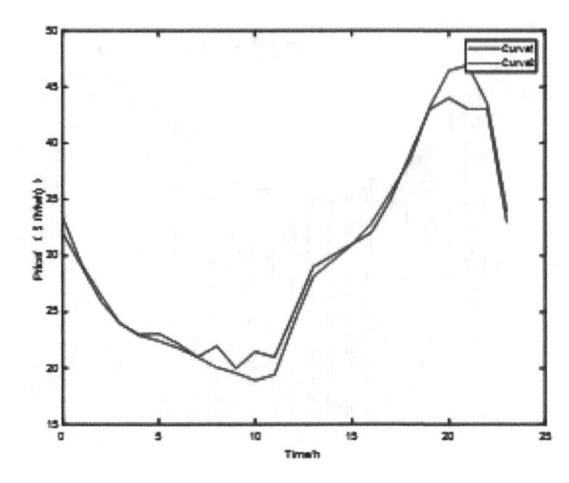

Figure 5. Real-time electricity price curve of sales.

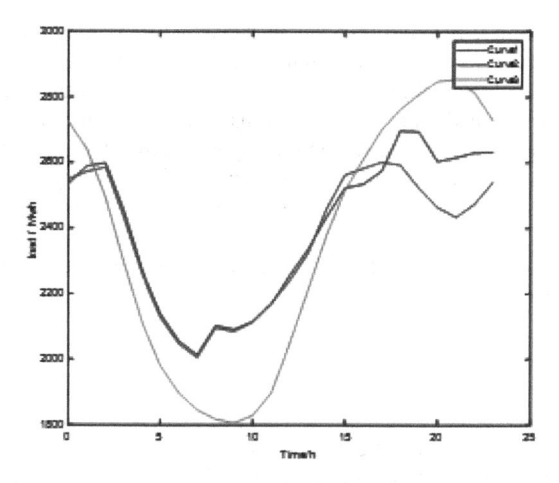

Figure 6. User load curve.

The comparison of the optimized load before and after the optimization is shown in Figure 6 through the above optimized real-time electricity price to guide the user to respond to the demand. It can be seen that the demand response model can guide the user to complete peak cutting and valley filling, and optimize the user load curve. Curve 1 is the load curve after considering the user behavior (that is, the degree of user participation). Besides, curve 2 is the load curve in which the user fully participates in the demand response (that is, the degree of participation is 100%). Moreover, curve 3 is the user's original load curve. As can be seen from Figure 5, whether or not the user behavior is taken into account, the implementation of the demand response to the user will optimize the user load curve, reduce the peak period and increase the power consumption during the trough period, making the curve relatively flat. When the electricity seller makes the decision of purchasing and selling electricity, it is assumed that all the users participate, and the load curve shows the law of curve 2 if the user behavior is not taken into account. In the actual transaction, it will produce the deviation between the actual electricity load of the user and the planned purchase load of the electricity seller, resulting in the electricity seller can only trade in the spot market to make up for the deviation and increase the electricity purchase cost of the electricity seller though the load optimization effect is better. If the user behavior is taken into account, the electricity seller will avoid the above errors when making the decision of purchasing and selling electricity. As a consequence it can reduce the loss of the seller.

Table 2. Parameter comparison.

Strategy of buying and selling electricity	Medium-and long-term electricity purchase ratio	Medium-and long-term electricity purchase / (MW·h)	Revenue from e-commerce sellers ($)
Regardless of demand response	0.4523	28904	34569
Regardless of user behavior	0.5075	32857	39654
User behavior	0.4963	30758	36589

As can be seen from the above table, e-commerce sellers make strategic decisions on the purchase and sale of electricity considering that the response of users' demand can increase their revenue and increase their market share. It can also concluded that without considering the user behavior (that is, when the user fully participates in the demand response), it is a relatively ideal state relative to taking into account the user behavior through the above analysis. However, it is impossible for the user to achieve the complete demand response in the actual transaction in the electricity market. From Table 2, taking into account the differences in user behavior and not taking into account user behavior, it can also be concluded that the power seller's power purchase cost increases and revenue decreases when considering the degree of user participation. It also verifies the load law described above. The sellers regard the maximum response value of users as the transaction volume in the spot market because when formulating the strategy of purchasing and selling electricity. As a consequence, it can reduce the cost of purchasing electricity. However, the effect of demand response can not reach the ideal state expected by the sellers in the specific implementation stage. As a consequence, taking into account the user behavior is closer to the actual transaction situation in the electricity market.

5 CONCLUSION

This paper constructs a decision-making model of power sellers considering user behavior choice and dynamic demand response, and uses cplex in matlab to obtain the optimal solution that meets the constraint conditions based on the regret matching mechanism and the electricity price-market share model. The example analysis verifies the following conclusions:

1)The response of users' demand is considered, the load curve of users is improved, the purchase of electricity in the spot market is reduced, and the income of the sellers is increased in the decision-making of purchasing and selling electricity.

2)The response of users' demand is considered, the load curve of users is improved, the purchase of electricity in the spot market is reduced, and the income of the sellers is increased in the decision-making of purchasing and selling electricity.

REFERENCES

[1] Nojavan S , Nourollahi R , Pashaei-Didani H , et al. Uncertainty-based electricity procurement by retailer using robust optimization approach in the presence of demand response exchange[J]. International Journal of Electrical Power & Energy Systems, 2019, 105(FEB.):237–248.
[2] J. Hou et al., "An Energy Imbalance Settlement Mechanism Considering Decision-Making Strategy of Retailers Under Renewable Portfolio Standard," in IEEE Access, vol. 7, pp. 118146–118161, 2019, doi: 10.1109/ACCESS.2019.2936459.
[3] Wang Y , Wang D , Zhang H , et al. Optimal bidding of price-maker retailers with demand price quota curves under price uncertainty[J]. IEEE Access, 2020, PP(99):1–1.
[4] W. Xu, P. Zhang and D. Wen, "Decision-making model of Electricity Purchasing for Electricity Retailers Based on Conditional Value-atRisk in Day-ahead Market," 2020 12th IEEE PES Asia-Pacific Power and Energy Engineering Conference (APPEEC), Nanjing, China, 2020, pp. 1–5, doi: 10.1109/APPEEC48164.2020.9220431.
[5] J. Campos Do Prado and W. Qiao, "A Stochastic Bilevel Model for an Electricity Retailer in a Liberalized Distributed Renewable Energy Market," in IEEE Transactions on Sustainable Energy, vol. 11, no. 4, pp. 2803–2812, Oct. 2020, doi: 10.1109/TSTE.2020.2976968.
[6] H. Pan, H. Gao, W. Ma and J. Liu, "Dynamic Bidding Strategy for Electricity Retailers Considering Multi-type Demand Response," 2020 IEEE Sustainable Power and Energy Conference (iSPEC), 2020, pp. 1127–1132, doi: 10.1109/iSPEC50848.2020.9350983.
[7] FENG Xiaofeng, XIE Tiankuo, GAO Ciwei, et al. A Demand Side Response Strategy Considering Long-term Revenue of Electricity Retailer in Electricity Spot Market[J].Power System Technology, 2019, 43(08): 2761–2769.
[8] HUANG Guori, SONG Yihang, CAI Hao, GAO Ciwei, CHEN Tao. Combinatorial Optimization Strategy of Power Sales Company Considering Transferable load[J]. lectric Power Construction, 2020, 41(11):126–134.
[9] XIAO Yong, WANG Yan, QIAN Bin, JIANG Jing, GAO Ciwei. Modeling Method of Load Combination Optimization for Electricity Retailer Considering Coordination of Power Generation and Consumption[J]. Automation of Electric Power Systems, 2020, 44(20):148–156.

[10] GUO Manlan, CHEN Haoyong, XIAO Wenping, N Zhifeng, SHEN Na, HUANG Zhaowen, LIU Fenghua, ZHANG Lirong.Operational Optimization Model for Electricity Suppliers under the Internet of Things Environment[J]. Smart Power, 2020, 48(11):80-85+91.

[11] TAN XiANDong, CHEN Yuchen, LI Yang, JING Jiangbo, JIANG Ning, WANG Zijian, SHEN Yunwei.esearch on Optimization of TOU Considering Load Development and User Behavior[J]. Electric Power, 2018, 51(07):136–144.

[12] LI Hongzhu, CAO Renzhong, ZHANG Xinyu. Demand Response Model Based on Real-time Pricing Under Household Power Background[J]. Proceedings of the CSU-EPSA, 2018, 30(11):26-31+51.

[13] Wang Y, Zhang Ning, Kang Chongqing, Xi Weiming, Huo Molin.Electrical Consumer Behavior Model: Basic Concept and Research Framework[J].TRANSACTIONS OF CHINA ELECTROTECHNICAL SOCIETY, 2019, 34(10):2056–2068.

[14] Wu Geng. Influence of Demand Response on Supply Adequacy of Distributed Generation System Considering Behaviors of Users[J]. Automation of Electric Power Systems, 2018, 42(08):119–126.

[15] LI Chunyan, XU Zhong, MA Zhiyuan. Optimal Time-of-use Electricity Price Model Considering Customer Demand Response[J]. Proceedings of the CSU-EPSA, 2015, 27(03):11–16.

[16] H. P. KHOMAMI, M. H. JAVIDI. Energy management of smart microgrid in presence of renewable energy sources based on real-time pricing[C]//2014 Smart Grid Conference (SGC) . Tehran, Iran: IEEE, 2014, pp:1–6.

Theoretical studies on the electronic structure and optical properties of NaTaO₃ doped with different concentrations of cerium and nitrogen

G.W. Pang, D.Q. Pan, C.X. Liu, L.Q. Shi, X.D. Wang, L.Z. Liu, J.B. Liu, L. Ma, L.L. Zhang & B.C. Lei

Xinjiang Laboratory of Phase Transitions and Microstructures in Condensed Matter, College of Physical Science and Technology, Yili Normal University, Yining, China

ABSTRACT: In this paper, density functional theory is used to investigate the effect of single doping and co-doping of Ce and N atoms at different concentrations on the electronic structure and optical properties of NOT system. The calculation results reveal that Ce concentration of 0.375 at% provides the best stability in the co-doped systems. Doping elements have a significant influence on the chemical properties of NOT and can lead to lattice distortion, which is beneficial for the separation of photogenerated electron-hole pairs. After doping, the band gap of the system is reduced. Consequently, the energy required for the electronic transition from valence band to conduction band is reduced, which causes a red shift in the absorption band edge. The absorption spectrum of the co-doped system appears in the visible range. When the Ce concentration in the co-doped system is 0.125 at% and 0.25 at%, a stronger absorption peak appears in the visible range, indicating that these two systems have the best response to visible light.

1 INTRODUCTION

Owing to its unique advantages of mild reaction conditions and direct conversion of solar energy into chemical energy, photocatalysis has garnered considerable researcher attention [1–4]. Recently, several researchers have focused on finding an efficient photocatalytic material that can respond to visible light [5, 6]. Fujishima and Honda [7] were the first to discover the photocatalytic properties of TiO_2, which had opened the door for photocatalysis research. In 1998, $NaTaO_3$ (Referred to as NTO in the text) was found to be completely hydrolyzed under the action of ultraviolet light [8]. Such tantalate-based semiconductors with perovskite-type structure have attracted global attention as a new type of photocatalytic material [9–11], because their photocatalytic activity are much higher than that of TiO_2 under undoped condition. Further, NiO-doped NTO exhibits significantly improved photocatalytic activity [12]. However, since the intrinsic forbidden gap width of NTO is extremely large (4.1 eV [13]), it can only absorb ultraviolet (UV) light irradiation, which accounts for just 4% of solar energy at the Earth's surface. Therefore, people have done a lot of research on how to improve the photocatalytic activity and visible light response of NTO by doping. Since metal ions are effective acceptors of electrons, they can capture electrons in the conduction band, which can reduce the recombination of photogenerated electrons and photogenerated holes on the semiconductor surface, thus affecting the absorption and catalytic activity of light. [14] The reported doping of common metal ions mainly includes alkali metal (Sr, Ba, Ca) [15, 16] and lanthanide (La, Pr, Nd, Eu.) [17] For example, Su et al. [18] Prepared NTO crystal doped with Eu^{3+} by solid-state method. Eu^{3+} is uniformly incorporated into the lattice of NTO, which can adjust the photocatalytic activity of NTO. It shows the performance of photocatalytic water desorption of hydrogen in both UV and visible light regions. Kudo Al found [19] when the doping concentration of La is 2%, the quantum efficiency can be improved to 56% under the 270nm UV irradiation. However, although the quantum efficiency has been improved, the response light region has not been expanded to the visible light region. Therefore, people try to modify NOT with non-metallic doping. The doping of NOT with nonmetallic ions usually leads to the increase of O hole concentration or the substitution of nonmetal ions for part of O^{2-}, which leads to the narrowing of the band gap of NOT and the broadening of the response range of visible light. In recent years, the doping of N, S, C and other nonmetallic ions into NOT has become the focus of scientific research. For instance, Liu Darui [20] found that after doping S in NOT, O^{2-} ions are replaced by S^{2-} ions to form Ta-S-Ta bonds. The photoresponse range of the doped sample extends to the visible light region. Hu [21] prepared N-doped NOT by hydrothermal and solid-phase methods. Due to the hybridization of O 2p and N 2p, the band gap of the sample is reduced, and the photocatalytic activity of the catalyst under visible light is significantly enhanced. Although single doping of metal or single doping with non-metal can improve the performance of the material to a certain extent, but, when the valence of the doping element and the bulk element are different, the impurity state occupied by the impurity element can act as the recombination center of electrons and holes. Thereby reducing the photocatalytic performance [22] Compared with single-doping, co-doping can greatly improve the stability of the system, which is mainly due to the strong Coulomb interaction between the positive and negative charges carried by the anion and cation, and it can effectively inhibit the electron-hole recombination [23] As proof, Tian et al. co-doped NOT with non-metallic elements N/C and metallic element Fe it is observed that co-doping can reduce the band gap to a larger extent than single doping, causing an obvious red shift in the absorption band edge and improving the photocatalytic efficiency in the visible region.

In this study, we attempt to modify NOT by co-doping with non-metal N and metal Ce. It is expected that the co-doped systems has both responsiveness to visible light and excellent photocatalytic activity. After investigation, lanthanides have excellent performance in doping modification. Therefore, we will modify NTO with doping different concentrations of Ce and N analyzed the electronic structures and optical properties of the systems in detail to provide a theoretical basis for the experimental preparation of NOT

2 COMPUTATIONAL DETAILS

In the article, the first-principles method based on density functional theory (DFT) is used to calculate the electronic and structural properties by using CASTEP [24] software in Materials Studio 2017. The Perdew-Burke-Ernzerhof (PBE) [25] functional under the generalized gradient approximation (GGA) is chosen as the exchange correlation functional. K lattice points are selected in the Brillouin zone with $3 \times 3 \times 3$ grid. The plane wave truncation energy E_{cut} is 410 eV, and the self-consistent convergence accuracy is 5.0×10^{-6} eV/atom. Here, the electronic configurations of five atoms are discussed: Na $(2p^63s^1)$, Ta $(5d^36s^2)$, $O(2s^22p^6)$, $Ce(4f^15d^16s^2)$, and $N(2s^22p^3)$. The space group of perovskite NOT is P12/M1 (No. 10). The experimental values of lattice constants [26] are a = c = 3.8895 Å b = 3.8855 Å $\alpha = \gamma = 90°$ and $\beta = 90.367°$. Researchers have tested three kinds of large and small supercells [27] ($2 \times 2 \times 2$, $2 \times 2 \times 3$, $3 \times 3 \times 3$) all parameters were shown in Table 1:

Table 1. Lattice constant, volume and energy band values of different size intrinsic NOT supercell systems.

Model	a/Å	b/Å	c/Å	Cell volume/Å³	Energy
2×2×2	8.006575	8.006575	8.006575	513.263421	-88546.45
2×2×3	8.003908	8.003908	12.003978	769.005338	-132819.71
3×3×3	12.001187	12.001185	12.001187	1728.512569	-298844.10

It was found that the calculated results are basically consistent. Therefore, to save computational resources, we chose $2 \times 2 \times 2$ supercell model (40 atoms per supercell). The pure NOT model is shown in Figure 1(a). Considering the boundary effect, N and Ce replace O and Ta atoms in the body center of NOT to form doped models of N@O and Ce@Ta, respectively, which were shown in Figure 1(b-c) Subsequently, one, two, and three Ta atoms in the supercell are replaced with Ce atoms to form the three doped models in Figure 1(df), called $Na_8Ce_xTa_{8-x}O_{24}$ (x=1, 2, 3), respectively. For convenience, they are abbreviated as N@O&xCe@xTa (x = 1, 2, 3).

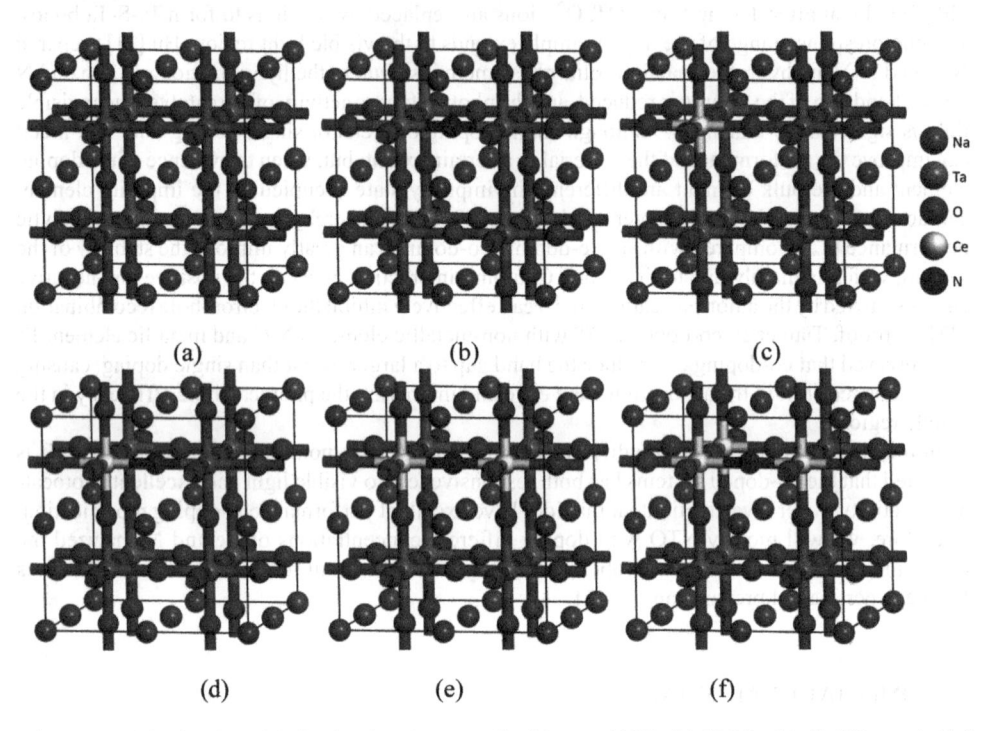

Figure 1. Calculated model for $2 \times 2 \times 2$ supercell: (a) pure NOT, (b) N@O, (c) Ce@Ta, and (d-f) N@O&xCe@xTa (x = 1, 2, 3).

3 RESULTS AND DISCUSSION

3.1 *Structural optimization*

Table 2. Comparison of lattice parameters a, b, c, c/a and cell volume of pure NOT, N@O, Ce@Ta, and N@O&xCe@xTa (x = 1, 2, 3) from this work.

Model	a/Å	b/Å	c/Å	c/a	Cell volume/Å^3
NOT	8.276	8.277	8.276	1.000	566.932
N@O	8.299	8.282	8.282	0.998	569.217
Ce@Ta	8.355	8.355	8.355	1.000	586.166
N@O&Ce@Ta	8.566	8.288	8.286	0.967	588.317
N@O&2Ce@2Ta	8.384	8.383	8.499	1.014	597.365
N@O&3Ce@3Ta	8.663	8.379	8.402	0.970	610.454

Table 2 shows the lattice constants and cell volumes of undoped, single-doped, and co-doped NOT. It was evident that the unit cell volume of single-doped and co-doped systems were increased as compared to that of pure NOT because the radius of the dopant Ce atom was slightly larger than that of the Ta atom ($R_{Ta} = 2.09$Å, $R_{Ce} = 2.70$Å); As the doping concentration of Ce increases, the cell volume of Ce-doped crystals also increases. After the Ta atom were replaced by the Ce atom, due to the different properties of Ce atoms and Ta atoms, the bond lengths of Ce-O and Ta-O change, and the adjacent lattice points were subjected to stress, which causes the periodic potential field in the crystal to change. In turn, the crystal lattice was distorted. Figure 2 compares the cell diagram of pure NOT and N@O&3Ce@3Ta after geometric optimization. It was clear that the lattice of the doped systems were distorted (the distorted position was shown in Figure 2(b)), which was consistent with the analysis of cell parameters.

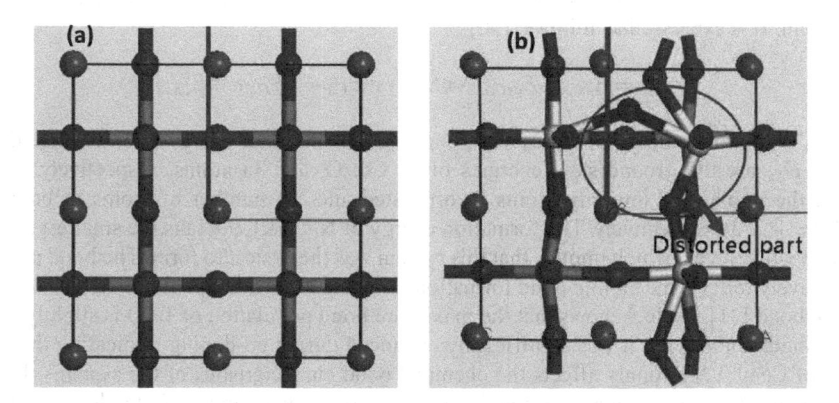

Figure 2. Schematic diagram of the lattice structure after geometric optimization: (a) Intrinsic NOT (b) N@O&3Ce@3Ta.

Table 3. Binding energys energies, formation energies, and bond populations of various models.

Model	E_b/eV	E/eV	E_f/eV	Ta-O(max)	Ta-O(min)	Ce-O(max)	Na-O(max)
NOT	−36.615	−22045.473	/	0.71	0.71	/	−0.06
N@O	−36.433	−21877.017	1.591	0.72	0.67	/	−0.05
Ce@Ta	−36.238	−22970.085	2.941-	0.72	0.67	0.47	−0.05
N@O&Ce@Ta	−36.149	−22802.074	4.086	0.95	0.37	0.51	−0.04
N@O&2Ce@2Ta	−56.563	−23726.611	7.201	0.94	0.41	0.52	−0.01
N@O&3Ce@3Ta	−77.228	−24655.994	5.326	0.95	0.32	0.57	−0.01

Table 3 shows that compared to pure NOT, the lattice defect of the doped systems changes the total energy of the crystal. As the doping concentration of Ce increases, the total energy of the doping systems decreases gradually. Further, the N@O&3Ce@3Ta system exhibits the lowest total energy, indicating that it was the most stable. When the crystal is in a stable state, the total energy is different from when n atoms are in a free state. And the difference between the two is defined as the binding energy of the crystal [28]:

$$E_b = E_0 - E_n \qquad (1)$$

E_0 is the total energy of the crystal, and E_n represents the sum of the energies of the n atoms composing the crystal in their free state. The more negative the binding energy is, the stronger the binding strength is and the more stable the structure is. According to the definition of binding energy, for NOT, its binding energy can be expressed as [29]:

$$E_b(NaTaO_3) = \frac{E_t(NaTaO_3) - nE_{isolate}(Na) - nE_{isolate}(Ta) - 3nE_{isolate}(O)}{n} \qquad (2)$$

$E_t(NaTaO_3)$ is the total energy of $NaTaO_3$ the supercell, $E_{isolate}(Na)$ and $E_{isolate}(Ta)$ are the total energy of Na and Ta atoms respectively, $E_{isolate}(O)$ is the total energy of a single isolated O atom, n is the number of NOT formula in the adopted supercell, and the total energy of a single isolated Na, Ta, and O atom is directly output by CASTEP. According to the data in the table 3, compared with pure NOT, the binding energy of the doping systems decreases gradually with the increase of the co-doping concentration of N and Ce, among which the binding energy of the N@O&3Ce@3Ta doping system was the smallest, indicating that the system was the most stable, which was Consistent with the total energy analysis. The formation energy signifies the degree of difficulty in the formation of a system. The smaller the formation energy, the easier the formation of a system. It is expressed as follows: [30]

$$E_f = E_{defect} - E_{perfect} - nE_N - mE_{Ce} + nE_O + mE_{Ta} \qquad (3)$$

$E_{perfect}$ is the total energy of the system, and E_{defect} is the total energy after doping. E_N, E_{Ce}, E_O, and E_{Ta} are the ground state energies of N, Ce, O, and Ta atoms, respectively. n and m indicate the number of impurity atoms incorporated and the number of atoms to be replaced ($n = 1 m = 1, 2, 3$), respectively. The formation energy of N@O&Ce@Ta is the smallest among all the co-doped systems, which implies that this system was the easiest to form. The bond population quantitatively reflects the atomic bond formation. The larger the bond population, the stronger the covalent bond [31]. Table 3 shows that the maximum bond population of Ta-O is slightly changed during single doping, but it was significantly enhanced during co-doping, indicating that the co-doping of Ce and N strongly affects the chemical bond characteristics of the systems. However, the difference between the maximum and minimum bond population of Ta-O also increases due to co-doping, which can lead to an increase in lattice distortion. This result was consistent with the analysis of lattice constants, and the lattice distortion destroys the original crystal symmetry of NOT. The negative charge centers are not coincident, but a local potential difference was generated, which was beneficial to hinder the recombination of photogenerated electron-hole pairs. Therefore, the photocatalytic performance of the material is expected to be improved. Furthermore, it can be seen from Table 3 that the Na-O bond always maintains the ionic bond property, while the Ta-O bond always maintains the covalent bond property, indicating that TaO_3 forms a covalent bond with isolated Na atom in the crystal. This implies that Na^+ and TaO_3 primarily bond through ionic interactions, which is consistent with the results of Zhurova [32].

3.2 Electronic structures

Figure 3(a) shows the energy band structure of pure NOT The calculated band gap width is 1.678 eV, which is much smaller than the experimental value of 4.1 eV. This is attributed to the fact that the DFT method underestimates the band gap width of the system. However, we are interested in the relative variation trend, so it does not affect the analysis of the system. Figure 3(b) displays the N@O band diagram, which reveals that the energy band gap is significantly reduced by 20% as compared to that of pure NOT. In addition, the energy required for the electronic transition from valence band to conduction band is also reduced. This is because of N atom, which has the same valence state as the O atom is introduced as an impurity. The holes in the vicinity of the Fermi level are mutually repulsive, and the hole carriers are localized at the top of the valence band. At the same time, the energy levels were split, and multiple deep acceptor levels were formed near the Fermi level. Therefore, the quantity of adjacent atoms increase in the vicinity of the Fermi level, and the repulsive force also increases, leading to system instability and lattice distortion, which is consistent with the analysis of lattice constant Figure 3(c) shows the Ce@Ta energy band diagram. Since the number of valence electrons in Ce is less than that in Ta, doping introduces redundant carrier holes, and the forbidden gap width also decreases slightly. Further, the distribution of energy levels becomes dense. These changes significantly impact the generation, transition, and composite behavior of electrons. To give you an idea, the energy required for the electrons to transition from the valence band to the conduction band is reduced, so the absorption band edge is red-shifted, thereby extending the response range of the material to visible region, which is consistent with the results of optical properties. Figure 3(d-f) show the energy band diagram of N@O&xCe@xTa ($x = 1, 2, 3$), Among them, the Fermi levels of N@O&xCe@xTa ($x = 1, 2$) systems pass through the valence band, indicating the characteristics of N-type semiconductor. Further, the width of the band gap was reduced. Although the reduction was not large, obviously, the number of energy levels near the Fermi level was far more than that of pure non-energy levels. There was a significant increase in the number of active electrons, i.e., their transition probability has also increased.

Furthermore, the introduction of dopants increases the lattice defects, resulting in some irregular structures inside and on the surface. This irregularity is closely related to the surface electronic state. The surface electronic states are generally different from the bulk energy states in semiconductors. Such electronic states act as a trap for capturing charge carriers, which help to suppress the recombination of photogenerated electrons and holes and improves the photocatalytic efficiency of the material [33].

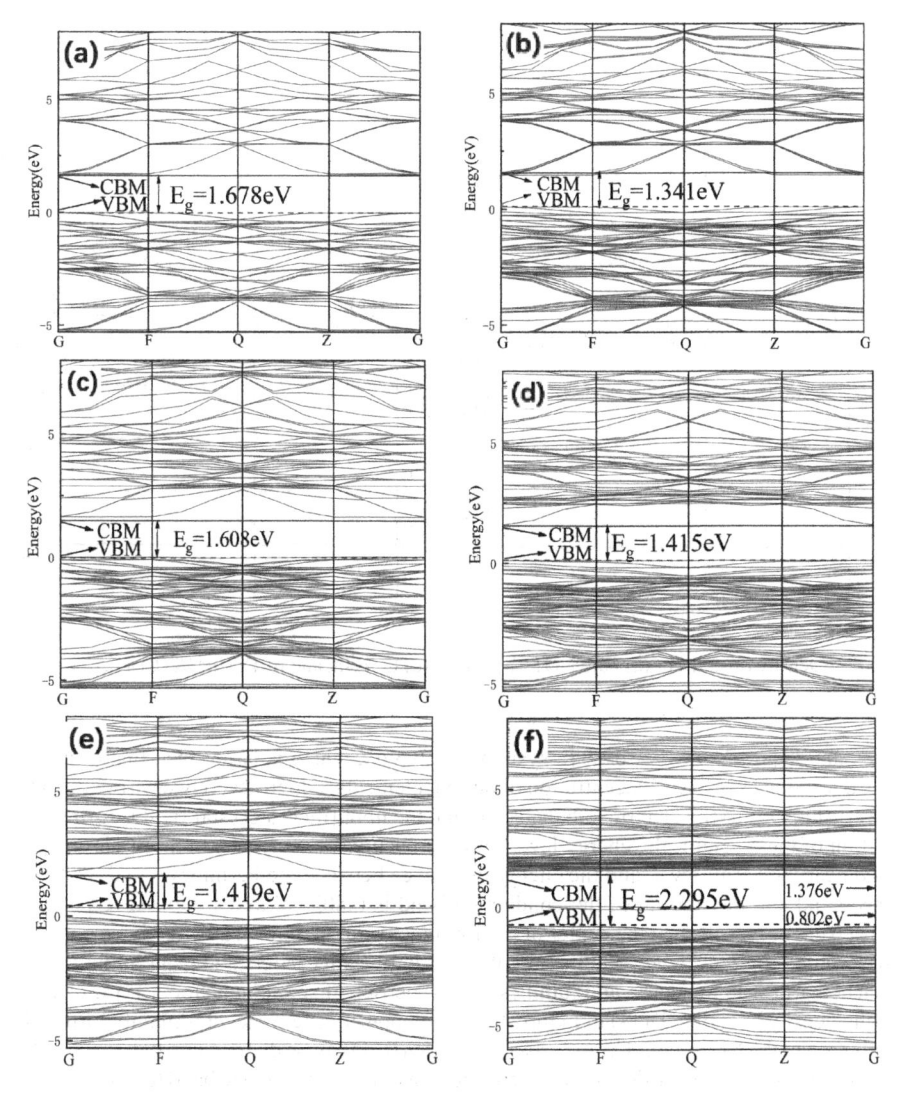

Figure 3. Band structures of (a) pure NOT, (b) N@O, (c) Ce@Ta, and (d-f) N@O&xCe@xTa ($x = 1, 2, 3$).

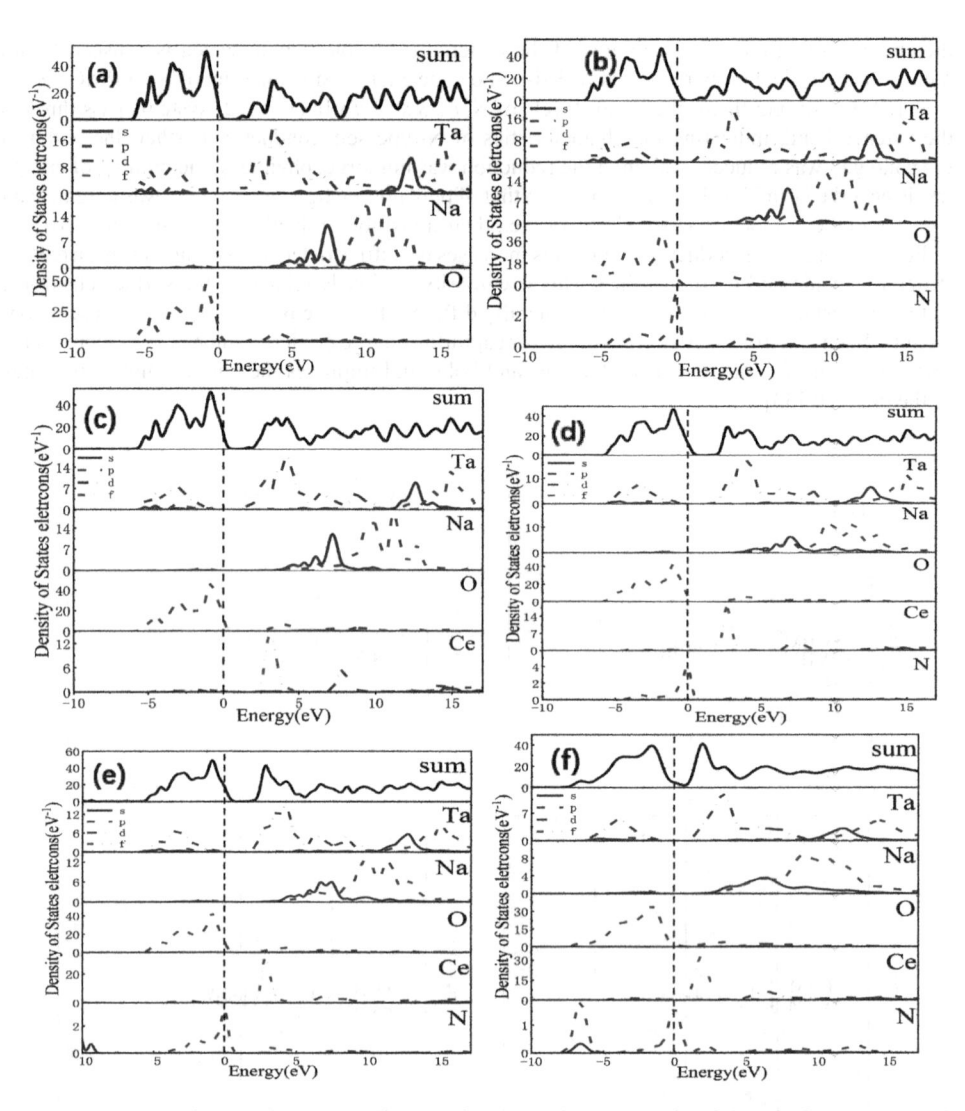

Figure 4. Density of states of (a) pure NOT, (b) N@O, (c) Ce@Ta, and (d-f) N@O&xCe@xTa ($x = 1, 2, 3$).

The band gap of N@O&3Ce@3Ta system is 2.295 eV, which is larger than that of pure NOT. This is due to the excessive concentration of doping elements, which severely destroys the integrity of the crystal. Due to the electric imbalance around the impurities, a crystal field is formed. Under the action of the crystal field, energy level splitting occurs. Concurrently, impurity levels [34] are formed near the Fermi level, and the energy difference between the Fermi level and the top of the valence band and the bottom of the conduction band is 0.802 eV and 1.376 eV, respectively. In fact, deep energy levels near the center of the forbidden gap often form an effective recombination center, which promotes the recombination of non-equilibrium carriers and plays an important role in enhancing the photoelectric and luminescent properties of semiconductors [35] It can be seen from the density of states (DOS) in Figure 4 that the impurity energy level is mainly formed by the contribution of the N-2p state. The increased forbidden band can be divided into two small forbidden bands, which become the bridge of the electronic transition from valence band to conduction band. A small amount of energy can cause a transition to the impurity level and then to the conduction band. Due to the reduced energy required, NOT can respond to low-energy visible light, i.e., its photocatalytic efficiency is improved.

Compared to pure NOT, the energy required for electronic transition from valence band to the conduction band of all co-doped systems decreases, which is conducive to the red shift of the edge of the absorption band. The band gap of N@O&Ce@Ta system is the smallest, indicating that its photocatalytic activity is the best.

Figure 4(a) shows the DOS of pure NOT. It is clear that near the Fermi level, the valence band is mainly contributed by Ta-5d and O-2p states, and the conduction band is mainly contributed by Ta-5d and Na-3s states. The strong interaction between Ta-5d and O-2p states (Ta-5d and Na-3s states) results in a peak at -2.874 eV (7.360 eV). Figure 4(b) shows the DOS of N@O It is evident that doping of N has a minor effect on the atomic states in the intrinsic system. However, due to the contribution of the 2p state, the valence band of the system moves up, and the band gap width decreases, i.e., the energy required for electronic transition decreases, which extends the response range of the system to visible region. Figure 4(c) shows the DOS of Ce@Ta It can be seen that the introduction of Ce mainly affects the conduction band of the system. The density peak near the Fermi level is formed by the superposition of Ta-5d and Ce-4f states, which slightly lowers the conduction band of the system and the band gap width decreases, which is consistent with the band analysis. Figure 4(d-f) show the DOS of three systems: O&xCe@xTa (x=12,3). Here, the N-2p state exhibits a peak at the Fermi level, but the performance is quite different. The Fermi levels in N@O&xCe@xTa (x=1,2) systems passes through the valence band, while the N@O&3Ce@3Ta system exhibits an impurity level at the Fermi level because the Ce-4f state slowly reaches the Fermi level as the Ce concentration increases. The energy levels are close, which is inductive to the N-2p state and repellent to the O-2p state. Therefore, while reaching the maximum concentrationof Ce doping in this paper, the Ce-4f state pushes the O-2p state away from the top of the valence band. therefore the N-2p state may be the contributor of generating of impurity level.

3.3 *Optical properties*

In order to research the optical properties of NOT before and after doping with N and different concentrations of Ce, the dielectric function and absorption spectrum of NOT before and after doping are calculated and studied. In this paper, the scissors operator is used for correction when calculating the optical properties, and the scissors operator is taken The value is 2.422 eV. Figure 5 shows the real part of the dielectric function (dielectric constant) of the systems before and after doping as a function of incident light energy. The dielectric constant characterizes the system dielectric polarization ability under an external field as wells as its charge binding capacity. In the absence of light, the static dielectric constant of the system is the intersection of the curve and the vertical axis. The static dielectric constants of pure NOT N@O, Ce@Ta, and N@O&xCe@xTa ($x = 1, 2, 3$) are 4.667, 4.955, 5.894, 4.801, 6.163 and 8.977, respectively. It can be seen that the static permittivity of the doped systems increases in varying degrees compared to that of the pure system, which implies that doping enhances the polarization ability of the systems and the binding capacity of the charge carriers. Further, it facilitates a faster migration of the photogenerated carriers in the systems.

Since this article hopes to split water in the visible light range to produce hydrogen, the absorption spectrum of the visible light range is expanded, as shown in Figure 6, which shows the absorption spectrum of NOT before and after doping. It can be seen from the figure 6 that the edges of the absorption band of all doped systems move to the low-energy direction, causing red shift, thereby expanding the response range to the visible light region of systems. The wavelength at the edge of the absorption band represents the relative change trend of the forbidden band width in the systems. After doping, the forbidden band width is reduced to varying degrees, which is consistent with the red shift at the edge of the spectral absorption band. Ce@Ta and N@O&xCe@xTa ($x = 1, 2, 3$) compared with before doping, the absorption peak position and absorption capacity of the four doping systems have changed significantly. The absorption peaks appear at 2.71, 3.55, 3.12, 3.36 eV, and the absorption capacity of the co-doped system increases with the increase of Ce concentration in the co-doped systems. Among them, the absorption capacity of the N@O&3Ce@3Ta system has the most obvious increase, which is three times that of the Ce@Ta single-doped system, indicating that the system has the best photocatalytic performance.

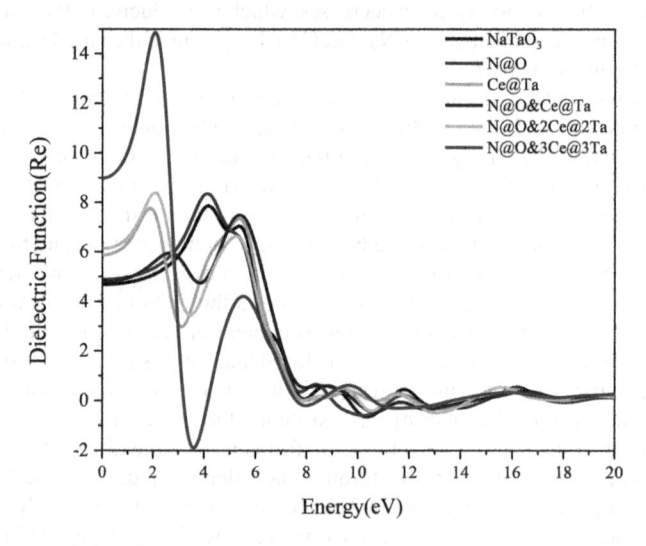

Figure 5. Real part of the dielectric function of (a) pure NOT, (b) N@O, (c) Ce@Ta, and (d-f) N@O&xCe@xTa ($x = 1, 2, 3$).

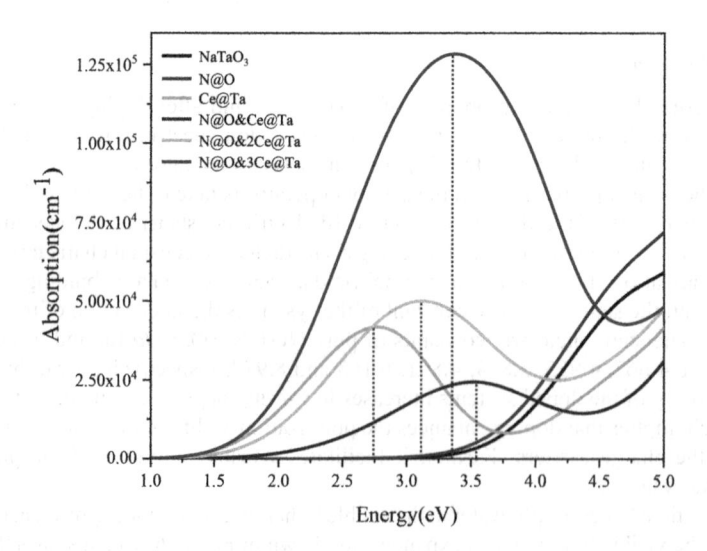

Figure 6. Optical absorption spectra of (a) pure NOT, (b) N@O, (c) Ce@Ta, and (d-f) N@O&xCe@xTa ($x = 1, 2, 3$).

4 CONCLUSIONS

In this study, the electronic structure and optical properties of N and Ce single-doped and co-doped NOT were calculated using first-principles methods, and then compared with those of pure NOT. The main results are summarized as follows.

- Among all the doped systems, the formation energy of N@O&Ce@Ta is the smallest, indicating that it is the easiest to form. Bond group analysis shows that the introduction of impurities strongly affects the chemical bonding properties of the crystal and causes lattice distortion, which is beneficial to hinder the recombination of photogenerated electron-hole pairs, thereby improving the photocatalytic activity of the system.
- From the analysis of the electronic structure, it can be seen that the band gaps of all systems are reduced to varying degrees after doping. This means that the energy required for the electronic transition is reduced, which is conducive to the red shift at the edge of the absorption band. The reduction of the gap is mainly attributed to the contribution of the Ce-4f state and the N-2p state.
- The analysis of optical properties shows that compared with pure NOT, the static dielectric constants of all doped systems are increased, indicating that the polarization ability and charge binding ability of the system have been enhanced. Therefore, photogenerated carriers exhibit faster internal migration in the system.
- The optical absorption spectrum of the system shows that compared with pure NOT, the edges of the absorption band of all doped systems were red-shifted, and Ce @ Ta, N @ O & xCe @ xTa ($x = 1, 2, 3$) four systems emerge absorption peak in the visible light region. The absorption intensity of each system in the co-mixing system increases with the increase of Ce concentration in the co-mixing of N and Ce. Among them, N@O&3Ce@3Ta exhibits the strongest absorption peak, which is three times that of Ce@Ta single-doped system, indicating that its photocatalytic performance is the best among all doped systems.

ACKNOWLEDGMENT

This work was supported by the Xinjiang research projects for colleges and Universities (Grant No.XJEDU2021Y044) and the National Natural Science Foundation of China(Grant No.11664042.

REFERENCES

[1] Sheng X, Liu Z, Zeng R, Chen L, Feng X, Jiang L. (2017) Enhanced Photocatalytic Reaction at Air–Liquid–Solid Joint Interfaces. J AM CHEM SOC. 36: 12402–12405.
[2] Feng X, Yin L, Wang Z, Jin Y, Huang W. (2017) Surface Reconstruction-Induced Site-Specific Charge Separation and Photocatalytic Reaction on Anatase TiO_2 (001) Surface. J PHYS CHEM C. 121: 9991–9999.
[3] Li F, Han T, Wang H, Zheng X, Wan J, Ni B. (2017) Morphology evolution and visible light driven photocatalysis study of Ti^{3+} self-doped TiO_{2-x} nanocrystals. J MATER RES. 32: 1563–1572.
[4] Zhao Shiqiang, Iocozzia, James, WangYang, Cui Xun, Li Zhen. (2017) Noble metal-metal oxide nanohybrids with tailored nanostructures for efficient solar energy conversion, photocatalysis and environmental remediation. J Energy & Environmental Science Ees. 10: 402–434.
[5] Kovacic, M Kusic, H Fanetti, M Stangar, U L Valant, M Dionysiou, D D Bozic, A L. (2017) TiO_2-SnS_2 nanocomposites: solar-active photocatalytic materials for water treatment. J Environmental science and pollution research international. 24: 19965–19979.
[6] Opoku F, Govender KK, Sittert CGCE, Govender PP. (2017) Recent Progress in the Development of SemiconductorBased Photocatalyst Materials for Applications in Photocatalytic Water Splitting and Degradation of Pollutants. J Advanced Sustainable Systems. 1: 1700006.
[7] Fujishima A, Honda K. (1972) Electrochemical Photolysis of Water at a Semiconductor Electrode. J NATURE. 238: 37–38.
[8] Hideki, Kato, And, Akihiko, Kudo. (1998) New tantalate photocatalysts for water decomposition into H_2 and O_2. J CHEM PHYS LETT. 295: 487–492.
[9] Ding, Q Liu, Y Chen, T Wang, X Feng, Z Wang, X Dupuis, M Li, (2020) Unravelling the water oxidation mechanism on $NaTaO_3$-based photocatalysts. J MATER CHEM A. 8: 6812–6821.
[10] Basaleh AS, Shawky A, Zaki ZI. (2021) Visible light-driven photodegradation of ciprofloxacin over sol-gel prepared Bi_2O_3-modified La-doped $NaTaO_3$ nanostructures. J Ceramics International 47:19205–19212.

[11] Tang ZK, Valentin CD, Zhao X, Liu LM, Selloni A. (2019) Understanding the Influence of Cation Doping on the Surface Chemistry of NaTaO$_3$ from First Principles. J *ACS CATAL*. 9: 10528–10535.
[12] Yeh MY, Li JH, Chang SH, Lee SY, Huang H. (2019) Facile hydrothermal synthesis of NaTaO$_3$ with high photocatalytic activity. J *MOD PHYS LETT B*. 14n15: 1940046.
[13] Zhang J, Zhang D. (2020) Electronic structure and optical properties of anion-doped monoclinic NaTaO$_3$ by first-principles calculations. J *CAN J PHYS*. 393–395: 80–83.
[14] Paz Y. (2019) Transient IR spectroscopy as a tool for studying photocatalytic materials. J *Journal of Physics: Condensed Matter*. 31: 503004.
[15] Mizutani, S Karimata, I An, L Sato, T Kobori, Y Onishi, H Tachikawa, T. (2019) Charge Carrier Dynamics in Sr-Doped NaTaO$_3$ Photocatalysts Revealed by Deep UV Single-Particle Microspectroscopy. J *The Journal of Physical Chemistry C*. 123: 12592–12598.
[16] Longjie An Yohan Park Youngku Sohn Hiroshi Onishi. (2015) Effect of Etching on Electron–Hole Recombination in Sr-Doped NaTaO$_3$ Photocatalysts. J *PHYS CHEM C*. 119: 28440–28447.
[17] Nakanishi H, Iizuka K, Takayama T, Iwase A, Kudo A. (2016) Highly Active Alkaline Earth Metals and Lanthanum-Doped NaTaO$_3$ Photocatalysts for CO$_2$ Reduction to Form CO Using Water as an Electron Donor. J *CHEMSUSCHEM*. 10: 112–118.
[18] Su Y, Peng L, Guo J, Huang S, Lv L, Wang X. (2014) Tunable Optical and Photocatalytic Performance Promoted by Nonstoichiometric Control and Site-Selective Codoping of Trivalent Ions in NaTaO$_3$. J *The Journal of Physical Chemistry C*. 118: 10728–1039.
[19] Kudo A, Kato H. (2000) Effect of lanthanide-doping into NaTaO$_3$ photocatalysts for efficient water splitting. J *Chemical Physics Letters*. 331:373–377.
[20] Liu Darui. (2018) The effect of nonmetallic S doping on the photocatalytic performance of NaTaO$_3$ under visible light. J *Journal of Inorganic Materials*. 33: 409–415.
[21] Hu CC, Huang HH, Huang YC. (2017) N-doped NaTaO$_3$ synthesized from a hydrothermal method for photocatalytic water splitting under visible light irradiation. J *ENERGY CHEM*. 26: 515–521.
[22] Paz Y. (2019) Transient IR spectroscopy as a tool for studying photocatalytic materials. J *Journal of Physics: Condensed Matter*. 31: 503004.
[23] Hu Xiaoying, Tian Hongwei, Song Lijun, Zhu Pinwen, Qiao Liang. (2012) First-principles study of P-type ZnO co-doped with Li-N and Li-2N. J *Acta Phys. Sin*. 61: 369–373.
[24] Varns R, Strange P. (2008) Stability of gold atoms and dimers adsorbed on graphene. J *Journal of Physics Condensed Matter*. 20: 225005.
[25] Perdew JP, Burke K, Ernzerhof M. (1998) Perdew, Burke, and Ernzerhof Reply. J *PHYS REV LETT*. 80: 891–891.
[26] Li H, Shi X, Liu X, Li X. (2020) Synthesis of novel, visible-light driven S,N-doped NaTaO$_3$ catalysts with high photocatalytic activity. J *APPL SURF SCI*. 508: 145306.
[27] Ding B (2013) First-principles study on the electronic structure and related properties of doped NaTaO$_3$ (Jinan: Shan Dong University) p 48.
[28] Fang Junxin, Lu Dong. (1981) Solid State Physics.: Shanghai Science and Technology Press.
[29] Emery AA, Wolverton C. (2017) High-throughput DFT calculations of formation energy, stability and oxygen vacancy formation energy of ABO$_3$ perovskites. J *SCI DATA*. 4: 170153.
[30] Wang D, Yang L, Cao J. (2021) First-principles study on the magnetic properties of IB group transition metal-doped MoS$_2$. J *MOD PHYS LETT B*. 35: 2141002.
[31] Segall MD, Pickard CJ, Shah R, Payne MC. (1996) Population analysis in plane wave electronic structure calculations. J *MOL PHYS*. 54: 16317–16320.
[32] Zhurova EA, Ivanov Y, Zavodnik V, Tsirelson V. (2000) Electron density and atomic displacements in KTaO$_3$. J *Acta Crystallographica*. 56: 594–600.
[33] Yu Zehua. (2007) Preparation of Visible Light Response Photocatalyst Pt/TiO$_2$ and Its Degradation of Phenol. Harbin University of Science and Technology: Harbin.
[34] Zheng, W.L, Z.W. Li and X.F. Wang. (2009) Energy level of hydrogen-like impurities in InAs quantum ring. J *Spectroscopy and Spectral Analysis*. 29: 607–610.
[35] Xia, J.B. (1984) Transition element impurity levels in silicon. J *Journal of Physics*. 33: 1418–1426.

Research on shale gas fracturing technology

X.T. Gao
Drilling and Production Technology Research Institute of Liaohe Oilfield, Panjin, China

ABSTRACT: Our country shale gas resource base is strong, the shale gas reservoir of efficient and reasonable development became more and more important. Because of low permeability reservoirs of low permeability and large seepage resistance, poor connectivity, sometimes the single well production capacity of horizontal well is low, which can't satisfy the demand of economic development. Shale gas is widely distributed and has great development potential. However, due to the characteristics of low porosity, low permeability and strong heterogeneity of shale reservoir, compared with conventional oil and gas resources, it is more difficult to exploit. In most cases, effective stimulation measures should be taken for shale reservoir.

1 SHALE GAS FRACTURING CONSTRUCTION TECHNOLOGY

With the deepening of shale gas development, conventional fracturing has been unable to meet the requirements of large-scale commercial development. Large-scale hydraulic fracturing of horizontal well has become the key technology of shale gas development. In addition, the "factory" fracturing repeated fracturing and fracture monitoring technology have also been widely used in shale gas fracturing construction.

1.1 Horizontal well staged fracturing

Drillable plug fracturing is a commonly used horizontal well fracturing technology in shale gas development.

Drillable plug fracturing technology is a combination of hydraulic pumping, perforation and plug operation as well as fast drilling plug fracturing technology. The technology is suitable for casing completions. The process is as follows: run the drillable plug frac string – set the plug – remove the plug and lift the string – point the perforating gun at the intended location – pull the string out of the wellbore – fracture. Repeat to the next stage. Because the drillable plug staging technology is a combination of perforating and fracturing, all plugs can be drilled in a short period of time, significantly increasing operational efficiency compared to conventional fracturing methods while reducing fracturing fluid damage to the reservoir.

1.2 Repeated fracturing

When the initial fracturing failure or proppant crushing results in a significant decrease in gas production from a shale gas well, refracturing can reopen or reorient the fracture to re-establish the flow path from the reservoir to the wellbore and restore or increase the production of the gas well. This is a low-cost method for increasing the production of the gas well. Shale gas Wells have been shown to produce more after refracturing than during the initial refracturing period.

1.3 "Factory" fracturing

The fracturing mode adopts circulating fracturing fluid system, which is suitable for cluster well group development, and is a new and efficient comprehensive development mode. The biggest characteristic of "factory" fracturing is continuous operation, which can greatly improve the utilization rate of fracturing equipment, while reducing the frequency of equipment relocation and installation, and reducing the labor intensity of workers. The development practice shows that this fracturing mode has a significant effect in shale gas development, and makes a great contribution to improving fracturing efficiency and reducing shale gas development cost.

1.4 Fracture monitoring technology

Fracture monitoring is an important means to evaluate the effect of fracturing and an important way to optimize the fracturing process and its design. Through the fracture monitoring, the fracture orientation can be predicted, the reconstructed volume and the drainage area can be calculated, which provides a reference for the later production prediction and new well layout. At present, the methods of fracture monitoring mainly include conventional chemical tracer method, physical tracer method, microseismic monitoring and inclinometer monitoring. Among them, microseismic monitoring is widely used, which can be divided into the same well monitoring and adjacent well monitoring. The principle of this method is to record microseismic events during fracture initiation and closure by placing multiple geophones close to each other, so as to calculate the volume of fracturing and predict the production after fracturing.

2 FRACTURING FLUID SYSTEM

2.1 Water skiing fracturing fluid system

Slippery water is a new liquid system developed for the reconstruction of shale gas reservoirs. It uses a very small amount of thickening drag reducer to reduce friction. The amount is generally less than 0.2%, and the amount of high-efficiency drag reducer can be reduced to Below 0.018%, this type of liquid system mainly relies on pumping displacement to carry sand instead of liquid viscosity, and is suitable for formations with no water sensitivity, well-developed natural fractures in the reservoir, and high brittleness. Its advantages include: ① suitable for fractured reservoirs; ② increase the probability of shear fracture formation, which is conducive to the formation of network fractures, which can greatly increase the fracture volume and improve the fracturing effect; ③ use a small amount of thickener to reduce resistance, Less damage to the formation and less proppant consumption; ④ Under the premise of the same scale of operation, the cost of slick water fracturing is 40%–60% lower than that of conventional gel fracturing.

2.2 Clear water fracturing fluid

In the clear water fracturing fluid, water occupies the vast majority. At the same time, a certain amount of proppant is added, as well as a little drag reducers, surfactants, clay stabilizers, etc. As a clean fracturing fluid, clear water fracturing fluid is mainly suitable for formations with weak water sensitivity, well-developed natural fractures in the reservoir, and high brittleness. Compared with conventional jelly fracturing fluid, clear water fracturing fluid has better fracturing effect, less damage to the reservoir and low cost. At present, clean water fracturing fluid has been widely used in shale gas fracturing. However, the clean water fracturing fluid also has its shortcomings, such as poor proppant carrying capacity due to low viscosity.

2.3 Composite fracturing fluid system

The composite fracturing fluid is mainly composed of high-viscosity jelly and low-viscosity slippery water. The proppant uses ceramsite of different particle sizes, which is suitable for shale gas reservoirs with high clay content and strong plasticity. The high-viscosity jelly guarantees a certain sand-carrying capacity and the width of artificial cracks. The low-viscosity slippery water has viscous fingering phenomenon in the jelly liquid and has a good joint-making ability, and finally makes the alternate injection of large and small particles support The agent has a lower settling velocity and a higher fracture conductivity.

3 FRACTURING DESIGN OPTIMIZATION

3.1 *Fracturing pump rate optimization*

Shale gas reservoirs are diverted by slick water fracturing, forming complex fracture networks. The poor proppant-carrying capacity of slick water ensures the fracturing fluid transportation and migration and branch method to form fracture network. The fracturing pump rate is generally required to be larger than $10m^3/min$, and the formation of higher net pressure in the formation makes the fracturing propping agent reach the depth of the formation is another important reason. The experimental conditions are as follows: slick water, 8% proppant ratio, 40/70 mesh proppant. According to the experimental results, under the condition of the same proppant ratio, the pump rate increases, the proppant settling speed decreases, and the proppant carrying capacity is relatively improved.

3.2 *Proppant ratio optimization*

The design of proppant-liquid ratio not only seeks to match the width of the supporting joint with the width of the crack, but also to increase the flow friction of the mortar in the crack, so as to enhance the net pressure of the main crack, so as to realize the purpose of turning the crack or opening the natural crack. Foreign fracturing experience of Barnett J shale shows that there is an approximately positive correlation between the amount of fluid used and the post-fracturing effect, while the influence of proppant-liquid ratio is not obvious, which fully indicates that the design of proppant-liquid ratio must be combined with the development degree of natural fractures in the reservoir. If the design is reasonable, the diversion of fractures can be realized, otherwise there will be no diversion. Therefore, the proppant-liquid ratio and the post-pressure effect are not regular.

Shale gas formation permeability is very low, for shale formation, produce more cracks, longer communicate more formation volume is more important, support layer of diverting capacity is relatively minor, so construction proppant is higher than the main consideration should be given the choice of proppant than homework is helpful for fracturing fluid conveying, and cracks in proppant dike heaped up speed. Avoid proppant plugging

The fracture width and the number of natural fractures opened mainly control the possibility of proppant plugging near the wellbore of the fracture. Fracture width depends on flow rate, fracturing fluid viscosity, formation brittleness, local in-situ stress, and the presence of effective fracture isolation. In general, the initial proppant concentration is 24–40kg/m^3, which can be increased by 40kg/m^3 per step after the pressure stabilizes. The upper limit of the slick water proppant depends on the proppant size, typically 300kg/m^3 for 70/100 mesh proppant and 240kg/m^3 for 40/70 mesh proppant. Proppant slug controlled fluid loss technology is commonly used in shale gas fracturing, with slug proppant concentrations ranging from 60 kg/m^3 to 180kg/m^3.

3.3 *Proppant adding mode*

Shale is fractured reservoir, the conventional hydraulic fracturing naturally inhibit crack propagation of cracks formed in the main based. For shale reservoir fracturing reconstruction, it is necessary to make full use of natural fractures, through multiple artificial fractures, to communicate with more natural fracture systems and increase the discharge area. At the same time, the fluid loss of shale is mainly the fluid loss of many beddings and natural fractures. In the process, we must consider how to use the many beddings and natural fractures instead of controlling it. In addition, slick water has a low viscosity and is easy to filter out. To achieve continuous and effective support of the fracture network in shale gas reservoirs, slug injection technology is needed to supplement the filtration loss of fracturing fluid and slow down the formation of proppant banks near the well. Embankment height.

The proppant levee peaks are scoured by multi-stage slugs, and the proppant levee formed is smoother, which is beneficial to reduce the risk of proppant blockage in construction and increase the success rate of construction. The optimization of slug volume and slug stage is mainly to prevent early proppant plugging. The optimization principle is to ensure the continuous fracture support profile.

4 DEVELOPMENT DIRECTION OF SHALE GAS FRACTURING TECHNOLOGY

4.1 *Improve the efficiency of shale gas fracturing*

High production cost is an important factor restricting the development of shale gas. How to improve the fracturing efficiency and reduce the cost is still an important direction for the development of shale gas fracturing technology in the future. The results show that the majority of production in horizontal well development comes from only about 30% of the effective intervals, especially in highly heterogeneous horizontal Wells, because most of the frac stages do not appear in the "sweet spots". Realizing the number of fracturing stages is an important way to improve the efficiency of shale gas fracturing.

4.2 *To improve the seepage conditions of shale gas*

Improving the seepage condition of shale gas is the key to increase shale gas production. The introduction of Hiway high-speed channel fracturing technology has revolutionized the conventional fracturing concept, where proppant insertion reduces fracture conductivity during conventional fracturing. Hiway high-speed channel fracturing technology, which integrates well completion technology, sand filling technology and fluid control engineering, forms an open high-speed seepage channel in the proppant filled area through intermittent alternate injection of proppant and high-concentration gel fracturing fluid, which greatly improves the seepage state of shale gas. The technology has been used in several parts of the world. The results show a 15% increase in production compared to conventional fracturing. Compared with vertical Wells, horizontal Wells greatly improve the seepage conditions of shale gas. However, complex structure Wells have greater advantages, and new technologies dominated by complex structure Wells will gradually become an effective means to develop unconventional oil and gas resources in the future.

5 CONCLUSION

Shale gas fracturing technology has experienced the stages of exploration, rapid development and large-scale popularization and application, and is gradually mature, which has effectively promoted the development process of shale gas. It calculates the fracture profile under different fracturing schemes, predicts construction risks, and analyzes the influence of controllable parameters on actual construction, determines the best scheme, and avoids construction risks. The degree of fracture development is an important factor that affects shale gas production. How to obtain more man-made fractures is the first consideration in fracturing design. At present, the selection of various fracturing parameters mainly focuses on how to generate main fractures, while effectively communicating natural fractures and generating fracturing induced fractures, that is, generating an effective fracture network and sealing natural fractures when necessary. If large particle size and high proppant ratio proppants need to be tailed, the sedimentation proppant bank of the proppant in the low-viscosity liquid must be studied to prevent proppant plugging accidents.

REFERENCES

[1] Research on Shale Gas Fracturing Technology and Fracturing Fluid Performance and Its Development StatusJB TaoCI AmpT Polytechnic
[2] The Key and the Countermeasures Research of Shale Gas Fracturing Technology, Y Tang,B Wang,F Zeng,J Wang.
[3] Shale Gas Fracturing Platform Design and Matching Technology, GQ Jiang

Dynamic characteristics analysis and pressure regulating system research of a tire

B. Zheng, H. Li, Y.C. Jiang & X.C. Zhang
School of Intelligent Manufacturing, Panzhihua University, Panzhihua, China

ABSTRACT: Tire is one of the important parts of automobiles, and it is also the only part of automobiles in contact with the road; it has an important influence on the safety, driving performance, and operation stability of automobile driving. In this paper, SolidWorks is used to establish a three-dimensional solid model of the tire, which is imported into ANSYS Workbench to carry out modal analysis of the tire, extract the first six natural frequencies and modal shapes, and find out the area with a large amplitude. Based on the results of modal analysis, the harmonic response of the tire is analyzed, and the displacement frequency response curve and acceleration frequency response curve of the tire in three directions of X axis, Y axis and Z axis are obtained. It is determined that the resonance frequency and region are mainly concentrated in the region between the tire and the rim. On this basis, a set of tire pressure regulating system is designed.

1 INTRODUCTION

Vehicle tires not only play the role of bearing the whole vehicle mass, but also play the role of shock absorption and cushioning [1, 2]. Tire is one of the important parts of the vehicle that have a direct impact on the driving performance, comfort, and safety of vehicle. At present, many scholars at home and abroad have carried out relevant research on tires. Based on ABAQUS finite element analysis software, Zhao established a three-dimensional finite element model of non-pneumatic plastic tire with a spoke plate. Then, the static load, steady rolling, and cornering rolling conditions were considered in the model calculation. The distribution of equivalent stress of tire carcass and contact stress of tread and the radial and lateral stiffness characteristic curves of the tire were obtained [3]. Zang used the method of implicit analysis and explicit analysis to study the dynamic characteristics of tire [4].

In this paper, a car tire is taken as the research object. First, the three-dimensional solid model of the tire is established in SolidWorks. The tire 3D solid model is saved as the format of x_ t. Second, the finite element analysis is opened and carried out in ANSYS Workbench. The first six natural frequencies and modal shapes of the tire are obtained by modal analysis in ANSYS Workbench. The harmonic response of the tire is analyzed in ANSYS Workbench. The displacement frequency response curve and acceleration frequency response curve of the tire in X, Y, and Z directions are obtained. The motion law and stress condition of the tire are analyzed, and the results are discussed and analyzed. It provides a theoretical basis for tire structure design and optimization.

2 THREE-DIMENSIONAL MODEL OF A TIRE

The structure of the tire studied in this paper is more complex. In the process of establishing its three-dimensional solid model, it should be appropriately simplified, such as deleting some small holes and threaded holes that have little influence on the dynamic analysis results. The simplified three-dimensional solid model of a tire was established in SolidWorks, as shown in Figure 1.

Figure 1. Simplified three dimensional solid model of a tire.

3 MODAL ANALYSIS OF TIRE

3.1 *Modal analysis theory*

Modal analysis is used to solve the vibration characteristics of the structure under the condition of free vibration. The main parameters are natural frequency and modal shape.

According to Newton's classical mechanics, the dynamic equation of the object can be depicted as

$$[M]\{\ddot{x}(t)\} + [C]\{\dot{x}(t)\} + [K]\{x(t)\} = \{F(t)\} \quad (1)$$

where $[M]$ is the mass matrix, $[C]$ is the damping matrix, $\{x\}$ is the displacement vector, $\{\dot{x}\}$ is the velocity vector, $\{\ddot{x}\}$ is the acceleration vector, and $\{F(t)\}$ is the external load.

In modal analysis, the external load $F(t)$ can be set to zero, and the damping matrix $[C]$ can also be set to zero [6]. The dynamic equation can be written as follows

$$[M]\{\ddot{x}\} + [K]\{x\} = 0 \quad (2)$$

The formula of simple harmonic motion is represented as

$$\{x(t)\} = \{\varphi_i\} \sin(\omega_i t) \quad (3)$$

The equation of motion without damping can be obtained as

$$([K] - \omega_i^2)[M]\{\varphi_i\} = \{0\} \quad (4)$$

Because of the amplitude of each node is not all zero. The formula can be obtained from (4).

$$\left|[K] - \omega_i^2[M]\right| = 0 \quad (5)$$

Where, ω_i is the ith natural frequency. $\{\varphi_i\}$ is the ith mode vibration shape.

3.2 Finite element modeling of a tire

Considering the complexity of the tire, higher mesh accuracy is used in the mesh generation, and the local mesh is refined. The tire finite element model has 377548 nodes and 198136 elements. The finite element model is shown in Figure 2.

Figure 2. Finite element model of a tire.

3.3 Modal analysis and results

Combined with the actual working condition of the tire, the fixed support constraint is added to the inner surface of the six threaded holes of the rim. Because the high-order modes have little effect on the vibration of the tire, only the first six natural frequencies and vibration modes of the tire are extracted in ANSYS. The modal diagram and description of the tire are shown in Figure 3 and Table 1.

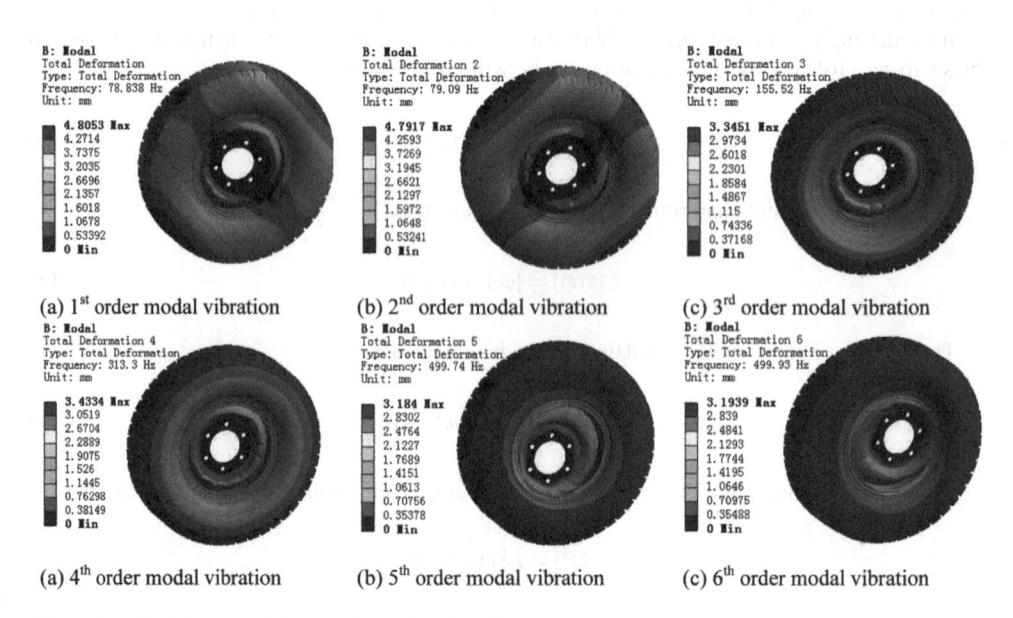

(a) 1st order modal vibration　　(b) 2nd order modal vibration　　(c) 3rd order modal vibration

(a) 4th order modal vibration　　(b) 5th order modal vibration　　(c) 6th order modal vibration

Figure 3. First six natural frequencies and modal vibration.

Table 1. First six natural frequencies and modal vibration.

Order	Natural Frequency /Hz	Modal description
1st	78.84	The overall performance of the tire is first-order bending.
2nd	79.09	The overall performance of the tire is first-order bending.
3rd	155.52	The overall performance of the tire is second-order bending.
4th	313.3	Tire twist on the XOY surface.
5th	499.74	The overall performance of the tire is second-order bending.
6th	499.93	The overall performance of the tire is second-order bending.

According to Figure 3 and Table 1, when the frequency of the external excitation is the same as or close to the natural frequency of the tire, it will lead to resonance, bending deformation, or torsional deformation. The resonance can be avoided by changing the size of the area.

4 HARMONIC RESPONSE ANALYSIS OF A TIRE

4.1 *Harmonic response analysis theory*

Harmonic response analysis is used to solve the structural response under periodic loads and the dynamic characteristics under different frequency loads [5].

The motion equation of the structure under harmonic load can be depicted as

$$[M]\{\ddot{x}\} + [C]\{\dot{x}\} + K[x] = \{F(t)\} \sin(\theta t) \quad (6)$$

where $[M]$ is the structural mass matrix, $[C]$ is the structural damping matrix, $[K]$ is the structural stiffness matrix, $\{f(t)\}$ is the load function varying with time, $\{x\}$ is the displacement vector, $\{\dot{x}\}$ is the velocity vector, and $\{\ddot{x}\}$ is the acceleration vector.

The displacement response of the node is represented as

$$\{x\} = \{A\} \sin(\theta t + \psi) \quad (7)$$

where $\{A\}$ is the displacement amplitude vector, and Ψ is the phase angle of the displacement response lag excitation load.

4.2 *Harmonic response analysis results and discussion*

The common harmonic response analysis method is the modal superposition method, which is based on the superposition of the modal shapes obtained from modal analysis multiplied by the coefficients.

Combined with the actual working condition of the tire, the maximum stress of 100 N is set on the inner surface of the small hole of the tire, and the direction is opposite to the Z axis. According to the results of modal analysis, the maximum frequency of harmonic response analysis is set to 600 Hz, and the solution interval is set to 300. Based on the first six natural frequencies in the abovementioned modal analysis, the displacement frequency response curve and acceleration frequency response curve of tire in X, Y, and Z directions are shown in Figure 4 and Figure 5, respectively.

(a) Displacement frequency response curve of the X direction.

(b) Displacement frequency response curve of the Y direction.

(c) Displacement frequency response curve of the Z direction.

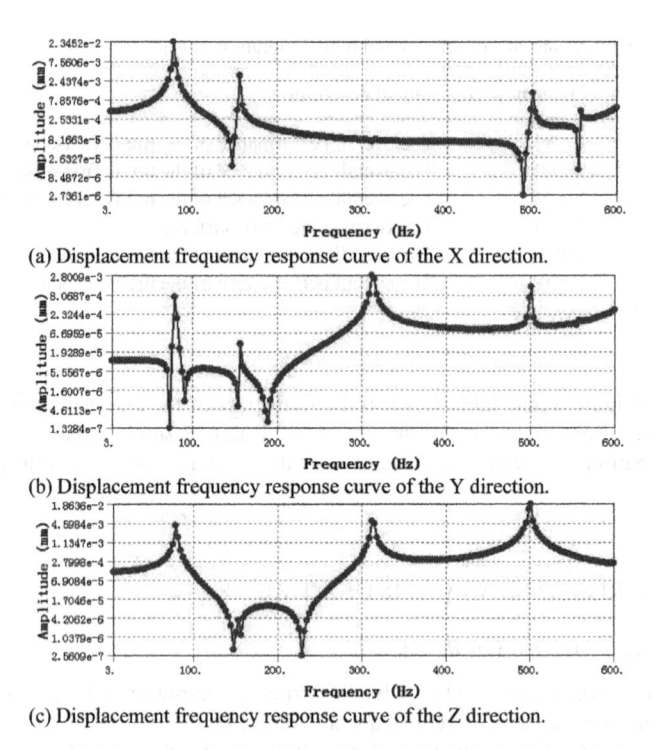

Figure 4. Displacement frequency response curve of harmonic analysis results.

(a) Acceleration frequency response curve of the X direction.

(b) Acceleration frequency response curve of the Y direction.

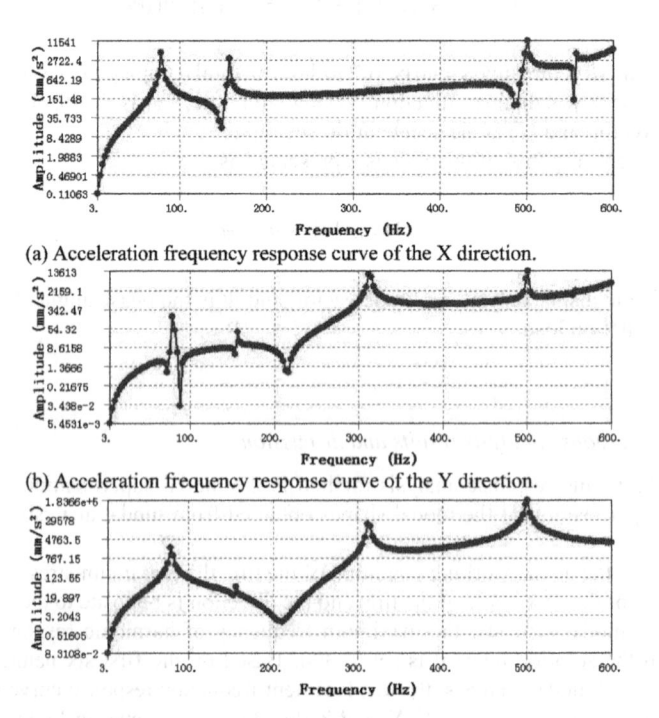

Figure 5. Acceleration frequency response curve of harmonic analysis results.

Figure 4 shows the harmonic response analysis results of a tire in X, Y, and Z directions. According to the displacement frequency response curve in the X direction, the vibration displacement reaches the maximum value when the frequency of the tire's big end position is 78 Hz, and then it decreases gradually. According to the displacement frequency response curve in the Y direction, when the tire frequency is 310 Hz, the vibration displacement reaches the maximum value and then decreases gradually. According to the Z direction displacement frequency response curve, when the tire frequency increases gradually, its vibration displacement also increases at any time. It can be seen from Figure 5 that the acceleration frequency response curve of a tire is similar to the displacement frequency response curve. Therefore, in order to improve the service life of a tire, the abovementioned frequency region should be avoided.

5 RESEARCH ON THE TIRE PRESSURE REGULATING SYSTEM

On the basis of modal analysis and harmonic response analysis, a tire pressure regulating system is designed to improve its dynamic characteristics. The overall structure of tire pressure regulating system is shown in Figure 6.

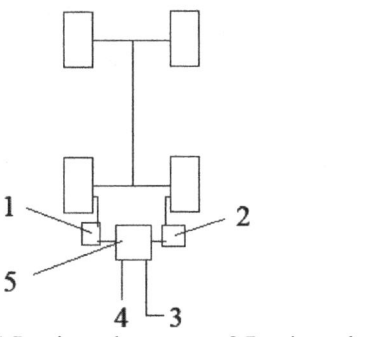

1-Rotating seal components 2-Rotating seal components 3-Air compressor 4-Exhaust pipe 5-Pressure regulating control valve

(a) Overall structure.

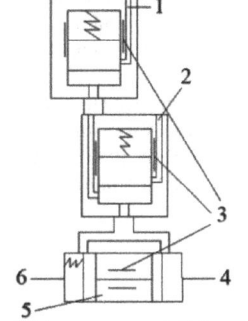

1-Air compressor 2-Exhaust pipe 3-Solenoid valve 1, 2, and 3 (from top to bottom) 4-Right wheel air pressure 5-Balance valve 6-Left wheel air pressure

(b) Pressure regulation control valve structure.

Figure 6. Tire pressure regulating system.

Figure 6 shows only the pressure regulation of the rear wheel, which is the same as that of the front wheel. When the tire pressure regulating system does not work, the double pistons are in the middle position under the action of the spring, and the left and right tire pressures are consistent. It is conducive to the driving stability of the vehicle. While the car deviates to the left, solenoid valve 3 is powered on to make the double pistons move to the left. At the same time, solenoid valve 2 is powered on for a period of time to make the piston move upward to deflate the right wheel, increase the driving resistance of the right wheel, and promote the balance and stability of the vehicle, or inflate the left wheel to reduce its driving resistance. In case of emergency braking, solenoid valve 2 is powered on to deflate the two wheels and shorten the braking distance. After braking, solenoid valve 2 is powered off and solenoid valve 1 is powered on to inflate the tire through the air reservoir or air compressor.

6 CONCLUSION

In this paper, the three-dimensional modeling and modal analysis of an automobile tire are carried out, and the natural frequencies of the tire are obtained. Based on the harmonic response analysis results, the displacement frequency and acceleration frequency response curve are obtained. It is determined that the resonance frequency and region are mainly concentrated in the region between the tire and rim. On this basis, a tire pressure regulating system is designed to improve its dynamic characteristics.

ACKNOWLEDGMENTS

This work was supported by Educational Commission of Sichuan Province Project No.15ZB042, Vanadium and Titanium Resources Comprehensive Utilization Key Laboratory of Sichuan Province No.2019FTSZ08 and Innovation and Entrepreneurship for College Students No.S202011360036 and 2019cxcy025.

REFERENCES

[1] Wang, J. Li, L., Sun, L., Sha, C. X. (2016) Finite element analysis of 245/70 R16 tire, Tire Industry, 36: 520–528.
[2] Zheng,B., Zhong,F., Zhang,J. D., Jiang,Y., Fan, X. P. (2020) Modal and harmonic response analysis of drum brake," Journal of Chinese Agricultural Mechanization, 41:117–122.
[3] Zhao,P., Yang,W. M. (2012) FEA for ground-contact behavior of plastic tire with wheel spoke, Plastics, 41:94–96+83.
[4] Zheng, B., Yin,G. F. (2019) Finite element analysis and optimization design for brake shoe of agricultural dump truck," Journal of Chinese Agricultural Mechanization, 40:85–90.
[5] Zheng,B., Zhang, J. D., Yin,G. F. (2019) Thermal-structure coupling characteristic simulation analysis for drum brake, Journal of Chinese Agricultural Mechanization, 40:119–124.

Experimental analysis of energy absorbing effect of elastomeric energy absorbing device for suspension type crossing frame

F.H. Meng, Y.J. Xia & Y. Ma
China Electric Power Research Institute, Beijing, China

ABSTRACT: For improving the impact resistance performance of the suspension crossing frame, the elastomeric energy absorption device suitable for the crossing frame is designed based on the load-bearing characteristics of the suspension crossing frame under the accident state, and the energy absorption effect verification test is carried out. Firstly, based on the analysis of the load-bearing characteristics of the suspension crossing frame under accident conditions, several different types of energy absorption principles are compared, and finally the elastic energy absorption device is selected for design. Then, in order to verify the energy absorption effect of the energy absorption device under real working conditions, the small-scale vertical state test of the bearing cable of the suspension type crossing frame is carried out, and the energy absorption effect of the energy absorption device is tested. The test results show that under the conditions of 0.5T, 1.0T, 1.5T and 0.5m impact height, the maximum energy absorption effect of the elastomeric energy absorption device can reach 52.59%, which can effectively improve the impact resistance of the suspension type crossing frame.

1 INTRODUCTION

In the construction process of extra-high voltage line (UHV) crossing, large crossing distance, large height difference and complex terrain are very common. The suspension type crossing frame has the advantages of simple structure, convenient installation, convenient transportation, no terrain restrictions, low cost and low construction cost, which is suitable for crossing construction under complex terrain conditions. In the current construction of crossing lines, the use of suspension type crossing frame accounts for more than 80%. As the safety requirements of important crossing construction (high-speed railway, expressway and important power lines) are gradually increasing, once the safety accident of crossing construction occurs, it will cause serious social impact. For example, during the construction of a UHV line crossing in Tianjin, a broken wire accident occurred during the process of hanging porcelain bottles on the wire, and the wire fell on the crossing sealing network, which caused great safety risk to the running 220kV power line and Beijing Shanghai expressway. Therefore, much requirements are put forward for the safety of the suspension type crossing frame.

The traditional method to improve the impact resistance of suspension type crossing frame is to increase the cross-sectional area of the bearing cable, so as to increase the breaking force of the fiber rope and improve its safety factor. However, this method significantly increases the overall weight of the suspension type crossing frame, resulting in the increase of construction difficulty and cost. The energy absorbing suspension type crossing frame cannot significantly increase its own weight, but also improve the impact resistance of the frame. Therefore, it is a useful technique to improve the impact resistance of the suspension type crossing frame.

Considering that the bearing cable is the key bearing structure of the suspension type crossing frame, its bearing capacity determines the impact resistance of the suspension type crossing frame. According to the stress characteristics and energy dissipation characteristics of the bearing cable in the process of bearing conductor impact, selecting the appropriate energy absorption principle and energy dissipation mechanism can effectively improve the impact resistance of the suspension type crossing frame. At present, the widely used energy dissipation mechanisms include friction energy absorption, metal deformation energy absorption, viscous energy absorption and viscoelastic energy absorption. Viscous energy absorbing devices mainly include hydraulic buffer and liquid viscous damper. Viscous damper is made of the principle that the fluid produce throttling resistance when passing through the orifice. It is a kind of damper related to the piston movement speed and widely used in high-rise buildings, bridges, seismic transformation of building structures, anti-vibration of industrial pipeline equipment, military industry and other fields [1–3], effectively alleviating the impact and damage of earthquake on building structures [4,5]. As a safety protection device to reduce rigid collision during operation, hydraulic buffer is widely used in various fields such as lifting transportation, metallurgy, port machinery and railway vehicles [6–8]. As a viscoelastic energy absorbing material, polymer thermoplastic elastomer can provide damping, shock absorption, sound insulation, impact resistance and other functions, especially applied in sports and work body protection, military and police body protection, explosion-proof, airborne and airdrop protection, automobile anti-collision, electronic and electrical products impact protection, etc. For example, the anti-collision energy absorption device applied in the field of rail transit is a typical viscoelastic material, which has a certain reference significance for the design of energy absorption device for suspension type crossing frame.

This paper designs the elastomer energy absorption device based on the load-bearing characteristics of the suspension type crossing frame in the accident state. For verifying the effectiveness of the energy absorbing device under real working conditions, the small-scale vertical state test of the load-bearing cable in the suspension type crossing frame is carried out.

2 MECHANICAL CHARACTERISTICS OF SUSPENSION TYPE CROSSING FRAME

Suspension type crossing frame is mainly composed of bearing cable (including insulated section and non-insulated section), temporary beam and insulation net. The bearing cable is the key bearing part of suspension type crossing frame, the temporary beam is the supporting part, and the insulation net is the direct bearing part of suspension type crossing frame, as shown in Figure 1.

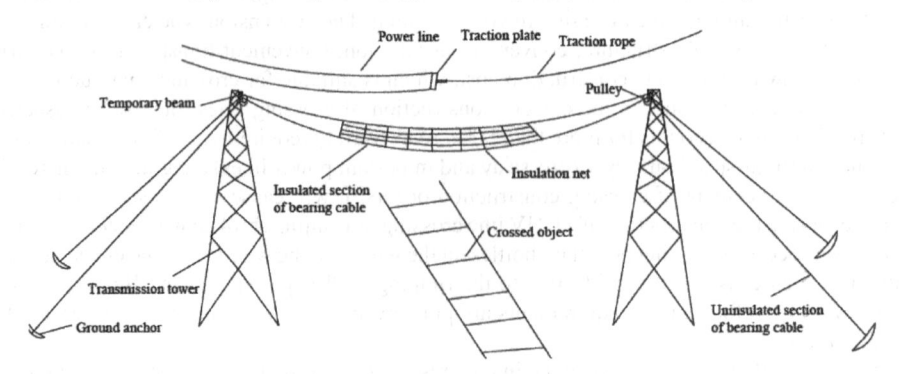

Figure 1. Schematic diagram of suspension type crossing frame.

2.1 Tension response law of bearing cable

For improving the bearing capacity of the suspension type crossing frame, it is necessary to analyse its stress characteristics in combination with the bearing conditions of the frame under accident conditions, which can be used as the basis for the research of energy absorption device.

Although a variety of coupling factors have certain influence on the peak tension, the tension time history response of the bearing cable under dynamic impact has certain regularity. According to the actual situation of the project, the dynamic response law of the spanning frame under the typical construction scheme can be analyzed. Combined with the change of the tension dissipation time history, the appropriate energy dissipation mechanism can be selected, which has certain representativeness. Taking the ±800kV Jiuhu line of Gansu Section 7 of Changji Guquan project undertaken by Tianjin power transmission and Transformation Engineering Co., Ltd. as an example, the dynamic simulation is carried out, and the time history response law of the bearing cable under the condition of line running accident and line breaking accident is obtained respectively, as shown in Figure 2 and Figure 3.

Based on the above simulation results, it can be seen that the tension change of the bearing cable of the frame under the accident state reaches the peak after about 1.0s -1.5s, and then after about 2s continuous vibration energy dissipation, the whole tension change process takes a short time and has instantaneous effect. The tension change amplitude of the bearing cable under the line breaking accident condition is large, and the maximum tension reaches 93kN.

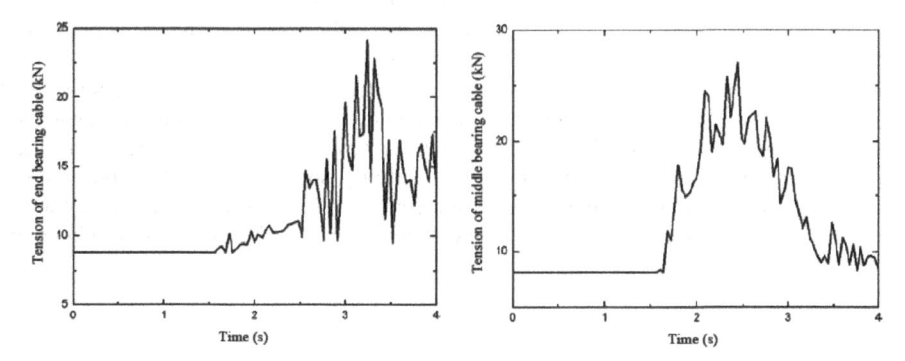

Figure 2. Time history curve of tension response of bearing cable under line running accident.

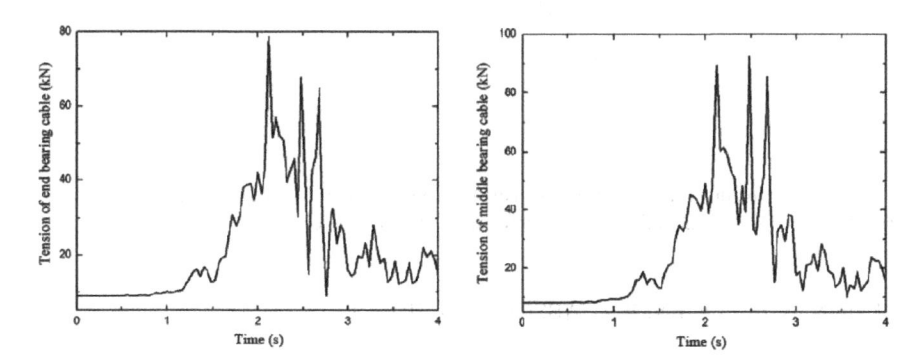

Figure 3. Time history curve of tension response of bearing cable under broken line accident.

2.2 *Installation position and connection of energy absorbing device*

In case of accident, the bearing cable and insulated net rope are the key load-bearing structures of suspension type crossing frame, so it has important engineering practical value to improve their impact resistance. The energy absorption device used in suspension type crossing frame can effectively reduce the peak tension of load-bearing cable and insulated wire rope in accident state, which can improve the impact resistance of suspension type crossing frame.

The installation positions of the energy absorption device of suspension type crossing frame can be roughly divided into two types: (1) installed on the bearing cable in the crossing gear; (2) installed on the bearing cable pulling line of non-crossing gear, as shown in Figure 4.

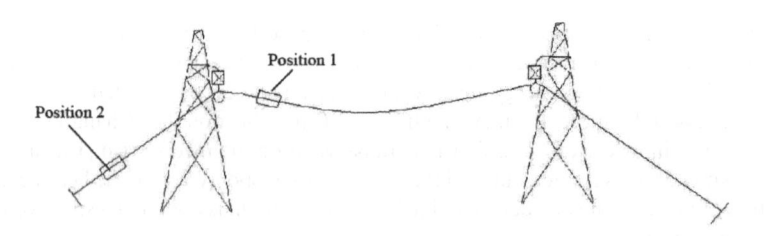

Figure 4. Schematic diagram of installation position of energy absorption device of suspension type crossing frame.

The main body of energy absorption device for suspension type crossing frame is metal structure, which needs to bear the impact action under accident state and absorb certain impact energy, so its structure and performance determine the larger concentrated mass. In the practical application of the project, the installation position 1 has a great potential safety hazard, so the installation position 2 is preferred as the installation position.

During installation, the temporary crossbeam is fixed on the iron towers on both sides, the bearing cable turns to the ground through the special pulley, and is anchored on the ground anchor by the energy absorption device through the suspension type crossing frame, and the net sealing cable is arranged on the bearing cable with a fixed spacing, so as to realize the protection of the crossed object. Both ends of the energy absorption device for suspension type crossing frame use eyebolts as connectors, which can be connected with ropes through shackles, as shown in Figure 5.

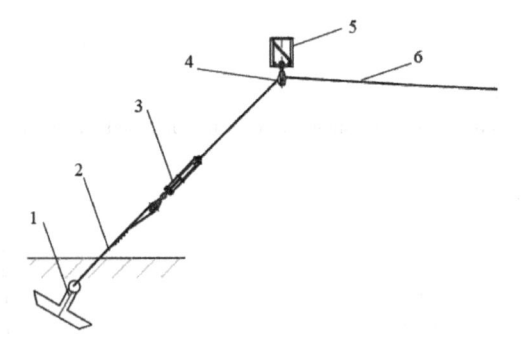

Figure 5. Connection diagram of energy absorber: 1 - ground anchor, 2 - ground anchor stay wire, 3 - energy absorption device, 4 - special pulley, 5 - temporary cross arm, and 6 - bearing cable.

3 DESIGN INDEX OF ENERGY ABSORBING DEVICE

3.1 Energy absorption mode of suspension type crossing frame

The energy absorption principle of the energy absorption device for the whole structure can be described from the perspective of energy, and the energy equation of the structure at any time when it is impacted is [9]:

Traditional seismic structure

$$E_{in} = E_e + E_c + E_k + E_h \tag{1}$$

Energy dissipation structure

$$E'_{in} = E'_e + E'_c + E'_k + E'_h + E_d \tag{2}$$

where, E_{in} is the input traditional structure during the earthquake, E'_{in} is the total energy of the energy dissipation structure system, E_e and E'_e are the strain energy of the system, E_c and E'_c are the viscous damping energy of the system, E_k and E'_k are the kinetic energy of the system, E'_h and E_h are the hysteretic energy of the system, E_d is the energy dissipated or absorbed by the energy dissipation (damping) device or energy dissipation element.

For the bearing cable structure of suspension bridge, the viscous damping energy consumption of the structural system can be ignored, the total impact energy is mainly consumed from two aspects: the hysteretic energy consumption of the bearing cable structure system and the energy consumption of the installed damping device.

The commonly used energy absorption methods include friction energy absorption, metal deformation energy absorption, viscous energy absorption and viscoelastic energy absorption. The effect of friction energy absorption and metal deformation energy absorption is related to the relative displacement of the two ends of the energy absorption device. The energy absorption effect of viscous and viscoelastic energy absorption methods is related to the relative velocity at both ends of the energy absorption device. Among them, friction energy absorption and metal deformation energy absorption belong to mechanical energy absorption. Viscoelastic energy absorption has higher requirements on energy absorption materials, which is affected by environmental temperature. The advantage of viscoelastic energy absorption is that compared with mechanical energy absorption, its overall structure size and mass are smaller. Compared with viscoelastic energy absorption, viscoelastic energy absorption has stable energy absorption effect, which is less affected by environment, but the overall structure size and mass have increased.

For the suspension type crossing frame, the energy absorption mode of its energy absorption device should consider the following aspects:

- The impact load of the bearing cable under the accident condition is large, and the limit load of the energy absorbing device is high;
- China has a vast territory, and the temperature difference between Northeast China and Guangdong is large. According to the universality of different construction environment temperature, the energy absorption effect of energy absorption device is less affected by temperature;
- The design of the energy absorption device of the suspension type crossing frame should be as light as possible, which can reduce the construction cost and the operation intensity of the construction personnel.

Based on the above analysis, friction energy absorption and metal deformation energy absorption belong to mechanical energy absorption methods, and their overall structure size and mass are large, which are not suitable for the energy absorption device of suspension type crossing frame. Viscous energy absorption has the characteristics of large bearing capacity and low influence of temperature on energy absorption effect, but its quality is large. Viscoelastic energy absorption structure is small in size and mass, which can be selected as the energy absorption mode of the suspension type crossing frame.

3.2 *Design index of energy absorbing device*

The design index of energy absorption device is shown in Table 1.

Table 1. Design index of energy absorption device.

NO.	Index item	Index value
1	Starting tension	15 kN-35 kN
2	Mass	<150 kg
3	Peak tension in energy absorption stage	120 kN
4	Ultimate tension at failure	403 kN
5	Energy absorption effect	The peak tension of bearing cable is reduced by 10%

4 DESIGN OF ELASTOMER ENERGY ABSORBING DEVICE

The elastic energy absorption device for suspension type crossing frame is mainly composed of elastic component, front pull plate, end pressing plate 1, middle pressing plate, end pressing plate 2, rear pull plate, pull rod, support rod, lifting ring, etc. The front pulling plate, the end pressing plate 1, the end pressing plate 2, the rear pulling plate, the pull rod and the support rod constitute the reverser, and the elastomer is assembled in the reverser through certain pre compression, as shown in Figure 6.

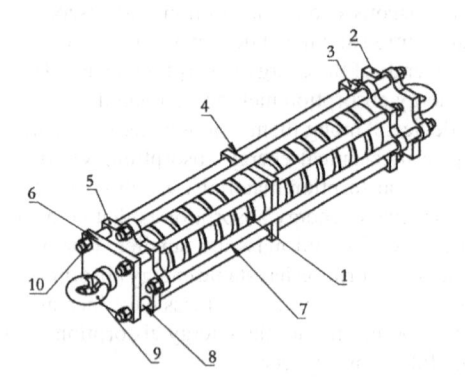

Figure 6. Structure of elastomer energy absorption device: 1-elastomer assembly, 2-front pulling plate, 3-end pressing plate 1, 4-middle pressing plate, 5-end pressing plate 2, 6-rear pulling plate, 7-pull rod, 8-bracket rod, 9-lifting ring, and 10-M20 hexagon nut.

When an accident occurs, the tension of the bearing cable increases, and the process of tension increasing is a time-varying process. When the tension of the bearing cable is less than the start threshold of the elastic energy absorption device, the elastic energy absorption device can be regarded as a rigid connecting rod. When the tension of the bearing cable reaches the starting threshold, the cable acts on the two connecting rings respectively. The pull rod of the elastomer energy absorption device drives the end pressing plate 2 to move against the support rod. The end pressing plate 2 and the end pressing plate 1 compress the deformation of the elastomer assembly, and play the role of buffering and absorbing kinetic energy through the slow reduction of the compression speed of the elastomer assembly.

4.1 Static pressure test

In order to test the performance of the elastomer, TPEE elastomer is selected to test the static pressure performance of the press. The elastomer test assembly and static pressure curve are shown in Figure 7.

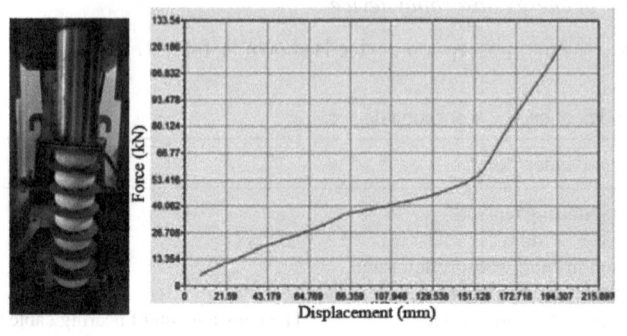

Figure 7. Elastomer test assembly and static pressure curve.

Table 2. Static pressure data of elastomer components.

Number of elastomers	Compression stroke (mm)	Maximum static load (kN)
7	197	120

According to the analysis of static pressure data in Table 2, when the maximum pressure is not more than 120kN, the compression stroke is 197mm, minus the displacement of 29mm when the trigger force is 16kN, the effective compression stroke is 168mm in the load range of 16kN-120kN. In addition, it can be seen from the curve that when the compression stroke is less than 150 mm, the load changes slightly with the increase of stroke, and the load is relatively gentle.

4.2 Drop weight impact test

Before the drop weight test, the internal rod of the elastomer assembly is used to pre-tighten the threads at both ends, and a pre-tightening force of 16kN is applied to the elastomer. The drop weight impact test and impact curve of elastomer components are shown in Figure 8.

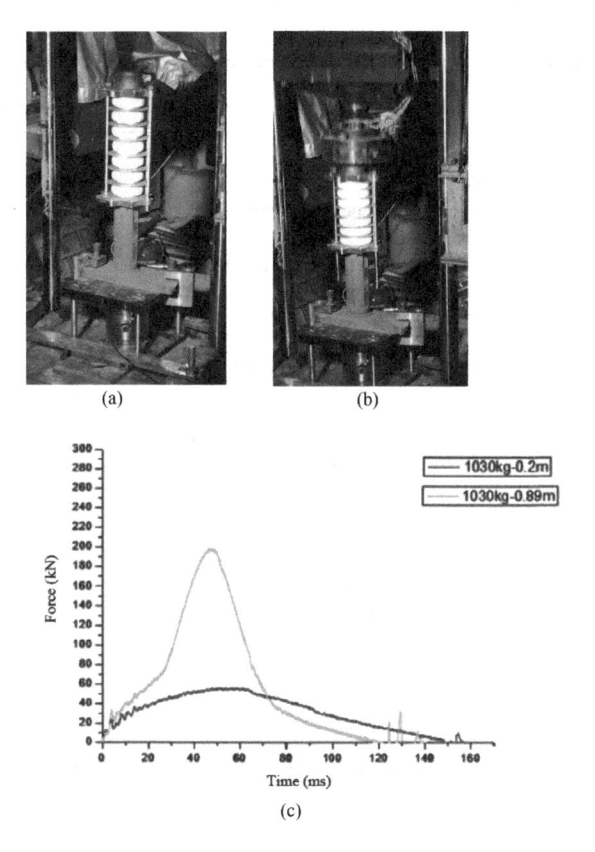

Figure 8. Drop weight impact test and impact curve of elastomer components: (a) Before drop hammer, (b) After drop hammer, and (c) Drop weight test curve.

Table 3 is the analysis of drop hammer impact data. It can be seen from the data and curve that when the maximum force of the buffer is 120kN, the corresponding stroke is 158mm, which is smaller than that under static pressure, but the difference is slightly, which is consistent with the characteristics of the elastomer. Therefore, the characteristics of the elastomer under impact can be basically evaluated by the static characteristics.

Table 3. Drop weight impact data of elastomer components.

NO.	Drop weight (kg)	Hammer height (m)	Impact velocity (m/s)	Energy (kJ)	Stroke (mm)	Maximum force (kN)
1	1030	0.20	2.00	2.06	/	57.0
2	1030	0.89	4.17	8.98	260	198.7

4.3 Parameters of elastomer energy absorbing device

According to the failure limit tension of 403kN, the maximum load of each of the four sliding bars is 100.75kN. The selected material is 45 # chromium plated bar after heat treatment, with tensile strength of 600MPa, yield strength of 355MPa and elongation of 1.6%. For the sake of safety, according to the yield strength calculation, the minimum cross-sectional area of single rod is $A = F/\sigma_s = 283.8mm^2$. The parameters of elastomer energy absorption device are shown in Table 4.

Table 4. Design parameters of elastomer energy absorption device.

NO.	Index item	Design value
1	Starting tension	16 kN
2	Mass	81.7 kg
3	Peak tension in energy absorption stage	120 kN
4	Buffer stroke	320mm
5	Theoretical maximum energy absorption	21kJ
6	Ultimate tension at failure	403 kN

The physical diagram and theoretical load curve of the elastomer energy absorption device as shown in Figure 9.

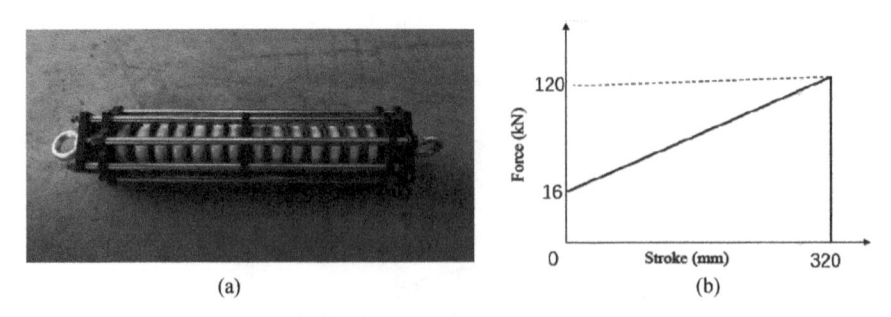

Figure 9. Elastic energy absorbing device and its theoretical load curve: (a) Elastomer energy absorbing device and (b) Theoretical load curve.

5 SMALL SCALE VERTICAL IMPACT TEST OF BEARING CABLE

The span of suspension type crossing frame is large, the time period of building true test environment is long, and the cost is high. Considering the test efficiency and economy, the small-scale vertical impact test is carried out to test the energy absorption effect of the energy absorption device.

5.1 Test installation and layout

The impact test of energy absorption device is composed of test frame, UHMWPE fiber rope, wireless tension sensor, wireless load release device, energy absorption device and standard configuration block in series. The connection arrangement is shown in Figure 10.

5.2 Test conditions

The small-scale vertical impact test of the bearing cable of the suspension type crossing frame is carried out. The impact test of 0.5T, 1T and 2T heavy objects falling from the height of 0.5m is implemented, respectively. The impact test without and with energy absorption device is carried out. The test conditions are shown in Table 5.

The energy absorption device is connected in series with the rope and installed on the gantry. In the initial state, the rope bears the weight of the energy absorption device, and the weight is suspended on the wireless release device, as shown in Figure 11. After starting the wireless load release device, the heavy object is decoupled from the wireless load release device, and the heavy object acts on the rings of the energy absorption device through the suspension rope to generate impact force. The test data are recorded. The state after impact is shown in Figure 11 (b).

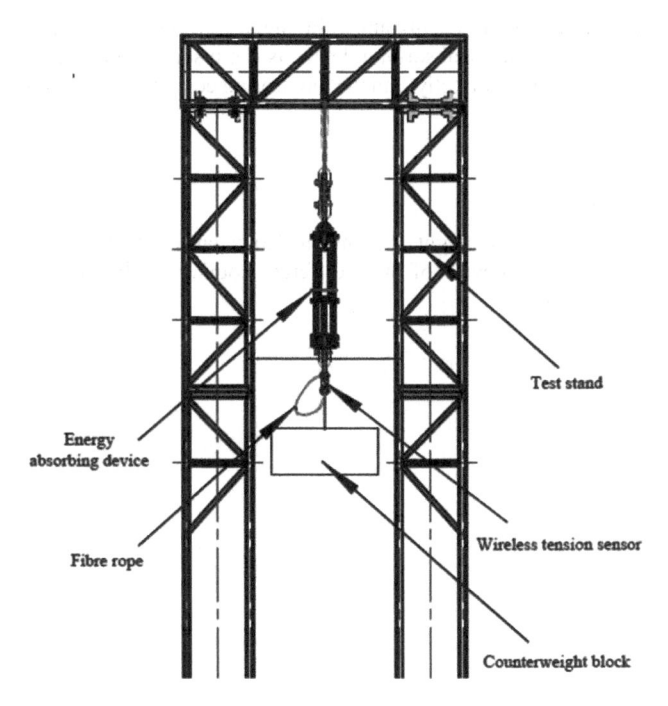

Figure 10. Test layout.

Table 5. Shock test condition of energy absorption device.

NO.	Drop height /m	French weight /T	Energy absorbing device
1		0.5	
2		1	without
3		1.5	
4	0.5	0.5	
5		1	with
6		1.5	

Figure 11. Working state of elastomer energy absorption device: (a) Initial installation state and (b) impact completion state.

The impact test process without energy absorbing device is similar to the above process. In the initial installation state, the energy absorbing device is not installed, and then the heavy object is released to impact the fibre rope directly, and the monitoring equipment records the tension change data of the rope.

5.3 Test results and analysis

Under the condition of 0.5m drop height, the tension change law of the impact rope with or without the elastic energy absorption device of three different counterweights of 0.5T, 1T and 1.5T is recorded, as shown in Figure 12.

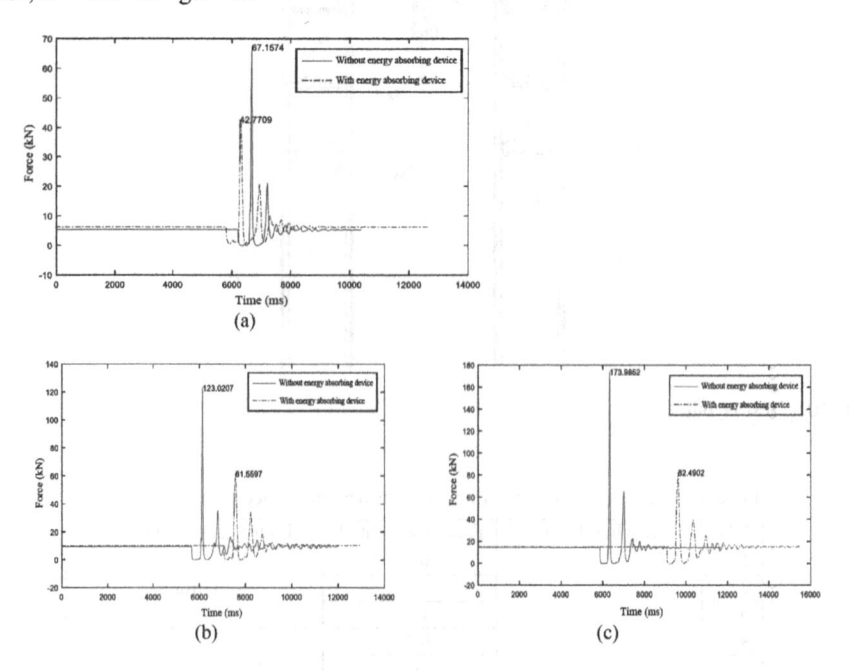

Figure 12. Tension curve of rope under different impact: (a) under impact of 0.5T counterweight, (b) under impact of 1T counterweight and (c) under impact of 1.5T counterweight.

It can be seen from the test results that when there is no elastic energy absorption device, the impact tension of the rope increases instantaneously, and gradually dissipates after the first wave crest. With the increase of counterweight mass, the peak value of tension of load-bearing cable increases. When the elastomer energy absorption device is installed, the peak load decreases. When the drop height is 0.5m, with the increase of counterweight mass, the energy absorption effect is more obvious. The energy absorption effect is shown in Table 6.

Table 6. Test data of elastomer energy absorption device.

NO.	Drop height /m	Counterweight mass/T	Energy absorbing device	Peak tension /kN	Energy absorption effect
1	0.5	0.5	with	42.77	36.31%
2			without	67.16	
3		1	with	61.56	49.96%
4			without	123.02	
5		1.5	with	82.49	52.59%
6			without	173.99	

From Table 6, it can be found from that the energy absorption effect of the elastomer energy absorption device can reach 52.59%, and it is significantly improved with the increase of impact force. Therefore, the elastomer energy absorption device as an important component of the whole frame can play a more anti-impact role.

6 CONCLUSION

In recent years, the impact resistance performance of power transmission equipment has been greatly improved in the field of practical engineering.

In this paper, based on the analysis of the bearing characteristics of the suspension crossing under accident conditions, several different types of energy absorption methods are compared and analysed. Finally, the viscoelastic energy absorption method is selected as the energy absorption method for the suspension crossing, and the elastomer energy absorption device is designed. Through the small-scale vertical state test of the bearing cable of the suspension type crossing frame, the energy absorption effect of the elastomer energy absorption device is verified. Based on the obtained results and their interpretations, the main findings are presented below:

- Under the conditions of 0.5T, 1.0T, 1.5T load and 0.5m impact height, the maximum energy absorption effect of the elastomer energy absorption device can reach 52.59%, which can effectively improve the impact resistance of the suspension type crossing frame.
- The energy absorption effect of the elastomer energy absorption device is significantly improved with the increase of impact force.

Finally, the findings of this study must be extended and validated to field tests on a real and full-scale suspension type crossing frame.

ACKNOWLEDGMENTS

Authors gratefully acknowledge the financial support of Science and Technology Research Project of State Grid (Research on Key Technologies for Improving Impact Resistance of Suspension Spanning Frame, Grant No. 5442GC180003).

REFERENCES

[1] Peng C, Chen Y Q 2013 Discovery for a new system, using fluid viscous damper and tuned mass damper (TMD) together in high-rise structure for wind-induced vibration reduction *Building Structure*. **s2** 720–25
[2] Lu G C, Hu L T 2005 Analysis of parametric sensitivity of fluid viscous dampers for Xihoumen Bridge *World Bridge*. **2** 45–47
[3] Zhao J D, Wang J C 2010 Research on buffer of asteroid lander based on semi-active control *J. Vibration & Shock*. **29** 78–80
[4] Zhang J 2008 The Application of viscous fluid damper in strengthening engineering *Earthquake Resistant Engineering and Retrofitting*. **30** 40–42
[5] Fu W J, Zhu C M, Zhang C Y 2003 Experimental analysis and dynamic simulation of hydraulic buffer used in elevator system *J. Vibration & Shock*. **22** 80–87
[6] Hao P F, Zhang X W 2003 Dynamic characteristic analysis of small hydraulic buffer *J. Chinese Mechanical Engineering*. **3** 155–158
[7] Cai W J, Wang P 2006 The structure and analysis on damping characteristic of a hydraulic damper *Machine Tool & Hydraulics*. **39** 149–153
[8] Hoback A. S 1996 Optimization of singular problems Struct. Opt. 12 94–97
[9] Zhou Y 2013 *Design theory and application of metal energy dissipation structure* (Wuhan: Wuhan University of Technology Press) p 203
[10] Hou C Y 2008 Fluid dynamics and behaviour of nonlinear viscous fluid dampers *J. Structural Engineering*. **134** 56–63.

Effect of alloy elements on corrosion resistance of low alloy steel in marine atmosphere

G.P. Liu, L.C. Yan, K.W. Gao, C.Y. Lang & Z. Ji
School of Materials Science and Engineering, University of Science and Technology Beijing, Beijing, China

ABSTRACT: Using data mining technology, the corrosion rate model was established through the support vector regression (SVR) algorithm, and the influence of Mn, P, Cu, Cr four alloy elements on the corrosion resistance of low alloy steel was studied. The results showed that the increase of Mn, Cu and Cr elements significantly improves the corrosion resistance of the alloy, while the increase of P element reduces the corrosion resistance of the alloy. Under the premise that the total content of alloying elements does not exceed 5 wt.%, the alloys of 1.4 wt.% Mn, 0.1 wt.% P, 1.4 wt.% Cu and 2.1 wt.% Cr have the best corrosion resistance and predict corrosion rate is 0.0205 mm/a.

1 INTRODUCTION

The marine atmosphere mainly refers to the atmospheric environment in the ocean and coastal areas. Due to the evaporation of seawater, an atmosphere with high salt content is formed, which has a strong corrosive effect on metals [1]. The marine atmospheric corrosion of low-alloy steel is that it is in a humid gas environment and a thin water film is formed on its surface. This is the first step of atmospheric corrosion of steel; and then through the corrosion of the thin water film, the dry/wet alternate of the micro-battery and electrochemical corrosion process is essentially the cathode and anode reaction at the corrosion interface; finally, the dissolution of anode iron occurs, resulting in the loss of the steel matrix [2–5].

Early research on weathering steel found that after adding certain alloying elements to carbon steel, the corrosion weight loss under atmospheric exposure conditions was significantly reduced. This is the initial intuitive understanding of the relationship between alloying elements and the corrosion resistance of weathering steel [6, 7]. Later, researchers continued to study the corrosion resistance mechanism of alloying elements and weathering steel, showing that the addition of certain alloying elements can significantly improve the atmospheric corrosion resistance of weathering steel [8].

Corrosion is a slow process of change, and the factors affecting corrosion are complex and changeable. Therefore, it is still an important challenge to predict the long-term development trend of material corrosion in the natural environment. This paper uses data mining technology to establish a corrosion rate model through the support vector regression algorithm to study the influence of Mn, P, Cu, and Cr on the corrosion resistance of low-alloy steels, and provide guidance for the design and preparation of high-corrosion-resistant low-carbon steels.

2 CORROSION RATE MODEL

2.1 Low alloy steel data set

The composition and environmental parameter data set of low alloy steel is shown in Table 1.

Table 1. Low alloy steel composition and environmental parameter data set.

feature		data range	description
material	C	0.0010~0.4200 wt.%	
	Si	0.0030~2.9800 wt.%	
	Mn	0.0030~1.5000 wt.%	
	P	0.0002~1.4800 wt.%	
	S	0.0001~0.0310 wt.%	
	Al	0.0010~3.0900 wt.%	
	Cu	0.0090~2.9700 wt.%	
	Cr	0.0013~9.0300 wt.%	
	Ni	0.0030~9.0600 wt.%	
	Mo	0.0050~0.4000 wt.%	
	Ti	0.0010~1.0010 wt.%	
environment	Temp.	12~32°C	average temperature
	R.H.	30.2~87.0%	relative humidity
	v-Cl	0.36~740.00 mg NaCl/$m^2 \cdot d$	Chloride deposition rate
	v-SO_2	0.1~130.0 mg SO_2/$m^2 \cdot d$	SO_2 deposition rate
	Year	0.5,1,2,3,4,5a	exposure period
Target attribute	Vcorr	0.0008~0.7568 mm/a	exposure period

2.2 Support vector regression (SVR) model

Use the training set to train the SVR model, and determine the C value and gamma value of the SVR model through grid search. Grid search is a parameter adjustment method. Through exhaustive search, that is, in all candidate parameter selections, through looping, try every possibility, and use the best performing parameter as the final result; it can be automatically adjusted Parameter, as long as the parameters are entered, the optimized results and parameters can be given. C is the penalty coefficient, which is the tolerance for errors. The higher the C, the less tolerable errors, and the easier to overfit; the smaller the C, the easier to underfit. If C is too large or too small, the generalization ability becomes poor. Gamma determines the distribution of data after mapping to the new feature space. The larger the gamma, the fewer the support vectors; the smaller the gamma value, the more support vectors.

3 RESULTS AND ANALYSIS

3.1 Influence of single elementcontent on corrosion resistance

The content of Mn, P, Cu, and Cr elements remained unchanged. The other components and environmental variables were set to the median of the data, and the corrosion time was set to 1 year. The predicted results are shown in Figure 1.

It can be seen from Figure 1 that after the other components and environmental variables are set to the median, the predicted corrosion rates are all above 0, indicating that this is more consistent with the actual situation. The addition of Mn element improves the corrosion resistance of the alloy to a certain extent; the addition of P element content accelerates the corrosion, and the corrosion rate reaches the maximum at about 1.5wt.%; the increase of Cu element content makes the corrosion resistance of the alloy improved The increase reached the minimum at about 1.5wt.%, and the subsequent increase in element content increased the corrosion rate; the corrosion rate of Cr added with less than 2wt.% increased with the increase in composition, and the Cr greater than 2wt.% reflected. In order to achieve strong corrosion resistance, the corrosion rate decreases with the increase of Cr content.

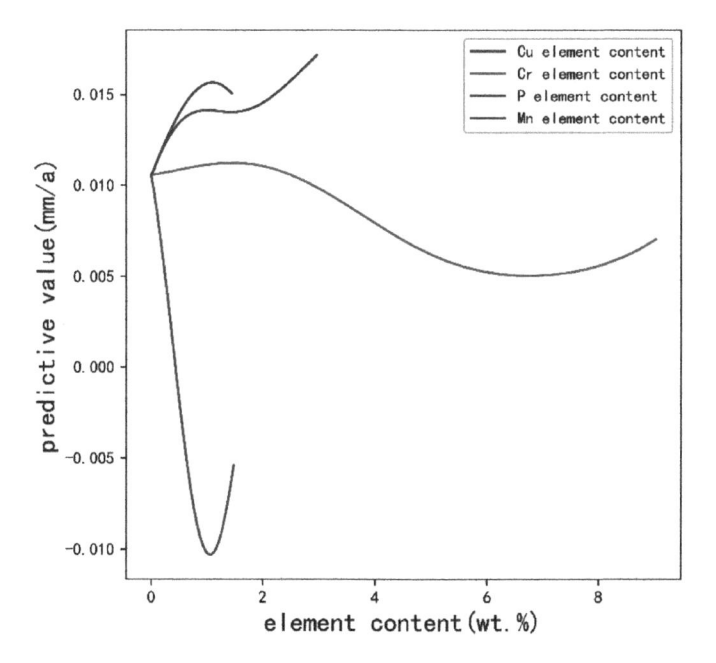

Figure 1. The SVR model predicts the influence of four component changes on the corrosion rate.

3.2 Optimum alloy element content

Establish a new element optimization data set, change the composition of the four alloy elements Mn, P, Cu, and Cr at the same time, and use the SVR model to predict the corrosion rate of the optimization data set to find the alloy composition with the best corrosion resistance. In order to ensure that the final element content is within the low alloy steel composition range, only the added element content should not exceed 5wt.%, that is, the total content of the four alloying elements should not exceed 5wt.%.

The box diagram shown in Figure 2 predicts the effect of changes in the content of the four elements on the corrosion resistance of the alloy. From this, we can see that the Mn content has little effect on the corrosion rate, and the corrosion rate with the addition of 1.4wt.% Mn is slightly lower than the corrosion rate when a small amount of Mn is added. The content of P element has obvious influence on the corrosion rate. With the increase of the content of P element, the overall corrosion rate shows an upward trend, which shows that the element P has a greater adverse effect on the corrosion resistance of the alloy. The addition of Cu element content reduces the corrosion rate of the alloy. When the content reaches 2.5wt.%, the effect of continuous addition of Cu on the corrosion resistance of the alloy weakens, and with the increase of Cu content, the minimum corrosion rate starts to increase instead. The higher the Cr element content, the lower the overall corrosion rate. Due to the large range of Cu and Cr elements and the restriction that the total content of the four alloy components does not exceed 5wt.%, there are relatively few calculation data for Cu and Cr when the content is high, resulting in a series of abnormal value.

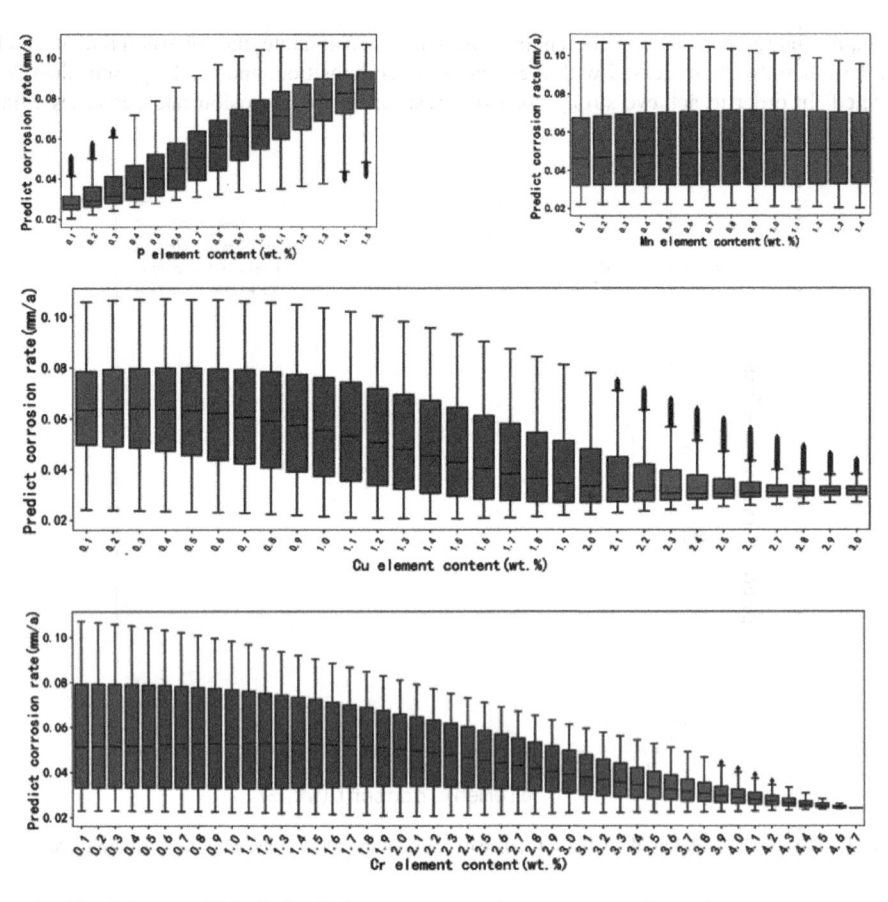

Figure 2. The influence of Mn, P, Cu, Cr four component changes on corrosion resistance.

Table 2 shows the 10 results with the smallest predicted corrosion rate and the composition of the corresponding alloy elements. It can be seen from the predicted results that the predicted minimum corrosion rate is 0.0205mm/a. Due to the limit of added elements and no more than 5wt.%, the higher the content of Mn and the lower the content of P, the higher the corrosion resistance; Cu and Cr The ratio of element content has a greater impact on the corrosion resistance of the alloy; the alloys with the lowest corrosion rate range are 1.4 wt.% Mn, 0.1 wt.%P, 1.3 to 1.5 wt.% Cu and 2.0 to 2.2 wt.% Cr.

Table 2. Optimization of the alloy composition with the best corrosion resistance.

Mn(wt.%)	P(wt.%)	Cu(wt.%)	Cr(wt.%)	Corrosion rate(mm/a)
1.4	0.1	1.4	2.1	0.0205
1.4	0.1	1.5	2.0	0.0205
1.4	0.1	1.3	2.2	0.0205
1.4	0.1	1.6	1.9	0.0207
1.3	0.1	1.4	2.2	0.0207
1.4	0.1	1.2	2.3	0.0207
1.4	0.1	1.4	2.0	0.0207
1.4	0.1	1.5	1.9	0.0207
1.3	0.1	1.3	2.3	0.0207
1.3	0.1	1.5	2.1	0.0207

4 CONCLUSIONS

(1) The increase of Mn, Cu and Cr elements significantly improves the corrosion resistance of the alloy, while the increase of P element reduces the corrosion resistance of the alloy.
(2) Under the premise that the total content of alloying elements does not exceed 5 wt.%, the alloys of 11.4 wt.% Mn, 0.1 wt.% P, 1.3~1.5 wt.% Cu and 2.0~2.2 wt.% Cr have the best Corrosion resistance, its predicted corrosion rate is 0.0205mm/a.

REFERENCES

[1] Zhenyao Wang, Guocai Yu, Wei Han. Investigation of atmospheric corrosivity of natural environment in my country[J]. Corrosion and protection, 2003, 8:323–326.
[2] Yuzhen Lin, Junde Yang. Corrosion and corrosion control principles[M]. Beijing: China Petrochemical Press, 2007.
[3] Singh J K, Singh D D N. The nature of rusts and corrosion characteristics of low alloy and plain carbon steels in three kinds of concrete pore solution with salinity and different pH[J]. Corrosion Science, 2012, 56:129–142.
[4] Yingying Zhen, Yan Zhou, Jia Wang. Research progress on corrosion behavior of carbon steel under rust layer in marine environment[J]. Corrosion Science and Protection Technology, 2011, 23(1):93–98.
[5] Yamashita M, Uchida H. Recent Research and Development in Solving Atmospheric Corrosion Problems of Steel Industries in Japan[J]. Hyperfine Interactions, 2002, 139–140(1):153–166.
[6] Morcillo M, Chico B, Diaz I, et al. Atmospheric corrosion data of weathering steels. A review[J]. Corrosion Science, 2013, 77(12):6–24.
[7] Lihong Liu, Huibin Qi, Yanping Lu, et al. General Situation of Research on Atmospheric Corrosion Resistant Steel[J]. Corrosion Science and Protection Technology, 2003, 2:86–89.
[8] Quancheng, Jianshen Wu, Wenlong Zheng, et al. Effect of the secondary distribution of alloying elements on the atmospheric corrosion resistance of weathering steel[J]. Material protection, 2001, 34(4): 4–5.

Game testing based on artificial intelligence emotional cognition

J.H. Cheng & J.Q. Li
Faculty of Humanities and Arts, Macau University of Science and Technology, Macau, China

ABSTRACT: Artificial intelligence is developing rapidly, and AI technology is playing an important role in various IT fields. Compared with the traditional game testing work, adding more automated and intelligent systems to the testing will greatly reduce the human and material investment. It is the entry point of this paper to try to apply new artificial intelligence techniques to find new models to help game testing work. Through the analysis and comparison of existing algorithms, theories and models, the feasibility of some current mainstream AI models and algorithms in game testing sessions is inferred and identified, in an attempt to find the model basis and algorithmic foundation for the new approach. Game testing is different from ordinary software testing. In addition to the testing of product functions, more attention should be paid to the entertainment and user experience of the game. The current conventional experience testing uses in-game test play and user surveys. In emotional cognition, artificial intelligence excludes human factors, and through big data and its own learning and computing capabilities, it can obtain more objective test results, while also saving the investment of human and material costs. This paper analyzes three aspects of AI in software testing, game testing and the current state of technology of emotion cognition, and explores the feasibility of game testing based on AI emotion cognition.

1 INTRODUCTION

In recent years, with the rapid development of computer and Internet technology, artificial intelligence technology has also continued to mature. In software testing, the application of artificial intelligence in software testing aims to write more effective test cases instead of humans, so as to better detect potential errors in the program. In this case, a large amount of input and output data is provided by the tester for training in AI techniques, which ultimately allows the AI to plan and execute tests according to specific needs and provide a reasonable summary of the test results [1].

In the gaming sector, game developers can use AI findings to begin creating immersive and intelligent games - a complex game that can be built automatically that can change and respond to player feedback, and where the game characters inside the game do not simply respond to the player, but can evolve over time.

Testing in games is an extremely important part of game software development and typically takes around 1/3 of the development cycle, with testing taking place throughout thedevelopment process. According to the game development process, testing can be broadly divided into unit testing, module testing, overall testing and product testing [2].

Product testing refers to the market research activities carried out by developers around the product, according to their own purposes, using professional technical means and research methods, and often accompanied by some market activities [3]. The main purpose of the market research method is to test players' experience of the game in the early stages, so as to make changes to the bad experience and improve the players' experience of the game, and thus improve the reputation and profitability after the release.

Traditional game product testing usually involves launching an internal test version, inviting some players to try it out, then collecting feedback from them and making changes to any bad feedback. This type of testing not only increases the investment in human and material resources during testing, but also bears a greater risk of leakage.

In terms of emotional cognition, AI can build corresponding models through data combining human behaviour and emotions, which can be used in monitoring the emotions represented by a person's behaviour in the environment.

Based on the current status and experience of AI in software testing, game development and emotion cognition, this paper further investigates the feasibility of using AI techniques in game testing to predict the emotions of players' experiences and provide player emotion feedback for game testing efforts.

2 ARTIFICIAL INTELLIGENCE AND SOFTWARE TESTING

Test cases play a decisive role in the entire software testing process. In software testing, which is designed to find errors and then execute the program, the writing of test cases is the key to finding bugs or errors in the program. In today's world of automated testing tools, the writing of complex test cases is still dominated by humans, except for simple data tests which can be generated automatically by computers. Automated testing tools can only automate the output of human-written test cases to execute programs and produce results.

As can be seen in Figure 1, AI has covered many areas of software testing, from the requirements analysis phase to test execution and closure. This is the current market contribution to artificial intelligence in software testing.

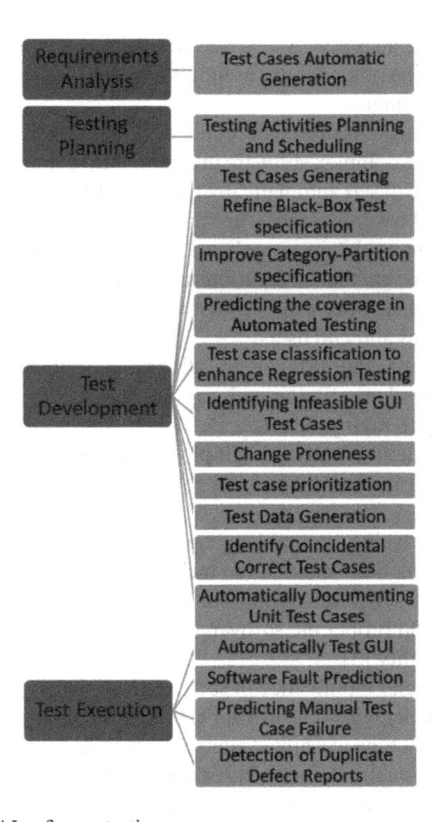

Figure 1. Current scope of AI software testing.

The emerging trend of artificial intelligence in the software testing industry is very promising and AI will drive the industry to great achievements. This is the future and companies are already investing in the industry. The following are key expected contributions to AI in the software testing sector in the near future (4 to 8 years) based on their research analysis and predictive studies.

Artificial intelligence software testing will become a separate industry and will play an important role in IT. They expect that AI software testing will replace quality assurance and tester engineers. Quality checking teams and tester engineers will play a new role in tuning and monitoring AI results.

Artificial intelligence will drive software testing and will cover all phases of testing from test preparation to planning, execution and reporting, without the need for human intervention and errors.

The AI software testing industry will produce more accurate results and will shorten the software development lifecycle compared to traditional testing techniques. The challenge of building software solutions to meet deadlines in particular, as they may not be able to meet overwhelming software requirements, is why AI will bridge this gap and alleviate this challenge by reducing the required testing time.

Artificial intelligence will eventually have dedicated tools to effectively test new technologies such as cloud computing, the Internet of Things, big data and other future technologies. Combining new technologies will bring innovation to AI software testing, as AI will play the role of integrator in generating the test data needed for a specific product.

They expect to have dedicated software and hardware solutions that can run AI deep learning as well as other AI algorithms and techniques to achieve more accurate test results in a competitive time frame.

Artificial intelligence will also play a key role in testing customer requirements by applying predictive analytics to examine other similar products and services to better understand which new features are required by customers.

Artificial intelligence software testing will reduce time to market and increase the efficiency of organisations producing more complex software in a competitive time frame. Artificial intelligence allows complex data to be analysed automatically using intelligent techniques and algorithms.

Artificial Intelligence will cover most software product testing in all areas including: application development, web development, database applications, mobile applications, the gaming industry, real-time critical applications, embedded solutions and more.

The new AI software testing tools will be innovative, agile and intelligent. They will provide better results for beneficiaries and end users.

By using AI algorithms and techniques, organisations and businesses will improve the customer experience, enhance their product offering and improve the quality of the services provided, and will bring software stability to their products.

Artificial intelligence predictive analytics will play an important role in discovering all possible test cases and will make software products more robust, more reliable and will exceed customer expectations.

Machine learning, deep learning, natural language processing and other areas of artificial intelligence are seen as leading edge for most of the technologies around them. As they highlight and discuss, introducing AI into software testing will unleash the power of intelligent software test automation and will drive and propel the software development and testing industry in a new era of driving innovation and agility [3].

As an application of artificial intelligence in software testing, the aim is to write more effective test cases instead of human beings, so as to find errors in the program. Based on this, they hope that by using the theory of machine learning, testers will provide a large amount of input and output data to train the AI, and eventually the AI will learn to automatically generate test cases, execute tests and analyse the test results according to specific needs. This will greatly reduce the manual testing workload and significantly improve efficiency [4].

3 ARTIFICIAL INTELLIGENCE AND EMOTIONAL COGNITION

In medical psychology, one can classify human emotions into hundreds of types, and there is no universal way to categorise them. However, all these complex emotions consist of four basic emotions: happiness, sadness, anger and fear, and all emotions can vary in intensity. According to medical psychology, for artificial intelligence to perceive emotions, it is necessary to quantify them and model them.

For example, the emotion model mentioned in 'An emotion-based approach to decision making and self learning in autonomous robot control' is based on four emotions: sadness, loneliness, disgust and fear, while the sensory inputs are energy, friendship, cleanliness and brightness. Like the sensory inputs, these four emotions are also real values between 0 and 1. The relationship between sensory input and emotion depends on some basic rules.

When the level of sensory energy is low, the emotion of sadness is obtained.
When the level of feeling friendship is low, the emotion of loneliness is obtained.
When the cleanliness of the environment is low, the emotion of disgust is obtained.
When the brightness of the environment is low, the emotion of fear is obtained.

In this model, all sensory inputs and emotions are mapped to three fuzzy sets S, M and L. The output is determined in fuzzy inference and the fuzzy if-then rule is described as follows.

IF inputi is Ai.1 and...and inputm is Ai,m,
THEN emi is Bi,1 and...and emm is Bi, m, [5]

Where Ai, j and Bi. j are the functions of the jth input and output at rule i, respectively, and m and n are the number of inputs and outputs, respectively. Figure 2 shows the rules used in this model. In addition, a weighted average method was used to deblur the images to obtain accurate sentiment values.

Taking into account the effect of recent emotional history, the temporal thresholds and the increments of dominant emotion are given as follows.

$emdom(t+1) = emdom(t) + \Delta e, t > tthr$

where tthr and positive Δe are the time threshold and the increment, respectively. This means that if a certain emotion has been the dominant emotion for a long period of time, then it will increase in intensity faster and faster.

Input	Value	Output	Value
Energy	S	Sad	L
	M		M
	L		S
Friendship	M	Lonely	M
	L		S
	S		L
Cleanness	M	Disgust	M
	L		S
	S		L
Brightness	M	Fear	M
	L		S

Figure 2. Fuzzy inference rules for emotion models.

A computational model of artificial intelligence emotional cognition is investigated based on the theory of emotional intuition and formal Bayesian rules are used to analyse the inference classification of emotional cognition [6]. The model represents an approach to emotion detection with a rational information integration process. The observer weighs up the available data on the behaviour displayed by the subject in different emotions and combines them using the statistical principles of Bayesian inference [7].

Artificial intelligence can use the data to build corresponding models that can be used in monitoring the emotions represented by a person's behaviour in the environment.

4 THE USE OF ARTIFICIAL INTELLIGENCE EMOTIONAL COGNITION IN GAME TESTING

Game testing is essentially a kind of software testing, and the essence of game testing and software testing is basically the same in terms of test engineering. Game testing inherits the characteristics of software testing, but has its own unique characteristics [8].

The major difference between the two is that games need to be entertaining in the process of player experience, which is mainly reflected in the scene character design, functional design and interaction design of the game, so the testing of player emotion and game entertainment is an indispensable part of game testing.

Current game testing is mainly divided into traditional software testing and testing of the game itself. In software testing of games, game test engineers need to write test plans, test cases and test specifications to find out the shortcomings of the game functions. In the testing of the game format, a good test engineer needs to use a certain amount of trial data to find the shortcomings of the game design and make changes. The aim of all game testing is to find software problems and aspects of the game that are not sufficiently entertaining and to improve the playability of the game by making changes.

Reinforcement learning (RL) models offer the possibility to complement current scripted and automated solutions by learning directly from playing the game without human intervention. Modern RL algorithms are capable of exploring complex environments [8], as well as identifying vulnerabilities in game mechanics [10]. In the case of first-person shooters (FPS), for example, RL is particularly suited to modern FPS games, which can be said to consist of two main phases: navigation (finding targets, enemies, weapons, health, etc.) and combat (shooting, reloading, taking cover, etc.). Artificial intelligence techniques have been able to learn a single model of the entire game dynamic [11].

Artificial intelligence in games can be understood as all the 'thinking' done by the computer in the game, which makes the game behave similarly to human intelligence, or to the player's thinking and perception. The application of AI technology in the design and development of computer games can improve the playability of games, improve the game development process, and even change the way games are made [12].

In the game testing process, AI can be applied to game testing based on its capabilities in software testing, for testing the performance and implementation of features and problems with the design of game scenes and characters.

Conventional game testing relies on the use of human game testers, game test scripts and a priori knowledge of regions of interest to generate relevant test data. Using Deep Reinforcement Learning (DRL), they have introduced a self-learning mechanism into the game testing framework. Using DRL, the framework is able to explore or exploit game mechanics based on user-defined reinforcement reward signals. The result was increased test coverage and the discovery of unexpected game mechanics, exploits and bugs in a wide range of game types. In this paper, they demonstrate that DRL can be used to increase test coverage, find exploits and thus test out map difficulty as well as detect common problems that occur in FPS games [13].

In terms of testing player sentiment in games, AI's model of perception of sentiment is used as the basis for collecting previous game data and establishing the degree of player entertainment experience in different situations based on player behaviour in different scenarios and different game episodes, such as game duration, motivation and top-ups, and then in testing different scenarios, levels and episodes of the game, etc., for making This will help to predict the players' emotions in the different scenarios, levels and storylines of the game, to test the players' emotional and entertainment deficiencies in the game, and to provide the developers with valid data to help them modify the game's entertainment deficiencies.

5 SUMMARY

Artificial intelligence techniques are currently being used in three areas - software testing, game development and emotional awareness - to improve efficiency and accuracy.

The current main testing methods in terms of player sentiment and game entertainment testing are market research and internal testing data, which costs a lot of human and material investment costs and risks leaking game content.

This paper analyses the current state of artificial intelligence in software testing, game development and emotion perception, and demonstrates the feasibility of using AI techniques for player emotion prediction in game testing through experience in the use and application of AI techniques in these three areas. Testing with AI reduces the cost of human and material resources in testing and guarantees absolute confidentiality of the game before release. It also reduces the interference of human subjective factors and increases the accuracy of the test.

REFERENCES

[1] Qin Xiaoyan. Research and application of artificial intelligence software testing[J]. Information and Computer (Theory Edition), 2019, 31(24):66–67+70.

[2] Zhang Hongyan, Guo Haixin, Wang Dongfang. An introduction to game testing techniques[J]. Silicon Valley, 2009(22):42.

[3] H. Hourani, A. Hammad and M. Lafi, "The Impact of Artificial Intelligence on Software Testing," 2019 IEEE Jordan International Joint Conference on Electrical Engineering and Information Technology (JEEIT), Amman, Jordan, 2019, pp. 565–570, doi: 10.1109/JEEIT.2019.8717439.

[4] Ren ZH, Li XX, Gong XUN. Applications and challenges of artificial intelligence on software testing[J]. Computer Knowledge and Technology, 2018, 14(29):218–219.

[5] Changsheng Yu, and Li Xu. "An Emotion-Based Approach to Decision Making and Self Learning in Autonomous Robot Control." Fifth World Congress on Intelligent Control and Automation (IEEE Cat. No.04EX788). Vol. 3. IEEE, 2004. 2386–2390 Vol.3. Web.

[6] Fan Yuehong. Analysis of computational models for emotional cognitive reasoning of artificial intelligence[J]. Journal of Shanghai Normal University (Philosophy and Social Science Edition), 2020, 49(02):94–103.

[7] Ong, Desmond C. "Affective Cognition: Exploring Lay Theories of Emotion." Cognition 143 (2015): 141–162. Web.

[8] Zhang Jingfeng. What is the difference between game testing and general software testing [EB/OL].

[9] https://www.zhihu.com/question/23225355/answer/87917642 , 2016-02-24/2020-11-28.

[10] Baker, Bowen, et al. "Emergent tool use from multi-agent autocurricula." arXiv preprint arXiv:1909.07528 (2019).

[11] Harmer, Jack, et al. "Imitation learning with concurrent actions in 3d games." 2018 IEEE Conference on Computational Intelligence and Games (CIG). IEEE, 2018.

[12] Zhang Yukong. Artificial intelligence in computer games [J]. Science and Technology Information (Academic Research), 2007(18):244–245.

[13] J. Bergdahl, C. Gordillo, K. Tollmar and L. Gisslén, "Augmenting Automated Game Testing with Deep Reinforcement Learning," 2020 IEEE Conference on Games (CoG), Osaka, Japan, 2020, pp. 600–603, doi: 10.1109/CoG47356.2020.9231552.

[14] Lample, Guillaume, and Devendra Singh Chaplot. "Playing FPS games with deep reinforcement learning." arXiv preprint arXiv:1609.05521 (2016).

Nonlinear dynamic characteristics of mechanical vibration system with multi-gap asymmetric elastic constraint

F.W. Yin & G.W. Luo
School of Mechatronic Engineering, Lanzhou Jiaotong University, Lanzhou, China
Key Laboratory of System Dynamics and Reliability of Rail Transport Equipment of Gansu Province, Lanzhou, China

D.Q. Liu
Jilin Railway Technology College, Jilin, China

ABSTRACT: Based on a two-degree-of-freedom forced vibration system with multi-gap elastic constraints, and the variable step Runge-Kutta method was used for numerical calculation. Multi-parameter and multi-objective collaborative simulation analyses the global description of the dynamics of mechanical system in its parameter domain. The influence of system parameters on pattern types and transition characteristics of basic periodic and subharmonic vibrations is studied. The calculation results reveal the distribution characteristics of basic periodic impact groups of the vibration system with equal gap thresholds in the low frequency domain. Singularities, hysteretic and non-hysteretic transition regions induced by irreversibility of transition between adjacent basic periodic impact vibration are found. The coexistence of adjacent basic periodic vibration in hysteretic transition region and the pattern types and regularity characteristics of subharmonic vibration in non-hysteretic transition region are revealed. The pattern types and emergence rules of periodic impact vibration of system with unequal gap constraint are analyzed. The research provides scientific basis for dynamic design and collaborative optimization of mechanical systems with multiple elastic constraints.

1 INTRODUCTION

Dynamics and control of mechanical systems with piecewise smooth mechanical factors such as clearance, constraint and hysteresis are the common research hotspots in mechanical engineering and vibration engineering disciplines. Especially when the coupling of multiple non-smooth mechanical factors reflects the engineering reality, its dynamic behavior becomes very complex due to the interaction of many non-linear factors. For the dynamic research of multi-degree-of-freedom forced vibration system, domestic and foreign scholars pay more attention to symmetric constraints and unilateral constraint. However, there are a large number of non-linear factors in the engineering practice, such as constraints and control methods, whose dynamic behavior becomes extremely complex due to the influence of asymmetry. Shaw and Holmes [1] studied a single-degree-of-freedom vibration system and converted the traditional method of periodic impact response of the system into impact mapping. In [2–5], the periodic vibration stability, bifurcations and chaos of various impact vibration systems with clearance-constraint were studied. Nordmark [6] discovered the singularities of the impact Poincaré map and the induced grazing bifurcations. Luo studied the periodic vibration, bifurcations and chaos control of the vibration system [7], and analyzed the evolution law of the hysteretic and non-hysteretic transition regions of the vibration system with rigid constraints [8]. Wagg [9–10] analyzed the low-frequency vibration characteristics of a two-degree-of-freedom vibration system with multiple rigid constraints, and focused on the chattering-sticking vibration and rising phenomenon in the low-frequency domain. Yin and Shi [11–13] studied the effects of system parameters on the type and bifurcation of periodic impact vibration in mechanical vibration systems with clearance elastic and rigid constraints. Literature [14–16] has carried out further research on chattering-impact and complete chattering-impact vibration characteristics of vibration system.

A two-degree-of-freedom forced vibration system with three-elastic constraints was established. Diversity and bifurcation characteristics of periodic impact vibration of this kind of non-smooth vibration system are studied. The global description of system dynamics with unequal gap threshold constraints is studied. The pattern types and distribution characteristics of system periodic impact vibration under symmetrical and asymmetrical harmonic force are analyzed.

2 SYSTEM MECHANICAL MODEL

A mechanical model of a vibration system with multiple gaps asymmetrical elastic constraints is expressed in Figure 1. Mass M_1 and mass M_2 are connected by a linear spring K_1 and a linear damper C_1. And mass M_2 is connected to the base by a linear spring K_2 and a linear damper C_2. The harmonic excitation force $P_i \sin(\Omega T + \tau)$ acts on the corresponding mass $M_i (i=1,2)$, where P_i, Ω and τ represent the amplitude, frequency and initial phase respectively. Elastic constraints of the same stiffness K_0 are placed on the right constraint of mass M_1 and on the bilateral constraints of mass M_2. When the harmonic force is small, the system shows no-impac free vibration. At this time, with the increase of harmonic force, when the displacement of the mass is equal to the gap threshold of the corresponding constraint, the mass has a soft impact at its corresponding constraint [17, 18].

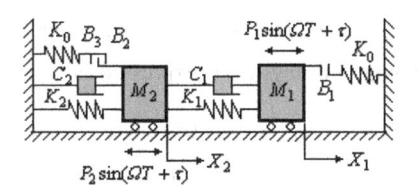

Figure 1. Mechanical model.

Defined Dimensionless Parameters, Variables and Time

$$\mu_m = \frac{M_2}{M_2 + M_1}, \mu_k = \frac{K_2}{K_2 + K_1}, \mu_c = \frac{C_2}{C_2 + C_1}, k_0 = \frac{K_0}{K_0 + K_1}, \quad (1)$$

$$f = \frac{P_2}{P_2 + P_1}, \delta_i = \frac{B_i K_1}{P_1 + P_2}, x_i = \frac{X_i K_1}{P_1 + P_2} (i = 1, 2, 3)$$

Dimensionless Differential Equation of Motion

$$\begin{bmatrix} 1 & 0 \\ 0 & \frac{\mu_m}{1-\mu_m} \end{bmatrix} \begin{Bmatrix} \ddot{x}_1 \\ \ddot{x}_2 \end{Bmatrix} + \begin{bmatrix} 2\zeta & -2\zeta \\ -2\zeta & \frac{2\zeta}{1-\mu_c} \end{bmatrix} \begin{Bmatrix} \dot{x}_1 \\ \dot{x}_2 \end{Bmatrix} + \begin{bmatrix} 1 & -1 \\ -1 & \frac{1}{1-\mu_k} \end{bmatrix} \begin{Bmatrix} x_1 \\ x_2 \end{Bmatrix}$$

$$+ \begin{Bmatrix} f_1(x_1) \\ f_2(x_2) \end{Bmatrix} = \begin{Bmatrix} 1-f \\ f \end{Bmatrix} \sin(\omega t + \tau) \quad (2)$$

in which $f_1(x_1)$ and $f_2(x_2)$ are expressed by

$$f_1(x_1)=\begin{cases}\dfrac{k_0}{1-k_0}(x_1-\delta_1), & x_1>\delta_1,\\ 0, & x_1\leq\delta_1,\end{cases} \quad f_2(x_2)=\begin{cases}\dfrac{k_0}{1-k_0}(x_2-\delta_2), & x_2>\delta_2,\\ 0, & -\delta_3\leq x_2\leq\delta_2,\\ \dfrac{k_0}{1-k_0}(x_2+\delta_3), & x_2<-\delta_3.\end{cases} \quad (3)$$

Based on dimensionless parameters, the range of some parameters of the system can be determined, i.e. $\mu_m \in (0,1) \mu_k \in (0,1), \mu_c \in (0,1) k_0 \in (0,1), f \in [0,1]$. The symbol p/n is used to describe the periodic vibration of system mass M_1, where n is the ratio of the vibration period of the system $T_n = 2n\pi/\omega$ to the period of the harmonic force $T_0 = 2\pi/\omega (n = 1,2,3,\ldots)$, p is the number of impacts on the elastic constraint of mass M_1 during the vibration period ($p = 0,1,2,3,\ldots$) At the same time, the symbol $n - p - q$ denotes the periodic vibration of mass M_2, and p (or q) represents the number of mass M_2 impact its left (or right) elastic constraint respectively. It is worth noting that the various periodic impact vibrations with $n = 1$ in the low-frequency domain are called the basic periodic impac vibration group. In order to calculate the values of n, p and q, the system state variable at the moment of impact at the B_1, B_2 and B_3 elastic constraints of the system is selected as Poincaré section σ_I^k ($k = 1,2,3$), and the system state variable at the moment of minimum displacement of the mass M_1 of the system within each period of harmonic force is selected as Poincaré section σ_{II}. Defined four kinds of Poincaré mapping

$$\sigma_I^1 = \{(x_1,\dot{x}_1,x_2,\dot{x}_2,t) \in R^4 \times T | x_1 = \delta_1, \dot{x}_1 > 0\};$$

$$\sigma_I^2 = \{(x_1,\dot{x}_1,x_2,\dot{x}_2,t) \in R^4 \times T | x_2 = \delta_2, \dot{x}_2 > 0\};$$

$$\sigma_I^3 = \{(x_1,\dot{x}_1,x_2,\dot{x}_2,t) \in R^4 \times T | x_2 = -\delta_3, \dot{x}_2 < 0\};$$

$$\sigma_{II} = \{(x_1,\dot{x}_1,x_2,\dot{x}_2,t)R^4 \times T | x_1 = x_{1\min}, \mod(t = 2\pi/\omega)\} \quad (4)$$

The impact Poincaré mapping of the system is expressed as

$$X^{(i+1)} = f(X^{(i)},\mu), \quad (5)$$

where $X^{(i)} = (\dot{x}_{1+}^{(i)}, x_2^{(i)}, \dot{x}_{2+}^{(i)}, \tau^{(i)})^T X^{(i+1)} = (\dot{x}_{1+}^{(i+1)}, x_2^{(i+1)}, \dot{x}_{2+}^{(i+1)}, \tau^{(i+1)})^T$, $X \in R^4$, μ are parameters, $\mu \in R^8$

3 WITH EQUAL GAP THRESHOLDS, VIBRATION CHARACTERISTICS OF A VIBRATION SYSTEM IN THE LOW-FREQUENCY DOMAIN

Dimensionless parameters of the system are selected $\mu_m = 0.35$, $\mu_k = 0.65$, $\mu_c = 0.65$, $\zeta = 0.1$, $f = 0.5$, $k_0 = 0.95$. Considering a two-degree-of-freedom forced vibration system with multiple elastic constraints with equal gap thresholds, the pattern type and bifurcation characteristics of impact vibration of the system in the (ω, δ) parameter plane are analyzed by multi-parameter and multi-objective collaborative simulation. Various periodic vibrations of the system are distinguished in the (ω, δ) parameter plane by different colors and symbols p/n (or $n - p - q$). The 70% black area marked 'C' in the figure represents chaos or long period multi-impact vibration. The $p/1$ (or $1-p-q, p \geq 6$ or $q \geq 6$) symbol indicates chattering-impact vibration. It can be seen from Figure 2(a) that in the parameter domain of gap threshold $\delta > 1.0$, mass M_1 mainly presents 1/1 periodic vibration, 1/2, 2/2 subharmonic vibration and 0/1 nonimpact periodic vibration. In the domain of gap threshold $\delta \leq 1.0$, mass M_1 shows 1/2, 2/2, 2/3, 2/4, 3/4, 4/4 subharmonic vibration and chaos in the high-frequency domain With the decrease of the frequency ω, a series of grazing contact occurs between the mass M_1 and its elastic constraint, resulting in the evolution of the $p/1$ vibration into $(p + 1)/1$ vibration due to the Grazing bifurcations. As a result, the number of impacts(p) of the mass M_1 increases gradually, showing the basic periodic impact vibration group $p/1$ ($p = 1, 2, 3, \ldots$) in the low-frequency domain. So far, when the number of impacts (p) becomes sufficient, the basic periodic vibration ($p/1$) of the system presents the characteristics of chattering-impact vibration At the same time, Figure 2(b) shows that in the large gap threshold(δ) domain, mass M_2 mainly presents 1-0-0 and 2-0-0 nonimpact periodic vibration. In the small gap threshold domain, when the frequency is $\omega \in (0, 1.0)$, the mass M_2 shows 1-1-0, 1-2-0, 1-3-0 basic periodic vibration group and 1-1-1, 1-1-2, 1-1-3, 1-1-4, 1-1-5 vibration group And with the increase of the frequency ω, the mass M_2 appears 1-2-1, 1-3-1 vibration. However, when the frequency (ω) is high, the mass M_2 shows 2-1-0, 2-1-1, 2-2-0, 2-2-1, 2-2-2, 4-2-2, 4-4-2 subharmonic vibration and chaos. Figures 2(c), (d), (e) and (f) are local details of Figure 2(a). Figure 2(c) starts with the lower left corner vertex (0,0), and the initial value of vibration is $X^{(0)} = (\dot{x}_{1+}^{(0)}, x_2^{(0)}, \dot{x}_{2+}^{(0)}, \tau^{(0)})^T = (0, 0, 0, 0)^T$, $x_1^{(0)} = 0.0$. Figure 2(d) starts with the top right corner vertex (0.7, 1.15), and the initial value of vibration is $X^{(0)} = (\dot{x}_{1+}^{(0)}, x_2^{(0)}, \dot{x}_{2+}^{(0)}, \tau^{(0)})^T = (0, 0, 0, 0)^T$, $x_1^{(0)} = 1.15$. In Figures 2(c) and 2(d), the horizontal coordinate $\vec{\omega}$ (or $\overleftarrow{\omega}$) and the vertical coordinate $\vec{\delta}$ (or $\overleftarrow{\delta}$) represent the incremental(or decrement) change of the frequency (ω) and the gap threshold (δ) during the numerical calculation. Figure 2(e) is an overlay of Figures 2(c) and (d) for the purpose of obtaining the hysteresis transition regions on the adjacent base periodic impact vibration boundary. The results show that with the decrease of the frequency (ω) or gap threshold (δ), the $p/1$ vibration is transformed into $(p + 1)/1$ vibration through Real-Grazing bifurcations, or subharmonic vibration or chaos is induced through Bare-Grazing bifurcations. On the boundary of Grazing bifurcations $G_{p/1}$, the $p/1$ vibration orbit of mass M_1 is in grazing contact with its constraint, i.e. the impact speed of the ($p + 1$) is zero. Through the boundary of Grazing bifurcations, the number of impacts increased from p to ($p + 1$) and the system maintained the original vibration period, forming a stable $(p + 1)/1$ vibration. This kind of Grazing bifurcation is defined as Real-Grazing bifurcation. Correspondingly, there are also Bare-Grazing bifurcation of $p/1$ vibration, but $(p + 1)/1$ vibration induced by Bare-Grazing bifurcations is unstable, and the system then presents subharmonic vibration or chaos. With the increase of the frequency (ω) or gap threshold (δ), the $(p + 1)/1$ vibration is transferred to $p/1$ vibration due to the saddle bifurcations, or $(2p + 2)/2$ vibration is generated due to the periodic doubling bifurcations. Except for singularities, the transition between adjacent basic periodic vibration $p/1$ and $(p + 1)/1$ is irreversible. It is irreversibility that causes a series of singularities and alternately distributed hysteretic and non-hysteretic transition regions on the boundary of adjacent basic periodic vibrations. Therefore, the singular point is not only the two-fold Grazing bifurcation point of $p/1$ vibration, but also the saddle-periodic doubling codimension-two bifurcation of $(p + 1)/1$ vibration. The hysteretic transition region is delineated by the saddle bifurcation boundary of $(p + 1)/1$ vibration and the Real-Grazing bifurcation boundary of $p/1$ vibration, as shown in the grey areas marked with HR_1, HR_2, and HR_3 in Figure 2(e). The adjacent basic periodic vibrations $p/1$ and $(p + 1)/1$ can coexist in the hysteretic transition region due to the difference of initial conditions of the system. The non-hysteretic transition region is defined by the doubling bifurcation boundary $PD_{(p+1)/1}$ of $(p + 1)/1$ vibration and the Bare-Grazing bifurcation boundary $G_{p/1}^b$ of $p/1$ vibration. There are pattern types $(2p + 1)/2$, $(3p + 1)/3$, $\ldots (np + 1)/n$ subharmonic vibrations, as shown in Figure 2(f).

In Figure 2(e), the 1/1 and 2/1 periodic motions coexist in the HR_1 hysteresis transition region due to the different initial conditions of the system. Select the system parameters as $\omega = 0.65$, $\delta = 0.25$. First, the phase space $\{x_1, \dot{x}_1 | -3 < x_1 < 3, -3 < \dot{x}_1 < 3, x_1, \dot{x}_1 \in R\}$ is divided into 40×40 grids and the midpoint of each grid is taken as the initial point for point mapping calculation. The fixed points of 1/1 vibration (-1.189489, 0.202811) and 2/1 vibration (-0.473204, 0.670069) are obtained by taking the phase plane as Poincaré section. Then divide the phase space into 800×800 cells, calculate the cell where the fixed point is located, and map the center points of the remaining cells to establish the mapping relationship between cells. The calculation results show that when the gap threshold $\delta = 0.285$, the number of cells that converge to 1/1 attractor is 441,668, accounting for 69.0% of the total number of cells, and the number of cells that converge to 2/1 attractor is 198,332, accounting for 31.0% of the total number of cells. When the gap threshold is $\delta = 0.25$, the number of cells that converge to 1/1 attractor is 188,620, accounting for 29.5% of the total cell,

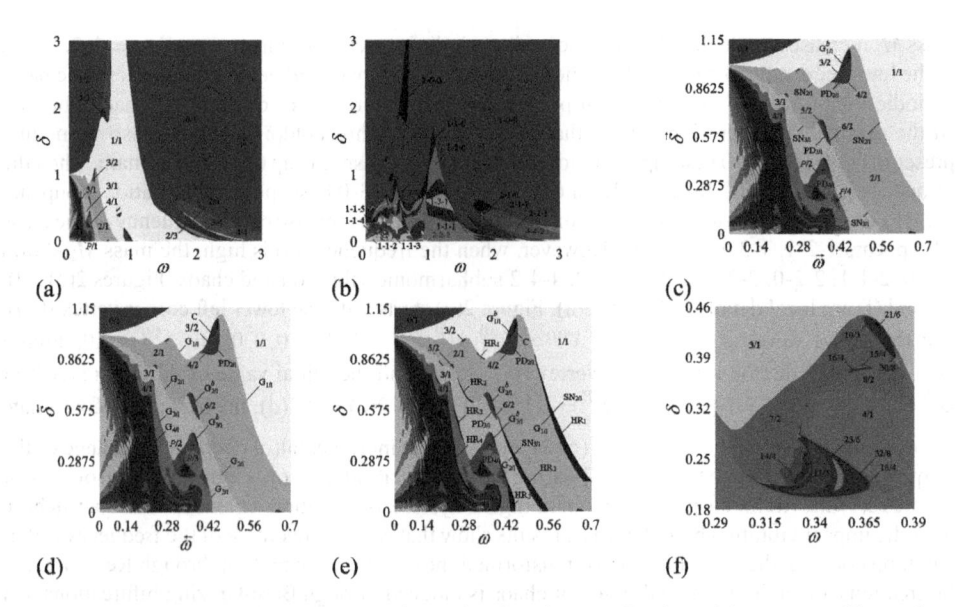

Figure 2. Pattern types and occurrence regions of various impact vibration of the system with equal gap threshold in the (ω, δ)-parameter plane: (a) global description of M_1, (b) global description of M_2, (c-d) local description of M_1, (e) hysteresis transition region of M_1, (f) non-hysteretic transition region of M_1.

and converge to 2/1 attractor is 451,380, accounting for 70.5%. It can be found in Figure 3 that the attraction basin of 1/1 vibration gradually reduces with the decrease of gap threshold(δ) and completely disappears at $\delta = 0.21$, while the attraction basin of 2/1 vibration gradually increases (Figure 3).

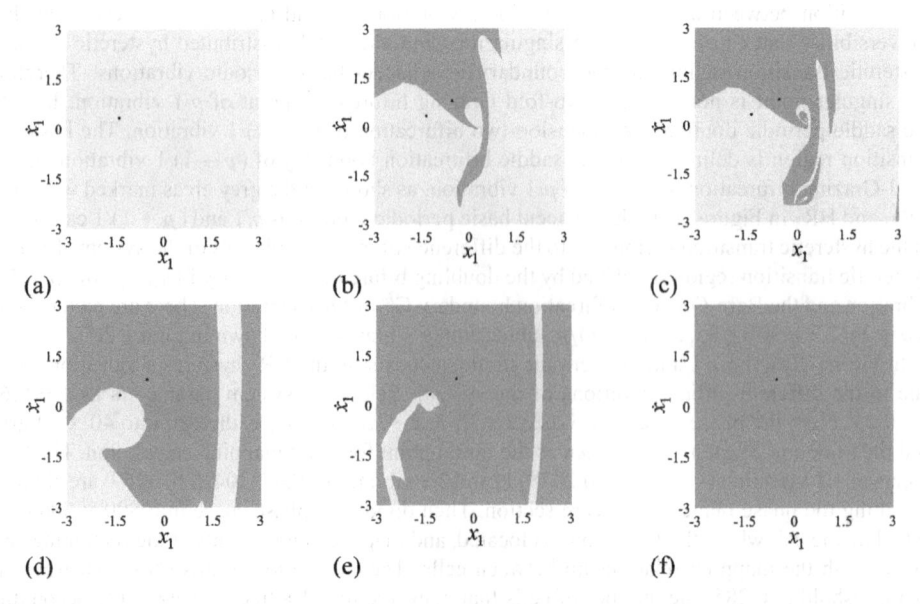

Figure 3. Evolution of the attraction basins of the elastic impact vibration obtained for $\omega=0.65$ and the gap threshold (a)δ=0.30,(b)δ=0.294,(c)δ=0.285,(d)δ=0.25,(e)δ=0.23,(f)δ=0.21. Here the 1/1 periodic vibration has a yellow basin and 2/1 periodic vibration has green basin.

4 DYNAMIC CHARACTERISTICS OF A VIBRATING SYSTEM WITH UNEQUAL GAP THRESHOLDS

In Figure 1, the right gap threshold (B_1) of mass M_1 and the left or right gap threshold ($B_2 = B_3$) of mass M_2 in the mechanical vibration system are set to different gap thresholds, i.e. $B_1 \neq B_2$ (and $B_2 = B_3$). In order to study the pattern types, occurrence regions and bifurcation characteristics of periodic impact vibration under unequal gap thresholds. Select a set of dimensionless parameters of the system, $\mu_m = 0.5$, $\mu_k = 0.5$, $\mu_c = 0.5$, $\zeta = 0.1$, $f = 0.5$, $k_0 = 0.95$. First, the gap threshold (B_1) between the mass M_1 and the right elastic constraint is set as a fixed value, and the gap threshold ($B_2 = B_3$) between the mass M_2 and the left or right elastic constraint is set as a variable for numerical calculation, the calculation results are shown in Figure 4. Then fix the gap thresholds ($B_2 = B_3$) of mass M_2, and take the gap thresholds (B_1) of mass M_1 as variables for numerical calculation (Figure 5). The fixed gap between the mass M_1 and the right constraint is $\delta_1 = 0.25$. In the smaller frequency (ω) domain, the mass M_1 exhibits basic periodic vibration group of 1/1, 2/1, 3/1, 4/1, 5/1, $p/1(p \geq 6)$. With the increasing frequency (ω), 3/3→2/3→2/2→4/4→2/2→1/2→2/4→1/2→0/1→C→1/3→0/1 vibration occurs in the larger gap threshold(δ_2) domain. In low frequency and small gap domain, the mass M_2 exhibits basic periodic vibration groups of 1-1-0, 1-2-0, 1-3-0, 1-4-0, 1-5-0, 1-p-0($p \geq 6$) as the gap(δ_2) decreases and 1-1-2, 1-2-2, 1-2-3, 1-2-4, 1-$p-q$ ($p \geq 6$ or$q \geq 6$) as the gap(δ_2) increases. When the frequency is $\omega \in (0.8, 1.5)$, the mass M_2 shows 1-1-1, 1-2-1, 1-3-1, 1-1-0 basic periodic vibration and 2-1-0, 2-2-0, 2-2-1, 2-2-3 subharmonic vibration with the increase of the gap (δ_2). With the increase of frequency (ω), the mass M_2 show 2-2-2, 2-3-1, 4-4-2 subharmonic vibration and chaos. When the fixed gap increases to $\delta_1 = 1.25$, in the lower frequency and larger gap domain, with the decrease of frequency(ω), the mass M_1 passes through a series of Grazing bifurcations, resulting in an increase in the number of impact and the period 1 remains unchanged, showing a $p/1$ ($p > 0$) basic periodic vibration group. At the same time, in the lower frequency and smaller gap threshold domain, the mass M_2 shows 1-1-0, 1-2-0, 1-3-0, 1-4-0, 1-5-0, 1-p-0($p \geq 6$) basic periodic vibration group with the decrease of gap (δ_2), and 1-1-1, 1-2-2, 1-3-3, 1-4-4, 1-5-5, 1-$p-p(p \geq 6)$ basic periodic

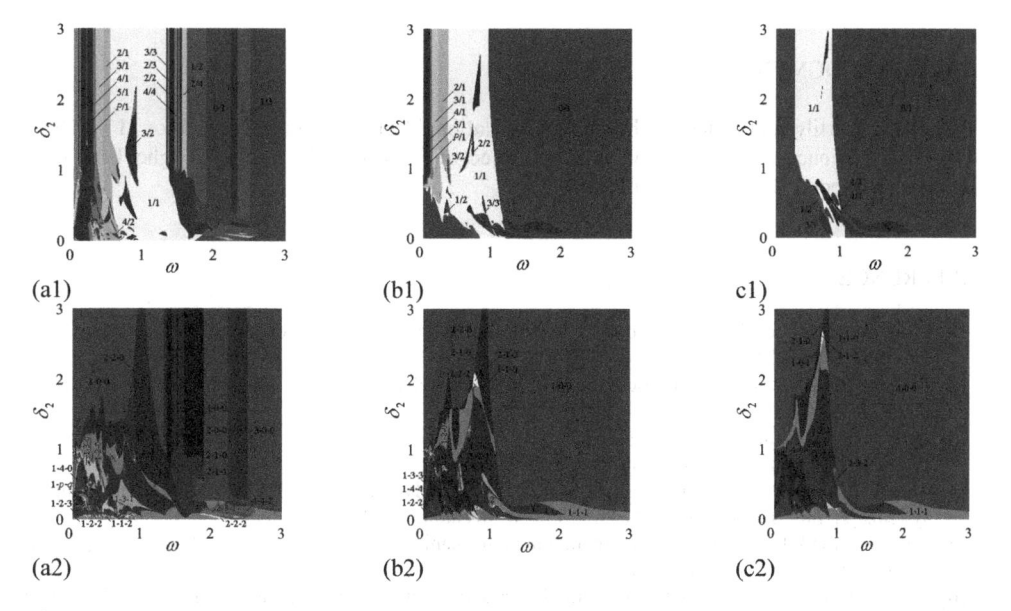

Figure 4. Pattern types and occurrence regions of impact vibration of the system with unequal gap threshold in the (ω, δ_2)-parameter plane: (a1) mass M_1, $\delta_1 = 0.25$, (b1) mass M_1, $\delta_1 = 1.25$, (c1) mass M_1, $\delta_1 = 2.05$, (a2) mass M_2, $\delta_1 = 0.25$, (b2) mass M_2, $\delta_1 = 1.25$, (c2) mass M_2, $\delta_1 = 2.05$.

vibration group with the increase of gap (δ_2). When the fixed gap is $\delta_1 = 2.05$, the mass M_1 mainly presents 0/1 non-impact vibration, 1/1 vibration, a small amount of 1/2, 2/2, 3/3, 4/4 periodic vibration and chaos in the (ω, δ_2)-parameter plane. Mass M_2 exhibits basic periodic vibration of 1-p − $p(p \geq 6)$ in low frequency and small gap domain. With the increase of the frequency, the mass M_2 shows 1-1-0, 1-0-1, 1-1-1, 1-2-1 basic periodic vibration, a small amount of 2-2-0, 2-1-0, 2-1-2, 2-3-2 periodic vibration and chaos. The calculation results show that with the increase of the fixed gap (δ_1), the pattern types and occurrence regions of the basic periodic impact vibration groups of the mass M_1 and the mass M_2 decrease. The mass M_2 shows complex dynamic characteristics in the low-frequency and small gap domain, and various basic periodic impact vibration groups emerge certain competition laws.

5 CONCLUSION

(1) The mechanical vibration system exhibits diverse and complex dynamic characteristics in the low frequency domain. In the low-frequency and small gap threshold, mass M_1 presents basic periodic impact vibration group ($p/1$) through a series of Grazing bifurcations, and causes the phenomenon of chattering-impact vibration. The irreversibility of the mutual transfer of the adjacent basic periodic vibration results in a series of singularities, hysteretic and non-hysteretic transition regions on the bifurcation boundary.
(2) When the fixed gap (δ_1) is small, the basic periodic vibration group ($p/1$) occurrence region of the mass M_1 is large in small frequency domain, and extends to the direction of gap (δ_2) increase. There is a certain competition law among various basic periodic impact vibration groups of mass M_2. With the increase of the fixed gap, the occurrence region of the basic periodic vibration group of mass M_1 is gradually reduced until it disappears, and the types of basic periodic vibration groups of mass M_2 are also reduced.
(3) The vibration system with equal and unequal gap threshold presents non-impact periodic vibration in the high frequency and large gap domain, and various basic periodic impact vibration groups in the low frequency domain are generated from Grazing bifurcations.

ACKNOWLEDGMENTS

The author heartily acknowledges the support by National Natural Science Foundation (11862011, 11672121), Science and Technology Planning Project of Gansu (18YF1WA059), Higher Education Research Project of Jilin(JGJX2020C118).

REFERENCES

[1] Shaw S W, Holmes P J. A periodically forced piecewise linear oscillator [J]. *Journal of Sound and Vibration*, 1983, 90(1): 129-155.
[2] Peterka F. Behaviour of impact oscillator with soft and preloaded stop[J]. *Chaos, Solitons and Fractals*, 2003, 18(1):79-88.
[3] Le Y, Miao P C. Catastrophe and quasi-periodic quasi-periodic paroxysms of a class of impact vibration systems [J]. *Journal of Vibration and Shock*, 2017, 36(07): 1-7+20.
[4] Zhang Y X, Luo G W. Detecting unstable periodic orbits and unstable quasiperiodic orbits in vibro-impact systems [J]. *International Journal of Non-Linear Mechanics*, 2017, 96: 12-21.
[5] Peterka F, Vacík J. Transition to chaotic motion in mechanical systems with impacts[J]. *Journal of Sound and Vibration*, 1992, 154(1):95–115.
[6] Nordmark A B. Non-periodic motion caused by grazing incidence in an impact oscillator [J]. *Journal of Sound and Vibration*, 1991, 145(1): 279–297.
[7] Luo G W, Lv X H. Controlling bifurcation and chaos of a plastic impact oscillator [J]. *Nonlinear Analysis: Real World Applications*, 2009, 10(3): 2047–2061.

[8] Luo G W, Zhu X F, Shi Y Q. Dynamics of a two–degree–of freedom periodically–forced system with a rigid stop: diversity and evolution of periodic–impact motions[J]. *Journal of Sound and Vibration*, 2015, 334: 338–362.

[9] Wagg D J, Bishop S R. Dynamics of a two degree of freedom vibro-impact system with multiple motion limiting constraints[J]. *International Journal of Bifurcation and Chaos*, 2004, 14(01):119–140.

[10] Wagg D J. Periodic sticking motion in a two-degree-of- freedom impact oscillator[J]. *International Journal of Non- Linear Mechanics*, 2005, 40(8):1076–1087.

[11] Yin F W, Luo G W, Tong C H. Diversity and Regularity of Periodic Vibro-impact of Vibration System with Clearance-Elastic Constraints[J]. *Journal of Vibration and Shock*, 2020,39(24):1–10.

[12] Yin F W, Luo G W. Relation between Dynamic Characteristics and Parameters of a Two-degree-of-freedom Vibro-impact System with Multi-elastic Constraints[J]. *IOP Conference Series: Materials Science and Engineering*, 2020, 746(1):012032 (7pp).

[13] Shi Y Q, Du S S, Yin F W. Dynamics of a two-degree-of-freedom vibration system with bilateral rigid stops[J]. *Journal of Vibration and Shock*, 2019,38(14):37–47.

[14] Nordmark A B, Piiroinen P T. Simulation and stability analysis of impacting systems with complete chattering[J]. *Nonlinear Dynamics,* 2009, 58():85–106.

[15] Csaba Hõs, Champneys A R. Grazing bifurcations and chatter in a pressure relief valve model[J]. *Physica D: Nonlinear Phenomena*, 2012, 241(22):0–0.

[16] Zhang Y L, Tang B B, Wang L. Dynamic analysis of impact vibration system with clearance under dynamic friction [J]. *Journal of Vibration and Shock*, 2017, 36(24): 58–3.

[17] Kundu S, Banerjee S , Ing J , et al. Singularities in soft-impacting systems[J]. *Physica D Nonlinear Phenomena*, 2012, 241(4):553–565.

[18] Jiang H, Chong A S E, Ueda Y. Grazing-induced bifurcations in impact oscillators with elastic and rigid constraints[J]. *International Journal of Mechanical Sciences*, 2017: S0020740317302874.

Kinematic and dynamic simulation of landing gear retraction mechanism in aircraft

Z.J. Sun
Engineering College, The Open University of China, Beijing, China

J.T. Dai & L. Han
Aeronautical Mechanism Department, Naval Aviation University Qingdao Branch, Qingdao, China

ABSTRACT: The performance of landing gear retraction mechanism in aircraft directly affects the safe operation of the aircraft. When landing gear retraction mechanism in a military aircraft was designed, the virtual prototype model was established by UG (Unigraphic) software, and the motion analysis and interference checking were carried on. Then the digital prototype model was established in ADAMS (Automatic Dynamic Analysis of Mechanical System) software, and the virtual simulation parameters were set to analyze the aerodynamic drag, mass, inertia forces and the load of actuator cylinder. Thus the kinematic and dynamic properties could be got to simulate real working situation, which could verify the correctness of the design and provide evidence for further optimal design of the machine.

1 INTRODUCTION

The landing gear system is one of the key devices in the aircraft. Among the various working components of the aircraft landing gear system, the landing gear retraction mechanism has relative higher failure probability, which is about 34.4% [1, 2], so its working performance directly affects the safe operation of the aircraft. When the landing gear retraction mechanism is designed, it is very important to analyze its kinematic and dynamic performance.

In order to reduce the aerodynamic drag, the landing gear in modern aircraft is usually retractable. The landing gear is put inside the aircraft after takeoff. While the landing gear is put down and locked reliably before landing. Although the landing gear retraction mechanism increases the weight, and complicates the structural design and use of the aircraft, it improves the efficiency of flight. Due to the relatively complex structure of the landing gear retraction mechanism, there're many non-linear factors, it is difficult to resolved more accurate dynamic results with the actual situation. In order to obtain better simulation results consisting with the actual work, a three-dimensional mathematical model of landing gear retraction mechanism in a military aircraft is established in UG software by virtual prototype technology [3]. Then using the interface of UG and ADAMS software, a digital prototype of landing gear retraction mechanism is simulated and calculated by ADAMS software instead of the original physical prototype test analysis. The impacts of aerodynamic drag, mass, inertia forces and the load of actuator cylinder on retractable motion are studied, thus the kinematic and dynamic properties are analyzed accordingly.

2 VIRTUAL PROTOTYPE ESTABLISHMENT AND RETRACTION MOTION ANALYSIS

By studying the composition and working principle of the landing gear retraction mechanism, the motion diagram is obtained. Then the motion transmission of the mechanism is analyzed, which lays the foundation for the structure design, kinematic and dynamic analysis of the mechanism [4]. Furthermore, the landing gear retraction mechanism is modelled, assembled in UG software, and motion simulation is applied. The rationality of the structure design is determined by static interference check and dynamic interference check.

2.1 Landing gear retraction mechanism analysis

The retraction mechanism generally adopts a linkage mechanism. Usually, as the aircraft's front landing gear is retracted forward into the front airframe, the aircraft's fairings are retracted accordingly. Therefore, the retracting of the landing gear and the opening and closing of the fairings constitute a spatial linkage motion. Structural analysis is the first step to analyze an actual mechanism. The landing gear retraction mechanism is simplified for convenient analysis. Since the opening and closing fairings are symmetrical, only one side of the fairing is researched, and some of the secondary components are omitted, the motion diagram of the landing gear retraction mechanism is shown in figure 1.

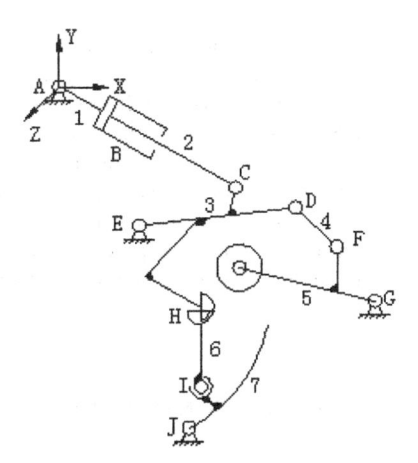

Figure 1. Motion diagram of landing gear retraction mechanism in a military aircraft. 1, 2-retraction cylinder; 3- upper support rod; 4-lower support rod; 5-buffer pillar (including wheel); and 6-pull rod; 7-fairing.

When the landing gear is retracted into the airframe, the state of the landing gear retraction mechanism is shown in figure 1. The retraction cylinder is driven, and the mechanism could complete a single input- output motion. When the landing gear moves down, the hinge point C will move downward, so the buffer pillar 5 will be lowered, and the fairing 7 will open accordingly.

2.2 Virtual prototype establishment

UG software could perform parametric three-dimensional modelling of complex surfaces and entities, and could reflect the assembly relationship of parts intuitively and accurately, and also provides sophisticated simulation module with analysis results in multiple formats. The results are provided to other simulation software for further analysis. The three-dimensional parts such as retraction cylinders, support rods, buffer pillar, pull rod and fairing are established in UG software. Then whole mechanism is assembled according to the assembly relationship between the parts. The assembly model is shown in figure 2.

Figure 2. Assembly model of landing gear retraction mechanism.

2.3 *Retraction motion analysis*

The motion simulation of the landing gear retraction mechanism is real-time, which could visualize the motion process of the three-dimensional model, and analyze the interference of the motion process. First, create motion pairs for the mechanism model according to figure 1, and then define the motion drive on the point B. The motion drive is uniform linear motion, and the speed of the retraction cylinder is 10mm/s. So the motion of the mechanism simulation could be analyzed, and the motion time could be obtained as 15.3s. Finally, because multiple components move at the same time, including the lifting of the buffer pillar and the linear motion of the retraction cylinder, the space movement is relatively messy, so it is necessary to conduct dynamic interference check between the various components during the retraction process. After dynamic interference check, it is showed that the model of the landing gear retraction mechanism is correct, and there is no interference between the components.

3 KINEMATIC AND DYNAMIC SIMULATION OF RETRACTION MECHANISM

Based on virtual prototype model of the landing gear retraction mechanism of a military aircraft in UG software, the load carried by the landing gear is analyzed. Further, a digital prototype model of the mechanism is established in ADAMS software. The model is loaded to simulates the effects of various loads, including aerodynamic drag of the landing gear, mass of the landing gear, the action of the retraction cylinder. Then the performance of the landing gear retraction mechanism is analyzed.

3.1 *Simulation environment settings*

To simulate the motion of the landing gear retraction mechanism, it is necessary to analyze the load and various factors that may affect the retraction motion at first. The landing gear retraction mechanism bears a wide variety of loads, and the calculations are also complicated. Therefore, the following loads of main influence are determined during the simulation process [5].

3.1.1 *Gravity*
The gravity acts on the gravity center, and its direction always points to the ground. When the aircraft is flying in a steady airflow, the gravity P_m is determined by the following equations:

$$P_m = n^u_{g,d} \cdot G_t \tag{1}$$

$$n_{g,d}^{lu} = 1 + 0.5KC_y^a \frac{\rho_0 VwS}{G_a} \quad (2)$$

$$K = 0.8 \frac{1 - e^{-\lambda}}{\lambda} \quad (3)$$

$$\lambda = 0.5 C_y^a \frac{\rho_H g L}{G_a/s} \quad (4)$$

G_t——gravity of the rotating part, N;
$n_{g,d}^{lu}$ ——overload when the landing gear is retracted, ≥ 2.0;
K——coefficient determined by Equation 3;
C_y^a ——derivative of the aircraft normal force coefficient to the angle of attack;
V——maximum flight speed allowed to retract the landing gear;
w——gust speed, 10m/s;
G_a——gravity of the aircraft taking off or landing, N;
S——surface area of wing, m²;
ρ_H——air density at flight altitude, $\rho_H = \rho_0 = 1.2 \text{kg}/\text{m}^3$;
g——acceleration of gravity, 9.8m/s²;
L——diffusion section length of gust, 30m.

Based on equations 2, 3, and 4, the overload of each component in figure 1 could be calculated when the landing gear is retracted, as shown in Table 1.

Table 1. Overload of each component when retracted.

Component	1	2	3	4	5	6	7
$n_{g,d}^{lu}$	2.13	2.36	2.63	2.08	3.14	2.33	2.96

3.1.2 Aerodynamic drag

The aerodynamic drag of each part of the landing gear acts on the pressure center and points in the direction of the airflow. The aerodynamic drag $P_{a,di}$ of each part is determined by the following equations:

$$P_{a,di} = C_{xi} \cdot q \cdot S_i \quad (5)$$

$$q = \frac{1}{2} \rho_0 V_{g,d}^{max^2} \quad (6)$$

C_{xi}——drag coefficient of each part in the landing gear;
q——speed pressure;
S_i——projected area of each part perpendicular to the airflow of the landing gear on the plane.

The aerodynamic drag acting on the component 5 and 7 is considered mainly. The drag coefficient C_{x0} of the circular cross-section buffer pillar varies with the wheel's width-to-diameter ratio. In this paper, the wheel's width-to-diameter ratio is selected as 5, so C_{x0}=0.76; the drag coefficient of the wheel is checked by reference [6], so $C_{xw} = 0.5$; the drag coefficient of the fairing is calculated according to the drag coefficient of board, so $C_{xh} = 1.28$. Finally, the converted aerodynamic drag torques acting on the rotational joints of component 5 and 7 are determined by the following equation:

$$M_{ad} = \sum P_{adi} \cdot b_i \quad (7)$$

P_{adi}——joint pneumatic force of each part changing with retraction;
b_i——lever arm of joint pneumatic force changing with retraction.

In the retraction motion, the aerodynamic drag torques on the rotating shaft of the fairing and buffer pillar (including wheel) corresponding to the upwind direction are shown in figure 3.

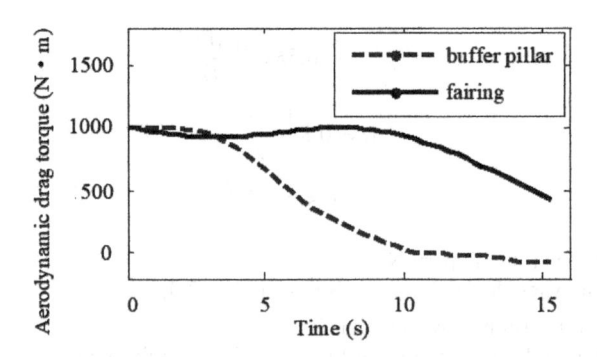

Figure 3. Aerodynamic drag torques of the fairing and buffer pillar.

3.1.3 *Inertial force*

If the landing gear needs to be retracted in short time, a large inertial force may appear. The direction of inertial torque on the axis of the landing gear is opposite to the angular acceleration. The inertial torque M_g is determined by the following equation:

$$M_g = -J \cdot \frac{d^2\varphi}{dt^2} \tag{8}$$

J——moment of inertia of the landing gear to the rotation shaft, $J = mr^2$;
m——mass of the rotating part of the landing gear;
r——distance from the gravity center of the rotating part to the rotation shaft;
$\frac{d^2\varphi}{dt^2}$——angular acceleration of the landing gear, it is related to the force of the retraction cylinder and the drag torque of the landing gear.

3.1.4 *Friction*

The additional load f_P caused by the total friction on the retraction cylinder is approximately calculated by the following equation:

$$f_P = (0.18 \sim 0.3) P_{aa} \tag{9}$$

P_{aa}——load of the retraction cylinder caused by gravity and aerodynamic force

3.2 *Kinematic and dynamic simulation*

ADAMS software is widely used for mechanical system's kinematic and dynamic simulation. The model in the UG software is imported into the ADAMS software, and is redefine the constraints, including rotational joints, moving joints, universal joints and ball joints. The Lagrangian equation method in the multi-rigid body dynamics theory is used to establish the dynamics equation for mechanical system, and the entire movement process of the landing gear retraction mechanism is simulated through the virtual prototype. Each force is load as below:

- The quality attributes of each part and gravitational acceleration. As the material of each rod in the landing gear retraction mechanism is high-strength steel, and the material of the tire is rubber, the material attributes of each part are set in ADAMS software. Then the gravitational acceleration is defined the magnitude and direction.
- During the normal flight of the aircraft, because of the positive wind and cross wind, the aerodynamic drag of the landing gear is from the heading and lateral direction. The aerodynamic drag torques of the fairing and buffer pillar is added in ADAMS software according to figure 3. The aerodynamic drag torque loaded on the right and left fairings is the same in size and opposite in direction.

- The inertial force of each part is loaded on the basis of the acceleration or angular acceleration of each part in ADAMS software.
- The friction coefficient is 0.3, then the additional friction caused by the total friction on the retraction cylinder is set in ADAMS software.

The motion time of the landing gear retraction mechanism is set to 15.3s. Then the time-varying angular displacements and velocities of the fairing and buffer pillar could be got, and the driving force of the retraction cylinder could also be got, as shown in figure 4.

As shown in figure 4(a), during the retraction process of the landing gear, while the angular displacement of the buffer pillar increases monotonically, the angular displacement of the fairing first decreases and then increases, but the angular displacement of the buffer pillar is always greater than the fairing's. This result is consistent with the requirements of the motion process of the mechanism. Because the fairings should be ensured closed reliably after the buffer pillar is retracted, and also should not interfere with the motion of the buffer pillar, so the fairings first open, then close again during the retraction process of the buffer pillar. As shown in figure 4(b), during the retraction process of the landing gear, while the angular velocity of the buffer pillar first increases and then decreases, the angular velocity of the fairing is a process of decrease-increase-decrease. This result is also consistent with the requirements of the motion process of the mechanism. Because the buffer pillar should be quickly retracted at first, and then the angular velocity gradually decreases during the retraction process; while the angular velocity of the fairing gradually decreases during the opening process, and increases during the closing process, and then decreases at the end of the retraction process. As shown in figure 4(c), during the retraction process of the landing gear, there're 3 crests and troughs in the curve of driving force of the retraction cylinder, and they appears when the fairing changes from opening to closing, the angular velocity of the buffer pillar begins to decrease, the angular velocity of the fairing begins to decrease respectively. Then certain hydraulic impact forces are formed according these times. Obviously, the simulation results in ADAMS software are completely consistent with the qualitative analysis and solution in the design process, thus verifying the correctness of the design.

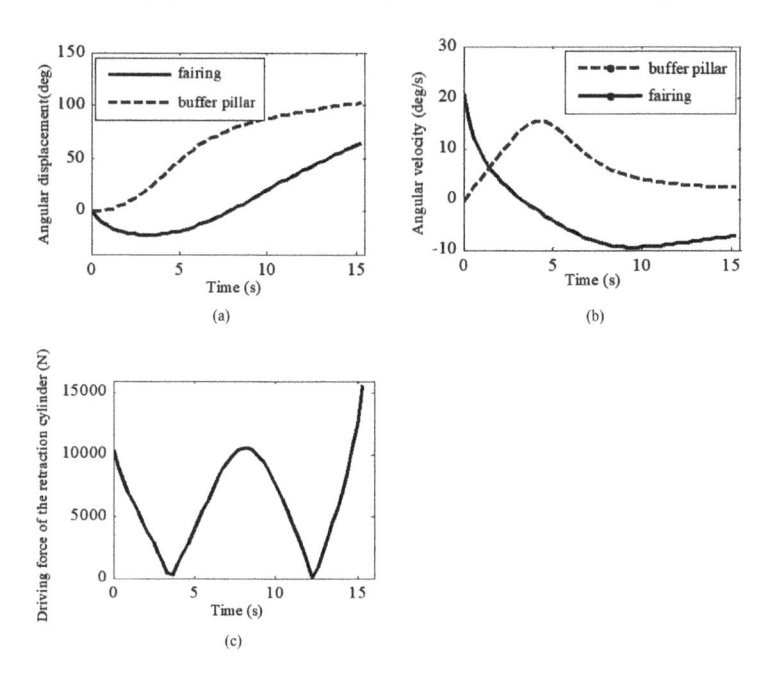

Figure 4. Output curves of simulation analysis: (a) Angular displacement of the fairing and buffer pillar, (b) Angular velocity of the fairing and buffer pillar, and (c) Driving force of the retraction cylinder.

4 CONCLUSIONS

A parametric virtual prototype of landing gear retraction mechanism in a military aircraft is established by UG software. The mechanism could be virtual assembled. Then, through the interface of UG software and ADAMS software, a digital prototype is established in ADAMS, the aerodynamic drag, mass, inertia forces and the load of actuator cylinder are analyzes. The digital prototype is loaded, and some environment parameters are set, then the kinematic and dynamic characteristics of the mechanism are obtained, which are closer to the real operating conditions. These simulation results verify the correctness of the design and also provide for the optimal design of the whole machine.

REFERENCES

[1] Swift K G, Raines M, Booker J D 2001. Advances in probabilistic design; manufacturing knowledge and applications. *Institution of Mechanical Engineering* 215 (Part B) pp 297~313
[2] Lin Zhu, Fanrang Kong, Chenglong Yin, etc. 2007. Dynamic Simulation of Retracting Mechanism of the Landing Gear of an Aircraft with Computer Simulation Techniques. *China Mechanical Engineering* 18 pp 26–29
[3] Xudong Shi, Fujun Ying, Yu Zhang, etc. 2021. Performance Simulation and Fault Analysis of Aircraft Landing Gear Extension and Retraction System. *Computer Applications and Software* 38 pp 65–71+184
[4] Kozhevnikov C H 1988 *Mechanism Reference Manual* (Beijing: China Machine Press) p97
[5] Lin Chen 2007 *Analysis on the Kinematic and Dynamic Performance of Landing-gear Retraction* (Nanjing: Master Thesis of Nanjing University of Aeronautics and Astronautics) pp 34–40
[6] Aircraft Design Manual Editor in Chief Committee 2002 *Aircraft Design Manual (Volume 14): Take-off and Landing System Design* (Beijing: Aviation Industry Press) p126

Experimental research on the forming process of large-curvature composite honeycomb sandwich structure

K.X. He, Q.Y. Qiu & H.J. Ye
Department of Composite Component, AVIC Composite Corporation Ltd., Beijing, China

ABSTRACT: Two process plans are used to cure a large-curvature composite honeycomb sandwich structure which is filled with foaming adhesive in a local area. By using experimental observation and X-ray inspection, it is found that there are many problems in the plan that the honeycomb core adheres to the outer layer, and the honeycomb core is filled with foaming adhesive and cured. The problem is the poor curing appearance of foaming adhesive, and the part was deformed in 9 months. The reason for the poor curing appearance is that the curing shrinkage of foaming adhesive is limited by outer layer. The outer layer hinders the export of nitrogen and reaction heat energy produced by foaming adhesive curing. The reason for deforming is that the incompletely curing shrinkage of foaming adhesive causes residual stress on the part. The slow release of residual stress over time leads to deformation of the part. The other plan is that the honeycomb core is filled with foaming adhesive and cured in a free state. A good foaming adhesive curing appearance is got. The part is no longer deformed.

1 INTRODUCTION

Fiber-reinforced resin-based composite materials have been widely used in many important engineering structures due to their high specific strength, high specific rigidity, and strong designability [1]. Composites are usually formed by heating. During the forming process, the material undergoes a series of physical and chemical changes, causing the composite to produce curing deformation and residual stress. The curing deformation can be divided into thermal stress, curing shrinkage stress, temperature gradient and degree of cure, pressure distribution and resin flow, tool-part interaction [2]. The occurrence of curing deformation has an extremely unfavorable effect on the shape accuracy of parts and the connection and matching of components. It will cause additional residual stress and poor sealing during assembly, resulting in reduced structural strength and fatigue life of the part, and even scrapping of the part. Residual stress will cause micro-cracks at the material matrix and interface, causing defects in the structure, affecting the proliferation of matrix cracks, interface debonding, layer-to-layer delamination damage during the service process of the material, affecting the performance of entire structure and service life, especially structural fatigue performance [3]. In order to reduce the curing deformation and residual stress, a large number of workers adjust the curing process and compensate mold through numerical simulation and process experiment [4–7].

The composite honeycomb sandwich structure can be composed of a variety of materials. The filling of foaming adhesive in the honeycomb core cells can not only prevent the instability of the honeycomb core but also improve the compressive strength of the honeycomb core. The curing deformation and residual stress of the multi-material structure are complicated. At present, most researches are before and after the composite is formed. For the deformation caused by the release of residual stress after being placed for a long time after forming, there are few findings in domestic and foreign studies.

2 FORMING PROCESS EXPERIMENT

2.1 *Composite shell model*

The structure of the composite shell is a honeycomb sandwich structure composed of a composite material outer layer, an Nomex honeycomb core, and a composite material inner layer. The big arc is irregular, and the small arc is regular. In a local area of the honeycomb core needs to be filled with foaming adhesive to form a foaming area. The honeycomb core is connected with the outer layer and the inner layer by adhesive film. The structure diagrams of the composite shell are shown in Figure 1 and Figure 2.

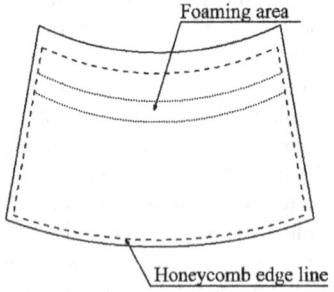

Figure 1. Front view of the composite shell.

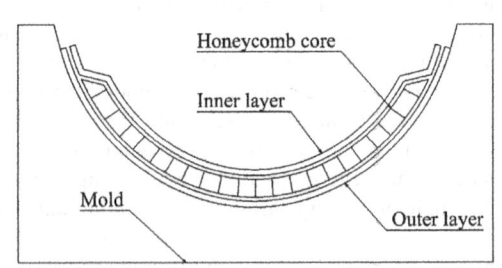

Figure 2. Profile of the composite shell.

2.2 *Plans of forming process*

In order to obtain a high-strength outer layer, the outer layer is first cured under a high pressure. Considering the extreme pressure that the honeycomb core can withstand, the inner layer is cured under a low pressure. According to the structural characteristics of the shell, two plans of forming process methods are developed. The plan A is that, firstly the outer layer and the honeycomb core adhere to each other with adhesive film, then the honeycomb core is filled with foaming adhesive and cured, and finally co-curing with the inner layer. The forming process is shown in Figure 3. The outer layer and the honeycomb core adhere to each other can fix the honeycomb core on the outer layer to prevent the honeycomb core from slipping and collapsing due to lateral high pressure when the inner layer is curing [8]. The part formed by this plan is referred to as Part 1 hereinafter. The plan B is that the honeycomb core is filled with foaming adhesive and cured in a free state, and then adhere the shaped honeycomb core to the outer layer and inner layer. The forming process is shown in Figure 4. The direct benefit of eliminating the process that the outer layer and the honeycomb core adhere to each other is to simplify the manufacturing process and reduce the manufacturing cost of the composite honeycomb sandwich structure. The part formed by this plan is referred to as Part 2 hereinafter.

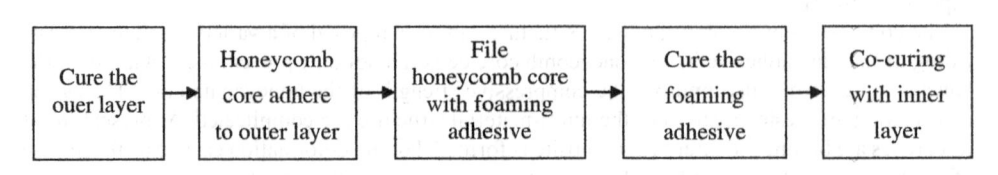

Figure 3. Forming process of Part 1.

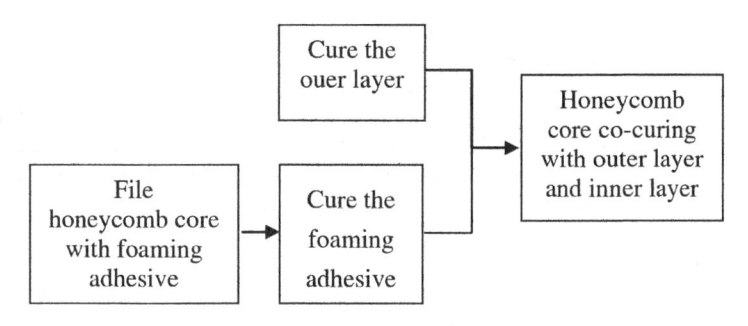

Figure 4. Forming process of Part 2.

3 EXPERIMENTAL RESULTS AND DISCUSSION

3.1 *Cracks in foaming adhesive and cause analysis*

Figure 5 (a-c) are the surface appearances of Part 1 after the foaming process is completed. It can be seen that the honeycomb core and foaming adhesive are cracked at local locations, the width of the crack is about 0.1mm, and the length of the crack runs through the L direction of the honeycomb core of the foaming area. There are many holes, bubbles and burn marks on the foaming adhesive surface, as shown in Figure 5 (a-c). The surface appearance of Part 2 when the foaming process is completed is shown in Figure 5 (d). It can be seen that the foamed surface is uniform and dense, without bubbles, holes, burn marks, and cracks.

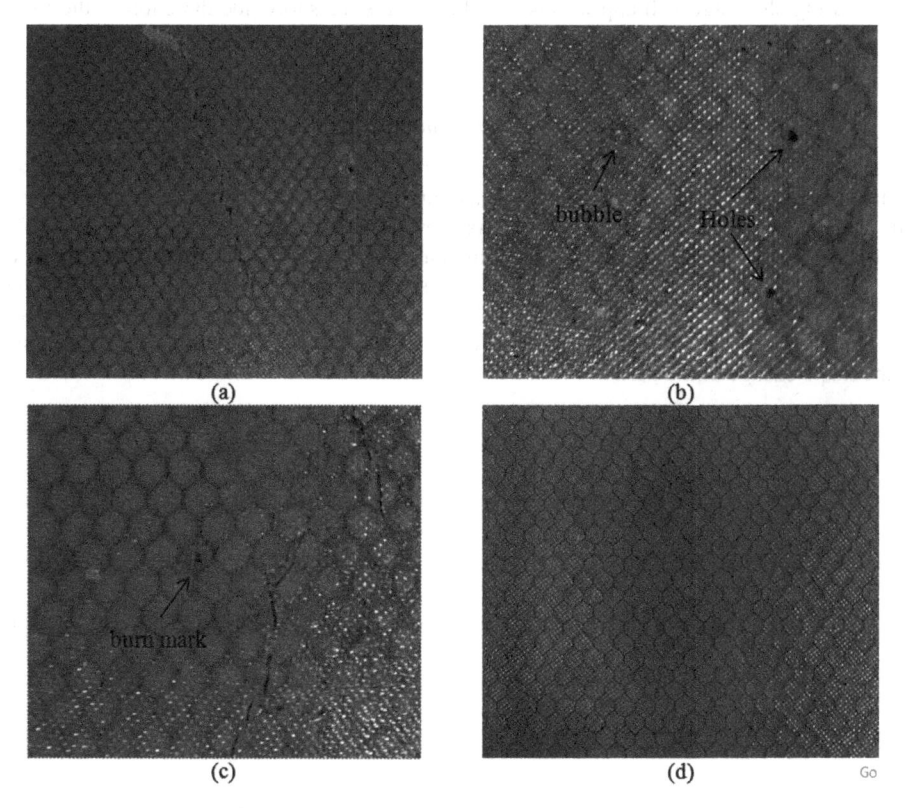

Figure 5. Surface appearances of foaming adhesive.

Foaming adhesive is heated and melted, from powdery solid state to viscous liquid, foaming agent is heated to break down nitrogen, so that the expansion volume of foaming adhesive increase, cyanate resins are heated or under the action of catalysts, curing to form a thiotriazinone structure. In the foaming process produce a large amount of reaction heat, the reaction heat needs to be exported in time, otherwise there will be excessive heat accumulation, resulting in burning.

The lower surface of the honeycomb core cell of Part 1 is closed by the outer layer, and the nitrogen generated is not smoothly discharged, so there are holes and bubbles. Due to the existence of the outer layer, the heat generated during the foaming process is not easily transferred out. The reaction heat energy accumulates in some places and causes the temperature high, resulting in slightly scorching. During the foaming process, the foaming adhesive undergoes curing shrinkage deformation. The reason for the curing shrinkage deformation is that the foaming adhesive changes from viscous liquid to solid, the volume of which is reduced. The lower surface of the honeycomb core is bound by the outer layer, resulting in the curing shrinkage of the foaming adhesive cannot be released. And the W direction of honeycomb core is the long side direction of the foaming adhesive filling area, the curing shrinkage strain is large. The cured foaming adhesive is a brittle substance, which is distributed around the brittle phase near the foaming area to form a stress concentration, thus producing cracks in the L direction of the honeycomb core.

Because the honeycomb core of Part 2 is in a free state, the upper and lower sides can release the same amount of curing shrinkage, so there is no external condition for cracks. In addition, the lower surface of the honeycomb core is not closed, and the air-permeable material is covered and attached to the mold, which releases the nitrogen and reaction heat generated during the foaming process, and obtains the part as shown in Figure 5 (d).

The following conclusions can be drawn. The honeycomb core adheres to the outer layer and then filled with foaming adhesive and cured. There is a non-negligible curing shrinkage. Cracks due to curing shrinkage will appear, and the direction is the short side direction of the foaming area.

3.2 Deformations of composite shell and cause analysis

The shape tracking measurement of Part 1 and Part 2 was carried out for 1 year. The measuring method is to place the shell freely on the measuring frame, and record the warpage value of the four corners of the shell every other month. As shown in Figure 6, it is found that Part 1 is slowly deformed. It is stable and no longer deformed after about 9 months. The warpage value has changed from 3mm~5.5mm initially to 12mm~16mm. As shown in Figure 7, Part 2 has no shape change over time. The warpage value is maintained at 3mm~5.8mm. It can be obtained that there is residual stress in Part 1 after forming, and the release of the residual stress over time has a great influence on the shape change of the part.

Figure 6. Curve of Part 1 warpage value.

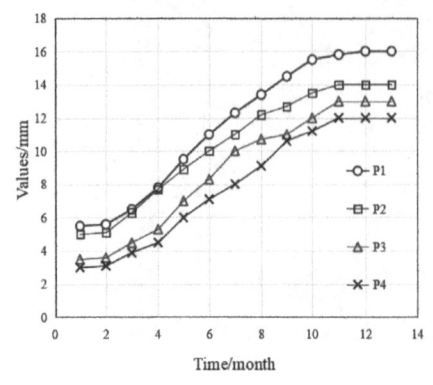

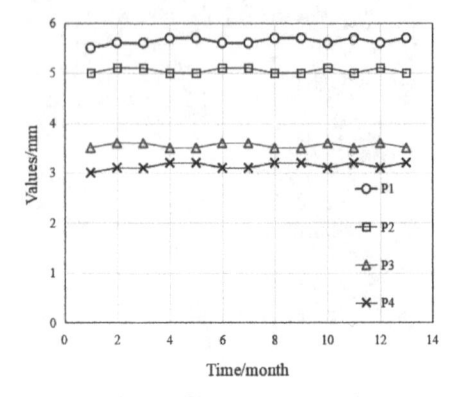

Figure 7. Curve of Part 2 warpage value.

The FDR-160 X-ray imaging inspection system is used to perform X-ray inspection on the two parts that have stopped deforming. Figure 8 shows the filling area of Part 1. It can be seen that the color of the cells is different, indicating that the foaming density inside the cells is uneven. The light areas have low density and the dark areas have high density. A large number of diamond-shaped cracks appear in the honeycomb core, and the crack positions are all honeycomb core node positions, and the direction is the L direction of honeycomb core. Figure 9 shows the filling area of Part 2. It can be seen that the foaming adhesive is very uniform after curing and adheres to the surface of the honeycomb core cell cavity. This is in line with the foaming characteristics of foaming adhesive. And there is no crack.

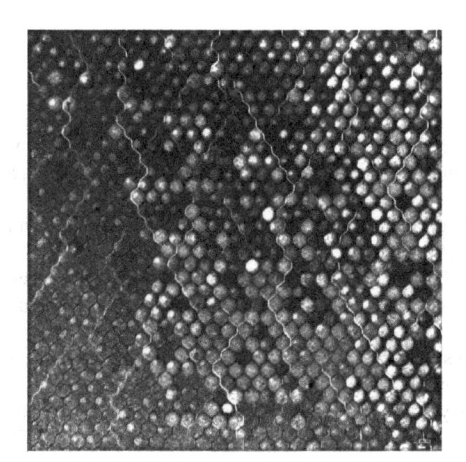

Figure 8. X-ray inspection of Part 1 foaming area.

Figure 9. X-ray inspection of Part 2 foaming area.

The reason for the above situation is that. Firstly, both parts have curing warpage deformation. On the one hand, the thermal expansion coefficient of the mold used in the test is much larger than that of the carbon fiber composite material used by the shell. On the other hand, because the structure is a curved structure, the thermal expansion coefficient of the fiber and the resin is quite different. So even if the composite material curved laminate is symmetrical, there is still curing warpage deformation [9], called spring-in. When Part 1 is co-cured, there is residual stress that has not been released. The residual stress is released over time by the honeycomb core and foaming adhesive cracking. The L-direction node position of the honeycomb core is a fragile area. And the W direction of honeycomb core is the long side direction of the foaming adhesive filling area, the curing shrinkage strain is large. So the fracture positions are all honeycomb core node positions and the direction is the L direction of honeycomb core. After the cracks are formed, the residual stress is released.When the cracks stop expanding, and the shell shape no longer changes.

There is no change in the shape of Part 2, and there is no crack in X-ray inspection. The following conclusions can be drawn.The plan B that the honeycomb core is filled with foaming adhesive and cured in a free state has the following advantages. When the foaming adhesive is cured, the curing shrinkage occurs equally on the upper and lower sides of the honeycomb core, eliminating cracks, curing shrinkage can be completely released, and eliminating residual stress, so that there is no shape change.

4 CONCLUSION

(1) There are some problems with the plan that the honeycomb core adheres to the outer layer and then filled with foaming adhesive and cured. As the outer layer limits the curing shrinkage of

the foaming adhesive, the foaming adhesive and the honeycomb crack. The crack direction is the short edge direction of the foaming area. The incompletely curing shrinkage of foaming adhesive causes residual stress on the part. The residual stress is released through the honeycomb core and foaming adhesive cracking over time. After the release, the shape of the part is greatly changed.

(2) The plan that the honeycomb core is filled with foaming adhesive and cured in a free state has the following advantages. When the foaming adhesive is cured, the curing shrinkage occurs equally on the upper and lower sides of the honeycomb core, eliminating cracks, curing shrinkage can be completely released, and eliminating residual stress, so that there is no shape change.

REFERENCES

[1] Hongwei Tian, Xianzhu Liang and Xiangchen Xue 2014 Response of laminated composite plates of CCF300/BA9916-2 under low-velocity impact loading J. Aeronautical Manufacturing Technology. 15 pp 74–77
[2] Wenli Cheng, Qiyan Qiu and Jing Chen 2012 Study on the cure-induced deformation mechanism and control of composite structures in auto clave process J.Materials Reports. 26 pp 410–414
[3] Anxin Ding, Shuxin Li, Aiqing Ni and Jihui Wang 2017 A review of numerical simulation of cure-induced distortions and residual stresses in thermoset composites J. Acta Materiae Compositae Sinica. 34 pp 471–485
[4] Qun Liang and Xiping Feng 2019 Residual stress and structural distortion analysis for the curing process of SRM composite case J. Journal of Solid Rocket Technology. 42 pp 628–634
[5] Wucherb, Lanif and Pardoent 2014 Tooling geometry optimization for compensation of cure-induced distortions of a curved carbon/epoxy C-spar J. Composites Part A:Applied Science and Manufacturing. 56 pp 27–5
[6] Juan Xu, Jianchuan Li, Jian Peng and Kai He 2013 Study on the effect of curing parameters on the curing deformation of cap-shaped stiffened composite panels J. FRP/CM. 04 pp 7–12
[7] Tian Jiang, Jifeng Xu, Weiping Liu, Jinrui Ye, Lijie Jia and Boming Zhang 2013 Simulation and verification of cure-induced deformation by stages for integrated composite structure J. Acta Materiae Compositae Sinica. 30 pp 61–66
[8] Wei Chen, Li Cheng, Hongjun Ye and Jing Chen 2017 Study on the forming process of composites filled with Nomex honeycomb sandwich structure J. FRP/CM. 07 pp 70–73
[9] Lijie Jia, Jinrui Ye, Weiping Liu, Changchun Wang and Boming Zhang 2013 Role of structural factors in process cure-induced deformation of the complex composites J. Acta Materiae Compositae Sinica. 30 pp 261–265

Deformation behavior of strip steel matrix under shot blasting impact descaling

S. Wang, H.G. Liu, R.C. Hao & Z.X. Feng
School of Automotive Engineering, Beijing Polytechnic, Beijing, China

X.C. Wang & Y.Z. Sun
School of Engineering Technology Research, University of Science and Technology Beijing, Beijing, China

ABSTRACT: In order to improve the descaling efficiency of shot blasting, the finite element model of impact descaling was established by ABAQUS software, and the deformation behavior of strip steel matrix under impact was studied. The diameter of the projectile in the model was set as 0.6 mm, and the impact velocity was set as 40 m/s. Then, the simulation calculation was carried out. Results show that: the scale is broken and peeled under high velocity impact, and the descaling area is roughly semicircular with "whisker" extending to all sides. The direct descaling area caused by impact is 0.214 mm^2, and the indirect descaling area is 0.939 mm^2. After the impact, the strip steel matrix presents the morphology similar to "meteorite crater", with stress concentration at the edge of the crater, and the maximum value of Mises stress is 734 MPa. The Z-coordinate value of the nodes and the strain value of the distance between adjacent nodes of the Z-axis at the impact crater are statistically analyzed. Results show that the function curves of both are "funnel-shaped". Different from the curve of the node Z-coordinate value, the curve of the strain value of the distance between adjacent nodes is not smooth, which is caused by the combined action of the projectile Z-axis impact and the metal plastic flow in XOY plane.

1 INTRODUCTION

Compared with the traditional pickling process, the shot blasting descaling technology does not use chemical acid solution, and has the advantages of pellet recycling, fast processing speed and no emissions [1,2]. The related scholars and technicians have carried out extensive research on the shot blasting descaling process. These studies focus on the impact velocity, particle size of the pellet on the surface roughness, as well as the impact of the pellet material, flow rate on the descaling effect [3–6], and so on. Obviously, these works are more focused on the detection of strip surface quality and morphology after shot blasting and descaling, while neglecting the evolution of scale failure and matrix deformation under high speed impact of projectile. This is because the impact velocity of the projectile is very fast (about 50 m/s) and the scale layer is very thin (about 20μm). It is obvious that the impact descaling is completed in an instant (10^{-6} s), which makes it difficult to observe the rupture and peeling of the scale layer under the impact of the projectile on industrial production lines.

In order to further explore the mechanism of shot blasting descaling, scholars used finite element software to do simulation, and obtained some research results. Zhi G [7] established the finite element model of descaling under two different conditions (wet and dry) and analyzed the descaling difference under different conditions, and found that wet descaling had better performance. Yujun R [8] used the finite element model to analyze the descaling effect at different speeds, and believed that the speed of 40–75 m/s could effectively remove the scale layer on the strip surface. Although some breakthroughs have been made, the models are generally simple and deviate from the actual industrial scene. For example, Zhi G' model does not set the scale layer, but only calculates the impact force of the projectile on the strip steel surface. Some models are not accurate enough, such as Yujun R' model scale layer is too thick (0.5 mm), which is very different from the real size (about 20μm).

In order to further analyze the descaling regularity of shot blasting, a more accurate finite element model was established in this paper, focusing on the impact descaling performance and the deformation of steel matrix, including:

1. The descaling effect at a speed of 40 m/s is simulated, and the descaling area is fitted.
2. The Z-axis coordinates and strain values of the impact crater of steel matrix are statistically analyzed.

2 FINITE ELEMENT MODELING

On the basis of the previous work [9], the finite element model of descaling was optimized, as shown in Figure 1. The finite element model consists of three parts: scale layer, strip steel matrix and projectile. The dimensions and properties of each part are as follows:

The scale layer is a semi-cylinder with diameter of 2 mm and thickness of 0.015 mm.

The steel matrix is a semi-cylinder with diameter of 2 mm and thickness of 0.985 mm. The steel matrix material property was set as elastoplastic material, and Johnson-Cook model [10] was selected. The equation of Johnson-Cook model is as follows:

$$\sigma = (A + B\varepsilon^{-pn})(1 + Cln\varepsilon^*)(1 - T^{*m}) \tag{1}$$

The parameters of the equation of Johnson-Cook model[11,12] are: $A=244.8$ MPa, $B=899.7$ MPa, $n=0.94$, $C=0.0391$. Due to the extremely long collision time, the influence of temperature can be ignored, that is, $T^*=0$.

The projectile is a half sphere, as a rigid body, with diameter of 0.6 mm. The material properties of each component of the finite element model are shown in Table 1.

Table 1. Simulation parameters.

	Density / (kg·m^{-3})	Modulus of elasticity /GPa	Poisson's ratio	Direct stress after cracking/MPa	Direct stress failure strain /%
projectile	7700	150	0.3	–	–
strip steel matrix	7800	200	0.288	–	–
scale layer	7750	210	0.29	180	0.08

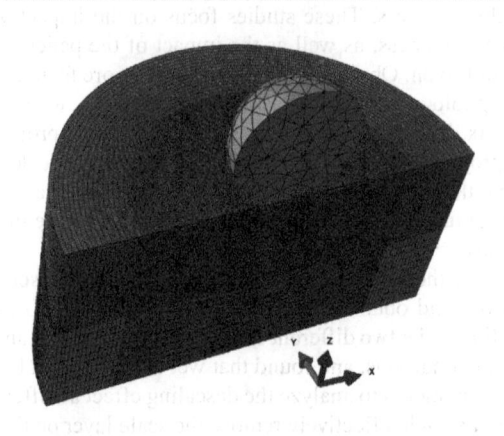

Figure 1. Finite element descaling model.

The coordinate system, material properties, cell type, analysis step, boundary constraint and other settings in the model are all the same as those in reference [9]. The model mesh was refined and optimized, including: the grid size of the scale layer and matrix mesh was set as 0.0075 mm; the grid size of the remaining unrefined areas was set as 0.05 mm, and the grid size of the projectile was set as 0.05 mm. The coordinate system of the finite element model was set, as indicated in Figure 1.

The impact speed was set as 40 m/s. The direction of speed was the negative direction of Z-axis, that is, along the vertical direction of the strip. The operation time was set as $t = 6 \times 10^{-6}$ s, and 200 time intervals were set in the "H-output" command for calculation.

3 RESULTS AND ANALYSIS

3.1 *Finite element calculation results*

The finite element model was simulated on the computer. Figure 2 is the image of the finite element model after the calculation. Figure 3 shows the scale rupture diagram observed along the negative direction of the Z-axis after calculation(the projectile andsteel matrix have been hidden). As can be seen from Figure 2, the projectile bounced off after impact, and the direction of its rebound velocity was roughly perpendicular to the XOY plane. The finite element calculation data shows that the velocity of the projectile core is located in the XOZ plane, that is, the velocity component in the Y-axis direction is zero. In the XOZ coordinate system, the velocity vector is located in the second quadrant at an angle of 1.04° from the positive direction of the Z-axis.

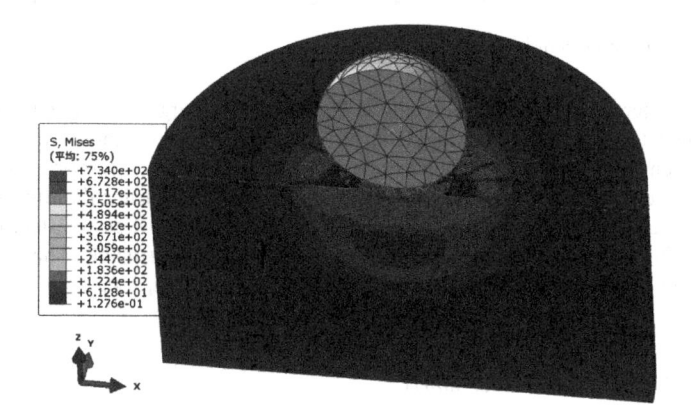

Figure 2. Finite element model after simulation operation.

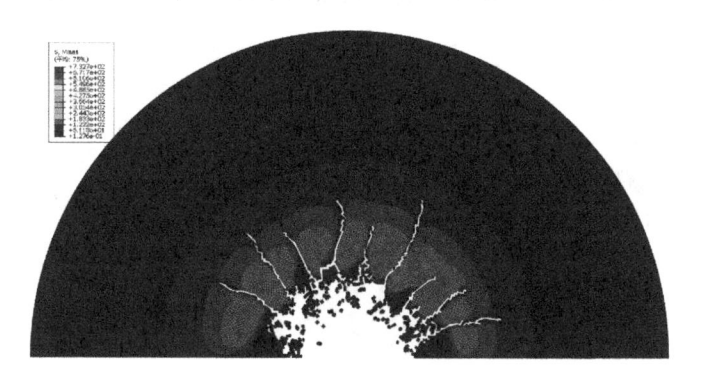

Figure 3. The descaling area obtained by simulation calculation.

The blank area of the scale layer component is the element group whose self-strain has reached critical strain failure and has been deleted from the model. The size of this blank area represents the descaling area. The descaling area is roughly semicircular, with "whiskers" extending in all directions in the XOY plane. In reference [9], direct descaling radius R_1, direct descaling area S_1, indirect descaling radius R_2 and indirect descaling area S_2 were specified. Using the same method, the relevant data calculated by the model are statistically analyzed in this paper. The fitting results show that: the direct descaling radius is $R_1 = 0.261$ mm, direct descaling area is $S_1 = 0.214$ mm^2, indirect descaling radius is $R_2 = 0.547$ mm, indirect descaling area is $S_2 = 0.939$ mm^2.

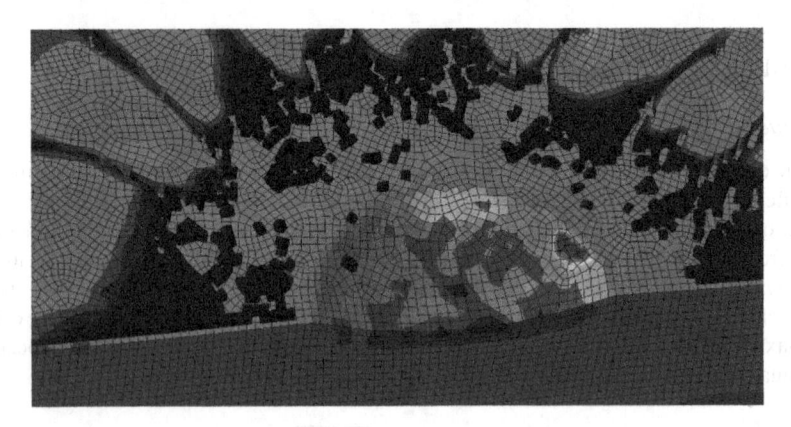

Figure 4. Enlargement view of impact crater.

The impact crater part below the projectile in Figure 2 was enlarged to obtain the local enlarged view of the impact crater as shown in Figure 4(the projectile has been hidden). As can be seen from Figure 4, the center of the impact crater is depressed, and the edge is semi-circular and slightly raised. In terms of Mises stress distribution, the further from the projectile impact position, the smaller the value of Mises stress. The results show that the Mises stress of steel matrix is only 0.1276 MPa, which indicates that the finite element model size is reasonable. The maximum Mises stress is located at the edge of the crater with a value of 734 MPa, located in the red region in Figure 4.

3.2 Deformation behavior of steel matrix

In order to further study the deformation behavior of the matrix, the Z-axis coordinate transformation of some nodes of the matrix crater is statistically analyzed. In the XOZ plane, there are a total of 35 nodes from the leftmost to the right of the matrix impact crater, as shown in Figure 5.

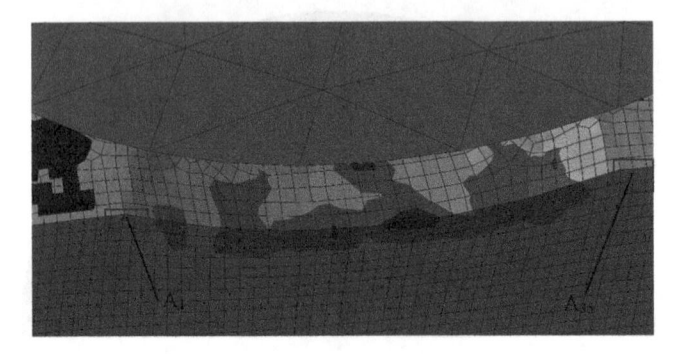

Figure 5. Impact crater and the node to be studied.

Label the leftmost node as A_1 (indicated by the red box in Figure 5), and from left to right, mark it as A_2, A_3, A_4 ..., all the way to the rightmost node A_{35} (indicated by the purple box in Figure 5). After statistical analysis of these nodes' coordinates, the Z-axis coordinate statistical graph of impact crater nodes is obtained, as shown in Figure 6.

The red dotted line in Figure 6 is the initial value of each node in the model Z-axis ($-15\mu m$). As can be seen from Figure 6, the node coordinates obtained after the model operation are a "funnel-shaped" as a whole, with a relatively flat curve that is smaller in the middle and larger at both ends, which is consistent with the appearance of impact crater. The data showed that, the Z-axis coordinate values of nodes A_1, A_2, A_{34} and A_{35} are larger than the initial coordinate values after the end of model operation. Another words, these nodes bulge after the projectile impact. The Z-axis coordinates of nodes A_3 and A_{33} are close to the initial values of coordinates ($-15.13\mu m$ and $-15.28\mu m$). The Z-axis coordinate values of the other nodes are all smaller than the initial coordinate values, among which the Z-axis coordinate value of node A_{18} is the smallest ($-39.99\mu m$), which is the lowest point of the impact crater.

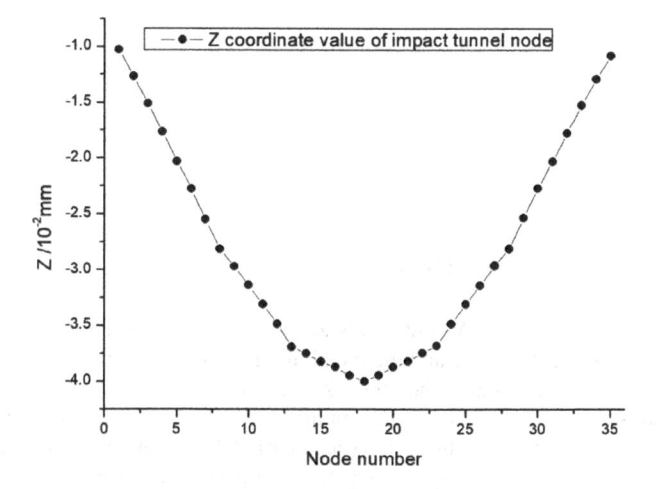

Figure 6. Z-axis coordinate numerical statistics of crater nodes.

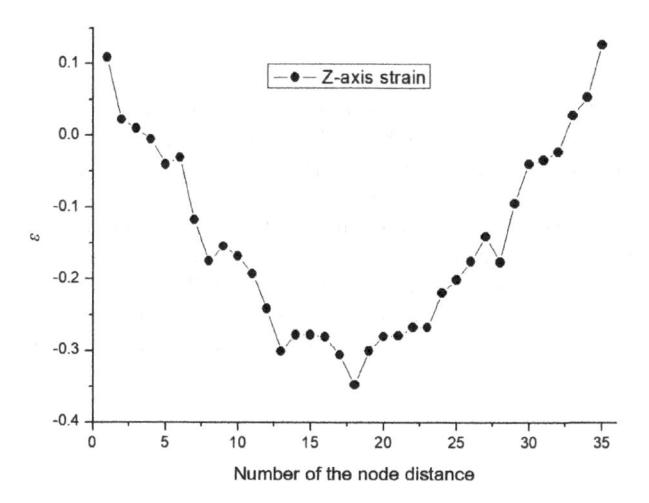

Figure 7. Strains statistics of distance between adjacent nodes.

Node A_1, A_2 ... A_{35} adjacent nodes along the negative side of the Z axis are named B_1, B_2...... B_{35} in turn. The distance between adjacent nodes ($A_1 B_1$, $A_2 B_2$...... $A_{35} B_{35}$) have been counted, and the statistical graph of strain was obtained as shown in Figure 7. As can be seen from Figure 7, the strain statistics of the impact crater in the direction of the Z-axis also show a "funnel-shaped" overall and symmetrical, but the local changes are more complex. The positive strain (tensile state) is $A_{35} B_{35}$ (the value is 0.1273), and the negative strain (compression state) is $A_{18} B_{18}$ (the value is -0.3474). The one whose strain is closest to zero is $A_4 B_4$, whose strain data is -5.363×10^{-3}.

As shown in Figure 8, under the high speed impact of the projectile, the metal is shaped to flow, and the flow direction is roughly shown by the arrow in the figure. Immediately below the projectile, a unit N bears the compressive stress caused by impact, as shown in Figure 8. In contrast, at a position far from the projectile, a unit M does not bear the impact force of the projectile in the direction of the Z-axis, and has a large degree of freedom in the direction of the Z-axis. However, the plastic flow of metal from the surrounding sides causes the unit M to undergo compressive stress along the X-axis. With the metal shaping flow, the unit M moves in the positive direction of the Z axis, and finally forms a convex part at the edge of the impact crater.

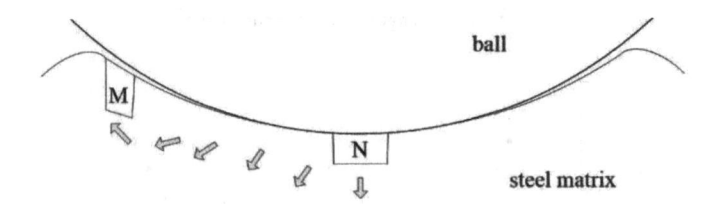

Figure 8. Projectile impact and metal plastic flow.

The units located between unit M and unit N are subjected to both the compression stress from the negative Z-axis of the projectile and the metal molding flow from the side. The former causes the negative strain accumulation of the above Z-axis, while the latter causes the positive strain accumulation. It is this dual effect that results in the "sawtooth" volatility of the curve in Figure 7. For example, the absolute value of the strain (negative) of $A_6 B_6$ and $A_8 B_8$ is greater than that of $A_7 B_7$, which indicates that, relatively, the negative strain accumulation caused by the projectile's Z-axis impact borne by $A_6 B_6$ and $A_8 B_8$ is more obvious. In contrast, $A_7 B_7$ has more obvious positive strain accumulation due to metal plastic flow.

4 CONCLUSION

The finite element descaling model was established and the deformation behavior of descaling and matrix was simulated. Under the high speed impact of projectile, the scale layer of strip steel surface ruptures and peels off, and the matrix deforms on a large scale, resulting in the "meteorite crater". The descaling area is roughly semicircular with "whiskers" extending from the center of the circle to all sides. The direct descaling area is 0.214 mm^2, and the indirect descaling area is 0.939 mm^2. The projectile impact in the Z-axis direction causes the element strain to increase negatively in the Z-axis direction. At the same time, metal plastic flow occurs in the matrix laterally (in the XOY plane) under the impact, resulting in a positive increase in the z-axis strain. Under the comprehensive action of the projectile impact in the Z axis direction and the metal plastic flow, the complex deformation of the base takes place and "meteorite crater" is formed.

ACKNOWLEDGMENTS

This work was supported by the Beijing City Board of Education Project(NO.KM202110858004).

REFERENCES

[1] Xin F, Mingnan D, Xiangpeng Y, et al. Review of acid-free descaling technology for hot rolled strip steel[J]. Baosteel Technology, 2019(01):7–14.
[2] Voges K C, Mueth A R. Method of producing rust inhibitive sheet metal through scale removal with a slurry blasting descaling cell: US, US8128460[P]. 2012.
[3] Zhiqiang Z. Brief discussion on influence of shot blasting process on surface quality of stainless steel sheet[J]. Tianjin metallurgy, 2014(5):18–20.
[4] Yongsheng Y, Wenmao X, Zhi L, et al. Effect of shot blasting process on descaling capacity and roughness of SUS430 hot-rolled coil[J]. Bao-steel technology, 2015(3):30–34.
[5] Xuchu L. Comparison of application of different shot blasting media in descaling process of stainless steel hot rolled sheet[J]. Shanxi metallurgy, 2012 (2):13–16.
[6] Jingquan M. Steel grit descaling for the raw material of the cold rolling stainless steel[J]. Shanxi metallurgy, 2010(1):28–30.
[7] Zhi G, Bu Y. Research on the wet shot blasting technology of strip steel surface topography and finite element simulation[C]// International Conference on Materials, Environmental and Biological Engineering. 2015.
[8] Yujun R, Jun G G, Jun L. Mechanism analysis about breakdown and feed of slurry abrasive in wet blasting. [J]. Metallurgical Equipment, 2014, 2: 013.
[9] Shang W, Quan Y, Xiaochen W, et al. Damage behavior of oxide scale at different blasting impact velocities [J]. Journal of Harbin Engineering University, 2018, 39(4):766–771.
[10] Johnson G R, Cook W H. A constitutive model and data for metals subjected to large strains, high strain rates and high temperatures[J]. Engineering Fracture Mechanics, 1983, 21:541–548.
[11] Li L, Xudong Z, Feng F, et al. Determination of parameters of Johnson-Cook models of Q235B steel[J]. Journal of Vibration and Shock, 2014, 33(9):153–158.
[12] Shang W, Quan Y, Yunhe R, et al. Tensile cracking behavior of oxide scale in hot-rolled steel[J]. Chinese Journal of Engineering, 2017, 39(10):1540–1545.

Author index

An, P. 509

Bai, X.F. 170
Bao, F.J. 383
Bao, W.B. 518

Cai, S.W. 455
Cai, Y.X. 343
Cao, J. 547
Cao, Q.X. 554
Cao, W. 15
Cao, Y.H. 315
Chang, Z.Y. 64
Chen, D.B. 554
Chen, G.Y. 75
Chen, L. 96, 212, 532
Chen, L.Q. 146
Chen, Q. 57
Chen, S.Y. 124, 219
Chen, X.H. 467
Cheng, G.X. 263
Cheng, J.H. 704
Cheng, Y.Q. 41, 242
Cong, F.L. 623
Cui, T. 124

Dai, J.T. 718
Dang, W. 547
Deng, D.W. 319
Deng, X. 304
Deng, X.Y. 3, 108
Dong, J.C. 299
Duan, N. 26
Dun, C.Y. 602

Fan, K. 589
Fang, R.Y. 583
Fang, S. 227
Fang, Y.H. 103
Feng, Z.X. 731
Fu, Y. 9

Gao, K.W. 699
Gao, W. 589

Gao, X.T. 675
Gao, Y. 177, 257
Guan, B.X. 602
Guan, S.Y. 351
Guo, J.L. 431
Guo, L. 331
Guo, S.R. 293
Guo, S.Y. 96, 212, 331, 532
Guo, Y. 602
Guo, Y.H. 26
Guo, Z.L. 31

Han, D.D. 299
Han, L. 718
Hao, D.H. 467
Hao, R.C. 731
He, J. 623
He, K.X. 725
He, W.L. 293
Hou, A.J. 315
Hou, Q.L. 20
Hou, X.D. 319
Hou, Y. 473
Hou, Z.Q. 193
Hu, J.G. 518
Hu, J.S. 315
Hu, K. 493, 610
Hu, Z.H. 610
Huang, D.F. 366
Huang, J. 455
Huang, W.J. 518
Huang, X.J. 84, 193
Huang, Y.J. 47, 137

Ji, J.R. 324
Ji, N. 137
Ji, Z. 90, 199, 699
Jia, C.C. 90, 199
Jia, N. 401
Jiang, T.C. 616
Jiang, Y.C. 680
Jin, C.F. 41, 242
Jin, C.P. 463
Jin, Z. 324

Kang, S.Z. 337
Kong, Z.G. 204
Kouser, T. 577

Lang, C.Y. 699
Lei, B.C. 664
Li, B. 487
Li, B.K. 96, 212, 532
Li, D.Y. 26
Li, G.L. 406
Li, H. 3, 108, 680
Li, H.Z. 146, 395
Li, J. 276
Li, J.Q. 704
Li, J.R. 351
Li, J.X. 96, 212
Li, J.Y. 431, 449, 504
Li, L. 415
Li, M.T. 64
Li, M.W. 324
Li, Q.F. 304
Li, S.X. 310
Li, W.H. 361
Li, X.Z. 119
Li, Y. 96, 304, 366, 406, 463, 532
Li, Y.A. 331
Liang, J. 124
Liang, X.B. 547
Lin, J.Q. 443
Lin, L. 623
Liu, B. 653
Liu, C. 635
Liu, C.G. 421
Liu, C.M. 152
Liu, C.S. 124
Liu, C.X. 664
Liu, D.D. 377
Liu, D.Q. 710
Liu, G.P. 699
Liu, H. 113, 487
Liu, H.G. 731
Liu, H.N. 431
Liu, J.B. 664
Liu, J.F. 337

Liu, J.L. 443
Liu, J.S. 487
Liu, L. 119
Liu, L.Z. 664
Liu, M. 653
Liu, P.Y. 421
Liu, Q.H. 642
Liu, S. 616
Liu, W.B. 242
Liu, W.H. 47, 137, 186
Liu, X.J. 395
Liu, X.X. 487
Liu, Y. 84, 193, 570
Liu, Y.F. 186
Liu, Y.L. 47, 186
Liu, Z. 90, 199
Liu, Z.D. 96, 212, 532
Liu, Z.S. 635
Lu, C.H. 271
Lu, D. 242
Lu, G.Z. 31, 130
Lu, W.H. 647
Lu, Y.J. 177, 257
Lu, Z.H. 9
Luo, G.W. 710
Lv, C.S. 248
Lv, L.Y. 463
Lv, S.Z. 389
Lv, Y.G. 41, 242, 310
Lv, Y.P. 629

Ma, H. 324
Ma, H.R. 212, 532
Ma, L. 664
Ma, M.E. 3, 108
Ma, W.F. 547
Ma, X.L. 343
Ma, Y. 509, 687
Ma, Z.L. 26
Mao, J.F. 493
Meng, F.H. 509, 687
Mi, T.W. 343

Ni, Y. 467
Nie, H.L. 547
Ning, T.B. 293

Pan, B.C. 629
Pan, D.Q. 664
Pang, G.W. 664
Pang, L. 146, 395
Pei, J.X. 395

Peng, C. 161
Peng, F. 635

Qi, B.W. 57
Qian, X. 146
Qiao, L. 377
Qiao, Y.A. 152
Qin, J. 635
Qiu, Q.Y. 725

Ren, E.X. 64
Ren, J.J. 547
Ren, Y. 319
Ruan, Z.G. 562

Shen, B. 421
Shi, C.F. 324
Shi, H.P. 595
Shi, L.Q. 664
Si, W.M. 119
Song, X.W. 219
Sun, J.H. 602
Sun, J.S. 20, 113
Sun, L.J. 276, 509
Sun, M. 219
Sun, S.X. 406
Sun, Y.M. 119
Sun, Y.N. 3, 108
Sun, Y.Z. 731
Sun, Z.J. 421, 718

Tan, Z.K. 653
Tang, H.M. 642
Tang, M.G. 642
Tang, S.Y. 152
Tantai, F.L. 20, 113
Tian, G. 271
Tian, G.F. 90, 199
Tian, H.F. 20, 113
Tian, M.Y. 467
Tian, Q.C. 263
Tian, Y. 361

Wang, B.T. 623
Wang, C. 20
Wang, D.Y. 64
Wang, H.W. 421
Wang, J.C. 152
Wang, J.D. 629
Wang, J.L. 602
Wang, J.W. 75
Wang, K. 547, 547
Wang, L. 242, 310

Wang, L.M. 366
Wang, L.Q. 248
Wang, M. 124, 219
Wang, M.L. 293
Wang, M.M. 75
Wang, P. 103
Wang, P.R. 642
Wang, R.Y. 284
Wang, S. 731
Wang, S.Y. 337
Wang, W. 20, 113
Wang, X. 319
Wang, X.B. 589
Wang, X.C. 731
Wang, X.D. 664
Wang, X.H. 629
Wang, X.P. 361
Wang, Y. 41, 299, 351
Wang, Y.F. 635
Wang, Y.N. 9
Wang, Y.Q. 653
Wang, Z.J. 31, 130, 554
Wang, Z.M. 493, 610
Wang, Z.P. 467
Wang, Z.W. 415
Wei, B.D. 9
Wei, M.W. 219
Wei, X.M. 170
Wen, N. 242, 310
Wu, B.F. 481
Wu, C.Y. 304
Wu, G.Q. 493, 610
Wu, J.W. 57
Wu, L.Q. 351
Wu, P. 26
Wu, P.P. 75
Wu, Q. 15
Wu, Q.H. 421
Wu, Q.P. 64
Wu, W. 343

Xia, Y.J. 509, 687
Xie, M.X. 70, 428
Xin, X.Y. 635
Xiong, H.P. 227
Xiong, Y.L. 577
Xu, R.C. 431
Xu, T. 518
Xu, X. 119
Xu, Y. 373, 463
Xu, Y.C. 75
Xu, Y.G. 589
Xue, G.Q. 642

Yan, F. 449
Yan, G.H. 595
Yan, L.C. 699
Yan, T.Y. 146
Yan, X.Q. 90, 199
Yang, D. 577
Yang, G. 421
Yang, G.H. 90, 199
Yang, G.S. 304
Yang, J. 170
Yang, S.G. 161
Yang, X.Q. 47, 137, 186
Yang, Y. 525, 583, 647
Yang, Z.L. 443
Yang, Z.Y. 421
Yao, T. 547
Ye, D.H. 47, 186
Ye, H.J. 725
Yi, J.Z. 248
Yin, F.W. 710
Yin, G.L. 124
Yin, Z.F. 130
Ying, Z.G. 562
You, H. 119
Yu, D.G. 293
Yu, L. 389, 583
Yu, L.S. 41
Yu, Q. 236

Yu, Q.F. 647
Yu, Q.Y. 227
Yu, T.Y. 3, 108
Yu, X. 146, 395
Yu, Z.Q. 547
Yuan, D.Y. 361

Zaheer, M. 137
Zhang, D. 518
Zhang, D.X. 26
Zhang, G.F. 319
Zhang, H. 421
Zhang, J.R. 9
Zhang, J.X. 473
Zhang, L.L. 664
Zhang, M.C. 227
Zhang, S.X. 389
Zhang, W.J. 15, 616
Zhang, X. 299, 337
Zhang, X.C. 455, 680
Zhang, X.L. 90, 199
Zhang, X.W. 324
Zhang, Y. 31, 236, 589
Zhang, Y.C. 204
Zhang, Y.F. 463
Zhang, Y.G. 473
Zhang, Y.X. 406
Zhang, Y.Z. 271
Zhang, Z.S. 31, 130

Zhang, Z.X. 113
Zhang, Z.Y. 331
Zhao, B. 57
Zhao, G.F. 284
Zhao, H.L. 455
Zhao, J. 653
Zhao, L.R. 271
Zhao, Q.Y. 20
Zhao, R.G. 47, 137, 186
Zhao, X. 437
Zhao, Z.J. 583
Zheng, B. 680
Zheng, C.M. 263
Zheng, D.W. 337
Zheng, H.F. 331
Zheng, S. 421
Zheng, X. 449
Zhou, D.Z. 299
Zhou, H. 554
Zhou, L. 219
Zhou, Q. 75
Zhou, W.H. 293
Zhou, X. 47, 137, 186
Zhu, L. 64
Zhu, P.P. 467
Zhuo, Y.C. 473
Zou, Y.L. 337
Zhang, W.H. 315

9781032127071